MUSEUM OF BROADCAST COMMUNICATIONS

ENCYCLOPEDIA OF
TELEVISION

MUSEUM OF BROADCAST COMMUNICATIONS

ENCYCLOPEDIA OF

TELEVISION

VOLUME 3
Q–Z

Editor
HORACE NEWCOMB

Photo Editor
CARY O'DELL

Commissioning Editor
NOELLE WATSON

FITZROY DEARBORN PUBLISHERS
CHICAGO AND LONDON

Library of Congress Cataloging-in-Publication Data

Encyclopedia of Television / editor, Horace Newcomb;
    p. cm.
  Includes bibliographical references and index.
  Contents: v. 3. entries Q–Z

British Library Cataloguing in Publication Data

Encyclopedia of Television
I. Newcomb, Horace, 1942

ISBN 1-884964-25-7
Set ISBN 1-884964-26-5

First published in the U.S.A. and U.K. 1997
Typeset by Acme Art, Inc.
Printed by Braun-Brumfield, Inc.

Cover photographs: *Hallmark Hall of Fame: Macbeth*; Astronaut Edwin E. Aldrin, Jr., Apollo 11 mission (courtesy of NASA); *The Sherlock Holmes Mysteries*, *The Jewel in the Crown* (courtesy of Goodman Associates); Benny Hill (courtesy of DLT Entertainment Ltd.); *I Love Lucy*, *The Steve Allen Show* (courtesy of Steve Allen); *The Muppet Show* (Jim Henson, creator); *I Spy*.
Covers designed by Peter Aristedes, Chicago Advertising and Design.

CONTENTS

ALPHABETICAL LIST OF ENTRIES

Volume 1

Volume 2

Volume 3

Q

QUATERMASS

British Science-Fiction Series

Years before the English Sunday supplements ever discovered the Angry Young Man, jazz, science fiction and other "marginal" art forms began to gather adherents among those who formerly might have quickly passed by them. Postwar British culture had entered a self-conscious period of transition, and science fiction suddenly seemed much more important to both pundits like Kingsley Amis, and readers in general, who made John Wyndham's novels (beginning with *The Day of the Triffids*, 1951) surprising best-sellers.

The 1950s were also a period of adjustment for the BBC, which lost its television monopoly midway through the decade with the dreaded debut of the Independent Television Authority (ITA)—the invasion of commercial TV. Classical works and theatrical adaptations suddenly seemed insufficient to secure the BBC's popular support. Perhaps not surprisingly, the corporation turned to science fiction: in 1953, the drama department put its development budget behind one writer, Nigel Kneale, who in exchange produced the script for the BBC's first original, adult science-fiction work. It was a serial to be produced and directed by Rudolph Cartier, and titled *The Quatermass Experiment*. The summer of that year its six half-hour episodes aired, and with them began a British tradition of science-fiction television which runs in various forms from *Quatermass* to *A Is for Andromeda* to *Blake's Seven*, and from *Doctor Who* to *Red Dwarf*. Kneale himself went on to adapt George Orwell's *Nineteen Eighty-Four* for Cartier's controversial 1954 telecast. Later in the decade, Kneale adapted John Osbourne's *Look Back in Anger* and *The Entertainer* for the screen.

Yet Kneale's first major project was quite possibly his most elegant as well. The story of *The Quatermass Experiment* is fairly simple: a British scientist, Professor Bernard Quatermass, has launched a rocket and rushes to the site of its crash. There he discovers that only one crew member, Victor Carroon, has returned with the ship. Carroon survived only as a host for an amorphous alien life form, which is not only painfully mutating Carroon's body, but preparing to reproduce. Carroon escapes and wreaks havoc upon London, until Quatermass finally tracks the now unrecognizably human mass to Westminster Abbey. There Quatermass makes one final appeal to Carroon's humanity.

Years before, H.G. Wells had inaugurated contemporary science fiction with *War of the Worlds'* warnings about Britain's failure to advance its colonial self-satisfaction. The

Quatermass Experiment's depiction of an Englishman's transformation into an alienated monster dramatized a new range of gendered fears about Britain's postwar and post-colonial security. As a result, or perhaps simply because of Kneale and Cartier's effective combination of science fiction and poignant melodrama, audiences were captivated.

With a larger budget and better effects, Kneale and Cartier continued the professor's story with *Quatermass II* (1955), an effectively disturbing story of alien possession and governmental conspiracies prefiguring *Invasion of the Body Snatchers* (1956). Perhaps fittingly, *Quatermass II* provided early counter-programming to the BBC's new commercial competition.

That same year, the small, struggling Hammer Films successfully released its film adaptation of *The Quatermass Experiment* in Britain. The next year the film (re-titled *The Creeping Unknown*) performed unexpectedly well in the lucrative U.S. market, providing the foundation for the company's subsequent series of Gothic horror films. Hammer released its film adaptation of the second serial (re-titled *The Enemy Within* for the United States) in 1957.

Kneale and Cartier's third serial in the series, *Quatermass and the Pit*, combined the poetic horror of the first and the paranoia of the second. In it, Quatermass learns that an archaeological discovery made during routine subway expansion means nothing less than humanity itself is

Quatermass

not what we have believed. The object discovered in that subway "pit" is an ancient Martian craft, and its contents indicate we were their genetically-engineered offspring. By the conclusion of the serial, London's inhabitants have been inadvertently triggered into a programmed "wilding" mode, and the city lies mostly in ruins. "We're all Martians!," became Quatermass' famous cry, and the serial's ample references to escalating racial and class tensions gives his words an ominous power.

It is this grim, elegant ending, filmed by Hammer in 1967 (and released in the United States as *Five Million Years to Earth*), that Greil Marcus used in his history of punk to describe the emotional experience of a Sex Pistols concert. If nothing else, Marcus' reference in *Lipstick Traces* (1989) suggests that Quatermass, like those repressed Martian memories, may return at the most curious moments. Even when more expected, the name may still operate as a certain sort of cultural code word: Brian Aldiss, in his extensive science-fiction history *Trillion Year Spree* (1986), uses "the Quatermass school" as if every reader should automatically understands its meaning.

But by the late 1970s, the BBC was no longer willing to commit itself to the budget necessary for Kneale's fourth and final Quatermass serial, simply titled *Quatermass*. Commercial television was ready, however, and in 1979, at the conclusion of a 75-day ITV strike, the four-part *Quatermass* debuted with John Mills starring as the now elderly professor in his final adventure.

Only the serial's opening sequence, involving Quatermass deriding a U.S.-U.S.S.R. "Skylab 2," displays the force of the earlier serials: a moment after Quatermass blurts out his words in a live television interview, the studio monitors are filled with the image of "Skylab 2" blowing to pieces. Subsequent episodes were less successfully provocative. Concerning a dystopic future Britain where hippie-like youth are being swept up by aliens, the serial's narrative was recognized as somewhat stale and unconvincing. Yet even in the late 1970s, despite the last serial's lukewarm reviews, *Quatermass* remained a source of fan preoccupation reminiscent of the commitment to *Star Trek*.

Unlike the three earlier serials, broadcast live but recorded on film, *Quatermass* was not adapted for the screen. It was simply edited and re-packaged as *The Quatermass Conclusion* for theatrical and video distribution abroad. Of the original serials, only *Quatermass and the Pit* has had a video release, although most of the first serial and all of the second have been preserved by the British Film Institute.

—Robert Dickinson

CAST

THE QUATERMASS EXPERIMENT

Professor Bernard Quatermass	Reginald Tate
Judith Carroon	Isabel Dean
John Paterson	Hugh Kelly
Victor Carroon	Duncan Lamont
James Fullalove	Paul Whitsun-Jones

QUATERMASS II

Quatermass	John Robinson
Paula Quatermass	Monica Grey
Dr. Leo Pugh	Hugh Griffiths
Captain John Dillon	John Stone
Vincent Broadhead	Rupert Davies
Fowler	Austin Trevor

QUATERMASS AND THE PIT

Quatermass	Andre Morrell
Dr. Matthew Roney	Cec Linder
Barbara Judd	Christine Finn
Colonel Breen	Anthony Bushell
Captain Potter	John Stratton
Sergeant	Michael Ripper
Corporal Gibson	Harold Goodwin
Private West	John Walker
James Fullalove	Brian Worth
Sladden	Richard Shaw

QUATERMASS

Quatermass	John Mills
Joe Kapp	Simon MacCorkindale
Clare Kapp	Barbara Kellerman
Kickalong	Ralph Arliss
Caraway	Paul Rosebury
Bee	Jane Bertish
Hettie	Rebecca Saire
Marshall	Tony Sibbald
Sal	Toyah Wilcox
Guror	Brewster Mason
Annie Morgan	Margaret Tyzack

PRODUCERS Rudolph Cartier (*The Quatermass Experiment; Quatermass II; Quatermass and the Pit*)
Verity Lambert, Ted Childs (*Quatermass*)

PROGRAMMING HISTORY

THE QUATERMASS EXPERIMENT

• BBC

6 30-minute episodes
18 July 1953–22 August 1953

QUATERMASS II

• BBC

6 c. 30-minute episodes
22 October 1955–26 November 1955

QUATERMASS AND THE PIT

• BBC

6 35-minute episodes
22 December 1958–26 January 1959

QUATERMASS

• ITV
4 60-minute episodes
24 October 1979–14 November 1979

FURTHER READING

Briggs, Asa. *The History of Broadcasting in the United Kingdom*, volume IV. Oxford: Oxford University Press, 1979.

Fulton, Roger. *The Encyclopedia of TV Science Fiction*. London: Boxtree, 1990.

Kneale, Nigel. *Quatermass*. London: Hutchinson, 1979.

———. *The Quatermass Experiment. Quatermass II. Quatermass and the Pit*. London: Penguin, 1960.

Leman, Joy. "Wise Scientists and Female Androids: Class and Gender in Science Fiction." In, Corner, John, editor. *Popular Television in Britain*. London: British Film Institute, 1991.

Marcus, Greil. *Lipstick Traces: A Secret History of the Twentieth Century*. Cambridge, Massachusetts: Harvard University Press, 1989.

Pirie, David. *A Heritage of Horror: The English Gothic Cinema 1946–1972*. London: Gordon Fraser, 1973.

See also Cartier, Rudolph; Lambert, Verity; Science-fiction Programs

QUENTIN DURGENS, M.P.

Canadian Drama Series

One of the first hour-long Canadian drama series produced by the CBC, *Quentin Durgens, M.P.* began as six half-hour episodes entitled *Mr. Member of Parliament* in the summer of 1965 as part of *The Serial*, a common vehicle for Canadian dramas. The program starred a young Gordon Pinsent as a naive rookie member of Parliament who arrives in Ottawa and quickly learns that the realities behind public service can be alternately humorous, overwhelming, and frustrating.

Consciously designed to be an absolutely distinctive Canadian drama series, *Quentin Durgens, M.P.* contrasted the private struggles and controversies faced by politicians with the more sedate, pompous image presented by Parliament. Many of its plots were inspired by real-life issues and situations. Pornography, violence in minor-league hockey, gender discrimination, and questions of religious tolerance were topics addressed among its episodes. In all of them, however, the inner workings of power, with its back-room deals and interpersonal struggles, remained the backbone of the series.

The regular series of *Quentin Durgens, M.P.* began in December 1966 as a winter season replacement, and followed the popular series *Wojeck* in a Tuesday 9:00 P.M. time slot. And like *Wojeck*, *Quentin Durgens* was hailed as an example of Canadian television, distinct and set apart from Hollywood drama. The show still carried its imprint as a serial with open narratives, unresolved psychological conflicts, and the freedom to construct stories around topical issues. Frequent allusions to actual social events and a great deal of subtext were interwoven in plots that juxtaposed rational and emotional behaviours. The result made for what its director and producer David Gardner called an "ironic drama." Documentary techniques grounded in the tradition of the National Film Board of Canada also added to the "behind-the-scene" feel of the series and reflected, according to Canadian television critic, Morris Wolfe, a Canadian tradition of "telling it like it is." Despite these claims, other Canadian television critics and historians such as Paul Rutherford have questioned the uniqueness of these "made-in-Canada" dramas, arguing instead that many of the characteristics attributed to Canadian drama series such as *Wojeck*, *Quentin Durgens, M.P.*, and *Cariboo Country* were already to be found in some American and, especially, British dramas.

Though *Quentin Durgens, M.P.* was part of a formidable line-up, it was never popular with Canadian viewers. With fewer funds and resources than *Wojeck*, the show had to be videotaped (on location and in the studio) for its initial two seasons. The flattened, taped images and sometimes awkward edits detracted from the documentary feel. Nor were its scripts consistently strong. Despite the increased support in its third season (after the end of *Wojeck*) when all 17 episodes were filmed and in colour, *Quentin Durgens* failed to hold the large audiences which *Wojeck* had won for the evening. Canadian viewers, it seemed, did not share the CBC's and producers' interest in developing a distinctive Canadian perspective. Parliamentary intrigues were not fascinating enough to attract a large following and *Quentin Durgens, M.P.* simply lacked the excitement of cop shows.

—Manon Lamontagne

CAST

Quentin Durgens, M.P.	Gordon Pinsent
His Secretary	Suzanne Levesque
Other Members of Parliament	Ovila Legere
	Franz Russell, Chris Wiggins

PRODUCERS   David Gardner, Ron Weyman, John Trent, Kirk Jones

Quentin Durgens, M.P.
Photo courtesy of the National Archives of Canada

PROGRAMMING HISTORY

Summer 1965 (*The Serial*)	Six Episodes
December 1966–January 1967	Eight Episodes
	Tuesdays 9:00-10:00
February 1967–April 1967	Ten Episodes
	Tuesdays 9:00-10:00
September 1968–January 1969	Seventeen Episodes
	Tuesdays 9:00-10:00

FURTHER READING

Miller, Mary Jane. *Turn Up the Contrast: CBC Drama Since 1952.* Vancouver: University of British Columbia Press, 1987.
Rutherford, Paul. *When Television Was Young: Primetime Canada 1952–1967.* Toronto: University of Toronto Press, 1990.
Wolfe, Morris. *Jolts: The TV Wasteland and the Canadian Oasis.* Toronto: James Lorimer, 1985.

See also Canadian Programming in English

QUIZ AND GAME SHOWS

Prior to the quiz show scandals in 1958 no differentiation existed between quiz shows and "game shows." Programs such as *Truth or Consequences* or *People Are Funny* that relied mainly on physical activity and had no significant quiz element to them were called quiz shows, as was as an offering like *The $64,000 Question* that emphasized factual knowledge. The scandals mark an important turning point because in the years following, programs formerly known as "quiz shows" were renamed "game shows." This change coincides with a shift in content, away from high culture and factual knowledge common to the big money shows of the 1950s. But the renaming of the genre also represents an attempt to

Quiz Kids
Photo courtesy of Rachel Stevenson

distance the programs from the extremely negative conno-tations of the scandals, which had undermined the legiti-macy of the high cultural values that quiz shows—the term and the genre—embodied. Thus, the new name, "game shows," removed the genre from certain cultural assump-tions and instead creates associations with the less sensitive concepts of play and leisure. Nevertheless, the historical and material causes for this re-naming still fail to provide a sufficient basis for a definition of this genre as a whole. John Fiske, in *Television Culture*, suggests more satisfactory defi-nitions and categories with which to distinguish among different types of shows.

One of the main appeals of quiz shows is that they deal with issues such as competition, success, and knowledge—central concerns for American culture. It makes sense, then, to follow Fiske in defining this genre according to its relation to knowledge. He begins by suggesting a basic split between "factual" knowledge and "human" knowledge. Factual knowledge can be further divided into "academic" and

"everyday" knowledge. Human knowledge consists of knowledge of "people in general" and of specific "individu-als." While Fiske does not clearly distinguish between the terms "game" and "quiz show," his categories reflect a sig-nificant difference in program type. All shows that deal with competitions between individuals or groups, and based pri-marily on the display of factual knowledge will be considered quiz shows. Shows dealing with human knowledge (knowl-edge of people or of individuals) or that are based primarily on gambling or on physical performances fall in the category of game shows. Thus, *The Gong Show* or *Double Dare* are not considered quiz shows since they rely primarily or com-pletely on physical talents; *Family Feud* and *The Newlywed Game* rely entirely on knowledge of people or of individuals and would therefore also be considered game shows. *Jeop-ardy!*, however, with its focus on academic, factual knowl-edge, is clearly a quiz show.

Many early television quiz shows of the 1940s were transferred or adapted from radio, the most prominent

University Challenge
Photo courtesy of the British Film Institute

among them, *Information Please, Winner Take All*, and *Quiz Kids*. These shows also provided a professional entry point for influential quiz show producers such as Louis Cowan, Mark Goodson, and Jack Barry. While a number of early radio and television quiz shows were produced locally and later picked up by networks, this trend ended in the early 1950s when increasing production values and budgets centralized the production of quiz shows under the control of networks and sponsors. Nevertheless, the relatively low production costs, simple sets, small casts, and highly formalized production techniques have continually made quiz shows an extremely attractive television genre. Quiz shows are more profitable and faster to produce than virtually any other form of entertainment television.

In the late 1940s and early 1950s most quiz shows were extremely simple in visual design and the structure of the games. Sets often consisted of painted flats and a desk for an expert panel and a host. The games themselves usually involved a simple question and answer format that displayed the expertise of the panel members. An important characteristic of early quiz shows was their foregrounding of the expert

knowledge of official authorities. A standard format (e.g. *Americana, Information Please*) relied on home viewers to submit questions to the expert panel. Viewers were rewarded with small prizes (money or consumer goods) for each question used and with larger prizes if the panel failed to answer their question. While this authority-centered format dominated the 1940s, it was slowly replaced by audience-centered quizzes in the early 1950s. In this period "everyday people" from the studio audience became the subjects of the show. The host of the show, however, remained the center of attention and served as a main attraction for the program (e.g. Bert Parks and Bud Collyer in *Break the Bank* and James McClain in *Doctor I.Q.*). At this point the visual style of the shows was still fairly simple, often recreating a simple theatrical proscenium or using an actual theater stage. The Mark Goodson-Bill Todman production *Winner Take All* was an interesting exception. While it also used charismatic hosts, it introduced the concept of a returning contestant who faced a new challenger for every round. Thus, the attention was moved away from panels and hosts and toward the contestants in the quiz.

A 1954 Supreme Court ruling created the impetus for the development of a new type of program when it removed "Jackpot" quizzes from the category of gambling and made it possible to use this form of entertainment on television. At CBS producer Louis Cowan, in cooperation with Revlon Cosmetics as sponsor, developed the idea for a new Jackpot quiz show based on the radio program *Take It or Leave It*. The result—*The $64,000 Question*—raised prize money to a spectacular new level and also changed the visual style and format of quiz shows significantly. *The $64,000 Question*, its spinoff *The $64,000 Challenge* and other imitations following between 1955 and 1958 (e.g. *Twenty-One, The Big Surprise*) all focused on high culture and factual, often academic, knowledge. They are part of television's attempts in the 1950s to gain respectability and, simultaneously, a wider audience. They introduced a much more elaborate set design and visual style and generally created a serious and ceremonious atmosphere. *The $64,000 Question* introduced an IBM sorting machine, bank guards, an isolation booth and neon signs, while other shows built on the same ingredients to create similar effects. In an effort to keep big money quiz shows attractive, the prize money was constantly increased and, indeed, on a number of shows, became unlimited. *Twenty-One* and *The $64,000 Challenge* also created tense competitions between contestants, so that audience identification with one contestant could be even greater. Consequently, the most successful contestants became celebrities in their own right, perhaps the most prominent among them being Dr. Joyce Brothers and Charles Van Doren.

This reliance on popular returning contestants, on celebrities in contest, also created, however, a motivation to manipulate the outcome of the quizzes. Quiz show sponsors in particular recognized that some contestants were more popular than others, a fact that could be used to increase audience size. They required and advocated the rigging of the programs to create a desired audience identification with these popular contestants.

When these practices were discovered and made public, the ensuing scandals undermined the popular appeal of big money shows and, together with lower ratings, led to the cancellation of all of these programs in 1958–59. Entertainment Productions Inc. (EPI), a production company founded by Louis Cowan, was particularly involved in and affected by the scandals. EPI had produced a majority of the big money shows and was also most actively involved in the riggings. Following the scandals, the networks used the involvement of sponsors in the rigging practices as an argument for the complete elimination of sponsor-controlled programming in prime-time television.

Still, not all quiz shows of the late 1950s were cancelled due to the scandals. A number of programs which did not rely on the huge prizes (e.g. *The Price is Right, Name that Tune*) remained on the air and provided an example for later shows. Even these programs, however, were usually removed from prime time, their stakes signif-

icantly reduced, and the required knowledge made less demanding. In the early 1960s very few new quiz shows were introduced, and most were game shows focusing less on high culture and more on gambling and physical games. Overall, the post-scandal era is marked by a move away from expert knowledge to contestants with everyday knowledge. *College Bowl* and *Alumni Fun* still focused on "academic" knowledge without reviving the spectacular qualities of 1950s quiz shows, but *Jeopardy!*, introduced by Merv Griffin in 1964, is the only other significant new program developed in the decade following the scandals. It re-introduces "academic" knowledge, a serious atmosphere, elaborate sets, and returning contestants, but offers only moderate prizes. The late 1960s were marked by even more cancellations (CBS cancelled all of its shows in 1967) and by increasing attempts of producers to find alternative distribution outlets for their products outside the network system. Their hopes were realized through the growth in first run syndication.

In 1970, the FCC introduced two new regulations, the Financial Interest and Syndication Rules (Fin-Sin) and the Prime Time Access Rule (PTAR), that had a considerable effect on quiz/game show producers and on the television industry in general. Fin-Syn limited network ownership of television programs beyond their network run and increased the control of independent producers over their shows. The producers' financial situation and their creative control was significantly improved. Additionally, PTAR gave control of the 7:00-7:30 P.M. time slot to local stations. The intention of this change was to create locally based programming, but the time period was usually filled with syndicated programs, primarily inexpensive quiz and tabloid news offerings. The overall situation of quiz/game show producers was substantially improved by the FCC rulings.

As a result, a number of new quiz shows began to appear in the mid-1970s. They were, of course, all in color, and relied on extremely bright and flashy sets, strong, primary colors, and a multitude of aural and visual elements. In addition to this transformation traditionally solemn atmosphere of quiz shows the programs were thoroughly altered in terms of content. Many of the 1970s quiz shows introduced an element of gambling to their contests (e.g. *The Joker's Wild, The Big Showdown*) and moved them further from a clear "academic" and serious knowledge toward an everyday, ordinary knowledge.

Blatant consumerism, in particular, began to play an important role in quiz shows such as *The Price is Right* and *Sale of the Century* as the distinctions between quiz and game shows became increasingly blurred in this period. As Graham points out in *Come on Down!!!*, quiz shows had to change in the 1970s, adapting to a new cultural environment that included flourishing pop culture and countercultures. Mark Goodson's answer to this challenge on *The Price is Right* was to create a noisy, carnival atmosphere that challenged cultural norms and assumptions represented in previous generations of quiz shows.

The same type of show remained prevalent in the 1980s, though most of them now apprear primarily in syndication and, to a lesser extent, on cable channels. Both *Wheel of Fortune* and a new version of *Jeopardy!* are extremely successful as syndicated shows in the prime time access slot (7:00-8:00 P.M.). In what may become a trend Lifetime Television has introduced two quiz shows combining everyday knowledge (of consumer products) with physical contests (shopping as swiftly—and as expensively—as possible). These shows, *Supermarket Sweep* and *Shop 'Til You Drop,* also challenge assumptions about cultural norms and the value of everyday knowledge. In particular they focus on "women's knowledge," and thus effectively address the predominantly female audience of this cable channel.

One future area of growth for quiz shows in the era of cable television, then, seems to be the creation of this type of "signature show" that appeals to the relatively narrowly defined target audience of specific cable channels. *Jeopardy!* and *Wheel of Fortune,* then, notable examples to be sure, remain as the primary representatives of the quiz show genre, small legacy for one of the more powerful and popular forms of television.

—Olaf Hoerschelmann

FURTHER READING

Barnouw, E. *A History of Broadcasting in the United States: Volume III—The Image Empire.* New York: Oxford University Press, 1970.

Boddy, W. *Fifties Television: The Industry and its Critics.* Urbana: University of Illinois Press, 1990.

Fabe, M. *TV Game Shows.* Garden City, New York: Doubleday, 1979.

Fiske, J. *Television Culture.* London: Routledge, 1987.

Graham, J. *Come on Down!!!: The TV Game Show Book.* New York: Abbeville Press, 1988.

Schwartz, D., S. Ryan, and F. Wostbrock. *The Encyclopedia of Television Game Shows.* New York: Zoetrope, 1987.

Shaw, P. "Generic Refinement on the Fringe: The Game Show." *Southern Speech Communication Journal* (Winston-Salem, North Carolina), 1987.

Stone, J., and T. Yohn. *Prime Time and Misdemeanors: Investigating the 1950s TV Quiz Scandal—A D.A.'s Account.* New Brunswick, New Jersey: Rutgers University Press, 1992.

See also Goodson, Mark, and Bill Todman; Griffin, Merv; Grundy, Reg; *I've Got A Secret*; Moore, Garry; Quiz Show Scandals; *Sale of the Century*; *$64,000 Dollar Challenge/$64,000 Question*

QUIZ SHOW SCANDALS

No programming format mesmerized televiewers of the 1950s with more hypnotic intensity than the "big money" quiz show, one of the most popular and ill-fated genres in U.S. television history. In the 1940s, a popular radio program had awarded top prize money of $64. The new medium raised the stakes a thousand fold. From its premiere on CBS on 7 June 1955, *The $64,000 Question* was an immediate sensation, racking up some of the highest ratings in television history up to that time. Its success spawned a spin-off, *The $64,000 Challenge,* and a litter of like-minded shows: *The Big Surprise, Dotto, Tic Tac Dough,* and *Twenty -One.* When the Q and A sessions were exposed as elaborate frauds, columnist Art Buchwald captured the national sense of betrayal with a glib name for the producers and contestants who conspired to bamboozle a trusting audience: the quizlings.

Broadcast live and in prime time, the big money quiz show presented itself as a high pressure test of knowledge under the heat of kleig lights and the scrutiny of fifty-five million participant-observers. Set design, lighting, and pure hokum enhanced the atmosphere of suspense. Contestants were put in glass isolation booths, with the air conditioning turned off to make them sweat. Tight close-ups framed faces against darkened backgrounds and spot lights illuminated contestants in a ghostly aura. Armed police guarded "secret" envelopes and impressive looking contrap-

Charles Van Doren
Photo courtesy of Wisconsin Center for Film and Theater Research

tions spat out pre-cooked questions on IBM cards. The big winners—like Columbia university student Elfrida Von Nardroff who earned $226,500 on *Twenty-One* or warehouse clerk Teddy Nadler who earned $252,000 on *The $64,000 Challenge*—took home a fortune in pre-inflationary greenbacks.

By the standards of the dumbed-down game shows of a later epoch, the intellectual content of the 1950s quiz shows was downright erudite. Almost all the questions involved some demonstration of cerebral aptitude—retrieving lines of poetry, identifying dates from history, and reeling off scientific classifications, the stuff of memorization and canonical culture. (Who wrote "Hope is a thing with feathers/it whispers to the soul"?) Since victors returned to the show until they lost, risking accumulated winnings on future stakes, individual contestants might develop a devoted following over a period of weeks. Among the famous for fifteen pre-Warhol minutes were opera buff Gino Prato, science prodigy Robert Strom, and ex-cop and Shakespeare expert Redmond O'Hanlon. Matching an incongruous area of expertise to the right personality was a favorite hook, as in the cases of Richard McCutchen, the rugged marine captain who was an expert on French cooking, or Dr. Joyce Brothers, not then an icon of pop psychology, whose encyclopedic knowledge of boxing won her (legitimately) $132,000.

If the quiz shows made celebrities out of ordinary folk, they also sought to engage the services of celebrities. Orson Welles claimed to have been approached by a quiz show producer looking for a "genius type" who guaranteed him $150,000 and a seven-week engagement. Welles refused, but bandleader Xavier Cugat won $16,000 as an expert on Tin Pan Alley songs in a rigged match against actress Lillian Roth on *The $64,000 Challenge*. "I considered I was giving a performance," he later explained guilelessly. Twelve-year-old Patty Duke won $32,000 against child actor Eddie Hodges, then the juvenile lead in *The Music Man* on Broadway. Hodges had earlier won the $25,000 grand prize on *Name That Tune* teamed with a personable marine flyer named John Glenn.

Far and away the most notorious quizling was Charles Van Doren, a contestant on NBC's *Twenty-One*, a quiz show based on the game of blackjack. Scion of the prestigious literary family and himself a lecturer in English at Columbia University, Van Doren was an authentic pop phenomenon whose video charisma earned him $129,000 in prize money, the cover of *Time* magazine, and a permanent spot on NBC's *Today*, where he discussed non-Euclidean geometry and recited seventeenth-century poetry. He put an all-American face to the university intellectual in an age just getting over its suspicion of subversive "eggheads."

From the moment Van Doren walked onto the set of *Twenty-One* on 28 November 1956 for his first face-off against a high-IQ eccentric named Herbert Stempel, he proved himself a telegenic natural. In the isolation booth,

Van Doren managed to engage the spectator's sympathy by sharing his mental concentration. Apparently muttering unself-consciously to himself, he let viewers see him think: eyes alert, hand on chin, then a sudden bolt ("Oh, I know!"), after which he delivered himself of the answer. Asked to name the volumes of Churchill's wartime memoirs, he mutters, "I've seen the ad for those books a thousand times!" Asked to come up with a biblical reference, he says self-depreciatingly, "My father would know that." Van Doren's was a remarkable and seductive performance.

Twenty-One's convoluted rules decreed that, in the event of a tie, the money wagered for points doubled—from $500 a point, to $1000 and so on. Thus, contestants needed to be coached not only on answers and acting but on the amount of points they selected in the gamble. A tie meant double financial stakes for each successive game with a consequent ratcheting up of the tension. By pre-game arrangement, the first Van Doren-Stempel face off ended with three ties; hence, the next week's game would be played for $2000 a point, and publicized accordingly.

On Wednesday, 5 December 1956, at 10:30 P.M., an estimated 50 million Americans tune in to *Twenty-One* for what host and co-producer Jack Berry calls "the biggest game ever played in the program." A pair of twin blondes escort the pair to their isolation booths. The first category is boxing and Van Doren blows it. Ahead sixteen points to Van Doren's zero, Stempel is given the chance to stop the game. Only the audience knows he's in the lead and, if he stops the game, Van Doren loses. At this point, on live television, Stempel could have reneged on the deal, vanquished his opponent, and won an extra $32,000. But he opts to play by the script and continue the match. The next category—movies—proves more Van Doren friendly. Asked to name Brando's female co-star in *On the Waterfront* Van Doren teases briefly ("she was that lovely frail girl") before coming up with the correct answer (Eva Marie Saint). Stempel again has the chance to ad-lib his own lines, but—in an echo of another Brando role—it is not his night. Asked to name the 1955 Oscar Winner for Best Picture, he hesitates and answers *On the Waterfront*. Stempel later recalled how that choice was the unkindest cut. The correct answer—*Marty*—was not only a film he knew well but a character he identified with, the lonesome guy wondering what he was gonna do tonight.

But another tie means another round at $2,500 a point. "You guys sure know your onions," gasps Jack Berry. The next round of questions is crucial and Van Doren is masterful. Give the names and the fates of the third, fourth, and fifth wives of Henry the Eighth. As Berry leads him through the litany, Van Doren takes the audience with him every step of the way. ("I don't think he beheaded her...Yes, what happened to her.") Given the same question, Stempel gets off his best line of the match up. After Stempel successfully names the wives, Berry asks him their fates. "Well, they all died," he cracks to gales of laughter. Van Doren stops the game and wins the round. Seemingly gracious in defeat, in reality steaming with resentment, Stempel says truthfully, "This all came so suddenly...Thanks for your kindness and courtesy."

The gravy train derailed in August and September of 1958 when disgruntled former contestants went public with accusations that the results were rigged and the contestants coached. First, a standby contestant on *Dotto* produced a page from a winner's crib sheet. Then, the still bitter Herbert Stempel, Van Doren's former nemesis on *Twenty-One,* told how he had taken a dive in their climatic encounter. The smoking gun was provided by an artist named James Snodgrass, who had taken the precaution of mailing registered letters to himself with the results of his appearances on *Twenty-One* predicted in advance. Most of the high-drama match ups, it turned out, were as carefully choreographed as the June Taylor Dancers. Contestants were drilled in Q and A before airtime and coached in the pantomime of nail-biting suspense (stroke chin, furrow brow, wipe sweat from forehead). The lucky few who struck a chord with audiences were permitted a good run before a fresh attraction took their place; the patsies were given wrist watches and a kiss off.

By October 1958, as a New York grand jury convened by prosecutor Joseph Stone investigated the charges and heard closed-door testimony, quiz show ratings had plummeted. For their part, the networks played damage control, denying knowledge of rigging, canceling the suspect shows, and tossing the producers overboard. Yet it was hard to credit the Inspector Renault-like innocence of executives at NBC and CBS who claimed to be shocked that gambling was *not* going on in their casinos. A public relations flack for *Twenty-One* best described the implied contract: "It was sort of a situation where a husband suspects his wife, but doesn't want to know because he loves her."

Despite the revelations and the grand jury investigation, the quiz show producers, Van Doren, and the other big money winners steadfastly maintained their innocence. Solid citizens all, they feared the loss of professional standing and the loyalty of friends and family as much as the retribution of the district attorney's office. Thus, even though there was no criminal statute against rigging a quiz show, the producers and contestants called to testify before the New York grand jury mainly tried to brazen it out. Nearly one hundred people committed perjury rather than own up to activities that, though embarrassing, were not illegal. Prosecutor Joseph Stone lamented that "nothing in my experience prepared me for the mass perjury that took place on the part of scores of well-educated people who had no trouble understanding what was at stake."

When the judge presiding over the New York investigations ordered the grand jury report sealed, Washington smelled a cover-up and a political opportunity. Through October and November 1959, the House Subcommittee on Legislative Oversight, chaired by Oren Harris (D-Arkansas), held standing-room-only hearings into the quiz show scandals. A renewed wave of publicity recorded the now repentant testimony of network bigwigs and star contestants whose minds, apparently, were concentrated powerfully by federal intervention. At one point, committee staffers came upon possible communist associations in the background of a few witnesses. The information was turned over to the House Committee on Un-American Activities, a move that inspired one wiseacre to suggest the networks produce a new game show entitled *Find That Pinko!*

Meanwhile, as newspaper headlines screamed "Where's Charlie?", the star witness everyone wanted to hear from was motoring desperately through the back roads of New England, ducking a congressional subpoena. Finally, on 2 November 1959, with tension mounting in anticipation of Van Doren's appearance to answer questions (the irony was lost on no one), the chastened professor fessed up. "I was involved, deeply involved, in a deception," he told the Harris Committee. "The fact that I too was very much deceived cannot keep me from being the principal victim of that deception, because I was its principal symbol." In another irony, Washington's made-for-TV spectacle never made it to the airwaves due to the opposition of House Speaker Sam Rayburn, who felt that the presence of television cameras would undermine the dignity of Congress.

The firestorm that resulted, claimed *Variety,* "injured broadcasting more than anything ever before in the public eye." Even the sainted Edward R. Murrow was sullied when it was revealed that his celebrity interview show, CBS's *Person to Person,* provided guests with questions in advance. Perhaps most significantly in terms of the future shape of commercial television, the quiz show scandals made the networks forever leery of "single sponsorship" programming. Henceforth, they parceled out advertising time in fifteen, thirty, and sixty-second increments, wrenching control away from single sponsors and advertising agencies.

The fallout from the quiz show scandals can be gauged as cultural residue and written law. To an age as yet unschooled in credibility gaps and modified, limited hang-outs, the mass deception served as an early warning signal that the medium, and American life, might not always be on the up and up. As if to deny that possibility, Congress promptly made rigging a quiz show a federal crime. A televised exhibition may be fixed; a game show must always be upright.

—Thomas Doherty

FURTHER READING

Anderson, Kent. *Television Fraud: The History and Implications of the Quiz Show Scandals.* Westport, Connecticut: Greenwood, 1978.

Karp, Walter. "The Quiz-show Scandal." *American Heritage* (New York), May-June 1989.

Real, Michael. "The Great Quiz Show Scandal: Why America Remains Fascinated." *Television Quarterly* (New York), Winter 1995.

Stone, Joseph, and T. Yohn. *Prime-time and Misdemeanors: Investigating the 1950s TV Quiz Scandal: A D.A.'s Account.* New Brunswick, New Jersey: Rutgers University Press, 1992.

See also Quiz and Game Shows; *$64,000 Question/ $64,000 Challenge*

R

RACISM, ETHNICITY AND TELEVISION

Until the late 1980s whiteness was consistently naturalized in U.S. television—social whiteness, that is, not the "pinko-grayishness" that British novelist E.M. Forster correctly identified as the "standard" skin-hue of Europeans. This whiteness has not been culturally monochrome. Irish, Italians, Jews, Poles, British, French, Germans, and Russians, whether as ethnic entities or national representatives, have dotted the landscape of TV drama, providing the safe spice of white life, entertaining trills and flourishes over the *basso ostinato* of social whiteness.

In other words, to pivot the debate on race and television purely on whether and how people of color have figured, on or behind the screen or in the audience, is already to miss the point. What was consistently projected, without public fanfare, but in teeming myriads of programs, news priorities, sportscasts, movies, and ads, was the naturalness and normalcy of social whiteness. Television visually accumulated the heritage of representation in mainstream U.S. science, religion, education, theatre, art, literature, cinema, radio, and the press. According to television representation, the United States was a white nation, with some marginal "ethnic" accretions that were at their best when they could simply be ignored, like well-trained and deferential maids and doormen. This was even beyond being thought a good thing. It was axiomatic, self-evident.

Thus, American television in its first two generations inherited and diffused—on an hourly and daily basis—a mythology of whiteness that framed and sustained a racist national self-understanding. Arguably all the more powerfully for seemingly being so integral, so...inevitable.

There is a second issue. Insofar as the televisual hegemony of social whiteness has been critiqued, either on television itself, or on video, or in print, it has most often tended to focus on African-American issues. Yet, in reviewing racism and ethnicity in U.S. television, we need not downplay four centuries of African-American experience and contribution in order to recognize as well the importance of Native American nations, Chicanos and other Latinos, and Asian-Americans in all their variety. Thus, in this essay, attention will be paid so far as research permits to each one of these four groupings, although there will not be space to treat the important sub-groupings (Haitians, Vietnamese, etc.) within each. The discussion will commence with representation, mainstream and alternative, and then move on to employment patterns in the TV industry, broadcast and

cable. The conclusion will introduce the so far under-researched question of racism, ethnicity and TV audiences. Before doing so, however, a more exact definition is needed of racism in the U.S. context.

Firstly, racism is expressed along a connected spectrum, from the casual patronizing remark to the sadism of the prison guard, from avoidance of skin-contact to the starving of public education in inner cities and reservations, or to death-rates among infants of color higher than in some Third-World countries. Racism does not have to take the form of lynching, extermination camps or slavery to be systemic, virulent—yet simultaneously dismissed as of minor importance or even as irrelevant by the white majority.

Secondly, racism may stereotype groups differently. Class is often pivotal here. Claimed success among Asian-Americans and Jews is attacked just as is claimed inability to make good among Latinos and African-Americans. Multiple Native American nations with greatly differing languages and cultures are squashed into a generic "Indian" left behind by history. Gender plays a role too: white stomachs will contract at supposedly truculent and violence-prone men of color, but ethnic minority women get attributed with pli-

In Living Color *(Keenan Ivory Wayans)*

1333

Broken Arrow

The Courtship of Eddie's Father

ancy—even, for white males, to presuming their special eagerness for sexual dalliance.

Thirdly, racism in the United States is binary. You are either a person of color or you are not. People of mixed descent are not permitted to confuse the issue, but belong automatically to a minority group of color. Ethnic minority individuals whose personal cultural style may be read as emblematic of the ethnic majority's, are quite often responded to as betrayers, and thus either warmly as the "good exception" by the white majority, or derisively as "self-hating" by the minority.

Lastly, as Entman (1990) and others have argued, racist belief has changed to being more supple, and "modern" racism has shed its biological absolutism. In the "modern" version the Civil Rights Movement won, racial hatred is past, and talented individuals now make it.

Therefore—triumphantly—continuing ethnic minority poverty is solely the minority's overall cultural/attitudinal fault. There are many other dimensions to racism, such as the economic. Indeed, race relations in U.S. life still closely resemble the depth and width of the Grand Canyon, but rarely its beauty.

Mainstream Representation

In discussing mainstream representation, it is vital to note two issues. One is the importance of historical shifts in the representation of these issues, especially since the mid-

1980s, but also at certain watermark junctures before then. The second is the importance of taking into account the entire spectrum of what television provides, including ads (perhaps 20% of TV content), weathercasting, sitcoms, documentaries, sports, MTV, non-English-language programming, religious channels, old films, breaking news, and talk shows. Too many studies have zeroed in on one or other format and then taken it as representative of the whole. Here we will try to engage with the spectrum, although space and available research will put most of the focus on whites and blacks in mainstream television news and entertainment.

Historically, as MacDonald has shown, U.S. television perpetuated U.S. cinema, radio, theatre and other forms of public communication and announced people of color overwhelmingly by their absence. It was not that they were malevolently stereotyped or denounced. They simply did not appear to exist. If they surfaced, it was almost always as wraiths, silent black butlers smiling deferentially, Chicano field-hands laboring sweatily, Indian braves whooping wildly against the march of history. Speaking parts were rare, heavily circumscribed, and typically an abusive distortion of actual modes of speech. But the essence of the problem was virtual non-existence.

Thus, the TV industry collaborated to a marked degree with the segregation that marks the nation, once legally and residentially, now residentially. Programs and advertisements that might have inflamed white opinion in the South

Webster

Brooklyn Bridge

were strenuously avoided, partly in accurate recognition of the militancy of some opinions that might lead to boycotts of advertisers, but partly yielding simply to inertia in defining that potential as a fact of life beyond useful reflection.

The programs shunned were rarely in the slightest degree confrontational, or even suggestive of horrid interracial romance. The classic case was *The Nat "King" Cole Show*, which premiered on NBC in November 1956, and which was eventually taken off for good in December of the following year. A Who's Who of distinguished black as well as white artists and performers virtually gave their services to the show, and NBC strove to keep it alive. But it could not find a national sponsor, at one point having to rely on no less than 30 sponsors in order to be seen nationwide. Cole himself explicitly blamed the advertising agencies' readiness to be intimidated by the White Citizens Councils, the spearhead of resistance to desegregation in southern states.

This was not the only occasion that African Americans were seen on the TV screen in that era. A number of shows, notably *The Ed Sullivan Show*, made a point of inviting black performers on to the screen. Yet entertainment was only one thin slice of the spectrum. Articulate black individuals, such as Paul Robeson, with a clear critique of the racialization of the United States, were systematically excluded from ex-

pressing their opinion on air, in his case on the pretext he was a Communist (and thus apparently deprived of First Amendment protections).

This generalized absence, this univocal whiteness, was first really punctured by TV news coverage of the savage handling of Civil Rights demonstrations in the latter 1950s and early 1960s. Watching police dogs, fire-hoses and billy-clubs unleashed against unarmed and peaceful black demonstrators in Montgomery, Alabama, and seeing white parents—with their own children standing by their side—spewing obscenities and racially charged curses at Dr. King's march through Cicero, Illinois, and hurling rocks at the marchers: these TV news images and narratives may still have portrayed African Americans as largely voiceless victims, but they were nonetheless able to communicate their dignity under fire, whereas their white persecutors communicated their own monstrous inhumanity. The same story repeated itself in the school desegregation riots in New Orleans in 1964 and Boston in 1974.

U.S. television since then made sporadic attempts to address these particular white-black issues, with such shows as *Roots*, *The Cosby Show*, and *Eyes on the Prize*, and through a proliferation of black newscasters at the local level, but all the while cleaving steadfastly to three traditions. These are, firstly, the continuing virtual invisibility of Latinos, Native Americans, and Asian Americans. Indeed, some studies indicate that for decades Latinos have hovered around 1 to 2% of characters in TV drama, very substantially less than their percentage of the public. Hamamoto (1994), similarly, charges that "By and large, TV Asians are inserted in programs chiefly as semantic markers that reflect upon and reveal telling aspects of the Euro-American characters." Secondly, the tradition of color-segregating entertainment changed but little. Even though from the latter 1980s black shows began to multiply considerably, casts have generally been white or black (and never Latino, Native or Asian). Thirdly, the few minority roles in dramatic TV have frequently been of criminals and drug addicts. This pattern has intensively reinforced, and seemingly been reinforced by, the similar racial stereotyping common in "reality TV" police shows and local TV news programs. The standard alternative role for African Americans has been comic actor (or stand-up comic in comedy shows). Ramírez-Berg (1990), commenting upon the wider cinematic tradition of Latino portrayal, has identified the bandit/greaser, the mixed-race slut, the buffoon (male and female), the Latin lover and the alluring Dark Lady, as six hackneyed tropes. (If Latinos are given more TV space, will the first phase merely privilege the audience with negative roles in a wider spectrum?) Let us examine, however, some prominent exceptions.

Roots (1977; *Roots: The Next Generations*, 1979) confounded the TV industry's prior expectations, with up to 140 million viewers for all or part of it, and over 100 million for the second series. For the first time on U.S. television some of the realities of slavery—brutality, rape, enforced de-culturation—were confronted over a protracted period, and through individual characters with whom, as they fought to escape or survive, the audience could identify. Against this historic first was the individualistic focus on screenwriter Alex Haley's determined family, presented as "immigrant-times-ten" fighting an exceptionally painful way over its generations toward the American Dream myth of all U.S. immigrants. Against it, too, was the emphasis on the centuries and decades before the 1970s, which the ahistorical vector in U.S. culture easily cushions from application to the often devastating here and now. Nonetheless, it was a signal achievement.

The Cosby Show (1984–92) was the next milestone. Again defeating industry expectations, the series scored exceptionally high continuing ratings right across the nation. The show attracted a certain volume of hostile comment, some of it smugly supercilious. The fact it was popular with white audiences in the South, and in South Africa, was a favorite quick shot to try to debunk it. Some critics claimed it fed the mirage that racial injustice could be overcome through individual economic advance, others that it primly fostered Reaganite conservative family values. Both were indeed easily possible readings of the show within contemporary U.S. culture. Yet critics often seemed to think a TV text could actually present a single monolithic meaningfulness or set up a firewall against inappropriate readings.

Most critics missed the oasis-in-a-desert dimension of the series for black viewers, representing a functional black family quietly confident in being black. The critics appeared oblivious of the lilywhite wasteland that had preceded the show. Most also missed the gate-opening function of *The Cosby Show*. Their eyes were seemingly so set on an overnight revolution in TV's racial discourses that they could not acknowledge the pivotal difference made within the industry by a show that combined being financially successful and never demeaning African Americans.

Herman Gray, one of the few critics to acknowledge this industrial role of the show in opening the gate to a large number of black television shows and to new professional experience and openings for many black media artists, is also correct in characterizing *The Cosby Show* as assimilationist. It hardly ever directly raised issues of social equity, except in interpersonal gender relations. Nonetheless, in the context of the nation's and the industry's history, the show could have been exquisitely correct—and never once have hit the screen.

By way of response to Gray's reading, two further complicating dimensions are worth comment. The new job-openings were valuable, but were often in the gang-exploitation genre of *New Jack City*. One step forward, one to the side. And "assimilation"—in the sense of showing how African Americans share many values common across the United States—is still a novel message, Lesson 1, to far too many of the majority. In turn, the African American specificities that continue to contribute so much to the nation are Lessons 2 and following for many citizens of every ethnic background.

Eyes on the Prize (1987; 1990) is much more straightforward to discuss. A brilliant documentary series on the American Civil Rights Movements from 1954 to 1985, it too marked a huge watershed in U.S. television history. Partly, its achievement was to bring together historical footage with movement participants, some very elderly, who could supply living oral history. Partly, too, its achievement was that producer-director Henry Hampton consistently included in the narrative the voices of segregationist foes of the movement, on the ground that the story was theirs too. This gave the opportunity for self-reflexion within the white audience rather than easy self-distancing.

However, the series was on PBS and thus never drew the kind of audience *Roots* did. The public appetite for documentaries was also at something of a low toward the end of the century, as opposed to Europe and Russia, where the documentary form was much more popular. *Eyes'* influence would be bound to be slower, though significant, through video rentals and college courses. Its primary significance for present purposes is its demonstration of what could be done televisually, but was never contemplated to be undertaken by the commercial TV companies.

In 1996, PBS screened a similar four-part series, *Chicano!*, by documentarist Hector Galan on the Chicano social movements in the southwest, a story much less known even than the civil rights movements.

These then were turning points, not in the sense of an instantaneous switch, but in terms of setting a high water mark that expanded the definition of the possible in U.S. TV. The other turning point was the proliferation, mostly locally, of black and other ethnic minority group individuals as newscasters. Although newscasters rarely had the clout to write their own bulletin scripts, let alone decide on news priorities for reporting or investigation, they had the cachet of a very public, trusted role. To that extent, this development did carry considerable symbolic prestige for the individuals concerned.

Only as time went on and racial news values and priorities remained the same or similar despite the change in faces, did the limits of this development begin to become more apparent. At about the same time, most news bulletins, especially locally, were deteriorating into infotainment, with lengthy weather and sports reports incorporated into the half hour. Perhaps television news over the longer term will be increasingly vacated of its traditional significance in the United States, and will become more a reaffirmation of community and localism, with ethnic minority newscasters as a rather indeterminate entity within the endeavor.

Alternative Representations

Alternative representation became somewhat more frequent after *The Cosby Show*'s success. In part this change was also due to the steadily declining price of video-cameras and editing equipment, to support from federal and state arts commissions, and to developments in cable TV, especially public access, which opened up more scope for independent video-makers to develop their own work, some of which could be screened locally and even nationally.

MacDonald, however, goes so far as to forecast cable TV's multiple channels as an almost automatic technical solution to the heritage of unequal access for African Americans. The "technological fix" he envisages would not of itself address the urgent national need for dialogue on race and whiteness in television's public forum. Nor does it seem to bargain with the huge costs of generating mostly new product for even a single cable channel.

All in all, though, the emergence of a variety of shows such as *Frank's Place*, *A Different World*, *In Living Color*, and of cable and UHF channels such as Black Entertainment Television (BET), Univision, and Telemundo, together with leased ethnic group program-slots in metropolitan areas, did begin to change the standard white face of television at the margins, even though the norm remained.

These new developments were often contradictory. The often cheap-shot satirization of racial issues on *In Living Color*, the question Gray and others raise concerning BET programming as often simply a black reproduction of white televisual tropes, the role of black sitcoms and stand-up comics as a new version of an older tradition in which blackness is acceptable as farce, are all conflicted examples.

Another contradictory example is Univision, effectively dominated by Mexico's near-monopoly TV giant Televisa.

Its entertainment programs are mostly a secondary market for Televisa's products, and while they are certainly popular, they have had little direct echo of Chicano or other Latino life in the United States. Its news programs have been dominated by Cuban political expatriates, whose obsession with the Castro regime and whose frequent avoidance of Chicanos and Mexican issues have often raised hackles within the largest Latino group. At the same time, as Rodríguez (1996) has shown, Univision's news program has cultivated—for commercial reasons of mass appeal—a pan-ethnic Spanish that over time may arguably contribute to a pan-Latino U.S. cultural identity, rather than the Chicano, Caribbean, Central and South American fragments that constitute the Latino minority.

It is difficult to summarize a sense for the profusion of single features and documentaries, either generated by video-artists of color, or on ethnic themes, scattered as they are over multiple tiny distributors or self-distributed. Suffice it to say that distribution, cable channels notwithstanding, is the hugest single problem that such work encounters. (Sources of information on these videos include Asian-American CineVision, the Black Filmmakers Foundation, and National Video Resources, all in New York City, and Facets Video in Chicago.)

In examining alternatives, finally, we need to take stock of some of the mainstream alternatives to segregated casts, such as one of the earliest, *Hawaii Five-O*, and the later *Miami Vice* and *NYPD Blue*. The first was definitely still within the Tonto tradition insofar as the ethnic minority cops were concerned ("Yes boss" seemed to be the limit of their vocabulary). *Miami Vice*'s tri-ethnic leads were less anchored in that tradition, although Edward James Olmos as the police captain often approximated Captain Dobey in *Starsky and Hutch*, apparently only nominally in charge. *NYPD Blue* carried over some of that tradition as regarded the African American lieutenant's role, but actually starred Latinos in two of the three key police roles in the second series. (One was played by an Italian American, in a continuing variation on "blackface" seemingly popular with casting directors.) A central issue, however, raised once more the question of "modern" racism. A repetitive feature of the show was the skill of the police detectives in pressuring people they considered guilty to sign confessions and not to avail themselves of their legal rights.

Two comments are in order. One is that a police team is shown at work, undeflected by racial animosity, strenuously task-driven. It is a theme with its roots in many World War II movies, though in them ethnicity was generally the focus rather than race. The inference plainly to be drawn was that atavistic biases should be laid aside in the face of clear and present danger, with the contemporary "war" being against the constant tide of crime.

A second issue is that a vastly disproportionate number of prisoners, in relation to their percentage of the nation, are African Americans and Latinos. On *NYPD Blue* we see firm unity among white, black and Latino police professionals in defining aggressive detection and charge practices as legitimate and essential, even though it is procedures like those that, along with racially differential sentencing and parole procedures, have often helped create that huge imbalance in U.S. jails. A war is on, and hard-headed, loyal cops, regardless of their race, in the firing line know it.

Within the paradigm of "modern" racism, co-opting ethnic minority individuals into police work made a great deal of sense (the security industry was living proof). Any TV reference was extremely rare to the fierce racial tensions often seething between police officers. How much had changed? It was like the energetically gyrating multi-racial perpetual-party dancers of MTV: a heavily sugared carapace clamped on a very sour reality.

The Television Industry and Race Relations

Except for a clutch of public figures led by Bill Cosby, CNN's Bernard Shaw, talk-show hosts Oprah Winfrey and Geraldo, and moderately influential behind-the-camera individuals such as Susan Fales, Charles Floyd Johnson, and Suzanne de Passe, and local newscasters, the racial casting of television organizations has been distinctly leisurely in changing. Cable television has the strongest ratio of minority personnel, but this should be read in connection with its lower pay-scales and its minimal original production schedules. Especially in positions of senior authority, television is still largely a white enterprise.

The Federal Communications Commission's (FCC) statistics are often less than helpful in determining the true picture, and represent a classic instance of bureaucratic response to the demand to collect evidence by refusing to focus with any precision on the matter in hand. According to the FCC's 1994 figures the two seeming top categories (Officials and Managers, and Professionals) showed percentages of 13.5 and 18.5 ethnic minority employees. Moderately encouraging it would seem in the latter case, against a national percentage of around 25% people of color, until the tiresome question is posed as to what roles are covered by those categories. At that juncture, fog descends. Only at the time of writing is more careful research, sponsored by the Radio and Television News Directors Association, about to delve into those ragbag categories and disaggregate them.

What can be said from FCC statistics is that Sales Workers positions were only 13.3% occupied by people of color as of 1994. This sounds rather a low-level job, until it is recalled that this is the prime category from which commercial station managers are recruited. Public TV had 11.1% for that year. The statistic does not bode well for the future. By contrast, the Laborers category in commercial TV in 1994 was 56.1% non-majority.

NTIA data show that ownership of commercial TV stations was in ethnic minority hands in just 31 out of 1155 cases across the United States in 1994. Six were in California, six in Texas, three in Michigan, two in Illinois, and one each in Colorado, Florida, Louisiana, Maine, Minnesota, Mississippi, Missouri, New Mexico, New York, Oregon, Pennsylvania, Virginia, Washington, D.C., and Wisconsin.

The question then at issue is how far this absence from positions of TV authority determines the mainstream representation patterns surveyed above. Abstractly conceived, if no customary formats or tropes were changed, and none of the legal, financial and competitive vectors vanished, a television executive stratum composed entirely of ethnic minority individuals would likely proceed to reproduce precisely the same patterns.

But this is abstract, and only helps to shed light on the pressures to conform faced by the few ethnic minority individuals scattered through the TV hierarchy. Sociologically, were their executive numbers to increase even to within hailing distance of their percentage of the nation, a much wider internal dialogue would be feasible concerning the very limits of the possible in television. We come back, in a sense, to Cosby.

Since the proportion of black and Latino viewers was higher than the national average, and since between them they accounted in 1995 for at least $300 billion consumer spending a year, the economic logic of advertising by the mid-1990s seemed to point toward increasing inclusiveness in TV. How this clash between economic logic and inherited culture would work out remained to be seen.

Audience and Spectatorship

We come to the most complex question of all, namely how viewers process televisual content related to race and ethnicity. It has already been argued that decades of daily programs have mostly underwritten the perception of the United States as at core a white nation with a white culture, rather than a pluricultural nation beset by entrenched problems of ethnic inequity. Television fare has obviously not been a lone voice in this regard; nor has it been anything resembling a steady opposition voice. This judgment obviously transcends interpretations of particular programs or even genres. It is sufficiently loose in formulation to leave its plausible practical consequences open to extended discussion. Yet given the ever greater dominance of television in U.S. culture, TV's basic vision of the world can hardly be dismissed as impotent.

It was a vision likely to reassure the white majority that it had little to learn or benefit from people of color. Rather, TV coverage of immigration and crime made it much easier to be afraid of them. George Bush's manipulation of the Willie Horton case for his 1988 campaign commercial had even the nation's vice president and president-to-be drawing on, and thus endorsing, the standard tropes of local TV news.

Naturally, not all of the white majority were to be found clicked into position behind that vision. However, it was ever harder to muster a coherent and forward-looking public debate about race, whiteness and the nation's future, given TV's continuing refusal, in the main, to step up to the plate. It was not the only agency with that responsibility, nor the unique forum available. But TV was and is crucial to any solution. Should the conclusion be that TV's dependence on so many other national forces—advertisers, corporations, government—have reduced its generals to the power-level of Robert Burns' "wee, cowering, timorous" fieldmouse?

The detailed analysis of audience reception of particular shows or series is a delicate business, linking as it will into the many filaments of social and cultural life for white audiences and for audiences of color. It is, though, a sour comment on audience researchers that so little has been done to date to explore how TV is appropriated by various ethnic minority audiences, or how majority audiences handle ethnic themes. Commercial research has been content simply to register viewer levels by ethnicity; academic research, with a scatter of exceptions has rarely troubled to explore ethnic diversity in processing TV, despite the outpouring of ethnographic audience studies in the 1980s and 1990s. Truly, as W.E.B. DuBois forecast in 1903, the color line has been the problem of the twentieth century.

—John D.H. Downing

FURTHER READING

Abernathy-Lear, Gloria. "African Americans' Criticisms Concerning African American Representations on Daytime Serials." *Journalism Quarterly* (Urbana, Illinois), 1994.

Bobo, Jacqueline. *Black Women as Cultural Readers.* New York: Columbia University Press, 1995.

Campbell, Christopher P. *Race, Myth and the News.* Thousand Oaks, California: Sage, 1995

Corea, Ash. 1995. "Racism and the American Way of Media." In, Downing, John, Ali Mohammadi, and Annabelle Sreberny-Mohammadi, editors. *Questioning the Media: A Critical Introduction.* Thousand Oaks, California: Sage, 1990; 2nd edition, 1995.

Cosby, Camille O. *Television's Imageable Influences: The Self-Perceptions of Young African Americans.* Lanham, Maryland: University Press of America, 1994.

Dates, Jannette L., and William Barlow, editors. *Split Image: African Americans in the Mass Media.* Washington, D.C.: Howard University Press, 1993.

Downing, John. "The Cosby Show and American Racial Discourse." In, Van Dijk, Teun A., and Geneva Smitherman-Donaldson, editors. *Discourse and Discrimination.* Detroit, Michigan: Wayne State University Press, 1988.

Drinnon, Richard. *Facing West: The Metaphysics of Indian-Hating and Empire-Building.* Minneapolis: University of Minnesota Press, 1980.

Entman, Robert. "Modern Racism and the Images of Blacks in Local Television News." *Critical Studies in Mass Communication* (Annandale, Virginia), 1990.

Essed, Philomena. *Understanding Everyday Racism: An Interdisciplinary Theory.* Thousand Oaks, California: Sage, 1991.

Fiske, John. *Media Matters: Everyday Culture and Political Change.* Minneapolis: University of Minnesota Press, 1994.

Gray, Herman. *Watching Race: Television and the Struggle for "Blackness."* Minneapolis: University of Minnesota Press, 1995.

Hamamoto, Darrell Y. *Monitored Peril: Asian Americans and the Politics of TV Representation.* Minneapolis: University of Minnesota Press, 1994.

hooks, bell. *Black Looks: Race and Representation.* Boston: South End Press, 1992.

Jhally, Sut, and Justin Lewis. *Enlightened Racism: The Cosby Show, Audiences, and the Myth of the American Dream.* Boulder, Colorado: Westview, 1992.

MacDonald, J. Fred. *Blacks and White TV: Afro Americans in Television Since 1948.* Chicago: Nelson-Hall, 1983; 2nd edition, 1992.

Navarrete, Lisa, and Charles Kamasaki. *Out of the Picture: Hispanics in the Media.* Washington D.C.: National Council of La Raza, 1994.

Ramírez-Berg, Charles. "Stereotyping in Films in General and of the Hispanic in Particular." *The Howard Journal of Communication* (Washington, D.C.), 1990.

Rodríguez, America. "Objectivity and Ethnicity in the Production of the Noticiero Univision." *Critical Studies in Mass Communication* (Annandale, Virginia), 1996.

West, Cornel. *Race Matters.* Boston: Beacon Press, 1993.

See also Allen, Debbie; *Amen; Amos 'n' Andy;* Berg, Gertrude; *Beulah,* Black Entertainment Network; Cosby, Bill; *Cosby Show, Different World, A; Ed Sullivan Show; Eyes on the Prize;* Family on Television; *Frank's Place; Flip Wilson Show; Goldbergs; Good Times;* Haley, Alex; Hemsley, Sherman; Hooks, Benjamin Lawson; *I Spy; Jeffersons; Julia;* Music on Television; *Nat "King" Cole Show;* National Asian Americans in Telecommunications Association; Parker, Everett C.; Pryor, Richard; Reid, Tim; Riggs, Marlon; *Room 222; Roots; 227;* Social Class and Television; Telemundo; Univision; Waters, Ethel; Wilson, Flip; Winfrey, Oprah; *Women of Brewster Place, Zorro*

RADIO CORPORATION OF AMERICA

U.S. Radio Company

In 1919, General Electric (GE) formed a privately owned corporation to acquire the assets of the wireless radio company American Marconi from British Marconi. The organization, known as the Radio Corporation of America (RCA), was formally incorporated on 17 October of that year. Shortly thereafter, American Telephone and Telegraph (AT and T) and Westinghouse acquired RCA assets and became joint owners of RCA. In 1926, RCA formed a new company, the National Broadcasting Company (NBC), to oversee operation of radio stations owned by RCA, General Electric, Westinghouse and AT and T.

In the early 1930s, the Justice Department filed an antitrust suit against the company. In a 1932 consent decree, the organization's operations were separated and GE, AT and T, and Westinghouse were forced to sell their interests in the company. RCA retained its patents and full ownership of NBC. Shortly after becoming an independent company, RCA moved into new headquarters in the Rockefeller Center complex in New York City, into what later became known as Radio City.

While other American companies were cutting back on research expenditures during the depression years, David Sarnoff, president of RCA since 1930, was a staunch advocate of technological innovation. He expanded RCA's technology research division, devoting increased resources to television technology. Television pioneer Vladimir Zworykin was placed in charge of RCA's television research division. RCA acquired competing and secondary patents related to television technology, and once the organization felt that the technology had attained an appropriate level of refinement, it pushed for commercialization of the new medium.

In 1938, RCA persuaded the Radio Manufacturers Association (RMA) to consider adoption of its television system for standardization. The RMA adopted the RCA version, a 441-line, 30-pictures-per-second system, and presented the new standard to the FCC on 10 September 1938. Upon the recommendation of the RMA, the Federal Communications Commission (FCC) scheduled formal hearings to address the adoption of standards. The hearings, however, did not take place until January 1940.

In the interim, RCA began production of receivers and initiated a limited schedule of television programming from the New York transmitters of the National Broadcasting Corporation (NBC) basing their service upon the RMA-RCA standards. The service was inaugurated in conjunction with the opening of the New York World's Fair on 30 April 1939 and continued throughout the year. At the commission's hearing addressing standards on 15 January 1940, opposition to the proposed RMA standards emerged. The two strongest opponents of the standard were DuMont Laboratories and Philco Radio and Television. One of the criticisms voiced by both organizations was the assertion that the 441-line standard did not provide sufficient visual detail and definition. Given the lack of a clear industry consensus, the Commission did not act on the proposed RMA standards.

Despite the absence of official approval, RCA continued to employ the RMA standards and announced plans in early 1940 to increase production of television receivers, cut the price to consumers by one-third, and double their programming schedule. While some commentators saw this as a reasonable and progressive action, the Commission perceived it as a step towards prematurely freezing the standards in place, and as a consequence, scheduled another set of public hearings for 8 April 1940. At these hearings, opponents argued that the action taken by RCA was stifling research and development into other alternative standards. As a result of the hearings, the Commission eliminated commercial broadcasting until further development and refinement had transpired. Furthermore, the Commission asserted that commercialization of broadcasting would not be permitted until there was industry consensus and agreement on one common system. To marshal industry wide support for a single standard, the RMA formed the National Television System Committee (NTSC). The NTSC standards, a 525-line, 60-fields-per-second system, were approved by the FCC in 1941.

Several years later, RCA also became a major participant in the establishment of color television standards. In 1949, the organization proposed to the FCC that its dot sequential color system, which was compatible with existing black and white receivers, be adopted as the new color standard. Citing shortcomings in the compatible systems offered by RCA and other organizations, the FCC opted to formally adopt an incompatible color system offered by the Columbia Broadcasting System as the color standard. RCA appealed this decision all the way to the Supreme Court, while simultaneously refining their color system. A second NTSC was formed to examine the color issue. In 1953, the FCC reversed itself and endorsed a modified version of the RCA dot sequential system compatible color system offered by the NTSC.

In the 1950s, RCA continued the military and defense work in which it had been heavily engaged during World War II. In the late 1950s and early 1960s, the company became involved with both satellite technology and the space program. During the 1960s, RCA began to diversify as the company acquired such disparate entities as the publishing firm Random House, and the car rental company Hertz. Throughout the 1970s and early 1980s, RCA began to divest itself of many of its acquired subsidiaries. In June 1986, RCA was acquired by General Electric, the organization that had originally established it as a subsidiary. GE retained the brand name RCA, established NBC as a relatively autonomous unit, and combined the remainder of RCA's businesses with GE operations.

—David F. Donnelly

FURTHER READING

Barnum, Frederick O. *"His Master's Voice" in America: Ninety Years of Communications Pioneering and Progress.*. Camden, N.J.: Victor Talking Machine Company, Radio Corporation of America, and General Electric, 1991

Bilby, Kenneth M. *The General: David Sarnoff and the Rise of the Communications Industry.* New York: Harper and Row, 1986.

Graham, M. *RCA and the VideoDisc: The Business of Research* Cambridge: Cambridge University Press, 1986.

Lewis, Thomas S. W. *Empire of the Air: The Men Who Made Radio.* New York: Edward Burlingame Books, 1991.

Lyons, Eugene. *David Sarnoff, A Biography.* New York: Harper and Row, 1966.

Sarnoff, David. *Looking Ahead; The Papers of David Sarnoff.* New York: Wisdom Society for the Advancement of Knowledge, Learning, and Research in Education, 1968

The Wisdom of Sarnoff and the World of RCA. Beverly Hills: Wisdom Society for the Advancement of Knowledge, Learning, and Research in Education, 1967.

See also National Broadcasting Company; Sarnoff, David; Sarnoff, Robert; Silverman, Fred; Tartikoff, Brandon; Tinker, Grant; United States: Networks

RADIO TELEVISION NEWS DIRECTORS ASSOCIATION

U.S. Professional Organization

RTNDA (Radio Television News Directors Association) is the trade organization representing broadcast news professionals in the United States. Founded in 1946 when radio was the dominant broadcast news medium, the association now serves all electronic media, with the bulk of its membership comprised of local television news professionals. Its primary focus is on the needs of broadcast news managers; while membership is open to all electronic journalists as well as students, educators, suppliers, and other interested parties, only members who exercise significant editorial supervision of news programming are allowed to vote.

Among the organization's services to members are a monthly magazine, *RTNDA Communicator*, and an annual convention held in the fall featuring training sessions, notable speakers, technology demonstrations, and an exhibit area for suppliers of news products and services. RTNDA also produces a variety of specialty publications for members, including a weekly fax sheet of late breaking developments, targeted newsletters focussing on such areas as TV production and radio reporting, and a monthly newsletter covering legal issues. Other ongoing member services include a resource catalog of related books and tapes; one-day training sessions held throughout the year in different parts of the country; industry research projects that examine pertinent issues such as salaries, staff size, and profitability; and a biweekly Job Bulletin of available personnel and positions.

The number and scope of RTNDA services reflect the dramatic changes experienced by the broadcast news industry in recent years. Among such developments have been the growing profitability and expansion of local television news; the emergence of new outlets such as Cable News Network, C-SPAN, and online information services; and advances in the technology of news gathering, particularly in live remote broadcast capabilities and satellite transmission. In addition, local TV news operations, unlike their newspaper counterparts, are generally locked in fierce three-way competition with other local news programs in the same market. The pressure to maximize ratings often puts the news manager in the precarious situation of having to decide between news values and entertainment values. The nature of a commercial medium such as television generally makes such conflict unavoidable.

Through its ongoing activities and services, RTNDA strives to set and promote professional standards for electronic journalists. The RTNDA Code of Ethics is published in each issue of the organization's monthly magazine. The code states that "the responsibility of radio and television journalists is to gather and report information of importance and interest to the public accurately, honestly and impartially," and provides guidelines for fair, balanced reporting that respects the dignity and privacy of subjects and sources, avoiding deception, sensationalism, and conflicts of interest.

RTNDA honors professional excellence through its Edward R. Murrow Awards in the areas of spot news coverage, feature reporting, series, investigative reporting, and overall newscast (awarded separately for small and large market stations). The organization's top honor is the Paul White Award, given each year to an individual for lifetime achievement in the field of broadcast journalism. RTNDA also sponsors the Radio Television News Directors Foundation, a nonprofit organization that engages in research, education, and training activities in four principal areas: journalistic ethics, impact of technology on electronic news gathering, the role of electronic journalism in politics and public policy, and cultural diversity in the profession.

—Jerry Hagins

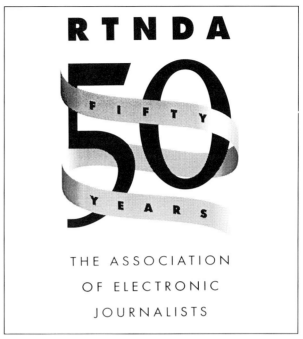

Courtesy of RTNDA

FURTHER READING

Cook, Philip S., Douglas Gomery, and Lawrence W. Lichty. *The Future of News.* Washington, DC: Woodrow Wilson Center Press, 1992.

Fields, Howard. "RTNDA at 40: Major Lobbying Role." *Television-Radio Age* (New York), 18 August 1986.

Jacobs, Jerry. *Changing Channels.* Mountain View, California: Mayfield Publishing, 1990.

Kaniss, Phyllis. *Making Local News.* Chicago: University of Chicago Press, 1991.

McManus, John H. *Market-Driven Journalism.* Thousand Oaks, California: Sage Publications, 1994.

"RTNDA and the State of Electronic Journalism." *Broadcasting* (Washington, D.C.), 12 December 1988.

See also News, Local and Regional

RANDALL, TONY

U.S. Actor

Tony Randall, an Emmy Award-winning television and film actor, is most noted for his role as the anal-retentive Felix Unger in the ABC sitcom *The Odd Couple*. A popular guest on numerous variety and talk shows, Randall has been connected with all three major broadcast networks, as well as with PBS.

Randall began his career in radio in the 1940s, appearing on such shows as the *Henry Morgan Program* and *Opera Quiz*. From 1950-52, Randall played Mac on the melodramatic TV serial *One Man's Family*. He then went on to play Harvey Weskit, the brash, overconfident best friend of Robinson Peepers (Wally Cox) in the live sitcom *Mr. Peepers* (1952-1955). After finding a niche in films, including numerous roles in romantic comedies, Randall won the part of Felix Unger in the ABC television version of *The Odd Couple* (1970-75).

Although the Broadway and film versions of *The Odd Couple* became established hits with different stars (Randall, however, did play Felix in a Chicago production), Randall lent numerous additions to the Felix character. Drawing upon his interest in opera, Randall had Felix become an opera lover. Randall also added the comedic honking noises that accompanied Felix's ever-present sinus attacks. Much like Jack Klugman's close connection to the Oscar Madison role, Randall became synonymous with Unger.

Despite low ratings for the series, ABC, the third-place network, allowed *The Odd Couple* a five season run. In 1975, Randall won an Emmy as lead actor for his role as Felix. A popular guest on numerous variety shows, Randall was present on two Emmy Award-winning variety show episodes in 1970 (*The Flip Wilson Show*) and 1971 (*The Sonny and Cher Show*). Randall's frequent appearances as a guest on the *Tonight Show* won him a role playing himself in Martin Scorsese's *King of Comedy* (1983).

Beginning in 1976, Randall starred in the CBS sitcom *The Tony Randall Show*. Randall played Walter Franklin, a judge who deliberated over his troubled family as much as he did over the cases presented to him in his mythical Philadelphia courtroom. In 1981, Randall returned to television playing Sidney Shorr in NBC's *Love, Sidney*, a critically-acclaimed yet commercially unsuccessful sitcom canceled in 1983. The series did attract some criticism from the religious and culturally conservative communities. In *Sidney Shorr*, the made-for-television movie which preceded the series, Randall's character was presented as homosexual. In the series this was simply dropped.

Randall reprised his Felix Unger role in a 1993 TV-movie version of *The Odd Couple*. He has also hosted the PBS opera series *Live from the Met*.

—Michael B. Kassel

TONY RANDALL. Born Leonard Rosenberg in Tulsa, Oklahoma, U.S.A., 26 February 1920. Educated at Northwestern University, Chicago; Columbia University, New York; the Neighborhood Playhouse School of the Theatre, New York City, 1938-40; and the Officer Candidate School at Fort Monmouth, New Jersey. Married: 1) Florence Gibbs (died, 1992); 2) Heather Harlan, 1995; one child. Served as private and first lieutenant, U.S. Army Signal Corps, 1942-46. Announcer and actor in radio soap operas; New York debut as stage actor, *A Circle of Chalk*, 1941; various theater and radio work, 1947-52; television actor, from 1952; continues to star in film, on stage, and on television. Member: Actors' Equity Association; Screen Actors Guild; American Federation of Television and Radio Artists; Association of the Metropolitan Opera Company; founder and artistic director of the National Actors' Theatre in New York City. Recipient: Emmy Award, 1975. Address: c/o National Actors' Theatre, 1560 Broadway, Suite 409, New York City, New York 10036, U.S.A.

Tony Randall

TELEVISION SERIES

1949–55	*One Man's Family*
1952–55	*Mr. Peepers*
1970–75	*The Odd Couple*
1976–78	*The Tony Randall Show*

1981–82   *Love, Sidney*

MADE-FOR-TELEVISION MOVIES

1978   *Kate Bliss and the Ticker Tape Kid*
1981   *Sidney Shorr: A Girl's Best Friend*
1984   *Off Sides*
1985   *Hitler's SS: Portrait in Evil*
1986   *Sunday Drive*
1988   *Save the Dog*
1989   *The Man in the Brown Suit*
1993   *The Odd Couple: Together Again*

TELEVISION SPECIALS (selection)

1956   *Heaven Will Protect the Working Girl* (host)
1960   *Four for Tonight* (co-star)
1960   *So Help Me, Aphrodite*
1962   *Arsenic and Old Lace*
1967   *The Wide Open Door*
1969   *The Littlest Angel*
1977   *They Said It with Music: Yankee Doodle to Ragtime* (co-host)
1981   *Tony Randall's All-Star Circus* (host)
1985   *Curtain's Up* (host)
1987   *Walt Disney World Celebrity Circus* (host)

FILMS

Oh Men, Oh Women, 1957; *Will Success Spoil Rock Hunter?*, 1957; *The Mating Game*, 1959; *Pillow Talk*, 1959; *Let's Make Love*, 1960; *Lover Come Back*, 1962; *Send Me No Flowers*, 1964; *The Brass Bottle*, 1964; *Fluffy*, 1965; *Bang, Bang, You're Dead*, 1966; *Hello Down There*, 1969; *Everything You Always Wanted to Know about Sex...*, 1972; *Huckleberry Finn*, 1974; *Scavenger Hunt*, 1979; *Foolin' Around*, 1980; *The King of Comedy*, 1983; *My Little Pony*, 1986; *That's Adequate*, 1989; *Gremlins 2: The New Batch* (voice), 1990; *Fatal Instinct*, 1993.

STAGE (selection)

Circle of Chalk, 1941; *Candida*, 1941; *The Corn Is Green*, 1942; *The Barretts of Wimpole Street*, 1947; *Anthony and Cleopatra*, 1948; *Caesar and Cleopatra*, 1950; *Oh Men, Oh Women*, 1954; *Inherit the Wind*, 1955-56; *Oh Captain*, 1958; *UTBU*, 1966; *Two Into One*, 1988; *M. Butterfly*, 1989; *A Little Hotel on the Side*, 1992; *Three Men on a Horse*, 1993; *The Government Inspector*, 1994; *The Odd Couple*, 1994.

RADIO

I Love a Mystery; Portia Faces Life; When a Girl Marries; Life's True Story.

PUBLICATION

Which Reminds Me, with Michael Mindlin. New York: Delacorte Press, 1989.

See also *Odd Couple*

RATHER, DAN

U.S. Broadcast Journalist

In a career in journalism that is now in its fifth decade, Dan Rather has established himself as a crucial figure in broadcast news. Anchor of the *CBS Evening News* since 1981, Rather has enjoyed a long and sometimes colorful career in broadcasting. Rather has interviewed every United States President from Dwight D. Eisenhower to Bill Clinton, and international leaders from Nelson Mandela to Boris Yeltsin. In 1990, he was the first American journalist to interview Saddam Hussein after Iraq's invasion of Kuwait. Rather's hard-hitting journalistic style has sometimes been as much discussed as the content of his reporting, particularly in the case of well-publicized contretemps with Richard Nixon and George Bush.

Rather began his career in journalism in 1950 as an Associated Press reporter in Huntsville, Texas. He subsequently worked as a reporter for United Press International, for KSAM Radio in Huntsville, for KTRH Radio in Houston, and at the *Houston Chronicle*. He became news director of KTRH in 1956 and a reporter for KTRH-TV in Houston in 1959. He was news director at KHOU-TV, the CBS affiliate in Houston, before joining CBS News in 1962 as chief of the southwest bureau in Dallas.

In 1963, Rather was appointed chief of CBS' southern bureau in New Orleans, responsible for coverage of news events in the South, Southwest, Mexico and Central America. He reported extensively on southern racial strife, becoming well acquainted with Dr. Martin Luther King, Jr. On 22 November 1963 in Dallas, Rather broke the news of the death of President John F. Kennedy. A few weeks after the assassination, he became CBS' White House correspondent.

Rather attracted notice in 1974 for an exchange with Richard Nixon. At a National Association of Broadcasters convention in Houston, Rather was applauded when he stood to ask a question, drawing Nixon's query, "Are you running for something?" Many saw Rather's quick retort, "No, sir, Mr. President. Are you?" as an affront to Presidential dignity.

A year later, Rather was selected to join the roster of journalists on CBS' *60 Minutes*, and in 1981, after lengthy negotiations with the network, Rather became the successor to Walter Cronkite, anchoring the *CBS Evening News*. During Rather's tenure, he has sometimes been associated with striking, even bizarre, moments of news coverage. For one week in September 1986, Rather concluded his nightly broadcast with the solemn, ominous-sounding, single-word sign-off "Courage." The line, seen as an attempt to respond to or replace audience familiarity with Cronkite's "And that's the way it is," attracted widespread media coverage and more than a little satire. In October 1986, Rather was attacked outside the CBS building by thugs reportedly demanding "What's the frequency, Kenneth?", and he subsequently appeared on the air with a swollen and bruised face. In September 1987, Rather walked off the *CBS Evening News* set in protest over the network's decision to allow U.S. Open tennis coverage to cut into the broadcast. His action on this occasion left CBS with a blank screen for more than six minutes. This moment was recalled in an explosive live interview Rather conducted with Vice President George Bush in January 1988. When Rather pressed Bush about his contradictory claims regarding his involvement in the Iran-Contra scandal, the vice president responded by asking Rather if he would like to be judged by those minutes resulting from his decision to walk off the air.

Connie Chung joined Rather on the *CBS Evening News* in a dual anchor format in 1993 amid constant speculation that he did not approve of the appointment. When Chung left the *Evening News* spot in 1995, he did not seem displeased. Rather also continues to anchor and report for the CBS News broadcast *48 Hours* (which premiered in 1988). He was the first network journalist to anchor an evening news broadcast and a prime-time news program at the same time, a practice which has since been adopted by other networks.

Rather's career reflects the passing of the era in which one anchor, Walter Cronkite, was unproblematically "the most trusted man in America." Along with Tom Brokaw and Peter Jennings, Rather is one of a triumvirate of middle-aged white male anchors who dominate the U.S. national nightly news. The three network news broadcasts continue to be locked in a tightly contested ratings race, and these highly paid anchors are decidedly valuable properties, the "stars" of television news.

—Diane M. Negra

DAN RATHER. Born in Wharton, Texas, U.S.A., 31 October 1931. Educated at Sam Houston State College, Huntsville, Texas, B.A. in journalism 1953; attended University of Houston and South Texas School of Law. Married: Jean Goebel; children: Dawn Robin and Daniel Martin. Journalism instructor, Sam Houston State College; worked for the *Houston Chronicle*; news writer, reporter, and news director, CBS radio affiliate KTRH, Houston, mid-late 1950s; director of news and public affairs, CBS television affiliate KHOU, Houston, late 1950s to1961; chief, CBS' southwestern bureau, Dallas, 1962–64; CBS White House correspondent, 1963; chief, CBS' London bureau, 1965–66; war correspondent, Vietnam, 1966; returned to position as CBS White House correspondent, 1966–74; anchor-correspondent, *CBS Reports*, 1974–75; correspondent and co-editor, *60 Minutes*, 1975–81; anchor, *Dan Rather Reporting*, CBS Radio Network, since 1977; anchor and managing editor, *CBS Evening News with Dan Rather*, since 1981; anchor, *48*

Dan Rather

Hours, since 1988; anchored numerous CBS news specials. Recipient: Texas Associated Press Broadcasters' Awards for spot news coverage, 1956, 1959; numerous Emmy Awards. Address: CBS News, 524 West 57th Street, New York City, New York 10019, U.S.A.

TELEVISION

1974–75	*CBS Reports*
1975–81	*60 Minutes*
1981–	*CBS Evening News with Dan Rather*
1988–	*48 Hours*

PUBLICATIONS

The Camera Never Blinks: Adventures of a TV Journalist, with Mickey Herskowitz. New York: Morrow, 1977.

The Camera Never Blinks Twice: Further Adventures of a Television Journalist. New York: Morrow, 1994.

FURTHER READING

Corliss, Richard. "Broadcast Blues." *Film Comment* (New York), March-April, 1988.

Goldberg, Robert, and Gerald Jay Goldberg. *Anchors: Brokaw, Jennings, Rather, and the Evening News*. New York: Birch Lane, 1990.

Jones, Alex S. "The Anchors: Who They Are, What They Do, The Tests They Face." *The New York Times,* 27 July 1986.

Matusow, Barbara. *The Evening Stars.* Boston: Houghton Mifflin, 1983.

Westin, Av. *Newswatch: How TV Decides the News.* New York: Simon and Schuster, 1982.

Zelizer, Barbie. "What's Rather Public About Dan Rather: TV Journalism and the Emergence of Celebrity." *Journal of Popular Film and Television* (Washington, D.C.), Summer 1989.

See also Anchor; Columbia Broadcasting System; News, Network; *60 Minutes*

RATINGS

Ratings are a central component of the television industry, almost a household word. They are important in television because they indicate the size of an audience for specific programs. Networks and stations then set their advertising rates based on the number of viewers of their programs. Network revenue is thus directly related to the ratings. The word "ratings," however, is actually rather confusing because it has both a specific and a general meaning. Specifically, a rating is the percentage of all the people (or households) in a particular location tuned to a particular program. In a general sense, the term is used to describe a process (also referred to as "audience measurement") that endeavors to determine the number and types of viewers watching TV.

One common rating (in the specific sense) is the rating of a national television show. This calculation measures the number of households—out of all the households in the United States that have TV sets—watching a particular show. There are approximately 92.4 million households in the United States and most of them have TV sets. In order to simplify the example, assume that there are 100,000,000 households. If 20,000,000 of them are watching NBC at 8:00 P.M. NBC's rating would be 20 (20,000,000/100,000,000=20). Another way to describe the process is to say that one rating point is worth 1,000,000 households.

Ratings are also taken for areas smaller than the entire nation. For example, if a particular city (Yourtown) has 100,000 households, and 15,000 of them are watching the local news on station KAAA, that station would have a rating of 15. If Yourtown has a population of 300,000 and 30,000 people are watching KAAA, the station's rating would be 10. And because television viewing is becoming less and less of a group activity with the entire family gathered around the living-room TV set, some ratings are expressed in terms of people rather than households.

Many calculations are related to the rating. Sometimes people, even professionals in the television business, confuse them. One of these calculations is the share. This figure reports the percentage of households (or people) watching a show out of all the households (or people) *who have the TV set on.* So if Yourtown has 100,000 households but only 50,000 of them have the TV set on and 15,000 of those are watching KAAA, the share is 30 (15,000/50,000=30). Shares are always higher than ratings unless, of course, everyone in the country is watching television.

Another calculation is the cume, which reflects the number of different persons who tune in a particular station or network over a period of time. This number is used to show advertisers how many different people hear their message if it is aired at different times such as 7:00 P.M., 8:00 P.M., and 9:00 P.M.. If the total number of people available is 100, five of them view at 7:00, those five still view at 8:00, but three new people watch, and then two people turn the TV off, but four new ones join the audience at 9:00, the cume would be 12 (5+3+4=12). Cumes are particularly important to cable networks because their ratings are very low. Two networks with ratings of 1.2 and 1.3 can not really be differentiated, but if the measurement is taken over a wider time span, a greater difference will probably surface.

Average quarter hours (AQH) are another measurement. This calculation is based on the average number of people viewing a particular station (network, program) for at least five minutes during a fifteen-minute period. For example, if, out of 100 people, ten view for at least five minutes between 7:00 and 7:15, seven view between 7:15 and 7:30, eleven view between 7:30 and 7:45, and four view between 7:45 and 8:00, the AQH rating would be 8 (10+7+11+4=32/4=8).

Many other calculations are possible. For example, if the proper data has been collected, it is easy to calculate the percentage of women between the ages of 18 and 34, or of men in urban areas, who watch particular programs. Networks and stations gather as much information as is economically possible. They then try to use the numbers that present their programming strategies in the best light.

The general ratings (audience measurement) process has varied greatly over the years. Audience measurement started in the early 1930s with radio. A group of advertising interests joined together as a non-profit entity to support ratings known as Crossleys, named after Archibald Crossley, the man who conducted them. Crossley used random numbers from telephone directories and called people in about 30 cities to ask them what radio programs they had listened to the day before his call. This method became known as the recall method because people were remembering what they had listened to the previous day. Crossleys existed for about 15 years but ended in 1946 because several for-profit commercial companies began offering similar services that were considered better.

One of these, the Hooper ratings, was begun by C. E. Hooper. Hooper's methodology was similar to Crossley's, except that respondents were asked what programs they were listening to at the time of the call—a method known as the coincidental telephone technique. Another service, the Pulse, used face-to-face interviewing. Interviewees selected by random sampling were asked to name the radio stations they had listened to over the past 24 hours, the past week, and the past five midweek days. If they could not remember, they were shown a roster containing station call letters to aid their memory. This was referred to as the roster-recall method.

Today the main radio audience measurement company is Arbitron. The Arbitron method requires people to keep diaries in which they write down the stations they listen to at various times of the day. In these diaries, they also indicate demographic features—their age, sex, marital status, etc.—so that ratings can be broken down by sub-audiences.

The main television audience measurement company is the A.C. Nielsen company. For many years Nielsen used a combination of diaries and a meter device called the Audimeter. The Audimeter recorded the times when a set was on and the channel to which it was tuned. The diaries were used to collect demographic data and list which family members were watching each program. Nielsen research in some markets still uses diaries, but for most of its data collection, Nielsen now attaches Peoplemeters to TV sets in selected homes. Peoplemeters collect both demographic and channel information because they are equipped with remote control devices. These devices accommodate a number of buttons—one for each person in the household and one for guests. Each person watching TV presses his or her button, which has been programmed with demographic data, to indicate viewing choices and activities.

There are also companies that gather and supply specialized ratings. For example, one company specializes in data concerning news programs and another tracks Latino viewing.

All audience measurement is based on samples. As yet there is no economical way of finding out what every person in the entire country is watching. Diaries, meters, and phone calls are all expensive, so sometimes samples are small. In some cases no more than .004 percent of the population is being surveyed. However, the rating companies try to make their samples as representative of the larger population as possible. They consider a wide variety of demographic features—size of family, sex and age of head of household, access to cable TV, income, education—and try to construct a sample comprising the same percentage of the various demographic traits as in the general population.

In order to select a representative sample, the companies attempt to locate every housing unit in the country (or city or viewing area), mainly by using readily available government census data. Once all the housing units are accounted for, a computer program is used to randomly select the sample group in such a way that each location has an equal chance of being selected. Company representatives then write or phone people in the households that have been

selected trying to secure their cooperation. About 50% of those selected agree to participate. People are slightly more likely to allow meters in their house and to answer questions over the phone than they are to keep diaries. Very little face-to-face interviewing is now conducted because people are reluctant to allow strangers into their houses. When people refuse to cooperate, the computer program selects more households until the number needed for the sample have agreed to volunteer.

Once sample members have agreed to participate, they are often contacted in person. In the case of a diary, someone may show them how to fill it out. In other cases the diary and instructions may simply be sent in the mail. For a meter, a field representative goes to the home (apartment, dorm room, vacation home, etc.) and attaches the meter to the television set. This person must take into account the entire video configuration of the home—multiple TV sets, VCRs, satellite dishes, cable TV, and anything else that might be attached to the receiver set. The field representative also trains family members in the use of the meter.

People participating in audience measurement are usually paid, but only a small amount, such as fifty cents. Ratings companies have found that paying people something makes them feel obligated, but paying them a large amount does not make them more reliable.

Ratings companies try to see that no one remains in the sample very long. Participants become weary of filling out diaries or pushing buttons and cease to take the activities seriously. Soliciting and changing sample members is expensive, however, so companies do keep an eye on the budget when determining how to update the sample.

Once the sample is in order, the data must be collected from the participants. For phone or face-to-face interviews, the interviewer fills in a questionnaire and the data is later entered into a computer. For meters, the data collected is sent over phone lines to a central computer. People keeping diaries mail them back to the company and employees then enter the data into a computer. Usually only about 50% of diaries are useable; the rest are never mailed back or are so incorrectly filled out that they can not be used.

From the data collected and calculated by the computer, ratings companies publish reports. These vary according to what was surveyed. Nielsen covers commercial networks, cable networks, syndicated programming, public broadcasting, and local stations. Other companies cover more limited aspects of television. Reports on each night's prime-time national commercial network programming, based on Nielsen Peoplemeters, are usually ready about twelve hours after the data is collected. It takes considerably longer to generate a report based on diaries. The reports dealing with stations are published less frequently than those for prime-time network TV. Generally station ratings are undertaken four times a year—November, February, May, and July—periods that are often referred to as "Sweeps." The weeks of the Sweeps are very important to local stations because the numbers produced then determine advertising rates for the following three months. Most reports give not only the total ratings and shares but also information broken down into various demographic categories—age, sex, education, income. The various reports are purchased by networks, stations, advertisers, and any other companies with a need to know audience statistics. The cost is lower for small entities, such as TV stations, than for larger entities, such as commercial networks. The latter usually pay several million dollars a year to receive a ratings service.

While current ratings methods may be the best yet devised for calculating audience size and characteristics, audience measurement is far from perfect. Many of the flaws of ratings should be recognized, particularly by those employed in the industry who make significant decisions based on ratings.

Sample size is one aspect of ratings that is frequently questioned in relation to rating accuracy. Statisticians know that the smaller the sample size the more chance there is for error. Ratings companies admit to this and do not claim that their figures are totally accurate. Most of them are only accurate within two or three percent. This was of little concern during the times when ratings primarily centered around three networks, each of which was likely to have a rating of 20 or better. Even if CBS' 20 rating at 8:00 P.M. on Monday was really only 18, this was not likely to disturb the network balance. In all likelihood CBS' 20 rating at 8:00 Tuesday evening was really a 22, so numbers evened out. Now that there are many sources of programming, however, and ratings for each are much lower, statistical inaccuracies are more significant. A cable network with a 2 rating might actually be a 4, an increase that might double its income.

Audience measurement companies are willing to increase sample size, but doing so would greatly increase their costs, and customers for ratings do not seem willing to pay. In fact, Arbitron, which had previously undertaken TV ratings, dropped them in 1994 because they were unprofitable.

As access to interactive communication increases, it may be easier to obtain larger samples. Wires from consumer homes back to cable systems could be used to send information about what each cable TV household is viewing. Many of these wires are already in place. Consumers wishing to order pay-per-view programming, for example, can push a button on the remote control that tells the cable system to unscramble the channel for that particular household. Using this technology to determine what is showing on the TV set at all times, however, smacks of a "Big Brother" type of surveillance. Similarly, by the 1970s a technology existed that enabled trucks to drive along streets and record what was showing on each TV set in the neighborhood. This practice, perceived as an invasion of privacy, was quickly ended.

Sample composition, as well as sample size, is also seen as a weakness in ratings procedures. When telephone numbers are used to draw a sample, households without telephones are excluded and households with more than one phone have a better chance of being included. For many of the rating samples, people who do not speak either English or Spanish are eliminated. Perhaps one of the greatest difficulties for ratings companies is caused by those who eliminate themselves from the sample by refusing to cooperate. Although rating services make every attempt to replace these people with others who are similar in demographic characteristics, the sample's integrity is somewhat downgraded. Even if everyone originally selected agreed to serve, the sample can not be totally representative of a larger population. No two people are alike, and even households with the same income and education level and the same number of children of the same ages do not watch exactly the same television. Moreover, people within the sample, aware that their viewing or listening habits are being monitored, may act differently than they ordinarily do.

Other problems rise from the fact that each rating technique has specific drawbacks. Households with Peoplemeters may suffer from "button pushing fatigue," thereby artificially lowering ratings. Additionally, some groups of people are simply more likely to push buttons than others. When the Peoplemeter was first introduced, sports viewing soared and children's program viewing decreased significantly. One explanation held that men, who were watching sports intently, were very reliable about the button pushing, perhaps, in some cases, out of fear that the TV would shut off if they didn't push that button. Children, on the other hand, were confused or apathetic about the button, therefore underreporting the viewing of children's programming. Another theory held that the women of the household had previously kept the diaries and though not always aware of what their husbands were actually viewing, were much more conscious of what their children were watching. Under the diary system, in this explanation, sports programming was underrated.

But diaries have their own problems. The return rate is low, intensifying the problem of the number of uncooperative people in the sample. Even the diaries that are returned often have missing data. Many people do not fill out the diaries as they watch TV. They wait until the last minute and try to remember details—perhaps aided by a copy of *TV Guide*. Some people are simply not honest about what they watch. Perhaps they do not want to admit to watching a particular type of television or a particular program.

With interviews, people can be influenced by the tone or attitude of the interviewer or, again, they can be less than truthful about what they watched out of embarrassment or in an attempt to project themselves in a favorable light. People are also hesitant to give information over the phone because they fear the person calling is really a sales person.

Beyond sampling and methodological problems, ratings can be subject to technical problems—computers that go down, meters that function improperly, cable TV systems that shift the channel numbers of their program services without notice, station antennas struck by lightning.

Additionally, rating methodologies are often complicated and challenged by technological and sociological change. Videocassette recorders, for example, have presented difficulties for the ratings companies. Generally, programs are counted as being watched if they are recorded. However, many programs that are recorded are never watched, and some are watched several times. In addition, people replaying tape often zip through commercials, destroying the whole purpose of ratings. And ratings companies have yet to decide what to do with sets that show four pictures at once.

Another major deterrent to the accuracy of ratings is that fact that electronic media programmers often try to manipulate the ratings system. Local television stations program their most sensational material during ratings periods. Networks preempt regular series and present star-loaded specials so that their affiliates will fare well in ratings and can therefore adjust their advertising rates upward. Cable networks show new programs as opposed to reruns. All of this, of course, negates the real purpose of determining which electronic media entities have the largest regular audience. It simply indicates which can design the best programming strategy for Sweeps week.

Because of the possibility for all these sampling, methodological, technological, and sociological errors, ratings have been subjected to numerous tests and investigations. In fact, in 1963, the House of Representatives became so skeptical of ratings methodologies that it held hearings to investigate the procedures. Most of the skepticism had arisen because of a cease-and-desist order from the Federal Trade Commission (FTC), requiring several audience measurement companies to stop misrepresenting the accuracy and reliability of their reports. The FTC charged the rating companies with relying on hearsay information, making false claims about the nature of their sample populations, improperly combining and reporting data, failing to account for non-responding sample members, and making arbitrary changes in the rating figures.

The main result of the hearings was that broadcasters themselves established the Electronic Media Rating Council (EMRC) to accredit rating companies. This group periodically checks rating companies to make sure their sample design and implementation meets preset standards that electronic media practitioners have agreed upon, to determine whether or not interviewers are properly trained, to oversee the procedures for handling diaries, and in other ways to assure the ratings companies are compiling their reports as accurately as possible. All the major rating companies have EMRC accreditation.

The EMRC and other research institutions have continued various studies to determine the accuracy of ratings. Some of the findings include: people who cooperate with rating services watch more TV, have larger families, and are younger and better educated than those who will not cooperate; telephone interviewing gets a 13% higher cooperation rate than diaries; Hispanics included in the ratings samples watch less TV and have smaller families than Hispanics in general.

Both electronic media practitioners and audience measurement companies want their ratings to be accurate, so both groups undertake testing to the extent they can afford it. In 1989, for example, broadcasters initiated a study to conduct a thorough review of the Peoplemeter. The result was a list of recommendations to Nielsen that included changing the amount of time people participate from two years to one year to eliminate button pushing fatigue, metering all sets including those on boats and in vacation homes, and simplifying the procedures by which visitors log into the meter.

Still, the weakest link in the system, at present, seems to be how the ratings are used. Networks tout rating superiorities that show .1 percent differences, differences that certainly are not statistically significant. Programs are canceled because their ratings fall one point. Sweeps weeks tend to become more and more sensationalized. At stake, of course, are advertising fees that can translate into millions of dollars. Advertisers and their agencies need to remain vigilant so that they are not paying rates based on artificially stimulated ratings that bear no resemblance to the programs in which the sponsor is actually investing.

At this time all parties in the system seem invested in some form of audience measurement. So long as the failures and inadequacies of these systems are accepted by these major participants, the numbers will remain a valid type of "currency" in the system of television.

—Lynne Schafer Gross

FURTHER READING

Beeville, Hugh Malcolm. *Audience Ratings: Radio, Television, and Cable.* Hillsdale, New Jersey: Erlbaum, 1985; revised edition, 1988.

Buzzard, Karen. *Electronic Media Ratings.* Boston, Massachusetts: Focal, 1992.

Clift, Charles III, and Archie Greer, editors. *Broadcast Programming: The Current Perspective.* Washington, D.C.: University Press of America, 1981.

Dominick, Joseph R., and James E. Fletcher. *Broadcasting Research Methods.* Boston, Massachusetts: Allyn and Bacon, 1985.

Gross, Lynne S. *Telecommunications: An Introduction to Electronic Media.* Madison, Wisconsin: Brown and Benchmark, 1995.

Webster, James G., and Lawrence W. Lichty. *Ratings Analysis: Theory and Practice.* Hillsdale, New Jersey: Erlbaum, 1991.

See also A.C. Nielsen Company; Advertising; Advertising, Company Voice; Cost-Per-Thousand/Cost-Per-Point; Demographics; Market; Nielsen, A.C.; Programming; Share

REAGAN, RONALD

U.S. Actor/Politician

Ronald Reagan lived in the public eye for more than fifty years as an actor and politician. He appeared in fifty-three Hollywood movies, from *Love Is in the Air* (1937) to *The Killers* (1964). Never highly touted as an actor, his most acclaimed movie was *Kings Row* (1942), while his favorite role was as George Gipp in *Knute Rockne—All American* (1940). He served as president of the Screen Actors Guild from 1947 to 1952 and again in 1959 where he led the fight against communist infiltration in the film industry and brokered residual rights for actors.

Reagan made his debut on television 7 December 1950 as a detective on the CBS *Airflyte Theater* adaptation of an Agatha Christie novel. After a dozen appearances over the next four years on various shows, Reagan's big television break came when Taft Schreiber of MCA acquainted him with *General Electric Theater*. Reagan hosted this popular Sunday evening show from 1954 to 1962, starring in thirty-four episodes himself. Reagan was one of the first movie stars to see the potential of television, and, as host, he introduced such Hollywood notables as Joan Crawford, Alan Ladd, and Fred Astaire to their television debuts. He also became a goodwill ambassador for General Electric (G.E.), plugging G.E. products, meeting G.E. executives, and speaking to G.E. employees all over the country. This proved fine training for his future political career as he honed his speaking skills, fashioned his viewpoints, and gained exposure to middle-America.

In 1965, Reagan began a two-season stint as host of *Death Valley Days*, which he had to relinquish when he announced his candidacy for governor of California, in January 1966. During his terms as governor of California (1966–74), Reagan made frequent televised appearances on *Report to the People*.

The hinge between Reagan's acting and political careers swung on a nationally televised speech, "A Time for Choosing," on 27 October 1964. This speech for Barry Goldwater, which David Broder hailed as "the most successful political debut since William Jennings Bryan electrified the 1896 Democratic convention with his 'Cross of Gold' speech," brought in over one million dollars for the Republican candidate and marked the beginning of Reagan's reign as the leading conservative for the next twenty-five years.

By 1980, the year Reagan was elected president for the first of his two terms, more people received their political information from television than from any other source. Reagan's experience as an actor on the screen and on television gave him an enormous advantage as politics moved fully into its television era. His mastery of the television medium earned for him the title, "the great communicator." He perfected the art of "going public," appealing to the American public on television to put pressure on Congress to support his policies. The rhetoric of this "prime-time president" suited television perfectly. Whether delivering a State of the Union address, eulogizing the crew of the *Challenger*, or speaking directly to the nation about

Ronald Reagan

his strategic defense initiative, he captured the audience's attention by appealing to shared values, creating a vision of a better future, telling stories of heroes, evoking memories of a mythic past, exuding a spirit of "can-do" optimism, and converting complex issues into simple language that people could understand and enjoy.

He understood that television is more like the oral tradition committed to narratival communication than like the literate tradition committed to linear, factual communication. As Denton puts it, in video politics "how something is said is more important than what is said." Reagan surmounted his numerous gaffes and factual inaccuracies until the Iran-Contra affair, when it became apparent that his style could not extricate him from the suspicion that he knew more than he was telling the American public.

His administration also greatly expanded the Office of Communication to coordinate White House public relations, stage important announcements, control press conferences, and create visual productions such as *That's America*, shown at the 1984 Republican convention. Image management and manipulation increased in importance because of television. Reagan's aides perfected a new political art form—the visual press release—whereby Reagan could take credit for new housing starts while visiting a construction site in Fort Worth, Texas, or announce a new welfare initiative during a visit to a nursing home.

Ronald Reagan was an average television actor but a peerless television politician. Both Reagan and his staff set the standard by which future administrations will be judged. As Schmuhl argues in *Statecraft and Stagecraft*, Ronald Reagan represented not only the rhetorical presidency, but the theatrical presidency as well.

—D. Joel Wiggins

RONALD (WILSON) REAGAN. Born in Tampico, Illinois, U.S.A., 6 February 1911. Eureka College, Illinois, B.A. in economics and sociology 1932. Married: 1) Jane Wyman, 1940 (divorced, 1948); children: Maureen and Michael; 2) Nancy Davis, 1952; children: Patti and Ron. Served in U.S. Army Air Force, 1942–45. Wrote sports column for Des Moines, Iowa newspaper; sports announcer, radio station WOC, Davenport, Iowa, 1932–37; in films, 1937–1964; contract with Warner Brothers, 1937; first lead role in big-budget film was in *Kings Row*, 1942; president, Screen Actors Guild, 1947–52, and 1959; in television, 1953–66, starting as host of *The Orchid Awards*, 1953–54; governor of California, 1966–74; U.S. president, 1980–88.

TELEVISION SERIES

1953–54	*The Orchid Awards* (host)
1953–62	*General Electric Theater* (host and program supervisor)
1965–66	*Death Valley Days* (host)

MADE-FOR-TELEVISION MOVIE

1964	*The Killers* (released as theatrical feature due to violent content)

FILMS

Love Is in the Air, 1937; *Hollywood Hotel*, 1937; *Swing Your Lady*, 1938; *Sergeant Murphy*, 1938; *Accidents Will Happen*, 1938; *The Cowboy from Brooklyn*, 1938; *Boy Meets Girl*, 1938; *Girls on Probation*, 1938; *Brother Rat*, 1938; *Going Places*, 1939; *Secret Service of the Air*, 1939; *Dark Victory*, 1939; *Code of the Secret Service*, 1939; *Naughty but Nice*, 1939; *Hell's Kitchen*, 1939; *Angels Wash Their Faces*, 1939; *Smashing the Money Ring*, 1939; *Brother Rat and a Baby*, 1940; *An Angel from Texas*, 1940; *Murder in the Air*, 1940; *Knute Rockne—All American*, 1940; *Tugboat Annie Smith Sails Again*, 1940; *Santa Fe Trail*, 1940; *The Bad Men*, 1941; *Million Dollar Baby*, 1941; *Nine Lives Are Not Enough*, 1941; *International Squadron*, 1941; *Kings Row*, 1941; *Juke Girl*, 1942; *Desperate Journey*, 1942; *This Is the Army*, 1943; *Stallion Road*, 1947; *That Hagen Girl*, 1947; *The Voice of the Turtle*, 1947; *John Loves Mary*, 1949; *Night Unto Night*, 1949; *The Girl From Jones Beach*, 1949; *It's a Great Feeling*, 1949; *The Hasty Heart*, 1950; *Louisa*, 1950; *Storm Warning*, 1951; *Bedtime for Bonzo*, 1951; *The Last Outpost*, 1951; *Hong Kong*, 1952; *She's Working Her Way Through College*, 1952; *The Winning Team*, 1952; *Tropic Zone*, 1953; *Law and Order*, 1953; *Prisoner of War*, 1954; *Cattle Queen of Montana*, 1954; *Tennessee's Partner*, 1955; *Hellcats of the Navy*, 1957; *The Young Doctors* (narrator), 1961; *The Killers*, 1964.

PUBLICATIONS

Where's the Rest of Me?, with Richard Hubler. New York: Dyell, Sloan and Pierce, 1965.

The Reagan Wit, edited by Bill Adler. New York: Thornwood, 1981.

Ronald Reagan: An American Life. New York: Simon and Schuster, 1990.

FURTHER READING

Barilleaux, Ryan J. *The Post-modern Presidency: The Office after Ronald Reagan*. New York: Praeger, 1988.

Cannon, Lou. *Reagan*. New York: Putnam's, 1982.

Deaver, Michael, with Mickey Herskowitz. *Behind the Scenes: In Which the Author Talks About Ronald and Nancy Reagan. . . and Himself*. New York: William Morrow, 1987.

Denton, Robert E., Jr. *The Primetime Presidency of Ronald Reagan*. New York: Praeger, 1988.

Erickson, Paul D. *Reagan Speaks: The Making of an American Myth*. New York: New York University Press, 1985.

Gold, Ellen Reid. "Ronald Reagan and the Oral Tradition." *Central States Speech Journal* (West Lafayette, Indiana), 1988.

Jamieson, Kathleen Hall. *Eloquence in an Electronic Age: The Transformation of Political Speechmaking*. New York: Oxford University Press, 1988.

Kernal, Samuel. "Going Public." *Congressional Quarterly Press* (Washington, D.C.), 1986.

Kiewe, Amos, and Davis W. Houck. *A Shining City on a Hill: Ronald Reagan's Economic Rhetoric, 1951–1989*. New York: Praeger, 1991.

Leamer, Laurence. *Make-believe: The Story of Nancy and Ronald Reagan*. New York: Harper and Row, 1983.

McClelland, Doug. *Hollywood on Ronald Reagan: Friends and Enemies Discuss our President, the Actor*. Winchester, Massachusetts: Faber and Faber, 1983.

McClure, A.F., C.D. Rice, and W.T. Stewart, editors. *Ronald Reagan: His First Career; A Bibliography of the Movie Years*. Studies in American History, vol. 1. Lewiston, New York: Edwin Mellen, 1988.

Pearce, Barnett, and Michael Weiler. *Reagan and Public Discourse in America*. Tuscaloosa: University of Alabama Press, 1992.

Schmuhl, Robert. *Statecraft and Stagecraft: American Politics in the Age of Personality*. Indiana: University of Notre Dame Press, 1990.

Stuckey, Mary E. *Getting into the Game: The Pre-presidential Politics of Ronald Reagan*. New York: Praeger, 1989.

———. *Playing the Game: The Presidential Rhetoric of Ronald Reagan*. New York: Praeger, 1990.

———. *The President as Interpreter-in-Chief*. Chatham, New Jersey: Chatham House, 1991.

Thomas, Tony. *The Films of Ronald Reagan*. Secaucus, New Jersey: Citadel, 1980.

See also *General Electric Theater*; U.S. Presidency

REALITY PROGRAMMING

Reality programming is an expansive television industry label which includes both syndicated and "on-net" (network) programs such as "tabloid" television newsmagazine shows (*Entertainment Tonight, Hard Copy, A Current Affair, Inside Edition, Day One, Dateline NBC*), video-verite (*COPS*) re-created crime or rescue programs (*Top Cops, Rescue 911, America's Most Wanted, Unsolved Mysteries, Real Stories of the Highway Patrol*), and family amateur video shows (*America's Funniest Home Videos, America's Funniest People*). While the corpus of programs grouped under this generic rubric is admittedly varied, the one consistent characteristic which underscores each of these genres is a visible reference to, and dramatization of, "real" events and occupations.

As a program form which purports to exhibit the actual or "the real," reality programming is evocative of non-fictional genres, particularly mainstream television news. Many of the formal conventions of television journalism—such as the style of electronic news gathering, the use of anchors and stand-up shots of reporters on location—are variously found within reality programs. Most importantly, it is reality programming's involvement in the immediacy of the scene or the event which tends to evince the naturalism of television news. Additionally, such similarities are found not only on conventional, but structural levels. For instance, syndicated "tabloid" newsmagazines or crime shows are often competitively time-shifted into the "prime access" scheduling time-slots immediately succeeding local or regional news.

Nonetheless, "the real" in reality programming is a highly flexible concept. Rather than solely relying upon the use of actual documentary or "live" footage for its credibility, reality programming often draws upon a mix of acting, news footage, interviews and re-creations in a highly simulated pretense towards the "real." Admittedly, mainstream television news is also involved in the recreation of reality, rather than simply recording actual events. And yet, "reality" is dramatized on reality programming to an extent quite unlike conventional television news, and this dramatization is often geared towards more promotional, rather than informational, ends. Tabloid newsmagazines, for instance, make liberal use of flashy graphics, creative editing and increased use of music beds in an effort to "hype" the story, often to the point where there is little difference between the promotional trailers for the upcoming report and the actual story itself. In essence, the effectivity of reality programs lies in their ability to dramatize "the real" by drawing upon popular memory and forms, specifically the popular forms of commodity culture.

In addition to a reliance upon an actual or fabricated "real," much reality programming (particularly of the "law and order" or tabloid genre) is concerned with defining moral boundaries within society. These programs tend to accentuate moral or criminal threats to everyday life, and their narrative structure follows classical lines of contrasting

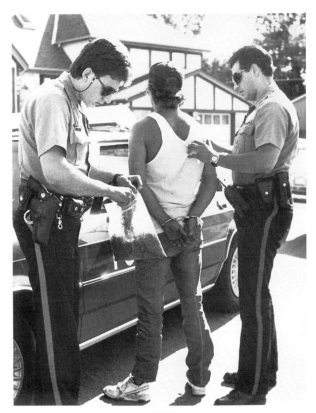

COPS
Photo courtesy of FOX

victims and heroes against criminals and deviants. Criminality and deviance are posed as constant and random factors of everyday life, and their existence demands moral response and redress. It is this heightened emphasis upon moral or criminal disorder which accounts for much of reality programming's disrepute as sensational, excessive, and indulgent of vulgar tastes.

Coupled with the tendency towards moral polarity in reality programming is an emphasis upon the subjective or personal. Reality programming expresses social or moral dilemmas in emotional terms; and it is the emotional affectivity of a program which acts as the key support for its "truthfulness" or credibility. Stress is laid less upon the social, political or historical context of an event, than on its individual and immediate ramifications, particularly in terms of how someone feels or responds to the reported event. In this respect, it is no longer a supposedly neutral objectivity which acts to establish the authenticity of "reality," but rather an appeal towards subjective identification, wherein a distanced or impartial reasoned analysis is replaced with the "closeness" of feeling and sensation. One feature which is emphasized within all types of reality programming—tabloid newsmagazines, crime and rescue shows, and family amateur video programs—is the proximity of the depicted "reality" to the experiences of the audience. In

1352   ENCYCLOPEDIA OF TELEVISION

other words, the adulterous affair on *Inside Edition*, the senseless mugging on *COPS*, or the hapless pratfall on *America's Funniest Home Videos* could all possibly happen to the viewer. Additionally, subjective involvement is further established through participatory strategies which encourage audiences to "interact" with the program itself. For instance, audiences of *Hard Copy* are offered 1-900 numbers in order to place phone-in votes at the end of the program—"Burt or Loni? Whom do you believe? (Callers must be 18 years or older)." *America's Most Wanted* asks it viewers to assist in the capture of suspected fugitives profiled on the show by calling a toll-free hotline. And studio audience members of *America's Funniest Home Videos* vote for the prize-winning "funniest" video shown during the program.

Despite, or even perhaps due to, reality programming's emphasis upon moral conflict, its accentuation of the subjective, and its use of a simulated "real," the genre has experienced wide financial success since its inception in the late 1980s. Emerging during a period of intensified competition for viewers and advertising revenues, early reality-based shows such as the tabloid newsmagazine *A Current Affair* (which debuted in 1986), the video-verite "true crime" series *COPS* (1988) or the re-created "manhunt" series *America's Most Wanted* (1988)—all productions developed by Fox Television—have proven to be long-lasting and solid ratings performers. Similarly, during the 1988–89 season, each of the "Big Three" networks launched at least one weekly reality series (NBC's *Unsolved Mysteries*, ABC's *Funniest Home Videos* and CBS' *Rescue 911*), each of which still enjoys consistent financial viability.

Producers attribute the longevity of such programs to their ability to tell "good stories" and the fact that they are free from the capriciousness of actors or scripts. There are, however, more pragmatic reasons for the genre's success. Such programs are inexpensive to produce, particularly when compared to the production costs of network drama (typically $1 million per hour) or other conventional newsmagazines. While Paramount's *Entertainment Tonight* (which has served as the programming model for other reality-based magazines since its inception in 1980), has one of the highest weekly production budgets at $500,000 to $600,000 per week, the production costs of other tabloid newsmagazines such as King World's *Inside Edition* typically range from $250,000 to $400,000 per week. Production costs for reality-based crime and rescue series are considerably lower at the $150,000 to $250,000 week range. This factor of cost is crucial for countries such as Canada, where both public and private broadcasters have always been dependent upon the availability of inexpensive American shows for their programming schedules, much to the demise of an indigenous product. It may be argued, then, that reality programs are especially attractive to countries outside of the United States. Because of their low cost, each country can create its own version of the programs, which then qualify as indigenous productions and therefore enjoy the privileges of state support. For example, the

Canadian program *Battle Against Crime*, produced by MacBac Productions, is modeled in part upon the video-verite style of Barbour-Langley's *COPS*.

An additional economic incentive is the proven syndication record of reality shows. While relatively strong on network schedules, such programs have also found prosperity when launched as either syndicated first-run series or half-hour strips aired during prime access or fringe time-slots. Reality programs are generally sold on a "cash-plus-barter" basis, meaning that in addition to receiving cash for license fees, syndicators reserve the right to sell one or two minutes of national advertising time while local stations sell the remaining minutes themselves. Much of the success of these syndicated shows is due to the ease with which they can be shifted into compatible schedules. Both tabloid newsmagazines (*A Current Affair, Hard Copy*) and law enforcement and rescue shows (*COPS, Top Cops, Rescue 911*) have done well in prime-access spots, acting as a lead-in or lead out from local newscasts with whom they share similarities in structure and content. The cop and rescue genre, however, evidences more flexibility in its ability to be sold for further programming in strip syndication. While conventional industry wisdom once held that first-run reality programs were too deadline-oriented and time-sensitive to be launched in repeat sales, the cop and rescue sub-genres are not limited to the same temporal constraints as newsmagazines.

Audiences for reality shows tend to fit conventional expectations with regard to the gender of viewers; men in the 18-49 age group are the predominant viewers of the crime and rescue sub-genre, and women in the 18-49 age group comprise the audience for the tabloid shows. An interesting variable is the audience for the family amateur video programs. Besides consistently garnering high weekly ratings, *America's Funniest Home Videos* is also atypical when defined as a "family" oriented program; it appeals foremost to men and children, rather than women.

While reality programs have earned relatively strong ratings, and their advertising time is inexpensive in comparison to programs garnering similar audience numbers, advertisers have often been wary of the genre. This is especially the case for the tabloid newsmagazine shows, sometimes termed "trash TV" for their excessive style and sensational stories. Unwilling to associate their product with programs considered exploitative or in ill-taste, many advertisers have refused to buy air-time on such programs. In response, reality shows have attempted to unburden themselves of the "trash TV" stigma. Paramount's *Hard Copy*, originally sold as *Tabloid*, changed its name after adverse media attention threatened advertiser support.

Such negative connotations do not appear to pertain to the crime, rescue or manhunt sub-genres. Producers of these programs claim this is due to the fact that they are perceived, and pitched as, "pro-social," as offering a form of public service. Supposedly, these shows are designed to foster a solid consensual ground of moral and social certitude. In their appeals to viewer identification, and the participatory strategies of toll-free numbers used to report criminal activity, they presumably offer an engagement with the social authority of the state. And yet, as the Canadian media scholar Graham Knight has argued, the moral and political consensus established by these programs is directed less towards collectivist and statist ends, than it is geared towards an individualist and conservative populism.

This last point demonstrates the importance of situating the historical emergence of reality programming within a specific political and cultural climate. Much of the controversy surrounding the presence of reality programs concerns the blurring between reality and representation, wherein the ability to determine what is real and what is not is increasingly brought under question. In this respect, the controversy and confusion surrounding reality programming's mutation of fictional and non-fictional genres may be indicative of wider cultural and political shifts within society. The genre's violation of conventional distinctions between reality and representation can be seen as symptomatic of a culture in which the lines drawn between culture and commerce, the private and the public, and around categories of social identities have become muddled at best. Hence, and in a quite contradictory way, the moral preoccupations of reality programming may also be read as attempts to re-assert

social and moral order and to provide a simulated relief from the assault upon conventional cultural values.

—Beth Seaton

FURTHER READING

Glynn, Kevin. "Tabloid Television's Transgressive Aesthetic." *Wide Angle* (Athens, Ohio), April 1990.

Goodwin, Andrew. "Reality Programmes." *Sight and Sound* (London), January 1993.

Knight, Graham. "Reality Effects: Tabloid Television News." *Queen's Quarterly* (Toronto), Spring 1989.

Mellencamp, Patricia. *High Anxiety: Catastrophe, Scandal and Comedy.* Bloomington, Indiana: Indiana University Press, 1992.

Nichols, Bill. *Blurred Boundaries: Questions of Meaning in Contemporary Culture.* Bloomington, Indiana: Indiana University Press, 1994.

Scholle, David. "Buy Our News: Tabloid Television and Commodification." *Journal of Communication Inquiry* (Iowa City, Iowa), Winter 1993.

See also *America's Funniest Home Videos; America's Most Wanted;* Tabloid Television

THE RED SKELTON SHOW

U.S. Comedy/Variety

The Red Skelton Show, which premiered on 30 September 1951, was not only one of the longest running variety series on television, but also one of the first variety shows to make the successful transition from radio to television. Despite his popularity as an entertainer in nightclubs, vaudeville, radio and 26 feature films, Skelton was unsure of the new medium. Consequently, he continued his weekly radio broadcasts while simultaneously working on the first two seasons of his television show.

The series originally aired in a half-hour format on NBC. Despite an outstanding first year in which his show was ranked fourth in the Nielsens and won two Emmy Awards, the series' ratings toppled in its second season. When NBC canceled the show, it was immediately picked up by CBS, and *The Red Skelton Show* became a Tuesday night staple from 1954 to 1970.

The format of the series was similar to Skelton's radio program. Each show began with Skelton performing a monologue based on topical material, followed by a musical interlude. He would then perform in a series of blackout sketches featuring one or more of his characters. The sketches were a mixture of new material and old routines (including his popular "Guzzler's Gin") perfected over the years in vaudeville and in nightclubs. At the end of the program, Skelton would become serious and express his gratitude to his audience for their love and laughter. His signature closing line became "Good night and may God bless."

The Red Skelton Show, unlike other variety series, did not rely on guest stars every week. Skelton had a strong group of support players, most of whom had worked with him on his radio program. They included Benny Rubin, Hans Conried, Mel Blanc, and Verna Felton.

Most of Skelton's characters were first developed for radio and worked equally well on television. Among the best known were Junior the Mean Widdle Kid (who was famous for his expression, "I Dood It"), country boy Clem Kadiddlehopper, Sheriff Deadeye, boxer Cauliflower McPugg, drunkard Willy Lump-Lump, and con man San Fernando Red. Skelton had a reputation for his extensive use of "headware." Each character had his own specific hat, which Skelton used as a means to find the center of each personality.

The only television addition to his repertoire of characters was Freddie the Freeloader, a hobo who never spoke. A special "silent spot" featuring the hobo character was added to the program, and provided Skelton the opportunity to demonstrate his talents as a pantomimist.

Skelton's forte was his use of slapstick. He seemed oblivious to physical punishment and often ended his vaudeville act by falling off the stage into the orchestra pit. One of his most popular pieces was created for his premiere show. At the end of his monologue, while Skelton was taking a bow, two hands reached out from under the curtain, grabbed him by the ankles, and swept him off the stage.

Many stars got their start on *The Red Skelton Show*. Johnny Carson, one of Skelton's writers, was called upon to fill in for the star when, in 1954, Skelton injured himself during a rehearsal. The Rolling Stones made one of their earliest American appearances on the show in 1964.

Critics often chastised Skelton for breaking into laughter at his own material on the air. But, no matter how many times he succumbed to his giggles, took another pratfall, mugged for the camera, or made asides to the audience, his popularity only increased.

Although the series remained among the top 20 rated shows, CBS canceled it in 1970, citing high production costs. But it was also the case that Skelton's main audience was very young viewers and speculation suggested that the network wanted to increase its audience share of young adults. The next season, Skelton returned to NBC in a half-hour format on Monday night, but the new show lasted only one season.

During the run of his variety series, Skelton was also able to demonstrate his dramatic abilities. He played the punch-drunk fighter, Buddy McCoy, in *Playhouse 90*'s *The Big Slide* (CBS, 1956) for which he was nominated for an Emmy Award as Best Actor.

—Susan R. Gibberman

The Red Skelton Show

REGULAR PERFORMERS
Red Skelton
David Rose and His Orchestra
Carol Worthington (1970–71)
Chanin Hale (1970–71)
Jan Arvan (1970–71)
Bob Duggan (1970–71)
Peggy Rea (1970–71)
Brad Logan (1970–71)
The Burgundy Street Singers (1970–71)

PRODUCERS   1951–70: Nat Perrin, Cecil Barker, Freeman Keyes, Ben Brady, Gerald Gardner, Bill Hobin, Seymour Berns; 1970–71: Guy Della Cioppa, Gerald Gardner, Dee Caruso

PROGRAMMING HISTORY

• NBC

| September 1951–June 1952 | Sunday 10:00-10:30 |
| September 1952–June 1953 | Sunday 7:00-7:30 |

• CBS

September 1953–June 1954	Tuesday 8:30-9:00
July 1954–September 1954	Wednesday 8:00-9:00
September 1954–December 1954	Tuesday 8:00-8:30
January 1959–June 1961	Tuesday 9:30-10:00
September 1961–June l962	Tuesday 9:00-9:30
September 1962–June 1963	Tuesday 8:30-9:30
September l963–June 1964	Tuesday 8:00-9:00
September 1964–June 1970	Tuesday 8:30-9:30

• NBC

| September 1970–March 1971 | Monday 7:30-8:00 |
| June 1971–August 1971 | Sunday 8:30-9:00 |

FURTHER READING

Abramson, M. "The Red Skelton Story." *Cosmopolitan* (New York), September 1956.

Busch, N. F. "Red Skelton: Television's Clown Prince." *Reader's Digest* (Pleasantville, New York), March 1965.

Chassler, S. "Helter Skelton." *Colliers* (New York), 29 March 1952.

"Clown of the Year." *Newsweek* (New York), 17 March 1952.

"Invincible Red: Tormented Skelton is Top U.S. Clown." *Life* (New York), 21 April 1961.

Jennings, D. "Sad and Lonely Clown." *Saturday Evening Post* (Philadelphia, Pennsylvania), 2 June 1962.

Marx, Arthur. *Red Skelton.* New York: E. P. Dutton, 1979.

Pryor, Thomas M. "Impromptu Comic: In TV, Red Skelton is a Free-Wheeling Clown." *New York Times*, 2 March 1952.

"Rubber Face on TV." *Life* (New York), 22 October 1951.

"Still Fighting for Laughs." *Look* (New York), 2 April 1957.

See also Skelton, Red; Variety Programs

REDMOND, PHIL

British Producer

Phil Redmond is the most well-known drama producer in Britain, and his name is familiar in most households as the creator of the long-running children's school drama *Grange Hill*, and the soap opera *Brookside*. Redmond rose from a council estate childhood in north Liverpool to become a media celebrity and owner of a large private production company. As for most working-class children, a career in the media lay outside his reach, and in 1968 he left his local comprehensive school to train as a quantity surveyor in the building trade. However, by 1972, he had abandoned this, having resolved instead to become a writer, and to take a university degree in social studies to help him in the task. The course had a profound effect on his career, and his writing and programs continually draw on forms of social observation.

The producer's career in television began as a scriptwriter for comedy programs, but his major breakthrough came in 1978 when his proposals for a new children's drama series were adopted by Anna Home at the BBC. What set *Grange Hill* apart from other high school dramas was the program's realism, and its interweaving of serious moral and social issues, such as bullying, teenage sex, and heroin addiction, into the story lines. The program's unsentimental approach to schooling and controversial subject matter has frequently provoked complaints from pressure groups. Despite the objections, however, the series has always been hugely popular with young people, and successive generations of school students have grown up with the program and enjoyed exposure to the problems of the "real" world.

Redmond wrote over thirty episodes for *Grange Hill* in its first four seasons, but his ambitions were driving him toward becoming a producer in his own right and following up the opportunities created by the advent of the fourth channel in Britain. He approached the head of Channel 4, Jeremy Isaacs, and its commissioning editor for fiction, David Rose, and succeeded in convincing them that they should adopt his proposals for *Brookside*, a twice-weekly soap opera focusing on social issues based around family life on a new private housing estate. Channel 4 brought a new style of television production to Britain by commissioning independent production companies to make programs. In 1981, Redmond secured a £4 million investment from Channel 4 to establish his own company, Mersey Television, and to begin work on *Brookside*. Much of the money was spent purchasing and fitting-out the real Liverpool housing estate that was to serve both as the production and company base.

The development of Redmond's soap opera is of considerable importance to the history of the British television institution. Since its launch in 1982, *Brookside* has provided Channel 4 with by far its most popular program, and has played a major role in establishing the viability of the channel. The setting up of Mersey Television in Liverpool to produce the program represents a considerable innovation, for it has created not only the largest independent produc-

Phil Redmond
Photo courtesy of the British Film Institute

tion company in Britain, with over one hundred full-time jobs for the local workforce, but has also significantly extended the opportunities for television production outside London. With his production base secure, Redmond has continued to maintain an anti-metropolitan stance and, going against the industry's received wisdom, has championed the cause of regional television.

Redmond has always contended that the audience of popular drama will respond positively to challenging subject matter. With *Brookside* he was to prove his point. After a slightly shaky start, the program's realist aesthetics, pioneering single-camera video production on location, and engaging major social issues such as unemployment, rape, drugs, and lesbian politics has won over an up-market audience group not normally interested in soaps. The program has helped to raise the stakes of production design, and has added a new seriousness to popular drama. A new generation of realist drama programs, including top shows such as *EastEnders* and *Casualty*, have followed *Brookside*'s example and explored contemporary social problems.

Redmond's success as a producer necessarily stems as much from his shrewd business instincts as his ability to generate creative ideas. His early training as a surveyor instilled in him a respect for the kind of strict budget control and resource management that underpins the whole *Brookside* operation. The permanent locations do not just contribute to the realist look of the program, but are a way of reducing production overheads. He has been equally adroit in marketing the program and creating media events out of the dramatic sensations that are introduced into the storylines from time to time.

Redmond's wider business activities provide a conspicuous example of the new entrepreneurial spirit that pervades broadcasting in Britain following deregulation. In 1991, he was at the centre of the £80 million consortium bid for the new ITV franchise in North West England, which had been held by Granada since 1956. Though the bid was unsuccessful, the additional premises that had been acquired to substantiate it have strengthened the power-base of Mersey Television and enabled it to extend its production. In 1990, the output of *Brookside* was increased to three episodes a week. In 1995, Redmond successfully bid for a new youth soap opera, and *Hollyoaks* was introduced into Channel 4's early evening schedule. Currently, the company's annual turnover is more than £12 million.

Redmond is also active in helping to formulate new training policy for the television industry. He is particularly concerned with the vocational opportunities for new entrants and, as Honorary Professor of Media Studies at the Liverpool John Moores University, he is helping to develop a media degree program with close industry links.

—Bob Millington

PHIL REDMOND. Born in Liverpool, Lancashire, England, 1949. Began career as a television scriptwriter, contributing to *Z Cars* and other series; established reputation with the realistic school series *Grange Hill*, BBC; subsequently moved into independent television, setting up Mersey Television and creating *Brookside* soap opera for Channel 4.

TELEVISION SERIES

1978–93	*Grange Hill*
1981	*Going Out*
1982–	*Brookside*
1990–91	*Waterfront Beat*

FURTHER READING

Geraghty, Christine. *Women and Soap Opera*. London: Polity, 1990.

———. "*Brookside*: No Common Ground." *Screen* (London), 1983.

Redmond, Phil. *Brookside, The Official Companion*. London: Weidenfeld and Nicolson, 1987.

Tunstall, Jeremy. *Television Producers*. London: Routledge, 1993.

Vahimagi, Tise, editor. *British Television: An Illustrated Guide*. Oxford: Oxford University Press, 1994.

See also British Programming; British Programme Production Companies; *Brookside*; *Grange Hill*

REDSTONE, SUMNER

U.S. Media Mogul

Sumner Redstone has become one of the most powerful media moguls of the late 20th century. In his capacity as owner and chief executive officer of Viacom, Inc., Redstone controls Hollywood's Paramount Pictures television and motion picture factory; a handful of cable TV networks including MTV, the Movie Channel, Showtime, Nickelodeon, and VH1; several radio and TV stations; and a TV production and syndication business that owns the lucrative syndication rights to *Roseanne*, *A Different World*, *I Love Lucy*, *Perry Mason*, *The Twilight Zone*, and *The Cosby Show*. Viacom has also produced such prime-time fare as *Matlock* and *Jake and the Fatman*.

Redstone's father Michael first sold linoleum from the back of a truck, later became a liquor wholesaler, and finally purchased two nightclubs and set up one of the original drive-in movie operations in the United States. By the time Redstone graduated from Harvard in 1943, his father was concentrating on the movie industry. One of a number of struggling owners in the fledgling drive-in business, he was unable to book first-run films because the vertically integrated Hollywood giants promoted their own movie theaters.

Redstone graduated first in his class from the prestigious Boston Latin School, and then finished Harvard in less than three years. Upon graduation, he was recruited by Edwin Reischauer, a future United States ambassador to Japan, for an ace U.S. Army intelligence unit that would become famous for cracking Japan's military codes. After three years of service, during which he received two Army commendations, Redstone entered Harvard Law School.

After graduating from Harvard Law in 1947, he began to practice law, first in Washington, D.C., and then in Boston, but soon was lured into the family movie theater business. Two decades later, Redstone was president and chief executive officer of the family firm, National Amusements, Inc. Indeed, even with his move to Viacom, Redstone has continued in the movie exhibition business. At the end of the 20th century, National Amusements operates more than 800 screens in a dozen states across the United States.

Redstone is a physically tough individual. In 1979, he survived a Boston hotel fire by clinging to a third-floor window with one severely burned hand. Doctors never expected him to live through 60 hours of surgery, but he did.

Medical experts told him he would never walk again, yet Redstone began to exercise daily on a treadmill and to play tennis regularly, wearing a leather strap that enabled him to grip his racquet. Those who know the Boston tycoon say that his recovery spurred his ambition to succeed in the motion picture and later television business.

As he recovered from his burns, Redstone used his knowledge of the movie business to begin selectively acquiring stock in Hollywood studios. In a relatively short time, he made millions of dollars buying and selling stakes in 20th Century-Fox, Columbia Pictures Entertainment, MGM/UA Entertainment, and Orion. At first, Viacom represented simply another stock market investment, but soon Redstone realized that the company needed new management and, in 1987, he resolved to take over and run the operation.

Redstone's acquisition proved difficult. The company had rebuffed an earlier takeover attempt by financier Carl Icahn, and Viacom executives had sought to buy and protect their own company. Redstone became embroiled in a bitter, six-month corporate raid which forced him to raise his offer three times. Upon final acquisition, rather than break up Viacom and sell off divisions to pay for the deal as his bankers advised, Redstone slowly and quietly built the company into one of the world's top TV corporations.

Redstone hired former Home Box Office chief executive Frank Biondi to build on Viacom's diversity. For example, by the mid-1990s, Viacom had expanded its MTV music network far beyond its original base in the United States to reach more than 200 million households in approximately 80 countries in Europe, Latin America, and Asia. Redstone felt that his networks needed a Hollywood studio to make new products, and in 1993 he decided to acquire Paramount. He soon found himself in a battle with QVC Network, Inc., and in time joined forces with video rental empire Blockbuster Entertainment to cement the deal.

Owning more than two-thirds of Viacom's voting stock in 1995 meant that Redstone controlled a vast media empire second only to that of Rupert Murdoch. Through the mid-1990s, *Forbes* ranked Redstone among the richest persons in the United States, with a net worth in excess of $4 billion. Yet Redstone has never "gone Hollywood." As the 20th century ends, he continues to operate his collection of enterprises, not from Paramount's sprawling studio on Melrose Avenue in Hollywood, but from his longtime National Amusements, Inc., headquarters in Dedham, Massachusetts.

—Douglas Gomery

SUMNER MURRAY REDSTONE. Born Sumner Murray Rothstein in Boston, Massachusetts, U.S.A., 27 May 1923. Harvard University, B.A. 1944, LLB. 1947. Married: Phyllis Gloria Raphael, 1947; children: Brent Dale and Shari Ellin. Served as 1st Lt., U.S. Army, 1943–45. Admitted to the Massachusetts Bar, 1947; instructor of law and labor management, University of San Francisco, 1947; law secretary, U.S. Court of Appeals for 9th Circuit, San Francisco, 1947–48; admitted to U.S. Court of Appeals (1st and 9th Cir-

Sumner Redstone
Photo courtesy of Broadcasting and Cable

cuits), 1948; special assistant to U.S. Attorney General, Washington, D.C., 1948–51; admitted to U.S. Court of Appeals (8th Circuit), 1950; admitted to Washington, D.C. Bar, 1951; partner in firm of Ford, Bergson, Adams, Borkland and Redstone, Washington, D.C., 1951–54; admitted to U.S. Supreme Court, 1952; executive vice president, Northeast Drive-In Theatre Corporation, 1954–68; president, Northeast Theatre Corporation; assistant president, Theatre Owners of America, 1960–63, president, 1964–65; chair of the board, National Association of Theatre Owners, 1965–66; chair of the board, president, and chief executive officer, National Amusements, Inc., Dedham, Massachusetts, from 1967; chair of the board, Viacom International, Inc. and Viacom, Inc., New York City; professor, Boston University Law School, 1982, 1985–86. Charitable work includes: chair, metropolitan division, North East Combined Jewish Philanthropies, Boston, 1963; trustee, Children's Cancer Research Foundation; chair, American Cancer Crusade, State of Massachusetts, 1984-86; vice president and member of executive committee, Will Rogers Memorial Fund; board of directors, Boston Arts Festival; board of overseers, Dana Farber Cancer Center and the Boston Museum of Fine Arts; member, presidential advisory committee on arts, John F. Kennedy Center for Performing Arts; board of directors, John F. Kennedy Library Foundation, 1984–86. Member: American Bar Association; National Association of Theatre Owners; Theatre Owners of America; Motion Picture Pioneers; Boston Bar Association; Massachusetts Bar Association; Harvard Law School Association; American Judicature Society. Recipient: Army Commendation Medal; William J. German Human Relations Award, American Jewish Committee Entertainment and Communication Division, 1977; Silver Shingle Award, Boston University Law School, 1985; Man of the Year, Entertainment Industries Division of United Jewish Appeal Federation, 1988; Variety New England Humanitarian Award, 1989; Pioneer of the Year, Motion Picture Pioneers, 1991. Home address: 98 Baldpate Hill Road, Newton, Massachusetts 02159-2825, U.S.A. Office address: National Amusements, Inc., 200 Elm Street, Dedham, Massachusetts 02026-4536, U.S.A.

FURTHER READING

Auletta, Ken. "The Last Studio in Play." *The New Yorker*, 4 October 1993.

Bart, Peter. "Owners Take Over the Asylum: Murdochian Moguls Become Hands-on." *Variety* (Los Angeles), 26 February 1995.

Gallese, Liz Roman. "'I Get Exhilarated by It.'" *Forbes* (New York), 22 October 1990.

Greenwald, John. "The Man with the Iron Grasp." *Time* (New York), 27 September 1993.

Landler, Mark. "The MTV Tycoon: Sumner Redstone Is Turning Viacom into the Hottest Global TV Network." *Business Week* (New York), 21 September 1992.

———. "Sumner at the Summit." *Business Week* (New York), 28 February 1994.

Lenzer, Robert. "Late Bloomer." *Forbes* (New York), 17 October 1994.

Matzer, Marla. "Winning Is the Only Thing." *Forbes* (New York), 17 October 1994.

Stern, Christopher. "Ready to Take On the World" (interview). *Broadcasting and Cable* (Washington, D.C.), 20 September 1993.

"Sumner Redstone: A Drive to Win." *Broadcasting* (Washington, D.C.), 14 November 1988.

See also Cable Networks; Music Television (MTV); Syndication

REES, MARIAN

U.S. Producer

After graduating with honors in sociology from the University of Iowa, Marian Rees moved to Los Angeles in 1952, where she began her television career as a receptionist-typist at NBC. By 1955, she had joined the Norman Lear-Bud Yorkin company, Tandem Productions, and in 1958, served as an associate producer of the much-honored *An Evening with Fred Astaire*. She continued to advance in the organization, and by the early 1970s, served as associate producer of the pilots of *All in the Family* and *Sanford and Son*. In 1972, however, she was told by Tandem that she would be happier elsewhere, and was given two weeks' notice. It was a stunning blow, but as she told an interviewer in 1986, she used the firing to grow.

Rees assumed a new position at the independent production company, Tomorrow Entertainment, where she broadened her knowledge of development, pre-production, and post-production. At Tomorrow, Rees was associated with a variety of quality productions, including *The Autobiography of Miss Jane Pittman*. She then spent two years as vice president of the NRW Company, where she was the executive producer of *The Marva Collins Story*, a *Hallmark Hall of Fame* presentation starring Cicely Tyson. In 1982, Rees formed her own company, Marian Rees Associates. Anne Hopkins joined the company as a partner, and has continued to work with Rees ever since.

In order to fund her first independent productions, Rees initially mortgaged her home and car, facing demands for financial qualification far more extensive than would have been required for a man. She pressed for months to gain

Marian Rees
Photo courtesy of Marian Rees

network approval for her first production, *Miss All-American Beauty,* but resistance continued, and she finally learned that the male executive she had to convince simply did not want to trust a woman. Finally, with funds running extremely low, approval for the project came from CBS. Rees completed the production under budget, and her company at last found itself on solid footing.

In the succeeding years, Rees has garnered 11 Emmy Awards and 30 additional nominations. In 1992, just ten years after her company began, she saw her film for NBC, *Miss Rose White,* garner four Emmys out of ten nominations, a Golden Globe nomination, and the Humanitas Award. Seven of her productions have been aired as part of the *Hallmark Hall of Fame* series.

Rees has remained faithful to her vision of excellence, even in times of financial difficulty. She examines potential stories to ascertain whether they speak to her personally, and make her proud to be associated with the final product. These same concerns are reflected in the meticulous attention given to each project once it is in production. While filming *Miss Rose White* in spring 1992 in Richmond, Virginia, for example, both Rees and Hopkins supervised details at every stage, and personally examined each location shot for authenticity. Such care has meant that their work is usually focused on a single film at a time. Only once since the company was started have they broadcast more than two productions in any given year. Rees and Hopkins form a remarkable team, taking considerable risks, and always delivering quality products, a task made more difficult in the U.S. television industry at the end of the decade.

A champion for women's rights in the U.S. television industry throughout her career, Marian Rees served two terms as president of Women in Film. Her service to her profession also includes board membership at the American Film Institute and the Producer's Guild of America, where she now serves as vice president. "Producer" may be an easy title to acquire in the modern television age. Few earn it, and certainly none deserve it more than Marian Rees.

—Robert S. Alley

MARIAN REES. Worked in live television, New York City, from 1950s; associate producer, Tandem Productions, 1955–72; executive, Tomorrow Entertainment, First Artists Television, EMI Television, and NRW Company's features division, 1972–82; founder, Marian Rees Associates, 1982; producer, numerous made-for-television movies. Member: Women in Film (twice elected president); board of directors, American Film Institute; Producers Guild of America (vice president, 1996).

TELEVISION SERIES (selection)

1971–79	*All in the Family*
1972–77	*Sanford and Son*

MADE-FOR-TELEVISION MOVIES (selection)

1979	*Orphan Train*
1981	*The Marva Collins Story*
1981	*Angel Dusted*
1982	*Miss All-American Beauty*
1983	*Between Friends*
1984	*License to Kill*
1984	*Love Is Never Silent*
1986	*Christmas Snow*
1986	*Resting Place*
1987	*The Room Upstairs*
1987	*Foxfire*
1988	*Little Girl Lost*
1989	*The Shell Seekers*
1989	*Home Fires Burning*
1990	*Decoration Day*
1992	*Miss Rose White*
1995	*In Pursuit of Honor*
1995	*When the Vows Break*

TELEVISION SPECIAL

1958	*An Evening with Fred Astaire*

See also *All in the Family; Hallmark Hall of Fame; Sanford and Son*

REID, TIM

U.S. Actor/Producer

Tim Reid is an accomplished television actor and producer whose critically acclaimed work has, unfortunately, often failed to meet with sustained audience acceptance. As an African American, Reid has tried to choose roles and projects that help effect a positive image for the black community. Through both his acting and writing, Reid has provided important insights regarding black/white relationships and bigotry.

Being a part of show business was one of Reid's childhood dreams. Not content with simply being an actor, Reid

hoped to play a vital role behind the scenes, as well. Like many young actors, he began his career as a stand-up comedian, working with Tom Dreesen as part of the comedy duet "Tim and Tom." It was during this experience that Reid began exploring the dynamics of black/white relationships. In 1978, after performing in various episodic series, Reid received the role of Venus Flytrap in Hugh Wilson's *WKRP in Cincinnati.* From the beginning, Reid made it clear to Wilson that he was not interested in playing just another "jive-talking" black character. Wilson agreed, eventually giving Reid control over his character's development, which culminated in a story that revealed a much deeper character than the Flytrap persona first presented.

It was during *WKRP* that Reid gained experience as a writer, contributing several scripts to the series. One episode, "A Family Affair," dealt with the underlying tones of bigotry that plague even the best of friends. Reid also worked closely with Hugh Wilson on the script "Venus and the Man," in which Venus helped a young black gang member decide to return to high school. Teacher's organizations applauded the effort, and scenes from the show were reproduced, in comic book form, in *Scholastic* magazine.

After *WKRP*, Reid landed a recurring role in the detective drama *Simon and Simon*, for which he also wrote a number of scripts. In 1987, Reid joined forces with Wilson to co-produce one of television's finest half-hour programs—*Frank's Place*—which starred Reid as a Boston professor who took over his deceased father's bar in a predominately black section of New Orleans. While critics raved about the rich writing (Wilson won an Emmy for the *Frank's Place* script "The Bridge"), acting and photography, the series was canceled after its first season. Reid feels this was due to the constant schedule changes which afflicted the series (a problem he and Wilson experienced previously with *WKRP*), as well as CBS' overall dismal ratings at the time.

In 1989, Reid became executive producer of *Snoops*, a drama in which he starred with his wife, Daphne Maxwell Reid, as a sophisticated husband-and-wife detective team in the tradition of the *Thin Man* series. Just as with *Moonlighting* and *Remington Steele*, *Snoops* placed character development over mystery. Once again, despite quality scripting and performances, the show failed to find an audience. Reid has continued to appear in a variety of series, including ABC's *Sister, Sister*, a disappointing sitcom that pales in comparison to Reid's previous work.

—Michael B. Kassel

TIM REID. Born in Norfolk, Virginia, U.S.A., 19 December 1944. Educated at Norfolk State College, B.B.A. 1968. Married: Daphne Maxwell, 1982; children: Tim II, Tori LeAnn, Christopher Tubbs. Marketing representative for Dupont Corporation, 1968–71; actively involved in anti-drug movement, since 1969; stand-up comedian, Tim and Tom Comedy Team, 1971–75; actor in series television, from 1976; founded Timalove Enterprises, 1979; creator, producer, anti-drug video *Stop the Madness*, 1986; co-founded, with Black Entertainment Television, United Image Entertainment Enterprises, 1990, also co-chair; organizer and sponsor, Annual Tim Reid Celebrity Tennis Tournament, Norfolk State University campus. Member: Writers Guild of America; Screen Actors Guild; board of directors, Phoenix House of California; board of trustees, Norfolk State University, Commonwealth of Virginia; board of directors, National Academy of Cable Programming; AFTRA; life member, NAACP. Recipient: Emmy Award; Critics Choice Award, 1988; NAACP Image Award, 1988; Viewers for Quality Television Best Actor in a Comedy Award,

Tim Reid
Photo courtesy of Tim Reid

1988; National Black College Alumni Hall of Fame, 1991. Address: United Image Entertainment, 1640 South Sepulveda Boulevard, #311, Los Angeles, California 90025-7510, U.S.A.

TELEVISION SERIES

1976	*Easy Does It...Starring Frankie Avalon*
1977	*The Marilyn McCoo and Billy Davis, Jr. Show*
1977	*The Richard Pryor Show*
1978–82	*WKRP in Cincinnati*
1983	*Teachers Only*
1983–87	*Simon and Simon*
1987–88	*Frank's Place* (also co-executive producer)
1989–90	*Snoops* (also co-creator, executive producer)
1994–	*Sister, Sister* (also creator, producer)

MADE-FOR-TELEVISION MOVIES

1979	*You Can't Take It With You*
1990	*Perry Mason: The Case of the Silenced Singer*
1991	*Stephen King's It*
1991	*The Family Business*
1992	*You Must Remember This*
1994	*Race to Freedom: The Underground Railroad*
1995	*Simon and Simon: In Trouble Again*

FILMS

Dead Bang, 1989; *The Fourth War*, 1990; *Once Upon a Time...When We Were Colored* (director), 1995.

FURTHER READING

Gray, Herman. *Watching Race: Television and the Struggle for "Blackness."* Minneapolis: University of Minnesota Press, 1995.

See also Comedy, Workplace; Dramedy; *Frank's Place*; Pryor, Richard; Racism, Ethnicity, and Television

REINER, CARL

U.S. Comedian/Writer/Producer

Carl Reiner is one of the few true Renaissance persons of 20th-century mass media. Known primarily for his work as creator, writer, and producer of *The Dick Van Dyke Show*—one of a handful of classic sitcoms by which others are measured—Reiner has also made his mark as a comedian, actor, novelist, and film director. From Reiner's "Golden Age" TV connection with Sid Caesar to his later film work with Steve Martin, the Emmy Award-winning Reiner has touched three generations of American comedy.

According to Vince Waldron's *Official "Dick Van Dyke Show" Book*, Reiner began his career as a sketch comedian in the Catskill Mountains. After serving in World War II, he landed the lead role in a national touring company production of *Call Me Mister*, which he later reprised on Broadway. Reiner's big break came in 1950 when producer Max Leibman, whom he had met while working in the Catskills, cast Reiner as a comic actor in Sid Caesar's *Your Show of Shows*. Drawn to the creative genius of the show's writers, which included Mel Brooks and Neil Simon, Reiner ended up contributing ideas for many of the series' sketches. The experience undoubtedly provided Reiner with a good deal of fodder for his later *Dick Van Dyke Show*. While he never received credit for his writing efforts on *Your Show of Shows*, in 1955 and 1956 he received his first two of many Emmy Awards, these for his role as supporting actor. In 1957, Reiner conquered another medium when he adapted one of his short stories into *Enter Laughing*, a semi-autobiographical novel focusing on a struggling actor's desire to break into show business. In 1963 the book became a hit play.

By the summer of 1958, after Caesar's third and final series was canceled, Reiner spent the summer preparing for what many consider his greatest accomplishment—writing the first thirteen episodes of *Head of the Family*, a sitcom featuring the exploits of fictional New York comedy writer Rob Petrie. Originally intended as an acting vehicle for himself, Reiner's pilot failed to sell. However, Danny Thomas Productions' producer Sheldon Leonard liked the idea and said it had potential if it were recast—which was Leonard's nice way of saying, "Keep Reiner off camera." When Reiner's Rob Petrie was replaced with TV newcomer Dick Van Dyke—who had just enjoyed a successful Broadway run in *Bye, Bye Birdie*—*The Dick Van Dyke Show* was born.

As with *Enter Laughing*, Reiner's sitcom was autobiographical. Like Petrie, Reiner was a New York writer who lived in New Rochelle. Like Petrie, Reiner spent part of his World War II days at Camp Crowder in Joplin, Missouri, a fact that was brought out in several flashback episodes. Even Petrie's 148 Bonny Meadow Road address was an allusion to Reiner's own 48 Bonny Meadow Road home.

Perhaps it was this realism that contributed to the series' timelessness, making it a precursor for such sophisticated and intelligent sitcoms as *The Mary Tyler Moore Show* and

Carl Reiner

The Bob Newhart Show. Just as with these later works, Reiner's series placed character integrity over raw laughs. By being the first to combine both the home and work-life of the series' main character, Reiner also provided interesting insights regarding both sedate suburbia and urbane New York. *The Dick Van Dyke Show* also serves as an early example of the "co-workers as family" format, which has become a staple relationship in modern sitcoms.

Carl Reiner was one of the first "*auteur*" producers," with his first thirteen episodes becoming the bible upon which consequent episodes were based. He continued to write many of the series' best episodes, as well as portray recurring character Alan Brady, the egomaniacal star of the variety program for which Petrie and crew wrote. After a tough first season in 1961, Leonard was able to convince CBS executives, who had canceled the series, to give it a second chance. The series became a top hit in subsequent years, enjoying five seasons before voluntarily retiring. The reruns have never left the air, and it, along with *I Love Lucy*, comprises some of the most-watched programs in syndication history. Those series, along with *The Mary Tyler Moore Show*, have also become the flagship programs of classic TV powerhouse Nick at Nite.

While many view *The Dick Van Dyke Show* as the culmination of Reiner's career, his films cannot be ignored. After directing *Enter Laughing* in 1967, Reiner went on to do several critically acclaimed films such as *The Comic* (1969), a black comedy which starred Dick Van Dyke as an aging silent-film comedian, and *Where's Poppa?* (1970). Reiner also directed the wildly successful George Burns vehicle *Oh, God!* (1977). Reiner is also significant for his role as straight man in "The 2,000 Year Old Man" recordings, which he began with Mel Brooks in 1960.

In the 1970s, Reiner and Van Dyke re-entered television with *The New Dick Van Dyke Show*. While Reiner had hoped to break new ground, he became frustrated with the network's family standard provisions that hampered its sophistication. It was not until 1976 that Reiner returned to series television as actor and executive producer of the short-lived ABC sitcom *Good Heavens*.

Just as *The Dick Van Dyke Show* represented a departure from the standard sitcom fare of the 1960s, *Saturday Night Live* and its most famous guest-host Steve Martin were forging their own late-1970s humor. Once again on the cutting edge, Reiner joined forces with Martin as the "wild and crazy" comedian made the transition to film, with Reiner directing *The Jerk* (1979), *The Man with Two Brains* (1983), and *All of Me* (1984).

In a 1995 episode of the NBC comedy series, *Mad about You*, Reiner reprised his role as Alan Brady. In the fictional world of the newer sitcom, *The Dick Van Dyke Show* is "real," as is the Brady character. Reiner's performance drew on the entire body of his work, from his days with Sid Caeser through his work as writer, director, and producer, and the portrait he presented in this new context echoed with references to the television history he has lived and to which he has so fully contributed.

—Michael B. Kassel

CARL REINER. Born in the Bronx, New York, U.S.A., 20 March 1922. Educated at the School of Foreign Service, Georgetown University, 1943. Married: Estelle Lebost, 1943; children: Robert, Sylvia, and Lucas. Served in the U.S. Army, attached to Major Maurice Evans' special services unit, 1942–46. Worked in Broadway shows, 1946–50; character actor and emcee, television show *Your Show of Shows*, 1950–54; appeared in *Caesar's Hour*, 1954–57; appeared in short-lived *Sid Caesar Invites You*, 1958; emcee, *Keep Talking*, 1958–59; writer, actor, and producer, various TV series, from 1960; director and star,

numerous motion pictures, since 1959. Recipient: numerous Emmy Awards, since 1965. Address: c/o George Shapiro, Shapiro West, 141 El Camino Drive, Beverly Hills, California 90212, U.S.A.

TELEVISION SERIES

1950–54	*Your Show of Shows*
1954–57	*Caesar's Hour*
1956–63	*The Dinah Shore Chevy Show*
1958–59	*Keep Talking*
1961–66	*The Dick Van Dyke Show* (producer and writer)
1971–74	*The New Dick Van Dyke Show* (producer and writer)
1976	*Good Heavens* (actor and producer)

TELEVISION SPECIALS

1967	*The Sid Caesar, Imogene Coca, Carl Reiner, Howard Morris Special*
1968	*The Fabulous Funnies* (host)
1969	*The Wonderful World of Pizzazz* (co-host)
1970	*Happy Birthday Charlie Brown* (host)
1984	*Those Wonderful TV Game Shows* (host)
1984	*The Great Stand-Ups: 60 Years of Laughter* (narrator)
1987	*Carol, Carl, Whoopi, and Robin*

FILMS (selection)

Happy Anniversary, 1959; *The Gazebo*, 1960; *Gidget Goes Hawaiian*, 1961; *It's A Mad, Mad, Mad, Mad World*, 1963; *The Russians Are Coming!*, 1966; *Enter Laughing* (director), 1967; *Where's Poppa?*, 1970; *Heaven Help Us* (co-producer), 1976; *Oh, God!* (director), 1977; *The End*, 1978; *The One and Only* (director), 1978; *The Jerk* (director), 1979; *Dead Men Don't Wear Plaid*, 1982; *The Man with Two Brains* (co-director), 1983; *All of Me* (director), 1984; *Summer Rental* (director), 1985; *Summer School* (director), 1987.

STAGE

Call Me Mister, 1947–48; *Inside U.S.A.*, 1948–49; *Alive and Kicking*, 1950; *Enter Laughing*, 1963.

See also Caesar, Sid; *Dick Van Dyke Show*

REITH, JOHN C.W.

British Media Executive

John Reith, the founding director general of the British Broadcasting Corporation (BBC) from 1922 to 1938, was aptly designated by *The New York Times* as "the single most dominating influence on British broadcasting." Reith developed strong ideas about the educational and cultural public-service responsibilities of a national radio service,

ideas subsequently pursued by many broadcasting systems around the world.

Reith was born the fifth son of a Scottish minister and trained in Glasgow as an engineer. After service in World War I, where he was severely wounded (his face carried the scars), and a growing boredom with engineering, he answered a 1922 advertisement for a post at the new BBC, then a commercial operation. He knew nothing of radio or broadcasting and did not even own a receiver. He was hired and a year later was promoted to managing director.

Learning on the job, Reith soon defined public-service broadcasting as having four elements, which he described in his book *Broadcast over Britain* (1924). Such a system, he argued, operated on a public-service rather than commercial motive, offered national coverage, depended upon centralized control and operation rather than local outlets, and developed high-quality standards of programming. He held broadcasting to high moral—almost religious—standards and rather quickly identified the BBC (which became a public corporation early in 1927) with the political establishment just as he also insisted on BBC operational independence from any political pressures.

Reith directed the expanding BBC operations from Broadcasting House, the downtown London headquarters he initiated, which opened in 1932 and remains a landmark. His primary interest was in radio, however, and the BBC was slow to cooperate with John Logie Baird and other TV experimenters. With the development of effective all-electronic television, Reith's BBC inaugurated the world's first regularly public schedule of television broadcasts from November 1936 until Britain entered World War II in September 1939.

Reith felt increasingly underutilized at the BBC by the late 1930s; the system he had built and the key people he had selected were all doing their jobs well and the system hummed relatively smoothly. He was both revered and somewhat feared in the organization he had shaped. In a mid-1938 managerial coup, however, Reith was eased out as director general by the BBC's Board of Governors (acting in consort with the government), which had grown weary with his self-righteous inflexibility within the organization as well as his political stance. He left the BBC after 16 years with considerable bitterness which remained for the rest of his life.

Reith's remaining three decades were a disappointment to him and others. After a brief period (1938-40) heading Imperial Airways as it became the British Overseas Airways Corporation (BOAC, the government-owned predecessor of British Airways), he held a number of minor cabinet posts in wartime and post-war governments, and served as chair of several companies. Reith's strong views, conviction that he was nearly always right, and dour personality made it difficult for him to readily get along in the rapidly-changing postwar British scene. He wrote an autobiography, *Into the Wind* (1949), and complained he had never been "fully stretched." Indeed, he saw his entire life as one of failure. He argued strongly in the House of Lords against the inception of commercial television in 1954. He felt the BBC had long since given way to social pressures and lowered its standards. It was no longer his child.

Reith was an obsessive keeper of diaries all his life—excerpts published in 1975 showed him to be a man with strong convictions, powerful hatreds, considerable frustration, and an immense ego.

—Christopher H. Sterling

JOHN C(HARLES) W(ALSHAM) REITH. Born Stonehaven, Grampian, Scotland, 1889. Attended Glasgow Academy; Gresham's School, Holt. Served in World War I. Engineer, Coatbridge; first general manager, BBC, 1922; director-general, 1927–38, pioneering public-service broadcasting; chair, Imperial Airways, 1938; elected member of Parliament, Southampton, 1940; appointed minister of works and buildings, 1940–42; chair, Commonwealth Telecommunications Board, 1946–50. Annual Reith lectures inaugurated in his honour, 1948. Knighted, 1927; created Baron Reith of Stonehaven, 1940. Died 1971.

PUBLICATIONS

Broadcast over Britain. London: Hodder and Stoughton, 1924.

Into the Wind (autobiography). London: Hodder and Stoughton, 1949.

Wearing Spurs. London: Hutchinson, 1966.

The Reith Diaries, edited by C.H. Stewart. London: Collins, 1975.

FURTHER READING

Allighan, Garry. *Sir John Reith.* London: Stanley Paul, 1938.

BBC Yearbook (1928–34), *Annual* (1935–37) and *Handbook* (1938). London: BBC, 1928–38.

Boyle, Andrew. *Only the Wind Will Listen: Reith of the BBC.* London: Hutchinson, 1972.

Briggs, Asa. *The History of Broadcasting in the United Kingdom: The Birth of Broadcasting.* Oxford: Oxford University Press, 1961.

———. *And the Golden Age of Wireless.* Oxford: Oxford University Press, 1965.

———. *Governing the BBC.* London: BBC, 1979.

McIntyre, Ian. *The Expense of Glory: A Life of John Reith.* London: Harper Collins, 1993.

Milner, Roger. *Reith: The B.B.C. Years.* Edinburgh: Mainstream Publishing, 1983.

See also British Television; Public Service Television

RELIGION ON TELEVISION

Religion is uncommon in American television. It does appear, however, through two primary avenues. First, consistent with traditions developed in the radio era, there have been a variety of religious programs on the air. Second, there are occasions when religion has appeared in general entertainment program offerings.

Religious programs have been a fixture of television from its earliest years. The pattern was established in radio, where certain sectarian organizations—both national and local—receive free air time (called "sustaining time") for productions intended to elucidate consensual "broad truths" about religion. Programs produced by the National Council of Churches, the United States Catholic Conference, the New York Board of Rabbis, and the Southern Baptist Convention, received such air play without competition until the 1970s when an entirely new type of religious television developed.

These newer programs, which came to be called Televangelism, first emerged nationally after changes in federal policy began to allow use of domestic satellite transmission for the creation of alternative "networks." A number of new and existing television "ministries" capitalized on the situation. These were largely outside the religious mainstream, representing independent, non-denominational, conservative Fundamentalist or Pentecostal organizations. Among the earliest programs were Rex Humbard's *Cathedral of Tomorrow*, *Oral Roberts and You*, Pat Robertson's *700 Club*, and Jim and Tammy Bakker's *PTL Club*.

From the mid–1970s until a series of scandals struck three prominent programs ten years later, televangelism was a force on television and in the world of religion. Early on, this new religious broadcasting was feared to have negative consequences for conventional religion by drawing members and financial support away from churches. After academic studies confirmed that audiences for these programs tended to be small and made up of already-religious, church-going people, that controversy faded.

Televangelism's role in politics has been a more persistent issue. Fundamentalist minister Jerry Falwell used his *Old Time Gospel Hour* program as a platform for political influence through the founding of the Moral Majority, a conservative think-tank, and the Liberty Lobby, a political organization. Falwell withdrew from politics at the time of the scandals, but Pat Robertson used his position as host of *The 700 Club* to launch his own political career, culminating in a run for the presidency in 1988, and the founding of his own political organization, the Christian Coalition, shortly thereafter. Several televangelism ministries also founded and developed their own universities, such as Falwell's Liberty University, Oral Roberts University, and Robertson's CBN University, which was renamed Regent University in 1990.

Robertson's is the singular case which typifies the evolution of modern televangelism from its roots in "Bible Belt" fundamentalist radio toward an altogether conventional television presentation. While other televangelists continued to hold to more traditional "worship and preaching" production, *The 700 Club* evolved a sophisticated "Christian talk show" format. At the same time, its Christian Broadcasting Network (CBN) evolved into the Family Channel, a widely-carried cable service featuring "family-oriented" re-runs and motion pictures.

Another lasting legacy of televangelism has been its impact on sustaining-time or "public service" religion. The conventional churches and church organizations saw their air time gradually erode as "paid time" televangelism rose to prominence. By the mid-1990s virtually no national or network-based sustaining-time religion persisted. A number of these organizations participated in the founding of their own cable network, the Faith and Values Channel (originally the Vision Interfaith Satellite Network), in 1988.

Religion appears in entertainment programs more rarely. In the 50-year history of television in the United States, fewer than two dozen series or pilots have featured religious persons in leading or title roles. The majority of these were Roman Catholic, with only nine non-Catholic examples. Four of these were pilots which were not developed into regularly appearing series. The Catholic programs include some of the most memorable: *Father Murphy*, in which the main character pretends to be a priest; *Sarge*, featuring a former detective who becomes a priest; and *The Father Dowling Mysteries*, in which a priest becomes a detective.

For reasons that are not entirely clear, the programs featuring non-Catholic characters have been less successful: *Bridget Loves Bernie*, a sitcom which turned on the theme of a religiously-mixed marriage; *Keep the Faith*, a pilot featuring two rabbis; *St. Peter*, a pilot about a young Episcopalian priest in Greenwich Village; *Almost Heaven*, the antics of a group of deceased souls trying to find their way to heaven; *Steambath*, adapted from a stage play by the same name; *Great Bible Adventures*, a 1966 pilot, and; *Greatest Heros of the Bible*. Jewish or Muslim characters are rarely depicted, except in a Biblical or period drama.

The presentation of religious characters and themes holds that religion be as general and conventional as possible, so as to avoid potential controversy. For example, whereas Roman Catholics are most often identified as such, Protestant characters are not identified by denomination. And, religiosity is most often limited to the most obvious and innocuous external signifiers, such as place of domicile (a convent, for instance) or dress (nun's habit, yarmulke, or Roman collar). One of the most overtly religious programs in this general sense was *Highway to Heaven*, starring Michael Landon. Landon portrays Jonathan Smith, an angel whose assignment is to help ordinary mortals through difficult times. This show built on the gentle persona developed by Landon in *Little House on the Prairie*, and during the mid-1980s was successful with both adults and children. More explicit religious activities are rarely presented and superficial when they are (group prayers on *M*A*S*H*, perfunctory table grace on *The Simpsons*).

The Hour of Power with Robert Schuller
Photo courtesy of Crystal Cathedral Ministries

Of the 2000 made-for-television movies and miniseries produced between 1964 and 1986, fewer than 30 dealt with religious matters or included religious main characters. Nine of these were historical (usually Biblical): *A.D., The Day Christ Died, Jesus of Nazareth, Mary and Joseph, Masada, The Nativity, Samson and Delilah, The Story of David*, and *The Story of Jacob and Joseph*. Four were profiles of historical Catholic figures, three involved Protestant characters, two were Jewish in theme or character.

Religion began to find its way into some prominent series in the early 1990s, this time not as a major theme, but as a significant element nonetheless. A "new age" or "seeker" religiosity was a fairly common theme of *Northern Exposure*, frequently introduced by the character Chris. *Picket Fences* regularly dealt with religious themes and ideas, and a born-again Christian joined the firm in *L.A. Law* during this time. *Touched by an Angel* repised themes found in *Highway to*

Heaven. Religious awakening and interest was also a theme of *thirtysomething*.

Christy, a series based on a novel by Katherine Marshall, was hailed as a religiously-attuned program during its short run in 1994. It thus followed in the footsteps of such earlier period pieces as *The Waltons* and *Little House on the Prairie*, where religion was portrayed as a more obvious and natural dimension of "earlier times."

Religious places are rarely depicted, at least in use. Early programs such as *The Goldbergs* and *Leave It to Beaver* did show families attending church or synagogue as did *The Simpsons* in the 1990s. However, these were the exceptions. Religion is most frequently shown in connection with rites of passage, specifically in connection with births, deaths and—most frequently—weddings. There have been hundreds of weddings shown on daytime serials alone.

—Stewart M. Hoover and J. Jerome Lackamp

FURTHER READING

Abelman, Robert. "How Political Is Religious Television?" *Journalism Quarterly* (Urbana, Illinois), Summer 1988.

Armstrong, Ben. *The Electric Church*. Nashville, Tennessee: Thomas Nelson, 1979.

Bluem, A. William. *Religious Television Programs; A Study of Relevance*. New York: Hastings House, 1969.

Bruce, Steve. *Pray TV: Televangelism in America*. London: Routledge, 1990.

Diekema, David A. "Televangelism and the Mediated Charismatic Relationship." *Social Science Journal* (Fort Collins, Colorado), April 1991.

Ellens, J. Harold. *Models of Religious Broadcasting*. Grand Rapids, Michigan: Eerdmans, 1974.

Erickson, Hal. *Religious Radio and Television in the United States, 1921–1991: The Programs and Personalities*. Jefferson, North Carolina: McFarland, 1992.

Ferre, John, editor. *Channels of Belief: Religion and American Commercial Television*. Ames, Iowa: Iowa State University Press, 1990.

Gunter, Barrie, and Rachel Viney. *Seeing Is Believing: Religion and Television in the 1990s*. London: John Libbey, 1994.

Hadden, Jeffrey, and Charles Swann. *Prime Time Preachers*. Reading, Massachusetts: Addison-Wesley, 1981.

Hoover, Stewart M. *Mass Media Religion: The Social Sources of the Electronic Church*. Beverly Hills, California: Sage, 1988.

Horsfield, Peter. *Religious Television: The American Experience*. New York: Longman Press, 1984.

Medved, Michael. *Hollywood vs. America: Popular Culture and the War on Traditional Values*. New York: Harper Collins, 1992.

Peck, Janice. *The Gods of Televangelism*. Lexington, Massachusetts: Greenwood Press, 1993.

Schmidt, Rosemarie, and Joseph F. Kess. *Television Advertising and Televangelism: Discourse Analysis of Persuasive Language*. Amsterdam and Philadelphia, Pennsylvania: J. Benjamins, 1986.

Schultze, Quentin. *Televangelism and American Culture*. Grand Rapids, Michigan: Baker, 1991.

Skill, Thomas, and James Robinson. "The Portrayal of Religion and Spirituality on Fictional Network Television." *Review of Religious Research* (New York), 1994.

Wolfe, Kenneth M. *The Churches and the British Broadcasting Corporation: The Politics of Broadcast Religion*. London: SCM Press, 1984.

Wolff, Rick. "'Davey and Goliath': The Response of a Church-produced Children's Television Program to Emerging Social Issues." *Journal of Popular Film and Television*, Fall 1990.

See also Billy Graham Crusades; Christian Broadcasting Network/The Family Channel; *Man Alive*; Parker, Everett C.; Robertson, Pat

REMOTE CONTROL DEVICE

The Remote Control Device (RCD), available in over 90% of U.S. households, has become a central technological phenomenon of popular culture. Though many cartoons, anecdotal accounts, and even television commercials trivialize the RCD, they also reflect its ubiquity and importance in everyday life. For better or for worse, the RCD has permanently altered television viewing habits by allowing the user to exercise some of the functions once the exclusive province of program and advertising executives. The RCD has altered viewing styles by increasing activities such as "zapping" (changing channels during commercials and other program breaks), "zipping" (fast forwarding through pre-recorded programming and advertising) and "grazing" (the combining of disparate program elements into an individualized programming mix).

Although wired RCDs existed in the "Golden Age" of radio, their history is more directly tied to the television receiver manufacturing industry and, more recently, to the diffusion of videocassette recorders (VCRs) and cable television. Zenith Radio Corporation engineer Robert Adler developed the "Space Command," the first practical wireless RCD in 1956. Although other manufacturers would offer both wired and wireless RCDs from the mid-1950s on, the combination of high cost (RCDs typically were available only on more expensive "high end" receivers), technological limitations, and, most critically, the limited number of channels available to most viewers made the RCD more a novelty than a near standard feature of television receivers until the 1980s.

The rapid increase in the number of video distribution outlets in the 1980s was instrumental in the parallel mass diffusion of RCDs. The RCD, in essence, was the necessary tool for the use of cable, VCRs, and more complex television receivers. Without the RCD, the popularity and impact of these programming conduits would have been much less. In the 1990s, a converging television/telecommunications industry redefined the RCD as a navigational tool whose design is essential to the success of interactive consumer services.

While some industry figures see the RCD as a key to the success of future services, the same elements that allow viewers to find and use specific material from the many channels and services available also enables them to avoid content that they find undesirable. Of particular concern are two gratifications of RCD use that have emerged from both academic and industry studies: advertising avoidance and "getting more out of television." These gratifications are symptomatic of a generation of "restless viewers" who

Zenith print ad for remote control television (c. 1957)
Courtesy of Zenith Electronics Corporation

challenge many of the conventional practices of the television industry.

The industry has coped with the RCD "empowered" viewer by implementing changes in programming and advertising. Examples include "seamless" and "hot switch" scheduling where one program immediately segues into the following program, the reduction or elimination of opening themes, shorter and more visually striking commercials, and an increase in advertising/program integration. Although not solely a result of RCD diffusion, the ongoing economic consolidation of the world television/telecommunications industry; the continuing shift of costs to the television viewer/user through cable, pay-per-view, and emerging interactive services; and the increased emphasis on integrated marketing plans that treat traditional advertising spots as only one element of the selling process can all be regarded in part as reactions to restless and RCD-wielding television viewers.

—Robert V. Bellamy Jr.

FURTHER READING

Bellamy, Jr., R.V., and J.R. Walker. *Grazing on a Vast Wasteland: The Remote Control and Television's Second Generation.* New York: Guilford, 1995.

Ferguson, Douglas A. "Channel Repertoire in the Presence of Remote Control Devices, VCRs, and Cable Television." *Journal of Broadcasting and Electronic Media* (Washington, D.C.), Winter 1992.

Walker, J.R., and R.V. Bellamy, Jr., editors. *The Remote Control in the New Age of Television.* Westport, Connecticut: Praeger, 1993.

See also Zapping

RERUNS/REPEATS

A television program that airs one or more times following its first broadcast is known as a rerun or a repeat. In order for a program to be rerun it must have been recorded on film or videotape. Live telecasts, obviously, can not be rerun. The use of reruns is central to the programming and economic strategies of television in the United States and, increasingly, throughout the world.

In the early days of U.S. television, most programming was live. This necessitated the continuous production of new programs which, once aired, were gone. Certain program formats, such as variety, talk, public affairs, quiz, sports, and drama, dominated the airwaves. With the exception of variety and drama, each of these formats is inexpensive to produce, so the creation of live weekly or daily episodes worked fairly well for broadcasters. Even the production costs for variety shows could be reduced over time with the repeated use of sets and costumes.

Production of dramatic programming, however, was more expensive. Most dramatic series were "anthologies"—a different story was programmed each week, with different characters, and often times, different talent. The costs involved in creating each of these plays was considerable and could rarely be reduced, as in the case of variety programs, by repeated use of the durable properties. Because of the expense, dramatic programs decreased, and the number of other less expensive types of programs increased during the first decade of television.

During the early 1950s, however, several weekly prime-time series, most notably *I Love Lucy*, began filming episodes instead of airing live programs. This allowed producers to create fewer than 52 episodes a year, yet still present weekly episodes throughout the year. They could produce 39 new episodes and repeat 13 of those, usually during the summer months when viewership was lower. While some expenses, for additional payments to creative personnel, are involved

in airing reruns, the cost is almost 75% less than that incurred in presenting a new first-run episode. The practice proved so successful that by the end of the 1950s there was very little live entertainment programming left on U.S. television, and the television industry, which had been well established in New York, had shifted its center to Hollywood, the center of U.S. film production.

By the 1970s most network prime-time series were producing only 26 new episodes each year, repeating each episode once (the 26/26 model). And by the 1980s, the standard prime-time model was 22/22 with specials or limited series occupying the remaining weeks.

But the shift to film or videotape as the primary form of television production also turned out to have benefits far exceeding the reduction of production costs and modifications of the programming schedule. Reruns and repeats are not used merely to ease production schedules and cut costs. By contractual arrangement episodes usually return to the control of the producer after two network showings. They may then be licensed for presentations by other television distributors. This strategy is financially viable only after several years of a successful network run, when enough episodes of a television program are accumulated to make the series valuable to other programmers. It does lead to the possibility, however, that reruns of a program can be in syndication forever and almost anywhere. A common industry anecdote claims—and it may be true—that *I Love Lucy* is playing somewhere in the world at any given moment of the day.

The development of the rerun system, particularly as it supports syndication, has become the economic foundation on which the American television industry does business. Because networks, the original distributors of television programs, rarely pay the full production costs for those programs, independent producers and/or studios must create programs at a deficit. That deficit can only be recouped if the program goes into syndication (not a foregone outcome). If the program is sold into syndication the profits may be huge, sufficient to pay off the cost of deficit financing for the original production and to support the development of other series and the programming of less successful programs that may never be syndicated. This entire system is dependent on a sufficient market for rerun programs, a market traditionally composed of independent television stations and the international television systems, and on an economical means of reproduction.

Initially, film was more desirable than videotape as a means of storing programs because film production contracts called for lower residual payments—the payments made to performers in the series when episodes are repeated. Programs produced on film were under the jurisdiction of the Screen Actors Guild (SAG), which required lower residual payments than did the American Federation of Television and Radio Artists (AFTRA), which oversaw programs produced on videotape. By the mid-1970s residual costs for film and taped performances evened out and more and more programs are now produced on or transferred to videotape for syndication.

In addition to their use in prime time, reruns are scheduled in all other dayparts by the networks. Several unions have petitioned the Federal Communications Commision (FCC) in an attempt to restrict network use of reruns, claim that the use of reruns results in a loss of jobs because it leads to less original production. All of these attempts have failed.

With the tremendous growth of television distribution outlets throughout the world in the 1980s—growth often founded on the expansion of cable television systems and the multi-channel environment—additional markets for reruns of old network series were created. So long as these venues continue to increase, the financial basis for American television production will continue to be stable. And as more and more countries establish large programming systems of their own, the amount of material available for second, third and continuing airings will continue to grow.

—Mitchell E. Shapiro

FURTHER READING

Boddy, William. *Fifties Television: The Industry and Its Critics.* Urbana, Illinois: University of Illinois Press, 1990.

"Brits Bank on Rerun Bonanza with U.S. Help." *Variety* (Los Angeles), 28 September 1992.

Eastman, Susan T. *Broadcast/Cable Programming: Strategies and Practices.* Belmont, California: Wadsworth Publishing, 1981; 4th edition, 1993.

Godfrey, Donald G. *Reruns on File: A Guide to Electronic Media Archives.* Hillsdale, New Jersey: Lawrence Erlbaum, 1992.

Moore, Barbara. "The *Cisco Kid* and Friends: The Syndication of Television Series from 1948 to 1952." *Journal of Popular Film and Television* (Bowling Grccn, Ohio), 1980.

Nelson, Jenny L. "The Dislocation of Time: A Phenomenology of Television Reruns." *Quarterly Review of Film and Video* (Chur, Switzerland), October 1990.

Robins, J. Max. "Rerun Resurrection: Webs Favor Old Shows, Newsmags, to Summer Startups." *Variety* (Los Angeles, 27 June 1994.

Shales, Tom. "The Re Decade." *Esquire* (New York), March 1985.

Simon, Ronald. "The Eternal Rerun: Oldies But Goodies." *Television Quarterly* (New York), 1986.

Story, David. *America on the Rerun: TV Shows That Never Die.* Secaucus, New Jersey: Carol, 1993.

Williams, Phil. "Feeding Off the Past: The Evolution of the Television Rerun." *Journal of Popular Film and Television* (Washington, D.C.), Winter 1994.

See also Prime Time Access Rule; Programming; Syndication

RESIDUALS

Residuals are payments made to actors, directors, and writers involved in the creation of television programs or commercials when those properties are rebroadcast or distributed via a new medium. These payments are also called "re-use fees" or "royalties." For example, when a television series goes into syndication, the writers, actors and directors who work on a particular episode are paid a percentage of their original fee each time that episode is re-broadcast. This also includes re-use through cable, pay television, and videocassette sales.

Residuals have played an important part in the history of broadcasting unions. In the early days of live radio in the United States, actors had to perform twice, once for the Eastern time zone, and again three hours later for the Pacific time zone. As recording technology developed, networks recorded the first performance and simply replayed it three hours later. In 1941, the American Federation of Radio Artists (AFRA) insisted that the actors be compensated for the rebroadcast.

With the development of television in the 1950s, a new distribution outlet was created for motion pictures. The Hollywood unions representing actors, writers, and directors feared that they would lose job opportunities if television stations broadcast preexisting movies instead of paying for new programming. In 1951, the American Federation of Musicians negotiated residual payments with film producers for the broadcast of movies on television. The next year, the Screen Actors Guild (SAG) conducted a strike against

Monogram Pictures to force the company to make residual payments when its movies were broadcast on television. In the mid-1950s, SAG was also able to negotiate residual payments from the emerging television networks for reruns and from advertisers for the reuse of television commercials. By 1960, residuals had become standard practice throughout the film and television industry.

When new distribution markets emerge, unions such as the American Federation of Television and Radio Artists (AFTRA), the Writers Guild of America (WGA), the Directors Guild (DGA), and SAG negotiate for residuals in those markets as well. In the 1970s and 1980s, unions negotiated residuals for cable, videocassette, pay per view, and even in-flight movies on airplanes. In the mid-1990s, unions were fighting for residuals in new markets such as CD-ROMs and computer networks.

Residuals are a lucrative source of income, and thus a major source of contention between unions and producers. As Archie Kleingartner and Alan Paul point out in *Labor Relations and Residual Compensation in The Movie and Television Industry* (1992), residuals have played a major role in 18 strikes by the various unions. Low-paid actors working in television commercials often earn four times as much from residuals as they do from their initial fees. Series actors, who are paid much more for their initial services, still earn about 30% of their income from residuals. In 1990, total residual payments exceeded 337 million dollars, not counting residuals from television commercials.

Unions negotiate residuals with the Alliance of Motion Picture and Television Producers (AMPTP), which represents most studios and independent producers. In 1995, the three biggest television networks (ABC, CBS, and NBC) separated from AMPTP over a dispute regarding the status of the newer Fox network. The three older networks wanted Fox to pay the same residual rates that they pay, while Fox argued that it was not technically a network by FCC standards. At this time, the unions negotiate with the three networks and AMPTP separately.

Residuals are an important source of compensation for actors, writers, and directors whose works are distributed in an ever wider array of foreign and domestic markets. They are a major factor in the continuing strength of the various unions over the years.

—Matt Jackson

FURTHER READING

Gilbert, Robert W. "'Residual Rights' Established by Collective Bargaining in Television and Radio." *Law and Contemporary Problems* (Durham, North Carolina), 1958.

Kleingartner, Archie, and Alan Paul. *Labor Relations and Residual Compensation in the Movie and Television Industry.* (Working Paper Series No. 224). Los Angeles, California: UCLA Institute of Industrial Relations, 1992.

Mittleman, Shel. "Residuals Under the Guild Agreements—WGA, DGA, IATSE, SAG and AFM: Accommodating the New Media." In, *Reel of Fortune: A Discussion of the Critical Business and Legal Issues Affecting Film and Television Today.* Los Angeles, California: University of California at Los Angeles, The Twelfth Annual UCLA Entertainment Symposium, 1987.

Paul, Alan, and Archie Kleingartner. "Flexible Production and the Transformation of Industrial Relations in the Motion Picture and Television Industry." *Industrial and Labor Relations Review* (Ithaca, New York) 1994.

Prindle, David F. *The Politics of Glamour: Ideology and Democracy in the Screen Actors Guild.* Madison, Wisconsin: University of Wisconsin Press, 1988.

Spring, Greg. "Unions Approve Film, TV Production Pacts." *Los Angeles Business Journal* (Los Angeles), 22 May 1995.

See also Syndication; Unions; Writing for Television

REYNOLDS, GENE

U.S. Actor/Producer/Director

From a child movie actor in *Boy's Town*, Gene Reynolds grew into a respected producer-director identified with thoughtful television dramas reflecting complex human situations. The programs Reynolds is associated with often possess an undercurrent of humor to entertain, but without softening socially significant story lines.

As producer-director of *Room 222* (1969-74), Reynolds found a supportive, kindred spirit in the series' creator James L. Brooks. Exploring life among high school teachers, administrators, and students, their program featured African-American actor Lloyd Haynes as a revered, approachable teacher. A lighter touch in dialogue and situations helped keep the stories attractive to casual viewers. Still, the central characters were involved each week in matters of personal and social import such as drugs, prejudice, self-worth, and dropping out of school.

Again aligning himself with a congenial, creative associate for a TV version of the novel and motion picture *M*A*S*H*, Reynolds sought out respected "comedy writer with a conscience" Larry Gelbart. Together they fleshed out a sensitive, probing, highly amusing, and wildly successful series about the foibles and aspirations of a military surgical team in the midst of warfare. Raucous, sometimes ribald comedy acted as counterpoint to poignant human dilemmas that are present when facing bureaucratic tangles amid willful annihilation. Though intended as comedy-drama commentary on the devastating absurdities of war in general, and the Vietnam conflict in particular, Reynolds and Gelbart pushed the time period of their show back to Korea in the 1950s in order to be acceptable to the network and stations, and to a deeply-divided American public. Gelbart left the series early on, and Reynolds eventually became executive producer, turning the producer's role over to Burt Metcalf. The ensemble cast only grew stronger as new actors replaced departing ones through the decade. The acclaimed series earned awards from all sectors during its 11-year run (1972-83), including the Peabody Award in 1975, Emmy Awards for outstanding comedy series in 1974; Emmys many other seasons for outstanding writing, acting, and direction; Emmys twice for best directing by Gene Reynolds (1975, 1976); and the Humanitas Prize.

The public voted, too; their sustained viewing kept the program among the top-ranked five or ten programs every year *M*A*S*H* aired. The concluding two-and-one-half-hour "farewell" episode (February 28, 1983) still stands as the single-most-watched program in American TV history, attracting almost two out of every three homes in America (60.3 rating). More than 50 million families tuned in that evening to watch the program.

Reynolds left *M*A*S*H* in 1977. He teamed up again with James L. Brooks and Alan Burns, all as executive producers of *Lou Grant*. This series explored the combative turf of a major metropolitan newspaper. It dealt with the constitutional and ethical issues found in pitting journalists against politicians, corporate executives, courts, and the general public. Reynolds' creative team avoided cliché-driven plots, focusing instead on complex, unresolved issues and depicting their impact on a mix of vulnerable personalities. The series (1977–82) received critical acclaim, including Peabody, Emmy, and Humanitas Awards, for exploring complicated challenges involving media and society.

Gene Reynolds' modus operandi for producing a television series is to thoroughly research the subject area by extended visits to sites—schools, battlefields (Vietnam to replicate Korean field hospitals), and newspaper offices. There he interviews at length those engaged in career positions. He and his creative partners regularly returned to those sites armed with audiotape recorders to dig for new story ideas, for points of view, for technical jargon and representative phrases, and even for scraps of dialogue that would add verisimilitude to the words of studio-stage actors recreating an incident. Reynolds and his associates always strive for accuracy, authenticity, and social significance. They present individual human beings caught up in the context of controversial events, but affected by personal interaction.

A thoughtful, serious-minded creator with a quiet sense of humor, Gene Reynolds' ability to work closely with colleagues earns the respect of both actors and production crews. He often directs episodes, regularly works with writers on revising scripts, and establishes a working climate on the set that invites suggestions from the actors for enhancing dialogue and action.

Reynolds directed pilots for potential TV series and movies for television, including *People Like Us* (1976), *In Defense of Kids* (1983), and *Doing Life* (1986). In 1995,

Gene Reynolds
Photo courtesy of Gene Reynolds

having served actively in organizations and on committees in the creative community for many years, he was elected president of the Directors Guild of America.

—James A. Brown

GENE REYNOLDS. Born in Cleveland, Ohio, U.S.A., 4 April 1925. Married: Bonnie Jones. Began career as film actor, debut in *Thank You, Jeeves*, 1936; producer and director of numerous television series, from 1968. Recipient: five Emmy Awards, Directors Guild of America Award, Peabody Award.

TELEVISION (producer)

1968–70	*The Ghost and Mrs. Muir* (pilot)
1969–74	*Room 222* (executive producer)
1972	*Anna and the King*
1972–83	*M*A*S*H* (also director)
1973–74	*Roll Out*
1975	*Karen*
1977–82	*Lou Grant*

MADE-FOR-TELEVISION MOVIES (selection)

1976	*People Like Us* (producer, director)
1983	*In Defense of Kids* (director)

1986   *Doing Life* (director)

FILMS

Thank You, Jeeves (actor) 1936; *In Old Chicago* (actor), 1937; *Boys Town* (actor), 1938; *They Shall Have Music* (actor), 1939; *Edison, the Man* (actor), 1940; *Eagle Squadron* (actor), 1942; *The Country Girl* (actor), 1954; *The Bridges at Toko-Ri* (actor), 1955; *Diane* (actor), 1955.

See also *Lou Grant; M*A*S*H; Room 222*

RICH MAN, POOR MAN

U.S. Miniseries

One of the first American television miniseries, *Rich Man, Poor Man* aired on ABC from 1 February to 15 March 1976. Adapted from the best-selling 1970 Irwin Shaw novel, *Rich Man, Poor Man* was a limited twelve-part dramatic series consisting of six two-hour prime-time made for television movies. The televised novel chronicled the lives of the first-generation immigrant Jordache family. The story focused on the tumultuous relationship between brothers, Rudy (Peter Strauss) and Tom Jordache (Nick Nolte), as they suffered through twenty years (1945–65) of conflict, jealousy, and heartbreak.

The serial was enormously successful, leading the weekly ratings and ending as the second highest rated how for the 1976–77 television season. Along with its enormous audience popularity, it also garnered critical praise, reaping 20 Emmy nominations and winning four—two for acting achievement, one for directing, and one for musical score.

The success of *Rich Man, Poor Man* hinged on its employment of several innovative techniques. The narrative struck a unique combination which contained both the lavish film-style production values of prestigious special event programming while relying upon the "habit viewing" characteristic of a weekly series. Also, by utilizing historical backdrops like McCarthyism, the Korean War, campus riots, and the Black Revolution, *Rich Man, Poor Man* suggested larger circumstances than those usually found in a traditional soap opera. However, the limited series also liberally applied a range of risqué melodramatic topics including adultery, power struggles, and alcoholism. Another inventive concept introduced by *Rich Man, Poor Man* was the use of multiple, revolving guest stars throughout the series. While the three principal cast members were relatively unknown at the time, shuffling better known actors throughout the six-part series was a way to maintain interest and achieve some form of ratings insurance on the six-million dollar venture.

By invigorating the concept of adapting novels into television miniseries, *Rich Man, Poor Man* began a rapid proliferation of similar prime-time programming, including a sequel. The continuation, *Rich Man, Poor Man—Book II*, was a twenty-one part weekly series that aired in the fall of 1976. Although the sequel was not as successful as its predecessor, the idea of extended televised adaptations of popular novels quickly became a component of network schedules. In the season following the debut of *Rich Man, Poor Man*, all major networks scheduled at least one miniseries, including an adaptation of Harold Robbins' *The Pirates* and Alex Haley's historical epic *Roots*.

Although eclipsed by the record-breaking 1977 miniseries *Roots* (aired 1 January through 30 January on ABC), *Rich Man, Poor Man* nonetheless has staked a spot in television history. It helped to create a special niche for televised novels as an economically viable miniseries genre that can still be found in such offerings as *North and South* and *Lonesome Dove*.

—Liza Treviño

CAST—BOOK I

Rudy Jordache	Peter Strauss
Tom Jordache	Nick Nolte
Julie Prescott Abbott Jordache	Susan Blakely
Axel Jordache	Edward Asner
Mary Jordache	Dorothy McGuire
Willie Abbott	Bill Bixby
Duncan Calderwood	Ray Milland
Teddy Boylan	Robert Reed
Virginia Calderwood	Kim Darby
Sue Prescott	Gloria Grahame
Asher Berg	Craig Stevens
Joey Quales	George Maharis
Linda Quales	Lynda Day George
Nichols	Steve Allen
Smitty	Norman Fell
Teresa Sanjoro	Talia Shire
Marsh Goodwin	Van Johnson
Irene Goodwin	Dorothy Malone
Kate Jordache	Kay Lenz
Sid Gossett	Murray Hamilton
Arnold Simms	Mike Evans
Al Fanducci	Dick Butkus
Clothilde	Fionnula Flanagan
Brad Knight	Tim McIntire
Bill Denton	Lawrence Pressman
Claude Tinker	Dennis Dugan
Gloria Bartley	Jo Ann Harris
Pete Tierney	Roy Jenson
Lou Martin	Anthony Carbone
Papadakis	Ed Barth
Ray Dwyer	Herbert Jefferson, Jr.
Arthur Falconetti	William Smith
Col. Deiner	Andrew Duggan

Rich Man, Poor Man

Pinky	Harvey Jason
Martha	Helen Craig
Phil McGee	Gavan O'Herlihy
Billy	Leigh McCloskey
Wesley	Michael Morgan

PRODUCERS   Harve Bennett, Jon Epstein

PROGRAMMING HISTORY   9 Episodes

• ABC

February 1976–March 1976	Monday 10:00-11:00
May 1977–June 1977	Tuesday 9:00-11:00

CAST—BOOK II

Senator Rudy Jordache	Peter Strauss
Wesley Jordache	Gregg Henry
Billy Abbott	James Carroll Jordan
Maggie Porter	Susan Sullivan
Arthur Falconetti	William Smith
Marie Falconetti	Dimitra Arliss
Ramona Scott	Penny Peyser

Scotty	John Anderson
Charles Estep	Peter Haskell
Phil Greenberg	Sorrell Brooke
Annie Adams	Cassie Yates
Diane Porter	Kimberly Beck
Arthur Raymond	Peter Donat
Claire Estep	Laraine Stephens
Senator Paxton	Barry Sullivan
Kate Jordache	Kay Lenz
John Franklin	Philip Abbott
Max Vincent	George Gaynes
Al Barber	Ken Swofford
Senator Dillon	G. D. Spradlin

PRODUCERS   Michael Gleason, Jon Epstein

PROGRAMMING HISTORY   21 Episodes

• ABC

September 1976–March 1977	Tuesday 9:00-10:00

See also Adaptation; Miniseries

RIGG, DIANA

British Actor

After shooting her first twelve episodes in the role of Mrs. Emma Peel in *The Avengers*, Diana Rigg made one of those discoveries most likely to madden newly-minted stars: her weekly salary as the female lead in an already highly successful series was £30 less than what *The Avengers*' lowly cameraman earned. Rigg had not even been the first choice to replace the popular Honor Blackman as Steed's accomplice; the first actress cast had been sacked after two weeks. The role then fell to Rigg, whose television résumé at the time consisted only of a guest appearance on *The Sentimental Agent* and a performance of Donald Churchill's *The Hothouse*.

Rigg's stage experience, however, was already solid. After joining the Royal Shakespeare Company (RSC) in 1959, the same year as Vanessa Redgrave, Rigg had steadily amassed a strong string of credits, including playing Cordelia to Paul Schofield's Lear. Years later, Rigg described the rationale for her turn to television: "The trouble with staying with a classical company is that you get known as a 'lady actress.' No one ever thinks of you except for parts in long skirts and blank verse."

Rigg's salary complaints were quickly satisfied, and American audiences, who had never been exposed to Blackman's *Avengers* episodes until the early 1990s, quickly embraced Rigg's startlingly assertive (but always upper-class) character. Peel's name may have been simply a play upon the character's hoped-for "man appeal," but Rigg's embodiment of the role suggested a much more utopian representation of women; like Peel—and Rigg's own persona in interviews—women can be intelligent, independent, and sexually confident. After three seasons and an Emmy nomination, Rigg left the series in 1968, claiming "Emma Peel is not fully emancipated." Still, she resisted publicly associating herself with feminism; to the contrary, Rigg flippantly claimed to find "the whole feminist thing very boring."

Following Blackman into Bond films (in 1964 Blackman had been *Goldfinger*'s Pussy Galore), Rigg's presence in *On Her Majesty's Secret Service* (1969) as the tragic Mrs. James Bond added intertextual interest to the film. Paired with the unfamiliar George Lazenby as Bond, it was Rigg who carried the film's spy genre credentials, even though her suicidal, spoiled character displayed few of Peel's many abilities. But the British spy genre had already begun to collapse, followed by the rest of the nation's film industry, and Rigg's career as a movie star never soared.

Rigg did not immediately return to series television. In fact, she publicly attributed her problems on film to having learned to act for television only too well—she had become too "facile" before film cameras, a trait necessitated by the grueling pace of series production. Apparently her stage skills remained unaffected, and Rigg went on to assay a wide range of both classical and contemporary roles as a member of the RSC, the National Theatre, and Broadway. But while Rigg has originated

Diana Rigg
Photo courtesy of Diana Rigg

the lead roles in such stylish works as Tom Stoppard's *Jumpers* (1972), the stage work she performed for television broadcast tended to fit more snugly into familiar anglophilic conventions. In the United States, her television appearances in the 1960s included *The Comedy of Errors* (1967) and *Women Beware Women* (1968) for *N.E.T. Playhouse*; in the 1980s, they included *Hedda Gabler*, *Witness for the Prosecution*, Lady Dedlock in a multi-part adaptation of *Bleak House* (1985), and Laurence Olivier's *King Lear* (1985).

During the decade between, however, NBC attempted to capitalize upon what Rigg jokingly called her "exploitable potential" following *The Avengers*. After one failed pilot, the network picked up *Diana* (1973-74), a *Mary Tyler Moore Show*-inspired sitcom, and Rigg returned to series television as a British expatriate working in New York's fashion industry. As if to acknowledge the sexual daring of her first series, Rigg's character became American sitcoms' first divorcee, just as Moore's character had been initially conceived. But the comic actress television critics had once praised as wry and deliberately understated did not appear; in *Diana*, Rigg appeared rather bland, and the series provided no Steed for verbal repartee. (Perhaps even more damning, *Diana* showed few traces of *The Avengers*' always dashing fashion sense.) NBC programmed *Diana* during what had once been *The Avengers*' time slot, but the sitcom shortly disappeared.

Only a year later, Rigg successfully played off both her previous roles and her sometimes bawdy public persona in a sober religious drama, *In This House of Brede* (1975). Portraying a successful businesswoman entering a convent, Rigg's combination of restraint and technique seemed quintessentially British, and earned her a second Emmy nomination.

In recent years, however, Rigg's range of roles seems more limited to one-dimensional versions of the days when she masqueraded as *The Avengers*' "Queen of Sin." Of course middle age has, as for many other women, resulted in a narrowed range of options, particularly in film. Still, Rigg carries a coolly sexual charge: she has taken on a range of "ageless" stage roles (including *Medea*, for which she won a 1994 Tony Award), as well as more and more character roles on television. Most often these latter roles are villainous to some degree, whether in bodice rippers (*A Hazard of Hearts*, 1987), light comedy (*Mrs. 'arris Goes to Paris*, 1992), or edgy comedy like the Holocaust farce *Genghis Cohn* (1994).

In 1990, Rigg impressed American audiences as the star of an Oedipal nightmare, *Mother Love*, a multi-part British import presented as part of the PBS series, *Mystery!* Rigg had also succeeded Vincent Price in hosting *Mystery!* in 1989. In a sense, Rigg has become that "lady actress" she had once entered television to avoid: ensconced in finely tailored suits and beaded gowns, her performance as host displays all the genteel, ambassadorial authority of a woman now entitled to be addressed as Dame Rigg (Dame Commander, Order of the British Empire, 1994).

—Robert Dickinson

DIANA RIGG. Born in Doncaster, Yorkshire, England, 20 July 1938. Attended Fulneck Girls' School, Pudsey; Royal Academy of Dramatic Art, London. Married: 1) Menachem Gueffen, 1973 (divorced, 1974); 2) Archibald Stirling, 1982; child: Rachel. Began career as stage actor, making debut with RADA during the York Festival at the Theatre Royal, York, 1957; made London stage debut, 1961; member, Royal Shakespeare Company (RSC), 1959–64; made London debut with RSC, Aldwych Theatre, 1961; toured Europe and the United States with RSC, 1964; made television debut as Emma Peel in *The Avengers*, 1965; film debut, 1967; joined National Theatre Company, 1972; has since continued to appear in starring roles both on screen and on stage; director, United British Artists, since 1982; vice president, Baby Life Support Systems, since 1984. Companion of the Order of the British Empire, 1988; Dame Commander of the Order of the British Empire, 1994. Chair: Islington Festival; MacRoberts Arts Centre. Recipient: *Plays and Players* Award, 1975, 1979; Variety Club Film Actress of the Year Award, 1983; British Academy of Film and Television Arts Award, 1989; *Evening Standard* Drama Award, 1993; Tony Award, 1994. Address: London Management, 235 Regent Street, London W1A 2JT, England.

TELEVISION SERIES (selection)

1965–67	*The Avengers*
1973–74	*Diana*
1989–	*Mystery!* (host)

MADE-FOR-TELEVISION MOVIES (selection)

1975	*In This House of Brede*
1982	*Witness for the Prosecution*
1986	*The Worst Witch*
1987	*A Hazard of Hearts*
1989	*Mother Love*
1994	*Genghis Cohn*
1995	*The Haunting of Helen Walker*
1996	*Chandler and Co.*

TELEVISION SPECIALS (selection)

1964	*The Hothouse*
1968	*Women Beware Women*
1981	*Hedda Gabler*
1985	*King Lear*
1985	*Bleak House*
1986	*Masterpiece Theatre: 15 Years*
1992	*The Laurence Olivier Awards 1992* (host)

FILMS (selection)

A Midsummer Night's Dream, 1968; *The Assassination Bureau*, 1969; *On Her Majesty's Secret Service*, 1969; *Married Alive*, 1970; *Julius Caesar*, 1970; *The Hospital*, 1971; *Theatre of Blood*, 1973; *A Little Night Music*, 1977; *The Serpent Son*, 1979; *Hedda Gabler*, 1980; *The Great Muppet Caper*, 1981; *Evil Under the Sun*, 1982; *Little Eyolf*, 1982; *Held in Trust*, 1986; *Snow White*, 1986.

STAGE (selection)

The Caucasian Chalk Circle, 1957; *Ondine*, 1961; *The Devils*, 1961; *Becket*, 1961; *The Taming of the Shrew*, 1961; *Madame de Tourvel*, 1962; *The Art of Seduction*, 1962; *A Midsummer Night's Dream*, 1962; *Macbeth*, 1962; *The Comedy of Errors*, 1962; *King Lear*, 1962; *The Physicists*, 1963; *Twelfth Night*, 1966; *Abelard and Heloise*, 1970; *Jumpers*, 1972; *'Tis Pity She's a Whore*, 1972; *Macbeth*, 1972; *The Misanthrope*, 1974; *Pygmalion*, 1974; *Phaedra Britannica*, 1975; *The Guardsman*, 1978; *Night and Day*, 1979; *Colette*, 1982; *Heartbreak House*, 1983; *Little Eyolf*, 1985; *Antony and Cleopatra*, 1985; *Wildfire*, 1986; *Follies*, 1986; *Love Letters*, 1990; *All for Love*, 1991; *Berlin Bertie*, 1992; *Medea*, 1992.

PUBLICATIONS

No Turn Unstoned. Garden City, New York: Doubleday, 1982.
So to the Land. London: Headline, 1994.

FURTHER READING

Jenkins, Henry. *Textual Poachers: Television Fans and Participatory Culture.* New York and London: Routledge, 1992.
Nathan, David. "Heavy-Duty Lightweight." (London) *Times*, 20 April 1991.
Rogers, Dave. *The Avengers.* London: ITV Books, 1983.
Story, David. *America on the Rerun: TV Shows That Never Die.* New York: Citadel Press, 1993.

See also *Avengers*

RIGGS, MARLON

U.S. Filmmaker

Before his death in 1994, African American filmmaker, educator and poet Marlon Riggs forged a position as one of the more controversial figures in the recent history of public television. He won a number of awards for his creative efforts as a writer and video producer. His theoretical-critical writings appeared in numerous scholarly and literary journals and professional and artistic periodicals. His video productions, which explored various aspects of African-American life and culture, earned him considerable recognition, including Emmy and Peabody awards. Riggs will nonetheless be remembered mostly for the debate and contention that surrounded the airing of his highly charged video productions on public television stations during the late 1980s and early 1990s. Just as art-photographer Robert Mapplethorpe's provocative, homoerotic photographs of male nudes caused scrutiny of government agencies and their funding of art, Marlon Riggs' video productions similarly plunged public television into an acrimonious debate, not only about funding, but censorship as well.

Riggs' early works received little negative press. His production *Ethnic Notions* aired on public television stations through the United States. This program sought to explore the various shades of mythology surrounding the ethnic stereotyping of African Americans in various forms of popular culture. The program was well-received and revolutionary in its fresh assessment of such phenomenon as the mythology of the Old South and its corresponding caricatures of black life and culture.

The video *Color Adjustment*, which aired on public television stations in the early 1990s, was an interpretive look at the images of African-Americans in fifty years of American television history. Using footage from shows like *Amos 'n' Andy*, *Julia*, and *Good Times*, Riggs compared the grossly stereotyped caricatures of blacks contained in early television programming to those of recent, and presumably more enlightened, decades.

By far the most polemical of Riggs' work was his production, *Tongues Untied*. This fifty-five minute video, which "became the center of a controversy over censorship" as reported *The Independent* in 1991, was aired as part of a series entitled *P.O.V. (Point of View)*, which aired on public television stations and featured independently produced film and video documentaries on various subjects ranging from personal reflections on the Nazi holocaust to urban street life in contemporary America.

Tongues Untied is noteworthy on at least three accounts. First, Riggs chose as his subject urban, African-American gay men. Moving beyond the stereotypes of drag queens and comic-tragic stock caricatures, Riggs offered to mainstream America an insightful and provocative portrait of a distinct gay sub-culture—complete with sometimes explicit language and evocative imagery. Along with private donations, Riggs had financed the production with a $5,000 grant from

Marlon Riggs
Photo courtesy of Signifyin' Works/Andy Stern

the National Endowment for the Arts (NEA), a federal agency supporting visual, literary and performing artists. News of the video's airing touched off a tumult of debate about the government funding of artistic creations that to some were considered obscene. While artists argued the basic right of free speech, U.S. government policy makers, especially those of conservative bent, engaged in hotly contentious debate regarding the use of taxpayer money for the funding of such endeavors.

The second area of consternation brought on by the *Tongues Untied* video concerned the area of funding for public broadcasting. The *P.O.V.* series also received funding from the NEA, in the amount of $250,000, for its production costs. Many leaders of conservative television watch-dog organizations labeled the program as obscene (though many had not even seen it). Others ironically heralded the program's airing, in the hope that American taxpayers would be able to watch in dismay how their tax dollars were being spent.

Lastly, the question of censorship loomed large throughout the debate over the airing *Tongues Untied*. When a few frightened station executives decided not to air the program, the fact of their self-censorship was widely reported in the press. As mentioned, *Tongues Untied* was not the first *P.O.V.* production to be pulled. Arthur Kopp of People for the American Way noted in *The Independent* "the most insidious censorship is self-censorship. . . It's a frightening sign when television executives begin to second guess the far right and pull a long-planned program before it's even been attacked."

Riggs defended *Tongues Untied* by lambasting those who objected to the program's language and imagery by stating in a 1992 *Washington Post* interview, "People are far more sophisticated in their homophobia and racism now . . .they say 'We object to the language, we have to protect the community'. . . those statements are a ruse."

Tongues Untied was awarded Best Documentary of the Berlin International Film Festival, Best Independent Experimental Work by the Los Angeles Film Critics, and Best Video by the New York Documentary Film Festival.

Before his death Riggs began work on a production entitled *Black Is, Black Ain't.* In this video presentation Riggs sought to explore what it meant to be black in America, from the period when "being Black wasn't always so beautiful" to the 1992 Los Angeles riots. This visual reflection on gumbo, straightening combs, and Creole life in New Orleans was Riggs' own personal journey. It also unfortunately served as a memorial to Riggs. Much of the footage was shot from his hospital bed as he fought to survive the ravaging effects of AIDS. The video was finished posthumously and was aired on public television during the late 1990s.

—Pamala S. Deane

MARLON RIGGS. Born in Ft. Worth, Texas, U.S.A., 3 February 1957. Graduated from Harvard University, magna cum laude, B.A. in history 1978; University of California at Berkeley, M.A. in journalism 1981. Taught documentary film, Graduate School of Journalism, University of California, Berkeley, from 1987; produced numerous video documentaries, from 1987. Honorary doctorate, California College of Arts and Crafts, 1993. Recipient: Emmy Awards, 1987 and 1991; George Foster Peabody Award, 1989; Blue Ribbon, American Film and Video Festival, 1990; Best Video, New York Documentary Film Festival, 1990; Erik Barnouw Award, 1992. Died in Oakland, California, 5 April 1994.

TELEVISION DOCUMENTARIES

1987	*Ethnic Notions*
1988	*Tongues Untied*
1989	*Color Adjustment*
1992	*Non, Je Ne Regrette Rein (No Regret)*
1994	*Black Is, Black Ain't*

PUBLICATIONS (selection)

"Black Macho Revisited: Reflections of a Snap! Queen." *Black American Literature Forum* (Terre Haute, Indiana), Summer 1991.

"Notes of a Signifying Snap! Queen." *Art Journal* (New York), Fall, 1991.

Grundmann, R. "New Agendas in Black Filmmaking: An Interview with Marlon Riggs." *Cineaste* (New York), 1992.

FURTHER READING

Becquer, M. "Snap-Thology and other Discursive Practices in *Tongues Untied.*" *Wide Angle—A Quarterly Journal of Film History, Theory and Criticism* (Athens, Ohio), 1991.

Berger, M. "Too Shocking to Show." *Art in America* (New York), July 1992.

Creekmur, Corey K., and Alexander Doty. *Out in Culture: Gay, Lesbian, and Queer Essays on Popular Culture.* Durham, North Carolina: Duke University Press, 1995.

Harper, Phillip Brian. "Marlon Riggs: The Subjective Position of Documentary Video." *Art Journal* (New York), Winter 1995.

Maslin, Janet. "Under Scrutiny: TV Images of Blacks." *The New York Times*, 29 January 1992.

Mercer, Kobina. "Dark and Lovely Too: Black Gay Men in Independent Film." In, Gerver, Martha, with others, editors. *Queer Looks: Perspectives on Lesbian and Gay Film and Video.* New York: Routledge, 1993.

Mills, David. "The Director with Tongue Untied; Marlon Riggs, A Filmmaker Who Lives Controversy." *Washington* (D.C.) *Post,* 15 June 1992.

Prial, Frank J. "TV Film About Gay Blacks is Under Attack." *The New York Times*, 25 June 1991.

Scott, Darieck. "Jungle Fever? Black Gay Identity Politics, White Dick, and the Utopian Bedroom." *GLQ: A Journal of Lesbian and Gay Studies* (Yverdon, Switzerland), 1994.

See also Public Service Broadcasting; Racism, Ethnicity, and Television

RINTELS, DAVID W.

U.S. Writer/Producer

Writer-producer David W. Rintels has worked in a variety of dramatic television forms, including series, made-for-television movies, and miniseries. He began his television career in the early 1960s, writing episodes for the critically acclaimed CBS courtroom drama series *The Defenders.* He continued his series involvement writing episodes for *Slattery's People* (1964–65), a CBS political drama, and be-

came head writer for the ABC science-fiction series *The Invaders* (1967–68) before concentrating his energies on writing and producing made-for-television movies and miniseries. His work has been honored with two Emmy Awards for outstanding writing (*Clarence Darrow*, 1973, and *Fear On Trial*, 1975); Writers Guild of America Awards for outstanding script ("A Continual Roar of Musketry," Parts I and II of the series *The*

Senator, 1970, *Fear On Trial*, 1975, and *Gideon's Trumpet*, 1980); and a cable ACE Award for writing (*Sakharov*, 1984). Rintels' credits also include the sole story and joint screenplay for the feature film *Scorpio* (1972).

Rintels' television work in the genres of fictional history (using novelistic invention to portray real historical figures and events) and historical fiction (placing fictional characters and events in a more or less authentic historical setting) has been praised by *Los Angeles Times* television critic Howard Rosenberg, who noted that Rintels' "fine record for using TV to present history as serious entertainment is probably unmatched by any other present dramatist." Some critics have argued, however, that while his faithfulness to historical detail and accuracy is commendable, his use of lengthy expository sequences has, on occasion, diminished the stories' dramatic power.

Following his involvement as an episode writer for *The Defenders*, the Emmy Award-winning drama series featuring a father and son legal team defending people's constitutional rights, Rintels returned to the subject of the courts in *Clarence Darrow* (NBC, 1973), and *Gideon's Trumpet* (CBS, 1980), a *Hallmark Hall of Fame* production which he both wrote and produced. The latter, based on Anthony Lewis' book, was the real-life story of Clarence Earl Gideon (played by Henry Fonda). A drifter with little education, Gideon was arrested in the early 1960s for "breaking and entering." The U.S. Supreme Court held that Gideon was entitled to an attorney, although he could not afford to pay for one.

Rintels also frequently focused on the political sphere, and especially on idealistic individuals who become ensnared in the nefarious webs woven by those seeking power or influence. In "A Continual Roar of Musketry," he developed the character of Hayes Stowe, an idealistic U.S. senator (played by Hal Holbrook).

In the 1975 CBS docudrama *Fear on Trial*, starring George C. Scott and William Devane, Rintels told the story of John Henry Faulk, a homespun radio personality who wrote a book about the blacklisting in television in the 1950s. Upon its publication, Faulk suddenly found his own name appearing in the AWARE bulletin—a blacklisting sheet created by two Communist-hunting businessmen who proclaimed themselves protectors of the entertainment industry.

Washington: Behind Closed Doors (1977), a twelve-and-one-half-hour ABC miniseries co-written (with Eric Bercovici) and co-produced by Rintels, was a provocative examination of the Nixon Administration, including a striking psychological portrait of Nixon, fictionalized as President Richard Monckton. Played to perfection by Jason Robards, Nixon is described by Michael Arlen as "nervous and disconnected . . . insecure, vengeful, riddled with envy, and sublimely humorless." Although loosely based on *The Company*—the rather amateurish first novel of Nixon insider John Erlichman—the Rintels and Bercovici script transcended Erlichman's unidimensional characterizations to bring to the small screen "an intelligent and well-paced scenario of texture and character." Yet working in the genre of historical fiction was not without its pitfalls. In a foreshadowing of the heated debate surrounding Oliver Stone's

David W. Rintels
Photo courtesy of David W. Rintels

1995 feature film *Nixon*, Arlen questioned the production's mixing of fiction with fact:

> . . . there should be room in our historical narratives for such a marvelously evocative (though not precisely factual) interpretation as Robards' depiction of Nixon-Monckton's strange humorous humorousless-ness, where an actor's art gave pleasure, brought out character, and took us closer to truth. At the same time, for major television producers . . . to be so spaced out by the present Entertainment Era as to more or less deliberately fool around with the actual life of an actual man, even of a discredited President . . . seems irresponsible and downright shabby.

Rintels turned his attention to political repression abroad in *Sakharov* (HBO, 1984), the moving story of the courageous Soviet scientist Andrei Dmitrievich Sahkarov, played by Jason Robards, and his second wife Yelena G. Bonner, played by Glenda Jackson. *Sakharov* chronicles the 1975 Nobel Peace Prize winner's painful journey into dissent, and his outspoken advocacy of human rights. Because so much information about affairs in the Soviet Union was cloaked in secrecy, it would have been tempting to invent much of Sakharov's tale. Rintels, however, was loath to do this. Rather, in order to present the personal side of Sakharov, he compiled information from extensive interviews with his children and their spouses, who had emigrated to the United States, and with Yelena Bonner's mother. He also drew upon Sakharov's own accounts and those of his friends, and on reports from journalists stationed in

Moscow. As the story unfolded for Rintels, he decided to use, as a primary framing device, Sakharov's "growing awareness—through his personal relationship with Yelena—of his moral duty." Rintels was careful to avoid painting the Soviet bureaucrats and security police as "evil" in simplistic melodramatic terms in order to glorify Sakharov. The script attempted to explain why the Soviet officials perceived Sakharov as an internal threat, and was circumspect regarding his motivations when the facts (or lack thereof) warranted.

In two recent efforts—*Day One* (AT and T Presents/CBS, 1989), and *Andersonville* (TNT, 1996)—Rintels has examined America at war. *Day One* was a three-hour drama special detailing the history of the Manhattan Project to build an atomic bomb during World War II. Based on Peter Wyden's book *Day One: Before Hiroshima and After*, the program was written and produced by Rintels, and won an Emmy Award for Outstanding Drama Special. The story began with the flight of top European scientists, who feared Nazi Germany was progressing toward developing an atomic bomb, to the United States. Near its conclusion, a lengthy, balanced, and soul-searching debate transpires among scientists, military leaders, and top civilian government officials, including President Truman, regarding whether to drop the bomb on Japan without prior notice, or to invite Japanese officials to a demonstration of the bomb in hopes that they would surrender upon seeing its destructive power. Throughout the piece, Rintels explores the symbiotic relationship that developed between the two key players in the Manhattan Project—the intellectual scientist and project leader, J. Robert Oppenheimer, and the military leader charged with overall coordination of the effort, General Leslie R. Groves.

Andersonville, a four-hour, two-part drama written and produced by Rintels, recounts the nightmare of the Civil War Confederate prison camp in Southwest Georgia—a 26-acre open-air stockade designed for 8,000 men, which at peak operation contained 32,000 Union Army prisoners of war. Of the 45,000 Union soldiers imprisoned there between 1864 and 1865, nearly 13,000 died, mostly from malnutrition, disease, and exposure. Not only were the Confederate captors cruel; there also existed in the camp a ruthless gang of prisoners, the Raiders, who intimidated, beat, and even killed their fellow prisoners for their scraps of food. The other prisoners eventually revolted against the Raiders, placing their six ring leaders on trial and hanging them with the Confederates' blessing. Rintels places the blame for the squalid conditions in the camp both on the camp's authoritarian German-Swiss commandant, Henry Wirz, the only person tried and executed for war crimes following the Civil War, and on larger forces that were the products of a devastating four-year war: shortages of food, medicine, and supplies that plagued the entire Confederacy, and forced it to choose between supplying its own armies or the Union prisoners. To Rintels, the Andersonville camp, unlike the Nazi concentration camps, seemed less the result of a conscious evil policy than the tragic result of a brutal war.

In addition to his creative work, Rintels has also been active in the politics of television. As president of the Writers Guild of America (1975–77), he coordinated the successful campaign, led by the Guild and producer Norman Lear, to have the courts overturn the FCC's 1975 "Family-Viewing" policy, which designated the first two hours of prime time (7:00-9:00 P.M.) for programs that would be suitable for viewing by all age groups. Rintels and Lear argued that the FCC policy violated the First Amendment, forcing major script revisions of more adult-oriented programs appearing before 9:00 P.M. and the rescheduling of series such as *All in the Family.*

Since the early 1970s, Rintels has been a vocal critic of television networks' timidity in their prime-time programming. In 1972, he condemned commercial television executives for rejecting scripts dealing with Vietnam draft evaders, the U.S. Army's storing of deadly nerve gas near large cities, anti-trust issues, and drug companies' manufacture of drugs intended for the illegal drug market. In a 1977 interview, Rintels criticized the bulk of prime-time entertainment television: "That's the television most of the people watch most of the time—seventy-five to eighty million people a night. And it is for many people a source of information about the real world. But the message they are getting is, I think, not an honest message."

—Hal Himmelstein

DAVID W. RINTELS. Born in Boston, Massachusetts, U.S.A., 1939. Educated at Harvard University, B.A. magna cum laude 1959. Journalist, *Boston Herald,* 1959–60; news director, WVOX-Radio, New Rochelle, New York, 1959; researcher, National Broadcasting Company, 1961; television writer, since the early 1960s. Member: Writers Guild of America, West, president, 1975–77; chair, Committee on Censorship and Freedom of Expression. Recipient: ACE Award, George Foster Peabody Award, 1970; Silver Gavel Award from the American Bar Association, 1971; Writers Guild of America Awards, 1970, 1975, 1980; Emmy Awards, 1973, 1975.

TELEVISION

1961–75	*The Defenders*
1964–65	*Slattery's People*
1965–68	*Run for Your Life*
1967–68	*The Invaders*
1970–71	*The Senator*
1970–71	*The Young Lawyers*

MADE-FOR-TELEVISION MOVIES

1973	*Clarence Darrow*
1975	*Fear on Trial*
1980	*Gideon's Trumpet*
1980	*The Oldest Living Graduate*
1981	*All the Way Home*
1982	*The Member of the Wedding*
1984	*Choices of the Heart*
1984	*Mister Roberts*
1984	*Sakharov*
1985	*The Execution of Raymond Graham*
1989	*Day One* (also producer)
1990	*The Last Best Year* (also producer)

1992	*A Town Torn Apart*
1994	*World War Two: When Lions Roared*
1995	*My Antonia*

TELEVISION MINISERIES

| 1977 | *Washington: Behind Closed Doors* (co-producer, co-writer) |
| 1996 | *Andersonville* |

FILMS

Scorpio, (co-writer), 1972; *Not without My Daughter*, 1992.

FURTHER READING

Arlen, Michael J. "The Air: Getting the Goods on Pres. Monckton." *The New Yorker*, 3 October, 1977.

Nordheimer, Jon. "How the Ordeal of Sakharov was Recreated for Cable TV." *The New York Times*, 17 June 1984.

Rintels, David W. "Not for Bread Alone." *Performance 3* (New York), July/August 1972.

Rosenberg, Howard. "Civil War POWs' Tale of Horror." *Los Angeles Times*, 1 March 1996.

See also *Defenders*; Writing for Television

RISING DAMP

British Situation Comedy

Rising Damp, the classic Yorkshire Television situation comedy series set in a run-down northern boarding house, was originally screened on ITV between 1974 and 1978, and has continued to be revived on British television at regular intervals ever since, always attracting large audiences (many of whom were no doubt lodgers at one time or another in similarly seedy houses). Created by writer Ernie Chappell, the series depicted the comic misadventures and machinations of Rupert Rigsby, the embittered down-at-heel landlord, who constantly spied on the usually very innocent private lives of an assortment of long-suffering tenants.

The success of *Rising Damp* depended largely upon the considerable comic talent of its star, Leonard Rossiter, who played the snooping and sneering Rigsby. Rossiter had first demonstrated his impeccable comic timing in the same role (though under the name Rooksby) in the one-off stage play *Banana Box*, from which the television series was derived. Rossiter rapidly stamped his mark upon the money-grubbing, lecherous, manneristic landlord, making him at once repulsive, vulnerable, paranoid, irrepressible, ignorant, cunning, and above all hilarious. Sharing his inmost fears and suspicions with his cat Vienna, he skulked about the ill-kempt house, bursting in on tenants when he thought (almost always mistakenly) that he would catch them *in flagrante*, and impotently plotting how to seduce university administrator Miss Jones, the frustrated spinster who was the reluctant object of his desire.

Rigsby's appalling disrespect for the privacy of his lodgers and his irrepressible inquisitiveness were the moving force behind the storylines, bringing together the various supporting characters who otherwise mostly cut lonely and inadequate, even tragic, figures. The supporting cast was in fact very strong, with Miss Jones played in highly individualistic style by the respected stage actress Frances de la Tour; the confused, naive medical student Alan played by an ingenuous but appealing Richard Beckinsale; and Philip, the proud but smug son of an African tribal chief, played by Don

Warrington. Only Beckinsale had not appeared in the original stage play. Other lodgers later in the series were Brenda (Gay Rose) and Spooner (Derek Newark).

The frustrations and petty humiliations constantly suffered by the various characters, coupled with their dingy surroundings, could easily have made the series a melancholy affair, but the deft humour of the scripts, married to the inventiveness and expertise of the performers, kept the tone light, if somewhat hysterical at times, and enabled the writers to explore Rigsby's various prejudices (concerning sex, race, students, and anything unfamiliar) without causing offence. In this respect, the series was reminiscent of the techniques employed in *Steptoe and Son*, and by Johnny Speight and Warren Mitchell in the "Alf Garnett" series, though here there was less emphasis on invective and more on deliberately farcical comedy. One occasion on which the series did come unstuck was when fun was had at the expense of an apparently fictional election candidate named Pendry, who was described as crooked and homosexual. Unfortunately, there was a real Labour member of Parliament of the same name, and Yorkshire Television was obliged to pay substantial damages for defamation as a result.

The success of the television series led to a film version in 1980, but this met with mixed response, lacking the conciseness and sharpness of the television series and also lacking the presence of Beckinsale, who had tragically died of a heart attack at the age of 31 the previous year. Rossiter himself went on to star in the equally popular series *The Fall and Rise of Reginald Perrin* before his own premature death from heart failure in 1984.

—David Pickering

CAST

Rupert Rigsby	Leonard Rossiter
Alan Moore	Alan Beckinsale
Ruth Jones	Frances de la Tour
Philip Smith	Don Warrington
Spooner	Derek Newark
Brenda	Gay Rose

Rising Damp
Photo courtesy of the British Film Institute

PRODUCERS Ian MacNaughton, Ronnie Baxter, Len Luruck, Vernon Lawrence

PROGRAMMING HISTORY

• Yorkshire Television (ITV)

28 Episodes

2 September 1974	Pilot Episode
December 1974–January 1975	5 Episodes
November 1975–December 1975	8 Episodes
27 December 1976	Christmas Special
April 1977–May 1977	7 Episodes
April 1978–May 1978	6 Episodes

See also British Programming

RIVERA, GERALDO

U.S. Journalist/Talk-Show Host

The name of journalist and talk-show host Geraldo Rivera has become synonymous with more sensational forms of talk television. His distinctive style, at once probing, aggressive, and intimate, has even led, at times, to parody by a variety of print and broadcast mediums. He has seemed to contribute to this high-profile identification by playing himself (or a close approximation) in an episode of *thirtysomething*, a 1992 *Perry*

Mason TV movie, and the theatrical film *The Bonfire of the Vanities* (1990). Yet, ironically, his fear of going too far with his public image led him to turn down an offer to play the role of an over-the-top tabloid reporter in Oliver Stone's *Natural Born Killers* (1994). A master of self-promotion, Rivera's drive has taken his career in directions he may not have predicted. Despite having won ten Emmys and numerous journalism awards (including the Peabody), Rivera is still primarily known for the more public nature of both his personal life and his talk show.

Rivera was discovered while working as a lawyer for the New York Puerto Rican activist group the Young Lords. During the group's occupation of an East Harlem church in 1970, Rivera had been interviewed on WABC-TV local news and caught the eye of the station's news director Al Primo, who was looking for a Latino reporter to fill out his news team. In 1972, Rivera gained national attention with his critically acclaimed and highly rated special on the horrific abuse of mentally retarded patients at New York's Willowbrook School. He then went on to work for ABC national programs, first as a special correspondent for *Good Morning, America* and then, in 1978, for the prime-time investigative show *20/20*. But his brashness led to controversies with the network, and in 1985 he was fired after publicly criticizing ABC for canceling his report on an alleged relationship between John F. Kennedy and Marilyn Monroe.

Rivera was undaunted by his altercation with the network, and moved to boost his visibility with an hour-long special on the opening of Al Capone's secret vault in April 1986. The payoff for the audience was virtually nil since the vault contained only dirt, but the show achieved the highest ratings for a syndicated special in television history. Rivera wrote in his autobiography, "My career was not over, I knew, but had just begun. And all because of a silly, high-concept stunt that failed to deliver on its titillating promise."

The same high-concept approach became the base for Rivera's talk show *Geraldo*, which debuted in September 1987. The first guest was Marla Hanson, a model whose face had been slashed on the orders of a jilted lover. Many critics attacked the show—and Rivera—for his theatrics and "swashbuckling bravado," but *Geraldo* garnered a respectable viewership. However, Rivera has pointed out that it was his 1987 show, "Men in Lace Panties and the Women Who Love Them," which turned the talk format in a more sensational direction. The following year, he broke talk-show rating records with a highly publicized show on Nazi skinheads. During the show's taping, a brawl had broken out between two of the guests—a 25-year-old leader of the White Aryan Resistance Youth and black activist Roy Innis. A thrown chair hit Rivera square in the face, breaking his nose. The show was news before it even aired. The press jumped on this opportunity to use Rivera as an example of television's new extremes. A November 1988 cover of *Newsweek* carried a close-up of his bashed face next to a headline reading, "Trash TV: From the Lurid to the Loud, Anything Goes."

Geraldo continued throughout the late 1980s and early 1990s to capitalize on the sensational aspects of his reputation. He inserted himself into the talk-show narrative, often using his own exploits and bodily desires to fill out the issue at hand. In a show on plastic surgery, Rivera had fat sucked from his buttocks and injected into his forehead in a procedure to reduce wrinkles. A few years later, in another procedure, he had his eyes tucked on the show. The publication of his autobiography, *Exposing Myself*, in the fall of 1991 caused a major stir due to

Rivera's revelations of his numerous affairs. In 1993, Rivera tried to recoup his former role as a "serious" journalist by hosting a nightly news talk show on CNBC called *Rivera Live* as well as continuing the daytime *Geraldo*.

In a 1993 interview, Rivera offered an analysis of his own place in American life: "I'm so much a part of the popular culture now. I'm a punch line every night on one of the late-night shows.... I'm used as a generic almost in all the editorials and commentaries and certainly all the books about whether the news media has gone too far. It's just that, what is a review going to do to me? They either like me or don't like me, but I'm always interesting to watch." By mid-1996, Rivera was once again redesigning his programs to reduce elements that could be charged with an exploitative tone. With an outcome yet to be seen, he remains, as he says, "always interesting."

—Sue Murray

GERALDO RIVERA. Born Jerry Rivers, in New York City, New York, U.S.A., 4 July 1943. University of Arizona, B.S. 1965; Brooklyn Law School, J.D. 1969; postgraduate work at University of Pennsylvania, 1969; attended School of Journalism, Columbia University, New York, 1970. Married: 1) Edith Bucket "Pie" Vonnegut, 1971; 2) Sherryl Raymond, 1976; 3) C.C. Dyer, 1987; children: Gabriel Miguel and Isabella. Member, anti-poverty neighborhood law firm Harlem Assertion of Rights and Community Action for Legal Services, New York City, 1968–70; admitted to New York Bar, 1970; in television, from 1970, beginning at *Eyewitness News*, WABC-TV, New York City; host, numerous television specials and talk-shows. Member: Puerto Rican Legal Defense and Education Fund; Puerto Rican Bar Association. Recipient: Smith Fellowship, University of Pennsylvania, 1969; ten Emmy Awards; two Robert F. Kennedy Awards; Peabody Award; Kennedy Journalism Awards, 1973 and 1975. Address: Geraldo Investigative News Group, 555 West 57th Street, New York City, New York 10019, U.S.A.

TELEVISION SERIES

1970-75	*Eyewitness News,*
1973-76	*Good Morning, America*
1974-78	*Geraldo Rivera: Goodnight, America*
1975-77	*Good Morning, America*
1978-85	*20/20* (correspondent/senior producer)
1987-	*Geraldo* (host)
1991-92	*Now It Can Be Told*
1993	*Rivera Live*

MADE-FOR-TELEVISION MOVIE

1992	*Perry Mason: The Case of the Reckless Romeo*

TELEVISION SPECIALS (selection)

1986	*The Mystery of Al Capone's Vault*
1986	*American Vice: The Doping of a Nation*
1986	*American Vice: The Real Story of the Doping of a Nation*

1987	*Modern Love: Action to Action*
1987	*Innocence Lost: The Erosion of American Child-hood*
1987	*Sons of Scarface: The New Mafia*
1988	*Murder: Live from Death Row*

FILM

The Bonfire of the Vanities, 1990.

PUBLICATION

Exposing Myself, with Daniel Paisner. New York: Bantam, 1991.

FURTHER READING

Heaton, Jeanne Albronda, and Nona Leigh. *Tuning in Trouble: Talk TV's Destructive Impact on Mental Health.* San Francisco: Josey-Bass, 1995.

Leershen, Charles. "Sex, Death, Drugs and Geraldo." *Newsweek* (New York), 14 November 1988.

Levine, Art. "Blitzed: Ed Murrow, Meet Geraldo." *The New Republic* (Washington, D.C.), 9 January 1989.

Littleton, Cynthia. "Geraldo Takes the Pledge." *Broadcasting and Cable* (Washington, D.C.), 8 January 1996.

Livingstone, Sonia, and Peter Lunt. *Talk on Television: Audience Participation and Public Debate.* London: Routledge, 1994.

Munson, Wayne. *All Talk: The Talkshow in Media Culture.* Philadelphia, Pennsylvania: Temple University Press, 1993.

Priest, Patricia Joyner. *Public Intimacies: Talk Show Participants and Tell-All TV.* Creskill, New Jersey: Hampton, 1995.

Silverman, Art. "Network McNews: The Brave New World of Peter, Dan, Tom. . . and Geraldo." *ETC.: A Review of General Semantics* (San Francisco, California), Spring 1990.

Timberg, Bernard. "The Unspoken Rules of Television Talk." In, Newcomb, Horace, editor. *Television: The Critical View.* New York: Oxford University Press, 1994.

See also Talk Shows

ROAD TO AVONLEA

Canadian Family Drama

Road to Avonlea, one of English Canada's most successful dramatic series, aired on CBC (the Canadian Broadcasting Corporation network) for seven seasons, from 1990 to 1996. In addition to this domestic success, the series has been among the most widely circulated Canadian programs in international markets; it was sold in over 140 countries by the end of its domestic run. The series was both a popular and a critical success, and is a singular example of the adaptation of "national" Canadian fiction for the generic constraints of both domestic and international televisual markets. This singularity is evident in both the production context of the series and in its narrative development across the seven seasons. The program was produced by Sullivan Entertainment in association with the Disney Channel in the United States, and was supported with the participation of Telefilm Canada. Thus, from the beginning of its production run, the series was developed in relation to both domestic and international markets. In addition, the program was plotted in relation to the considerations of both a national broadcasting service and a specialty cable service.

The narrative was developed from the novels of Lucy Maud Montgomery, following the previous success of Sullivan Entertainment's miniseries adaptation of Montgomery's best-known novel, *Anne of Green Gables.* Set in the Atlantic province of Prince Edward Island in the first decades of the 20th century, *Avonlea* opens with the move of young Sara Stanley (Sara Polley) from Montreal to the small P.E.I. town of Avonlea to live with two aunts, Hetty King (Jackie Burroughs) and Olivia King (Mag Ruffman). Over the seven seasons, the narrative traces the coming of age of Sara and the other children of the town as well as the adjustments of the adults in the community to the increasing changes that 20th-century modernisation brings to rural island life. The series is situated simultaneously within period-costume drama and children's, or family, drama—on the CBC, the series ran in the 7:00 P.M. family hour.

The dramatic formula for the series was relatively stable. Episode plots built upon the development of the children's interrelationships and their increasing entrance into the "adult" world of family and community life. At the same time, the shape of the community was developed through the interactions of series regulars with "outsiders" who instigated disruptions into both family and kinship ties, and who served as indices of the invasive modernity encroaching on town life. The dramatic formula therefore intertwined the coming-of-age incidents and the character development of a traditional children's series with an idealised and nostalgic accounting of rural forms of community life. The fact that the series' narrative ends on the eve of World War I serves to reinforce this linking of childhood, family, and community in an earlier, more innocent period.

The episodic use of outsider characters also integrated well with the series development in relation to both domestic and foreign markets. Over the years the producers succeeded in recruiting for these roles a number of internationally-known Canadian guest stars (for example, Kate Nelligan, Colleen Dewhurst) and international guest stars (Michael York, Stockard Channing), a production decision which greatly aided in the international marketing of the series. *Road to Avonlea*, therefore, is a prime example of the adap-

Road to Avonlea
Photo courtesy of Sullivan Entertainment/Marni Grossman

tation of a national popular culture narrative to the constraints of the international television culture of the 1990s. At the same time, it demonstrates one possible strategy for series finance within relatively "small" national television industries.

—Martin Allor

CAST

Sara Stanley (1990–94) . . . . Sara Polley
Aunt Hetty King . . . . Jackie Burroughs
Janet King . . . . Lally Cadeau
Alec King . . . . Cedric Smith
Olivia King Dale . . . . Mag Ruffman
Jasper Dale . . . . R.H. Thompson
Felicity King . . . . Gema Zampogna
Felix King . . . . Zachary Bennett

Rachel Lynde . . . . Patricia Hamilton

PRODUCERS   Kevin Sullivan, Trudy Grant

PROGRAMMING HISTORY   91 Episodes

• CBC

January 1990–March 1996   Sunday 7:00-8:00

FURTHER READING

Miller, Mary Jane. "Will English Language Television Remain Distinctive? Probably." In, McRoberts, Kenneth, editor. *Beyond Quebec: Taking Stock of Canada.* Montreal, Quebec: McGill Queen's Press, 1995.

See also Canadian Programming in English

ROBERTSON, PAT

U.S. Religious Broadcaster

Pat Robertson is the leading religious broadcaster in the United States. His success has made him not only a television celebrity, but a successful media owner, a well-known philanthropist and a respected Conservative spokesman. Robertson experienced a religious conversion while running his own electronics company in New York, and became increasingly certain that God wanted him to buy a television station to spread the gospel. Robertson brought his family to Portsmouth, Virginia, in November 1959, with only $70 in his pocket, and a year later bought a $500,000 bankrupt UHF station in Portsmouth for a mere $37,000. He went on the air the following year with an evangelistic religious format. Robertson's decision to ask for 700 supporters to contribute $10 a month led to the 1963 birth of *The 700 Club*, his religious talk show. Robertson, an ordained minister of the Southern Baptist church, resigned his ordination in 1986 before his presidential bid. He also has authored several books, including *The Secret Kingdom* which contains his "Kingdom" principles for a healthy, wealthy life.

Robertson can claim to have built the popularity of the religious talk-show format, a format that has proved consistently popular over the last thirty years. The 1995 version of *The 700 Club* talk show is a mixture of news, in-depth feature reports on current ethical and moral issues like school prayer, the agenda of the New Christian Right, and Christian evangelism with a charismatic flavor. The program is an important indicator of what evangelicals and pentecostals believe about current moral and political issues.

Robertson was the first religious broadcaster to understand the importance of having his programs carried on a satellite transponder so that they could be down-loaded to the nation's 600 (rising to 800) cable television systems, coast-to-coast. The Annenberg/Gallup Survey of 1981 showed that one-quarter of his audience had some college education; that among religious broadcasts *The 700 Club* had the highest proportion of viewers between the ages of 30 and 50 (47%), the highest share in the Midwest (40%), and the greatest number of viewers who regularly attend church.

In 1986 Robertson celebrated the 25th anniversary of his network Christian Broadcasting Network (CBN). Thanks to his broadcasts, CBN's gift income was running in excess of $139 million a year, and CBN had branched out to include humanitarian arm Operation Blessing, and CBN University (founded in 1978 and renamed Regent University in 1990), and the now-defunct Freedom Council. Robertson then decided that he was to answer a higher call and run for the presidency against Republican vice president George Bush. After failing to win the Republican nomination, Robertson returned to *The 700 Club* in May 1988. Donations had fallen by 40%, but major staff reductions solved the financial crisis.

Pat Robertson
Photo courtesy of Broadcasting and Cable

With daily audiences for *The 700 Club* averaging one million households, Robertson's contribution to American broadcasting has been influential. His was the only religious broadcast to finance a Washington newsroom, and CBN has included a news element since 1980.

What will happen to CBN as Robertson begins to take less of an active role is uncertain. Robertson's combination of religious fervor and political rhetoric is unusual, and if he retires, there seems no one else likely to replace him.

—Andrew Quicke

PAT ROBERTSON. Born Marion Gordon Robertson in Lexington, Virginia, U.S.A., 22 March 1930. Washington and Lee University, B.A. 1950; Yale University, J.D. 1955; New York Theological Seminary, MDiv 1959. Married: Adelia Elmer; children: Timothy, Elizabeth, Gordon and Ann. Founder and president, Christian Broadcasting Network, Virginia Beach, from 1960; ordained minister, Southern Baptist Convention, 1961–86; author of numerous books, from 1972; on board of directors, National Broadcasters, from 1973; founder and president, CBN (now Regent) University, 1978; started relief organization Operation Blessing, 1978; founder and president, Continental Broadcasting Network, from 1979; co-founded Freedom Council foundation, 1981; member, Presidential Task Force on Victims of Crime, Washington, D.C., 1982; candidate for Republican nomination for U.S. president, 1988. ThD. (honorary), Oral Roberts University, 1983. Recipient: National Council of Christians and Jews Distinguished Merit citation; Knesset Medallion; Religious Heritage of America Faith and Freedom Award; Southern California Motion Picture Council Bronze Halo Award; Religion in Media's International Clergyman of the Year, 1981; International Committee for Goodwill's Man of the Year, 1981; Food for the Hungry Humanitarian Award, 1982; Freedoms Foundation George Washington Honor Medal, 1983. Address: Christian Broadcasting Network, CBN Center, 1000 Centerville Turnpike, Virginia Beach, Virginia 23463, U.S.A.

TELEVISION SERIES

1963–   *The 700 Club* (host)

PUBLICATIONS (selection)

The Secret Kingdom. Nashville: Nelson, 1982; revised edition. Dallas: Word, 1992.

Beyond Reason. New York: Morrow, 1984.

Answers to 200 of Life's Most Probing Questions. Virginia Beach, Virginia: CBN University Press, 1985.

America's Date with Destiny. Nashville: Thomas Nelson, 1986.

The New World Order. Dallas: Word, 1991.

The Turning Tide. Dallas: Word, 1993.

The End of the Age: A Novel. Dallas: Word, 1995.

FURTHER READING

Boston, Rob. *The Most Dangerous Man in America?: Pat Robertson and the Rise of the Christian Coalition.* Amherst, New York: Prometheus, 1996.

Donovan, John B. *Pat Robertson: The Authorized Biography.* New York: Macmillan and London: Collier Macmillan, 1988.

Harrow, David Edwin. *Pat Robertson: A Personal, Religious, and Political Portrait.* New York: Harper and Row, 1987.

Hertzke, Allen D. *Echoes of Discontent: Jesse Jackson, Pat Robertson, and the Resurgence of Populism.* Washington, D.C.: Congressional Quarterly Press, 1992.

Peck, Janice. *The Gods of Televangelism.* Crosskill, New Jersey: Hampton, 1993.

Straub, Gerard Thomas. *Salvation for Sale: An Insider's View of Pat Robertson.* Buffalo, New York: Prometheus, 1988.

See also Christian Broadcasting Network/The Family Channel; Religion on Television

ROBINSON, HUBBELL

U.S. Writer/Producer/Network Executive

Hubbell Robinson was active in American broadcasting as a writer, producer, and network programming executive for over 40 years. As the CBS executive who championed the 1950s anthology drama *Playhouse 90*, his efforts to develop high-quality programming that he described as "mass with class" contributed to CBS' long-lived reputation as the "Tiffany" network.

Robinson's broadcasting career began in 1930, when he became the first head of the new radio department at the advertising agency Young and Rubicam. In the era of early commercial broadcasting, when corporate clients sought new radio programs to sponsor, many advertising agencies helped develop program genres, such as the soap opera, that encouraged habitual listening. At Young and Rubicam, Robinson created and wrote scripts for General Foods' soap opera *The Second Mrs. Burton.* The program's success was based, according to Robinson, on "four cornerstones": simple characterizations, understandable predicaments, the centrality of the female characters, and the soap opera's philosophical relevance.

During the late 1930s and early 1940s, Young and Rubicam became an important radio program provider, simultaneously producing *The Jack Benny Show*, Fred Allen's *Town Hall Tonight*, and *The Kate Smith Hour*, among others. As did other radio executives at the agency, Robinson wrote many scripts and commercials, in addition to producing programs.

By the time Robinson joined CBS Television in 1947, his extensive background in radio programming had pre-

Hubbell Robinson
Photo courtesy of Broadcasting and Cable

pared him well for the new medium. Indeed, in his autobiography, *As It Happened,* then-CBS chairman William Paley referred to Robinson as "the all-around man in our programming department." As executive vice president in charge of television programming at CBS, Robinson championed and oversaw the development of such popular programs as *I Love Lucy, You'll Never Get Rich* (with Phil Silvers as Sergeant Bilko), and *Gunsmoke.*

However, according to Paley, "Culturally, [Robinson's] interests were levels above many of his colleagues.... His special flair was for high-quality programming." Robinson organized and championed the 90-minute dramatic anthology series, *Playhouse 90,* which featured serious dramas written by Paddy Chayevsky, Reginald Rose, and Rod Serling, among others. During its run from 1956 to 1961, *Playhouse 90*'s plays included *Requiem for a Heavyweight, A Sound of Different Drummers, The Miracle Worker,* and *Judgment at Nuremberg.* Robinson was credited with bringing serious television drama to its peak with *Playhouse 90.*

For Paley and others at CBS, however, the anthology drama format was a drawback: its lack of continuity from week to week did not seem to encourage regular television viewing habits. But the networks' increasing reliance on filmed episodic programs was disparaged by many admirers of live anthology drama. Referring to critics' concerns that network programming quality was declining, Robinson openly criticized the television industry's "willingness to settle for drama whose synonym is pap." Paley, on the other hand, expressed concern that as a network executive, Robinson "may have lacked the common touch." Still, it was Robinson's stance that helped CBS deal with federal regulators when questions were raised about whether or not CBS programs served the (loosely defined) public interest.

Robinson returned to CBS briefly from 1962 to 1963, and later joined ABC as executive producer of the *Stage 67* series and the on-location series *Crisis!* from 1966 to 1969. In the early 1960s, he was credited with helping erode stereotyping of African Americans on television by distributing a memorandum calling for producers to cast them in a greater variety of roles. Robinson's contributions as a producer and programmer spanned the crucial decades of radio's maturity and television's early growth. As the executive responsible for the programming of both popular and innovative television programs in the 1950s, he helped CBS establish and maintain its reputation as the network with the highest ratings and best programming, a reputation that endured for several decades.

—Cynthia Meyers

HUBBELL ROBINSON. Born in Schenectady, New York, U.S.A., 16 October 1905. Graduated from Phillips Exeter Academy, 1923; Brown University, B.A. 1927. Married: 1) Therese Lewis, 1940 (divorced, 1948); 2) Margaret Whiting (divorced); 3) Vivienne Segal (legally separated, 1962). Drama critic, *Exhibitors Herald,* 1927; reporter, *Schenectady Union Star,* Albany Knickerbocker Press, 1929; radio producer, Young and Rubicam, 1930, vice president and radio director, 1942; vice president and program director, ABC radio, New York City, 1944–45; vice president, Foote, Cone, and Belding advertising agency, 1946; vice president and program director, CBS, 1947–56; executive vice president, CBS-TV, 1956-59; organized Hubbell Robinson Productions, 1959; senior vice president, television programs, CBS, 1962–63; executive in charge of various productions, ABC-TV, 1966–69; contributing critic, *Films in Review,* 1971–74; film critic, CATV Channel 8, New York City, 1969–72. Recipient: Emmy Awards, 1958 and 1959; two TV Digest Awards, 1960; Producers Guild Award, 1962; Fame Award, 1967; Television Academy's Salute Award, 1972. Died in New York City, 4 September 1974.

TELEVISION SERIES (executive producer)

1956–61	*Playhouse 90*
1966–69	*Crisis!*
1967	*Stage 67*

RADIO

The Second Mrs. Burton; The Jack Benny Show; Fred Allen's Town Hall Tonight; The Kate Smith Hour.

FURTHER READING

Boddy, William. *Fifties Television: The Industry and Its Critics.* Urbana: University of Illinois Press, 1990.

Metz, Robert. *CBS: Reflections in a Bloodshot Eye.* Chicago: Playboy, 1975.

Paley, William S. *As It Happened.* Garden City, New York: Doubleday, 1979.

See also Anthology Drama; "Golden Age" of Television; *Playhouse 90*

THE ROCKFORD FILES

U.S. Detective Drama

The *Rockford Files* is generally regarded (along with *Harry O*) as one of the finest private eye series of the 1970s, and indeed of all time, consistently ranked at or near the top in polls of viewers, critics, and mystery writers. The series offered superbly-plotted mysteries, with the requisite amounts of action, yet it was also something of a revisionist take on the hard-boiled detective genre, grounded more in character than crime, and infused with humor and realistic relationships. Driven by brilliant writing, an ensemble of winning characters, and the charm of its star, James Garner, the series went from prime-time Nielsen hit in the 1970s, to a syndication staple with a loyal cult following in the 1980s, spawning a series of made-for-TV movie sequels beginning in 1994.

The show was created by producer Roy Huggins and writer Stephen J. Cannell. Huggins originally sketched the premise of a private eye who only took on closed cases (a conceit quickly abandoned in the series), at one point intending to introduce the character in an episode of the cop show *Toma*. Huggins assigned the script to Cannell—a professed aficionado of the hard-boiled detective tradition—who decided to have fun with the story by flouting the genre's clichés and breaking its rules. After the *Toma* connection crumbled, James Garner signed on to the project, NBC agreed to finance the pilot, and *The Rockford Files* was born.

Cannell was largely responsible for the character and the concept that finally emerged in the pilot script and the series. Jim Rockford did indeed break the mold set by television's earlier two-fisted chivalric P.I.s. His headquarters was a mobile home parked at the beach rather than a shabby office off Sunset Boulevard; in lieu of a gorgeous secretary, an answering machine took his messages; he preferred to talk, rather than slug, his way out of a tight spot; and he rarely carried a gun. (When one surprised client asked why, Rockford replied, "Because I don't want to shoot anybody.") No troubled loner, Jim Rockford spent much of his free time fishing or watching TV with his father Joe Rockford (Noah Beery, Jr.), a retired trucker with a vocal antipathy to "Jimmy's" chosen profession. Inspired by an episode of *Mannix* in which that tough-guy P.I. took on a child's case for some loose change and a lollipop, Cannell decided to make his creation "the Jack Benny of private eyes." Rockford always announced his rates up front: $200 a day, plus expenses (which he itemized with abandon). He was tenacious on the job, but business was business—and he had payments on the trailer.

For all of its ostensible rule-breaking, however, *The Rockford Files* hewed closely to the hard-boiled tradition in style and theme. The series' depiction of L.A.'s sun-baked streets and seamy underbelly rivals the novels of Raymond Chandler and Ross MacDonald. Chandler, in his essay "The Simple Art of Murder," could have been writing about Jim Rockford when he describes the hard-boiled detective as a poor man, a common man, a man of honor, who talks with the rude wit of his age. Rockford's propensity for wisecracks, his fractious relationship with the police, and his network of shady underworld connections, lead straight back to Dashiell Hammett by way of Chandler and Rex Stout. As for his aversion to fisticuffs, Rockford was not a coward, but a pragmatist, different only by degree (if at all) from Philip Marlowe; when violence was inevitable, he was as tough as nails. Most tellingly of all, he shared the same code as his L.A. predecessors Marlowe and Lew Archer: an unwavering sense of morality, and an almost obsessive thirst for the truth. Thus, despite his ostensible concern for the

The Rockford Files

bottom line, in practice Rockford ended up doing as much or more charity work as any fictional gumshoe (as in "The Reincarnation of Angie," when the soft-hearted sleuth agrees to take on a distressed damsel's case for his "special sucker rate" of $23.74).

Ultimately—perhaps inevitably—all of Cannell's generic revisionism served to make his hero more human, and the stories that much more realistic. Jim Rockford could be the Jack Benny of private eyes precisely because he was the first TV private eye—perhaps the first literary one—to be created as a fully credible human being, rather than simply a dogged, alienated purveyor of justice. *The Rockford Files* was as much about character and relationships as it was about crime and detection. The presence of Rockford's father was more than a revisionist or comic gimmick. Although "Rocky" and Jim's wrangling was the source of much humor, that humor was credible and endearing; their relationship was the emotional core of the show, underlining Jim's essential humanity—and subtly, implicitly, sketching in a history for the detective. By the same token, a tapestry of supporting and recurring characters gave Rockford a life beyond the case at hand: L.A.P.D. Sergeant Dennis Becker (Joe Santos), Jim's buddy on the force, served a stock genre function as a source of favors and threats, but their friendship, which played out apart from the precinct and the crime scene, added another dimension of character; likewise, Jim's attorney and sometimes girlfriend Beth Davenport (Gretchen Corbett) further fleshed out the details of his personal life, and served as an able foil for Becker and his more ill-tempered superiors (in the process imparting a dash of seventies feminism to the show); and Angel Martin (Stuart Margolin), Rockford's San Quentin cellmate, the smallest of small-time grifters, the weasel's weasel, at once hilarious and pathetic, evoked Rockford's prison past, evinced his familiarity with L.A.'s seamier side, and balanced Rocky's hominess with an odious measure of sleaze. These regular members of the Rockford family, and a host of distinctive recurring characters—cops, clients, crooks, con-men, ex-cons—helped create, over time, a web of relationships that grounded *Rockford*, investing it with a more intense and continuing appeal than would a strict episodic focus on crime and detection.

As the preceding might suggest, *The Rockford Files* was underlined with a warmth not usually associated with the private eye genre. Much of the show's distinctiveness was its emphasis on humor, exploiting Garner's comic gifts (and his patented persona of "reluctant hero") and the humor of the protagonist's often prickly relationships with his dad, Becker, Angel, and his clients. In later seasons the series occasionally veered into parody—especially in the episodes featuring dashing, wealthy, virtuous detective Lance White (Tom Selleck), and bumbling, pulp-fiction-addled, would-be private-eye Freddie Beamer (James Whitmore, Jr.)—and even flirted with self-parody, as the show's signature car chases became more and more elaborate and (sometimes) comical (as when Rockford is forced to give chase in a VW bug with an enormous pizza adorning the top). Even so, the series was faithful to its hard-boiled heritage. Yet the series also brought a contemporary sensibility to the hard-boiled tradition's anti-authority impulses, assailing political intrigue, official corruption, and bureaucratic absurdity with a distinctly post-Watergate cynicism.

Rockford's most profound homage to the detective tradition was first-rate writing, and a body of superbly-realized mysteries. Cannell and Juanita Bartlett wrote the bulk of the series' scripts, and most of its best, with writer-producer David Chase (*I'll Fly Away, Northern Exposure*) also a frequent contributor of top-notch work. Mystery author Donald Westlake, quoted in *The Best of Crime and Detective TV*, captures the series' central strengths in noting that "the complexity of the plots and the relationships between the characters were novelistic." John D. MacDonald, critiquing video whodunits for *TV Guide*, proposed that in terms of "believability, dialogue, plausibility of character, plot coherence, *The Rockford Files* comes as close to meeting the standards of the written mystery as anything I found." During its run the series was nominated for the Writers Guild Award and the Mystery Writer's of America "Edgar" Award, in addition to winning the Emmy for Outstanding Drama Series in 1978.

The Rockford Files ran for five full seasons, coming to a premature end in the middle of the sixth, when Garner left the show due to a variety of physical ailments brought on by the strenuous demands of the production. Yet *Rock-ford* never really left the air; not only has the series remained steadily popular in syndication and on cable, three of a projected six made-for-television reunion movies aired on CBS between 1994 and 1996 (the first scoring blockbuster ratings). In addition, a loyal cult following celebrates the series on the *Rockford Files* Web site, and Internet discussion groups. The show's rather rapid canonization as a touchstone of the private eye genre is evinced by its conscious imitation or outright quotation in subsequent series including *Magnum, P.I., Detective in the House*, and *Charlie Grace*.

The Rockford Files marked a significant step in the evolution of the television detective, honoring the traditional private eye tale with well-crafted mysteries, and enriching the form with what television does best: fully-developed characters and richly-drawn relationships. In musing on the hard-boiled detective whose tradition he helped shape, Raymond Chandler wrote, "I do not care much about his private life." In *Rockford*, Cannell and company embraced and exploited their detective's private life. Television encourages, even demands this intimacy. For all the gritty realism of Spade and Marlowe's mean streets, they were, in their solitary asceticism, figures of romantic fantasy. Jim Rockford was no less honorable, no less resolute in his quests; he was, however, by virtue of his trailer, his dad, his gun in the cookie jar, just that much more real.

—Mark Alvey

CAST

Jim Rockford	James Garner
Joseph "Rocky" Rockford	Noah Beery Jr.
Detective Dennis Becker	Joe Santos
Beth Davenport (1974–78)	Gretchen Corbett
Evelyn "Angel" Martin	Stuart Margolin
John Cooper (1978–79)	Bo Hopkins
Lieutenant Alex Diehl (1974–76)	Tom Atkins
Lieutenant Doug Chapman (1976–80) . . .	James Luisi
Lance White (1979–80)	Tom Selleck

PRODUCERS Meta Rosenberg, Stephen J. Cannell, Charles Floyd Johnson, Juanita Bartlett, David Chase

PROGRAMMING HISTORY 114 Episodes

• NBC

September 1974–May 1977	Friday 9:00-10:00
June 1977	Friday 8:30-9:30
July 1977–January 1979	Friday 9:00-10:00
February 1979–March 1979	Saturday 10:00-11:00
April 1979–December 1979	Friday 9:00-10:00
March 1980–April 1980	Thursday 10:00-11:00
June 1980–July 1980	Friday 9:00-10:00

FURTHER READING

Chandler, Raymond. *The Simple Art of Murder*. New York: Houghton Mifflin, 1950.

Collins, Max, and John Javna. *The Best of Crime and Detective TV.* New York: Harmony, 1988.

Grillo, Jean. "A Man's Man and a Woman's Too." *N.Y. Daily News, TV Week* (New York), 10 June 1979.

Kane, Hamilton T. "An Interview with Stephen J. Cannell." *Mystery* (Glendale, California), January 1981.

MacDonald, John D. "The Case of the Missing Spellbinders." *TV Guide* (Radnor, Pennsylvania), 24 November 1979.

Martindale, David. *The Rockford Phile.* Las Vegas: Pioneer, 1991.

Randisi, Robert J. "The Best TV Eyes of the 70s." *Mystery* (Glendale, California), January 1981.

Robertson, Ed. *"This is Jim Rockford . . .": The Rockford Files.* Beverly Hills: Pomegranate, 1995.

Torgerson, Ellen. "James Garner Believes in Good Coffee— And a Mean Punch." *TV Guide* (Radnor, Pennsylvania), 2 June 1979.

Vallely, Jean. "The James Garner Files." *Esquire* (Chicago, Illinois), July 1979.

Wicking, Christopher, and Tise Vahimagi. *The American Vein.* New York: Dutton, 1985.

See also Cannell, Stephen; Detective Programs; Garner, James; Huggins, Roy

RODDENBERRY, GENE

U.S. Writer/Producer

Gene Roddenberry, who once commented, "No one in his right mind gets up in the morning and says, 'I think I'll create a phenomenon today,'" is best known as the creator and executive producer of *Star Trek*, one of the most popular and enduring television series of all time.

A decorated B-17 pilot during World War II, Roddenberry flew commercially for Pan American Airways after the war while taking college writing classes. Hoping to pursue a career writing for the burgeoning television industry, Roddenberry resigned from Pan Am in 1948 and moved his family to California. With few prospects, he followed in his father's and brother's footsteps and joined the Los Angeles Police Department, where he served for eight years. During his career as a police officer, the LAPD was actively involved with Jack Webb's *Dragnet* series. The LAPD gave technical advice on props, sets, and story ideas based on actual cases, many of which were submitted by police officers for $100 in compensation. Roddenberry submitted treatments based on stories from friends and colleagues.

Roddenberry's first professional television work was as technical advisor to Frederick Ziv's *Mr. District Attorney* (1954). The series also gave him his first professional writing work. In addition to creating episodes for *Mr. District Attorney*, Roddenberry also wrote the science-fiction tale "The Secret Weapon of 117," which was broadcast on the syndicated anthology series *Chevron Hall of Stars* (6 March 1956). As he gained increasing success in his new career, he decided to resign from the LAPD in 1956 to pursue writing full time.

Roddenberry continued working on Ziv TV's new series, *The West Point Story* (CBS, 1956-57; and ABC, 1957-58), and eventually became the show's head writer. For the next few years, he turned out scripts for such series as *Highway Patrol* (syndicated), *Have Gun, Will Travel* (CBS), *The Jane Wyman Theater* (NBC), *Bat Masterson* (NBC), *Naked City* (ABC), *Dr. Kildare* (NBC), and *The Detectives* (ABC/NBC). Even at this furious pace, Roddenberry continued to develop ideas for new series.

The first series created and produced by Roddenberry was *The Lieutenant* (NBC, 1963–64). Set at Camp Pendleton, *The*

Lieutenant examined social questions of the day in a military setting. Coincidentally, the show featured guest performances by actors who later played a large role in *Star Trek*: Nichelle Nichols, Leonard Nimoy, and Majel Barrett, whom he later married. Casting director Joe D'Agosta and writer Gene L. Coon also worked with Roddenberry on *Star Trek*.

A life-long fan of science fiction, Roddenberry developed his idea for *Star Trek* in 1964. The series was pitched to the major studios, and finally found support from Desilu

Gene Roddenberry
Photo courtesy of the Academy of Motion Pictures Arts and Sciences

Studios, the production company formed by Lucille Ball and Desi Arnaz. The original $500,000 pilot received minor support from NBC executives, who later commissioned an unprecedented second pilot. The series premiered on 8 September 1966.

Like *The Lieutenant*, *Star Trek* episodes comment on social and political questions in a military (albeit futuristic) setting. Roddenberry described *Star Trek* as a "*Wagon Train* to the stars" because, like that popular series, its stories focused on the "individuals who traveled to promote the expansion of our horizons." *Star Trek* was the first science-fiction series to depict a peaceful future, and Roddenberry often credited the enduring success of the series to the show's positive message of hope for a better tomorrow. It was also the first series to have a multicultural cast. *Star Trek*, which received little notoriety during its three-year run, was canceled after the third season due to low ratings. However, it gained worldwide success in syndication.

In addition to producing the *Star Trek* feature films, Roddenberry continued to write and produce for television, but without the same degree of success. His pilot for *Assignment: Earth* (NBC) was incorporated as an episode of *Star Trek* (29 March 1968). Later pilots included *Genesis II* (CBS, 23 March 1973), *The Questor Tapes* (NBC, 23 January 1974), *Planet Earth* (ABC, 23 April 1974), and *Spectre* (21 May 1977). Roddenberry also served as executive consultant on an animated *Star Trek* series (NBC, 1974–75). A second *Star Trek* series, *Star Trek: The Next Generation*, premiered as a syndicated series in 1987 and had a successful seven-year run.

Roddenberry was the first television writer to be honored with his own star on the Hollywood Walk of Fame (on 4 September 1985). Known affectionately to *Star Trek* fans as "The Great Bird of the Galaxy," Roddenberry died on 24 October 1991. With the permission of Roddenberry's widow, Majel Barrett, the producer's ashes were carried aboard a 1992 flight of the space shuttle *Columbia*. In 1993, Roddenberry was posthumously awarded NASA's Distinguished Public Service Medal for his "distinguished service to the nation and the human race in presenting the exploration of space as an exciting frontier and a hope for the future."

—Susan R. Gibberman

GENE (EUGENE WESLEY) RODDENBERRY. Born in El Paso, Texas, U.S.A., 19 August 1921. Educated at Los Angeles City College; University of Miami; Columbia University; University of Southern California. Married: Majel Leigh Hudec (Majel Barrett), 1969; children: Darleen, Dawn Alison, and Eugene Wesley. Served in U.S. Army Air Force, World War II. Pilot for Pan American Airways, 1946–49; worked for Los Angeles Police Department, 1949–51; television scriptwriter, 1951–62; wrote first science-fiction script, "The Secret Defense of 117," episode for *Chevron Theater*, 1952; created and produced several television series. D.H.L., Emerson College, 1973; D.Sc., Clarkson College, 1981. Recipient:Distinguished Flying Cross; Emmy Award; Hugo Award. Died in Santa Monica, California, 24 October 1991.

TELEVISION SERIES

1952	*Chevron Theatre: "The Secret Defense of 117"* (writer)
1955–58	*Jane Wyman Theater* (writer)
1955–59	*Highway Patrol* (writer)
1956–58	*The West Point* Story (writer)
1957–63	*Have Gun, Will Travel* (writer)
1958–63	*Naked City*
1959–61	*Bat Masterson*
1959–62	*The Detectives*
1961–66	*Dr. Kildare*
1963–64	*The Lieutenant* (creator and producer)
1966–69	*Star Trek* (creator and producer)
1973–74	*Star Trek* (animated show)
1987–91	*Star Trek: The Next Generation* (executive producer)

MADE-FOR-TELEVISION MOVIES (pilots; producer)

1973	*Genesis II*
1974	*Planet Earth*
1974	*The Questor Tapes*
1975	*Strange New World*
1977	*Spectre* (director)

FILMS

Pretty Maids All in a Row (producer and writer), 1971; *Star Trek: The Motion Picture* (producer), 1979; *Star Trek II: The Wrath of Khan* (executive consultant), 1982; *Star Trek III: The Search for Spock*, 1984; *Star Trek IV: The Voyage Home*, 1986; *Star Trek V: The Final Frontier*, 1989.

PUBLICATIONS

The Making of "Star Trek", with Stephen E. Whitfield. New York: Ballantine, 1968.

Star Trek: The Motion Picture. New York: Pocket, 1979.

The Making of "Star Trek: The Motion Picture", with Susan Sackett. New York: Pocket, 1980.

Star Trek: The First Twenty-Five Years, with Susan Sackett. New York: Pocket, 1991.

Gene Roddenberry: The Last Conversation: A Dialogue with the Creator of Star Trek, with Yvonne Fern. Berkeley: University of California Press, 1994.

FURTHER READING

Alexander, David. *Star Trek Creator: The Authorized Biography of Gene Roddenberry*. New York: ROC, 1994.

Barret, Majel. *The Wit and Wisdom of Gene Roddenberry*. New York: Harper Collins, 1995.

Engel, Joel. *Gene Roddenberry: The Myth and the Man Behind Star Trek*. New York: Hyperion, 1994.

Paikert, Charles. "Gene Roddenberry: American Mythmaker." *Variety* (Los Angeles), 2 December 1991.

Van Hise, James. *The Man Who Created Star Trek: Gene Roddenberry*. Las Vegas: Movie Publisher Services, 1992.

See also *Star Trek*

ROGERS, FRED MCFEELY

U.S. Children's Television Host/Producer

Fred McFeely Rogers, better known to millions of American children as Mr. Rogers, is the creator and executive producer of the longest-running children's program on public television, *Mister Rogers' Neighborhood*. While commercial television most often offers children animated cartoons, and many educational programs employ the slick, fast-paced techniques of commercial television, Rogers' approach is as unique as his content. He simply talks with his young viewers. Although his program provides a great deal of information, the focus is not upon teaching specific facts or skills, but upon acknowledging the uniqueness of each child and affirming his or her importance.

Rogers did not originally plan to work in children's television. Rather, he studied music composition at Rollins College in Florida, receiving a bachelor's degree in 1951. He happened to see a children's television program, and felt it was so abysmal that he wanted to offer something better. While he worked in television, however, he also pursued his dream of entering the ministry, continuing his education at Pittsburgh Theological Seminary. In 1962, Rogers received a Bachelor of Divinity degree, and was ordained by the United Presbyterian Church with the charge to work with children and their families through the mass media.

Rogers began his television career at NBC, but joined the founding staff of America's first community-supported television station, WQED in Pittsburgh, as a program director in 1953. His priority was to schedule a children's program; however, when no one came forward to produce it, Rogers assumed the task himself, and in April 1954, launched *The Children's Corner*. He collaborated with on-screen hostess Josie Carey on both the scripts and music to produce a show that received immediate acclaim, winning the 1955 Sylvania Award for the best locally-produced children's program in the country. Rogers and Carey also created a separate show with similar material for NBC network distribution on Saturday mornings. With only a meager budget, their public television show was not a slick production, but Rogers did not view this as a detriment. He wanted children to think that they could make their own puppets, no matter how simple, and create their own fantasies. The important element was to create the friendly, warm atmosphere in the interactions of Josie and the puppets (many of whom are still a part of *Mister Rogers' Neighborhood*), which has become the hallmark of the program.

In 1963, the Canadian Broadcasting Corporation (CBC) in Toronto provided Rogers another opportunity to pursue his ministerial charge through a fifteen-minute daily program called *Misterogers*. This was his first opportunity to develop his on-camera style: gentle, affirming, and conversational. The style is grounded in Rogers' view of himself as an adult who takes time to give children his undivided attention rather than as an entertainer.

Fred Rogers
Photo courtesy of Family Communications, Inc.

Rogers returned to Pittsburgh in 1964, acquired the rights to the CBC programs, and lengthened them to thirty minutes for distribution by the Eastern Educational Network. When production funds ran out in 1967 and stations began announcing the cancellation of the show, an outpouring of public response spurred the search for new funding. As a result of support by the Sears, Roebuck Foundation and National Educational Television, a new series entitled *Misterogers' Neighborhood* began production for national distribution. Currently there are 700 episodes in the library, and, since 1979, Rogers has produced a few new segments each year, adding freshness and immediacy to the series.

Mister Rogers' Neighborhood is unique because it provides a warmth and intimacy seldom found in mass media productions. The show is designed to approximate a visit between friends, and is meticulously planned in consultation with psychologists at the Arsenal Family and Children's Center. The visit begins with a model trolley which travels through a make-believe town to Rogers' home. He enters, singing "Won't you Be My Neighbor?," an invitation for the viewer to feel as close to him as to an actual neighbor. He also creates a bond with his audience by speaking directly to the camera, always in an inclusive manner about things of interest to his viewers. As he speaks, he changes from his sport coat to his trademark cardigan sweater and from street shoes to tennis shoes to further create a relaxed, intimate atmosphere.

The pacing of the program also approximates that of an in-depth conversation between friends. Rogers speaks slowly, allowing time for children to think about what he has said and to respond at home. And psychologists studying the show verify that children do respond. He also takes time to examine objects around him or to do simple chores such as feed his fish. Although he invites other "neighbors," such as pianist Van Cliburn, to share their knowledge, the warm rapport also allows him to tackle personal subjects, such as fears of the dark or the arrival of a new baby.

Recognizing the importance of play as a creative means of working through childhood problems, he also invites children into the Neighborhood of Make Believe. Because Rogers wants children to clearly separate fantasy from reality, this adjacent neighborhood can only be reached via a trolley through a tunnel. The Neighborhood of Make Believe is populated by a number of puppets who are kindly and respectful but not perfect. King Friday XIII, for example, is kind but also somewhat pompous and authoritarian.

Human characters also inhabit this neighborhood and engage the puppets on an equal level. Since Rogers is the puppeteer and voice for most of the puppets, it is difficult for him to interact in this segment. This movement away from "center stage," however, is a conscious choice. His lack of visible participation underscores the separation between the reality he creates in his "home" and these moments of fantasy. The trolley then takes the children back to Rogers' home, and the visit ends as he changes back into his street clothes and leaves the house, inviting the children back at a later date.

In 1971, Rogers formed Family Communications, Inc., a nonprofit corporation of which he is president, to produce *Mister Rogers' Neighborhood* and other audiovisual, educational materials. Many of these productions, such as the prime- time series *Mister Rogers Talks with Parents* (1983), and his books *Mister Rogers Talks with Parents* (1983) and *How Families Grow* (1988), are guides for parents. He has also recorded six albums of children's songs. However, these activities are viewed as educational endeavors rather than profit-generating enterprises, and most of the funding for his productions still comes from grants.

Fred Rogers has succeeded in providing something different for children on television, and in acknowledgment of his accomplishments has received two Peabody Awards, a first for non-commercial television. Rather than loud, fast-paced animation or entertaining education, he presents a caring adult who visits with children, affirming their distinction and value, and understanding their hopes and fears.

—Suzanne Hurst Williams

FRED MCFEELY ROGERS. Born in Latrobe, Pennsylvania, U.S.A., 20 March 1928. Educated at Dartmouth College, 1946; Rollins College, Winter Park, Florida, B.A. in music 1951; Pittsburgh Theological Seminary, Bachelor of Divinity 1962. Married: Sara Joanne Byrd, 1952; children: James Byrd and John Frederick. Assistant television producer and network floor director, NBC, 1951–53; program director, producer, writer, and performer, WQED, Pittsburgh,

1953–62; producer and television host, Canadian Broadcasting Corporation, Toronto, Ontario, 1963–64; producer and host, PBS show *Mister Rogers' Neighborhood*, since 1967; producer and host, *Old Friends, New Friends*, 1979–81; producer of videocassettes, CBS, 1987-88. Founder and president, Family Communications, Inc., 1971. Member: Esther Island Preserve Association; Luxor Ministerial Association; board of directors, McFeely Rogers' Foundation; honorary chair, National PTA, 1992–94. Numerous honorary degrees. Recipient: Peabody Awards, 1969 and 1993; Emmy Awards, 1980 and 1985; Ohio State Awards, 1983 and 1986; ACT Award, 1984; Christopher Award, 1984; Educational Press Association of America's Lamplighter Award, 1985; Children's Book Council Award, 1985; Gold Medal at the International Film and TV Festival, 1986; Parent's Choice Award, 1987-88; PBS Award in recognition of 35 years in public television, 1989; Eleanor Roosevelt Val-Kill Medal, 1994; Joseph F. Mulach, Jr. Award, 1995. Address: 4802 5th Avenue, Pittsburgh, Pennsylvania 15213, U.S.A.

TELEVISION SERIES

1954–1961	*Children's Corner*
1963–67	*Misterogers*
1967–	*Mister Rogers' Neighborhood*
1979–81	*Old Friends, New Friends*

TELEVISION SPECIAL (selection)

| 1994 | *Fred Rogers' Heroes* |

RECORDINGS

Won't You Be My Neighbor?, 1967; *Let's Be Together Today*, 1968; *Josephine, The Short-Necked Giraffe*, 1969; *You Are Special*, 1969; *A Place of Our Own*, 1970; *Bedtime*, 1992; *Growing*, 1992.

PUBLICATIONS (selection)

Mister Rogers Talks with Parents. New York: Berkley Books, 1983.

The New Baby (Mister Rogers' First Experiences Books). New York: Putnam, 1985.

Making Friends (Mister Rogers' First Experiences Books). New York: Putnam, 1987.

Mister Rogers: How Families Grow. New York: Berkley Books, 1988.

You Are Special. New York: Penguin Books, 1994.

FURTHER READING

Barringer, Felicity. "Mister Rogers Goes to Russia." *The New York Times,* 21 September 1987.

Berkvist, Robert. "Misterogers Is a Caring Man." *The New York Times,* 16 November 1969.

Blau, Eleanor. "Rogers Has New TV Series on School." *The New York Times,* 20 August 1979.

Briggs, Kenneth A. "Mr. Rogers Decides It's Time to Head for New Neighborhoods." *The New York Times,* 8 May 1975.

Collins, Glenn. "TV's Mr. Rogers—A Busy Surrogate Dad." *The New York Times,* 19 June 1983.

Fischer, Stuart. "Children's Corner." *Kids TV: The First Twenty-Five Years.* New York: Facts on File Publications, 1983.

"Fred M(cFeely) Rogers." In, Moritz, Charles, editor. *Current Biography.* New York: H. W. Wilson, 1970.

"Fred McFeely Rogers." *Broadcasting and Cable* (Washington, D.C.), 26 July 1993.

"The Man Kids Believe." *Newsweek* (New York), 12 May 1969.

McCleary, Elliott H. "Big Friend to Little People." *Today's Health* (New York), August 1969.

O'Connor, John J. "An Observer Who Bridges the Generation Gap." *The New York Times,* 23 April 1978.

———. "Mr. Rogers, a Gentle Neighbor." *The New York Times,* 15 February 1976.

"TV: On Superheroes." *The New York Times,* 4 February 1980.

Ziaukas, Tim. "Kid Video." *Pittsburgh* (Pittsburgh, Pennsylvania), July 1986.

See also Children and Television

ROGERS, TED

Canadian Media Executive

The founder and chief executive officer of Rogers Communications, Inc., Ted Rogers has become Canada's undisputed new media mogul. A tireless worker, over the last 30 years Rogers has ceaselessly expanded his business undertakings by plunging headlong into each new communication technology. He has compared his corporate machinations to the likes of Rupert Murdoch's News Corporation and Time-Warner, Inc., maintaining that only by building Canadian companies of comparable size and diversity can Canadians be assured of a distinctive voice at the forefront of the electronic highway.

Established in 1967, Rogers Communications has grown into one of Canada's largest media conglomerates. Rogers Communications is the largest cable television business in Canada, with 50 systems that embrace close to 35% of all Canadian cable subscribers. As a broadcaster and television content provider, Rogers Communications owns over 40 radio stations, CFMT in Toronto (a multicultural television station), YTV (a youth-oriented specialty cable channel), the Canadian Home Shopping Channel, and a 25% stake in Viewers Choice Canada, a pay-per-view cable service. It also owns a chain of video stores. In telecommunications, Rogers holds a major stake in Unitel Communications, Inc., a long-distance telephone company, and owns 80% of Cantel Communications, Inc., a Canada-wide cellular phone service. As a result of its 1994 takeover of Maclean-Hunter Ltd., Rogers Communications is the majority share holder of the Toronto Sun Publishing Corporation, which publishes five newspapers across Canada, and is also the owner of 191 periodicals in Canada, Britain, the United States, and Europe. In 1993, Rogers Communications generated revenues of $1.34 billion; the addition of the assets from Maclean-Hunter will bring the revenues of Rogers Communications close to $3 billion.

Rogers' interest in broadcasting continues a family tradition. His father, Edward Samuel Rogers, was the first amateur radio operator in Canada to successfully transmit a signal across the Atlantic. He later invented the radio tube that made it possible to build "batteryless" alternating cur-

rent (AC) receiving sets and in the 1920s founded Rogers Majestic Corporation to build them. Until then neither radio receivers nor transmitters could utilize existing household wiring or power lines, and the batteries that powered

Ted Rogers
Photo courtesy of Ted Rogers

radio receivers were cumbersome, highly corrosive, and required frequent changing. Rogers' "batteryless radio" greatly increased the popularity of broadcasting. The elder Rogers also established CFRB (for Canada's First Radio Batteryless), a commercial radio station in Toronto that grew to command Canada's largest listening audience. In 1935, Rogers was granted the first Canadian license to broadcast experimental television. He died eight years later at the age of 38, when Ted Rogers was five. After Edward Samuels' death, the Rogers family lost control of CFRB.

In 1960, while still a student at Osgoode Hall Law School in Toronto, Ted Rogers bought all the shares in CHFI, a small 940-watt Toronto radio station that pioneered the use of FM (frequency modulation) at a time when only 5% of the Toronto households had FM receivers. By 1965, he was in the cable TV business. In the 1970s he bought out two competitors—Canadian Cablesystems and Premier Cablevision—both larger than his own operation and, by 1980, Rogers Communications had taken over UA-Columbia Cablevision in the United States, to become for a time the world's largest cable operator, with over million subscribers.

Rogers has since sold his stake in U.S. cable operations to concentrate on the Canadian market. His forays into long-distance and cellular telephony, his ownership of cable services such as the Home Shopping Network and specialty channels such as YTV, and the acquisition of Maclean-Hunter's publishing interests make Rogers a key player in the unfolding of the information superhighway.

While the Canadian Radio-Television and Telecommunication Commission (CRTC) has generally given its assent to Rogers' corporate maneuvers, there are many who believe that the commission has neither the regulatory tools nor the will to adequately monitor or control the activities of Rogers and other large cable operators, especially in regards to pricing and open network-access. While cable rates rose an average of 80% between 1983 and 1993, Rogers was busy adding to its corporate empire and up-grading its technical infrastructure ($1 billion over the past five years). Rogers Communications has paid no dividend to its share-holders since 1980 and has posted profits only three times in the last ten years. It is hard not to conclude that cable subscribers are bearing the costs of Rogers' grand corporate scheme to lead Canada into the information age. As smaller cable operators tremble at the prospect of competition from direct-to-home satellites and telephone companies, Ted Rogers has ensured that Rogers Communications is well positioned for life after the era of local cable monopolies.

—Ted Magder

TED (EDWARD SAMUEL) ROGERS. Born in Toronto, Ontario, Canada, 27 May 1933. Educated at the Upper Canada College, Toronto; University of Toronto, Trinity College, B.A. 1956; Osgoode Hall Law School, LL.B. 1961. Married: Loretta Anne Robinson, 1963; children: Lisa Anne, Edward Samuel, Melinda Mary, and Martha Loretta. Read law for Tory, Tory, DesLauriers and Binnington; called to bar of Ontario, 1962; founder, Rogers Communications, 1967; president and chief executive officer. Director: Toronto-Dominion Bank, Canada Publishing Corporation, Hull Group, Wellesley Hospital, Junior Achievement of Canada. Address: P.O. Box 249, Toronto-Dominion Centre, Suite 2600, Commercial Union Tower, Toronto, Ontario M5K 1J5, Canada.

FURTHER READING

Dalglish, Brenda. "Shifting Ground: Changes in Canada and U.S. Rulings Give Rogers Second Thoughts on His Bid for Maclean Hunter." *Maclean's* (Toronto), 7 March 1994.

———. "King of the Road." *Maclean's* (Toronto), 21 March 1994.

Fotheringham, Allan. "The Revenge of Mila Mulroney." *Maclean's* (Toronto), 14 February 1994.

Newman, Peter C. "Life in the Fast Lane." *Maclean's* (Toronto), 21 March 1994.

———. "The Ties that Bind: Ted Rogers Past is Shaping His Future." *Maclean's* (Toronto), 21 February 1994.

See also Canadian Production Companies

ROOM 222

U.S. High School Drama

Room 222 was a half-hour comedy-drama that aired on ABC from 1969 to 1974. While seldom seen in syndication today, the show broke new narrative ground that would later be developed by the major sitcom factories of the 1970s, Grant Tinker's MTM Enterprises and Norman Lear's Tandem Productions. Mixing dramatic elements with traditional television comedy, *Room 222* also prefigured the "dramedy" form by almost two decades.

The series was set at an integrated high school in contemporary Los Angeles. While the narrative centered around a dedicated and student-friendly African-American history teacher, Pete Dixon (Lloyd Haynes), it also depended upon an ensemble cast of students and other school employees. The optimistic idealism of Pete, guidance counselor Liz McIntyre (Denise Nichols), and student-teacher Alice Johnson (Karen Valentine) was balanced by the experienced, somewhat jaded principal, Seymour Kaufman (Michael Constantine). These characters and a handful of other teachers would spend each episode arguing among themselves about the way in which to go about both educating their students and acting as surrogate parents.

A season and a half before Norman Lear made "relevant" programming a dominant genre with the introduction of programs like *All in the Family* and *Maude, Room 222* was using the form of the half-hour comedy to discuss serious contemporary issues. During its five seasons on the air, the show included episodes that dealt with such topics as racism, sexism, homophobia, dropping out of school, shoplifting, drug use among both teachers and students, illiteracy, cops in school, guns in school, Vietnam war veterans, venereal disease, and teenage pregnancy.

Most importantly, *Room 222* served as a prototype of sorts for what would become the formula that MTM Enterprises would employ in a wide variety of comedies and dramas during the 1970s and 1980s. When Grant Tinker set up MTM, he hired *Room 222*'s executive story editors James L. Brooks and Allan Burns to create and produce the company's first series, *The Mary Tyler Moore Show.* This series eschewed issue-oriented comedy, but it picked up on *Room 222*'s contemporary and realistic style as well as its setting in a "workplace family." Treva Silverman, a writer for *Room 222*, also joined her bosses on the new show, and Gene Reynolds, another *Room 222* producer, produced *The Mary Tyler Moore Show* spin-off *Lou Grant* several years later.

Room 222 was given a number of awards by community and educational groups for its positive portrayal of important social issues seldom discussed on television at the time. It won an Emmy for Outstanding New Series in 1969.

—Robert J. Thompson

CAST

Pete Dixon	Lloyd Haynes
Liz Mcintyre	Denise Nicholas
Seymour Kaufman	Michael Constantine
Alice Johnson	Karen Valentine
Richie Lane (1969–71)	Howard Rice
Helen Loomis	Judy Strangis
Jason Allen	Heshimu
Al Cowley (1969–71)	Pendrant Netherly
Bernie (1970–74)	David Jollife
Pam (1970–72)	Te-Tanisha
Larry (1971–73)	Eric Laneuville

PRODUCERS Gene Reynolds, William D'Angelo, John Kubichan, Ronald Rubin

PROGRAMMING HISTORY 112 Episodes

• ABC

Room 222

September 1969–January 1971	Wednesday 8:30-9:00
January 1971–September 1971	Wednesday 8:00-8:30
September 1971–January 1974	Friday 9:00-9:30

FURTHER READING

Eisner, Joel, and David Krinsky. *Television Comedy Series: An Episode Guide to 153 TV Sitcoms in Syndication.* Jefferson, North Carolina: McFarland, 1984.

Feuer, Jane, Paul Kerr, and Tise Vahimagi, editors. *MTM-'Quality Television.'* London: British Film Institute, 1984.

MacDonald, J. Fred. *Blacks and White TV: Afro-Americans in Television since 1948.* Chicago: Nelson-Hall, 1992.

Newcomb, Horace, and Robert Alley. *The Producer's Medium: Conversations with Creators of American TV.* New York: Oxford University Press, 1983.

Tinker, Grant, and Bud Rukeyser. *Tinker in Television: From General Sarnoff to General Electric.* New York: Simon and Schuster, 1994.

See also Brooks, James L.; Burns, Allan; Tinker, Grant

ROOTS

U.S.Miniseries

Roots remains one of television's landmark programs. The twelve-hour miniseries aired on ABC from 23 to 30 January 1977. For eight consecutive nights it riveted the country. ABC executives initially feared that the historical saga about slavery would be a ratings disaster. Instead, *Roots* scored higher ratings than any previous entertainment program in history. It averaged a 44.9 rating and a 66- audience share for the length of its run. The seven episodes that followed the opener earned the top seven spots in the ratings for their week. The final night held the single-episode ratings record until 1983, when the finale of *M*A*S*H* aired on CBS.

The success of *Roots* had lasting impact on the television industry. The show defied industry conventions about black-oriented programming: executives simply had not expected that a show with black heroes and white villains could attract such huge audiences. In the process, *Roots* almost single-handedly spawned a new television format — the consecutive-night miniseries. (Previous miniseries, like the 1976 hit, *Rich Man, Poor Man*, had run in weekly installments.) *Roots* also validated the docudrama approach of its executive producer, David Wolper. The Wolper style, blending fact and fiction in a soap-opera package, influenced many subsequent miniseries. Finally, *Roots* was credited with having a positive impact on race relations and expanding the nation's sense of history.

Based on Alex Haley's best-selling novel about his African ancestors, *Roots* followed several generations in the lives of a slave family. The saga began with Kunta Kinte (LeVar Burton), a West African youth captured by slave raiders and shipped to America in the 1700s. Kunta received brutal treatment from his white masters and rebelled continually. An older Kunta (John Amos) married and his descendants carried the story after his death. Daughter Kizzy (Leslie Uggams) was raped by her master and bore a son, later named Chicken George (Ben Vereen). In the final episode, Kunta Kinte's great-grandson Tom (Georg Stanford Brown) joined the Union Army and gained emancipation. Over the course of the saga, viewers saw brutal whippings and many agonizing moments, rapes, the forced separations of families, slave auctions. Through it all, however, *Roots* depicted its slave characters as well-rounded human beings, not merely as victims or symbols of oppression.

Apprehensions that *Roots* would flop shaped the way that ABC presented the show. Familiar television actors like Lorne Greene were chosen for the white, secondary roles, to reassure audiences. The white actors were featured disproportionately in network previews. For the first episode, the writers created a conscience-stricken slave captain (Ed Asner), a figure who did not appear in Haley's novel but was intended to make white audiences feel better about their historical role in the slave trade. Even the show's consecutive-night format allegedly resulted from network apprehen-

Roots

sions. ABC programming chief Fred Silverman hoped that the unusual schedule would cut his network's imminent losses—and get *Roots* off the air before sweeps week.

Silverman, of course, need not have worried. *Roots* garnered phenomenal audiences. On average, 80 million people watched each of the last seven episodes. Over 100 million viewers, almost half the country, saw the final episode, which still claims one of the highest Nielsen ratings ever recorded, a 51.1 with a 71 share. A stunning 85% of all television homes saw all or part of the miniseries. *Roots* also enjoyed unusual social acclaim for a television show. Vernon Jordan, former president of the Urban League, called it "the single most spectacular educational experience in race relations in America." Today, the show's social effects may appear more ephemeral, but at the time they seemed widespread. Over 250 colleges and universities planned courses on the saga, and during the broadcast, over 30 cities declared "Roots" weeks.

The program drew generally rave reviews. Black and white critics alike praised *Roots* for presenting African-American characters who were not tailored to suit white audiences. The soap-opera format drew some criticism for its emphasis on sex, violence, and romantic intrigue. A few critics also complained that the opening segment in Africa was too Americanized—it was hard to accept television regulars like O.J. Simpson as West African natives. On the whole, however, critical acclaim echoed the show's resounding popular success. *Roots* earned over 30 Emmy Awards and numerous other distinctions.

The program spawned a 1979 sequel, *Roots: The Next Generations.* The sequel did not match the original's ratings, but still performed extremely well, with a total audience of 110 million. Overall, *Roots* had a powerful and diverse impact—as a cultural phenomenon, an exploration of black history, and the crown jewel of historical miniseries.

—J.B. Bird

PRODUCER   Stan Margulies

Adapted for Television by William Blinn

CAST

Kunta Kinte (as a boy)	LeVar Burton
Kunta Kinte (*Toby*: adult)	John Amos
Binta	Cicely Tyson
Omoro	Thalmus Rasula
Nya Boto	Maya Angelou
Kadi Touray	O. J. Simpson
The Wrestler	Ji-Tu Cumbuka
Kintango	Moses Gunn
Brimo Cesay	Hari Rhodes
Fanta	Ren Woods
Fanta (later)	Beverly Todd
Capt. Davies	Edward Asner
Third Mate Slater	Ralph Waite
Gardner	William Watson
Fiddler	Louis Gosett, Jr.
John Reynolds	Lorne Greene
Mrs. Reynolds	Lynda Day George
Ames	Vic Morrow
Carrington	Paul Shenar
Dr. William Reynolds	Robert Reed
Bell	Madge Sinclair
Grill	Gary Collins
The Drummer	Raymond St. Jacques
Tom Moore	Chuck Connors
Missy Anne	Sandy Duncan
Noah	Lawrence-Hilton Jacobs
Ordell	John Schuck
Kizzy	Leslie Uggams
Squire James	Macdonald Carey
Mathilda	Olivia Cole
Mingo	Scatman Crothers
Stephen Bennett	George Hamilton
Mrs. Moore	Carolyn Jones
Sir Eric Russell	Ian McShane
Sister Sara	Lillian Randolph
Sam Bennett	Richard Roundtree
Chicken George	Ben Vereen
Evan Brent	Lloyd Bridges
Tom	Georg Stanford Brown
Ol' George Johnson	Brad Davis
Lewis	Hilly Hicks
Jemmy Brent	Doug McClure
Irene	Lynne Moody
Martha	Lane Binkley
Justin	Burl Ives

PROGRAMMING HISTORY

• ABC

January 1977
   Eight consecutive nights at   9:00-11:00, or 10:00-11:00
September 1978
   Five consecutive nights at   8:00-11:00, or 9:00-11:00

FURTHER READING

Adams, Russell L. "An Analysis of the Roots Phenomenon in the Context of American Racial Conservatism." *Presence Africaine: Revue Culturelle du Monde Noir/Cultural Review of the Negro World* (Paris), 1980.

Blayney, Michael Steward. "Roots and the Noble Savage." *North Dakota Quarterly* (Grand Forks, North Dakota), Winter 1986.

Bogle, Donald. "*Roots* and *Roots: The Next Generations.*" *Blacks in American Film and Television: An Encyclopedia.* New York: Garland, 1988.

Brooks, Tim, and Earle Marsh. *The Complete Directory to Prime-Time Network TV Shows: 1946–Present.* New York: Ballantine, 1979; 5th edition, 1992.

Gray, Herman. *Watching Race: Television and the Struggle for "Blackness."* Minneapolis: University of Minnesota Press, 1995.

Gray, John. *Blacks in Film and Television, A Pan-African Bibliography of Films, Filmmakers, and Performers.* New York: Greenwood, 1990.

Haley, Alex. *Roots.* Garden City, New York: Doubleday, 1976.

Journal of Broadcasting, special issue on *Roots* (Washington, D.C), 1978.

Kern-Foxworth, Marilyn. "Alex Haley." In, *Dictionary of Literary Biography.* Detroit: Gale, 1985.

Tucker, Lauren R., and Hemant Shah. "Race and the Transformation of Culture: The Making of the Television Miniseries *Roots.*" *Critical Studies in Mass Communication* (Annandale, Virginia), December 1992.

"Why *Roots* Hit Home." *Time* (New York) 14 February 1977.

Winship, Michael. *Television.* New York: Random House, 1988.

Woll, David. *Ethnic and Racial Images in American Film and Television.* New York: Garland, 1987.

See also Adaptation; Haley, Alex; Miniseries; Racism, Ethnicity, and Television

ROSE, REGINALD

U.S. Writer

Reginald Rose was one of the outstanding television playwrights to emerge from the golden age of television drama anthology series. Like his acclaimed contemporaries—Paddy Chayefsky, Tad Mosel, and Rod Serling, for example—Rose takes a place in history at the top of the craft of television writing. In addition to other accolades,

Rose was nominated for six Emmy awards during his career, and won three. Although most of Rose's fame derives from his teleplays for the live drama anthologies, he wrote a number of successful screen and stage plays, and went on to create and write scripts for *The Defenders* at CBS, as well as winning recognition for the revived *CBS Playhouse* in the late 1960s.

Rose's first teleplay to be broadcast was "The Bus to Nowhere", which appeared on *Studio One* (CBS) in 1951. It was the 1954–55 season, however, that gave Rose his credentials as a top writer: that year has been referred to as "*the* Reginald Rose season" at *Studio One*. His contributions included the noted plays "12:32 A.M.", "An Almanac of Liberty", "Crime in the Streets", as well as the play that opened the season and became perhaps Rose's most well-known work, "Twelve Angry Men". In addition to winning numerous awards and undergoing transformation into a feature film, "Twelve Angry Men" undoubtedly established Rose's reputation almost immediately as a major writer of drama for television.

What distinguished Rose's teleplays from those of his colleagues such as Chayefsky and Serling was their direct preoccupation with social and political issues. Although the other writers were perhaps equally concerned with the larger social dimensions of their work, they concentrated on the conflicts that emerge in private life and the domestic sphere, and the problems of society as a whole remained implicit in their writing. Rose, in contrast, tackled controversial social issues head-on.

In one of his most well-known and contentious plays, "Thunder on Sycamore Street" (*Studio One*, 1953), Rose attempts to confront the problem of social conformity. In this story, an ex-convict moves to an up-scale neighborhood in an attempt to make a new beginning. When his past is discovered, one of his neighbors organizes a community march to drive the ex-convict out of his new home. Rose deals directly with the issues of mob anger and difference from the norm, issues of general concern in a time when the pressures of conformity were overwhelming, and the memory of fascism still prevalent. This play was controversial from the outset since the central character was originally written to be an African American. Rose was forced, under pressure from *Studio One* sponsors fearful of offending (and losing) audiences in the south, to change him into an ex-convict. This, perhaps more than anything, is indicative of his ability to touch on the most sensitive areas of American social life of that time.

Reginald Rose
Photo courtesy of Reginald Rose

Although Rose kept his sights directed at the scrutiny of social institutions and mechanisms, his characters were as finely drawn as those writers who focused on domestic struggles. The tension created by exhausting deliberations within the confined closeness of the jury room in which "Twelve Angry Men" occurs is exemplary in this regard. The remake of this powerful drama and Paddy Chayefsky's *Marty* into successful feature films marked the breakthrough of the television drama aesthetic into Hollywood cinema. Rose was responsible in part for the creation of this new approach. This gritty realism that became known as the "slice of life" school of television drama was for a time the staple of the anthology shows and reshaped the look of both television and American cinema.

—Kevin Dowler

REGINALD ROSE. Born in New York City, New York, U.S.A., 10 December 1920. Studied at City College (now of the City University of New York), New York, 1937–38. Married: 1) Barbara Langbart, 1943 (divorced); children: Jonathan, Richard, Andrew and Steven; 2) Ellen McLaughlin, 1963; children: Thomas and Christopher. Served in U.S. Army, 1942–46. Writer in television, from 1951, starting with CBS, eventually working for all the major networks; wrote CBS-TV's *Studio One* episode "Twelve Angry Men," 1954; wrote and co-produced *Twelve Angry Men* film version, 1957, and wrote stage version, 1964; writer of films, from 1956; author of books, from 1956; wrote CBS pilot for series "The Defender," as episode of *Studio One*, 1957; wrote Emmy-nominated "The Sacco-Vanzetti Story", NBC-TV's *Sunday Showcase*, 1960; president, Defender Productions, from 1961; created series and with others wrote *The Defenders*, 1961–65; wrote Emmy-nominated "Dear Friends" for *CBS Playhouse*, 1967; wrote multiple-award-winning CBS mini-series, *Escape from Sobibor*, 1987. President of Reginald Rose Foundation. Recipient: Emmy Awards, 1954, 1962, 1963 (with Robert Thom), 1968; Edgar Allan Poe Award, 1957; Berlin Film Festival Golden Berlin Bear Award, 1957; Writers Guild of America Award, 1960; Writers Guild of America Laurel Award, 1958 and 1987. Address: Defender Productions, c/o Philip Plumber, 105-58 Flatlands 5th Street, Brooklyn, New York 11236, U.S.A.

TELEVISION SERIES (various episodes)

1948–55	*Philco Television Playhouse-Goodyear Playhouse*
1948–58	*Studio One*
1951	*Out There*
1954–55	*Elgin Hour*
1955–57	*The Alcoa Hour-Goodyear Playhouse*
1956–61	*Playhouse 90*
1959–60	*Sunday Showcase*
1961–65	*The Defenders* (creator and writer)
1967	*CBS Playhouse*
1975	*The Zoo Gang* (creator and writer)
1977	*The Four of Us* (pilot)

TELEVISION MINISERIES

1979	*Studs Lonigan*
1987	*Escape from Sobibor*

MADE-FOR-TELEVISION MOVIES

1982	*The Rules of Marriage*
1986	*My Two Loves* (with Rita Mae Brown)

FILMS

Crime in the Streets, 1956; *Dino*, 1957; *Twelve Angry Men* (also co-produced), 1957; *Man of the West*, 1958; *The Man in the Net*, 1958; *Baxter!*, 1972; *Somebody Killed Her Husband*, 1978; *The Wild Geese*, 1978; *The Sea Wolves*, 1980; *Whose Life Is It, Anyway?* (with Brian Clark), 1981; *The Final Option*, 1983; *Wild Geese II*, 1985.

STAGE

Black Monday, 1962; *Twelve Angry Men*, 1964; *The Porcelain Year*, 1965; *Dear Friends*, 1968; *This Agony, This Triumph*, 1972.

PUBLICATIONS

Six Television Plays. New York: Simon and Schuster, 1957.
The Thomas Book. New York: Harcourt, 1972.

FURTHER READING

Hawes, William. *The American Television Drama: The Experimental Years*. University, Alabama: University of Alabama Press, 1986.
Sturken, Frank. *Live Television: The Golden Age of 1946–1958 in New York*. Jefferson, North Carolina: McFarland, 1990.
Wilk, Max. *The Golden Age of Television: Notes from the Survivors*. New York: Dell, 1977.

See also *Defenders*; *Playhouse 90*; *Studio One*; Writing for Television

ROSEANNE

U.S. Actor/Comedienne

Roseanne (née Roseanne Barr, formerly Roseanne Arnold) is the star of the situation comedy *Roseanne,* for several years the most highly rated program on American television and the centerpiece of ABC comedy programming. She is also one of the more controversial and outspoken television stars of the 1980s and 1990s. Her public statements, appearances on celebrity interview shows, and feature articles about her life in magazines and tabloid newspapers often overshadow her work on the television show.

Roseanne's career did not begin in network dramatic television. In the mid-1980s, she starred in two HBO comedy specials and in the feature film *She-Devil* with Meryl Streep. When she did create the series character, it was based on her own comic persona, a brash, loudmouthed, working-class mother and wife who jokes and mocks the unfairness of her situation and who is especially blunt about her views of men and sexism. Her humor aggressively attacks whomever and whatever would denigrate fat poor women—husbands, family and friends, the media, or government welfare policies. She has often stated that her life experiences are the basis for the TV character and her comedy. Critics have described the persona as a classic example of the "unruly" woman who challenges gender and class stereotypes in her performances.

Roseanne's published self-disclosures, in her two autobiographies, provide a detailed public record of her life. She grew up in Salt Lake City in a working class Jewish family she has defined as "dysfunctional," a description that includes assertions of having been sexually molested by family members. A high-school dropout, she reports getting married while still in her teens in order to get away from her family. She worked as a waitress, and, according to *People* magazine, began her comedy by being rude to her customers. Her career as a standup comic began in Denver, where her club appearances gained a following among the local feminist and gay communities. She toured nationally on the comedy club circuit and made well-received appearances on late-night talk shows before starring in her own comedy specials on HBO. In 1986, the Carsey-Werner Company approached her with a proposal for developing a situation comedy based on the standup routines. The show would be an antidote to the upper-middle-class wholesomeness of the previous Carsey-Werner hit, *The Cosby Show.* The popularity of her sitcom, which first aired in the fall of 1988, has broadened the audience for Roseanne as a public persona and greatly increased her power within show business (she has been compared to Lucille Ball in this regard). But there have been missteps.

One highly publicized gaffe was her off-key performance of the national anthem at a professional baseball game, a performance that ended with a crude gesture. Still, the resulting flurry of outraged criticism from public officials and in the media did not diminish the popularity of the show. In another exercise of industry clout, she threatened to move *Roseanne* to a different network when ABC decided to cancel the low-rated *The Jackie Thomas Show,* which starred her then-husband Tom Arnold. The threat created real jitters among network executives until it was discovered that she did not own the rights to the show—only Carsey-Werner could make such a decision. Roseanne has also pushed boundaries by having her series take a number of risks by raising issues of gender, homosexuality, and family dysfunction. The forthrightness of these dramatic moments is rare in prime-time sitcoms. Despite their frankness, the series continues to appeal to a wide segment of the viewing audience.

The show's treatment of such charged issues is consistent with Roseanne's stated political and social views. While she does not write the scripts (for a time, Arnold was heavily involved in writing), she retains a good deal of artistic control. Many of the plots draw on aspects of Roseanne's life prior to her success, or refer to contemporaneous events in her "real" life. Other episodes may include entire dialogues proposed by Roseanne to address specific themes or issues. The show occasionally strays from the sitcom formula of neatly tying up all the plotlines by the end of the episode. As Kathleen Rowe notes, one year saw Darlene (Sara Gilbert), the younger daughter character, going through an early adolescent depression that continued for the entire season.

After eight years, the program continues to be extremely popular, now in syndication as well first-run, and some critics have argued that it has improved over its earlier seasons. Most recently, Roseanne herself has had a good deal more media exposure about her personal life—cosmetic surgery, divorce, remarriage, pregnancy—than about her political views or her career as an actor. In almost every case she seems able to turn such public discussions into more authority and control within the media industries, and her position as a major figure in that context seems assured for some time to come.

—Kathryn Cirksena

Roseanne

ROSEANNE BARR (Roseanne Arnold; Roseanne). Born in Salt Lake City, Utah, U.S.A., 3 November 1952. Married: 1) Bill Pentland, 1974 (divorced, 1989); children: Jessica, Jennifer, Brandi, and Jake; 2) Tom Arnold, 1990 (divorced, 1994); 3) Ben Thomas, 1994. Cocktail waitress in Denver and comedy performer in local clubs, including the Comedy Store in Los Angeles, California, 1985; appeared in or starred in several TV specials; star of television series *Roseanne,* since 1988; co-executive producer, *The Jackie Thomas Show,* 1993–94; starred in motion pictures since 1989. Recipient: Cable Ace Award, 1987; Best Comedy Special, 1987; Emmy Award, 1993. Address: Full Moon and High Tide Productions, 4024 Radford Avenue, Dressing Rooms 916-917, Studio City, California 90614, U.S.A.

TELEVISION SERIES

1988–	*Roseanne*
1993–94	*The Jackie Thomas Show* (co-producer)
1994	*Tom* (co-producer)

MADE-FOR-TELEVISION MOVIES

1991	*Backfield in Motion*
1993	*The Woman Who Loved Elvis*

TELEVISION SPECIALS

1985	*Funny*
1986	*Rodney Dangerfield - It's Not Easy Bein' Me*
1987	*Dangerfield's*
1987	*On Location: The Rosanne Barr Show*
1990	*Mary Hart Presents Love in the Public Eye*
1992	*The Rosey and Buddy Show* (voice; co-producer)
1992	*Class Clowns*

FILMS

She-Devil, 1989; *Look Who's Talking Too* (voice), 1990; *Freddy's Dead*, 1991; *Even Cowgirls Get the Blues*, 1994.

PUBLICATIONS

Roseanne: My Life as a Woman. New York: Harper and Row, 1989.
"What Am I, a Zoo?" *The New York Times*, 31 July 1989.
"I Am an Incest Survivor: A Star Cries Incest." *People* (New York), 7 October 1991.

My Lives. New York: Ballantine Books, 1994.

FURTHER READING

Cole, Lewis. "Roseanne." *The Nation* (New York), 21 June 1993.
Klaus, Barbara. "The War of Roseanne." *New York Times*, 22 October 1990.
Murphy, Mary, and Frank Swertlow. "The Roseanne Report." *TV Guide* (Radnor, Pennsylvania), 4 January 1992.
Rowe, Kathleen. *The Unruly Woman: Gender and the Genres of Laughter.* Austin: University of Texas Press, 1995.
Van Buskirk, Leslie. "The New Roseanne—The Most Powerful Woman in Television." *US* (New York), May 1992.
Wolcott, James. "On Television: Roseanne Hits Home." *The New Yorker* (New York), October 1992,

See also Comedy, Domestic Settings; Family on Television; Gender and Television; *Roseanne*

ROSEANNE

U.S. Domestic Comedy

*R*oseanne evolved from the stand-up comedy act and HBO special of its star and executive producer, Roseanne (formerly Roseanne Barr Arnold). In the act, Roseanne deemed herself a "domestic goddess" and dispensed mock cynical advice about child-rearing: "I figure by the time my husband comes home at night, if those kids are still alive, I've done my job." *Roseanne,* the program, built a working-class family around this matriarchal figure and became an instantaneous hit when it premiered in 1988 on ABC.

Roseanne's immediate success may well have been in reaction to the dominant 1980s domestic situation comedy, *The Cosby Show.* Like *The Cosby Show, Roseanne* starred an individual who began as a stand-up comic, but the families in the two programs were polar opposites. Where *The Cosby Show* portrayed a loving, prosperous family with a strong father figure, *Roseanne's* Conner family was discordant, adamantly working class and mother-centered.

The Conner family included Roseanne, her husband Dan (John Goodman), sister Jackie (Laurie Metcalf), daughters Darlene (Sara Gilbert) and Becky (Lecy Goranson, replaced in the fall of 1993 by Sarah Chalke), son D.J. (Michael Fishman). Over the years the household expanded to include Becky's husband Mark (Glenn Quinn) and Darlene's boyfriend David (Johnny Galecki) and, in 1995, a new infant for Roseanne and Dan, Jerry Garcia Conner (Cole and Morgan Roberts).

The Conners were constantly facing money problems as they both worked in blue-collar jobs—in factories, hanging sheetrock, running a motorcycle shop, and even-

tually owning their own diner where they served "loose-meat" sandwiches. Their parenting style was often sarcastic, bordering on scornful. Once, when the kids left for school, Roseanne commented, "Quick. They're gone. Change the locks." But caustic remarks such as these were always balanced by scenes of affection and support so that the stability of the family was never truly in doubt. Much as in its working-class predecessor, *All in the Family*, the Conner family was not genuinely dysfunctional, despite all the rancor.

Roseanne often tested the boundaries of network standards and practices. One episode dealt with the young son's masturbation. In others, Roseanne frankly discussed birth control with Becky and explained her choice to have breast reduction surgery. The program also featured gay and lesbian characters, which made ABC nervous—especially when a lesbian character kissed Roseanne. The network initially refused to air that episode until Roseanne, the producer, demanded they do so.

Controversy attended the program off screen as well. During its first season there were well-publicized squabbles among the producing team, which led to firings and Roseanne assuming principal control of the program. Subsequently, Roseanne battled ABC over its handling of her then-husband Tom Arnold's sitcom, *The Jackie Thomas Show.* Dwarfing these professional controversies has been the strife in Roseanne's publicly available personal life. Among the events that have been chronicled in the tabloid press are her tumultuous marriage to and divorce from Arnold (amid accusations of spousal abuse), the reconcilia-

Roseanne

tion with the daughter she put up for adoption (which was forced by a tabloid newspaper's threat to reveal the story), her charges of being abused as a child, struggles with addic-tions to food and other substances, and a misfired parody of the national anthem at a baseball game (1990).

—Jeremy G. Butler

CAST

Roseanne Conner	Roseanne
Dan Conner	John Goodman
Becky Conner (1988–92, 1995–96) . . .	Lecy Goranson
Becky Conner (1993–95; 1996)	Sarah Chalke
Darlene Conner	Sara Gilbert
D.J. (David Jacob) Conner (pilot)	Sal Barone
D.J. Conner	Michael Fishman
Jackie Harris	Laurie Metcalf
Crystal Anderson (1988–92)	Natalie West
Booker Brooks (1988–89)	George Clooney
Pete Wilkins (1988–89)	Ron Perkins
Juanita Herrera (1988–89)	Evalina Fernandez
Sylvia Foster (1988–89)	Anne Falkner
Ed Conner (1989–)	Ned Beatty
Bev Harris (1989–)	Estelle Parsons
Mark Healy (1990–)	Glenn Quinn
David Healy (1992–)	Johnny Galeki
Grandma Nanna (1991–)	Shelley Winters
Leon Carp (1991–)	Martin Mull
Bonnie (1991–92)	Bonnie Sheridan
Nancy (1991–)	Sandra Bernhard
Fred (1993–95)	Michael O'Keefe
Andy	Garrett and Kent Hazen
Jerry Garcia Conner	Cole and Morgan Roberts

PRODUCERS   Marcy Carsey, Tom Werner, Roseanne

PROGRAMMING HISTORY

• ABC

October 1988–February 1989	Tuesday 8:30-9:00
February 1989–September 1994	Tuesday 9:00-9:30
September 1994–March 1995	Wednesday 9:00-9:30
March 1995–May 1995	Wednesday 8:00-8:30
May 1995–September 1995	Wednesday 9:30-10:00
September 1995–	Wednesday 8:00-8:30

FURTHER READING

Arnold, Roseanne. *My Lives.* New York: Ballantine, 1994.

Dresner, Zita Z. "Roseanne Barr: Goddess or She-devil." *Journal of American Culture* (Bowling Green, Ohio), Summer 1993.

Dworkin, Susan. "Roseanne Barr: The Disgruntled Housewife as Stand-up Comedian." *Ms.* (New York), July-August 1987.

Givens, Ron. "A Real Stand-up Mom." *Newsweek* (New York), 31 October 1988.

Klaus, Barbara. "The War of the Roseanne: How I Survived Three Months in the Trenches Writing for TV's Sitcom Queen." *The New York Times*, 22 October 1990.

Lee, Janet. "Subversive Sitcoms: Roseanne as Inspiration for Feminist Resistance." *Women's Studies: An Interdisciplinary Journal* (Claremont, California), 1992.

Mayerle, Judine. "Roseanne—How Did You Get Inside My House? A Case Study of a Hit Blue-Collar Situation Comedy." *Journal of Popular Culture* (Bowling Green, Ohio), Spring 1991.

Rich, Frank. "What Now My Love." *The New York Times*, 6 March 1994.

Rowe, Kathleen. *The Unruly Woman: Gender and Genres of Laughter.* Austin: University of Texas Press, 1995.

Volk, Patricia. "Really Roseanne." *The New York Times Magazine*, 8 August 1993.

See also Comedy, Domestic Settings; Family on Television; Gender and Television; Roseanne

ROSENTHAL, JACK

British Writer

As one of British television's most successful dramatists, Jack Rosenthal has received BAFTA Awards for *The Evacuees, Bar Mitzvah Boy, P'tang Yang Kipperbang,* and *Ready When You Are, Mr. McGill,* an Emmy Award for *The Evacuees,* and the Prix Italia for *Spend, Spend, Spend,* and *The Knowledge.* He has written for the big screen with *The Chain* and *The Knowledge,* and has also authored five plays for the live stage, notably *Smash!*

Rosenthal learned the craft of writing for the medium of television in the 1960s, at a time when television drama in Britain (particularly on the BBC) was still dominated by writers schooled in theatrical conventions and overly concerned with being taken seriously. This resulted in a preoccupation with adaptations of theatrical successes, revivals of classics (e.g., Shakespeare, Dickens), and writing that exploited literary rather than visual resources. Independent television in the late 1950s was looking to develop more popular forms of drama to attract wider audiences, and brought in Sydney Newman from Canada, who fostered new dramatists and initiated new series. It was against this background that Rosenthal started work in Granada, where he served his apprenticeship by creating more than 150 scripts for the popular TV soap *Coronation Street.* The experience of writing for a popular genre prepared him for originating such comedy serials as *The Dustbinmen, The Lovers,* and *Sadie, It's Cold Outside.* His growing reputation in the 1970s as a reliable professional writer led to his being entrusted with the prestigious single play: a form that Rosenthal himself prefers because of the freedom it offers the artist to explore his own vision.

Rosenthal was born in Manchester of Jewish parents, and drew on his experiences to write *Bar Mitzvah Boy* and *The*

Evacuees. But his interest lies in observing the interactions of individuals in diverse social networks, and the Jewish community is merely one of the many institutions that he explores: schools (*P'tang Yang Kippperbang*), taxi drivers (*The Knowledge*), the army (*Bootse and Snudge*), fire fighters (*London's Burning*), and TV drama (*Ready When You Are, Mr. McGill*). He is also interested in the common experiences that many face at particular moments in life: moving (*The Chain*), growing up (*Bar Mitzvah Boy, P'tang Yang Kippperbang*), falling in love (*The Lovers*), and forgetfulness and old age (*A Day to Remember*).

The strength of Rosenthal's comedy lies in its closeness to tragedy; from another perspective, the petty cruelties of the stepmother in *The Evacuees* could have blighted the lives of the children, but both plot and psychological insight combine to restore harmony and recognize the cruelty as misplaced possessiveness. So too, in *A Day to Remember*, the terror and pain of short-term memory loss, attendant on a stroke in old age, are contained and balanced by the comic presentation of the gaps and imperfections that beset the middle-aged. If the comic vision is shown as perceptive about the frailties of the human condition, it is not sentimentalized. The insight that comes through comedy is one that is often painfully achieved. The schoolboy hero of *P'tang Yang Kipperbang* is only able to kiss his first love; he enters upon adult sexuality by recognizing the fantasy element of that anticipated delight. To fulfil his desire means abandoning private fantasy and entering the real world in which people are both less than we would wish and more diverse than we could expect. Similarly, when the aspirant cabby in *The Knowledge* finally achieves his ambition to be a London taxi driver, he discovers his girlfriend, the initial driving force behind his application, has fallen for somebody else. He neglected her to focus on the discipline of acquiring "the knowledge" (learning by heart the streets and landmarks of London by perpetually driving around them). Knowledge of chaps rather than maps turns out to be that which is most difficult to acquire.

Although the comedy of Jack Rosenthal is invariably rooted in a recognizable social setting which has been carefully researched, the characters are not deeply explored. The story is, instead, focused on the themes: in *Another Sunday and Sweet FA,* the frustrations of refereeing a football match provide the opportunity for a comic disquisition on the competing claims of power and justice; in *P'tang Yang Kipperbang,* imagination and reality struggle for an accommodation; in *The Chain,* the seven deadly sins provide the motivation for Fortuna's wheel of house-hunting. If there is a thread which underlies most of Rosenthal's work, it is that our desire as individuals to do good in order to be liked and admired is at variance with our role as social beings to impose order, our order, on others. Wisdom comes when we learn to accommodate these competing demands and accept responsibility for fulfilling our desires.

—Brendan Kenny

JACK MORRIS ROSENTHAL. Born in Manchester, England, 8 September 1931. Attended Colne Grammar School; Sheffield University, B.A. in English language and literature; University of Salford, M.A. 1994. Married: Maureen Lip-

Jack Rosenthal
Photo courtesy of Jack Rosenthal

man, 1973; one son and one daughter. Writer for television; subsequently consolidated reputation with comedy series and one-off dramas, several of which were pilots for series. Commander of the Order of the British Empire, 1994. D.Litt., University of Manchester, 1995. Recipient: British Academy of Film and Television Arts Writer's Awards; Emmy Award; Prix Italia; Royal Television Society Writer's Award, 1976; British Academy of Film and Television Arts Best Play Award, 1976, 1977. Address: William Morris Agency, 31–32 Soho Square, London W1V 5DG, England.

TELEVISION SERIES

1960–	*Coronation Street*
1962–63	*That Was the Week That Was*
1965	*Pardon the Expression*
1969–70	*The Dustbinmen*
1970–71	*The Lovers*
1975	*Sadie, It's Cold Outside*
1994	*Moving Story*

TELEVISION SPECIALS

1963	*Pie in the Sky*
1963	*Green Rub*
1968	*There's a Hole in Your Dustbin, Delilah*

1972	*Another Sunday and Sweet FA*
1974	*Polly Put the Kettle On*
1974	*Mr. Ellis Versus the People*
1974	*There'll Almost Always Be an England*
1975	*The Evacuees*
1976	*Ready When You Are, Mr. McGill*
1976	*Bar Mitzvah Boy*
1977	*Spend, Spend, Spend*
1979	*Spaghetti Two-Step*
1979	*The Knowledge*
1982	*P'tang Yang Kipperbang*
1985	*Mrs. Capper's Birthday*
1986	*Fools on the Hill*
1986	*London's Burning*
1986	*A Day to Remember*
1989	*And a Nightingale Sang*
1989	*Bag Lady*
1991	*Sleeping Sickness*
1992	*'Bye, 'Bye, Baby*
1993	*Wide-Eyed and Legless*

FILMS

Lucky Star, 1980; *Yentl*, with Barbra Streisand, 1983; *The Chain*, 1985.

STAGE (selection)

Smash!, 1981.

PUBLICATIONS

The Television Dramatist, with others. London: Elek, 1973.
Three Award Winning Television Plays: Bar Mitzvah Boy, The Evacuees, Spend, Spend, Spend. London: Harmondsworth, 1978.
First Loves: Stories (anthology). London: Hamilton, 1984.
The Chain, with *The Knowledge*, and *Ready When You Are, Mr. McGill*. Boston: Faber and Faber, 1986.

See also *Coronation Street*; *That Was the Week That Was*

ROUTE 66

U.S. Drama

*R*oute 66 was one of the most unique American television dramas of the 1960s, an ostensible adventure series that functioned, in practice, as an anthology of downbeat character studies and psychological dramas. Its 1960 premiere launched two young drifters in a Corvette on an existential odyssey in which they encountered a myriad of loners, dreamers and outcasts in the small towns and big cities along U.S. Highway 66 and beyond. And the settings were real; the gritty social realism of the stories was enhanced by location shooting that moved beyond Hollywood hills and studio backlots to encompass the vast face of the country itself. *Route 66* took the anthology on the road, blending the dramaturgy and dramatic variety of the *Studio One* school of TV drama with the independent filmmaking practices of the New Hollywood.

Route 66 was the brainchild of producer Herbert B. Leonard and writer Stirling Silliphant, the same creative team responsible for *Naked City*. The two conceived the show as a vehicle for actor George Maharis, casting him as stormy Lower East Side orphan Buz Murdock, opposite Martin Milner as boyish, Yale-educated Tod Stiles. When Tod's father dies, broke but for a Corvette, the two young men set out on the road looking for "a place to put down roots." Maharis left the show in 1963 in a dispute with the show's producers, and was replaced by Glenn Corbett as Linc Case, a troubled Vietnam vet also seeking meaning on the road.

Like *Naked City*, which producer Leonard had conceived as an anthology with a cop-show pretext, the picaresque premise of *Route 66* provided the basis for a variety of weekly encounters from which the stories arose. Episodes emphasized the personal and psychological dramas of the various troubled souls encountered by the guys in their stops along the highway. Guest roles were filled by an array of Hollywood faces, from fading stars like Joan Crawford and Buster Keaton, to newcomers such as Suzanne Pleshette, Robert Duvall, and Robert Redford. The show's distinct anthology-style dimension was symptomatic of a trend *Variety* dubbed "the semi-anthology," a form pioneered by *Wagon Train* and refined by shows like *Bus Stop* and *Route 66*. The series' nomadic premise, and its virtual freedom from genre connections and constraints, opened it up to a potentially limitless variety of stories. While the wandering theme was hardly new in a television terrain overrun with westerns, for a contemporary drama the premise was quite innovative. *Route 66* was consistent in tone to the rest of TV's serious, social-realist dramas of the period, but unencumbered by any predetermined dramatic arena or generic template—as against the likes of *The Defenders* (courtroom drama), *Dr. Kildare* (medical drama), *Saints and Sinners* (newspaper drama) or *Mr. Novak* (blackboard drama). Indeed, the show's creators met initial resistance from their partner/distributor Screen Gems for this lack of a familiar "franchise," with studio executives arguing that no one would sponsor a show about two "bums." Of course, Chevrolet proved them wrong.

Perhaps even more startling for the Hollywood-bound telefilm industry was the program's radical location agenda. Buz and Tod's cross-country search actually was shot across the country, in what *Newsweek* termed "the largest weekly mobile operation in TV history." Remarkably, by the end of its four-season run, the *Route 66* production caravan had traveled to twenty-five states—as far from L.A. as Maine and Florida—as well as Toronto. The show's stark black and white photography and spectacular locations provided a powerful backdrop to its downbeat stories, and yielded a photographic and geographical realism that has never been duplicated on American television.

The literate textures and disturbing tones of *Route 66*'s dramas were as significant as its visual qualities. The wandering pretext provided both a thematic foundation and a narrative trajectory upon which a variety of psychological dramas, social-problem stories, and character studies could be played out. The nominal series "heroes" generally served as observers to the dramas of others: a tormented jazz musician, a heroin addict, a washed-up prizefighter, migrant farm workers, an aging RAF pilot (turned crop-duster), a runaway heiress, Cajun shrimpers, a weary hobo, an eccentric scientist, a small-time beauty contest promoter, drought-stricken ranchers, Cuban-Basque jai-alai players, a recent ex-con (female and framed), a grim Nazi-hunter, a blind dance instructor, a dying blues singer—each facing some personal crisis or secret pain.

The show's continuing thread of wandering probed the restlessness at the root of all picaresque sagas of contemporary American popular culture. The search that drove *Route 66* was both a narrative process and a symbolic one. Like every search, it entailed optimism as well as discontent. The unrest at the core of the series echoed that of the Beats—especially Kerouac's *On the Road*, of course—and anticipated the even more disaffected searchers of *Easy Rider*. The show's rejection of domesticity in favor of rootlessness formed a rather startling counterpoint to the dominant prime-time landscape of home and family in the sixties, as did the majority of the characters encountered on the road. The more hopeful dimension of *Route 66* coincided with the optimism of the New Frontier circa 1960, with these wandering samaritans symbolic of the era's new spirit of activism. Premiering at the dawn of a new decade, *Route 66* captured in a singular way the nation's passage from the disquiet of the fifties to the turbulence of the sixties, expressing a simultaneously troubled and hopeful vision of America.

Despite its uniqueness as a contemporary social drama, and its radical break from typical Hollywood telefilm factory practice, *Route 66* has been largely forgotten amid the rhetoric of 1960s TV-as-wasteland. When the series is cited at all by television historians, it is as the target of CBS-TV president James Aubrey's attempts to inject more "broads, bosoms, and fun" into the series ("the Aubrey dictum"). Aubrey's admitted attempts to "lighten" the show, however, only serve to underscore its dominant tone of seriousness. What other American television series of the 1960s could have been described by its writer-creator as "a show about a statement of existence, closer to Sartre and Kafka than to anything else"? (*Time*, 1963). Silliphant's hyperbole is tempered by critic Philip Booth, who suggested in a *Television Quarterly* essay that the show's literacy was "sometime spu-

Route 66

rious," and that it could "trip on its own pretensions" in five of every ten stories. Still, Booth wrote, of the remaining episodes, four "will produce a kind of adventure like nothing else on television, and one can be as movingly universal as Hemingway's 'A Clean, Well-Lighted Place.'"

How often *Route 66* matched the power of Hemingway (or the existential insight of Sartre) is debatable. That it was attempting something completely original in television drama is certain. Its footloose production was the antithesis of the claustrophobic stages of the New York anthologies of old, yet many of its dramatic and thematic concerns—even certain of its stories—echoed those of the intimate character dramas of the *Philco Playhouse* era. Indeed, one of Aubrey's CBS lieutenants, concerned with the show's "downbeat" approach to television entertainment, protested to its producers that *Route 66* should not be considered "a peripatetic *Playhouse 90*"—capturing, willingly or not, much of the show's tenor and effect. *Route 66* was trying to achieve the right mix of familiarity and difference, action and angst, pathos and psychology, working innovative elements into a commercial package keyed to the demands of the industry context. Even with its gleaming roadster, jazzy theme song, obligatory fistfights and occasional romantic entanglements, *Route 66* was far removed indeed (both figuratively and geographically) from the likes of *77 Sunset Strip*.

In 1993 the Corvette took to the highway once more in a nominal sequel, a summer series (on NBC) that put

Buz's illegitimate son at the wheel with a glib Generation-X partner in the passenger seat. Although the new *Route 66* lasted only a few weeks, by reviving the roaming-anthology premise of the original, it evidenced television's continuing quest for narrative flexibility (and Hollywood's inherent penchant for recycling). From *The Fugitive* to *Run For Your Life* to *Highway to Heaven* to *Quantum Leap* to *Touched by an Angel*, television has continued to exploit the tradition of the wandering samaritan, to achieve the story variety of an anthology within a series format. *Route 66* established the template in 1960, launching a singular effort at contemporary drama in a non-formulaic series format. That the series mounted its dramatic agenda in a Corvette, on the road, is to its creators' everlasting credit.

—Mark Alvey

CAST

Tod Stiles . . . . Martin Milner

Buz Murdock (1960–63) . . . . George Maharis

Linc Case (1963–64) . . . . Glenn Corbett

PRODUCERS Herbert B. Leonard, Jerry Thomas, Leonard Freeman, Sam Manners

PROGRAMMING HISTORY 116 Episodes

• CBS

October 1960–September 1964   Friday 8:30-9:30

FURTHER READING

Barnouw, Erik. *Tube of Plenty.* New York: Oxford University Press, 1990.

Bergreen, Laurence. *Look Now, Pay Later.* New York: Mentor, 1980.

Booth, Philip. "*Route 66*—On the Road Toward People." *Television Quarterly* (New York), Winter 1963.

Castelman, Harry, and Walter Podrazik. *Watching TV: Four Decades of American Television.* New York: McGraw-Hill, 1982.

Chandler, Bob Chandler. "Review of *Route 66.*" *Variety* (Los Angeles), 12 October 1960.

Dunne, John Gregory. "Take Back Your Kafka." *The New Republic* (Washington, D.C.), 4 September 1965.

"The Fingers of God." *Time* (New York), 9 August 1963.

"Have Camera, Will Travel." *Variety* (Los Angeles), 12 October 1960.

"The Hearings that Changed Television." *Telefilm* (New York), July-August, 1962.

Jarvis, Jeff. "The Couch Critic." *TV Guide* (Radnor, Pennsylvania), 12 June 1993.

Jenkins, Dan. "Talk About Putting a Show on the Road!" *TV Guide* (Radnor, Pennsylvania), 22 July 1961.

"A Knock Develops on *Route 66.*" *TV Guide* (Radnor, Pennsylvania), 26 January 1963.

"Rough Road." *Newsweek* (New York), 2 January 1961.

Seldes, Gilbert. "Review of *Route 66.*" *TV Guide* (Radnor, Pennsylvania), 10 February 1962.

See also Silliphant, Sterling

ROWAN AND MARTIN'S LAUGH-IN

U.S. Comedy Variety

Rowan and Martin's Laugh-In was the NBC comedy-variety program which became an important training ground for a generation of comic talent. If *The Smothers Brothers Comedy Hour* captured the political earnestness and moral conscience of the 1960s counterculture, *Laugh-In* snared its flamboyance, its anarchic energy, and its pop aesthetic, combining the black-out comedy of the vaudeville tradition with a 1960s-style "happening."

In an age of "sit-ins," "love-ins" and "teach-ins," NBC was proposing a "laugh-in" which somehow bridged generational gaps. Originally a one-shot special, *Laugh-In* was an immediate hit and quickly became the highest-rated series of the late 1960s. In a decade of shouted slogans, bumper stickers, and protest signs, *Laugh-In* translated its comedy into discrete one-liners hurled helter-skelter at the audience in hopes that some of them would prove funny. Many of them became catch-phrases: "Sock it to me," "Here come de judge," "You bet your sweet bippy," and "Look that up in your *Funk and Wagnalls.*" In this frenetic and fragmented series, comic lines were run as announcements along the bottom of the screen, printed in lurid colors on the bodies of bikini-clad go-go girls, and shouted over the closing credits. The humor was sometimes topical, sometimes nonsensical, sometimes "right on" and sometimes right of center, but it largely escaped the censorship problems which besieged the Smothers Brothers. Its helter-skelter visual style stretched the capabilities of television and video-tape production, striving for the equivalent of the cutting and optical effects Richard Lester brought to the Beatles movies.

Laugh-In broke down the traditional separation of comedy, musical performance, and dramatic interludes which had marked most earlier variety shows and decentered the celebrity host from his conventional position as mediator of the flow of entertainment. Dan Rowan and Dick Martin, successful Las Vegas entertainers, sought to orchestrate the proceedings but were constantly swamped by the flow of sight-gags and eccentric performances which surrounded them. Similarly, guest stars played no privileged role here. For a time, everyone seemed to want to appear on *Laugh-In*, with guests on one memorable episode including Jack Lemmon, Zsa Zsa Gabor, Hugh Hefner, and presidential candidate Richard Nixon. But no guest appeared for more than a few seconds at a time, and none received the kind of screen time grabbed by the program's ensemble of talented young clowns.

The comic regulars—Gary Owens' over-modulated announcer, Ruth Buzzi's perpetually-frustrated spinster, Arte Johnson's lecherous old man, Goldie Hawn's dizzy blonde, Jo Anne Worley's anti-Chicken Joke militant, Henry Gibson's soft-spokenly banal poet, Lily Tomlin's snorting telephone operator, Pigmeat Markham's all-powerful Judge, and countless others—dominated the program. Many of these comics moved almost overnight from total unknowns to household names and many became important stars for the subsequent decades. Not until *Saturday Night Live* would another television variety show ensemble leave such a firm imprint on the evolution of American comedy. These recurring characters and their associated shtick gave an element of familiarity and predictability to a program which otherwise depended upon its sense of the unexpected.

While *Laugh-In* lacks the satirical bite of later series such as *Saturday Night Live*, or *In Living Color* or of *That Was the Week That Was* (to which it was often compared by contemporary critics), *Laugh-In* brought many minority and female performers to mainstream audiences, helping to broaden the composition of television comedy. Its dependence upon stock comic characters and catch-phrases was clearly an influence on the development of *Saturday Night Live*, which by comparison, has a much more staid visual style and more predictable structure. Unfortunately, *Laugh-In's* topicality, even its close fit with 1960s aesthetics, has meant that the program has not fared well in re-runs, being perceived as dated almost from the moment it was aired. However, the on-going success of *Laugh-In* alums such as Hawn, Tomlin, or even gameshow host Richard Dawson point to its continued influence.

—Henry Jenkins

REGULAR PERFORMERS

Dan Rowan
Dick Martin
Gary Owens
Ruth Buzzi
Judy Carne (1968–70)
Eileen Brennan (1968)
Goldie Hawn (1968–70)
Arte Johnson (1968–71)
Henry Gibson (1968–71)
Roddy-Maude Roxby (1968)
Jo Anne Worley (1968–70)
Larry Hovis (1968, 1971–72)
Pigmeat Markham (1968–69)
Charlie Brill (1968–69)
Dick Whittington (1968–69)
Mitzi McCall (1968–69)
Chelsea Brown (1968–69)

Rowan and Martin's Laugh-In
Photo courtesy of George Schlatter Productions

Alan Sues (1968–72)
Dave Madden (1968–69)
Teresa Graves (1969–70)
Jeremy Lloyd (1969–70)
Pamela Rodgers (1969–70)
Byron Gilliam (1969–70)
Ann Elder (1970–72)
Lily Tomlin (1970–73)
Johnny Brown (1970–72)
Dennis Allen (1970–73)
Nancy Phillips (1970–71)
Barbara Sharma (1970–72)
Harvey Jason (1970–71)
Richard Dawson (1971–73)
Moosie Drier (1971–73)
Patti Deutsch (1972–73)
Jud Strunk (1972–73)
Brian Bressler (1972–73)
Sarah Kennedy (1972–73)
Donna Jean Young (1972–73)
Tod Bass (1972–73)
Lisa Farringer (1972–73)
Willie Tyler and Lester (1972–73)

PRODUCERS George Schlatter, Paul W. Keyes, Carolyn Raskin

PROGRAMMING HISTORY 124 Episodes

• NBC

January 1968–May 1973   Monday 8:00-9:00

FURTHER READING

Castleman, Harry, and Walter J. Podrazik. *Watching TV: Four Decades of American Television.* New York: McGraw-Hill, 1982.

Rowan, Dan. *A Friendship: The Letters of Dan Rowan and John D. McDonald, 1967–1974.* New York: Knopf, 1986.

Waters, Harry R. "Laugh-In." *Newsweek* (New York), 8 February 1993.

See also Variety Programs

THE ROYAL CANADIAN AIR FARCE

Canadian Satirical Review

On 9 December 1973 the first radio show by *The Royal Canadian Air Farce* comedy troupe was broadcast coast-to-coast on CBC Radio and CBC Stereo. Now in its 24th year of political commentary, social satire and general nonsense, the *Air Farce*, a Canadian institution, moved into television in the fall of 1993 with a weekly series on CBC Television. Like the radio show, *Air Farce* is topical, on the edge of controversy, and performed in front of a live audience. The group consists of Roger Abbot, Don Ferguson, Luba Goy and John Morgan. Dave Broadfoot, who was a member for fifteen years before going out on his own, makes frequent guest appearances. Two non-performing writers, Rick Olsen and Gord Holtam, have been with the troupe for 20 years.

In 1992, the group became the first Canadian inductees into the International Humour Hall of Fame. The editors of *Maclean's* (Canada's national news magazine) chose the *Air Farce* for the 1991 Honour Roll of Canadians who make a difference. The group has won fifteen ACTRA Awards (Association of Canadian Television and Radio Artists) for radio and television writing and performing and a Juno Award (Canadian recording award) for Best Comedy Album. In 1993, the four current members of the Farce were each awarded Honorary Doctor of Law degrees by Brock University in St. Catharines.

The *Air Farce* keeps in touch with Canadians and ensures that their humour remains relevant by performing and recording in all ten provinces and two territories. "We're reluctant to give up radio," Ferguson told *Toronto Star* journalist Phil Johnson. "Radio allows us to showcase new acts and characters." They generally play in halls which hold 2,000 or 2,500, even when taping for radio. This creates the need for more visual interest. "I did [former Prime Minister] Brian Mulroney for 20 years—the worst years of my life I might add," Ferguson has told *Globe and Mail* columnist Liam Lacey. "On-stage, I'd have a long walk over to the microphone, so I'd start from the side of the stage with just the chin first, and then the stuckout bum would follow. The audiences would be roaring before I reached the microphone. Then we'd edit all that out, and cut to the voice."

When the *Farce* first tried a television show in 1981, it was shot in advance and produced with canned laughter. The lack of live performance and topicality destroyed the spontaneity that is at the heart of the *Farce* and the show failed. Then in 1993, a New Year's Eve special was made, raking in two million viewers, almost 10% of the entire Canadian population. Network honcho Ivan Fecan approved a series. It became one of the top 20 Canadian shows, and one of the CBC's top five.

Rather than leaning towards a particular point of view, the Farce points fingers at all parties. Skewered politicians and media figures regularly show up in person to do sketches on the show. Individual performers don't even know how the other members of the group vote and wouldn't dream of discussing it. As Liam Lacey wrote in noting indirect governmental support of the *Farce* in the form of the CBC: "One would be hard-pressed to imagine another country in the world where purveyors of official disrespect would be regarded with such widespread affection." Dave Broadfoot used to say, "Do you know what they'd call us in the Soviet Union? Inmates."

—Janice Kaye

REGULAR PERFORMERS
Roger Abbot
Don Ferguson
Luba Goy
John Morgan
Dave Broadfoot

FURTHER READING
Turbide, Diane. "The Air Farce is Flying High." *Maclean's* (Toronto), 26 February 1996.

See also Canadian Programming in English

ROYALTY AND ROYALS ON TELEVISION

The relationship between television and the royalty of the United Kingdom and other states has always been uneasy, albeit generally mutually respectful, as the perceived dangers to both sides have been immense. With television audiences of grand royal occasions and major documentaries running into many millions around the globe, the impact of a mishandled interview could have serious political repercussions for any monarchy, as well as huge public relations problems for television networks anxious not to outrage public opinion.

The idea that members of the British royal family might allow themselves to be seen on television in any other capacity than at the end of a long-range lense in the course of a formal state occasion or fleetingly in newsreel footage was once considered unthinkable. In the early days, immediately after World War II, television was regarded by many in the establishment as too trivial to be taken seriously, and it was argued that it was inappropriate for heads of nations to appear. In Britain, Sir Winston Churchill was in the vanguard of those who considered television a vulgar plaything and beneath the dignity of the crown.

The crunch came in 1953, when it was suggested that television cameras be allowed to film the coronation of Elizabeth II. Churchill, the Archbishop of Canterbury, the Earl Marshal, and various members of the British cabinet strongly opposed the idea, but to their surprise the 26-year-old Princess Elizabeth, in a decision subsequently hailed for its sagacity, insisted upon the rest of the nation being able to witness her enthronement via television—and the cameras were allowed in. The resulting broadcast, expertly narrated by the BBC's anchorman Richard Dimbleby, was a triumph, bringing the monarchy into the "television age" and cementing the image of Elizabeth II as a "people's monarch."

Following the 1953 coronation experiment, it became accepted that the television cameras would be permitted to film grand royal occasions, including weddings, the state opening of Parliament, and the trooping of the colour, as well as jubilee celebrations, visits by the royal family to local businesses, and so forth. Coverage of royal events, however, remained a sensitive area in broadcasting, and many rows erupted when it was felt cameras had intruded too far, or conversely (and increasingly in recent years), that too much deference had been shown. Certain presenters, including ITV's Alistair Burnet and BBC's Raymond Baxter, specialized in coverage of royal stories or spectacles—but found they had to tread a very thin line between being accused of sycophancy or else of gross insensitivity.

The Queen is sheltered from more intrusive interrogation on television by necessity: the constitutional imperative that the monarch should not comment personally on the policies of her government because of the implications this might have in terms of party politics means that Buckingham Palace, in concert with the government of the day, closely controls the style and content of all broadcasts in which she appears. In 1969 an attempt was made for the first time, in the joint BBC and ITV production, *Royal Family*, to portray the Queen as a private person rather than as a constitutional figurehead. The program attracted an audience of 40 million in the United Kingdom alone, and similarly large audiences have watched her celebrated annual Christmas broadcasts, which have over the years become more relaxed in tone, inspiring further occasional documen-

The Prince and Princess of Wales
Photo courtesy of Camera Press Ltd.

taries inviting the cameras "behind the scenes" (though, again, only under strict direction from the palace).

There is more leeway with other members of the royal family; however, this has been exploited with increasing vigor since the 1980s, in response to changing public attitudes toward royalty. Prince Philip's hectoring manner during rare appearances on chat shows did little to endear television audiences, and he was henceforth discouraged from taking part in such programs. Princess Anne developed a similarly tempestuous relationship with the media as a whole, though she was better received after her good works for charity won public recognition. Prince Andrew came over as bluff and hearty, and Prince Edward was considered affable enough—though there were adverse comments about loss of dignity in 1987 when the three youngest of the Queen's children attempted to sound a populist note by appearing in a special *It's a Knockout* program for charity (royal guests stormed out of press meetings when the questioning became hostile and the experiment was not repeated).

After years of carefully treading the line between deference and public interest, television's relationship with the royals was stretched to the limit in the 1990s during the furor surrounding the break-up of several royal marriages, notably that of the heir-apparent, Prince Charles (whose wedding to Lady Diana Spencer had been seen by 700 million people worldwide in 1981). A notorious interview with Princess

Diana that was broadcast on *Panorama,* when it was becoming clear that the rift was irreparable (though many still hoped the marriage could be saved), provoked howls of protest from many quarters—not least from the palace itself. Charles was given his own program in which to tell his side of the story, but only succeeded in drawing more fire upon himself and his family. For many viewers both interviews were enthralling, though to others they were distasteful and reflected badly both on the individuals themselves and on the institution of the monarchy.

Other monarchies have experienced not dissimilar difficulties in their relations with television and other organs of the media. For a number of years, the Rainiers of Monaco, for instance, seemed to live their lives in the constant glare of the cameras. Some, however, have protected themselves by insisting that the cameras remain at a discreet distance (as in Japan, where the emperor is only rarely filmed), despite the demands imposed by unflagging public interest.

Television's fascination with royalty has expressed itself in other forms beside coverage of contemporary royals, notably in the field of drama. The BBC in particular won worldwide acclaim in the late 1960s and 1970s for lavish costume series dealing with Henry VIII, Elizabeth I, Edward VII and, rather more controversially, Edward VIII. More recently a documentary series in which Prince Edward delved into the lives of some of his royal ancestors was also well received.

—David Pickering

See also Parliament, Coverage by Television; Political Processes and Television

RULE, ELTON

U.S. Media Executive

Elton Rule took the ABC TV network from a struggling operation in 1968 to top of the television network world a decade later. Under Rule's leadership, ABC-TV expanded its number of affiliates from 146 to 214 stations, and revenues increased from $600 million to $2.7 billion. The "alphabet network" began turning a profit in 1972; by 1976 it was the highest rated network in prime time; a year later Rule was presiding over a television empire that was collecting more money for advertising time than any media corporation in the world.

The key to this extraordinary success was Rule's ability to find top programming. During the 1970s Rule helped introduce such innovations as the made-for-television movie, the miniseries, and *Monday Night Football.* One of his first moves as network president was to sign the Hollywood producer Aaron Spelling, who through the 1970s added a string of top-ten hits to ABC's line-up, including *Mod Squad, Family, Starsky and Hutch, Love Boat,* and *Charlie's Angels.* Rule pioneered the presentation of made-for-television movies as a regular part of network schedules, billing them as ABC's Movie of the Week, and producing such early hits as *Brian's Song* and *That Certain Summer.* In 1974 Rule approved the miniseries, *QB VII.* Three years later a week of *Roots,* from Alex Haley's best-selling book, set ratings records, earned Rule wide acclaim, and generated for ABC vast sums of advertising dollars.

During the 1970s, Rule made ABC the leading sports network, centered around *Monday Night Football* and the Olympics. Rule must also be credited with making the ABC news division the industry leader. He moved sports producer Roone Arledge over to head a languishing network operation, approved hiring reporters from major newspapers, and expanded the locus of the network's foreign news bureaus. By the mid-1980s ABC News was the leading broadcast journalism operation in the United States.

When Rule retired in January 1984, he was properly hailed as a corporate savior. Through the remainder of the 1980s he bought and sold television stations, becoming a multi-millionaire. He is remembered—and heralded—for creating a television network empire, an economic, political,

Elton Rule
Photo courtesy of Broadcasting and Cable

social, and cultural force second to none in the history of television.

—Douglas Gomery

ELTON (HOERL) RULE. Born in Stockton, California, U.S.A., 13 June 1916. Graduated from Sacramento College, 1938. Married: Betty Louise Bender; children: Cindy Rule Dunne, Christie, James. Served in the U.S. Army Infantry, 1941–45. Worked at KROY, Sacramento, 1938–41; radio sales account executive, 1946–52; assistant sales manager, KECA-TV (now KABC-TV), 1952; general sales manager, 1953–1960; general manager, 1960–61; vice-president and general manager, 1961–68; president, California Broadcasters Association, 1966–67; president, ABC-TV, 1968–70; group vice-president, ABC, 1969–72; president, ABC division, 1970–72; director, ABC, 1970–84; president, chief executive officer, and member of executive committee, ABC, 1972–83; vice chair ABC, 1983–84; president, chair, investment funds with I. Martin Pompadur; co-chair, National Center of Film and Video Preservation. Member: advisory board, Institute of Sports Medicine and Athletic Trauma, Lenox Hill Hospital, 1973–84; board of visitors, University of California, Los Angeles, School of Medicine, 1980–84. Recipient: Purple Heart; Bronze Star with Oak Leaf Cluster; International Radio and TV Society Gold Medal Award, 1975; Academy of TV Arts and Sciences Governor's Award, 1981. Died in Beverly Hills, California, 5 May 1990.

FURTHER READING

Brown, Les. *Televi$ion: The Business Behind the Box*. New York: Harcourt Brace Jovanovich, 1971.

Goldenson, Leonard H. *Beating the Odds*. New York: Scribner's, 1991.

Gunthar, Marc. *The House That Roone Built: The Inside Story of ABC News*. Boston: Little Brown, 1994.

Quinlan, Sterling. *Inside ABC: American Broadcasting Company's Rise to Power*. New York: Hastings's House, 1979.

Williams, Huntington. *Beyond Control: ABC and the Fate of the Networks*. New York: Atheneum, 1989.

See also American Broadcasting Company; United States: Networks

RUMPOLE OF THE BAILEY

British Legal/Mystery Comedy

Rumpole of the Bailey, a mix of British courtroom comedy and drama, aired on Thames Television in 1978. The program made a successful transatlantic voyage and is popular on the American Public Broadcasting Service as part of the *Mystery!* anthology series.

All episodes feature the court cases of Horace Rumpole (Leo McKern), a short, round, perennially exasperating, shrewd, lovable defense barrister. His clients are often caught in contemporary social conflicts: a father accused of devil worshiping; the Gay News Ltd. sued for blasphemous libel; a forger of Victorian photographs who briefly fooled the National Portrait Gallery; a pornographic publisher. His deep commitment for justice leads him to wholeheartedly defend hopeless cases and the spirit of the law, as opposed to his fellow barristers who stubbornly defend the letter of the law. Rumpole is given to frequent oratorical outbursts from the *Oxford Book of English Verse* and manages to aim the elegant passages at upper-class hypocritical trumpeters, buffoons and other barristers, and prosecution inspiring justices. He comments on the phenomenon of "judgitis [pomposity] which, like piles, is an occupational hazard on the bench." His suggested cure is "banishment to the golf course."

Rumpole is married to Hilda (played at various times by Joyce Heron, Peggy Thorpe-Bates, and Marion Mathie), to whom he refers as "She Who Must Be Obeyed." Even though Hilda—whose father was head of chambers—aspires for a more prestigious position for her husband and a bit more luxurious life-style for herself, she continues to support

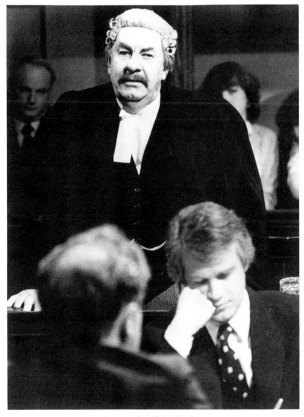

Rumpole of the Bailey
Photo courtesy of DLT Entertainment Ltd.

her husband's brand of justice rather than that sought by egotistical or social climbing royal counsels. Rumpole revels in lampooning his fellow colleagues, whom he believes to be a group of twits. They include the dithery and pompous Claude Erskine-Brown (Julian Curry), the full-of-himself Samuel Ballard (Peter Blythe), and the variety of dour judges who preside in court—the bumbling Justice Guthrie Featherstone (Peter Bowles), the blustering "mad bull" Justice Bullingham (Bill Fraser), the serious and heartless Justice Graves (Robin Bailey), and the almost kindly Justice "Ollie" Oliphant (James Grout). Among Rumpole's colleagues he favors the savvy and stylish Phillida Neetrant Erskine-Brown (Patricia Hodge), one feminist voice of the series who is married to Claude, and the endearing Uncle Tom (Richard Murdoch), an octogenarian waiting to have the good sense to retire, who, in the meantime, practices his putting in chambers.

John Mortimer, the creator of the Rumpole stories, has exclusive rights in writing the television series. Mortimer draws upon both his 36 years of experiences as Queen's Counsel and his life with his father, a blind divorce lawyer. Much like Rumpole, Mortimer adores good food, enjoys a bottle of claret before dinner, loves Dickens, and fights for liberal causes.

His series, then, in addition to the quick-witted dialogue among characters, is distinguished by its social commentary. Specifically, the program is a cleverly entertaining vehicle for tweaking the legal profession and the general state of British mores and manners. In chambers and during court cases, Rumpole provides viewers with grumbling commentaries and under-the-breath critiques of pomposity and the all-too-frequent soulless application of strict legalism. Yet, even though these comments on various social issues such as gay rights, censorship, and the treatment of children in court are quite serious, Mortimer never allows the issues to get in the way of the story. Meticulous attention to detail, well-written scripts, and top-notch actors are the factors that contribute to fine television without the formula-driven action/adventure genres typically associated with drama programming.

All these aspects of the program's charm are enhanced by the superb casting of Leo McKern. Each actress and actor appears uniquely qualified for a specific role, but McKern is the very embodiment of the fictional Rumpole. Robert Goldberg, a television critic from *The Wall Street Journal,* compares this match with other strokes of casting genius: "Every once in a while a character and an actor fit together so precisely that is becomes hard to imagine one without the other (Sean Connery and James Bond, Jeremy Brett and Sherlock Holmes)." McKern's jowls, bulbous nose, the erratic eyebrows, were made to fit the eccentric, irrepressibly snide barrister who is, in Goldberg's words, as "lovable as a grumpy old panda."

Rumpole of the Bailey is a cherished series in U.S. television. According to WGBH's senior producer Steven Ashley, *Rumpole* has solid ratings and continues to be regarded as one of the most popular titles in the *Mystery!* schedule despite stiff competition from commercial networks for the Thursday night 9:00 P.M. time slot. Approximately 300 public television stations carry the Rumpole series on an ongoing basis, representing 95% of all PBS stations. In the San Francisco Bay Area, some of the show's more active fans have formed the "Rumpole Society" with over 450 members: they feature principal actors or John Mortimer as guest speakers at their annual fete, and have visited the Rumpole studios in London.

—Lynn T. Lovdal

CAST

Horace Rumpole	Leo McKern
Guthrie Featherstone	Peter Bowles
Erskine-Brown	Julian Curry
Phillida	Patricia Hodge
George Frobisher	Moray Watson
Uncle Tom	Richard Murdoch
Hilda Rumpole	Joyce Heron
	Peggy Thorpe-Bates
	Marion Mathie
Justice Bullingham	Bill Fraser
Fiona Allways	Rosalyn Landor
Henry	Jonathan Coy
Diane	Maureen Derbyshire
Marigold Featherstone	Joanna Van Gysegham
Nick Rumpole	David Yelland
Liz Probert	Abigail McKern
Judge Graves	Robin Bailey
Samuel Ballard	Peter Blythe

PRODUCERS Irene Shubik, Jacqueline Davies

PROGRAMMING HISTORY 44 Episodes

• BBC1

As an installment of *Play for Today*   16 December 1975

• Thames

April 1978–May 1978	6 Episodes
May 1979–June 1979	6 Episodes
December 1980	Special: Rumpole's Return
October 1983–November 1983	6 Episodes
January 1987–February 1987	6 Episodes
November 1988–December 1988	6 Episodes
October 1991–December 1991	6 Episodes
October 1992–December 1992	6 Episodes

FURTHER READING

Gussow, Mel. "The Man Who Put Rumpole on the Case." *The New York Times,* 13 April 1995.

Mortimer, John Clifford. *The Best of Rumpole* (short stories). New York: Viking, 1993.

———. *The First Rumpole Omnibus.* Harmondsworth, England, and New York: Penguin, 1983.

See also British Programming; McKern, Leo

RUSHTON, WILLIAM

British Author/Actor/Artist

A versatile cartoonist, broadcaster, author, and actor, William Rushton's range of talent emerged early, while a student at Shrewsbury School. There he edited the school magazine, *The Salopian*, and regularly illustrated its issues. The public school friendships and joint contributions for *The Salopian* led to the idea of a satirical publication, *The Private Eye*, co-founded by Rushton and first published in 1962. With its comprehensive attack on the establishment, who were presented as running England in the manner of a private club, *The Private Eye* pioneered a style of satire that was to become fashionable in the early 1960s.

In 1962, Rushton moved on to television to take part in BBC's satirical program, *That Was the Week That Was (TW3)*. Under director Alasdair Milne and producer Ned Sherrin, the crew put together their best work to express doubts about the old order in Britain. In an even more practical step, *The Private Eye* team, upset by the possibility of Sir Alec Douglas Home's further career in politics, posted Rushton to run against him in the Kinross by-election. Rushton's failed candidacy and his Macmillan impersonation on *TW3* made his name, but the irreverent show, anchored by David Frost, deeply divided the public, and the resulting controversy led to its removal from television screens.

In the 1964–65 season, Rushton co-hosted the follow-up to *TW3*, called *Not So Much a Programme, More a Way of Life*. The show had less clear direction, and was at its most successful when it approached the impertinence of *TW3*. Even this milder satirical program, however, faced political criticism that put an end to its existence.

The success of *TW3* opened the way to the cinema for Rushton. Director Clive Donner incorporated three of the show's presenters into *Nothing But the Best* (1964). The film featured a young opportunist, and provided a brash criticism of affluent Britain through a mocking celebration of its values. Rushton also played a role in *Those Magnificent Men in Their Flying Machines* (1965), a humorous take on the early days of aviation.

The slightly overweight Rushton, who describes his hobbies as "gaining weight, losing weight and parking," served as presenter for *Don't Just Sit There* (1973), a BBC series on healthy living. He also took part in the television show *Up Sunday* (1975–78), and entertained the viewers in *Celebrity Squares* (1979–80), a popular game show based on the idea of the American syndicated *Hollywood Squares*.

As a stage actor, Rushton had made his debut in Spike Milligan's *The Bed-Sitting Room* in Canterbury in 1961. After a number of smaller parts, he returned to stage in a full-length role in Eric Idle's play *Pass the Butler* (1982). This witty black comedy, written by a member of the offbeat *Monty Python* team, played successfully in Britain.

Rushton has written and illustrated a number of books, such as *William Rushton's Dirty Book* (1964), *Superpig* (1976), *The Filth Amendment* (1981), and *Marylebone Versus*

William Rushton
Photo courtesy of William Rushton

the Rest of the World (1987). He has also provided illustrations and cartoons for many others, including a number of children's books.

After his early success in the 1960s, Rushton continued to work for *The Private Eye* for decades. He also took on a number of smaller roles in films, plays, and television shows. Known particularly for his humorous cartoons and funny personal presentations, he is a fine performer, a versatile and interesting artist for whom television has provided a continuing opportunity for comic invention.

—Rita Zajacz

WILLIAM GEORGE RUSHTON. Born in Chelsea, London, England, 18 August 1937. Attended Shrewsbury School, Shropshire. Married: Arlene Dorgan, 1968; children: Tobias, Matthew, and Sam. After National Service, worked as solicitor's articled clerk; freelance cartoonist and satirist; co-founder and editor, *Private Eye*, 1961; stage debut, 1961; made television debut as one of *That Was the Week That Was* team, 1962; comic performer on radio, film, and television, appearing on numerous panel shows. Address: Wallgrave Road, London SW5 0RL, England.

TELEVISION SERIES

1962–63   *That Was the Week That Was*
1964–65   *Not So Much a Programme, More a Way of Life*
1969–72   *Up Pompeii!*
1975–78   *Up Sunday*
1979–80   *Celebrity Squares*
1980   *Rushton's Illustrated*

FILMS

It's All Over Town, 1963; *Nothing But the Best*, 1964; *Those Magnificent Men in Their Flying Machines*, 1965; *The Mini-Affair*, 1968; *The Bliss of Mrs. Blossom*, 1968; *The Best House in London*, 1969; *Monte Carlo or Bust/ Those Daring Young Men in Their Jaunty Jalopies*, 1969; *Flight of the Doves*, 1971; *The Adventures of Barry McKenzie*, 1972; *Keep It Up Downstairs*, 1975; *The Chiffy Kids*, 1976; *Adventures of a Private Eye*, 1977; *Adventures of a Plumber's Mate*, 1978; *The Blues Band*, 1981; *The Magic Shop*, 1982; *Consuming Passions*, 1987.

RADIO

I'm Sorry I Haven't a Clue, 1976–; *Trivia Test Match.*

STAGE

The Bed-Sitting Room, 1961; *Gulliver's Travels*, 1971, 1979; *Pass the Butler*, 1982; *Tales from a Long Room*, 1988.

PUBLICATIONS (selection)

William Rushton's Dirty Book, 1964; *How to Play Football: The Art of Dirty Play*, 1968; *The Day of the Grocer*, 1971; *The Geranium of Flüt*, 1975; *Superpig*, 1976; *Pigsticking: A Joy for Life*, 1977; *The Reluctant Euro*, 1980; *The Filth Amendment*, 1981; *W. G. Grace's Last Case*, 1984; *Willie Rushton's Great Moments of History*, 1985; *The Alternative Gardener: A Compost of Quips for the Green-Fingered*, 1986; *Marylebone Versus the Rest of the World*, 1987; *Spy Thatcher* (editor), 1987; *French Letters*, *Every Cat in the Book*, 1993.

FURTHER READING

Marnham, Patrick. *The Private Eye Story. The First 21 Years.* London: Andre Deutsch, 1982.
Murphy, Robert. *Sixties British Cinema.* London: British Film Institute Publishing, 1992.

See also *That Was the Week That Was*

RUSSELL, KEN

British Filmmaker

Ken Russell, a British filmmaker, is best known in the United States as director of such feature films as *Women in Love* (1969), *The Music Lovers* (1970), *Tommy* (1975), and *Altered States* (1980). Although his television work is less well known outside the United Kingdom, it has had a major impact on the development of the television genre of fictional history—described by historian C. Vann Woodward as the portrayal of "real historical figures and events, but with the license of the novelist to imagine and invent." Russell's special province in the genre (a psychobiographical form which he terms the "biopic"), has been music composers and other artists such as dancers and poets. His imaginative interpretations of the lives of artists have, on occasion, outraged both critics and the general public.

After a brief career as a ballet dancer, and later as a successful commercial photographer, Russell turned his attention to film directing. On the basis of a portfolio of three low-budget short films, he was hired by the British Broadcasting Corporation (BBC) in 1959, at the age of 32, to work as a director on its arts series *Monitor*. Most of the *Monitor* pieces (10- 15-minute short subjects) focused on contemporary artists working in British music, dance, and literature. Russell noted that, at the time, there was no real experimental film school in Britain, except for *Monitor*. *Monitor* producer Huw Wheldon, who later became managing director of BBC-TV, encouraged experimentation, within limits, and Russell took full advantage of this.

The two most important productions from Russell's *Monitor* period were *Elgar* (1962) and *The Debussy Film* (1965). *Elgar*, Russell's attempt to counter British music critics' negative assessments of the British composer Edward Elgar, was his first full-length *Monitor* film, lasting 50 minutes. It also marked the celebration of the 100th *Monitor* program. In *Elgar*, Russell advanced the idea of using actors to impersonate historical characters, which he had introduced the previous year on *Monitor* in the short film *Portrait of a Soviet Composer*, on the life of Sergei Prokofiev. Prior to this, the BBC had prohibited the use of actors in the portrayal of historical personages. In the Prokofiev film, Russell used an actor to show the composer's hands, a so-called "anonymous presence." In *Elgar*, Russell took the concept a step further, allowing Elgar to be seen (but still not heard). Five different actors, mostly amateurs, portrayed the composer at various stages of his life. Most of the scenes with the actors were shot in medium-shot. According to Russell, the viewer was "not aware of a personality; just a figure." Russell skillfully combined silent footage of the actors, stock footage of English life at the turn of the century, and photographs of Elgar and his family, all of which were enhanced by Elgar's compositions. Russell focused his interpretation on Elgar's reverence for the English countryside—his "return to the strength of the hills" (a theme of great importance in Russell's own life). That theme would reemerge in many subsequent Russell biopics. *Elgar* was extremely popular with the audience, in large measure because of Russell's romantic use of Elgar's music; the show was repeated at least three times. As Baxter points out, this work launched Russell's national reputation.

After an unsuccessful feature film, *French Dressing,* Russell returned to the BBC to direct *The Debussy Film: Impressions of the French Composer* (1965). Here, Russell broke through the BBC's last remaining prohibition against using actors in speaking roles in historical drama. According to Russell, as quoted in Phillips, Wheldon thought the film "a bit esoteric," and insisted on beginning the film "with a series of photographs of Debussy along with a spoken statement assuring viewers they were about to see a film based on incidents in Debussy's life and incorporating direct quotations from Debussy himself." The BBC feared that viewers might believe they were watching newsreels of real people. To circumvent this potential problem, Russell created an intriguing "film-within-a-film," in which the framing story depicts a French film director coming to England to shoot a film on Debussy. In the script, actors were clearly identified as actors playing the various historical figures. Russell, and writer Melvyn Bragg (who would collaborate with Russell on several films and later become the editor and presenter of *The South Bank Show*), conceived Debussy as "a mysterious, shadowy character"—an unpredictable and sensual dreamer. This is accentuated by Russell's evocative use of macabre physical comedy.

Isadora Duncan: The Biggest Dancer in the World (1966) is the most celebrated and least factual of Russell's BBC biopics. The film used a mix of classical music and popular tunes (from Beethoven to *Bye, Bye, Blackbird*), and featured a nude dance, suicide attempts, and wild parties to depict Isadora's sensational life and her death wish. Excerpts from Leni Riefenstahl's *Olympia* were intercut with original footage, Hanke reports, to convey the "ideal of German perfection" Duncan sought to emulate. Isadora was at once "sublime" and "vulgar," if not grotesque. Interestingly, some of Russell's more hostile critics have accused the director of the same tendencies.

Song of Summer (BBC, 1968) chronicles the last years of the life of composer Frederick Delius, who, blind and crippled with syphilis, is living in a French village with his wife, Jelka, and his amanuensis, Eric Fenby. Fenby, who advised Russell on the film, is portrayed as a young man who sacrificed his own career out of love and respect for Delius. In the end, according to Russell, as quoted in Phillips, Fenby feels "robbed of his own artistic vision." The ultimate irony, says Russell, is that much of Delius' music is second-rate. In *Song of Summer*, Russell is able to express an understanding and even compassion for a composer whose basic personality and music he clearly dislikes. The theme, evident in *Isadora*, of what Hanke refers to as "the artist's unfortunate need to debase himself and his art," re-emerges here. As in *Elgar*, Russell highlights the artist's obsession with nature. According to Hanke, in *Song of Summer*, Russell exhibited his "ability to work in a restrained manner if the subject matter calls for it."

Ken Russell
Photo courtesy of the Academy of Motion Pictures Arts and Sciences

The last film Russell would make for the BBC, the infamous *The Dance of the Seven Veils: A Comic Strip in Seven Episodes on the Life of Richard Strauss* (1970), exhibited no such restraint. The complete title reveals Russell's intention to create a satirical political cartoon on the life of the German composer, whom Russell saw as a "self-advertising, vulgar, commercial man . . . [a] crypto-Nazi with the superman complex underneath the facade of the distinguished elderly composer." Although, according to Russell, "95 percent of what Strauss says in the film he actually did say in his letters and other writings," many critics and viewers found Russell's treatment of the venerated composer itself to be vulgar. Hanke's assessment is that in the film, Russell contends that Strauss "betrayed himself and his art through his lack of personal responsibility," which included his currying favor with the Nazis during World War II. The most objectionable sequences in the film were Strauss conducting *Der Rosenkavalier,* and exhorting his musicians to play ever louder to drown out the screams of a Jew being tortured in the audience by SS men, who were carving a Star of David on his chest with a knife; and the playing of Strauss' *Domestic Symphony* over shots of Strauss and his wife making love, their climax being mirrored by the orchestra. The film concludes with Russell himself portraying a wild-haired orchestra conductor bowing and walking away from the camera as his director's credit appears on the screen (perhaps signaling his own farewell to the BBC). The film aired once, leading to mass protests and questions raised in Parliament. As Russell put it, "all hell broke loose." Huw Wheldon, head of BBC-TV, defended Russell. At the same time, the BBC tried to placate critics, including Strauss' family and his publisher, by presenting a roundtable discussion in which music critics and conductors denounced both Russell and the film. By the time *The Dance of the Seven Veils* aired on the BBC, Russell's feature film *Women in Love* had assured him a reputation in feature-film circles, and the BBC experience convinced him it was time to abandon the small screen.

Russell would return to television, but not to the BBC. He was in fact eager to do so, as he felt the medium would allow him to make more "personal and optimistic films." In 1978, Russell directed *Clouds of Glory* for British independent television's Grenada-TV. This was actually two one-hour episodes. The first, *William and Dorothy*, was a biopic on the love of William Wordsworth for his sister Dorothy. Their relationship was understated in the film; neither William nor Dorothy ever explicitly verbalized its incestuous nature. The second episode, *The Rime of the Ancient Mariner*, was a biopic on the lurid life of Samuel Taylor Coleridge. Lines from the title poem are recited over various scenes, accompanied by the music of British composer Ralph Vaughn Williams. In one fantasy sequence, Coleridge, the opium addict, buries an anchor in his estranged wife's breast—a reference to the albatross in the poem—as he attempts to rid himself of her.

In 1988, Russell directed and starred in *Ken Russell's ABC of British Music*, a special episode of London Weekend Television's *The South Bank Show*, hosted by Melvyn Bragg. This light-hearted treatment of a serious subject finds Russell, dressed in a variety of humorous costumes, running through the letters of the alphabet in a carnival barker's voice, extolling "neglected geniuses" of British classical and pop music and "bulldozing a few sacred cows at the same time." One of the most inventive moments comes with the letter "U": "U is for . . . ucch . . . critics." Here we see a dream-like video sequence of six midgets carrying a coffin through a field while, in their munchkin voices, they babble their condemnation of Elgar and Delius.

—Hal Himmelstein

KEN (KENNETH ALFRED) RUSSELL. Born in Southampton, Hampshire, England, 3 July 1927. Attended Pangbourne Nautical College, 1941–44; Walthamstow Art School; International Ballet School. Married: 1) Shirley Ann Kingdon, 1957 (divorced, 1978); five children; 2) Vivian Jolly, 1984; children: Molly and Rupert. Served in Merchant Navy, 1945, and Royal Air Force, 1946–49. Dancer, Ny Norsk Ballet, 1950; actor, Garrick Players, 1951; photographer, 1951–57; amateur film director; documentary film-maker, BBC, 1958–66; debut as professional film director, 1963; established reputation on television with series of biographical films about great composers for the arts program *Omnibus*, from 1966, and the *South Bank Show*, from 1983; freelance film director, also staging opera and directing pop videos, since 1966. Recipient: Screen Writers Guild Awards, 1962, 1965, 1966, 1967; Guild of Television Producers and Directors Award, 1966; Desmond Davis Award, 1968; Emmy Award, 1988. Address: Peter Rawley International Creative Management, 8899 Beverly Boulevard, West Hollywood, California 90048-2412, U.S.A.

TELEVISION SERIES

1993	*Lady Chatterley*

TELEVISION DOCUMENTARIES

1959	*Poet's London*
1959	*Gordon Jacob*
1959	*Variations on a Mechanical Theme*
1959	*Robert McBryde and Robert Colquhoun*
1959	*Portrait of a Goon*
1960	*Marie Rambert Remembers*
1960	*Architecture of Entertainment*
1960	*Cranks at Work*
1960	*The Miners' Picnic*
1960	*Shelagh Delaney's Salford*
1960	*A House in Bayswater*
1960	*The Light Fantastic*
1961	*Old Battersea House*
1961	*Portrait of a Soviet Composer*
1961	*London Moods*
1961	*Antonio Gaudi*
1962	*Pop Goes the Easel*
1962	*Preservation Man*
1962	*Mr. Chesher's Traction Engines*
1962	*Lotte Lenya Sings Kurt Weill*
1962	*Elgar*
1963	*Watch the Birdie*
1964	*Lonely Shore*
1964	*Bartok*
1964	*The Dotty World of James Lloyd*
1965	*The Debussy Film: Impressions of the French Composer*
1965	*Always on Sunday*
1966	*The Diary of a Nobody*
1966	*Don't Shoot the Composer*
1966	*Isadora Duncan: The Biggest Dancer in the World*
1967	*Dante's Inferno*
1968	*Song of Summer*
1970	*The Dance of the Seven Veils: A Comic Strip in Seven Episodes on the Life of Richard Strauss*
1978	*Clouds of Glory, Parts I and II*
1983	*Ken Russell's View of the Planets*
1984	*Elgar*
1984	*Vaughan Williams*
1988	*Ken Russell's ABC of British Music*
1989	*Ken Russell—A British Picture*
1990	*Strange Affliction of Anton Bruckner*
1992	*The Secret Life of Sir Arnold Bax*
1995	*Classic Widows*

FILMS (director)

Amelia and the Angel, 1957; *Peep Show*, 1958; *Lourdes*, 1958; *French Dressing*, 1963; *Billion Dollar Brain*, 1967; *Women in Love*, 1969; *The Music Lovers* (also producer), 1970; *The Devils* (also writer and co-producer), 1971; *The Boy Friend* (also writer and producer), 1971; *The Savage Messiah* (also producer), 1972; *Mahler* (also writer), 1974; *Tommy* (also writer and co-producer), 1975; *Lisztomania* (also writer), 1975; *Valentino* (also co-writer), 1977; *Altered States*, 1980; *Crimes of Passion*, 1984; *Gothic*, 1986; *Aria* (episode), 1987; *Salomé's Last Dance*, 1988; *The Lair of the White Worm*, 1988; *The Rainbow*, 1989; *Whore*, 1991; *The Russia House* (actor), 1991.

STAGE (operas)

The Rake's Progress, 1982; *Die Soldaten*, 1983; *Madame Butterfly*, 1983; *La Bohème*, 1984; *Faust*, 1985; *Princess Ida*, 1992; *Salomé*, 1993.

PUBLICATIONS

A British Picture: An Autobiography. London: Heinemann, 1989.

Fire Over England: British Cinema Comes Under Friendly Fire. London: Hutchinson, 1993.

The Lion Roars: Ken Russell on Film. Winchester, Massachusetts: Faber and Faber, 1993.

FURTHER READING

Atkins, Thomas. *Ken Russell*. New York: Monarch, 1976.

Baxter, John. *An Appalling Talent: Ken Russell*. London: Joseph, 1973.

Dempsey, Michael. "The World of Ken Russell." *Film Quarterly* (Berkeley), Spring 1972.

————. "Ken Russell, Again." *Film Quarterly* (Berkeley), Winter 1977–78.

Farber, Stephen. "Russellmania." *Film Comment* (New York), November-December 1975.

Fisher, Jack. "Three Paintings of Sex: The Films of Ken Russell." *Films Journal* (New York), September 1972.

Gilliatt, Penelope. "Genius, Genia, Genium, Ho Hum." *The New Yorker*, 26 April 1976.

Gomez, Joseph. "*Mahler* and the Methods of Ken Russell's Films on Composers." *Velvet Light Trap* (Madison, Wisconsin), Winter 1975.

————. *Ken Russell: The Adaptor as Creator*. London: Muller, 1976.

Hanke, Ken. *Ken Russell's Films*. Metuchen, New Jersey: Scarecrow, 1984.

Jaehne, Karen. "Wormomania: Ken Russell's Best Laid Planaria." *Film Criticism* (Meadville, Pennsylvania), 1988.

Kolker, Robert. "Ken Russell's Biopics: Grander and Gaudier." *Film Comment* (New York), May-June 1973.

Phillips, Gene D. *Ken Russell*. Boston: Twayne, 1979.

Rosenfeldt, Diane. *Ken Russell: A Guide to Reference Sources*. Boston: Hall, 1978.

Woodward, C. Vann. *The Future of the Past*. New York: Oxford University Press, 1989.

Yacowar, M. "Ken Russell's *Rabelais*." *Literature/Film Quarterly* (Salisbury, Maryland), 1980.

See also Bragg, Melvyn; British Programming; Wheldon, Huw

RUSSIA

Russia was the largest and the culturally predominant republic of the U.S.S.R., and the history of Russian television up to the disintegration of that country in 1991 is inseparable from that of Soviet television. Moreover, in spite of the changes that have taken place since then, Russian television remains the principal inheritor of the traditions (as well as the properties) of its Soviet predecessor.

Regular television broadcasting began in Moscow in 1939, though the service was interrupted for the duration of World War II (1941–1945). Broadcasting was always given a high priority by the Soviet authorities, and television expanded rapidly in the post-war years, so that by the late 1970s there were two general channels that could be received over most of the country and two further channels (one local and one educational) in certain large cities. There were also television stations in the constituent republics and studios in most large cities. Apart from a gradual extension of the coverage of the two national channels until the first, at least, could be received in virtually the whole of the country, this situation remained little changed until 1991.

Because of its size the Soviet Union was a pioneer of satellite transmission: by the mid-1980s both national channels were broadcast in four time-shifted variants to eastern parts of the country, while the first channel was among the earliest television programs to be made available world wide. Regular colour transmissions began in 1967, using the SECAM system.

Administratively, television was the responsibility of the All-Union Committee for Television and Radio (generally known as Gosteleradio), the chairman of which was a member of the Council of Ministers and of the Central Committee of the Soviet Communist Party. Equivalent committees existed in the constituent republics, with the exception, owing to a quirk of the system, of Russia itself. Only in May 1991, after sustained pressure from the Russian parliament, did a separate Russian oganisation start its own television transmissions; its programs, broadcast for six hours per day on the second channel, were in the summer of that year a focus of opposition to President Gorbachev. Broadcasting was financed out of the state budget, the receiving licence having been replaced in 1962 by a notional addition to the retail price of television sets.

The social, political and economic upheavals that accompanied the collapse of the Soviet system have led to major changes in Russian television. The period since 1991 has been characterised by a rapid growth of commercialisation and a continuing debate concerning the rôle of the state in owning, financing and controlling the content of the electronic media. There has also been continuous disagreement between the executive and legislative branches of power over which of them should exercise control over broadcasting. Up to now this has invariably been resolved in favour of the former, and the entire structure of Russian television has in effect been put into place by a series of presidential decrees.

One aspect of the involvement of the state in television is the Federal Service for Television and Radio, a regulatory body with relatively few powers, whose principal function is to issue licences to broadcasting organisations. There are in addition two broadcasting companies wholly owned by the state: the All-Russian State Television and Radio Company (RTR), the organisation founded in 1991, and Peterburg—piatyj kanal (St. Petersburg—the fifth channel), converted into a state company in 1993. A third state company, Ostankino, which was created out of the former Gosteleradio when the Soviet Union disintegrated, was abolished in 1995. Its functions were taken over by Obshchestvennoe rossiiskoe televidenie (Russian Public Television, known as ORT), owned 51% by the state and 49% by a consortium of banks and private companies. ORT produces its own news bulletins, but otherwise is essentially a commissioning company. Publicly-owned broadcasting organisations continue to exist in each of the regions of Russia. The proliferation of state companies and the rapid inflation from 1992 onwards has meant that allocations from the state budget have covered an ever smaller proportion of the costs of these companies: for both ORT and RTR this had declined to 25% by 1995. The shortfall is made up by revenue from advertising.

In the commercial sector two companies, NTV and TV6, aspire to national coverage, though at present their programs can be seen in certain large cities only; both commenced operations in 1993. There are also several hundred local stations, and cable television has started to appear in certain large cities. There has been little or no foreign investment in Russian television; CNN were involved in TV6 when it started up, but subsequently withdrew from the operation. NTV is owned by a consortium of banks which also owns the daily newspaper Segodnia and the main television listings journal Sem' dnei and can be said to be part of Russia's first media conglomerate. An interesting feature is the growth of independent production companies, the oldest of which, ViD and ATV, date back to 1990, when they were "semi-detached" outgrowths of Gosteleradio. These now provide programs for the various broadcasting companies, especially ORT and NTV.

The changes since 1991 have had an equally profound effect on programs and their content. In Soviet times television was first and foremost an instrument of propaganda, serving the interests of party and state, and this purpose was reflected in all news bulletins and political programs. The main evening news program, Vremia (Time), was shown simultaneously on all channels and often ran far beyond its allotted forty minutes (a cavalier attitude towards the published schedules is characteristic of both Soviet and Russian television). All programs were in effect, if not formally, subject to censorship, and caution usually prevailed: the popular student cabaret KVN was taken off the air in the 1970s for being too daring, and a high proportion of the non-political programs consisted of high culture (opera, ballet and classical drama), films made for the Soviet cinema and sport, all of which could be guaranteed in advance to be inoffensive.

Because of its importance as a means of propaganda, the effects of glasnost' were felt more slowly in television than in the print media. By the late 1980s, however, a certain liberalisation could be discerned: KVN returned to the screens, and previously taboo topics began to be discussed in programs such as Vzgliad (View) and Do i posle polunochi (Before and After Midnight). These were followed by a range of lively and innovatory productions originated by ATV (see above), as well as by attempts to liven up news presentation, though as late as the 1990–91 season all of these programs were liable to suffer cuts imposed by the censors or even to disappear altogether; the suspension of Vzgliad in January 1991 was a particular cause célèbre. In the circumstances it is not surprising that the removal of all restrictions after the collapse of the August 1991 putsch led to a brief flowering of creative talent (and the emergence of long-forbidden programs) that may prove to have been something of a golden age of Russian television.

The 1990s have seen a gradual westernisation of Russian television with the appearance of genres hitherto eschewed. Among these are game-shows, such as Pole chudes (Field of Miracles), which is based on Wheel of Fortune and which is one of Ostankino/ORT's most popular programs; talk-shows, such as Tema (Theme) and My (We), which likewise have clear ancestral links with their American counterparts, and soap operas. These are almost invariably imported from the United States (Santa Barbara), Mexico (Los Ricos también lloran, Simplemente María and others), Brazil and elsewhere; home-grown versions have been few in number and short-lived. A number of British and U.S. crime series have also been imported (for example, The Sweeney and Moonlighting). One genre to which Russian television has remained immune is situation comedy, though in the area of satire it is worth mentioning NTV's Kukly (Puppets), which uses the format of the British Spitting Image and which has occasionally succeeded in annoying the authorities. Films made in the United States and other Western countries are now widely shown, though in the 1995–96 season, presumably in response to complaints from viewers, there has been a marked increase in the number of Russian/Soviet films being broadcast. Religious programs of various types, most connected with the Russian Orthodox Church, but some originating with certain strands of western Protestantism, are now transmitted, but literature, classical music and serious drama have disappeared almost totally from the screens.

This westernisation has by no means met with universal approval, though it is not only a reaction to Soviet isolationism, but also a response to commercial pressures. All channels are now dependent on income from advertising, and while the relationship between audience ratings and the prices charged for advertisements is not as sophisticated as in the West, there is a requirement to show programs which will attract viewers. Advertising is lightly regulated and takes many forms, including spots between and during programs and sponsorship. It tends to be unpopular, partly because of the unfamiliar intrusiveness, but mainly because a high proportion of the advertisements are for foreign goods which are not widely available or (especially from 1992 to 1994) for disreputable financial institutions which subsequently collapsed. Nevertheless, while some companies prefer to re-cycle advertisements previously used in their older markets, the best Russian-produced examples of the form will bear comparison with anything shown in the West. A noteworthy, even notorious example is the sequence of advertisements produced in 1994 for the now-defunct MMM, which featured the fictional Lionia Golubkov and his "family". The rapid growth of advertising has led to widespread allegations of corruption, particularly in connection with Ostankino/ORT, and the murkier side of Russian television received prominence in March 1995 with the still unsolved murder of Vladislav List'ev, originator and presenter of several popular programs and director-general-designate of ORT.

Commercial pressures have not entirely succeeded in supplanting political pressures, though the latter are incomparably subtler than in Soviet times. Nevertheless, in both areas the long-established Soviet practice of "telephone law" (whereby a person in power uses that instrument to convey his or her wishes/instructions) continues to prevail. Ostankino and its successor ORT have had a reputation for being "pro-presidential", but this is principally due to the perceived slant of their news coverage. Indeed, certain programs produced for these channels by independent production companies have been accused, somewhat contradictorily, of giving opponents of the president too much air time, and it is generally considered that the demagogic nationalist Vladimir Zhirinovskii largely owes his political career to television. In general, state-owned companies (including and perhaps especially ORT) are more likely to come under political pressures, particularly during periods of heightened tension, such as the run-up to elections, while commercial companies retain more freedom of manoeuvre. One of the principal concerns of NTV has been to build up a reputation for independence and lack of bias in its news programs.

The outside observer can occasionally discern signs of the growth of informal power networks involving politicians and businessmen with media interests, and this development, together with the subtle combination of public and private patronage and political and commercial pressures, suggests that post-Soviet television in Russia may end up following most closely the French or Italian patterns, albeit that there is no evidence that anyone has deliberately set out to achieve this result. If, however, the reaction against all forms of westernisation which became noticeable in the mid-1990s continues, there may well be a partial retreat towards Soviet models, although any "re-sovietisation" of Russian television, with its implied enhancement of the rôle of the state, will inevitably encounter serious financial obstacles. Whatever happens, it is difficult to see how television in Russia can escape the effects of that country's continuing political and economic instability.

—J.A. Dunn

FURTHER READING

Dunn, J.A. "A Pot of Boiling Milk." *Rusistika* (Rugby, England), December 1993.

———. "The Rise, Fall and Rise(?) of Soviet Television." *Rusistika* (Rugby, England), December 1991.

Graffy, Julian, and Geoffrey A. Hosking, editors. *Culture and Media in the USSR Today.* London: MacMillan, with the School of Slavonic and East European Studies, University of London, 1989.

McNair, Brian. "From Monolith to Mafia: Television in Post-Soviet Russia." *Media, Culture and Society* (London), July 1996.

———. *Glasnost, Perestroika and the Soviet Media.* London: Routledge, 1991.

Mickiewicz, Ellen. *Split Signals: Television and Politics in the Soviet Union.* Oxford and New York: Oxford University Press, 1988.

Paasilinna, Reino. *Glasnost and Soviet Television, Research Report 5.* Helsinki: Ylesradio (Finnish Broadcasting Company), 1995.

Seifert, Marsha, editor. *Mass Culture and Perestroika in the Soviet Union.* Oxford and New York: Oxford University Press, 1991.

S

ST. ELSEWHERE

U.S. Serial Medical Drama

St. Elsewhere was one of the most-acclaimed of the upscale serial dramas to appear in the 1980s. Along with shows like *Hill Street Blues, L.A. Law,* and *thirtysomething, St. Elsewhere* was a result of the demographically-conscious programming strategies that had gripped the networks during the years when cable TV was experiencing spectacular growth. Often earning comparatively low ratings, these shows were kept on the air because they delivered highly desirable audiences consisting of young, affluent viewers whom advertisers were anxious to reach. In spite of its never earning a seasonal ranking above 49th place (out of about 100 shows), *St. Elsewhere* aired for six full seasons on NBC from 1982 to 1988. The series was nominated for 63 Emmy Awards and won 13.

Set in a decaying urban institution, *St. Elsewhere* was often and aptly compared to *Hill Street Blues,* which had debuted a season and a half earlier. Both shows were made by the independent production company MTM Enterprises, and both presented a large ensemble cast, a "realistic" visual style, a profusion of interlocking stories, and an aggressive tendency to break traditional generic rules. While earlier medical dramas like *Dr. Kildare, Ben Casey,* and *Marcus Welby, M.D.* featured godlike doctors healing grateful patients, the staff of Boston's St. Eligius Hospital exhibited a variety of personal problems and their patients often failed to recover.

St. Elsewhere's content could be both controversial and surprising. In 1983, for instance, it became the first prime-time series episode to feature an AIDS patient. Six years before *NYPD Blue* began introducing nudity to network television, *St. Elsewhere* had shown the naked backside of a doctor (Ed Flanders) who'd dropped his trousers in front of his supervisor (Ronny Cox) before leaving the hospital and the show. It was also not uncommon for principal characters to die unexpectedly, which happened on no fewer than five occasions during the run of the series.

As a medical drama, *St. Elsewhere* dealt with serious issues of life and death, but every episode also included a substantial amount of comedy. The show was especially noted for its abundance of "in jokes" that made reference to the show's own ancestry. In one episode, for example, an amnesia patient comes to believe that he is Mary Richards from *The Mary Tyler Moore Show,* MTM Enterprises' first production. Throughout the episode the patient makes oblique references to MTM's entire program history. Later, in the series' final episode, a scene from the last installment of *The Mary Tyler Moore Show* is restaged, and the cat that had appeared on the production logo at the end of every MTM show for eighteen years, dies as the final credits roll.

St. Elsewhere proved to be a fertile training ground for many of its participants. At the start of the 1992–93 season, creators John Falsey and Joshua Brand had a critically-acclaimed series on each of the three major networks: *Northern Exposure* (CBS), *I'll Fly Away* (NBC), and *Going to Extremes* (ABC). Writer-producer Tom Fontana became the executive producer of *Homicide: Life on the Street* with Baltimore-based film director Barry Levinson. Other *St. Elsewhere* producers and writers went on to work on such respected series as *Moonlighting, China Beach, L.A. Law, Civil Wars,*

St. Elsewhere

NYPD Blue, ER, and *Chicago Hope.* Actor Denzel Washington, virtually unknown when he began his role as Dr. Phillip Chandler, had become a major star of feature films by the time *St. Elsewhere* ended its run.

St. Elsewhere also exerted a significant creative influence on *ER,* the hit medical series that debuted on NBC in 1994. While the pacing of *ER* is much faster, both the spirit of the show and many of its story ideas have been borrowed from *St. Elsewhere.*

—Robert J. Thompson

CAST

Dr. Donald Westphall	Ed Flanders
Dr. Mark Craig	William Daniels
Dr. Ben Samuels (1982–83)	David Birney
Dr. Victor Ehrlich	Ed Begley, Jr.
Dr. Jack Morrison	David Morse
Dr. Annie Cavanero (1982–85)	Cynthia Sikes
Dr. Wayne Fiscus	Howie Mandel
Dr. Cathy Martin (1982–86)	Barbara Whinnery
Dr. Peter White (1982–85)	Terence Knox
Dr. Hugh Beale (1982–83)	G.W. Bailey
Nurse Helen Rosenthal	Christina Pickles
Dr. Phillip Chandler	Denzel Washington
Dr. V. J. Kochar (1982–84)	Kavi Raz
D. Wendy Armstrong (1982–84)	Kim Miyori
Dr. Daniel Auschlander	Norman Lloyd
Nurse Shirley Daniels (1982–85)	Ellen Bry
Orderly Luther Hawkins	Eric Laneuville
Joan Halloran (1983–84)	Nancy Stafford
Dr. Robert Caldwell (1983–86)	Mark Harmon
Dr. Michael Ridley (1983–84)	Paul Sand
Mrs. Ellen Craig	Bonnie Bartlett
Dr. Elliot Axelrod (1983–98)	Stephen Furst
Nurse Lucy Papandrao	Jennifer Savidge
Dr. Jaqueline Wade (1983–88)	Sagan Lewis
Orderly Warren Coolidge (1984–88) . . .	Byron Stewart
Dr. Emily Humes (1984–85)	Judith Hansen
Dr. Alan Poe (1984–85)	Brian Tochi
Nurse Peggy Shotwell (1984–86)	Saundra Sharp
Mrs. Hufnagel (1984–85)	Florence Halop
Dr. Roxanne Turner (1985–87)	Alfre Woodard
Ken Valere (1985–86)	George Deloy
Terri Valere (1985–86)	Deborah May
Dr. Seth Griffin (1986–88)	Bruce Greenwood
Dr. Paulette Kiem (1986–88)	France Nuyen
Dr. Carol Novino (1986–88)	Cindy Pickett
Dr. John Gideon (1987–88)	Ronny Cox

PRODUCERS Bruce Paltrow, Mark Tinker, John Masius, John Falsey, Joshua Brand

PROGRAMMING HISTORY

• NBC

October 1982–August 1983	Tuesday 10:00-11:00
August 1983–May 1988	Wednesday 10:00-11:00
July 1988–August 1988	Wednesday 10:00-11:00

FURTHER READING

Barker, David. "*St. Elsewhere*: The Power of History." *Wide Angle* (Athens, Ohio), 1989.

Feuer, Jane, Paul Kerr, and Tise Vahimagi, editors. *MTM-"Quality Television."* London: British Film Institute, 1984.

Paisner, Daniel. *Horizontal Hold: The Making and Breaking of a Network Television Pilot.* New York: Birch Lane Press, 1992.

Schatz, Thomas. "St. Elsewhere and the Evolution of the Ensemble Series." In, Newcomb, Horace, editor. *Television: The Critical View.* New York: Oxford University Press, 1976; 4th edition, 1987.

Tartikoff, Brandon, and Charles Leerhsen. *The Last Great Ride.* New York: Random House, 1992.

Thompson, Robert J. *Good TV: The St. Elsewhere Story.* New York: Syracuse University Press, 1996.

Tinker, Grant, and Bud Rukeyser. *Tinker in Television: From General Sarnoff to General Electric.* New York: Simon and Schuster, 1994.

Turow, Joseph. *Playing Doctor: Television, Storytelling, and Medical Power.* New York: Oxford University Press, 1989.

See also *Marcus Welby, M.D.*; *Medic*; Melodrama; Workplace Programs

SALANT, RICHARD S.

U.S. Media Executive

Richard S. Salant started in television in 1952, as vice president and general executive of CBS. The Harvard-educated lawyer worked in government and private practice for 12 years before switching industries. His corporate experience was fueled by his lifetime commitment to such issues as freedom of the press; ethics in news production; and the relationship of government, corporate broadcast management, and news production. His longevity in the industry

stemmed from such intangible qualities as skillful conflict resolution that minimized public debate, the ability to isolate issues from complex events, and verbal clarity in articulating his position.

After a decade as vice president for CBS, with no experience or training as a journalist, Salant became president of the CBS news division in 1961. His appointment was greeted with reservation. He moved to corporate management in 1964, as vice president for corporate affairs and special assistant to the president of CBS, then returned to again head the news division from 1966 to 1979. Because of the strength of his advocacy for the division during both tenures in this position, reservations regarding his commitment and ability abated.

Utilizing his legal background, from 1953 through 1959 Salant represented CBS in Washington, D.C., in congressional hearings and forums pertaining to broadcast regulation and rights. He learned the structure of the industry for his speeches and testimony on issues such as subscription television, UHF-VHF allocations, monopoly rulings, coverage of house hearings by broadcasters, and the barriers constructed to free expression by Section 315 (the Equal Time Provision) of the Communications Act. He argued that Congress's ban on cameras and microphones as unacceptable journalistic tools placed broadcasters as second class citizens, and Section 315 prevented the free pursuit and airing of information. From his participation in the complex discussions of these legal issues, Salant slowly derived the position that news should be based on information the public needs to know to participate in a democratic system, not on what they would like to know. Small experiences supported his thinking, such as when he discovered, while in Washington, that CBS cooperated with the CIA by providing outtakes of news stories. He stopped the practice in 1961.

Salant had a passion for the potential of television news; in 1961 he brought a meticulous set of policies to the news division so that the ethics and credibility of news remained unscathed. These ranged from the sweeping change that separated sports and other entertainment projects from the news division, to detailed guidelines for editing interviews. His directives banished music and sound effects from any news or documentary program. They stopped the involvement of news personnel in entertainment ventures. They both limited the use of and marked all all occurrences of simulations. Salant published these directions in the *CBS News Standards Handbook*, which all new employees still were required to read at the end of the 1990s. Employees also signed an affidavit agreeing to comply with the guidelines.

In 16 years as president, Salant looked at small and large policies for their potential contribution toward building a credible image in the public eye. He spoke out against the news division creating "personalities" to market programs. He was especially concerned for the potential harm of docudramas, which, if not consistently marked and explained as fictionalizations, might be taken as news products by the public. Most troubling to Salant was the network's lack of supervision over news emanating from CBS owned stations. Integrity and credibility came in a package under the CBS name, and the package extended, in his view, to the local level.

Salant's continuous examination of broadcast ethics and news judgment set the pace for other networks and the industry. When Fred Friendly resigned as president of CBS News in 1966 because network executives declined to preempt regular

Richard S. Salant
Photo courtesy of Broadcasting and Cable

daytime programming in order to air the Senate Foreign Relations Committee Hearings on Vietnam, Salant reiterated the importance of news judgment under the criterion of selective coverage. Congress, Washington, and the president would not, he argued, dominate airways with a selective coverage policy. The networks were responsible for alternative ways of reporting, such as evening news specials, half-hour news summaries, and the provision of alternative voices.

Salant realized that his background in the CBS corporate arena would always cast doubt on his decisions. His record of wrestling more broadcasting time for news in prime time as well as daytime eventually changed that. In fact, Salant's inside knowledge of CBS helped the news division move from 15-minutes to a 30-minute newscast. Under his guidance, CBS started a full-time election unit, created additional regional news bureaus outside New York and Washington, launched *60 Minutes*, started a regular one-hour documentary series called *CBS Reports*, produced many investigative and controversial documentaries, and covered the Watergate Affair with more than 20 1-hour specials on the events.

These accomplishments were not Salant's most difficult. He succeeded, with great pain, in insulating news division personnel from the wrath of corporate criticism and deflected movements against the division's autonomy. When CBS President William S. Paley vehemently objected to Cronkite's *Evening News* report on Watergate, the first by a network, and demanded the story never appear again, Salant defied Paley, airing a second part, but reduced the number of issues covered. Although this action is open to multiple interpretations, his decisions in 1973 are clearer. He supported CBS News journalists in a protest against Paley's call for the elimination of instant specials after Presidential speeches or news conferences.

Salant continually addressed the volatile connection between news and corporate management in a pragmatic manner. He did not see the relationship as strictly adversarial, nor did he see it as polarized between two opposing sides. Every conflict was a path toward new strategies to apply in the future. Salant's brilliance as division president was grounded in the attitude and communication skills he brought to conflicts. He diverted escalating personal attacks and swung discussions back to issues.

Not everyone appreciated this strategy. When Friendly resigned, Salant referred to it as a misunderstanding, and explained CBS' strategy on the congressional hearings. When local affiliates called for less Watergate coverage, and when they demanded Dan Rather's reassignment after talking back to the President at a news conference, Salant did denounce defiance and arrogance in any news division. But he turned the argument so that affiliates had to examine the central issue as a matter of news judgment: network news needed its independence, even if it was dependent on affiliates.

In one of the most widely discussed controversies of his tenure, the findings reported in the CBS documentary, *The Selling of the Pentagon* (1971), put Salant in a difficult and complex position. The government called congressional hearings and subpoenaed CBS documents, accusing the news division of manipulative editing and false claims. Again, Salant simplified the matter, accusing the government of infringing on the freedom of speech. He argued that a network has the right to be wrong and, even when wrong, the right not to be judged by the government. To support this view he pointed to an issue with ramifications for the entire television industry: the government had the power to jeopardize free speech by its power to intimidate affiliates that carried controversial programs. Even in the midst of his defense, however, Salant was not afraid to criticize CBS or network news, and his attitude provided credibility to his position. After the confrontation with Congress, when CBS did something questionable—such as paying H. R. Haldeman $50,000 for an interview on *60 Minutes*—an admission of wrongdoing was forthcoming.

Upon mandatory retirement from CBS, Salant immediately went to NBC, serving two uneventful years as a vice president and general advisor in the network. Only one Salant proposal received extensive coverage. He recommended development of a one-hour evening news program, from 8:00 to 9:00 P.M., freeing the earlier prime-time slot for local news, and saving networks the expense of an hour of dramatic programming. Salant finished his career as pres-

ident and chief executive officer of the National News Council. This independent body, recommended in 1973 by a Twentieth Century Fund panel on which Salant served, was created in 1983 to make non-binding decisions on complaints brought against the press or by the press. Faced by a hostile industry that wanted no monitor looking at its work, the council disbanded after one year. This attitude on the part of the industry was discouraging to Salant, especially considering the increased government attacks on media credibility that also functioned to maintain government credibility. Potentially, the council could do what Salant did at CBS, protect news standards and press freedom. But the networks had changed radically. By the mid-1980s news was a profit center, noted Salant, and these larger issues were irrelevant. Although Salant did not succeed in having the standards of broadcast journalism maintained, he set historical precedent with CBS news programming.

—Richard Bartone

RICHARD SALANT. Born in New York City, New York, U.S.A., 14 April 1914. Educated at Harvard College, A.B. 1931–35; Harvard Law School, 1935–38. Married: 1) Rosalind Robb, 1941 (divorced, 1954), children: Rosalind, Susan, Robb, and Priscilla; 2) Frances Trainer, 1955, child: Sarah. Served in U.S. Naval Reserve, 1943–46. Worked for U.S. Attorney General's committee on administrative procedure, 1939–41; worked for Office of the Solicitor General, U.S. Department of Justice, 1941–43; associate, Roseman, Goldmark, Colin and Kave, 1946–48, partner, 1948–51; vice president, special assistant to the president, CBS, Inc., 1952–61, 1964–66; president, CBS news division, 1961–64, 1966–79; member, board of directors, CBS, Inc., 1964–69; vice chair, NBC, 1979–81; senior adviser, 1981–83; president and chief executive officer, National News Council, 1983–84. Died 16 February 1993.

PUBLICATIONS

"TV News's Old Days Weren't all that Good." *Columbia Journalism Review* (New York), March-April, 1977.

"When the White House Cozies up to the Home Screen." *The New York Times,* 23 August 1981.

"Clearly Defining Fact from Fiction on Docudramas." *Stamford Advocate* (Stamford, Connecticut), 24 February 1985.

"CBS News's 'West 57th': A Clash of Symbols." *Broadcasting* (Washington, D.C.), 28 October 1985.

FURTHER READING

Auletta, Ken. *Three Blind Mice: How the TV Networks Lost Their Way.* New York: Random House, 1991.

Bleiberg, Robert M. "Salanted Journalism." *Barron's* (New York), 21 June 1976.

Brown, Les. "Salant Defends Coverage of Watergate." *The New York Times,* 16 May 1974.

———. "Salant Talks about His Plans for NBC News." *The New York Times,* 17 April 1979.

Collins, Thomas. "Salant: Calm before the Storm." *Newsday* (Hempstead, New York), 3 March 1971.

"Dick Salant and the Purity of the News." *Broadcasting* (Washington, D.C.), 19 September 1977.

Friendly, Fred. *Due to Circumstances beyond our Control . . .* New York: Random House, 1967.

Gardella, Kay. "CBS News Prexy Salant Declares War on Past." *New York News,* 7 June 1967.

Gates, Gary Paul. *Air Time: The Inside Story of CBS News.* New York: Harper and Row, 1978.

Gould, Jack. "Salant, C.B.S.'s Man Behind the 'Selling of the Pentagon.'" *The New York Times,* 1 March 1971.

Hammond, Charles Montgomery, Jr. *The Image Decade: Television Documentary 1965–1975.* New York: Hastings House, 1981.

Leonard, Bill. *In the Storm of the Eye: A Lifetime at CBS News.* New York: Putnam's, 1987.

Midgley, Leslie. *How Many Words Do You Want?* New York: Birch Lane, 1989.

"Never a Newsman, Always a Journalist." *Broadcasting* (Washington, D.C.), 26 February 1979.

Shayon, Robert Lewis. "The Pragmatic Mr. Salant." *Saturday Review* (New York), 11 March 1961.

Williamson, Lenora. "Salant Tackles Problems at National News Council." *Editor and Publisher* (New York), 4 June 1983.

See also Columbia Broadcasting System; Cronkite, Walter; News, Network; Paley, William S.; *Selling of the Pentagon*; *60 Minutes*; Stanton, Frank

SALE OF THE CENTURY

Australian Game Show

S*ale of the Century* is the most successful game show ever produced and shown on Australian television. The series began on the Nine Network early in 1980, and apart from the short four-week summer break each year, has been transmitted in the same prime- time access slot of 7:00 P.M. five nights a week ever since. Apart from the historical ratings dominance of the Nine Network in the Australian television market place, the reasons for the success of *Sale* have much to do with the format of the program, its pace, and its prizes. The game consists of three rounds in which three contestants compete for the right to buy luxury prizes at low prices. The first to sound a buzzer gains the opportunity to answer a general knowledge question. Each contestant begins with a bankroll of $25, receiving $5 for a correct answer and losing $5 for an incorrect one.

At the end of each round, the contestant with the highest score is offered the opportunity to buy a luxury item such as a colour TV set with some of the points. At the end of the program, the overall winner goes to a panel where he or she tries to guess the location of a particular prize behind a set of panels. Whether lucky or not, the contestant returns to the next episode of *Sale*. From time to time, the producers have varied the format as *Celebrity Sale of the Century,* using television personalities and other celebrities as contestants, playing either for home viewers or charity.

The program succeeds because it is a blend of general knowledge, luck, and handsome prizes. The question-and-answer format, combined with the time factor, draws in the home viewer while guesses at the panels and whether to buy items offered by the compere involve luck and risk. This combination gives *Sale of the Century* a pace and interest that make it a bright, attractive game show.

Sale of the Century originally ran on NBC, the American television network, from 1969 to 1973. The Australian-based Grundy Organisation had since 1961 been a very

frequent licensee/producer of American game show formats, but it had decided in the early 1970s to develop or

Sale of the Century
Photo courtesy of Grundy Television

buy-in formats of its own. Grundy bought the format for *Sale of the Century* in 1979, and later the same year sold the program to the Australian Nine Network. By this time, the Grundy Organisation was the biggest program packager in Australian television, and had decided that the only way to continue to expand was to internationalise its operation. However, because of differing licensing arrangements, Grundy was aware that many of the American game-show-format licence rights were not available to the company in other territories—hence the decision to buy format copyrights on programs such as *Sale*. The outstanding rating success of *Sale* in the Australian television market made it easier to sell the format elsewhere. Thus, since 1982, the company has re-versioned *Sale of the Century* in five other territories: Hong Kong (RTV, 1982); United States (NBC, 1982/1988); United Kingdom (Sky, 1989/1991); New Zealand (TVNZ, 1989/1993); and Germany (Telos/DSP, 1990/1993).

Some of the program's hosts in different countries have included Tony Barber (Australia), Joe Garagiola (U.S.), Jack Kelly (U.S.), Steve Parr (New Zealand), Nicholas Parsons (U.K.), Jim Perry (U.S.), and Glen Ridge (Australia).

—Albert Moran

PROGRAMMING HISTORY

• Nine Network

3,460 Episodes
July 1980–   Weeknight 7:00-7:30

See also Australian Production Companies; Australian Programming; Quiz and Game Shows

SALHANY, LUCIE

U.S. Broadcasting Executive

Lucille S. (Lucie) Salhany became the first woman to manage an American broadcast television network when she was appointed chair of FOX Broadcasting Company in January 1993. The company, a subsidiary of Rupert Murdoch's FOX Incorporated, is the fourth national television network to be formed in the United States, after ABC, CBS, and NBC. Salhany resigned from FOX in July 1994 and became the president and chief executive officer of the nascent United Paramount Network (UPN) where she supervised the broadcast inauguration of the network in January 1995.

In the history of American television broadcasting there had been no previous female managers who had shattered the "glass ceiling" barrier to the senior executive suite. Salhany started her career in programming at the station level in Cleveland in 1967, and by 1979 she had become vice president for programming for the Taft Broadcasting Company. She moved to Paramount Domestic Television in Los Angeles as president in 1985 and supervised the production of *Entertainment Tonight, The Arsenio Hall Show, Hard Copy,* and *Star Trek: The Next Generation.* The latter program, a revival of the original television classic, was to become one of the most successful syndicated programs in international broadcast history.

Salhany had acquired an insider's knowledge of television broadcast programming at the station-level, and used this expertise to craft a number of series that were highly salable in syndication. At the time of the premiere of *The Arsenio Hall Show,* the program introduced a number of innovations in talk show form and content—not the least of which was the replacement of the traditional host-at-a-desk with comfortable sofas allowing greater interaction between host and guest.

Salhany was recruited from Paramount by FOX Broadcasting CEO Barry Diller to manage Twentieth Television—the production and distribution arm of the network—at a time when the parent company was becoming a formidable competitor to the traditional big three networks. Salhany's open management style was well received by FOX

Lucie Salhany
Photo courtesy of Lucie Salhany

station affiliates, and she was selected by Rupert Murdoch as head of FOX Broadcasting after Diller's departure. However, the rapid growth of the network came to a halt during the 1993–94 season as the number of viewers declined and efforts to reach older viewers were not successful. Salhany had championed the *Chevy Chase Show* in the late night market, and her tenure at FOX was jeopardized when the program proved to be a brief and expensive failure. Murdoch had increasingly taken over hands-on management of his broadcast operations, and when he proposed that Salhany report to him through an intermediary she resigned and moved back to Paramount as they were about to launch their UPN network.

Lucie Salhany is a perceptive television executive who understands the intricacies of affiliate programming needs and network production operations. Her success in syndicated programming at Paramount and in operations at Twentieth Television enabled her to rise into the rarefied, but often-tenuous, environment of senior network management.

—Peter B. Seel

LUCIE (LUCILLE) S. SALHANY. Married, two children. Began television career as program director, WKBF-TV, Cleveland, Ohio, U.S.A., 1967; program manager, WLVI-TV, Boston, from 1975; vice president of television and cable programming, Taft Broadcasting Co., 1979–85; president, Paramount Domestic Television, Paramount Pictures, 1985–91; chair, Twentieth Television (a division of FOX Broadcasting Company), 1991; chair, FOX Broadcasting Company, 1993–94; named president and chief executive officer, United Paramount Network, 1994. Honorary degree: doctorate of humane letters, Emerson College, Boston, 1991. Member: board of directors, Fox Inc.; board of directors, Emerson College; executive committee, University of California, Los Angeles, School of Theater, Film and Television; board, Academy of Television Arts and Sciences; board, Hollywood Supports. Recipient: American Jewish Committee's Sherrill C. Corwin Human Relations Award (first woman recipient), 1995; American Women in Radio and Television's Silver Satellite Award, 1995. Address: United Paramount Network, 5555 Melrose Avenue, Los Angeles, California, 90038-3197, U.S.A.; 11800 Wilshire Boulevard, Los Angeles, California 90025, U.S.A.

FURTHER READING

"Fox's Passionate Pro." *Broadcasting and Cable* (Washington, D.C.), 1 March 1993.
Harris, Kathryn. "Madame Chairman." *Forbes* (New York), 5 August 1991.
Tobenkin, D. "Fox's New Team at the Top." *Broadcasting and Cable* (Washington, D.C.), 11 July 1994.
"The View from Atop Twentieth" (interview). *Broadcasting* (Washington, D.C.), 21 January 1992.

See also Diller, Barry; FOX Broadcasting Company; Murdoch, Rupert

SANDFORD, JEREMY

British Writer

Jeremy Sandford is the writer of *Cathy Come Home* and *Edna the Inebriate Woman*; his oeuvre may be one of the smallest, yet most famous, in the history of British television drama. *Cathy Come Home* is surely the most talked about television play ever, an iconic text in the radical canon of the 1960s *Wednesday Play*, which has become overshadowed by the association with its director, Ken Loach, and producer, Tony Garnett.

After more or less disappearing from television, Sandford surfaced in 1980 with a play commissioned for the series *Lady Killers,* and then in 1990, as the homeless population in Britain began once again to be a topic of public debate, with a documentary for the BBC, *Cathy, Where Are You Now?*

When *Cathy* was reshown in 1993 as part of a season commemorating the setting up of the housing charity Shelter, Sandford wrote to the *Independent,* taking issue with a claim that doubts had been raised over the accuracy of the homelessness and family separation statistics given at the end of the play. "I work as a journalist as well as an author," he wrote, "and it would be professional suicide to be inaccurate." Sandford has never wholly identified himself as a television dramatist. At one time a poet and artist, he had nursed an early ambition to be a professional musician, and played the clarinet in an RAF band during his national service. One of his first plays, *Dreaming Bandsmen,* broadcast by BBC Radio in 1956 and later staged in Coventry, seemed to confirm his early reputation as a surrealist, but at the same time he was recording radio documentaries about working-class life in the East End, and it was as a journalist and activist that he began writing about homelessness in the early 1960s. As he told an interviewer in 1990, he had always sought to play his role on the stage of life rather than simply reflecting it. Thus, not only did he submerge himself in the nether world of the down-and-out for his research on *Edna,* but went on to arm himself with his written work as part of an active crusade on behalf of the dispossessed. A special showing of *Cathy* was arranged for Parliament, and Sandford himself toured the country screening and talking about both plays at public meetings.

Homelessness, itinerancy, and housing policy have been particular obsessions of Sandford. His Anglo-Irish grandmother, Lady Mary Carbery, was a member of the Gypsy Lore Society, and he has campaigned on behalf of gypsies as well as editing their newspaper, *Romano Drum*. A play about gypsies, *Till the End of the Plums*, was to complete a trilogy about the homeless but was never produced.

Born of wealthy parents (his father owned a private printing press) and educated at Eton and Oxford, Sandford was brought up in a stately Herefordshire home. In the late 1980s, after a long association with the alternative communities of folk festivals and camps, he moved into a large country house and opened it up as a study centre for New Age travellers.

A further play, *Smiling David*, about the case of a Nigerian drowned in a Leeds river and the agencies implicated in the events, was commissioned for radio and broadcast in 1972, but never made it to the television screen. Sandford's oft-remarked status as a documentarist and social advocate rather than a natural television dramatist is emphasised by the fact that the scripts for *Cathy* and *Edna* are published in a series of political and social treatises. His polemical and factual writing, such as *Down and Out in Britain,* which accompanied *Edna,* far exceeds the amount he has written for television. However, the importance of his two major works in defining the cultural role of television drama in Britain as an intrinsic part, rather than mere mirror, of socio-political actuality, cannot be ignored. *Cathy Come Home* remains a landmark in this sense. Sandford's exchange with Paul Ableman in the pages of *Theatre Quarterly* over the ethics of fictional form in *Edna the Inebriate Woman* set the agenda for a debate about the aesthetics and politics of drama-documentary that was to dominate television drama criticism through the 1970s and 1980s.

—Jeremy Ridgman

JEREMY SANDFORD. Attended Eton Public School, Berkshire; Oxford University. Married: 1) Nell Dunn, 1956 (divorced, 1986); three sons; 2) Philippa Finnis, 1988. Worked initially as a journalist; established reputation as socially-committed writer for television and radio with *Cathy Come Home,* 1966; editor, *Romano Drum* (gypsy newspaper); director, Cyrenians; executive, Gypsy Council; sponsor, Shelter. Recipient: Screen Writers Guild of Great Britain Awards, 1967 and 1971; Prix Italia for Television Drama, 1968; Critics' Award for Television Drama, 1971. Address: Hatfield Court, Hatfield, Leominster, Herefordshire HR6 0SD, England.

TELEVISION PLAYS

1966	*Cathy Come Home*
1971	*Edna the Inebriate Woman*
1980	*Don't Let Them Kill Me on Wednesday (Lady Killers)*

TELEVISION DOCUMENTARY

1990	*Cathy, Where Are You Now?*

RADIO

Dreaming Bandsmen, 1956; *Smiling David,* 1972.

STAGE

Dreaming Bandsmen, 1956.

PUBLICATIONS (selection)

Dreaming Bandsmen. Coventry: BBC Radio and Belgrade Theatre, 1956.
Cathy Come Home. London: BBC, 1966.
Synthetic Fun, with Roger Law. Harmondsworth: Penguin, 1967.
Edna the Inebriate Woman. London: BBC, 1971.
Down and Out in Britain. London: Owen, 1971; revised edition, London: New English Library, 1972.
In Search of the Magic Mushrooms. London: Owen, 1972.
"Edna and Cathy: Just One Huge Commercial" (Production Casebook No. 10). *Theatre Quarterly* (London), April-June 1973.
Gypsies. London: Secker and Warburg, 1973.
Tomorrow's People. London and New York: Jerome, 1974.
Smiling David. London: Calder and Boyars, 1974.
Prostitutes. London: Secker and Warburg, 1975; revised edition, London: Abacus, 1977.
Virgin of the Clearways. London: Boyars, 1978.
Songs from the Roadside, Sung by Romani Gypsies in the West Midlands. Clun: Redlake Press, 1995.

FURTHER READING

Ableman, Paul. "Edna and Sheila: Two Kinds of Truth." *Theatre Quarterly* (London), July-September 1972.
Banham, Martin. "Jeremy Sandford." In, Brandt, G.W. editor. *British Television Drama.* Cambridge: Cambridge University Press, 1981.
Dunn, Elizabeth. "Gimme Shelter." *Sunday Telegraph* (London), 8 July 1990.
Rosenthal, Alan. *The New Documentary in Action: A Casebook in Film Making.* Berkeley: University of California Press, 1971.
Shubik, Irene. *Play for Today: The Evolution of Television Drama.* London: Davis and Poynter, 1975.
Worsley, T.C. *Television: The Ephemeral Art.* London: Alan Ross, 1970.

See also *Cathy Come Home;* Garnett, Tony; Loach, Ken; *Wednesday Play*

SANDRICH, JAY

U.S. Director

The career of Jay Sandrich, a leading director of American situation comedies, covers much of the first few decades of the sitcom. His programs have been characterized by wit, a supportive working environment, and care for his actors.

The son of film director Mark Sandrich, Jay Sandrich began his television work in the mid-1950s as a second assistant director with Desilu Productions, learning to direct television on *I Love Lucy, Our Miss Brooks,* and *December Bride.* Later he worked on both *The Danny Thomas Show* and *The Dick Van Dyke Show.* In 1965, Sandrich put in his only stint as a producer, serving as associate producer for the first season of the innovative comedy *Get Smart.* He enjoyed the experience but vowed to stick to directing in future. He told Andy Meisler of *Channels* magazine, "I really didn't like producing. I liked being on the stage. I found that, as a producer, I'd stay up until four in the morning worrying about everything. As a director, I slept at night."

In 1971, he signed on as regular director for the relationship-oriented, subtly feminist *Mary Tyler Moore Show,* beginning a long-term partnership with the then fledgling MTM Productions. Directing two-thirds of the episodes in the program's first few seasons, he won his first Emmys and worked on the pilot for the program's spin-off, *Phyllis.* In an interview for this encyclopedia, he spoke glowingly of the MTM experience: "[MTM chief] Grant [Tinker] created this wonderful atmosphere of being able to have a lot of fun at your work—plus you were working next door to people who were interesting and bright. And there was this feeling of sharing talent."

Sandrich went on to work as a regular director on the satirical *Soap* and eventually created another niche for himself as the director of choice for *The Cosby Show* from 1985 to 1991. Meisler's article painted an appealing portrait of the director's relationship with the star and with other *Cosby* production personnel, quoting co-executive producer Tom Werner on the show's dynamics: "Although we're really all here to service Bill Cosby's vision, the show is stronger because Jay challenges Bill and pushes him when appropriate." Sandrich was proud of the program's pioneering portrayal of an upper-class black family, and of its civilized view of parent-child relations.

During and following *Cosby*'s run, Sandrich directed pilots and episodes for a number of successful programs, including *The Golden Girls, Benson, Night Court,* and *Love and War.*

Although he ventured briefly into the field of feature films, directing *Seems Like Old Times* in 1980, Sandrich decided quickly that he preferred to remain in television. "The pace is much more interesting," he explained. "In features you sit around so much of the time while lighting is going on, and then you make the picture, and you sit around for another year developing projects. I like to work. I like the immediacy of television." Asked whether there was a Jay Sandrich type of program, Sandrich ruminated, "I don't know if there is, but I like more human-condition shows, not really wild and farcy, although *Soap* gave me really a bit of everything to do....

Jay Sandrich
Photo courtesy of Jay Sandrich

Basically, I like men-women shows.... I go more for shows that have more love than anger in them." Certainly most of his programs have lived up to this inclination.

For many of his colleagues, Sandrich has defined the successful situation-comedy director. "I think it was Jay who first made an art form of three-camera film," said producer Allan Burns (quoted in Meisler), referring to the shooting technique most often used for sitcoms. Although he was modest about his own accomplishments, and quick to note that good writing is the starting point for any television program, Sandrich asserted that he cherishes his role as director in a medium often viewed as the domain of the producer.

"If there's a regular director every week," he stated, "[television] should be a major collaboration between the director and the producer—if the director's any good—because he is the one who sets the style and the tone of the show. He works with the actors. And a good director, whether he is rewriting or not, he is always making suggestions ... and in many cases knows the script a little bit better than the producer because he's been seeing each scene rehearsed and understands why certain things work and why they don't.... So when it's a regular director on a series, I think it's not a producer's medium. It is the creative team [that shapes a series]."

In his early 60s at this writing, Sandrich still worked frequently but denied that he was any longer the king of pilots for

American comedies. "I think Jimmy Burrows is the king," he said of his former protégé. "He's gotten so many shows on the air. No, I think I'm the dowager queen or something by now."

—Tinky "Dakota" Weisblat

JAY SANDRICH. Born in Los Angeles, California, U.S.A., 24 February 1932. Educated at University of California Los Angeles, B.A. 1953. Married 1) Nina Kramer, 1953 (divorced, 1974); two sons and one daughter; 2) Linda Green, 1984. Started career as second assistant director, *I Love Lucy,* Desilu Productions, 1955, then first assistant director, *I Love Lucy* and *The Danny Thomas Show;* director, MTM Productions, from 1971; currently director, primarily for television. Recipient: Emmy Awards, 1971, 1973, 1985, and 1986; DGA Awards, 1975, 1984, 1985, and 1986. Address: c/o Creative Artists Agency, 9830 Wilshire Boulevard, Beverly Hills, California 90212, U.S.A.

TELEVISION SERIES (selection)

1965–70	*Get Smart* (producer)
1967–70	*He and She*
1971–77	*The Mary Tyler Moore Show*
1972–78	*The Bob Newhart Show*
1975–77	*Phyllis*
1976–78	*The Tony Randall Show*
1977–79	*Soap*
1979–86	*Benson*
1984–92	*The Cosby Show*
1985–92	*The Golden Girls*
1988–95	*Empty Nest*
1992–95	*Love and War* (pilot only)
1993–94	*Thea*
1995	*The Office*
1995–96	*The Jeff Foxworthy Show*
1996	*London Suites*

FILMS

Seems Like Old Times, 1980; *For Richer, For Poorer,* 1992.

FURTHER READING

Kuney, Jack. *Take One: Television Directors on Directing.* New York: Greenwood, 1990.

Meisler, Andy. "Jay Sandrich: Ace of Pilots." *Channels* (New York), October 1986.

Ravage, John W. *Television: The Director's Viewpoint.* Boulder, Colorado: Westview, 1978.

See also *Cosby Show; Danny Thomas Show; Dick Van Dyke Show;* Director, Television; *Get Smart; I Love Lucy; Mary Tyler Moore Show; Our Miss Brooks;* Tinker, Grant

SANFORD AND SON

U.S. Domestic Comedy

The 1972 NBC television program *Sanford and Son* chronicled the adventures of Fred G. Sanford, a cantankerous widower living with his grown son, Lamont, in the notorious Watts section of contemporary, Los Angeles, California. Independent producers Norman Lear and Bud Yorkin licensed the format of a British program, *Steptoe and Son,* which featured the exploits of a cockney junk dealer, and created *Sanford and Son* as an American version. *Sanford and Son, The Jeffersons* and *Good Times,* all produced by Lear and Yorkin, featured mostly black casts—the first such programming to appear since the *Amos 'n' Andy* show was canceled in a hailstorm debate in 1953.

The starring role of *Sanford and Son* was portrayed by actor-comedian Redd Foxx. Foxx (born John Elroy Sanford) was no newcomer to the entertainment industry. His racy nightclub routines had influenced generations of black comics since the 1950s. Born in St. Louis, Missouri, Foxx began a career in the late 1930s performing street acts. During the 1950s he achieved a measure of success as a nightclub performer and recorder of bawdy joke albums. By the 1960s he was headlining in Las Vegas. In 1969, he earned a role as an aging junk dealer in the motion picture *Cotton Comes to Harlem,* a portrayal that brought him to the attention of Lear and Yorkin.

It was Foxx's enormously funny portrayal of sixty-five-year-old Fred G. Sanford that quickly earned *Sanford and Son* a place among the top-ten watched television programs to air on NBC television. He was supported by Lamont, his thirtyish son, and a multi-racial cast of regular and occasional characters who served as the butt of Sanford's often bigoted jokes and insults. Fred's nemesis, the "evil and ugly" Aunt Esther (portrayed by veteran actor LaWanda Page), often provided the funniest moments of the episode, as she and Fred traded jibes and insults. The trademark routine of the series occurred when Fred feigned a heart attack by clasping his chest in mock pain. Staggering drunkenly he would threaten to join his deceased wife Elizabeth, calling out "I'm coming to join you, Elizabeth!"

Though enormously successful, Foxx became dissatisfied with the show, its direction, and his treatment as star of the program. In a *Los Angeles Times* article, he stated, "Certain things should be yours to have when you work your way to the top." At one point he walked off the show complaining that the white producers and writers had little regard or appreciation of African-American life and culture. In newspaper interviews he lambasted the total lack of black writers or directors. Moreover, Foxx believed that his efforts were not appreciated, and in 1977 he left NBC for his own variety show on ABC. The program barely lasted one season.

Sanford and Son survived some five years on prime-time television. It earned its place in television history as the first successful, mostly black cast television sitcom to appear on American network, primetime television in twenty years since the cancellation of *Amos 'n' Andy*. It was an enormously funny program, sans obvious ethnic stereotyping. "I'm convinced that *Sanford and Son* shows middle class America a lot of what they need to know," Foxx said in a 1973 interview. "The show. . . doesn't drive home a lesson, but it can open up people's minds enough for them to see how stupid every kind of prejudice can be." After Foxx left the show permanently, a pseudo-spin-off, called *Sanford Arms*, proved unsuccessful and lasted only one season.

—Pamala S. Deane

CAST

Fred Sanford	Redd Foxx
Lamont Sanford	Demond Wilson
Grady Wilson (1973–77)	Whitman Mayo
Aunt Esther (1973–77)	LaWanda Page
Woody Anderson (1976–77)	Raymond Allen
Bubba Hoover	Don Bexley
Janet Lawson (1976–77)	Marlene Clark
Roger Lawson (1976–77)	Edward Crawford
Donna Harris	Lynn Hamilton
Officer Swanhauser (1972)	Noam Pitlik
Officer Hopkins ("Happy") (1972–76)	Howard Platt
Aunt Ethel (1972)	Beah Richards
Julio Fuentes (1972–75)	Gregory Sierra
Rollo Larson	Nathaniel Taylor
Melvin (1972)	Slappy White
Officer Smith ("Smitty") (1972–76)	Hal Williams
Ah Chew (1974–75)	Pat Morita

PRODUCER   Norman Lear

PROGRAMMING HISTORY   136 Episodes

• NBC

January 1972–September 1977	Friday 8:00-8:30
Also April 1976–August 1976	Wednesday 9:00-9:30

FURTHER READING

Bogel, Donald. *Blacks, Coons, Mulattos, Mammies and Bucks: An Interpretive History of Blacks in American Film.* New York: Garland, 1973.

———. *Blacks in American Television and Film.* New York: Garland, 1988.

Friedman, Lester D. *Unspeakable Images: Ethnicity and the American Cinema.* Urbana: University of Illinois Press, 1991.

Sanford and Son

Gray, Herman. *Watching Race: Television and the Struggle for "Blackness."* Minneapolis: University of Minnesota Press, 1995.

MacDonald, J. Fred. *Blacks and White TV: Afro-Americans in Television Since 1948.* Chicago: Nelson-Hall, 1993.

Marc, David, and Robert J. Thompson. *Prime Time, Prime Movers: From I Love Lucy to L.A. Law, America's Greatest TV Shows and People Who Created Them.* Boston: Little, Brown, 1992.

Taylor, Ella. *Prime Time Families: Television Culture in Postwar America.* Berkeley: University of California Press, 1990.

See also *Amen; Amos 'n' Andy;* Comedy, Domestic Settings; *Good Times;* Lear, Norman; Racism, Ethnicity and Television; *227*

SARNOFF, DAVID

U.S. Media Executive

A pioneer in radio and television, David Sarnoff was an immigrant who climbed the rungs of corporate America to head the Radio Corporation of America (RCA). Born 27 February 1891, in Uzlian, in the Russian province of Minsk, Sarnoff's early childhood years were spent studying to be a rabbi, but when he emigrated to the United States in

1900, he was forced to work to feed his mother, ailing father, and siblings.

Learning early the value of self-promotion and publicity, Sarnoff falsely advanced himself both as the sole hero who stayed by his telegraph key for three days to receive information on the *Titanic's* survivors and as the prescient prophet of broadcasting who predicted the medium's rise in 1915. While later described by others as the founder of both the Radio Corporation of American (RCA) and the National Broadcasting Company (NBC), Sarnoff was neither. These misconceptions were perpetuated because Sarnoff's later accomplishments were so plentiful that any myth was believable. Indeed, his foresight and corporate savvy led to many communication developments, especially television.

Sarnoff began his career at age nine, selling Yiddish-language newspapers shortly after arriving in New York. To better his English, he picked up discarded English-language newspapers. By the time he was ten, he had a fairly passable vocabulary. He also soon had his own newsstand. During the day he attended grade school, while at night he enrolled in classes at the Educational Alliance, an East Side settlement house. At age 15, with his father's health deteriorating, Sarnoff was forced to seek a full-time job.

He became a messenger for the Commercial Cable Company, the American subsidiary of the British firm that controlled undersea cable communication. The telegraph key lured him to the American Marconi Company a few months later, where he was hired as an office boy. Once there, he began his corporate rise, including the job of being Marconi's personal messenger when the inventor was in town. With Marconi's endorsement, Sarnoff became a junior wireless telegraph operator and, at age 17, volunteered for wireless duty at one of the company's remote stations. There he studied the station's technical library and took correspondence courses. Eighteen months later, he was appointed manager of the station in Sea Gate, New York. He was the youngest manager employed by Marconi. After volunteering as a wireless operator for an Arctic seal expedition, he became operator of the Marconi wireless purchased by the John Wanamaker department stores. At night he continued his studies.

Then, on the evening of 14 April 1912, he heard the faint reports of the *Titanic* disaster. One of a number of wireless operators reporting the tragedy, Sarnoff would later claim he was the only one remaining on air after President Taft ordered others to remain silent. Another probably spurious claim was Sarnoff's assertion he wrote his famous "Radio Music Box Memo" in 1915. The version so often cited was actually written in 1920, when others were also investigating and predicting broadcasting.

As his career thrived, Sarnoff's personal life also grew. On 4 July 1917, he married Lizette Hermant, following a closely supervised courtship. Their 54-year marriage survived Sarnoff's occasional philanderings and proved the bedrock of his life. They had three sons: Robert, Edward, and Thomas. Robert succeeded his father as RCA's presi-

David Sarnoff
Photo courtesy of the David Sarnoff Research Center

dent. In 1919, when British Marconi sold its American Marconi assets to General Electric (GE) to form RCA, Sarnoff came on board as commercial manager. Under the tutelage of Owen D. Young, RCA's chair, Sarnoff was soon in charge of broadcasting as general manager of RCA and was integral in formation of NBC in 1926. Again as Young's protégé, he negotiated the secret contracts with American Telephone and Telegraph (AT and T) that led to NBC's development. With acquisition of AT and T's broadcasting assets, RCA had two networks, the Red and the Blue, and they debuted in a simulcast on 15 November 1926.

In 1927 Sarnoff was elected to RCA's board and during the summer of 1928, he became RCA's acting president when General James G. Harbord, RCA's president, took a leave of absence to campaign for Herbert Hoover. His eventual succession to that position was assured. During the end of the decade Sarnoff negotiated successful contracts to form Radio-Keith-Orpheum (RKO) motion pictures, to introduce radios as a permanent fixture in automobiles, and to consolidate all radio manufacturing by the Victor company under RCA's banner. On 3 January 1930, the 39-year-old Sarnoff became RCA's president.

The next two years were pivotal in Sarnoff's life as the Department of Justice sued GE and RCA for monopoly and restraint of trade. Sarnoff led industry efforts to combat the government's suits that would have destroyed RCA. The result was a consent decree in 1932 calling for RCA's divestiture from GE and the licensing of RCA's patents to competitors. When GE freed RCA, Sarnoff was at the helm and, for nearly the next three decades, he would oversee numerous communications developments, including television.

Sarnoff's interest in television began in the 1910s, when he became aware of the theory of television. By 1923, he was convinced television would be the next great step in mass communication. In 1929 Westinghouse engineer Vladimir Zworykin called on Sarnoff to outline his concept of an electronic camera. Within the year, Sarnoff underwrote Zworykin's efforts, and Zworykin headed the team developing electronic television. As the Depression deepened, Sarnoff bought television patents from inventors Charles Jenkins and Lee De Forest, among others, but he could not acquire those patents held by Philo Farnsworth. These he had to license, and in 1936, RCA entered into a cross licensing agreement with Farnsworth. This agreement solved the technological problems of television, and establishing television's standards became Sarnoff's goal.

The Federal Communications Commission (FCC) would set those standards, but within the industry, efforts to reach consensus failed. Other manufacturers, especially Philco, Dumont and Zenith, fought adoption of RCA's standards as the industry norm. In 1936, the Radio Manufacturers Association (RMA) set up a technical committee to seek agreement on industry standards, an action blessed actively by Sarnoff and silently by the FCC. For more than five years the committee would fight over standards. Sarnoff told the RMA, standards or not, he would initiate television service at the opening of the New York World's Fair on 20 April 1939, and he did. Skirmishes continued for the next two years over standards, but finally in May 1941 the FCC's National Television System Committee (NTSC) set standards at 525 lines, interlaced, and 30 frames per second. But rapid television development stalled as World War II intervened. Sarnoff's attention then turned to devices, including radar and sonar, that would help win the war.

During World War I Sarnoff had applied for a commission in naval communications, only to be turned down, ostensibly because his wireless job was considered essential to the war effort. Sarnoff suspected anti-Semitism. Now as head of the world's largest communication's firm, Sarnoff was made a brigadier general and served as communication consultant to General Dwight Eisenhower. After the war, with the death of RCA chair of the board, General J.G. Harbord in 1947, General Sarnoff, as he preferred to be called, was appointed chair and served in that capacity until his death in 1971.

After the war, RCA introduced monochrome television on a wide scale to the American population, and the race for color television with CBS was on. CBS picked up its pre-war experiments with a mechanical system, which Sarnoff did not see initially as a threat because it was incompatible with already approved black-and-white standards. When CBS received approval for its system in 1951, Sarnoff challenged the FCC's decision in the courts on the grounds it contravened the opinions of the industry's technical leaders and threatened the public's already $2-billion investment in television sets. When the lower court refused to block the FCC ruling, Sarnoff appealed to the Supreme Court, which affirmed the FCC action as a proper exercise of its regulatory power.

Sarnoff counterattacked through an FCC-granted authority for RCA to field-test color developments. Demonstrations were carefully set for maximum public exposure, and they were billed as "progress reports" on compatible color. By then, the Korean War intervened in the domestic color television battle and blunted introduction of CBS' sets on a large scale. Monochrome still reigned, and Sarnoff continued pressing the compatibility issue. In 1953 CBS abandoned its color efforts as "economically foolish" in light of the 25 million incompatible monochrome sets already in use. The FCC was forced to reconsider its earlier order and, on 17 December 1953, voted to reverse itself and adopt standards along those proposed by RCA. During the 1950s and 1960s Sarnoff's interests included not only television but also satellites, rocketry, and computers.

At the same time he was battling CBS over color, Sarnoff's feud with Edwin Howard Armstrong over FM radio's development and patents continued. Sarnoff and Armstrong, once close friends, were hopelessly alienated by the end of World War II. Their deadly feud lasted for years, consumed numerous court challenges and ended in Armstrong's suicide in 1954.

Sarnoff died in his sleep 12 December 1971, of cardiac arrest. At his funeral he was eulogized as a visionary who had the capacity to see into tomorrow and to make his visions work. His obituary began on page one and ran nearly one full page in the *New York Times* and aptly summed up his career in these words: "He was not an inventor, nor was he a scientist. But he was a man of astounding vision who was able to see with remarkable clarity the possibilities of harnessing the electron."

—Louise Benjamin

DAVID SARNOFF. Born near Minsk, Russia, 27 February 1891. Attended public schools, Brooklyn, New York, U.S.A.; studied electrical engineering at Pratt Institute. Married: Lizette Hermant, 1917; three sons. Joined Marconi Wireless Company, 1906–19, telegraph operator, 1908, promoted to chief radio inspector and assistant chief engineer, when Marconi was absorbed by Radio Corporation of America (RCA), 1919–70, commercial manager; elected general manager, RCA, 1921, vice president and general manager, 1922, executive vice president, 1929, president, 1930, chair of board, RCA, 1947–70; oversaw RCA's manufacture of color television sets and NBC's color broadcasts. Received 27 honorary degrees, including doctoral degrees from Columbia University and New York University. Died in New York City, 12 December 1971.

PUBLICATION

Looking Ahead: The Papers of David Sarnoff. New York: McGraw-Hill, 1968.

FURTHER READING

Benjamin, Louise. "In Search of the Sarnoff 'Radio Music Box' Memo." *Journal of Broadcasting and Electronic Media* (Washington, D.C.), Summer 1993.

Bilby, Kenneth M. *The General: David Sarnoff and the Rise of the Communications Industry.* New York: Harper and Row, 1986.

"David Sarnoff of RCA Is Dead; Visionary Broadcast Pioneer." *The New York Times,* 13 December 1971.

Dreher, Carl. *Sarnoff, An American Success.* New York: Quadrangle/New York Times Book, 1977.

Lyons, Eugene. *David Sarnoff, A Biography.* New York: Harper and Row, 1966.

Sobel, Robert. *RCA.* New York: Stein and Day, 1986.

The Wisdom of Sarnoff and the World of RCA. Beverly Hills, California: Wisdom Society for the Advancement of Knowledge, Learning and Research in Education, 1967.

See also American Broadcasting Company; Color Television; Columbia Broadcasting System; Farnsworth, Philo; Goldenson, Leonard; National Broadcasting Company; Paley, William S.; Radio Corporation of America; Sarnoff, Robert; United States: Networks; Zworykin, Vladimir

SARNOFF, ROBERT

U.S. Media Executive

Robert Sarnoff, eldest son of broadcasting mogul David Sarnoff, followed in his father's professional footsteps through his career at NBC and the Radio Corporation of America (RCA). Contemporaries attributed the son's corporate promotions to nepotism, and constantly drew comparisons between his executive performance and style and that of his father. During his years as company head, Robert Sarnoff practiced decision-making by consensus, displayed an obsession with corporate efficiency, and constantly sought to implement modern management techniques. David Sarnoff's aggressive, imperial, dynamic manner of command often overshadowed his son's practical, yet increasingly mercurial, character.

After a short stint in the magazine business, Robert Sarnoff joined NBC as an accounts executive in 1948—at a time when David Sarnoff had recently assumed chairmanship of electronics giant RCA, the parent company of NBC. Robert Sarnoff served in a variety of positions over the next few years, working his way up the business ladder. As vice president of NBC's film unit, he oversaw the development of *Project XX* and *Victory at Sea*—the latter a pioneer in the documentary series format that traced the naval campaigns of World War II through compilation footage. Passing as educational programming, the series was well attuned to Cold War patriotism and earned Sarnoff a Distinguished Public Service Award from the U.S. Navy.

NBC television programming strategies during the first half of the 1950s were largely determined by the flamboyant Pat Weaver. RCA funded Weaver's extravagant experiments in the medium since it wished to establish NBC's reputation as a "quality" network and was realizing a return on its investment through increased sales of television receivers. By mid-decade, however, RCA policy was modified: NBC was now expected to achieve economic self-sufficiency and advertising sales parity with arch-rival CBS. Weaver was first promoted to NBC chair in 1955, and then forced to resign from the company several months later. In turn, Robert Sarnoff ascended to fill that vacant positions.

Sarnoff assumed leadership of the network's financial interests and general policy decisions. Robert Kintner, who had shown a propensity for budget-conscious scheduling at ABC, took over as head of NBC-TV programming and was elevated to the rank of NBC president in 1958. Together, the "Bob and Bob Show" (as it was known in the industry) stabilized network operations and routinized programming. Sarnoff established a clear chain of command by streamlining NBC's staff, increasing middle management positions, and delegating more operating responsibilities to department heads. In order to cut overheads, in-house production was curtailed, and links with several dependable suppliers of filmed programming were created. Program development

Robert Sarnoff
Photo courtesy of the David Sarnoff Research Center

and series renewal became subject to ratings success and spot advertising sales. Toward the end of the decade, westerns, action shows, sitcoms, and quiz shows were regular prime-time features. Gone, for the most part, were the costly "spectaculars" and live dramas of the Weaver years. NBC profits improved steadily.

Sarnoff's most public phase came in the late 1950s and early 1960s when he defended NBC programming policies against critics in the press and in Congress. The public interest was best served by popular programming, Sarnoff's reasoning went. He espoused the benefits of a "well-rounded schedule," but clearly practiced a policy of programming to majority tastes. Sarnoff insisted that competition for advertisers, audiences, and affiliate clearance would ensure that the networks would remain receptive to the multiple demands of the market. Ratings were the economic lifeblood of the medium; "high brow" interests would have to remain secondary to "mass appeal" shows in the NBC schedule. Critics who lamented the disappearance of "cultural" programming were elitist, he claimed. Neither the Federal Communications Commission nor Congress should interfere in network operations or establish program guidelines, according to Sarnoff, since this would encourage political maneuvering and obstruct market forces. More effective industry self-regulation and self-promotion, spearheaded by the networks, would ensure that recent broadcasting transgressions (symbolized by the quiz show scandals and debates over violence on television) would not reoccur.

Sarnoff's agenda did not dismiss "public service" programming entirely. Kintner had turned NBC's news department into a commercially viable operation, most notably with *The Huntley–Brinkley Report*. During these years, NBC undertook various educational projects, including *Continental Classroom* (the first network program designed to provide classes for college credit) and several programs on art history (a particular passion of Sarnoff). Sarnoff extolled television's ability to enlighten through its capacity to channel and process the diverse fields of information, knowledge, and experience that characterized the modern age. He touted television's ability to generate greater viewer insight into the political process, and is credited with bringing about the televised "Great Debates" between Kennedy and Nixon during the 1960 presidential campaign.

In general, NBC's public service record during the Sarnoff years was disappointing. NBC did, however, become a serious ratings and billings competitor to CBS. In marked contrast to the dismal results of the previous decade, the network's color programming in the 1960s helped to dramatically boost color set sales and, consequently, RCA coffers.

On the first day of 1966, again thanks largely to his father's influence, Robert Sarnoff became president of RCA. Two years later he assumed also the role of chief executive officer. David Sarnoff remained chairman of the board until 1970, when ill health forced him to relinquish that position to his son. At RCA Robert Sarnoff inherited, and exacerbated, problematic developments that would

result in his forced resignation in 1975. The younger Sarnoff continued to diversify the corporation, but with some ill-chosen investments that yielded poor returns. Most significantly, he over-committed company resources to an abortive attempt to achieve competitiveness in the mainframe computer market. During Sarnoff's tumultuous time at RCA he continued to oversee operations at NBC. There he found little solace, as the network lost ground to CBS and ABC in the early 1970s. NBC's weakened performance contributed to declining RCA stock prices—a state of affairs that resulted in Robert Sarnoff's displacement from the company that had been synonymous with the Sarnoff name over the previous half century.

—Matthew Murray

ROBERT SARNOFF. Born in New York City, New York, U.S.A., 2 July 1918. Educated at Harvard University, B.A. 1939; Columbia Law School, 1940. Worked in office of Coordinator of Information, Washington, D.C., 1941; U.S. Navy, 1942; assistant to publisher Gardner Cowles, Jr., 1945; staff member, *Look*, 1946; president, NBC, 1955–58; board of directors, RCA, 1957; chair of board, NBC, 1958; chair of board, chief executive officer, NBC, 1958–65; president, RCA, 1966; chief executive officer, 1968; chair of board, 1970–75. Member: Television Pioneers, 1957 (president, 1952–53); International Radio and Television Society; Broadcasters Committee for Radio Free Europe; American Home Products, Inc.; director, Business Committee for the Arts; chair and former president of council, Academy of Television Arts and Sciences; vice president and member of board of directors, Academy of Television Arts and Sciences Foundation.

PUBLICATIONS

"What Do You Want From TV?" *Saturday Evening Post* (Philadelphia, Pennsylvania), 1 July 1961.
"A View from the Bridge of NBC." *Television Quarterly* (New York), Spring 1964.

FURTHER READING

Bilby, Kenneth. *The General: David Sarnoff and the Rise of the Communications Industry.* New York: Harper and Row, 1986.
Kepley, Jr., Vance. "From 'Frontal Lobes' to the 'Bob-and-Bob' Show: NBC Management and Programming Strategies, 1949-65." In, Balio, Tino, editor. *Hollywood in the Age of Television.* Cambridge, Massachusetts: Unwin Hyman, 1990.
"Sarnoff of NBC: The Decision Makers, Part 5." *TV Guide* (Radnor, Pennsylvania), 2 February 1963.
Sobel, Robert. *RCA.* New York: Stein and Day, 1986.

See also Kintner, Robert; National Broadcasting Company; Radio Corporation of America; Sarnoff, David; United States: Networks; *Victory at Sea*; Weaver, Sylvester "Pat"

SATELLITE

Television could not exist in its contemporary form without satellites. Since 10 July 1962, when NASA technicians in Maine transmitted fuzzy images of themselves to engineers at a receiving station in England using the Telstar satellite, orbiting communications satellites have been routinely used to deliver television news and programming between companies and to broadcasters and cable operators. And since the mid-1980s they have been increasingly used to broadcast programming directly to viewers, to distribute advertising, and to provide live news coverage.

Arthur C. Clarke, a British engineer turned author, is credited with envisioning the key elements of satellite communications long before the technical skill or political will to implement his ideas existed. In 1945 he published a plan to put electronic relay stations—a radio receiver and re-transmitter—into space at 23,000 miles above the earth's equator. At this altitude, the satellite must complete a full rotation around the earth every 24 hours in order to sustain orbit (countering the pull of the earth's gravity). Given the rotation of the earth itself, that keeps the satellite at the same relative position. This "geosynchronous orbit" is where several hundred communications satellites sit today providing telephone and data communications, but mostly, relaying television signals. Television is currently the largest user of satellite bandwidth.

An "uplink" transmitter on earth, using a "dish" antenna pointed toward the satellite, sends a signal to one of the satellite's "transponders." The transponder amplifies that signal and shifts it to another frequency (so as not to interfere with the incoming signal) to be transmitted back to earth. A "downlink" antenna and receiver on earth then captures that signal and sends it on its way. The essential advantage of the satellite is that the uplink and downlink may be 8000 miles apart. In practice, satellite communications is more efficient over a shorter distances than that, but the advantages over terrestrial transmissions—cable, fiber optics, and microwave—are profound, particularly across oceans. As with Direct Broadcast Satellites (DBS), satellites can transmit to an unlimited number of ground receivers simultaneously, and costs do not increase with distance.

Each satellite has a distinct "footprint," or coverage area, which is meticulously shaped and plotted. In 1971, the first communications satellites carrying "spot beam" antennas were launched. A spot beam antenna can be steered to focus the satellite's reception and transmission capabilities on a small portion of the earth, instead of the 40% of the earth's surface a wider antenna beam could cover. Spot coverage is crucial in international broadcasting, when neighboring countries may object to signal "spillover" into their territory.

Communications satellites since the 1960s have received uplink signals in a range of frequencies (or "bandwidth") near six GHz (gigahertz, or a billion cycles per second) and downlinked signals near four GHz. This range of frequencies is known as "C-band." Each range of frequen-

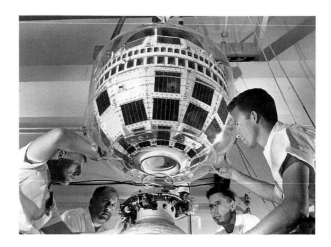

Technicians attaching the Telestar satellite to
a Delta rocket for launch
Photo courtesy of AT and T

cies is subdivided into specific channels, which, in the case of C-band, are each from 36 to 72 MHz wide. A single analog television transmission may occupy enough bandwidth to fully utilize a single 36 MHz channel. Hundreds or thousands of voice or data signals requiring far less bandwidth would fit on the same channel. In the 1980's a new generation of satellites using bandwidths of 11 to 12 GHz (uplink) and 14 GHz (downlink) came into use. The "Ku-band" does not require as much power to be transmitted clearly, thereby permitting the use of small (and less expensive) earth stations for uplink and downlink. With the introduction of the Ku-band, television entered the era of live news—satellite news gathering (SNG)—as Ku-band satellites made it easy to uplink television signals with a portable dish from the scene of a breaking news story. Television news has also made some use of another satellite technology, remote sensing, using pictures taken by satellites to illustrate or verify news stories.

In the late 1970s, with the satellite distribution of Home Box Office, home satellite dishes, or "television receive only" (TVRO), became popular for people out of reach of cable television. Later, direct satellite broadcasting (DBS) to small home dishes became possible through the use of these higher frequencies. Since 1988 DBS has been heavily used in Europe, and it is rapidly gaining popularity in the United States. Overuse of the C- and Ku-bandwidths and the desire for even greater signal strength is leading to new satellites that use other areas of the radio spectrum. A typical communications satellite launched in the early 1990s has a mix of C- and Ku-band transponders, and is capable of relaying over 30,000 voice or data circuits and four or more television transmissions. Telephony and television use roughly equivalent portions of available satellite capacity, but the demand for DBS has led to a number of satellites dedicated to TV transmission.

Like other communications technologies, the satellite industry has embraced digitalization and signal compression as a means of maximizing the use of limited bandwidth. By converting analog signals to digital signals, less bandwidth is required, and digital signals can be broken into smaller pieces for transmission through bits of available bandwidth, and reassembled at the point of reception. Compression eliminates otherwise redundant portions of a television transmission, allowing for a signal to be sent using far less bandwidth. Encryption, or scrambling, of satellite television signals is now becoming common to ensure that only customers who have bought or rented a decoder can receive transmissions. Even inter-company television feeds via satellite, such as daily feeds to broadcasters from television news agencies, are being encrypted to prevent unauthorized use. Typical television transmissions via satellite in the 1990s are digital, and are often compressed and encrypted. Compression technology is expected to considerably increase the number of DBS services available.

Some developing countries have demonstrated success in using satellite delivered television to provide useful information to portions of their populations out of reach of terrestrial broadcasting. In 1975, an experimental satellite communications project called SITE (Satellite Instructional Television Experiment) was used to bring informational television programs to rural India. The project led to Indian development of its own satellite network. China has also embarked on a ambitious program of satellite use for development, claiming substantial success in rural education.

STAR-TV, controlled by media mogul Rupert Murdoch, transmits television programming over much of Asia and has forced governments worldwide to reevaluate their stance on issues of national sovereignty and control of incoming information. STAR-TV reaches over 50 countries and potentially half of the world's population—far more than any other satellite television service (though it is technically not DBS, still requiring larger dishes). A slew of contentious political and cultural issues have resulted. Murdoch dropped BBC World Service Television from his STAR-TV program lineup as a concession to the Chinese government. Other governments have complained about the unrestricted importation of news presented from an Anglo-American viewpoint, though their concerns about political consequences are often couched in terms of protecting local culture. Reports of disruptions to local cultures stemming from international satellite broadcasting are widespread.

In all these instances satellite technology has called into question conventional notions of the nation state. Geographic borders may be insufficient definitions of culture and nationality in an era of electronic information, beamed from multiple sources into the sky, and down again into almost any location.

—Chris Paterson

FURTHER READING

Akwule, R. *Global Telecommunications: Their Technology, Administration, and Policies*. New York: Free Press, 1992.

Breeds, John, editor. *The Satellite Book: A Complete Guide to Satellite TV Theory and Practice*. Cricklade, Wilts, England: Swift Television Publications, 1994.

Chippindale, Peter. *Dished!: The Rise and Fall of British Satellite Broadcasting*. London: Simon and Schuster, 1991.

Clarke, Neville, and Edwin Riddell. *The Sky Barons*. London: Methuen, 1992.

Collins, Richard. "The Language of Advantage: Satellite Television in Western Europe." *Media, Culture and Society* (London), July 1989.

———. *Satellite Television in Western Europe*. London: John Libbey, 1992.

Frederick, Howard. *Global Communication and International Relations*. Belmont, California: Wadsworth, 1993.

Goldstein, I. "Broadcasting International Crisis: Retrospect and Prospects" *Journal of International Communications*, 1994.

Grant, August, editor. *Communication Technology Update*. Austin, Texas: Technology Futures, 1995.

Lacy, Stephen. "Use of Satellite Technology in Local Television News." *Journalism Quarterly* (Urbana, Illinois), Winter 1988.

Long, Mark *World Satellite Almanac*. Indianapolis, Indiana: Howard W. Sams, 1987.

Miller, M., B. Vucetic, and L. Berry, editors. *Satellite Communications: Mobile and Fixed Services*. Boston, Massachusetts: Kluwer Academic Publishers, 1993.

Stewart, M. LeSueur. *To See the World: The Global Dimension in International Direct Television Broadcasting by Satellite*. Dordrecht, Netherlands: M. Nijhoff, 1991.

Tefft, Sheila. "Satellite Broadcasts Create Stir Among Asian Regimes." *Christian Science Monitor* (Boston, Massachusetts), 8 December 1993.

Weid, Denis von der. *Development, Democracy, and Outer Space*. Geneva: United Nations Non-Governmental Liaison Services, 1992.

See also Ancillary Markets; Association of Independent Television Stations; British Sky Broadcasting; Cable Networks; Cable News Network; Channel One; Copyright Law and Television; Communication Satellite Corporation; Development Communication; Digital Television; Direct Broadcast Satellite; Distant Signal; European Broadcast Union; European Commercial Broadcasting Satellite; European Union: Television Policy; Federal Communications Commission; First People's Television Broadcasting in Canada; Geography and Television; Home Box Office; Midwest Video Case; International Telecommunication Union; Knowledge Network; Medical Video; Microwave; Movies on Television; Murdoch, Rupert; Narrowcasting; National Cable Television Association; News Corporation, Ltd.; Olympics and Television; Pay Cable; Pay Television; Pay-Per-View Cable; Public Access Television; Scrambled Signal; Star-TV; Space Program and Television; Telcos; Television Technology; Translators; Turner Broadcasting Systems; United States: Cable; U.S. Policy: Telecommunications Act of 1996

SATURDAY NIGHT LIVE

U.S. Comedy Variety Program

Saturday Night Live first aired on 11 October 1975 on NBC and has continued since to hold that spot in the line-up despite major cast changes, turmoil in the production offices and variable ratings. A comedy-variety show with an emphasis on satire and current issues, the program has been a staple element of NBC's dominance of late-night programming since its inception.

The program was developed by Dick Ebersol with producer Lorne Michaels in 1975 as a result of NBC's search for a show for its Saturday late night slot. The network had long enjoyed dominance of the weekday late night slot with *The Tonight Show* and sought to continue that success in the unused weekend time period. With the approval of Johnny Carson, whose influence at the network was strong, Ebersol and Michaels debuted their show, which was intended to attract the 18-to-34 age demographic.

The regulars on the show have always been relative unknowns in the comedy field. The first cast (the Not Ready for Prime Time Players) included Chevy Chase, Dan Aykroyd, John Belushi, Jane Curtin, Gilda Radner, Laraine Newman and Garrett Morris, all of them from the New York and Toronto comedy scenes. Featuring a different guest host each week (comedian George Carlin was the first) and a different musical guest as well, the programs reflected a non-traditional approach to television comedy from the start. The cast and writers combined the satirical with the silly and non-sensical, not unlike *Monty Python's Flying Circus*, one of Michaels' admitted influences.

The program was produced live from NBC's studio 8-H for 90 minutes. This difficult schedule and pressure-filled production environment has resulted in some classic comedy sketches and some abysmally dull moments over the years. Creating comedy in such a situation is difficult at best and the audience was always aware when the show was running dry (usually in the last half hour). But this sense of the immediate and the unforeseen also gave the show its needed edge. By returning to TV's live roots, *Saturday Night Live* gave its audiences an element of adventure with each program. It acquainted the generations who never experienced live television programming in the 1950s with the sense of theatre missing from pre-recorded programming.

For the performers, crew and writers, the show was a test of skill and dedication. The show has undergone several major changes since its beginning. The most obvious of these were the cast changes. *SNL*'s first "star," Chevy Chase, left the show in the second season for Hollywood. Aykroyd and Belushi followed in 1979. The rest of the original cast, including Bill Murray who replaced Chase, left when Lorne Michaels decided to leave the show after the 1979–80 season. Michaels' departure created wide-spread doubt about the viability of the show without him and his cast of favorites. Jean Doumanian was chosen as producer and her tenure lasted less than a year. With the critics attacking the show's diminished satirical edge and the lackluster replacement performers, NBC enticed Ebersol to return as producer in the spring of 1981. Ebersol managed to attract some of the original staff for the 1981–82 season, particularly writer Michael O'Donoghue. With the addition of Eddie Murphy, the show began to regain some of its strength, always based in its focus on a young audience and the use of relevant material.

Michaels rejoined the show as producer in 1985 and oversaw a second classic period of *Saturday Night Live*. With talented performers such as Dana Carvey, Jon Lovitz, Jan Hooks and Phil Hartman, the program regained much of its early edge and attitude. But the nature of the the the program is that the people who make it funny (the performers and writers) are the ones who tend to move on after a few years of the grind of producing a weekly live show. As the program moved into the 1990s, this trend still affected the quality. But Michaels' presence established a continuity which reassured the network and provided some stability for the audience.

From the beginning, *Saturday Night Live* provided America with some of its most popular characters and catch-phrases. Radner's Roseanne Roseannadana ("It's always something") and Emily Litella ("Never mind"), Belushi's Samurai, Aykroyd's Jimmy Carter, Murphy's Mr. Robinson, Billy Crystal's Fernando ("You look mahvelous"), Martin Short's Ed Grimley, Lovitz's pathological liar, Carvey's Church Lady ("Isn't that special?") and Carvey and Kevin Nealon's Hans and Franz have all left marks on popular culture. The program's regular news spot has been done by Chase, Curtin, Aykroyd, Nealon and Dennis Miller, among others and, at its best, provided sharp comic commentary on current events. It was particularly strong with Miller as the reader.

Saturday Night Live has seen many of its cast members move on to success in other venues. Chase, Aykroyd, Murray, Murphy and Crystal have all enjoyed considerable movie success. Short, Lovitz, Carvey, Jim Belushi, Adam Sandler, Chris Farley and Joe Piscopo have been mildly successful in films. Curtin, Julia Louis-Dreyfuss, Hooks and Phil Hartman moved on to other television shows.

As a stage for satire, few other American programs match *Saturday Night Live*. As an outlet for current music, the show has featured acts from every popular musical genre and has hosted both old and new artists (from Paul Simon, the Rolling Stones and George Harrison to R.E.M. and Sinead O'Connor). Due to its longevity, *SNL* has crossed generational lines and made the culture of a younger audience available to their elders (and the opposite is also true). Ultimately, *Saturday Night Live* must be considered one of the most distinctive and significant programs in the history of U.S. television.

—Geoffrey Hammill

Third-season cast of Saturday Night Live

ANNOUNCER
Don Pardo (1975–81, 1982–)

REGULAR PERFORMERS
Chevy Chase (1975–76)

Albert Brooks (1975–76)
Jim Henson's Muppets (1975–76)
John Belushi (1975–79)
Dan Aykroyd (1975–79)
Gilda Radner (1975–80)

Garrett Morris (1975–80)
Jane Curtin (1975–80)
Laraine Newman (1975–80)
Gary Weis (1976–77)
Bill Murray (1977–80)
Don Novello (1978–80, 1985–86)
Paul Shaffer (1978–80)
Al Franken (1979–80, 1988–)
Tom Davis (1979–80, 1988–)
Denny Dillon (1980–81)
Gilbert Gottfried (1980–81)
Gail Matthius (1980–81)
Joe Piscopo (1980–84)
Ann Risley (1980–81)
Charles Rocket (1980–81)
Eddie Murphy (1981–84)
Robin Duke (1981–84)
Tim Kazurinsky (1981–84)
Tony Rosato (1981–82)
Christine Ebersole (1981–82)
Brian Doyle-Murray (1981–82)
Mary Gross (1981–85)
Brad Hall (1982–84)
Gary Kroeger (1982–85)
Julia Louis-Dreyfus (1982–85)
Jim Belushi (1983–85)
Billy Crystal (1984–85)
Christopher Guest (1984–85)
Harry Shearer (1984–85)
Rich Hall (1984–85)
Martin Short (1984–85)
Pamela Stephenson (1984–85)
Anthony Michael Hall (1985–86)
Randy Quaid (1985–86)
Joan Cusack (1985–86)
Robert Downey, Jr. (1985–86)
Nora Dunn (1985–90)
Terry Sweeney (1985–86)
Jon Lovitz (1985–90)
Damon Wayans (1985–86)
Danitra Vance (1985–88)
Dennis Miller (1985–90)
Dana Carvey (1986–93)
Phil Hartman (1986–94)
Jan Hooks (1986–91)
Victoria Jackson (1986–92)
A. Whitney Brown (1986–91)
Kevin Nealon (1986–91, 1993–)
Mike Myers (1989–)
G. E. Smith and the Saturday Night Live Band (1989–)
Chris Farley (1990–)
Chris Rock (1990–93)
Julia Sweeney (1990–94)
Ellen Cleghorne (1991–)
Siobhan Fallon (1991–93)
Tim Meadows (1991–)

Adam Sandler (1991–)
David Spade (1991–)
Rob Schneider (1991–94)
Melanie Hutshell (1991–94)
Beth Cahill (1991–93)
Sarah Silverman (1993–94)
Norm MacDonald (1993–)
Jay Mohr (1993–)
Michael McKean (1994–)
Chris Elliott (1994–)
Janeane Garofalo (1994–)
Mark McKinney (1994–)
Laura Kightlinger (1994–)
Molly Shannon (1994–)
Morwenna Banks (1994–)

PRODUCERS   Lorne Michaels, Jean Doumanian, Dick Ebersol

PROGRAMMING HISTORY

• NBC

October 1975–	Saturday 11:30-1:00 A.M.
October 1979–March 1980	Wednesday 10:00-11:00
March 1980–April 1980	Friday 10:00-11:00

FURTHER READING

Barol, Bill, and Jennifer Foote. "Saturday Night Lives!" *Newsweek* (New York), 25 September 1989.

Beatts, Anne, and John Head, editors. *Saturday Night Live.* New York: Avon, 1977.

Cader, Michael, editor. *Saturday Night Live: The First Twenty Years.* Boston, Massachusetts: Cader Books, 1994.

Corliss, Richard. "Party on, Wayne—from TV to Movies." *Time* (New York), 2 March 1992.

Cullingford, Elizabeth Butler. "Seamus and Sinead: From 'Limbo' to *Saturday Night Live* by Way of Hush-A-Bye-Baby." *Colby Quarterly* (Waterville, Maine), March 1994.

Greenfield, Jeff. "Live from New York." *The New York Times,* 19 April 1993.

Harkness, John. "Out of the Joke Box." *Sight and Sound* (Leicestershire, England), March 1994.

Hill, Doug, and Keff Weingrad. *Saturday Night: A Backstage History of Saturday Night Live.* New York: Beech Tree, 1986.

Myers, Mike, and Robin Ruzan. *Wayne's World: Extreme Close-Up.* New York: Hyperion, 1991.

Partridge, Marianne, editor. *Rolling Stone Visits Saturday Night Live.* Garden City, New York: Dolphin, 1979.

Saltzman, Joe. "The Agony and the Ecstasy: Live Television Comedy." *USA Today Magazine* (New York), November 1987.

See also Variety Programs

SAUNDERS, JENNIFER

British Actor

Since the early 1980s, Jennifer Saunders has been a popular and influential figure in British television comedy. Her success stems from her involvement as both a performer in, and writer of, several comedy shows which have been heralded as innovative by critics and received as hugely entertaining by audiences.

Saunders established her career as part of a double act with Dawn French on the live comedy circuit in the late 1970s. She and French, who have remained collaborators on many projects since, made their initial impact while on tour in 1981 with the Comic Strip, a group consisting of several young comedians performing an alternative, innovative form of comedy. The group were rapidly transferred to television, appropriately making their debut on Channel Four's opening night in November 1982. Throughout the 1980s, the original members appeared in *The Comic Strip Presents...* in which they wrote, directed, and performed a series of narratives satirising a variety of genre themes. The programme set a precedent for the so-called alternative comedy of the 1980s, won critical approval, and was awarded a Golden Rose at the Montreux Festival.

Saunders and French's role within this group was particularly significant in that the two succeeded in providing much more complex and interesting female characters than had hitherto been offered by television comedy. They placed their characters in opposition to the traditional representations of women in British television comedy—such as the sexual accessories of *The Benny Hill Show*; the domesticated, subservient wife of *The Good Life*; and the nag of *Fawlty Towers*. Saunders and French's very presence in *The Comic Strip Presents...* was a timely intrusion into a realm of comedy that had previously been the exclusive domain of male performers, from *Monty Python* to the double acts of the 1970s: *Morecombe and Wise*, and *Little and Large*.

The autonomy women were gaining was confirmed in *French and Saunders*. This show, the first series of which was screened on the BBC in 1987, presented the pair as partners combining stand-up and sketches. *French and Saunders* offered a uniquely feminine version of British comedy (unique, with the notable exception of *Victoria Wood: As Seen on TV*, first screened in 1985). Their writing and acting focused directly, and with hilarious results, on female experience. Many of the scenes worked to reinforce the centrality of women's talk and to parody the position and representations of women in the media.

It was out of a *French and Saunders* sketch that Saunders conceived of and developed her most prolific work, *Absolutely Fabulous*. Saunders has written and starred in three six-part series of *Absolutely Fabulous* (BBC, 1992, 1994, and 1995) which have achieved uniformly high viewing figures as well as critical acclaim. In some respects a domestic sitcom, *Absolutely Fabulous* satirises the matriarchal household of fashion P.R. executive, Edina Monsoon

Jennifer Saunders
Photo courtesy of Jennifer Saunders

(Saunders), and the women around her, including her unruly best friend, Patsy (Joanna Lumley), and long-suffering daughter, Saffron (Julia Sawalha). Because *Absolutely Fabulous* remains an unusual example of a peak-time situation comedy written by women, with a predominantly female cast and a specific address to a female audience, it provides rare viewing pleasures of self-recognition and humour to women. In addition to having feminist concerns at the core of its structure and themes, it stresses the artificiality surrounding "womanliness," and celebrates gender as a complex social and cultural construction.

In terms of her writing and performance, Saunders helped to raise the profile of female comedians in television, leading the way for others, such as Jo Brand, and Dawn French in her solo series, *Murder Most Horrid*. Saunders took on her first non-comedy role for a BBC drama, *Heroes and Villains* (1995), a period piece based on the true life of Lady Hester Stanhope, an eccentric 19th-century traveller. As well as revealing a further talent for dramatic acting, the show crystallises Saunders' TV persona, and arguably her role in British television, as an independent and powerful woman.

—Nicola Foster

JENNIFER SAUNDERS. Born in Sleaford, Lincolnshire, England, 12 July 1958. Attended Central School of Speech and Drama. Married: Adrian Edmondson; children: Ella, Beattie, and Freya. Formed cabaret partnership with comedian Dawn French, the Comedy Store, London; appeared in the *Comic Strip* series, early 1980s, and subsequently in *French and Saunders* sketch show and, without French, in *Absolutely Fabulous*. Address: Peters Fraser and Dunlop, Fifth Floor, The Chambers, Chelsea Harbour, Lots Road, London SW10 0XF, England.

TELEVISION SERIES

1982–92	*The Comic Strip Presents* (*Five Go Mad in Dorset, Five Go Mad on Mescalin, Slags, Summer School, Private Enterprise, Consuela; Mr. Jolly Lives Next Door, Bad News Tour, South Atlantic Raiders, G.L.C., Oxford, Spaghetti Hoops, Le Kiss, Wild Turkey, Demonella, Jealousy, The Strike*)
1985	*Happy Families*
1985–86	*Girls on Top* (also co-writer)
1987–	*French and Saunders*
1992–95	*Absolutely Fabulous*

FILM

The Supergrass, 1985.

PUBLICATION

Absolutely Fabulous. New York: Pocket, 1995.

See also *Absolutely Fabulous*; British Programming; French, Dawn; Wood, Victoria

SAWYER, DIANE

U.S. Broadcast Journalist

Diane Sawyer, co-anchor on ABC News *PrimeTime Live*, is one of broadcast journalism's most prominent and successful female presences. Sawyer began her career as a weather reporter on a Louisville, Kentucky, television station. In 1970, she took a job at the White House on the staff of Presidential Press Secretary Ron Ziegler. She continued her career as a press aide during the Nixon administration until 1974, and then assisted the former President with the preparation of his memoirs. The transition to broadcast journalism was made in 1978, when she joined CBS News as a reporter in the Washington bureau. When Sawyer accepted the job of State Department correspondent for CBS News (1978-81), she began a career as a popular figure in television journalism: she was the co-anchor of *CBS Morning News* (from 1981), the co-anchor of *CBS Early Morning News* (1982-84), and the first woman on the network's flagship public affairs program, *60 Minutes* (1984-89). Sawyer left this successful position at CBS to sign a multi-year contract to co-anchor *PrimeTime Live* on ABC News with Sam Donaldson in 1989.

In addition to her impressive professional resume, Sawyer is known for a variety of individual characteristics. Her intelligent reporting and tenacious coverage of the Three Mile Island crisis assisted her in garnering heavy journalistic assignments which at the time were considered a challenge to male colleagues working in early morning news. At *CBS Morning News* she earned a reputation for skilled reporting as well as her ability to help increase ratings. Her commanding delivery helped edge the network's program closer to its rivals in the Nielsen ratings. Her presence and teamwork with Bill Kurtis gave CBS its first healthy ratings in this time slot in three decades. High-profile assignments as correspondent of *60 Minutes* established her as a national figure: viewers admired her equally for her personality and her talents as an investigative reporter. Sawyer's skill contributed to *PrimeTime Live's* success and its distinct style.

In the fall of 1994, Sawyer signed a $7 million contract, making her one of the highest paid women in broadcast news. Though one critique characterized her as "the warm ice maiden," such views may reflect forms of professional

Diane Sawyer
Photo courtesy of Diane Sawyer

jealously. Margo Howard, entertainment critic of *People* magazine, writes "...she got to the top with a formidable blend of smarts, drive, [warmth], and earnestness." Another characterization as "a girl who is one of the boys" points to Sawyer's authoritative, intelligent, enterprising manner.

A frequently referred to aspect of Sawyer's work is her willingness to move between two styles—that of a tabloid journalist and the "legitimate" journalist. Diligent reporting pieces coexist with celebrity interviews, such as her coverage of the Iranian hostage crisis and her interview with Michael Jackson and Lisa Marie Presley. Her "softball" questions to Tonya Harding during the 1994 Olympics, her low-camp interview with Marla Maples (asking whether Donald Trump was "really the best sex" she ever had), and her brief, heavily promoted and news-free encounter with Boris Yeltsin in the Kremlin during the 1987 coup contribute to the "tabloid" label.

Though the critiques are valid to some degree, Sawyer's distinctive personality has helped *PrimeTime Live* move toward unqualified success and produce millions of dollars in profits for ABC. All four major networks have sought her services, and she has become a "brand name," a person the viewers remember, and a television personality who can deliver ratings. She remains one of the most visible news figures in U.S. television prime-time hours.

—Lynn T. Lovdal

DIANE SAWYER. Born in Glasgow, Kentucky, U.S.A., 22 December 1945. Educated at Wellesley College, Wellesley, Massachusetts, B.A. 1967. Married: Mike Nichols, 1988. Reporter, WLKY-TV, Louisville, Kentucky, 1967–70; ad-ministrator, White House press office, 1970–74; researcher for Richard Nixon's memoirs, 1974–78; general assignment reporter, then U.S. State Department correspondent, CBS News, 1978–89; ABC News, since 1989. Member: Council on Foreign Relations. Recipient: two Peabody Awards; Robert F. Kennedy Award; nine Emmy Awards. . Address: ABC News, 147 Columbus Avenue, New York City, New York 10023-5904, U.S.A.

TELEVISION

1978–81	*CBS Evening News* (correspondent)
1981–84	*CBS Morning News* (co-anchor)
1982–84	*CBS Early Morning News* (co-anchor)
1984–89	*60 Minutes* (correspondent and co-editor)
1989–	*PrimeTime Live* (co-anchor)

FURTHER READING

Auletta, Ken. "Promise Her the Moon." *The New Yorker* (New York), 14 February 1994.

Exley, Frederick. "If Nixon Could Possess the Soul of this Woman, Why Can't I? The Decade's Last Piece about Diane Sawyer." *Esquire* (New York), December 1989.

Unger, Arthur. "Diane Sawyer: 'The Warm Ice Maiden'" (interview). *Television Quarterly* (New York), Spring 1992.

Zoglin, Richard. "Star Power: Diane Sawyer, with a New Prime-time Show and a $1.6 Million Contract, Is Hot. But Are Celebrity Anchors Like Her Upstaging the News?" *Time* (New York), 7 August 1989.

See also News, Network; *60 Minutes*

SCALES, PRUNELLA

British Actor

Prunella Scales is an established star of British situation comedy, although she has also won praise in a wide range of other productions, including drama for television and stage. Television viewers are most likely to associate her, however, with the classic John Cleese comedy *Fawlty Towers*, in which she played the unflappable Sybil to Cleese's appallingly inept hotelier Basil Fawlty.

As Sybil Fawlty, the archetypal gossipy and battle-hardened nagging wife who in her husband's eyes was more of a hindrance than a help (though in truth she spent much of her time smoothing, with carefully rounded vowels, the ruffled feathers of guests her husband had offended), Scales was deemed perfect. Employing all the skills she had acquired from her early experience in repertory theater and subsequently with the Royal Shakespeare Company and other leading troupes, she easily countered the manic ranting of her screen husband, ensuring that life—such as it was—could carry on at Fawlty Towers. When not seeing to her monstrous coiffure, her Sybil took desultory pleasure in providing her husband with new irrita-tions, usually guaranteed to send him into paroxysms of helpless rage. As a mark of the degree to which the performances of Scales and Cleese were essential to the success of the series—widely judged a classic of television comedy—an attempt to make a U.S. version under the title *Amanda's*, with a cast headed by Bea Arthur of *Golden Girls* fame, was a total failure (even though, in desperation, some episodes were duplicated word for word).

Scales had previously performed as bus conductress Eileen Hughes in *Coronation Street*, and also as co-star of the series *Marriage Lines*, a relatively conventional husband-and-wife situation comedy in which she was paired with Richard Briers. As Kate Starling in the latter production, she charted the ups and downs experienced by typical newly-weds in the 1960s, wrestling with a range of more or less mundane financial and domestic problems (later complicated by the arrival of their baby).

In the wake of the huge success of *Fawlty Towers*, Scales enjoyed further acclaim from critics and audiences alike in the role of the widowed Sarah in Simon Brett's *After Henry*, a compassionate and often hilarious comedy that was equally successful as a series for radio and subsequently on television. When not contemplating the future course of her life as the widowed mother of a teenage daughter, she indulged in entertaining sparring with "mother," played by the redoubtable Joan Sanderson.

Other highlights of Scales' career have included her performance as Elizabeth Mapp in the television version of E. F. Benson's Edwardian *Mapp and Lucia* stories, in which she was cast opposite the equally distinguished Geraldine McEwan. Another triumph was her enthralling impersonation of Queen Elizabeth II in a much-acclaimed television version of Alan Bennett's celebrated play, *A Question of Attribution*, which concerned the relationship between the monarch and her art adviser Anthony Blunt, who was fated to be exposed as a spy for Communist Russia. On the stage, meanwhile, she added another monarch to her list of credits when she impersonated Queen Victoria in her own one-woman show.

Considered one of the most technically proficient actresses of stage and screen of her generation, as well as an accomplished occasional director, Scales has continued to divide her time between television and the theater throughout her career, sometimes appearing in partnership with her real-life husband actor Timothy West. In 1996, in recognition of her skills, she was invited to share some of her secrets concerning acting as part of a short series of master classes on the art of comedy performance.

—David Pickering

Prunella Scales
Photo courtesy of Prunella Scales

PRUNELLA SCALES (Prunella Margaret Rumney Illingworth). Born in Sutton Abinger, Surrey, England, 22 June 1932. Attended Moira House, Eastbourne; trained for stage at the Old Vic Theatre School, London, and the Herbert Berghof Studio, New York. Married: Timothy West, 1963; children: Samuel and Joseph. Started in repertory theater in Huddersfield, Salisbury, Oxford, Bristol Old Vic, and elsewhere; performed in theater seasons at Stratford-upon-Avon and Chichester Festival Theatre, 1967–68; also acted on London stage; had greatest success on television as Sybil in *Fawlty Towers*, 1975; subsequently appeared in numerous sitcoms and plays; also teaches and directs theater. Companion of the British Empire, 1992. Address: Jeremy Conway, 18-21 Jermyn Street, London SW1Y 6HP, England.

TELEVISION SERIES

1963–66	*Marriage Lines*
1975, 1979	*Fawlty Towers*
1985–86	*Mapp and Lucia*
1988, 1990	*After Henry*
1994	*The Rector's Wife*
1995	*Searching*

TELEVISION SPECIALS

1953	*Laxdale Hall*
1954	*What Every Woman Wants*
1954	*The Crowded Day*
1959	*Room at the Top*
1962	*Waltz of the Toreadors*
1976	*Escape from the Dark*
1977	*The Apple Cart*
1979	*Doris and Doreen*
1982	*A Wife Like the Moon*
1982	*Grand Duo*
1982	*Outside Edge*
1983	*The Merry Wives of Windsor*
1985	*Absurd Person Singular*
1987	*The Index Has Gone Fishing*
1987	*What the Butler Saw*
1991	*A Question of Attribution*
1994	*Fair Game*
1995	*Signs and Wonders*

FILMS

Hobson's Choice, 1953; *The Hound of the Baskervilles*, 1978; *The Boys from Brazil*, 1978; *The Wicked Lady*, 1982; *Wagner*, 1983; *The Lonely Passion of Judith Hearne*, 1987; *Consuming Passions*, 1988; *A Chorus of Disapproval*, 1989; *Howards End*, 1992; *Second Best*; *Wolf*; *An Awfully Big Adventure*.

RADIO

After Henry.

SCHAFFNER, FRANKLIN

U.S. Director

Franklin Schaffner, one of several prominent directors during U.S. television's "Golden Age," worked in such prestigious anthology series as *Studio One* (CBS); *The Kaiser Aluminum Hour* (NBC), *Playhouse 90* (CBS); *The DuPont Show of the Week* (NBC), the Edward R. Murrow series, *Person to Person* (CBS), and the dramatic series *The Defenders* (CBS). Schaffner later became known as an "actor's director," but his television work is known primarily for his unique use of the camera.

Schaffner attended Franklin and Marshall College in Lancaster, where he majored in government and english. A prize-winning orator, Schaffner appeared in several university productions and also worked part-time as an announcer at local radio station WGAL. His plans to attend Columbia Law School were interrupted by his enlistment in the U.S. Navy during World War II, during which he served with amphibious forces in Europe and North Africa and, later, with the Office for Strategic Services in the Far East.

After the war, Schaffner first sought work as an actor. He was eventually hired as a spokesperson and copywriter for Americans United for World Government, a peace organization. During this period, Schaffner met ABC Radio Vice President Robert Saudek and worked as a writer for Saudek's radio series *World Security Workshop*. For that series, Schaffner wrote "The Cave," which was the series' final broadcast (8 May 1947). Schaffner's experience on this series encouraged him to pursue a career in broadcasting.

Schaffner was hired as an assistant director on the radio documentary series *The March of Time* for $35 per week. His work brought him to the attention of Robert Bendick, director of television news and special events for CBS. Bendick hired Schaffner in April 1948 as director of Brooklyn Dodgers baseball as well as other sporting events and public service programs. Schaffner's experience with the spontaneity and immediacy of live special events made him a logical choice as one of three directors for the 1948 Democratic and Republican political conventions held in Philadelphia.

By 1949, Schaffner was ready for the challenge of directing live dramatic programs. After directing *Wesley* (CBS), a situation comedy produced by Worthington Miner about a precocious twelve-year-old and his family, Schaffner alternated directing assignments with Paul Nickell on Miner's live anthology series *Studio One* (CBS). On the series, Schaffner directed adaptations of classics as well as original productions, including the series' first color telecast, *The Boy Who Changed the World* (18 October 1954). At a time when other networks used static cameras, Schaffner utilized a moving camera with long, graceful tracking shots. In addition to hiding the limitations of the studio set, Schaffner's camera work drew audiences into the action of the play. In *Twelve Angry Men* (20 September 1954), Schaffner designed a 360$ shot that required orchestrated moves of the set's walls during the shot. Schaffner won the 1954 Emmy for Best Direction for his work on *Twelve Angry Men*.

While working on the *Studio One* series, Schaffner drew on his news and public affairs experience to serve as producer and studio director for the Edward R. Murrow interview program, *Person to Person* (CBS, 1953-1961). Although the initial episodes utilized static camera set-ups for the remote interviews, Schaffner later incorporated tracking cameras

Franklin Schaffner
Photo courtesy of Wisconsin Center for Film and Theater Research

that moved with guests to show their home and activities. Schaffner worked on the series until 1957, when more of his work originated from Los Angeles.

Schaffner utilized his news experience once again for *A Tour of the White House with Mrs. John F. Kennedy* (NBC, 14 February 1962). Schaffner's moving camera and unique camera angles provided viewers with an intimate look at the White House renovation. He won a 1962 Directorial Achievement Award from the Directors Guild of America for his work on the program.

One of Schaffner's best-known works is the production of *The Caine Mutiny Court Martial* for *Ford Star Jubilee* (CBS, 19 November 1955), which was broadcast from the new CBS state-of-the-art facilities at Television City in Los Angeles. The static action of the play is kept moving by Schaffner's mobile camera and dramatic crane shots. Schaffner was awarded two Emmys for his work on the teleplay: one for best director and another for best adaptation (with Paul Gregory). The show was originally broadcast in color, but only black-and-white kinescopes survive.

After years as a director of live television dramas, Schaffner directed various episodes of the dramatic series *The Defenders* (CBS, 1961–65), produced by Herbert Brodkin and written by Reginald Rose. The series originated as a two-part episode on *Studio One* in 1957, directed by Robert Mulligan. Schaffner used film editing to create montages of busy New York scenes and unusual camera angles to concentrate on the characters. Schaffner won his fourth Emmy for his work on the series.

Schaffner left television to direct and produce feature films. His film work includes *Planet of the Apes* (1968), *Patton* (1970, for which he received the Oscar and Directors Guild awards for Best Director), *Nicholas and Alexandra* (1971), *Papillon* (1973), and *The Boys from Brazil* (1978). In 1977, Schaffner's alma mater, Franklin and Marshall College, established the Franklin J. Schaffner Film Library and presented the director with an honorary Doctor of Humane Letter. Schaffner died of cancer on 2 July 1989.

—Susan R. Gibberman

FRANKLIN J. SCHAFFNER. Born in Tokyo, Japan, 30 May 1920. Graduated from Franklin and Marshall College, Lancaster, Pennsylvania, U.S.A., 1942; studied law at Columbia University, New York, New York. Married Helen Jean Gilchrist, 1948; children: Jenny and Kate. Served in U.S. Navy, 1942–46. Began television career as assistant director, *March of Times* documentary series, 1947–48; television director, CBS, including such programs as *Studio One, Ford Theater* and *Playhouse 90*, 1949–62; formed Unit Four production company with Worthington Miner, George Roy Hill, and Fielder Cook, 1955; directed *Advise and Consent* on Broadway, 1960; signed three-picture deal for 20th Century Fox and directed first feature, 1961; TV counselor to President Kennedy, 1961–63; president, Gilchrist Productions, 1962–68; president, Franklin Schaffner Productions,

1969–89. Member: Directors Guild of America, president 1987–89; National Academy of Television Arts and Sciences; Phi Beta Kappa; board member: Center Theater Group of the Music Center, Los Angeles; Academy of Motion Pictures Arts and Sciences, National Council of the Arts, Presidential Task Force on the Arts and Humanities; chairman, executive committee, American Film Institute. Recipient: Sylvania Award, 1953, 1954; Emmy Awards, 1954, 1955, 1962; Best Direction Award, Variety Critics Poll, 1960; Trustee Award (shared with Jacqueline Kennedy) for documentary *Tour of the White House*, American Academy of Television Arts and Sciences, 1962; Oscar for Best Director, 1970; Directors Guild Award, 1970. Died in Santa Monica, California, 2 July 1989.

TELEVISION SERIES (selection; director)

1949	*Wesley*
1949–56	*Studio One*
1950–51	*Ford Theater*
1953–61	*Person to Person*
1955–56	*Ford Star Jubilee*
1956–57	*Kaiser Aluminum Hour* (also producer)
1957	*Producer's Showcase*
1957–60	*Playhouse 90*
1959	*Ford Startime*
1961–65	*The Defenders*
1962–64	*DuPont Show of the Week* (also producer)

TELEVISION SPECIAL

1962	*A Tour of the White House with Mrs. John F. Kennedy*

FILMS (director)

A Summer World (incomplete), 1961; *The Stripper*, 1963; *The Best Man*, 1964; *The War Lord*, 1965; *The Double Man* (also actor), 1967; *The Planet of the Apes*, 1968; *Patton*, 1970; *Nicholas and Alexandra* (also producer), 1971; *Papillon* (also co-producer), 1973; *Islands in the Stream*, 1977; *The Boys from Brazil*, 1978; *Sphinx* (also executive producer), 1981; *Yes, Giorgio*, 1982; *Lionheart*, 1987; *Welcome Home*, 1989.

RADIO

World Security Workshop; The March of Time.

STAGE (director)

Advise and Consent, 1960.

PUBLICATIONS

"The TV Director: A Dialog," with Fielder Cook. In, Bluem, William A., and Roger Manvell, editors. *The Progress in Television*. New York: Focal, 1967.

"Interview," with Gerald Pratley. *Cineaste* (New York), Summer 1969.

"Interview," with R. Feiden. *InterView* (New York), March 1972.

"Interview," with R. Applebaum. *Films and Filming* (London), February 1979.

"Interview," with D. Castelli. in *Films Illustrated* (London), May 1979.

Worthington Miner: Interviewed by Franklin J. Schaffner. Metuchen, New Jersey: Scarecrow Press, 1985.

FURTHER READING

Cook, B. "The War Between Writers and the Directors: Part II: The Directors." *American Film* (Washington, D.C.), June 1979.

"Franklin J. Schaffner." *Kosmorama* (Copenhagen), Autumn 1977.

Geist, Kathe. "Chronicler of Power." *Film Comment* (New York), September/October 1972.

Kim, Erwin. *Franklin J. Schaffner.* Metuchen, New Jersey: Scarecrow Press, 1985.

Lightman, Herb. "On Location with *Islands in the Stream.*" *American Cinematographer* (Los Angeles), November 1976.

Sarris, Andrew. "Director of the Month—Franklin Schaffner: The Panoply of Power." *Show* (Hollywood), April 1970.

"TV to Film: A History, a Map and a Family Tree." *Monthly Film Bulletin* (London), February 1983.

Wilson, David. "Franklin Schaffner." *Sight and Sound* (London), Spring 1966.

See also *Defenders;* "Golden Age" of Television; *Person to Person; Playhouse 90; Studio One; Tour of the White House with Mrs. John F. Kennedy*

SCHORR, DANIEL

U.S. Broadcast Journalist

Daniel Schorr is an American television newsman whose aggressive investigative style of reporting made him, at various times in his career, the bane of the KGB, U.S. presidents from Dwight D. Eisenhower to Gerald Ford, CIA chiefs, television executives, and his fellow TV newsmen and women. In 1976 he, himself, became "the story" when he published a previously suppressed congressional report on CIA assassinations.

Schorr was born and brought up in New York City and did his apprenticeship in print journalism on his high school and college newspapers. During his college years he also worked on a number of small New York City papers, among them the *New York Journal-American*. Drafted in World War II, he served in Army Intelligence. Following the war he became a stringer for a number of U.S. newspapers and the Dutch News agency ANETA. His radio reports on the floods in Holland brought him to the attention of Edward R. Murrow, who hired him for CBS News in l953.

In 1955, Schorr was assigned to open the first CBS bureau in Moscow since 1947. His refusal to cooperate with Soviet censors soon earned him their disapproval, and when he returned home for a brief period at the end of 1957 the Soviets refused to permit him to return. For the next few years Schorr was a roving diplomatic correspondent. In 1959, he provoked the first in a long series of incidents that aroused the ire of various presidents. Schorr's report of the impending resignation of Secretary of State John Foster Dulles so irked President Eisenhower that he denied the report, only to have it confirmed by his press secretary a week later.

During the Kennedy administration, the President asked CBS to transfer Schorr, then the station's correspondent in West Germany, because he felt that Schorr's interpretations of American policy were pro-German. During the 1964 election, Schorr's report that the Republican-nominee

Senator Barry Goldwater had formed an alliance with certain right-wing German politicians and was thinking of spending some time at Adolf Hitler's famous Berchtesgaden retreat caused a furor, and Schorr was ordered to make a "clarification."

In 1966, Schorr returned to the United States without a formal assignment. He created his own beat, however, by

Daniel Schorr
Photo courtesy of National Public Radio

investigating the promise and reality of the "Great Society" for the *CBS Evening News.* In this role he turned in excellent reports on poverty, education, pollution, and health care. His interest in health care led to a provocative 1970 contribution to the documentary series, *CBS Reports.* The program, "Don't Get Sick in America," appeared as a book that same year from Aurora Publishers.

Schorr's muckraking reporting during the Nixon administration earned him a prominent place on Nixon's so-called "enemies list." In addition, his subsequent reporting on the Watergate scandal garnered him Emmys for outstanding achievement within a regularly scheduled news program in 1972, 1973, and 1974.

Following Nixon's resignation, Schorr was assigned to cover stories involving possible criminal CIA activities at home and abroad. He soon achieved a scoop based on a tip he received about an admission by President Ford regarding CIA assassination attempts. The comment had come in an off-the-record conversation with the editors of *The New York Times.* Schorr's report forced the Rockefeller Commission investigating the CIA to broaden its inquiry, and prompted an exclamation from former CIA chief Richard Helms, referring to him as "Killer Schorr."

Commenting on his journalistic method, more akin to print journalism than conventional television journalism, Schorr has said, "My typical way of operating is not to stick a camera and a microphone in somebody's face and let him say whatever self-serving thing he wants to say, but to spend a certain amount of time getting the basic information, as though I was going to write a newspaper story.... [I] may end up putting a mike in somebody's face, but it is usually for the final and hopefully embarrassing question."

Soon after making these remarks, Schorr found himself at the center of a huge controversy involving both journalistic ethics and constitutional issues. Schorr came into possession of the Pike Congressional Committee's report on illegal CIA and FBI activities. Congress, however, had voted not to make the report public. In hopes of being able to publish the report, Schorr contacted Clay Felker of the *Village Voice,* who agreed to pay him for it and to publish it. To Schorr's suprise, instead of supporting him, many of his colleagues and editorialists around the country excoriated him for selling the document. Making matters worse was Schorr's initial reaction, which was to shift suspicion from himself as the person who leaked the documents to his CBS colleague Lesley Stahl.

Schorr managed to turn opinion around when, after being subpoenaed to appear before a House Ethics Committee, he eloquently defended himself on the grounds that he would not reveal a source. While this put off the congressional bloodhounds, it certainly didn't satisfy some of the wolves at CBS, among whom was Chairman William S. Paley, who wanted Schorr fired. Schorr and CBS news executives resisted until the story of the internal dissension over Schorr's conduct broke during an interview he did with Mike Wallace on *60 Minutes.* As a result, Schorr resigned

from CBS News in September 1976. A year later, he wrote about it in his autobiographical account, *Clearing the Air.*

Subsequently Schorr toured on the lecture circuit, taught journalism courses, and wrote a syndicated newspaper column. In 1979, hoping to give his new Cable News Network (CNN) instant journalistic credibility, Ted Turner hired Schorr as a commentator. However, in 1985 CNN refused to renew his contract. Schorr commented at the time that he had been "forced out" because, "they wanted to be rid of what they considered a loose cannon." Since 1985, Schorr has been a senior news analyst for National Public Radio. His reporting and commentary are heard on *All Things Considered* and *Weekend Edition.*

Schorr represents the traditions of investigative print journalism transfered to the world of TV reporting. His work, though it has sometimes overstepped boundaries, is in vivid contrast to the often image-conscious attitudes of contemporary TV news.

—Albert Auster

DANIEL SCHORR. Born in New York City, U.S.A., 31 August 1916. Educated at the College of the City of New York, B.S. 1939. Married: Lisbeth Bamberger, 1967; children: Jonathan and Lisa. Served in U.S. Army, stationed at Camp Polk, Louisiana, and at Fort Sam Houston, Texas, 1943–45. Worked as a stringer for the Bronx *Home News,* the *Jewish Daily Bulletin,* and several metropolitan dailies, 1930s; assistant editor, Jewish Telegraphic Agency, 1939; worked for the *New York Journal-American,* 1940; New York news editor, ANETA (Dutch news agency), 1941–43, 1945–48; freelance journalist, 1948–53; Washington correspondent and special assignments, CBS News, Latin America and Europe, 1953–55; reopened CBS Moscow Bureau, 1955; roving assignments, United States and Europe, 1958–60; chief, CBS News Bureau, Germany, Central Europe, 1960–66; CBS News Washington correspondent, 1966–76; Regents professor, University of California at Berkeley, 1977; columnist, *Des Moines Register-Tribune* Syndicate, 1977–80; senior Washington correspondent, CNN, 1979–85; senior analyst, National Public Radio, since 1985. Member: American Federation of Radio-TV Artists; New York City Council on Foreign Relations. Recipient: Emmy Awards, 1972–74; Peabody Award for lifetime of uncompromising reporting of highest integrity, 1992; inducted into the Society of Professional Journalists Hall of Fame, 1991.

RADIO

National Public Radio shows, from 1985.

PUBLICATIONS

Don't Get Sick in America! Nashville: Aurora, 1970.
Clearing the Air. Boston: Houghton Mifflin, 1977.
"Introduction." *Taking the Stand: The Testimony of Lieutenant Colonel Oliver L. North.* New York: Pocket, 1987.

FURTHER READING

Boyer, Peter J. *Who Killed CBS? The Undoing of America's Number One News Network.* New York: Random House, 1988.

Carter, Bill. "Daniel Schorr Wins Top duPont-Columbia Journalism Award." *The New York Times*, 26 January 1996.

Smith, Sally Bedell. *In All His Glory: William S. Paley, the Legendary Tycoon and His Brilliant Circle.* New York: Simon and Schuster, 1990.

See also Cable News Network; Columbia Broadcasting System; News, Network

SCIENCE PROGRAMS

When most people consider the history and development of scientific television programming in the United States, they are quick to mention the popular 1950's show, *Watch Mr. Wizard*. This program was indeed one of the first attempts to bring science to the general public through the medium of television. Forty three years later, in 1994, Don Herbert, creator of the *Mr. Wizard* series, launched a new show entitled *Teacher to Teacher with Mr. Wizard*. The enduring image of Herbert as "Mr. Wizard" is a testament to the presence of science oriented programmmg throughout the history of television.

Early growth in the area of scientific television programmmg closely paralleled increasing public awareness of the impact of science and technology on everyday life in an era more completely defined by mass communication. As issues of science and public policy became intertwined, television was seen as the perfect vehicle through which to develop a "public understanding of science" (Tressel 25). Over the years, scientific television programming evolved to serve three primary goals—to entertain, to educate, and, ultimately, to bridge the gap between the general public and the scientific community. In order to service such goals, however, sustainable funding had to be secured.

Scientific television was a key element in the National Science Foundation's (NSF) early initiatives to promote a public understanding of science. Through station by station syndication, the NSF funded several short programs which aired on commercial television. In the 1970s, *Closeups*, produced by Don Herbert, introduced children to scientific concepts through everyday objects (Tressel 26). During this same period, Herbert also developed a syndicated scientific news report aimed at adults entitled *How About* (Tressel 32). Most recently, syndication has facilitated the entry of independently funded and produced scientific programs into commercial formats.

In the realm of public television, the NSF invested in the series *Nova*. The controversial subject matter engaged in early *Nova* programs tested the NSF's funding procedures, however. In an attempt to balance the interests of a free press against those of the scientific community, the NSF established a grant approval system mediated by "outside advisors," most often experts in the field addressed in the program (Tressel 27). With "balanced, objective, and accurate" programming in mind, the "outside advisor" has become a standard feature of most scientific television production regardless of funding sources (Tressel 27).

The success of *Nova* sparked an ongoing relationship between the NSF and public broadcasting, one which positioned public television at the forefront of scientific programming. This coalition continues to be responsible for the development of several science based specials, such as *The Mind*, and a myriad of children's shows, including *3-2-1 Contact* and *Square One TV* (Tressel 29-31). In many ways, the Public Broadcasting System (PBS) has forged its identity around science programs and shows every indication of continuing its commitment to scientific television in the future.

Alongside the ongong efforts of the NSF, today's multifaceted television market has led to the development of scientific programming in unanticipated arenas, most notably cable. Cable networks have capitalized on the entertainment value of science and technology to become prolific purveyors of scientific television shows, such as *Beyond 2000* (Discovery Channel) and *Science and Technology Week* (CNN). Science programs have become a staple ingredient on education oriented cable channels such as The Discovery Channel. Recently, cable has also directed its attention back toward the scientific community with the development of professional programing such as Lifetime's *Medical Television* (Barinage, 1307).

Closely paralleling respective funding sources, current scientific television programs can be divided into three basic categories: commercial programming, children's programming, and PBS programming. These categories often overlap. For example, many children's science programs are produced by and aired on PBS. While such categories are useful in providing basic understanding of the focus of certain programs, they are by no means a definitive description of their content.

Most commercial science programming is developed by either network or syndicated sources. The majority of programs target adult audiences, and the topics of the episodes vary greatly. Most of the programs in this category are series, with each episode focusing on a specific topic, such as new technology, the universe, aeronautics, zoology, and genetic engineering. A few, such as the *NASA Space Films* (1990), are dedicated to one specific topic. Almost all entries in this category include a focus on "science and technology" in their program description. In addition to several already mentioned, programs in this category include: *Sci-Tech TV* (1994), *World of Discovery* (ABC: 1990–1994), *A View of the World* (1993), *Quantum*

(1993), *The Science Show* (1990-1993), *Omni: Visions of To-morrow* (1985), *Eye on Science* (CBS: 1981–1985), *Introducing Biology* (1980's), and *Universe* (1979).

Programming for children is a rapidly growing genre of science television. Since the implementation of the Children's Television Act of 1990, programmers have been required to air a certain amount of educational material during day-time slots when children are prime viewers. Several shows, such as Walt Disney's *Bill Nye the Science Guy*, (1993–94) are a direct response to this act. Other science programs targeting children and/or teens include *Beakman's World (1992)*, *Timehoppers* (1992), *The Voyage of the Mimi* (no date available), *Newton's Apple* (1982–1988) and, of course, *Watch Mr. Wizard* (1951).

Science television programming produced and aired by PBS also encompasses a wide range of topics. Series such as *NOVA* and *Nature* consist of single episodes focusing on areas as diverse as general science, nature, medicine and teclnology. Other similar programs include *Future Quest* (1993), and *The Infinite Voyage* (1987).

Clearly, cable is positioned to become a front runner in future scientific programming by virtue of its resources, funding, and widespread distribution. While PBS has traditionally set the standard in science television, its leadership may be weakened by the continued assault on federal funding of public broadcasting.

New technology will also undoubtedly play a role in the future development of scientific television programs. Following a trend set by science museums, scientific television will likely move toward interactve programming. Likewise, the anticipated profusion of cable channels may lead to high degrees of specialization in programming, such as an "all biology channel".

In the final analysis, the future of science television lies with the audience itself, as tbe first generations of viewers raised on science-based children's programming reach maturity and reach for the remote control.

—Joanna Ploeger-Tsoulos and Robbie Shumate

FURTHER READING

Banks, Jane. "Science as Fiction: Technology in Prime Time Television." *Critical Studies in Mass Communication* (Annandale, Virginia), March 1990.

Barinaga, Marcia. "Science Television: Colleagues on Cable." *Science*, 1991.

Bib Television Programming Source Books. 1994–1995 ed.

Goldsmith, Donald. "Two Years in Hollywood: An Astronomer in Television Land." *Mercury*, March-April 1991.

Hornig, Susanna. "Television's 'Nova' and the Construction of Scientific Truth." *Critical Studies in Mass Communication* (Annandale, Virginia), March 1990.

Jerome, Fred. "A Retreat—And an Advance—for Science on TV." *Issues in Science and Technology* (Washington, D.C.), Summer, 1988.

Jones, Glyn. "New Directions for Science on TV." *New Scientist* (London), 13 October 1990.

Tressel, George. "Science on the Air: NSF's Role." *Physics Today* (New York),1990.

See also *Ascent of Man*; Attenborough, David; Cousteau, Jaques; Educational Television; *Hitchhiker's Guide to the Galaxy*; *Nature of Things*; Open University; Suzuki, David; *Watch Mr. Wizard*; Wild Kingdom, Mutual of Omaha's; Wildlife and Nature Programs

SCIENCE-FICTION PROGRAMS

Although not one of television's predominant genres in terms of overall programming hours, science fiction nonetheless spans the history of the medium, beginning in the late 1940s as low-budget programs aimed primarily at juvenile audiences and developing, by the 1990s, into a genre particularly important to syndication and cable markets. For many years, conventional industry wisdom considered science fiction to be a genre ill-suited to television. Aside from attracting a very limited demographic group for advertisers, science fiction presented a problematic genre in that its futuristic worlds and speculative storylines often challenged both the budgets and narrative constraints of the medium, limitations especially true in television's first decades. Over the years, however, producers were to discover that science fiction could attract an older and more desirable audience, and that such audiences, though often still limited, were in many cases incredibly devoted to their favorite programs. As a consequence, the eighties and nineties saw a tremendous increase in science-fiction programming in the U.S., especially in markets outside the traditional three broadcast networks.

As a children's genre in the late 1940s and early 1950s, science-fiction programs most often followed a serial format, appearing in the afternoon on Saturdays or at the beginning of prime time during the weeknight schedule. At times playing in several installments per week, these early examples of the genre featured the adventures of male protagonists working to maintain law and order in outer space. These early "space westerns" included *Buck Rogers* (ABC 1950–51), *Captain Video and His Video Rangers* (DuMont 1949–54), *Flash Gordon* (Syndicated 1953), *Space Patrol* (ABC 1951–52), and *Tom Corbett, Space Cadet* (CBS/ABC/NBC 1950–52). Each series pitted its dynamic hero against a variety of intergalactic menaces, be they malevolent alien conquerors, evil mad scientists, or mysterious forces of the universe. All of these programs were produced on shoe-string budgets, but this did not stop each series from equipping its hero with a fantastic array of futuristic gadgetry, including

Captain Midnight

The Prisoner

Star Trek

The X-Files

radio helmets, ray-guns, and Captain Video's famous "decoder ring." Viewers at home could follow along with their heroes on the quest for justice by ordering plastic replicas of these gadgets through popular premium campaigns. Of these first examples of televised science fiction, *Captain Video* was particularly popular, airing Monday through Friday in half-hour (and later, fifteen-minute) installments. One of the first "hits" of television, the program served for many years as a financial lynchpin for the struggling DuMont network, and left the air only when the network itself collapsed in 1954.

As was typical of much early programming for children, Captain Video concluded each episode by delivering a lecture on moral values, good citizenship, or other uplifting qualities for his young audience to emulate. Such gestures, however, did not spare Captain Video and his space brethren from becoming the focus of the first of many major public controversies over children's television. In a theme that would become familiar over the history of the medium, critics attacked these shows for their "addictive" nature, their perceived excesses of violence, and their ability to "over-excite" a childish imagination. In this respect, early science fiction on television became caught up in a larger anxiety over children's culture in the fifties, a debate that culminated with the 1954 publication of Dr. Fredric Wertham's *Seduction of the Innocent,* an attack on the comic book industry that eventually led to a series of Congressional hearings on the imagined links between popular culture and juvenile delinquency.

Science-fiction programming aimed at older audiences in early television was more rare, confined almost entirely to dramatic anthology series such as *Lights Out* (NBC 1949–52), *Out There* (CBS 1951–52), and *Tales of Tomorrow* (ABC 1951–53). As with other dramatic anthologies of the era, these programs depended heavily on adaptations of pre-existing stories, borrowing from the work of such noted science-fiction writers as Jules Verne, H.G. Wells, and Ray Bradbury. *Tales of Tomorrow* even attempted a half-hour adaptation of Mary Shelly's *Frankenstein.* When not producing adaptations, these anthologies did provide space for original and at times innovative teleplays. Interestingly, however, as science fiction became an increasingly important genre in Hollywood during the mid-late-1950s, especially in capturing the burgeoning teenage market its presence on American television declined sharply. One exception was *Science Fiction Theater* (1955–57), a syndicated series that presented speculative stories based on contemporary topics of scientific research.

Science fiction's eventual return to network airwaves coincided with the rising domestic tensions and cold war anxieties associated with the rhetoric of the Kennedy administration's "New Frontier." As a response to the Soviet launch of Sputnik, for example, CBS' *Men Into Space* (1959–60) participated in the larger cultural project of explicitly promoting interest in the emerging "space race" while also celebrating American technology and heroism that had been threatened by the Soviets' success. Other series were more complex in their response to the social and technological conflicts of the New Frontier era. In particular, *The Twilight Zone* (CBS 1959–64) and *The Outer Limits* (ABC 1963–65), programs that would become two of the genre's most celebrated series, frequently engaged in critical commentary on the three pillars of New Frontier ideology—space, suburbia, and the superpowers.

Hosted and for the most part scripted by Rod Serling, a highly acclaimed writer of live television drama in the fifties, *The Twilight Zone* was an anthology series that, while not exclusively based in science fiction, frequently turned to the genre to frame allegorical tales of the human condition and America's national character. Some of the most memorable episodes of the series used science fiction to defamiliarize and question the conformist values of post-war suburbia as well as the rising paranoia of Cold War confrontation. Of these, "The Monsters are Due on Maple Street" was perhaps most emblematic of these critiques. In this episode, a "typical" American neighborhood is racked with suspicion and fear when a delusion spreads that the community has been invaded by aliens. Neighbor turns against neighbor to create panic until at the end, in a "twist" ending that would become a trademark of the series, the viewer discovers that invading aliens *have* actually arrived on earth. Their plan is to plant such rumors in every American town to tear these communities apart thus laying the groundwork for a full-scale alien conquest.

More firmly grounded in science fiction was *The Outer Limits,* an hour-long anthology series known primarily for its menagerie of gruesome monsters. Much more sinister in tone than Serling's *Twilight Zone, The Outer Limits* also engaged in allegories about space, science, and American society. But in an era marked by the almost uniform celebration of American science and technology, this series stood out for its particularly bleak vision of technocracy and the future, using its anthology format to present a variety of dystopic parables and narratives of annihilation. Of the individual episodes, perhaps most celebrated was Harlan Ellison's award-winning time-travel story, "Demon with a Glass Hand," an episode that remains one of the most narratively sophisticated and willfully obtuse hours of television ever produced.

While *The Twilight Zone* and *The Outer Limits* remain the most memorable examples of the genre in this era, science-fiction television of the mid-1960s was dominated, in terms of total programming hours, by the work of producer Irwin Allen. Allen's series, aimed primarily at juvenile audiences on ABC, included *Voyage to the Bottom of the Sea* (ABC 1964–68), *Lost in Space* (CBS 1965–68), *Time Tunnel* (ABC 1966–67), and *Land of the Giants* (ABC 1968–70). Each series used a science-fiction premise to motivate familiar action-adventure stories. Of these, *Lost in Space* has been the most enduring in both syndication and national memory. Centering on young Will Robinson and his friend the Robot, the series adapted the "Swiss Family Robinson" story to outer space, chronicling a wandering family's adventures as they tried to return to earth.

Many other television series of the sixties, while not explicitly science fiction, nevertheless incorporated elements of space and futuristic technology into their storyworlds. Following the success of *The Flintstones,* a prime time animated series about a prehistoric family, ABC premiered *The Jetsons* (1962–63), a cartoon about a futuristic family of the next century. The sitcom *My Favorite Martian* (CBS 1963–66), meanwhile, paired an earthling newspaper reporter with a Martian visitor, while *I Dream of Jeannie* (NBC 1965–70) matched a NASA astronaut with a beautiful genie. The camp hit *Batman* (ABC 1966–68) routinely featured all manner of innovative "bat" technologies that allowed its hero to outwit Gotham City's criminals. Also prominent in this era was a cycle of spy and espionage series inspired by the success of the James Bond films, each incorporating a variety of secret advanced technologies. Of this cycle, the British produced series, *The Prisoner* (CBS 1968–69), was the most firmly based in science fiction, telling the Orwellian story of a former secret agent stripped of his identity and trapped on an island community run as a futuristic police state.

By far the most well-known and widely viewed science-fiction series of the 1960s (and probably in all of television) was *Star Trek* (NBC 1966–69), a series described by its creator, Gene Roddenberry, as *"Wagon Train* in space." Although set in the 23rd century, the world of *Star Trek* was firmly grounded in the concerns of sixties America. Intermixing action-adventure with social commentary, the series addressed such issues as racism, war, sexism, and even the era's flourishing hippie movement. A moderately successful series during its three-year network run, *Star Trek* would become through syndication perhaps the most actively celebrated program in television history, inspiring a whole subculture of fans (known variously as "trekkies" or "trekkers") whose devotion to the series led to fan conventions, book series, and eventually a commercial return of the Star Trek universe in the 1980s and 1990s through motion pictures and television spin-offs.

Like *Star Trek,* the BBC produced serial *Doctor Who* also attracted a tremendous fan following. In production from 1963 to 1989, *Doctor Who* stands as the longest running continuous science-fiction series in all of television. A time-travel adventure story aimed primarily at children, the series proved popular enough in the United Kingdom to inspire two motion pictures pitting the Doctor against his most famous nemesis-the Daleks *(Doctor Who and the Daleks* (1965) and *Daleks: Invasion Earth 2150 AD* (1966). The series was later imported to the United States, where it aired primarily on PBS affiliates, and quickly became an international cult favorite.

While most television science fiction in the 1950s and 1960s had followed the adventures of earthlings in outer space, increasing popular interest in Unidentified Flying Objects (UFOs) led to the production, in the late 1960s and into the 1970s, of a handful of programs based on the premise of secretive and potentially hostile aliens visiting the earth. *The Invaders* (ABC 1967–68) chronicled one man's struggle to expose an alien invasion plot, while *UFO* (Syndicated 1972) told of a secret organization dedicated to repelling an imminent UFO attack. Veteran producer Jack Webb debuted *Project UFO* (NBC) in 1978, which investigated, in Webb's characteristically terse style, unexplained UFO cases taken from the files of the United States Air Force. Such series fed a growing interest in the early seventies with all manner of paranormal and extraterrestrial phenomena, ranging from Erich von Daniken's incredibly popular speculations on ancient alien contact in *Chariots of the Gods* to accounts of the mysterious forces in the "Bermuda Triangle." Such topics from the fringes of science were the focus of the syndicated documentary series, *In Search Of* (Syndicated 1976), hosted by Star Trek's Leonard Nimoy.

For the most part however, science fiction once again went into decline during the 1970s as examples of the genre became more sporadic and short-lived, many series running only a season or less. Series such as *Planet of the Apes* (CBS 1974) and *Logan's Run* (CBS 1977–78) attempted to adapt popular motion pictures to prime time television, but with little success. A much more prominent and expensive failure was the British series, *Space: 1999* (Syndicated 1975). Starring Martin Landau and Barbara Bain, the program followed a group of lunar colonists who were sent hurtling through space when a tremendous explosion drives the moon out of its orbit. The series was promoted in syndication as the most expensive program of its kind ever produced, but despite such publicity, the series went out of production after only 48 episodes.

Two of the more successful science-fiction series of the era were *The Six Million Dollar Man* (ABC 1975–78) and its spin-off *The Bionic Woman* (ABC/NBC 1976–78). The "six million dollar man" was Lt. Steve Austin, a test pilot who was severely injured in a crash and then reconstructed with cybernetic limbs and powers that made him an almost superhuman "bionic man." Austin's girlfriend, also severely injured (in a separate incident) and rebuilt (by the same doctors) debuted her own show the following season (complete with a "bionic" dog). The moderate success of these two series sparked a cycle of programs targeted at children featuring superheros with superpowers of one kind or another, including *The Invisible Man* (NBC 1975–76), *Gemini Man* (NBC 1976) *Man From Atlantis* (NBC 1977–78), *Wonder Woman* (ABC/CBS 1976–79), and *The Incredible Hulk* (CBS 1978–82).

Also moderately successful in the late 1970s were a pair of series designed to capitalize on the extraordinary popularity of George Lucas' 1977 blockbuster film, *Star Wars.* Both *Battlestar Galactica* (ABC 1978–80), starring *Bonanza's* patriarch Lorne Greene, and *Buck Rogers in the 25th Century* (NBC 1979–81) spent large amounts of money on the most complex special effects yet seen on television, all in an attempt to recreate the dazzling hardware, fast-paced space battles, and realistic aliens of Lucas' film. Less successful in riding *Star Wars'* coat-tails was the parodic sitcom, *Quark* (NBC 1978), the story of a garbage scow in outer space.

In England, the 1970s saw the debut of another BBC produced series that would go on to acquire an international audience. *Blake's Seven* (BBC 1978–81) was created by Terry Nation, the same man who introduced the Daleks to the world of *Doctor Who* in the early 1960s. Distinguished by a much darker tone than most television science fiction, *Blake's Seven* followed the adventures of a band of rebels in space struggling to overthrow an oppressive regime.

Alien invasion was once again the theme on American television in 1983, when NBC programmed a high-profile mini-series that pitted the earth against a race of lizard-like creatures who, though friendly at first, were actually intent on using the earth's population for food. *V* (NBC 1984–85) proved popular enough to return in a sequel miniseries the following year, which in turn led to its debut as a weekly series in the 1984–85 season. More provocative was ABC's short-lived *Max Headroom* (ABC 1987), television's only attempt at a subgenre of science fiction prominent in the eighties known as "cyberpunk." "Max," who through commercials and a talk-show became a pop cult phenomenon in his own right, was the computerized consciousness of TV reporter Edison Carter. Evoking the same "tech noir" landscape and thematic concerns of such cinematic contemporaries as *Blade Runner, Robocop,* and *The Running Man,* Max and Edison worked together to expose corporate corruption and injustice in the nation's dark, cybernetic, and oppressively urbanized future.

Less weighty than Max, but certainly more successful in their network runs, were two series that, while not necessarily true "science fiction," utilized fantastic premises and attracted devoted cult audiences. *Beauty and the Beast* (CBS 1987–90) was a romantic fantasy about a woman in love with a lion-like creature who lived in a secret subterranean community beneath New York City, while *Quantum Leap* (NBC 1989–93) followed Dr. Sam Beckett as he "leapt" in time from body to body, occupying different consciousnesses in different historical periods. The series was less concerned with the "science" of time travel, however, than with the moral lessons to be learned or taught by seeing the world through another person's eyes.

By far the most pivotal series in rekindling science fiction as a viable television genre was *Star Trek: The Next Generation* (Syndicated 1987–94), produced by Paramount and supervised by the creator of the original *Star Trek,* Gene Roddenberry. Already benefiting from the tremendous built-in audience of *Star Trek* fans eager for a spin-off of the old series, Paramount was able to bypass the networks and take the show directly into first-run syndication, where it quickly became the highest rated syndicated show ever. In many ways, *Next Generation* had more in common with other dramatic series of the 1980s and 1990s than it did with the original series. In this new incarnation, *Star Trek* became an ensemble drama structured much like *Hill St. Blues* or *St. Elsewhere,* featuring an expanded cast involved in both episodic and serial adventures. Broadcast in conjunction with a series of cinematic releases featuring the original *Star Trek* characters, *Next Generation* helped solidify *Star Trek* as a major economic and cultural institution in the eighties and nineties. After a seven-year run, Paramount retired the series in 1994 to convert the *Next Generation* universe into a cinematic property, but not before the studio debuted a second spin-off, *Star Trek: Deep Space Nine* (Syndicated 1993), which proved to be a more claustrophobic and less popular reading of the *Star Trek* universe. A third spin-off, *Star Trek: Voyager* (Syndicated 1995–), served as the anchor in Paramount's bid to create their own television network in 1995.

The success of the *Star Trek* series in first-run syndication reflected the changing marketplace of television in the 1980s and 1990s. As the three major networks continued to lose their audience base to the competition of independents, cable, and new networks such as Fox, Warner Brothers, and UPN, the entire industry sought out new niche markets to target in order to maintain their audiences. The *Star Trek* franchise's ability to deliver quality demographics and dedicated viewership inspired a number of producers to move into science fiction during this period. These series ranged from the literate serial drama, *Babylon 5* (Syndicated 1994), to the bizarre police burlesque of *Space Precinct* (Syndicated 1994–). Also successful in syndication were "fantasy" series such as *Highlander* (Syndicated 1992–) and *Hercules: The Legendary Journeys* (Syndicated 1994–).

For the most part, the three major networks stayed away from science fiction in the 1990s, the exceptions being NBC's *Earth 2* (1994–95) and *Seaquest DSV* (1993), the latter produced by Steven Spielberg's Amblin Entertainment. By far the most active broadcaster in developing science fiction in the 1990s was the FOX network, which used the genre to target even more precisely its characteristically younger demographics. FOX productions included *Alien Nation* (1989–91), *M.A.N.T.I.S.* (1994–95), *Sliders* (1995), *VR.5* (1995), and *Space: Above and Beyond* (1995–96). FOX's most successful foray into science fiction, however, was *The X-Files* (1993–). A surprise hit for the network, *The X-Files* combined horror, suspense, and intrigue in stories about two FBI agents assigned to unsolved cases involving seemingly paranormal phenomena. Although the series originally centered on a single "spook" of the week for each episode, it eventually developed a compelling serial narrative line concerning a massive government conspiracy to cover up evidence of extraterrestrial contact. Like so many other science-fiction programs, the series quickly developed a large and organized fan community.

By the early 1990s, television science fiction had amassed a sizable enough program history and a large enough viewing audience to support a new cable network. The Sci-Fi Channel debuted in 1992, scheduling mainly old movies and television re-runs, but planning to support new program production in the genre sometime in the future.

—Jeffrey Sconce

FURTHER READING

Bellafante, Ginia. "Out of this world." *Time* (New York), 3 April 1995.

Coe, Steve. "Networks Take a Walk on Weird Side: Programmers Tap into Taste for Fantasy, Sci-Fi and the Bizarre." *Broadcasting and Cable* (Washington, D.C.), 30 October 1995.

Fulton, Roger. *The Encyclopedia of TV Science-Fiction.* London: Boxtree, 1995.

Jenkins, Henry. *Textual Poachers: Television Fans and Participatory Culture.* New York: Routledge, 1992.

Lentz, Harris M. *Science Fiction, Horror and Fantasy Film and Television Credits.* Jefferson, North Carolina: McFarland, 1994.

Littleton, Cynthia. "First-Run Faces Unreality: Fantasy, Sci-Fi and the Unexplained Have Proved Fertile Field for Syndication." *Broadcasting and Cable* (Washington, D.C.), 30 October 1995.

Menagh, Melanie, and Stephen Mills. "A Channel for Science Fiction." *Omni* (New York), October 1992.

Okuda, Denise, Debbie Mirek, and Doug Drexler. *The Star Trek Encyclopedia.* New York: Pocket Books, 1994.

Peel, John. *Island in the Sky: The Lost in Space Files.* San Bernardino, California: Borgo, 1986.

Phillips, Mark, and Frank Garcia. *Science Fiction Television Series: Episode Guides, Histories, and Casts and Credits for 62 Prime Time Shows, 1959 Through 1989.* Jefferson, North Carolina: McFarland, 1996.

Rigelsford, Adrian, and Terry Nation. *The Making of Terry Nation's Blake's 7.* London: Boxtree, 1995.

Schow, David. *The Outer Limits: The Official Companion.* New York: Ace Science Fiction Books, 1986.

Sconce, Jeffrey. "The 'Outer Limits' of Oblivion." In, Spigel, Lynn, and Michael Curtin, editors. *The Revolution Wasn't Televised: Sixties Television and Social Conflict.* New York: Routledge, 1995.

Spigel, Lynn. "From Domestic Space to Outer Space: The 1960s Fantastic Family Sit-Com." In, Penley, Constance, editor, with others. *Close Encounters: Film Feminism, and Science Fiction.* Minneapolis: University of Minnesota Press, 1991.

Tulloch, John, and Henry Jenkins. *Science Fiction Audiences: Watching Dr. Who and Star Trek.* London: Routledge, 1995.

Van Hise, James. *New Sci Fi TV from The Next Generation to Babylon 5.* Las Vegas, Nevada: Pioneer, 1994.

———. *Sci Fi Tv: From The Twilight Zone to Deep Space Nine.* New York: Harper Paperbacks, 1993.

White, Matthew, and Jaffer Ali. *The Official Prisoner Companion.* London: Sidgwick and Jackson, 1988.

Wright, Gene. *The Science Fiction Image: The Illustrated Encyclopedia of Science Fiction in Film, Television, Radio and the Theater.* New York: Facts on File, 1983.

Zicree, Marc Scott. *The Twilight Zone Companion.* New York: Bantam, 1982.

See also *Captain Video and His Video Rangers*; *Dark Shadows*; *Doctor Who*; *Max Headroom*; *Hitchhiker's Guide to the Galaxy*; Nation, Terry; Pertwee, Jon; *Prisoner*; Roddenberry, Gene; Serling, Rod; *Star Trek*; Troughton, Patrick; *Twilight Zone*

SCOTLAND

Scotland is a small country located on the periphery of Europe. Its television service reflects many of the key issues surrounding broadcasting in minority cultures. Politically part of the multi-nation state of the United Kingdom along with the other "Celtic" countries of Wales and Northern Ireland, Scotland's legal, educational, and religious institutions remain separate from those of England, the dominant partner. Its broadcasting systems, like much of its cultural organisation, display a mixture of autonomy and dependence which reflects Scotland's somewhat anomalous position.

Scotland's current programming reflects the evolution of Britain's broadcasting ecology, offering viewers a choice of four channels and a mix of British networked television and Scottish national and local productions. A brief history of its development sets in context both the present state of television in Scotland and some of the prevailing debates about its nature.

The first television service in Scotland was introduced by the British Broadcasting Corporation (BBC) in 1952.

To a large extent, the constitution and character of this new medium was determined by its existing radio system. John Reith, the architect of the BBC, and himself a Scot, was determined that the BBC should provide an essentially British service. The consequent emphasis placed on the centralisation of public-service broadcasting led to a downgrading of other forms of more local production, as well as to the BBC's oxymoronic categorisation of Scotland as a "national region." This decision was not simply an organisational choice but, as McDowell suggests, reflected the dominant ideological belief in the superiority of "metropolitan culture." The BBC's early television broadcasts consisted of largely the same programmes as those of London. What was produced in Scotland received considerable criticism in terms of its nature and quality; the Pilkington Report of 1962 noted that the few programmes produced by BBC Scotland often "failed to reflect distinctive Scottish culture."

The arrival of independent—or commercial—television in Scotland offered a new source of programming.

Like the BBC, the independent companies broadcast a mix of network provision and more local, opt-out, productions. Franchises were awarded to Scottish Television, covering central Scotland; Border Television, covering the Scottish and English borders; and Grampian Television, serving the North of Scotland. The enthusiasm of some for the new medium can be gauged by the notorious comment of Scottish's first proprietor, Canadian magnate Roy Thomson, that an independent franchise was "a license to print money." In these early years, perhaps unsurprisingly, Scottish programme schedules, too, were heavily criticised for their poor quality and parochial outlook. The 1970s and 1980s saw both the BBC and STV upping the level of their local programming, improving its quality and diversity, and beginning to form a stronger presence on the network through programmes such as the long-running police drama *Taggart* and the popular soap *Take the High Road*.

Recent years have brought significant changes, diversifying the type and origins of programmes produced in Scotland. The introduction of Channel 4 in 1982 and quotas for independent production in the 1990 Broadcasting Act have led to the emergence of numerous independent companies, as is the case across the United Kingdom as a whole. While they have undoubtedly broadened the production base and often pioneered innovative forms of programming, the vast majority of these companies are relatively small and powerless in their ability to affect broadcast policy.

In the 1990s, extensive lobbying has brought governmental support of £9.5 million for the production of television programmes in Scotland's minority indigenous language, Gaelic. Unquestionably a welcome move, it nonetheless demonstrates (as does the support of *Sianal Pedwar Cymru,* the Welsh Channel 4), that it is easier to gain recognition for linguistic than for cultural differences.

These moves in television are indicative of wider cultural shifts. For some years debate has been growing over Scotland's constitutional position in the United Kingdom, manifested in some quarters by demands for political change in the form of self-government or independence. More widespread, however, has been a transformation in cultural activity in Scotland over the past two decades—most notably in literature, but also in theatre, music, and film—which many see as a form of cultural nationalism.

This climate of cultural and political contention has led to a new attention to questions of representation and national identity. In one of the most significant interventions, *Scotch Reels (1982),* critics Colin McArthur and Cairns Craig exposed and deconstructed the dominant representations of Scottishness, identifying two central rhetorics which have informed representations of Scotland—the associated discourses of tartanry and kailyard. While tartanry harks back to a romantic celebration of lost Scottish nationhood and draws on the emblems of a vanished (and imagined) premodern Highland way of life, kailyard celebrates the virtues of small- town life through couthy homilies. These discourses are seen to run through heterogeneous productions from Hollywood cinema and *Brigadoon* to indigenous programmes such as *Dr. Finlay's Casebook* and *The White Heather Club.*

This deconstruction of what Murray Grigor terms "Scotch Myths" has become widely circulated, and indeed parodying the clichés of Scottishness has become something of a trope in contemporary Scottish television productions (although it has yet to penetrate a Hollywood increasingly enamoured of Highland heroes such as *Braveheart* and *Rob Roy*). Scottish television offers its audiences antiheroes like Ian Pattison's comic creation *Rab C. Nesbitt,* a gloriously loud-mouthed Glaswegian drunkard and member of the underclass, who exaggerates to comic excess accepted notions of nationality and class. A more sophisticated and ambiguous demonstration of this parodic process is to be found in BBC Scotland's police series *Hamish Macbeth.* Set in a picturesque Highland village populated by bizarre characters, it simultaneously sends up the stereotypes of Highland life, while embracing their more marketable forms.

Much of the debate about television in Scotland, in academic and popular circles, has concerned itself with analysing and often attacking the dominant images of Scottishness which have been produced, while comparatively little attention has been paid to questions of production and policy. In Scotland questions of cultural identity and diversity, and independence and control, reverberate through television production at both a symbolic and material level.

— Jane Sillars

FURTHER READING

Caughie, J. "Questions of Representation." In, Dick, E., editor. *From Limelight to Satellite: A Scottish Film Book.* London: British Film Institute and Scottish Film Council, 1990.

McArthur, C., editor. *Scotch Reels: Scotland in Cinema and Television.* London: British Film Institute, 1982.

McDowell, W.H. *A History of the BBC in Scotland.* Edinburgh: Edinburgh University Press, 1992.

Nairn, T. *The Break-up of Britain.* London: Verso, 1977.

Sendall, B., and J. Potter. *A History of Independent Television, Volumes 1-4.* London: MacMillan, 1982.

See also British Television; British Programming; British Production Companies; Channel Four; Ireland; Wales

SCRAMBLED SIGNALS

Scrambled signals refers to the encryption of satellite data streams by cable television program providers to prevent the unauthorized reception of their signals by home satellite dish owners. Program providers scramble the signals they beam up to satellites which distribute their programming to local or regional cable operators.

With the relaxation of satellite broadcast and reception regulations by the FCC in 1979, and the tremendous reduction in the cost of satellite receiving equipment due to advances in technology, a booming market developed for home satellite dish receivers in the early 1980s. These satellite dishes were known as television receive only satellite earth stations—TVRO. Essentially, TVRO dish owners were able to intercept, free of charge, cable television programming distributed over C-band satellites. Though most early adopters of TVRO dishes were located in rural areas where cable television was unavailable, cable system operators were nevertheless concerned about the actual and potential loss of subscribers who opted to receive cable programming for "free". When Congress passed the Cable Communications Policy Act of 1984, which specified that it was indeed lawful to receive unencrypted satellite signals for private viewing, cable system operators convinced program suppliers to scramble their satellite uplink feeds. Though they sought to protect the system operators (their clients) by scrambling, program suppliers also realized the profit potential in selling programming directly to the TVRO owners.

By early 1985, therefore, most major program suppliers (led by HBO and Showtime) had begun scrambling. As a result, TVRO owners were required to purchase a signal descrambler and pay a monthly fee to receive scrambled programming. Though many TVRO owners worried that they would have to deal with several different encryption systems, the industry adopted M/A-Com's (later purchased by General Instrument Corporation) Videocipher II as the standard for scrambling. Though the industry was confident that the Videocipher II (VC-II) would reduce satellite programming "theft," the system was quickly plagued with problems.

The process of scrambling includes several steps. Program providers scramble their programs on earth and then beam them to a satellite. The maker of the descrambler receives instructions from the programmers as to which subscribers have paid for what programming, information which it too beams to the satellite. The satellite transmits both the program and the subscriber information back to earth where a TVRO owner's dish picks up the signals and sends them to the decoder. The decoder includes various computer chips which contain the information necessary to descramble the programming. Problems arose for Videocipher II when an enormous black market developed for altered descramblers. To receive free programming, dish owners could simply purchase a descrambler with one of the chips in the unit replaced, enabling the unit to descramble all programming. Industry sources estimated that 600-800,000 VC-II units had been illegally altered, and that approximately 5,120 of the 6,404 equipment dealers were somehow involved in the selling of pirated units. And after six years of program scrambling, it was estimated that only 10% of the three million dish owners were paying subscribers.

To correct this flaw (and to protect their near monopoly status), General Instrument released an updated version of the descrambler called Videocipher II Plus in late 1991. Also known as VC-RS (for "renewable security"), the new units replaced the multiple chips in the unit with a single chip. Any effort to copy or replace the chip would disable the unit entirely. More importantly, the units include a renewable encryption system through the use of a "TYPass" smart-card (similar to a credit card). Should a breach in security occur, the encryption information on the cards can be changed quickly and inexpensively. Major programmers switched to the upgraded system with due speed, as HBO became the first programmer to shut off its consumer Videocipher II data stream on 19 October 1992. Other programmers quickly followed suit. Furthermore, HBO's satellite transmissions to Europe, Latin America, and elsewhere use the VC-RS technology.

Though the scrambling of signals has primarily been the concern of cable programmers and operators, the broadcast networks (CBS, NBC, and ABC) also began to scramble the transmission of programs to their affiliates (in 1986, 1988, and 1991 respectively). Defending his network's move to scramble such transmissions, CBS Vice President Robert McConnell contends that network feeds are "private property", and he encourages viewers instead to watch their local affiliates for local news, weather, and commercials. Though obviously directed at protecting the advertising revenues of its affiliates, such justifications ignore the lack of "local" reception for many rural satellite dish owners.

Although VC-RS is currently the de facto industry standard for the scrambling of C-band satellite programming signals, the imminent move to Ku-band satellite transmissions (such as SkyPix's EchoSphere system), digital television, and the introduction of digital video encryption and compression technologies (such as GI's DigiCipher) means that scrambling technologies for television transmissions will continue to change as program providers and cable system operators seek to maintain a firm control of any "illegal" reception.

—Jeffrey P. Jones

FURTHER READING

Hsiung, James C. "C-band DBS: An Analysis of the US Scrambling Issue." *Telecommunications Policy* (Guildford, England), March 1988.

Sims, Calvin. "A New Decoder to Foil Satellite-TV Pirates." *New York Times* (New York), 31 January 1990.

"Unscrambling Pay TV's New Descramblers." *Discover* (Los Angeles, California), May 1986.

See also Cable Networks; Distant Signal; Pay Television; Pay-Per-View; Satellite; United States: Cable

SECOND CITY TELEVISION

Canadian Comedy Program

econd City Television (SCTV) was a popular comedy television show originating from Canada that ran in the late 1970s and early 1980s in a variety of incarnations. Pulling much of its talent and ideas from the Chicago and Toronto Second City comedy clubs, the show became an important pipeline for comedians, especially Canadians, into the mainstream of the U.S. entertainment market. Popular performers who moved from *SCTV* into U.S. television and movies included John Candy, Martin Short, Dave Thomas, Catherine O'Hara, Andrea Martin, Rick Moranis, Harold Ramis, Robin Duke, Tony Rosato, Joe Flaherty, and Eugene Levy. Their training in live improvisational comedy meant that they could appear in a variety of capacities, but primarily work as writers and performers.

SCTV's early opening credit sequence set the tone for the show. As the announcer said, "*SCTV* now begins its programming day," a number of television sets were thrown out of an apartment building's windows, smashing on the pavement below. Using impersonations of well-known celebrities and ongoing original characters, *SCTV* presented a parody of every aspect of television, including programs, advertising, news, and network executives. In effect, *SCTV* was a cross between a spoof of television and a loose parodic soap opera about the running of the fictional Melonville television station. The station's personnel included the owner, Guy Caballero (Flaherty), the station manager, and Moe Green (Ramis), to be replaced by Edith Prickley (Martin), whose sister, Enda Boil (also Martin), advertised her Organ Emporium with husband, Tex (Thomas), in a parody of cheap late-night commercials. Other recurring figures were the bon vivant and itinerant host, Johnny LaRue (Candy), and the endearingly inept Ed Grimley (Short). Over the years, the *SCTV* programming line-up included the local news, read by Floyd Robertson (Flaherty) and Earl

Second City Television
Photo courtesy of Broadcasting and Cable

Camembert (Eugene Levy), "Sunrise Semester," "Fishin' Musician," and "The Sammy Maudlin Show," hosted by Maudlin (Flaherty) and his sidekick, William B. (Candy), with regular guest appearances from Bobby Bittman (Levy) and Lola Heatherton (O'Hara). Other spoofs included Yosh and Stan Shmenge's polka show (Levy and Candy), Count Floyd's "Monster Chiller Horror Theatre," whose host was played by the news anchor Floyd Robertson (Flaherty), the ersatz children's show "Captain Combat" (Thomas), "Farm Film Report" (Flaherty and Candy), and the improvised editorials of Bob and Doug Mackenzie's "Great White North" (Moranis and Thomas).

SCTV's trademark was the use of complex intertextual references to produce original hybrid comic sketches. A parody of *The Godfather* (Francis Ford Coppola, 1972) became the story of the Mafia-like operations of television networks. "Play It Again, Bob" took *Play It Again, Sam* (Woody Allen, 1972) and paired Woody Allen (Moranis) with Bob Hope (Thomas). Brooke Shields (O'Hara) and Dustin Hoffman (Martin) were guests on the "Farm Film Report," where they "blew up real good." In the station owner's attempt to capture a youth audience, *SCTV* tried to mimic *Saturday Night Live*, with guest host Earl Camembert, a ridiculously over-enthusiastic studio audience, and set-ups based around humorless references to drug use. *SCTV's* continual use of *mise en abyme* devices produced an intricate, layered text, in addition to a knowing fan culture. Further, this program, with its markedly satirical view of television and North American culture in general, was an important contribution to the notion that Canadian humour is ironic, self-deprecatory, and parodic.

The show's history began in 1976, when Andrew Alexander, Len Stuart, and Bernie Sahlins produced the first half-hour episodes, called *Second City TV*, for Global Television Network in Toronto, where it ran for two seasons. Filmways Productions acquired the syndication rights for the U.S. market in 1977.

A deal was struck in 1979 with the Canadian Broadcasting Corporation (CBC) and Allarcom Ltd. in which the show would move to Edmonton for broadcast on the national CBC network. In 1981, NBC bought the program, shifted it to a 90-minute format, and moved the show back to Toronto. At NBC, it became part of the "Late Night Comedy Wars" between the renamed *SCTV Network 90* on Fridays from 12:30 A.M. to 2:00 A.M., ABC's *Fridays* on the same night from 12:30 A.M. to 1:30 A.M., and NBC's *Saturday Night Live*. When NBC did not renew *SCTV Network 90* in 1983, Cinemax took it over. Over the years, *SCTV* produced 72 half-hour shows, 42 90-minute shows, and 18 45-minute shows, as well as numerous spin-offs and specials. With 13 Emmy nominations, *SCTV* won two for best writing. The show has since been re-edited and repackaged into a half-hour "best of" format for syndication.

—Charles Acland

CAST

Guy Caballero	Joe Flaherty
Moe Green	Harold Ramis
Edith Prickley	Andrea Martin
Earl Camembert	Eugene Levy
Floyd Robertson	Joe Flaherty
Count Floyd	Joe Flaherty
Dr. Tongue	John Candy
Bruno	Eugene Levy
Johnny LaRue	John Candy
Bob MacKenzie	Dave Thomas
Doug MacKenzie	Rick Moranis
Tex Boil	Dave Thomas
Edna Boil	Andrea Martin
Mayor Tommy Shanks	John Candy
The Schmenge Brothers	John Candy and Eugene Levy
Perini Scleroso	Andrea Martin
Ed Grimley	Martin Short
Lin Ye Tang	Dave Thomas
Sammy Maudlin	Joe Flaherty
William B.	John Candy
Bobby Bittman	Eugene Levy
Lola Heatherton	Catherine O'Hara
Big Jim McBob	John Candy
Billy Saul Hurok	Joe Flaherty
Harry, the Guy with the Snake on His Face	John Candy
Rockin' Mel Slurp	Eugene Levy
Jackie Rogers, Jr.	Martin Short
Rusty van Reddick	Martin Short

PRODUCERS Andrew Alexander, Ben Stuart, Bernie Sahlins

PROGRAMMING HISTORY 72 Half-Hour Programs; 42 90-Minute Programs; 18 45-Minute Programs

• Global Television Network

1976–78

• CBC

1979–80

• NBC

1981–83   12:30-2:00 A.M.

• Cinemax Cable

1983–84   Various Times

FURTHER READING

McCrohan, Donna. *The Second City: A Backstage History of Comedy's Hottest Troupe.* New York: Perigree, 1987.

See also Canadian Programming in English

SECONDARI, JOHN H.

U.S. Documentary Producer

John Secondari played a major role in the early growth of television news at ABC during the 1960s. As executive in charge of the network's first regular documentary series, Secondari forged a coherent house style that featured a heavy emphasis on visualization and dramatic voice-over narration. He later carried these qualities over to a series of occasional historical documentaries that earned him wide recognition and numerous national broadcasting awards.

Born in Rome in 1920, Secondari was educated in the United States and served in the army during World War II. Afterward, he worked in Europe first for CBS and then as the chief of information for the Marshall Plan in Italy. He quit in 1951 to devote himself to fiction writing on a full-time basis. Over the next six years he authored four books, one of which was turned into the popular Hollywood feature film *Three Coins in the Fountain*. During this period he also wrote scripts for television anthology dramas such as *The Alcoa Hour* and *Playhouse 90*. Both his background as a fiction writer and his fondness for Italy would figure prominently in his documentary career at ABC.

Secondari joined the network's Washington news bureau in 1957 and started producing documentaries toward the end of the decade. At the time, ABC's news operation was tiny by comparison to its rivals and its output was therefore quite limited. In the early 1960s, as television news expanded rapidly and as network news competition escalated, the smallest of the three major networks relied heavily on its documentary unit in order to sustain its stature as a *bona fide* news organization. ABC's major contribution to prime-time information fare during this period was the weekly *Bell and Howell Close-Up!* series, which Secondari took charge of shortly after its launch in 1960.

Underfunded by comparison to his network rivals and lacking a seasoned staff of broadcast newsworkers, Secondari nevertheless mounted a creditable series and even made some significant contributions during documentary's television heyday. He accomplished this in part by tapping freelance contributors such as producers Robert Drew and Nicholas Webster. Drew's *cinema verite* style offered dramatic glimpses of Castro's Cuba, the Kennedy White House, and the cockpit of an X-15. Similarly, Webster provided first-person accounts of racism in New York City, the school system in Moscow, and the revolving door in America's penal system. In these and many other *Close-Up!* documentaries, the camera escorted the protagonist through the routines and challenges of everyday life. The style emphasized intimacy and visual dynamism, qualities explicitly requested by the series sponsor Bell and Howell, a major manufacturer of amateur motion picture equipment. The same qualities could be seen in the output of regular staff members in the ABC documentary unit. A critic for *Variety* once commented on the house style of each network's flagship series by noting that *CBS Reports* could

John H. Secondari
Photo courtesy of Broadcasting and Cable

be described as the *Harper's* of television documentary, *NBC White Paper* as the *Atlantic*, and *Bell and Howell Close-Up!* as the *Redbook*. Indeed, the emphasis on dramatic visualization at ABC was accompanied by a commitment to florid voice-over narration that sometimes seemed excessive. Several critics noted that at the end of "Comrade Student" (a profile of Soviet schools), Secondari's commentary turned self-consciously propagandistic. Similarly, a documentary about the Italian Communist Party—on which he collaborated with his wife, Helen Jean Rogers—closes with a paean to the spirit of republican Rome that reputedly dwells in the souls of all Italians and serves as the last bulwark against leftist revolution.

This penchant for the dramatic continued to mark Secondari's work as he moved to historical topics with a series entitled the *Saga of Western Man*. Co-produced with Rogers, it began in 1963 with each episode focusing on a particular year, person, or incident that Secondari believed had significantly influenced the progress of Western civilization. Using the camera "as if it were the eyes of someone who had been present in the past," Secondari transported the viewer to historical locations while voice-over narrators read authentic journal entries or letters from the period.

For example, Secondari outfitted historical ships in Spain and put to sea with his camera crew in order to capture the sensations of Columbus' transoceanic voyage. These historical reenactments were then edited together with close-up shots scanning the canvases of period paintings. Meanwhile, the audio track featured music and dramatic readings from the navigation logs of Columbus done by actor Frederic March. These techniques—which were also being developed by NBC producers Lou Hazam and George Vicas—generated widespread critical acclaim and numerous awards for the series, thereby encouraging ABC to sign on for a second season. By year's end, however, some critics began to complain that the method was wearing thin. *The Saga of Western Man* was scaled back and continued on an occasional basis until the end of the 1960s when Secondari and Rogers left ABC to form their own production company.

Secondari died in 1975 at the age of 55. In all, he garnered some twenty Emmy and three Peabody awards. Perhaps most important, however, was his contribution to the development of the historical television documentary. Secondari's style not only anticipated the later efforts of such producers as Ken Burns, but also laid the groundwork for the emergence of the television docudrama in the 1970s.

—Michael Curtin

JOHN H. SECONDARI, Born in Rome, Italy, 1 November 1919. Fordham University, New York, U.S.A., B.A. 1939; Columbia University, M.S. in Journalism 1940. Married: 1) Rita Hume, 1948 (died); 2) Helen Jean Rogers, 1961. Enlisted in U.S. Army, 1941; appointed to staff of Cavalry School; commanded a reconnaissance unit and a tank company in combat in France, Germany, and Austria; served on staff of General Mark Clark in Vienna; left Army with rank of captain, 1946. Worked as a newspaper reporter for the Rome *Daily American*, 1946; foreign corespondent for CBS, 1948; deputy chief, information division of the Economic Cooperation Administration's Special Mission to Italy, 1948–51; freelance writer, 1951–56; chief, ABC's Washington news bureau, 1956; executive producer, ABC's special projects division, 1960–68; formed own production company, 1968. Recipient: *Radio Television Daily*'s Television Writer of the Year, 1963; Italy's Guglielmo Marconi World Television Award, 1964; 20 Emmy Awards; three Peabody Awards. Died 8 February 1975.

TELEVISION SERIES

1957–58	*Open Hearing* (moderator)
1960–63	*Bell and Howell Close-Up!*
1963–66	*The Saga of the Western Man* (co-producer)

TELEVISION SPECIALS (selection)

1958	*Highlights of the Coronation of Pope John XXIII*
1960	*Japan: Anchor in the East*
1960	*Korea: No Parallel*
1963	*Soviet Women*
1963	*The Vatican*
1970	*The Golden Age of the Automobile*
1970	*The Ballad of the Iron Horse*
1972	*Champions*

PUBLICATIONS (selection)

Coins in the Fountain (novel). Philadelphia, Pennsylvania: Lippincott, 1952.
Temptation for a King (novel). Philadelphia, Pennsylvania: Lippincott, 1954.
Spinner of the Dream (novel). Boston, Massachusetts: Little, Brown, 1955.

FURTHER READING

Bluem, A. William. *Documentary in American Television.* New York: Hastings House, 1965.
Curtin, Michael. *Redeeming the Wasteland: Television Documentary and Cold War Politics.* New Brunswick, New Jersey: Rutgers University Press, 1995.
Einstein, Daniel. *Special Edition: A Guide to Network Television Documentary Series and Special News Reports, 1955–1979.* Metuchen, New Jersey: Scarecrow, 1987.
Hammond, Charles M. *The Image Decade: Television Documentary: 1965–1975.* New York: Hastings House, 1981.

See also Documentary; Drew, Robert; *Tour of the White House with Mrs. John F. Kennedy*

SEE IT NOW

U.S. Documentary Series

See It Now (1951–58), one of television's earliest documentary series, remains the standard by which broadcast journalism is judged for its courage and commitment. The series brought radio's premier reporter, Edward R. Murrow, to television, and his worldly expertise and media savvy helped to define television's role in covering and, more importantly, analyzing the news.

The genesis of *See It Now* was a series of record albums that Murrow created during the late 1940s with Fred W. Friendly, a former radio producer at a Rhode Island station. The *I Can Hear It Now* records, which interwove historical events and speeches with Murrow narration, became such a commercial success that the partnership developed a radio series for CBS that also creatively used taped actualities. The weekly *Hear It Now* was modeled on a magazine format, with a variety of "sounds" of current events, such as artillery fire from Korea and an atom smasher at work, illuminated by Murrow and other expert columnists.

After his World War II experience, Murrow had assiduously avoided television, having been overheard stating, "I wish goddamned television had never been invented." Friendly was eager to test the new technology, and in 1951 the team agreed to transfer the *Now* concept yet again, this time emphasizing the visual essence of the medium, and calling their effort *See It Now*. Murrow never desired to anchor the evening newscast, and he did not want *See It Now* to be a passive recitation of current events, but an active engagement with the issues of the day. To implement this vision, Murrow and Friendly radically transformed the fundamental nature of news gathering on television.

Unlike other news programs that used newsreel companies to record events, *See It Now* maintained its own camera crews to coordinate filming on location, using 35mm- cameras to record the most striking images. Murrow and Friendly also deviated from standard practice by mandating that all interviews would not be rehearsed and there would be no background music to accompany the visuals. Although *See It Now* relied on CBS correspondents around the world, Murrow, serving as editor-in-chief, and Friendly, as managing editor, organized the first autonomous news unit, whose ranks included reporter-producers Joe Wershba and Ed Scott; director Don Hewitt; production manager Palmer Williams; and former newsreel cameramen Charlie Mack and Leo Rossi.

"This is an old team trying to learn a new trade," intoned Murrow to inaugurate *See It Now* on 18 November 1951. Murrow, as in all the programs that followed, was ensconced in Studio 41, exposing the tricks of the electronic trade—the monitors, the microphones, and the technicians all in view. To underscore this new technological undertaking, Murrow summoned up a split screen of the Brooklyn Bridge in New York City and the Golden Gate Bridge in San Francisco, the first live coast-to-coast transmission.

See It Now was the first news magazine series on television, alternating live studio commentary with reports from such seasoned correspondents as Howard K. Smith and Eric Sevareid. The series was initially scheduled in the intellectual ghetto of Sunday afternoon. By its third outing, *See It Now* gained a commercial sponsor, Alcoa (the Aluminum Company of America), which sought prestige among opinion makers to offset antitrust troubles. As the half-hour series became the most influential news program on television, it moved into prime time, first on Sunday evenings, and then for three years on Tuesday evenings at 10:30 P.M.

See It Now established its voice by covering the campaign rituals throughout the 1952 presidential year. Two early pieces were also emblematic of what Murrow and Friendly wanted to accomplish for the new venture: simulated coverage of a mock bomb attack on New York City, a segment that addressed the tensions of the nuclear age, and

See It Now
Photo courtesy of Washington State University Libraries

a one-hour report on the realities from the ground of the Korean War during the 1952 Christmas season. The later special evoked the frustrations and confusions of everyday soldiers, and was described by one critic as "the most graphic and yet sensitive picture of war we have ever seen."

Despite the laudatory reviews and the respectability that *See It Now* brought to television news, a question plagued the partnership: how to cover the anti-Communist hysteria that was enveloping the nation. The team first searched for what Friendly called "the little picture," an individual story that symbolized a national issue. In October 1953, Murrow and reporter Wershba produced "The Case of Milo Radulovich," a study of an Air Force lieutenant who was deemed a security risk because his father, an elderly Serbian immigrant, and sister supposedly read subversive newspapers. Because of the report, for which Murrow and Friendly used their own money to advertise, the secretary of the Air Force reviewed the case and retained Radulovich in the service. In "Argument in Indianapolis," broadcast one month later, *See It Now* investigated an American Legion chapter that refused to book its meeting hall to the American Civil Liberties Union. Again, Murrow and staff succeeded in documenting how McCarthyism, so-called because of the demagogic tactics of Senator Joseph McCarthy, had penetrated the heartland.

Having reported discrete episodes on the Cold War, Murrow and Friendly decided to expose the architect of the paranoia, McCarthy himself. On 9 March 1954, *See It Now* employed audiotapes and newsreels to refute the outrageous half-truths and misstatements of the junior senator from

Wisconsin. In his tailpiece before the signature "Good Night and Good Luck," Murrow explicitly challenged his viewers to confront the nation's palpable fears. A month later, McCarthy accepted an invitation to respond, and his bombastic rhetoric, calling Murrow "the leader and cleverest of the jackal pack," coupled with the later failure of his televised investigation into the Army, left his career in a shambles. The McCarthy program also produced fissures in the relationship between Murrow and the network. Again, CBS did not assist in promoting the broadcast; but this time CBS executives suggested that Murrow had overstepped the boundaries of editorial objectivity. In the process, he had become controversial and, therefore, a possible liability to the company's business opportunities.

Provocative programs, targeting the most pressing problems of the day, continued during the 1954-55 season. Murrow conducted an interview with J. Robert Oppenheimer, the physicist who was removed as advisor to the Atomic Energy Commission because he was accused of being a Soviet agent. See It Now documented the effects of the Brown v. Board of Education desegregation decision on two southern towns. Murrow, a heavy smoker, examined the link between cigarettes and lung cancer. By the end of the season, Alcoa, stung by See It Now's investigation into a Texas land scandal where it was expanding operations, ended its sponsorship. Because of the profitability of other entertainment shows, most notably the bonanza in game shows, CBS also decided that See It Now should yield its regular timeslot and become a series of specials. Many insiders thought the series should be retitled See It Now and Then.

During the final three seasons of specials, the tone of See It Now became softer. Despite exclusive interviews with Chinese Premier Chou En-lai and Yugoslavian strongman Marshal Josip Tito, the most memorable programs were almost hagiographic profiles of American artists, including Louis Armstrong, Marian Anderson, and Danny Kaye. Controversy for Murrow was now reserved for outside the studio; his 1958 speech to radio- and news-directors was an indictment of the degrading commercialism pervading network television. The final broadcast, "Watch on the Ruhr," on 7 July 1958, surveyed the mood of postwar Germany. After See It Now's demise, CBS News made sure to split the Murrow-Friendly team; Murrow hosted specials, the most significant being Harvest of Shame, and left the network in 1961; Friendly was named executive producer of Now's public affairs successor, CBS Reports.

Murrow and Friendly invented the magazine news format, which became the dominant documentary form on network television. The most esteemed inheritor of its legacy, 60 Minutes, was conceived by integral See It Now alumni: Don Hewitt (as 60 Minutes's executive producer), Palmer Williams (as managing editor), and Joe Wershba (as producer). See It Now was also a seminal force in how most television documentaries conveyed a national issue: illuminating the individual story, immediately and directly, so that it resonates with deeper implications. If Murrow and

Friendly established the model for the documentary for both form and content, they also tested the limits of editorial advocacy. Although the series of McCarthy programs have been lionized as one of television's defining moments, Murrow and Friendly exposed as well the inherent tension between the news and the network and sponsor. How to deal with controversy in a commercial medium has remained controversial ever since.

—Ron Simon

HOST
Edward R. Murrow

PRODUCERS   Fred W. Friendly, Edward R. Murrow

PROGRAMMING HISTORY

• CBS

November 1951–June 1953	Sunday 6:30-7:00
September 1953–July 1955	Tuesday 10:30-11:00
September 1955–July 1958	Irregular Schedule

FURTHER READING

Barnouw, Erik. Tube of Plenty: The Evolution of American Television. New York: Oxford University Press, 1975; revised edition, 1990.

Bliss, Edward J. Now the News: The History of Broadcast Journalism. New York: Oxford University Press, 1975.

————, editor. In Search of Light: The Broadcasts of Edward R. Murrow, 1938-1964. New York: Knopf, 1967.

Boyer, Peter J. Who Killed CBS? The Undoing of America's Number One News Network. New York: Random House, 1988.

Cloud, Stanley, and Lynne Olson. The Murrow Boys. Boston: Houghton Mifflin, 1996.

Friendly, Fred W. Due to Circumstances Beyond Our Control... New York: Vintage, 1967.

Gates Gary Paul. Air Time: The Inside Story of CBS News. New York: Harper and Row, 1978.

Halberstam, David. The Powers That Be. New York: Knopf, 1979.

Kendrick, Alexander. Prime Time: The Life of Edward R. Murrow. Boston: Little, Brown, 1969.

Matusow, Barbara. The Evening Stars. New York: Ballantine, 1983.

Murrow, Edward R., and Fred W. Friendly, editors. See It Now. New York: Simon and Schuster, 1955.

O'Connor, John E., editor. American History/American Television: Interpreting the Video Past. New York: Frederick Ungar, 1983.

Paley, William S. As It Happened. Garden City, New York: Doubleday, 1979.

Persico, Joseph E. Edward R. Murrow: An American Original. New York: McGraw-Hill, 1988.

Reeves, Thomas C. The Life and Times of Joe McCarthy. New York: Stein and Day, 1982.

Smith, Sally Bedell. *In All His Glory.* New York: Simon and Schuster, 1990.

Sperber, A. M. *Murrow: His Life and Times.* New York: Freundlich, 1986.

See also Censorship; Columbia Broadcasting System; Documentary; Friendly, Fred W.; Hewitt, Don; Murrow, Edward R.; Paley, William S.; Stanton, Frank

SEINFELD

U.S. Situation Comedy

Jerry Seinfeld, American standup comedian and author of the best-selling book *SeinLanguage* (1993), is now best known as the eponymous hero of *Seinfeld*, a sitcom that has been a great success for NBC for the last five years. Yet hero, for the show's fans in the United States and around the world, is not the right word for Jerry in *Seinfeld.* Nor would it describe the show's other main characters, Elaine, George, and (Cosmo) Kramer, all thirtysomething and leading the single life in New York. The program's distinctiveness lies in being a comedy made out of trivia and minutiae, a bricolage of casual incidents and situations of everyday metropolitan life, all of which belie any conventional notion of "heroism," any notion, indeed, of distinction. We see Jerry in his apartment, with bizarre neighbour Kramer constantly dropping in, and Elaine and George visiting, or in the café where they are all regular customers, or at Elaine's office where she worked as a publisher, until she lost her job. (She has since worked in a series of situations, usually as personal assistant to eccentric, bizarre individuals.) Seinfeld himself, in an interview, suggested that *Seinfeld* was adding something new to television comedy, some new representation of the quotidian that might be influencing other TV and film culture. He cited some of the coffee shop conversation between the John Travolta and Samuel Jackson characters in Quentin Tarantino's *Pulp Fiction*, and Tarantino in turn has admitted to being a big fan of *Seinfeld.*

Seinfeld does not mix seemingly trivial conversation and incidents with sudden unnerving violence as does *Pulp Fiction*, whose main characters, gangsters, create a world of shattering absurdity. Jerry, Elaine, George, and Kramer instead lead a life of quiet absurdity. They appear always to be relentlessly superficial. Even to say they are friends would be too kind. If they do help each other, it is out of self-interest only. They create a comic world out of the banally cruel and amoral, of trivial lies, treachery, and betrayal. In their relations with each other, with anyone else they encounter, or with their families, they rarely find it in themselves to act out of altruism, kindness, generosity, support, courage, caring, sharing, concern, neighbourliness, sense of human community, trust. Like comedy through the ages, they say the unsayable, do the undoable, as they casually ignore sanctioned morality and recognised correctness. Watching someone being operated on, they pass callous remarks, and accidentally pop a chocolate ball into the body.

George in particular is freely given to making trouble and then denying all responsibility; to boasting, deceiving, lying. We wait for him to do disgusting things, expecting, hoping, he'll do them. He tries to get money out of a hospital when someone falls to his death from the hospital's window onto his car. He makes love on his parents' bed and leaves behind a used condom. He sells his father's beloved old clothes to a shop, saying his father had died and this was his dearest wish. He hopes an artist will die so his paintings will go up in value.

Jerry and a girlfriend, who can't make love in his apartment because his parents are visiting, entwine themselves in the flickering darkness when they go to see *Schindler's List* and consequently miss most of the film. Their behaviour is reported to Jerry's Jewish parents by another acquaintance, the treacherous Newman. Much of *Seinfeld* involves similar comic humiliation, and so recalls and reprises a long Jewish tradition of humour that has flourished this century in vaudeville, radio, then film and television: in the figure of the *schlemiel* (think of Woody Allen), making comedy out of failure, ineptitude, defeat, minor disaster.

In *Seinfeld* disasters multiply for each character, except for the mysterious Kramer, a trickster figure, who like trickster figures through the ages always gets out of daily work, is a renowned sexual reptile, generally out-tricks every adversary, and ignores the havoc he insists on causing. In *Seinfeld* Kramer functions as pure sign of folly, misrule, turning the world upside-down at every chance.

Elaine is Jerry's former girlfriend. With George she has a relationship of uneasiness, if not sharp mutual dislike. Elaine is sassy and spunky, but her spunkiness usually emerges as irritability and impatience (especially in restaurants or waiting to see a film). She picks arguments with almost everyone she encounters, including any boyfriend. In matters of romance, Elaine constantly self-destructs. So, too, do Jerry and George, usually quickly allowing a trivial difference or unfounded suspicion to end a relationship. Once Jerry insisted that he and Elaine make love again, but he can't get it up, and here Elaine emerges as similar to the irrepressible female carnival figures of early modern Europe (as discussed by Natalie Zemon Davis in her famous essay "Women on Top"), overturning men's power and self-image.

Seinfeld also recalls a long comic tradition of farce that descends from Elizabethan drama. In the plays and the jigs following, the audience was presented with a contestation of ideals and perspectives. Whatever moral order is realized in the play is placed in tension with its parody in the closing jig. There the clown dominated as festive Lord of Misrule, creating, for audiences to ponder, not a definite conclusion but an anarchy of values, a play of play and counterplay. Similarly, *Seinfeld* continuously presents an absurd mirror image of other television programs that, like Shakespeare's romances, hold out hope for relationships despite every obstacle that tries to rend lovers, friends, kin, neighbours apart, obstacles that create amidst the comedy sadness, pathos, and intensity.

The possible disadvantage of a genre like absurdist farce is repetition and sameness, comic action turning into ritualised motion. Seinfeld himself comments that in *Seinfeld*, "You can't change the basic situation or the basic characters." Nevertheless, he rejected the suggestion that even the show's devotees think the characters are becoming increasingly obnoxious and the jokes forced (*TV Week*, 4 March 1995). While some contemporary satirical comedy such as *Married...With Children* may have fatally succumbed to this danger, *Seinfeld* remains one of the most innovative and inventive comedies in the history of American television.

—John Docker

CAST

Jerry Seinfeld . . . . Himself
Elaine Benes . . . . Julia Louis-Dreyfus
George Costanza . . . . Jason Alexander
Kramer . . . . Michael Richards

PRODUCERS   Larry David, Jerry Seinfeld

PROGRAMMING HISTORY

• NBC

May 1990–July 1990	Thursday 9:30-10:00
January 1991–February 1991	Wednesday 9:30-l0:00
April 1991–June 1991	Thursday 9:30-10:00
June 1991–December 1991	Wednesday 9:30-10:00
December 1991–January 1993	Wednesday 9:00-9:30
February 1993–August 1993	Thursday 9:30-10:00
August 1993–	Thursday 9:00-9:30

FURTHER READING

Davis, Natalie Zemon. *Society and Culture in Early Modern France.* Stanford, California: Stanford University Press, 1975.

Seinfeld

Docker, John. *Postmodernism and Popular Culture: A Cultural History.* Melbourne, Australia: Cambridge University Press, 1994.

Johnson, Carla. "Luckless in New York: The Schlemiel and the Schlimazel in Seinfeld." *Journal of Popular Film and Television* (Washington, D.C.), Fall 1994.

Radway, Janice A. *Reading the Romance.* Chapel Hill: University of North Carolina Press, 1984.

Rapping, Elayne. "The *Seinfeld* Syndrome." *The Progressive* (Madison, Wisconsin), September 1995.

"Sein of the Times?" (interview). *TV Week* (Australia), 4 March 1995.

Wiles, David. *Shakespeare's Clown: Actor and Text in the Elizabethan Playhouse.* Cambridge: Cambridge University Press, 1987.

See also Comedy, Domestic Settings

SELLERS, PETER

British Comedian and Actor

While the late actor Peter Sellers is primarily known for his roles in film comedies such as the *Pink Panther* series, he first became a British celebrity as a member of the cast of *The Goon Show*, a satirical BBC radio series.

Originally aired in 1951, the show teamed Sellers with fellow comedians Spike Milligan and Harry Secombe. The program was a shocking departure for listeners accustomed to urbane humor from the BBC—the Goons combined a

Peter Sellers

zany blend of odd characters in sketches that poked fun at every aspect of English society. Sellers used mimicry skills honed as a stand-up comedian in London striptease bars to create a number of distinctive characters with equally memorable names: Grytpype Thynne, Bluebottle, Willum Cobblers, and Major Bloodnok. The show acquired a cult following with BBC audiences around the world, and helped launch Sellers' film career.

Goon Show influences can be traced to equally-eccentric British television progeny such as *Monty Python's Flying Circus* and *The Benny Hill Show*. The Goons, led by Sellers, created a distinctive media genre that combined Kafkaesque humor with hilariously stereotypical English characters. This new genre paved the way for the Pythons and others to follow in the 1960s and 1970s.

In 1979, Peter Sellers appeared in Hal Ashby's production of *Being There*, a film version of Jerzy Kosinski's satirical novel on the cultural influence of television. In the film, Sellers played Chauncey Gardiner, a none-too-bright gardener who was for-

cibly thrust into the outside world after the death of his benefactor. Sheltered in his employer's home, Chauncey's world-view was entirely shaped by the television shows he watched on sets scattered throughout the house. After being cast from this TV-defined Eden, Chauncey and his childlike innocence were challenged by the harsh realities of the outside world at every turn. In one memorable scene, he was menaced by members of an inner-city street gang as he urgently pressed a TV remote control to make them "go away." In another scene, Sellers kissed a passionate female character played by Shirley MacLaine as he mimiced a televised love scene that he was watching over her shoulder.

Being There reflected Kosinski's jaundiced view of the influence of television on modern culture, and the tendency to confuse actual events with their symbolic media representations. In Kosinski's sardonic world, the innocent jabberings of a moronic child-man were mistaken as profound wisdom—at the end of the film Chauncey was feted as a presidential candidate.

This story resonated with Peter Sellers at first reading, and he pursued Kosinski for seven years for the film rights. During the making of the motion picture, Sellers became Chauncey Gardiner—so much so that friends were alarmed at his 24-hour-a-day transformation. The result was one of Sellers' funniest and most poignant screen roles. He was an innocent man cast adrift in a world full of duplicitous people and contrived mediated images. The film, like Kosinski's novel, was one of the most trenchant indictments of the role of television in society yet mounted in fictional form. The film was a fitting end to a career built on Sellers' own unique mimicry skills. He contrived a number of quirky illusory personas—a diverse world that included such memorable characters as Grytpype Thynne, Jacques Clouseau, and Chauncey Gardiner.

—Peter B. Seel

PETER (RICHARD HENRY) SELLERS. Born in Southsea, Hampshire, England, 8 September 1925. Attended St. Aloysius College, London. Married: 1) Anne Howe, 1951 (divorced, 1964); children: Michael and Sarah; 2): Britt Ekland, 1964 (divorced, 1969); child: Victoria; 3) Miranda Quarry, 1970 (divorced, 1974); 4) Lynne Frederick, 1977. Served in Royal Air Force, 1943–46. Began career in revue at the age of five; worked as drummer in dance band; entertainment director of holiday camp, 1946–47; vaudeville comedian, first at the Windmill Theatre, London, 1948, then on vaudeville circuit, 1949–56; made film debut, 1951; performer, *The Goon Show* and other radio programmes, 1948–59; achieved international stardom in *Pink Panther* film series. Commander of the Order of the British Empire, 1966. Recipient: British Academy Best British Actor Award, 1959; San Francisco International Film Festival Golden Gate Award for Best Fiction Short, 1960; San Sebastian Award for Best British Actor, 1962; Teheran Film Festival Best Actor Award, 1973; *Evening News* Best Actor of the Year Award, 1975. Died in London, 24 July 1980.

TELEVISION SERIES (selection)

1956	*The Idiot Weekly, Price 2d*
1956	*A Show Called Fred*
1956	*Son of Fred*
1957	*Yes, It's the Cathode Ray Tube Show*
1963	*The Best of Fred* (compilation)

FILMS

Penny Points to Paradise, 1951; *London Entertains*, 1951; *Let's Go Crazy*, 1951; *Down Among the Z Men*, 1952; *Super Secret Service*, 1953; *Orders Are Orders*, 1954; *John and Julie*, 1955; *The Ladykillers*, 1955; *The Case of the Mukkinese Battlehorn*, 1955; *The Man Who Never Was*, 1955; *The Smallest Show on Earth*, 1957; *Death of a Salesman*, 1957; *Cold Comfort*, 1957; *Insomnia is Good for You*, 1957; *The Naked Truth*, 1958; *Up the Creek*, 1958; *Tom Thumb*, 1958; *Carlton-Browne of the F.O.*, 1958; *The Mouse That Roared*, 1959; *I'm All Right, Jack*, 1959; *Battle of the Sexes*, 1960; *Two-Way Stretch*, 1960; *The Running, Jumping and Standing Still Film* (also producer), 1960; *Never Let Go*, 1961; *The Millionairess*, 1961; *The Road to Hong Kong*, 1961; *Mister Topaze* (also director), 1961; *Only Two Can Play*, 1962; *Waltz of the Toreadors*, 1962; *Lolita*, 1962; *The Dock Brief*, 1963; *Heavens Above*, 1963; *The Wrong Arm of the Law*, 1963; *The Pink Panther*, 1963; *Dr. Strangelove; or, How I Learned to Stop Worrying and Love the Bomb*, 1964; *The World of Henry Orient*, 1964; *A Shot in the Dark*, 1964; *What's New Pussycat?*, 1965; *The Wrong Box*, 1966; *After the Fox*, 1966; *Casino Royale*, 1967; *The Bobo*, 1967; *Woman Times Seven*, 1967; *The Party*, 1968; *I Love You, Alice B. Toklas*, 1968; *The Magic Christian*, 1969; *Hoffman*, 1970; *There's a Girl in My Soup*, 1970; *A Day at the Beach*, 1970; *Simon, Simon*, 1970; *Where Does it Hurt?*, 1972; *Alice's Adventures in Wonderland*, 1972; *The Blockhouse*, 1973; *The Optimist*, 1973; *Soft Beds and Hard Battles*, 1973; *Ghost in the Noonday*, 1974; *The Great McGonagall*, 1974; *The Return of the Pink Panther*, 1974; *Murder by Death*, 1976; *The Pink Panther Strikes Again*, 1976; *Revenge of the Pink Panther*, 1978; *Being There*, 1979; *The Prisoner of Zenda*, 1979; *The Fiendish Plot of Dr. Fu Manchu*, 1980; *The Trail of the Pink Panther*, 1982.

RADIO

Show Time, 1948; *Ray's a Laugh*, 1949; *The Goon Show*, 1951.

RECORDINGS (selection)

I'm Walking Backwards for Christmas; *The Ying Tong Song*; *Any Old Iron*; *A Hard Day's Night*; *Goodness Gracious Me*; *Bangers and Mash*; *The Best of Sellers*; *Songs for Swingin' Sellers*.

STAGE

Brouhaha, 1958.

PUBLICATION (selection)

The Book of the Goons, with Spike Milligan. London: Robson, 1974.

FURTHER READING

Braun, Eric. "Authorized Sellers." *Films* (London), August 1982.

Evans, Peter. *Peter Sellers: The Mask Behind the Mask*. New York: New American Library, 1980.

Lewis, Roger. *The Life and Death of Peter Sellers*. London: Century, 1994.

McGillivray, D. "Peter Sellers." *Focus on Film* (London), Spring 1974.

McVay, D. "The Man Behind." *Films and Filming* (London), May 1963.

Miller, M. "Goonery and Guinness." *Films and Filming* (London), January 1983.

Peary, Gerald. "Peter Sellers." *American Film* (Washington, D.C.), April 1990.

Sellers, Michael, with Sarah and Victoria Sellers. *P.S. I Love You: Peter Sellers, 1951–80*. New York: Dutton, 1981.

Sinoux, J. "Bye Bye Birdie—mun-num." *Positif* (Paris), February 1981.

Sylvester, D. *Peter Sellers*. New York: Proteus, 1981.

Thomson, D. "The Rest Is Sellers." *Film Comment* (New York), September-October 1980.

Walker, A. *Peter Sellers: The Authorized Biography*. London: Weidenfeld and Nicholson, 1981.

THE SELLING OF THE PENTAGON

U.S. Documentary

The *Selling of the Pentagon* was an important documentary aired in prime time on CBS on 23 February 1971. The aim of this film, produced by Peter Davis, was to examine the increasing utilization and cost to the taxpayers of public- relations activities by the military-industrial complex in order to shape public opinion in favor of the military. The subject was not new, and had been heavily discussed in the press and debated in Congress. The junior senator from Arkansas, J. William Fulbright, had first raised the subject in a series of four widely-publicized speeches in the Senate in December 1969. In November 1970, Fulbright published his book *The Pentagon Propaganda Machine*, and this formed the core around which the network constructed its version of the Senator's ideas. While the controversial nature of the subject-matter was clearly understood by the producers, and a strong reaction was anticipated, the virulence and direction of this reaction could not have been foreseen. In the end, the furor surrounding *The Selling of the Pentagon* would serve as a significant benchmark in evaluating the First Amendment Rights of the broadcast media.

The documentary, narrated by Roger Mudd, concentrated on three areas of Pentagon activity to illustrate its theme of public manipulation: direct contacts with the public, Defense Department films, and the Pentagon's use of the commercial media—the press and television. From the opening sequence of "firepower display" at Armed Forces Day in Fort Jackson, South Carolina, culminating in the last "mad minute" when all the weapons on display were fired simultaneously, through the middle section which showed clips of the anti-Communist film *Red Nightmare*, to the closing section which detailed how the media are "managed" by the Pentagon, the documentary unveiled a massive and costly public-relations effort to improve the public perception of the military. However, these facts, while open to some subjective interpretation, were not the real cause of the dispute.

The real issues of contention centered around how the producers had "reconstructed" several key interviews and speeches shown in the documentary. The first controversial sequence involved a lecture by Army Colonel John A. McNeil, which began with Mudd's voice-over noting that "The Army has a regulation stating 'Personnel should not speak on the foreign policy implication of U.S. involvement in Vietnam.'" McNeil was then shown delivering what appeared to be a six-sentence passage from his talk, which

made him seem to be contravening official military regulations. In fact, the sequence was reconstructed from several different passages over a wide range of pages, and taken out of context in places.

The second of the controversial interview sequences was with Assistant Secretary of Defense Daniel Henkin on the reasons for the public displays of military equipment at state fairs and shopping centers. Again, many of Henkin's answers were taken out of context and juxtaposed, making him appear, in television critic Martin Mayer's words, "a weasler and a fool." Henkin, in keeping with government policy, had made his own tape recording of the interview, and was therefore able to demonstrate how skillful editing had distorted what he had actually said.

The complaints about the show began only 14 minutes after it went on the air with phone calls to the network. The outcry in subsequent days was centered around two main sources: Representative F. Edward Hebert, chair of the House Armed Services Committee; and Representative Harley O. Staggers, chair of the House Committee on Interstate and Foreign Commerce and its Special Subcommittee on Investigations. On 23 March 1971, CBS ran the documentary again, and this time followed it by 20 minutes of critical remarks by the vice president, Spiro T. Agnew, Representa-

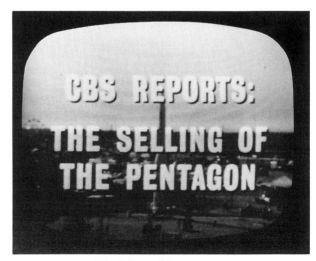

The Selling of the Pentagon
Photo courtesy of Broadcasting and Cable

tive Hebert, and Secretary of Defense Melvin Laird, with a rebuttal by CBS News president Richard Salant. This did not satisfy the politicians, and on 7 April, Representative Staggers caused subpoenas to be issued to CBS demanding the record of the production of the documentary.

The next move was up to CBS, and on the afternoon of 20 April, the network responded to the first executive session of the Special Subcommittee on Investigations through its deputy general counsel, John D. Appel. CBS disputed Representative Staggers' comment that "the American public has a right to know and understand the techniques and procedures which go into the production and presentation of the television news documentaries upon which they must rely for their knowledge of the great issues and controversies of the day." The network had voluntarily submitted the film and complete script of *The Selling of the Pentagon*, but refused to supply the outtakes, draft notes, payments to persons appearing, and other material that had been subpoenaed.

The Federal Communications Commission (FCC) refused to become involved in the case, and the subcommittee held a series of hearings which included testimony from Assistant Secretary Henkin, and Col. John A. McNeil (who had in the interim filed a $6 million lawsuit against the network). On 24 June, at the subcommittee's third meeting, the star witness was Dr. Frank Stanton, the president of CBS. Stanton claimed that he had "a duty to uphold the freedom of the broadcast press against Congressional abridgment," and pointed out the differences between print and broadcast journalism. He noted that these issues would not arise with the print media, but "because broadcasters need government licenses while other media do not, the First Amendment permits such an intrusion into the freedom of broadcast journalism, although it admittedly forbids the identical intrusion into other press media." There was a provocative exchange between Representative Springer over the definition of "the press," with the Congressman trying to prove, with the aid of a 1956 edition of Webster's *New Collegiate Dictionary*, that broadcasting was not part of "the press." Stanton testified for more than four hours, and in the end he refused to submit to the subcommittee's subpoena.

In the midst of the furor concerning *The Selling of the Pentagon*, an even more important First Amendment issue was thrust upon the public scene. On 13 June, *The New York Times* published the first installment of the series of what became known as *The Pentagon Papers*. This case moved rapidly through the courts, and on 30 June the Supreme Court, by a vote of six to three, allowed the unrestrained publication of the documents.

It was against this background that on 28 June the subcommittee voted unanimously to refer the entire case to its parent Committee on Interstate and Foreign Commerce. On 1 July, the full committee voted 25 to 13 to report the matter to the House, with a recommendation that the network and Stanton be cited for contempt. Stanton could

not help but notice the contrast between the two decisions: "This action is in disappointing contrast to the Supreme Court's ringing reaffirmation yesterday of the function of journalism in a free society."

On 8 July, Staggers made his bid for House support with a floor speech and a letter to members of Congress. On 13 July, in a surprisingly heated debate, the issue came to a head. In the end one of the committee members, Representative Hastings Keith, introduced a motion to recommit the resolution to the committee, which was asked to report back to the floor with legislation that would more adequately express the intent of Congress, and give authority to the FCC to move in a constitutional way that would require the networks to be as responsible for the fairness and honesty of their documentaries as for quiz shows and other programs. After a roll call vote, the resolution was approved 226 to 181, effectively negating the contempt citations. Staggers commented: "The networks now control this Congress." Stanton, as was to be expected, was extremely pleased by what he felt was "the decisive House vote."

What was the final outcome? Was the vote really that decisive? On 15 July, Representative Keith followed through on his promise and introduced legislation that would have prohibited broadcasters from staging an event, or "juxtaposing or rearranging by editing," without indicating to the public that this had occurred. The proposed legislation never made it to the floor. The final outcome was a victory of sorts for CBS specifically, and broadcast journalism in general, for never in modern history had the House failed to sustain the vote of one of its committees to cite for contempt.

The Selling of the Pentagon was a milestone in the development of the television documentary, not so much for what it contained, but because it represented a clear statement that the networks could not be made to bend to government control in the technological era.

—Garth S. Jowett

NARRATOR
Roger Mudd

PRODUCER   Peter Davis

PROGRAMMING HISTORY

• CBS
23 February 1971

FURTHER READING
Fulbright, J. William. *The Pentagon Propaganda Machine*. New York: Liveright, 1970.
Irvine, Reed J. "The Selling of 'The Selling of the Pentagon,'" *National Review* (New York), 10 August 1971.
Jowett, Garth S. "'The Selling of the Pentagon': Television Confronts the First Amendment." In, O'Connor, John, editor. *American History/American Television; Interpreting the Video Past*. New York: Ungar, 1983.

Mayer, Martin. *About Television*. New York: Harper and Row, 1972.

Rogers, Jimmie N., and Theodore Clevenger, Jr. "'The Selling of the Pentagon': Was CBS the Fulbright Propaganda Machine?" *Quarterly Journal of Speech* (Falls Church, Virginia), October 1971.

Smith, F. Leslie. "'Selling of the Pentagon' and the First Amendment." *Journalism History* (Northridge, California), Spring 1975.

See also Columbia Broadcasting Company; Stanton, Frank; Vietnam on Television; Wallace, Mike

SERLING, ROD

U.S. Writer/Producer

Rod Serling was perhaps the most prolific writer in American television. It is estimated that during his twenty-five year career, from the late 1940s to 1975, over 200 of his teleplays were produced. This staggering body of work for television has ensured Serling's place in the history of the medium. His emphasis on character (psychology and motivation), the expedient handling of incisive, direct and forceful and painfully penetrating dialogue, alongside his moralizing subtext, placed him in a unique position to question humankind's prejudices and intolerance as he saw it.

Following army service Serling entered Ohio's Antioch College as a student under the GI bill, where he began writing radio and television scripts, selling a number while still an undergraduate. Upon leaving college he went to work as a continuity writer for a Cincinnati television station, WLWT-TV, and then began writing a regular weekly series of live dramas for the anthology show *The Storm*, produced by Robert Huber for WKRC-TV Cincinnati. Turning freelance in 1952, Serling sold scripts to such network anthologies as *Lux Video Theatre*, *Hallmark Hall of Fame*, *The Doctor*, *Studio One*, and *Kraft Television Theatre*. It was for the latter show that Serling wrote "Patterns" (ABC, 12 January 1955), a powerful drama about corporate politics and big business power games. It was an instant success with both the viewers and critics, winning him his first of six Emmy Awards (for Best Original Teleplay Writing), as well as a Sylvania Award for Best Teleplay.

He followed this with, among others, an adaptation of Ring Lardner's "The Champion"/*Climax* (1955), "The Rack"/*U.S. Steel Hour* (1955), "Incident in an Alley"/*U.S. Steel Hour* (1955), "Noon on Doomsday"/*U.S. Steel Hour* (1956), and "Forbidden Area"/*Playhouse 90* (1956). "Forbidden Area" was his first for *Playhouse 90* (an adaptation of a Pat Frank story) and was also that show's premiere episode. But it was *Playhouse 90*'s second presentation that brought him his greatest success: "Requiem for a Heavyweight" (CBS, 11 October 1956). This compelling yet overlong story of a boxer who knows that he's washed up but does not know anything else than the world of the ring projected Serling to the top ranks of the TV writing elite and brought him a gallery of awards, including another Emmy (for Best Teleplay Writing), a Harcourt-Brace Award, another Sylvania Award (for Best Teleplay Writing), a Television-Radio Writers' Annual Award, a Writers Guild of America Award, and the first ever George Foster Peabody Award for writing. *Playhouse 90* and CBS promptly signed him to a contract and he became one of the show's chief writers (among such distinguished names as Horton Foote and Reginald Rose). Serling's next *Playhouse 90*, "The Comedian" (CBS, 14 February 1957), based on Ernest Lehman's story about an egomaniacal entertainer, gave him his third Emmy for Best Teleplay Writing.

But then, from 1958, his conflicts with networks and sponsors over censorship of his work became increasingly intense. "I can recall the blue-penciling of a script of mine called 'A Town Has Turned to Dust'", he said in a 1962 *TV Guide* interview, "in which a reference to a 'mob of men in masks and sheets' was cut because of possible affront to

Rod Serling

Southern institutions". Eventually these censorship battles led to Serling making a transition from live drama to filmed series television, and his own *The Twilight Zone*.

Stemming from a Serling-scripted *Westinghouse-Desilu Playhouse* entry called "The Time Element" in November 1958, Serling created, executive-produced, hosted and (for the most part) wrote the half-hour science-fantasy anthology *The Twilight Zone*, networked by CBS from 1959 to 1964. The series not only created a whole new programming genre for television, it also offered Serling an opportunity to say things he could never get away with in more conventional dramatizations. The weekly tales remain memorable for allowing the viewer to enter "the middle ground between light and shadow, between science and superstition" which lay "between the pit of man's fears and the summit of his knowledge".

The Twilight Zone added two more Emmy awards (Outstanding Writing Achievement in Drama) to Serling's already impressive collection of tributes. His sixth and final Emmy came during *Twilight Zone's* run for the 1963 *Bob Hope Presents the Chrysler Theatre* segment "It's Mental Work" (also for Outstanding Writing Achievement in Drama, Adaptation). But it was with *The Twilight Zone* that Serling reached the peak of his success, for most of what followed after this period would be below Serling's personal standard.

In the fall of 1965 CBS premiered Serling's *The Loner*, a half-hour, post-Civil War Western about a wandering, introspective cowboy in search of life's meaning, starring Lloyd Bridges. The story behind *The Loner* went back almost five years to the time when Serling believed that his *Twilight Zone* would not be renewed by CBS and, as an alternative, he came up with a one-hour pilot script about a character he called *The Loner*, heading west after the Civil War. CBS turned it down. However, around the same time, *The Twilight Zone* was given the go-ahead for another season and *The Loner* script was shelved. When in early 1965 CBS was looking for a half-hour western for their Saturday night schedules, independent producer William Dozier, remembering Serling's *The Loner* proposal from his CBS days, sold the package (now consisting of Serling as writer, Bridges as star, and Dozier as producer) to the network. The series of 26 episodes (14 of them by Serling) opened to poor ratings and lukewarm reviews. When CBS demanded more "action" (meaning less character and motivation, and more "running gun battles") Serling refused to comply, causing a rift between the writer and the network. *The Loner* left the schedules in April 1966.

For the next few years Serling occupied himself with various projects and programs. He served a two-year term as president of the National Academy of Television Arts and Sciences, hosted TV entertainment shows (*The Liar's Club*, 1969; *Rod Serling's Wonderful World of . . .* , 1970), and turned, once again, to screenplay work with adaptations of novels for *Planet of the Apes* (1968; based on the novel by Pierre Boulle) and *The Man* (1972; from the novel by Irving Wallace, which had actually started out as a telefilm). Not unlike other 1950s TV writers, Serling had based his earliest screenplays on his own television work: *Patterns* (UA, 1956),

The Rack (MGM, 1956), *Incident in an Alley* (UA, 1962), and *Requiem for a Heavyweight* (Columbia, 1962).

In 1969 he was approached by producer Aaron Spelling to write a pilot for a series called *The New People* (ABC, 1969–70), featuring an assorted group of young Americans stranded on a South Pacific atoll. Serling delivered his script but later commented on the *Lord of the Flies* theme that "it may work, but not for me". NBC's horror-fantasy anthology *Night Gallery* (1970–73) was to occupy his time during the early 1970s, following the pilot TV-movie (NBC, 1969), adapted from his short-story collection (*The Season to Be Wary*), published in 1967. Based on the three stories (one directed by the young Steven Spielberg), the Mystery Writers of America presented him with their special Edgar Award for the suitably suspenseful scripts. Also known as *Rod Serling's Night Gallery* (he acted as host and sometime contributor), the series failed to come anywhere close to his *Twilight Zone* sense of "seriousness", as Serling had hoped, and the show quickly deteriorated, according to Tim Brooks and Earle Marsh, into "the supernatural equivalent of *Love, American Style*". There were, however, two Serling episodes that remain outstanding for their sense of compassion and morality: "They're Tearing Down Tim Riley's Bar" and "The Messiah on Mott Street"; both nominated for Emmys.

After *Night Gallery* was cancelled in 1973, he retreated to Ithaca College, in upstate New York, and taught writing. Teaching the art of writing sustained him more than anything else during the last few years of his life. *The Twilight Zone*, in constant reruns, remains a cultural milestone to Serling's art and craft and practice.

—Tise Vahimagi

ROD SERLING. Born Edward Rodman Serling in Syracuse, New York, U.S.A., 25 December 1924. Educated at Antioch College, Yellow Springs, Ohio, B.A. 1950. Married: Carolyn Kramer, 1948; two daughters. Served as paratrooper in U.S. Army during World War II. Worked as writer for WLW-Radio, Cincinnati, Ohio, 1946–48, WKRC-TV, Cincinnati, 1948–53; freelance writer, from 1953; producer, television series *The Twilight Zone*, 1959–64, and *Night Gallery*, from 1969; taught at Antioch College, 1950s, and Ithaca College, 1970s. Honorary degrees: D.H.L., Emerson College, Boston, Massachusetts, 1971, and Alfred University, New York, 1972; Litt.D., Ithaca College, 1972. President, National Academy of Television Arts and Sciences, 1965–66; member of the council, Writers Guild of America West, 1965–67. Recipient: six Emmy Awards; Sylvania Awards, 1955 and 1956; Christopher Awards, 1956 and 1971; Peabody Award, 1957; Hugo Awards, 1960, 1961, and 1962. Died 28 June 1975.

TELEVISION PLAYS (selection; writer)

1953	"Nightmare at Ground Zero" *Suspense*
1953	"Old MacDonald Had a Curve" *Kraft Television Theatre*
1954	"One for the Angels" *Danger*
1955	"Patterns" *Kraft Television Theatre*

1955–56	*U.S. Steel Hour*
1956	"Requiem for a Heavyweight" *Playhouse 90*
1956	"Forbidden Area" *Playhouse 90* (from Pat Frank's novel)
1957	"The Comedian" *Playhouse 90*
1959	"The Lonely" *Twilight Zone*
1959	"Time Enough at Last" *Twilight Zone*
1965–66	*The Loner,* 14 Episodes
1966	*The Doomsday Flight*
1970	"A Storm in Summer" *Hallmark Hall of Fame*
1971	"Make Me Laugh" *Night Gallery*

TELEVISION SERIES (producer)

| 1959–64 | *The Twilight Zone* |
| 1970–73 | *Night Gallery* |

FILMS (writer)

Patterns, 1956; *Saddle the Wind* (with Thomas Thompson), 1958; *Requiem for a Heavyweight,* 1962; *The Yellow Canary,* 1963; *Seven Days in May,* 1964; *Assault on a Queen,* 1966; *Planet of the Apes* (with Michael Wilson), 1968; *A Time for Predators,* 1971.

STAGE

The Killing Season, 1968.

PUBLICATIONS

Stories from the Twilight Zone. New York: Bantam, 1960.
More Stories from the Twilight Zone. New York: Bantam, 1961.
New Stories from the Twilight Zone. New York: Bantam, 1962.
Requiem for a Heavyweight (novel). New York: Bantam, and London: Corgi, 1962.
From the Twilight Zone (short stories). New York: Doubleday, 1962.
Night Gallery (short stories). New York: Bantam, 1971.
Night Gallery 2 (short stories). New York: Bantam, 1972.
Rod Serling's Night Gallery Reader. Greenberg, Martin H., Carol Serling, and Charles G. Waugh, editors. New York: Dembner Books, 1987.

FURTHER READING

Engel, Joe. *Rod Serling: The Dreams and Nightmares of Life in the Twilight Zone.* Chicago: Contemporary Books, 1989.
Sander, Gordon. *Serling: The Rise and Twilight of Television's Last Angry Man.* New York: Penguin, 1992.

See also Anthology Drama; "Golden Age" of Television; *Playhouse 90; Twilight Zone*

SEVAREID, ERIC

U.S. Journalist

Eric Sevareid was one of the earliest of a group of intellectual, analytic, adventureous, and sometimes even controversial newspapermen, hand-picked by Edward R. Murrow as CBS radio foreign correspondents. Later Sevareid and others of this elite band of broadcast journalists, known as "Murrow's Boys," distinguished themselves in television. From 1964 until his retirement from CBS in 1977, he carried on the Murrow tradition of news analysis in his position as national correspondent for *The CBS Evening News.* There, his somber, eloquent commentaries were either praised as lucid and illuminating, or criticized for sounding profound without ever reaching a conclusive point.

Sevareid's image as a scholarly commentator on the *CBS Evening News* was belied by an early career in which he was something of a swashbuckler. Sevareid was working at the *New York Herald Tribune's* Paris office when his writing abilities caught the eye of Edward R. Murrow, who offered him a job. Later Sevareid would say of those early years, "We were like a young band of brothers in those early radio days with Murrow." In his final 1977 *CBS Evening News* commentary, Sevareid referred to Murrow as the man who "invented me."

As one of "Murrow's Boys" during World War II, Sevareid "scooped the world" with his broadcast of the news of the French surrender in 1940. He joined Murrow in covering "The Battle of Britain"; he was lost briefly after parachuting into the Burmese Jungle when his plane developed engine trouble while covering the Burmese-China theater; he reported on Tito's partisans; and he landed with the first wave of American troops in Southern France, accompanying them all the way to Germany.

In 1946, after reporting on the founding of the United Nations, Sevareid wrote *Not So Wild a Dream,* which appeared in 11 printings and became a primary source on the lives of the generation of Americans who had lived through the Depression and World War II. For the 1976 edition of the book, he wrote, "It was a lucky stroke of timing to have been born and lived as an American in this last generation. It was good fortune to be a journalist in Washington, now the single news headquarters in the world since ancient Rome. But we are not Rome; the world is too big, too varied."

Always considering himself a writer first, Sevareid felt uneasy behind a microphone and even less comfortable with television; nevertheless, he did such early Sunday news ghetto programs as *Capitol Cloakroom* and *The American Week,* and served as host and science reporter on the CBS series *Conquest.* As head of the CBS' Washington bureau from 1946 to 1959, Sevareid was an early critic of McCarthyism, and, in one of the few even mildly critical comments he ever made about Murrow, he observed that he came to the issue rather late.

Serving as CBS' roving European Correspondent from 1959 to 1961, Sevareid contributed stories to *CBS Reports* as well as serving as moderator of series such as *Town Meeting of the World*, *The Great Challenge*, *Where We Stand*, and *Years of Crisis*. In addition, he also appeared in every presidential election coverage from 1948 to 1976. However, one of Sevareid's scoops of those years, his 1965 exclusive interview with Adlai Stevenson shortly before his death, for which he won a New York Newspaper Guild Page One award, was not broadcast over CBS, but instead appeared in *Look* magazine.

From 1963 until his retirement Sevareid appeared on the *CBS Evening News with Walter Cronkite*. During that period his Emmy and Peabody award-winning two-minute commentaries, with their penchant to elucidate rather than advocate, inspired those who admired him to refer to him as the "Grey Eminence." On the other hand, those who were irked by his tendency to overemphasize the complexity of every issue nicknamed him "Eric Severalsides." Sevareid himself said that as he had grown older his tendency was toward conservatism in foreign affairs and liberalism in domestic politics. Despite this, after a trip to South Vietnam in 1966 he commented that prolonging the war was unwise and a negotiated settlement was advisable. His commentary on the resignation speech of President Richard M. Nixon ("Few things in his presidency became him as much as his manner of leaving the presidency") was hardly as perceptive.

Beside keeping alive the Murrow tradition of news commentary at CBS, Sevareid, in keeping with another Murrow tradition, interviewed noted individuals such as West German Chancellor Willy Brandt, novelist Leo Rosten, and many others on the series *Conversations with Eric Sevareid*. In something of a spoof of this tradition he also did a conversation with King George III (played by Peter Ustinov) entitled *The Last King in America*.

After his retirement, Sevareid continued to be active as a CBS consultant and narrator of shows such as *Between the Wars* (Syndicated, 1978), a series on American diplomacy between 1920 to 1941, *Enterprise* (PBS, 1984), a series on American business, and *Eric Sevareid's Chronicle* (Syndicated, 1982). His final appearance, before his death in 1992, was on the 1991 CBS program *Remember Pearl Harbor*. Needless to say, Sevareid's presence at CBS was a link to the Murrow tradition, long after Murrow himself and many of his "Boys" left the network, and after that tradition ceased to have significant practical relevance at CBS News.

—Albert Auster

ERIC SEVAREID. Born in Velva, North Dakota, U.S.A., 26 November 1912. Educated at the University of Minnesota, B.A. in political science 1935; studied at London School of Economics, and Alliance Française, Paris. Married: 1) Lois

Eric Sevareid

Finger, 1935 (divorced, 1962); two sons; 2) Belén Marshall, 1963; one daughter; 3) Suzanne St. Pierre. Worked as teenager as copy boy for the *Minneapolis Journal*; worked during college as freelancer for the *Minneapolis Star*; served on staff of the *Minneapolis Journal*, 1936–37; reporter, Paris edition of the *New York Herald Tribune,* 1938; recruited to join CBS radio by Edward R. Murrow, 1939; traveled with French army and air force for CBS, 1939–40, became first to report France's capitulation to Germany; assigned to CBS News Bureau in Washington, D.C., 1941–43; served as war correspondent in China, 1943–44, London, 1945; served as chief Washington, D.C., correspondent for CBS, 1946–59; worked as European correspondent, 1959–61; moderator, numerous CBS News programs, 1961–64; served as commentator for *The CBS Evening News*, from 1963; national correspondent, CBS News, from 1964; hosted interview series, *Conversations With Eric Sevareid*, from 1977; consultant, CBS News, from 1977; reported on numerous presidential conventions. Received numerous honorary degrees. Recipient: Peabody Awards, 1950, 1964, and 1976; Emmy Awards, 1973, 1974, 1977; two Overseas Press Club Awards; Harry S. Truman Award, 1981; numerous other awards. Died 10 July 1992.

TELEVISION

1957–58   *Conquest* (host and science reporter)

1963–77   *CBS Evening News* (commentator)
1964–77   *CBS Evening News* (national correspondent)
1977   *Conversations With Eric Sevareid*

TELEVISION SPECIAL

1959   *CBS Reports: Great Britain—Blood, Sweat and Tears Plus Twenty Years*

PUBLICATIONS

Canoeing With the Cree. New York: Macmillan, 1935.
Not So Wild a Dream. New York: Knopf, 1946.
In One Ear. New York: Knopf, 1952.
Small Sounds in the Night. New York: Knopf, 1956.
Candidates 1960, editor. New York: Basic Books, 1959.
This is Eric Sevareid. New York: McGraw-Hill, 1964.

FURTHER READING

Fensch, Thomas, editor. *Television News Anchors: An Anthology of Profiles of the Major Figures and Issues in United States Network Reporting.* Jefferson, North Carolina: McFarland, 1993.
Gates, Gary Paul. *Air Time: The Inside Story of CBS News.* New York: Harper and Row, 1978.
McCabe, Peter. *Bad News at Black Rock: The Sell-out of CBS News.* New York: Arbor, 1987.
Schoenbrun, David. *On and Off the Air: An Informal History of CBS News.* New York: Dutton, 1989.
Schroth, Raymond A. *The American Journey of Eric Sevareid.* South Royalton, Vermont: Steerforth, 1995.

See also Columbia Broadcasting System; Cronkite, Walter; Murrow, Edward R.; News, Network

SEX

Australian Talk Show

S*ex*, also known as *Sex with Sophie Lee*, was a "lifestyle" show launched in Australia in 1992. Produced by Tim Clucas for the Nine Network, the show went to a second series in 1993 with a new presenter, the comedian Pamela Stevenson. *Sex* can be seen as the first show on Australian TV to try to modernize sexual attitudes and make sex a vital topic of mainstream public discussion in the HIV era; or it can be seen as an attempt by commercial television to consumerize sex itself, making sexual preference into supermarket choice, and use public education as an excuse for exploitative television.

The show was launched to phenomenally high ratings (a 32 share), largely on the lure of its presenter Sophie Lee's own reputation for sexiness. But the early episodes succeeded in mixing straightforward advice about common problems, with some noteworthy firsts for prime-time television, especially by showing human reproductive organs, both male and female, on screen. Most notably, even though its own format comprised traditional magazine-style journalistic and "expert" segments, linked by a studio anchor in glamorous evening-wear, *Sex* crossed one of television's most policed generic boundaries: characters (fiction) can have sex while people (fact) can only talk about it. The presentation of ordinary people being sexual on screen, and the screening of sexualized bodies (even if only in bizarre slow-motion "reconstruction" mode) was enough to give the show an unsettling, innovative feel, and to ensure that *Sex* provoked widespread discussion in the press and popular magazines as well as rating highly. Not all reaction was positive; for instance, General Motors president Holden announced that the giant car company would not advertise during *Sex* because it wanted its products to be associated with "wholesome" topics.

Sophie Lee became progressively disenchanted with the lack of control she had over the items she was contracted to introduce, segments which began to interpret "sex" in terms of ratings-potential rather than public utility. She left the show at the end of its first season, to be replaced by Pamela Stephenson, the Australian-born comedian best-known for the 1970s BBC series *Not the Nine O'Clock News*. Stevenson recorded her links for *Sex* in a studio in Los Angeles, clearly regarding it as her brief to supply the "nudge, nudge, wink, wink" element. After the departure of Sophie Lee, without anyone on or behind the screen to argue for the show's importance in changing public attitudes

Sex
Photo courtesy of TCN Channel Nine

to sex, the series slid from interesting experiment to unstylish exploitation, and was canceled by the Nine Network after two seasons, to be replaced by safer lifestyle shows about money, home improvement, tourism, and gardening.

—John Hartley

HOSTS

Sophie Lee, 1992
Pamela Stevenson, 1993

PRODUCER   Tim Clucas

PROGRAMMING HISTORY   20 Episodes

• Nine Network

May 1992-July 1992	Thursday 8:30-9:00
February 1993-May 1993	Thursday 9:30-10:00

See also Lee, Sophie

SEXUAL ORIENTATION AND TELEVISION

When the freeze on television broadcast licenses was lifted by the Federal Communications Commission in 1952, television stations proliferated throughout the United States. In the same period, the FCC set regulation standards for the mass production of television receivers, making them relatively inexpensive to produce and affordable for the middle-class American public. "Television," previously a phenomenon related primarily to an East Coast, upper-class definition, quickly became an economically profitable industry catering to perceived middle-class tastes.

Throughout the 1950s and 1960s, the television broadcast networks implicitly constructed the mainstream viewing public as replications of the idealized middle-class nuclear family, defined as monogamous, heterosexual couples with children. In response, the overwhelming trend was to provide programming targeted toward this consumer group. To a large degree, of course, this construction stemmed from the larger context of American society in which the ideals of heterosexuality and family dominated the overall hierarchy of sexual orientation.

The assumptions were even more fundamental with this new medium, however, because the mode of distribution of programming and the measure of economic success were significantly different for television broadcasting than for most other forms of popular culture. In those contexts consumers had to actively purchase a product: a movie ticket, a record or a book. Economic success and popularity were determined by the number of sales of the cultural product. Within the setting of American broadcasting, however, the programming was distributed free of charge to anyone with a television set capable of receiving the broadcast signal. The networks generated profits through advertising, selling the viewing audience to commercial sponsors as potential targets for commercial messages. In this mode of distribution, a network's success was determined by the number of viewers it attracted, not the number of programs sold. This interaction among the networks, advertisers and the viewing audience developed into a very complex economic relationship.

Until the early 1970s and the introduction of demographic measurements, the networks quantified a mass audience as an index of a program's popularity to set commercial rates for advertisers. Since most television use by the American public has been and continues to be in a domestic environment, the networks and advertisers easily assumed that the viewing audience mirrored, in its values, the idealized middle-class nuclear family of the 1950s. Given this institutional construction of the television viewer, the networks produced and broadcast a plethora of programs built around the values and concerns of the contemporary nuclear family. Series such as *I Love Lucy*, *Father Knows Best*, *Leave It to Beaver*, and *The Donna Reed Show* developed scripts explicitly exploring gender and sexual roles in the context of the 1950s. For example, *Father Knows Best* often defined appropriate and inappropriate gender behavior as Jim and Margaret Anderson negotiated their marital and implied (hetero)sexual relationship. Explicit discussion of sexual behavior was forbidden. In addition, the Anderson children were groomed for heterosexuality on a weekly basis as they entered into the adolescent dating arena. In the context of the series, same sex romantic attraction was not offered as a viable or legitimate option for offspring Betty, Bud and Kitten. Nor did episodes deal with many heterosexual options outside of conventional coupling, limited to traditional heterosexual norms.

Even series which were not located in the contemporary family milieu of the 1950s or 1960s reinforced a narrow range of heterosexual choices. In a series such as *Gunsmoke* with its surrogate family, traditional heterosexual coupling was the status quo. What sexual tension existed in the series surfaced between Marshall Matt Dillon and saloon owner Miss Kitty, not between Matt and his deputy sidekick Chester. Even between Matt and Miss Kitty overt sexuality was seldom displayed in the series. After all, how was the wild expanse of the Western prairie to be tamed if the product of sexuality was pleasure rather than population growth? Given the baby boom mentality of the 1950s and 1960s, the sexual orientation of *Gunsmoke*'s characters and their sexuality replicated the dominant values of American society, at least as they were perceived by network programmers and advertisers.

This perception of sexuality began to shift slightly by the early 1970s as pleasure became a more acceptable foundation for sexual activity. Even so, sexual orientation continued to be overwhelmingly defined as heterosexual, although an occasional gay or lesbian character began to make an appearance.

Several factors account for this cultural breakthrough. At this time, the Prime Time Access Rule forced the networks out of the business of program production. As a result, the networks began to license programming from independent production companies such as Norman Lear's Tandem Productions and MTM Enterprises. These independents were willing to address subject material, including explicit sexual pleasure and homosexuality, that had previously been ignored by the networks.

Additionally, the networks and advertisers began to shift their conception used to market the viewing audience. In the ratings competition between NBC and CBS during this same period, reliance on undifferentiated mass numbers gave way to the first wave of demographic marketing directed at a younger, urban, rather than older, rural, audience. These young, urban viewers, at least in the perception of the networks and advertisers, were less inclined to take offense at potentially controversial topics. In conjunction with the moxie of independent program producers, sexuality, including explicitly gay characters, began to surface in some programs.

Images of gay men and lesbians began to appear in fictional programming during the early 1970s for another reason as well. Culturally, gay men and lesbians became more visible in American society after the Stonewall Riots in June 1969, a date now celebrated as a watershed moment of the modern gay rights movement. As gays and lesbians entered the struggle for social acceptance and legitimization within mainstream discourse, the emergence of gay characters became part and parcel of this burgeoning social consciousness. In response to a newfound possibility of representation, gay activist groups such as the National Gay Task Force, formed in 1973, attacked any outright negative mainstream media images of gay men and lesbians.

Initially, single-episode gay characters, at best self-destructive and at worst evil, were used as narrative plot devices to create conflict among the regular characters of a prime-time series. This was not an acceptable representation for most gay activists. The first major conflict between gay activists and the networks occurred over just such a depiction in "The Other Martin Loring," an episode of *Marcus Welby, M.D.* during the 1973 broadcast season. The confrontation focused on the dilemma of a closeted gay man worried about the effect of his homosexuality on his family life. Welby's advice and the resolution to the narrative conflict finally rested upon the repression of sexual desire. As Kathryn Montgomery points out in *Target: Prime Time: Advocacy Groups and the Struggle Over Entertainment Television* (1989), this initial conflict had little effect on preventing the broadcast of the episode. However, it did open the door for continued discussion between gay activists and the networks concerning subsequent representations.

Indeed, the networks began to solicit advice about gay representation before programming went into actual pro-

duction. By 1978, the National Gay Task Force provided the networks a list of positive and negative images which it considered to be of greatest importance. From the negative perspective, the organization wanted to eliminate stereotypically swishy gay men and butch lesbians as characters as well as inhibit the portrayals of gay characters as child molesters, mentally unbalanced or promiscuous. In contrast, positive images would include gay characters within the mainstream of the television milieu. These images would reflect individuals performing their jobs well, who were personable and comfortable about their sexual orientation. Additionally, the NGTF asked to see more gay couples, more lesbian portrayals and instances where gayness was incidental rather than the focus of a narrative controversy centered on sexual preference.

As one manner of achieving these positive goals, gay activists suggested that continuing regular gay or lesbian characters be used within a series format, expanding beyond the plot function of a "problem" that needed to be solved and eliminated. However, the inclusion of a recurring gay character created problems of its own. Story editors and script writers had to maintain a delicate balance between creating gay characters who were too extreme in their behavior and therefore offensive to heterosexual mainstream viewers, or characters so innocuous that they become nearly indistinguishable in their gayness. Several series, beginning with *Soap* and *Dynasty* and more recently *Doctor, Doctor* and *Melrose Place*, have included regular gay characters as part of their narrative foundation, with varying degrees of success. Often within these series, the gay character is isolated from any connection to a larger gay community and lacks any presentation of overt sexuality. While it has certainly been acceptable for heterosexual individuals and couples to engage in displays of affection, it has been untenable, until recently, for gay characters to exhibit similar behavior.

Despite this glaring drawback, gay characters as series regulars have functioned differently in the narrative context than in one-shot episodic appearances. For the most part, recurring gay characters have been comfortable with their sexual identity. (The possible exception is Steven Carrington, oil heir apparent in *Dynasty*, who fluctuated in his sexual orientation from season to season.) While a series regular's gayness could still initiate some problems in a series, his or her sexuality, however, was no longer an outside problem. Rather, the series regular could provide a narrative position whereby sexual "otherness" could be used to discuss and critique the dominant representation of both homosexuality and heterosexuality. Contextually, adaptation to, rather the elimination of, homosexuality became the narrative strategy.

Despite *Dynasty*'s wavering on the subject of homosexuality, early installments of the series illustrate this narrative shift. The gay subplots of this prime-time soap opera often performed a pivotal role in exposing the contradictions of heterosexual patriarchy. An excellent example occurred when Blake Carrington, the series' patriarchal figure, stood trial for the death of son Steven's gay lover. The courtroom

Billy Crystal (left) with Richard Mulligan in Soap

setting of this particular subplot created an ideological arena in which Steven could critique his father's homophobia, patriarchal dominance and sense of socially constructed gender roles from an explicitly gay perspective. As can be seen by this example, a gay man or lesbian who appears as part of the regular constellation of a series' cast naturalizes gayness within the domain of mainstream broadcast narratives, thus allowing that sexual otherness a cultural voice of its own. In some instances of this process of naturalization, these fictional gay characters face many of the same problems that their heterosexual counterparts encounter. This has not necessarily meant that their sexual orientation has been ignored, but has been woven together with other concerns to create multi-dimensional, sometimes contradictory characters that reflect some of the experience of gay men and lesbians in American society.

Since 1973, the broadcast networks, program producers and gay activists have maintained an ongoing working relationship with each other. The Alliance for Gay and Lesbian Artists in the Entertainment Industry, an internal industry activist organization, has provided an important connection with outside gay activists. Often, gay men or lesbians within production companies have alerted activists to potential problems with plot lines or characters. Many producers and scriptwriters now elicit opinions from gay and lesbian activists in the preproduction process, thereby circumventing costly confrontations once a production is under way. Also, Network Broadcast Standards and Practices departments have internalized many of the activist's concerns and criticisms, thus pressuring program producers to eliminate potential trouble spots from scripts. The activists have also learned to praise producers, directors and scriptwriters creating appropriate gay-themed programming with positive reinforcement such as yearly awards and congratulatory telegrams, letters and e-mail messages. Because of this de-facto system of checks and balances, antagonistic confrontations seldom arise between gay activists and the television broadcast industry.

The gay activists' success in dealing with the networks and program producers has also activated a strong response from religious and political conservatives since the mid-1970s. As Gitlin argues in *Inside Prime-Time* (1983), these conservative social forces have regarded the social inroads made by gay men and lesbians as a threat to their own social power and deeply embedded patriarchal values including traditional conceptions of the family, gender roles and heterosexuality. Any positive representation of homosexuality (or even bisexuality) is taken to undermine the legitimacy of these traditional values. The conservative far right has been dominated by religious fundamentalist whites males such as Jerry Falwell and Donald Wildmon as well as white anti-feminists such as Phyllis Schlafly. Indeed, Wildmon heads the American Family Association, a formidable advocacy organization which monitors the television broadcasting industry's presentation of sexuality with a bible-thumping fervor.

In contrast to the gay activists who have been more than willing to confront the networks and program producers directly about the representation of sexual orientation, the AFA has employed an indirect approach. Providing members with postcards pre-addressed to advertisers, the AFA has often threatened a boycott of consumer products manufactured by companies placing commercials within the broadcast of objectionable programming. While the direct, preemptive approach of the gay activists appears so far to have been more successful with the commercial networks than the post-broadcast method used by the AFA, the latter organization's efforts have produced some effect. For one thing, advertisers who have come under fire from the AFA have begun to consider placement of a commercial in potentially objectionable programming less lucrative than they might have previously.

As a response to advertisers' reluctance to place commercials in programs that include a positive discussion of homosexuality, the networks' Broadcast Standards and Practices departments have codified some of the AFA's concerns about sexual orientation as a means to counter any negative criticism from conservative advocacy groups. The positive portrayal of any physically romantic or sexual interaction between gay or lesbian characters, for example, has generally been exorcised from programming content. In addition, any gay-themed script must include at least one character who presents a critique of homosexuality to provide a balanced discussion of the subject. As a side note, the Gay and Lesbian Alliance Against Defamation, formed in the mid-1980s, has appropriated AFA's practice of sending out pre-addressed postcards. GLAAD has also urged individuals to send them to advertisers, praising their bravery in placing commercials in gay-themed programming.

At times, program producers and the networks have ended up at the center of a cultural tug of war between gay activists and conservative religious fundamentalists. Perhaps the best illustration of this predicament occurred in the summer of 1977. ABC had scheduled *Soap* for the fall lineup. The series was created by Susan Harris as a satire on both the nuclear family and the overdrawn angst of daytime television drama. One of the regular characters was Jodie Dallas, a swishy gay man. In addition, the heterosexual characters engaged in a number of extra-marital affairs, hardly reinforcing traditional monogamy. ABC pre-

viewed the initial episodes of the series for local affiliates and gay activists. Some disgruntled station owners alerted the National Council of Churches about the risqué content of the show. Also, the conservatives felt the inclusion of Jodie Dallas condoned homosexuality. As a result of the conservative backlash, some affiliates refused to carry *Soap*. Conservative forces picketed stations which did air the satire. Under threat of a product boycott, several potential sponsors backed out of buying time in the series. Gay activists were not pleased with the premise of the Dallas character either. He was too much the gay stereotype. In addition, Dallas was not particularly satisfied with his sexual orientation and planned a sex change operation.

In an attempt to appease both sides, *Soap*'s producers adjusted the series after the first few episodes. Dallas' stereotypical elements were modified, nearly neutering the character in the process. In comparison to the other characters, his behavior became less explicitly sexual. Even so, he became more affirmative about his sexual orientation, dropping any desire to change his gender. Ironically, the more stable, less sexually outrageous Jodie Dallas seemed to address conservative concerns about homosexuality as well. Without the overt presentation of Jodie's sexual desire, apparently religious conservatives believed the series did not condone homosexuality as strongly.

Throughout the 1980s and into the 1990s, opposing gay and conservative advocacy groups have continued to pressure networks, program producers and advertisers on the representational boundaries regarding sexual orientation. As in the case of *Soap*, gay and lesbian characters have usually appeared in a highly diluted form, nominally gay, perhaps with a political stance, but lacking sexuality. Only in a very few instances have these limits been successfully challenged, most notably in an episode of *Roseanne*, a domestic situation comedy and *Serving in Silence: The Margarethe Cammermeyer Story*, a made-for-television movie. In both instances, the cultural and economic clout of their respective production companies provided the impetus to include moments of intimacy and sexuality for lesbian characters. During the spring of 1994, Roseanne, as reigning prime-time diva and executive producer of her series, threatened to withhold an episode from ABC if it did not air with its lesbian kiss intact. The network initially balked, but eventually broadcast the unedited episode rather than lose potential commercial profits from a top-ten series. The combined talents of Barbra Streisand, as executive producer, and Glenn Close, as additional executive producer and star, added production muscle to *Serving in Silence*. With their involvement, NBC gave a green light to the movie, which dealt with both Cammermeyer's fight to be reinstated into the military as an open lesbian and her blossoming romantic relationship with her lover Diane. With Streisand's and Close's involvement providing an aura of quality and legitimacy, this production opened the cultural space for moments of physical intimacy as integral narrative elements. *Roseanne* and *Serving in Silence* have been hallmarks in the presentation of gay and lesbian experience in American television broadcasting, and in 1995

Serving in Silence received the Emmy award for "Outstanding Movie Made for Television."

While gay men and lesbians inside and outside the television industry have applauded these cultural steps forward, the gains are by no means secure, especially outside of the commercial networks where gay activists have less social and economic power. In the American social context of the 1990s, the struggle between gay rights activists and anti-gay rights advocates has reached a crescendo. Both sides have confronted each other over the legitimacy of sexual orientation in the political and legislative arenas, with neither side winning any clear legal victories. However, a conservative shift has occurred in the political arena which could drastically affect gay and lesbian representation in non-commercial American Public Broadcasting. Because the Federal government economically supports non-commercial broadcasting, funding for the Corporation for Public Broadcasting can be reduced or eliminated altogether based on the agendas of powerful political interests. Therefore, proactive intervention—techniques used by groups such as GLAAD with network representatives, program producers and advertisers—have not worked as well in the non-commercial broadcast setting.

Once the bastion of liberal tolerance and a cultural podium for marginal social groups, the Corporation for Public Broadcasting has increasingly come under attack from conservative forces in Congress for precisely those reasons. Conservatives have threatened to eliminate funding and privatize CPB in response to the use of Federal tax dollars to produced non-traditional programming, especially programming targeted to the gay community. Special programming such as Marlon Riggs' *Tongues Untied*, an exploration of gay African-American men's experiences with both homophobia and racism, and Masterpiece Theatre's production of *Armistead Maupin's Tales of the City*, a narrative set in the 1970s San Franciscan milieu of sexual experimentation, have been specific targets of conservatives. Both productions contained a fair amount of frank, adult language about sexuality and a modicum of nudity. Indeed, many PBS affiliates refused to air either program or, if they did broadcast the offerings, censored the material radically. *Tales of the City* generated enough controversy that conservative forces were able to pressure CPB to withdraw funding for the sequel, *More Tales of the City*.

As the social and political struggle over legitimization of gay rights accelerates in the mid-1990s, the inclusion and representation of gay men and lesbians in entertainment television programming will continue to be a point of cultural conflict. Driven by the economic demands placed on network broadcasting as it competes with the relaxed standards on cable channels, programming will probably broaden the parameters of acceptable content. Thus, the economic demands of commercial television will create an atmosphere for further presentation of alternatives to monogamous heterosexual orientation. Also, the gay community has gained more interest from advertisers as a demographic social group with relatively more disposable income to spend. Indeed, some manufacturers of products such as clothing, alcohol, and travel have begun to produce print ads directly targeting gay men and lesbians. Similar advertising in television programming which attracts a gay audience is probably not far behind. In contrast, the strong shift to the conservative right in the political arena has already imposed government regulations on funding for the arts. The Federal government has placed limits on the range of appropriate subject matter for grants from the National Endowment for the Arts, National Endowment for the Humanities, and even the Corporation for Public Broadcasting. It is not outside the realm of possibility that conservative political forces will also attempt to regulate commercial television programming content. Given the larger context, issues about sexual orientation are hardly going to disappear in the near future. If anything, the number of confrontations over sexual orientation and the intensity of those conflicts will only increase.

—Rodney A. Buxton

FURTHER READING

Barnouw, Eric. *Tube of Plenty: The Evolution of American Television.* Oxford, New York, Toronto, and Melbourne: Oxford University Press, 1975; revised edition, 1990.

Brown, Les. *Television: The Business Behind the Box.* New York: Harcourt Brace, 1971.

Buxton, Rodney. "'After It Happened...': The Battle to Present AIDS in Television Drama." In, Newcomb, Horace, editor. *Television: The Critical View.* New York and Oxford: Oxford University Press, 1994.

Cowan, Geoffrey. *See No Evil: The Backstage Battle Over Sex and Violence on Television.* New York: Simon and Schuster.

Doty, Alexander. *Making Things Perfectly Queer: Interpreting Mass Culture.* Minneapolis, Minnesota: University of Minnesota Press, 1993.

"Dream Consumers," *The Austin American-Statesman* (Austin, Texas), 12 March 1989.

Gitlin, Todd. *Inside Prime Time.* New York: Pantheon Books, 1983.

Howes, Keith. *Broadcasting It: An Encyclopaedia of Homosexuality on Film, Radio and TV in the UK, 1923–1993.* London, New York: Cassell, 1993.

Leo, John R. "The Familialism of 'Man' in American Television Melodrama." *South Atlantic Quarterly* (Durham, North Carolina), Winter 1989.

Montgomery, Kathryn. *Target: Prime Time: Advocacy Groups and the Struggle over Entertainment Television.* New York and Oxford: Oxford University Press, 1989.

Parish, James Robert. *Gays and Lesbians in Mainstream Cinema: Plots, Critiques, Casts and Credits for 272 Theatrical and Made-for-Television Hollywood Releases.* Jefferson, North Carolina: McFarland, 1993.

See also Advocacy Groups; Family and Television; Gender and Television; *Pee-wee's Playhouse*; Racism, Ethnicity, and Television; Randall, Tony; *Soap*; *Starsky and Hutch*

SHARE

Share is an audience measurement term that identifies the percentage of television households with sets in use which are viewing a particular program during a given time period. If the total TV audience is represented by a pie, the audience for each program is a slice or share of that pie. The slices are not equal, however, since audience share varies widely according to the relative popularity of each program. Share is a comparative tool; it allows station and network executives to determine how well their programs are doing when compared with competing programs on other broadcast or cable channels.

Share is closely associated with rating, another measurement term. Both terms are derived from the same estimates of audience size, but the percentage quotient is calculated differently. Share measures the percentage of TV viewers who are actually watching a particular program, while the rating for a program calculates the percentage of *all* television households—both those using TV and those not using TV.

For example, station WXXX airs *Jeopardy!* at 7 P.M. Sample data estimate that 10,000 or 10 percent of the city's 100,000 TV households are viewing that program. Some 40,000 households are viewing other programs, but another 50,000 are not using their TV sets. Since 10,000 of the 50,000 active viewers (20%) are watching *Jeopardy!*, that program has a share of 20 even though its rating (the percentage of TV households) is only 10.

Electronic media trade journals generally report both rating and share. Rating is expressed first and is given to the nearest tenth of a percent. Share follows and is rounded to the nearest whole percent. For example, an audience estimate for *60 Minutes* may report a 13.0/28; i.e., 13% of the total TV households (the rating) and 28% of the viewing audience (the share).

If every television household was using TV during a given time period, the share and the rating would be equal. But since this never happens, the share for any program is always greater than its rating because different divisors are used to calculate the two equations.

The gap between share and rating is greatest during periods of very light viewing. An early morning newscast with a share of 30 and a rating of only 3 is competing very well against other programs in the same time block even though the total number of viewers for all programs is small.

Share is useful as a comparative tool during virtually any portion of the day, however. When a program gains share,

it usually does so at the expense of competing programs since the total audience for television during any given day-part is relatively stable.

Share can also be used to illustrate programming trends. One network may average its share of successive programs to illustrate its dominance on a particular weekday night. A new broadcast or cable network may average its share across an entire season to illustrate its increasingly competitive position over a previous season.

Share can be used to demonstrate industry trends. For example, the combined share of ABC, CBS, and NBC for the 1980-81 programming year was 90. This meant that 90% of the viewing audience was watching one of these three networks. The remaining 10% of the audience was distributed among independent stations, public television, and the few cable networks then operation. By 1993–94, combined network share had dropped to 60, primarily because the cable networks collectively had captured one-third of the network viewers. Some industry observers predict that network share will continue to decline; others believe the erosion of network share has been halted and possibly reversed. A study of network share measures the competition between traditional broadcasters and their new technology competitors.

Unless otherwise specified, share refers to the total universe of television households. Share can be used in demographics breakouts, however. A morning talk show may have a 2.2/20 for women 18 to 34 years of age. That would be the rating and share for this particular demographic grouping.

—Norman Felsenthal

FURTHER READING

Beville, Hugh Malcolm. *Audience Ratings: Radio, Television, and Cable*. Hillsdale, New Jersey: L. Erlbaum, 1988.

Rust, Roland T., and Naras V. Eechambadi. "Scheduling Network Television Programs: A Heuristic Audience Flow Approach to Maximizing Audience Share." *Journal of Advertising* (Provo, Utah), Spring 1989.

Webster, James C., and Lawrence W. Lichty. *Ratings Analysis: Theory and Practice*. Hillsdale, New Jersey: L. Erlbaum, 1992.

See also A.C. Nielsen Company; Audience Research, Industry and Market Approaches; Cost Per Thousand; Demographics; Market; Programming; Ratings

SHAW, BERNARD

U.S. Broadcast Journalist

As principal Washington anchor for the Cable News Network (CNN), Bernard Shaw has built a reputation for asking difficult questions and upholding unfaltering journalistic ethics. His style and professionalism have

enabled him to land interviews with world leaders. His most visible, sensational, and some would say, impressive moment as a journalist came in 1991. In Baghdad, Iraq, to complete a follow-up interview with Iraqi President

Saddam Hussein, Shaw was one of three CNN reporters who worked during a major attack by the Allied Forces. With his colleagues, Shaw brought unprecedented live coverage of the Allied Forces' bombing. On 16 January 1991, more than one billion homes watched Shaw and his colleagues deliver around-the-clock coverage of Operation Desert Storm.

Shaw's coverage of the war earned him numerous national and international journalism prizes, including the Eduard Rhein Foundation's Cultural Journalistic Award, a George Foster Peabody Award, and a cable ACE Award for best newscaster of the year. Shaw's receipt of the Rhein Foundation Award was the first time this honor had been bestowed on a non-German.

Live coverage was not new for Shaw; he also presented live broadcasts of the events surrounding the student revolt in China's Tiananmen Square until CNN was forced by the Chinese government to discontinue coverage. His coverage of the uprising earned him and CNN considerable recognition. His awards for coverage of Tiananmen Square included a cable ACE for best news anchor and an Emmy for anchoring the single most outstanding news event. CNN won a Golden ACE, an Alfred I. duPont-Columbia University Silver Baton, and a Peabody for its coverage of China.

Shaw is best known for his political reporting at CNN. Through the 1990s, he has been anchor of *The International Hour, The World Today,* and *Inside Politics.* He has covered debates, primaries, conventions, and the hoopla of presidential campaigning.

In 1988, while moderating a presidential debate between George Bush and Michael Dukakis, Shaw asked Dukakis if he would change his mind about opposing the death penalty if his own wife were raped and killed. Political analysts credit Shaw's question and Dukakis' off-guard response with portraying Dukakis as unemotional. Dukakis' campaign never recovered from the backlash of his reaction to Shaw's question.

Shaw is a graduate of the University of Illinois, which established the Bernard Shaw Endowed Scholarship Fund to honor his career and assist promising young men and women who share his interests and integrity. Shaw is a major benefactor of that fund.

—John C. Tedesco

BERNARD SHAW. Born in Chicago, Illinois, U.S.A., 22 May 1940. Educated at the University of Illinois, Chicago, 1963–68. Married: Linda Allston, 1974; children: Amar Edgar and Anil Louise. Served in U.S. Marine Corps, Oahu, Hawaii, 1959–63. Reporter, WNUS, Chicago, 1963; news writer, WFLD, Chicago, 1965; reporter, WIND, 1966–68; White House reporter, Westinghouse Broadcasting Company, 1968–71; reporter, CBS News, 1971–74; correspondent, CBS News, 1974–77; Latin American bureau chief and correspondent, ABC, 1977–78;

Bernard Shaw
Photo courtesy of CNN

Capitol Hill correspondent, ABC, 1978–79; CNN news anchor, 1980. Honorary degrees: Marion College, 1985; University of Chicago, 1993; Northeastern University, 1994. Member: Society of Professional Journalists, National Press Club, Sigma Delta Chi. Recipient: International Platform Association's Lowell Thomas Electronic Journalist Award, 1988; Awards for Cable Excellence (ACE) from the National Academy of Cable Programming, 1988, 1990, 1993, and 1994; Emmy Awards, 1989 and 1992; National Association of Black Journalists, Journalist of the Year Award, 1989; gold medal, International Film and TV Festival, 1989; Peabody Award, 1990; Congress of Racial Equality, Dr. Martin Luther King, Jr. Award for Outstanding Achievement, 1993; University of Kansas, William Allen White Medallion for Distinguished Service, 1994. Address: Principal Washington Anchor, CNN America, Inc., CNN Building, 820 First Street NE, 11th Floor, Washington, D.C. 20002., U.S.A.

TELEVISION

1980   *CNN News*
1989   *The World Today*

FURTHER READING

Kellner, Douglas. *The Persian Gulf TV War.* Boulder, Colorado: Westview, 1992.

Smith, Perry M. *How CNN Fought the War: A View from the Inside.* New York: Carol, 1991.

Whittemore, Hank. *CNN, The Inside Story.* Boston: Little, Brown, 1990.

Wiener, Robert. *Live from Baghdad: Gathering News at Ground Zero.* New York: Doubleday, 1992.

See also Anchor; Cable News Network

SHEEN, FULTON J.

U.S. Religious Broadcaster

Widely known by his Roman Catholic ecclesiastical title, Bishop Sheen established a very successful niche for religious programming in U.S. television's early days with his *Life Is Worth Living* program. Sheen's show originally aired on the DuMont network on Tuesday evenings in 1952 and then moved to ABC where it remained until Sheen withdrew it in 1957. The shows—really half-hour talks by Sheen—proved very popular and ultimately were carried on 123 ABC television stations and another 300 radio stations.

Life Is Worth Living followed a simple format. Sheen would choose a topic and, with only a blackboard for a prop and his church robes for costuming, would discuss the topic for his allotted 27 minutes. He spoke in a popular style, without notes but with a sprinkling of stories and jokes, having spent up to 30 hours preparing his presentation. Because the program was sponsored by the Admiral Corporation rather than the Catholic Church, Sheen avoided polemics and presented a kind of Christian humanism. In his autobiography he noted that the show was not "a direct presentation of Christian doctrine but rather a reasoned approach to it beginning with something that was common to the audience." He covered topics as diverse as art, science, aviation, humor, communism, and philosophy.

Like many others in its early days, Sheen had moved into television from radio. As a professor at the Catholic University of America, he began commuting from Washington, D.C., in 1928 to broadcast on WLWL in New York. Two years later he became the first regular speaker on *The Catholic Hour*, a sustaining time program on NBC radio, sponsored by the National Council of Catholic Men. In 1940, he made his television debut presiding at New York City's first televised religious service.

After several years off, Sheen attempted to come back to television a number of times, but without the success that had greeted *Life Is Worth Living*. He hosted a series on the life of Christ in the 1950s; in 1964, he worked on *Quo Vadis, America?*; and he revived the format of *Life Is Worth Living*, now called *The Bishop Sheen Program*. Television had changed and his lecture style no longer commanded audience loyalty. He ended his long career in broadcasting with numerous guest appearances on television talk shows during the 1960s and 1970s.

Broadcasting was never Sheen's full-time occupation. He left The Catholic University of America in 1950 to become the national director of the Society for the Propagation of the Faith, a fund-raising office for missionaries, a position he held until Pope Paul VI named him Bishop of Rochester, New York, in 1966.

Sheen's importance for television lies in two areas. First, he pioneered a nonsectarian style of religious programming and found commercial sponsors for his message. By doing this he both adapted to and helped to shape commercial

Fulton J. Sheen

broadcasting's attitudes toward religious shows. The need to develop audiences meant that only those programs with the widest possible appeal would find a place in mainstream or network programming. Second, Sheen provided a role model (if not an ideal) for the next generation of ministers interested in television—the televangelists. Many of the later stars of cable religious television acknowledge that the widespread acceptance of Sheen's *Life Is Worth Living* inspired their own forays into television. They too hoped to escape the "Sunday morning ghetto" of religious programming for a place in the mainstream.

—Paul A. Soukup

FULTON JOHN SHEEN. Born in El Paso, Illinois, U.S.A., 8 May 1895. Graduated from St. Viator College, Bourbonnais, Illinois, 1917, M.A. 1919; studied at St. Paul Seminary, Minnesota, ordained 20 September 1919; University of America, Washington, D.C., S.T.B and J.C.B. Catholic; University of Louvain, Belgium, Ph.D. 1923; Collegio Angelico, Rome, D.D. 1924; made *Agrege en Philosophie* at Louvain. Served in St. Patrick's parish, Peoria, Illinois, 1924–26; instructor in religion, Catholic University of America, 1926, remaining affiliated with university until 1950; preacher, WLWL Radio in New York, 1928; became regular preacher on NBC radio program *The Catholic Hour*, 1930; made papal chamberlain and given rank of monsignor, 1934; presided over New York's first televised religious service, 1940; director, U.S. activities for the Society for Propagation of the Faith, 1950–66; consecrated as auxiliary bishop of the New York archdiocese, 11 June 1951; began long-running television program *Life Is Worth Living*, 1952; bishop, Rochester, New York, 1966–69; made titular archbishop of Newport, Wales, 1969. Died in New York, 10 December 1979.

TELEVISION SERIES

1952–57	*Life Is Worth Living*
1955–57	*Mission to the World*

1961–68   *The Bishop Sheen Program*
1964   *Quo Vadis, America?*

RADIO
The Catholic Hour, from 1930.

PUBLICATIONS (selection)
Peace of Soul. New York: Whittlesey House, 1949.
Three to Get Married. New York: Appleton-Century-Croft, 1951.
Life Is Worth Living. New York: McGraw Hill, 1953.
The Priest Is Not His Own. New York: McGraw Hill, 1963.
Missions and the World Crisis. Milwaukee, Wisconsin: Bruce, 1964.

That Tremendous Love. New York: Harper and Row, 1967.
Treasure in Clay: The Autobiography of Fulton J. Sheen. Garden City, New York: Doubleday, 1980.

FURTHER READING
Griffin, W. "Foreword." In, Sheen, F. J. *The Electronic Christian: 105 readings from Fulton J. Sheen.* New York: Macmillan, 1979.
Noonan, D. P. *The Passion of Fulton Sheen.* New York: Dodd, Mead, 1972.

See also Religion on Television

SHERLOCK HOLMES

Mystery (Various National Productions)

Sherlock Holmes, the fictional character created by Sir Arthur Conan Doyle is, perhaps, the most popular literary character adapted to the performing arts. The adventures of Sherlock Holmes have been transformed for the dramatic stage (*Sherlock Holmes*, 1899, and *The Crucifer of Blood*, 1978), the musical stage (*Baker Street*, 1965), ballet (*The Great Detective*, 1953), film, radio and television. On television, the character has appeared in specials, series, parodies, animation, made-for-television films, and even in a recurring role-playing game by the android Data (Brent Spiner) on *Star Trek: The Next Generation*.

The actors who have undertaken the role for television include Ronald Howard (son of film actor Leslie Howard), Alan Napier, Peter Cushing, Christopher Lee, Frank Langella, Tom Baker (later the Doctor in *Doctor Who*), Edward Woodward, Charlton Heston, Roger Moore, Leonard Nimoy, Peter O'Toole (as the voice of the detective in the Australian animated *Sherlock Holmes and the Baskerville Case*), and Jeremy Brett. Even Basil Rathbone, who portrayed the character in 14 feature films and eight years on the radio, played Holmes on the small screen. Comic actors such as Milton Berle, *Monty Python*'s John Cleese, Larry Hagman, and Peter Cook have all played the master sleuth in television parodies.

Sherlock Holmes was the first fictional character adapted for television. *The Three Garridebs*, a trial telecast, was broadcast on 27 November 1937 from the stage of New York City's Radio City Music Hall by the American Radio Relay League. The live presentation was augmented with filmed footage to link scenes together. Louis Hector played the detective, and William Podmore played his associate, Dr. Watson.

Until 1951, Holmes' appearances on television were limited to a variety of special broadcasts, including the hour-long parody, *Sherlock Holmes in the Mystery of the Sen Sen Murder*, on the 5 April 1949 episode of NBC's *Texaco Star Theatre*. The satire featured Milton Berle and Victor Moore as Holmes and Watson, and a guest appearance by Basil Rathbone as Rathbone of Scotland Yard.

The first television series of Sherlock Holmes adventures was produced in the United Kingdom. Vandyke Pictures intended for its half-hour adaptation of *The Man with the Twisted Lip*, starring John Longden as Holmes and Campbell Singer as Watson, to be the first of a six-episode series. However, the pilot did not impress executives, and only the one episode was broadcast (in March 1951). Three months later, the BBC aired its own pilot, an adaptation of *The Mazarin Stone*, with Andrew Osborn as Holmes and

The Sherlock Holmes Mysteries

Philip King as Watson. In late 1951, the BBC produced the first television series of Sherlock Holmes adventures, but with a new producer and new actors (Alan Wheatley as Holmes and Raymond Francis as Watson). Six of Arthur Conan Doyle's stories were adapted to the 35-minute format by C.A. Lejeune, a film critic for *The Observer.*

Basil Rathbone who, for many years gave what was considered the definitive portrayal of Holmes, reprised his role as the detective in a half-hour live presentation for the 26 May 1953 episode of CBS' *Suspense.* The episode, *The Adventures of the Black Baronet,* was adapted by Michael Dyne from an original story by crime novelist John Dickson Carr and Adrian Conan Doyle, son of the character's creator. The episode was intended as a pilot for an American series, but it was not selected for programming by any network.

The first and only American television series of Sherlock Holmes adventures finally aired in syndication in the fall of 1954. The 39 half-hour original stories were produced by Sheldon Reynolds and filmed in France by Guild Films. Ronald Howard starred as Holmes and Howard Marion Crawford starred as Watson. The series' associate producer, Nicole Milinaire, is considered to be the first woman to attain a senior production role in a television series.

Since 1954, American adaptations of the Holmes stories have been limited to various made-for-television films (e.g., *The Return of the World's Greatest Detective* with Larry Hagman as Holmes, *Sherlock Holmes in New York* with Roger Moore as Holmes, and *The Hound of The Baskervilles*) or televised stage plays (Frank Langella's *Sherlock Holmes* and *The Crucifer of Blood* with Charleton Heston).

In addition to producing made-for-television Holmes films in Britain, the BBC continued to produce other series of Holmes adventures. A 1965 series of 12 adaptations was produced by David Goddard and featured Douglas Wilmer who, *The Times* noted, bore an "uncanny resemblance" to the sleuth in the original book illustrations by Sydney Paget. A 1968 series starring Peter Cushing dispensed with many of the conventions invented by other actors for the character, such as the meerschaum pipe, the deer-stalker cap, and the phrase, "Elementary, my dear Watson." The series aspired to be true to the character as written in the novels. In an attempt to capitalize on Cushing's popular work in 1950s

and 1960s horror films, the BBC series accentuated the elements of horror and violence in the original stories.

In 1984, Britain's Granada Television mounted the most popular series to date. Shown under various titles (*The Adventures of Sherlock Holmes, The Return of Sherlock Holmes,* and *The Casebook of Sherlock Holmes*) in Britain, the series was broadcast in the U.S. as part of PBS' *Mystery!* series. Critics have praised the high quality of the series' productions, including an authentic-looking Baker Street, and Jeremy Brett's performance as Holmes has been ranked as among the finest portrayals of the detective.

The appeal of the character has not been limited to English-speaking countries. An original teleplay, *The Longing of Sherlock Holmes* (*Touha Sherlocka Holmese*), in which Holmes is tempted to commit the perfect crime, was produced for Czechoslovakian television in 1972. In 1983, Russian television produced a series of five 80-minute adaptations of Conan Doyle's stories featuring leading Soviet actors Vassily Livanov and Vitaly Solomin as Holmes and Watson.

—Susan R. Gibberman

FURTHER READING

Bunson, Matthew E. *Encyclopedia Sherlockiana: An A-to-Z Guide to the World of the Great Detective.* New York: Macmillan, 1994.

Collins, Max Allan, and John Javna. *The Best Crime and Detective TV: Perry Mason to Hill Street Blues, The Rockford Files to Murder, She Wrote.* New York: Harmony, 1988.

deWall, Ronald Burt. *The World Bibliography of Sherlock Holmes and Dr. Watson: A Classified and Annotated List of Materials Relating to Their Lives and Adventures.* New York: Bramhall House, 1974.

Eyles, Allen. *Sherlock Holmes: A Centenary Celebration.* New York: Harper and Row, 1986.

Haining, Peter. *The Television Sherlock Holmes.* London: Virgin, 1994.

"Shadowing a Sleuth: Holmes Stalks the City by Television— Football Pictures are Clear." *New York Times,* 28 November 1937.

See also British Programming; Detective Programs

SHORE, DINAH

U.S. Musical Performer/Talk-Show Host

Dinah Shore ranks as one of the important on-air musical stars of the first two decades of television in the United States. Indeed, from 1956 through 1963, there were few TV personalities as well-known as she was. More than any song she sang, Shore herself symbolized cheery optimism and southern charm, most remembered for blowing a big kiss to viewers at the end of her 1950s variety show. As hostess, she sometimes danced and frequently partici-

pated in comedy skits, but was best loved as a smooth vocalist reminiscent of a style associated with the 1940s.

Shore pioneered the prime-time color variety show when *The Dinah Shore Chevy Show* started in October 1956 on NBC and ran on Sunday nights until the end of the 1963 season. Sponsored by General Motors, then the largest corporation in the world, Shore helped make the low-priced Chevrolet automobile the most widely selling car up to that point in history.

Shore represented a rare woman able to achieve major success hosting a TV variety show. In the late 1950s, her enthusiasm and lack of pretension proved so popular that she was four times named to the list of the "most admired women in the world." Her desire to please showed in her singing style, which some purists dismissed as sentimental, but through her recording career she did earn nine gold records. Shore made listeners and later viewers feel good, and beginning with her first broadcasts on radio in the late 1930s and then on television, she was able to remain a constant presence in American broadcasting for more than 50 years.

When Fanny Rose Shore was old enough to go to school, in her hometown of Nashville, Tennessee, she found herself taunted for being Jewish in the decidedly non-Jewish world of a segregated Deep South. Undeterred, Shore logged experience on Nashville radio while in college, on her hometown's WSM-AM, best known as the home of the Grand Ole Opry. But Shore was no hillbilly singer, no typical Southern belle. She took a degree in sociology at Vanderbilt University, putting herself through college with her radio earnings. Her show's theme song was the Ethel Waters blues-inspired "Dinah," and Shore changed her name accordingly. The success of her local radio show, *Our Little Cheerleader of Song,* enabled Shore to move to New York City to try to make it in Tin Pan Alley, then the center of the world of pop music.

Shore, by her own admission, did not have the vocal equipment of Ella Fitzgerald or Billie Holiday, and never chose to reveal as much of herself in music as did her other idol, Peggy Lee. However, she was persistent. During the late 1930s, having auditioned unsuccessfully for such band leaders as Benny Goodman and Tommy and Jimmy Dorsey, Shore finally hooked up with the Xavier Cugat band. Through the 1940s, she sold one million copies of "Yes, My Darling Daughter," and that recording success was followed quickly by such hits as "Blues in the Night," "Shoo Fly Pie," "Buttons and Bows," "Dear Hearts and Gentle People," and "It's So Nice to Have a Man Around the House." During the World War II, Shore sang these songs for the troops in Normandy and for shows at other Allied bases in Europe.

In 1950, Shore made a guest appearance on Bob Hope's first NBC television special. A year later, NBC assigned her a regular TV series which ran until 1956 on Tuesday and Thursday nights from 7:30-7:45 P.M., Eastern time, following 15 minutes of network news. This led, in time, to her Sunday night series. RCA and NBC corporate chief David Sarnoff loved Shore's conservative vocal choices and middle-brow sensibilities. In retrospect, Shore's famed signature theme song, the catchy Chevrolet jingle, "See the USA in your Chevrolet," accompanied by her sweeping smooch to the audience, were so theatrically commercial they made Ed Sullivan seem subversive and Pat Boone look like an rock star. Shore did best when she played the safe 1950s non-threatening "girl next door," with no blond (she was born a brunette) hair out of place, no joke offensive to anyone. The outcast of Nashville finally fit in.

Dinah Shore

The Dinah Shore Chevy Show rarely entered the top 20 ratings against CBS' *General Electric Theater,* hosted by Ronald Reagan, which regularly won the time slot. Reagan had a better lead-in from Ed Sullivan. Still, Shore won Emmy Awards for Best Female Singer (1954-55), Best Female Personality (1956–57), and Best Actress in a Musical or Variety Series (1959).

After the *Chevy Show,* Shore went on to host three daytime television programs: the 90-minute talk show *Dinah!* (1970–74), *Dinah's Place* (1970–74), and *Dinah and Friends* (1979–84). Her TV career ended in 1991 on cable TV's Nashville Network with *A Conversation with Dinah.* By then she was better known as Hollywood heart throb Burt Reynolds' "older" girl friend, and sponsor of a major golf tournament for women.

—Douglas Gomery

DINAH SHORE. Born Frances Rose Shore in Winchester, Tennessee, U.S.A., 1 March 1917. Educated at Vanderbilt University, Nashville, Tennessee, B.A. 1939. Married: 1) George Montgomery, 1943 (divorced, 1962); one daughter and one son; 2) Maurice Fabian Smith, 1963 (divorced, 1964). Singer, WNEW, New York, 1938; sustaining singer, NBC, 1938; signed contract with RCA-Victor, 1940; starred in *Chamber Music Society of Lower Basin Street,* NBC radio program, 1940; joined Eddie Cantor radio program, 1941; starred in own radio program for General Foods,

1943; entertained American troops in European theater of operations, 1944; hosted radio program for Procter and Gamble; starred in TV show for Chevrolet, 1956–63; hosted numerous variety and talk shows. Recipient: Emmy Awards, 1954, 1956, 1959, 1973, 1974, and 1976. Died in Beverly Hills, California, 24 February 1994.

TELEVISION SERIES

1951–57	The Dinah Shore Show
1956–63	The Dinah Shore Chevy Show
1970–74	Dinah!
1974–80	Dinah's Place
1976	Dinah and Her New Best Friends
1979–84	Dinah and Friends

1989–91	A Conversation with Dinah

FILMS

Thank Your Lucky Stars, 1943; Up In Arms, 1944; Belle of the Yukon, 1944; Follow the Boys, 1944; Make Mine Music (voice only), 1946; Till the Clouds Roll By, 1946; Fun and Fancy Free (voice only), 1947; Aaron Slick from Punkin Crick, 1952; Oh, God!, 1977; Health, 1979.

PUBLICATION

Someone's in the Kitchen with Dinah. Garden City, New York: Doubleday, 1971.

See also Dinah Shore Chevy Show

SILLIPHANT, STIRLING

U.S. Writer

Stirling Silliphant was one of the most important and prolific writers of television drama in the 1960s, remembered particularly for his work on Naked City and Route 66. Although he had early success in the 1950s with a spate of feature films, and went on to even greater big-screen achievements in the late 1960s and 1970s, Silliphant maintained a constant presence in television throughout his writing career, and in the 1980s focused most of his attention on television movies, historical miniseries, and novels.

Silliphant's passage between big-screen and small-screen writing marked his work very early on. He began his association with the movies as a publicist, first for Disney, and later 20th Century-Fox. Silliphant left that end of the business in 1953 to package an independent feature, The Joe Louis Story (honing his rewrite skills on the script). In 1955 he transformed a rejected screenplay into the novel Maracaibo (which was adapted by another writer and filmed three years later), and within the next three years saw five feature scripts produced, including Jacques Tourneur's Nightfall and Don Siegel's The Lineup. During the same period he aimed his typewriter at television, generating dozens of scripts for such anthologies as General Electric Theater, Alcoa-Goodyear Theatre, Suspicion, Schlitz Playhouse, and Alfred Hitchcock Presents, as well as two episodes of Perry Mason.

Silliphant was completing his sixth feature script (Village of the Damned) when independent producer Herbert B. Leonard (Adventures of Rin Tin Tin, Circus Boy) hired him to write the pilot for Naked City, a half-hour series based on the 1948 "semi-documentary" feature The Naked City. With a resume composed almost exclusively of anthologies and features, Silliphant's proclivity for self-contained stories was consistent with Leonard's vision of the series as a character-oriented dramatic anthology with a police backdrop, as opposed to a police procedural in the Dragnet mold. Silliphant wrote thirty-one of Naked City's first thirty-nine episodes, remembered today as taut, noirish thirty-minute thrillers offering both character drama and gunplay. Can-

celled after one season in its original form, the series was resurrected as an hour-long show in 1960.

In the interim, Silliphant remained busy with scripts for crime series like Markham, Tightrope, and The Brothers Brannagan, as well as an unsold private eye pilot, Brock Callahan. When Naked City was resurrected at a sponsor's behest for the 1960 season in the longer form, Silliphant was already collaborating with Leonard on another series-anthology hybrid, Route 66. (A third Leonard-Silliphant project for 1960 called Three-Man Sub—a sort of underwater Mediterranean variation on Route 66—did not sell.) Although he did write the pilot script and served as "executive story consultant" for the new version of Naked City, Silliphant would provide fewer scripts for the show because of his intense involvement with Route 66; still, the writing remained first-rate. The all-New York production offered a fascinating mix of action and Actor's Studio, yielding three seasons of compelling urban tragedy. The series was nominated for an Emmy in the Outstanding Drama category every year of its run.

Route 66 proved to be a critical and commercial hit, despite early concerns from Screen Gems studio about its premise: two young drifters searching for meaning on the highways of America. Filmed on location across the United States, the wide-ranging backdrops and visual realism of Route 66, and its mix of psychological drama, social commentary, romance, action, and big-name guest stars, all underlined by strong writing and supervision from Silliphant (and story editor Howard Rodman), paved the way for a four-year run. Spending much of this time writing and observing on the road, Silliphant would go on to write some three-fourths of Route 66's 116 episodes. Silliphant calls those four years the most intensive period of writing in his career, and the site of some of his best work.

Naked City was cancelled in 1963, and Route 66 a year later, but the "writing machine" (as one producer dubbed Silliphant in a Time magazine profile) did not pause. During the mid-1960s Silliphant freelanced for *Bob Hope Presents the Chrysler Theater*, *Mr. Novak*, and *Rawhide* before signing on as writer-creator of another nomadic adventure series, *Maya*, in 1967 (this time, two teens on an elephant wandering India). That same year Silliphant made a triumphant return to features, winning an Academy Award for his adaptation *In the Heat of the Night*. Even with this big-screen success (followed up with films like *Marlowe*, *Charly*, and *The New Centurions*), Silliphant did not abandon television. Despite a 1960 interview in which he eschewed the growing plague of "hyphenated billing"—alleging that the "miasma of memos and meetings" inevitably curtailed the insight and blunted the creativity of writer-producers—by 1971 Silliphant was one, serving as executive producer of the mystery series *Longstreet*. Notable as part of the 1970s-era cycle of "gimmick" detective series (*Cannon*, *Ironside*, *McCloud*), *Longstreet*—the story of a blind insurance investigator—was otherwise unremarkable. A year later the writer attempted to mount yet another picaresque series entitled *Movin' On*, this time concerning a pair of itinerant stock car racers (not to be confused with the 1974 series about truckers); the pilot aired as a TV movie, but the series did not sell. *Longstreet*'s cancellation after one season effectively ended Silliphant's involvement in the continuing series form—but not his television career.

Although he did pen several TV movies and his first miniseries, *Pearl* (based on his novel) during the 1970s, Silliphant concentrated most of his efforts on features. He produced *Shaft* in 1971 (and wrote the 1973 sequel *Shaft in Africa*); in 1972 he helped launch the popular cycle of disaster movies by scripting *The Poseidon Adventure*, followed by *The Towering Inferno* and *The Swarm*; and turned out successful thrillers like *Telefon* and *The Enforcer* (Clint Eastwood's third "Dirty Harry" film). A few more features followed in the 1980s, but for the most part Silliphant settled back into television, scripting a succession of made-for-TV movies (and unsold pilots), and epic miniseries such as *Mussolini: The Untold Story*, and *Space*. True to form, the fertile author also found time during the decade to publish three adventure novels featuring roving adventurer John Locke.

Silliphant's writing career is remarkable not only for its sheer volume of output, its duration, and its spanning of television and feature work, but also for the very fact that he kept an active hand in television after achieving big-screen success, and that, he considered television to be the medium most conducive to the writer's vision. Silliphant has charged that his *In the Heat of the Night* script was inferior to many of his *Naked City* teleplays. "As a matter of fact," he declared to writer William Froug, "I can think of at least twenty different television scripts I've written which I think are monumental in comparison." Truth be told, the bulk of Silliphant's features—most of which are adaptations—have tended toward formula, while the passion for character and ideas comes through most strongly in the television work.

Silliphant has repeatedly pronounced *Naked City* and *Route 66* as the best of his writing. It is difficult to disagree. These two series are surely Silliphant's finest achievements, and

rank among the most original and well-written dramas ever created for the medium. A *Variety* columnist observed in a 1962 review of *Route 66* that Silliphant "composes poetry which is often raw and tenuous, so it requires delicacy of treatment." As this suggests, Silliphant's "poetry" carried some risk. John Gregory Dunne cited Silliphant as a prime purveyor of television "pseudo-seriousness" in a 1965 article, and Silliphant himself has admitted a proclivity for the overwrought phrase. But with the right director and actors, no writing for the screen has been more powerful. And if the intense demands of series writing—and writing on the road, at that—occasionally failed to limit a slight propensity for pretension that sometimes overwhelmed characterization or credibility, by and large Silliphant's scripts for *Naked City* and *Route 66* yielded moving renderings of troubled relationships and tortured psyches. Even his more purple moments speak to the ambitions he had for television as a dramatic form.

In 1968 *TV Guide* critic Dick Hobson, bemoaning the exodus of writing talent from the medium, lamented, "What became of writer Stirling Silliphant, whose *Naked City*'s and *Route 66*'s were once a repertory theater of contemporary life and times?" Seven years earlier, these selfsame programs were overlooked by a Federal Communications Commission member branding television a "vast wasteland," and critics bemoaning the demise of live drama. Meanwhile, Silliphant's "poetry" was living (broadcast) proof of television's capacity for brilliant writing and provocative drama, and thirty years later, the writing machine was still writing.

—Mark Alvey

STIRLING DALE SILLIPHANT. Born in Detroit, Michigan, U.S.A., 16 January 1918. Educated at the University of Southern California, B.A. magna cum laude 1938. Married: Tiana Du Long, 1974; one daughter and two sons (one deceased). Served as lieutenant, U.S. Navy, 1943–46. Publicity director, 20th Century-Fox, New York City, 1946–53; screenwriter, independent producer for various Hollywood studios, 1953–80s; moved to Thailand, where he continued to work on movie and TV projects, 1980s. Member: Writers Guild of America West, Mystery Writers Association, and Authors League. Recipient: Academy Award, Edgar Allen Poe Award, 1967; Golden Globe Awards, 1968 and 1969; Writer of the Year Award from the National Theater Owners, 1972; Image Award from the NAACP, 1972; Writer of the Year Award from the National Theater Owners, 1974. Died, in Thailand, 26 April 1996.

TELEVISION SERIES (writer; selection)

1958–63	*Naked City*
1960–64	*Route 66*

TELEVISION SERIES (contributing writer; selection)

1953–62	*General Electric Theater*
1955–65	*Alfred Hitchcock Presents*
1957–60	*Alcoa-Goodyear Theater*

1963–67   *Bob Hope Presents the Chrysler Theater*

MADE-FOR-TELEVISION MOVIES (selection)
1978   *Pearl*
1981   *Fly Away Home*
1985   *Space*
1985   *Mussolini: The Untold Story*
1987   *The Three Kings*

FILMS (selection)
In the Heat of the Night, 1967; *Marlowe, 1969; Charly*, 1969; *Shaft*, 1971; *The Poseidon Adventure*, 1972; *The New Centurions*, 1972; *The Liberation of L. B. Jones*, 1973; *Murphy's War*, 1974; *The Towering Inferno*, 1974; *The Killer Elite*, 1975; *The Enforcer*, 1976; *Telefon*, 1977; *The Swarm*, 1978; *Circle of Iron*, 1978; *When Time Ran Out*, 1980; *Over the Top*, 1986.

PUBLICATIONS
Maracaibo (novel). New York: Farrar, Straus, 1953.
"Lo, The Vanishing Writer." *Variety* (Los Angeles), 6 January 1960.
The Slender Thread (novel). New York: New American Library, 1966.
Pearl (novel). New York: Dell, 1978.
Steel Tiger (novel, John Locke Adventures). New York: Ballantine, 1983.

Bronze Bell (novel, John Locke Adventures). New York: Ballantine, 1985.
Silver Star (novel, John Locke Adventures). New York: Ballantine, 1986.
"Stirling Silliphant" (interview). *American Film* (Washington, D.C.), March 1988.

FURTHER READING
Dunne, John Gregory. "Take Back Your Kafka." *The New Republic* (Washington, D.C.), 4 September 1965.
"The Fingers of God." *Time* (New York), 9 August 1963.
Froug, William. *The Screenwriter Looks at the Screenwriter.* New York: MacMillan, 1972.
Hobson, Dick. "TV's Disastrous Brain Drain." *TV Guide* (Radnor, Pennsylvania), 15 June 1968.
"Route 66." *Variety* (Los Angeles), 7 November 1962.
"Silliphant Deplores that Bum Literary Rap Pinned on VidPix Writer." *Variety* (Los Angeles), 15 April 1959.
Silliphant, Stirling. Collected Papers. University of California, Los Angeles, Special Collections.
Stempel, Tom. *Storytellers to the Nation: A History of American Television Writing.* New York: Continuum, 1992.

See also *Naked City*; Police Programs; *Route 66*; Writing for Television

SILVERMAN, FRED

U.S. Media Executive and Producer

Fred Silverman devoted his life to programming television. He is the only person to have held key programming positions at all of the three traditional networks in the United States and today he owns the Fred Silverman Company, which produces programs for those networks. What makes Silverman unique in the history of American network television is that he raced through network jobs while still in his thirties and that his career mysteriously waned after having waxed so splendidly for so long.

Fred Silverman graduated with a Master's degree from Ohio State University (his master's thesis analyzed programming practices at ABC) and went to work for WGN-TV in Chicago to oversee children's programs. Soon, however, he moved to the network level. He assumed responsibility for daytime programming at CBS, where he later took charge of all of CBS Entertainment programming. During his tenure at CBS, Silverman remade the Saturday morning cartoon lineup and, in so doing, remade the ratings—from third to first. He also helped devise the programming strategy that brought *All in the Family, The Mary Tyler Moore Show* and *The Waltons* to CBS. With the success of the CBS schedule assured, Silverman moved on. In 1975, he became head of ABC Entertainment.

From 1975 to 1978, Fred Silverman took ABC from ratings parity with the other networks to ratings dominance over them. Among the shows and mini-series he was responsible for programming were *Rich Man, Poor Man, Roots, Charlie's Angels* and *Starsky and Hutch*. Silverman made the "third" network a ratings power, and, as some of these program selections suggest, is credited with creating what critics called "jiggle TV," the type of television that features beautiful, scantily clad, frolicking women. In short, he bore partial responsibility for programming both acclaimed and reviled. But he demonstrated at ABC the same touch he had at CBS—an almost unerring sense of what the public, in great numbers, would watch on television. In 1977, a *Time* magazine cover story referred to Silverman as the "man with the golden gut," ostensibly referring to his unfailing programming instincts. At the height of his power at ABC, Silverman left to take on the presidency of NBC.

It was there, however, that whatever abilities brought him fame at the other two networks seemed to abandon Fred Silverman. Some of his program selections were disastrous, (*Supertrain* and *Hello, Larry*, an ill-conceived effort starring McLean Stevenson, formerly of *M*A*S*H*). Also, without the success he had enjoyed earlier, his mercurial behavior was less tolerable. After three difficult years, he was replaced at NBC by Grant Tinker. Fred Silverman's eighteen-year run with the networks was over.

Silverman left programming to make programs, but he did not enjoy immediate success. The first years for the Fred Silverman Company were difficult, particularly because the former program buyer was now forced to try to sell programming to many of the persons he had alienated at the networks. But in 1985, Silverman and partner Dean Hargrove produced the first *Perry Mason* movie with Raymond Burr. It was wildly successful and established the formula that would drive Silverman's comeback in television. He took identifiable television stars from the recent past and recast them in formulaic dramas. Andy Griffith in *Matlock* and Carroll O'Connor in *In the Heat of the Night* are but two examples. Silverman also used his programming acumen to push for favorable time slots for his shows. Because Silverman has enjoyed great success with his production company, some industry observers have called him the Nixon of television.

Throughout his career in network television, Silverman was considered a hero in the industry because he could devise program schedules that delivered strong ratings. But during the latter stages of his network years, some industry observers saw a danger in so much television programming having the imprimatur of one individual. Moreover, his critics often looked beyond the bottom line and lamented the content of the programming used to build Silverman's various ratings empires. His work at ABC has been particularly criticized because of messages regarding sex and violence in the programs. Television programming has been criticized for appealing to the lowest common denominator in its quest for raw numbers of viewers and more than once, Silverman has been targeted as the chief instrument of that appeal. Indeed, columnist Richard Reeves observed in 1978 that Silverman had probably done more to lower the standards of the viewing audience than any other individual.

Of Silverman's comeback, this much can be said—he returned to his roots. His productions, using familiar faces and formulas which have enjoyed prior television success, can be seen as part of a larger pattern. It has been suggested that one current programming trend is to look back to a time when network television was at its peak. In the face of a complex and mercurial telecommunications landscape, those involved in broadcasting seek comfort from a time more stable. Many of the programs meeting this need are revivals, retrospectives, or old faces in new attire. One need look no further than the "new" *Burke's Law, Columbo,* or Dick Van Dyke in *Diagnosis Murder.* Silverman has capitalized on this tendency and has very probably become its leading practitioner. In a time when the term "auteur," or author, is being applied to television producers, the career of Fred Silverman suggests that an auteur could just as easily be the programmer as the program producer. For better or worse, few individuals have had as profound an impact on television programming for as long as Fred Silverman.

—John Cooper

FRED SILVERMAN. Born in New York City, New York, U.S.A., 1937. Studied at Syracuse University, New York; Television and Theater Arts at Ohio State University, Athens, M.A. Worked for WGN-TV, Chicago, 1961-62; worked for WPIX-TV, New York City; director of daytime programs, then vice president of

Fred Silverman
Photo courtesy of the Fred Silverman Company

programs, CBS-TV, New York City, 1963–75; president, ABC Entertainment, New York City, 1975–78; president and chief executive officer, NBC, New York City, 1978–81; president, Fred Silverman Company, Los Angeles, from 1981. Address: Fred Silverman Company, 12400 Wilshire Boulevard, Suite 920, Los Angeles, California 90025, U.S.A.

TELEVISION SERIES (executive producer)

1985–94	*Perry Mason* (movies)
1986–95	*Matlock*
1987–93	*Jake and the Fatman*
1988–95	*In the Heat of the Night*
1989, 1990–91	*Father Dowling Mysteries*
1994–	*Diagnosis Murder*

FURTHER READING

Bedell, Sally. *Up the Tube: Prime-time TV and the Silverman Years.* New York: Viking Press, 1981.

Reeves, Richard. "The Dangers of Television in the Silverman Era." *Esquire* (New York), 25 April 1978.

See also American Broadcasting Company; *Charlie's Angels*; Columbia Broadcasting System; *Mary Tyler Moore Show*; National Broadcasting Company; *Perry Mason*; Programming; *Rich Man, Poor Man*; *Roots*; *Starsky and Hutch*; Tartikoff, Brandon; United States: Networks

SILVERS, PHIL

U.S. Actor/Comedian

Phil Silvers was one of the great stars for CBS television during the late 1950s. Already a minor star on the vaudeville stage and in motion pictures, Silvers created, with writer-producer Nat Hiken, a pioneering television situation comedy, *You'll Never Get Rich*. In this satirical look at life in the U.S. Army, Silvers played Sergeant Ernest Bilko, the con-man with a heart of gold.

You'll Never Get Rich premiered on CBS-TV at the beginning of the 1955–56 TV season, and soon became a hit. For three years, as CBS took command of the prime-time ratings race, *You'll Never Get Rich* was a fixture in the 8:00 P.M. Tuesday night time slot. Between 1955 and 1958 the show was highly rated, and its success spelled the end of Milton Berle's Tuesday night reign on rival NBC.

As played by Silvers, Bilko was an army lifer, a motor pool master sergeant at isolated Fort Baxter located near the fictional army small town Roseville, Kansas. The show was a send up of army life (or of any existence within any confined and rigid society) and loved by ex-GIs of World War II and the Korean conflict, a generation still close to its own military experiences, and willing to laugh at them. With little to do in the U.S. Army of the Cold War era and stuck in the wide open spaces of rural Kansas, Ernest "Ernie" Bilko spent most of his time planning and trying one elaborate scam after another. Always, predictably, they failed. Bilko was never able to make that one big score. But the comedy came in the trying.

His platoon, played by a cast of wonderful ex-burlesque comics and aspiring New York actors, reluctantly assisted him. His right hand henchmen, the corporals Barbella and Henshaw, were ever by his side. The remainder of the group, following the pattern of numerous World War II films, seemed to have a man from every ethnic group: the brassy New Yorker, Private Fender, the Italian city boy, Private Paparelli, the high strung country lad, Private Zimmerman, and the loveable slob, Private Doberman. Others who manned the platoon included black actors in a rare, racially integrated TV situation comedy telecast in the 1950s.

If Silvers was the show's star, Nat Hiken, one of television's first writer-producers, was its creator-auteur. Hiken had first written for Fred Allen's hit radio show, then moved to television to help pen Milton Berle's *Texaco Star Theater*. His scripts provided a mine of comic gems for Bilko and company. Possibly the funniest was "The Case of Harry Speakup," in which a Bilko scheme backfires and he is forced to help induct a chimpanzee into the army. Only Bilko could run such a recruit past army doctors and psychiatrists, have him pass an IQ test and receive a uniform, be formally sworn in as a private, and then moments later honorably discharged. No bureaucracy has ever been spoofed better than was the Cold War U.S. Army in this 26-minute comic masterpiece.

Nat Hiken did more than write wonderfully funny scripts. As a producer he had an eye for talent. Guests on *You'll Never*

Phil Silvers

Get Rich included a young Fred Gwynne in "The Eating Contest" (first telecast on 15 November 1955), a youthful Dick Van Dyke in "Bilko's Cousin" (first telecast on 28 January 1958), and Alan Alda in his first significant TV role in "Bilko, the Art Lover" (first telecast on 7 March 1958).

You'll Never Get Rich shot up in the ratings, and less than two months after the premiere was renamed—not surprisingly, *The Phil Silvers Show*, with "You'll Never Get Rich" thereafter relegated to the subtitle. So popular was this show that in September 1957, as it started its second season, it inspired one of television's first paperback collections of published scripts.

Yet, as would be the case for television since the 1950s, the Bilko magic fell out of prime-time favor almost as swiftly as it had seized the public's fascination. The end began in 1958 when CBS switched *The Phil Silvers Show* to Friday nights and moved Bilko and company to Camp Fremont in California. A year later the show was off the schedule, and since then has functioned as staple in syndication around the world. Phil Silvers had had his four year run in television's spotlight.

He would find it again—briefly—in the 1963–64 television season when CBS tried *The New Phil Silvers Show*, a knockoff of the earlier program. Here, Silvers played Harry

Grafton, a plant foreman, trying (unsuccessfully) to get rich. It lasted but a single season and thereafter Silvers filled out his career doing occasional TV specials.

But Silvers—and Nat Hiken—should always be remembered for their pioneering work with *You'll Never Get Rich*. This show hardly dates at all; its comic speed, invention, and ensemble performances rank it among television's greatest comic masterworks.

—Douglas Gomery

PHIL SILVERS. Born Phillip Silversmith in Brooklyn, New York, U.S.A., 11 May 1912. Married: 1) Jo Carroll Dennison (divorced); 2) Evelyn Patrick (divorced); five daughters. Started career as vaudeville singer; became comedian in burlesque, then on Broadway; made screen debut in *The Hit Parade*, 1940; gained fame for television show *The Phil Silvers Show*, CBS, 1955–59. Recipient: Tony Awards, 1952 and 1972; Emmy Awards, 1955 and 1956. Died in Los Angeles, California, 1 November 1985.

TELEVISION SERIES

1955–59   *You'll Never Get Rich* (became *The Phil Silvers Show*, 1955)
1963–64   *The New Phil Silvers Show*

MADE-FOR-TELEVISION MOVIES

1975   *The Deadly Tide*
1975   *All Trails Lead to Las Vegas*
1977   *The New Love Boat*
1978   *The Night They Took Miss Beautiful*
1979   *'Hey Abbott!'*
1979   *Goldie and the Boxer*

FILMS

The Hit Parade, 1940; *Strike Up the Band*, 1940; *Pride and Prejudice*, 1940; *Ball of Fire*, 1941; *The Penalty*, 1941; *The Wild Man of Borneo*, 1941; *Ice Capades*, 1941; *Tom, Dick and Harry*, 1941; *Lady Be Good*, 1941; *You're in the Army Now*, 1941; *Roxie Hart*, 1942; *All Through the Night*, 1942; *Tales of the Night*, 1942; *My Gal Sal*, 1942; *Footlight Serenade*, 1942; *Just Off Broadway*, 1942; *Coney Island*, 1943; *A Lady Takes a Chance*, 1943; *Cover Girl*, 1944; *Four Jills in a Jeep*, 1944; *Something for the Boys*, 1944; *Take It or Leave It*, 1944; *Billy Rose's Diamond Horseshoe*, 1945; *A Thousand and One Nights*, 1945; *If I'm Lucky*, 1946; *Summer Stock*, 1950; *Top Banana*, 1952; *Lucky Me*, 1956; *40 Pounds of Trouble*, 1962; *It's a Mad, Mad, Mad, Mad World*, 1963; *A Funny Thing Happened on the Way to the Forum*, 1966; *A Guide for the Married Man*, 1967; *Follow that Camel*, 1967; *Buona Sera, Mrs. Campbell*, 1968; *The Boatniks*, 1970; *The Strongest Man in the World*, 1975; *Won Ton Ton: The Dog Who Saved Hollywood*, 1975; *Murder By Death*, 1976; *The Chicken Chronicles*, 1976; *Racquet*, 1978; *There Goes the Bride*, 1979; *The Cheap Detective*, 1979; *The Happy Hooker Goes to Washington*, 1980; *Hollywood Blue*, 1980.

STAGE (selection)

Yokel Boy, 1939; *High Button Shoes*, 1947; *Top Banana*, 1952; *Do Re Mi*, 1960, 1962; *How the Other Half Lives*, 1971; *A Funny Thing Happened on the Way to the Forum*, 1971–72.

PUBLICATION

This Laugh Is on Me: The Phil Silvers Story, with Robert Saffron. Englewood Cliffs, New Jersey: Prentice-Hall, 1973.

FURTHER READING

Everitt, David. "Kingmaker of Comedy." *Television Quarterly* (New York), Summer 1990.
———. "The Man Behind the Chutzpah of Master Sgt. Ernest Bilko." *New York Times*, 14 April 1996.
Hamamoto, Darrell Y. *Nervous Laughter: Television Situation Comedy and Liberal Democratic Ideology.* New York: Praeger, 1989.
Javna, John. *The Best of TV Sitcoms: Burns and Allen to the Cosby Show, The Munsters to Mary Tyler Moore.* New York: Harmony Books, 1988.
Marc, David. *Comic Visions: Television Comedy and American Culture.* Boston, Massachusetts: Unwin-Hyman, 1989.
———. *Demographic Vistas: Television in American Culture.* Philadelphia: University of Pennsylvania Press, 1984.

See also Comedy, Workplace; *Phil Silvers Show*; Workplace Programs

THE SIMPSONS

U.S. Cartoon Situation Comedy

The Simpsons, longest-running cartoon on American prime-time network television, chronicles the animated adventures of Homer Simpson and his family. Debuting on the FOX network in 1989, critically acclaimed, culturally cynical and economically very successful, *The Simpsons* helped to define the satirical edge of prime-time television in the early 1990s and was the single most influential program in establishing FOX as a legitimate broadcast television network.

The Simpsons' household consists of five family members. The father, Homer, is a none-too-bright safety inspector for the local nuclear power plant in the show's fictional location, Springfield. A huge blue beehive hairdo characterizes his wife, Marge, often the moral center of the program.

Their oldest child, Bart, a sassy 10-year-old and borderline juvenile delinquent, provided the early focus of the program. Lisa, the middle child, is a gifted, perceptive-but-sensitive saxophone player. Maggie is the voiceless toddler, observing all while constantly sucking on her pacifier. Besides *The Simpsons* clan, other characters include Moe the bartender; Mr. Burns, the nasty owner of the Springfield Nuclear Power Plant; and Ned Flanders, *The Simpsons'* incredibly pious neighbor. These characters and others, and the world they inhabit, have taken on a dense, rich sense of familiarity. Audiences now recognize relationships and specific character traits that can predict developments and complications in any new plot.

The Simpsons is the creation of Matt Groening, a comic strip writer/artist who until the debut of the program was mostly known for his syndicated newspaper strip "Life in Hell." Attracting the attention of influential writer-producer and Gracie Films executive James L. Brooks, Groening developed the cartoon family as a series of short vignettes featured on the FOX variety program *The Tracey Ullman Show* beginning in 1987. A Christmas special followed in December 1989, and then *The Simpsons* became a regular series.

Despite its family sitcom format, *The Simpsons* draws its animated inspiration more from Bullwinkle J. Moose than Fred Flintstone. Like *The Bullwinkle Show*, two of the most striking characteristics of *The Simpsons* are its social criticism and its references to other cultural forms. John O'Connor, television critic for *The New York Times*, has labeled the program "the most radical show on prime time" and indeed, *The Simpsons* often parodies the hypocrisy and contradictions found in social institutions such as the nuclear family (and nuclear power), the mass media, religion and medicine. Homer tells his daughter Lisa that it is acceptable to steal things "from people you don't like." Reverend Lovejoy lies to Lisa about the contents of the Bible to win an argument. Krusty the Clown, the kidvid program host, endorses dangerous products to make a quick buck. Homer comforts Marge about upcoming surgery with the observation that "America's health care system is second only to Japan's . . . Canada's . . . Sweden's . . . Great Britain's. . . well, all of Europe."

The critical nature of the program has been at times controversial. Many elementary schools banned Bart Simpson T-shirts, especially those with the slogan, "Underachiever, and Proud of It." U.S. President George Bush and former U.S. Secretary of Education William Bennett publicly criticized the program for its subversive and anti-authority nature.

In addition to its ironic lampoons, it is also one of the most culturally literate entertainment programs on prime time. Viewers may note references to such cultural icons as *The Bridges of Madison County*, Ayn Rand, Susan Sontag and the film *Barton Fink*, in any given episode. These allusions extend far beyond explicit verbal notations. Cartoon technique allows free movement in *The Simpsons*, and manipu-

The Simpsons

lation of visual qualities, often mimicking comic strip perspectives and cinematic manipulation of space creates an extraordinary sense of time, place, and movement. On occasion *The Simpsons* has reproduced the actual camera movements of the films it models. At other times the cartoonist's freedom and ability to visualize internal psychological states such as memory and dream have produced some of the program's most hilarious moments.

The unique nature of *The Simpsons* reveals much about the nature of the television industry. Specifically, the existence of the show illustrates the relationship of television's industrial context to its degree of content innovation. It was a program that came along at the right place, the right time, and appealed to the right demographic groups. Groening has said that no other network besides FOX would have aired *The Simpsons*, and in fact conventional television producers had previously turned down Groening's programming ideas. The degree of competition in network television in the late 1980s helped to open the door, however. Network television overall found itself in an increased competitive environment in this period because of cable television and VCRs. The FOX network, specifically, was in an even more precarious economic position than the Big Three. Because FOX was the new, unestablished network, attempting to build audiences and attract advertisers, the normally restrictive nature of network television gatekeeping may have been loosened to allow the program on the air. In addition, the championing of *The Simpsons* by Brooks, an established producer with a strong track record, helped the program through the industrialized television filters that might have watered down the program's social criticism. Finally, the fact that the program draws young audiences especially attractive to advertisers also explains the network's willingness to air such an unconventional and risky program. The "tween" demographic, those between the ages of 12 and 17, is an especially key viewing group for *The Simpsons* as well as a primary consumer group targeted by advertisers.

The Simpsons was a watershed program in the establishment of the FOX network. The cartoon has been the FOX program most consistently praised by television critics. It was the first FOX program to reach the Top 10 in ratings, despite the network's smaller number of affiliates compared to the Big Three. When FOX moved *The Simpsons* to Thursday night in 1990, it directly challenged the number one program of the network establishment at the time, *The Cosby Show*. Eventually, *The Simpsons* bested this powerful competitor in key male demographic groups. The schedule change, and the subsequent success, signaled FOX's staying power to the rest of the industry, and for viewers it was a powerful illustration of the innovative nature of FOX programming when compared to conventional television fare.

The Simpsons is also noteworthy for the enormous amount of merchandising it sparked. Simpsons T-shirts, toys, buttons, golf balls and other licensed materials were everywhere at the height of Simpsonsmania in the early 1990s. At one point retailers were selling approximately one million Simpsons T-shirts per week.

The Big Three networks attempted to copy the success of the prime-time cartoon, but failed to duplicate its innovative nature and general appeal. Programs like *Capital Critters*, *Fish Police* and *Family Dog* were all short-lived on the webs.

—Matthew P. McAllister

CAST (voices)

Homer Simpson . . . . Dan Castellaneta
Marge Simpson . . . . Julie Kavner
Bartholomew J. "Bart" Simpson . . . . Nancy Cartwright
Lisa Simpson . . . . Yeardley Smith
Mrs. Karbappel . . . . Marcia Wallace
Mr. Burns
Principal Skinner
Ned Flanders
Smithers
Otto the School Bus Driver
   (and Others) . . . . Harry Shearer
Moe
Apu
Chief Wiggins
Dr. Nick Riviera . . . . Hank Azaria

PRODUCERS   Larina Adamson, Sherry Argaman, Joseph A. Boucher, James L. Brooks, David S. Cohen, Jonathan Collier, Gabor Csupo, Greg Daniels, Paul Germain, Matt Groening, Al Jean, Ken Keeler, Harold Kimmel, Jay Kogen, Colin A.B.V. Lewis, Jeff Martin, Ian Maxtone-Graham, J. Michael Mendel, George Meyer, David Mirkin, Frank Mula, Conan O'Brien, Bill Oakley, Margo Pipkin, Richard Raynis, Mike Reiss, David Richardson, Jace Richdale, Phil Roman, David Sachs, Richard Sakai, Bill Schultz, Mike Scully, David Silverman, Sam Simon, John Swartzwelder, Ken Tsumura, Jon Vitti, Josh Weinstein, Michael Wolf, Wallace Wolodarsky

PROGRAMMING HISTORY

• FOX

December 1989–August 1990	Sunday 8:30-9:00
August 1990–	Thursday 8:00-8:30

FURTHER READING

Berlant, Lauren. "The Theory of Infantile Citizenship." *Public Culture: Bulletin of the Society for Transnational Cultural Studies* (Chicago), Spring 1993.

Coe, Steve. "Fox Hoping *Simpsons* Will Boost Slow Start." *Broadcasting* (Washington, D.C.), 8 October 1990.

Corliss, Richard. "*Simpsons* Forever!" *Time* (New York), 2 May 1994.

Elder, Sean. "Is TV the Coolest Invention Ever Invented? Subversive Cartoonist Matt Groening Goes Prime Time." *Mother Jones* (Boulder, Colorado), December 1989.

Freeman, Mike. "Fox Affils Deal for Radical Dude: *Simpsons* Pricing Appears to Remain Apace of Big-Ticket '80s Sitcoms." *Broadcasting and Cable* (Washington, D.C.), 1 March 1993.

Henry, Matthew. "The Triumph of Popular Culture, Situation Comedy, Postmodernism and *The Simpsons*." *Studies in Popular Culture* (Louisville, Kentucky), October 1994.

Larson, Mary Strom. "Family Communication on Prime-time Television." *Journal of Broadcasting and Electronic Media* (Washington, D.C.), Summer 1993.

McConnell, Frank. "'Real' Cartoon Characters: *The Simpsons*." *Commonweal* (New York), 15 June 1990.

Ozersky, Josh. "TV's Anti-families: Married....With Malaise." *Tikkun* (Oakland, California), January-February 1991.

Rebeck, Victoria A. "Recognizing Ourselves in *The Simpsons*." *Christian Century* (Chicago), 27 June 1990.

Waters, Harry F. "Family Feuds." *Newsweek* (New York), 23 April 1990.

Zehme, Bill. "The Only Real People on TV." *Rolling Stone* (New York), 28 June 1990.

See also Brooks, James L.; Cartoons; Family on Television; FOX Broadcasting Company

SIMULCASTING

Simulcasting is a term used to describe the simultaneous transmission of a television and/or radio signal over two or more networks or two or more stations. The most obvious example would be a major address by the President of the United States which might be carried simultaneously by three television networks (ABC, CBS, NBC), one or more cable networks (CNN, CNBC), and several radio networks.

The term has taken a different meaning during various periods in broadcasting. Initially, the term was applied to the simultaneous transmission of important events over two or more radio outlets. Later, it referred to the simultaneous transmission of programs on radio and television. This occurred during the 1960s when some of the most popular radio programs became television programs. but the audio portion was still simulcast on radio. This practice was short-lived, however, as the number of homes with TV sets increased and radio shifted to a more music-based programming.

The very slow growth in FM radio during the 1950s and 1960s was due, in part, to the simulcasting of radio programming over co-owned AM and FM stations. In 1964, the FCC (Federal Communications Commission) acted to force the independence of FM stations by severely restricting the number of hours that AM and FM stations could simulcast during any given broadcast day, although protests by radio station owners delayed implementation of the rule until 1 January 1967. (Ironically, the FCC removed the restrictions on AM-FM simulcasting a quarter of a century later so that struggling AM stations could simulcast the programming of their stronger FM sister stations.)

Simulcasting of musically-oriented programs by television and FM stations occurred on an occasional basis during the 1970s and 1980s. Sometimes these programs included opera or other classical presentations; on other occasions rock concerts were simulcast. The improved sound fidelity

and stereo capability of newer television sets have diminished the need for such audio-enhancement simulcasting although some TV/FM simulcasting still occurs.

Currently the term simulcasting is most relevant to the development and adaptation of high definition television (HDTV). Both broadcasters and regulators realize that newer more advanced forms of television transmission will have to be phased in gradually since viewers with standard television receivers would not be willing to accept the immediate obsolescence of their current TV sets.

Proposals now under consideration would require television stations to simulcast two separate signals. One standard, or NTSC, analog signal suitable for reception by current receivers would be transmitted over the channels currently allocated to television stations; a second high definition, digital signal suitable for newer, more advanced receivers would be transmitted over a separate channel.

This type of simulcasting would require the allocation of additional broadcast frequencies to those television stations that transmit the second signal. While broadcasters have expressed an interest in acquiring second channels for various uses including HDTV, they resist the stipulation that links the second channel to the mandated simulcasting of NTSC and HDTV signals. Another point of controversy involves the time frame for simulcasting: i.e. how long would such transmissions be required and what requirement, if any, would have broadcasters return unused frequencies to the FCC for reallocation once simulcasting ended.

While this particular utilization of simulcasting is still under discussion the traditional simulcasting of major events by one or more television and/or cable outlets is a well established practice and one not likely to end in the near future.

—Norman Felsenthal

See also Music on Television; Public Television

THE SINGING DETECTIVE

British Serial Drama

The Singing Detective (1986) is a six-part serial by one of British television's great experimental dramatists, Dennis Potter. Produced for the BBC by Kenith Trodd and directed by Jon Amiel, it revolves around the personal entanglements—real, remembered, and imagined—of the thriller author, Philip Marlow (played by Michael Gambon), who is suffering from acute psoriasis and from the side-effects associated with its treatment. The result is a complex, multi-layered text which weaves together, in heightened, anti-realist form, the varied interests and themes of the detective thriller, the hospital drama, the musical and the autobiography.

A first level of narrative centres on Marlow in his hospital bed. Set in the present, this narrative includes his fantasies and

hallucinations. The second narrative is played out in Marlow's mind as he mentally re-writes his story *The Singing Detective*, with himself as hero, set in 1945. The third narrative, also set in 1945, consists of memories from his childhood as a nine-year-old boy in the Forest of Dean and in London, told through a series of flashbacks. The fourth area of narrative involves Marlow's fantasy about a conspiracy between his wife, Nicola, and a supposed lover, set in the present.

There are obvious parallels between the story and Potter's own personal history. Like Marlow, Potter was born and brought up in the Forest of Dean at about the same time as Marlow was a wartime evacuee, and like Marlow he stayed in Hammersmith with relations who had difficulty with his strong Gloucestershire accent. Two key incidents in *The Singing Detective* are based on real-life incidents in childhood—Potter's mother, a pub pianist, being kissed by a man, and Potter's writing a four-letter word on the blackboard when his precocious facility as a young writer made him unpopular with other schoolchildren.

The serial is explicitly concerned with psychoanalysis: the spectator is constructed both as detective and as psychoanalyst in a drama which Potter saw as "a detective story about how you find out about yourself." The text is rich in Freudian imagery and symbolism, and also deals with psychoanalytical technique as Dr. Gibbons attempts to involve a linguistically skeptical Marlow in the talking cure. Marlow's neurosis and paranoia are explicitly linked to his repression of painful childhood memories, notably his mother's adultery, her eventual suicide and the mental breakdown of a fellow pupil after a beating by a teacher. At this level, for Potter the story was about paranoia—"one man's paranoia and the ending of it".

But *The Singing Detective* does not offer a straightforward case of autobiographical drama—for Potter, the serial was "one of the least autobiographical pieces of work I've ever attempted"—nor does it lead to conventional psychological or psychoanalytical resolution. He translates basic concerns, instead, to a more complex level where the narrative and generic dimensions of the text endlessly merge and overlap, fusing past and present, fantasy and "reality", challenging the organic conventions of realist drama and mixing the stabilities of popular television with the textual instabilities of modernism and postmodernism.

The Singing Detective is thus not only the serial that the TV viewer is watching, but the fiction that Marlow is rewriting in his head. Although his name is not unfamiliar in the genre, Marlow is no conventional focus for identification: he is obstreperously unlikeable and contradictory and his illness has been hideously disfiguring. More important, he is sometimes not the major "focaliser" of the narrative at all, but is repeatedly displaced by other themes and discourses in the process of a drama in which "character" itself rapidly becomes an unstable entity. The same character, for example, can appear in different narratives, played by the same actor; characters from one narrative can appear in another, or a character may lip-synch the lines of another character from a different narrative, or, in true Brechtian-Godardian style, characters may feel free to comment on their role, or to speak directly to the camera.

Questions of time and its enigmas, past and present, are also rendered complex. In narrative 1, in the present, Marlow is reconstructing two pasts: the book he wrote a long time ago, which was itself set in the past, and a part of his childhood, also set in 1945. The main enigmas in his text are set in that year. In narrative 2, who killed the busker, Sonia, Amanda, Lilli and Mark Binney? And why? In narrative 3, who shat on the table? Why did Mrs. Marlow commit suicide? Although narratives 1 and 2 usually (but not always) follow story chronology, in narrative 3, it is not really clear what the actual chronology of the young Philip's life might be. In terms of narrative frequency, *The Singing Detective* is further marked by a high degree of repetition—of words, events, and visual images—as the same event, or part of it, is retold, re-worked, or recontextualised.

The final shoot-out in the hospital thus merges narratives 1 and 2 by uniting past (1945) with the present time of its reconstruction (1986), i.e. its reconstruction in Marlow's head rather than in his book itself. The "villain" who is killed is not just one of the characters but also the sick author himself, thus liberating the singing detective and ensuing an ending for narrative 2. Although it does not resolve any of the enigmas posed by this second narrative, the "dream" of the "sick" Marlow allows the Marlow who is "well" to get up and walk out of the hospital, concluding narrative 1. As he walks away down a long corridor on Nicola's arm, bird sounds from the Forest of Dean (narrative 3) are heard; past and present are again combined, if, typically, not reconciled.

The Singing Detective thus refuses any simple reading, and even contests the traditional definition of television "reading" altogether. It is witty, comic, and salacious, and yet also savage, bleak and nihilistic. It is blunt and populist, and yet arcane and abstruse. Its key themes are language and communication, memory and representation, sexual and familial betrayal and guilt, the transition from childhood to adulthood, the relationships between religion, knowledge and belief, the processes of illness and of dying. Whilst its themes are resonant, its main enduring claim on critical attention lies in its thoroughgoing engagement with the textual politics of modernism. Its swirl of meanings and enigmas render it British prime-time television's most sustained experiment with classic post-Brechtian strategies for anti-realism, reflexivity, textual deconstruction, and for the encouragement of new reading practices on the part of the TV spectator.

—Phillip Drummond and Jane Revell

CAST

Philip Marlow	Michael Gambon
Raymond Binney/Mark Binney/Finney	Patrick Malahide
Nurse Mills/Carlotta	Joanne Whalley
Dr. Gibbon	Bill Paterson
Philip Marlow at Ten	Lyndon Davies
Nicola	Janet Suzman
Mrs. Marlow/Lili	Alison Steadman
Mr. Marlow	Jim Carter
Schoolteacher/Scarecrow	Janet Henfrey
Mark Binney (age of 10)	William Speakman

PRODUCERS John Harris, Kenith Trodd

PROGRAMMING HISTORY 6 episodes of 60–80 minutes

• BBC

16 November–21 December 1986

FURTHER READING

Cook, John R. Dennis Potter: A Life on Screen. Manchester and New York: Manchester University Press, 1995.

Corliss, Richard. "Notes from The Singing Detective: Dennis Potter Makes Beautiful Music from Painful Lives." Time (New York), 19 December 1988.

Fuller, Graham. "Dennis Potter." *American Film* (Washington, D.C.), March 1989.

———, editor. *Potter on Potter*. London and Boston: Faber and Faber, 1993.

Gilbert, W. Stephen. *Fight and Kick and Bite: The Life and Work of Dennis Potter*. London: Hodder and Stoughton, 1995.

Potter, Dennis. *Seeing the Blossom: Two Interviews and a Lecture*. London and Boston: Faber and Faber, 1994.

———. "An Interview with Dennis Potter: An Edited Transcript of Melvyn Bragg's Interview with Dennis Potter, Broadcast on the 5th of April, 1994." London: Channel 4 Television, 1994.

Stead, Peter. *Dennis Potter*. Bridgend, England: Seren, 1993.

Wyver, John. "Arrows of Desire" (interview). *New Statesman and Society* (London), 24 November 1989.

See also British Programming; *Pennies from Heaven*; Potter, Dennis; Trodd, Kenith

SISKEL AND EBERT

U.S. Movie Review Program

Siskel and Ebert represents the first and most popular of the movie review series genre that emerged on television in the mid-1970s. The lively series focuses on the give-and-take interaction and opinions of its knowledgeable and often contentious co-hosts, Gene Siskel, film critic of the *Chicago Tribune* and Roger Ebert, film critic of the *Chicago Sun-Times*. Syndicated to approximately 180 markets across the United States, as of this writing, the spirited pair reach a potential 95% of the country on a weekly basis.

Developed from an idea credited to producer Thea Flaum of PBS affiliate WTTW in Chicago, the original series, *Opening Soon at a Theater Near You*, was broadcast once a month to a local audience beginning in September 1975. Using brief clip of movies in current release, the rival critics debated the merits of the films making simple yes or no decisions for positive and negative review. On those not so rare occasions when the two disagreed, sparks might fly, which delighted viewers. An additional element of interest featured Spot the Wonder Dog jumping on to a balcony seat and barking on cue to introduce the film designated "dog of the week."

After two seasons, the successful series was retitled *Sneak Previews* and appeared biweekly on the PBS network. By its fourth season, the show became a once-a-week feature on 180 to 190 outlets and achieved status as the highest rated weekly entertainment series in the history of public broadcasting. Based on their success, in 1980, WTTW made plans to remove the show from PBS and sell it commercially as a WTTW production. The two stars indicate they were offered a take-it-or-leave-it contract, which they declined. They left the series in 1981 to launch *At the Movies* for commercial television under the banner of Tribune Entertainment, a syndication arm of the *Chicago Tribune*. Basically utilizing the same format as *Sneak Previews*, the new series made some minor adjustments including the replacement of the black-and-white Wonder Dog with Aroma the skunk which ultimately was removed to make room for commercials. At WTTW, *Sneak Previews* replaced Siskel and Ebert with New York based critics Jeffrey Lyons and Neal Gabler. In time, the PBS offering would settle on Lyons and Michael Medved as its hosts and the show remained on air through the 1995–1996 season.

Citing contractual problems with Tribune Entertainment, in 1986, Siskel and Ebert departed *At the Movies* for Buena Vista Television, a subsidiary of the Walt Disney Company, and created a new series entitled *Siskel and Ebert and the Movies*. The order of the names was decided by the flip of a coin and the show title was eventually shortened to *Siskel and Ebert*. Ebert also suggested the Romanesque thumbs up-thumbs down rating system, which has since become a distinctive Siskel-Ebert trademark. Their former show, *At the Movies*, acquired Rex Reed and Bill Harris as hosts, and added news of show business to the format. Harris left the series in 1988 and was replaced by Dixie Whatley, former co-host on *Entertainment Tonight*, and the series continued into 1990.

Of all the different series and co-hosts in this genre, the Siskel-Ebert partnership has remained the most celebrated. In twenty years of offering responsible commentary in an unedited spontaneous fashionthe two critics have reviewed more than 4,000 films and have compiled an impressive list of firsts and show milestones. In his defense of television film critics in the May/June 1990 issue of *Film Comment*, Ebert, the only film critic to have won a Pulitzer Prize for criticism, points out that *Siskel and Ebert* was the first national show to discuss the issue of film colorization, the benefits of letterbox video dubbing and the technology of laser disks. They have provided an outlet for the ongoing examination of minority and independent films, attacked the MPAA rating system as de facto censorship and protested product placement, i.e., incidental advertising, within films. And, in May 1989, extolling the virtues of black-and-white cinematography, they videotaped their show in monochrome—the first new syndicated program to do so in twenty-five years.

Siskel and Ebert's influence with audiences is also notable. Their thumbs-up reviews are credited with turning films such as *My Dinner with Andre* (1981), *One False Move* (1992) and *Hoop Dreams* (1994) into respectable box-office hits. Thumbs down reviews have had the opposite effect but many filmmakers feel that ultimately it is up to the public to choose what films they see and many directors and producers speak to the benefits that exposure on *Siskel and Ebert* can provide. Notwithstanding, there have been occasional disgruntled feelings. As reported in the *Los Angeles Times* (10 December 1995), screenwriter Richard LaGravanese used "Siskel" as the name for one of the "bad guys" in his film *The Ref* after a negative review of his previous work, *The Fisher King*.

Both Siskel and Ebert agree their animated dialogue is crucial to the show's success and more compelling than criticism from a solitary voice. They view their disagreements as those of two friends who have seen a movie and have a difference of opinion. But, they have had more intense moments, as evidenced in a pre-Oscar special broadcast in 1993—when an angry Ebert took exception to Siskel's revelation of the significant plot twist that concludes the film *The Crying Game*.

Through the years, the television industry has recognized Siskel and Ebert with six national Emmy nominations and one local Emmy (1979). In 1984, the pair were among the first broadcasters initiated into the National Association of Television Programming Executives (NATPE) Hall of Fame. They also received NATPE's Iris Award for their achievement in nationally syndicated television. The Hollywood Radio and Television Society named them Men of the Year in 1993. As Richard Roeper wrote in the *Chicago Sun-Times* (15 October 1995) on the occasion of their twentieth anniversary, "Siskel and Ebert took serious film criticism and made it palatable to a mass audience—and in so doing, became celebrities themselves, as recognizable as most of the movie stars whose films they review."

—Joel Sternberg

HOSTS

Gene Siskel and Roger Ebert

PROGRAMMING HISTORY

Syndicated   Various Times

FURTHER READING

Arlidge, Ron. "'At the Movies' Rates Higher than Improving 'Sneak.'" *Chicago Tribune*, 11 November 1982.

Anderson, John. "Why is 'Movies' So Successful? It's Simple." *Chicago Tribune TV Week*, 1-7 September 1985.

Bellafante, Ginia. "Pro Thumb Wrestling." *Time* (New York), 5 April 1993.

Ebert, Roger. "All Stars or, Is There a Cure for Criticism of Film Criticism." *Film Comment* (New York), May/June 1990.

Roeper, Richard. "Thumbs Up! 20 Years in the Balcony." *Chicago Sun-Times*, 15 October 1995.

Turan, Kenneth. "Rating TV's Movie Critics." *TV Guide* (Radnor, Pennsylvania), 18-24 March 1989.

Zoglin, Richard. "'It Stinks!' ' You're Crazy!'" *Time* (New York) 25 May 1987.

See also Movies on Television

THE SIX WIVES OF HENRY VIII

British Historical Drama Serial

The Six Wives of Henry VIII, first broadcast by the BBC in 1970, became one of its most celebrated historical drama serials. The nine-hour, six-part series went on to be shown in some 70 countries and attracted no less than seven major awards, winning plaudits both for the quality of the performances and for its historical authenticity.

Towering over the series was the gargantuan figure of Henry himself, played by the hitherto unknown Australian actor Keith Michell, who earned an award for Best Television Actor as a result of his efforts. Michell, who started out as an art teacher, owed the role to Laurence Olivier, who had been impressed by Michell while on tour in Australia and had brought him back to England in order to advance his career. The faith the BBC put in the young actor was more than amply rewarded; Michell went to extraordinary lengths to vitalize the larger-than-life character of the king.

The series was neatly split into six episodes, each one dealing with one of the six wives and tracing their varied experiences and sometimes bloody ends at the hands of one of England's most infamous rulers. The wives themselves were played by Annette Crosbie, Dorothy Tutin, Anne Stallybrass, Elvi Hale, Angela Pleasance, and Rosalie Crutchley, all respected and proven stars of stage and screen. Annette Crosbie, playing Catherine of Aragon, collected a Best Actress Award for her performance.

Michell, though, was always the focus of attention. The task for the actor was to portray Henry at the different stages of his life, beginning with the athletic 18-year-old monarch and culminating in the oversize 56-year-old tyrant plagued by a variety of physical ailments. Playing the aging Henry in the later episodes proved the most demanding challenge. Michell, who boasted only half the girth of the real king, spent some four hours each day getting his make-up on, and was then unable to take any sustenance except through a straw because of the padding tucked into his cheeks. The impersonation was entirely convincing, however, and critics hailed the attention to detail in costume and sets. No one, it seemed, twigged that Henry's mink robes were really made of rabbit fur, or that the fabulous jewels studding his hats and coats were humble washers and screws sprayed with paint.

The lavishness of the costumes and settings and the brilliance of Michell and his co-stars ensured the success of the series, though some viewers expressed reservations. In particular, it was felt by some critics that the underlying theme of the lonely and essentially reasonable man beneath the outrageous outer persona was perhaps rather predictable, and further that Michell—who admitted to admiring Henry's excesses—had a tendency to reduce Henry to caricature (a fault more clearly evident in the film *Henry VIII and His Six Wives* that was spawned by the television series in 1972).

Whatever the criticisms, the success of *The Six Wives of Henry VIII* brought stardom to Michell and also did much to establish the BBC's cherished reputation for ambitious and historically authentic costume drama, consolidated a year later by the equally-acclaimed series *Elizabeth R*, starring Glenda Jackson as Henry's daughter.

—David Pickering

The Six Wives of Henry VIII
Photo courtesy of BBC

CAST

Henry VIII	Keith Michell
Catherine of Aragon	Annette Crosbie
Anne Boleyn	Dorothy Tutin
Jane Seymour	Anne Stallybrass
Anne of Cleves	Elvi Hale
Catherine Howard	Angela Pleasance
Catherine Parr	Rosalie Crutchley
Duke of Norfolk	Patrick Troughton
Lady Rochford	Sheila Burrell
Thomas Cranmer	Bernard Hepton
Thomas Cromwell	Wolfe Morris
Sir Thomas Seymour	John Ronane

NARRATOR
Anthony Quayle

PRODUCERS   Ronald Travers, Mark Shivas, Roderick Graham

PROGRAMMING HISTORY   Twelve 90-Minute Episodes

- BBC-2

1 January–5 February 1970

FURTHER READING
Trewin, J.C., editor. *The Six Wives of Henry VIII*. New York: Ungar, 1972.

See also British Programming; Miniseries

60 MINUTES

U.S. News Magazine Show

In 1967 Don Hewitt conceived of his new program, *60 Minutes,* as a strategy for addressing issues given insufficient time for analysis in two minutes of the *Evening News* but not deemed significant enough to justify an hour-long documentary. *60 Minutes* was born, then, in an environment of management tension and initial ambiguity regarding its form. Bill Leonard, CBS vice president for News Programming, supported the new concept, but Richard Salant, president of the News Division, argued it countered that unit's commitment to the longer form and risked taking the hard edge off television journalism. In the end Salant acquiesced.

Hewitt's direction remained flexible and uncertain, with design for the program possibly including any number of "pages" and "chapters" lasting one to twenty minutes, and spanning breaking news, commentary, satire, interviews with politicians and celebrities, feature stories, and letters to the editor. CBS proclaimed the groundbreaking potential of this magazine form, announcing that no existing phrase could describe the series' configuration, and that any attempt to gauge (or predict) demographic appeal based on comparisons with traditional public affairs programming was a limited prospect. Yet, by the spring of 1993 the series' success was so established within the history of network programming that CBS and *60 Minutes* had competition from six other prime-time magazine programs.

From September 1966 through December 1975, network management shifted the scheduling position of *60 Minutes* seven times. Its ratings were very low according to industry standards, although slightly higher than those of *CBS Reports* when aired in the same time slot, but critical response remained positive. In today's competitive environment, where "unsuccessful" programs are quickly removed from the schedule, the series would not remain on the air. But in the early 1970s the CBS News Division sought a more engaging weekly documentary form.

Almost three decades later Hewitt flippantly claimed *60 Minutes* destroyed television by equating news with the profit motive; news organizations sought money in magazine and entertainment news programs, reducing their long-standing, and expensive, commitments to breaking news. But Hewitt set the groundwork. His blunt statements suggesting that success depends on marketing, and his continuous refinements of the product often generated controversy. Audiences must experience stories in the pit of their stomach, the narrative must take the viewer by the throat, and, noted Hewitt, when a segment is over it's not significant what they have been told—"only what they remember of what you tell them." Hewitt predicted high ratings if *60 Minutes* packaged stories, not news items, as "attractively as Hollywood packages fiction." Such stories require drama, a simplified structure, a narrative maximizing conflict, a quick editing pace, and issues filtered through personalities. Although the series profiled celebrities, politicians, and popular or well-known people in numerous fields, the stress on personality meant that a human being would be positioned in the story in a manner inviting the public to "identify with" or "stand against."

The *60 Minutes* correspondents narrated and focused these "mini-dramas." Several of the show's journalists had established positions as personalities before *60 Minutes*, but with the program's growing success and significance, the correspondents reached international celebrity status, becoming crusaders, detectives, sensitive and introspective guides through social turmoil, and insightful probers of the human psyche. A confrontational style of journalism, pioneered by Mike Wallace, grew and was embraced by a more confrontational society. In the 1970s certain correspondents seemed to speak for a public under siege by institutional greed and deceit.

Through it all Hewitt remained sensitive to balancing the series at any one time with varying casts. Wallace's role remained consistent as the crusading detective, played, as the series began, opposite Harry Reasoner's calm, analytical and introspective persona. As correspondents were added—Morley Safer, Dan Rather, Ed Bradley, Diane Sawyer, Meredith Vieria, Steve Kroft, and Lesley Stahl—Hewitt developed complimentary personas. The correspondents became part of his "new form" of storytelling, allowing the audience to watch their intimate involvement in discovering information, tripping up an interviewee, and developing a narrative. As a result, the correspondents are often central to

60 Minutes

Hewitt's notion of stories as morality plays, the confrontation of vice and virtue.

The most explosive segments of *60 Minutes*, for example, accuse companies, government agencies, or organizations of massive deceit, of harming public welfare. Correspondents, often in alliance with an ex-employee or group member, have confronted the Illinois Power Company, Audi Motors, the Worldwide Church of God, tobacco companies, Allied Chemical Corporation, the U.S. Army, adoption agencies and land development corporations. Smaller entities and individuals, such as owners of fraudulent health spas, used-car dealers, or clothing manufacturers, often put faces and names on compelling images of deceit. Because of these investigative segments, the series was the focus of consistent examination by the press concerning such issues as journalism ethics and integrity. *60 Minutes* has been taken to task for having correspondents or representatives use false identities to generate stories, establishing sting operations for the camera, confronting the person under inquiry by surprise, and revealing new documents without prior notice to a cooperative interviewee in order to increase the shock value of the information. By raising these issues the series focused attention on emerging techniques of broadcast journalism. But even when stories relied on more thoughtful critical analysis they could shake the foundations of institutions and have strong and lasting effects. Morley Safer's 1993 story arguing that the contemporary art world is filled with "junk" sparked more than two years of defense and response from different members of the art community.

In spite of widespread knowledge of these strong techniques, individuals still subject themselves to interviews, offering the audience an opportunity to anticipate who will win the battle. Indeed, part of the appeal of *60 Minutes* is whether the possibility of getting a corporate perspective across is worth the risk encountered by company representatives when facing the penetrating (aggressive) questioning and fact-finding by the correspondent. The consequences and repercussions of appearing on the program can be severe. Stark revelations by eyewitnesses have lead to extensive damage and bankruptcy of companies, even to death threats. One person, after disclosing odometer tampering in the automotive industry, had his house blown up.

The high stakes involved in such public confrontations led Herb Schmertz, former vice president of the Mobil Oil Corporation, to write a guide for corporate America instructing companies and individuals how to prepare and withstand an interview by *60 Minutes'* correspondents. But public figures still appear, seeking to enhance their position or rectify a situation. In doing so they risk unexpected changes in the direction of public opinion, as demonstrated by Ross Perot's drop in approval ratings after raising questionable topics in his interview.

The series continues to establish historical markers regarding legal issues of press freedom, and some cases have set precedents for legal aspects of broadcast journalism. One reason for this continuing involvement is that for each segment, the outtakes, transcribed interviews, editors' notes, and relevant documents are archived and entered into a database at CBS. Following the segment entitled "The Selling of Col. Herbert," for example, Col. Anthony Herbert initiated a defamation suit against producer Barry Lando. The suit was dismissed after ten years, but not before the Supreme Court decision giving Herbert's lawyers the right to "direct evidence" about the editorial process. Specifically, they were given access to film outtakes and editors' notes that could establish malicious intent by illustrating the producer's "state of mind." Dr. Carl Galloway's slander suit against Dan Rather and *60 Minutes* went to court after Rather left the show to anchor the *Evening News,* but when Rather, and the series' production process, were scrutinized on the witness stand, the examination raised questions about the power of editing to construct specific images of an individual.

In these and other cases, *60 Minutes* continues, intentionally and unintentionally, to be at the center of struggles concerning the rights of the press. Risks taken by the series have the potential to harm the image and credibility of CBS as well as that of the program, and such concerns have conditioned CBS and the broadcast industry to a rapid response to legal challenges.

But *60 Minutes* has also become one of most analyzed programs concerning television's effect on viewer behavior. When a story endorsed moderate consumption of red wine to prevent heart disease, sales of red wine jumped significantly. Although the use and gradual discontinuation of Alar on apple crops received moderate coverage by the press, *60 Minutes* addressed the issue of this use of the cancer-causing agent in 1989. The story, and other media reports contributing to what became a national hysteria, cost the agriculture industry over 100 million dollars. The series' scrutiny of companies even led to tangible effects on their stocks. During one two-year period, stocks rose an average of 14% for companies negatively profiled on *60 Minutes.* Market insiders, aware of the upcoming story, bought to increase shares, knowing that the market had previously responded to the companies' problems.

Critics, researchers, and the public continue to investigate the reasons behind the longevity of *60 Minutes* as a popular culture phenomenon. The series' timeliness, its bold stand on topics, its confrontations with specific individuals, all provides audiences with the pleasure of knowing accountability does exist. For some, the program compels with its crusades, as in the case of Lenell Geter, freed from life imprisonment after his case was explored and analyzed. For others, the appeal comes with vigorous self-defense, as when Senator Alfonso D'Amato (Republican, New York) poured out his wrath in a 30-minute response to claims that he misused state funds.

Point/Counterpoint, a program feature from 1971 to 1979, illustrated that two opposing positions can remain unreconciled, and served, in three-minute debates between left- and right-wing critics, to agitate viewer emotions with ideological battles. The segment's popularity probably explains why, in 1996, Hewitt added a similar "commentator" section, resurrecting the art of speaking what the public may think but dare not say with such force. And the series' perennial "light" moment, "A Few Minutes with Andy Rooney" confirms the value of personal opinion on otherwise mundane matters.

60 Minutes is also able to generate news about itself and thus keep the series attractive by humanizing its trials and tribulations. For over two decades the producers, correspondents, and Hewitt have played out issues in public. Twice, producer Marion Goldin quit the program after accusing the unit of sexism. Hewitt charged Rooney with hypocrisy for criticizing CBS owner, Lawrence Tisch, on air instead of quitting. Wallace has been reprimanded for using hidden cameras to tape a reporter who agreed to help him with a story. And when the series dropped to number 13 in the 1993–94 Nielsen ratings (after being first for two years), the drop became a "story." Hewitt and others blamed CBS, Inc., for losing affiliates in urban areas and for allowing the FOX network win the bid for Sunday afternoon football, *60 Minutes'* long-time lead-in program.

When *Dateline NBC,* a similar news magazine, was programmed opposite *60 Minutes* in the spring of 1996, the press covered the move as a battle for the hearts and minds of the audience. But for several months before the direct competition, Hewitt began to revamp the series, adding brief, hard news segments, announcing production of new stories throughout the summer, adding a "Commentary" section, and tracking down new and unfamiliar topics. Although the series has been criticized for following compelling stories broken by magazines such as *The Nation*, instead of breaking news, the strategy meets Hewitt's mandate to impact a large audience. Entering its fourth decade, then, *60 Minutes* continues to shift strategy and change in form. The one constant is that the program's producers still believe in validating its journalistic integrity through its popularity on American television.

—Richard Bartone

REPORTERS

Mike Wallace
Harry Reasoner (1968–70, 1978–91)
Morley Safer (1970–)
Dan Rather (1975–81)
Andrew Rooney (1978–)
Ed Bradley (1981–)
Diane Sawyer (1984–89)
Meredith Vieira (1989–91)
Steve Kroft (1989–)
Leslie Stahl (1991–)

PRODUCER   Don Hewitt

PROGRAMMING HISTORY

• CBS

September 1968–June 1971	Tuesday 10:00-11:00
January 1972–June 1972	Sunday 6:00-7:00
January 1973–June 1973	Sunday 6:00-7:00
June 1973–September 1973	Friday 8:00-9:00
January 1974–June 1974	Sunday 6:00-7:00
July 1974–September 1974	Sunday 9:30-10:30
September 1974–June 1975	Sunday 6:00-7:00
July 1975–September 1975	Sunday 9:30-10:00
December 1975–	Sunday 7:00-8:00

FURTHER READING

Campbell, Richard. *"60 Minutes" and the News: A Mythology for Middle America*. Urbana: University of Illinois Press, 1991.

Coffey, Frank. *"60 Minutes": 35 Years of Television's Finest Hour*. Los Angeles: General Publishing Group, 1993.

Fury, Kathleen, editor. *Dear 60 Minutes*. New York: Simon and Schuster, 1984.

Goodman, Walter. "How *60 Minutes* Holds its Viewers' Attention." *The New York Times*, 22 September 1993.

Hewitt, Don. *Minute by Minute*. New York: Random House, 1985.

Madsen, Axel. *"60 Minutes": The Power and the Politics of America's Most Popular TV News Show*. New York: Dodd, Mead, 1984.

Moore, Donovan. "*60 Minutes*." *Rolling Stone* (New York), 12 January 1978.

Punch, Counterpunch: "60 Minutes" vs. Illinois Power Company. Washington, D.C.: Media Institute, 1981.

Reasoner, Harry. *Before the Colors Fade*. New York: Knopf, 1981.

Rosenberg, Howard. "Child Abuse: A Compound Travesty." *Los Angeles Times*, 21 May 1986.

———. "*60 Minutes*: Time Out for a Correction." *Los Angeles Times*, 23 August 1993.

"The *60 Minutes* Team Tells: The Toughest Stories We've Ever Talked." *TV Guide* (Radnor, Pennsylvania), 19-25 January 1991.

"*60 Minutes*" Verbatim: Who Said What to Whom—The Complete Text of 114 Stories with Mike Wallace, Morley Safer, Dan Rather, and Andy Rooney. New York: Arno Press, 1980.

Shales, Tom. "Still Ticking at 25: The Great Granddaddy of Magazine Shows." *The Washington* (D.C.) *Post*, 13 November 1993.

Shaw, David. "Alar Panic Shows Power of Media to Trigger Fear." *Los Angeles Times*, 13 September 1994.

Spragens, William C. *Electronic Magazines: Soft News Programs on Network Television*. Westport, Connecticut: Praeger, 1995.

Stein, Harry. "How *60 Minutes* Makes News." *The New York Times*, 6 May 1979.

Wallace, Mike, and Gary Paul Gates. *Close Encounters*. New York: Morrow, 1984.

See also Columbia Broadcasting Company; Documentary; Hewitt, Don; News, Network; Sawyer, Diane; Wallace, Mike

THE $64,000 QUESTION/THE $64,000 CHALLENGE

U.S. Quiz Shows

The premiere of *The $64,000 Question* as a summer replacement in 1955 marked the beginning of the big money quiz shows. Following a Supreme Court ruling in 1954 that exempted "Jackpot" quizzes from charges of illegal gambling, Louis G. Cowan, the creator and packager of the program, Revlon, its main sponsor, and CBS were able to

bring this new type of quiz show on the air. Based on the popular 1940s radio quiz show *Take It Or Leave It* with its famous $64 question, *The $64,000 Question* increased the prize money to an unprecedented, spectacular level. It also added public appeal with a security guard and a "trust officer" who monitored questions and prizes, and its fairly elaborate

The $64,000 Question

set design, which included an "isolation booth" for the contestants. Intellectual "legitimacy" was further claimed through the employment of Professor Bergen Evans as "Question Supervisor." With its emphasis on high culture, academic knowledge, and its grave, ceremonious atmosphere, *The $64,000 Question* represented an attempt to gain more respectability for the relatively new and still despised television medium, while at the same time appealing to a large audience.

Each contestant began his or her quest for fortune and fame by answering a question in an area of expertise for $64. Each subsequent correct answer doubled their prize money up to the $4,000 level. After this stage contestants could only advance one level per week and were asked increasingly elaborate and difficult questions. They were allowed to quit the quiz at any level—and keep their winnings—but missing a question always eliminated the contestant. Nevertheless, contestants were guaranteed the $4,000 from the first round, and, if they missed a question after having reached the $8,000 level, they received an additional consolation prize—a new Cadillac. At this level, candidates were also moved from the studio floor to

the "Revlon Isolation Booth," a shift designed to intensify the dramatic effects at the higher levels of the quiz.

Besides its use of such spectacular features, the appeal of *The $64,000 Question* was also strongly grounded in the audience's identification with returning contestants. Thus, many of the early competitors were transformed from "common people" into instant superstars. Policeman Redmond O'Hanlon, a Shakespeare expert, and shoemaker Gino Prato, an opera fan, are among the noted examples. The popularity of these and other contestants proved the viability of "the serialized contest," a concept that *The $64,000 Question* and many imitators (e.g., *Twenty-One*; *The Big Surprise*) followed.

Due to the immense success of *The $64,000 Question* (at one point in the 1955 season it had an 84.8% audience share), CBS and Cowan created a spin-off, *The $64,000 Challenge*. This program allowed those contestants from *The $64,000 Question* who had won at least $8,000 to continue their quiz show career. The format was changed into a more overt contest; two candidates competed against each other in a common area of expertise. As a minimum prize, contestants were guaranteed the amount at which they beat their opponents. Additionally, the $64,000 limit on winnings was removed, making the contests even longer and more spectacular.

The combination of these two shows allowed the most successful candidates to become virtual television regulars, as in the case of Teddy Nadler, who had accumulated $252,000 by the time *The $64,000 Challenge* was canceled. These programs held top rating spots until *Twenty–One* found a format and a contestant, Charles Van Doren, which were even more appealing to the audience.

The need for regular contestants to appear over long periods of time, one of the central factors in the popularity of the big prize game shows, also proved to be an central factor in their downfall with the quiz show scandal of 1958. The sponsors of the programs implicitly expected and sometimes explicitly demanded that popular contestants be supplied with answers in advance, enabling them to defeat unpopular competitors and remain on the show for extended periods. Although no allegations against Entertainment Productions, Inc., and CBS were ever substantiated, Barnouw points out in *The Image Empire* that their production personnel claimed that Revlon had frequently tried to influence the outcome of the quizzes. Ultimately, both shows were canceled due to public indignation and waning ratings in the wake of the scandals.

One of the most significant results of the quiz show scandal and the involvement of sponsors in it was the shift in the power to program television. The scandal was used as an argument by the networks to completely eliminate sponsor-controlled programming in prime-time broadcasting and to take control of program production themselves.

—Olaf Hoerschelmann

THE $64,000 QUESTION

EMCEE
Hal March

ASSISTANT
Lynn Dollar

AUTHORITY
Dr. Bergen Evans

PROGRAMMING HISTORY

• CBS

June 1955–June 1958	Tuesday 10:00-10:30
September 1958–November 1958	Sunday 10:00-10:30

THE $64,000 CHALLENGE

EMCEE
Sonny Fox (1956)
Ralph Story (1956–58)

PRODUCERS   Steve Carlin, Joe Cates

PROGRAMMING HISTORY

• CBS

April 1956–September 1958	Sunday 10:00-1:30

FURTHER READING

Barnouw, E. *A History of Broadcasting in the United States: Volume III—The Image Empire.* New York: Oxford University Press, 1970.

Boddy, W. *Fifties Television: The Industry and its Critics.* Urbana: University of Illinois Press, 1990.

Schwartz, D., S. Ryan, and F. Wostbrock. *The Encyclopedia of Television Game Shows.* New York: Zoetrope, 1987.

See also Quiz and Game Shows; Quiz Show Scandals

SKELTON, RED

U.S. Comedian

It was not until 1986, a full fifteen years after his weekly television show had ended, that "one of America's clowns" received his overdue critical praise. Only then did the critics realize what the public had long known. Regardless of his passion for corny gags and slapstick comedy, Red Skelton was a gifted comedian. He is one of the few performers to succeed in four entertainment genres—vaudeville, radio, film, and television. To honor his lifetime achievements, Skelton received the Academy of Television Arts and Sciences Governor's Emmy Award in 1986 and the critical praise he deserved.

Skelton's youth was characterized by poverty and a fascination for vaudeville. It was the influence of vaude-

ville great Ed Wynn that led Skelton to perfect his own comedy routines. The basics of Red's vaudeville act consisted of pantomimes, pratfalls, funny voices, crossed eyes, and numerous sight gags that would serve to identify Skelton throughout his entertainment career. It was also during this period that Red began developing various comedy characters.

His radio show, which ran from 1941 to 1953, provided the opportunity to present his comedy to a mass audience. The limitations of the sound medium also made it necessary for him to further develop the characters he would later bring to television: Freddie the Freeloader; Clem Kadiddlehopper, the country bumpkin; Willy Lump Lump, the drunk; Cauliflower McPugg, the boxer; The Mean Widdle Kid; San Fernando Red, the con man.

In conjunction with his radio show, Skelton also enjoyed film success, most notably in *Whistling in the Dark* (1941), *The Fuller Brush Man* (1948), *A Southern Yankee* (1948), and *The Yellow Cab Man* (1950). Regardless of his vaudeville, radio, and film success, it would be television that would bring him his greatest fame and endear him to his largest audience.

The Red Skelton Show began in 1951 on NBC as a comedy-variety show. Skelton co-produced this initial show, which was a half-hour program on Sunday evenings. In its first year, the show finished fourth in the ratings and received the Emmy Award for Best Comedy Show. Unlike other radio comedians Skelton's comedy act entailed more than his voice, and television provided the opportunity to fully display the showmanship talents he had begun in vaudeville.

In 1953, the show moved to CBS on Tuesday nights and received a second Emmy Award for Outstanding Writing Achievement in Comedy in 1961, and expanded to an hour-long show the following year. In 1964, the show made the Nielsen Top Twenty, where it stayed until its end in 1970.

The show consisted of Skelton's opening monologue, performances by guest stars, and comedy sketches which included his various characters. Perhaps the most unique part of the show (and for all of television) was "The Silent Spot," a mime sketch that often featured his character Freddie the Freeloader. The only regulars on the show were Skelton and the David Rose Orchestra. *The Red Skelton Show* set the precedent for future comedy-variety shows, such as *The Carol Burnett Show*.

According to CBS, the show's 1970 cancellation was due to rising production costs and the network's desire to appeal to more upscale advertisers (the show finished seventh in its final season). The following year, Skelton returned to NBC with a half-hour comedy variety show which included a cast of regulars. The show's premiere featured then Vice-President Spiro Agnew. This time, unfortunately, the uneven comedy failed to match Skelton's previous success. Its cancellation marked the end of Skelton's television career, a run of twenty-one straight years which also included guest appearances on other television series and involvement with thirteen television specials. The only television performer with a longer stay was Ed Sullivan (twenty-four years as host of *The Ed Sullivan Show*).

Following his departure from television, Skelton maintained a low profile and performed at resorts, clubs, and casinos. In the early 1980s a series of superb performances at Carnegie Hall received critical praise and briefly thrust him back into the public spotlight. The new-found interest in Skelton resulted in three comedy specials for Home Box Office (HBO).

Since his TV show was seldom rerun and is not syndicated, it is easy to forget his popularity. Based on longevity and audience size, *The Red Skelton Show* was the second most popular show in TV history (*Gunsmoke* is first). As Groucho Marx once said, Red Skelton is "the most unacclaimed clown in show

Red Skelton

business." Marx noted that by using only a soft, battered hat as a prop, Red could entertain with a dozen characters.

—Robert Lemieux

RED SKELTON. Born Richard Red Skelton in Vincennes, Indiana, U.S.A., 18 July 1913. Married: Lothian Toland (third wife). Joined medicine show at age 10; later appeared in show-boat stock, minstrel shows, vaudeville, burlesque, and circuses; began appearing on radio in 1936; starred in long-running *The Red Skelton Show* on television. Recipient: Emmy Awards, 1951, 1956, 1960/61; ATAS Hall of Fame and Governor's Award, 1986.

TELEVISION SERIES
1951–53, 1953–70,
   1970–71   *The Red Skelton Show*

MADE-FOR-TELEVISION MOVIE
1956   *The Big Slide*

TELEVISION SPECIALS (selection)
1954   *The Red Skelton Revue*
1959   *The Red Skelton Chevy Special*
1960   *The Red Skelton Timex Special*

1966	*Clown Alley* (host, producer)
1982	*Red Skelton's Christmas Dinner*
1983	*Red Skelton's Funny Faces*
1984	*Red Skelton: A Royal Performance*

FILMS

Having Wonderful Time, 1938; *Seein' Red*, 1939; *Broadway Buckaroo*, 1939; *Flight Command*, 1940; *Lady Be Good*, 1941; *The People vs. Dr. Kildare*, 1941; *Dr. Kildare's Wedding Day*, 1941; *Whistling in the Dark*, 1941; *Whistling in Dixie*, 1942; *Ship Ahoy*, 1942; *Maisie Gets Her Man*, 1942; *Panama Hattie*, 1942; *DuBarry Was a Lady*, 1943; *Thousands Cheer*, 1943; *I Dood It*, 1943; *Whistling in Brooklyn*, 1943; *Bathing Beauty*, 1944; *Ziegfeld Follies*, 1944; *Radio Bugs* (voice only), 1944; *The Show-Off*, 1946; *Merton of the Movies*, 1947; *The Fuller Brush Man*, 1948; *A Southern Yankee*, 1948; *Neptune's Daughter*, 1949; *The Yellow Cab Man*, 1950; *Three Little Words*, 1950; *The Fuller Brush Girl*, 1950; *Watch the Birdie*, 1951; *Duchess of Idaho*, 1950; *Excuse My Dust*, 1951; *Texas Carnival*, 1951; *Lovely to Look At*, 1952; *The Clown*, 1952; *Half a Hero*, 1953; *The Great Diamond Robbery*, 1953; *Susan Slept Here*, 1954; *Around the World in 80 Days*, 1956; *Public Pigeon No. 1*, 1957; *Ocean's Eleven*, 1960; *Those Magnificent Men in Their Flying Machines*, 1965.

RADIO

The Red Skelton Show, 1941–53.

PUBLICATION

I Dood It. n.p., 1943.

FURTHER READING

Adir, Karen. *The Great Clowns of American Television*. Jefferson, North Carolina: McFarland, 1988.

Davidson, Bill. "I'm Nuts and I Know It." *Saturday Evening Post* (Philadelphia, Pennsylvania), 17 June 1967.

Jennings, Dean. "Sad and Lonely Clown." *Saturday Evening Post* (Philadelphia, Pennsylvania), 2 June 1962.

Marx, Arthur. *Red Skelton*. New York: Dutton, 1979.

Rosten, Leo. "How to See RED—SKELTON—That Is." *Look* (New York), 23 October 1951 and 6 November 1951.

Shearer, Lloyd. "Is He a Big Laugh!" *Collier's* (New York), 15 April 1950.

See also *Red Skelton Show*, Variety Programs

SKIPPY

Australian Children's Program

Before the international sales success of Australian soap operas such as *Neighbours* and *Home and Away* in the late 1980s and 1990s, *Skippy* was the most successful series ever made in Australia. It had sales in over 100 overseas markets and was syndicated on U.S. television. In addition, in a lucrative deal the series' central figure of Skippy, the bush kangaroo, was licensed to the U.S. breakfast food giant, Kellogg's.

Skippy was produced by Fauna Productions, a partnership formed by film producer-director Lee Robinson and former film actor John McCallum, with a Sydney lawyer as the third partner. Robinson had had an extensive background in Australian documentary filmmaking, and had created the position of Australian and Pacific film correspondent for the *High Adventure* series on U.S. television, hosted by newsman and explorer Lowell Thomas. Ever the internationalist, in the 1950s, Robinson had produced a series of feature films in Australia, in partnership with actor Chips Rafferty, which combined familiar Hollywood narrative structures; drawn from such genres as the western, they used exotic locations, flora, and fauna, and were based in different parts of the Pacific.

McCallum, although born in Australia, had spent most of his professional life in Britain, where he had worked extensively on stage and in film. He returned to Australia to take a senior executive position with J. C. Williamson and Company, the largest theatrical group in Australia and New Zealand, where he became involved with the latter's comedy feature, *They're a Weird Mob*. McCallum and Robinson, who had both been production managers on the film, briefly considered producing a spin-off television series. However, they followed the advice of the international distributor, Global, about what would sell well in the world market, and finally decided on *Skippy*.

The genre that they settled on for *Skippy* was a family/children's series with a child and an animal at its centre, in a familiar vein that stretched from *Lassie* to *Flipper*. The "difference" in the Australian series was the fact that it featured native flora and fauna. Skippy was a bush kangaroo (a universal symbol of Australia), and the series was set in a national park north of Sydney that featured bushland, waterways, and ocean shores. The series concerned ranger Matt Hammond (Ed Devereaux), his son Sonny (Garry Pankhurst), the latter's pet kangaroo, his brother (Ken James), and two other junior rangers played by Tony Bonner and Liza Goddard. Altogether three different kangaroos played Skippy.

Airing between 1968 and 1970, *Skippy* resulted in 91 half-hour episodes together with one feature film, *Skippy and the Intruders*. The series was produced on film and in colour, even though Australian television had not yet moved to a colour transmission system, and was sold to the Packer-owned Nine Network, where it first aired in February 1968. With high production values, the program was costly to produce and an initial financial risk for the packaging company Fauna. However, they soon achieved sufficient overseas sales to maintain their cash flow, and the series eventually achieved very high sales. In the meantime, Fauna had become bored producing *Skippy*, and had embarked on a new series, *Barrier Reef,* which featured the reef off the northeast coast of Australia, the largest coral formation in the world.

In the 1990s, *Skippy* has had to share international recognition with other Australian series, most especially *Neighbours*, but there is still strength in the former's format. In 1991, the Nine Network licensed the format from Fauna and produced a spin-off series, *The Adventures of Skippy*, which ran to 39 half-hour episodes and was again produced on film and in colour. Set in an animal sanctuary near the Gold Coast and featuring a different group of children and adults, this second series did, however, preserve both the theme song and the kangaroo character from the original.

–Albert Moran

CAST

Matt Hammond	Ed Devereaux
Sonny Hammond	Gary Pankhurst
Mark Hammond	Ken James
Clancy	Liza Goddard
Jerry King	Tony Bonner

PRODUCERS   John McCallum, Lee Robinson, Joy Cavill, Dennis Hill

PROGRAMMING HISTORY   91 Half-hour Episodes

• Nine Network

February 1968–November 1968	7:00-7:30
January 1969–November 1969	7:00-7:30
February 1970–May 1970	7:00-7:30

See also Australian Programming

SMITH, HOWARD, K.

U.S. Journalist

Howard K. Smith, an outspoken, often controversial television newsman, developed a career that spanned the decades from his sober analytic foreign news reporting at CBS as one of "Murrow's Boys," to years as co-anchor and commentator on *ABC Evening News*. Smith's career also saw his transformation from CBS' "resident radical" to his persona "Howard K. Agnew," a sobriquet granted by critics for his support of conservative Republican Vice President Spiro T. Agnew's bitter 1969 attack on TV news.

In 1940 he joined United Press as their correspondent in London and Copenhagen, and in 1941 joined CBS news, where he replaced William L. Shirer as CBS' Berlin correspondent. The last American correspondent to leave Berlin after war was declared, he reached safety in Switzerland with a manuscript that decribed conditions in Germany, which became the basis for his best selling book *Last Train from Berlin.*

During the war Smith accompanied the Allied sweep through Belgium, Holland and into Germany. He was on hand when the Germans surrendered to the Russians under Marshal Zhukov in 1945, and then covered the Nuremburg trials. In 1946 he succeeded Murrow as CBS' London correspondent, where he spent the next 11 years covering Europe and the Middle East.

In 1949 Smith published *The State of Europe*, advocating a planned economy and the Welfare State for postwar Europe. Perhaps for this reason, and to some extent because of his radical past, he was named as a communist supporter in *Red Channels*, a McCarthyite document purporting to uncover Communist conspiracy in the media industries. He hardly suffered from these accusations,

however, since both Murrow and his overseas posting protected him. Indeed in 1957, Smith returned to the United States and in 1960 was named chief of the CBS Washington Bureau, where he hosted programs such as *The Great Challenge, Face the Nation*, and the Emmy-award-winning *CBS Reports* documentary "The Population Explosion." He served as the moderator of the first Kennedy-Nixon presidential debate.

Howard K. Smith

As a Southerner, Smith was more and more drawn to the battle over civil rights, and in 1961 he narrated a *CBS Reports* special, "Who Speaks for Birmingham?" His final commentary included a quote from Edmund Burke, "All that is necessary for the triumph of evil is for good men to do nothing." The quote was cut from the program. In a showdown with the company chairman, William S.Paley, Smith resigned after Paley supported his executives over Smith and his alleged "editorializing."

Shortly thereafter Smith signed with ABC News and began doing a weekly news show, *Howard K. Smith—News and Comment.* Smith's program made creative use of film, graphics, and animation, and explored controversial topics such as illegitimacy, disarmament, physical fitness, the state of television and the "goof-off Congress." The program won critical approval and generally high ratings. However, in 1962, Smith was again the center of controversy over his broadcast of a program entitled "The Political Obituary of Richard Nixon."

This program followed Nixon's loss of the California governer's election in 1962. In his review of Nixon's career, Smith included an interview with Alger Hiss, whom Nixon, as a member of the House Un-American Activities Committee, had investigated for his alleged membership in and spying for the Communist Party, and whose conviction for perjury in 1950 had helped launch Nixon's national political career. For balance Smith also included Murray Chotiner, a Nixon supporter and campaign advisor. The result was an avalanche of telephone calls to ABC criticizing Smith for permitting a convicted perjurer and possible spy to appear on the program. Smith's sponsor quickly ended support of the show and it was cancelled. Some historians have contended that Smith's documentary enabled Nixon to regain some of the sympathy he had lost after the disastrous temper tantrum at his self-titled "last press conference."

Following the cancellation of his show, Smith covered news for ABC-TV's daily newscast and hosted the network's Sunday afternoon public affairs program *Issues and Answers.* In 1966, he became the host of the ABC documentary program *Scope.* Until then, *Scope* had been a general documentary show dealing with many topics. In 1966 the decision was made to devote all its programs to the Vietnam War. Between 1966 and its cancellation in 1968, the program dealt with seldom touched issues of the war, such as the experience of African American soldiers, North Vietnam, and the air war.

Unlike many other newsmen, who became progressively disillusioned with the war, Smith became more and more hawkish as the war progressed. Among other things he advocated bombing North Vietnam's dike system, bombing Haiphong, and invading Laos and Cambodia. Indeed, in one of his commentaries shortly after the Tet Offensive, Smith said "There exists only one real alternative: that is to escalate, but this time on an overwhelming scale."

Smith's conservative drift on foreign affairs was also reflected in his domestic views. He was vociferous in his support of Vice President Spiro T. Agnew's 1969 "Des Moines speech," in which the vice president accused the TV networks' producers, newscasters, and commentators of a highly selective and often biased presentation of the news. Smith concurred and in salty language criticized network newsmen as, among other things, "conformist," adhering to a liberal "party line," for "stupidity," and, at least in some cases lacking "the depth of a saucer."

In March 1969 when Av Westin took over as head of ABC News, he immediately installed Smith as the co-anchor of *ABC Evening News*, with Frank Reynolds. In 1971 he was teamed with the newly arrived former CBS newsman Harry Reasoner, and given additional duties as commentator. Smith's support of the Vietnam War and Vice President Agnew's attacks on TV news stood him in good stead with President Nixon, who granted him the unique privilege of an hour-long solo interview in 1971 titled, *White House Conversation: The President and Howard K. Smith.* Despite this, when evidence grew of Nixon's involvement in the Watergate scandal, Smith was the first major TV commentator to call for his resignation.

In 1975 Smith relinquished his co-anchor role on the *ABC Evening News* but stayed on as commentator. Following the 1977 arrival of Roone Arledge as head of ABC News, Smith found himself being used less and less. In l979, he resigned from ABC, denouncing Arledge's evening newscast featuring Peter Jennings, Max Robinson, Frank Reynolds, and Barbara Walters as a "Punch and Judy Show." Since his retirement Smith has been inactive in television and radio. He was one of the last of TV newsmen who saw their role as not merely reporting the news but analyzing and commenting on it passionately.

—Albert Auster

HOWARD K(INGSBURY) SMITH. Born in Ferriday, Louisiana, U.S.A., 12 May 1914. Educated at Tulane University, New Orleans, Louisiana, 1936; Heidelberg University, 1936; Rhodes Scholar, Merton College, Oxford, 1939. Married: Benedicte Traberg Smith, 1942; one daughter and one son. Worked as reporter for the *New Orleans Item-Tribune*, 1936–37; worked for United Press International (UPI), Copenhagen, 1939; worked for UPI, Berlin, 1940; correspondent, CBS News radio, Berlin, 1941; European correspondent, CBS News, 1941–46; chief European correspondent, CBS News, 1946–57; correspondent, Washington, D.C., 1957; chief correspondent and general manager, CBS News, Washington, D.C., 1961; reporter and anchor, ABC television and radio networks, 1961–75; ABC news commentator, from 1975; host, *ABC News Closeup*, from 1979. Recipient: Peabody Award, 1960; Emmy Award, 1961; Paul White Memorial Award, 1961; duPont Commentator Award, 1962; Overseas Press Club Award, 1967; special congressional honoree for contribution to journalism; numerous other awards.

TELEVISION SERIES

1959	*Behind the News with Howard K. Smith*
1960–81	*Issues and Answers*
1960–63	*Face the Nation* (moderator)

1960–62   *Eyewitness to History* (narrator)
1961–62   *CBS Reports* (narrator)
1962–63   *Howard K. Smith—News and Comment*
1966–68   *ABC Scope*
1969–75   *ABC Evening News* (co-anchor)
1979   *ABC News Closeup*

FILM

The Best Man (cameo), 1964.

PUBLICATIONS

Last Train from Berlin. New York: Knopf, 1942.
The State of Europe. New York: Knopf, 1949.
Washington, D.C.: The Story of our Nation's Capital. New York: Random House, 1967.

Events Leading up to my Death: The Life of a Twentieth Century Reporter. New York: St. Martin's Press, 1996.

FURTHER READING

Bliss, Edward. *Now the News: The Story of Broadcast Journalism.* New York: Columbia University Press, 1991.
Fensch, Thomas, editor. *Television News Anchors: An Anthology of the Major Figures and Issues in United States Network Reporting.* Jefferson City, North Carolina: McFarland, 1993.
Gunther, Marc. *The House that Roone Built: The Inside Story of ABC News.* Boston: Little, Brown, 1994.

See also American Broadcasting Company; Arledge, Roone; Murrow, Edward R.; News, Network

THE SMOTHERS BROTHERS COMEDY HOUR

U.S. Comedy Variety Program

The *Smothers Brothers Comedy Hour*, starring the folk-singing comedy duo Tom and Dick Smothers, premiered on CBS in February 1967. A variety show scheduled opposite the top rated NBC program, *Bonanza*, the *Comedy Hour* attracted a younger, hipper, and more politically engaged audience than most other video offerings of the 1960s. The show's content featured irreverent digs at many dominant institutions such as organized religion and the presidency. It also included sketches celebrating the hippie drug culture and material opposing the war in Vietnam. These elements made *The Smothers Brothers Comedy Hour* one of the most controversial television shows in the medium's history. Questions of taste and the Smothers' oppositional politics led to very public battles over censorship. As CBS attempted to dictate what was appropriate prime-time entertainment fare, the Smothers tried to push the boundaries of acceptable speech on the medium. The recurring skirmishes between the brothers and the network culminated on 4 April 1969, one week before the end of the season, when CBS summarily threw the show off the air. Network president Robert D. Wood charged that the Smothers had not submitted a review tape of the upcoming show to the network in a timely manner. The Smothers accused CBS of infringing on their First Amendment rights. It would be twenty years before the Smothers Brothers again appeared on CBS.

In their earliest days, however, the network and the brothers got along quite well. The Smothers began their association with CBS in a failed situation comedy called *The Smothers Brothers Show* which ran for one season in 1965–66. The show featured straight man Dick as a publishing executive and slow-witted, bumbling Tom as his deceased brother who had come back as an angel-in-training. The sitcom format did not prove to be appropriate for Tom and Dick's stand-up brand of comedy. CBS, feeling that the

brothers still had potential, decided to give them another try in a different program format.

Considering how contentious *The Smothers Brothers Comedy Hour* became, it is worth noting that, in form and style, the show was quite traditional, avoiding the kinds of experiments associated with variety show rival, *Rowan and Martin's Laugh-In.* The brothers typically opened the show with a few minutes of stand-up song and banter. The show's final segment usually involved a big production number, often a costumed spoof, featuring dancing, singing and comedy. Guest stars ran the gamut from countercultural icons like the Jefferson Airplane and the Doors to older generation, "Establishment" favourites like Kate Smith and Jimmy Durante. Nelson Riddle and his orchestra supplied musical accompaniment, and the show had its own resident dancers and singers who would have been as comfortable on *The Lawrence Welk Show* as on the Smothers' show.

The show was noteworthy for some of the new, young talent it brought to the medium. Its corral of writers, many of whom were also performers, provided much of the energy, and managed to offset some of the creakiness of the format and the older guest stars. Mason Williams, heading the writing staff, achieved fame not so much for his politically engaged writing, but for his instant guitar classic, "Classical Gas." Bob Einstein wrote for the show and also played the deadpan and very unamused cop, Officer Judy. He went on to greater fame as Super Dave. Finally, the as yet unknown Steve Martin cut his comedic teeth as a staff writer for the show.

What also raised *The Smothers Brothers Comedy Hour* above the usual fare of comedy variety was the way the Smothers and their writers dealt with some of their material. Dan Rowan of *Laugh-In* noted that while his show used politics as a platform for comedy, the Smothers used comedy as a platform for politics. A recurring political sketch during the 1968 presidential year tracked regular cast member, the

The Smothers Brothers

lugubrious Pat Paulsen, and his run for the nation's top office. Campaigners for Democratic contender Hubert Humphrey apparently worried that write-in votes for Paulsen would take needed votes away from their candidate.

Another *Comedy Hour* regular engaged in a different kind of subversive humour. Comedienne Leigh French created the recurring hippie character, Goldie O'Keefe, whose parody of afternoon advice shows for housewives, "Share a Little Tea with Goldie," was actually one long celebration of mind-altering drugs. "Tea" was a countercultural code word for marijuana, but the CBS censors seemed to be unaware of the connection. Goldie would open her sketches with salutations such as "Hi(gh)—and glad of it!"

While Goldie's comedy was occasionally censored for its pro-drug messages, it never came in for the suppression that focused on other material. One of the most famous instances was the censorship of folk singer Pete Seeger. Seeger had been invited to appear on the Smothers' second season premiere to sing his anti-war song, "Waist Deep in the Big Muddy." The song—about a gung-ho military officer during World War II who attempts to force his men to ford a raging river only to be drowned in the muddy currents—was a thinly veiled metaphor for President Lyn-

don Johnson and his Vietnam policies. The censoring of Seeger created a public outcry, causing the network to relent and allow Seeger to reappear on the *Comedy Hour* later in the season to perform the song.

Other guests who wanted to perform material with an anti-war message also found themselves censored. Harry Belafonte was scheduled to do a calypso song called "Don't Stop the Carnival" with images from the riotous 1968 Chicago Democratic Convention chromakeyed behind him. Joan Baez wanted to dedicate a song to her draft-resisting husband who was about to go to prison for his stance. In both cases, the network considered this material "political," thus not appropriate for an "entertainment" format. Dr. Benjamin Spock, noted baby doctor and anti-war activist, was prevented from appearing as a guest of the show because, according to the network, he was a "convicted felon."

Other material that offended the network's notions of good taste also suffered the blue pencil. One regular guest performer, comedian David Steinberg, found his satirical sermonettes censored for being "sacrilegious." Even skits lampooning censorship, such as one in which Tom and guest Elaine May played motion-picture censors trying to find a more palatable substitution for unacceptable dialogue, ended up being censored.

The significance of all this censorship and battles between the Smothers and CBS is what Bert Spector has called a "clash of cultures." The political and taste values of two generations were colliding with each other over *The Smothers Brothers Comedy Hour*. The show, appearing at a pivotal moment of social and cultural change in the late 1960s, ended up embodying some of the turmoil and pitched conflict of the era. The Smothers wanted to provide a space on prime-time television for the perspectives of a disaffected and rebellious youth movement deeply at odds with the dominant social order. CBS, with a viewership skewed to an older, more rural, more conservative demographic, could only find the Smothers' embrace of anti-establishment politics and lifestyles threatening.

In the aftermath of the show's cancellation, the Smothers received a great deal of support in the popular press, including an editorial in *The New York Times* and a cover story in the slick magazine *Look*. Tom Smothers attempted to organize backing for a free speech fight against the network among Congressional and Federal Cummunications Commission members in Washington, D.C. While they were unsuccessful in forcing CBS to reinstate the show, the Smothers did eventually win a suit against the network for breach of contract.

In the years following their banishment from CBS, the Smothers attempted to recreate their variety show on the other two networks. In 1970, they did a summer show on ABC, but were not picked up for the fall season. In 1975 they turned up on NBC with another variety show which disappeared at mid-season. Then, finally, twenty years after being shown the door at CBS, the brothers were welcomed back for an anniversary special in February 1988. The success of the special, which re-introduced stalwarts Goldie O'Keefe (now a yuppie) and Pat Paulsen, led to another short-lived and uncontroversial run of *The Smothers Brothers Comedy Hour* on CBS. Most recently, in 1992, the Smothers re-edited episodes of the original *Comedy Hour* and ran them on the E! cable channel, providing introductions and interviews with the show's guests and writers to explain the show's controversies.

—Aniko Bodroghkozy

REGULAR PERFORMERS
Tom Smothers
Dick Smothers
Pat Paulsen
Leigh French
Bob Einstein
Mason Williams (1967–69)
Jennifer Warnes (1967–69)

John Hartford (1968–69)
Sally Struthers (1970)
Spencer Quinn (1970)
Betty Aberlin (1975)
Don Novello (1975)
Steve Martin (1975)
Nino Senporty (1975)

DANCERS
The Louis Da Pron Dancers (1967–68)
The Ron Poindexter Dancers (1968–69)

MUSIC
The Anita Kerr Singers (1967)
Nelson Riddle and His Orchestra (1967–69)
The Denny Vaughn Orchestra (1970)

PRODUCERS (1967–1969)   Saul Ilson, Ernest Chambers, Chris Bearde, Allen Blye

PROGRAMMING HISTORY

• CBS
February 1967–June 1969   Sunday 9:00-10:00

• ABC
July 1970–September 1970   Wednesday 10:00-11:00

• NBC
January 1975–May 1975   Monday 8:00-9:00

FURTHER READING

Bodroghkozy, Aniko. "The Smothers Brothers Comedy Hour and the 1960s Youth Rebellion." In, Spigel, Lynn, and Michael Curtin, editors. *The Revolution Wasn't Televised: Sixties Television and Social Conflict*. New York: Routledge, 1997.

Carr, Steven Alan. "On the Edge of Tastelessness: CBS, the Smothers Brothers, and the Struggle for Control." *Cinema Journal* (Champaign, Illinois), Summer 1992.

Hendra, Tony. *Going too Far*. New York: Doubleday, 1987.

Kloman, William. "The Transmogrification of the Smothers Brothers." *Esquire* (New York), October, 1969.

Metz, Robert. *CBS: Reflections in a Bloodshot Eye*. Chicago: Playboy, 1975.

Spector, Bert. "A Clash of Cultures: The Smothers Brothers vs. CBS Television." In, O'Connor, John E., editor. *American History, American Television*. New York: Frederick Ungar, 1983.

See also Columbia Broadcasting Company

SOAP

U.S. Serial Comedy

Soap was conceived by Susan Harris as a satire on the daytime soap operas. The show combined the serialized narrative of that genre with aspects of another U.S. television staple, the situation comedy, and was programmed in weekly, half-hour episodes. Harris, Paul Witt and Tony Thomas had formed the Witt/Thomas/Harris company in 1976 and *Soap* was their first successful pitch to a network. They received a good response from Marcy Carsey and Tom Werner at ABC and Fred Silverman placed an order for the series. Casting began in November 1976, at which point director Jay Sandrich became involved. The producers and director created an ensemble of actors, several of whom had had considerable success on Broadway. They produced a one-hour pilot by combining two half-hour scripts and developed a "bible" for the show that outlined the continuing comical saga of two families, the Tates and the Campbells, through several potential years of their stories.

In the spring of 1977 *Newsweek* reviewed the new TV season and characterized *Soap* as a sex farce that would include, among other things, the seduction of a Catholic priest in a confessional. The writer of the piece had never seen the pilot and his story was completely in error. However, that did not deter a massive protest by Roman Catholic and Southern Baptist representatives condemning the show. Later the National Council of Churches entered the lists against *Soap*. Refusing to listen to reason, the religious lobby sought to generate a boycott of companies that sponsored *Soap*. In the summer, when the producers quite properly denied requests by church groups to have the pilot sent to them for viewing, the religious groups insisted they were denied opportunity to see an episode. That was simply not true. *Soap* was in production in late July in Hollywood and each week any person walking through the lobby of the Sheraton-Universal Hotel could have secured tickets for the taping. The tapings were always open to the public and any priest or preacher could have easily gone to the studio stage for that purpose.

This combination of irresponsible journalism and misguided moral outrage by men of the cloth resulted in a dearth of sponsors. The campaign, led by ecclesiastical executives, sought to define and enforce a national morality by the use of prior censorship. It almost worked. Costs for advertising spots in the time slot for *Soap* were heavily discounted in order to achieve full sponsorship for the premiere on 13 September 1977. Only the commitment to the series by Fred Silverman prevented its demise. Some ABC affiliates were picketed and a few decided not to air it. Other stations moved it from 9:30 P.M. to a late-night time slot. A United Press International story for 14 September reported a survey of persons who had watched the first episode of *Soap* carried out by University of Richmond (Virginia, U.S.) professors and their students. They discovered that 74% of viewers found *Soap* inoffensive, 26% were offended, and half of

Soap

those offended said they were planning to watch it the next week. The day after the premiere Jay Sandrich, who had directed most of the *Mary Tyler Moore Show* episodes, stated, "If people will stay with us, they will find the show will grow." Still, producer Paul Witt believes the show never fully recovered from the witch-hunting mentality that claimed banner headlines across the country.

In spite of these difficulties, all three of the producers recall the "joy of doing it." It was their first hit, and arguably one of the most creative efforts by network television before or after. The scripts and acting were calculated to make audiences laugh—not snicker—at themselves. Indeed, in its own peculiar way it addressed family values. In one of the more dramatic moments in the series, for example, Jessica Tate, with her entire family surrounding her, confronted the threat of evil, personified by an unseen demon, and commanded the menacing presence to be gone. She invoked the family as a solid unit of love and informed the demon, "You have come to the wrong house!"

Perhaps *Soap* was not quite the pace-setting show one might have hoped for, since nothing quite like it has been seen since. In content, it had some characteristics of another pioneer effort, Norman Lear's *Mary Hartman, Mary Hart-*

man. But the differences between the two were greater than the similarities and each set a tone for what might be done with television, given freedom, imagination and talent.

Soap was a ratings success on ABC and a hit in England and Japan. In spite of the concerted attacks it was the 13th most popular network program for the 1977–78 season. *Eight is Enough* was rated 12th. Soap ended, however, under suspicion that resistance from ad agencies may have caused ABC to cancel at that point. The series may still be seen in syndication in various communities and for several years has been available on home video.

—Robert S. Alley

CAST

Chester Tate	Robert Mandan
Jessica Tate	Katherine Helmond
Corrine Tate (1977–80)	Diana Canova
Eunice Tate	Jennifer Salt
Billy Tate	Jimmy Baio
Benson (1977–79)	Robert Guillaume
The Major	Arthur Peterson
Mary Dallas Campbell	Cathryn Damon
Burt Campbell	Richard Mulligan
Jodie Dallas	Billy Crystal
Danny Dallas	Ted Wass
The Godfather (1977–78)	Richard Libertini
Claire (1977–78)	Kathryn Reynolds
Peter Campbell (1977)	Robert Urich
Chuck/Bob Campbell	Jay Johnson
Dennis Phillips (1978)	Bob Seagren
Father Timothy Flotsky (1978–79)	Sal Viscuso
Carol David (1978–81)	Rebecca Balding
Elaine Lefkowitz (1978–79)	Dinah Manoff
Dutch (1978–81)	Donnelly Rhodes
Sally (1978–79)	Caroline McWilliams
Detective Donahue (1978–80)	John Byner
Alice (1979)	Randee Heller
Mrs. David (1979–81)	Peggy Hope
Millie (1979)	Candace Azzara
Leslie Walker (1979–81)	Marla Pennington
Polly Dawson (1979–81)	Lynne Moody
Saunders (1980–81)	Roscoe Lee Brown
Dr. Alan Posner (1980–81)	Allan Miller
Attorney E. Ronald Mallu (1978–81)	Eugene Roche
Carlos "El Puerco" Valdez (1980–81)	Gregory Sierra
Maggie Chandler (1980–81)	Barbara Rhoades
Gwen (1980–81)	Jesse Welles

PRODUCERS Paul Junger Witt, Tony Thomas, Susan Harris, J.D. Lobue, Dick Clair, Jenna McMahon

PROGRAMMING HISTORY 83 30-Minute Episodes; 10 60-Minute Episodes

• ABC

September 1977–March 1978	Tuesday 9:30-10:00
September 1978–March 1979	Thursday 9:30-10:00
September 1979–March 1980	Thursday 9:30-10:00
October 1980–January 1981	Wednesday 9:30-10:00
March 1981–April 1981	Monday 10:00-11:00

See also Advocacy Groups; Harris, Susan; Sexual Orientation and Television; Silverman, Fred; Thomas, Tony; Witt, Paul Junger

SOAP OPERA

The term "soap opera" was coined by the American press in the 1930s to denote the extraordinarily popular genre of serialized domestic radio dramas, which, by 1940, represented some 90% of all commercially-sponsored daytime broadcast hours. The "soap" in soap opera alluded to their sponsorship by manufacturers of household cleaning products; while "opera" suggested an ironic incongruity between the domestic narrative concerns of the daytime serial and the most elevated of dramatic forms. In the United States, the term continues to be applied primarily to the approximately fifty hours each week of daytime serial television drama broadcast by ABC, NBC, and CBS, but the meanings of the term, both in the United States and elsewhere, exceed this generic designation.

The defining quality of the soap opera form is its seriality. A serial narrative is a story told through a series of individual, narratively linked installments. Unlike episodic television programs, in which there is no narrative linkage between episodes and each episode tells a more or less self-contained story, the viewer's understanding of and pleasure in any given serial installment is predicated, to some degree, upon his or her knowledge of what has happened in previous episodes. Furthermore, each serial episode always leaves narrative loose ends for the next episode to take up. The viewer's relationship with serial characters is also different from those in episodic television. In the latter, characters cannot undergo changes that transcend any given episode, and they seldom reference events from previous episodes. Serial characters do change across episodes (they age and even die), and they possess both histories and memories. Serial television is not merely narratively segmented, its episodes are designed to be parceled out in regular installments, so that both the telling of the serial story and its reception by viewers is institutionally regulated. (This generalization obviously does not anticipate the use of the video tape recorder to "time shift" viewing).

Soap operas are of two basic narrative types: "open" soap operas, in which there is no end point toward which

All My Children

the action of the narrative moves; and "closed" soap operas, in which, no matter how attenuated the process, the narrative does eventually close. Examples of the open soap opera would include all U.S. daytime serials (*General Hospital, All My Children, The Guiding Light*, etc.), the wave of prime-time U.S. soaps in the 1980s (*Dallas, Dynasty, Falcon Crest*), such British serials as *Coronation Street, EastEnders,* and *Brookside*), and most Australian serials (*Neighbours, Home and Away, A Country Practice*). The closed soap opera is more common in Latin America, where it dominates prime-time programming from Mexico to Chile. These *telenovelas* are broadcast nightly and may stretch over three or four months and hundreds of episodes. They are, however, designed eventually to end, and it is the anticipation of closure in both the design and reception of the closed soap opera that makes it fundamentally different from the open form.

In the United States, at least, the term "soap opera" has never been value-neutral. As noted above, the term itself signals an aesthetic and cultural incongruity: the events of everyday life elevated to the subject matter of an operatic form. To call a film, novel, or play a "soap opera" is to label

it as culturally and aesthetic inconsequential and unworthy. When in the early 1990s the fabric of domestic life amongst the British royal family began to unravel, the press around the world began to refer to the situation as a "royal soap opera," which immediately framed it as tawdry, sensational, and undignified.

Particularly in the United States, the connotation of "soap opera" as a degraded cultural and aesthetic form is inextricably bound to the gendered nature of its appeals and of its target audience. The soap opera always has been a "woman's" genre, and, it has frequently been assumed (mainly by those who have never watched soap operas), of interest primarily or exclusively to uncultured working-class women with simple tastes and limited capacities. Thus the soap opera has been the most easily parodied of all broadcasting genres, and its presumed audience most easily stereotyped as the working-class "housewife" who allows the dishes to pile up and the children to run amuck because of her "addiction" to soap operas. Despite the fact that the soap opera is demonstrably one of the most narratively complex genres of television drama whose enjoyment requires con-

As the World Turns

General Hospital

Guiding Light

siderable knowledge by its viewers, and despite the fact that its appeals for half a century have cut across social and demographic categories, the term continues to carry this sexist and classist baggage.

What most Americans have known as soap opera for more than half a century began as one of the hundreds of new programming forms tried out by commercial radio broadcasters in the late 1920s and early 1930s, as both local stations and the newly-formed networks attempted to marry the needs of advertisers with the listening interests of consumers. Specifically, broadcasters hoped to interest manufacturers of household cleaners, food products, and toiletries in the possibility of using daytime radio to reach their prime consumer market: women between the ages of eighteen and forty-nine.

In 1930, the manager of Chicago radio station WGN approached first a detergent company and then a margarine manufacturer with a proposal for a new type of program: a daily, fifteen-minute serialized drama set in the home of an Irish-American widow and her young unmarried daughter. Irna Phillips, who had recently left her job as a speech teacher to try her hand at radio, was assigned to write *Painted Dreams*, as the show was called, and play two of its three regular parts. The plots Phillips wrote revolved around morning conversations "Mother" Moynihan had with her

daughter and their female boarder before the two young women went to their jobs at a hotel.

The antecedents of *Painted Dreams* and the dozens of other soap operas launched in the early 1930s are varied. The soap opera continued the tradition of women's domestic fiction of the nineteenth century, which had also been sustained in magazine stories of the 1920s and 1930s. It also drew upon the conventions of the "woman's film" of the 1930s. The frequent homilies and admonitions offered by "Mother" Moynihan and her matriarchal counterparts on other early soap operas echoed those presented on the many advice programs commercial broadcasters presented in the early 1930s in response to the unprecedented social and economic dislocation experienced by American families as a result of the Great Depression. The serial narrative format of the early soap opera was almost certainly inspired by the primetime success of *Amos 'n' Andy*, the comic radio serial about "black" life on the south side of Chicago (the show was written and performed by two white men), which by 1930 was the most popular radio show to that time.

In the absence of systematic audience measurement, it took several years for broadcasters and advertisers to realize the potential of the new soap opera genre. By 1937, however, the soap opera dominated the daytime commercial radio schedule and had become a crucial network programming strategy for attracting such large corporate sponsors as Procter and Gamble, Pillsbury, American Home Products, and General Foods. Most network soap operas were produced by advertising agencies, and some were owned by the sponsoring client.

Irna Phillips created and wrote some of the most successful radio soap operas in the 1930s and 1940s, including *Today's Children* (1932), *The Guiding Light* (1937), and *Woman in White* (1938). Her chief competition came from the husband-wife team of Frank and Anne Hummert, who were responsible for nearly half the soap operas introduced between 1932 and 1937, including *Ma Perkins* (1933) and *The Romance of Helen Trent* (1933).

On the eve of World War II, listeners could choose from among sixty-four daytime serials broadcast each week. During the war, so important had soap operas become in maintaining product recognition among consumers that Procter and Gamble continued to advertise Dreft detergent on its soap operas—despite the fact that the sale of it and other synthetic laundry detergents had been suspended for the duration. Soap operas continued to dominate daytime ratings and schedules in the immediate post-war period. In 1948 the ten highest-rated daytime programs were all soap operas, and of the top thirty daytime shows all but five were soaps. The most popular non-serial daytime program, *Arthur Godfrey*, could manage only twelfth place.

As television began to supplant radio as a national advertising medium in the late 1940s, the same companies that owned or sponsored radio soap operas looked to the new medium as a means of introducing new products and exploiting pent-up consumer demand. Procter and Gamble, which established its own radio soap opera production subsidiary in 1940, produced the first network television soap opera in 1950. *The First Hundred Years* ran for only two and demonstrated some of the problems of transplanting the radio genre to television. Everything that was left to the listener's imagination in the radio soap had to be given visual form on television. Production costs were two to three times that of a radio serial. Actors had to act and not merely read their lines. The complexity and uncertainty of producing fifteen minutes of live television drama each weekday was vastly greater than was the case on radio. Furthermore, it was unclear in 1950 if the primary target audience for soap operas—women working in the home— could integrate the viewing of soaps into their daily routines. One could listen to a radio soap while doing other things, even in another room; television soaps required some degree of visual attention.

By the 1951–52 television season, broadcasters had demonstrated television's ability to attract daytime audiences, principally through the variety-talk format. CBS led the way in adapting the radio serial to television, introducing four daytime serials. The success of three of them, *Search for Tomorrow*, *Love of Life* (both produced by Roy Winsor), and *The Guiding Light*, established the soap opera as a regular part of network television daytime programming and CBS

as the early leader in the genre. *The Guiding Light* was the first radio soap opera to make the transition to television, and one of only two to do so successfully (The other was *The Brighter Day*, which ran for eight years). Between its television debut in 1952 and 1956 *The Guiding Light* was broadcast on both radio and television.

By the early 1960s, the radio soap opera—along with most aspects of network radio more generally—was a thing of the past, and "soap opera" in the United States now meant "television soap opera." The last network radio soap operas went off the air in November 1960. Still, television soap operas continued many of the conventions of their radio predecessors: live, week-daily episodes of fifteen minutes, an unseen voice-over announcer to introduce and close each episode, organ music to provide a theme and punctuate the most dramatic moments, and each episode ending on an unresolved narrative moment with a "cliffhanger" ending on Friday to draw the audience back on Monday.

The thirty-minute soap opera was not introduced until 1956, with the debut of Irna Phillips's new soap for Procter and Gamble and CBS, *As the World Turns*. With an equivalent running time of two feature films each week, *As the World Turns* expanded the community of characters, slowed the narrative pace, emphasized the exploration of character, utilized multiple cameras to better capture facial expressions and reactions, and built its appeal less on individual action than on exploring the network of relationships among members of two extended families: the Lowells and the Hughes. Although it took some months to catch on with audiences, *As the World Turns* demonstrated that viewers would watch a week-daily half-hour soap. Its ratings success plus the enormous cost savings of producing one half-hour program rather than two fifteen-minute ones persuaded producers that the thirty-minute soap opera was the format of the future. The fifteen-minute soap was phased out, and all new soap operas introduced after 1956 were at least thirty-minutes in length.

CBS' hegemony in soap operas was not challenged until 1963. None of the several half-hour soaps NBC introduced in the wake of *As the World Turns'* popularity made the slightest dent in CBS' ratings. However, in April of 1963 both NBC and ABC launched soaps with medical settings and themes: *The Doctors* and *General Hospital*, respectively. These were not the first medical television soaps, but they were the first to sustain audience interest over time, and the first soaps produced by either network to achieve ratings even approaching those of the CBS serials. Their popularity also spawned the sub-genre of the medical soap, in which the hospital replaces the home as the locus of action, plot lines center on the medical and emotional challenges patients present doctors and nurses, and the biological family is replaced or paralleled by the professional family as the structuring basis for the show's community of characters.

The therapeutic orientation of medical soaps also provided an excellent rationale for introducing a host of contemporary, sometimes controversial social issues, which Irna

Phillips and a few other writers believed soap audiences in the mid-1960s were prepared to accept as a part of the soap opera's moral universe. *Days of Our Lives* (co-created for NBC in 1965 by Irna Phillips and Ted Corday, the first director of *As the World Turns*) presented Dr. Tom Horton (played by film actor Macdonald Carey) and his colleagues at University Hospital with a host of medical, emotional, sexual, and psychiatric problems in the show's first years, including incest, impotence, amnesia, illegitimacy, and murder as a result of temporary insanity. This strategy made *Days of Our Lives* a breakthrough hit for NBC, and it anchored its daytime line-up through the late 1960s.

Medical soaps are particularly well-suited to meet the unique narrative demands of the "never-ending" stories American soap operas tell. Their hospital settings provide opportunities for the intersection of professional and personal dramas. They also allow for the limitless introduction of new characters as hospital patients and personnel. The constant admission of new patients to the medical soap's hospitals facilitates the admission to the soap community of a succession of medical, personal, and social issues which can be attached to those patients. If audience response warrants, the patient can be "cured" and admitted to the central cohort of community members. If not, or if the social issue the patient represents proves to be too controversial, he or she can die or be discharged—both from the hospital and from the narrative. Such has been the appeal (to audiences and writers alike) of the medical soap, that many non-medical soaps have included doctors and nurses among their central characters and nurses' stations among their standing sets. Among them has been *As the World Turns, The Guiding Light, Search for Tomorrow,* and *Ryan's Hope.*

The latter half of the 1960s was a key period in the history of U.S. daytime soap operas. By 1965 both the popularity and profitability of the television soap opera had been amply demonstrated. Soaps proved unrivaled in attracting female viewers aged between eighteen and forty-nine—the demographic group responsible for making most of the non-durable good purchasing decisions in U.S. Production costs were a fraction of those for primetime drama, and once a new soap "found" its audience, broadcasters and advertisers knew that those viewers would be among television's most loyal. For the first time CBS faced competition for the available daytime audience. With the success of *Another World* (another Irna Phillips vehicle launched in 1964), *Days of Our Lives* and *The Doctors,* by 1966 NBC had a creditable line-up across the key afternoon time-slots.

This competition sparked a period of unprecedented experimentation with the genre, as all three networks assumed that audiences would seek out a soap opera "with a difference." As the network with the most to gain (and the least to lose) by program innovation, ABC's new soaps represented the most radical departures from the genre's thirty-five-year-old formula. Believing that daytime audiences would also watch soaps during primetime, in September 1964 ABC introduced *Peyton Place,* a twice-weekly

half-hour prime-time serial based on the best-selling 1957 novel by Grace Metalious and its successful film adaptation. Shot on film and starring film actress Dorothy Malone, *Peyton Place* was one of ABC's biggest primetime hits of the 1964-65 television season and made stars of newcomers Mia Farrow and Ryan O'Neal. The show's ratings dropped after its first two seasons, however, and in terms of daytime soap longevity its run was relatively brief: five years.

In 1966 ABC launched the most unusual daytime soap ever presented on American television. *Dark Shadows* was an over-the-top gothic serial, replete with a spooky mansion setting, young governess (lifted directly from Henry James's *The Turn of the Screw*), and two-hundred-year-old vampire. Broadcast in most markets in the late afternoon in order to catch high-school students as well as adult women, *Dark Shadows* became something of a cult hit in its first season, and it did succeed in attracting to the soap opera form an audience of teenage viewers (male and female) and college students who were not addressed by more mainstream soaps. The show was too camp for most of those mainstream soap viewers, however, and it was canceled after five years.

ABC's most durable innovations in the soap opera genre during this period, however, took the form of two new mainstream soap operas, both created by Irna Phillips's protégé, Agnes Nixon. Nixon, who had apprenticed to Phillips for more than a decade as dialogue writer for most of her soaps and head writer of *The Guiding Light,* sold ABC on the idea of new soap that would foreground rather than suppress class and ethnic difference. *One Life to Live,* which debuted in 1968, centered initially on the family of wealthy WASP newspaper owner Victor Lord, but established the Lords in relation to three working-class and ethnically "marked" families: the Irish-American Rileys, the Polish-American Woleks, and after a year or two, the Jewish-American Siegels. Ethnic and class difference was played out primarily in terms of romantic entanglements.

Where most soap operas still avoided controversial social issues, Nixon exploited some of the social tensions then swirling through American society in the late 1960s. In 1969 *One Life to Live* introduced a black character who denied her racial identity (only to proudly proclaim it some months and dozens of episodes later). The following year when a teenage character is discovered to be a drug addict, she is sent to a "real life" treatment center in New York, where the character interacts with actual patients.

Some of this sense of social "relevance" also found its way into Nixon's next venture for ABC, *All My Children,* which debuted in 1970. It was the first soap opera to write the Vietnam War into its stories, with one character drafted and (presumably) killed in action. Despite an anti-war speech delivered by his grieving mother, the political force of the plot line was blunted by the discovery that he was not really killed at all.

Even before *One Life to Live* broke new ground in its representation of class, race, and ethnicity, CBS gestured (rather tentatively, as it turns out) in the direction of social realism in response to the growing ratings success of NBC and ABC's soaps. *Love Is a Many Splendored Thing* had been a successful 1955 film, with William Holden playing an American journalist working in Asia who falls in love with a young Eurasian woman, played by Jennifer Jones. Irna Phillips wrote the soap opera as a sequel to the film, in which the couple's daughter moves to San Francisco and falls in love with a local doctor. *Love Is a Many Splendored Thing* debuted on 18 September 1967, its inaugural story (indeed, its very premise) concerning the social implications of this interracial romance. After only a few months, CBS, fearing protests from sponsors and audience groups, demanded that Phillips write her Eurasian heroine out of the show. She refused to do so and angrily resigned. Rather than cancel the show, however, CBS hired new writers, who refocused it on three young, white characters (played by Donna Mills, David Birney, and Leslie Charleson).

What the replacement writers of *Love Is a Many Splendored Thing* did in a desperate attempt to save a wounded show, Agnes Nixon did in a very premeditated fashion some thirty months later in *All My Children*. As its name suggests, *All My Children* was, like many radio and tv soaps before it, structured around a matriarch, the wealthy Phoebe Tyler (Ruth Warwick), but to a greater degree than its predecessors, it emphasized the romantic relationships among its "children." Nixon realized that after nearly two decades of television soaps, many in the viewing audience were aging out of the prime demographic group most sought by soap's sponsors and owners: women under the age of fifty. *All My Children* used young adult characters and a regular injection of social controversy to appeal to viewers at the other end of the demographic spectrum. It was a tactic very much in tune with ABC's overall programming strategy in the 1960s, which also resulted in *The Flintstones* and *American Bandstand*. *All My Children* was the first soap opera whose organizational structure addressed what was to become the form's perennial demographic dilemma: how to keep the existing audience while adding younger recruits to it.

The problem of the "aging out" of a given soap opera's audience was particularly acute for CBS, whose leading soaps were by the early 1970s entering their second or third decade (*Search for Tomorrow, Love of Life, The Guiding Light, As the World Turns, Secret Storm*, and *The Edge of Night* were all launched between 1951 and 1957). Consequently a troubling proportion of CBS' soap audience was aging out of the "quality" demographic range.

Thus for the first time CBS found itself in the position of having to respond to the other networks' soap opera innovations. As its name rather baldly announces, *The Young and the Restless* was based upon the premise that a soap opera about the sexual intrigues of attractive characters in their twenties would attract an audience of women also in their twenties. Devised for CBS by another of Irna Phillips's students, William Bell, and launched in 1973, *The Young and the Restless* is what might be called the first "Hollywood" soap. Not only was it shot in Hollywood (as some other soaps already were), it borrowed something of the "look" of

a Hollywood film (particularly in its use of elaborate sets and high-key lighting), peopled Genoa City with soap opera's most conspicuously attractive citizens, dressed them in fashion-magazine wardrobes, and kept its plots focused on sex and its attendant problems and complications. The formula was almost immediately successful, and *The Young and the Restless* has remained one of the most popular soap operas for more than twenty years. It is also the stylistic progenitor of such recent "slick" soaps as *Santa Barbara* and *The Bold and the Beautiful*.

The early 1970s saw intense competition among the three networks for soap opera viewers. By this time, ABC, CBS, and NBC all had full slates of afternoon soap operas (at one point in this period the three networks were airing ten hours of soaps every weekday), and the aggregate daily audience for soap operas had reached twenty million. With a four-fold difference in ad rates between low-rated and high-rated soaps and the latter having the potential of attracting $500,000 in ad revenue each week, soap operas became driven by the Nielsen ratings like never before.

The way in which these ratings pressures affected the writing of soap opera narratives speaks to the genre's unique mode of production. Since the days of radio soap operas, effective power over the creation and maintenance of each soap opera narrative world has been vested in the show's head writer. She (and to a greater degree than in any other form of television programming, the head writers of soap operas have been female) charts the narrative course for the soap opera over a six-month period and in doing so determines the immediate (and sometimes permanent) fates of each character, the nature of each intersecting plot line, and the speed with which each plot line moves toward some (however tentative) resolution. She then supervises the segmentation of this overall plot outline into weekly and then daily portions, usually assigning the actual writing of each episode to one of a team of script writers ("dialoguers" as they are called in the business). The scripts then go back to the head writer for her approval before becoming the basis for each episode's actual production.

The long-term narrative trajectory of a soap opera is subject to adjustment as feedback is received from viewers by way of fan letters, market research, and, of course, the weekly Nielsen ratings figures, which in the 1970s were based on a national sample of some 1200 television households. Looking over the head writer's shoulder, of course, is the network, whose profitability depends upon advertising revenues, and the show's sponsor, who frequently was (and, in the case of four soaps today, still is) the show's owner.

By the early 1970s, head writers were under enormous pressure to attain the highest ratings possible, "win" the ratings race against the competition in the show's time slot, target the show's plots at the demographic group of most value to advertisers, take into account the production-budget implications of any plot developments (new sets or exterior shooting, for example), and maintain audience interest every week without pauses for summer hiatus or reruns. These pressures—and the financial stakes producing them—made soap opera head writers among the highest paid writers in broadcasting (and the most highly paid women in the industry), but they also meant that, like the manager of a baseball team, she became the scapegoat if her "team" did not win.

If the mid– and late–1960s were periods of experimentation with the soap opera form itself, the early 1970s launched the era of incessant adjustments within the form—an era that has lasted to the present. Although individual soap operas attempted to establish defining differences from other soaps (in the early 1970s *As the World Turns* was centered on the extended Hughes family; *The Young and the Restless* was sexy and visually striking; *The Edge of Night* maintained elements of the police and courtroom drama; *General Hospital* foregrounded medical issues; etc.), to some degree all soap opera meta-narratives over the past twenty-five years have drawn upon common sets of tactical options, oscillating between opposed terms within each set: fantasy versus everyday life, a focus on individual character/actor "stars" versus the diffusion of interest across the larger soap opera community, social "relevance" versus more "traditional" soap opera narrative concerns of family and romance, an emphasis on one sensational plotline versus spreading the show's narrative energy across several plotlines at different stages of resolution, attempting to attract younger viewers by concentrating on younger characters versus attempting to maintain the more adult viewer's interest through characters and plots presumably more to her liking.

At any given moment, the world of any given soap opera is in part the result of narrative decisions that have been made along all of these parameters, mediated, of course, by the history of that particular soap opera's "world" and the personalities of the characters who inhabit it. Any head writer brought in to improve the flagging ratings of an ongoing soap is constrained in her exercise of these options by the fact that many of the show's viewers have a better sense of who the show's characters are and what is plausible to happen to them than she does. And being among the most vocal and devoted of all television viewers, soap opera fans are quick to respond when they feel a new head writer has driven the soap's narrative off-course.

Despite the constant internal adjustments being made in any given soap opera, individual shows have demonstrated remarkable resilience and overall soap operas exhibit infinitely greater stability than any primetime genre. With the exception of several years in the late 1940s when Irna Phillips was in dispute with Procter and Gamble, *The Guiding Light* has been heard or seen every weekday since January 1937, making it the longest story ever told. Of the ten currently running network soap operas (1995), eight have been on the air for more than twenty years, five for more than thirty years, and two (*The Guiding Light* and *As the World Turns*) survive from the 1950s.

Although long-running soap operas have been canceled (*Love of Life* and *Search for Tomorrow* were both canceled in the 1980s after thirty-year runs) and others have come and gone, the incentive to keep an established soap going is considerable in light of the expense and risk of replacing it with a new soap opera, which can take a year or more to "find" its audience. Viewers who have invested years in watching a particular soap are not easily lured to a new one, or, for that matter, to a competing soap on another network. In the mid-1970s, rather than replacing failing half-hour soaps with new ones, NBC began extending some of its existing soaps to a full hour (*Days of our Lives* and *Another World* were the first to be expanded in 1976). Eight of the ten currently running soap operas are one hour in length.

In the 1980s, despite daytime soap operas' struggles to maintain audience in the face of declining overall viewership, the soap opera became more "visible" in the United States as a programming genre and cultural phenomenon than at any point in its history. Soap operas had always been "visible" to its large and loyal audience. By the 1980s some fifty million persons in the United States "followed" one or more soap operas, including two-thirds of all women living in homes with televisions. As a cultural phenomenon, however, for thirty years the watching of soap operas had for the most part occurred undetected on the radar screen of public notice and comment. Ironically, soap opera viewing became the basis for a public fan culture in the late 1970s and early 1980s in part because more and more of the soap opera audience was unavailable during the day to watch. As increasing numbers of soap opera viewing women entered the paid workforce in the 1970s, they obviously found it difficult to "keep up" with the plots of their favorite soaps. A new genre of mass-market magazine emerged in response to this need. By 1982 ten new magazines had been launched that addressed the soap opera fan. For the occasional viewer they contained plot synopses of all current soaps. For them and for more regular viewers, they also featured profiles of soap opera actors, "behind-the-scenes" articles on soap opera production, and letters-to-the-editor columns in which readers could respond to particular soap characters and plot developments. *Soap Opera Digest*, which began in 1975, had a circulation of 850,000 copies by 1990 and claimed a readership of four million. Soap opera magazines became an important focus of soap fan culture in the 1980s—a culture that was recognized (and exploited) by soap producers through their sponsorship or encouragement of public appearances by soap opera actors and more recently of soap opera "conventions."

Soaps and soap viewing also became more culturally "visible" in the 1980s as viewer demographics changed. By the beginning of the decade, fully thirty percent of the audience for soap operas was made up of groups outside the core demographic group of eighteen- to forty-nine-year-old women, including substantial numbers of teenage boys and girls (up to fifteen percent of the total audience for some soaps) and adult men (particularly those over fifty). Under-reported by the Nielsen ratings, soap opera viewing by some three million college students was confirmed by independent research in 1982.

The 1980s also was the decade in which the serial narrative form of the daytime soap opera became an important feature of primetime programming as well. The program that sparked the primetime soap boom of the 1980s was *Dallas*. Debuting in April 1978, *Dallas* was for its first year a one-hour episodic series concerning a wealthy but rough-edged Texas oil family. It was the enormous popularity of the "Who Shot J.R.?" cliffhanger episode at the end of the second season (21 March 1980) and the first episode the following season (21 November 1980—the largest audience for any American television series to that time) that persuaded producers to transform the show into a full-blown serial.

Dallas not only borrowed the serial form from daytime soaps, but also the structuring device of the extended family (the Ewings), complete with patriarch, matriarch, good son, bad son, and in-laws—all of whom lived in the same Texas-sized house. The kinship and romance plots that could be generated around these core family members were, it was believed by the show's producers, the basis for attracting female viewers, while Ewing Oil's boardroom intrigues would draw adult males, accustomed to finding "masculine" genres (westerns, crime, and legal dramas) during *Dallas's* Sunday 10:00 P.M. time slot. By 1982 *Dallas* was one of the most popular programs in television history. It spawned direct imitators (most notably *Dynasty* and *Falcon Crest*), and a spin-off (*Knot's Landing*). Its success in adapting the daytime serial form to fit the requirements of the weekly one-hour format and the different demographics of the primetime audience prompted the "serialization" of a host of primetime dramas in the 1980s—the most successful among them *Hill Street Blues, St. Elsewhere,* and *L.A. Law.*

Dallas and *Dynasty* were also the first American serials (daytime or primetime) to be successfully marketed internationally. *Dallas* was broadcast in fifty-seven countries where it was seen by 300 million viewers. These two serials were particularly popular in western Europe, so much so that they provoked debates in a number of countries over American cultural imperialism and the appropriateness of state broadcasting systems spending public money to acquire American soap operas rather than to produce domestic drama. Producers in several European countries launched their own direct imitations of these slick American soaps, among them the German *Schwarzwaldklinik* and the French serial *Chateauvallon.*

But even as soap opera viewing came out of the closet in the 1980s and critics spoke (usually derisively) of the "soapoperafication" of primetime, daytime soaps struggled to deal with the compound blows struck by continuing changes in occupational patterns among women, the transformation of television technology (with the advent of the video tape recorder, satellite distribution of programming, and cable television), and the rise of competing, and less expensive, program forms. Between the early 1930s and the beginning of the 1970s, broadcasters and advertisers could count on a stable (and, throughout much of this period, expanding) audience for soap operas among what industry trade papers always referred to as "housewives": women working in the home, many of them

caring for small children. But with the end of the post-war "baby boom," American women joined the paid workforce in numbers unprecedented in peacetime. In 1977 the number of daytime households using television ("HUTs" in ratings terminology) began to decline and with it the aggregate audience for soap operas. Although daytime viewing figures have fluctuated somewhat since then, the trend over the past twenty years is clear: the audience for network programming in general and daytime programming specifically is shrinking.

In large measure the overall drop in network viewing figures is attributable to changes in television technology, especially the extraordinarily rapid diffusion of the video tape recorder in the 1980s and, at the same time, an explosion in the number of viewing alternatives available on cable television. The penetration of the video tape recorder into the American household has had a paradoxical impact on the measurement of soap opera viewing. Although the soap opera is the genre most "time-shifted" (recorded off the air for later viewing), soap opera viewing on video tape does not figure into audience ratings data, and even if it did, advertisers would discount such viewership, believing (accurately) that most viewers "zip" through commercials.

The wiring of most American cities for cable television in the 1970s and 1980s has meant the expansion of program alternatives in any given time period in many markets from three or four channels to more than fifty. In the 1960s and 1970s, daytime television viewers were limited in the viewing choices in many time slots to two genres: the game show and the soap opera. By the 1990s, network soaps were competing not only against each other and against game shows, but also against an array of cable alternatives, including one cable channel (Lifetime) targeted exclusively at the soap opera's core audience: women between the ages of eighteen and forty-nine.

For the three commercial networks, dispersed viewership across an increasingly fragmented market has meant lower ratings, reduced total advertising revenue, reduced advertising rates, and reduced profit margins. Although soap operas actually gained viewership in some audience segments in the 1980s—men and adolescents, in particular—these are not groups traditionally targeted by the companies whose advertising has sustained the genre for half a century. As they scrambled to staunch the outflow of audience to cable in the early 1990s, the networks and independent producers (who supply programming both to the networks and in syndication to local broadcasters) turned to daytime programming forms with minimal start-up costs and low production budgets, especially the talk show. In many markets soap operas' strongest competition comes not from other soaps but from Montel Williams, Ricki Lake, Jerry Springer, or another of the dozens of talk shows that have been launched since 1990.

It is impossible here to set the history of serial drama in U.S. broadcasting in relation to the history of the form in the dozens of other countries where it has figured prominently—from China and India to Mexico and Brazil—except to say that the form has proven to be extraordinarily malleable and responsive to a wide variety of local institutional and social requirements. However, it may be instructive to contrast briefly the British experience with the serial drama with that surveyed above in the United States

The tradition of broadcast serial drama in Britain goes back to 1940s radio and *The Archers*, a daily, fifteen-minute serial of country life broadcast by the BBC initially as a means of educating farmers about better agricultural practices. The British television serial, on the other hand, grows out of the needs of commercial television in the late 1950s. Mandated to serve regional needs, the newly chartered "independent" (commercial) television services were eager to capture the growing audience of urban lower-middle class and working-class television viewers. In December 1960, Manchester-based Granada Television introduced its viewers to *Coronation Street*, a serial set in a local working-class neighborhood. The following year it was broadcast nationwide and has remained at or near the top of the primetime television ratings nearly ever since.

Coronation Street's style, setting, and narrative concerns are informed by the gritty, urban, working-class plays, novels, and films of the 1950s—the so-called "angry young man" or "kitchen sink" movement. Where U.S. daytime serials were (and still are) usually disconnected from any particular locality, *Coronation Street* is unmistakably local. Where U.S. soaps usually downplay class as an axis of social division (except as a marker of wealth), *Coronation Street* began and has to some degree stayed a celebration of the institutions of working-class culture and community (especially the pub and the cafe)—even if that culture was by 1960 an historical memory and *Coronation Street*'s representation of community a nostalgic fantasy.

In part because of the regionalism built into the commercial television system, all British soap operas since *Coronation Street* have been geographically and, to some degree, culturally specific in setting: *Crossroads* (1964–88) in the Midlands, *Emmerdale Farm* (1972–) in the Yorkshire Dales, *Brookside* (1982–) in Liverpool, and the BBC's successful entry in the soap opera field *EastEnders* (1985–) in the East End of London. All also have been much more specific and explicit in their social and class settings than their American counterparts, and for this reason their fidelity to (and deviation from) some standard of social verisimilitude has been much more of an issue than has ever been the case with American soaps. *Coronation Street* has been criticized for its cozy, insulated, and outdated representation of the urban working-class community, which for decades seemed to have been bypassed by social change and strife.

Still, by American soap opera standards, British soaps are much more concerned with the material lives of their characters and the characters' positions within a larger social structure. *EastEnders*, when it was launched in 1985 the BBC's first venture into television serials in twenty years, was designed from the beginning to make contemporary material and social issues part of the fabric of its grubby East End community of pensioners, market traders, petty criminals, shopkeepers, the homeless, and the perennially unemployed.

Internationally, the most conspicuous and important development in the soap opera genre over the past twenty years has not involved the production, reception, or export of American soap operas (whether daytime or primetime), but rather the extraordinary popularity of domestic television serials in Latin America, India, Great Britain, Australia, and other countries, and the international circulation of non-U.S. soaps to virtually every part of the world *except* the United States. With their *telenovelas* dominating primetime schedules throughout the hemisphere, Latin American serial producers began seriously pursuing extra-regional export possibilities in the mid-1970s. Brazil's TV Globo began exporting telenovelas to Europe in 1975. Within a decade it was selling soap operas to nearly 100 countries around the world, its annual export revenues increasing five-fold between 1982 and 1987 alone. Mexico's Televisa exports serials to fifty-nine countries, and its soap operas have topped the ratings in Korea, Russia, and Turkey. Venezuelan serials have attracted huge audiences in Spain, Italy, Greece, and Portugal. Latin American soap operas have penetrated the U.S. market but, thus far, only among its Spanish-speaking population: serials comprise a large share of the primetime programming on Spanish-language cable and broadcast channels in the United States.

Although Australian serials had been shown in Britain for some years, they became a major force in British broadcasting with the huge success of Reg Grundy Productions' *Neighbours* in 1986. For most of the time since then, it has vied with either *EastEnders* or *Coronation Street* as Britain's most-viewed television program. *Neighbours* has been seen in more than twenty-five countries and has been called Australia's most successful cultural export.

The global circulation of non-U.S. serials since the 1970s is, in part, a function of the increased demand for television programming in general, caused by the growth of satellite and cable television around the world. It is also due, particularly in western and eastern Europe, to a shift in many countries away from a state-controlled public service television system to a "mixed" (public and commercial) or entirely commercial model. The low production cost of serials (in Latin America between $25,000 and $80,000 an episode) and their ability to recover these costs in their domestic markets mean that they can be offered on the international market at relatively low prices (as little as $3000 per episode) in Europe. Given the large audiences they can attract and their low cost (particularly in relation to the cost of producing original drama), imported serials represent good value for satellite, cable, and broadcast services in many countries.

Ironically, American producers never seriously exploited the international market possibilities for daytime soap operas until the export success of Latin American serials in the 1980s, and now find themselves following the lead of TV Globo and Venezuela's Radio Caracas. NBC's *The Bold and the Beautiful*, set in the fashion industry, is the first U.S. daytime soap to attract a substantial international following.

Derided by critics and disdained by social commentators from the 1930s to the 1990s, the soap opera is nevertheless the most effective and enduring broadcast advertising vehicle ever devised. It is also the most popular genre of television drama in the world today and probably in the history of world broadcasting: no other form of television fiction has attracted more viewers in more countries over a longer period of time.

—Robert C. Allen

FURTHER READING

Allen, Robert C. *Speaking of Soap Operas*. Chapel Hill: University of North Carolina Press, 1985.

———, editor. *To Be Continued: Soap Operas Around the World*. London: Routledge, 1995.

Ang, Ien. *Watching Dallas: Soap Opera and the Melodramatic Imagination*. London: Methuen, 1985.

Buckingham, David. *Public Secrets: EastEnders and Its Audience*. London: British Film Institute, 1987.

Cantor, Muriel G., and Suzanne Pingree. *The Soap Opera*. Beverly Hills, California: Sage, 1983.

Cassata, Mary, and Thomas Skill. *Life on Daytime Television*. Norwood, New Jersey: Ablex, 1983.

Dyer, Richard, with others. *Coronation Street*. London: British Film Institute, 1981.

Geraghty, Christine. *Women and Soap Operas*. Cambridge: Polity Press, 1991.

Hobson, Dorothy. *Crossroads: The Drama of a Soap Opera*. London: Methuen, 1982.

Intintoli, Michael. *Taking Soaps Seriously: The World of Guiding Light*. New York: Praeger, 1984.

Modleski, Tania. *Loving with a Vengeance: Mass-Produced Fantasies for Women*. Hamden, Connecticut: Archon Books, 1982.

Nochimson, Martha. *No End to Her: Soap Opera and the Female Subject*. Berkeley: University of California Press, 1992.

Silj, Alessandro. *East of Dallas: The European Challenge to American Television*. London: British Film Institute, 1988.

Williams, Carol Traynor. *"It's Time for My Story": Soap Opera Sources, Structure, and Response*. Westport, Connecticut: Praeger, 1992.

See also Nixon, Agnes; Phillips, Irna; Telenovela; Teleroman

SOCIAL CLASS AND TELEVISION

Social class has been a neglected factor in research on American television programs and audiences. Only a few studies specifically focus on the portrayal of class in television programming though some additional information can be gleaned from incidental remarks relevant to class in studies on other topics. Class has seldom been considered in audience research either, although media researchers from the British cultural studies tradition, through their applications of ethnographic audience research, have recently directed more attention to this topic.

Research on class content has focused on drama programming. News, talk shows, and most other genre remain unexamined. Several studies have examined sex role portrayals in television commercials, but little exists on the matter of class, except frequency counts of occupations used in studies of gender. A wide range of writers, from television critics to English professors to communications researchers, have examined the texts of single drama programs or of small numbers of drama series, selected for their prominence in the television landscape.

Woven into the textual analysis of some of these analyses are remarks on class, but only a few studies have concentrated on the class-related messages of particular programs. In a 1977 *Journal of Communication* article Lynn Berk argued that Archie Bunker exemplified the equation of bigotry with working class stupidity, a stereotype no longer applied to race but still acceptable in characterizing the working class. Robert Sklar in his 1980 book, *Prime Time America*, was more hopeful about two Gary Marshall shows of the mid-1970s, when a number of working-class characters populated prime time. The Fonz and Laverne and Shirley retained their dignity in their everyday struggles against class biases. In a 1986 *Cultural Anthropology* article George Lipsitz examined seven ethnic working class TV sitcoms from the 1950s and found sentimental images of ethnic families combined with themes promoting consumption.

While textual studies focus on in-depth analysis of particular shows, other researchers have compiled demographic portraits across all television drama programming at a given point in time. They categorize fictional characters by sex, race, age, occupations, and occasionally the evaluative tone of these portrayals. Only a few of these studies extend beyond occupation to discuss social class specifically. But data on occupations can be used as a measure of the class distribution of television characters.

Many such studies have been done since the 1950s. Collectively, they provide a series of snapshots over time. The overall results of studies from the 1950s to the 1980s have revealed a repeated under-representation of blue collar and over-representation of white collar characters. Professionals and managers predominate. Central characters were even more likely than peripheral characters to be upper-middle-class white males. The movement of working-class people to the periphery of television's dramatic worlds produces what Gerbner called "symbolic annihilation", i.e. they are invisible background in the dominant cultural discourse. Over-representation of those at the top or at least in the upper middle class, simultaneously gives the impression that those not among these classes are deviant.

Textual criticism gives depth, demographic surveys, breadth to the understanding of television. An approach which provides some aspects of both metods is genre study, the close examination of many shows within a given genre. Sitcoms, and particularly domestic sitcoms, have been studied in this way. Ella Taylor's *Prime Time Families* (1989) is a good example of this type of work. Only a small number of such studies, however, address social class in more than a cursory fashion. The most extensive genre studies of class are Richard Butsch's "Class and Gender in Four Decades of Television Situation Comedies" (in *Critical Studies in Mass Communication),* and Butsch and Lynda Glennon's 1982 and 1983 essays in the *Journal of Broadcasting,* and the report on *Television and Behavior,* published by the U.S. Department of Health and Human Services. These studies found remarkable consistencies in domestic situation comedies over four decades, from 1946 to 1990. Working-class families were grossly and persistently under-represented compared to their proportion of the nation's population. For over half the forty years, there was only one working-class series on the air, out of an average of 14 domestic sitcoms broadcast annually. From 1955 to 1971 not one new working-class domestic sitcom appeared. Middle-class families headed by professional/managerial fathers predominated.

Butsch found that the portrayals themselves are strikingly persistent. The prototypical working-class male is incompetent and ineffectual, often a buffoon, well-intentioned but dumb. In almost all working-class series, the male is flawed, some more than others: Ralph Kramden, Fred Flintstone, Archie Bunker, Homer Simpson. He fails in his role as a father and husband, is lovable but not respected. Heightening this failure is the depiction of working-class wives as exceeding the bounds of their feminine status, being more intelligent, rational, and sensible than their husbands. In other words gender status is inverted, with the head of house, whose occupation defines the families social class, demeaned in the process. Class is coded in gendered terms. Working-class men are de-masculinized by depicting them as child-like; their wives act as mothers. Some writers fail to note that these male buffoons are almost always working class. They miss the message about class, and instead define it as a message about gender. These results indicate the importance of accounting for class along with gender.

In middle-class domestic situation comedies the male buffoon is a rarity. When a character plays the fool it is the dizzy wife, like Lucy Ricardo in *I Love Lucy.* In most middle-class series, however, both parents are mature, sensible, and competent, especially when there are children in the series. It is the children who provide the antics and humor. They are, appropriately, child-like. Nor are sex roles inverted in these series. The man is appropriately "manly," and the woman "womanly." The family as a whole represents an orderly, well functioning unit, in contrast to the chaotic scenes in the working class families. The predominance of middle-class series, combined with persistently positive treatment, equated the middle-class family with the American family ideal.

The Nanny

Reinforcing the middle-class ideal was an exaggerated display of affluence and upward mobility. Maids and other household help were far more prevalent than in the real world. Even working-class families were upwardly mobile, moving to the suburbs or having the father promoted to foreman or starting his own business.

In his 1992 article "Social Mobility in Television Comedies" (in *Critical Studies in Mass Communication),* Lewis Freeman found that upward mobility in sitcoms of 1990–1992 was achieved through self-sacrifice and reliance, reinforcing the ethic of individualism which makes each person responsible for his or her socio-economic status. Thus one's status is an indicator of one's ability, character and moral worth. However, as if to temper desires of the audience, the economic benefits of upward mobility were counter-balanced by the personal consequences. The economic rewards disrupted relations with family and friends.

Sari Thomas and Brian Callahan argue in "Allocating Happiness: TV Families and Social Class" (*Journal of Communication*) that portrayals in the late 1970s showed working class families who were sympathetic and supportive of each other and the characters generally "good" people. The middle class was portrayed this way too, but less so. Both contrast to portrayals of the rich who were often depicted as unsympathetic and unsupportive of each other, and as "bad" or unhappy people. The contrasts between classes convey the moral that money does not buy happiness.

Dynasty

Married . . . with Children

Rarely has class been considered a variable in research seeking to identify specific effects resulting from television viewing. This research tradition has concentrated on generalizations about psychological processes rather than on group differences. In a major bibliography of almost 3,000 studies of audience behavior only seven articles on television effects and thirteen on use patterns examined class differences. Joseph Klapper's classic summary of effects research, *The Effects of Mass Communication* (1960), did not even mention class as a factor. The few studies that have considered class found that there were no class differences in children's susceptibility to violence on television, in contrast to the usual stereotype of working-class children being more likely to be led into such behavior.

Studies of family television use patterns have looked more broadly at people's behavior with the television set. But even in these class is often peripheral. Books on television audiences seldom include social class as a topic in their indexes. One traditional research technique, however, has been to distinguish class differences in television use, usually with an evaluative preference for the patterns established in "higher" classes. Ira Glick and Sidney Levy's *Living with Television* (1962) firmly established the tradition from their 1950s market surveys. The working-class family tended to use TV as a continuing background, with children and parents doing other things while the TV was on. They did not plan viewing, but watched whatever was available at the time they had to watch. They were defined

as indiscriminate users, the term suggesting an unhealthy habit. Middle-class families tended to turn on the TV for a specific program and then turn it off. They planned a schedule of activities, including when and what to watch on television. The middle-class pattern was defined as intellectually superior and as approved child-rearing practice. Other researchers adopted this description of working class viewers, confirming popular critics prejudices about the working class, and favoring of the middle class. Recent family communication research has continued to distinguish these class differences, but has avoided the evaluative tone.

Buried within the 1950s and 1960s sociological literature on working-class lifestyle are a few ethnographic observations on working-class uses of and responses to television. These have confirmed the working-class pattern of using the TV as filler and background to family interaction. They also revealed distinctive responses to program content. Working-class men preferred shows featuring a character sympathetic to working-class values. They identified with working-class types even when those types were written as peripheral characters or villains. They contradicted the notion of working-class viewers as passive and gullible.

These results are consistent with effects research which indicated that audiences tend to reject as unrealistic television portrayals that they can compare to their own experience. Thus working-class viewers would not be likely to accept stereotypic portrayals of their class such as described above. Indirect evidence suggests that working-class viewers tended to perceive Archie Bunker as winning arguments with his college-educated son-in-law. In a recent study of soap operas and their viewers (*Remote Control: Television Audiences and Cultural Power*), working-class women viewers of daytime serials rejected the affluent long suffering heroines in favor of villainesses who transgressed feminine norms and thus cast off middle class respectability.

British researchers have given more attention to class [e.g., Piepe]. Cultural studies in particular have popularized the methods of talking with working-class viewers about their reactions to television. Studies of British working-class viewers have painted a more complicated picture of working class viewing than popular stereotypes, encouraged by the portrayals of class on television, would suggest. As with the earlier American studies, working people construct their own alternative readings of television programs.

This wide range of studies over decades provide consistent evidence that working-class viewers are not the passive dupes with their eyes glued to the screen, that popular television criticism has concocted. Nor are they the bumbling, ineffectual clowns often constructed in television comedies. Rather, they use television to their advantage, and interpret content to suit their own needs and interests.

—Richard Butsch

FURTHER READING

(Content)

Berk, Lynn. "The Great Middle American Dream Machine." *Journal of Communication* (New York), Summer 1977.

Butsch, Richard. "Class and Gender in Four Decades of Television Situation Comedies." *Critical Studies in Mass Communication* (Annandale, Virginia), December 1992.

———, and Lynda Glennon. "Social Class Frequency Trends in Domestic Situation Comedy, 1946–1978." *Journal of Broadcasting* (Washington, D.C.), 1983.

Freeman, Lewis. "Social Mobility in Television Comedies." *Critical Studies in Mass Communication* (Annandale, Virginia), December 1992.

Gerbner, George, Larry Gross, Suzanne Jeffries-Fox, Marilyn Jackson-Beeck, and Nacy Signorielli. "Cultural Indicators: Violence Profile No. 9." *Journal of Communication* (New York), 1978.

Glennon, Lynda, and Richard Butsch. "The Family as Portrayed on Television 1946–1978." In, David Pearl, editor, with others. *Television and Behavior: Technical Reviews* (vol.2). Washington D.C., U.S. Dept. of Health and Human Services, 1982.

Greenberg, Bradley, editor. *Life on Television: Content Analysis of US TV Drama*. Norwood, New Jersey: Ablex, 1980.

Lipsitz, George. "The Meaning of Memory." *Cultural Anthropology* (Washington, D.C.), 1986.

Sklar, Robert. "The Fonz, Laverne, Shirley and the Great American Class Struggle." In, *Prime Time America*. New York: Oxford University Press, 1980.

Steeves, H. L., and M. C. Smith. "Class and Gender on Prime Time Television Entertainment." *Journal of Communication Inquiry*, 1987.

Taylor, Ella. *Prime Time Families*. Berkeley, California: University of California Press, 1989.

Thomas, Sari. "Mass Media and the Social Order." In, G. Gumpert, and R. Cathcart, editors. *Intermedia: Interpersonal Communication in a Media World*. New York: Oxford University Press, 1986.

Thomas, Sari, and Brian Callahan. "Allocating Happiness: TV Families and Social Class." *Journal of Communication* (New York), Summer, 1982.

(Audiences)

Blum, Alan. "Lower Class Negro Television Spectators." In, Shostak, Arthur B., and William Gomberg, editors. *Blue Collar World*. New York: Prentice Hall, 1964.

Bryce, Jennifer. "Family Time and Television Use." In, Lindlof, Thomas, editor. *Natural Audiences*. Newbury Park, California: Sage, 1987.

Gans, Herbert. *The Urban Villagers*. New York: Free Press, 1962.

Glick, Ira, and Sidney Levy. *Living With Television*. Chicago: Aldine, 1962.

Jordan, Amy. "Social Class, Temporal Orientation, and Mass Media Use within the Family System." *Critical Studies in Mass Communication* (Annandale, Virginia), December, 1992.

Seiter, Ellen, Hans Borchers, Gabriele Kreutzner, and Eva-Maria Warth, editors. "Don't Treat Us Like We're So Stupid and Naive: Toward an Ethnography of Soap Opera Viewers." In, *Remote Control: Television Audiences and Cultural Power*. London: Routledge, 1989.

Turner, Graeme. "Audiences." In, *British Cultural Studies*. London: Routledge, 1992.

See also Family and Television; Gender and Television; Racism, Ethnicity and Television

SOCIETY FOR MOTION PICTURE AND TELEVISION ENGINEERS

While the emergence of motion pictures and television is typically linked to the rise of commercial culture and mass entertainment, the extent of industry growth cannot be adequately explained without acknowleging the extensive benefits that came from technical standardization.

Incorporated in July 1916, the Society for Motion Picture Engineers (SMPE) sought to act as a professional forum for its members, and to publish technical findings "deemed worthy of permanent record." The impact of the society, however, extended far beyond the research reports published in SMPE's *Journal* and *Transactions*. With film pioneer Francis Jenkins installed as its charter president, the society took as its first task the development of a 35mm format—the standard upon which the the motion picture and telefilm industries were built. Subsequent SMPE interventions codified two-color cinematography (November 1918), three-color technicolor (August 1935), and optical sound recording technologies (September 1938, October 1930). Although the organization began as a professional association for technical specialists, its public actions worked as an antidote to the high-risk economic and methodological instabilities that accompanied the introduction of each new film/television technology.

Research interests in television predated by decades the formal addition of "Television" to the society's name in 1950 (SMPTE). Groundbreaking work was published on alternative delivery systems ("Radio Photographs, Radio Movies and Radio Vision" by C.F. Jenkins, May 1923), on vacuum tube imaging devices ("Iconoscopes and Kinescopes" by V.K. Zworykin, May 1937), and on RCA's field test of a comprehensive broadcasting system in New York (R.R. Beal, August 1937). While this pre-war flurry of engineering interest in television may suggest a proactive and determining influence, subsequent actions demonstrate just how provisional SMPE's recommendations were. For example, although the *Journal* published standards for CBS' new high-resolution color television system in April 1942, the subsequent combination of coercion and economic clout lead the government to opt for an inferior system in 1947. The FCC favored RCA/NBC's less developed alternative, thereby forcing engineers to impose color information onto the limited black-and-white bandwidth of NTSC—a system that had itself been hastily (and some would say prematurely) adopted in 1941. Or again, despite the open-ended, forward-thinking proposals put forth by Jenkins for theatrical television, pay per view, and set-licencing subsidies in 1923, the harsh regulatory realities of the FCC licensing freeze from 1948 to 1952 effectively deferred development of alternative delivery technologies for decades. A three network oligopoly would dominate for almost thirty years as a result of the freeze; enabled by economic and regulatory collusion rather than engineering wisdom.

While such actions demonstrate the provisional nature of the society's recommendations—SMPTE is not a government regulatory body like the FCC, but an association of professionals representing a wide range of proprietary corporations—subsequent breakthroughs mark key points in the history of television technology. Standards for the eventual victor in the color television race (NTSC) were finally published in April 1953. Engineers from Ampex disseminated information on the first commercially successful videotape recorder in April 1957—an event that led to the precipitous death of the kinescope, initiated intense competition among VTR developers in the years that followed, and altered forever the way viewers see liveness (live-on-tape). Although American industry lagged behind foreign compet-

Courtesy of SMPTE

itors in the race for viable "digital" video systems, SMPTE began to disseminate engineering standards for a spate of new digital television recording formats developed in Europe and Japan starting in December 1986.

The international battle over high-definition television (HDTV) demonstrates the strategic role a standardizing organization can take in the international arena. NHK in Japan had produced and begun marketing an HDTV system in the early 1980s—long before American corporations entered the fray with working prototypes. European corprations soon offered a competing system. U.S. broadcasters, however, resisted HDTV development given the tremendous costs involved in changing-over from current transmission systems. Eventually, however, SMPTE worked on and proposed a third HDTV system. Unlike the analog systems from NHK and Europe, SMPTE's late start allowed them to propose an all digital sytem. When the FCC started competitive trials between three American-centered consortia—and then cancelled the trial before rendering a verdict on the winner—the implications were clear. Government intervention meant that the U.S. would produce a single "consensus" HDTV system. The resulting "grand alliance" minimized the risk of losing an expensive R and D race, and affirmed SMPTE's all digital lead. The foreign trade journalists howled at the prospect of what many now considered—given America's late HDTV entry and government muscle—the odds-on international favorite. Engineering standards, then, can be political footballs used for economic leverage and technological nationalism. They also frequently provide a demilitarized zone for manufacturers; especially for those corporations that wait on the sidelines to apply the lessons of the proprietary risk-takers; that wait, in short, until the corporations that are first off the technological runway go down in flames. Japanese equipment manufacturers—Sony and Matsushita—stood on the sidelines and watched pioneers Ampex and RCA in the 1960s. Computerized video and HDTV now show that the process works in other directions as well.

SMPTE's future influence will depend upon how well it comes to grips with several substantive changes. It must respond to the technological "convergence" blurring boundaries between film and electronic media; it must continue to demonstrate the value of common technical ground within the proprietary world of mulitnational corporations; and it must engage a membership that increasingly lies outside of the confines of engineering. As studios are reduced to computerized desktops, and practitioners with technical backgrounds cross-over into creative capacities (and vice versa), technological discourses will become no less important or problematic. Given the inevitable capital-intensive nature of electronic media—and the public shift to paradigms of decentralization, entrepeneurial imperative, and market volatility—issues of standardization and technological "order" will be more crucial to the future of television than ever.

—John Thornton Caldwell

FURTHER READING

Boddy, William. *Fifties Television*. Urbana, Illinois: University of Illinois Press, 1990.

Bordwell, David, Kristin Thompson, and Janet Staiger. *Classical Hollywood Cinema*. New York: Columbia University Press, 1985.

Gilder, George. *Life After Television*. New York: Norton, 1992.

Milestones in Motion Picture and Television Technology. White Plains, New York: SMPTE, 1991.

Winston, Brian. *Misunderstanding Media*. Cambridge, Massachusetts: Harvard University Press, 1986.

See also Jenkins, Charles Francis; Standards; Television Technology; Zworykin, Vladimir

SOME MOTHERS DO 'AVE 'EM

British Comedy Series

Some Mothers Do 'ave 'em was a hugely popular British comedy series, broadcast by the BBC in the 1970s. Initially considered unlikely to succeed, the series triumphed through the central performance of Michael Crawford as the hapless Frank Spencer and became one of the most popular comedy series of the decade, attracting a massive family audience.

Frank Spencer was the ultimate "loser", unemployable, unable to cope with even the simplest technology, and the victim of his surroundings. Every well-meaning attempt that he made to come to terms with the world ended in disaster, be it learning to drive, getting a job, or realizing some long-cherished dream. What saved him, and kept the story comic, was his innocence, his dogged persistence, and his outrage at the injustices he felt he had suffered.

The theme of the naive innocent comically struggling in an unforgiving world is an old one, but in this incarnation the most obvious antecedents for the slapstick Spencer character were such silent movie clowns as Charlie Chaplin's tramp and, some three decades later, British cinema's Norman Wisdom. Writer Raymond Allen insisted, however, that he based the character on himself, quoting as his qualifications as the original Frank Spencer his outdated dress sense, complete lack of self-confidence, and overwhelming inability to do anything right. As proof of the character's origins, Allen recalled how he had bought himself a full-length raincoat to wear to the first rehearsals of the series in London—and was dismayed to see Crawford acquire one virtually the same as the perfect costume to play the role. The mac, together with the beret and the ill-fitting tanktop jumper, quickly became visual trademarks of the character.

It was Michael Crawford (really Michael Dumble Smith), complete with funny voice and bewildered expres-

Some Mothers do 'ave 'em
Photo courtesy of BBC

sion, who turned Frank Spencer into a legend of British television comedy, employing the whole battery of his considerable comic skills. Disaster-prone but defiant, the little man at odds with a society judging people solely by their competence and ability to fit in, he turned sets into battlefields as he fell foul of domestic appliances, motor vehicles, officials, in-laws, and just about anyone or anything else that had the misfortune to come into his vicinity.

Some Mothers Do 'ave 'em was essentially a one-joke escapade, with situations being set up chiefly to be exploited for the admittedly often inventive mayhem that could be contrived from them. What kept the series engaging, however, was the pathos that Crawford engendered in the character, making him human and, for all the silliness of many episodes, endearing. In this Crawford was ably abetted by Michelle Dotrice, who played Frank Spencer's immensely long-suffering but steadfastly loyal (if occasionally despairing) girlfriend, and later wife, Betty.

In the tradition of the silent movie stars, Crawford insisted on performing many of the hair-raising and life-threatening stunts himself, teetering in a car over lofty cliffs, dangling underneath a helicopter, and risking destruction under the wheels of a moving train in a way that would not have been tolerated by television companies and their insurers a few years later. The professionalism that he displayed in pulling off these stunts impressed even those who balked

at the show's childish humour and overt sentimentalism. It is not so surprising that Crawford himself, after six years in the role, was able to escape the stereotype that threatened to obscure his talent and to establish himself as a leading West End and Broadway musical star.

—David Pickering

CAST

Frank Spencer . . . .   Michael Crawford
Betty   . . . .   Michele Dotrice

PRODUCER   Michael Mills

PROGRAMMING HISTORY   19 Half Hour Episodes; 3 Fifty-Minute Specials

• BBC

February 1973–March 1973	7 Episodes
November 1973–December 1973	6 Episodes
25 December 1974	Christmas Special
25 December 1975	Christmas Special
October 1978–December 1978	6 Episodes
25 December 1978	Christmas Special

See also British Programming

SONY CORPORATION

International Media Conglomerate

An innovative Japanese consumer-electronics company founded by Masaru Ibuka and Akio Morita in 1946, Sony started out manufacturing heating pads, rice cookers, and other small appliances, but soon switched to high technology, bringing out Japan's first reel-to-reel magnetic tape recorder in 1950 and then its first FM transistor radio in 1955. Sony's later innovations in consumer electronics included the Trinitron color television picture tube (1968), the Betamax videocassette recorder (1975), the Walkman personal stereo (1979), the compact disc player (1982), the 8mm video camera (1985), and the Video Walkman (1988).

Sony's success in marketing its products worldwide rested on distinctive styling and "global localization," a practice that retained product development in Japan, while disbursing manufacturing among plants in Europe, the United States, and Asia. To maintain quality control, Sony dispatched large numbers of Japanese managers and engineers to supervise these plants.

Under the leadership of Norio Ohga, who joined the company 1959 and ran Sony's design center, Sony pursued the course of marrying Japanese consumer electronics with American entertainment software. After purchasing CBS Records for $2 billion in 1987, Sony initiated the Japanese invasion of Hollywood by acquiring Columbia Pictures Entertainment (CPE) from Coca-Cola for $3.4 billion in

1989. The following year, Sony's Japanese rival Matsushita Electric Industrial Company, the largest consumer-electronics company in the world, purchased MCA for $6.9 billion. The two takeovers led to charges that the Japanese were about to dominate American popular culture, but the controversy soon died out when it became apparent that Sony and Matsushita would have to stay aloof from production decisions if their studios were to compete effectively.

In 1989, the year Sony acquired CPE, Sony generated over $16 billion in revenues from the following categories: (1) video equipment other than TV—$4.3 billion; (2) audio equipment—$4.2 billion; (3) TV sets—$2.6 billion; (4) records—$2.6 billion; and (5) other products—$2.5 billion. The CPE acquisition, which included two major studios—Columbia Pictures and TriStar Pictures—home video distribution, a theater chain, and an extensive film library, brought in an additional $1.6 billion in revenues.

By becoming vertically-integrated, Sony hoped to create "synergies" in its operations, or stated another way, Sony wanted to stimulate the sales of hardware by controlling the production and distribution of software. The company may have been reacting to the so-called "format wars" of the 1970s when Sony's Betamax lost out to Matsushita's VHS video tape recorder. Industry observers believed that the greater availability of VHS software in video stores naturally led consumers to choose VHS machines over Betamaxes. Sony would not make the same mistake again and found a way to protect itself as it contemplated introducing the 8-millimeter video and high definition television systems it had in development.

To strengthen CPE as a producer of software, Sony spent an added $1 billion and perhaps more to acquire and refurbish new studios and to hire film producers Peter Guber and Jon Peters to run the company, which it renamed Sony Entertainment. Sony performed reasonably well under the new regime until 1993, but afterwards, Columbia and TriStar struggled to fill their distribution pipelines. Virtually all of Sony's hits had been produced by independent producer affiliates and when these deals lapsed, Sony lagged behind the other majors in motion picture production and market share. Some industry observers claimed Sony lacked "a clear strategy" for taking advantage of the rapid shifts in the entertainment business. After top production executives left Columbia and TriStar in 1994, Sony took a $3.2 billion loss on its motion picture business, reduced the book value of its studios by $2.7 billion, and announced that "it could never hope to recover its investment" in Hollywood.

—Tino Balio

FURTHER READING

Lardner, James. "Annals of Law; The Betamax Case, Part 1." *The New Yorker* (New York), 6 April 1987.

———. "Annals of Law; The Betamax Case, Part 2." *The New Yorker* (New York), 13 April 1987.

Lyons, Nick. *The Sony Vision.* New York: Crown, 1976.

Morita, Akio. *From a 500-dollar Company to a Global Corporation: The Growth of Sony.* Pittsburgh, Pennsylvania: Carnegie-Mellon University Press, 1985.

Morita, Akio, with Edwin M. Reingold, and Mitsuko Shimomura. *Made in Japan: Akio Morita and Sony.* New York: Dutton, 1986.

See also Betamax Case; Camcorder; Home Video; Time Shifting; Videocassette; Videotape

SOUL TRAIN

U.S. Music-Variety Show

Soul Train, the first black-oriented music variety show ever offered on American television, is one of the most successful weekly programs marketed in first-run syndication and one of the longest running syndicated programs in American television history. The program first aired in syndication on 2 October 1971 and was an immediate success in a limited market of seven cities: Atlanta, Cleveland, Detroit, Houston, Los Angeles, Philadelphia and San Francisco. Initially, syndicators had difficulty achieving their 25-city goal. However, *Soul Train'*s reputation as a "well produced" and "very entertaining" program gradually captured station directors' attention. By May 1972, the show was aired in 25 markets, many of them major cities.

The show's emergence and long-standing popularity marks a crucial moment in the history of African-American television production. Don Cornelius, the show's creator, began his career in radio broadcasting in Chicago in November 1966. At a time when African Americans were systematically denied media careers, Cornelius left his $250-a-week job selling insurance for Golden State Mutual Life to work in the news department at WVON radio for $50 a week. It was a bold move, and clearly marked his committed optimism. By seizing a small opportunity to work in radio broadcasting, Cornelius was able to study broadcasting first hand. His career advancement in radio included employment as a substitute disc jockey and host of talk shows. Radio broadcasting techniques informed Cornelius' vision of the television program *Soul Train.*

By February 1968, Cornelius was a sports anchorman on the black-oriented news program, "A Black's View of the News" on WCIU-TV, Channel 26, a Chicago UHF TV station specializing in ethnic programming. Cornelius pitched his idea for a black-oriented dance show to the management of WCIU-TV the following year. The station agreed to Cornelius' offer to produce the pilot at his own expense in exchange for studio space. The name *Soul Train* was taken from a local promotion Cornelius produced in 1969. To create publicity, he hired several Chicago entertainers to perform live shows at up to four high schools on the same day. The caravan performances from school to school reminded the producer of a train.

Cornelius screened his pilot to several sponsors. Initially, no advertising representatives were impressed by his idea for black-oriented television. The first support came from Sears, Roebuck and Company, which used *Soul Train* to advertise phonographs. This small agreement provided only a fraction of the actual cost of producing and airing the program. Yet, with this commitment, Cornelius persuaded WCIU-TV to allow the one-hour program to air five afternoons weekly on a trial basis. The program premiered on WCIU-TV on 17 August 1970, and within a few days youth and young adult populations of Chicago were talking about this new local television breakthrough. The show also had the support of a plethora of Chicago-based entertainers. As an independent producer of the program, Cornelius acted as host, producer and salesman five days a week. He worked without a salary until the local advertising community began to recognize the program as a legitimate advertising vehicle, and *Soul Train* began to pay for itself.

The *Soul Train* format includes guest musical performers, hosts, and performances by the *Soul Train* dancers. Set in a dance club environment, the show's hosts are black entertainers from music, television and the film industries. The dancers are young women and men, fashionably dressed, who dance to the most popular songs on the Rhythm and Blues, Soul, and Rap charts. The show includes a game called "*The Soul Train* Scramble," in which the dancers compete for prizes. The program's focus on individual performers, in contrast to the ensemble dancing more common in televisual presentation, has been passed down to many music variety shows such as *American Bandstand, Club MTV,* and *Solid Gold.*

The television show's success can be linked to the increasing importance of black-oriented radio programs taking advantage of FM stereo sound technology. With that support soul and funk music exploded in popularity across the nation. Black record sales soared due to the increased radio airplay, and the opportunity to view popular performances without leaving home became the appeal of *Soul Train.*

The popularity of the show in Chicago prompted Cornelius to pursue national syndication of the program. One of the nation's largest black-owned companies, the Johnson Products Company, agreed to support the show in national syndication. Sears, Roebuck and Company increased their advertising support. In 1971 Cornelius moved the production of the *Soul Train* to Hollywood. The show continued to showcase musical talent and to shine the spotlight on stand up comedians. The program's presentation of vibrant black youth attracted viewers from different racial backgrounds and ethnicities to black entertainment. The show has been credited with bringing 1970s black popular culture into the American home.

In 1985, the Chicago-based Tribune Entertainment company became the exclusive distributor and syndicator of *Soul Train.* In 1987, the Tribune company helped to launch the "*Soul Train* Music Awards." This program is a live two-hour television special presented annually in prime-time syndication and reaches more than 90% of U.S. television households. The *Soul Train* Music Awards represent the ethos of the *Soul Train* program, which is to offer exposure for black recording artists on national television.

—Marla L. Shelton

PRODUCER   Don Cornelius

PROGRAMMING HISTORY
Syndicated, Various Times
1971–

Don Cornelius
Photo courtesy of Don Cornelius

FURTHER READING

Meisler, Andy. "For *Soul Train*'s Conductor, Beat Goes On: A TV Show Sticks with its Niche and a Time-Tested Formula." *The New York Times,* 7 August 1995.

Reynolds, J.R. "Big Draw." *Billboard* (New York), 29 April 1995.

Rule, Sheila. "Off the *Train.*" *The New York Times,* 20 October 1993.

See also Music on Television; Racism, Ethnicity, and Television

SOUTH AFRICA

The South African television service, launched in 1976, is among the youngest in Africa, but by far the most advanced on the continent. Propped by the country's large economy and high living standards among the minority populations, South Africa's television industry developed rapidly to become one of the first satellite-based broadcasting systems on the continent, with the most widely-received national service.

The industry is dominated by a state organization, the South African Broadcasting Corporation (SABC), which was established in 1936 by an act of Parliament. The corporation however concentrated on radio broadcasting during its first 40 years of operation, as the racist National Party in power during most of this period opposed the introduction of television under the pretext of preserving cultural sovereignty. The launching of the communication satellite Intelsat IV in 1972 by Western countries ushered in new fears about the dangers of uncontrolled reception of international television via cheap satellite dishes. The South African government, fearing imperialism, swiftly resolved to introduce a national television service as an anti-imperial device.

Between 1976 and 1990, the SABC-TV service was state-controlled and heavily censored, and functioned as an arm of the government. SABC was banned from broadcasting pictures or voices of opposition figures, and its editorial policy was dictated through an institutional censoring structure.

The blackout on politically-dissenting voices was discontinued in 1990 as the corporation purged itself of racial bias, and shifted its focus to public service broadcasting. Since then, SABC-TV has balanced its programs to reflect the country's cultural and political diversity and embraced a policy of affirmative action in staff recruitment.

At inception, SABC-TV operated four national television channels: TV-1, TV-2, TV-3, and TV-4. This configuration was revised in a 1992 restructuring program; TV-1 retained its autonomy and the rest were merged into a new multicultural channel called Contemporary Community Values Television (CCV-TV). The two national channels now compete for audiences and advertising with M-Net, a highly successful privately-owned pay channel.

TV-1, the largest and most influential, was directed at the minority white population, with all programs broadcast in Afrikaans and English. Since mid-1986, the channel's 18-hour daily programming has been relayed through a transponder on an Intelsat satellite to 40 transmitting stations with an ERP of 100Kw, and 42 stations with an ERP range of between 1Kw and 10Kw. These transmissions are augmented by 63 gap fillers and an estimated 400 privately-owned low power transmitters, enabling the channel to be received by three quarters of the country's population.

The CCTV channel broadcasts in nine local languages via fourteen 100Kw terrestrial stations, nine 1 to 10Kw stations, and 33 gap fillers. The channel's programming is received by 64% of the country's population.

SABC's domination of radio and television has enabled it to develop advanced products and services for its audience. The corporation offers simulcasting of dubbed material on television with the original sound track on radio Teledata, a teletext service initially established as a pilot project on spare TV-1 signal capacity, has been expanded to a 24-hour service with over 180 pages of news, information, and educational material. Selected material from the Teledata database is also copied onto TV-1 outside program transmission to provide an auxiliary service that is available on all TV sets countrywide.

The Electronic Media Network, widely known by its acronym, M-Net, is South Africa's only private television channel. Founded by a consortium of newspaper publishers in October 1986 to counter the growing threat that the commercially-driven SABC-TV posed to the newspaper industry, M-Net has grown into the most successful pay-TV station in the world outside the United States. Its nearly 850,000 subscribers (1995 estimate) received 120 hours a week of entertainment,

documentaries, film, series, and miniseries. The large national audience is accessed through a number of leased or rented SABC terrestrial reception facilities.

The subscription service is offered on an internationally-patented decoder originally developed from the American Oak Systems decoder technology. M-Net's subscriber management subsidiary, Multichoice Ltd., markets the programming services to individual subscribers across Southern Africa. It also markets the Delta 9000 Plus decoders to pay-TV operations elsewhere; by 1994, it was marketing the technology to Pelepiu pay-TV system in Italy. Another of its subsidiaries, M-Net International, has been actively seeking subscribers in tropical and northern Africa after successful operations in Namibia, Lesotho, and Swaziland. Through the use of two transponders on C-band satellites, the channel has a footprint covering the entire African continent and parts of the Middle East. During 1994, Multichoice Ltd. signed an agreement with a private TV station in Tanzania to relay programming across the country via satellite. At the same time, M-Net International began broadcasting across Africa on a channel shared with the BBC World Service Television. Plans were also afoot to extend rebroadcast rights to sub-Sahara African countries, and to expand satellite services and individual subscriptions.

Three small regional television stations are operated in the former homelands of Bophuthatswana, Transkei, and Ciskei. The Bophuthatswana television, Bop-TV, is a commercial operation that is aired via 18 small transmitters (all with ERP below 1Kw), and relay stations in Johannesburg and Pretoria. The Transkei Broadcasting Corporation operates a television service which competes with the pay service of M-Net Transkei. M-Net Transkei is a scrambled service except between 3:00 P.M. and 5:00 P.M. when its signal is unscrambled. The Rhena Church of South Africa runs two private TV stations in Ciskei and Transkei, which broadcast in English via two small stations. Plans were underway in 1994 to install two 1Kw transponders.

Since the early 1980s, South Africa has been considering venturing into satellite communications. The first involvement in satellite-aided broadcasting came in mid-1986 when a transponder was fitted on an Intelsat satellite to relay TV-1 to terrestrial transmitting stations. In early 1992, the C-band satellite service was upgraded from a hemispherical beam to a zonal beam to enhance the establishment of cellular transmitters in remote areas of the country. At the same time, the transmission standards were upgraded from B-MAC to PAL System 1. Together with the introduction of transmissions in the Ku-Band range, these modifications are expected to provide television coverage to the entire country. The Ku-Band satellite service is also expected to be utilized in telecommunication applications.

With over 150 production houses, South Africa has the largest broadcasting production industry on the continent. Local productions, from SABC teams and independent production houses, account for about 50% of airtime of SABC-TV and between 10 and 30% on M-Net. Both organizations have laid heavy emphasis on Afrikaans language productions. However, independent producers, brought together by the Film and Television Foundation (FTF), have in the past lobbied for higher local content quotas. However, such proposals have been contested by M-Net on the grounds that pay-TV service is customer-driven. The FTF suggests that where a broadcaster is unable to offer local content quotas, a levy should be introduced on the turnover to finance local productions.

—Nixon K. Kariithi

FURTHER READING

Bourgault, Louise Mahon. *Mass Media in Sub-Saharan Africa.* Bloomington: Indiana University Press, 1995.

Hachten, William A. *Mass Communication in Africa: An Annotated Bibliography.* Madison, Wisconsin: University of Wisconsin, 1971.

Mytton, Graham. *Mass Communication in Africa.* London: Edward Arnold, 1983.

Nixon, Rob. *Homelands, Harlem, and Hollywood: South African Culture and the World Beyond.* New York: Routledge, 1994.

Prinsloo, Jeanne, and Costas Criticos, editors. *Media Matters in South Africa.* Durban, South Africa: Media Resource Centre, 1991.

Wilcox, Dennis L. *Mass Media in Black Africa: Philosophy and Control.* New York: Praeger, 1975.

SOUTH KOREA

In the past half century, television broadcasting has been introduced in the majority of Western nations. In the 1950s, when television broadcasting evolved into the dominant electronic medium in the West, some Asian countries established their own television services. Korea, the fourth adopter in Asia, began television broadcasting on 12 May 1956 with the opening of HLKZ-TV, a commercially operated television station. HLKZ-TV was established by the RCA Distribution Company (KORCAD) in Seoul with 186-192 MHz, 100-watt output, and 525 scanning lines.

Korean television celebrated its 40th birthday in 1996 and a great deal has changed in the past four decades. In 1956 there were only 300 television sets in Korea, but that number has climbed to an estimated 6.27 million by 1980 and television viewing has become the favorite form of entertainment or amusement for the mass audience. As of 1993, Koreans owned nearly 11.2 million television sets, a penetration rate of nearly 100 percent.

The early 1960s saw a phenomenal growth in television broadcasting. On 31 December 1961 the first full-scale television station, KEWS-TV, was established and began operation under the Ministry of Culture and Public Information. The second commercial television system, MBC-TV, following the first commercial television, TBC-TV, made its debut in 1969. The advent of MBC-TV brought significant development to the television industry in Korea and after 1969 the television industry was characterized by furious competition among the three networks.

The 1970s were highlighted by government intervention into the media system in Korea. In 1972, President Park's government imposed censorship upon media through the Martial Law Decree. The government revised the Broadcasting Law under the pretext of improving the quality of television programming. After the revision of the law, the government expanded its control of media content by requiring all television and radio stations to review programming before and after transmission. Although the government argued that its action was taken as a result of

growing public criticism of broadcasting media practices, many accused the government of wanting to establish a monopoly over television broadcasting.

The 1980s were the golden years for Korea's television industry. Growth was phenomenal in every dimension: the number of programming hours per week rose from 56 in 1979 to nearly 88.5 in 1989; the number of television stations increased from 12 in 1979 to 78 by 1989; and the number of television sets grew from 4 million in 1979 to nearly 6 million in the same period. In 1981 another technological breakthrough happened, the introduction of color television. Color broadcasting, however, occasioned a renewal of strong competition among the networks.

As the decade progressed, more controversial entertainment programming appeared, prompting the government to establish a new broadcasting law. With the Broadcasting Law of 1987, the Korean Broadcasting Committee was established to oversee all broadcasting in the country. The most important feature of this law was that it guaranteed freedom of broadcasting. However, one of its main provisions required that television stations allocate at least 10% of their broadcasting hours to news programming, 40% to cultural/educational programming, and 20% to entertainment programming. At the time of the imposition of these new regulations, the three networks broke new ground by successfully broadcasting the 1988 Seoul Olympics. The coverage of the 24th Olympiad was the product of technological prowess and resourceful use of manpower by the Korean broadcasting industry.

Since the early 1980s, the structure of the Korean television industry has remained basically unchanged. The government ended the 27-year-long freeze on new commercial licenses by granting a license to SBS-TV in 1990. This breakthrough paved the way for competition between the public and the private networks.

Another technological breakthrough took place in the beginning of the 1990s with the introduction of cable television. In 1990, the government initiated an experimental multi-channel and multi-purpose cable television service. In addition, Korea launched its first broadcasting/communication satellite, Mugungwha, to 36,000 km above the equator in 1995. The development of an integrated broadband network is expected to take the form of B-ISDN immediately after the turn of the century.

The decade of the 1990s is likely to be a period of great technological change in the Korean broadcasting industry, which will make broadcasting media even more important than in the past. In this decade the Korean broadcasting industry will maximize the service with new technological developments such as DBS, satellites, and interactive cable systems, all of which will allow Korea to participate fully in the information society.

Regulation of Broadcasting

The aim of the latest Broadcasting Act, legislated on 1 August 1990, is to strive for the democratic formation of public opinion and improvement of national culture, and to contribute to the promotion of broadcasting. The act consists of six chapters: (1) General Provisions; (2) Operations of the Broadcasting Stations and Broadcasting Corporations; (3) The Broadcasting Commission; (4) Payment and Collection of the Television Reception Fee; (5) Matters to be observed by the Broadcasting Stations; and (6) Remedy for Infringement.

In the article on the definition of terms, "broadcast" is defined as a transmission of wireless communication operated by a broadcast station for the purpose of propagating to the general public, news, comments and public opinion on politics, economy, society, culture, current events, education, music, entertainment, etc. Accordingly, cable television is not subject to this act.

Article Three of the act states: (1) The freedom of broadcast programming shall be guaranteed, and (2) No person shall regulate or interfere with the programming or operation of a broadcasting station without complying with the conditions as prescribed by this act or other acts.

Regarding the operation of broadcasting stations, it is prescribed that no person may hold stocks or quotas of the same broadcasting corporation, including stocks or quotas held by a persona having a special relation, in excess of one-third of the total stocks or quotas.

No broadcasting corporation may concurrently operate any daily newspaper or communication enterprise under the control of the Registration of Periodicals. Inflow of foreign capital is also prohibited. That is, no broadcasting corporation shall receive any financial contribution on the pretext of donation, patronage, or other form of foreign government or organization, except a contribution from a foreign organization having an objective of education, physical training, religion, charity or other international friendship, which is approved by the Minister of Information.

Any person who has a television set in order to receive a television broadcast, shall register the television set and pay the reception fee of 2500 Won (about $3) a month. Black-and-white television sets are not subject to the reception fee.

An Overview of Television Programming in Korea

Currently, the four networks (KBS-1TV, KBS-2TV, MBC-TV, and SBS-TV) offer four hours of daytime broadcasting beginning at 6:00 A.M., then resume broadcasting from 5:30 P.M. to midnight. There is no broadcasting between 10:00 A.M. and 5:30 P.M. on weekdays. However, the four networks operate an additional 7.5 hours on Saturday and Sunday.

A typical programming schedule for Korean television networks begins at 6:00 A.M. with either a "brief news report" or "a foreign-language lesson" (English or Japanese). Early morning programs offer daily news, information, and cultural/educational programs. Each network begins its evening schedule at 5:30 P.M. with an afternoon news brief, followed by a time slot reserved for network children's programming. After this another news brief at 7:00 P.M. introduces prime-time. The four networks fill the next three hours with programs ostensibly suitable for family viewing, including dramas, game shows, soap operas, variety shows, news magazines, situation comedies, occasional sports, and specials. Traditionally, networks also broadcast 40 to 50 minutes of "Nine O'Clock News" during prime-time. This news broadcast attracts many viewers and produces extremely high ratings. Over the course of the evening, each network also provides brief reports and sports news. Late evening hours are usually devoted to imported programs, dramas, movies, and talk shows. Weekend programming is similar to weekday programming except that it is designed to attract specific types of viewers who are demographically desirable to advertisers.

In its early years Korean television networks depended heavily on foreign imports, most from the United States, for their programming. Overall, imported programs averaged approximately one-third of the total programming hours in 1969. In 1983, 16% of programming originated outside the country. By 1987 imported programming had decreased to 10%, though in March 1987, the networks did still broadcast programs such as *Love Boat, Hawaii 5-0, Mission Impossible,* "Weekend American Movies," and cartoons.

In addition to watching imported television programs on Korean television networks, many Koreans also watch AFKN-TV, which is an affiliate of the American Forces Radio and Television Service, the second largest of five networks managed by the Army Broadcasting Service. AFKN has been broadcasting for 39 years as an information and entertainment medium for 60,000 United States military personnel, civilian employees, and dependents. AFKN-TV also plays a significant role for many young Koreans. No one is quite sure of the size of the Korean "shadow audience" for AFKN-TV. However, it is watched by so many ordinary people that all Korean newspapers and most television guides carry AFKN-TV along with Korean program schedules.

Research by Drs. Won-Yong Kim and Jong-keun Kang has mapped the "cultural outlook" of Korean television. Their sample includes all prime-time dramatic programming on three Korean television networks aired during 1990. It demonstrates that the world of Korean prime-time television significantly under-represents children and adolescents. It grossly over-represents adult groups, however—those who are between the ages of 20 and 39, who constitute one-third of the Korean population, comprise 56.7% of the fictional population. In sum, age distribution in the world of Korean television is bell shaped as compared to the diagonal line of the Korean population.

Another significant difference between Korean prime-time drama and reality is that farmers and fishermen, who constitute 25% of the population, make up only 7.4% of television characters. Social class distribution among characters reveals that nearly half of all television characters appear in the "lower" part of a three-way classification.

With regard to violence, among 49 characters who are involved in violence, 44.9% commit violence and 55.1% suffer it. Among them, mostly adult groups of both sexes are involved with violence. Children and adolescents of both sexes are never involved in violence and young female adults are the most frequent victims in all age groups.

Although these findings show somewhat different patterns between Korea and other countries, they are not strictly comparable with each other, due to the differences among their media systems.

The Korean Television Audience

According to Media Service Korea, each household in Seoul has an average of 1.6 television sets. A poll conducted by KBS shows that Korean television viewers watch an average of a little over three hours on weekdays, 4.5 hours on Saturday, and about 5.5 hours on Sunday. When broken down by demographic information, men watch more television than women. On weekends there were no differences in television viewing among age groups.

In terms of ratings, the most popular time slot is between the hours of 9:00 P.M. and 10:00 P.M. and the highest-rated program is the 9:00 P.M. evening news. Approximately 70% of the adult audience watches the news program every night. The second highest rated time slot is between the hours of 7:00 and 8:00 A.M. The average ratings are 31 points on weekdays and 20 points on weekends.

Korean adults frequently watch news and comedy programs, while teenagers watch comedy programs more frequently and people in the 30-50 age group watch the news more. Men tend to watch more sports, but women tend to watch soap operas and movies.

In terms of information provided by audiences with reference to their stated uses and gratifications, the motive for watching television is most often described as intentions: "to get information" and "to understand other opinions and ways of life," "to get education and knowledge," and "to relax." Another study done by the KBS Broadcasting Culture Center indicates that many viewers considered watching television as a news providing function. Others thought of it as a "craving for refreshment," a "social relation function," or "identification." The motives for watching television news are cited variously as a way to "get information from around the world," a practice done "out of habit" or with the intent "to listen to expert opinions and commentary." For soap operas, the stated reasons for viewing include "because they are interesting," "to kill time," and for some "they seem useful." People watch comedy "to alleviate stress" and "to have fun."

Television ratings and audience viewing information is studied by most broadcasting companies as well as research firms and in Korea, ratings have been measured by diary and people-meter. Currently, a people meter is generally used for gathering ratings and Media Service Korea is engaged in the business of providing the people meter ratings.

—Won-Yong Kim

FURTHER READING

Chang, Won-Ho. *Mass Communication and Korea: Toward a Global Perspective for Research*. Seoul, Korea: Nanam Publishing House, 1990.

Kang, Jong-Geun. *Cultural Indicators: The Korean Cultural Outlook Profile* (Ph.D. dissertation, University of Massachusetts, 1988).

Kang, Jong-Geun, and Michael Morgan. "Cultural Clash: U.S. Television Programs in Korea." *Journalism Quarterly* (Urbana, Illinois), 1988.

Kim, Kyu, Won-Yong Kim, and Jong-Geun Kang. *Broadcasting in Korea*. Seoul, Korea: Nanam Publishing House, 1994.

Lee, Jae Won. "Korea." In, Bank, M. B., and J. Johnson, editors. *World Press Encyclopedia*. New York: Facts on File, 1982.

SPACE PROGRAM AND TELEVISION

The American Space Program and the American television industry contributed mightily to each other's growth. Space missions have matched Hollywood productions for drama, suspense and excitement, and have consistently pulled in some of the medium's largest audiences.

America's first astronauts were among television's first celebrity heroes. Some television journalists, such as Walter Cronkite and ABC's Jules Bergman (1930–1987), became famous for their chronicalling of the space program. The 69-year-old Cronkite even applied to become an astronaut in 1986 (as part of NASA's short-lived "journalist in space" program).

The Soviet Union's Sputnik satellite launch in 1957 was one of the earliest big stories for television news, then growing rapidly in popularity and influence. With the framing of the Sputnik story as crisis, an affront to American superiority and a military threat, the U.S. government justified a strong response, a crash program to beat the Soviets to space. Unfortunately, several of the earliest uncrewed U.S. rocket tests did just that—crash— further heightening the crisis atmosphere as each major attempt was anxiously reported on the 15-minute national evening newscasts.

Eventually American satellites were launched successfully, and in 1959 seven military pilots were chosen for the astronaut corps. Television, egged on by the print press, elevated the astronauts to hero status, as celebrated as Hollywood's leading stars. Publicists from NASA, the new civilian space agency, worked to fuel that perception. They schooled the seven in on-camera behavior and prohibited military uniforms, to the astronauts' discomfort but to the benefit of the program's all-civilian image.

Immediately after the triumphant sub-orbital flight of Alan Shepard in May 1961 (following the orbital flight of Cosmonaut Gagarin), Vice President Johnson, with Defense Secretary Robert McNamara and NASA Administrator Webb, sent a report to President John F. Kennedy justifying the eventual forty billion dollar investment in a moon landing program. "The orbiting of machines is not the same as orbiting or landing of man...," they wrote; "It is man in space that captures the imagination of the world." So from its inception, the crewed space program had at its core a propaganda objective—capturing the world's imagination. With Johnson's report as ammunition and the political goal of justifying massive government projects and fulfilling his vision of a "new frontier", Kennedy went before Congress to challenge the nation to land a man on the moon before 1970.

The remaining five Mercury space flights (1961–1963) and ten Gemini flights (1965–1966) were covered virtually from launch to splashdown by adoring TV networks. Each mission promised new accomplishments, such as Ed White's first American spacewalk. For television news it was a welcome reprieve from the 1960s morass of assassination, war, and inner-city unrest. A favorite theme of television— the "horse-race"—here between the Soviet and American space programs, was prominent. However, by 1965 it was apparent that the Soviet's had no hope of putting someone on the moon, a fact that rarely entered the "space race" discourse.

The ideal marriage of space and television was not merely the result of political and ideological agendas nor technical and logistical circumstance, but of more resonant connections between the program and American cultural mythology. The space program was a Puritan narrative, with its crew-cut NASA technocrats tirelessly striving toward the Moon, and a Western narrative, with lone heroes conquering a formidable new frontier (from mostly western facilities). And as the parallel narrative to the Vietnam war, it offered a reassuringly benign, yet pow-

Astronaut Edwin E. Aldrin, Jr., Apollo 11 mission
Photo courtesy of NASA

erful government, while simultaneously reinforcing cold war fear (and the need for military spending) in demonstrating the awesome power of rockets.

In 1967 three astronauts died in an early Apollo program test and the theme of astronaut as hero was tragically revived, and the public reminded of the risks of conquering space. But the first of the Apollo flights (1968–1972) were enormously successful, including the Christmas, 1968, first lunar orbits by Apollo 8. The astronaut's reading from the Book of Genesis while in lunar orbit made for stirring television, but firmly anchored the NASA TV spectacular as a believers-only enterprise. In July 1969, the space TV narrative reached its climax, as the networks went on the air nearly full time to report the mission of Apollo 11, the first lunar landing. 528 million people around the world—but not in the Soviet Union—marvelled at Apollo 11 on TV.

As with other Apollo missions providing TV coverage from the spacecraft, informal visits with the astronauts were highly scripted, using cue cards. Second moon walker Edwin Aldrin suggested the United States Information Agency scripted Apollo Eight's Bible reading and Neil Armstrong's first words from the lunar surface. Whether Armstrong said, "That's one small step for man," or "a man", as he intended (with the article "a" lost to static), has never been resolved. The blurry black-and-white images of Armstrong jumping onto the lunar surface and the short surface explorations by Armstrong and Aldrin are widely regarded as television's first, and perhaps greatest, example of unifying a massive worldwide audience in common wonder and hope.

After the Apollo 11 television spectacular, coverage of the following moon missions became increasingly brief and critical. Under considerable pressure to begin cutting back, NASA eliminated the last three planned Apollo missions, terminating the program with Apollo 17 in 1972. NASA actually paid the networks to cover the last Apollo mission (NASA official Chris Kraft, Jr., quoted in Hurt, 282). Coverage was spectacular nonetheless, from the nail-biting return of the explosion-crippled Apollo 13 spacecraft, to the lengthy moon walks and moon buggy rides of the last Apollos, covered live with color cameras. Such a part of American culture was NASA of the 1960s that it routinely provided technical assistance and advice to Hollywood, as with the many permutations of *Star Trek*, or provided entire series storylines, as with *I Dream of Jeanie*. Footage from NASA's massive film library appears in all manner of productions.

Television coverage of the long-duration Skylab missions (1973–1974) provided entertaining images of astronaut antics in weightlessness, but was overshadowed by the Watergate hearings. Watergate signalled an end of the trust of government and hero worship characterizing the 1960s space program. NASA could no longer sell its heroes and expensive programs to the public. The heroism of ex-astronauts was often dismantled by the same media which had constructed it, as astronauts were exposed in shady business deals or dysfunctional lives, criticized for making commercials, or doubted in new corporate and political roles. Television could not accept the astronaut as human.

Interest in space was occasionally revived in the 1970s by spectacular NASA accomplishments. In 1976 America enjoyed the extraordinary experience of seeing live pictures of the Martian surface as they arrived from the Viking lander—a visual thrill rivalling coverage of Apollo 11. In subsequent years the Voyager and Pioneer spacecraft had close encounters with the outer planets of the solar system, sending back dazzling images. But television coverage outside of regular newscasts was minimal. As Johnson's report had predicted, television reported the accomplishments of NASA's incredible robot explorers, but reserved its greatest excitement for crewed missions.

Between the last Skylab mission and the first Space Shuttle orbital mission in 1981, the only crewed space flight was Apollo-Soyuz in 1975, an odd public relations stunt intended as a tangible demonstration of detente with the Soviet Union. The orbital link up of three astronauts with two cosmonauts was entertaining if unimpressive by lunar mission standards, but NASA public relations was heavy handed. The mission was highly scripted and choreographed for a potential international television audience of a half billion. As Walter Cronkite noted, "This is one mission when we can truly say the television picture is as important as the mission itself, because that is the picture of detente." It was important in the Soviet Union, where it was the first space mission on live television.

The first Space Shuttle test landings over California were covered live, with NASA providing remarkable pictures from chase planes as Enterprise (named after pressure from

Star Trek fans) separated from its 747 mother plane and glided to Earth. Coverage of the long delayed first Shuttle space flight in 1981 was as abundant as in 1960s missions, and occasionally reminiscent of 1960s coverage for its cold war rhetoric—including the breathless reporting of a Soviet spy ship lurking off the coast as the Shuttle Columbia returned from orbit.

Coverage of the space shuttle rapidly diminished, and live coverage of missions had ended long before the 25th shuttle mission on 28 January 1986. On that day the shuttle Challenger, with a crew of seven including school teacher and media darling Christine McAuliffe, exploded after lift off like a daytime fireworks display. As President Reagan would speculate and the media would faithfully repeat, TV became America's "electronic hearth," a common gathering place to seek understanding and solace. Television was unprepared for such a tragedy, with speechless anchors, an unfortunate tendency to repeat the videotape of the explosion constantly, and irresponsible speculation about the possibility of survivors. But as shared national tragedy, it was an event like none other.

Thanks in part to television, the history of the American space program and its role in American life (including the dramatic acceleration of technological development which resulted, to which the television industry itself owes much), has never been completely written. Television presented fleeting spectacles, devoid of analysis, perspective, and retrospective. Because America saw the Space Program as television program, there was little demand for deeper analysis in journalism and literature. Only since the 1970s have writers and scholars attempted to specify the place of the space program in American culture. While television may have obscured issues, it presented such unforgettable images that few people who witnessed Apollo 11, Viking, or Challenger on TV have forgotten where they watched. But with nearly seventy space shuttle missions to date, the space program has now become too ordinary for television.

—Chris Paterson

FURTHER READING

Aldrin, Edwin, and Wayne Warga. *Return to Earth.* New York: Random House, 1973.

Atwill, William. *Fire and Power: The American Space Program as Postmodern Narrative.* Athens, Georgia: University of Georgia Press, 1994.

Axthelm, Pete. "Where Have All the Heroes Gone?" *Newsweek* (New York), 6 August 1979.

Carpenter, M. Scott, with others. *We Seven.* New York: Simon and Schuster, 1962.

Collins, Michael. *Carrying the Fire.* New York: Ballentine, 1974.

Hurt, Harry, III. *For All Mankind.* New York: Atlantic Monthly Press, 1988.

Kauffman, James. *Selling Outer Space.* Tuscaloosa: University of Alabama Press, 1994.

Life in Space. Alexandria, Virginia: Time-Life, 1983.

Martin, Ann Ray. "Getting the Picture." *Newsweek* (New York), 28 July 1975.

O'Toole, Thomas, and Jim Schefter. "The Bumpy Road that Led Man to the Moon." *The Washington Post* (Washington, D.C.), 15 July 1979.

Vladimirov, Leonid. *The Russian Space Bluff.* London: Tom Stacey Ltd., 1971.

Wolfe, Tom. *The Right Stuff.* New York: Farrar, Straus, Giroux, 1979.

See also Cronkite, Walter; Satellite

SPAIN

Five national channels now serve a Spanish population of 39.4 million and a television audience of 29.2 million. Of these five, two, TVE-1 and TVE-2, are state-owned, financed by subsidy and advertising. Antena-3 and Telecino are private channels, financed by advertising; Canal+ is private and financed by subscription.

Eight regional channels also contribute to the Spanish television environment: TV-3 and Canal 33 (financed by advertising and subsidy of the Catalan government); Canal Sur (financed by advertising and subsidy of the Andalusian government); Telemadrid (property of the Madrid regional government, financed by advertising and bank loans); Canal 9 (financed by advertising and subsidy of the Valencian government); TVG (financed by advertising and subsidy of the Galician government); ETB-1 and ETB-2 (financed by advertising and subsidy of the Basque government). Projects for cable television in the year 2000 speculate that 3 million TV households will be connected with 1 million subscribers. By that year, Spain will have exceeded nine decades of broadcasting.

In 1908, the Spanish government enacted a law that gave the central state the right to establish and exploit "all systems and apparatuses related to the so-called 'Hertzian telegraph,' 'ethereal telegraph, 'radiotelegraph,' and other similar procedures already invented or that will be invented in the future." Scattered experiments in radiowave communication evolved into regular broadcasts by 1921 with such events as Radio Castilla's program of concerts from the Royal Theater of Madrid. In 1924, the first official license for radio was granted, and all experimental stations were ordered to cease broadcasting and request state authorization. The first "legal" radio broadcast began in Barcelona and, like most radio programs that preceded the Spanish Civil War (1936-39), it was started up by private investors to make a profit. The broadcasting law of 1934 defined radio as "an essential and exclusive function of the state" and was amended in 1935 to confirm that all "sounds and images already in use or to be invented in the future" would be established and exploited by the state.

The government of the Second Republic (1931-39) kept centralized control over spectrum allocation and the diffusion of costly high-power transmitters, while it encouraged independent operators to install low-power transmitters for local radio. Radio spread with investments in urban zones, and only one significant private chain, the Union Radio, showed signs of economic concentration. The conditions of the Spanish Civil War (1936-39) halted the growth of independent radio when broadcasters were transformed into voices of military propaganda on both sides of the conflict. The leader of the fascist insurgents, Francisco Franco, ordered the nationalization of all radio stations under the direction of the new state, and the existing collection of transmitters merged into a state-controlled network called Radio Nacional de Espana. Use of the distinct idioms of Basque, Catalan, Galician was outlawed, and new laws aimed at the press gave the Ministry of the Interior full power to suppress communication which "directly, or indirectly, may tend to reduce the prestige of the Nation or Regime, to obstruct the work of the government of the new State, or sow pernicious ideas among the intellectually weak."

The first public demonstration of television took place in Barcelona in 1948 as part of a promotion by the multi-national communications firm Philips. Experiments continued until October 1956, when the first official TV broadcast appeared on an estimated six hundred television sets in Madrid—the program consisted of a mass conducted by Franco's chaplain, a speech by the Minister of Information and Tourism commemorating the twenty year regime, and a French language documentary. Much of the early programming came from the U.S. Embassy, but there were also live transmissions of variety and children's shows, and a news program was started in 1957. By 1958 there were approximately thirty thousand TV sets in Madrid. From the beginning, Television Espanola (TVE) was supported by advertising, although it also received subsidies derived from a luxury tax on television receivers. In 1959, TVE reached Barcelona via terrestrial lines, where a second studio was soon installed. At the end of the decade, there were fifty thousand sets in use. Through Eurovision, Spanish viewers joined European viewers in an audience of some fifty million, and one of the first images they shared was the historic meeting in Madrid between Franco and Eisenhower. By 1962, TVE claimed its sole VHF channel covered 65% of the Spanish territory and was viewed regularly by one percent of the population.

Television was a strictly urban phenomenon at this time, and there were only two production centers, one in Madrid and one in Barcelona. Transmissions originated from Madrid and were relayed in one direction to the rest of the territory. In 1964, a modern studio and office building were erected in Madrid to commemorate the 28th anniversary of the regime, and a year later, a second channel (TVE-2, UHF) with production studios located in Madrid and Barcelona, began testing. In 1965, the luxury tax on television sets was eliminated, making advertising the major resource for TVE-I and TVE-2. Estimates put yearly advertising investment in television at $1 million by the early 1960s, while airtime increased from 28 to 70 hours a week between 1958 and 1964, rising to 110 hours in 1972. Advertising income for TVE multiplied one-hundred times between 1961 and 1973, reaching estimated totals of over $100 million.

In the early 1970s, new regional centers were constructed in Bilbao, Oviedo (Asturias), Santiago de Compostela (Galicia), Valencia, and Seville (Andalusia). The entire system was finally united with radio in 1973 and was placed under the management of one state-owned corporation, Radio Television Espanola (RTVE). The regional circuit was wired into a highly centralized network in which all regional broadcasts were obliged to pass through Madrid. The only centers with the capacity to produce programs of any length were those in Barcelona and the Canary Islands. Though the records of RTVE management during the Franco dictatorship are unreliable, one study for 1976 reported that the Barcelona center contributed 3% of the total broadcast hours, followed by the center at Las Palmas in the Canary Islands at 2.9%. The rest transmitted a negligible amount of 1.8 to 1.85% of the total. The one way flow from the center to the regions was an effect of the Franco regime's centralism, which kept the regional centers (other than Barcelona and Las Palmas) from connecting with Madrid.

Television in Spain changed radically in the years following the death of Francisco Franco in 1975. In 1980, the government enacted a reform statute which established norms to ensure that a plurality of political parties would control RTVE. The Statute of RTVE also stipulated that broadcasting should be treated as an essential public service and that it should defend open and free expression. The Statute called for the upgrading of the regional circuit with a view to this becoming the basis for a network of television stations operated by regional governments, whose recognition in the constitution of 1978 was part of the reorganization of Spain as a "State of the Autonomies." The parliaments of the newly formed autonomous governments of the Basque Country and Catalonia founded their own television systems—the Basques in May 1982, the Catalans a year later. These actions resulted in the most decisive change in the broadcast structure since radio was nationalized during the Spanish Civil War, as they contravened existing laws that gave the central state the right to control all technology using the electromagnetic spectrum. In response, the central government enacted the Third Channel Law in 1984 in order to regulate the establishment of any additional networks in the regions.

The Third Channel Law was designed to stabilize the process of decentralization of the television industry, and it was based in the principle of recognition for the cultures, languages, and communities within the Spanish territory—suppressed during the forty-year Franco dictatorship. The law stipulated that regional networks remain under the state's control and within the RTVE infrastructure. Parliaments in Catalonia, the Basque Country, and Galicia resisted control by the central state and set up technical structures that ran parallel to, but separate from, the national network. Despite ongoing legal battles between the central state and the regions over rights of access to regional airwaves and rights of ownership of the infrastructure, eleven autonomous broadcast companies have been founded, six of which were broadcasting regularly by 1995. In 1989, the directors of these systems agreed to merge into a national federation of autonomous broadcasters, known as the Federation of Autonomous Radio and Television Organizations (FORTA).

Between 1975 and 1990, Spanish television emerged from a system of absolute state control to a regulated system in which both privately- and publicly-owned channels compete for advertising sales within national and regional markets. This structure was completed with the development of the 1988 law and technical plan for private television. The law furnished three licenses for the bidding of private corporations, a three-phase framework for the extension of universal territorial coverage, and restrictions on legal ownership to promote multiple partnerships, rather than monopoly control, and to limit foreign ownership. The technical plan created an independent public company, Retevision, to manage the network infrastructure, abolishing RTVEs economic and political control over the airwaves. Today all broadcasters must pay an access fee to use the public infrastructure. Regular transmissions from the private companies began in 1990.

The signals of state-owned Television Espanola cover 98.5% of the territory with its first channel and 94.7% with its second. Privately owned stations, Antena-3 and Telecinco, cover 80% of the territory, as does the subscription service Canal+, which has 1 million subscribers. On the regional scale, TV-3 and Canal 33 cover Catalonia with Catalan language programs, having significant spillover into contiguous regions and parts of France, reaching beyond their official audience of 5.8 million. Canal Sur covers the Andalusian audience of 6.7 million.

Telemadrid, owned by the regional government of Madrid, reaches an official audience of 4.8 million. Valencia's Canal 9's 3.7 million viewers can watch programs in Valenciano, a language similar to Catalan. Signals of TVG in Galicia spill over into northern Portugal and parts of Asturias in Spain, taking Galician language programming to more than the region's 2.6 million viewers. ETB-I and ETB-2 cover the Basque Country, and parts of surrounding provinces to reach beyond the official audience of 2 million; notably ETB-I broadcasts in the Basque language (Euskera), while ETB-2 does so in Spanish.

Ninety-eight percent of Spanish households have a television set, 86% have a color receiver (in contrast, 76% have a radio). In 1980, only one percent of Spanish households had a VCR, today 42% of them do, and in over 10% of them a video is watched each day. On average, Spaniards watch about three and a half hours of television a day, mostly in the afternoon and late evening hours. They are shown films 25.9% of the time, followed by series (15.4%), kids programs (12%), news (11.2%), musicals and variety shows (10%), sports (9.2%), and game shows and other programs (16.3%). Since 1993, the categories of programs they watched most often were soccer, sitcoms, reality TV shows, tabloid interview shows, and films or teleseries. The largest audience in every yearly account watches a soccer match on TVE. In 1993, the second and third largest audiences watched live broadcasts of political debates between the Spanish president and the opposition leader on private TV channels. A reality-TV show on TVE-I, *Quien sabe donde*, based on the American sensationalist format of true crime and human curiosities, consistently ranks among the top five most watched programs. Also among the leading formats is *Lo que necesitas es amor*, which spotlights "plain folks" and their concerns and pleasures about intimacy and sexuality. The American films or teleseries most watched in 1994 were *Pretty Woman, Scarlett, Doctor Quinn*, and *Police Academy Two*; a Spanish teleseries bettered the American competition once in 1994, though with a smaller audience than *Pretty Woman*. Spaniards also like to watch a situation comedy called *Farmacia en guardia*, about neighborhood life that centers around a family-run pharmacy. These preferences vary in each of the six regions where regional broadcasters compete with national programming.

The period 1990 to 1994 shows a trend of equalization of audience shares among the major national networks, with decreases in TVE-1 and TVE-2 probably caused by increases in Antena-3 and Telecinco. TVE-2's decline began with the establishment of the regional systems, though in the most recent "war over audiences," TVE-2 lost significant numbers to the private channels. On the regional scale, the companies of the autonomous communities have retained a stable audience, though the aggregate figures hide the dominance of the Catalan (TV-3 and Canal 33) and Madrid (Telemadrid) systems within FORTA. Figures for municipal and local television stations (there are over 100 in Catalonia alone) are as they are as yet insignificant on the national register.

TVE-1, Telecinco, and Antena-3 attract over 70% of the advertising investments made in commercial television in Spain. Antena-3 rose to the top of the ratings in 1994, an advance that translated into a 65% increase in its advertising revenues over figures for 1992. TVE's subsidy has not helped it overcome the growing debt of the company, despite its stable position in the market. In contrast, the private firms have been profitable. One reason for this is the presence of foreign and finance capital in their ownership structure, which support the firms with larger film and video libraries and easier access to foreign currencies. This support became increasingly important following the inflationary spiral that was initiated when European financial markets destabilized in June 1992—by the end of 1994, the value of the peseta had fallen about 30% against the values of the dollar and the deutsche mark, and the trend continued into 1995.

Telecinco is owned by Silvio Berlusconi (25%), the Leo Kirch Group of Munich (25%), Radiotelevision Luxembourg (19%), the French investor Jacques Hachuel (10%), the Bank of Luxembourg (8%), and Spanish investors, who hold the remainder. The ownership of Antena-3 TV is more complicated with Grupo Zeta and Renvir holding 25% each, the bank Banesto with 10%, the French company, Bouygues, with 15%, and Invacor and Corpoban sharing 25%. The Spanish investor, Antonio Asensio, controls nearly 70% of Grupo Zeta, while the banks which helped him finance this control, Banco Central Hispano and Banesto, hold about 12.5% each. Banesto's media holdings were being divested in 1995 after its president was arrested and charged with fraud and illegal trading. The British conglomerate Cable and Wireless is also a major shareholder of Bouygues Telecom. Antonio Asensio also has indirect holdings of Antena-3 through his investment shell companies Renvir, Corpoban, and Invacor. Canal+ has remained stable since its founding: 25% belongs to the Spanish media conglomerate, PRISA, 25% to Canal Plus France, with about 42% divided among Spanish banks. PRISA owns the largest daily newspaper in Spain, *El Pais*, as well as a leading popular commercial radio station. PRISA also has holdings in Britain (*The Independent*), Portugal, France, Germany (with Bertelsmann in the German pay-TV service, Premiere), and Mexico (*La Prensa*).

In anticipation of the enactment of a 1995 cable regulation, foreign and national firms are forming large consortia. Among the national firms positioning themselves for the future cable market are the leading banks, the largest electrical power companies, the national phone company, the national network Retevision, construction firms, regional press groups, the regional governments, and the private TV operators. Among foreign investors are Time Warner, US West, Sprint, TCI, Bell Atlantic, Cable and Wireless, and the various investors active in the commercial television market. Notable aspects of the draft legislation include municipal control over the demarcation of markets within cities, protection of intellectual property rights, and the stipulation that operators must carry and pay for the terrestrial output of all national and regional channels.

Audiovisual production from the U.S. accounts for practically all the imported programs on the public and private networks. Estimates for 1993 are that one out of every five programs on TVE-1, TVE-2, and Telemadrid is from the United States, the rest are Spanish. For Telecinco and Antena-3, two out of every five programs are from the United States, the rest are Spanish. These ratios show an improvement over 1990 figures when imports took up 40% of the program schedule on TVE-1, 33% on Andalusia's Canal Sur, 34% on Catalonia's TV-3, 35% on Galicia's TVG, and 39% on the Basque ETB-1. In 1990, Telemadrid showed twice as many U.S. programs as it did Spanish ones, while a ratio of one to one could be seen on Valencia's Canal 9, the Basque ETB-2, and the two private channels.

Language is a key characteristic of the Spanish TV culture. The regional firms in the Basque Country, Galicia, Catalonia, and Valencia were founded with the objective of fomenting the language and culture in the regions. In Galicia, 99% of the people understand Gallego, but only 14% actually prefer to watch TV in Gallego. Estimates are that 95% of the people in Catalonia understand Catalan, though only a third of the Catalans watch programs exclusively in the idiom. Up to 90% of the people in Valencia understand Valenciano, a linguistic cousin of Catalan, but 12% like TV only in Valenciano. In the Basque County, as many as half of the people claim to understand Euskera, but only one-fifth of the Basques show strong preferences for their TV in Euskera. These figures are dwarfed by the scale of the national population, where practically 100% of the people understand Spanish. Despite the linguistic, territorial, and financial limitations affecting the regional networks, they manage to retain a stable audience of viewers because of the political and cultural history of centralism in Spanish communication. Both for the managers and audiences of these systems, the presence of the local idiom alongside Spanish recalls the multilingual identity of the regions and helps sustain a sense of place as Spain positions itself within the European Union and opens its borders to globalized audiovisual production.

—Richard Maxwell

FURTHER READING

Bustamante, Enrique. "TV and Public Service in Spain: A Difficult Encounter." *Media, Culture and Society* (London), January 1989.

"Democracy by Television." *Economist* (London), 20 May 1989.

Maxwell, R. *The Spectacle of Democracy: Spanish Television. Nationalism, and Political Transition.* Minneapolis: University of Minnesota Press, 1994.

Report by the Think Tank on the Audiovisual Policy in the European Union. Luxembourg: Office for Official Publications of the European Communities, 1994.

Villagrasa, J. M. "Spain: the Emergence of Commercial Television." In, Silj, A., editor. *The New Television in Europe.* London: John Libbey, 1992.

SPANISH INTERNATIONAL NETWORK

The Spanish International Network (SIN) was the first Spanish language television network in the United States. From its inception in 1961, SIN was the U.S. subsidiary of Televisa, the Mexican entertainment conglomerate, which today holds a virtual monopoly on Mexican television, and is the world's largest producer of Spanish language television programming.

From the point of view of a U.S. entrepreneur in the early 1960s, the U.S. Spanish speaking population was so small and so poor a community that it was not considered a viable advertising market. The 1960 Census counted 3.5 million Spanish surnamed U.S. residents. The vast majority of this population were Mexican immigrants and Mexican-Americans living in the United States. (Large scale immigration from Puerto Rico, Cuba and other Latin American countries had not yet begun.) Spanish language advertising billed through the U.S. advertising industry amounted to $5 million dollars annually, less than one-tenth of one percent of all advertising expenditures at that time. From the perspective of a Latin American entrepreneur, however, this U.S. Latino audience was one of the wealthiest Spanish language markets in the world.

SIN was founded by Emilio Azcárraga, the "William Paley of Mexican broadcasting." Azcárraga was an entrepreneurial visionary, and owner of theaters and recording companies, who first built a radio, then a television empire in Mexico, before expanding it north of the border. SIN began with two television stations, KMEX, Los Angeles and KWEX, San Antonio, and from the beginning had national ambitions. In fulfilling these aims SIN pioneered the use of five communications technologies, the UHF band, cable television, microwave and satellite interconnections and repeater stations. All these applications contributed to rapid growth in the 1960s and 1970s, and by 1982 SIN could claim it was reaching 90% of the Spanish speaking households in the United States with 16 owned and operated UHF stations, 100 repeater stations and 200 cable outlets.

In these first decades, virtually every broadcast hour of each SIN affiliate was *Televisa* programming produced in Mexico: *telenovelas* (soap operas), movies, variety shows and sports programming. The vertical integration of Emilio Azcárraga's transnational entertainment conglomerate gave tremendous economic advantages to early U.S. Spanish language television. The performers under contract to Azcárraga's theaters and recording companies also worked for his television network. In other words, SIN programming had covered costs and produced a profit in Mexico, before it was marketed in the United States.

After 1981, and the start of satellite distribution of its programming, SIN began producing programs in the United States. The network created a nightly national newscast, the *Noticiero Univisión*, and national public service programming such as voter registration drives. It also provided coverage of U.S. national events such as the Tournament of Roses parade and the Fourth of July celebrations. The larger network-owned stations also began airing two hours a day of locally produced news and public affairs programming. This programming represented a limited recognition by SIN that the United States and Mexican television audiences had different needs and interests. Moreover, it was an attempt to modify the SIN audience profile from that of a "foreign" or "ethnic" group interested only in Mexican programming, to that of a more "American" community participating in the same national rituals as the mainstream consumer market. Perhaps SIN's most enduring contribution to U.S. culture was its leading institutional role in the creation of a commercially viable, panethnic, national Hispanic market.

The entrepreneurial financial and marketing acumen displayed by Emilio Azcárraga (and since 1972 by his son and heir Emilio Azcárraga Milmo) in the creation and development of SIN, were matched by his legal skills in maneuvering around U.S. communications law. The Communications Act of 1934 simply and explicitly bars "any alien or representative of any alien . . . or any corporation directly or indirectly controlled by . . . aliens" from owning U.S. broadcast station licenses. For Azcárraga and his SIN associates, perhaps the most salient part of this law is what it did *not* address. It does not prohibit the importation or distribution of foreign broadcast signals, or programming. In other words, U.S. law does not limit foreign ownership of broadcast networks; it does bar foreign ownership of the principal means of dissemination of the programming, the broadcast station. On paper, and in files of the Federal Communications Commission (FCC), none of the SIN stations or affiliates was owned by Emilio Azcárraga or *Televisa*. Rather, the foreign ownership prohibition was avoided by means of a time-honored business stratagem known, in Spanish, as the "*presta nombre*," which translates literally to "lending a name," or in colloquial English, a "front." SIN stations were owned by U.S. citizens with long professional and familial ties to Azcárraga and *Televisa*, with Azcárraga retaining a 25% interest (the limit permitted by law) in the SIN network.

Though long a subject of criticism by Latino community leaders and would-be U.S. Spanish language television entrepreneurs, the foreign control of SIN was not successfully challenged until the mid-1980s when a dissident shareholder filed a complaint with the Federal Communications Commission. In January 1986 the FCC ordered the sale of SIN. The FCC action was met with much excited anticipation by U.S. Latino groups who felt that for the first time since its creation 25 years earlier, there was a possibility that U.S. Spanish television would be controlled by U.S. Latino interests.

Several U.S. Latino investor groups were formed, but ultimately the bid (for $301.5 million) of Hallmark, Inc., of Kansas City, Missouri, the transnational greeting card company, received FCC approval. Hallmark changed the network's name to Univision, pledging to keep the network broadcasting in Spanish. Under the terms of the sale, *Televisa*, in addition to cash, was given a guaranteed U.S. customer (the new network, Univision, was given a right of first refusal for all *Televisa* programming), free advertising (for its records and tapes division) on Univision for two years, and 37.5% of the profits of its former stations for two years. After a quarter century, SIN, the Spanish International Network, ceased to exist as a corporate entity, leaving a significant cultural and economic legacy: a commercially viable U.S. Spanish language television network, and a new U.S. consumer group, the Hispanic market.

—America Rodriguez

FURTHER READING

All About the SIN Television Network. New York: SIN Television Network, 1984.

Arrarte, Anne Moncreiff. "And Galavision Makes Three." *Advertising Age* (New York), 12 February 1990.

Conference on Telecommunications and Latinos. Stanford, California: Stanford Center for Chicano Research, 1985.

"Hispanic Broadcasting Comes of Age" (special section). *Broadcasting* (Washington, D.C.), 3 April 1989.

Navarrete, Lisa, and Charles Kamasaki. *Out of the Picture: Hispanics in the Media: State of Hispanic America, 1994.* Washington, D.C.: Policy Analysis Center, Office of Research Advocacy and Legislation: National Council of La Raza, 1994.

Sobel, Robert. "Where's Spanish TV Going." *Television-Radio Age* (New York), 23 November 1987.

"Spanish Spending Power Growing Dramatically, But Consumers Retain Special Characteristics." *Television-Radio Age* (New York), 10 December 1984.

Stilson, Janet. "New SIN Prexy Sets Sights Skyward: Aims to Double TV Web Revenue." *Variety* (Los Angeles), 24 September 1986.

See also Univision

SPECIAL/SPECTACULAR

The television special is, in many ways, as old as television itself. Television specials are (usually) one-time only programs presented with great network fanfare and usually combining music, dance, and comedy routines (or "bits") presented in a variety format. When television was still new, specials were common, in that weekly, ongoing shows were expensive to produce and not yet proven as tools for securing long-term viewer loyalty. Hence, early television schedules did contain many one-time presentations, such as "The Damon Runyon Memorial Fund" (1950, TV's first telethon hosted by Milton Berle), the "Miss Television USA Contest" (1950, won by Edie Adams), "Amahl and the Night Visitors" (1951, the first *Hallmark Hall of Fame* program), and the "Ford 50th Anniversary Show" (1953, featuring duets between stage stars Mary Martin and Ethel Merman).

But the TV special entered its greatest and most prolific phase in 1954 when genius programmer Sylvester "Pat" Weaver conceptualized what he called television "spectaculars." These one of a kind, one-night broadcasts were Weaver's attempt to bring new and larger audiences and prestige to the television medium and to his network, NBC.

Breaking with the format of television at that time, the spectaculars regularly pre-empted the normal network program schedule of sponsored weekly shows. Weaver's move was a controversial gamble—to forgo sponsorship by single companies (basically money in the bank for the network) on these nights in order to regain air time on the Mondays, Saturdays and Sundays of every fourth week for the presentation of his spectaculars. Instead, following his trademark "magazine" formula for sponsorship, Weaver sold different segments of each spectacular to different sponsors, in the process laying the foundation for the future of multiple sponsorship and commercials on all of U.S. television.

In creating his spectaculars, Weaver drew on the talents of three producers—Fred Coe, Max Liebman, and Albert McCleery. Coe created his works for *Producer's Showcase*, airing on Mondays, Liebman for his series *Max Liebman Presents* on Saturdays, and Albert McCleery on Sundays for *Hallmark Hall of Fame*. Under Weaver and his team of producers the spectacular could be a musical extravaganza (such as *Peter Pan*, with Mary Martin repeating her Broadway triumph), or a play (such as Coe's *Our Town* with Paul Newman, Eva Marie Saint and Frank Sinatra), or a dramatic film (such as Olivier's *Richard III*).

In time, spectaculars became known by the less hyperbolic term "special" and generally they were shortened in length; most lasting only one hour as opposed to the ninety minutes to three hours sometimes taken by NBC. For the most part, specials took on a lighter tone, becoming variety oriented, with the emphasis on music, dance and elaborate production numbers. This era of the special saw the presentation of such benchmark television offerings as *Astaire Time* with Fred Astaire and Barrie Chase (1960), *Julie and Carol at Carnegie Hall* with Julie Andrews and Carol Burnett (1962), *My Name Is Barbra* starring Streisand (1964), and *Frank Sinatra: A Man and His Music* (1964).

These types of programs continued successfully into the late 1960s and 1970s featuring such diverse talents as Carol Channing, Bill Cosby, Elvis Presley, Liza Minnelli, Lily Tomlin, Shirley MacLaine, Bette Midler, Ann-Margaret, Olivia Newton-John, Tom Jones and Carol Burnett, who often paired herself with the other performers such as Beverly Sills, Dolly Parton or Julie Andrews. Throughout this period, stars

of contemporary television programs such as Lynda Carter, Cheryl Ladd and Ben Vereen also headlined occasional hour-long specials, frequently with substantial ratings success.

As the weekly variety show all but disappeared from network television (*The Carol Burnett Show*, TV's last successful variety show, ceased in 1978), the trend also signaled the beginning of the decline for the television music-dance special. As audiences began to select their musical entertainment from other media, or in shorter forms such as the music video, the hour-long, star centered special began to appear dated. At the same time, the shows were proving too expensive to produce in relation to their ratings.

Currently, with the exception of such yearly traditions as award shows, Christmas specials, pageants such as Miss USA, annual NBC installments by the unsinkable Bob Hope, and NBC's *This Is...* series (which have so far spotlighted Michael Bolton and Garth Brooks, among others), the television special/spectacular is now the domain of channels other than ABC, CBS, FOX, or NBC. PBS, for example, at times presents films of Broadway musicals and pay-cable stations such as HBO, the site of Barbra Streisand's most recent concert special, will air the highly touted entertainment event. Increasingly, pay-per-view is becoming the purveyor of the made-for-television extravaganza, having so far offered audiences the musical talents of David Hasselhoff and an extremely popular and profitable concert by the country music duo the Judds. In the world of 50-channel television, then (not to speak of the 500 channel universe), it is difficult to know what events might qualify as "special," harder still to identify the truly "spectacular."

—Cary O'Dell

FURTHER READING

Bailey, Robert Lee. *An Examination of Prime Time Network Television Special Programs: 1948-1966.* New York: Arno, 1979.

Terrace, Vincent. *Television Specials: 3,201 Entertainment Spectaculars, 1939-1993.* Jefferson, North Carolina: McFarland, 1995.

See also Coe, Fred; *Peter Pan*; Programming; Weaver, Sylvester (Pat)

SPEIGHT, JOHNNY

British Writer/Producer

Johnny Speight is the creator of the BBC series *Till Death Us Do Part*, upon which the U.S. series *All in the Family* (CBS) was based. As controversial in its time and place as was *All in the Family*, Speight's creation spawned a generation of relevant, hard-hitting sitcoms both in the United States and England.

A former factory worker and jazz musician, Speight began writing for television in 1956. In 1966, after serving

as head writer for the *Arthur Haynes Show*, Speight launched *Till Death Us Do Part*. The series revolved around the different values and beliefs held by blue-collar bigot Alf Garnett and his liberal son-in-law Mike. Originally committed to shows about the family itself, Speight maneuvered *Till Death* to more relevant social issues. Norman Lear, who was working in feature films at the time, saw the series, and, with partner Bud Yorkin, he optioned the series for their company Tandem Productions. The resulting hit was *All in the Family*, which debuted on CBS in 1971.

Speight's more controversial episodes prompted the Conservative Central Office to ask for advance copies of the *Till Death* scripts. When Speight refused, the matter was soon dropped. In 1968, Speight produced a BBC movie version of the series, and, in 1972, he also penned a short-run revival of the series. During that run, the series reached 24 million viewers, making it the most popular show in Britain.

Speight has written several plays, including *If there Weren't Any Blacks You Would Have to Invent Them*, which was produced in seventeen countries. He has also won numerous awards.

—Michael B. Kassel

JOHNNY SPEIGHT. Born in Canning Town, London, England, 2 June 1920. Attended St. Helen's Roman Catholic School, London. Married: Constance Beatrice Barrett, 1956; children: one daughter and two sons. Worked in a factory, then as a jazz drummer and insurance salesman; writer, BBC radio and television, from 1956; created the sitcom *Till Death Us Do Part*. Recipient: Screenwriters Guild Award, 1962, 1966, 1967, 1968; Prague Festival Award, 1969; *Evening Standard* Award, 1977; Pye Award, for television writing, 1983. Address: Fouracres, Heronsgate, Chorleywood, Hertfordshire, England.

TELEVISION SERIES (selection)

1960–66	*Arthur Haynes Show*
1966–75	*Till Death Us Do Part*
1969	*Curry and Chips*
1972	*Them*
1973	*Speight of Marty*
1979	*The Tea Ladies* (with Ray Galton)
1980	*Spooner's Patch* (with Ray Galton)
1982	*The Lady Is a Tramp*
1985	*In Sickness and in Health*
1989	*The 19th Hole*

TELEVISION SPECIALS

1961	*The Compartment*
1962	*Playmates*
1963	*Shamrot*
1965	*If there Weren't any Blacks You'd Have to Invent Them*
1967	*To Lucifer a Sun*
1970	*The Salesman*
1975	*For Richer...For Poorer*

FILMS (writer)

French Dressing, 1964; *Privilege*, 1967; *Till Death Us Do Part*, 1968; *The Alf Garnett Saga*, 1972; *The Secret Policeman's Third Ball*, 1987.

FILMS (actor)

The Plank, 1967; *The Undertakers*, 1969; *Rhubarb*, 1970.

RADIO (writer)

The Edmondo Ross Show, 1956–58; *The Morecambe and Wise Show*, 1956–58; *The Frankie Howerd Show*, 1956–58; *Early to Braden*, 1957–58; *The Deadly Game of Chess*, 1958; *The April 8th Show (Seven Days Early)*, 1958; *The Eric Sykes Show*, 1960–61.

STAGE (writer)

Mr. Venus, 1958; *The Art of Living* (with others), 1960; *The Compartment*, 1965; *The Knacker's Yard*, 1962; *Playmates*, 1971; *If there Weren't any Blacks You'd Have to Invent Them*, 1965; *The Picture* (with others), 1967; *The Salesman*, 1970; *Till Death Us Do Part*, 1973; *The Thoughts of Chairman Alf*, 1983.

PUBLICATIONS

It Stands to Reason: A Kind of Autobiography. Walton-on-Thames: Hobbs, 1973.

The Thoughts of Chairman Alf: Alf Garnett's Little Blue Book; or, Where England Went Wrong: An Open Letter to the People of Britain. London: Robson Books, 1973.

Pieces of Speight. London: Robson Books, 1974.

The Garnett Chronicles: The Life and Times of Alf Garnett, Esq. London, 1986.

See also *Till Death Us Do Part*

SPELLING, AARON

U.S. Producer

Aaron Spelling is one of television's most prolific and successful producers of dramatic series and made-for-television films. Spelling began his career as a successful student playwright at Southern Methodist University, where he won the Eugene O'Neill Award for original one-act plays in 1947 and 1948. After graduating in 1950 and spending a few years directing plays in the Dallas area, and then trying less than successfully to make his way on Broadway, Spelling moved to Hollywood. There he initially found work as an

actor and later as a scriptwriter for such anthology and episodic series as *Dick Powell's Zane Grey Theater*, *Playhouse 90*, *Wagon Train*, and *The Jane Wyman Theater*. Within a few years, Spelling had become a producer at Four Star Studio Productions, where he created *The Lloyd Bridges Show* (1962–63), *Burke's Law* (1963–66), *Honey West* (1965–66), and helped develop *The Smothers Brothers Show* (1967–75).

Spelling's first really successful series, *Mod Squad* (1968–73), was produced after he left Four Star and formed a partnership with Danny Thomas. During its five-year run, *Mod Squad* earned six Emmy Award nominations, including one for outstanding dramatic series of the 1969–70 season. In 1972, Spelling formed a new partnership with Leonard Goldberg, which lasted until 1977 and produced such hits as *The Rookies*, *Starsky and Hutch* (1972–76), and *Charlie's Angels* (1976–81).

Spelling's series featuring both wealthy crime fighters and regular cops continued in the 1980s with *Hart to Hart* (1979–84), *Matt Houston* (1982–85), *Strike Force* (1981–82), *T.J. Hooker* (1982–87), and *McGruder and Loud* (1985). But Spelling also ventured into new genres with his innovative hour-long comedy, *Love Boat,* and the prime-time serial *Dynasty.* Reminiscent of the 1960s anthology comedy, *Love, American Style,* Spelling's *Love Boat* turned the three separate comedy stories into three intertwined storylines. Intercutting three separate plots in short scenes which recapitulated and advanced each storyline plot was a brilliant strategy that enabled the series to appeal to different sets of viewers, each of whom might be attracted to a particular plotline, within a format that was admirably suited to the fragmented and distracted way that most people view television. Another Spelling innovation which first appeared in *Love Boat* was the ritualized introductory sequence that formally presented the multiple plots in each week's episode as well as the series' main characters.

In 1980s television, Spelling was king. In 1984, Spelling's seven series on ABC accounted for one-third of the network's prime-time schedule, leading some critics to rename ABC "Aaron's Broadcasting Company." Spelling's 18-year exclusive production deal with ABC ended in 1988, but his ability to create hit series did not; in the 1990s, he introduced *Beverly Hills 90210* and *Melrose Place.*

Among the recurring thematic features that have characterized Spelling's productions over the years are socially relevant issues such as the disaffected militant youth of the 1960s, institutional discrimination against women, racism, and homophobia; altruistic capitalism; conspicuous consumption and valorization of the wealthy; the optimistic, moralistic maxims that people can be both economically and morally successful; good ultimately triumphs over evil; the grass often looks greener but rarely is; and the affirmation of the "caring company" work family (e.g., in *Hotel*) as well as the traditional kinship family. Stylistically, his productions have included high-key lighting, gratuitous displays of women's bodies, heavily orchestrated musical themes, lavish sets, and what Spelling himself thinks is the most important element in television—"style and attention to detail."

One Spelling series which stands out as truly anomalous among this auteur's prime- time and movie ventures is *Family* (ABC, 1976–80). Spelling and Mike Nichols co-produced this weekly hour-long drama, which many consider to be his best work. During the four years that this serious portrayal of an upper middle-class suburban family was in first run, it won four Emmy Awards for the lead performers and was twice nominated for outstanding drama series.

"Innovator," "over-achiever," "spin doctor," "angel," "king of pap," "ratings engineer," "TV's glitzmeister," win-

Aaron Spelling
Photo courtesy of Spelling Television, Inc.

ner of six NAACP Awards—whatever other labels Spelling's critics and admirers have used to describe this prolific, successful producer, one which certainly describes the unique signature Aaron Spelling has left on four decades of television is that of television auteur.

—Leah R. Vande Berg

AARON SPELLING. Born in Dallas Texas, U.S.A., 22 April 1923. Educated at the Sorbonne, Paris, 1945–46; Southern Methodist University, Dallas, Texas, B.A. 1950. Married: 1) Carolyn Jones, 1953 (divorced, 1964); 2) Carole Gene Marer, 1968; one daughter and one son. Served in U.S. Air Force, 1942-45, decorated with Bronze Star Medal, Purple Heart with oak leaf cluster. Actor, from 1953, appearing in 50 television shows and 12 films; began career as a writer after selling script to *Zane Grey Theater,* worked in production, Four Star, 1956–65; co-owner, with Danny Thomas, Thomas-Spelling Productions, 1968–72; co-president, Spelling-Goldberg Productions, 1972–77; president, Aaron Spelling Productions, Inc., Los Angeles, 1977–86, chair and chief executive officer, since 1986. Member: board of directors, American Film Institute; Writers Guild of America; Producers Guild of America; Caucus of Producers, Writers and Directors; Hollywood Radio and TV Society; Hollywood TV Academy of Arts and Sciences; Academy of Motion Picture Arts and Sciences. Recipient: Eugene O'Neill Awards, 1947 and 1948; six National Association for the Advancement of Colored People (NAACP) Image Awards;

named Man of the Year by the Publicists Guild of America, 1971; named Man of the Year by Beverly Hills chapter of B'Nai B'rith, 1972, 1985; named Humanitarian of the Year, 1983; named Man of the Year by the Scopus Organization, 1993. Address: Spelling Television Inc., 5700 Wilshire Boulevard, Los Angeles, California 90036-3659, U.S.A.

TELEVISION SERIES (selection; producer)

1956–62	*Dick Powell's Zane Grey Theater* (writer only)
1959–60	*Johnny Ringo*
1959–61	*The duPont Show with June Allyson*
1963–65	*Burke's Law*
1965–66	*Amos Burke—Secret Agent*
1968–73	*The Mod Squad*
1975–79	*Starsky and Hutch*
1976–81	*Charlie's Angels*
1976–80	*Family*
1977–86	*The Love Boat*
1978–84	*Fantasy Island*
1981–89	*Dynasty*
1983–88	*Hotel*
1984–85	*Finder of Lost Loves*
1985–87	*The Colbys*
1986	*Life with Lucy*
1989	*Nightingales*
1990–	*Beverly Hills 90210*
1992–	*Melrose Place*
1994	*Winnetka Road*
1994–95	*Models, Inc.*
1995–	*Savannah*
1995	*Malibu Shores*

MADE-FOR-TELEVISION MOVIES

1976	*The Boy in the Plastic Bubble*
1977	*Little Ladies of the Night*
1981	*The Best Little Girl in the World*
1993	*And the Band Played On*

FILMS (selection; producer)

Mr. Mom, 1983; *'night, Mother*, 1986; *Surrender*, 1987; *Cross My Heart*, 1987; *Soapdish*, 1991.

PUBLICATION

Spelling, Aaron, with Jefferson Graham. Aaron Spelling: A Prime Time Life. New York: St.Martin's, 1996.

FURTHER READING

Bark, E. "Aaron Spelling." *The Dallas* (Texas) *Morning News*, 12 March 1989.

Budd, M., S. Craig, and C. Steinman. "*Fantasy Island*: Marketplace of Desire." *Journal of Communication* (Philadelphia, Pennsylvania), 1983.

Carson, T. "The King of Pap." *Village Voice* (New York), 12 April 1994.

Coe, S., and D. Tobenkin. "Aaron Spelling: TV's Overachiever." *Broadcasting and Cable* (Washington, D.C.) 23 January 1995.

Davis, I. "He's baaaaaack! TV's King of Jiggle, Aaron Spelling." *Los Angeles Magazine*, April 1991.

Friedman, D. "Fox Sees Spelling as an Angel." *Variety* (Los Angeles), 6 January 1988.

Goldenson, Leonard. *Beating the Odds*. New York: Scribners, 1991.

Grover, Ronald. "Is Aaron Spelling Still in His Prime Time?" *Business Week* (New York), 17 April 1989.

Marc, D., and R.J. Thompson. *Prime Time, Prime Movers*. Boston: Little Brown, 1992.

Marc, D. "TV Auteurism." *American Film* (Washington, D.C.), November 1981.

Pesman, Sandra. "Finding '90s on 'Winnetka Rd'" (interview). *Advertising Age* (New York), 11 April 1994.

Quill, G. "Spin Doctor Spelling Aspires to the Heights of Hyperbole." *The Toronto Star*, 26 August 1992.

Schwichtenberg, C. "A Patriarchal Voice in Heaven." *Jump Cut* (Berkeley, California), 1984.

Seiter, E. "The Hegemony of Leisure: Aaron Spelling Presents *Hotel*." In, Drummond, P. and R. Paterson, editors. *Television in Transition*. London: British Film Institute, 1985.

Swartz, M. "Aaron Spelling: Entertainment's King of the Jiggle." *Texas Monthly* (Austin), September 1994.

———. "Aaron Spelling: The Trash TV Titan." *Texas Monthly* (Austin, Texas), September 1994.

Thompson, H. "Messing with *Texas*: TV Glitzmeister Aaron Spelling Tries to Wake up Michener's Epic Snooze." *Texas Monthly* (Austin, Texas), August 1994.

Thompson, R. "*Love Boat*: High Art on the High Seas." *Journal of American Culture* (Bowling Green, Ohio), 1983.

See also *Beverly Hills 90210*; *Charlie's Angels*; *Dynasty*; Melodrama; *Starsky and Hutch*

SPIN-OFF

The spin-off is a television programming strategy that constructs new programs around characters appearing in programs already being broadcast. In some cases the new venue is created for a familiar, regular character in the existing series (e.g. *Gomer Pyle, U.S.M.C.* from *The Andy Griffith Show*). In others, the existing series merely serves as an introduction to, and promotion for, a completely new program (*Mork and Mindy* from *Happy Days*).

The most famous examples of the spin-off surround the work of producer Norman Lear and that of the pro-

Diff'rent Strokes

The Facts of Life

ducers working at MTM Productions during the 1970s. A list of the originating programs with their spin-offs reads like a genealogy of popular television comedy. Thus, *All in the Family* begat *Maude*, which begat *Good Times*, and *The Jeffersons*, which begat *Checking In. All in the Family* also begat *Gloria*, which lasted only one season and begat nothing.

The Mary Tyler Moore Show begat *Phyllis*, *Rhoda*, and *Lou Grant*, and though none of these "offspring" engendered specific shows of their own, their producers went on to create numerous programs with the distinctive style of these earlier works.

Other prolific sources of spin-offs were *The Danny Thomas Show*, the source of *The Andy Griffith Show*, which led to *Gomer Pyle, U.S.M.C.,* and *Mayberry, R.F.D.* From *Happy Days* the list includes *Laverne and Shirley, Joannie Loves Chachi, Mork and Mindy,* and *Out of the Blue.* As should be clear from these lists, a spin-off is no guarantee of success. For every *Wanted: Dead or Alive* (from *Trackdown*), there is a *Beverly Hill Buntz* (from *Hill Street Blues*).

The existence of spin-offs can lead to puzzling problems when one considers the relations among programs across the schedule. The long-running prime-time serial, *Knots Landing*, for example, was a spin-off of *Dallas*, the most famous example of that genre. During the famous 1985-86 season of *Dallas*, the season that was "dreamt" by Pamela Ewing (Victoria Principal), various events on *Knots Landing* occurred in response to Bobby Ewing's (Patrick Duffy)

"death." Yet no one on *Knots Landing* troubled to explain how the history of their own fictional world might be altered by the fact that a "year in the life of *Dallas*" never occurred.

In any instance, spin-offs attest to television's constant demand for new, if not always different, material. This demand often leads to mindless repetition and the most meager attempts to cash in on previous success. While spin-offs may lead to new sources of creativity in their own right, the result of applying this strategy is often no more than a program that temporarily fills a time slot.

Indeed, it should be noted that spin-offs often result from producers' financial arrangements. Successful producers frequently contract for future commitments from studios or networks. New shows constructed around proven, popular characters offer obvious advantages in these arrangements. Similarly, the existence of a successful program offers the producer and the network a ready-made billboard for advertising new work. Characters from the new production may appear in no more than a single episode of the ongoing program in order to be introduced to a large audience.

A final version of the spin-off is related to variations on a program franchise or formula, variations that often cross national boundaries. It is important to remember that *All in the Family* and *Sanford and Son*, two of the most highly acclaimed shows produced by Norman Lear, were copies of British productions, *Till Death Us Do Part* and *Steptoe and Son*, respectively. Currently, the most prominent examples in the United States are the international versions of *Wheel of Fortune.* Licensed by the parent company, Merv Griffin Productions, to producers in other countries, some form of *Wheel* is popular from France to Taiwan, from Norway to Peru. In each country small variations are created to express particular cultural expectations and attitudes. Because game shows are cheaply and easily produced, this type of the spin-off concept is likely to expand.

—Horace Newcomb

SPITTING IMAGE

British Puppet/Satire Programme

The premiere of *Spitting Image* opened with a puppet caricature of Israel's prime minister Menachem Begin wearing a magician's outfit. With a flourish, he produced a dove of peace from his top hat, then announced, "For my first trick . . ."—and wrung its neck.

This was the first of many outrages perpetrated on the British public, who were either offended or delighted each Sunday evening from 1984 to 1992. *Spitting Image* was roundly condemned for its lampooning of the Royal family: the queen was portrayed as a harried housewife, beset by randy, dullard children and screaming grandkids. Britain's most cherished figure, the Queen Mother, appeared as a pleasant, if somewhat boozy great-grams.

The Conservative leadership was a constant target: Margaret Thatcher's puppet was a needle-nosed Reagan groupie who consulted with Hitler on immigration policy and sold off England's infrastructure to baying packs of yuppies and her eventual successor, John Major, was portrayed as a dull, totally grey man who ate nothing but peas. The opposition Labour leaders, including Neil Kinnock as "Kinnochio," were pilloried for their inability to challenge decades of Tory rule.

In spite of its detractors, over 12 million viewers (a quarter of England's adult population) watched *Spitting Image* on Central Independent Television, a subsidiary of ITV. Its spin-off records, books, comics and videos sold in the million. It won an International Emmy for "Outstanding Popular Arts" program in the 1985–86 season, and a franchised edition appeared on Moscow television.

Spitting Image originated with Peter Fluck and Roger Law, who first met at Cambridge School of Art. They became involved in the liberal politics favored by art students, through which they met another student, Peter Cook. In 1961, Cook fronted England's flowering of political satire by starring with Dudley Moore in the revue "Beyond the Fringe," which inspired the TV program *That Was the Week That Was.* Cook employed Law as an illustrator for his projects such as the satire magazine *Private Eye* and a political comic strip in the *Observer* newspaper. Fluck and Law built separate careers in magazine illustration, and Law took two commissions in the music business that

Spitting Image

Spitting Image

Spitting Image

yielded classic album covers: The Jimi Hendrix Experience as Hindu deities for "Axis: Bold as Love," and "The Who Sell Out," for which Roger Daltrey posed sitting in a bathtub filled with baked beans.

Peter and Roger each began working with sculpted caricatures, creating several images that appeared in London's *Sunday Times Magazine,* where Law had become an artistic director and reporter. In 1975, they formed a partnership, spooneristically named Luck and Flaw, to turn out their 3-D portraits for outlets like *The New York Times Sunday Magazine,* Germany's *Stern,* international editions of *Time,* and the *National Lampoon.* The work proved barely profitable until 1981, when Martin Lambie-Nairn invited them to lunch.

Lambie-Nairn was a graphic designer at London Weekend Television. He thought a political television program using puppets or animation might be a good investment, and he proposed to front Fluck and Law the capital for a pilot episode (thus the credit at the end of each episode: "From an original lunch by Martin Lambie-Nairn."). The pilot took two years to complete.

The pair quickly decided the show should use puppets, which, like the Muppets, required two operators, for the face and one arm. Jim Henson, in fact, turned down an offer to collaborate on the puppet workshop. The first puppet designs were bogged down by expensive, heavy electronics needed just to make their eyes move. After several months without any film being shot, Fluck cobbled together a simple mechanism using steel cable and air bulbs. They also picked

up Tony Hendra of the *National Lampoon* (and later of *Spinal Tap*) as a writer, and their producers: Jon Blair, a producer of current affairs programming, and John Lloyd of the *Not the Nine O'Clock News. Spitting Image,* the pilot's title, exhausted the resources of several backers, including computer executive Clive Sinclair, before it was completed at a cost of 150,000 pounds, a record for a light entertainment program.

In its first season, *Spitting Image* focused exclusively on politics, and played to mediocre ratings. For the next round, Fluck and Law were obliged to caricature entertainment and sports figures as well, and the show's fortunes immediately improved. They worked out a schedule in which they spent the off-season stockpiling non-topical segments such as music video parodies (in one, Barry Manilow was all nose; another showed off Madonna's singing belly button). Each episode had a window of six minutes for fresh political commentary, written and taped the night before its broadcast.

The *Spitting Image* parodies reached a status not unlike that of *Mad* magazine in the early 1960s, when many of those whom the show caricatured took it as a sign that they had "made it." While Thatcher has only commented, "I don't ever watch that program," members of the House of Commons had tapes of each show delivered to them the following Monday, and former Tory Defense Minister Michael Heseltine tried to purchase his puppet.

The commercial broadcaster Central Television gave *Spitting Image* few censorship problems. BBC radio, however, refused to play their first spin-off record, with a Prince Andrew imitator boasting "I'm Just a Prince Who Can't Say No." "The Chicken Song," however, a single that parodied the singalong ditties that infest pub jukeboxes and vacation discotheques every summer, reached number one on the charts.

The influence of U.S. politics on the British scene was apparent in frequent lampoons of Ronald Reagan. American news outlets excerpted a video with Ron and Nancy as Leaders of the Pack, singing "Do Do Ron Ron." The befuddled Reagan also appeared in a serial thriller, "The President's Brain is Missing," and was featured prominently in the Spitting Image-produced video for Genesis' song, "Land of Illusion." In September 1986, NBC aired a two-part original *Spitting Image* special in which the secret arbiters of fame, including Bill Cosby and Ed McMahon, hatch a clandestine plot to have an over-muscled Sylvester Stallone elected president.

Spitting Image projects continue to appear on both sides of the Atlantic. American VCRs can play a compilation of their music videos, a puppet production of "Peter and the Wolf," and a mock documentary, "Bumbledown: the Life and Times of Ronald Reagan" (a double-feature with the musical, "The Sound of Maggie!"). Most recently, the group has collaborated with American cable channel, Comedy Central, to illustrate a book by Glenn Eichler, *Bill and Hillary's 12-Step Recovery Guide*. The book is promoted through a series of commercial cutaways on the cable channel, featuring the puppet Clinton family.

—Mark R. McDermott

CAST

Puppets by . . . . Peter Fluck, Roger Law
Voices by . . . . Chris Barrie, Steve Nallon, Enn Reitel,
. . . . . Harry Enfield, Pamela Stephenson, Jon Glover,
. . . . Jan Ravens, Jessica Martin, Rory Bremner,
. . . . Kate Robbins, Hugh Dennis

PRODUCERS David Frost, Jon Blair, John Lloyd, Geoffrey Perkins, David Tyler, Bill Dare

PROGRAMMING HISTORY 89 30-minute episodes; 3 Specials

• ITV

26 February 1984–17 June 1984
6 January 1985–24 March 1985
5 January 1986–9 February 1986
30 March 1986–4 May 1986
14 September 1986
28 September 1986–2 November 1986
1 November 1987–6 December 1987
17 April 1988
29 October 1988
6 November 1988–11 December 1988
6 May 1989
11 June 1989–16 July 1989
12 November 1989–17 December 1989
13 May 1990–24 June 1990
11 November 1990–6 December 1990
10 November 1991–15 December 1991
8 April 1992
12 April 1992–17 May 1992
4 October 1992–8 November 1992

FURTHER READING

Heller, Steven. "Spitting Images." *Print* (New York), May-June, 1986.

Iyer, Pico, and John Wright. "Stringing Along." *Time* (New York), 28 April 1986.

Law, Roger, with Lewis Chester, and Alex Evans. *A Nasty Piece of Work: The Art and Graft of Spitting Image*. London: Booth-Clibborn Editions, 1992.

"Major Threat." *The Nation* (New York), 5 April 1985.

O'Neil, Thomas. *The Emmys: Star Wars, Showdowns, and the Supreme Test of TV's Best*. New York: Penguin, 1992.

Robinson, Andrew. "*Spitting Image*: Lampoon from Limehouse." *Sight and Sound* (London), Spring 1984.

Waters, Harry F. "Stringing Up the Celebrated." *Newsweek* (New York), 1 September 1986.

Wolf, Matt. "Ace Puppeteers Pull No Punches with Clintons." *Chicago Tribune*, 30 August 1995.

See also British Programming

SPONSOR

Television in the United States is a profit-maximizing set of entities, an industry whose success is largely measured by its ability to deliver viewers to advertisers. The lure of television is its programs; commercial broadcasters seek shows of optimal value (be it in terms of ratings generated or demographics attracted) in order to maximize advertising revenue. The sponsor—the organization, corporation, institution, or other entity willing to pay the broadcaster revenue in exchange for the opportunity to advertise on television—stands at the center of program strategies. This situation requires recognition of the complex interrelationship between television networks and advertisers, two industries whose differing responsibilities and sometimes conflicting needs produce the programming that draws the audience to the advertisement. In U.S. television, the economic and industrial systems supporting these arrangements have their beginnings in radio broadcasting.

The emergence of radio in the early 1930s as an astonishingly effective means of delivering consumers to producers attracted an array of enthusiastic advertisers, and soon

"Speedy" for Alka-Seltzer
Photo courtesy of Bayer Corporation

the radio schedule was dominated by shows named for their sponsors—the *Chase and Sanborn Hour*, the *Cliquot Club Eskimos*, and the *Maxwell House Concert*, for example. Produced for their clients by such advertising agencies as J. Walter Thompson and Young and Rubicam, the single-sponsored program was a staple of commercial broadcasting; it was an article of faith that if a listener identified a show with its sponsor, he or she was more likely to purchase the advertised product.

Although agency involvement in television was little more than tentative prior to 1948, advertisers soon embraced the new medium with great fervor; *Pabst Blue Ribbon Bouts*, *Camel Newsreel*, and the *Chesterfield Supper Club* were testimony to the steadfast belief in sponsor identification. However, as program costs soared in the early 1950s, it became increasingly difficult for agencies to assume the financial burdens of production, and even the concept of single sponsorship was subject to economic pressure.

By the 1952–53 season, television's spiraling costs (an average 500% rise in live programming budgets from 1949 to 1952) threatened to drive many advertisers completely out of the market. Many sponsors turned to a non-network syndication strategy, cobbling together enough local station buys across the country to approximate the kind of national coverage a network usually provided. Television executives—most notably Sylvester L. "Pat" Weaver at NBC—countered sponsor complaints by championing the idea of participation advertising, or the "magazine concept." Here, advertisers purchased discrete segments of shows (typically one- or two-minute blocks) rather than entire programs. Like magazines, which featured advertisements for a variety of products, the participation show might, depending on its length, carry commercials from up to four different sponsors. Similarly, just as a magazine's editorial practice was presumably divorced from its advertising content, the presence of multiple sponsors meant that no one advertiser could control the program.

Even as agencies relinquished responsibility for production, they still maintained some semblance of control over the content of the programs in which their clients advertised, a censorship role euphemistically referred to as "constructive influence." As one advertising executive noted, "If my client sells peanut butter and the script calls for a guy to be poisoned eating a peanut butter sandwich, you can bet we're going to switch that poison to a martini." Still, this type of input was mild compared with the actual melding of commercial and editorial content, a practice all but abandoned by the vast majority of agencies by 1953.

Despite Madison Avenue's initially hostile reaction, participation advertising ultimately became television's dominant paradigm for two reasons. One was purely cost; purchasing 30- to 60-minute blocks of prime time was prohibitively expensive to all but a few advertisers. More importantly, participations were the ideal promotional vehicle for packaged-goods companies manufacturing a cornucopia of brand names. While it is true that the magazine concept opened up television to an array of low-budget advertisers, and thus expanded the medium's revenue base, it was companies like Procter and Gamble that catalyzed the trend (ironically, given that Procter and Gamble today has operational control over two soap operas, the last vestige of single-sponsored shows on television). Further, back-to-back recessions in the mid-1950s provided an impetus for the producers of recession-proof goods to scatter their spots throughout the schedule; their subsequent sales success solidified the advent of participation on the schedule. Without the economic rationale of single sponsorship, most advertisers chose to circulate their commercials through many different shows rather than rely on identification with a single program.

By 1960, sponsorship was no longer synonymous with control—it now merely meant the purchase of advertising time on somebody else's program. While sponsor identification remained important to such advertisers as Kraft and Revlon, most sponsors prized circulation over prestige; as a result, fewer agencies offered advertiser-licensed shows to the networks. The quiz scandals of 1958-59, often identified as the causative factor in network control of program procurement, were in actuality only a coda.

Ironically, it was the networks' assumption of programming control that resulted in a narrower and more conservative conception of program content, with a greater reliance on established genres and avoidance of technical or narrative experimentation. In an effort to provide shows that would offend no sponsor, network television's attempts to be all things to all advertisers drained the medium of its youthful vigor, plunging it into a premature middle age. By appealing to target audiences—at least in the early 1950s—advertisers were in many ways more responsive and innovative than the networks.

While the vestiges of single sponsorship remain in, of all places, public television—*Mobil Masterpiece Theater*, for example—advertisers still wield enormous, if indirect, influence on program content. For example, in 1995 Procter and Gamble, the nation's largest television advertiser, announced that it would no longer sponsor daytime talk shows whose content the company considered too salacious. Today's marketers believe they can influence programs through selective breeding, bankrolling the content they support and pulling dollars from topics they do not.

—Michael Mashon

FURTHER READING

Allen, Robert C., editor. *Channels of Discourse, Reassembled.* Chapel Hill: University of North Carolina Press, 1992.

Barnouw, Erik. *The Sponsor: Notes on a Modern Potentate.* New York: Oxford University Press, 1978.

Bellaire, Arthur. *TV Advertising: A Handbook of Modern Practice.* New York: Harper, 1959.

Boddy, William. *Fifties Television: The Industry and Its Critics.* Urbana: University of Illinois Press, 1990.

Cantor, Muriel B. *Prime-Time Television: Content and Control.* Beverly Hills, California: Sage, 1980.

Ewen, Stuart, and Elizabeth Ewen. *Channels of Desire.* Minneapolis: University of Minnesota Press, 1992.

Kepley, Vance. "The Weaver Years at NBC." *Wide Angle* (Athens, Ohio), April 1990.

See also Advertising; Advertising, Company Voice; Advertising Agency; "Golden Age" of Television; Programming; Sustaining Program

SPORTS AND TELEVISION

The history of sports on U.S. television is the history of sports on *network* television. Indeed, that history is closely related to the development and success of the major television networks. "Television got off the ground because of sports," reminisced pioneering television sports director Harry Coyle. He continued, "Today, maybe, sports need television to survive, but it was just the opposite when it first started. When we (NBC) put on the World Series in 1947, heavyweight fights, the Army-Navy football game, the sales of television sets just spurted."

With only 190,000 sets in use in 1948, the attraction of sports to the networks in its early period was not adver-tising dollars. Instead, broadcasters were looking toward the future of the medium, and aired sports as a means of boosting demand for television as a medium. They believed their strategy would eventually pay off in advertising revenues. But because NBC, CBS and DuMont manufactured and sold receiver sets, their more immediate goal was to sell more of them. Sports did indeed draw viewers, and although the stunning acceptance and diffusion of television cannot be attributed solely to sports, the number of sets in use in the U.S. reached ten and a half million by 1950.

Technical and economic factors made sports attractive to the fledgling medium. Early television cameras were heavy and cumbersome and needed bright light to produce even a passable picture. Boxing and wrestling, contested in confined, very well-lit arenas and baseball and football, well-lit by the sun and played out in a familiar, well-defined spaces, were perfect subjects for the lens. Equally important, because sporting events already existed there were no sets to build, no writers and actors to hire. This made sports inexpensive to produce, a primary concern when the audience was small and not yet generating large advertising revenues.

The first televised sporting event was a college baseball game between Columbia and Princeton in 1939, covered by one camera providing a point of view along the third base line. But the first network sports broadcast was NBC's *Gillette Cavalcade of Sports*, which premiered in 1944 with the Willie Pep vs. Chalky White Featherweight Championship bout. Sports soon became a fixture on prime-time network programming, often accounting for one third of the networks' total evening fare. But in the 1950s, as television's other genres matured and developed their own large and loyal (and approximately 50% female) followings, sports began to disappear from network prime-time, settling into a very profitable and successful niche on weekends. This, too, would change, like so much else in television, with alterations in the technology and economics of the medium.

Gillette Cavalcade of Sports stayed on the network air for 20 years, a prime example of sporting events presented by a single sponsor. By the mid-1960s, however, televised sports had become so expensive that individual advertisers found it increasingly difficult to pay for sponsorship of major events by themselves. Still, the number of hours of sports on network television exploded as the audience grew and the multiplying ranks of spot-buying advertisers coveted these valuable minutes. This mutually beneficial situation persisted until well into the 1980s when the historically increasing amounts of advertising dollars began to decline, and networks experienced diminishing profit margins on sports.

But the economics of televised sports had begun to unravel earlier. In 1970, for example, the networks paid $50 million to broadcast the National Football League (NFL), $2 million for the National Basketball Association (NBA) and $18 million for major league baseball. In 1985 those figures had risen to $450 million, $45 million and $160 million respectively. These large increases were fueled by growing public interest in professional sports, in part as a result of more and better television coverage. But equally important, the networks saw the broadcasting of big time sports as the hallmark of institutional supremacy in broadcasting. Major league sports meant major league broadcasting—not an unimportant issue for the networks now challenged by VCR, the newly empowered independent stations, and cable. Many of these cable channels were themselves carrying sports (WGN, WTBS, and HBO, for example), and one, ESPN, offered nothing but sports. Seemingly unconcerned, the CBS, NBC, and ABC attitude could be described as "Who cares about Australian Rules Football?" (a high point of early ESPN programming).

But rising fees for rights to major sporting events were not, in themselves, bad for the networks. They could afford them, and the cable and independent channels could not. But increasing rights fees, accompanied by falling ratings, proved to be disastrous. From 1980 to 1984, broadcasts of professional football lost 7% of their viewership (12% among men 18 to 34 years old) and baseball lost 26% of its viewers, showing a 63% decline among young males. Nonsports programming on cable, home video use, and the independents took many of these viewers. In addition, sports on the competing channels further diluted the remaining sports audience. To make up for falling revenues on all its programming as they began to lose audience, the networks began to raise the price of advertising time on sports shows to cover the huge rights fees contractually owed to the sports leagues.

Advertisers balked. Not only were they unwilling to pay higher prices for smaller audiences, but the once attractive male audience was becoming less desirable as working women came to control even larger amounts of consumer capital. Rather than pay what they saw as inflated rates for a smaller and now less prized set of viewers, many advertisers bought commercial time away from sports altogether, feeling they could reach their target audiences more efficiently through other types of shows. Car manufacturers turned to prime-time drama to reach women, who were increasingly making car-buying decisions; beer makers were turning to MTV to attract young women and young men.

Finally, in order to make the most of their expensive contracts with the major sports leagues, the networks began broadcasting more sports. But spots on sports shows would have been easier to sell had there been fewer of them on the market. The three networks together showed 1,500 hours of sports in 1985, double what they programmed in 1960. With about eight minutes of commercials an hour, the addition of even relatively few hours of programming had a noticeable effect on the supply-and-demand balance of the commercial spot market.

It was during this same period that superstations WTBS and WGN, and premium channel HBO began national, cable-fed sports programming. ESPN was launched in 1979 and by mid-1980 reached 4 million homes. By 1986, 37 million households subscribed. The glut of sports on television was abetted even more by crucial court decisions affecting intercollegiate competition. Universities, desirous of their own access to broadcast riches, successfully challenged the National Collegiate Athletic Association (NCAA) and, at times their own regional athletic conferences, to be free of what they considered restrictive television contracts and broadcast revenue-sharing agreements. College basketball and football, once local or regional in appeal, began appearing on the television dial in a complex array of syndication packages and school-centered or conference-centered television networks.

While the history of televised sports may have been directly related to network television, the current and future

NBC broadcasting a baseball game in the early 1950s

states of the genre certainly are not. There are more televised sports today than ever before and they continue to draw a large total audience, but it is an audience fragmented among many available choices. Sports on television, then, is decreasingly likely to originate on a national network. Despite the Super Bowl's annually growing audience and increases in the price of a 30-second spot ($1.3 million for some aired during the 1995 San Francisco/San Diego mismatch), it remains a television anomaly, unique as a television and cultural event. Ratings for individual television sports programs generally continue to decline in the 1990s. The 1993 World Series, for example, had a cumulative rating for all its games of slightly more than 16, surpassing only the 1989 Series interrupted by the Loma Prieta earthquake. Game 1 of the 1993 contest between Toronto and Atlanta was the lowest rated World Series game ever recorded by Nielsen. When CBS' four-year, $1.06 billion deal with major league baseball ended with that Series, the best network deal that the leagues could make was an arrangement with both ABC and NBC tying baseball's income to the amount of advertising sold. Baseball was forced into the business of selling advertising time for the networks (and therefore, for itself). Hockey's ratings on its ABC and ESPN/ESPN2 telecasts,

never big, also declined, from 0.9 in 1992 to 0.8 in 1993. The pattern is the same for football, basketball and the Olympics.

The Industrial Benefits of Televised Sports

This does not mean, however, that viewers no longer watch sports on television. But they have ceased to watch the big marquee events in numbers as great as they once did. Now cable channels and local independent stations have joined the networks as primary outlets for sports programming, and sports remains valuable and attractive to programmers for several distinct reasons:

Except for the big ticket events like the NCAA Basketball Championships, the Super Bowl, the World Series, the National Basketball Association (NBA) Championships and the major college football bowl games, televised sports generally produce smaller audiences than prime-time network programming. Of course, for independents and cable channels sports contests may often draw their biggest viewership. But regardless of the size of the audience for a sports telecast, it is audience composition that is important—this is the demographically attractive audience for advertisers who want to reach males, 18 to 49 years old. Certain sports also bring with them even more narrowly defined audience segments. The products advertised during a bowling match, a game of golf or an auto race, make it immediately clear that a particular demographic group is being targeted. And advertising rates for these events are usually well below those charged on more general-interest programming.

When the new technologies began to divide the television audience, huge rights fees to big sports leagues became a burden for the networks. But for cable channels, local broadcasters and even for certain events on the networks, sports are often cheaper to buy and air than much first-run programming. This is why many regional sports networks such as the Boston Red Sox Baseball Network have developed. This is why team- and conference-centered ad hoc networks (groups of stations that come together in a network for specific programming like the Pac-10 Football Network or the Big East Game of the Week) have grown in number. And this is why ballgames or boxing matches are programmed on cable channels such as HBO, WTBS, WGN, and TNT.

Sports is the only programming that has successfully attracted large audiences on a weekend day. This creation of regularized audience behavior enables the medium to maintain its role as a familiar aspect of "everyday" life.

Sports also link the medium into a system of cross-promotion. Newspapers, radio stations, even stations in competition with a channel airing the weekend's big game all provide free advertising in the course of their usual sports coverage.

Ultimately, the reason sports are popular with broadcasters is that they are popular with viewers. Even the 1993 "low rated" World Series drew an average of fourteen and a half million households a game. The National Hockey League's (NHL) tiny 0.8 rating per game translates into three-quarters of a million homes. And, as FOX Television's 1994 acquisition of the pro football rights from CBS makes clear, the networks still see the ownership of the rights to major league sports as tantamount to being in the Big Time. Owner Rupert Murdoch dismissed the $350 million loss for FOX Television in its first year of NFL broadcasts as "an investment" in altering audience perceptions of his low-rated fourth network.

The Appeals of Televised Sports

But why are these contests of skill, originally designed to test the abilities of the participants and then to delight those who attend, so popular from a distance, on an illuminated iridescent screen? A range of possible appeals may be involved in gathering the large audiences for sporting events.

Viewers identify with their team, their favorite players, those warriors who carry the good name of their city, college, conference, nation, ethnic heritage, or other characteristic, into battle. Sports offer real heroes and villains, as opposed to the fictional characters of televised drama and comedy. Fans become familiar with those real individuals and their teams, following them, learning about them, living and

dying with them, or, in the immortal words of *ABC Wide World of Sports*, experiencing with them the "joy of victory and the agony of defeat."

Sports on television is live television, it is history in the making, it is being "up close and personal" (again, thanks to ABC) as possibly momentous events unfold. To thrill in the victory of a favorite, to join the excitement of the moment in an exhilarating game or to learn more about the teams, players or games on television are among possible satisfactions that are obviously specific to sports on television.

The Aesthetics of Televised Sports

And no doubt a fourth reason why people watch televised sports is that the contests often make great television. Carlton Fisk's famous 1975 World Series homer, the American hockey victory over the Soviet Union team at the Lake Placid Olympics and the camera's sad attention to Thurman Thomas in the last quarter of the 1994 Super Bowl, its focus on the individual miscues that had led to a fourth straight Buffalo Bills defeat, are only three examples of the wonder that can be sports on television.

But what, specifically, makes an individual sporting event "good television?" As *Channels* writer Julie Talen wrote, "All sports are not created equal. The most popular sports on TV are those best served by the medium's limitations." What she means is that even if there are 20 cameras and 40 microphones at an event, the viewer still receives one picture and one set of sounds. Together these must convey a sense of what is happening in the actual contest. *Monday Night Football*'s long-time director, Chet Forte, argued, "It's impossible to blow a football game. . .Football works as a flattened sport. Its rectangular field fits on the screen far more readily than, for example, golf's far-flung woods and sand traps. The football moves right or left on the screen and back again. Its limited repertoire—kick, pass, and run—sets it apart from, say, baseball, where the range of possibilities for the ball and the players at any given moment is enormous." And CBS' top football director, Sandy Grossman, says, "The reason (the gridiron) is easier to cover is because every play is a separate story. There's a beginning, a middle, and an end, and then there's 20 or 30 seconds to retell it or react to it."

There are, in other words, certain characteristics of the different sports that make them better dramatic and visual matches for television, and in doing so, render them more popular with audiences.

The camera, and therefore the fans' attention, is repeatedly redirected to a specific starting point for each new play, serve, or pitch. This is what CBS' Grossman above called a "separate story." Therefore, football and baseball are better than hockey and soccer in providing a discrete starting point.

Tension can be sustained and viewer interest maintained if something crucial can occur at any moment. Any pitch can result in a home run or a fine running catch by the center fielder. Any pass can produce a touchdown or an interception. In contrast, the first three quarters of a basketball game usually serve only to set up the last three minutes and much of soccer's action happens at mid-field, yards and yards away from the goal (and a potential exciting save or game-changing score).

Baseball has innings, football has time-outs and quarters. Those covering and those watching the event can establish a rhythm that allows for the more-or-less natural insertion of commercials and visits to the refrigerator. Soccer has continuous action, as does hockey, which makes commercial insertion more complex.

Cameras and viewers have to be able to follow the object of interest on the field and on the small screen, respectively. Basketballs and footballs are big while hockey pucks and golf balls are quite small.

Television is a visual medium; it lives by the pictures it offers its viewers. Baseball and football offer spectacle—big, full, beautiful stadiums, lovely playing surfaces, the blimp, cheerleaders (football) and the bullpen (baseball). Golf presents the manicured scenery of country club settings and the occasional glimpse at windswept Scottish headlands. Tennis, by contrast, has a small rectangular court and bowling has a skinny lane of wood (though each has the beginning-middle-end story structure so desireable to directors).

Nothing adds to visual variety like physical action, people moving and competing. Basketball is ballet above the rim. In football there are incredible tests of strength and aggression. Tennis demands action defined by precision and endurance.

Fans follow players as well as teams and the camera is well versed in the close-up. Roone Arledge of ABC called this "sports as soap opera." Baseball gives us the tight shot of the pitcher's anxiety as he holds the runners on first and third or zooms in on the concentration in the basketball player's eyes as she shoots two from the charity stripe with the game on the line. In hockey it's much more difficult to provide close-up, personal video images because the players wear helmets and skate at 30 miles an hour, but fans can still be attached to individual personalities, waiting for the grudge-induced fistfight on the ice. And as the celebrity status of Arnold Palmer or Jack Nicklaus, or the glamorous intensity of a John McEnroe, a Martina Navratilova, or a Jimmy Conners attest, even the more sedate sports create a cult-like status for their superstars.

Of course, television prefers sports with wide interest because it assures more viewers and ad revenue; but this is a plus for sports fans as well. Surely many fans watch games between teams they would not typically follow. The outcome might affect their home-town favorites or they want to see that scrappy second baseman they've read so much about?

Televised Sports and "Real World" Sports

Fans may watch televised sports for many of these reasons, but this involvment is not without its costs. Here the difference between sports and television's other forms of programming becomes clearer. That is, unlike soap operas and situation comedies, sports exist apart from television. Major league baseball, for example, was born before radio was invented and developed its rules, traditions, nature and character apart from television. Moreover, sports are played in front of and for paying customers. This produces two important tensions. First, what have sports lost and gained from their wedding to television? Second, what have fans lost and gained?

The gains might be obvious. The leagues and athletes have prospered. More and more teams and tournaments are played in more and more cities and fill more and more television screens. Television has helped create tremendous interest and excitement for the public, turning the Super Bowl, for example, into something akin to a national celebration.

The losses, however, might be less obvious. Trying to explain dips in television ratings and attendance at games in the 1993 NFL season, for example, sports reporter Bud Geracie of the San Jose *Mercury News* wrote, "Terry Bradshaw (former NFL quarterback and Hall of Fame inductee) says that although Dallas is 'as good as any team that's ever played, the league as a whole isn't fun to watch.' Is this a temporary lull in the action, or a permanent condition? Is this (1993) NFL season the product of fluky misfortunes, or is it the beast born of parity... . The NFL wanted parity—and took measures to achieve it—and you can't argue with the logic or the success of the concept. The NFL wished to maximize the number of teams in play-off contention late in the season, thereby maximizing fan interest, TV ratings, revenues and the rest. This is what the NFL bosses sought, and this is what they got. What they seem to have lost in the process is the big game. 'There are no big games between 5-4 teams vying for wildcard spots,' said Bob Costas of NBC."

Television has also been instrumental in changing sports in other not-so-obvious ways, for example in the alteration, even the destruction, of traditional college sports conferences. In February 1994 four schools, the University of Texas, Texas A and M, Texas Tech and Baylor, left the 80 year old Southwest Conference to join another regional conference, the Big Eight. One goal was to cash in on ABC's promise to pay the newly expanded league between $85 million and $90 million for the next five years, with the promise of an additional $10 million if this new football "super-conference" developed a play-off.

Other schools in the former Southwestern Conference were left behind. Bubba Thornton, alumnus and track coach of one of the jilted schools, Texas Christian University, lamented in a *Sports Illustrated* interview, "What the Southwest Conference was about was small towns and big cities, Texans against Texans, wives and girlfriends dressing up, bragging rights, the Methodist preacher talking Sunday morning about beating the Christians [Church of Christ], all the things that keep you going. We were about tradition all these years instead of instant gratification and egos. This decision will come back to haunt us."

Such a view might be attributed to no more than nostalgia, a common aspect of sports in any medium. And certainly different critics' lists might vary. But here are several other "concessions" that fans and the games themselves have made to television: 1) games moved to awkward times of day to satisfy television schedules, ignoring fans who've bought tickets; 2) giant video screens in arenas and stadiums; 3) alteration of game rules, as in the creation of the "TV time-out" for television commercials; 4) free agency for players and consequent moves to the "highest bidder;" 5) pro teams moving to better "markets;" 6) wild-card games designed to increase playoff partipants; 7) expanded playoffs; 8) the 40-second shot clock in the NBA; 9) the designated hitter in the American League; 10) over-expansion in the professional leagues; 11) salary caps; 12) umpire and officials strikes; 13) recruiting abuses as college teams chase television riches; 14) the playing of World Series games at night in freezing October weather (Game 7 of the 1994 Series was scheduled for October 30); 15) electric lights in Wrigley Field; and 16) players strikes and lock-outs.

The 1994 World Series fell victim to baseball's labor problems, but at the root of the dispute that also killed the last half of that season was the inability of the sports' owners to resolve "revenue disparities" between the small and large television market teams. The 1994-95 National Hockey League season lost nearly half its contests as well as its All-Star Game to precisely the same dispute among its franchise owners.

Still, the future of sports on television is certainly one that promises more contests on the screen and more transformation of both the games and the medium. It's widely accepted, for example, that one reason the networks paid such large rights fees to the professional sports leagues throughout the 1980s was to keep them out of the hands of pay-per-view television programmers. In the short term the strategy was successful. But virtually every cable system in America offers at least boxing on a pay basis and the Sports Channel is, for all intents and purposes, pay television. What will happen to competitiveness, franchise stability, and scheduling as individual teams become star attractions? What will happen to the look of the broadcasts and the nature of the games if television "tickets" rather than advertising become the basis of program support? What will changes in the economics of the sports-television marriage mean to the teams, the medium and to the fans? What technological innovations (in covering the games and in distributing them) will we see and what might be their impact?

Answers to these questions are neither immediate nor obvious. But as a sports reporter might put it, one thing is certain—sports will continue to be closely intertwined with devleopments in television. Major events will continue to serve as national rituals. And audiences will continue to follow favorite teams and celebrity players, watching from a distance as the skills, the strength, the speed, and the tactics of athletes, coaches, and owners are pitted against one another on the screen in the home.

—Stanley J. Baran

FURTHER READING

Bellamy, Robert V., Jr. "Impact of the Television Marketplace on the Structure of Major League Baseball." *Journal of Broadcasting and Electronic Media* (Washington, D.C.), Winter 1988.

Bennett, Randall W. "Telecast Deregulation and Competitive Balance: Regarding NCAA Division I Football." *American Journal of Economics and Sociology* (New York), April 1995.

Berkman, Dave. "Long before Arledge...Sports and TV: The Earliest Years: 1937–1947—As Seen by the Contemporary Press." *Journal of Popular Culture* (Bowling Green, Ohio), Fall 1988.

Bierman, Jeffrey A. "The Effect of Television Sports Media on Black Male Youth." *Sociological Inquiry* (Haverford, Pennsylvania), Fall 1990.

Bryant, J., D. Brown, P.W. Comisky, and D. Zillman. "Sports and Spectators: Commentary and Appreciation." *Journal of Communication* (New York), 1983.

Chandler, J.M. *Television and National Sports: The United States and Britain.* Urbana, Illinois: University of Illinois Press, 1988.

Cox, Phillip M., II. "Flag on the Play? The Siphoning Effect on Sports Television." *Federal Communications Law Journal* (Los Angeles, California), April 1995

Gantz, Walter. "Men, Women, and Sports: Audience Experiences and Effects." *Journal of Broadcasting and Electronic Media* (Washington, D.C.), Spring 1991.

Hocking, J.E. "Sports and Spectators: Intra-audience Effects." *Journal of Communication* (New York), 1982.

McKay, J. *My Wide World.* New York: Macmillan, 1973.

Messner, Michael A. "Separating the Men From the Girls: The Gendered Language of Televised Sports." *Gender and Society* (Newbury Park, California), March 1993.

Neal-Lunsford, J. "Sport in the Land of Television: The Use of Sport in Network Prime-time Schedules 1946–50." *Journal of Sport History* (Radford, Virginia), 1992.

O'Neil, T. *The Game Behind the Game: High Pressure, High Stakes in Television Sports.* New York: Harper and Row, 1989.

Patton, P. *Razzle Dazzle: The Curious Marriage of Television and Professional Football.* Garden City, New York: Dial Press, 1984.

Powers, R. *Supertube: The Rise of Television Sports.* New York: Coward-McCann, 1984.

Rader, B.G. *In Its Own Image: How Television Has Transformed Sports.* New York: Free Press, 1984.

Rushin, Steve. "1954–1994: How We Got Here." *Sports Illustrated,* 16 August 1994.

Spence, J. *Up Close and Personal: The Inside Story of Network Television Sports.* New York: Atheneum, 1988.

See also Arledge, Roone; Australian Programming; Canadian Programming in English; Canadian Programming in French; *Hockey Night in Canada*; Grandstand; Ohlmeyer, Don; Olympics and Television

SPORTSCASTERS

The history of sportscasting, like almost everything else on television, is rooted in radio. Radio's first generation of great sportscasters—Graham McNamee, Ted Husing, Tommy Cowan, Harold Arlin, Ford Frick, and Grantland Rice—transformed the airwaves into an "arena of the mind" in which hyperbole would become honored as an art. McNamee, regarded as the first well-known play-by-play announcer, was unapologetic about sacrificing accuracy for excitement. His emphasis on enthusiasm, of course, lives on in the television performance of John Madden.

The radio days of sportscasting are notable as a period when sporting events would be "re-created" from bare-bones wire service reports. Announcers located in studios sometimes hundreds of miles away from the game site used sound effects and imaginative language to manufacture the impression that they were actually on location and describing the play-by-play action as it unfolded on the field of play. Perhaps the most notable conjurer of the illusion of sport re-creation landed his first job in the entertainment industry at WOC in Davenport, Iowa, as a football announcer. The year was 1932 and Ronald "Dutch" Reagan, who 48 years later would ride his oratorical skills into the White House as the 40th president of the United States, was paid $5 a game for his services. Many of the second generation of distinguished radio sportscasters—Mel Allen, Red Barber, Jack Brickhouse, Clem McCarthy, Lindsay Nelson, and Bill Stern—would later be prominent voices in television's first decades as a mass medium. Two of this group, Allen and Barber, were the first broadcasters to be enshrined in the Baseball Hall of Fame.

Sports programming played a central role in transforming television into a mass medium in the late 1940s, stimulating much of the initial demand for this expensive new technology. As much as 30% of the prime-time schedule was devoted to sports programming during this period.

In urban centers, many Americans' first television experience was watching an athletic contest on a set prominently displayed at the local tavern. Although roller derby and bowling also played well on the small screen, the sports best suited to the limitations of primitive television sets were boxing and professional wrestling. During this period when most television programs were sponsored by a single company, *Gillette Cavalcade of Sports* stayed on the air for 50 years, the longest continuous run of any television boxing show. The prototypical wrestling announcer, Dennis James, would become one of the first sportscasters to become known solely for his performances on television.

Sportscasting in the 1950s basically followed the radio pattern of enthusiasm. Sportscasters, and their employers, conceived of their role as being ambassadors of the game. As exemplified by the partisan commentary of Chicago Cub's announcers Bert Wilson and Harry Caray, many operated as boosters for a franchise. Point men in the team's public relations efforts, these sportscasters are often identified as the beloved "voice" of their organizations.

Howard Cosell

In the 1960s, television sports would be revolutionized by the advent of instant-replay technology. Introduced on 31 December 1963, during the Army-Navy football game, instant replay would figure prominently in enabling the phenomenal rise in popularity of televised football during the 1960s. Unseating boxing as the supreme made-for-TV sport, slow-motion replay technology made chaotic combat on the gridiron into an aesthetic experience in which even the most grotesque display of brutality (for example, the snapping of Joe Theismann's leg) became a thing of beauty, a kind of improvised ballet of violent masculinity. In 1964, immediately after the advent of instant replay technology, CBS paid an unprecedented $28 million dollars for television rights for NFL games and instantly recouped its investment with two $14 million sponsorship contracts—one with Ford Motor Company, the other with Philip Morris.

Establishing Sunday afternoons, once considered a "cultural ghetto," as a showcase for the masculine melodrama of professional football would set the stage for premiering what would eventually become the single most significant regularly-scheduled special event of the television year—the Super Bowl. Known then as the World Championship Game, the first match-up between the top teams in the AFL and NFL was played on 15 January 1967, and did not attract enough paying customers to fill the Los Angeles Coliseum game site—even with a local blackout of televised game coverage. The game was carried by both CBS and NBC. In pre-game promotions, both networks emphasized the excellence of their sportscasters. CBS offered its regular NFL announcer/analyst staff of Ray Scott, Jack Whitaker, Frank Gifford, and Pat Summerall, while NBC featured Curt Gowdy and Paul Christman. However, *The New York Times* found that the much ballyhooed competition between the two sportscasting staffs "proved to be malarkey . . . never were two networks more alike." Despite such negative reviews, the Super Bowl quickly caught on. In 1972, Super Bowl VI established a new record for the largest audience ever to watch an American TV program, a record since surpassed by several other Super Bowl telecasts.

An account of sportscasting in the 1960s and 1970s would not be complete without acknowledging the influence of Roone Arledge, the man who first penned the lines (on the back of an airline ticket), "the thrill of victory, the agony of defeat." In the early 1960s, as the architect of what was destined to become the longest running sports program on television, Arledge hired a young Baltimore announcer named James K. McManus to host *ABC's Wide World of Sports*. After McManus changed his name to Jim McKay, he would go down in broadcast history as the man who first informed the world about the tragic terrorist attack that took the lives of ten Israeli athletes at the 1972 Olympics in Munich, West Germany. As an ABC vice-president, Arledge played an instrumental role in the adoption and refinement of instant replay. Then, after being promoted to president of ABC Sports, Inc., Arledge would author yet another American institution, *Monday Night Football*. When Arledge's salary reached the one-million-dollar mark in 1975 he was considered the television industry's highest-paid executive.

Arledge's philosophy of sportscasting was established early on when, in 1961, he adopted a policy of not signing contracts that included the traditional announcer-approval clause. This policy made ABC the first network to allow critical commentary to accompany the play-by-play, a clear break with the obligatory boosterism of sportscasting's past. This philosophy is perhaps most clearly embodied in Arledge's underwriting of Howard Cosell's stormy career. Arledge designed the innovative narration of *Monday Night Football* around Cosell's quasi-journalistic commentary. Thanks to the Cosell factor, the bantering of the most memorable stars of *Monday Night Football*—Cosell, Frank Gifford, and Don Meredith—was both controversial and tremendously successful. Often humorous and sometimes rancorous disputes between Cosell and Meredith, refereed to a certain extent by Gifford, made even the most lopsided contest entertaining. While the public loved it when Meredith won an argument with a well-placed zinger, Cosell, near the end of his life, got the last laugh. In 1994, Cosell joined the elite ranks of Jim McKay, Lindsay Nelson, Curt Gowdy, Chris Schenkel, and Pat Summerall when the National Academy of Television Arts and Sciences recognized his

accomplishments with a Lifetime Achievement Award. In the post-Cosell era, the mantle of the sportscaster that Americans love to hate is now worn by Brent Musburger.

Although former Miss America Phyllis George is generally credited with breaking sportscasting's gender barrier in 1975 when she joined *The NFL Today* on CBS, at least one woman performed as a color commentator in the 1950s. Her name was Myrtle Power and she was signed by CBS after achieving brief celebrityhood as a baseball expert on *The $64,000 Question*. While women have continued to make inroads into sportscasting, it has not been without struggle. Perhaps the most confounding obstacle involves female access to male locker rooms, the subject of much controversy during the 1980s and 1990s. Jocko Maxwell, an African American who went to work as a sports announcer in 1935 for WHOM in Jersey City, New Jersey, is recognized as the person who first crossed sportscasting's color line. Recently, several African Americans have distinguished themselves as announcers and commentators, the most noteworthy and notorious being Irv Cross, Bryant Gumbel, Jayne Kennedy, Joe Morgan, Ahmad Rashad, and O. J. Simpson. Even so, racism, like sexism, is still alive in the world of television sportscasting as demonstrated in 1988 in a racist comment by Jimmy "the Greek" Snyder. A well-known odds-maker, Snyder's career as a game analyst on CBS was curtailed when it was reported that he attributed the competitive excellence of African American athletes to selective breeding.

With the launching of ESPN in 1979, sportscasting entered a new era. ESPN quickly became the first profitable basic cable network and is currently available in over 59 million U.S. television households making it the largest cable network. As the dominant national supplier of sports-related programming to cable systems, ESPN has had a profound impact on both the economic and stylistic dimensions of sportscasting. Largely because of ESPN's "March Madness" coverage, NCAA basketball's annual 64-team national championship tournament has joined the World Series and the Super Bowl as a mega-event on the television calendar. ESPN coverage is also responsible for cultivating a larger following for women's basketball and men's baseball at the collegiate level.

Stylistically, the ESPN era in sportscasting has been marked by what media critic Leslie Savan calls "the ironic reflex." This style, associated with the postmodern turn in popular culture, appeals to the "Bud Bowl" generation—a generation that pays more attention to the Super Bowl commercials than to the game itself. Rightfully or wrongfully, this generation believes that being conscious of the contrivances of television's commercialism makes it somehow immune to its manipulation. NBC's Bob Costas surely ranks as the most sophisticated and celebrated member of the new wave of hyper-cool sportscasters. Earning five Sports Emmy Awards between 1987 and 1994, Costas's postmodern credentials were validated when NBC scheduled his interview show, *Later with Bob Costas*, in the time slot immediately following the nightly performance of television's king of irony, David Letterman.

Three of ESPN's most familiar sportscasters put very different spins on the postmodern style. Chris Berman impersonates Cosell's "He...could...go...all...the...way!" when narrating football highlights on *Sportscenter*, the terrifically popular sports news show that appears daily on ESPN. However, Berman is best known as a punster whose nicknames for baseball players (e.g., Roberto "Remember the" Alomar, Greg Gagne "with a spoon," Wally "Absorbine" Joyner) command their own page on the World Wide Web. In contrast to Berman's playfulness, Keith Olbermann, another regular on *Sportscenter*, is the epitome of postmodern cynicism. As a sportscaster with an attitude, Olbermann carries on the Cosell tradition of opinionated commentary, but inflects it with the Costa-esque smirk or the Groucho-Marxian raised eyebrow. On ESPN's college basketball beat, Dick Vitale takes enthusiasm over the edge, the excessiveness of his hyperkinetic performance making Vitale a parody of himself.

And, yet, amidst all the winks, nudges, insincerity and excess of contemporary sportscasting, there is still a profitable place for good, old-fashioned enthusiasm. In 1993, after FOX outbid CBS for the rights to broadcast NFC football, John Madden negotiated a contract with Rupert Murdoch that would pay Madden $30 million over four years. With that deal, Madden, a sportscaster who emotes the sincere enthusiasm of sportscasting's pre-ESPN days, became the highest paid sportscaster of all time.

—Jimmie L. Reeves

FURTHER READING

Alexander, Sue. "Gender Bias in British Television Coverage of Major Athletic Championships." *Women's Studies International Forum* (Oxford), November-December 1994.

Catsis, John R. *Sports Broadcasting*. Chicago: Nelson-Hall, 1996.

Cosell, Howard, with Peter Bonventre. *I Never Played the Game*. New York: Morrow 1985.

Farrell, Thomas B. "Media Rhetoric as Social Drama: The Winter Olympics of 1984." *Critical Studies in Mass Communication* (Annandale, Virginia), June 1989.

Madden, John, with Dave Anderson. *One Size Doesn't Fit All*. New York: Villard, 1988.

Merrill, Sam. "Roone Arledge." In, Golson, G. Barry, editor. *The Playboy Interview: Volume II*. New York: Perigee, 1983.

Messner, Michael A. "Separating the Men from the Girls: The Gendered Language of Televised Sports." *Gender and Society* (Newbury Park, California), March 1993.

O'Neil, Terry. *The Game behind the Game: High Pressure, High Stakes in Television Sports*. New York: Harper and Row, 1989.

Powers, Ron. *Supertube: The Rise of Television Sports*. New York: Coward-McCann, 1984.

Rader, Benjamin G. *In Its Own Image: How Television Has Transformed Sports*. New York: Free Press, and London: Collier Macmillan, 1984.

Savan, Leslie. *The Sponsored Life: Ads, TV, and American Culture*. Philadelphia, Pennsylvania: Temple University Press, 1995.

Spence, Jim. *Up Close and Personal: The Inside Story of Network Television Sports*. New York: Atheneum, 1988.

Sullivan, David B. "Commentary and Viewer Perception of Player Hostility: Adding Punch to Televised Sports." *Journal of Broadcasting and Electronic Media* (Washington, D.C.), Fall 1991.

See also Sports and Television

SPRIGGS, ELIZABETH

British Actor

Elizabeth Spriggs is among Britain's most established and well-loved character actors. An Associate Artiste with the Royal Shakespeare Company, her illustrious work in the theatre has run parallel with her lengthy and successful career in television. Work in the two media converged with her characterisation of Sonia in Wesker's *Love Letters on Blue Paper,* a role she originally created for television and then transferred to the stage, winning her the West End Managers Award for 1978.

Her versatility is revealed by both her skill at adapting her style for television, resisting the tendency of many actors with a theatrical background to "play to the gallery," and her work in a diverse set of television genres. Listed among her credits are the particularly noteworthy roles of the long-suffering and self-sacrificing wife and mother, Connie Fox, in the drama series *Fox*; Harvey Moon's no-nonsense and strong-willed mother in the situation comedy series *Shine on Harvey Moon*; the God-fearing gossip, May, in the critically acclaimed and highly popular drama, *Oranges Are Not the Only Fruit*; and the wayward and wonderfully funny nurse, Sairey Gamp, in the much-praised BBC adaptation of *Martin Chuzzlewit*.

While to a great extent subject to the standard type casting of older actresses, Spriggs takes the crones, gossips, and suffering matriarchs and transforms them with her engagingly strong and rooted presence. In doing so, she imbues the usual fare with additional weight and dimension.

Although there has been interest, particularly within feminist television criticism, in analysing the representations of older female characters and the contributions of actresses to these characterisations, most of the attention has been paid to the soap opera genre. The wider terrain remains largely unexplored and unevaluated within television studies.

—Nicola Strange

ELIZABETH SPRIGGS. Born in England1929. Educated at the Royal School of Music. Married: 1) Marshall Jones; 2) Murray Manson. Stage actor with the Bristol Old Vic and the Birmingham Repertory, 1958; joined the Royal Shakespeare Company, 1962; joined the National Theatre Company, 1976; numerous appearances on television and in motion pictures. Recipient: SWETM Best Supporting Actress Award, 1978.

TELEVISION SERIES

1982	*Shine On Harvey Moon*
1992–93	*The Young Indiana Jones Chronicles*

TELEVISION PLAY

1978	*Love Letters on Blue Paper*

TELEVISION MINISERIES

1976	*The Glittering Prizes*
1980	*Fox*
1990	*Oranges Are Not the Only Fruit*
1994	*Middlemarch*
1995	*Martin Chuzzlewit*

MADE-FOR-TELEVISION MOVIES

1979	*Julius Caesar*
1982	*Merry Wives of Windsor*
1984	*The Cold Room*
1989	*Young Charlie Chaplin*
1992	*The Last Vampyre*

Elizabeth Spriggs
Photo courtesy of Elizabeth Spriggs

FILMS

Work is a Four-Letter Word, 1967; *Three into Two Won't Go*, 1969; *An Unsuitable Job for a Woman*, 1981; *Richard's Things*, 1981; *Lady Chatterly's Lover*, 1981; *Going Undercover*, 1988; *Oranges Are Not the Only Fruit*, 1989; *Impromptu*, 1991; *Hour of the Pig*, 1993; *Sense and Sensibility*, 1995; *The Secret Agent*, 1996; *Paradise Road*, 1997.

STAGE (selection)

Cleopatra, 1958; *The Cherry Orchard*, 1958; *The Beggar's Opera*, 1963; *The Representative*, 1963; *Victor*, 1964; *Marat/Sade*, 1965; *The Comedy of Errors*, 1965; *Timon of Athens*, 1965; *Hamlet*, 1965; *The Governor's Lady*, 1965–66; *The Government Inspector*, 1965–66; *Henry IV*, 1966; *Henry V*, 1966; *All's Well That Ends Well*, 1966; *Macbeth*, 1966; *Romeo and Juliet*, 1966; *Julius Caesar*, 1968; *The Merry Wives of Windsor*, 1968; *A Delicate Balance*, 1969; *Women Beware Women*, 1969; *Twelfth Night*, 1970; *London Assurance*, 1970; *The Winter's Tale*, 1970; *Twelfth Night*, 1970; *Major Barbara*, 1970; *Much Ado About Nothing*, 1972; *Blithe Spirit*, 1976; *The Country Wife*, 1977; *Volpone*, 1977; *Love Letters on Blue Paper*, 1978.

SPY PROGRAMS

Although individual series have enjoyed enormous popularity and cult followings, the spy genre overall has never been as successful nor as ubiquitous in American television as westerns, medical dramas, and detective programs. Nevertheless, espionage-themed programs can boast a number of firsts, most notably the first African-American lead character in a regular dramatic series (*I Spy*); the first female action lead character in an hour-long American dramatic series (*The Girl From U.N.C.L.E.*); and the first Russian lead character in an American dramatic series (*The Man From U.N.C.L.E.*), the latter appearing less than three years after the Cuban Missile Crisis.

Except during the so-called "spy craze" period of the mid-1960s when it seemed that every action/adventure show borrowed elements from James Bond, spies as television action heroes have been far outnumbered by the more traditional figures of policemen and private investigators. Even when they do appear, television spies (or "secret agents") are often presented as international crime fighters rather than as true undercover operators, with the emphasis on justice and law enforcement rather than on clandestine activities. As a result, there are few "pure" spy programs and most of the long-running ones can be classed in other genre categories, including westerns (*The Wild, Wild West*), situation comedy (*Get Smart*), and science fiction (*The Avengers* and *The Prisoner*).

The boundaries between the spy and other television genres is extremely fluid, and the elements of the typical spy program are variable and not easily defined. On television, spies and detectives have a great deal in common. Both are tough, sometimes world-weary individuals who live and work on the edges of normal society. Their antagonists are rich, powerful, clever, and often apparently "respectable." In both genres, because of the wealth and resources of the villain, the heroes must use extra-legal means in order to triumph. Before they do, they must progress through various narrative situations including the assignment of the case/mission, investigation of the crime, abduction by the villain, interrogation and/or torture, at least one long, complicated chase and a final shoot-out or brawl.

The average secret agent tends to be more cerebral and sophisticated than the average detective and if not wealthy himself, at least comfortable with the trappings of wealth. Money is not an important incentive, however. The secret agent's motives are personal and philosophical, a dedication to certain moral or political ideals, or simply a taste for the game of espionage itself. In its focus on the "game"—the hero's intellectual ability to decipher clues, solve complex mysteries, and outmaneuver the bad guys—the television spy plot may resemble the classical detective story. Indeed, the chess metaphor appears often in each.

Nevertheless, there are several subtle differences that distinguish the television secret agent from the detective and these can be seen in the transformation of Amos Burke, the title character of the popular series *Burke's Law*. For the first two years of the series' run, Burke (played by Gene Barry) was a Los Angeles chief of detectives who also happened to be a millionaire. In solving his homicide cases, Burke was chauffeured around in a silver Rolls Royce. His cases were typical whodunits involving the rich and glamorous portrayed by large casts of guest stars.

Then in 1965, in order to cash in on the spy craze, Captain Amos Burke, detective, became Amos Burke—Secret Agent. Since he was already suave, sophisticated, witty, and charming, no character tinkering was needed. However, several important changes were made. Burke left the L.A.P.D. to work for a U.S. government intelligence agency, with his only contact, a mysterious character called simply, "The Man" (played by Carl Benton Reid). Burke's operating milieu subsequently expanded from the confines of the Los Angeles area to include the entire world. No longer a local millionaire sleuth, Burke became a continent-hopping agent and his quarry changed from small time murderers to international criminals whose schemes and machinations had global consequences.

These changes, then, define the essential elements of the television spy series: (1) The active presence of a government or quasi-government agency in the life of the protagonist. The agency is shown to be involved in clandestine and/or espionage activities. (2) Villains who are often foreign, usu-

The Man from U.N.C.L.E.

ally eccentric, and whose crimes have larger political consequences. Most commonly, these villains desire either to take over the world or to destroy it. (3) An expansion of the plot setting beyond local and even national boundaries to include a variety of countries and exotic locales.

Since James Bond appeared on the literary and later, cinematic scene, spy stories have also incorporated a number of stylistic motifs of his creator, Ian Fleming. These include ironic humor; the use martial arts techniques for self-defense; a preoccupation with expensive clothes, cars, food, accommodations and leisure pursuits; the presence of beautiful women either as agents, antagonists, or innocent bystanders caught up in the plot, and a fascination with weaponry and high technology.

The importance of these motifs should not be underestimated. For example, *Honey West* was essentially a series about a female detective similar to the later *Remington Steel*. Yet, critics have always categorized it as a spy program simply because of its stylistic trappings, most notably, Honey West's pet ocelot and the one-piece black jumpsuit worn by the star, Anne Francis, so reminiscent of the wardrobe of *The Avengers'* Emma Peel (Diana Rigg). On the other hand, series like *Tightrope* and the later *Wiseguy*, which both feature lead characters working undercover, are not considered spy programs because the international reach of the enemy crime syndicates is not emphasized, and because the heroes appear and function as police officers.

The primary reason why spy shows are so few and far between on television is that the genre does not adapt well to the production and aesthetic needs of medium. In their book, *The Spy Story* (1987), John G. Cawelti and Bruce A. Rosenberg delineate two subcategories of spy fiction, both of which can be applied to spy stories on television.

The first, originating with James Buchan (*The Thirty-Nine Steps*) and other "clubmen" writers and re-invented by Ian Fleming, consists of colorful, imaginative adventures with roving, honorable heroes, dastardly villains and exotic settings. By comparison, the second subcategory, identified with Eric Ambler, Grahame Green and more recently, John Le Carré, contains tales of espionage more realistically presented. Concerned with corruption, betrayal and conspiracy, these stories feature a grayer mood, more circumscribed settings and ordinary protagonists who seem, at first glance, not much different than the people they oppose. The plotting is complicated and subtle, and the endings are often downbeat, leaving the agent sadly disillusioned or dead. The chief difference between the two subcategories is the moral base of the narrative. In the first group, good and evil is rendered in stark black and white. In the second, the morality is ambiguous.

As with their literary equivalents, television spy stories may be similarly divided into the romantic and the realistic, although as one might expect, there is considerable overlap. Both types present problems in adapting to the television medium.

The romantic spy adventure, while meeting the aesthetic needs of the medium for simplicity in storytelling, escapist interest and fast paced excitement, requires foreign locations, numerous props, expensive wardrobes, and other production details that can severely strain a limited television budget. On the other hand, although the realistic espionage story is likely to be less expensive to produce, the difficult themes, depressive mood, and often unattractive characters do not lend themselves to the medium, particularly to the demands of a weekly network series.

As a result, to be produced for television, both types of spy stories must be "domesticated", both literally and figuratively. For the romantic spy program, elements of the so-called Bond formula of "sex, snobbery, and sadism" must be toned down to small screen standards. The intensity of torture sequences may be tempered by the use of outlandishly humorous devices and Perils-of-Pauline-style narrow escapes. Weapons may fire sleep inducing darts (*The Man From U.N.C.L.E.*) or the hero may not carry a gun at all (as in the quasi-espionage series, *MacGyver*).

Location shoots must also be kept to a minimum. Both *The Man* and *The Girl From U.N.C.L.E.* filmed on the MGM backlot used an ingenious swish-pan technique to get from one location to another. *I Spy* traveled overseas but filmed a number of episodes in each country it visited. *The Prisoner* was shot at an actual resort village at the Hotel Portmeiron in North Wales. *Adderly* was fortunate enough to find Canadian locations that could mimic the landscape

of the Soviet Union and other European countries. More recently, series like *The Scarecrow and Mrs. King* confine themselves to U.S. settings, saving stories set in foreign locales for season finales and sweeps weeks.

Several realistic, even dyspeptic, espionage series like *Danger Man*, *Callan*, and *Sandbaggers* enjoyed healthy runs in the United Kingdom but only one of these, *Danger Man*, ever crossed the Atlantic to be seen in the States. To make the plotlines and characters of realistic spy programs more appealing to American audiences, television producers have employed a number of different strategies. For example, *Danger Man* was retitled *Secret Agent* and a snazzy Johnny Rivers song was added to the opening and closing credits. Both *I Led Three Lives* in the 1950s and *The Equalizer* in the 1980s exploited anxieties that were close to home for the audience, mining Red Scare paranoia in the case of the earlier show and fears of urban crime in the later.

Another strategy used by creators of realistic spy programs is to make the central character morally certain. Although he was often surrounded by double-crossing colleagues and double agents in *Secret Agent*, John Drake's (Patrick McGoohan) own loyalty was never in question. In *The Equalizer*, Edward Woodward, who earlier played a lonely, cold-blooded assassin in *Callan*, returned as Robert McCall, a retired CIA operative. McCall clearly had a past career similar to Callan's, but now deeply regretted it. To expiate his past sins, McCall became the self-styled Equalizer of the title, dedicating his life and skills to protecting the weak and innocent free of charge. McCall was also given a family—an estranged son, a dead wife and a daughter whose existence he discovered during the run of the series.

Surrounding the usually isolated secret agent with family, colleagues, and friends is yet another television strategy for domesticating both strains of the genre. Humor and a fraternity boy camaraderie between Kelly Robinson (Robert Culp) and Alexander Scott (Bill Cosby) leavened *I Spy*'s sometimes bleak Cold War ideology, while the developing romance between the two lead characters (Bruce Boxleitner and Kate Jackson) kept interest high between chases in *The Scarecrow and Mrs. King*. In *Under Cover*, an intensely realistic series which featured plotlines drawn directly from recent world events, the husband and wife agents (Anthony John Denison and Linda Purl) were forced to juggle the dangerous demands of their profession with the everyday problems of home and family life. Finally, those spy stories that, for whatever reason, could not be domesticated, such as adaptations of bestselling spy thrillers, generally ended up on cable or PBS, or on network television as TV movies and miniseries.

The history of the spy on television reflects this continuing tension between the genre and the medium, and between romantic and realistic tendencies. Whenever public interest in foreign affairs is on the rise, spy programs of both types proliferate, with fictional villains reflecting the country's current political enemies.

The first regular spy series appeared on U.S. television in the early 1950s. A handful, including an early series also called

I Spy (hosted by Raymond Massey) and *Behind Closed Doors* (hosted by Bruce Gordon) were anthologies. Others, like *Biff Baker* (Alan Hale Jr.) and *Hunter* (the first of four series called *Hunter*, this one starring Barry Nelson), featured gentlemen amateurs caught up in foreign intrigue through chance or patriotism. The rest, which usually had the word "danger" in their titles (*Doorway to Danger, Dangerous Assignment, Passport to Danger*) were undistinguished half-hour series about professional agents battling Communists. These series lasted, with only three exceptions, a year or less.

Those exceptions were *I Led Three Lives, Foreign Intrigue* and *Five Fingers. I Led Three Lives* was an enormously popular hit series based on the real-life story of FBI undercover agent Herbert Philbrick who infiltrated the American Communist Party. A favorite of J. Edgar Hoover (who considered it a public service), the show reportedly was taken so seriously by some viewers that they wrote the producers to report suspected Communists in their neighborhood. *Foreign Intrigue*, a syndicated series boasted colorful European locations but replaceable stars (five in four years played four various wire service correspondents and a hotel owner) who stumble across international criminals. Only the last, *Five Fingers*, starring David Hedison as double agent Victor Sebastian, even hinted at the cool, hip style that was to be the hallmark of spy shows in the sixties.

An interesting oddity during this period was an adaptation of Ian Fleming's *Casino Royale* for the anthology series, *Climax*, in which the British James Bond is transformed into an American agent, "Jimmy" Bond (Barry Nelson) confronting a French Communist villain named Le Sheef (originally Le Chiffre). After a tense game of baccarat, Le Sheef (played by a sleepwalking Peter Lorre) captures Bond, confines him in a hotel bathtub, and rather bizarrely tortures him by twisting his bare toes with pliers.

There is no doubt that the mid-1960s was the high water mark for the spy genre. Spies were everywhere—in books, on records, on the big screen and the little screen, and their images were emblazoned on countless mass produced articles from toys to toiletries. Most were hour-long color shows which featured pairs or teams of professional agents of various races, genders and cultural backgrounds. The pace was fast, the style, cool, with lots of outrageous villains, sexual innuendo, technical gadgetry, and tongue-in-cheek humor. A third subcategory of the genre, the spy "spoof", developed during this time (*Get Smart*, created by Mel Brooks and Buck Henry, is the quintessential example) but there was so much humor in the "serious" shows that it was often difficult to distinguish spoofs from the real thing.

By 1968, the high spirits had soured and the spy craze came to a fitting end with the unsettlingly paranoiac series, *The Prisoner*, created and produced by its star, ex-Secret Agent Patrick McGoohan. Still, many of the shows of this period, including *The Man From U.N.C.L.E., The Avengers, I Spy, The Wild, Wild West, Mission: Impossible* and even *The Prisoner*, have enjoyed continued life in periodic film and television revivals and in cult fan followings throughout the world.

The decade of the 1970s saw a few sporadic attempts to breathe new life into a moribund genre. All the spy series introduced during this period featured gimmicky characters who worked for organizations identified by acronyms. Among the gimmicks were an agent with a photographic memory (*The Delphi Bureau*), agents fitted with electronic devices connected to a computer (*Search*), an agent accompanied by a giant assistant with a steel hand filled with gadgets (*A Man Called Sloane*) and a superhuman cyborg (*The Six Million Dollar Man*). With the exception of the last, which appealed primarily to children, all were quickly cancelled.

The beginning of the next decade saw several "return" movies of 1960s' favorites like *Get Smart, The Wild, Wild West*, and *The Man From U.N.C.L.E.* as well as quality television adaptations of John Le Carré's *Tinker Tailor Soldier Spy* and *Smiley's People* by the BBC (shown on PBS in the U.S.). This eventually led to a mini-revival in spy programs in the mid-1980s, which included serious, gritty series like *The Equalizer* and adaptions of bestselling spy novels including Le Carre's *A Perfect Spy*, Len Deighton's *Game Set Match*, Ken Follett's *Key to Rebecca* and Robert Ludlum's *The Bourne Identity*. As with Amos Burke in the sixties, action series like *The A-Team* began to boost their ratings by injecting espionage elements into their formulas.

However, unlike their predecessors of twenty years previous, the spies of the 1980s were less fantastic and more pragmatic, with believable technology and a post-modern sensibility. Even romantic adventure series like *Airwolf* and *Scarecrow and Mrs. King* were given a realistic edge. Indeed, this trend toward intense realism reached its culmination in *Under Cover*, a series so realistic that it was cancelled by a nervous ABC network after less than a month on the air. In January 1991, a two-part episode of *Under Cover*, in which Iraq planned to fire a virus-carrying missile at Israel, was pulled from the schedule when the war in Kuwait broke out.

For the 1994–95 season, the fledgling Fox network offered two spy series, *Fortune Hunter*, a James Bond clone, and a revival of *Get Smart* starring an aging Don Adams and Barbara Feldon. Both series were cancelled after extremely abbreviated runs.

—Cynthia W. Walker

FURTHER READING

Cawleti, John G., and Bruce A. Rosenberg. *The Spy Story.* Chicago, Illinois: University of Chicago Press, 1987.

Harper, Ralph. *The World of the Thriller.* Cleveland, Ohio: Press of Case Western Reserve University, 1969.

McCormick, Donald, and Kathy Fletcher. *Spy Fiction: A Connoisseur's Guide.* New York: Facts on File, 1990.

Meyers, Richard. *TV Detectives.* San Diego, California: A.S. Barnes, 1981.

Snelling, O.F. *007 James Bond: A Report.* New York: New American Library, 1964.

See also *Avengers; Get Smart; I Spy, Man from U.N.C.L.E./ The Girl from U.N.C.L.E.; Mission: Impossible; Prisoner; Tinker Tailor Soldier Spy*

STANDARDS

Recorded video signals are rather complex and tightly structured. The standard unit of video is a frame. Similar to film, motion video is created by displaying progressive frames at a rate fast enough for the human eye and brain to perceive continuous motion. The basic means by which video images are recorded and displayed is a scanning process. When a video image is recorded by most cameras, a beam of electrons sweeps across the recording surface in a progressive series of lines. This basic technology is simple enough, widely understood, and, after a certain point, easily manufactured. The concept can be applied and the effect of a video image can be achieved, however, in various ways, with varying rates of electronic activity. Line frequencies and scanning rates are flexible, determined in part by a level of user (producer and viewer) satisfaction, in part by concerns of equipment manufacturers and broadcasters. Consequently, not all video or television systems are alike. The variations among them are defined in terms of "standards."

In the United States industry-wide agreement on engineering standards for television did not come until 1941, when the Federal Communications Commission (FCC) decided to adopt a black-and-white standard (postponing the issue of color). The FCC accepted the National Television System Committee (also referred to as the National Television Standards Committee and referred to as NTSC) recommendations, and set line frequency at 525 per frame scanned at a rate of approximately 30 frames per second (29.97 to be exact). In 1953, corporate interests (CBS and RCA/NBC) agreed to another proposal which allowed the NTSC to establish color television standards; these standards were compatible with those already set for black-and-white transmission.

These standards are not, however, uniformly accepted elsewhere. There are presently three world standards for transmitting a color video signal. The NTSC recommendations accepted by the FCC as a national standard for the United States in 1953 are used in several other countries including Canada, Chile, Costa Rica, Cuba, El Salvador, Guatemala, Honduras, Japan, Mexico, Panama, Philippines, Puerto Rico, South Korea, and Taiwan.

PAL (phase alternating line) and SECAM (sequential couleur a memoire) are the two other major worldwide television standards. PAL is a modified form of NTSC, and specifies a different means of encoding and transmitting color video designed to eliminate some NTSC problems, specifically a shift in chroma phase (hue). PAL uses 625 lines per frame (versus NTSC's 525) scanned at a rate of 25 frames per second (versus NTSC's 29.97), and operates at a 50Hz frequency (versus NTSC's 60Hz frequency). The PAL system is standard in more countries than NTSC or SECAM, including Argentina, Australia, Belgium, Brazil, China, Denmark, Finland, Great Britain, India, Indonesia, Ireland, Italy, Norway, New Zealand, the Netherlands, Portugal, Spain, Sweden, Switzerland, and Turkey.

SECAM is a video color system developed by the French; though it differs from PAL, it too uses 625 lines per frame, scanned at a rate of 25 frames per second, and operates at a 50Hz frequency. SECAM is used in France, as well as several other countries, including Egypt, Germany, Greece, Haiti, Iran, Iraq, North Korea, Poland, and parts of the former Soviet Union.

There are enough differences between these three standards so that a videotape recorded using PAL will not play on a VCR set up for NTSC or SECAM, and vice versa. NTSC, PAL and SECAM are thus incompatible with each other. Standards converters can convert video from one standard to another, but the resultant image is often poor. Digital standards converters can provide better quality converted video. Productions intended to be broadcast or released in different video standards are often shot on film, which can be converted to any video standard with reasonably good quality.

Recent developments in high-definition television (HDTV) have closed the gap between the technical quality of broadcast television and motion pictures. HDTV doubles the current broadcast NTSC number of scanning lines per frame—from 525 to 1050 or 1125, depending on the specific system—with a fourfold improvement in resolution (and a change to a wide-screen format).

—Eric Freedman

FURTHER READING

Browne, Steven E. *Video Editing: A Postproduction Primer.* Boston: Focal, 1989; 2nd edition, 1993.

Caruso, James R., and Mavis E. Arthur. *Video Editing and Post Production.* Englewood Cliffs, New Jersey: Prentice Hall, 1992.

Inglis, Andrew F. *Video Engineering.* New York: McGraw Hill, 1993.

Lenk, John D. *Lenk's Video Handbook: Operations and Troubleshooting.* New York: McGraw Hill, 1991.

Zettl, Herbert. *Television Production Handbook.* London: Pitman, 1961; 5th edition, Belmont, California: Wadsworth, 1992.

See also Society for Motion Picture and Television Engineers

STANDARDS AND PRACTICES

Standards and Practices is the term most American networks use for what many, especially in the creative community, refer to as the "network censors." Standards and Practices Departments (known as Program Practices at CBS) are maintained at each of the broadcast and many of the cable networks. The concept came about as a direct outgrowth of the trusteeship model: broadcasters were said to have a responsibility to the pubic interest as a result of their having access to a scarce resource. Another factor was the fear of propaganda, deemed to have been so effective in World War I. The most important consideration, however, was the unprecedented reality that radio, and later television content, came into the home, unforeseen, often unbidden, and sometimes unwelcome. Historically, therefore, lest an offended audience demand government intervention, Standards and Practice's charge has been to review all non-news broadcast matter, including entertainment, sports and commercials, for compliance with legal, policy, factual, and community standards.

The broadcasters' insistence on setting and maintaining their own standards goes back to the very beginning of the medium in 1921, when engineers were instructed to use an emergency switch in the event that a performer or guest used language or brought up topics which were held to be unsuitable. One early memoir describes the use of the button when a distinguished ballerina launched into a discussion of birth control. During radio's first decade, taboos also included any mention of price or even the location of a sponsoring store. Later, the networks would have an organist at the ready in a standby studio. A noted incident is said to have occurred in 1932 when a major administration spokesman was reporting on the government's progress in dealing with the Great Depression. He allegedly used the word "damn," a light went on in the standby studio, and the nation heard organ arpeggios for the length of time it took to be assured that he wouldn't do that again.

By the late 1930s, the networks had established so-called continuity acceptance procedures to assure that their advertising policies and federal law were adhered to. Later, as the role of radio in American life became more clearly understood, a body of written policy was articulated, generally on a case by case basis, to guide not only advertisers and their agencies but also programmers and producers in entertainment and other programming.

More than 67 % of all television stations subscribed to the NAB (National Association of Broadcasters) Code adopted in 1950 (a similar radio code had been in operation since 1935). In addition to provisions which addressed historic concerns respecting the "advancement of education and culture," responsibility toward children, community responsibility, and general program standards, the NAB Code also included advertising standards and time limits for non-program material defined as "billboards, commercials, promotional announcements and all credits in excess of 30 seconds per program." In

1982, in settlement of an anti-trust suit brought by the U.S. Department of Justice, the NAB and the federal government entered into a consent decree abolishing the time standards and the industry-wide limitations on the number and length of commercials they provided. The Code program standards had been suspended in 1976 after a federal judge in Los Angeles ruled that the Family Hour violated the First Amendment. After the demise of the Code, the networks, which had already developed their own written standards, took over the entire burden.

Standards, and the broadcasters' efforts to implement them, come to the fore whenever an apparent breach of the implicit obligation to respect the public trust occurs. The celebrated 1938 broadcast by Orson Welles' Mercury Theater of "The War of the Worlds" which simulated a radio broadcast interrupted by news reports describing the landing of Martians; the quiz show scandals of the 1950s; Congressional hearings into violence, and concern over the possible blurring of fact and fiction in early docudrama, are notable examples of perceived abuse which resulted in expanding the duties and enlarging Standards and Practices operations, generally throughout the industry. By 1985, one of the traditional network's department had no fewer than 80 persons on its staff. Each episode of every series was reviewed in script form and as it was recorded.

With the changes in ownership of the traditional networks, the emergence of the cable networks, and the deregulatory climate, there has been considerable relaxation of the process—not every episode is reviewed once a series is established—but the essential responsibilities of the editors remain the same. These include, in addition to compliance with the law, serving as surrogates for the network's affiliates who are licensed to be responsive to their local communities; reflecting the concerns of advertisers; and, most important, for their employers, the networks themselves, assuring that the programming is acceptable to the bulk of the mass audience. This involves serving as guardians of taste with respect to language, sexual and other materials inappropriate for children, and the suitability of advertising, especially of personal products.

Commercial clearance involves the close screening of more than 50,000 announcements a year, falling into about 70 different product categories. The Federal Trade Commission's statements in the early 1970s which not only permitted but virtually mandated comparative advertising, resulted in the establishment of courtlike procedures to adjudicate between advertisers making conflicting claims. By the mid-1980s, at least 25 % of all commercials contained comparisons to named competitor's products or services.

Critics contend, with some justification, that standards and practices is anachronistic paternalism at best, and most often a form of censorship; the networks claim the publisher's right to exercise their judgment as to what is appropriate for broadcast to the American public. The affiliated stations sometimes complain but are generally, though not always, satisfied that the network are sufficiently vigilant as their surrogates. Network and sales executives worry that the very process of vetting leads to pettifoggery and rigidity. Advertisers rail at the scrupulous insistence that all claims be substantiated, as the law requires. By far the most frequent complaints, however, are heard from the creative community which argues that the networks are too accommodating of the most conservative members of the audience and that only by "pushing the envelope" with respect to sex, violence, or language, can the medium advance.

Standards and Practices' primary purpose has always been to maintain the networks' most precious asset, its audience-in-being — the delivery of a significant share of television households, hour after hour, to the advertising community. Secondary purposes historically, have included protecting the networks' images as responsible and responsive institutions, as sources of reliable information and satisfying entertainment for the entire family, and even as precious national resources. In the final analysis, if the concern for not giving offense has contributed to blandness, it must also be credited for making a commercially supported national system possible.

—George Dessart

FURTHER READING

Adler, Keith. *Advertising Resource Handbook.* East Lansing, Michigan: Advertising Resources, Inc., 1989.

Barnouw, Erik. *A Tower in Babel: A History of Broadcasting,* volume 1. New York: Oxford University Press, 1970.

Broadcast Self-regulation. 2nd ed. Washington, D.C.: NAB Code Authority, 1977.

Dessart, George. "Of Tastes and Times: Some Challenging Reflections on Television's Elastic Standards and Astounding Practices." *Television Quarterly* (New York), 1992.

Henderson, Alice M., and Helaine Doktori. "How the Networks Monitor Program Content." In, Oskam, Stuart, editor. *Television as a Social Issue: The Eighth Applied Social Psychology Annual.* Newbury Park, California: Sage Publications, 1988.

Legal Guide to FCC Rules, Regulations and Policies. Washington, D.C.: National Association of Broadcasters, 1977.

Pember, Don R. *Mass Media Law.* Dubuque, Iowa: Brown Publishers, 1977; 5th edition, 1990.

"Program Standards for the CBS Television Network." In, Oskam, Stuart, editor. *Television as a Social Issue: The Eighth Applied Social Psychology Annual.* Newbury Park, California: Sage Publications, 1988.

Sensitive Theme Programming and the New American Mainstream. New York: Social Research Unit, Marketing and Research Services, ABC, n.d.

The Television Code, 22nd ed. Washington, D.C.: Code Authority, National Association of Broadcasters, 1981.

See also Censorship; United States: Networks

STANTON, FRANK

U.S. Media Executive

Frank Stanton is a distinguished broadcast executive known for the leadership he brought to CBS, Inc., during his twenty-five-year presidency (1946–71). His guidance gave CBS crucial stability during the company's critical growth period. More than just a corporate president, however, Stanton acquired a reputation as the unofficial spokesperson for the broadcasting industry. His opinions were routinely sought, his speeches repeatedly quoted, and his testimony before Congress recognized as a major part of any debate in the broadcasting field.

Stanton was fascinated with radio from his days in graduate school at Ohio State, chiefly by the question of why people reacted positively to certain radio shows but negatively to others. He used his doctoral research in the psychology department to answer this question, examining why and how people perceive various stimuli. He analyzed the audio and visual effectiveness of information transmission and established test procedures for making rough measurements of their effectiveness. His dissertation, "A Critique of Present Methods and a New Plan for Studying Radio Listening Behavior," caught the attention of CBS, and launched his career in the audience research department in 1935.

In 1937, Stanton began a collaboration with Dr. Paul Lazarsfeld of Columbia University. They devised a program analysis system nicknamed "Little Annie." While Stanton tends to downplay the importance of the machine, others have credited it with being the first qualitative measurement device. "Little Annie" determines the probability of a program's appeal by suggesting how large an audience that program would be likely to attract. The system was devised for radio, but continues to be used for television, reporting an accuracy rate of 85%.

Stanton was promoted to vice president of CBS in 1942, and in 1946, at the age of 38, to the presidency. In this position, he guided CBS through a period of diversification and expansion. He reorganized the company in 1951, creating separate administrations for radio, TV and CBS Laboratories, a plan that served as a model for other broadcast companies. He helped CBS expand its operations by decentralizing its administration and creating autonomous divisions with a range of new investments, including the purchase of the New York Yankees in 1964. CBS also bought the book publisher Holt, Rinehart and Winston, and Creative Playthings, manufacturer of high-quality educational toys. Diversification paid off for CBS; the company earned $1 billion in annual sales in 1969.

As president of CBS, Stanton concentrated on organizational and policy questions, leaving the entertainment programming and the discovering and nurturing of talent to chair, William S. Paley. Stanton was also responsible for the political issues growing out of the network's news department. He was instrumental in bringing about the 1960 Kennedy-Nixon televised presidential debate and is known for his efforts to repeal Section 315 of the Federal Communications Act, which requires networks to grant equal time to all political candidates. A staunch proponent of broadcast journalism and defender of broadcasting's First Amendment rights, he led campaigns before Congress and in the courts on behalf of the broadcast industry for access and protection equal to that of the printed press.

Stanton's greatest battle with the government occurred in 1971, and focused on just this parallel to print press rights. The controversy surrounded *The Selling of the Pentagon*, a CBS News documentary, which exposed the huge expenditure of public funds, partly illegal, to promote militarism. The confrontation raised the issue of whether television news programming deserved protection under the First Amendment. Against threat of jail, Stanton refused the subpoena from the House Commerce Committee ordering him to provide copies of the outtakes and scripts from the documentary. He claimed that such materials are protected by the freedom of the press guaranteed by the First Amendment. Stanton observed that if such subpoena actions were allowed, there would be a "chilling effect" upon broadcast journalism.

But long before this particular case, and long before Watergate or Vietnam, CBS was the first broadcasting network to seriously examine the negative side of Washington politics on television. One of the earliest of these explorations occurred on the news program *See It Now*, in which host Edward R. Murrow confronted U.S. Senator, Joseph McCarthy. The program was constructed using film clips of McCarthy's accusatory speeches and Murrow refuting his charges. McCarthy demanded, and was granted, time for a response, and in that blustery performance many observers see the downfall of McCarthyism. In retrospect, the two programs were among the most important in the history of television.

Documentaries, even of this immediate sort, however, had a more difficult time attracting sponsors than did entertainment programs and for this reason *See It Now* was canceled following the 1958 season. Appalled by what the broadcasting industry had become, Murrow spoke before the Television News Directors Association and delivered what was to become known as one of the most famous public tongue lashings in media history, aimed directly at Stanton and Paley. The relationship between Stanton and Murrow soured into accusations and name-calling and was widely reported in the press.

Stanton received the title of vice chair in 1972, one year before the mandatory retirement age of 65. Upon retiring Stanton still held $13 million worth of CBS stock and he remained a director of CBS and consultant to the corporation under a contract that lasted until 1987.

—Garth Jowett and Laura Ashley

Frank Stanton
Photo courtesy of Frank Stanton

FRANK STANTON. Born in Muskegon, Michigan, U.S.A., 20 March 1908. Educated at Ohio Wesleyan University, Delaware, Ohio, B.A. 1930; Ohio State University, Ph.D. 1935; diplomate from American Board of Professional Psychology. Worked in CBS research department (later CBS-TV), New York City, 1935–45, vice president, 1942, president, CBS Inc., 1946–71 (was cited by three committees of the House of Representatives for contempt of Congress for refusal to grant access to CBS News' "outtakes" in connection with the CBS broadcast of *The Selling of the Pentagon*, 1971), vice-chair, 1972–73, president emeritus, since 1973; chair, Rand Corporation, Santa Monica, California, 1961–67, trustee, 1957–78; U.S. Advanced Communications Info., Washington, 1964–73; chair, ARC, Washington, 1973–79, vice-chair, League of Red Cross Societies, Geneva, Switzerland, 1973–80; chair, visiting committee, Kennedy School of Government, 1979–85; chair (now retired), Broadcast International, Inc.; director, Capital Income Builder, Inc., Capital World Growth and Income Fund, Inc., Sony Music Entertainment, Inc. Member: founding member and chair, Center for Advanced Study in Behavioral Sciences, Stanford, California, 1953–60, trustee, 1953–71; Business Council, Washington, since 1956 (honorary); National Portrait Gallery Commission, Washington, since 1973; board of overseers, Harvard College, 1978–84; President's Committee on Arts and Humanities, Washington, 1983–90; honorary director and trustee, William Benton Foundation, Bryant Park Restoration

Corporation, Educational Broadcasting Corporation; emeritus trustee and director, Lincoln Center for the Performing Arts, Rockefeller Foundation, Carnegie Institution Washington. Recipient: Paul White Memorial Awards, Radio and TV News Directors Association, 1957 and 1971; Peabody Awards, 1959, 1960, 1961, 1964, and 1972; Trustees Awards, National Academy of Television Arts and Sciences, 1959 and 1972; Special Honor Award, AIA, 1967; International Directorate Award, National Academy of Television Arts and Sciences, 1980; named to TV Academy Hall of Fame, 1986, Market Research Council of New York, 1988. Address: 25 West 52nd Street, New York City, New York, 10019-4223, U.S.A.

FILMS

Some Physiological Reactions to Emotional Stimuli, 1932; *Factors in Visual Depth Perception*, 1936.

PUBLICATIONS

Editor, with Paul Lazarsfeld. *Radio Research, 1941*. New York: Duell, Sloan and Pearce, 1941.

Editor, with Paul Lazarsfeld. *Radio Research, 1942-43*. New York: Duell, Sloan and Pearce, 1943.

Editor, with Paul Lazarsfeld. *Communications Research, 1948–49*. New York: Harper, 1949.

FURTHER READING

Smith, Sally Bedell. *In All His Glory: The Life of William S. Paley, the Legendary Tycoon and His Brilliant Circle.* New York: Simon and Schuster, 1990.

See also Audience Research, Industry and Marketing Perspective; Columbia Broadcasting System; Murrow, Edward R.; Paley, William S.; *See It Now, Selling of the Pentagon*

STAR TREK

U.S. Science-Fiction Program

With the premiere of *Star Trek* on NBC in September 1966, few could have imagined that this ambitious yet often uneven science-fiction series would go on to become one of the most actively celebrated and financially lucrative narrative franchises in television history. Although the original series enjoyed only a modest run of three season and 79 episodes, the story world created by that series eventually led to a library of popular novelizations and comic books, a cycle of motion-pictures, an international fan community, and a number of spin-off series that made the *Star Trek* universe a bedrock property for Paramount Studios in the 1980s and 1990s.

Star Trek followed the adventures of the *U.S.S. Enterprise*, a flagship in a 23rd-Century interplanetary alliance known as "the Federation." The ship's five-year mission was "to seek out new life and new civilizations, to boldly go where no man has gone before," a mandate that series creator and philosophical wellspring Gene Roddenberry described as "*Wagon Train* in space." Each episode brought the crew of the *Enterprise* in contact with new alien races or baffling wonders of the universe. When not exploring the galaxy, the crew of the *Enterprise* often scrapped with the two main threats to the Federation's benevolent democratization of space, the Hun-like Klingons and the more cerebral yet equally menacing Romulans.

The program's main protagonists, Captain James T. Kirk (William Shatner), Mr. Spock (Leonard Nimoy), and Dr. Leonard McCoy (DeForest Kelly), remain three of the most familiar (and most parodied) characters in television memory. As commander of the *Enterprise*, the hyper-masculine Kirk engaged in equal amounts of fisticuffs and intergalactic romance, and was known for his nerves of steel in negotiating the difficulties and dangers presented by the

ship's mission. McCoy was the ship's cantankerous chief medical officer who, when not saving patients, gave the other two leads frequent personal and professional advice. Perhaps most complex and popular of the characters was Spock. Half-human and half-Vulcan, Spock struggled to maintain the absolute emotional control demanded by his Vulcan heritage, and yet occasionally fell prey to the foibles of a more human existence. In addition to the three leads, *Star Trek* featured a stable of secondary characters who also became central to the show's identity. These included the ship's chief engineer, Scotty (James Doohan), and an ethnically diverse supporting cast featuring Uhura (Nichelle Nichols), Chekov (Walter Koening), Sulu (George Takei), Yeoman Rand (Grace Lee Whitney), and Nurse Chapel (Majel Barrett).

Scripts for the original series varied greatly in quality, ranging from the literate time-travel tragedy of Harlan Ellison's "City on the Edge of Forever" and the Sophoclean conflict of Theodore Sturgeon's "Amok Time," to less inspired stock adventure plots, such as Kirk's battle to the death with a giant lizard creature in "Arena." With varying degree of success, many episodes addressed the social and political climate of late-1960s America, including the Vietnam allegory, "A Private Little War," a rather heavy-handed treatment of racism in "Let That Be Your Last Battlefield," and even an encounter with space hippies in "The Way to Eden."

NBC threatened to cancel *Star Trek* after its second season, but persuaded in some degree by a large letter-writing campaign by fans to save the show, the network picked-up the series for a third and final year. Canceled in 1969, *Star Trek* went on to a new life in syndication where it found an even larger audience and quickly became a major phenomenon within popular culture. Beginning with a network

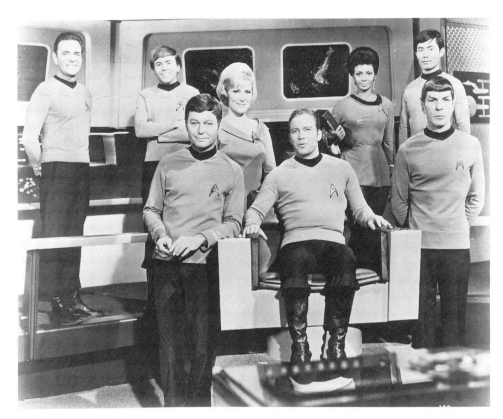

Star Trek

Star Trek: The Next Generation

of memorabilia collectors, fans of the show became increasingly organized, gathering at *Star Trek* conventions to trade merchandise, meet stars from the show, and watch old episodes. Such fans came to be known as "trekkies," and were noted (and often ridiculed) for their extreme devotion to the show and their encyclopedic knowledge of every episode. Through this explosion of interest, many elements of the *Star Trek* universe made their way into the larger lexicon of popular culture, including the oft heard line, "Beam me up, Scotty" (a reference to the ship's teleportation device), as well as Spock's signature commentary on the "illogic" of human culture. Along with Spock's distinctively pointed ears, other aspects of Vulcan culture also became widely popularized as television lore, including the Vulcan "mind-meld" and the Vulcan salute, "live long and prosper."

As "trekkie" culture continued to grow around the show during the seventies, a central topic of conversation among fans concerned rumors that the series might one day return to the airwaves. There was talk that the series might return with the original cast, with a new cast, or in a new sequel format. Such rumors were often fueled by a general sense among fans that the show had been unjustly canceled in the first place, and thus deserved a second run. Initially, Paramount did not seem convinced of the commercial potential of resurrecting the story world in any form, but by the late seventies, the studio announced that a motion picture version of the series featuring the original cast was under development. *Star Trek: The Motion Picture* premiered in 1979, and though it was a very clumsy translation of the series into the language of big-budget, big-screen science-fiction, it proved to be such a hit that Paramount developed a chain of sequels, including *Star Trek II: The Wrath of Kahn* (1982), *Star Trek III: The Search of Spock* (1984), and *Star Trek IV: The Voyage Home* (1986).

By the mid-1980, the *Star Trek* mythos had proven so commercially viable that Paramount announced plans for a new *Star Trek* series for television. Once again supervised by Roddenberry, *Star Trek: The Next Generation* debuted in first-run syndication in 1987 and went on to become one of the highest rated syndicated shows in history. Set in the 24th century, this series followed the adventures of a new crew on a new *Enterprise* (earlier versions of the ship having been destroyed in the movie series). The series was extremely successful at establishing a new story world that still maintained a continuity with the premise, spirit, and history of the original series. On the new *Enterprise*, the command functions were divided between a more cultured captain, Jean-Luc Picard (Patrick Stewart), and his younger, more headstrong "number one," Commander William Riker (Jonathan Frakes). Spock's character functions were distributed across a number of new crew members, including ship's counselor and Betazoid telepath, Deanna Troi (Marina Sirtis), the highly advanced android, Lt. Commander Data (Brent Spiner), who provided the show with "logical" commentary as ironic counter-point to the peculiarities of human culture, and finally, Lieutenant Worf (Michael

Dorn), a Klingon raised by a human family who struggled to reconcile his warrior heritage with the demands of the Federation. Other important characters included Lt. Geordi La Forge (LeVar Burton), the ship's blind engineer whose "vision" was processed by a high-tech visor, Dr. Beverly Crusher (Gates McFadden), the ship's medical officer and implicit romantic foil for Picard, and Wesley Crusher (Wil Wheaton), the dctor's precocious son.

Running for 178 episodes, *Star Trek: The Next Generation* was able to develop its characters and storylines in much more detail than the original series. As with many other hour-long dramas its era, the series abandoned a wholly episodic format in favor of more serialized narratives that better showcased the expanded ensemble cast. Continuing over the run of the series were recurring encounters with Q, a seemingly omnipotent yet extremely petulant entity, the Borg, a menacing race of mechanized beings, and Lor, Data's "evil" android brother. Other continuing stories included intrigue and civil war in the Klingon empire, Data's ongoing quest to become more fully human, and often volatile political difficulties with the Romulans. This change in the narrative structure of the series from wholly episodic to a more serialized form can be attributed in some part to the activities of the original series' enormous fan following. A central part of fan culture in the 1970s and 1980s involved fans writing their own *Star-Trek* based stories, often filling in blanks left by the original series and elaborating incidents only briefly mentioned in a given episode. *Star Trek: The Next Generation* greatly expanded the potential for such creative elaboration by presenting a more complex storyworld, one that actively encouraged the audience to think of the series as a foundation for imagining a larger textual universe.

Despite the show's continuing success, Paramount canceled *Star Trek: The Next Generation* after seven seasons to turn the series into a film property and make room for new television spin-offs, thus beginning a careful orchestration of the studio's Star Trek interests in both film and television. The cast of the original series returned to the theater for Star Treks 5 and 6, leading finally to *Star Trek: Generations,* in which the original cast turned over the cinematic baton to the crew of *Next Generation. Star Trek: Deep Space Nine* premiered in January of 1993 as the eventual replacement for *Next Generation* on television. In contrast to the usually optimistic and highly mobile structure of the first two series, *Deep Space Nine* was a much more claustrophobic reading of the Star Trek universe. Set aboard an aging space-station in orbit around a recently liberated planet, Bajor, the series generated its storylines from the aftermath of the war over Bajor and from a nearby "wormhole" that brought diverse travelers to the station from across the galaxy.

Hoping to compete with Fox and Warner Brothers in creating new broadcast networks, Paramount developed a fourth *Star Trek* series as the anchor for their United Paramount Network. *Star Trek: Voyager* inaugurated UPN in January 1995, serving as the network's first broadcast. Responding perhaps to the stagebound qualities and tepid reception of *Deep Space Nine, Voyager* opted for a premise that maximized the crew's ability to travel and encounter new adventures. Stranded in a distant part of the galaxy after a freak plasma storm, the *U.S.S. Voyager* finds itself seventy-five years away from earth and faced with the arduous mission of returning home.

Both *Deep Space Nine* and *Voyager* attracted the core fans of *Star Trek,* as expected, but neither series was as popular with the public at large as the programs they were designed to replace. Despite this, at century's end, there would seem to be every indication that the world of Star Trek will survive into the new millennium.

—Jeffrey Sconce

CAST

Captain James T. Kirk . . . . William Shatner
Mr. Spock . . . . Leonard Nimoy
Dr. Leonard McCoy . . . . DeForest Kelley
Yeoman Janice Rand (1966–67) . . . Grace Lee Whitney
Sulu . . . . George Takei
Uhura . . . . Nichelle Nichols
Engineer Montgomery Scott . . . . James Doohan
Nurse Christine Chapel . . . . Majel Barrett
Ensign Pavel Chekov (1967–69) . . . . . . Walter Koenig

PRODUCERS   Gene Roddenberry, John Meredyth Lucas, Gene L. Coon, Fred Freiberger

PROGRAMMING HISTORY   79 Episodes

• NBC

September 1966–August 1967	Thursday 8:30-9:30
September 1967–August 1968	Friday 8:30-9:30
September 1968–April 1969	Friday 10:00-11:00
June 1969–September 1969	Tuesday 7:30-8:30

FURTHER READING

Alexander, David, and Ray Bradbury. *Star Trek Creator: The Authorized Biography of Gene Roddenberry.* New York: Roc, 1994.

Asherman, Allan. *The Star Trek Compendium.* New York: Pocket, 1989.

Dillard, J. M., and Susan Sackett. *Star Trek, Where No One Has Gone Before: A History in Pictures.* New York: Pocket, 1994.

Gerrold, David. *The World of Star Trek.* New York, Ballantine, 1974.

Gibberman, Susan R. *Star Trek: An Annotated Guide to Resources on the Development, the Phenomenon, the People, the Television Series, the Films, the Novels, and the Recordings.* Jefferson, North Carolina: McFarland, 1991.

Jenkins, Henry. *Textual Poachers: Television Fans and Participatory Culture.* New York: Routledge, 1992.

Nemecek, Larry. *The Star Trek: The Next Generation Companion.* New York: Pocket, 1992.

Okuda, Michael, Denise Okuda, Debbie Mirek, and Doug Drexler. *The Star Trek Encyclopedia: A Reference Guide to the Future.* New York: Pocket, 1994.

Shatner, William, with Chris Kreski. *Star Trek Memories.* New York: Harper Collins, 1993.

Trimble, Bjo. *The Star Trek Concordance.* New York: Ballantine, 1976.

Tulloch, John, and Jenkins, Henry. *Science Fiction Audiences: Watching Doctor Who and Star Trek.* London and New York: Routledge, 1995.

Van Hise, James, and Hal Schuster. *Trek, The Unauthorized Story of the Movies.* Las Vegas, Nevada: Pioneer Books, 1995.

Whitfield, Stephen E., and Gene Roddenberry. *The Making of Star Trek.* New York, Ballantine, 1968.

See also Roddenberry, Gene; Science-fiction Programs

STAROWICZ, MARK

Canadian Broadcast Journalist/Producer

During his 25 years in radio and television with the Canadian Broadcasting Corporation, Mark Starowicz has produced a number of the more influential current affairs and documentary programs in Canadian broadcast history.

After beginning his career in newspaper journalism, Starowicz assumed the role of producer within the current affairs division of CBC radio at the age of 24. During the 1970s, Starowicz produced a total of five CBC radio programs, including *Radio Free Friday, Five Nights,* and *Commentary.* Starowicz received particular critical acclaim for his reworking of *As it Happens* (1973–76) and creation of *Sunday Morning* (1976-80), a three-hour weekend review.

CBC news programming chief Peter Herrndorf provided Starowicz's entry into television in 1979 by appointing him chair of a committee examining the corporation's news programming strategies. This resulted in the controversial move of *The National* news broadcast to 10:00 P.M. from its 11:00 P.M. slot, and the creation of *The Journal,* a current affairs and documentary program with Starowicz as executive producer. These decisions sought to take advantage of the larger audience numbers available at 10:00 P.M. (a difference between 10 and 4.5 million), and were part of CBC's strategy in the 1980s to invest its decreasing resources in its traditionally strong area of news and current affairs.

Despite Starowicz's lack of experience in television journalism, *The Journal* was a great success—both critically and in terms of viewership—and served to establish him as Canadian television journalism's new star. *The Journal* achieved an average 1.6 million viewers in its first year and comparable numbers during its ten-year run. Rather than decreasing the audience shares of its competitors, *The National* (22 minutes) and *The Journal* (38 minutes) combination actually increased the number of total viewers during the 10:00 P.M. time slot.

To deliver *The Journal*, Starowicz compiled a young staff, many of whom, like Starowicz, had previously only worked in radio. Hosts during the broadcast's life included Barbara Frum (formerly of *As It Happens)*, Mary Lou Finlay, Peter Kent, and Bill Cameron. Under Starowicz's leadership *The Journal* produced a total of 2,772 broadcasts between 1982 and 1992, consisting of 5,150 interviews and an amazing 2,200 documentaries. *The Journal* was notable for the depth with which it would develop stories, dedicating an entire broadcast to a single documentary if the subject required. For the interview segment of the show, Starowicz successfully re-invented the "double-ender" technique, (originally employed during the 1960s on CBS' *See it Now)*, wherein the anchor would interview guests that appear to the viewing audience to be projected on an in-studio screen. The high quality and volume of material were made possible by factors such as a staff of over 100, a budget of approximately $8(CDN) million per year (1980 dollars), and producer-reporter teams with as much as one month lead time for story preparation.

With the cancellation of *The Journal* in 1992, Starowicz accepted the position of executive producer of documentaries at the CBC. Starowicz oversees the weekly documentary prime-time series *Witness* (1990-), as well as CBC's new documentary unit. This one-hour broadcast consists of acquired, co-produced, and in-house documentaries dealing with a diverse array of often socially and politically charged issues. Although Starowicz's role as executive producer utilizes his capacity to orchestrate talent, he has produced and directed his own documentaries, including *The Third Angel* (1991) and *Red Capitalism* (1993), and is expected to contribute a minimum of two per year for *Witness*. Starowicz sees this series as an opportunity for the CBC to aggressively continue the strong documentary tradition in Canada started in the 1940s by John Grierson and the National Film Board. Significantly, Starowicz was able to get the CBC management to agree to the broadcasting of "point-of-view" documentaries, breaking free of the somewhat mythological pursuit of journalist "objectivity."

In numerous magazine and newspaper articles and public lectures, Starowicz has expressed his concern about the erosion of a uniquely Canadian sense of identity. He cites the absence of Canadian content in its own mass media and the dangers posed by U.S. cultural industries as key threats. To remedy this, he has proposed the introduction of a tax on U.S. media imports, continued public support for the CBC, the develop-

Mark Starowicz
Photo courtesy of Mark Starowicz

ment of a second public national network, and the extended financing of independent film and television production. Starowicz has argued that public television should not be produced for a small cultural and political elite, leaving private television to the "masses." Perhaps in illustration of his own goals, Starowicz is currently working on a co-production with the BBC on the history of public broadcasting.

—Keith C. Hampson

MARK STAROWICZ. Born in Worksop, United Kingdom, 8 September 1946. Educated at Loyola College High School, 1964; University of Grenoble, 1964; McGill University, B.A. 1968. Married: Anne, 1982; children: Caitlin-Elizabeth and Madeline Anne. Reporter, *Montreal Gazette*, 1964–68; editor, *McGill Daily*, 1968–69; reporter, *Toronto Star*, 1969–70; co-founder and writer, *The Last Post* magazine, 1969–73; producer, CBC Radio series, 1970–79; chair, Task Force to Reform CBC TV News and Current Affairs, 1979; executive producer, television program *The Journal*, 1982–92; executive documentary producer, CBC, since 1992. Member: Association of Toronto Producers and Directors. Recipient: Canadian Broadcasting League's Cybil Award, 1973; Ohio State Documentary Award, 1973; Anik Award, 1987; Gemini Award, 1987.

TELEVISION SERIES (producer)

1982–92	*The Journal*
1990	*Witness*

TELEVISION DOCUMENTARIES

1991	*The Third Angel*
1993	*Red Capitalism*
1994	*Romeo and Juliet in Sarajevo* (co-producer)
1994	*Escaping from History* (co-producer)
1994	*The Gods of Our Fathers* (co-producer)
1994	*The Tribal Mind* (co-producer)
1994	*The Bomb Under the World* (co-producer)
1994	*The Body Parts Business* (co-producer)

RADIO (producer)

Five Nights, 1970–73; *Radio Free Friday*, 1970–73; *Commentary*, 1970–73; *As It Happens*, 1973–76; *Sunday Morning*, 1976–80

FURTHER READING

Boone. "CBC Resignation Means There's Room at the Top for Starowicz." *Montreal Gazette*, 25 June 1994.

"CBC Producer Promoted." *Winnipeg Free Press*, 28 November 1992.

Quill. "Starowicz Stays with CBC-TV as the New Boss of Documentaries." *Toronto Star*, 26 November 1992.

Thosell. "Turning Your TV to the Final Nation State?" *Globe and Mail* (Toronto), 10 April 1993.

Underwood, Nora. "Twenty Years After." *Maclean's* (Toronto), 21 March 1988.

See also Canadian Programming in English; *National/ The Journal*

STARSKY AND HUTCH

U.S. Police Drama

At first glance, *Starsky and Hutch* (1975–79, ABC) seems of a piece with *Baretta*, *The Streets of San Francisco*, or even producer Aaron Spelling's own *Charlie's Angels*—one more post-1960s police series with street smarts and social cognizance, one that expresses at least a passing familiarity with youth culture. Yet on closer inspection, swarthy Dave Starsky (Paul Michael Glaser) and sensitive surfer Ken Hutchinson (David Soul), confirmed bachelors and disco-era prettyboys, seem to have taken the cop show maxim "Always watch your partner's back" well past their own private Rubicon.

The series was originally part of a logical progression by Spelling (with and without partner Leonard Goldberg) that traced the thread of the detective drama through the fraying social fabric at the end of the 1960s. Beginning with *The Mod Squad* (cops as hippies), this took him in logical sequence to *The Rookies* (cops as hippie commune), *S.W.A.T.* (cops as hippie commune turned collectivist cell/paramilitary cadre), and finally *Charlie's Angels* (ex-cops as burgeoning feminists/Manson Family pinups). This was before he jettisoned the cop show altogether and simply leached the raw hedonism out of 1960s liberalism—with *The Love Boat*, *Fantasy Island, Family* (sauteed in hubris), and ultimately, the neo-Sirkian *Beverly Hills 90210* and *Melrose Place*.

In this context, the freewheeling duo might seem the perfect bisecting point on a straight line between *Adam-12*'s Reed and Malloy and *Miami Vice*'s Crockett and Tubbs. *Butch Cassidy and the Sundance Kid* (1969) had ushered in the "buddy film" cycle, just then reaching its culmination with *All the President's Men*, and, in fact, the pair physically resemble no one so much as the high-gloss Redford and Hoffman assaying the golden boys of broadsheet expose, Woodward and Bernstein.

Yet viewed in retrospect, their bond seems at very least a curious one. Putting aside the ubiquitous costumes and leather, or Starsky's Coca-Cola-striped Ford Torino and Hutch's immense .357 Magnum handgun, which McLuhan or Freud might well have had a field day with, the drama always seems built around the specific gravity of their friendship. There is much of what can only be termed flirting—compliments, mutual admiration, sly winks, sidelong glances, knowing

Starsky and Hutch

smiles. They are constantly touching each other or indulging in excruciating cheek and banter—or else going "undercover" in various fey disguises. All of the women who pass between them—and their number is considerable, including significant ones from their past—are revealed by the final commercial break as liars or users or criminals or fatal attractions. And should one wind up alone with a woman, the other invariably retreats to a bar and drowns his sorrows. Following the inevitable betrayal, it is not uncommon for the boys to collapse sobbing into each other's arms.

This apparent secret agenda is perhaps best demonstrated in the opening credits themselves. Initially, these merely comprised interchangeable action sequences—Hutch on the prowl, Starsky flashing his badge. But by the second season, the action footage had been collapsed into a few quick images, followed by split-screen for the titles. To the left are three vertically stacked images: Hutch in a cowboy hat, both in construction outfits, and Starsky as Chaplin and Hutch in whiteface. Meanwhile, to the right, Starsky takes Hutch down in a full romantic clinch, the looks on their faces notably pained.

Next follows a series of quick clips: Starsky waits patiently while Hutch stops to ogle a bikini-clad dancer, and finally only gets his attention by blowing lightly on his cheek. Both gamble in a casino decked out in pinstripe Gatsby suits and fedoras, a la *The Sting*. Starsky, in an apron, fastidiously combs out a woman's wig, while Hutch sits dejectedly, shoulders squared, a dress pattern pinned around him. Hutch watches straight-faced while Starsky attempts the samba, festooned in thick bangles, flowing robes, and a Carmen Miranda headpiece. Each is then introduced individually—Soul shouting into the camera in freeze-frame, his mouth swollen in an enormous yawning oval, and Glaser as he ties a scarf foppishly to one side, frozen randily in mid-twinkle. Finally, a boiler-room explosion blows Starsky into Hutch's arms.

The entire sequence takes exactly one minute, with no single image longer than five seconds. And each scene is entirely explained away in context. Yet in the space of 60 seconds, these two gentlemen are depicted in at least four cases of literal or figurative transvestism, four cases of masculine hyperbole (encompassing at least two of the Village People), several prominent homosexual clichés (hairdresser, Carnival bacchanalian), a sendup of one of filmdom's most famous all-male couples, a wealth of Freudian imagery (including the pointed metaphor of fruit), two full-body embraces, two freeze-frames defining them in both homoerotic deed and dress, and one clearcut instance where the oral stimulation of a man prevails over the visual stimulation of a woman. This would seem to indicate a preoccupation on the part of someone with something. (And this doesn't even begin to address their dubiously named informant Huggy Bear—a flamboyant and markedly androgynous pimp.)

The tone of all this is uniformly playful, almost a parlor game for those in the know (not unlike *Dirty Harry*, whose most famous sequence—the bank robbery—is bookended on one side by Clint Eastwood biting into a hot dog, and on the other by a fire hydrant ejaculating over the attendant carnage). Meanwhile, the rather generic storylines consistently play fast and loose with gender.

Altogether, *Starsky and Hutch* is a fascinating digression for episodic television—especially considering that it was apparently conducted entirely beneath the pervasive radar of network censors.

—Paul Cullum

CAST

Detective Dave Starsky . . . . Paul Michael Glaser
Detective Ken Hutchinson (Hutch) . . . . . David Soul
Captain Harold Dobey . . . . Bernie Hamilton
Huggy Bear . . . . Antonio Fargas

PRODUCERS   Aaron Spelling, Leonard Goldberg, Joseph T. Naar

PROGRAMMING HISTORY   92 Episodes

• ABC

September 1975–September 1976 Wednesday 10:00-11:00
September 1976–January 1978   Saturday 9:00-10:00
January 1978–August 1978   Wednesday 10:00-11:00
September 1978–May 1979   Tuesday 10:00-11:00
August 1979   Tuesday 10:00-11:00

FURTHER READING

Collins, Max Allen. *The Best of Crime and Detective TV: Perry Mason to Hill Street Blues, The Rockford Files to Murder, She Wrote.* New York: Harmony Books, 1989.

Crew, B. Keith. " Acting Like Cops: The Social Reality of Crime and Law on TV Police Dramas." In, Sanders, Clinton R., editor. *Marginal Conventions: Popular Culture, Mass Media and Social Deviance.* Bowling Green, Ohio: Popular Culture Press, 1990.

Grant, Judith. "Prime Time Crime: Television Portrayals of Law Enforcement." *Journal of American Culture* (Bowling Green, Ohio), Spring 1992.

Hurd, Geoffrey. "The Television Presentation of the Police." In, Bennett, Tony, Susan Boyd-Bowman, Colin Mercer, and Janet Woollacott, editors. *Popular Television and Film.* London: British Film Institute, 1981.

Inciardi, James A., and Juliet L. Dee. "From the Keystone Cops to Miami Vice: Images of Policing in American Popular Culture." *Journal of Popular Culture* (Bowling Green, Ohio), Fall 1987.

Kaminsky, Stuart, and Jeffrey H. Mahan, editors. *American Television Genres.* Chicago: Nelson-Hall, 1985.

Marc, David. *Demographic Vistas: Television in American Culture.* Philadelphia: University of Pennsylvania Press, 1984.

See also Police Programs; Spelling, Aaron

STAR-TV

Asian Satellite Delivery Service

Star-TV is one of the most prominent regional satellite and cable television operations in the world. Its coverage footprint reaches from the Arab world to South Asia to East Asia. It carries global U.S. and British channels as well as channels in Mandarin and Hindi targeted at regional audiences defined by language and culture, and in so doing has helped define a new type of geo-cultural or geo-linguistic television market that stands between the U.S. dominated global market and national television markets. Fully acquired by Rupert Murdoch by the end of 1995 Star now forms a central part of his global media empire. In both news and culture, Star-TV is as challenging to some governments as United States imported programs and news have been. In 1995, Star-TV reached 53.7 million television households in 53 countries in English, Mandarin, and Hindi.

In April 1990, China's Long March III rocket launched a C-band satellite called AsiaSat-1. China International Trust and Investment Corp. (CITIC), Cable and Wireless of Britain and Hong-Kong's Hutchison Whampoa, jointly owned AsiaSat, making it Asia's first privately owned satellite. By picking up signals on parabolic dishes on the ground in an area under the satellite's footprint, regional broadcasting in Asia became possible.

In December 1990, Hong Kong granted a license to Hutchison Whampoa's satellite broadcasting arm, HutchVision, to begin a Direct Broadcast Satellite (DBS) service via AsiaSat. In a $300 million venture, the Satellite Television Asia Region operation, (Star-TV), began transmissions in August 1991. In July 1993, News Corp.'s Chairman Rupert Murdoch, already a power in Australia, Britain and America, bought into Star for $525 million (a 63.6% stake), forming a partnership with business tycoon Li Ka-shing, whose family owned the company. With the purchase, Murdoch's FOX studio and network had access to a successful Asian window in which to distribute programming. In July 1995, Rupert Murdoch's News Corp. paid $346 million to buy the remaining 36.4% of Star-TV.

Using AsiaSat for Star-TV created a problem, however, because the satellite was never meant to be used for broadcasting. Under the jurisdiction of the International Telecommunications Union (ITU), it was begun as a telecommunications satellite only. Little has been done about this situation, but criticism has developed in the scholarly community. In a 1992 paper for the International Communication Association, Seema Shrikhande asserted that, "Using telecommunications satellites for broadcasting goes against the ruling that national sovereignty includes the state's control over television within its borders and that satellite footprints should be tailored to national boundaries as far as possible." Following these assumptions, several countries have attempted to place restrictions on reception of Star-TV but have found them difficult to enforce.

Working with the idea of providing regionally-focused niche or genre-focused programming, Star-TV originally transmitted five channels 24-hours a day. These included MTV Asia (Viacom), British Broadcasting Corporation's World Service Television (WSTV), Prime Time Sports (a joint venture with the Denver-based Prime Network), entertainment and cultural programs through Star Plus, and a Mandarin Chinese-language channel. Subsequently MTV has withdrawn to offer its own wholly owned channels via satellite. Star-TV has replaced it with Channel V. More focused on Asian videos, Channel V has become quite popular and now competes favorably with MTV in the region.

In 1994, Murdoch removed the BBC World Service news from the northern part of Star's coverage area over China because its content offended news-sensitive China. In an earlier speech, Murdoch had said, "Advances in the technology of communications have proved an unambiguous threat to totalitarian regimes: Fax machines enable dissidents to bypass state-controlled print media; direct-dial telephone makes it difficult for a state to control interpersonal voice communication; and satellite broadcasting makes it possible for information-hungry residents of many closed societies to bypass state-controlled television channels." Despite this view, China subsequently demanded and received the removal of the BBC in order to permit reception of Star-TV in China.

Star-TV represents a very direct challenge to several Asian governments that have tended to restrict the inflow of information. Burma, Singapore, Saudi Arabia and Malaysia have made reception of Star essentially illegal. China requires a restrictive license for satellite reception dishes, although many individuals and cable systems continue to receive Star's offerings. India and Taiwan supposedly require licenses but permit both individuals and cable systems to receive it openly. Most other countries regulate redistribution via cable TV or apartment building antenna systems (SMATV) but are essentially open to Star and other satellite channels.

After its initial phase with five channels, Star has begun to target audiences more narrowly in terms of genres, language, and culture. For example, Star-TV is half-owner of its sixth channel, Zee TV, which offers Indian-produced Hindi-language programs. Zee reaches more than 25% of the total TV households in India and a significant viewership in the United Arab Emirates, Oman, Kuwait, Saudi Arabia and Pakistan. Another example of channel targeting by audience and culture comes as Star-TV has begun to refine several versions of its Channel V music channel for different regions of Asia.

In early 1996, Star-TV had 14 channels targeting various genres, languages and cultures, particularly in Hindi-speaking India and Mandarin Chinese-speaking areas. It expected to have a total of 30 channels in operation before the end of 1996, with most of the new ones concentrating on Japan, Indonesia, and other smaller markets. Its pattern as of 1996 in Japan and Indonesia is to start with one country-focused channel, including movies, sports, music and general entertainment, then to expand these into separate genre-based channels focused at that country.

Star's audience was originally concentrated in Taiwan, China and India, but has been steadily growing as its programming begins to target other cultures and languages as well. All of Star's channels are advertiser-driven; thus, they are free to viewers. This tends to give Star-TV a much larger audience than pay-TV operations. Star-TV takes programming from a number of sources, principally the United States, Hong Kong, China, Taiwan, India and Japan. It has tended to reduce non-Asian programming somewhat over time, responding to an apparent audience preference for what it calls "localized" programming. Star and similar regional operations add a new layer of complexity to discussions of concepts such as media imperialism, the globalization of culture, and the international flow of television. The system's emphasis on intra-regional cultural flows—across national borders but within language and cultural boundaries—assumes that audiences will respond to the cultural similarity or proximity of the programming. Given further satellite developments in other regions, Star-TV may be an example of one form of future television.

—Joseph Straubhaar

FURTHER READING

Beng, Y. S. "The Emergence of an Asian-Centred Perspective: Singapore's Media Regionalization Strategies." *Media Asia* (Singapore), 1994.

Bhatia, B. "Multi-channel Television Delivery Opportunities in the South Asia Region." *Media Asia* (Singapore), 1993.

Brauchelli, M. W. "Star Struck—A Satellite TV System Is Quickly Moving Asia into the Global Village." *Wall Street Journal* (New York), 10 May 1993.

Chan, J. M. "National Responses and Accessibility to Star TV in Asia." *Journal of Communication* (New York), 1994.

Fawthrop, T. "Chinese Shadows." *Index on Censorship* (London), 1994.

Greenwald, J. "Dish-Wallahs." *Wired* (San Francisco, California), May-June 1993.

Karp, J. "Cast Of Thousands." *Far Eastern Economic Review* (Hong Kong), 27 January 1994.

Khushu, O. P. "Satellite Communications in Asia: An Overview." *Media Asia* (Singapore), 1993.

Lau, T.-Y. "From Cable Television to Direct-Broadcast Satellite." *Telecommunication Policy.* September-October 1992.

Lull, J. *China Turned on: Television, Reform, and Resistance.* New York: Routledge, 1991.

Menon, V. "Regionalization: Cultural Enrichment or Erosion." *Media Asia* (Singapore), 1994.

Michaels, J. W., and N. Rotenier. "There Are more Patels out there than Smiths." *Forbes* (New York), 14 March 1994.

Nugent, P. "Down with the Dish." *Index on Censorship* (London), 1994.

Rajagopal, A. "The Rise of National Programming: The Case of Indian Television." *Media, Culture and Society* (London), 1993.

Scott, M. "News from Nowhere." *Far Eastern Economic Review* (Hong Kong), November 1991.

Thomas, P. "Informatization and Change in India—Cultural Politics in a Postmodern Era." *Asian Journal of Communication* (Singapore), 1993.

Wang, G. "Satellite Television and the Future of Broadcast Television in the Asia-Pacific." *Media Asia* (Singapore), 1993.

Waterman, D., and E. Rogers. "The Economics of Television Program Production in Far East Asia." *Journal of Communication* (New York), 1994.

See also Hong Kong; Murdoch, Rupert; News Corporation, Ltd.; Satellite

STATION AND STATION GROUP

A television station is an organization that broadcasts one video and audio signal on a specified frequency, or channel. A station can produce or originate its own programming, purchase individual programs from a program producer or syndicator, or affiliate with a "network" that provides a partial or complete schedule of programming. The term "station" is usually used to designate a local broadcast facility that includes origination or playback equipment and a transmitter, with the station being the last link between program producers and the viewer. Because the number of television channels available is limited, permission to operate a television station must be obtained from a governmental agency (in the United States, television stations are licensed by the Federal Communications Commission) and must operate within technical limitations to avoid interfering with signals from other television stations.

Television stations can be classified as "commercial" or "public" depending upon whether their source of funding is advertising revenue or government subsidy (although some stations rely upon both). Most television stations are divided into departments according to the primary functions of the station. The programming department is responsible for procuring or producing programming for the station and arranging the individual programs into a program schedule. The engineering department is responsible for the technical upkeep of station equipment, including transmitters, video recorders, switching equipment, and production equipment. The production department is responsible for producing local programs, commercial announcements, and other materials needed for broadcast. Many stations also have a news department that specializes in the production of news broadcasts. Commercial stations have a sales department responsible for selling commercial advertisements; many noncommercial stations have a similar "underwriting" department responsible for soliciting funds for the station. The promotions department is responsible for informing the audience about the program schedule using announcements on the station and in other media, such as newspapers and radio. Finally, many stations also have a business department responsible for collecting and distributing the revenues of the station. These departments are usually supervised by a station manager, general manager, or both.

An organization that owns or operates more than one station is known as a station group. There is a great deal of diversity in the manner in which groups operate individual stations. Some groups operate all the stations as a single unit, buying and scheduling programming for the station group as a unit in order to take advantage of economies of scale in negotiating the purchase price of programming or equipment. Other groups operate each station autonomously, with minimal group control over the daily operation of each station.

In the United States, the size of a station group is limited by federal regulations. As a result, the concentration of ownership of local television stations is extremely low,

with 1,181 commercial television stations being operated by more than 150 station groups, as of early 1996.

Many stations and stations groups are owned by companies with interests in other media. For example, the Tribune Company owns ten television stations, including WPIX (New York), KTLA (Los Angeles), and WGN (Chicago), as well as four radio stations, four daily newspapers (including the *Chicago Tribune*), and a television syndication company. Changes in broadcast ownership restrictions in the United States are expected to lead to larger station groups and increasing cross-ownership of broadcast and other media. (As of mid-1996, a station group was limited to 12 or fewer stations serving a maximum of 25% of the U.S. population.)

—August Grant

FURTHER READING

Broadcasting and Cable Yearbook. New Providence, New Jersey: R.R. Bowker, 1995.

Hilliard, R. L. *TV Station Operations and Management.* Newton, Massachusetts: Focal, 1989.

Smith, F.L., M. Meeske, and J.W. Wright, III. *Electronic Media and Government: The Regulation of Wireless and Wired Communication in the United States.* White Plains, New York: Longman, 1995.

See also Allocation; License; Ownership; United States: Networks

STEADICAM

In 1976 Cinema Products (CP), producer of motion picture support technologies, introduced Steadicam, a camera control device that profoundly influenced the look of both feature film and television in the years that followed. Developed by cinematographer Garrett Brown, the camera support mechanism was used on thousands of feature films world wide and earned an Oscar from the Academy of Motion Picture Arts and Sciences for technical achievement.

Steadicam wowed cinematographers and viewers alike with its apparent ability to "float" through space without physical constraints. At the center of this hand-held "revolution" was the patented use of gyroscopic motion to counter any irregularities in the camera operator's movement. For Steadicam was not just a body-brace that strapped a camera to an operator. It was a motorized, multi-directional, DC-powered mechanical arm that linked a padded vest on the operator's body with a sensitive "gimble" used for fingertip control of the camera head's pans and tilts. Without the gravity-bound lock of traditional camera supports (e.g. a tripod), Steadicam relied on the operator's physical skills to move nimbly through sets. Operators likened the task to the demands of ballet or long distance running.

Steadicam offered television directors and cinematographers benefits that were both logistical (speed of use, streamlined labor) and aesthetic (a film-look that was deemed dynamic and high-tech). The cinematic fluidity that became Steadicam's trademark was not limited to features. The device helped make exhibitionist cinematography a defining property of music videos after MTV emerged in 1981. Indeed, it became an almost obligatory piece of rental equipment for shoots in this genre. Most music videos, like primetime television, were shot on film and the Steadicam became a regular production component in both arenas. *Miami Vice*'s much celebrated hybridization of music video and the cop genre (1984-1989) made use of Steadicam flourishes even as it cloned music video segments within individual episodes. What critics of the show termed "overproduction" (stylized design, "excessively lensed" photography, and over-mixed sound tracks), fit well Cinema Product's pitch that Steadicam was "the best way to put production value on the screen." Postmodern stylization like that of *Miami Vice* defined American television in the 1980s, and Steadicam became a recognizable tool in primetime's menu of embellishment and "house looks," the signature visual qualities of individual production companies. ATAS

(Academy of Television Arts and Sciences), following AMPAS's lead, acknowleged Steadicam's impact on television with an Emmy.

Although orthodox production wisdom held that any given technique brought with it this type of distinct stylistic function, many practitioners in the early 1980s simply embraced CP's more pragmatic hype: that Steadicam was also a cost effective substitute for dolly or crane shots. Not only could the device preempt costly crane and dolly rentals, and the time needed to lay track across a set or location, but it cut to the heart of the stratified labor equation that producers imported to primetime from Hollywood. On scenes demanding Steadicam, the Director of Photography, the "A" camera operator, the focus-puller, and one or more assistants would merely stand aside as a single Steadicam operator executed lengthy moves that could previously consume inordinate amounts of program time. Steadicam was, then, not just a stylistic edge; it was also offered concrete production economies.

The popularity of steadicam was also affected by the growth of electronic field production. By the late 1980s CP had begun marketing its "EFP" version, a smaller variant better suited for 20-25-pound camcorder packages like the Betacam, and for the syndicated, industrial, and off-prime programming that embraced camcorders. At nearly 90 pounds loaded and at a cost of $40,000, the original Steadicam still represented a major investment. Steadicam EFP, by contrast, allowed tabloid and reality shows to move "showtime glitz" quickly into and out of their fragmentary exposes and "recreations." As channel competition heated up, and production of syndicated programming increased, Steadicam was but one stylistic tactic used to push a show above the "clutter" of look-alike programming. By the early 1990s, CP also marketed a "JR" version intended for the home market and "event videographers." At 2 pounds, and costing $600, CP hoped to tap into the discriminating "prosumer" market, a niche that used 8mm video and 3 pounds cameras. But video equipment makers were now building digital motion reduction systems directly into camcorders and JR remained a special interest resource.

While the miniaturization of cameras might imply a limited future for Steadicam, several trends suggest otherwise. HDTV (High Definition Television) cameras remain heavy armfuls, and Steadicam frequently becomes merely a component in more complicated camera control configurations. As a fluid but secure way of mounting a camera, that is, Steadicam is now commonly used at the end of cranes, cars, trucks, and helicopters—in extensions that synthesize its patented flourish into hybrid forms of presentational power.

While CP argued that the device made viewers "active participants" in a scene rather than "passive observers" it would be wrong to anthropomorphize the effect as a kind of human subjectivity. The Steadicam flourish is more like an out-of-body experience. A shot that races 6 inches above the

A steadicam and its operator
Photo courtesy of Jens Bogehegn

ground over vast distances is less a personal point-of-view than it is quadripedal or cybernetic sensation; more like a Gulf-war smart-bomb than an ontological form of realism. A stylistic aggression over space results, in part, because Steadicam worked to disengage the film/video camera from the operator's eyes; to dissociate it from the controlling distance of classical eye-level perspective. Video-assist monitors, linked to the camera's viewfinder by fiber-optic connections, made this optical "disembodiment" technically possible on the Steadicam and other motion control devices in the 1970s and 1980s, and liberated cameras to sweep and traverse diegetic worlds. Because running through obstruction-filled sets with a 90 pound apparatus myopically pressed to one's cornea could only spell disaster, operators quickly grasped the physical wisdom of using a flat LCD (liquid crystal display) video-assist monitor to frame shots. Yet the true impact of Steadicam, video-assist, and motion-control has less to do with how operators frame images, than with how film and television after 1980 turned the autonomous vision of the technologically disengaged eye into a stylistic index of cinematic and televisual authority.

An unheard of 75% of the scenes in *ER*—NBC's influential series that ranked number 1 or number 2 for all the 1994-95 season—were shot using the Steadicam. Many of these were included in the spectacular and complicated "one-er" sequences that defined the show, complicated flowing actions shot in one take with multiple moves and no cutaways. Citing these astonishing visual moments, trade magazine recognition confirmed that Steadicam's autonomous techno-eye now also provided a acknowledged programming edge.

—John Thornton Caldwell

FURTHER READING

Caldwell, John. *Televisuality: Style, Crisis, and Authority in American Television.* New Brunswick: Rutgers University Press, 1995.

Oppenheimer, Jean. "Lights, Camera: Eye-popping Cinematography, along With Sound and Editing, are Breaking New Ground in Hour Dramas." *The Hollywood Reporter* (Los Angeles, California), 6 June 1995.

STEPTOE AND SON

British Situation Comedy

Steptoe and Son was the most popular situation comedy in British television history, and one of the most successful. At the height of its fame in the early 1960s, it regularly topped the ratings and commanded audiences in excess of 20 million. In 1966, Labour Prime Minister Harold Wilson asked the BBC to delay the transmission of a repeat episode on election day until after the polls closed, because he was worried that many of his party's supporters would stay in to watch it rather than going out to vote.

Its creators, Ray Galton and Alan Simpson, were already well known and highly successful as the script writers for Tony Hancock. Indeed, it was Hancock's decision, the most disastrous of his career, to sever his links with Galton and Simpson which brought about the birth of *Steptoe and Son*. The BBC offered them a series of ten separate half-hour comedies, to be cast and produced according to their wishes, which they grabbed with alacrity, keen to produce more diverse material after such a long time working with the same star.

The most successful of these, transmitted in January 1962 under the banner title of *Comedy Playhouse*, was "The Offer", featuring a father and son firm of "totters", or rag-and-bone men. As soon as he saw it, head of Light Entertainment Tom Sloane knew it was a natural for a whole series. Galton and Simpson resisted at first, reluctant to commit themselves to another long-term venture, but were worn down by Sloane's persistence and the fact that he was clearly right.

The first series of *Steptoe and Son* was transmitted in June and July 1962 and consisted of five episodes. A further three series, of seven episodes each, followed in the next three years. The producer of all four series was Duncan Wood.

The basic plot line of *Steptoe and Son* is very simple and most episodes are in some way a variation on it. Albert Steptoe is an old-time, rag-and-bone man, a veteran of the Great War, who inherited the family business of the title from his father. He is a widower and lives with his son, Harold, and together they continue the business, with Harold doing most of the work. Albert is settled in his life and his lowly position in society, but Harold has dreams of betterment. He wants to be sophisticated and to enjoy the "swinging sixties". Above all he wants to escape from his father and make a life of his own, something which Albert

is prepared to go to any lengths to prevent. The comedy thus comes from the conflict of the generation gap and the interdependency of the characters. However hard he tries, we know that Harold will never get away. So, in his heart, does he, and that is his tragedy. Apart from anything else, his father is by far the smarter of the two.

The success of this formula was partly the result of the universality of the theme and partly the casting of the two leads. Galton and Simpson believed that they should cast straight actors rather than comedians and so signed up Wilfrid Brambell to play Albert and Harry H. Corbett as Harold. Between them, the writers and actors created two immortal characters and some extremely poignant drama, as well as the hilarious comedy. The television correspondent of *The Times* wrote in 1962: "*Steptoe and Son* virtually obliterates the division between drama and comedy."

A typical episode would see Albert ruining Harold's plans, whether it be in love, business or cultural pursuits. In "The Bird," Harold brings home a girl, only to find his father taking a bath in the main room. In "Sunday for Seven Days," Albert ruins Harold's choice of Fellini's *8 1/2* for an evening at the cinema. His father's generally uncouth behaviour frequently provokes Harold to utter the only catchphrase of the series: an exasperated "You dirty old man!"

In 1965, Galton and Simpson decided to stop writing the show while it was still an enormous success, although radio versions were produced in the following two years and the format was introduced to American television as *Sanford and Son*. However, with the arrival of colour television in Britain in 1967 and increased competition in comedy from the commercial network, the BBC decided in the early 1970s to bring back some of its top comedy successes of the middle 1960s. *Steptoe and Son* returned in 1970 for a further four series, a total of thirty episodes, between then and 1974.

The effectiveness of the show was in no way diminished. Indeed, the familiarity of the characters allowed the show to carry on where it had left off and achieve the same quality as before. Two feature films were also made of *Steptoe and Son*, though without the success of the television shows.

No more shows were made after 1974, but there is a footnote to the *Steptoe* story. Many programmes made on

Steptoe and Son
Photo courtesy of BBC

videotape were wiped by the BBC for purposes of economy in the early 1970s, including virtually all of the fifth and sixth series of *Steptoe and Son*. However, Ray Galton had made copies from the masters on the very first domestic video format, which became the only surviving copies. In 1990 he handed them to the National Film and Television Archive, which restored them to a viewable form and publicised the find with a theatrical showing. Although the technical quality was poor and they only played in black and white, the BBC transmitted a few of them to enormous success. The rest of the restored episodes were then transmitted, followed by all the black-and-white episodes from the 1960s, breaking the BBC's usual resistance to repeating black-and-white programmes.

Alas, the two leads were not around to witness the revival. Brambell died in 1985, following his screen son Corbett, who had died in 1983.

—Steve Bryant

CAST

Albert Steptoe . . . . Wilfrid Brambell
Harold Steptoe . . . . Harry H. Corbett

PRODUCERS   Duncan Wood, John Howard Davies, David Craft, Graeme Muir, Douglas Argent

PROGRAMMING HISTORY   55 30-minute episodes; 2 45-minute specials

• BBC

7 June 1962–12 July 1962	6 Episodes
3 January 1963–14 February 1963	7 Episodes
7 January 1964–14 February 1964	7 Episodes
4 October 1965–15 November 1965	7 Episodes
6 March 1970–17 April 1970	7 Episodes
2 November 1970–20 December 1970	8 Episodes
21 February 1972–3 April 1972	7 Episodes
24 December 1973	Christmas Special
4 September 1974–10 October 1974	6 Episodes
26 December 1974	Christmas Special

FURTHER READING

Burke, Michael. "You Dirty Old Man!" *The People* (London), 9 January 1994.

"How We Met: Ray Galton and Alan Simpson." *The Independent* (London), 11 June 1995.

See also Brambell, Wilfrid; Corbett, Harry H.; *Sanford and Son*

THE STEVE ALLEN SHOW (AND VARIOUS RELATED PROGRAMS)

U.S. Comedy Variety Program

One of the most famous ratings wars in television history began on 24 June 1956. That night NBC debuted *The Steve Allen Show* opposite the eighth anniversary program of what had become a television institution, *The Ed Sullivan Show* on CBS. The two hosts were markedly different. Sullivan was a rigorous master of ceremonies, known for enforcing strict conformity for both his guests and the members of his audience. Allen, too, served as host, but he was also innovative, funny and whimsical. Whereas Steve Allen liked to improvise and ad lib on his program, creating material and responding to guests and audience on the spot, *The Ed Sullivan Show* followed a strict format.

The appearances of Elvis Presley on the two programs serve to illustrate the differences between them. When Presley appeared on *The Ed Sullivan Show*, Sullivan instructed the camera operators to shoot the picture from the waist up only. On *The Steve Allen Show*, Presley appeared in a tuxedo and serenaded a bassett hound with his hit "You Ain't Nothing But a Hound Dog." Both strategies appeased nervous network censors, but each is emblematic of the show it served.

Relations between the two prominent hosts were not cordial and reached a low point in October 1956. Allen scheduled a tribute to the late actor, James Dean, for his 21 October program. When he learned that Sullivan planned his own tribute to Dean for his 14 October program, Allen charged that Sullivan had stolen his idea. Sullivan denied the charges and accused Allen of lying. Allen moved his segment to October 14 when both programs paid tribute to the late actor and showed clips from his last movie, *Giant.*

Much of Allen's work on *The Steve Allen Show* resembled previous performances on *The Tonight Show*, which he had hosted since 1954. He often opened the program casually, seated at the piano. He would chat with the audience, participate in skits, and introduce guests. Television critic Jack Gould considered the new program merely an expanded version of *The Tonight Show* and characterized it as "mostly routine stuff." Gould did concede that "more imagination could take the program far." *The Steve Allen Show* offered Allen a natural setting for what Gould termed his "conditioned social gift" of "creating spontaneous comedy in front of an audience in a given situation."

Allen also continued something else he had begun on *The Tonight Show*, discovering new talent. Andy Williams, Eydie Gorme and Steve Lawrence got their starts on *The Tonight Show*. And on the new show, Allen's man in the street interview segments launched the careers of comedians Bill Dana, Pat Harrington, Louis Nye, Tom Poston and Don Knotts. Dana played the timid Hispanic José Jiminez, and Harrington the suave Italian golfer Guido Panzino.

Characters created by Nye, Poston and Knotts were the best known of the group. Nye portrayed the effete and cosmopolitan Gordon Hathaway whose cry "Hi Ho Steverino" became a trademark of the program. Tom Poston was the sympathetic and innocent guy who would candidly answer any question but who could never remember his name. Probably the best remembered character was the nervous Mr. Morrison portrayed by Don Knotts. Often Morrison's initials were related to his occupation. On one segment he was introduced as K.B. Morrison whose job in a munitions factory was to place the pins in hand grenades. When asked what the initials stood for, Knotts replied, "Kaa Boom!" Invariably Allen would ask Knotts if he was nervous and always got the quick one word reply, "No!!!" Allen characterized the cast as the "happiest, most relaxed professional family in television."

The Steve Allen Show
Photo courtesy of Steve Allen

Allen became known for the outrageous. He conducted a geography lesson using a map of the world in the shape of a cube. He opened a program by having the camera shoot from underneath a transparent stage. Looking down at the camera, Allen remarked, "what if a drunk suddenly staggered into your living room and saw this shot?"

Although Allen won some of the ratings battles with Sullivan, he ultimately lost the war. In 1959 NBC moved *The Steve Allen Show* to Monday nights. The following year, it went to ABC for a fourteen week run. In 1961 Allen renamed the program *The Steve Allen Playhouse* and took it into syndication where it ran for three years.

—Lindsy E. Pack

THE STEVE ALLEN SHOW

REGULAR PERFORMER
Steve Allen

PROGRAMMING HISTORY

• CBS

December 1950–March 1951	Monday-Friday 7:00-7:30
July 1952–September 1952	Thursday 8:30-9:00

THE STEVE ALLEN SHOW
Comedy Variety

PROGRAMMING HISTORY

• NBC

June 1956–June 1958	Sunday 8:00-9:00
September 1958–March 1959	Sunday 8:00-9:00
March 1959	Sunday 7:30-9:00
April 1959–June 1959	Sunday 7:30-8:30
September 1959–June 1960	Monday 10:00-11:00

• ABC

September 1961–December 1961	Wednesday 7:30-8:30

REGULAR PERFORMERS
Steve Allen
Louis Nye
Gene Rayburn (1956–59)
Skitch Henderson (1956–59)
Marilyn Jacobs (1956–57)
Tom Poston (1956–59, 1961)
Gabe Dell (1956–57, 1958–61)
Don Knotts (1956–60)
Dayton Allen (1958–61)
Pat Harrington, Jr. (1958–61)
Cal Howard (1959–60)
Bill Dana (1959–60)
Joey Forman (1961)
Buck Henry (1961)

Jayne Meadows (1961)
John Cameron Swayze (1957–58)
The Smothers Brothers (1961)
Tim Conway (1961)
Don Penny (1961)

MUSIC
Les Brown and His Band (1959–61)

THE STEVE ALLEN COMEDY HOUR
Comedy Variety

REGULAR PERFORMERS
Steve Allen
Jayne Meadows
Louis Nye
Ruth Buzzi
John Byner

DANCERS
The David Winters Dancers

MUSIC
The Terry Gibbs Band

PROGRAMMING HISTORY

• CBS

June 1967–August 1967	Wednesday 10:00-1100

THE STEVE ALLEN COMEDY HOUR
Comedy Variety

PROGRAMMING HISTORY

• NBC

October 1980	Saturday 10:00-11:00
December 1980	Tuesday 10:00-11:00
January 1981	Saturday 10:00-11:00

REGULAR PERFORMERS
Steve Allen
Joe Baker
Joey Forman
Tom Leopold
Bill Saluga
Bob Shaw
Helen Brooks
Carol Donelly
Fred Smoot
Nancy Steen
Catherine O'Hara
Kaye Ballard
Doris Hess
Tim Lund
Tim Gibbon

MUSIC
Terry Gibbs and His Band

FURTHER READING

Allen, Steve. *Hi Ho Steverino! My Adventures in the Wonderful Wacky World of Television.* Fort Lee, New Jersey: Barricade, 1992.

———. *Mark It and Strike It: An Autobiography.* New York: Holt, 1960.

Gould, Jack. "To Meet Steve Allen." *New York Times*, 24 June 1956.

———. "Tribute to Actor Starts TV War," *New York Times*, 4 October 1956.

Shanley, J. P. "Trio of Thriving TV Bananas," *New York Times*, 10 November 1956.

"Steve Allen." *Current Biography Yearbook.* New York: H.W. Wilson, 1982.

See also Allen, Steve; *Tonight Show*

STREET LEGAL

Canadian Drama

When *Street Legal* completed its eighth and final season, one TV journalist called it "unblushingly sentimental, unblinkingly campy, unabashedly Canadian and completely addictive." The one-hour CBC drama series about a group of Toronto lawyers stands as a landmark event in Canadian broadcasting history. After taking two years to find its niche, it became extremely popular. In its last six seasons, it regularly drew about one million viewers, the benchmark of a Canadian hit.

The series debuted in 1987 with Maryke McEwen as executive producer. It experienced a rocky start, with good story ideas but weak execution, lacking style in directing, and consequently suffering low ratings. The theme music, however, was immediately identifiable—a distinctive, raunchy, rollicking saxophone piece by Mickey Erbe and Maribeth Solomon. At that time the show revolved around just three lawyers—Carrie Barr (played by Sonja Smits), Leon Robinovitch (Eric Peterson), and Chuck Tchobanian (C. David Johnson). Carrie and Leon were the committed, left-wing social activists and Chuck the motorcycle-riding, reckless, aggressive, 1980s lawyer.

From the third through the seventh seasons Brenda Greenberg was first senior producer, and then executive producer, with Nada Harcourt taking over for the final season. As CBC's director of programming in 1987, Ivan Fecan hired a Canadian script doctor at CBS, Carla Singer, to work with the producer on improving the show. It was after this that *Street Legal* began to find its niche, introducing aggressive, sultry, high-heeled, risk-taking Olivia Novak (played by Cynthia Dale) to contrast the niceness of the Carrie Barr character. Olivia became the most memorable and best-known, but other characters were also added. Alana (Julie Khaner) played a confident and compassionate judge, married to Leon, who confidently battled sexism in the workplace. Rob Diamond (Albert Schultz) handled the business affairs of the firm. In the fourth season, the first African-Canadian continuing character was introduced—crown prosecutor Dillon (Anthony Sherwood). He had a love affair with Carrie, and then with Mercedes (Alison Sealy-Smith), the no-nonsense black Caribbean secretary, and later joined the firm. New lawyer Laura (Maria Del Mar) clashed with

Olivia and romanced Olivia's ex-husband and partner, Chuck. Ron Lea played a nasty crown prosecutor called Brian Maloney, an in-joke to Canadians, who immediately connected him to the Conservative Prime Minister, lawyer Brian Mulroney. The enlarged ensemble cast allowed for more storylines and increased conflict.

The usual prime-time soap-opera shenanigans ensued, with ex-husbands and ex-wives reappearing, romances beginning and ending, children being born and adopted, promotions and firings, hirings and quittings, all against the backdrop of the Canadian legal system and the Toronto scene. The lawyers all wore gowns and addressed the court in Canadian legal terms, giving a different feeling from its American counterpart, *L.A. Law*, though the two shows were coincidentally developed and aired at the same time.

The issues dealt with were also definably Canadian as well as international. Leon fought an employment equity case for an RCMP candidate, as well as representing an African-Canadian nurse in front of the Human Rights Commission. Olivia became a producer of a Canadian movie. Chuck defended a wealthy Native cigarette smuggler on conspiracy to commit murder. Leon represented the survivors of a mine disaster and then ran for mayor of Toronto. Leon and Alana became involved with a Mexican refugee, eight months pregnant, who got in trouble with CSIS, the Canadian intelligence agency. Human-interest stories intertwined with the political issues and the characters' personal lives.

Street Legal represented a very important step in the Canadian television industry. Along with the CTV series *E.N.G.,* set in a Toronto television newsroom, the series established Canadian dramatic television stars. Cynthia Dale, who played vixen Olivia, has become nationally famous and has gone on to star in another series, as a Niagara Falls private eye in *Taking the Falls.* She has said that she gets letters from young girls who want to grow up to be just like Olivia. In one episode, when ogled and harassed by a construction worker as she passed his job site, Olivia knocked him off his sawhorse with her hefty briefcase. The scene was then inscribed into the new credit sequence.

The rest of the cast members have also gone on to other work, but the problem of a Canadian star system remains.

Street Legal
Photo courtesy of CBC

There are few series produced, even among all the networks, and often their stars will return to theatre or radio or, it has been noted, to auditioning again for TV parts. One reason *Street Legal* ended was that CBC could not afford to have two dramatic series on air at the same time, and the older program was supplanted by *Side Effects*, a medical drama. The show wrapped up with a two-hour movie in the spring of 1994 which drew a whopping 1.6 million viewers.

—Janice Kaye

CAST

Charles Tchobanian	C. David Johnson
Olivia Novak	Cynthia Dale
Dillon Beck	Anthony Sherwood
Alana Robinovitch	Julie Khaner
Rob Diamond	Albert Schultz
Laura Crosby	Maria Del Mar
Brian Maloney	Ron Lea
Leon Robinovitch	Eric Peterson
Mercedes	Alison Sealoy-Smith
Carrington Barr	Sonja Smits
Steve	Mark Saunders
Nick Del Gado	David James Elliott

PRODUCERS   Maryke McEwen, Brenda Greenberg, Nada Harcourt

PROGRAMMING HISTORY   126 episodes

• CBC

January 1987–March 1988	Tuesday 8:00-9:00
November 1988–March 1991	Friday 8:00-9:00
November 1991–March 1993	Friday 9;00-10:00
November 1993–March 1994	Tuesday 9:00-10:00

FURTHER READING

Miller, Mary Jane. "Inflecting the Formula: The First Seasons of *Street Legal* and *L.A. Law*." In, Flatery, David H. and, and Frank E. Manning, editors. *The Beaver Bites Back?: American Popular Culture in Canada*. Montreal, Quebec: McGill-Queen's University Press, 1993.

See also Canadian Programming in English

STUDIO

Studios are an integral part of independent television production, providing television programming created either by independent producers or, at times, the studio itself.

Studios have a long history with television. In 1944, three years before the FCC approved commercial broadcasting, RKO Studios announced plans to package theatrical releases and programming for television. Five years later, Paramount explored the profit potential of the new medium. By the early 1950s, Columbia and Universal-International had also started television subsidiaries. However, these early efforts were merely false starts. Low ad revenues and overall industry instability resulting from the 1948 anti-trust action against studio-owned theater chains made it difficult for studios to profit.

Toward the mid-1950s, after the networks successfully wrestled programming control away from commercial sponsors, however, studios provided the link between programming and a new breed of independent producers and syndicators. The most significant of these early studios—which began as an independent production company—was Desilu, founded in 1951 by Lucille Ball and Desi Arnaz. On the strength of its hit sitcom *I Love Lucy*, Desilu became a production empire that, by the late 1950s, rivaled the size and output of the largest motion picture studios. The company also solidified the position of the telefilm and independent producer's role in the medium. Under the leadership of Arnaz, Desilu hosted numerous successful independent producers, including Danny Thomas and Quinn Martin.

By this time, other studios were getting into the act, with Universal providing studio services for Jack Webb's Mark VII productions, and MCA's Revue Studios filming such series as *Alfred Hitchcock Presents* and *Leave it to Beaver*; although the Revue program's were quite diverse, they shared many studio qualities, including the same catalog of incidental and transitional music.

Warner Brothers studios became central to the rise of the action-oriented telefilm with its string of hit Westerns, including *Cheyenne*, *Sugarfoot*, and *Bronco Lane*. These shows were paired with a group of slick, contemporary detective shows such as *77 Sunset Strip* and *Hawaiian Eye*. In many ways Warner Brothers, the studio, was instrumental in discovering the techniques, the narrative strategies, and the modes of production needed for a large film studio to shift into the production of series television.

Another prolific 1960s independent producer/studio was Filmways, which began as a commercial production company. The studio's fortune grew when it joined with independent producer Paul Henning, creator and producer of such hits as *The Beverly Hillbillies*, *Green Acres*, and *Petticoat Junction*.

As the rural sitcom's corn-pone silliness gave way to the 1970s new age of relevance, Filmways was eclipsed by another major studio which also began as an independent—MTM Enterprises. Run on the fame of actress Mary Tyler Moore and the business-sense of her then-husband Grant Tinker, MTM became a major television studio that provided everything from writers and producers to stages and cameras. At the same time, the television divisions of Twentieth Century-Fox and Paramount Pictures were turning out such hits as *M*A*S*H* and *Happy Days*.

Today, producer/studios such as Desilu and MTM have faded, with most major television production provided by independents working in contractual relations with major studios such as Twentieth-Century Fox, Paramount, MCA-Universal and Warner Communication. For example, *The Simpsons*, which is independently produced by James L. Brooks' Gracie Films, is filmed by Twentieth Century-Fox (which, in the case of *The Simpsons*, farms out much of its animation to overseas production houses). In the sea of production logos flooding the end credits of most modern series, the final credit is usually that of a major film studio.

—Michael B. Kassel

FURTHER READING

Anderson, Christopher. *Hollywood/TV: The Studio System in the Fifties*. Austin: University of Texas Press, 1994.

Boddy, William. *Fifties Television: The Industry and Its Critics*. Urbana: University of Illinois Press, 1993.

Eastman, Susan Tyler, Sydney Head, and Lewis Klein. *Broadcast/Cable Programming Strategies and Practices*. Belmont, California: Wadsworth, 1981; 3rd edition, 1989.

STUDIO ONE

U.S. Anthology Drama

Studio One was one of the most significant U.S. anthology drama series during the 1950s. Like other anthology series of the time—*Robert Montgomery Presents, Goodyear Television Playhouse, Philco Theatre, Kraft Television Theatre*—the format was organised around the weekly presentation of a one-hour, live, television play. Several hours of live drama were provided by the networks per week, each play different: such risk and diversity is hard to come by today.

Writing about television, Stanley Cavell has argued, "What is memorable, treasurable, criticiseable, is not primarily the individual work, but the program, the format, not this or that day of *I Love Lucy*, but the program as such." While this admonition might admirably apply to the telefilm series that came later, the 1950s drama anthologies were premised on the fact that they were different every week. Yet the one-hour live format was one they had in common with each other, and because of that very fact they had to distinguish themselves from each other. They worked to develop a "house style," a distinctive reputation for a certain kind of difference and diversity, whether based on quality writing, attention to character over theme or, more typically, technical and artistic innovation which developed the form. A full assessment would necessarily consider each distinctive anthology series (and assess its "distinctiveness" from others) as a whole, and the failures and achievements of individual productions.

Studio One provides an emblematic continuity for the 1950s drama: it was the longest running drama anthology series, lasting ten years from 1948 to 1958, from the "big freeze" through the "golden age" to the made-in-Hollywood 90-minute film format: in all over 500 plays were produced. From the beginning *Studio One's* "house style" was foregrounded not only by the quality of its writers, but primarily by its production innovations, professionalism and experimentation within the limits of live production.

Studio One began as a CBS radio drama anthology show in the mid–1940s until CBS drama supervisor, Worthington Miner translated it to television. Its first production was an adaptation by Miner of "The Storm" (7 November 1948). Miner's control emphasised certain "quality" characteristics: adaptation (usually of classical works, e.g. *Julius Caesar*, 1948) and innovation ("Battleship Bismarck," 1949). *Studio One* adopted a serious tone under Miner, but also a pioneering spirit. For example, "Battleship Bismarck" made advanced use of telecine inserts, three-camera live editing within a confined and waterlogged set. Miner left to join NBC in 1952, but the show regained an even clearer sense of identity and purpose when Felix Jackson became the producer in 1953. Jackson used two directors, Paul Nickell and Franklin Schaffner, each with his own technical staff, who would alternate according to the material. Nickell would be given the more "sensitive" scripts, Schaffner the epics, the action. Both directors were committed to pushing the live studio drama to the limits. Nickell in particular has to stand as one of the greatest—and unsung—television directors: he never made the mistake of thinking a good TV drama has to look like a film.

By the mid–1950s the emphasis of production material had turned from adaptation to new works written for television, often giving attention to contemporary issues. *Studio One* followed this trend. Often the same writers, such as Reginald Rose, who had adapted for *Studio One,* now wrote originals. Rose worked as an adapter until 1954 when he wrote "12 Angry Men" (1954) and the controversial "Thunder on Sycamore Street" (1954). This story, about racial hatred, was modified to satisfy southern television station owners, replacing a black protagonist with a convict. By 1955 *Studio One* was receiving over 500 unsolicited manuscripts per week.

However, it was *Studio One's* technical innovation, rather than its coterie of writers, which made the series distinctive. Its chief rival in the ratings, Fred Coe's *Philco-Goodyear Theatre*, although it had a superior stable of writers (Paddy Chayefsky, Rod Serling, Horton Foote, Robert Alan Aurthur, Tad Mosel—most of who later worked for *Studio One*), could not match *Studio One's* technical daring. *Philco-Goodyear Theatre* developed a reputation for plays which explored the psychological realism of character, using many close-ups, but this was influenced by other factors. As Tad Mosel has said: "I think that began because the sets were so cheap, if you pulled back you'd photograph those awful sets. Directors began moving in to faces so you wouldn't see the sets. *Studio One* had much more lavish productions, they had more money."

After 1955 *Studio One* joined the general decline of the other New York based dramas. The formats began to favour 90-minute slots (such as CBS *Playhouse 90*), and drama shot on film, often in Hollywood. Eventually *Studio One* joined the drift to Hollywood and film. By 1957 the anthology was renamed *Studio One in Hollywood*—and the sponsors, Westinghouse, withdrew from the series.

Studio One's achievements have to be measured in terms of technical and stylistic superiority over their rival anthologies. With plays such as "Dry Run" and "Shakedown Cruise" (both set on a flooded submarine, built in the studio) and "Twelve Angry Men", they were the first to use four-walled sets, hiding the cameras behind flying walls, or using portholes to conceal cameras between shots. The freedom to innovate was in part due to CBS' policy of giving directors relative autonomy from network interference and the stability of the Schaffner-Nickell partnership, but it is also a pioneering quality which can be traced back to Worthington Miner and the late 1940s. Miner was quite clear that he wanted *Studio One* to advance the medium via its experimental storytelling techniques: "I was fascinated by the new medium and convinced that television was somewhere be-

Studio One: Wuthering Heights
Photo courtesy of Wisconsin Center for Film and Theater Research

tween drama and film ... a live performance staged for multiple cameras."

However, with the mature *Studio One* productions of the early and mid-1950s, one has the sense that the movements of the cameras were *not* subordinate to the requirements of the performance: quite the opposite. For example, "The Hospital" was an adaptation produced during the 1952 season, and directed by Schaffner. This play seems to achieve the impossible: it literally denies the existence of live studio time. Flashbacks and other interruptions could be achieved with some narrative jigging to allow for costume and scene changes. Still, unlike film, live studio time was real time, and the ineluctable rule of live drama was that the length of a performance was as long as it took to see it. But Schaffner had a reputation for thinking that nothing was impossible for live television. Most other anthologies of the period used a static three camera live studio set up, where two cameras were used for close-ups and the other for the two-shots. In such an arrangement the television camera acted as a simple, efficient, relay. Schaffner favoured instead a mobile mise-en-scene; his cameras were constantly on the move, with actors and props positioned and choreographed for the cameras.

This play concerns the drama of a local hospital, following the various staff and patients through typical medical crises. Although the transmitted play lasts 50 minutes, the story-time takes up only 18 minutes. Some scenes are therefore repeated during the three acts, using a different viewpoint, and requiring the actors to re-stage precisely their initial scenes. As some scenes are lengthened, or modified in the light of what we have seen before we gain a greater understanding of the events from each character's viewpoint. Whilst this would be relatively simple to achieve on film, for live drama it involved complex methods of panning and camera movement to capture and expand the chronicity of events and repeat them exactly as it had gone before. Schaffner achieves this by using several cranes to snake through the various sets as the scenes are played and repeated, often in a different order. Doing what seems technically impossible is therefore foregrounded in this drama, and the complexity of this achievement is emphasised by the ironic commentary of one of the hospital patients who, with head bandaged, is able to explain at the end, as the sponsors shout for their adverts, "Time? There is no time. Time is only an illusion." And *Studio One* could prove it.

—Jason J. Jacobs

SPOKESPERSON (1949–58)
Betty Furness

PRODUCERS   Herbert Brodkin, Worthington Miner, Fletcher Markle, Felix Jackson, Norman Felton, Gordon Duff, William Brown, Paul Nickell, Franklin Shaffner, Charles H. Schultz

PROGRAMMING HISTORY   466 Episodes

• CBS

November 1948–March 1949	Sunday 7:30-8:30
March 1949–May 1949	Sunday 7:00-8:00
May 1949–September 1949	Wednesday 10:00-11:00
September 1949–September 1958	Monday 10:00-11:00

FURTHER READING

Averson, Richard, and David Manning White, editors. *Electronic Drama: Television Plays of the Sixties.* Boston, Massachusetts: Beacon, 1971.

Brooks, Tim, and Earle Marsh, editors. *The Complete Directory to Prime Time Network TV Shows: 1946–Present.* New York: Ballentine, 1992.

Gianakos, Larry James. *Television Drama Series Programming: A Comprehensive Chronicle, 1947–1959.* Metuchen, New Jersey: Scarecrow, 1980.

Hawes, William. *The American Television Drama: The Experimental Years.* University, Alabama: University of Alabama Press, 1986.

Kindem, Gorham, editor. *The Live Television Generation of Hollywood Film Directors: Interviews with Seven Directors.* Jefferson, North Carolina: McFarland, 1994.

MacDonald, J. Fred. *One Nation Under Television: The Rise and Decline of Network TV.* New York: Pantheon, 1990.

Miner, Worthington. *Worthington Miner, Interviewed by Franklin Schaffner.* Metuchen, New Jersey: Scarecrow, 1985.

Skutch, Ira. *Ira Skutch: I Remember Television: A Memoir.* Metuchen, New Jersey: Scarecrow, 1989.

Stemple, Tom. *Storytellers to the Nation: A History of American Television Writing.* New York: Continuum, 1992.

Sturcken, Frank. *Live Television: The Golden Age of 1946–1958 in New York.* Jefferson, North Carolina: McFarland, 1990.

Wicking, Christopher, and Tise Vahimagi. *The American Vein: Directors and Directions in Television.* New York: Dutton, 1979.

Wilk, Max. *The Golden Age of Television: Notes from the Survivors.* New York: Dell, 1977.

See also Anthology Drama; "Golden Age" of Television; Miner, Worthington; Schaffner, Franklin

SUBTITLING

Subtitling is the written translation of the spoken language (source language) of a television program or film into the language of the viewing audience (the target language); the translated text usually appears in two lines at the foot of the screen simultaneously with the dialogue or narration in the source language.

This simultaneous provision of meaning in two different languages, one in oral and the other in written text, is thus a new form of language transfer created by film and further developed by television. It combines the two ancient forms of interlingual communication, i.e., "interpretation," involving speaking only, and "translation," involving writing only. The concept is sometimes used synonymously with "captioning." In terms of technical production and display on the screen, there is no difference between the two, although it is useful to reserve the term "caption" for the screen display of writing in the same language.

Subtitling is, together with dubbing, the main form of translation or "language transfer" in television, which is increasingly developing into a global medium in a world fragmented by about 5,000 languages. The scope of language transfer activity depends on the relative power of the television market of each country, its cultural, linguistic and communication environment, and audience preferences. Compared with North America, the countries of the European Union, for example, have a larger population, more TV viewers, TV households and program production. However, linguistic fragmentation has undermined their ability to effectively perform in the global market, and compete with the powerful, monolingual audiovisual economy of the United States. As a step toward the building of a "European single market," the Council of European Communities took measures in 1990 to overcome the "language barrier" by, among other means, promoting dubbing, subtitling, and multilingual broadcasting (see the text of the decision in Luyken, p. 208; Kilborn, p. 654). The deregulated market of Eastern Europe, too, is linguistically fragmented, and heavily dependent on imports. The annual total of foreign programs broadcast in Eastern Europe was estimated to be 19,000 hours in 1992 (Dries, p. 35). English has emerged as the largest source language in the world. Many countries prefer to import programs from the Anglophone audiovisual market in part because it is more economical to conduct language transfer from a single source language.

The ideal in subtitling is to translate each utterance in full, and display it synchronically with the spoken words on the screen. However, the medium imposes serious constraints on full text translation. One major obstacle is the limitations of the screen space. Each line, recorded on videotape, consists of approximately 40 characters or typographic spaces (letters, punctuation marks, numbers and word spaces) in the Roman alphabet, although proportional spacing (e.g. more space for "M" and less for "l") allows more room for words, which average five letters in English. Another constraint is the duration of a subtitle, which depends on the quantity and complexity of the text, the speed of the dialogue, the average viewer's reading speed (150 to 180 words per minute), and the necessary intervals between subtitles. Taking into account various factors, the optimum display time has been estimated to be four seconds for one line and six to eight seconds for two lines. As a result, the subtitler often presents the source language dialogue or narration in condensed form. Loss or change of meaning also happens because the written text cannot transfer all the nuances of the spoken language. Other problems relate to the reception process. Unlike the printed page, the changing screen does not allow the viewer to re-read a line, which disappears in a few seconds. Audiences have to divide the viewing time between two different activities, reading the subtitles and watching the moving picture, and constantly interrelating them. Thus, subtitling has created not only a new form of translation, but also new reading processes and reading audiences. This type of reading demands different literacy skills, which are individually and, often, effectively acquired in the process of viewing.

In spite of the limitations of subtitling, selectively outlined above, some broadcasters and viewers prefer it to dubbing in so far as it does not interfere with the source language. Although viewers of subtitled programs are not usually familiar with the source language, it is argued that they derive more authentic meaning by hearing the original speech. Preference for one or the other form of language transfer depends on the cultural, political, linguistic, and viewing traditions of each country as well as economic considerations such as audiovisual market size, import policies and the relative cost of each transfer method. It is known (Luyken, p. 181), for example, that Europe is divided into "subtitling countries" (e.g., Belgium, Cyprus, Finland, Greece, the Netherlands, Portugal, and Scandinavia) and "dubbing countries" (France, Germany, Italy and Spain). Dubbing is usually more expensive, more complex and time-consuming than subtitling or voice-over. Still, some of the economically troubled countries of Eastern Europe (Bulgaria, Czech Republic, Hungary, and Slovakia) dubbed the majority of their imported programs in 1992. In these countries, as in others, the professional community of actors supports the dubbing process as a source of employment.

Language transfer involves more than facilitating the viewer's comprehension of unfamiliar language. The European Commission has, for example, recommended subtitling as a means of improving knowledge of foreign languages within the European Union. Technological innovations are rapidly changing the production, delivery and reception of subtitles. Some satellite broadcasters provide multilingual subtitling by using a teletext-based system, which allows the simultaneous transmission of up to seven sets of subtitles in different languages. The viewer can choose any language by dialling the assigned teletext page. Subtitling has usually been a post-production activity but real-time subtitling for live broadcasting is already available. An interpreter watches a live broadcast, and provides simultaneous translation (interpretation) by speaking into a microphone connected to the headphone of a high-speed "audio typist." The interpreted text appears on the screen while it is keyed on the adapted keyboard of a computer programmed for formatting and boxing subtitles (Luyken, p. 64-65, 68). This kind of heavily mediated subtitling will no doubt be simplified when technological advance in voice recognition allows the direct transcription of the interpreted text. The demand for subtitling was growing in the mid-1990s outside North America, especially in Europe. In 1994, one company, the Subtitling International Group centered in Stockholm with branches in six capital cities, produced 26,000 hours of subtitles for cinema, video and television.

—Amir Hassanpour

FURTHER READING

Dries, Josephine. "Breaking Language Barriers Behind the Broken Wall." *Intermedia*, December/January, 1994.

Kilborn, Richard. "'Speak my Language': Current Attitudes to Television Subtitling and Dubbing," *Media, Culture and Society* (London), 1993.

Luyken, Georg-Michael. *Overcoming Language Barriers in Television: Dubbing and Subtitling for the European Audience.* Manchester: European Institute for the Media, 1991.

See also Closed Captioning; Dubbing; Language and Television; Voice-Over

SULLIVAN, ED

U.S. Variety Show Host

Anyone who watched television in America between 1948 and 1971 saw Ed Sullivan. Even if viewers did not watch his Sunday night variety show regularly, chances are they tuned in occasionally to see a favorite singer or comedian. Milton Berle may have been "Mr. Television" in the early years of TV, but for almost a quarter-century Sullivan was Mr. Sunday Night. Considered by many to be the embodiment of banal, middle-brow taste, Sullivan exposed a generation of Americans to virtually everything the culture had to offer in the field of art and entertainment.

Sullivan began as a journalist. It was his column in the New York *Daily News* that launched him as an emcee of vaudeville revues and charity events. This led to a role in a regular televised variety show in 1948. Known as the *Toast of the Town* until 1955, it became *The Ed Sullivan Show,* in September of that year. According to CBS president William S. Paley, Sullivan was chosen to host its Sunday night program because CBS could not hold anyone comparable to Berle. Ironically, Sullivan outlasted Berle in large measure because of his lack of personality. Berle came to be identified with a particular brand of comedy that was fading from popularity. On the other hand, Sullivan simply introduced acts, then stepped into the wings.

Ed Sullivan's stiff physical appearance, evident discomfort before the camera, and awkward vocal mannerisms (including the oft-imitated description of his program as a "reeeeeelly big shoe") made him an unlikely candidate to become a television star and national institution. But what Sullivan lacked in screen presence and personal charisma he made up for with a canny ability to locate and showcase talent. More than anything else, his show was an extension of vaudeville tradition. In an era before networks attempted to gear a program's appeal to a narrow demographic group, Sullivan was obliged to attract the widest possible audience. He did so by booking acts from every spectrum of entertainment: performers of the classics such as Itzhak Perlman, Margot Fonteyn and Rudolf Nureyev; comedians such as Buster Keaton, Bob Hope, Henny Youngman, Joan Rivers, and George Carlin; singers like Elvis Presley, Mahalia Jackson, Kate Smith, the Beatles, James Brown and Sister Sourire, the Singing Nun. Sports stars appeared on the same stage as Shakespearean actors. Poets and artists shared the spotlight with dancing bears and trained dogs. And then there were the ubiquitous "specialty acts" such as Topo Gigio, the marionette mouse with the thick Italian accent enlisted to "humanize" Sullivan, and Senor Wences, the ventriloquist who appeared over twenty times, talking to his lipstick-smeared hand and a wooden head in a box. Sullivan's program was a variety show in the fullest sense of the term. While he was not so notable for "firsts," Sullivan did seem to convey a kind of approval on emerging acts. Elvis Presley and many other performers had appeared on network television before ever showing up on the Sullivan program, but taking his stage once during prime time on Sunday night meant more than a dozen appearances on any other show.

Although Sullivan relented to the blacklist in 1950, apologizing for booking tap dancer and alleged Communist sympathizer Paul Draper, he was noted for his support of civil rights. At a time when virtually all sponsors balked at permitting black performers to take the stage, Sullivan embraced Pearl Baily over the objections of his sponsors. He also showcased black entertainers as diverse as Nat "King" Cole, Leontyne Price, Louis Armstrong, George Kirby, Richard Pryor, Duke Ellington, Richie Havens and the Supremes.

Ed Sullivan

Sullivan attempted to keep up with the times, booking rock bands and young comedians, but by the time his show was canceled in 1971 he had been eclipsed in the ratings by "hipper" variety programs like *Rowan and Martin's Laugh-In,* and *The Flip Wilson Show.* Sullivan became victim to his own age and CBS' desire to appeal to a younger demographic, regardless of his show's health in the ratings. He died in 1974.

Since *The Ed Sullivan Show* ended in 1971, no other program on American television has approached the diversity and depth of Sullivan's weekly variety show. Periodic specials drawing from the hundreds of hours of Sullivan shows, as well as the venue of the *Late Show with David Letterman,* continue to serve as tribute to Sullivan's unique place in broadcasting. Ed Sullivan remains an important figure in American broadcasting because of his talents as a producer and his willingness to chip away at the entrenched racism that existed in television's first decades.

—Eric Schaefer

ED(WARD VINCENT) SULLIVAN. Born in New York City, New York, U.S.A., 28 September 1902. Married: Sylvia Weinstein, 1930; one daughter. Covered high-school sports as a reporter, *Port Chester Daily Item;* joined *Hartford Post,* 1919; reporter and columnist, *New York Evening Mail,* 1920–24; writer, *The New York World,* 1924–25, and *Morning Telegraph,* 1925–27; sportswriter, *The New York Evening Graphic,* 1927–29, Broadway columnist, 1929–32; columnist, New York *Daily News*, from 1932; launched radio program over Columbia Station WCBS (then WABC), showcasing new talent, 1932; staged benefit revues during World War II; host, CBS radio program *Ed Sullivan Entertains*, from 1942; host, CBS television variety program *Toast of the Town* (later *The Ed Sullivan Show)*, 1948–71. Died October 1974.

TELEVISION SERIES

1948–71   *Toast of the Town* (became *The Ed Sullivan Show*, 1955)

FILMS (writer)

There Goes My Heart (original story), 1938; *Big Town Czar* (also actor), 1939; *Ma, He's Makin' Eyes at Me*, 1940.

RADIO

Ed Sullivan Show, 1932; *Ed Sullivan Entertains*, 1942.

FURTHER READING

Astor, David. "Ed Sullivan Was also a Syndicated Writer." *Editor and Publisher* (New York), 7 December 1991.

Barnouw, Eric. *Tube of Plenty: The Evolution of American Television.* New York: Oxford University Press, 1975; 2nd edition, 1990.

Bowles, Jerry. *A Thousand Sundays: The Story of the Ed Sullivan Show.* New York: Putnam's, 1980.

Falkenburg, Claudia, and Andrew Solt, editors; text by John Leonard. *A Really Big Show: A Visual History of the Ed Sullivan Show.* New York: Viking, 1992.

Harris, Michael David. *Always on Sunday/ Ed Sullivan: An Inside View.* New York: Signet, 1968.

Wertheim, Arthur Frank. "The Rise and Fall of Milton Berle." In, O'Connor, John E., editor. *American History/American Television: Interpreting the Video Past.* New York: Unger, 1983.

See also *Ed Sullivan Show, Steve Allen Show,* Variety Programs

SUPERSTATION

A superstation is an independent broadcast station whose signal is picked up and redistributed by satellite to local cable television systems. Within its originating market, the station can be received off the air using a home antenna. Once uplinked to a satellite, however, the station functions as a cable program service or cable "network."

The origins of modern superstations can be traced back to the start of distant signal importation by early cable (CATV) systems using microwave relays. At first the relays simply brought signals to communities too remote to receive them using rooftop or community antennas, but as cable systems began to penetrate television markets with one or more local stations, operators often would import the signals of popular, well-financed stations from major metropolitan areas to make their service more appealing to potential subscribers. In effect, the distant signals were combined with local signals to create a distinct cable programming package. All of today's superstations were carried by microwave at one time; however, the actual term "superstation" was not used until the late 1970s, shortly after Ted Turner's Atlanta station, WTBS, became the first independent station to be carried by satellite.

Not only was Turner's station the first satellite-delivered independent station (and the second satellite-delivered cable program service overall), it was an innovator in the type of programming that would be most successful on cable. As with many cable-only program services, the popularity of superstations stems largely from their numerous movie screenings and extensive sports coverage—program types available in much smaller quantities from the broadcast networks and their affiliates. Superstation status also gives an independent station an economic advantage when competing with other stations for the broadcast rights to popular syndicated series. The evolution of WTBS's successful program schedule represents an aggressive effort to acquire these sorts of programs.

The existence of WTBS dates back to 1968, when Turner purchased a failing UHF station. He quickly changed the fortunes of his new station (which he called WTCG during its early years) by using old movies and syndicated television series to counterprogram network affiliate stations, going after such audience segments as children and people not watching the news. By the early 1970s, Turner's station also offered local sports programming—first professional wrestling and later baseball, basketball, and hockey. As of 1972, WTCG had become popular enough in the Atlanta metropolitan area that its signal had begun to be carried by microwave to cable systems throughout Georgia and northern Florida. In 1976, when Turner uplinked

Courtesy of TBS

his signal to a communications satellite, WTCG's potential coverage was extended to locations as distant as Canada and Alaska. The station was renamed WTBS (for Turner Broadcasting System) in the late 1970s to reflect the scope of its new operations.

Within the next few years, the signals of other major-market independent stations began to be carried on satellite, as well. However, the stations that followed WTBS to satellite carriage represent a different category of superstation. WTBS is considered to be an "active" superstation because it pursues superstation status as part of day-to-day operations: programming targets a nationwide market more than a local market, and national advertising is sought. WTBS currently is the only active superstation.

"Passive" superstations, by contrast, traditionally have done little or nothing to acknowledge themselves as superstations. Satellite common carriers such as United Video, Inc. and EMI Communications Corp. retransmit the stations' signals without any formal consent, sometimes against the stations' wishes. In spite of their potential to be viewed thousands of miles away, passive superstations have continued to direct the greater portion of their programming and advertising toward local or regional markets. As with any cable program service, cable operators pay per-subscriber fees for the use of passive superstations' signals. However, the fees are paid to the common carriers, not to the stations.

As cable's popularity continues, passive superstations are giving more recognition to their own superstation status, often having an employee who functions as a liaison to the satellite carrier and possibly to the cable systems taking the service. Nonetheless, most continue to feel that their priorities remain with their local markets.

The five "passive" superstations currently in operation are: WOR and WPIX, New York; WSBK, Boston; WGN, Chicago; KTLA, Los Angeles; and KTVT, Dallas. It is worth noting that this group includes some of the country's most long-standing broadcast stations. Like WTBS, these stations have been extremely successful in counterprogramming other stations. All carry local sports, for example. WOR features Mets baseball; WPIX the Yankees; WSBK the Red Sox; WGN the Cubs; KTLA the Dodgers; and KTVT the Rangers. Other sports teams also are carried by all of these stations. Most also feature regularly scheduled movie programs, often with well-known hosts.

The popularity of independent stations as cable program services has surprised many, particularly those who have touted cable's potential to provide programming substantially different from that of broadcast television. This popularity indicates quite a lot about the economics of satellite-served cable, a new vehicle for television programming that has had to compete with the established and resource-laden broadcast networks. In many instances, the formula for success has been found in program schedules that are familiar to television audiences, but which nonetheless differ from those of the "big three"—a formula independent stations have been following for decades.

—Megan Mullen

FURTHER READING

Bibb, Porter. *It Ain't as Easy as it Looks: Ted Turner's Amazing Story*. New York: Crown, 1993.

Garay, Ronald. *Cable Television: A Reference Guide to Information*. New York: Greenwood, 1988.

Golbert, Robert, and Gerald Jay Goldberg. *Citizen Turner: The Wild Rise of an American Tycoon*. New York: Harcourt Brace, 1995.

Mair, George. *Inside HBO: The Billion Dollar War between HBO, Hollywood, and the Home Video Revolution*. New York: Dodd, Mead, 1988.

Moshavi, Sharon D. "Turner Superstation: Where Does It Go from here?" *Broadcasting* (Washington, D.C.), 29 October 1990.

Sherman, Barry L. *Telecommunications Management: The Broadcast and Cable Industries*. New York: McGraw-Hill, 1987.

Williams, Christian. *Lead, Follow or Get Out of the Way: The Story of Ted Turner*. New York: Times Books, 1981.

See also Cable Networks; Turner Broadcasting Systems; Turner, Ted; United States: Cable

SUSPENSE

U.S. Anthology Series

Suspense, an anthology drama featuring stories of mystery and the macabre, was broadcast live from New York on Tuesday evenings from 9:30-10:00 P.M. over CBS. The original series began on 1 March 1949 and continued for four seasons until August 1954. It was revived briefly between March and September 1964.

Suspense was based on the famous radio program of the same name and was one of many early television shows that had its origin in the older medium. The radio program began in 1942 and was broadcast weekly from Hollywood. Scripts were generally of high quality and featured at least one well-known stage or film performer. The famous broadcast of 1948 entitled "Sorry Wrong Number" starred Agnes Moorehead in a thrilling tale of an invalid woman who accidentally overhears a telephone conversation in which arrangements for her own murder are being discussed. For the rest of the program, she tries frantically to telephone someone for help. A stunning concept for the aural medium, the episode was later made into a film. In addition to such fine writing, the radio *Suspense* featured outstanding music by Bernard Herrmann and excel-

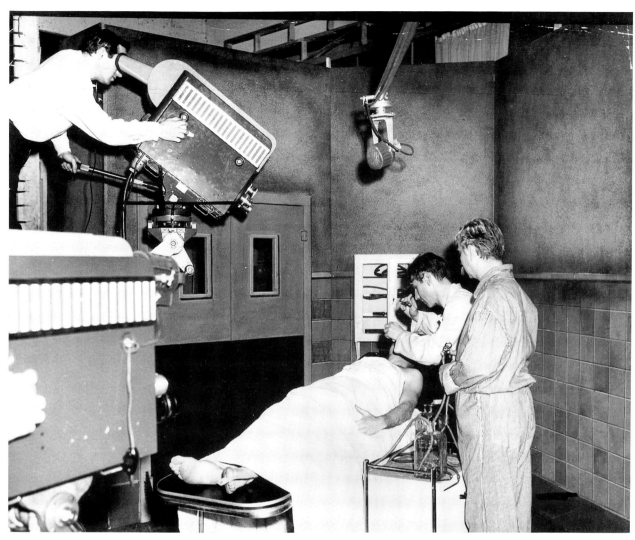

Suspense

Photo courtesy of Wisconsin Center for Film and Theater Research

lent production values. The program attracted a loyal following of listeners until September 1962. When it left the air, *Suspense* was the only remaining regularly scheduled drama on commercial network radio.

The television version of this popular show attempted to create the atmosphere of its radio predecessor by using the same opening announcement—"And now, a tale well calculated to keep you in. . . SUSPENSE!"—accompanied by the Bernard Herrmann theme played on a Hammond organ rather than by an orchestra. The television version, however, was not able to attain the generally high quality of the radio program. Part of the problem was the program's length. Thirty minutes hardly allowed sufficient time to develop characters of any subtlety. And the fact that the program was broadcast live from a New York studio severely restricted the mobility of its actions. It seemed too that writers sometimes offended public tastes by presenting subjects considered to be too violent for the conservative tastes of the early 1950s.

The first broadcast entitled "Revenge" was given a very negative review by *New York Times* radio and television columnist Jack Gould. He candidly stated that the program had more "corn than chill" and that the drab story about a man who stabs his wife while she is posing for a photograph gave actors "little opportunity for anything more than the most stereotyped portrayals." Gould noted that the most interesting thing about the program was its interspersing of live studio material with film to show exterior actions. Despite the interesting technique, Gould felt that the exteriors could have been dispensed with entirely without doing harm to the story.

He also complained of the excessive verbal explanation, and dialogue that was too simplified. He believed that the presence of pictures should free the dialogue from exposition and allow it to be more eloquent. As he put it, "with the pictures saying so much, the dialogue can afford to have more substance and be more subtle." His review concluded with a telling observation on the new medium: "The lesson of the first installment of 'Suspense' is that among all the mass media, television promises to demand a very high degree of compact and knowing craftsmanship for a mystery to be truly successful."

Gould continued to attend to the series, however, and became incensed about another episode entitled "Breakdown." Written by Francis Cockrell and Louis Polloch, the episode starred Ellen Violett and Don Briggs. The story focuses on a cruel and tyrannical office boss who breaks his neck in a plane crash and is taken for dead until just before his body is cremated.

Gould did not object so much to the story as to its mode of presentation. He was particularly upset by what he called "the unrelieved vividness of the details of death which no war correspondent would think of mentioning even in a dispatch from a battlefield." In closing Gould stated, "Both the sponsor, an auto accessories concern, and CBS should be thoroughly ashamed of themselves for their behavior last night. Mystery, murders, and suspense certainly have their place in any dramatic form. But a sustained and neurotic preoccupation with physical suffering for its own sake has nothing whatever to do with good theater. It is time for everyone concerned with 'Suspense' to grow up."

Most *Suspense* episodes were more conventional than "Breakdown." The program entitled "F.O.B. Vienna" of 28 April 1953 was fairly typical. It starred Walter Mathau and Jayne Meadows in the story of an American businessman who has accompanied a shipment of lathes to Austria and is trying to keep them out of the hands of Communists. The shipment ends up in Hamburg, and Mathau tracks it there with the help of Meadows who plays a newspaper reporter. At the last minute, he is able to destroy the shipment as the police arrive to round up the Communists. The ordinary script was not, in fact, very suspenseful and much of it cried for action impossible to depict within the confines of the studio.

A more successful broadcast was "All Hallows Eve" of 28 October 1952. Based on the story "Markheim" by Robert Louis Stevenson, this is the account of a man who murders his pawnbroker and is then visited by the devil, who urges him to kill the man's housekeeper in order to cover up his crime. In an attempt to atone for his utterly delinquent life, the man draws back at the last moment and tells the housekeeper to call the police because he has just murdered her master. Thwarted in his efforts to gain another soul, the devil disappears. Produced by Martin Manulis, this episode made excellent use of the pawnshop set. With its peculiar artifacts and many mirrors which reflect the face of the murderer as he thinks guiltily about his deed, the sense of confined space becomes central to the tale. Franchot Tone gave an outstanding performance as the main character. *Suspense* broadcast a number of other adaptations during its four years on the air. The program drew heavily on classic mystery and suspense offerings, including "The Suicide Club," "Dr. Jekyll and Mr. Hyde," by Stevenson, "The Mystery of Edwin Drood," and "The Signal Man" by Charles Dickens, and "The Cask of Amontillado" by Edgar Allen Poe.

On 26 May 1953, *Suspense* broadcast its only Sherlock Holmes story. "The Adventure of the Black Baronet" was written by Arthur Conan Doyle and John Dickson as an extension of the original Sherlock Holmes stories. The television adaptation was by Michael Dyne and starred Basil Rathbone as Holmes and Martyn Green as Dr. Watson. Jack Gould gave the program an unfavorable review, saying that much subtlety and brilliance of the Holmes character had been sacrificed by the compression of the story into thirty minutes. He added that Rathbone seemed unhappy with his part and that Martyn Greene was not as effective as Nigel Bruce who had played Dr. Watson to Rathbone's Holmes on the radio. The production was only one of many instances in which the television version of *Suspense* paled in comparison to its radio counterpart.

—Henry B. Aldridge

NARRATOR
Paul Frees

PRODUCERS   Robert Stevens, David Herlwell, Martin Manulis

PROGRAMMING HISTORY

• CBS

March 1949–June 1950	Tuesday 9:30-10:00
August 1950–August 1954	Tuesday 9:30-10:00
March 1964–September 1964	Wednesday 8:30-10:00

See also Anthology Drama

SUSSKIND, DAVID

U.S. Producer/Talk-Show Host

David Susskind was a key "mover-and-shaker" in the television industry during the medium's golden age and continued to take a high profile as a media personality long after the gold turned to waste, through some kind of reverse alchemy. In the process of leaving his mark on the histories of both live drama and television talk, Susskind would be honored with a Peabody, a Christopher, and 47 Emmy awards.

As Jack Gould observed in 1960, there were "virtually two Susskinds." One was a behind-the-scenes figure who was a major force, perhaps the major force, in the East-Coast branch of the television industry in the 1950s; the other

Susskind was the public man who would first achieve celebrityhood as the moderator-interviewer of *Open End*, a Sunday night discussion series aired by WNTA -TV in New York City. Some might say that his achievements were only surpassed by his arrogance. Described by his critics as "combative," "controversial," "blunt," "endearingly narcissistic," Susskind once aspired to be "the Cecil B. DeMille of television." As a self-styled "iconoclast" and "rebel," Susskind cultivated a reputation as a television insider who was an outspoken critic of the medium and its mediocrity. According to Susskind, "Ninety-five percent of the stuff shown on it [TV] is trash."

Susskind's ability to get things done, his genius as a logistician, was honed in the Navy during World War II. Serving as a communications officer aboard an attack transport, Susskind saw action at Iwo Jima and Okinawa. By the time he was discharged in 1946, he had given up his old ambition of landing a "job at Harvard as a teacher," and set his sights on show business. He actually went looking for his first job in his Navy uniform and quickly found a position as a press agent for Warner Brothers studios.

It was as an agent, that most despised, parasitic, and necessary of show business professionals, that the behind-the-scenes Susskind would first encounter success. After a brief stint as a talent scout for Century Artists, Susskind worked his way into the Music Corporation of America's television program department, where he managed such personalities as Jerry Lewis and Dinah Shore. In the early 1950s, he came to New York and joined Alfred Levy to form Talent Associates, Ltd., an agency that would represent creative personnel rather than actors and specialize in packaging programs for the infant television industry. The new firm's first package sale was the *Philco Television Playhouse*, a live, one-hour drama series on which Susskind would later find his first job as producer filling in for one of his clients, Fred Coe. After this heady experience, Susskind re-invented himself as a producer whose horizons extended far beyond the small screen, producing over a dozen movies and over a half-dozen stage plays in his forty-year career. As to television, in addition to serving as a producer on *The Kaiser Aluminum Hour*, *The DuPont Show of the Week*, and *Kraft Television Theatre* (among others), he was also the executive producer of *Armstrong Circle Theater*. During this period, Talent Associates, Ltd., also thrived. In 1959, Susskind's company contracted for nine million dollars in live shows, more than the combined efforts of the three major television networks.

From the 1960s to the 1980s, Susskind would come into his own. *Open End*, a forum which sometimes lasted for hours, went on the air in 1958. Called "Open Mouth" by Susskind's detractors, the show originally started at 11:00 P.M. and ran until the topics—or the participants—were exhausted. In 1961 the show was cut to two hours and went into syndication; in 1967 the title was changed to *The David Susskind Show*. Susskind's most significant interview, by far, was with Soviet Premier Nikita S. Krushchev. Broadcast in

David Susskind
Photo courtesy of Diana Susskind Laptook

October 1960, during the chilliest days of the Cold War, the interview dominated the headlines across the nation. Although station breaks featured a spot for Radio Free Europe depicting an ax-wielding communist soldier smashing a radio set, most observers scored the event as a propaganda coup for the impish Krushchev. As Jack Gould put it, "The televised tête-à-tête terminated in an atmosphere of Russian glee and Western chagrin."

In his twenty-nine years as a talk-show host and moderator, the abrasive Susskind would often rub a guest the wrong way resulting in what he termed "awkward moments." Tony Curtis even threatened to punch him "right on his big nose" after Susskind characterized Curtis as "a passionate amoeba." Susskind courted controversy by addressing such hot-button subjects as civil rights, abortion, terrorism, drugs, and a number of exotic or alternative lifestyles. His guests were as wide-ranging as his discussion topics. The roster of people who accepted invitations to appear on his show includes Harry S Truman, Richard M. Nixon, Robert F. Kennedy, Vietnam veterans, even a ski-masked professional killer.

Susskind continued to be intermittently involved as a producer of prestige programming, including *Hedda Gabler*

(1961), *The Price* (1971), *The Glass Menagerie* (1973), and *Eleanor and Franklin* (1976). It is ironic, yet somehow fitting, that the grand impresario who introduced millions of television viewers to Willy Loman would, himself, suffer the death of a traveling salesman. Susskind died alone in a hotel room of a heart attack at the age of 66 in 1987.

—Jimmie L. Reeves

DAVID SUSSKIND. Born in New York City, New York, U.S.A., 19 December 1920. Educated at University of Wisconsin; Harvard University, graduated with honors 1942. Married: 1) Phyllis Briskin, 1939 (divorced); 2) Joyce Davidson, 1966 (divorced, 1986); three daughters and one son. Served in U.S. Navy, 1943–46. Began career as a press agent; founder with Alfred Levy, Talent Associates Ltd.; hired by Music Corporation of America to produce *Philco Television Playhouse*; produced other early television programs; hosted own talk show for nearly thirty years; expanded production activities to Broadway and films; company purchased by Norton Simon, Inc., renamed Talent Associates-Norton Simon, for theatrical as well as film production, 1970; company sold to Time-Life Films, 1977. Recipient: Peabody Award; Christopher Award, numerous Emmy Awards. Died in New York City, New York, 22 February 1987.

TELEVISION SERIES (selection)

1947–58	*Kraft Television Theatre*
1948–55	*Philco Television Playhouse*
1950–63	*Armstrong Circle Theater*
1952–55	*Mr. Peepers*
1956–57	*Kaiser Aluminum Hour*
1954–56	*Justice*
1958–67	*Open End* (host)
1958–87	*The David Susskind Show* (formerly *Open End*; host)

1960–61	*Witness*
1962	*Festival of Performing Arts*
1963–64	*East Side/West Side*
1965–67	*Supermarket Sweep*
1965–70	*Get Smart*
1967–70	*He and She*
1967	*Good Company*

MADE-FOR-TELEVISION MOVIES (selection)

1960	*The Moon and Sixpence*
1967	*The Ages of Man*
1967	*Death of a Salesman*
1972	*Look Homeward, Angel*
1973	*The Bridge of San Luis Rey*
1973	*The Glass Menagerie*
1976	*Caesar and Cleopatra*
1976	*Truman at Potsdam*
1976	*Eleanor and Franklin: The White House Years*

FILMS (selection)

Edge of the City, 1957; *Raisin in the Sun*, 1961; *Requiem for a Heavyweight*, 1961; *All the Way Home*, 1963; *Lovers and Other Strangers*, 1969; *Alice Doesn't Live Here Anymore*, 1974; *Loving Couples*, 1980; *Fort Apache, The Bronx*, 1981.

STAGE (selection)

A Very Special Baby, 1959; *Rashomon*, 1959; *Kelly*, 1965; *All in Good Time*, 1965; *Brief Lives*, 1967.

FURTHER READING

Asinof, Eliot. *Bleeding Between the Lines*. New York: Holt, Rinehart, and Winston, 1979.

See also "Golden Age" of Television; Talk Shows

SUSTAINING PROGRAM

U.S. Programming Policy

In the United States broadcasting industries a program which does not receive commercial sponsorship or advertising support is known as a sustaining program. When the term was first used, sustaining programming included a wide variety of non-commercial programming offered by radio stations and networks to attract audiences to the new medium. Currently, most sustaining programming on commercial television is confined to public affairs, religious, and special news programs which are unsponsored.

At its inception radio programming was envisioned by many, including industry leaders (such as David Sarnoff, a guiding force behind the development of RCA and NBC), and government officials (such as then-Secretary of Commerce Herbert Hoover) as sustaining, i.e. provided by stations or networks as a public service. Since programming was

needed in order to sell radio transmitters and receivers, it was expected that the stations and networks established by manufacturers such as RCA would provide this programming and finance it from the profits on the sale of equipment. Programming provided by stations not associated with manufacturers was expected to be supported through endowments or municipal financing.

The vision of a commercial-free, public-service medium was short lived as AT and T began exploiting the commercial potential of radio in 1922. However, the public service responsibility of stations licensed to operate on scarce, public broadcast frequencies was affirmed in the Radio Act of 1927 and reaffirmed in the Communications Act of 1934 (Section 303), which states that the Federal Communications Commission (FCC) shall regulate the industry as required by "public convenience, interest, or necessity". The "public interest" standard was further delineated by the FCC in a 1946 document entitled *Public Service Responsibility of Broadcast Licensees,* commonly known as the "Blue Book". It states that devoting a reasonable percentage of broadcast time to sustaining programs is one criterion for operating in the public interest. Sustaining programming was deemed to be important, because it helped the station to maintain a balance in program content and provided time for programs not appropriate for sponsorship, programs serving minority interests or tastes, and non-profit and experimental programs. All licensees were expected to broadcast sustaining programs throughout the program schedule at times when the audience was expected to be awake. Thus, the importance of sustaining programming was firmly established before television began operation, and these standards were applied to the new medium.

Sustaining programming also became important in network affiliate contracts. In the early days of radio, NBC charged its affiliates for the sustaining programs they accepted and paid affiliates a small flat fee for broadcast of sponsored programs. In the early 1930s, William Paley, President of CBS, used sustaining programs to secure greater carriage of sponsored programs, offering the sustaining schedule free in return for an exclusive option on any part of the affiliate's schedule for sponsored programs. Thus, sustaining programming became a bargaining point in network affiliate contracts.

When experimental television was launched in the late 1930s, only sustaining programming was authorized by the FCC. The NBC schedule in 1939 included films supplied by outside sources; in-studio performances including interviews, musical performances, humorous skits, and educational demonstrations; and remote broadcasts, mostly of sporting events. Although NBC did not receive compensation to air these programs and shouldered much of the live and remote production costs, advertisers still had an influence on sustaining programming. In the January 1941, issue of *The Annals of the American Academy* David Sarnoff, then President of RCA and Chairman of the Board of NBC, wrote, "...invitations have been extended to members of the advertising industry to work with us in creating programs having advertising value, at no cost to the sponsors during this experimental period." When commercial operation was authorized in July 1941, NBC was prepared to convert many of its sustaining programs to commercially sponsored programs; however, World War II curtailed the development of television and of commercial. and sustaining programming.

As television regrouped after the war, sustaining programming became an important part of the industry's push to sell television receivers and transmitters. Since the financial strategy of many organizations was to use radio profits to provide funds for the fledgling television medium, a side effect of increased sustaining programming on television was the decrease in sustaining programming on radio as programs were dropped in favor of sponsored programming. Sustaining programming on television was varied, including dramatic series, educational programs, political events, and

Howdy Doody

public affairs programs. However, many programs (such as *The Howdy Doody Show)* which began as sustaining, quickly found sponsors once they became popular. As a result, the amount of sustaining programming on commercial television quickly diminished.

Further, after the freeze on the allocation of station licenses was lifted in 1950, channel space was allotted for educational stations. Industry leaders began to argue that much of the public service responsibility of broadcasting was being shouldered by these stations.

One of the more remarkable recent sustaining programs on commercial television was *Cartoon All-Stars to the Rescue* (an animated, anti-drug program), which was aired without advertisements in 1990 simultaneously on the ABC, CBS, NBC, FOX, Telemundo, Univision, Canadian Broadcasting Corp., CTV, Global Television (Canada), Televisa (Mexico), and Armed Forces Television networks; several hundred independent stations; plus the Black Entertainment Network, the Disney Channel, Nickelodeon, the Turner Broadcasting System, and the USA Network on cable. However, this program is the exception.

With the deregulatory push of the 1980s and the argument that non-profit, experimental, and minority programming is being provided by educational and public television, little regulatory attention is given to sustaining programming on commercial television. Currently, many programs which fulfill the FCC requirement for "public service" programming are sponsored and are, therefore, not sustaining.

—Suzanne Hurst Williams

FURTHER READING

Banning, William P. *Commercial Broadcasting Pioneer: The WEAF Experiment, 1922-1926.* Cambridge, Massachusetts: Harvard University Press, 1946.

Barnouw, Erik. *The Golden Web: A History of Broadcasting in the United States, Volume II—1933–1953.* New York: Oxford University Press, 1968.

"Cartoon Characters Enlisted in Anti-drug War." *Broadcasting,* 23 April 1990.

Federal Communications Commission. *Public Service Responsibility of Broadcast Licensees.* Washington: GPO, 7 March 1946.

Lichty, Lawrence W., and Malachi C. Topping. *American Broadcasting.* New York: Hastings House, 1975.

Sarnoff, David. *Looking Ahead: The Papers of David Sarnoff.* New York: McGraw-Hill, 1968.

———. "Possible Social Effects of Television." *The Annals of the American Academy* (New York), January 1941.

United States Congress. *Communications Act of 1934.* 73rd Congress, 2nd Session, S. Res. 3285. Washington: GPO, 1934.

———. *Radio Act of 1927.* 69th Congress, 2nd Session, H. Res. 9971. Washington: GPO, 1927.

See also Advertising; Advertising, Company Voice; Programming; Public Interest, Convenience, and Necessity; Sponsor; United States: Networks

SUZUKI, DAVID

Canadian Scientist/Television Personality and Host

A household name in English-speaking Canada, David Suzuki has almost single-handedly popularized some of the most complex scientific issues of our times largely through the medium of television. While students, teachers and heads of state continually laud his attempts to demystify contemporary science and nature, some in Canada's science community argue that Suzuki's work on environmental issues in particular is politically biased. Politics aside, Suzuki's awards of recognition clearly speak for themselves: Canada's most prestigious award, the Order of Canada; UNESCO's Kalinga Prize, and the United Nations Environmental Program Medal.

Such recognition, particularly awards bestowed to him in his native Canada, are in hindsight quite ironic. Growing up as a third-generation Japanese Canadian, Suzuki, his sisters and mother were placed in internment camps in 1942 by the Canadian government. After the war Suzuki and his family were forbidden by law to return to their Vancouver home.

On the faculty at the University of Alberta, Edmonton, Suzuki as a young academic began his illustrious television career by teaching science on campus TV. Some ten years later this experience, coupled with his scientific expertise, eventually landed Suzuki a host position on the weekly television program *Suzuki on Science*, broadcast by the Canadian Broadcasting Corporation (CBC). Suzuki would later extend his skills to radio where in 1975 he launched the CBC science affairs program *Quirks and Quarks*.

Although Suzuki continued on radio, his impact clearly remains in the sphere of Canadian public television. In 1974 he embarked upon his most successful broadcasting position, first as host of the CBC's television series *Science Magazine* . More importantly, five years later he became host of the well-established series, *The Nature of Things*. The longest-running science and nature television

series in North America, *The Nature of Things* is the CBC's top-selling international program. Established in 1960, the program is seen by viewers in over ninety countries, including on the Discovery Channel in the United States. The program's mandate is to cover a broad range of topics, including natural history and the environment, medicine, science and technology.

David Suzuki
Photo courtesy of CBC

It is widely recognized that *The Nature of Things*, as with Suzuki's work in general, surveys the scientific landscape though a critical, humanistic lens. Such an approach has increasingly lent itself to investigations of controversial contemporary issues of social importance. Suzuki's outspoken views on the clearcutting of old growth forests on Canada's west coast, for example, has gained him many friends (and enemies) in logging and environmentalist circles. Whatever one's opinion of his views, however, it would be safe to say that Suzuki remains the voice of popular science on the Canadian airwaves.

—Greg Elmer

DAVID SUZUKI. Born in Vancouver, British Columbia, Canada, 24 March 1936. Educated at Amherst College, Massachusetts, U.S.A., B.A. 1958; University of Chicago, Ph.D. 1961; postdoctoral research, the Rocky Mountain Biological Laboratory. Married: 1) Setsuko Joane Sunahara, 1958 (divorced, 1965), children: Tamiko, Laura, Troy; 2) Tara Elizabeth Cullis, 1972, children: Severn Cullis-Suzuki and Sarika Cullis-Suzuki. Held positions as research and teaching assistant, 1957–59; research associate, Oak Ridge National Laboratory, Tennessee, 1961; assistant professor, University of Alberta, 1962–63; assistant professor, University of British Columbia, Vancouver, 1963–69; professor, University of British Columbia, since 1969; television and radio host, various science programs; syndicated newspaper columnist, since 1989; author of numerous books and scientific articles. Recipient: E.W.R. Steacie Memorial Fellowship, 1969–71; Outstanding Japanese-Canadian of the Year Award, 1972; Order of Canada, 1976; Science Council of British Columbia Gold Medal, 1981; Biological Council of Canada Gold Medal, 1986; United Nations Environment Program Medal, 1985; UNESCO's Kalinga Award, 1986; Canadian Booksellers Association's Author of the Year, 1990. Address: c/o Sustainable Development Research Institute, University of British Columbia, Vancouver, British Columbia, V6T 2A9 Canada.

TELEVISION SERIES

1960–	*The Nature of Things*
1971–72	*Suzuki on Science*
1974–79	*Science Magazine*

TELEVISION SPECIALS

1977	*The Hottest Show on Earth* (also co-writer)
1977	Trouble in the Forest
1979	How Will We Keep Warm (Part 1 of *The Remarkable Society Series*)
1986	Fragile Harvest (narrator)

RADIO

Quirks and Quarks, 1975–79; *Earthwatch*, 1980; *Discovery with David Suzuki*, 1983–; *It's a Matter of Survival*, 1989.

PUBLICATIONS (selection)

An Introduction to Genetic Analysis, with A.J.F. Griffiths. New York: Freeman, 1976.

Metamorphosis. Toronto, Canada: Stoddart, 1987.

Genethics: The Ethics of Engineering Life, with Peter Knudson. Cambridge, Massachusetts: Harvard University Press, 1989.

See also *Nature of Things*; Science Programs

SWALLOW, NORMAN

British Producer/Media Executive

Norman Swallow's career in British broadcasting, from his joining the BBC in 1946 through to his continuing involvement in independent production, is that of a major pioneer of the British television documentary and, more broadly, a significant contributor to public service television.

Swallow went to school in Manchester and studied history at Oxford, before entering wartime military service. His first work for the BBC was in radio "drama-documentary" where he tackled a number of historical and social themes as a writer and producer. After moving to television, Swallow was a producer of the general election broadcast of 1951, which marked a decisive shift in television's treatment of elections, to their own distinctive form of extended national coverage and commentary. One year later he became the series director of *Special Enquiry*, a BBC-documentary series which concerned itself primarily with investigation into contemporary social issues. The series ran from 1952 to 1957 and was undoubtedly one of the most important innovations in television journalism of the period, acting as an influence upon a whole range of later work. In devising the series with his colleagues, Swallow was influenced both by the work of the 1930s British documentary film movement (particularly in films such as *Housing Problems,* 1935) and by the kind of feature journalism, making extensive use of location interviews, developed within BBC radio.

Special Enquiry started with a programme investigating life in the slum tenements of Glasgow. The program caused widespread and positive appreciation of the new series in the newspapers. It went on to engage with a variety of issues to do with housing, poverty, health, ageing, education, etc. As quoted in *Popular Television in Britain,* Swallow remarked on the response which the first programme caused, "we had many phone calls, even letters, from people who, because they know nothing about it, hadn't seen that sort of thing before, wouldn't believe it. They thought we were lying. That it was somehow fiction. So this was a television breakthrough."

One of the most controversial programmes in the series was entitled *Has Britain a Colour Bar?* and investigated racial prejudice against immigrants, taking the city of Birmingham as an example. Like all of the programmes, it consisted of a filmed report by an on-location investigative reporter (here Rene Cutforth), together with interview sequences. Following a convention of the period, interviews were often presented as direct-to-camera testimony, giving the series something of the feel of an "access programme" and linking it back to the precedent of direct address by ordinary people in the 1930s "classic" *Housing Problems.* The "colour bar" edition caused extensive public discussion, not least for the frankness with which racial prejudice was revealed in the speech of some of the participants, including trade union officials. There was also a powerful, and partly dramatized, scene in which a newly arrived immigrant looked for lodgings, to be repeatedly turned away by landladies, sometimes with the reason made perfectly clear. The *Daily Express* thought the programme to be "one of the most outspoken...ever screened."

At the time, Swallow was also the series producer of *The World Is Ours*, made in cooperation with the United Nations and produced within the BBC's new documentary department, headed by the distinguished filmmaker Paul Rotha. In 1960, Swallow became assistant editor of *Panorama* at a time when this series was establishing itself as the leading current-affairs programme on British television. Three years later he resigned to set up an independent company with Denis Mitchell, one the most brilliantly original documentary directors ever to work for British television. Together, the two did a series for Granada called *This England*, which further extended television's exploration of working-class life through a relaxed approach that kept commentary to a minimum. During this period Swallow made *A Wedding on Saturday*, a film about a wedding in a northern mining village, which won the Prix Italia in 1965.

Going back to the BBC in 1968, after a period of work which included the first Anglo-Soviet co-production, *Ten Days that Shook the World* (on the Russian revolution) for Granada, Swallow became series editor of the arts programme, *Omnibus.* During his first year, editions of this series included Ken Russell's much admired biographical film on Delius and Tony Palmer's pathbreaking programme on popular music, *All My Loving.* He went on to become the BBC's head of arts features before shifting northwards again, to rejoin Granada where, among other things, he worked on the 1985 series *Television*, an ambitious attempt at tracing the history and significance of the medium across the world.

Swallow has written extensively on the medium for newspapers and journals and his widely-cited book *Factual Television* remains one of the most thoughtful and sustained reflections on its subject by a practitioner. He was Television Advisor for the planning of the British Film Institute's Museum of the Moving Image, established in London's South Bank arts complex.

The career of Norman Swallow is both distinctive and representative. It is distinctive in his contribution (particularly in the shaping and supportive role of series editor)

both to the investigative documentary and to arts programming, where his interests, enthusiasms, and creative empathy have extended well beyond the confines of southern middle-class England. It is representative insofar as his ability to be both popular and serious, intellectually engaged, yet fully aware of the need to address a general audience, displays the best qualities of British public service television across four decades.

—John Corner

NORMAN SWALLOW. Born in Manchester, Lancashire, England, 17 February 1921. Attended Manchester Grammar School; Keble College, Oxford. Served in British Army, 1941–46. Began career as writer-producer of documentaries, BBC, 1948; producer of documentaries, from 1950; co-produced television coverage of the general election, 1951; produced monthly BBC program *Special Inquiry*, 1952–57; study tour of Middle East, India, Pakistan, and Ceylon (now Sri Lanka), 1956–57; assisted head of films for BBC, 1957; writer-producer for *On Target*, 1959; appointed chief assistant, BBC Television, 1960; assistant editor, *Panorama*, 1961; joined Denis Mitchell Films, 1963; head of arts features, BBC Television, 1972–74; executive producer, Granada Television, from 1974; freelance producer-director, since 1985. Recipient: Desmond Davis Award, 1977; Emmy Award, 1982.

TELEVISION SERIES (producer)

1952–57	*Special Inquiry*
1953–	*Panorama* (assistant editor)
1954–56	*The World Is Ours*
1959	*On Target* (also writer)
1968–72	*Omnibus*

TELEVISION SPECIALS

1964	*A Wedding on Saturday* (also writer)
1977	*The Christians*
1978	*Clouds of Glory*
1980	*This England* (co-producer)
1982	*A Lot of Happiness*
1986	*The Last Day* (also director)
1989	*Johnny and Alf Go Home*

PUBLICATIONS (selection)

"Documentary TV Journalism." In, Rotha, Paul, editor. *Television in the Making.* London: Focal, 1956.

Factual Television. London: Focal, 1966.

"Denis Mitchell." *The Listener* (London), 24 April 1975.

Eisenstein: A Documentary Portrait. London: Allen and Unwin, 1976.

FURTHER READING

Corner, John. "Documentary Voices." In, Corner, John, editor. *Popular Television in Britain: Studies in Cultural History.* London: British Film Institute, 1991.

See also British Programming; *Panorama*; Producer in Television

THE SWEENEY

British Police Drama

The Sweeney was the top-rated British police series of the 1970s, bringing a new level of toughness and action to the genre, and displaying police officers bending the rules to beat crime. The series was created by Ian Kennedy-Martin and produced by Ted Childs for Euston Films (a Thames Television subsidiary), and went out mid-week in prime time on ITV, the main commercial channel. In all, 54 episodes were made, and the programme ran for four seasons.

The Sweeney focused on the exploits of Jack Regan, a maverick detective inspector attached to the Flying Squad, the metropolitan police's elite armed-robbery unit, and featured John Thaw in the leading role. The programme, which derived its title from "Sweeney Todd," the Cockney rhyming slang for "Flying Squad," was a spin-off from the successful 1974 TV film, *Regan*, which had first introduced the protagonist. It had also established his professional relationships with his assistant, DS George Carter (played by Dennis Waterman), and his "governor," DCI Haskins (played by Garfield Morgan). Each episode in the series adopted the same basic narrative format—a three-act structure (with acts separated by adverts) preceded by a prologue that triggered the crime narrative. The first two acts were devoted to obtaining intelligence about a forthcoming robbery, often through tip-offs from informers or surveillance; the third involved the capture of the robbery gang, characteristically involving adrenalin-pumping action with car-chases, screaming tyres, spectacular smashes, and hand-to-hand fighting. The narrative was often further complicated through the addition of an anti-authority thread when Regan challenged Haskins' "rule-book" approach, or through the introduction of casual sex relationships when one of the detectives became involved with an available woman.

The programme's realism was considerable, and few other crime series have achieved so authentic an impression of the policing of London's underworld. To an extent, this was achieved by adopting the same visual style, fast action, and cynical outlook as contemporary rogue-cop films, such as *Dirty Harry* and *The French Connection*. Equally though, the programme relied on detailed inside-knowledge of the actual circumstances in which the Flying Squad operated and the sometimes rather dubious means used to secure prosecutions. The series' storylines frequently blurred the sharp distinctions that are normally drawn between good and evil characters in crime melodrama. Regan and Carter were shown inhabiting the same sleazy world as the criminals, mixing with low-life to obtain their leads, and adopting the same vernacular. Both law-enforcers and law-breakers indulged in womanising and heavy drinking, and used physical violence to achieve their objectives. The extent to which Regan was prepared to bend and break the rules to "nick villains" was well established in the pilot film, when he threatened a suspect with a longer sentence if he did not co-operate: "My sergeant is going to hit me, but I am going to say it's you." Throughout the series, however, the viewer's sense of Regan's integrity remained secure. Even though he might need to beat up suspects, strike deals with criminals and—on one occasion—burgle the office of the DCI to read his own personal file, such actions were legitimised in the narrative as the only means available to the serious crime-fighter to keep on top, and to cut through the dead weight of bureaucracy that continually threatened to impede the cause of justice.

Unsurprisingly, the series provoked fierce controversy, chiefly because of its potential to influence the public image of the police at a time of considerable social upheaval. However, the dark (if not confused) moral world that the series represented was difficult to fault on purely realistic grounds as, at the time of transmission, a prominent officer in the Squad was under investigation and was eventually imprisoned for corruption. Considered in wider cultural terms, the programme has been viewed as part of the general ideological shift to the right that occurred in the 1970s in Britain, as the post-war social-democratic consensus broke down. James Donald, notably, has argued that *The Sweeney* was fuelled by popular anxieties about law and order stimulated by the press campaign on mugging, and that episodes

The Sweeney

provided a "mapping fantasy" for the acting out of unconscious authoritarian urges.

The Sweeney had sold to 51 countries by 1985, and had also stimulated two successful feature films. It also established Dennis Waterman and John Thaw as household names with the British public. The series secured the reputation of Euston Films as a leading production company, and created an influential model in Britain not just for crime series on ITV, but for the production of cost-effective, high-quality drama in general. The lean and efficient production operation that Euston pioneered in *The Sweeney*, relying on short-term contracts and shooting wholly with 16mm-film, has been generally adopted across the industry; with the exception of soap opera, the great majority of drama projects today are manned by freelance crews and produced on film.

—Bob Millington

CAST

D.I. Jack Regan . . . . John Thaw
D.S. George Carter . . . . Dennis Waterman
D.C.I. Frank Haskins . . . . Garfield Morgan

PRODUCER   Ted Childs

PROGRAMMING HISTORY   53 50-minute Episodes; 1 77-minute Episode

• ITV

January 1975–March 1975	14 Episodes
September 1975–November 1975	13 Episodes
September 1976–December 1976	13 Episodes
September 1978–December 1978	14 Episodes

FURTHER READING

Alvarado, Manuel, and John Stewart. *Made for Television: Euston Films Ltd.* London: British Film Institute/Methuen, 1985.

Clarke, Alan. "This is Not the Boy Scouts." In, Bennett, Tony, Colin Mercer, and Janet Woollacott, editors. *Popular Culture and Social Relations.* Milton Keynes, England, and Philadelphia, Pennsylvania: Open University Press, 1986.

Donald, James. "Anxious Moments in *The Sweeney*." In, Alvarado, Manuel, and John Stewart. *Made for Television: Euston Films Ltd.* London: British Film Institute/Methuen, 1985.

Hurd, Geoffrey. "The Television Presentation of the Police." In, Bennett, Tony, with others, editors. *Popular Television and Film: A Reader.* London: British Film Institute/Open University Press, 1981.

See also British Programming; Thaw, John; Waterman, Dennis

SWITZERLAND

Switzerland, surrounded by Germany, Italy, France, Austria and the small country of Liechtenstein, is a multilingual and multicultural society. Due to its unique topography—a total of 41,293 square miles, most of it unpopulated mountain ranges—Switzerland is highly segmented. Nearly seven million inhabitants speak different languages and live in completely different surroundings. From industrialized cities such as Basel or Zurich to remote locations in closed-off valleys, they share a somewhat vague notion about what it means to be "Swiss." Still, commonalities have succeeded in overcoming the ever-present language barriers. So far, they have proven strong enough to keep Switzerland one of the few countries in Western Europe out of the European Union.

Television in Switzerland began in 1949 with an official delegation of Swiss technicians (and some staff members of General Electric) watching an experimental program, broadcast from Torino in Italy about 90 miles away. The first programs produced in Switzerland, in 1953, were received in Zurich only. By 1955, there were 8,600 television sets in Switzerland, 2,300 of them in public rooms and 6,300 in private households. In 1994 there were 2.6 million television licence fee holders in Switzerland.

Television, as developed in the 1950s and 1960s, was meant to be a tool of public communication and education. The technical objective was reception in all Swiss households (a goal still not attained, due to topography), but television broadcast had a political and social mission as well. "Audiovision," as it was termed, was supposed to play an important part in the national integration of different languages, regions, religions, generations and ways of living. Since there was, until 1992, only one network officially assigned with the mission to broadcast television programs, politicians of all parties kept an eye on content and on those responsible for developing and managing the broadcasting system.

The date 20 July 1953 marked the official beginning of Swiss broadcasting. Programming that night consisted of a demonstration of traditional Swiss woodcrafts and the recitation of a poem, "The Blind." Older Swiss citizens often remember broadcasts of live sports events that were viewed in crowded restaurants rather than at home. At the time, television was a social event.

Early viewers were especially interested in nature programs. And though educational programs rarely dealt with social problems, news and documentaries were something else. "Objectivity" was the key word during the 1970s and some television programmers, labeled as left-wing radicals by more conservative parties were constantly accused of undermining Swiss democracy.

When the French-speaking and Italian-speaking communities received their own television programs, news was still produced in one place, with different crews using the same facilities and sharing a single set until 1982. Heidi Abel began announcing programs in 1954, and went on to present many different kinds of programs. She finally found her place as a talk-show host covering the most sensitive topics with wit and courage.

Fiction programs, expensive to produce, did not develop for some time. Early production included Swiss plays, mostly comedies, that were adapted for the stage and televised, rather than true television productions. All other types of fiction required co-production with wealthier neighbors. Some miniseries and series have been developed, including *Die Sechs Kummerbuben, Heidi,* and *Die Direktorin.* The animated children's series *Pingu* achieved worldwide fame.

In the 1990s, family sitcoms based on American examples have become popular in all regions. In the German-speaking region, the popular program is *Fascht e Familie,* while in the French region, the favorite is *La petit Famille.* It is worth noting that some of the local stations have begun to produce experimental fiction. *The Eden Family,* for example, is a "dark" family sitcom, a parody of *The Addams Family* in which the characters live in a gay community.

The Swiss Broadcasting Company (SBC) is still organized as a private nonprofit association, not as a state institution. It is supported with licence fees paid every month. Advertising on television was introduced in 1965 and proved to be a most important additional source of income.

The system appears as complex as its political structure and its somewhat fragmented cultural identity. Radio and television stations are commercially or non-commercially organized. Yet the public broadcaster SBC (radio and tele-

vision) is still by far the biggest distributor of programs, beating other (foreign) stations in ratings. The SBC provides programs for a mainly German-speaking audience (64%) as well as the considerably smaller French-speaking (19%) and Italian-speaking (8%) communities. There is also a tiny Romansch-speaking audience in the east of the country (0.6% in 1990) counting on at least one weekly news-magazine being broadcast. There are four SBC television channels and nearly a dozen SBC radio channels altogether, all of them distributed terrestrially. 76% of all Swiss television and radio households are cabled.

A considerable number of small, local television-stations and/or text services were registered by 1995, most of them experimental and with very limited frequency ranges each. This domestic competition has been less influential than that caused by international developments such as the ongoing deregulation process in the European television market. More and more commercial television stations have emerged throughout Switzerland since the 1980s, changing viewing habits and taking a toll on the ratings. When legislation changed in 1992, allowing private television broadcasters to find (or at least search for) their specific segments in a more open market, those broadcasters were waiting in the wings, thus urging the public broadcaster SRG to develop market-oriented strategies as well.

—Ursula Ganz-Blaetller

FURTHER READING

Bonfadelli, Heinz, and Walter Hättenschwiler. "Switzerland: A Multilingual Culture Tries to Keep Its Identity." In, Becker, L., and K. Schönbach, editors. *Audience Responses to Media Diversification: Coping with Plenty.* Hillsdale, New Jersey: Lawrence Erlbaum, 1989.

Meier, Werner A., and Ulrich Saxer. "Switzerland." In, *The Media in Western Europe. The Euromedia Handbook.* London and Newbury Park, California: Sage, 1992.

SYKES, ERIC

British Comedy Actor

Eric Sykes, who cultivated his talent for comedy whilst serving in the army in World War II, worked as a writer on radio and a writer-performer on television through the 1950s before having his greatest success, the long running BBC sitcom *Sykes Versus TV* which debuted in 1960. The services had proved to be fertile ground for aspiring entertainers and many of Britain's favourite stars of the 1950s had discovered their performing skills whilst on wartime duty. Following the end of hostilities, these talents found themselves taking their acts on stage before getting the chance to do radio or television. Sykes was one such talent. He wrote comedy scripts as well as performing and eventually scripted one of radio's most popular comedies, *Entertaining Archie,* which was a prolific breeding ground for comic talent. His many appearances on TV were

usually comedy-variety specials and he developed a format for such one-offs which featured himself as a harassed producer struggling to put on a show and meeting with various obstacles.

But it was in 1960 that Sykes enjoyed his most enduring success. Comedy writer Johnny Speight collaborated with Sykes on the idea of a sitcom based loosely on Sykes existing stage persona. In the idea, Sykes would live in suburbia with his wife, getting involved in simple plots centering on everyday problems. However, Sykes soon realised that by making his partner his sister, rather than his wife, he would have more scope in storylines, with either or both of them able to get romantically entangled with other people. Comedy actor Hattie Jacques, who had worked with Sykes on the radio, was chosen as the sister and the first series, written by Speight, proved to be a success. The second series, written by Sykes and other writers from storylines suggested by Speight, consolidated that success. Subsequent series were all written by Sykes alone. The TV character Sykes was a proud, rather work-shy individual with somewhat childish habits; as if part of him hadn't grown up. His sister Hattie was formidable in stature but timid by nature, and was easily inveigled into her brother's schemes. It was a departure for a big woman to be portrayed on TV in this way but it was probably Hattie Jacques' radio career which had allowed her to formulate such characters, as her gentle voice belied her size allowing her to portray, on radio, small, timorous women.

The format was simple but enduring. Each week a single idea would be taken and every possible comedic situation of the theme would be exploited. For example, in one episode Sykes gets his toe stuck in the tap whilst having a bath and the entire programme revolves round efforts to free him; in another, highly memorable segment, Sykes and his sister accidentally get handcuffed together and spend the whole episode trying to do cope with ordinary domestic situations whilst remaining connected. By concentrating on this technique, Sykes was able to come up with seemingly endless storylines in which to place his characters.

The series was called simply *Sykes Versus TV* but each week bore a subtitle which began with "and", for instance "Sykes ..and a telephone," "Sykes.. and a Holiday" with the subtitle referring to that episodes particular theme to be milked. *Sykes* became the longest running sitcom of its time, continuing, with one notable seven-year break between 1965 and 1972, for 127 episodes until Hattie Jacques' death in 1980.

During the run of the sitcom Sykes also made a series of short dialogue-free films for the cinema, utilising the same structure as the TV show: that is one idea exploited to the limit, comedically. Most famous of these was called *The Plank* (1967) and just focused on the mishaps caused by a man carrying a large plank around—incidentally one of the Sykes episode also used this concept. Later he re-made two of these short films, *The Plank* and *Rhubarb* (1969) for television: *The Plank* (Thames 1979) and *Rhubarb, Rhubarb*

(Thames 1980). Subsequently Sykes, now a huge comedy star due to the success of the famous sitcom, appeared in specials and the odd series but never managed to recreate the popularity of *Sykes*. His long-lasting top flight career is even more remarking considering he has been dogged by hearing problems since 1952. The problems increased with the passing of time eventually leaving him completely deaf in one ear and with very poor hearing in the other.

—Dick Fiddy

ERIC SYKES. Born in Oldham, England, 4 May 1923. Served in British Army, World War II. Married: Edith. Began career as performer in military service; radio and television writer, 1950s; star of comedy series, *Sykes Versus TV*; star of short films based on television character. Recipient: OBE, 1986.

TELEVISION

1958–65,
    1972–80   *Sykes Versus TV*

TELEVISION SPECIALS (selection)

1971	*Sykes and a Big, Big Show*
1978	*Sykes and a Big, Big Show*
1979	*The Plank*
1980	*Rhubarb, Rhubarb*

FILMS (selection)

Watch Your Stern, 1960; *Invasion Quartet*; 1962; *Kill or Cure*, 1962; *Heaven's Above*, 1963; *One Way Pendulum*, 1965; *Those Magnificent Men in Their Flying Machines*, 1965; *Rotten to the Core*, 1965; *The Liquidator*, 1966; *The Spy with the Cold Nose*, 1966; *The Plank*, 1967; *Shalako*, 1968; *Monte Carlo or Bust*, 1969; *Rhubarb*, 1970; *The Alf Garnett Saga*, 1972; *Theatre of Blood*, 1973.

RADIO

Entertaining Archie (writer).

SYLVANIA WATERS

Australian Documentary

Sylvania Waters, a documentary television series which followed the lives of an Australian family, premiered on Australian television in 1992. A 12-part co-production by the Australian Broadcasting Commission (ABC) and the British Broadcasting Corporation (BBC), the controversial program chronicled the existence of couple Noeline Baker and Laurie Donaher and their largely adult offspring. The series took its name from the wealthy harbourside suburb in southern Sydney where Noeline and Laurie reside.

Billed as a "real-life" soap opera, *Sylvania Waters* was shot over a six-month period by a camera crew who lived with the Donaher/Bakers. According to an agreement struck with the

family, the crew was allowed to film "anywhere, at any time—except when family members were using the bathroom or making love". While ABC publicity for the documentary series emphasised the couple's new found wealth and luxurious lifestyle, the tightly edited result ruthlessly scrutinised the entrenched interpersonal conflicts which lay beneath the surface of the blended family's easygoing facade.

Like its 1978 British prototype, *The Family*, which brought instant infamy to the Wilkins family of Reading, and the 1973 U.S. program *An American Family*, which chronicled the lives of the Loud family in Santa Barbara, California, *Sylvania Waters* focused a national microscope on the values and behaviour of the Donaher/Baker family.

Noeline and Laurie's unwed status, Noeline's drinking problem, Laurie's racism, their materialism, and the family's routine domestic disputes, all became issues discussed widely in the Australian media.

A particularly passionate public debate erupted over the question of whether executive producer of *Sylvania Waters*, Paul Watson, who also produced *The Family* for the BBC, had chosen an Australian family which pandered to a British stereotype. Writing in *The Sydney Morning Herald*, popular cultural critic Richard Glover summed up these concerns when he wrote that the family was "hardly a surprising British choice: in Noeline and Laurie, every British preconception about the Aussies comes alive...Meet Australia's new ambassadors: a family whose members are variously materialistic, argumentative, uncultured, heavy drinking and acquisitive."

The debate intensified when the series screened in Britain and became the subject of widespread commentary in the press there. The tabloid newspaper *The Sun* headlined a story on the series "Meet Noeline. By Tonight You'll Hate Her Too," while *The Guardian* criticised "Noeline's bigotry and gruesome materialism." Critics of *Sylvania Waters* argued that this adverse publicity was proof that the producers of the series had effectively "set up" the Donaher/Baker family to feed British prejudices about Australians.

During the screening of the series, Noeline Baker, Laurie Donaher, and their extended family, also became the subject of intense media interest. While a number of family members claimed that the series had caused a family rift, they continued to give numerous press, radio, and television interviews and guest-hosted radio and television programs, both in Australia and in the United Kingdom.

On the level of genre, *Sylvania Waters* was also widely understood as representing a new trend dubbed "reality" television. This ambiguous term—generally identified by the use of unembellished documentary style footage of ordinary people for entertainment purposes—has been used to describe a number of programs which debuted in Australia in the early 1990s, including *Cops*, which showed footage of police arresting suspects, and *Hard Copy*, a current-affairs program which made frequent use of amateur video material.

—Catharine Lumby

EXECUTIVE PRODUCERS   Paul Watson, Pamela Wilson

PROGRAMMING HISTORY   12 Half-hour Episodes
21 July 1992–6 October 1992   Tuesday 9:30-10:00

FURTHER READING
Cunningham, Stuart. *Contemporary Australian Television.* Sydney: University of New South Wales Press, 1994.

See also Australian Programming

SYNDICATION

Syndication is the practice of selling rights to the presentation of television programs, especially to more than one customer, such as a television station, a cable channel, or a programming service such as a national broadcasting system. The syndication of television programs is a fundamental financial component of television industries. Long a crucial factor in the economics of the U.S. industry, syndication is now a worldwide activity involving the sales of programming produced in many countries.

A syndicator is a firm which acquires the rights to programs for purposes of marketing them to additional customers. The syndication marketplace in fact provides the bulk of programming seen by the public. For the internal U.S. market, for example, syndication is the source of the "reruns" often seen on network television, and of much material seen on cable networks. Internationally, large amounts of American television programming are sold through syndication for programming alongside material produced locally. Material not available in syndication includes current network prime time programs, live news programs and live coverage of sporting and other special events. Even current U.S. programs, however, may be syndicated in international markets, and American viewers may

sometimes see imported programs, usually from England, currently programmed in other countries.

The price for a syndicated television series is determined by its success with audiences and the number and type of "run" in which the program appears. A national run is the presentation of a film or program one time to a national audience. This notion of national run has been borrowed from the history of distributing theatrical films. Any number of theaters or communities may be included in the first run of a production. But as soon as any location receives a second presentation, the second national run has begun. Generally speaking, the cost of rights to present a television series declines as it is presented in later and later runs although, as indicated below, that rule does not always hold in the international market.

Repeated sales of television programs, both within the United States and throughout the world, has long been central to the profitability of the American television industry. Soon after U.S. television production shifted from live performance to film in the late 1950s, shrewd sales personnel realized that television products had additional life. Audiences would, in fact, watch the same program a second time, and perhaps return for repeated viewing. Moreover, many countries found it far more economical

Annual NAPTE syndicators convention
Photo courtesy of NAPTE

to purchase the syndicated rights to American television programs than to produce their own, opening a vast market for American products.

The cost of U.S. television programming in the international market place is generally based on whatever those markets will bear. Costs for programs in Europe are often far higher than in Africa or Latin America. No matter how small the syndication fee, however, the sales of programming produce additional income for their original production companies. In abstract economic terms this is an example of "public good theory," in which new profits are gained at no additional costs or at the marginal costs incurred in the marketing process.

Historically, syndication, whether domestic or international, served to underwrite the risky process of producing for American network television. From the late 1960s through the mid-1990s special regulations (the Financial Interest and Syndication Rules) governed relations between television networks and independent production companies. Under these rules ownership of the rights to the programs reverted to the producer/production company after a specified number of network runs. Profits

from any other sales, including syndication, generally benefited the production community. For this reason many production companies were willing to produce original programs at a loss, betting on the enormous income that might rise from successful syndication. Many "failed" programs could be created with the profits from one or two successfully syndicated shows.

One way of classifying television programs in the syndication marketplace is by the first national run of the program. If the first run of a program was as part of a national network schedule, then as the program is marketed for subsequent runs to other programmers, it is referred to as "off-network syndication." Thus a cable programmer who buys the rights to presentation of a situation comedy presented by NBC is buying off-network syndication. *Dallas*, presented in first run on CBS in the 1978 season, was heavily programmed throughout the world as an off-network syndication.

If a program is initially made to be sold to programmers other than the major networks, however, then the program is known as "first run syndication." An example would be the weekly program, *Star Search with Ed McMahon*, produced by Television Program Executives (T. P. E.) and Bob Banner Associates. Similarly, Paramount Television's *Star Trek: The Next Generation* and other *Star Trek* spin-offs are produced for first run syndication. On occasion, a television program originally developed for network programming will be shifted into the first run syndication mode. This is the case with *Baywatch,* a program that failed to attract a sufficient audience when programmed by NBC in 1989, and was canceled after a single season. It then went into production as a first run syndicated product and has become enormously successful in international markets.

First run syndication is often the origin of programs presented as programming "strips," that is, at the same time Monday through Friday. This is the case with *Entertainment Tonight,* another Paramount production, and also with numerous programs in the "tabloid TV" genre, game shows, and cartoons.

Barter syndication is a financial arrangement that supports a growing segment of the syndication marketplace. In barter syndication an advertiser purchases in advance all or some part of the advertising opportunities (commercial spots) in a syndicated program, no matter where the production is to be seen in any run. The advertiser benefits from the barter arrangement by insuring a friendly program environment for ads. The programmer—an independent station or a cable programmer—benefits because advertising slots are presold, assuring that the cost to acquire the program is at least partially covered. While this practice may reduce opportunities for the programmer to sell advertising time, the trade-off is considered a favorable one. The producer of the program also benefits because the prior purchase of advertising opportunities provides funds that may represent an important part of the production budget.

Increasingly, syndication is part of the worldwide television marketplace and the producers are not always part of the U.S. industry. Brazilian, Venezuelan, and Mexican *telenovelas* are programmed throughout the Spanish-speaking world and even in less predictable contexts such as India and Russia. British programming is seen in the United States and throughout Europe and the rest of the world. In these cases and many others, syndication is seen as an economic benefit. As in the American context, the profits generated by

syndication can be used to produce other material on a speculative basis and to bolster the production of the first-run production process. As television distribution channels proliferate throughout the world and the demand for product to fill those channels grows, it is likely that more and more producers in more and more contexts will create materials for sale to the syndication market.

—James E. Fletcher

FURTHER READING

Blumler, Jay G. "Prospects for Creativity in the New Television Marketplace: Evidence from Program Makers." *Journal of Communication* (New York), Autumn 1990.

Covington, William G., Jr. "The Financial Interest and Syndication Rules in Retrospect: History and Analysis." *Communications and the Law* (New York), June 1994.

Fletcher, James E. "The Syndication Marketplace." In, Alexander, Allison, James Owers, and Rod Carveth, editors. *Media Economics: Theory and Practice.* Hillsdale, New Jersey: Lawrence Erlbaum, 1993.

"Glossary of Syndication Terms." *Advertising Age* (New York), 16 April 1990.

Kaplar, Richard T. *The Financial Interest and Syndication Rules: Prime Time for Repeal.* Washington, D.C.: Media Institute, 1990.

Lazarus, P.N. "Distribution: A Disorderly Dissertation." In, Squires, J.E. *The Movie Business Book.* Englewood Cliffs, New Jersey: Prentice Hall, 1983.

"Syndication in the 1990s." *Advertising Age* (New York), 10 April 1995.

"What is Syndication." *Advertising Age* (New York), 16 May 1994.

Wildman, Steven S., and Stephen E. Siwek. *International Trade in Films and Television Programs.* Cambridge, Massachusetts: Ballinger, 1988.

See also Cable Networks; Financial Interest and Syndication Rules; International Television Program Markets; National Association of Television Programming Executives; Prime Time Access Rule; Programming; Superstation; Reruns/Repeats; Turner Broadcasting Systems

T

TABLOID TELEVISION

"Tabloid television" is the name often used to describe a group of journalistic program formats that achieved high visibility and great popularity during the middle to late 1980s and early 1990s. Generally used with derisive intonations, the label designates a loosely delineated collection of related genres rather than a singular cohesive one. It has typically been taken to include three primary types of popular journalism. The first is so-called "reality TV," which inserts minicams into a variety of ordinary scenarios like urban law enforcement, and extraordinary ones like spectacular accidents and rescues. Examples include *COPS, American Detective,* and *Rescue 911.* In "reality TV," however, post-hoc reenactments may substitute for "actual footage," and "actual footage" might itself be carefully orchestrated and edited in a variety of ways to match social expectations regarding the characteristics of cops and criminals, for example, and the conventions of television narrative. Tabloid television's second primary type includes unconventional newscasts and documentary programs such as *A Current Affair, Sightings* and *Unsolved Mysteries.* Each of these shows simultaneously embodies and violates television's established journalistic conventions. *A Current Affair,* for instance, copies the structure of the evening newscast, at times apparently only to parody it by transgressing norms of realistic representation or substituting mockery and laughter for high seriousness and reverentially solemn tones. The third primary type of tabloid television is the issue-oriented talk show, including *Donahue, Oprah* and *The Ricki Lake Show.* Like the other kinds of tabloid TV programs, these differ from "serious journalism" both in form and content. They typically value confrontation over "impartiality" and "objectivity" and include a multiplicity of contesting voices that challenges the traditional central role of the journalistic commentator or anchor. Additionally, they often deal with issues considered too "offensive" or "trivial" for serious journalism (such as marginalized sexual practices or the politics of romance and family life).

Tabloid television's explosion was abetted by a number of significant changes in American broadcasting that occurred during the 1980s. Among the most important of these were the expansion of cable television, a threefold increase in the number of independent broadcasting stations operating in the United States and the appearance of the FOX Network, owned by tabloid newspaper mogul Rupert Murdoch. One consequence of these industrial changes was an unprecedented level of demand for new programs designed specifically for syndication. Because of their relatively low production costs compared to fictional television, tabloid shows began to look increasingly attractive to producers of syndicated programming. Moreover, a long writers' strike in 1988 enhanced the value of "reality TV" and was directly responsible for tabloid style FOX Network shows like *COPS* and *America's Most Wanted.* These shows, produced with a minimum of narration or dialogue, were considered "writer proof," unaffected by unplanned production interruptions such as strikes.

The forms of tabloid television that emerged and became popular in the 1980s were not merely products of industrial dynamics and economics, though. They were also inevitably linked to the social context of the period, much of which was defined by Reaganism. As social historian Paul Boyer puts it, "Reaganism was a matter of mood and symbolism as much as of specific [government] programs." Assuming that the media do not "reflect" social history so much as they increasingly become an arena within which it is struggled over and played out, it is possible to find both congruence and dissonance between tabloid television and Reaganism.

Among the significant currents of meaning that Reaganism brought to the surface of American culture during the 1980s were those swirling around our collective anxieties over crime, drugs and, ultimately, race. For example, Reaganism helped popularize both a "war on drugs" and a politically successful "victims' rights" movement. The "war on drugs" saturated the electronic media with images of an urban battleground steeped in violent criminality that all-too-often struck at "innocent victims." Tabloid television played a significant role in both the circulation of images associated with the "drug war" and in the articulation of a populist sense of "victimhood." FOX's *America's Most Wanted,* for example, specialized in cinematically sophisticated reenactments of "actual crimes" followed by an open call for audience members to phone in whatever tips they might be able to provide the police that would help track down missing suspects or escaped fugitives. This premise implies not only a supportive stance toward police departments and crime victims, but also suggests that, in and of themselves, official institutions are incapable of ensuring social order. This was a premise that was extended in local as well as network broadcasting.

Thus, questions about the politics of these programs, which are quite contradictory and therefore difficult to assess, are unavoidable. On the one hand, the popularity of the shows indicates a level of popular distrust toward social institutions from which many people feel alienated. This distrust is often articulated as a class antagonism directed against "the system." Much crime-fighter tabloidism therefore appeals to the populist perception that only the people are capable of looking after their own interests, for "the system" is too often concerned with the narrow interests of the socially privileged. Thus, programs like *COPS,* whose minicams follow "the men and women of law enforcement" into dangerous situations, aren't interested in the upper echelons of police management and administration, but rather focus on the rank-and-file. In their emphasis upon the working conditions inhabited by "ordinary" cops, such programs resonate powerfully with a working class awareness that blue-collar folks inevitably labor under treacherous and difficult conditions and are poorly rewarded for it. As well, they appeal to a very real sense of vulnerability produced by a society in which the socially weak are far more likely to be criminally victimized than the powerful and the privileged.

On the other hand, these programs are part of a contemporary form of white racism that substitutes coded words and issues like "crime" and "drugs" for explicit ways of talking about race. As John Fiske has argued, this facilitates the exertion of racial power while enabling its agents to deny that race is involved at all. So, even though the individual criminals and suspects represented in these programs may often be white (albeit lower-class "white trash"), an emphasis on rampant urban disorder appeals to deeply rooted anxieties in the white imagination regarding people of color presumed to be "out of control" and therefore in need of stepped-up policing. One of the primary responses to these white anxieties in contemporary America has been a massive expansion of urban surveillance systems. Such systems have the two-fold aim of "visibilizing" especially nonwhite populations, and therefore making them available for social discipline, and of encouraging people to police themselves with greater circumspection and vigor. There is much justification for the view that reality-based "tabloid TV" is partly an extension of such surveillance practices. The case of Stephen Randall Dye, a fugitive who turned himself over to police after agonizing for two weeks over a story about him on *America Most Wanted,* provides anecdotal evidence in support of this position (Bartley, 1990).

Tabloidism's partial and populist distrust toward institutions of law and order is extended to the judicial system in the programs *Final Appeal* and *Trial and Error.* Like *America's Most Wanted,* these shows produce reenactments of crimes, but these are supplemented by further reenactments of the trials of the people accused and convicted of those crimes. Rather than supporting these convictions, *Final Appeal* and *Trial and Error* reexamine and question the validity of those criminal verdicts that have resulted in actual incarcerations. The voice-over narration

from *Trial and Error's* opening segment encapsulates the logic these programs follow:

> "Beyond a reasonable doubt." This is the guardian phrase that empowers juries to protect the innocent in America.... The most conservative estimates say that we wrongfully convict and imprison between six and seven thousand people every year. Two half-brothers were within sixteen hours of being executed when it was discovered that the prosecution's star witness was actually nowhere near the crime scene, and she'd only seen it in a dream. A couple in Southern California was convicted of a murder that never even occurred. The alleged victim was found alive and well and living in San Francisco years later....Witnesses sometimes lie, confessions are sometimes coerced, lawyers are sometimes incompetent, and sometimes juries make mistakes.

Final Appeal and *Trial and Error* ultimately question whether our courts ever operate "beyond a reasonable doubt." In doing this, they appeal to a form of popular skepticism that, at particular times and in particular contexts, turns against the judicial system and refuses its discursive power to produce authoritative truths. The first trial of Rodney King and the urban uprisings that answered its verdict provide the most obvious examples of this sort of popular skepticism erupting explosively and demonstrate that faith in American criminal justice is largely a consequence of one's position in American society. In turn, programs like *Final Appeal* and *Trial and Error* demonstrate one of the ways in which tabloid television is capable of tapping into widespread suspicions of officialdom shared by many people who occupy positions of social subordination.

The view that tabloid television circulates beliefs that appeal to a popular skepticism toward official truths receives anecdotal support from Dan Lungren, California's attorney general. Lungren has coined the term "Oprahization" to describe changes in American juries that many prosecutors feel have increased the difficulty of securing criminal convictions. Says Lungren, "people have become so set on the *Oprah* view, they bring that into the jury box with them" (Gregory, 1994). According to a professional jury consultant, "talk-show watchers . . . are considered more likely" than others "to distrust the official version" of events produced by prosecuting attorneys in courtrooms across the land (Gregory, 1994). Los Angeles District Attorney Gil Garcetti has gone so far as to pronounce that the criminal justice system is "on the verge of a crisis of credibility" due to these changes in the sensibilities of jurors (Gregory, 1994).

Talk shows, then, also appeal to a popular skepticism toward official truths. And like the other tabloid programs, their emergence and success bears no small relationship to Reaganism. In Elayne Rapping's words, "the people on these shows are an emotional vanguard, blowing the lid off the idea that America is anything like the place Ronald Reagan pretended to live in." It's no coincidence that tabloid talk shows achieved their highest visibility and popularity in the wake of Reagan, for Reaganism's widening of gaps between such groups as rich and poor, men and women, whites and Blacks brought social differences into clear definition and sharpened the conflicts around them (Fiske, 1994). If Reaganism entailed a widespread cultural repression of voices and identities representing social difference, Reaganism's repressed others returned with a vengeance on TV's tabloid talk shows, which invite the participation of people whose voices are often excluded from American commercial media discourse, such as African Americans, Latinos and Latinas, sex industry workers, "ordinary" women, blue and "pink" collar laborers, the homeless, the HIV positive, people living with AIDS, youths, gay men, lesbians, cross-dressers, transsexuals, convicted criminals, prison inmates, and other socially marginalized groups. This is not to say that tabloid talk shows have a political agenda of anti-racism, anti-sexism, anti-classism, or anti-homophobia, but rather that in opening themselves to the participation of a very broad range of voices, they necessarily encourage potentially progressive conflicts over cultural, racial and sexual politics. In particular, these shows often emphasize what we might call "the politics of normality." A number of prominent commentators such as Erving Goffman and Michel Foucault have examined the role of norms as instruments of power that facilitate the efficient identification of deviance, which is typically punished or subjected to "treatment" and social discipline. But tabloid talk shows are marked by a level of indiscipline that often disrupts the enforcement of norms and allows people who are disadvantaged by those norms to talk back against them.

The last genre of tabloid television includes unconventional newscasts and documentary programs like *A Current Affair* and *Sightings*. It is difficult to generalize about these programs, though often they utilize approaches to storytelling that violate the norms of mainstream journalistic practice in a number of ways. One is to disavow the seriousness of conventional journalism. For example, *A Current Affair,* one of the early definers of American television's tabloid style, was originally anchored by Maury Povich, a refugee from "serious" news whose style was playfully irreverent. This gave much offense to conventional journalists like Philip Weiss, who writes of Povich that "the rubber-faced lewdness his role calls for, the alacrity with which he moves through a half-dozen expressions and voices (from very soft to wired and mean) is a motility reminiscent of the veteran porn star." In his autobiography, Povich writes that his own scorn for the pretensions of the quality press shaped the agenda at *A Current Affair,* which he describes as a "daily fix of silliness, irony, and tub-thumping anger" infused with "an odor of disrespect for authority." He explains that "somehow the notion had come about that news was church business and had to be uttered with ponderous and humor-less reverence; instead news was a circus delivered by clowns and dancing bears and should be taken with a lot of serious skepticism."

The significance of *A Current Affair's* frequent disavowal of the seriousness of more traditional or "respectable" journalistic forms is suggested in Allon White's observation that "seriousness always has more to do with power than with content. The authority to designate what is to be taken seriously (and the authority to enforce reverential solemnity in certain contexts) is a way of creating and maintaining power." Official definitions of "serious journalism" such as those taught in university courses and circulated by the "respectable press" seemed to reinforce an established vision of "that information which the people *need,*" often as prescribed by a community of experts whose lives are quite removed from those of ordinary people. Consequently, analysts like Fiske argue that tabloid television's negotiated refusal of mainstream journalistic seriousness embodies an irreverent, laughing popular skepticism toward official definitions of truth that serve the interests of the socially powerful despite their constant appeals to "objectivity."

Besides mocking the seriousness of mainstream news, some tabloid programs like *Sightings* and *Unsolved Mysteries* confer seriousness upon issues that would likely be treated with laughing dismissal, if at all, in traditional newscasts. Thus, *Sightings* has featured stories about house hauntings, werewolves in the British countryside, and psychic detectives, while *Unsolved Mysteries* has delved into the paranormal terrain of UFO sightings and alien abductions. Popular interest and "belief" in such issues persists despite, or perhaps because of, official denials of their "truth" and "seriousness," and this antagonism between popular belief and official truth is part of the more general antagonism between the social interests of ordinary people and those of the powerful. *Sightings* opens each broadcast with a refreshing disclaimer that nicely encapsulates the difference between its attitude toward the process of informing and that which guides more conventional journalistic enterprises: "The following program deals with controversial subjects. The theories expressed are not the only possible interpretation. The viewer is invited to make a judgment based on all available information."

By transgressing certain norms of conventional journalism, tabloid television has drawn the scorn of a great many critics who feel that journalistic TV should address "loftier" issues in more "tasteful" and serious ways. And it has shown that television can be quite adept at speaking to a variety of forms of popular skepticism toward some of our social institutions and the versions of truth they pronounce.

—Kevin Glynn

FURTHER READING

Bartley, Diane. "John Walsh: Fighting Back." *The Saturday Evening Post* (Indianapolis, Indiana), April 1990.

Bernstein, Carl. "The Idiot Culture." *The New Republic* (Washington, D.C.), June 1992.

Beschloss, Steven. "TV's Life of Crime." *Channels* (New York), 24 September 1990.

Bird, S. Elizabeth. *For Inquiring Minds: A Cultural Study of Supermarket Tabloids*. Knoxville: University of Tennessee Press, 1992.

Boyer, Paul. "Introduction: Reaganism: Reflections on an Era." In, Boyer, Paul, editor. *Reagan as President*. Chicago: Ivan R. Dee, 1990.

Campbell, Richard. "Word vs. Image: Elitism, Popularity and TV News." *Television Quarterly* (New York), 1991.

Dahlgren, Peter, and Colin Sparks, editors. *Journalism and Popular Culture*. London: Sage, 1992.

Fiske, John. "Popularity and the Politics of Information." In, Dahlgren, Peter, and Colin Sparks, editors. *Journalism and Popular Culture*. London: Sage, 1992.

————. *Understanding Popular Culture*. Boston: Unwin Hyman, 1989.

————. *MediaMatters: Everyday Culture and Political Change*. Minneapolis: University of Minnesota Press, 1994.

Fiske, John, and Kevin Glynn. "Trials of the Postmodern." *Cultural Studies* (London), October 1995.

Foucault, Michel, and translated by Alan Sheridan. *Discipline and Punish: The Birth of the Prison*. New York: Vintage, 1979.

Glynn, Kevin. "Tabloid Television's Transgressive Aesthetic: *A Current Affair* and the Shows that Taste Forgot." *Wide Angle* (Athens, Ohio), April 1990.

————. "Reading Supermarket Tabloids as Menippean Satire." *Communication Studies* (West Lafayette, Indiana), Spring 1993.

Goffman, Erving. *Stigma: Notes on the Management of Spoiled Identity*. Englewood Cliffs, New Jersey: Prentice-Hall, 1963.

Gregory, Sophfronia Scott. "Oprah! Oprah in the Court." *Time* (New York), 6 June 1994.

Katz, Jack. "What Makes Crime 'News'?" *Media, Culture and Society* (London), 1987.

Katz, Jon. "Covering the Cops: A TV Show Moves in where Journalists Fear to Tread." *Columbia Journalism Review* (New York), February 1993.

Knight, Graham. "Reality Effects: Tabloid Television News." *Queen's Quarterly* (Kingston, Canada), March 1989.

Mellencamp, Patricia. *High Anxiety: Catastrophe, Scandal, Age, and Comedy*. Bloomington: Indiana University Press, 1992.

Morse, Margaret. "The Television News Personality and Credibility: Reflections on the News in Transition." In, Modleski, Tania, editor. *Studies in Entertainment: Critical Approaches to Mass Culture*. Bloomington: Indiana University Press, 1986.

Pauly, John J. "Rupert Murdoch and the Demonology of Professional Journalism." In, Carey, James, editor. *Media, Myths, and Narratives: Television and the Press*. Newbury Park, California: Sage, 1988.

Povich, Maury, with Ken Gross. *Current Affairs: A Life on the Edge*. New York: Berkeley, 1991.

Rapping, Elayne. "Daytime Inquiries." *The Progressive* (Madison, Wisconsin), October 1991.

Rose, Frank. "Celebrity Victims: Crime Casualties Are Turning into Stars on Tabloid TV." *New York* (New York), July 1989.

Sauter, Van Gordon. "In Defense of Tabloid TV." *TV Guide* (Radnor, Pennsylvania), 5 August 1989.

Seagal, Debra. "Tales from the Cutting-room Floor." *Harper's* (New York), November 1993.

Weiss, Philip. "Bad Rap for TV Tabs." *Columbia Journalism Review* (New York), May 1989.

See also *America's Most Wanted*; Donahue, Phil; Rivera, Geraldo; Talk Shows; Winfrey, Oprah

TAIWAN

The birth of the television era in Taiwan began when China Broadcasting Corporation (CBC) brought the Presidential Inauguration live to 50 television screens in May 1960. This event also marked the beginning of the extensive political influence of the three terrestrial broadcasting systems on all facets of life in the country. Taiwan Television Enterprise (TTV), the first network, was established in 1962 with a significant transfer of Japanese expertise and an initial 40% investment by the four leading Japanese electronic firms. China Television Company (CTV) was launched with exclusively domestic financing in 1969, and Chinese Television System (CTS) was transformed from an educational to a general broadcasting service in 1971. Two-and-a-half decades later, these three networks remain dominated by their stockholders which are, respectively, the Taiwan

Provincial Government, the political party Kuomintang, and the Ministries of Defense and Education. Ideological control, exercised by these major underwriters, remains apparent in both news and entertainment programming. In order to claim its political legitimacy over local Taiwanese politics, for example, the KMT government pronounced Mandarin as the official language in Taiwan and restricted the use of Fukienese to only 20% of television programming, despite the fact that it was used by the vast majority of the population in the 1960s.

Since the development of a political movement by the opposition party in the early 1980s, the KMT government has been under pressure to begin relaxation of its media monopoly. Opposition leaders fought for alternative voices with a massive wave of print media publications, followed by the creation of numerous underground radio broadcasting stations. Government crackdown on these activities proved ineffective when many opposition party members were voted into the legislature and the movement was backed by a significant number of intellectuals. In 1995, the Taipei city government, headed by a renowned Democratic Progress Party (DPP) leader, fought for a 30% share of TTV by threatening to block a signal license renewal. Ultimately, the attempt was dropped in exchange for a goodwill promise on the part of TTV to tone down its political partisanship. Furthermore, the legislature passed a regulation in 1996 which raised every terrestrial station's annual license fee from NT$60,000 to NT$10 million (exchange rate USD=NT$27.5), effective immediately.

These recent developments signal a passing of a television monarchy controlled by the three networks, which coincides with the emergence of the Fourth Channel, an abbreviated name for all underground cable systems and channels. This Fourth Channel surfaced as a powerful media alternative in 1994 with the official launch of TVBS and its landmark call-in program, *2100 All Citizens Talk*. A fourth official national television network is also in development, its license granted to People's Broadcasting Corporation, which consists largely of supporters of the opposition party, DPP. It is scheduled to be on air in February 1997, one year earlier than originally planned.

When the fourth channel begins programming, like the other broadcasters, it will turn to one of three types of sources for content: internal production by the networks, contracted domestic production by independent production companies, and foreign imports. The government ruled that foreign imports should not exceed 30% of the total daily programming hours and all foreign programs are required to use either Mandarin voice-over or Mandarin subtitles. CTS is particularly known for its effort in "localizing" its entertainment programming; the network wrote television history in 1994 when it first mixed Mandarin with Fukienese in its 8:00 P.M. prime-time drama series, *When Brothers Meet*. Instead of the neverending Romeo and Juliet-style of love and hate romance, this program established a brand new drama genre in which real-life conflicts were recreated in the context of real-life societal events. *When Brothers Meet* not only took the lead in the television prime-time ratings, it also began a continuing success in television drama for CTS.

With the exception of news, all television programs are subject to review by the Government Information Office (GIO). Even in newsrooms, however, self-censorship is practiced. Commercial air time—advertising—is limited to ten minutes per hour on terrestrial systems. Cable systems are limited to six minutes per hour, and coalition efforts are underway for some regional satellite broadcasters to unite in protesting the government's preferential treatment of the free-to-air terrestrials. In other areas, however, cable has its own advantages. Cigarette and liquor commercials are barred from free-to-air stations, yet in 1996 commercials for liquor have been allowed on cable after 9:00 P.M.

Such regulations are truly significant in economic terms. While 99.9% of the country receives broadcast television and 67% of the homes own at least two television sets, cable has penetrated 76% of the 5.6 million television households, according to Nielsen-SRT's second quarterly Media Index Report, released in July 1996. It is receivable in over 4.4 million homes and, since 1994, the channel share of all cable stations has surpassed the combined share of the three terrestrial systems. As of June 1996, cable homes or cable individuals spent two-thirds of their viewing time with cable. Certainly, the phenomenal cable growth in Taiwan from 18% of market penetration in 1991 to 50% in 1993 and the current 76% coincides with the economic well-being of the country.

Not surprisingly, the cable industry has been considered a highly lucrative market by both domestic and foreign investors. The Cable Law, however, passed in August 1993, explicitly outlawed foreign shareholding. Cross-media ownership is disallowed among newspaper owners, free-to-air broadcasters, and cable operators and programmers. Further regulations restrict any shareholder to no more than 10% of the total assets value.

Other regulations focus more precisely on cable systems. In the area of programming, for example, domestically-produced programs must represent at least 20% of the total programming hours. Nevertheless, in light of the fact that the Cable Law is designed exclusively to bring the system operators under control, cable programmers have often tested the limit of the law and frequently go their own way. The constant power struggles between system operators and cable program suppliers have left the GIO powerless most of the time.

In one area, however, the cable industry finally came under restriction in the fall of 1994 after severe protests by the U.S. copyright organizations. Cable operators engaged in extreme violations of copyright laws, airing literally everything from movies to sitcoms and variety shows without payment, which resulted in substantial revenue loss to the program copyright owners. Under threat from the U.S. government, authorities in Taiwan finally began an all-out effort to crack down on the illegal cable operators. The resulting rising costs for program purchases drove some operators out of business and contributed to a significant consolidation of cable systems in recent years.

Financial concerns also affect the terrestrial systems. Despite the fact that all three are financially dominated by the different government offices, they are essentially commercial rather than public stations. In 1995, they garnered 5% of the total NT$29.6 billion (or U.S.$1.1 billion) advertising revenues, with TTV slightly edging ahead of CTS by 3% and CTV by 6%. In the same year, television advertising revenues accounted for approximately 40% of the total advertising expenditures, topping newspapers by nearly 10%. With the significant cable growth, 90% of the top 300 advertisers replied in a 1995 survey that they were prepared to invest 15-20% of their advertising dollars in cable.

Essentially, the TV-advertising market has changed from a seller's market to a buyer's market. The three terrestrial networks are predicted to lose a quarter of net television advertising to other channels in 1997 and, by 2005, less than half the net total will go to the terrestrial systems. On the other hand, TV advertising is predicted to nearly double between 1995 and 2000 to U.S.$1.8 billion, and will almost triple to U.S.$2.7 billion in 2005. International advertisers dominate the top 20 list of largest advertisers in Taiwan. Ford leads the category with total annual billings of NT$1,592 million, followed by Proctor and Gamble with NT$1.103 million, Toyota with NT$1,005 million, and Mavibel, Kao, Matsushita, Hong Kong Shanghai Bank, AC Johnson, and Nestle among the biggest spenders.

These advertisers present their products in one of the most complex, multicultural media environments in the world. In a country with a population of 21 million, more than 180 satellite channels and 130 cable operators compete for audiences. A typical cable household receives 70 channels, all as part of the basic tier. In the movie category alone, more than 12 channels show movies originating from the United States, Spain, France, Italy, the United Kingdom, Russia, Japan, China, Hong Kong, and other countries.

In the face of this 70-channel environment, all regional satellite channels have made "channel localization" an integral part of their programming effort. They have created specific channel "identities" related to specific Asian countries and regions. Such localization has gone beyond the use of specific languages and has led regional broadcasters to produce "locally correct" cable content by teaming up with the local production entities or houses in the various Asian countries. The Discovery Channel, HBO, ESPN, MTV, and Disney are all prime examples of entities competing against these local cable channels and their localized content. Much of the programming effort by these "global" suppliers was, in fact, launched as an attempt to use the Taiwan market as a test for eventual programming in China.

The influx of new local and international cable channels is far from over. For every type of channel already in place, another is in formation. The Scholars' Corporation announced the launch of a five-channel package in May 1996; a very popular local channel, SanLi, is preparing for the release of its third channel; the Videoland Group is getting ready for its fourth channel; and the general-interest Super Channel, which came on the scene in October 1995, has added another channel devoted to sports.

The cable attraction has resulted in a large decline of viewership on the three terrestrial networks. Even the 7:00-8:00 P.M. news hour on the networks, dominant for almost three decades, is losing an audience share to cable. Individual ratings among viewers aged four and above have generally declined among all program genres.

On the other hand, almost every regional satellite channel and cable station has steadily gained viewership and momentum. Cable's niche programming orientation has led to the creation of many channels with clearly definable audience profiles. When analyzed within target audiences, some cable channel ratings even surpass those of the three networks. The current television climate may be summarized as follows: (1) A typical viewer spends an average of 2.2 hours daily watching television. Individuals with cable spend more time watching television than their non-cable counterparts. (2) "Program loyalty" has replaced "channel loyalty" in describing the viewer's logic of television choice. Viewers select specific programs and move among channels to do so. (3) Related to this development, a cable channel is recognized oftentimes because it carries a few popular programs. It is "programs" which define the character of any channel, not the channel itself, even for the 24-hour news channel. (4) Prime time on cable is virtually 23 hours a day; the only hour excluded is the 8:00-9:00 P.M. daily drama series time. (5) The new television ecology has gradually given rise to new sales and marketing concepts. Program suppliers can no longer emphasize the reality of "how many" viewers are watching; instead, it is the determination of "who" is watching that helps deliver the audience to the advertiser.

Behind this multi-channel, multicultural viewing environment is a series of questions baffling the policy-makers. The seemingly vast program choices conceal the reality that programming homogeneity still outweighs its heterogeneity. Not only are schedules for the three terrestrial networks similar across all dayparts, the same high level of repetition is also frequently observed within and among the cable channels. One hundred thirty cable operators spend a great deal of money to buy channels only to find that such operations are virtually the opposite of the principle of "natural monopoly" normally used to describe the cable industry. The government is busy making cable laws only to find that participants in the industry have invented new games which defy the regulations. While new channels continue to be rolled out on a monthly basis, new communication technologies such as the Internet are aggressively pursued and applied by many programmers to add to their marketing effort and competitive edge. The television market in Taiwan is far from saturated. It is instead loaded—with selection, repetition, excitement, energy, and challenges.

—Zoe Tan

FURTHER READING

Baum, Julian. "We Intercept This Broadcast: Taiwan Moves to Rein in Cable-TV Operators." *Far Eastern Economic Review* (Hong Kong), 29 July 1993.

———. "Untangling the Wires." *Far Eastern Economic Review* (Hong Kong), 27 October 1993.

"Boxing Clever." *The Economist* (London), 16 September 1995.

Nielsen SRG. *Asia Pacific Television Channels.* Shrub Oak, New York: Baskerville Communications, 1996.

———. *Asia Pacific Television Revenues.* Shrub Oak, New York: Baskerville Communications, 1996.

"Taiwan's Covert Cablers." *The Economist* (London), 20 November 1993.

"Tying Up Cable Television: Taiwan." *The Economist* (London), 18 September 1993.

TALK SHOWS

The television talk show is, on the face of it, a rather strange institution. We pay people to talk for us. Like the soap opera, the talk show is an invention of twentieth century broadcasting. It takes a very old form of communication, conversation, and transforms it into a low cost but highly popular form of information and entertainment through the institutions, practices and technologies of television.

The talk show did not originate overnight, at one time, or in one place. It developed out of forty years of television practice and antecedent talk traditions from radio, Chatauqua, vaudeville and popular theater. In defining the talk show it is useful to distinguish between "television talk" (unscripted presentational address) and "talk shows"—shows organized principally around talk. "Television talk" represents all the unscripted forms of conversation and direct address to the audience that have been present on television from the beginning. This kind of "live," unscripted talk is one of the basic things that distinguishes television from film, photography, and the record and book industries. Television talk is almost always anchored or framed by an announcer or host figure, and may be defined, in Erving Goffman's terms, as "fresh talk," that is, talk that appears to be generated word by word and in a spontaneous

manner. Though it is always to a degree spontaneous, television talk is also highly structured. It takes place in ritualized encounters and what the viewer sees and hears on the air has been shaped by writers, producers, stage managers and technical crews and tailored to the talk formulas of television.

Thus, though it resembles daily speech, the kind of talk that occurs on television does not represent unfettered conversation. Different kinds of television talk occur at different times of the broadcast day, but much of this talk occurs outside the confines of what audiences and critics have come to know as the "talk show." Major talk traditions have developed around news, entertainment, and a variety of social encounters that have been reframed and adapted for television. For example, talk is featured on game shows, dating or relationship shows, simulated legal encounters (*People's Court*) or shows that are essentially elaborate versions of practical jokes (*Candid Camera*). All of these shows feature talk but are seldom referred to as "talk shows."

A "talk show," on the other hand, is as a show that is quite clearly and self-consciously built around its talk. To remain on the air a talk show must adhere to strict time and money constraints, allowing time, for instance, for the ad-

Dick Cavett

Mike Douglas

Ricki Lake

Regis Philbin and Kathie Lee Gifford

vertising spots that must appear throughout the show. The talk show must begin and end within these rigid time limits and, playing to an audience of millions, be sensitive to topics that will interest that mass audience. For its business managers the television talk show is one product among many and they are usually not amenable to anything that will interfere with profits and ratings. This kind of show is almost always anchored by a host or team of hosts.

Host/Forms

Talk shows are often identified by the host's name in the title, an indication of the importance of the host in the history of the television talk show. Indeed, we might usefully combine the two words and talk about host/forms.

A good example of the importance of the host to the form a talk show takes would be *The Tonight Show*. *The Tonight Show* premiered on NBC in 1954 with Steve Allen as its first host. While it maintained a distinctive format and style throughout its first four decades on the air, *The Tonight Show* changed significantly with each successive host. Steve Allen, Ernie Kovacs, Jack Paar, Johnny Carson, and Jay Leno each took *The Tonight Show* in a significant new direction. Each of these hosts imprinted the show with distinctive personalities and management styles.

Though many talk shows run for only weeks or months before being taken off the air, once established, talk shows

and talk show hosts tend to have long runs. The average number of years on television for the thirty-five major talk show hosts listed at the end of this essay was eighteen years. Successful talk show hosts like Mike Wallace, Johnny Carson, and Barbara Walters bridge generations of viewers. The longevity of these "superstars" increases their impact on the forms and formats of television talk with which they are associated.

Television talk shows originally emerged out of two central traditions: news and entertainment. Over time hybrid forms developed that mixed news, public affairs, and entertainment. These hybrid forms occupy a middle ground position between news and entertainment, though their hosts (Phil Donahue, Oprah Winfrey, and Geraldo Rivera, for example) often got their training in journalism. Approximately a third of the major talk show hosts listed at the end of the essay came out of news. The other two-thirds came from entertainment (comedy in particular).

Within the journalistic tradition, the names Edward R. Murrow, Mike Wallace, Ted Koppel and Bill Moyers stand out. News talk hosts like Murrow, Koppel, and Moyers do not have bands, sidekicks, or a studio audience. Their roles as talk show hosts are extensions of their roles as reporters and news commentators. Their shows appear in evening when more adult and older aged viewers are watching. The morning host teams that mix "happy talk" and information also generally come from the news background. This format was pioneered by NBC's Sylvester "Pat" Weaver and host Dave Garroway with the *Today* show in the early 1950s. Hosts who started out on early morning news talk shows and went on to anchor the evening news or primetime interview shows include: Walter Cronkite, John Chancellor, Barbara Walters, Tom Brokaw, and Jane Pauley. Each developed a distinctive style within the more conversational format of his morning show.

Coming from a journalism background but engaging in a wider arena of cultural topics were hosts like Phil Donahue, Oprah Winfrey, and Geraldo Rivera. Mixing news, entertainment, and public affairs, Phil Donahue established "talk television," an extension of the "hot topic" live radio call-in shows of the 1960s. Donahue himself ran a radio show in Dayton, Ohio, before premiering his daytime television talk show. Donahue's Dayton show, later syndicated nationally, featured audience members talking about the social issues that affected their lives.

Within the field of entertainment/variety talk, it was the late-night talk show that assumed special importance. Late-night talk picked up steam when it garnered national attention during the talk show "wars" of the late 1960's and early 1970's. During this time Johnny Carson defended his ratings throne on *The Tonight Show* against challengers Joey Bishop, David Frost, Dick Cavett and Merv Griffin. Late-night talk show wars again received front page headlines when Carson's successors, Jay Leno, David Letterman, Chevy Chase, Arsenio Hall, Dennis Miller, and others engaged in fierce ratings battles after Carson's retirement. Within the United States these talk show wars assumed epic proportions in the press, and the impact that late night entertainment talk show hosts had over their audiences seemed, at times, to assume that of political leaders or leaders of state. In an age in which political theorists had become increasingly pessimistic about the possibilities of democracy within the public sphere, late-night talk show hosts became sanctioned court jesters who appeared free to mock and question basic American values and political ideas through humor. Throughout the 1960s, 1970s and 1980s Johnny Carson's monologue on the *Tonight* show was considered a litmus test of public opinion, a form of commentary on the news. Jay Leno's and David Letterman's comic commentary continued the tradition.

The ratings battle between Leno and Letterman in the early 1990s echoed the earlier battles between Carson, Dick Cavett, and Griffin. But it was not just comic ability that was demanded of the late night hosts. They had to possess a lively, quick-paced interview technique, a persistent curiosity arising directly from their comic world views, lively conversational skills, and an ability to listen to and elicit information from a wide range of show business and "civilian" guests. It was no wonder that a relatively small number of these hosts survived more than a few years on the air to become stars. Indeed, in all categories of the television talk show over four decades on the air, there were less than three

dozen news and entertainment talk show hosts who achieved the status of stars.

While entertainment/variety talk dominated late night television, and the mixed public affairs/entertainment audience participation talk shows with hosts like Phil Donahue and Oprah Winfrey increasingly came to fill daytime hours, prime time remained almost exclusively devoted to drama.

Talk Formats

While talk show hosts represent a potpourri of styles and approaches, the number of talk show formats is actually quite limited. For example, a general interest hard news or public affairs show can be built around an expert panel (*Washington Week in Review*), a panel and news figure (*Meet the Press*), a magazine format for a single topic (*Nightline*), a magazine format that deals with multiple topics (*60 Minutes*), or a one-on-one host/guest interview (Bill Moyers' *World of Ideas*). These are the standard formats for the discussion of hard news topics. Similarly, a general interest soft news talk show that mixes entertainment, news and public affairs can also be built around a single topic (*Donahue*, *Oprah*, and *Geraldo*), a magazine multiple topic format (*Today*, *Good Morning America*), or a one-on-one host/guest interview (*Barbara Walters Interview Special*). There are also special interest news/information formats that focus on such subjects as economics (*Wall Street Week*), sports (*Sports Club*), homemaking/fashion (*Ern Westmore Show*), personal psychology (*Dr. Ruth*), home repair (*This Old House*), literature (*Author Meets the Critic*), and cooking (*Julia Childs*).

Entertainment talk shows are represented by a similarly limited number of formats. By far the most prevalent is the informal celebrity guest/host talk show, which takes on different characteristics depending upon what part of the day it is broadcast. The late night entertainment talk show, with the publicity it received through the "talk show wars," grew rapidly in popularity among viewers during its first four decades on the air. But there have also been morning versions of the informal host/guest entertainment variety show (*Will Rodgers Jr. Show*), daytime versions (*The Robert Q. Lewis Show*), and special topic versions (*American Bandstand*). Some entertainment talk shows have featured comedy through satirical takes on talk shows (*Fernwood Tonight*, *The Larry Sanders Show*), monologues (*The Henry Morgan Show*), or comedy dialogue (*Dave and Charley*). Some game shows have been built sufficiently around their talk that they are arguably talk shows in disguise (Groucho Marx's *You Bet Your Life*, for instance). There are also a whole range of shows that are not conventionally known as "talk shows" but feature "fresh" talk and are built primarily around that talk. These shows center on social encounters or events adapted to television: a religious service (*Life is Worth Living*), an academic seminar (*Seminar*), a talent contest (*Talent Scouts*), a practical joke (*Candid Camera*), mating rituals (*The Dating Game*), a forensic event (*People's Court*), or a mixed social event (*House Party*). The line between "television talk" and what formally constitutes a talk show is often not easy to draw and shifts over time as new forms of television talk emerge.

How To Read a Television Talk Show

There are many approaches to understanding a television talk show. It may be viewed as a literary narrative, for instance, or as a social text. As literary texts, talk shows contain characters, settings, and even a loosely defined plot structure which re-enacts itself each evening in the talk rituals that take place in front of the camera. These narratives center on the host as the central recurring character who frames and organizes the talk. Literary analysis of talk shows is relatively rare, but Michael Arlen's essay on the talk show in *The Camera Age*, or Kenneth Tynan's profile of Johnny Carson in *The New Yorker*, are superb examples of this approach.

Talk shows can also be seen as social texts. Talk shows are indeed forums in which society tests out and comes to terms with the topics, issues and themes that define its basic values, what it means to be a "citizen," a participating member of that society. The "talk television" shows of Phil Donahue or Oprah Winfrey become microcosms of society as cutting-edge social and cultural issues are debated and discussed. By the early 1990s political and social analysts began to pay increasing attention to these forms of television and a number of articles were written about them.

Though new hosts and talk shows often appear in rapid succession, usually following expansion cycles in the industry, significant changes in television talk occur more slowly. These changes have traditionally come about at the hands of a relatively small number of influential talk show hosts and programmers and have occurred within distinct periods of television history.

Cycles of Talk: The History of the Television Talk Show

The term "talk show" was a relatively late invention, coming into use in the mid-1960s, but shows based on various forms of spontaneous talk were a staple of broadcasting from its earliest days. Radio talk shows of one kind or another made up 24% of all radio programming from 1927 to 1956, with general variety talk, audience participation, human interest, and panel shows comprising as much as 40-60% of the daytime schedule. Network television from 1949 to 1973 filled over half its daytime program hours with talk programming, devoting 15 to 20% of its evening schedule to talk shows of one kind or another. As the networks went into decline, their viewership dropping from 90% to 65% of the audience between the 1980s and the 1990s, talk shows were one form of programming that continued to expand on the networks and in syndication. By the summer of 1993 the television page of *USA Today* listed seventeen talk shows and local papers as many as twenty-seven. In all, from 1948 to 1993 over two hundred talk shows appeared on the air. These shows can be broken down into four cycles of televi-

sion talk show history corresponding to four major periods of television history itself.

The first cycle took place from 1948 to 1962 and featured such hosts as Arthur Godfrey, Dave Garroway, Edward R. Murrow, Arlene Francis, and Jack Paar. These hosts had extensive radio experience before coming to television and they were the founders of television talk. During this time the talk show's basic forms—coming largely out of previous radio and stage traditions—took shape.

The second cycle covers the period from 1962 to 1972 when the networks took over from sponsors and advertising agencies as the dominant forces in talk programming. A small but vigorous syndicated talk industry grew during this period as well. In the 1960s and early 1970s three figures established themselves on the networks as talk hosts with staying power: Johnny Carson, Barbara Walters, and Mike Wallace. Each was associated with a program that became an established profit center for their network and each used that position to negotiate the sustained status with the network that propelled them into the 1970s and 1980s as a star of television talk.

The third cycle of television talk lasted from 1970 to 1980. During this decade challenges to network domination arose from a number of quarters. While the networks themselves were initiating few new talk shows by 1969, syndicated talk programming exploded. Twenty new talk shows went on the air in 1969 (up to then the average number of new shows rarely exceeded five). It was a boom period for television talk—and the time of the first nationally publicized "talk show wars." New technologies of production (cheaper television studios and production costs), new methods of distribution (satellite transmission and cable), and key regulatory decisions by the FCC made nationally syndicated talk increasingly profitable and attractive to investors.

Talk show hosts like Phil Donahue took advantage of the situation. Expanding from 40 markets in 1974 to a national audience of 167 markets in 1979, Donahue became the nation's number one syndicated talk show host by the late 1970s. Other new talk show hosts entered the field as well. *Bill Moyers' Journal* went on the Public Broadcasting Service (PBS) in 1970, and William Buckley's *Firing Line*, which had appeared previously in syndication, went on PBS a year later. Both Moyers and Buckley, representing liberal and conservative viewpoints respectively, were to remain significant figures on public broadcasting for the next two decades. During this time independent stations and station groups, first-run syndication, cable and VCRs began to weaken the networks' once invincible hold over national audiences.

The fourth cycle of television talk took place in the period from 1980 to 1992, a period that has been commonly referred to as the "post-network" era. Donahue's success in syndication was emulated by others, most notably Oprah Winfrey, whose Donahue-style audience participation show went into national syndication in 1986. Winfrey set a new record for syndication earnings, grossing over a hundred million dollars a year from the start of her syndication. She became, financially, the most successful talk show host on television.

By the early 1980s the networks were vigorously fighting back. *Late Night with David Letterman* and Ted Koppel's *Nightline* were two network attempts to win back audiences. Both shows gained steady ratings over time and established Koppel and Letterman as stars of television talk.

Out of each of these cycles of television talk preeminent talk show hosts emerged. Following the careers of these hosts allows us to we see how talk shows are built from within by strong personalities and effective production teams, and shaped from without by powerful economic, technological, and cultural forces.

Paradigm Shifts in Late Night Entertainment: Carson to Letterman

Johnny Carson, for thirty years the "King of Late Night," and his successor, David Letterman, were in many ways alike. Their rise to fame could be described by the same basic story. A young man from America's heartland comes to the city, making his way through its absurdities and frustrations with feckless humor. This exemplary middle American is "square" and at the same time sophisticated, innocent, though ironic and irreverent. Straddling the worlds of common sense and show business, the young man becomes a national jester—and is so annointed by the press.

The "type" Johnny Carson and David Letterman represent can be traced to earlier archetypes: the "Yankee" character in early American theater and the "Toby" character of nineteenth century tent repertory. Carson brought his version of this character to television at the end of the Eisenhower and beginning of the Kennedy era, poking fun at American consumerism and politics in the late 1950s and 1960s.

Letterman brought his own version of this sharp-eyed American character to the television screen two decades later at the beginning of the Reagan era. By this time the "youth" revolts in the 1960s and 1970s were already on the wane, and Letterman replaced the politics of confrontation represented by the satire of such shows as *Saturday Night Live* and *SCTV* with a politics of accommodation, removal, and irony. His ironic stance was increasingly acknowledged as capturing the "voice" of his generation and, whether as cause or effect, Letterman became a generational symbol.

The shift from Carson to Letterman represented not only a cultural change but a new way of looking at television as a medium. Carson's camera was rooted in the neutral gaze of the proscenium arch tradition; Letterman's camera roamed wildly and flamboyantly through the studio. Carson acknowledged the camera with sly asides; Letterman's constant, neurotic intimacy with the camera, characterized by his habit of moving right up to the lens and speaking directly into it, represented a new level of self-consciousness about the medium. He extended the "self-referentiality" that Carson himself had promoted over the years on his talk show.

Indeed, Letterman represented a movement from what has been called a *transparent* form of television (the viewer taking for granted and looking *through* the forms of television: camera, lighting, switching, etc.) to an opaque form in which the technology and practices of the medium itself become the focus of the show. Letterman changed late night talk forever with his post-modern irreverence and mocking play with the forms of television talk.

Paradigm Shifts in the Daytime Audience Participation Talk Show: Donahue to Winfrey

When Oprah Winfrey rose to national syndication success in 1986 by challenging Phil Donahue in major markets around the country and winning ratings victories in many of these markets, she did not change the format of the audience participation talk show. That remained essentially as Donahue had established it twenty years before. What changed was the cultural dynamics of this kind of show and that in turn was a direct reflection of the person who hosted it.

The ratings battle that ensued in 1986 was between a black woman raised by a religious grandmother and strict father within the fold of a black church in the South against a white, male, liberal, Catholic Midwesterner who had gone to Notre Dame, and been permanently influenced by the women's movement. Just as Jackie Robinson had broken baseball's color barrier four decades earlier, Oprah Winfrey broke the color line for national television talk show hosts in 1986. She became one of the great "Horatio Alger" rags-to-riches story of the 1980s (by the early 1990s *People Weekly* was proclaiming her "the richest woman in show business," with an estimated worth of $200 million), and as Arsenio Hall and Bob Costas ended their six and seven year runs on television in the early 1990s, it became clear that Oprah Winfrey had staying power. She remained one of the few prominent talk show hosts of the 1980s to survive within the cluttered talk show landscape of mid–1990s.

Several factors contributed to this success. For one thing, Winfrey had a smart management team and a full-press national marketing campaign to catapult her into competition with Donahue. The national syndication deal had been worked out by Winfrey' representative, attorney-manager Jeffrey Jacobs, and thanks to King World's management, her marketing plan was a classic one. Executives at King World felt the media would pounce on "a war with Donahue" so they created one. The first step was to send tapes of Oprah's shows to "focus groups" in several localities to see how they responded. The results were positive. The next step was to show tapes to selected station groups—small network alliances of a half-dozen or more stations under a single owner. These groups would be offered exclusive broadcast rights. As the reactions began to come in, King World adjusted its tactics. Rather than making blanket offers, they decided to open separate negotiations in each city and market. The gamble paid off. Winfrey's track record proved her a "hot enough commodity" to win better deals through individual station negotiation.

To launch Winfrey on the air King World kicked off a major advertising campaign. Media publications trumpeted Oprah's ratings victories over Donahue in Baltimore and Chicago. The "Donahue-buster" strategy was tempered by Winfrey herself, who worked hard not to appear too arrogant or conceited. When asked about head-on competition with Donahue she replied that in a majority of markets she did not compete with him directly and that while Donahue would certainly remain "the king," she just wanted to be "a part of the monarchy." By the time *The Oprah Winfrey Show* went national in September 1986 it had been signed by over 180 stations—less than Donahue's 200-plus but approaching that number.

As well as refined marketing and advertising techniques, cultural issues also featured prominently in Winfrey's campaign. Winfrey's role as talk show host was inseparable from her identity as an African American woman. Her African American heritage and roots surfaced frequently in press accounts. One critic described her in a 1986 *Spy* magazine article as "capaciously built, black, and extremely noisy." These and other comments on her "black" style were not lost on Winfrey. She confronted with the issue of race constantly and was very conscious of her image as an African American role model.

When a *USA Today* reporter queried Winfrey bluntly about the issue of race in August of 1986, asking her "as someone who is not pencil-thin, white, nor blond," how she was "transcending barriers that have hindered many in television," Winfrey replied as follows:

> I've been able to do it because my race and gender have never been an issue for me. I've been blessed in knowing who I am, and I am a part of a great legacy. I've crossed over on the backs of Sojourner Truth, and Harriet Tubman, and Fannie Lou Hamer, and Madam C.J. Walker. Because of them I can now soar. Because of them I can now live the dream....

Winfrey's remarks represent the "double-voiced" identity of many successful African American public figures. Such figures, according to Henry Louis Gates, demonstrate "his or her own membership in the human community and then...resistance to that community." In the mid-1980s, then, the image of Oprah Winfrey as national talk show host played against both white and black systems of values and aesthetics. It was her vitality as a double-sign, not simply her role as an "Horatio Alger" figure, that made her compelling to a national audience in the United States.

Hosts like Letterman and Winfrey played multiple roles. They were simultaneously star performers, managing editors, entrepreneurs, cultural symbols, and setters of social trends. Of all the star performers who dot the landscape of television, the talk show host might have the most direct claim to the film director's status as *auteur*. Hosts like

Letterman and Winfrey had to constantly re-invent themselves, in the words of Kenneth Tynan, to sustain themselves within the highly competitive world of network television.

Conclusion

The talk show, like the daily newspaper, is often considered a disposable form. The first ten years of Johnny Carson's *Tonight* shows, for example, were erased by NBC without any thought of their future use. Scholars have similarly neglected talk shows. News and drama offered critics from the arts, humanities, and social sciences at least a familiar place to begin their studies. Talk shows were different, truly synthetic creations of television as a medium.

Nonetheless, talk shows have become increasingly important on television and their hosts increasingly influential. They speak to cultural ideas and ideals as forcefully as politicians or educators. National talk show hosts become surrogates for the citizen. Interrogators on the news or clown princes and jesters on entertainment talk shows, major television hosts have a license to question and mock—as long as they play within the rules. An investigation of the television talk show must, finally, delineate and examine those rules.

The first governing principle of the television talk show is that everything that occurs on the show is framed by the host who characteristically has a high degree of control over both the show and the production team. From a production point of view, the host is the managing editor; from a marketing point of view, the host is the label that sells the product; from an power and organizational point of view, the host's star value is the fulcrum of power in contract negotiations with advertisers, network executives, and syndicators. Without a "brand-name" host, a show may continue but it will not be the same.

A second principle of television talk show is that it is experienced in the present tense. This is true whether the show is live or taped "as-if-live" in front of a studio audience. Live, taped, or shown in "reruns," talk shows are conducted, and viewers participate in them, as if host, guest, and viewer occupy the same moment.

As social texts, television talk shows are highly sensitive to the topics of their social and cultural moment. These topics may concern passing fashions or connect to deeper preoccupations. References to the O.J. Simpson case on television talk shows in the mid-1990s, for example, reflected a preoccupation in the United States with domestic violence and issues of gender, race, and class. Talk shows are, in this sense, social histories of their times.

While it is host-centered, occurring in a real or imagined present tense, sensitive to the historical moment, and based on a form of public/private intimacy, the television talk show is also a commodity. Talk shows have been traditionally cheap to produce. In 1992 a talk show cost less than $100,000 compared to up to a million dollars or more for a prime time drama. By the early 1990s developments in video technology made talk shows even more economical to produce and touched off a new wave of talk shows on the air. Still, the rule of the marketplace prevailed. A joke on Johnny

Carson's final show that contained 75 words and ran 30 seconds was worth approximately $150,000—the cost to advertisers of a 30-second "spot" on that show. Each word of the joke cost approximately $2000. Though the rates of Carson's last show were particularly high, commercial time on television is always expensive, and an industry of network and station "reps," time buyers and sellers work constantly to negotiate and manage the cost of talk commodities on the television market. If a talk show makes money over time, its contract will be renewed. If it does not, no matter how valuable or critically acclaimed it may be, it will be pulled from the air. A commodity so valuable must be carefully managed and planned. It must fit the commercial imperatives and time limits of for-profit television. Though it can be entertaining, even "outrageous," it must never seriously alienate advertisers or viewers.

As we can see from the examples above, talk shows are shaped by many hands and guided by a clear set of principles. These rules are so well known that hosts, guests, and viewers rarely stop to think about them. What appears to be one of television's most unfettered and spontaneous forms turns out to be, on closer investigation, one of its most complex and artful creations.

—Bernard M. Timberg

MAJOR TALK SHOW HOSTS, 1948-94
Faye Emerson (1949–60), Arthur Godfrey (1948–61), Arlene Francis (1949–75), Dave Garroway (1949–61, 1969), Garry Moore (1951–77), Art Linkletter (1950–70), Steve Allen (1950–84), Ernie Kovacs (1951–61), Mike Wallace (1951–), Merv Griffin (1951–86), Edward R. Murrow (1952–59), Dinah Shore (1951–63, 1970–84, 1987–91), Jack Paar (1952–65, 1975), Mike Douglas (1953–82), Johnny Carson (1951–92), David Susskind (1958–87), Barbara Walters (1961–), David Frost (1964–65, 1969–73, 1977–), William Buckley, (1966–), Dick Cavett (1968–72, 1975, 1977–82, 1985–86, 1992–), Joan Rivers (1969, 1983–), Phil Donahue (1969–96), Bill Moyers (1971–), Tom Snyder (1973–82, 1994–), Geraldo Rivera (1974–), Ted Koppel (1980–), David Letterman (1980–), John Mclaughlin (1982–), Larry King (1985–), Oprah Winfrey (1986–), Sally Jesse Raphael (1986–), Arsenio Hall (1987–), Jane Pauley (1990–91), Jay Leno (1992–), Ricki Lake (1992–).
—Compiled by Robert Erler and Bernard Timberg

FURTHER READING
Carter, Bill. *The Late Shift: Letterman, Leno, and the Network Battle for the Night.* New York: Hyperion, 1994.
Corliss, Richard. "The Talk of our Town." *Film Comment* (New York), January-February 1981.
Donahue, Phil. *Donahue: My own Story.* New York: Simon and Schuster, 1979.
Downs, Hugh. *On Camera: My 10,000 Hours on Television.* New York: Putnam's, 1986.
Heaton, Jeanne Albronda, and Nona Leigh. *Tuning in Trouble: Talk TV's Destructive Impact on Mental Health.* San Francisco: Josey-Bass, 1995.
Himmelstein, Hal. *Television Myth and the American Mind.* New York: Praeger, 1995.
Hirsch, Alan. *Talking Heads: Political Talk Shows and Their Star Pundits.* New York: St. Martin's, 1991.
Latham, Caroline. *The David Letterman Story: An Unauthorized Biography.* New York: Franklin Watts, 1987.
Letterman, David. "Interview." *Playboy* (Chicago), October 1984.
Livingstone, Sonia, and Peter Lunt. *Talk on Television: Audience Participation and Public Debate.* London: Routledge, 1994.
Metz, Robert. *The Today Show.* Chicago: Playboy, 1977.
Munson, Wayne. *All Talk: The Talkshow in Media Culture.* Philadelphia, Pennsylvania: Temple University Press, 1993.
Priest, Patricia Joyner. *Public Intimacies: Talk Show Participants and Tell-All TV.* Creskill, New Jersey: Hampton, 1995.
Timberg, Bernard. "The Unspoken Rules of Television Talk." In, Newcomb, Horace, editor. *Television: The Critical View.* New York: Oxford University Press, 1994.
Tynan, Kenneth. *Show People.* New York: Simon and Schuster, 1979.

See also Allen, Steve; Carson, Johnny; *Dinah Shore Show*; Donahue, Phil; Downs, Hugh; Emerson, Faye; Francis, Arlene; Frost, David; *Garroway at Large*; Godfrey, Arthur; Griffin, Merv; King, Larry; Kovaks, Ernie; *Late Night with David Letterman/The Late Show with David Letterman*; Leno, Jay; Letterman, David; Moyers, Bill; Murrow, Edward R.; Paar, Jack; Pauley, Jane; *Person to Person*; Rivera, Geraldo; Shore, Dinah; *Steve Allen Show*; Susskind, David; *Tonight*; *Tonight Show*; Walters, Barbara; Wallace, Mike; Weaver, Sylvester "Pat"; Winfrey, Oprah

TARSES, JAY

U.S. Writer/Producer

Jay Tarses, a self-proclaimed outsider from the mainstream Hollywood television industry, achieved a reputation in the 1970s and 1980s as a "maverick" writer-producer—a maverick generally is described as brilliant, bold, outspoken, outrageous, and innovative. Tarses has been critically praised for introducing a bold new form of half-hour comedy series, often called character comedy or "dramedy," which achieved a radical stylistic break from the traditional sitcom formula. Tarses has had an ambivalent relationship with the three major networks, who often criticized—and frequently canceled—his shows for being too dark, inaccessible, and not "funny" enough for traditional sitcom audience expectations.

Beginning as a writer and actor with a Pittsburgh theater company, Tarses reportedly worked as a New York City truck driver for the *Candid Camera* series before beginning a career in advertising. In the late 1960s, he teamed with Tom Patchett as a stand-up comedy duo performing dry, semi-satirical material on the coffeehouse circuit. The Patchett-Tarses team turned to television writing, gaining credits on musical variety shows and assorted sitcoms prior to working on the writing staff of *The Carol Burnett Show*, for which they won an Emmy in 1972. The two went on to become collaborative executive producers for MTM Enterprises, where they achieved their first major impact on television history—as writers-producers for the original *Bob Newhart Show* (CBS 1972–78), in which Newhart played an introverted psychologist, surrounded by a circle of interesting and quirkily eccentric characters.

Building upon their success with Newhart, they developed *The Tony Randall Show* (ABC/CBS 1976–78), another MTM series, starring Randall as a widowed Philadelphia judge surrounded by his children, housekeeper, secretary, friends and legal associates. Apparently this sitcom was the site of great tension between the producers and the networks over the nature and style of the type of innovative "character comedy" that Tarses and Patchett were trying to introduce. During this period, they also produced several other short-lived and often-controversial series, including *We've Got Each Other* (CBS 1977–78), a domestic sitcom about the personal and professional lives of a professional couple, their colleagues and neighbors, and *Mary* (CBS 1978), a comedy/variety hour attempting to revive the televisual charisma of Mary Tyler Moore. However, *Mary* was a ratings disaster of such magnitude that it was canceled after three episodes, and its embarrassing failure "drummed us out of the TV business for a while," according to Tarses. During a hiatus from television during this time, the Patchett-Tarses team turned to writing screenplays, including two Muppet movies. The writing/producing team returned to television with the poorly-received *Open All Night* (ABC 1981–82), a sitcom about a convenience store with an ensemble of eccentric customers, and the notable *Buffalo Bill* (NBC 1983–84), about an unlikable, egomaniacal talk-show host, Bill Bittinger (played by Dabney Coleman), and his ensemble of television station coworkers.

During this period, Tarses split from Patchett and developed *The Faculty* (1985, canceled after one episode on ABC), about embattled high-school teachers, characterized by its black humor and mock documentary interviews. The ABC network reportedly asked Tarses to reshoot the pilot because they felt it was too dark and they wanted more emphasis on the students rather than the faculty; when he refused, the series was dropped.

Tarses achieved a critical comeback as producer and occasional writer/director of the controversial "dramedy"

The Days and Nights of Molly Dodd (NBC/Lifetime 1987–91). Originally produced for NBC, this series starred Blair Brown as a divorced woman living alone on New York City's Upper West Side, surrounded by an ensemble of quirky and likable characters representing her family, friends and lovers. After it was canceled by NBC, the series was picked up by the Lifetime cable network, which continued production of the series, reshaped to be aimed strategically at a female audience of a certain age, class and income level. The same year that *Molly Dodd* debuted, Tarses also introduced (on another network) *The "Slap" Maxwell Story* (ABC 1987–88), another critically-acclaimed "dramedy" about the professional and personal tribulations of an arrogant, provocative sports writer, played by Dabney Coleman.

In addition to writing and producing, Tarses has occasionally played cameo roles in his series—as a neighborhood cop in *Open All Night* and a garbage collector in *Molly Dodd*—as well as playing a writer for a cartoon studio in a 1984 MTM sitcom, *The Duck Factory*.

The dramatic/character comedies written and produced by Tarses have operated in what has been considered "uncharted territory" in the television industry. In terms of production style, they have generally not been shot as traditional sitcoms (four cameras, on videotape, in a studio before a live audience, with an added laugh track). Tarses has generally worked independent of the studio system, shooting in a cinematic style in warehouses or on location, and using a single 35mm-film camera. He has characterized his work as low budget, preferring to put his money into writing and actors rather than sets. Tarses' characters are distinguished as not always sympathetic or charismatic (an example is Bill Bittinger on *Buffalo Bill*). His dialogue is markedly low-key and "quirky," with a humor best described as biting and often darkly satirical, sometimes surreal, and written in a subtle comedic rhythm that eschews punch lines. Unlike traditional episodic sitcoms which attempt to solve problems in one episode, the narrative elements of Tarses' dramedies are serial, continuing from episode to episode.

Perhaps Tarses' two greatest contributions to the television industry have been his creativity in constantly pushing the limits of television style—both cinematically and narratively, and his willingness (often eagerness) to go to battle with the networks to champion the broadcasting of innovative and non-formulaic forms of narrative television at the expense of audience ratings. Tarses has increasingly refused to play the Hollywood programming "game", yet has produced what have been some of the freshest and most daring television series of the 1970s and 1980s.

—Pamela Wilson

JAY TARSES. Born in Baltimore, Maryland, U.S.A., 3 July 1939. Educated at Williams College, Williamstown, Massachusetts; Ithaca College, Ithaca, New York, B.F.A. Married Rachel Newdell, 1963; three children. Production assistant in New York for *Candid Camera*, 1963; worked in advertising and promotion, Armstrong Cork Company, Lancaster,

Pennsylvania; joined Tom Patchett in stand-up comedy team, playing the coffeehouse and college circuit, late 1960s; with Patchett, television writer, staff of *The Carol Burnett Show, The Bob Newhart Show,* and others; independent television producer since 1981. Recipient: Emmy Award, 1972; WGA Award, 1987. Address: Endeavor Agency, 350 South Beverly Drive, Suite 300, Beverly Hills, California 90212, U.S.A.

TELEVISION SERIES (with Tom Patchett; selection)

1967–79	*The Carol Burnett Show*
1972–78	*The Bob Newhart Show* (executive producer, writer)
1976–78	*The Tony Randall Show* (creator, executive producer, writer)
1977–78	*We've Got Each Other* (creator, executive producer, writer)
1978	*Mary* (creator, producer, writer)
1981–82	*Open All Night* (creator, producer, writer)
1983–84	*Buffalo Bill* (creator, executive producer, writer)

TELEVISION SERIES (creator, producer, writer, director)

1987–88, 1989–91	*The Days and Nights of Molly Dodd* (also actor)
1987–88	*The "Slap" Maxwell Story*
1992	*Smoldering Lust*

TELEVISION SERIES (actor)

1970–71	*Make Your Own Kind of Music* (also writer)
1981–82	*Open All Night*
1984	*The Duck Factory*

TELEVISION (pilots)

1977	*The Chopped Liver Brothers* (executive producer, actor; with Tom Patchett)
1985	*The Faculty* (executive producer, director, writer)
1990	*Baltimore*
1994	*Harvey Berger, Salesman*
1995	*Jackass Junior High*

FILMS (writer, with Tom Patchett)

Up the Academy, 1977; *The Great Muppet Caper,* 1981; *The Muppets Take Manhattan,* 1984.

FURTHER READING

Christensen, Mark. "Even Career Girls Get the Blues: Is Prime Time Ready for *The Days and Nights of Molly Dodd?*" *Rolling Stone* (New York), 21 May 1987.

———. "Jay Tarses" (interview). *Rolling Stone* (New York), 17 December 1987.

"The Humorous Days and Nights of Jay Tarses." *Broadcasting* (Washington, D.C.), 6 February 1989.

Kaufman, Peter. "The Quixotic Days and Nights of Jay Tarses" (interview). *The New York Times,* 6 June 1993.

Meisler, Andy. "Jay Tarses: 'I Don't Do Sitcoms'." *New York Times,* 2 August 1987.

Ross, Chuck. "What in the World is Molly Dodd?" *Television-Radio Age* (New York), 17 August 1987.

Wilson, Pam. "Upscale Feminine Angst: Molly Dodd, The Lifetime Cable Network and Gender Marketing." *camera obscura* (Berkeley, California), 1995.

See also Dramedy

TARTIKOFF, BRANDON

U.S. Media Executive/Producer

An independent producer and former president of Paramount Pictures, Brandon Tartikoff served from 1980 to 1991 as the youngest and most accomplished president of NBC's Entertainment division. During his tenure at NBC, Tartikoff developed a blockbuster Thursday night lineup which helped the ailing network rank number-one in primetime for the first time in 30 years.

Tartikoff, an admitted "child of television," confesses that he once dreamed of being the next Ed Sullivan, but his television career began at the local level. After undergraduate work in broadcasting at Yale, Tartikoff broke into the business at WTNH in New Haven, Connecticut. Driven to make it to the big leagues, he soon landed a job at the ABC owned-and-operated WLS in Chicago, the third largest market in the country. He worked under the tutelage of Lew Erlicht, his eventual rival.

In the mid-1970s, ABC President Fred Silverman was impressed by Tartikoff's high-camp promo for a series of "monkey-movies" dubbed "Gorilla My Dreams." Silverman recruited Tartikoff for manager of dramatic development at ABC. Three years later, the up-and-coming 30-year-old "boy wonder" of television was snatched by third-place NBC, where Silverman had become president in 1978. Tartikoff was named head of the entertainment division, where he stayed for the next 12 years, the longest any individual has held that position.

NBC's rating's breakthrough came in 1984, when Tartikoff happened to catch Bill Cosby doing a monologue on *The Tonight Show.* Convinced Cosby's family-based banter would make for an excellent sitcom, Tartikoff recruited the comedian and producers Tom Werner and Marcy Carsey. The resulting *Cosby Show* not only helped resurrect the failing sitcom format, but became the building block for a

Thursday night schedule which included *Family Ties*, *Cheers*, and *Night Court*.

Tartikoff was at the helm for the development of MTM Entertainment Inc., series *Hill Street Blues*, which exploded in popularity in its second season after receiving critical acclaim and an armload of Emmy awards in its first. And he shepherded as well *An Early Frost*, the first made-for-televison movie about AIDS. *Miami Vice* was also conceived under Tartikoff; according to executive producer Michael Mann, the head of entertainment presented him with a short memo which read: "MTV. Cops."

By 1991, when Tartikoff left NBC to head Paramount Pictures, the network had been ranked first in the ratings for six consecutive years. Tartikoff was replaced by Warren Littlefield. A series of organizational changes at Paramount and a near-fatal auto accident later led Tartikoff out of the studio arena and into the realm of independent production.
                                                    —Michael B. Kassel

BRANDON TARTIKOFF. Born on Long Island, New York, U.S.A., 13 January 1949. Educated at Yale University, New Haven, Connecticut, B.A. with honors 1970. Married: Lily Samuels, 1982; one daughter. Director of advertising and promotion, WTNH-TV, New Haven, 1971–73; programming executive for dramatic programming, WLS, Chicago, Illinois, 1973–76; manager, dramatic development, ABC, New York City, 1976–77; writer, producer, Graffiti; director of comedy programs, NBC entertainment, Burbank, California, 1977–78, vice president of programs, 1978–80, president, 1980–90; chair, NBC Entertainment Group, until 1991; chair, Paramount Pictures, 1991–92; independent producer, from 1992. Recipient: Tree of Life Award, Jewish National Foundation, 1986; Broadcaster of the Year, Television, Radio and Advertising Club of Philadelphia, 1986.

Brandon Tartikoff
Photo courtesy of New World Entertainment

PUBLICATIONS

"Tartikoff Talks" (interview). *Broadcasting* (Washington, D.C.), 4 June 1990.

The Last Great Ride, with Charles Leehrsen. New York: Turtle Bay Books, 1992.

Mandese, Joe. "Tartikoff Is Still One of TV's Idea Men" (interview). *Advertising Age* (New York), 5 July 1993.

"Brandon Tartikoff" (interview). *Mediaweek* (Brewster, New York), 23 January 1995.

FURTHER READING

Carter, Bill. "The Man who Owns Prime Time." *The New York Times Magazine*, 4 March 1990.

———. "Tartikoff's 11 Years at NBC: One for the Record Books." *The New York Times*, 6 May 1991.

Christensen, Mark, and Cameron Stauth. *The Sweeps: Behind the Scenes in Network TV*. New York: Morrow, 1984.

Coe, Steve. "Tartikoff Urges Networks to Take Risks." *Broadcasting* (Washington, D.C.), 20 May 1991.

Hammer, Joshua. "A TV King's Rough Passage." *Newsweek* (New York), 9 December 1991.

McClellan, Steve. "Grabbing the Grazers in a Crowded Field..." *Broadcasting* (Washington, D.C.), 1 February 1993.

Zoglin, Richard. "Return of the Slugger." *Time* (New York), 24 January 1994.

See also *Cheers*; *Cosby Show*; *Early Frost*; *Family Ties*; *Hill Street Blues*; *Miami Vice*; National Broadcasting Company; Programming; Silverman, Fred

TAXI

U.S. Situation Comedy

Taxi's television history is filled with contradictions. Produced by some of television comedy's most well-regarded talent, the show was canceled by two different net-

works. Despite winning fourteen Emmy Awards in only five seasons, the program's ratings were rock-bottom for its final seasons. Although it thrives in syndication and is still well-

Taxi

loved by many viewers, *Taxi* will be best remembered as the ancestral bridge between two of the most successful sit-coms of all time: *The Mary Tyler Moore Show* and *Cheers*.

In the mid–1970s, MTM Productions had achieved huge success with both popularity and critical appraisal. So

it was an unexpected move when four of the company's finest writers and producers, James L. Brooks, Stan Daniels, David Davis, and Ed Weinberger, jumped off the stable ship of MTM in 1978 to form their own production company, John Charles Walters Company. To launch their new venture, they looked back to an idea that Brooks and Davis had previously considered with MTM: the daily life of a New York City taxi company. From MTM head Grant Tinker they purchased the rights to the newspaper article that had initiated the concept and began producing this new show at Paramount for ABC. They brought a few other MTM veterans along for the ride, including director James Burrows and writer/producers Glen and Les Charles.

Although *Taxi* certainly bore many of the trademark signs of "quality television" as exemplified by MTM, other changes in style and focus distinguished this from an MTM product. After working on the middle-class female-centered worlds of *The Mary Tyler Moore Show*, *Rhoda*, and *Phyllis* for years, the group at John Charles Walters wanted to create a program focusing on blue-collar male experience. MTM programs all had clearly defined settings, but *Taxi'*s creators wanted a show that was firmly rooted in a city's identity—*Taxi'*s situations and mood were distinctly New York. Despite MTM Productions innovations in creating ensemble character comedy, there was always one central star around which the ensemble revolved. In *Taxi* Judd Hirsch's Alex Rieger was a main character, but his importance seemed secondary to the centrality of the ensemble and the Sunshine Cab Company itself. While *The Mary Tyler Moore Show* proudly proclaimed that "you're going to make it on your own," the destitute drivers of *Taxi* were doomed to perpetual failure; the closest any of them came to happiness was Rieger's content acceptance of his lot in life—to be a cabby.

Taxi debuted on 12 September 1978, amidst a strong ABC Tuesday night line-up. It followed *Three's Company*, a wildly-successful example of the type of show MTM "quality" sit-coms reacted against. *Taxi* used this strong position to end the season ninth in the ratings and garner its first of three straight Emmys for Outstanding Comedy Series. The show's success was due to its excellent writing, Burrows's award-winning directing using his innovative four-camera technique, and its largely unknown but talented cast. Danny DeVito's Louie DePalma soon became one of the most despised men on television—possibly the most unredeemable and worthless louse of a character ever to reside on the small screen. Andy Kaufman's foreign mechanic Latka Gravas provided over-the-top comedy within an ensemble emphasizing subtle character humor. But Kaufman sometimes also brought a demonic edge to the character, an echo of his infamous appearances on *Saturday Night Live* as a macho wrestler of women and Mighty Mouse lip-syncher. In the second season Christopher Lloyd's Reverend Jim Ignatowski was added to the group as television's first drugged-out '60s burn-out character. But Lloyd's Emmy-winning performance created in Jim more than just a storehouse of fried brain cells; he established a deep, complex humanity that moved far beyond mere caricature. The program launched successful movie careers for DeVito and Lloyd, as well as the fairly-notable television careers of Tony Danza and Marilu Hen-

ner; Kaufman's controversial career would certainly have continued had he not died of cancer in 1984.

In its third season ABC moved *Taxi* from beneath *Three's Company*'s protective wing to a more competitive Wednesday night slot; the ratings plummeted and *Taxi* finished the next two years in 53rd place. ABC canceled the show in early 1982 as part of a larger network push away from "quality" and toward the Aaron Spelling-produced popular fare of *Dynasty* and *The Love Boat*. HBO bid for the show, looking for it to become the first ongoing sitcom for the pay channel, but lost out to NBC, which scheduled the series for the 1982–83 season. Ironically, this reunited the show's executive producers with their former boss Tinker, who had taken over NBC. Tinker's reign at NBC was focused, not surprisingly, on "quality" programming which he hoped would attract viewers to the perennially last-place network. *Taxi* was partnered with a very compatible show on Thursday night—*Cheers*, created by *Taxi* veterans Charles, Burrows, and Charles. Although this line-up featured some of the great programs in television history—the comedies were sandwiched by dramas *Fame* and *Hill Street Blues*—the ratings were dreadful and *Taxi* finished the season in 73rd place. NBC was willing to stick by *Cheers* for another chance, but felt *Taxi* had run its course and canceled it at the end of the season. Had *Taxi* been given another year or two, it would have been part of one of the most successful nights on television, featuring *The Cosby Show* (co-created by *Taxi* creator Weinberger), *Family Ties*, *Hill Street Blues*, *L.A. Law*, and eventual powerhouse *Cheers*.

Taxi lives on in syndication, but its most significant place in television history is as the middle generation between *The Mary Tyler Moore Show* and *Cheers*. It served as a transition between the star-driven middle-class character comedy of MTM programs and the location-centered ensemble comedy inhabited by the losers of *Cheers* and *Taxi*. Considered one of the great sit-coms of its era, *Taxi* stands as a prime example of the constant tension in television programming between standards of "quality" and reliance on high ratings to determine success.

—Jason Mittell

CAST

Alex Rieger	Judd Hirsch
Bobby Wheeler (1978–81)	Jeff Conaway
Louie DePalma	Danny DeVito
Elaine Nardo	Marilu Henner
Tony Banta	Tony Danza
John Burns (1978–79)	Randall Carver
Latka Gravas	Andy Kaufman
"Reverend Jim" Ignatowski (1979–83)	Christopher Lloyd
Simka Gravas (1981–83)	Carol Kane

PRODUCERS   James L. Brooks, Stan Daniels, Ed Weinberger, David Davis, Glen Charles, Les Charles, Ian Praiser, Richard Sakai, Howard Gewirtz

PROGRAMMING HISTORY   111 Episodes

• ABC

September 1978–October 1980	Tuesday 9:30-10:00
November 1980–January 1981	Wednesday 9:00-9:30
February 1981–June 1982	Thursday 9:30-10:00

• NBC

September 1982–December 1982	Thursday 9:30-10:00
January 1983–February 1983	Saturday 9:30-10:00
March 1983–May 1983	Wednesday 9:30-10:00
June 1983–July 1983	Wednesday 10:30-11:00

FURTHER READING:

Feuer, Jane, Paul Kerr, and Tise Vahimagi, editors. *MTM-'Quality Television.'* London: British Film Institute, 1984.

Sorensen, Jeff. *The Taxi Book.* New York: St. Martin's, 1987.

Waldron, Vince. *Classic Sitcoms: A Celebration of the Best of Prime-Time Comedy.* New York: MacMillan, 1987.

See also Brooks, James L.; Burrows, James; Charles, Glen and Les; *Cheers*; Comedy, Workplace Settings; *Mary Tyler Moore Show*; Weinberger, Ed

TEASER

A teaser is a television strategy for attracting the audience's attention and holding it over a span of time. Typically a teaser consists of auditory or visual information, or both, providing the viewer a glimpse of what he or she can expect as programing continues. Teasers are used in several types of programming.

In news broadcasts, for example, a newscaster may address viewers in a fashion such as, "The state legislature gets ready for a showdown on taxes. Details when we return." The audience is being teased with information, and the purpose is to keep a viewer tuned to the station during a commercial. Similarly, teasers can also be used to keep a viewer tuned to a newscast. An anchor may begin a newscast with a tease for an upcoming story, like the state legislature story above, then shift the focus: "But first, we bring you our top story...."

According to Cohler, there are two types of news teasers. The first is best described as a headline, which contains the essential information about a story. In sports the headline may be, "Angels shut-out Pirates. Highlights when we return." The second type of teaser is more vague and leaves the reader wondering what exactly the news is about to report, as in the "showdown on taxes" example mentioned above.

For Yoakam and Cremer, there is little difference between "teasers" and "bumpers" since both are designed to promote upcoming stories. Thus a simple "We'll return in a moment" would qualify as a teaser as well as a bumper. So would a short video clip of a dramatic moment or a humorous exchange of words taken from the segment coming up after some commercials. Thus anything designed to get the attention of viewers and hold their attention through some span of time may be referred to as a teaser.

This is clearly the case in other types of programming. Daytime talk shows, for example, often open with provoca-tive summaries of their content, then cut to commercials. The teaser is designed to titillate the audience and entice it into returning.

Teasers for dramatic programming are similar. Short clips from the upcoming program can be used to highlight the most powerful or humorous moments. Bits of tense dialogue, jokes, tender moments can all be exerpted for use as an immediate promotion of the program at hand.

A related programming strategy uses the pre-commercial sequence to remind the audience of past events at the same time it pulls them into the current program. These summaries are often introduced with a voice-over announcement: e.g. "Previously on *Hill Street Blues*." In many cases (*Dallas* is a good example) the summary-teaser also serves as a prologue, indicating which stories, from the ever-growing collection of interrelated narratives, will be explored in the upcoming episode.

In the age of the remote control device a number of programs have abandoned teasers, plunging directly into the dramatic action of the narrative, sometimes without even an intervening commercial. Still, however, in some cases it is a prologue or a teaser, selected from the most powerful moments of previous and new material that is presented to the fickle audience. This strategy, it is hoped, prevents viewers from instantly changing the channel to "surf" between programs.

—Raul D. Tovares

FURTHER READING

Cohler, David Keith. *Broadcast Journalism: A Guide for the Presentation of Radio and Television News.* Englewood Cliffs, New Jersey: Prentice-Hall, 1985.

Yoakam, Richard D., and Charles F. Cremer. *ENG: Television News and the New Technology.* Carbondale, Illinois: Southern Illinois University Press, 1985.

TELCOS

Telephone companies (telcos) have always figured in the history of U.S. television. By the end of the 20th century they may attain the broadcast role they hoped for as long ago as the 1920s.

The earliest involvement of telephone companies in broadcasting dates to AT and T's interest in radio. Before World War I, AT and T was one among several companies actively experimenting with the hertzian waves with a view to controlling what seemed to be an imminent wireless communication era. AT and T's stake in the government-formed Radio Corporation of America (RCA) in the early 1920s seemed to guarantee the phone company a role in radio broadcasting, specifically with respect to developing the international market, selling transmitters, and providing anything that seemed to be telephony. Yet AT and T's definition of telephony broadened in that era: in 1922 it offered a special toll broadcasting service allowing people to use its "radio telephony" channels to send out their own programs—for a fee. At that time AT and T eschewed any interest in controlling content, although it did use its long distance lines to broadcast sports events, music, and certain other entertainment, avowing it desired only its rightful opportunity to transmit. Nevertheless, by 1924 the phone company had a regular radio programming schedule.

Its early control over broadcasting was broken up, however, by federal government (Federal Trade Commission) objections to the apparent growing monopoly power in radio. In 1926 a new structure was created to answer the monopoly charges, relegating the phone company to a role in transmission only while other companies involved in radio (General Electric, Westinghouse and RCA) would form the National Broadcasting Company and develop programming and an audience-oriented service.

AT and T, America's regulated, dominant national telephone carrier, operated in that capacity for several decades, conveying first radio and later television signals across the country, enabling the formation of national networks through its long distance links. The carriage fees it accumulated were enormous, and as the sanctioned, monopoly inter-state common carrier, AT and T had the business to itself, a monopoly role that was at times contested. In 1948, for example, the FCC debated procedures concerning inter-city video carriage. At that time the Commission espoused a rule reserving permanent microwave frequencies to common—not private—carriers. This rule thus sanctioned a *de facto* continued monopoly transmission role in television for AT and T. The company's first serious setbacks, however, did not occur until the mid-1970s.

In the 1970s regulatory liberalizations in two realms undermined AT and T's control of transmission services essential to television. First, communication satellites, an outgrowth of the U.S. space program, provided efficient and economical ways to transmit messages or signals over long distances. Although AT and T retained a major role for itself

in international satellite communication through provisions in the 1962 Communication Satellite Act, the stage was set for other companies to enter into satellite services. Ultimately, this development would provide crucial alternatives to television's (and cable television's) continued reliance on AT and T for transmission. In particular, telephone companies were unable to control domestic satellite services, which became the preferred and cost-effective method for broadcast and cable television networks to deliver their signals, thus ending their dependence on AT and T for interconnection. The successful launch of HBO nationwide on RCA's Satcom satellite in 1975 bypassed AT and T and illustrated a future independent of the telcos. The Public Broadcasting Service moved to satellite distribution of its signal in 1978, followed by the major television networks' migration from AT and T to satellites in the mid-1980s.

Second, skirmishes between telcos and the young cable television industry prompted the Federal Communcations Commision (FCC) and Congress to limit telcos' ability to own and operate cable television systems. The FCC ruled in 1970 that telcos could operate systems only in small, rural populations. In 1978, affirming that AT and T had abused its power in overcharging companies that wished to use its poles to establish cable television service, Congress enacted the Pole Attachment Act authorizing the FCC to "regulate the rates and conditions for pole attachments," and effectively removing the telcos' control over a key access and right-of-way issue and allowing cable television to expand under more favorable terms. Telephone company ability to enter into or otherwise control this new television medium clearly would be restrained. The cable television industry's insistence on this is in part reflected in a section of the 1984 Cable Communications Act that reiterated the 1970 telco-cable cross-ownership ban and explicitly forbade telephone companies from offering cable television services.

However, telephone companies' interest in video services never died. If the aforementioned two new communication technologies ultimately underscored telcos' limited hold on an expanding set of services, they can also be counted among the causes of a massive restructuring of the U.S. telephone system under the 1982 Modification of Final Judgment (MFJ), a federal court ruling that broke up AT and T's monopoly telephone service in the United States. The result of a long-standing inquiry into AT and T's vertical integration and possible abuse of power under antitrust laws, the MFJ separated competitive long-distance (interexchange) service from monopoly-provided local service. AT and T restructured, spinning off the "Baby Bells," regional companies restricted to the provision of local telephone service (local exchange companies). Both sets of companies, AT and T and other long-distance service providers (interexchange carriers), as well as the local service providers, again eyed the provision of video services as one among other competitive possibilities.

The MFJ put several restrictions on AT and T. The most notable was a seven-year restriction from 1984 (effective date of the MFJ), on entering into "electronic publishing." But in the late 1980s and 1990s AT and T, as well as several other telcos, quickly constructed a number of strategic liaisons with cable television, computer, software and even movie companies in order to position themselves for new video and multimedia services. Such liaisons built on the telephone companies' long standing interest in new media and their abortive history of attempting to provide teletext or videotext services in conjunction with publishers.

In 1988, amid the deregulatory fever of the 1980s, the FCC recommended lifting the cable-telco cross-ownership ban, but the requisite Congressional action was not forthcoming. Nevertheless, continued restructuring proceeded, allowing the convergence of what had been conceived as quite separate video, voice and data industries. In 1992 the FCC issued its "Video Dialtone" order allowing telcos (such as the "Baby Bells" or other local exchange companies) to provide the technological platforms for video service to subscribers. Essentially this also allowed them to enter the video services business, albeit without permitting them to directly own programming. One year later, in response to separate suits brought by telcos, several district courts began lifting the cable-telco cross-ownership ban. The first such suit was brought in 1993 (*Chesapeake and Potomac Telephone Co. of Virginia v. U.S., 830 F. Supp. 909*) by Bell Atlantic, a telco which also announced that same year a proposed merger with the largest cable company in the United States, TCI, a deal which later collapsed. Additionally, in the mid-1990s several telcos announced plans to provide video services as cable companies (which would allow them to own programming) rather than as telephone companies operating a video dialtone platform.

With new emphasis on creating a national information infrastructure, the role of telephone companies in providing an array of new services, including television, seems certain. Deregulating telcos and allowing them to offer video services, alone or in conjunction with already-established providers, has set the stage for a new television service and an entirely new set of corporate powers.

—Sharon Strover

FURTHER READING

Barnouw, E. *Tube of Plenty: The Evolution of American Television.* London: Oxford University Press, 1975.

Berniker, Mark. "Telcos Going their own Way into Video." *Broadcasting and Cable* (Washington, D.C.), 2 May 1994.

Horwitz, R. *The Irony of Regulatory Reform: The Deregulation of American Telecommunications.* London: Oxford University Press, 1989.

McAvoy, Kim. "Telco's Army Poised for Assault on TV Entry." *Broadcasting* (Washington, D.C.), 3 October 1988.

Oxley, Michael G. "The Cable-Telco Cross-Ownership Prohibition: First Amendment Infringement Through Obsolescence." *Federal Communications Law Journal* (Los Angeles, California), December 1993.

See also Cable Networks; Home Box Office; Radio Corporation of America; Satellite; Untied States: Cable

TELEFILM CANADA

Canadian Television and Film Development Corporation

Telefilm Canada is a Crown Corporation of the federal government. Its mandate is to support the development and promotion of television programs and feature films by the Canadian private sector. Telefilm is neither a producer nor a distributor and it is not equipped with a production studio; instead, it acts primarily as a banker and deals principally with independent Canadian producers. To this end, Telefilm invests over $100 million annually through a variety of funds and programs that encompass production, distribution and marketing, scriptwriting, dubbing and subtitling, festivals and professional development. Telefilm Canada also administers the official co-production treaties that exist with more than twenty countries, including France, Great Britain, Germany, Australia and New Zealand.

Until 1984, Telefilm Canada was known as the Canadian Film Development Corporation (CFDC). The CFDC began operations in 1968 with a budget of $10 million and a mandate to foster and promote the development of a feature film industry in Canada through the provision of loans, grants and awards to Canadian producers and filmmakers. Unlike the National Film Board of Canada (NFB), or the Canadian Broadcasting Corporation, the CFDC was expected to become a self-financing agency, interested as much (if not more) in the profitability of the films it supported as in their contribution to Canada's cultural life.

By 1971, the CFDC had exhausted its original budget and recouped barely $600,000, or roughly 9%, of its investments in 64 projects. In keeping with its commercial orientation, the CFDC contributed to a number of films that came to be referred to as "maple-syrup porn", movies like *Love Is a Four Letter World*. At the same time, the CFDC invested in a number of films that have come to be regarded as early Canadian classics, films such as *Goin' Down the Road*.

The federal government approved a second allotment of $10 million in 1971 and for the next six years the CFDC and industry representatives struggled to establish a clear set of corporate objectives. One option, which would have transformed the CFDC into something of an arts council for feature films, and brought it closer in line with the mandate of the NFB, was to rechannel its money into a system of grants that would provide for the production of a small number of Canadian films a year. The other option was to rechannel the CFDC's priorities toward the production of feature films with strong box-office potential, in particular films that would be attractive to the Hollywood majors.

This second option became viable after changes in tax regulations were accompanied by a change in the CFDC's financial practices. In 1974 the capital cost allowance for Canadian feature films was extended from 30% to 100%. In 1978, the CFDC shifted its focus from the provision of equity financing for low and medium-budgeted Canadian films, to the provision of bridge financing for projects that were designed to take advantage of the tax shelter. Both the number of productions and average budgets soared. Measured in terms of employment and total dollars spent, the tax-shelter boom was a success. But many of the films produced during this period were never distributed; many of the ones that did receive distribution were second-rate attempts at films that mimicked Hollywood's standard fare (notable examples include *Meatballs* and *Running*). By 1980, there was growing criticism of the direction taken by the CFDC, particularly from French-Canadian producers and filmmakers who benefited far less than their English-Canadian counterparts from the CFDC's shift in investment priorities. The tax-shelter boom came to a crashing halt in 1980.

The establishment of the Canadian Broadcast Program Development Fund in July 1983 dramatically shifted the CFDC's priorities from feature films to television programming. To reflect this shift in investment priorities the CFDC was renamed Telefilm Canada in February 1984. The Broadcast Fund has four overall objectives: a) to stimulate production of high quality, culturally relevant Canadian television programs in targeted categories, i.e. drama, children's, documentary and variety programming; b) to reach the broadest possible audience with those programmes through scheduling during prime time viewing hours; c) to stimulate the development of the independent production industry; d) to maintain an appropriate regional, linguistic and private/public broadcaster balance in the distribution of public funds. The fund had an initial budget of $254 million spread over five years. Since 1988, Telefilm has invested more than $60 million annually in television programming. On average its participation represents 33% of the total production budget.

The Broadcast Fund has been enormously successful in achieving its original objectives. Between 1986 and 1990, for example, the Fund helped finance close to $800 million in total production volume in 2,275 hours of original television programming, of which more than 1,000 hours consisted of dramatic programming exhibited during peak viewing hours. Some of these were *Anne of Green Gables*, the *Degrassi* series, *E.N.G.*, *Danger Bay*, *Love and War*, *Due South* and *The Boys of St. Vincent*. In terms of audience reach, viewing of Canadian programs in peak time has increased substantially. The Broadcast Fund has also played a crucial role in providing independent Canadian producers with the leverage to expand into export markets.

As a banker, Telefilm Canada is still a failure. It recoups only a small percentage of its annual investments. As a cultural agency and a support structure to Canada's independent producers, Telefilm has been remarkably successful, especially in terms of television programming. It is still the case that Canadians view far more foreign than domestic programming, but without Telefilm's presence there would be virtually no production of Canadian dramatic programming. In many respects, Canadian television is a function of Telefilm Canada.

—Ted Magder

FURTHER READING

Ayscough, Suzan. "Canadian Film Funder Tightens Purse Strings." *Variety* (Los Angeles), 30 March 1992.

———. "The Experiment That Spawned an Industry." *Variety* (Los Angeles), 16 November 1992.

———. "Factions Fracture Pic Funds." *Variety* (Los Angeles), 16 November 1992.

Kelly, Brendan. "Canada Funder Feels Sting of Budget Cuts." *Variety* (Los Angeles), 14 December 1992.

———. "Canadian Film Plan Rekindles Old Uproar." *Variety* (Los Angeles), 19 April 1993.

———. "More Homegrown Up There." *Variety* (Los Angeles), 24 April 1995.

Magder, Ted. *Canada's Hollywood: The Canadian State and Feature Films*. Toronto: University of Toronto Press, 1993.

Pendakur, Manjunath. *Canadian Dreams and American Control: The Political Economy of the Canadian Film Industry*. Detroit, Michigan: Wayne State University Press, 1990.

Posner, Michael. *Canadian Dreams: The Making and Marketing of Independent Films*. Vancouver, Canada: Donglas and McIntyre, 1993.

Wallace, Bruce, Joseph Treen, and Robert Enright. "A Campaign in Support of Entertainment." *Maclean's* (Toronto), 17 March 1986.

Winikoff, Kenneth. "They Always Get Their Film: The Canadian Government Has Sired a National Cinema, But Can a Film Industry Thrive When Every Taxpayer is a Producer?" *American Film* (Washington, D.C.), July 1990.

See also Canadian Programming in English

TELEMUNDO

U.S. Spanish-Language Network

Telemundo Group Inc. is the second largest Spanish-language television network in the United States. It reaches 86% of Hispanic households in 53 U.S. markets as well as over 19 countries in Latin America through its owned-and-operated stations, affiliates, and syndication.

Telemundo Group Inc. was formed in December 1986 by Saul Steinberg and Henry Silverman of Reliance Capital Group L. P., who were interested in moving into the Spanish-language market. They began by purchasing stations in Los Angeles, California, Miami, Florida, New York City, and Puerto Rico. In 1987, Telemundo began network broadcasting with *Noticiero Telemundo*, a world and national news program produced by Hispanic American Broadcasting Company in Miami. Later that year, *Deportes Telemundo*, a weekly two-hour round-up of sports highlights from around the world, premiered. Between 1988 and 1991, the network expanded both its station holdings, affiliates, and programming. Stations and affiliates in Houston, Dallas/Ft.Worth, McAllen/Brownsville, El Paso, Lubbock, and San Antonio, Texas; Albuquerque, New Mexico; Tucson and Phoenix, Arizona; and Yakima, Washington increased Telemundo's coverage to 78% of the Hispanic households in the United States. Meanwhile, U.S.-produced programming also expanded, including *Noticiero Telemundo/CNN*, a joint venture with CNN to produce a nightly national news program; *Cocina Crisco*, the first Spanish-language cooking show produced in the United States; *Angelica Mi Vida*, the first Spanish-language *telenovela* (soap opera) produced in this country, based on the lives of Hispanic Americans; *Cara a Cara*, a talk show starring Maria Laria; and *Ocurrio Asi*, a tabloid news program with sensational stories from the United States and Latin America. By 1991, Telemundo was producing 54% of its programming in the United States.

In 1992, Joaquin F. Blaya, a 22-year veteran of Spanish-language media and formerly president and chief executive officer of Univision Holdings, Inc., joined Telemundo as president and chief executive officer. Under his leadership, Telemundo continued its expansion. New programming targeting younger audiences and second-generation Hispanics, such as *Ritmo Internacional* and *Padrisimo*, was developed. In a joint venture among Telemundo, Univision, and Nielsen Media Research, Blaya also created the first nation-wide rating service focused on the Hispanic community's viewing habits. In another joint venture with Reuters and British Broadcasting Corporation World Television, in December 1994 he launched a 24-hour Spanish-language television news service called Telenoticias. By mid-1993, Telemundo filed for bankruptcy under Chapter 11 of the Bankruptcy Code. Through a subsequent financial restructuring, Apollo Advisors L.P. became the major shareholder in late 1994. In March 1995, corporate restructuring began with the naming of Roland A. Hernandez as president and chief executive officer. A native of Los Angeles, Hernandez is the founder and owner of Interspan Communications, which established KFWD-TV, the Telemundo affiliate in Dallas/Ft. Worth, Texas, and a member

Courtesy of Telemundo

of the board of directors of Telemundo Group Inc. One of his original moves was opening Telemundo's first West Coast production facility in Hollywood's famed Raliegh Studios in order to attract Spanish-speaking talent on both coasts so that Telemundo's programming would reflect the full cultural spectrum that is Hispanic America.

As of 1995, Telemundo Group Inc. consisted of six full-power owned-and-operated stations in Los Angeles, New York, Miami, San Francisco/San Jose, San Antonio, and Houston/Galveston. The company also operated low-power stations in five other markets, and one station in Puerto Rico. In addition, 18 full-power and 32 low-power stations were affiliated with Telemundo, and 614 cable systems carried Telemundo's signal. This represented a coverage of 53 markets and 86% of Hispanic households in the United States.

As of 1995, Telemundo produced 50% of its programming in the United States, at the network's production facilities in Hialeah, Florida, Los Angeles, California, and Puerto Rico. Programming consisted of *telenovelas*, movies, game shows, variety shows, sports programs, talk shows, and news programs. Movies and *telenovelas* represented the bulk of the imported programming. In April 1995, Telemundo added *Dando y Dando* to its game show line-up, along with *El Gran Juego de la Oca*. Also in 1995, *La Hora Lunatica* variety show was added to the noon hour. Telemundo capitalized on interest in sports with the weekend programming of *Boxeo*, *Futbol Telemundo*, and *Marcador Final*. Talk shows, *Lo Mejor de Sevcec* and *El y Ella*, are popular daytime programming. The world and national news program is *Telenoticias con Raul Peimbert*, which is seen not only in the United States but also in Latin America.

The future of Telemundo and Spanish-language television is unclear. Having emerged from bankruptcy in late 1994, Telemundo is still concerned with financial problems, as well as its competition with Univision. It does seem, however, that the Hispanic market in the United States and Latin America is growing, and is a prime target for advertisers and media. Given its fundamental objective (i.e., to make its programming relevant to the broadest base of Hispanic viewers in this country), it is to be expected that Telemundo will continue to move forward in meeting its many challenges.

—Patricia Constantakis-Valdés

FURTHER READING

Constantakis-Valdés, P. *Spanish-Language Television and the 1988 Elections: A Case Study of the 'Dual Identity' of Ethnic Media* (Ph.D. dissertation, University of Texas at Austin, 1993).

Subervi-Vélez, F.A., with C. R. Berg, P. Constantakis-Valdés, C. Noriega, D. I. Rios, and K. Wilkinson. "Mass Communication and Hispanics." In, Padilla, Felix, editor. *The Hispanic Almanac*. Houston: Arte Público Press, 1994.

Veciana-Suarez, A. *Hispanic Media, U.S.A.* Washington, D.C.: Media Institute, 1987.

———. *Hispanic Media: Impact and Influence.* Washington, D.C.: Media Institute, 1987.

TELENOVELA

The *telenovela* is a form of melodramatic serialized fiction produced and aired in most Latin American countries. These programs have traditionally been compared to English-language soap operas and even though the two genres share some characteristics and similar roots, the *telenovela* in the last three decades has evolved into a genre with its own unique characteristics. For example, *telenovelas* in most Latin American countries are aired in prime-time six days a week, attract a broad audience across age and gender lines, and command the highest advertising rates. They last about six months and come to a climactic close.

Telenovelas generally vary from 180 to 200 episodes, but sometimes specific *telenovelas* might be extended for a longer period due to successful ratings. The first *telenovelas* produced in Latin America in the 1950s were shorter, lasting between 15 and 20 episodes and were shown a few times a week. As they became more popular and more technologically sophisticated, they were expanded, becoming the leading genre in the daily prime-time schedule.

Unlike U.S. soap operas that tend to rely on the family as a central unit of the narrative, Latin American *telenovelas* focus on the relation between a romantic couple as the main motivator for plot development. During the early phases of their evolution in Latin America, until the mid-1960s, most *telenovelas* relied on conventional melodramatic narratives in which the romantic couple confronted opposition to their staying together. As the genre progressed in different nations at different rhythms it became more attuned to local culture. The Peruvian *telenovela Simplemente Maria,* for example, a version of the Cinderella story, dealt with the problems of urban migration. The Brazilian *telenovela, Beto Rockfeller,* presented the story of an anti-hero who worked as a shoe shop employee and pretended to be a millionaire getting simultaneously involved with two women, one rich and one poor. This *telenovela* appears to have led to the most dramatic changes in that nation's genre. It became an immediate hit in 1968. It introduced the use of colloquial dialogue. It presented social satire. And it offered new stylistic elements, such as the use of actual events in the plot, more natural acting, and improvisation.

The Globo network, Brazil's largest, which was only beginning to produce *telenovelas* in the late 1960s, soon took

El Vuelo del Aguila
Photos courtesy of Televisa

El Vuelo del Aguila
Photos courtesy of Televisa

the lead and imposed these new trends upon the *telenovela* market. Indeed, Globo owes its international recognition and economic powerhouse status to the *telenovela*. In the 1970s, Globo invested heavily in the quality of its *telenovelas*, using external locations traditionally avoided because of production costs. And Globo's export success forced other producers in the region to implement changes in production values and modernize their narratives to remain competitive. Mexico, for example, after dominating the international market for several years, had to adapt its *telenovelas* according to the influences of the main competitors, especially Argentina, Brazil and Venezuela.

There are important national distinctions within the genre in the areas of topic selection, structure and production values and there are also clear distinctions between the *telenovelas* produced in the 1960s and the 1990s, in terms of content as well as in production values. As Patricia Aufderheide has pointed out, recent *telenovelas* in Brazil "dealt with bureaucratic corruption, single motherhood and the environment; class differences are foregrounded in Mexican novelas and Cuba's novelas are bitingly topical as well as ideologically correct." In Colombia, recent *telenovelas* have dealt with the social violence of viewers' daily lives, but melodramatic plots that avoid topical issues are becoming more popular. In Brazil the treatment of racism is surfacing in *telenovelas* after being considered a taboo subject for several years.

The roots of the Latin American *telenovelas* go back to the radio soap operas produced in the United States, but they were also influenced by the serialized novels published in the local press. The origins of the melodramatic serialized romance date back to the sentimental novel in 18th-

El Vuelo del Aguila
Photos courtesy of Televisa

El Vuelo del Aguila
Photos courtesy of Televisa

century England, as well as 19th-century French serialized novels, the "feuilletons." In late 19th and early 20th centuries, several Latin American countries also published local writers' novels in a serialized form. However the proliferation of radionovelas that would later provide personnel as well as expertise to *telenovela* producers started in Cuba in the late 1930s. According to Katz and Wedell, Colgate and Sydney Ross Company were responsible for the proliferation of radionovelas in pre-Castro Cuba. In the beginning stages of *telenovelas* in Latin America, in the 1950s, Cuba was an important exporter of the genre to the region, providing actors, producers and also screenplays. U.S. multinational corporations and advertising agencies were also instrumental in disseminating the new genre in the region. Groups such as U.S. Unilever were interested in expanding their market to housewives by promoting *telenovelas* which contained their own product tie-ins. Direct influence of the United States on the growth and development of *telenovela* in the region subsides after the mid-1960s, and the genre slowly evolved in different directions in different countries. In the 1950s and early 1960s, *telenovelas* were primarily adaptations of novels and other literary forms, and only a few Latin American scriptwriters constructed original narratives. By the late 1960s local markets started producing their own stories, bringing in local influences, and shaping the narratives to particular audiences.

Presently the leading *telenovela* producers in the region are Televisa, Venevision, and Globo, the leading networks in Mexico, Venezuela and Brazil respectively. These networks not only produce *telenovelas* for the local market but also export to other Latin American nations and to the rest of the world. Televisa, for instance, is the leading supplier of *telenovelas* to the Spanish-speaking market in the United States. By 1988, Brazil had exported *telenovelas* to more than 128 countries. The more recent trend among *telenovela* producers in the region is to engage in co-productions with other nations, to guarantee better access to the international market.

—Antonio C. Lapastina

FURTHER READING

Allen, R., editor. *To Be Continued...: Soap Operas and Global Media Culture.* New York: Routledge, 1995.

Aufderheide, P. "Latin American Grassroots Video: Beyond Television." *Public Culture* (Chicago), 1993.

Katz, E., and G. Wedell. *Broadcasting in the Third World: Promise and Performance.* Cambridge, Massachusetts: Harvard University Press, 1977.

Lopez, Ana. "The Melodrama in Latin America." In, Landy, M. editor. *Imitations of Life: A Reader on Film and Television Melodrama.* Detroit, Michigan: Wayne State University Press, 1991.

Melo, Jose Marques de. *The Presence of the Brazilian Telenovelas in the International Market: Case Study of Globo Network.* Sao Paulo: University of Sao Paulo, 1991.

Rogers, E., and L. Antola. "Telenovelas: A Latin American Success Story." *Journal of Communication* (New York), 1985.

Singhal, Arvind. "Harnessing the Potential of Entertainment-education Telenovelas." *Gazette,* January 1993.

Straubhaar, Joseph. "The Development of the Telenovela as the Pre-eminent Form of Brazilian Popular Culture." *Studies in Latin American Popular Culture* (Las Cruces, New Mexico), 1982.

Vink, Nico. *The Telenovela and Emancipation: A Study of Television and Social Change in Brazil.* Amsterdam, Netherlands: Royal Tropical Institute, 1988.

See also Brazil; Mexico; Soap Opera; Teleroman

TELEROMAN

As a television genre, the weekly, prime-time *teleroman* can be defined as "A television program, fictitious in character with a realistic descriptive style which is comprised of a series of continuous episodes, diffused with fixed periodicity and characterized by a sequentiality which is either episodal, overlapping, or both" (author's translation).

The genre is generally recognized, both at home and abroad, as being specific to the French language television industry in Canada, located in the province of Quebec and intimately associated with Quebec society and its dominant francophone culture (82% of nearly 7 million inhabitants).

The term literally means "tele-novel," which strongly suggests its direct lineage with the modern, especially the nineteenth century, popular novel. The serial character of the *teleroman* makes it a descendant of Charles Dickens, Alexandre Dumas and Eugene Sue, whose works were published as series, one chapter or episode at a time, in the popular daily pennypress of their time. The upshot was of course to build customer loyalty for the supporting print media, a function not unlike that of the *teleroman* for the visual medium of television.

Next came the serial novel (the French feuilleton), a work of fiction written for the popular press. In this case authors, such as Honoré de Balzac, would write individual chapters which were then massively distributed and read at regular intervals; in other words, the "novel" was only produced in book form when each individual chapter had already been published. This new literature testifies to the technologies of modern mass communications in a liberal, urban, industrial, capitalist society. Because of its proximity to the United States, Quebec has benefitted and profited from these new technologies and even produced a cottage

The Plouffe Family
Photo courtesy of the National Archives of Canada

industry of popular serial novels, both within the pages of the popular press and between the covers of chapbooks.

With the advent of radio, both public and private, the serial novel became a permanent fixture of programming with such favorite *radioromans* (radio drama or radio-novel) as *La Pension Velder, Jeunesse dorée, La famille Plouffe* and the grandaddy of them all, *Un homme et son péché*. These of course developed under the far reaching shadow of the U.S. radio soap opera. While importing many of its basic characteristics, the Quebec radioroman showed the imprint of local cultural moorings, particularly in its reference to the history of this French speaking population on the North American continent dating back to the early seventeenth century (1604), its nationalistic fervor, its agrarian heritage and its forced adaptation to accelerated industrialization, urbanization and modernization.

There were no in-house writers for these radio plays; one could not earn a decent living writing radioromans or, for that matter, any type of novel. Still, many of the first telenovelists were radionovelists who were also established literary novelists. A literary profession of successful, independent novelists and telenovelists only emerged some ten years ago.

With the advent of television, classical and modern theatre (also prominent on radio—as in the United States), moved onto the small screen along with the radioroman. As elsewhere, theatre was shortlived on TV while the radioroman went on to become the *teleroman*. The *teleroman*, building on the loyal following of the radioroman by bringing "to life" the main characters of two of the best loved and most enduring radio productions, *Un homme et son péché* and *La famille Plouffe*, was able to experiment with new themes and new styles of writing. It thus adapted the century old popular novel to this modern medium without sacrificing tradition and its most endearing qualities.

As an indication not only of the rapid growth of the *teleroman*, but of the centrality of the position it holds within both the televison industry and the public discourse on television itself, one can cite the following figures. A recent repertoire lists nearly 600 titles of original works of fiction, including *teleromans*, produced by Quebecois screenwriters to the delight of tens of millions of television viewers from 1952 to 1992. A comparable feat is not to be found in any other French language television industry, including France's. Nor is the popularity of locally produced television fiction in Quebec to be equalled anywhere, particularly in terms of the loyalty that the *teleroman* commands. The "Who Killed JR" episode of *Dallas* set a new standard in American television market research with its 54 point market share, in the early 1980s, and it has rarely been challenged since. In Quebec a 50-point market share is considered the basic standard of a successful show with the yearly best-sellers, reaching the high 70s and low 80s.

Not surprisingly the *teleroman* has spawned some small but vibrant secondary commercial ventures and represents some notable investments by other communications industries. For example, a glossy magazine *Teleroman* is published four times a year with a readership of some 50,000. The well established television guides such as *TV Hebdo*, with nearly a million readers, often feature well known faces of actors or characters of the popular *teleroman* on its cover. Each year moreover, it devotes a special edition of the current lineup of best and least known *teleromans*. Every major daily newspaper publishes the weekly schedule of television programming and has a television critic whose main subject is the *teleroman*: its costs, production, writers, actors, characters, intrigues, and audience rates. Talk shows quite regularly invite authors, actors and TV characters to meet live studio audiences. Even "serious" public affairs television shows, magazines and newspapers give thoughtful attention to the phenomenon. Of course the *teleroman*, with its well-known and loved characters, is a bonanza for advertising agencies selling everything from sundries, to soft drinks, to automobiles; they are the spokespersons for industries; they appear on public announcements and telethons for the sick and the needy. But most importantly, these well-known and well-loved actors and characters have contributed to the birth and growth of a thriving, creative, French language Quebec-based advertising industry. Not too many years ago, this industry's main revenue was translating English language, Toronto or New York conceived, television commercials. Today French language advertisements for national Canadian and American brand names are conceived and produced in Quebec. The most eloquent product example is Pepsi, which failed miserably in the Quebec market until some 10 years ago when the company agreed to hire a local agency to build its campaign around a well-known fictitious comic figure. It has become a remarkable success story in its own right. Other examples abound and include, for example, campaigns by Bell Canada and General Motors.

Another commercial spinoff, besides the inevitable merchandizing of effigies on dolls, lunch boxes, and posters, is the phenomenon of "living museums." Here the sets, whether original or reconstructed, of *teleromans* such as *Un homme et son péché*, *Le temps d'une paix*, *Les filles de Caleb*, or *Cormoran* are rebuilt in their "natural" outdoor surroundings. These *teleromans* are historically grounded, either in a specific time frame such as the 1930s or 1940s, or in the lives of past public and semi-public figures. The actual historical site on which these sets are built, the authentic dwellings upon which they are grafted, even the now-permanent presence of actual descendants of the romanticized characters in these reconstructed settings, all lend a "museum" and educational quality to these commercial enterprises. The *teleroman* is thus much more than a television genre, it is also an industry in itself and a generator of economic activities in industrially related sectors.

One of the recurring themes in the *teleroman* is the city, and this city is Montreal, the largest French language city in North America. It is a character in its own right in the same manner as the London of Charles Dickens, Paris in the novels by Balzac and Zola, or New York and San Francisco for the modern American teleseries. The *teleroman* often looks and sounds like an indictment of the city with its wealth of social problems—anonymous violence, rackets, abused children, battered women, drug abuse, solitude, poverty, homelessness. But it is also an ode to the city's magnetism—riches, arts, adventure, beauty, fulfillment, empowerment, enlightenment, and above all, the chance for true love. The *teleroman* exudes both a sense of *déjà vu* and elsewhereism.

The *teleroman* focuses on the ordinary, even on the anti-hero who is allowed to fail, sometimes disastrously. It reaches into the banality of everyday life to gather the stuff out of which characters of flesh and blood appear on the television screen, live and evolve, cry and laugh, cheat and repent, love and hate, and sometimes disappear. The fact that ordinariness can be both enticing and serialized and still command loyalty from seasoned viewers of some forty years of television fiction, is the greatest hommage that can be paid to these writers, producers and actors. Such skill is attested to by the popularity, for example, of *Chambres en ville*, an exploration of the pains and joys of growing up as a teenager in Montreal.

Another remarkable feature of the Quebecois *teleroman* lies in its distinctive mixture of gendered world views. This particular mixture can be traced to the presence and influence of the women working in the *teleroman's* creative communities. Telenovelists include women such as former journalist Fabienne Larouche, former journalist and Quebec cabinet minister Lise Payette and her daughter Sylvie. Renowned women actors of both theatre and screen play lead roles in the *teleroman*. And women novelists whose best-selling novels have been adapted to the television genre, such as Arlette Cousture (*Les filles de Caleb*) and Francine Ouellet (*Au nom du père*) often contribute to the creative process.

The *teleroman*, like other works of fiction in many other societies, is a testimony to the creative use of technology, in this case a technology to transmit at a distance and in real time, images and sounds. Through the efforts and talents of many artists, professionals, and technicians a world of fiction is created. It is a world in which reality takes on certain meanings for a geographically, socially, historically and culturally designated community. That the *teleroman* succeeds in achieving this is not unique; what is unique is that it does so in a unique fashion. It thus contributes a small but original viewpoint, or narrative, to the accumulated human legacy of past efforts to give meaning to the lives of ordinary people.

—Roger de la Garde and Gisèle Tchoungui

See also Canadian Programming in French; *Family Plouffe/ La Famile Plouffe*; Soap Opera; Telenovela

TELETEXT

Teletext is the term commonly applied to electronic systems that transmit to specially equipped television receivers. This technology makes use of normal broadcast signals to distribute data to television sets or designated monitors. The data is provided in the form of "pages" made up of screens of colorful text and graphic information. The broadcasted data may contain information such as news headlines, sports scores, and traffic and financial reports.

Although the signals may sometimes be transmitted as a subcarrier on an FM radio signal, teletext systems usually

Examples of teletext
Photos courtesy of KSL-TV/Salt Lake City

Examples of teletext
Photos courtesy of KSL-TV/Salt Lake City

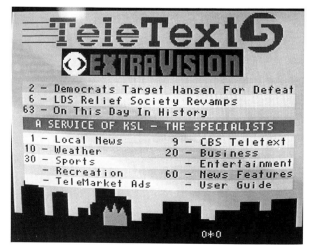

Examples of teletext
Photos courtesy of KSL-TV/Salt Lake City

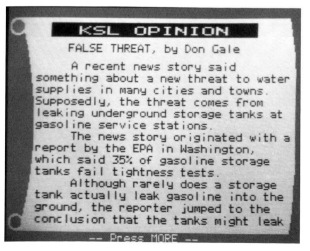

Examples of teletext
Photos courtesy of KSL-TV/Salt Lake City

transmit digital data by placing messages in the unused lines of the standard television signal. These unused lines are called the Vertical Blanking Interval or VBI: this can be seen as the black bar that appears when the vertical hold is defective on a television set. A special decoder is required to retrieve and display the text and graphic content: this decoder may be built into a television set or cable decoder or it may be installed as a peripheral device. The teletext information may be displayed in a superimposed format or full screen image, or it may be relegated to the lower part of the video display. Viewers are able to control the display of this information with a handheld remote control device. In some systems special printers may be connected to the sets which are able to reproduce provide hard copies of the broadcasted data.

In contrast to videotext services which are fully interactive, teletext has traditionally has been a one-way system. Because of the limitations of the broadcast signal, the user must access the information in order, page by page, as the station chooses to transmit it.

Teletext services have operated quite successfully in European markets since the late 1970s. In Great Britain, for example, several million sets have been equipped with decoders that allow users access to Ceefax, the British Broadcasting Corporation's teletext service, and to Oracle, a teletext service provided by Independent TV. However, teletext has had little success in the United States, in part because the Federal Communications Commission chose not to designate a national standard for teletext decoder devices. The resulting proliferation of incompatible equipment created confusion in the marketplace which ultimately contributed to the low adoption of teletext among U.S. television owners.

—Aviva Rosenstein

FURTHER READING

Alber, A. *Videotex-Teletext: Principles and Practices.* New York: McGraw Hill, 1985.

Mosco, Vincent. *Push-button Fantasies: Critical Perspectives on Videotex and Information Technology.* Norwood, New Jersey: Ablex, 1982.

Veith, Richard. *Televisions' Teletext.* New York: North-Holland, 1983.

TELETHON

A telethon is a marathon-length televised program devised for raising money for national and local charities or non-profit organizations. While *The Jerry Lewis Muscular Dystrophy Association Labor Day Telethon* is perhaps the most notable, numerous examples of the form have raised billions for various causes.

The first telethon, a 16-hour event broadcast by NBC and hosted by Milton Berle in 1949, raised $1,100,000 for the Damon Runyan Memorial Cancer Fund. Berle's pioneering effort set the tone for years to follow; a big-name star at the fore, a battery of telephone operators to collect pledges, and stage, film, and TV personalities appearing among impassioned pleas for donations. Jerry Lewis was one of the personalities to appear with Berle during the first telethon.

Without doubt, the Jerry Lewis MDA effort is the quintessential telethon. Hosted by Lewis since 1966, it is broadcast internationally, free of charge, by local stations signing on as part of the annual Labor Day "Love Network." The event, along with Lewis' off-key, emotional rendition

The MDA Labor Day Telethon
Photo courtesy of MDA

The MDA Labor Day Telethon
Photo courtesy of MDA

of the song "Walk On," have become synonymous with Labor Day itself.

Telethons began showing their age in the early 1990s, when various groups representing the disabled argued that telethons, with their accent on cures, paint a helpless and pathetic picture of people with disabilities. Lewis, a fervent campaigner for finding a cure for Muscular Dystrophy, has dismissed such complaints and continues his traditional approach. As of September 1995, the MDA telethon had raised over $1.4 billion for muscular dystrophy.

In addition to the MDA event, other annual telethons include those for Easter Seals, The Arthritis Foundation, United Cerebral Palsy, and the United Negro College Fund. On the local level, Public Television stations have borrowed from the form to raise money during their viewer pledge drives.

—Michael B. Kassel

FURTHER READING

"Charity Balls." *New Statesman and Society* (London), 1 June 1990.

Dempsey, John. "TV Stations Line Up for 'Flood Aid.'" *Variety* (Los Angeles), 2 August 1993.

"Jerry Lewis Criticized over Telethon's Approach." *New York Times*, 6 September 1992.

Johnson, Mary. "'Jerry's Kids.'" *The Nation* (New York), 14 September 1992.

Karpf, Anne. "'Give Us a Break, Not a Begging Bowl.'" *New Statesman* (London), 27 May 1988.

Kruger, Barbara. "Barbara Kruger on Television." *Artforum* (New York), November 1989.

Shapiro, Joseph P. "Disabling 'Jerry's Kids.'" *U.S. News and World Report* (Washington, D.C.), 14 September 1992.

See also Special/Spectacular

TELEVISION CRITICISM (JOURNALISTIC)

From the early 1900s, U.S. newspapers carried brief descriptions of distant reception of wireless radio signals and items about experimental stations innovating programs. After station KDKA in Pittsburgh inaugurated regular radio broadcast service in 1920, followed by hundreds of new stations, newspaper columns noted distinctive offerings in their schedules. In 1922 *The New York Times* started radio columns by Orrin E. Dunlap, Jr. From 1925 Ben Gross pioneered a regular column about broadcasting in the *New York Daily News*, which he continued for 45 years. Newspapers across the country added columns about schedules, programs, and celebrities during radio's "golden age" in the 1930s and 1940s. During those decades experiments in "radio with pictures" received occasional notice; attention to the new medium of television expanded in the late 1940s as TV stations went on the air in major cities, audiences grew, and advertisers and stars forsook radio for TV networks.

Chronicling those early developments were Jack Gould of *The New York Times* and John Crosby of the *New York Herald Tribune*, in addition to reviewer-critics of lesser impact in other metropolitan areas. From 1946 to 1972 Gould meticulously and even-handedly reported technical, structural (networks, stations), legal (Congress and Federal Communications Commission), economic (advertising), financial, and social aspects of TV as well as programming trends. Crosby began reviewing program content and developments in 1946 with stylistic vigor, offering a personalized judgment that could be caustic. As the medium matured in the 1960s and 1970s, Lawrence Laurent of *The Washington Post* joined the small group of influential media critics writing for major metro newspapers. He explored trends and causal relations and reported interrelations of federal regulatory agencies and broadcast corporations while also appraising major program successes and failures. On the West Coast where TV entertainment was crafted, the *Los Angeles Times'* Hal Humphrey and Cecil Smith covered the creative community's role in television, emphasizing descriptive reviews of individual programs and series. Other metro dailies and their early, influential program reviewer-critics included the *San Francisco Chronicle's* Terrance O'Flaherty, *Chicago Sun-Times'* Paul Malloy, and *Chicago Tribune's* Larry Wolters.

Meanwhile most newspapers carried popular columns about daily program offerings, reported behind-the-scenes information, and relayed tid-bits about TV stars. Some referred to this kind of column as "racing along in shorts," a series of brief items each separated by three dots. Complementing local columns were syndicated wire services, featuring a mix of substantive pieces and celebrity interviews. Among long-time syndicated columnists, in addition to *New York Times* and *Washington Post* columnists distributed nationally, were Associated Press' Cynthia Lowrey and Jay Sharbut.

Weekly and monthly magazines also published analyses of broader patterns and implications of television's structure, programming, and social impact. They featured critics such as *Saturday Review's* Gilbert Seldes and Robert Lewis Shayon, *Time's* "Cyclops" (John McPhee, among others), John Lardner and Jay Cocks in *Newsweek*, Marya Mannes in *The Reporter*, and Harland Ellison's ideosyncratic but trenchant dissections in *Rolling Stone*. Merrill Pannitt, Sally Bedell Smith, Neil Hickey, and Frank Swertlow offered serious analysis in weekly *TV Guide;* often multi-part investigative reports, those extended pieces appeared alongside pop features and interviews, plus think-pieces by specialists and media practitioners all wrapped around massive TV and cable local listings of regional editions across the country. Reporter-turned-critic Les Brown wrote authoritatively for trade paper *Variety*, then *The New York Times*, then as editor of *Channels of Communication* magazine. Weekly *Variety* published critical reviews of all new entertainment and documentary or news programs, both one-time-only shows and initial episodes of series; the newsmagazine's staff faithfully analyzed themes, topics, dramatic presentation, acting, sets and scenery, including complete listings of production personnel and casts. Reflecting shifting perspectives on the significance of modern mass media, Ken Auletta (*Wall Street Journal, New Yorker*) monitored in exhaustive detail the media mega-mergers of the 1980s and 1990s.

In the 1950s, TV columnists tended to be reviewers after the fact, offering comments about programs only after they aired, because almost all were "live." (Comedian Jackie Gleason quipped that TV critics merely reported accidents to eye witnesses.) They could also appraise continuing series, based on previous episodes. As more programs began to be filmed, following *I Love Lucy*'s innovation, and videotape was introduced in the late 1950s for entertainment and news-related programs alike, critics were able to preview shows. Their critical analyses in advance of broadcast helped viewers select what to watch. Producers and network executives could monitor print reviewers' evaluations of their product. Those developments increased print critics' influence, though their authority never approached New York drama critics' impact on Broadway's theatrical shows. Typically, many of a season's critically acclaimed new programs tend to be driven off the schedule by mass audience preferences for other less challenging or subtle programming. Praised, award-winning new series often find themselves cancelled for lack of popular ratings. Some might apply to television movie-critic Pauline Kael's aphorism about films; she cynically described the image industry as "the art of casting sham pearls before real swine."

Television critics often use a program or series as the concrete basis for examining broader trends in the industry. Analyzing a new situation comedy or action-adventure drama or documentary-like news magazine is more than an exercise in scrutinizing a 30- or 60-minute program; it serves as a paradigm representing larger patterns in media and society. The critical review traces forces that shape not only programming but media structures, processes, and public perceptions. Often reviewers not only lament failures but question factors influencing success and quality. They challenge audiences to support superior programming by selective viewing just as they challenge producers to create sensitive, authentic, depictions of deeper human values. Yet, Gilbert Seldes cautioned as early as 1956 that the critic must propose changes that are feasible in the cost-intensive mass media system; this would be "more intellectually honest and also save a lot of time" while avoiding pointless hostility and futility.

Over the decades studies of audiences and program patterns, and surveys of media executives, have generally discounted print media criticism as a major factor in program decision-making, particularly regarding any specific program content or scheduling. But critics are not wholly disregarded. Those published in media centers and Washington, D.C. serve as reminders to media managers of criteria beyond ratings and revenues. Critics in trade and metropolitan press are read by government agency personnel as well, to track reaction to pending policy moves. The insightful comments of critics come in many forms: courteous and cerebral (veteran John O'Connor since 1971 and Walter Goodman, *The New York Times*), stylistically sophisticated and witty (Tom Shales, *The Washington Post)*, sometimes abrasive (Ron Powers, Chicago *Sun-Times*), even cynical (Howard Rosenberg, *Los Angeles Times*). Each of these may illuminate lapses in artistic integrity or "good taste" and prod TV's creators and distributors to reflect on larger aesthetic and social implications of their lucrative, but ephemeral, occupations. Those published goadings enlighten readers, serve as a burr under the saddle of broadcasters/creators, and provide an informal barometer to federal law-makers and regulators.

At the same time television criticism published in print media serves the publisher's primary purpose of gaining readership among a wide and diverse circulation. That goal puts a premium on relevance, clarity, brevity, cleverness, and attractive style. The TV column is meant to attract readers primarily by entertaining them, while also informing them about how the system works. And at times columns can inspire readers to reflect on their use of television and how they might selectively respond to the medium's showcases of excellence, plateaus of mediocrity, and pits of meretricious exploitation and excess. Balanced criticism avoids blatant appeals and gratuitous savaging of media people and projects. The critic serves as a guide, offering standards or criteria for judgment along with factual data, so readers can make up their own minds. A test of successful television criticism is whether readers enjoy reading the articles as they grow to trust the critic's judgment because they respect his or her perspective. The critic-reviewer's role grows in usefulness as video channels proliferate; viewers innundated by dozens of cable and over-air channels can ensure optimum use of leisure viewing time by following critics' tips about what is worth tuning in and what to avoid.

Reflecting the quality of published television criticism in recent years, distinguished Pulitzer Prizes have been awarded to Ron Powers (1973), William Henry III (1980, *Boston Globe),* Howard Rosenberg (1985), and Tom Shales (1988). Early on, influential *Times* critic Jack Gould set the standard when in 1957 he won a special George Foster Peabody Award for his "fairness, objectivity and authority." Prerequisites for proper critical perspective outlined by Lawrence Laurent three decades ago remain apt today: sensitivity and reasoned judgment, a renaissance knowledge, coupled with exposure to a broad range of art, culture, technology, business, law, economics, ethics, and social studies all fused with an incisive writing style causing commentary to leap off the page into the reader's consciousness, possibly influencing their TV behavior as viewers or as professional practitioners.

—James A. Brown

FURTHER READING

Adkins, Gale. "Radio-Television Criticism in the Newspapers: Reflections on a Deficiency." *Journal of Broadcasting* (Washington, D.C.), Summer 1983.

Laurent, Lawrence. "Wanted: the Complete Television Critic." In, *The Eighth Art*. New York:Holt, Rinehart and Winston, 1962.

Orlik, Peter B. *Critiquing Radio and Television Content*. Boston: Allyn and Bacon, 1988.

Rossman, Jules. "The TV Critic Column: Is It Influential?" *Journal of Broadcasting* (Washington, D.C.) Fall, 1975.

Seldes, Gilbert. *The Public Arts*. New York: Simon and Schuster, 1956.

Shayon, Robert Lewis. *Open to Criticism*. Boston: Beacon Press, 1971.

Smith, Ralph Lewis. *A Study of the Professional Criticism of Broadcasting in the United States*. New York: Arno Press, 1973.

Watson, Mary Ann. "Television Criticism in the Popular Press." *Critical Studies in Mass Communication* (Annandale, Virginia), March, 1985.

See also Television Studies

TELEVISION NORTHERN CANADA

In 1983, after a decade of lobbying, the Canadian federal government established a Northern Broadcasting Policy which laid out the principles for the development of Northern native-produced programming for communities North of the 55th latitude line. With the policy was an accompanying program vehicle called the Northern Native Broadcast Access Program. This program was given $40.3 million to be distributed over a four-year period to 13 regional Native Communications Societies in the North in order to produce 20 hours of regional radio and/or five hours of television per-week in First Peoples' languages, reflecting their specific cultural perspectives. Although funding has eroded over time, the policy and programs are still operational.

One of the key problems identified quite soon after the Program was initiated was that of program distribution via satellite. Transponder rental costs were prohibitive and it became apparent early in the implementation process that a dedicated Northern transponder would be the solution to the negotiation problems involved in piggybacking on existing distribution services, such as that of the Canadian Broadcasting Corporation's Northern Services or CANCOM (a commercially-based Northern distribution service).

In January 1987, in response to the issue of distribution, Canadian aboriginal and Northern broadcasters met in Yellowknife, Northwest Territories to form a non-profit consortium with the goal of establishing a Pan-Northern television distribution service. In 1988, the Canadian government gave the organizers $10 million to establish Television Northern Canada (TVNC). In 1991, Canada's regulatory agency, the Canadian Radio-television and Telecommunications Commission (CRTC) approved TVNC's application for a native television network license to serve Northern Canada for the purpose of broadcasting cultural, social, political, and educational programming. On 21 January 1992, TVNC began broadcasting.

TVNC's network is owned and programmed by 13 aboriginal broadcast, government and education organizations in Northern Canada. Members include: the Inuit Broadcasting Corporation (Ottawa, Iqaluit) the Inuvialuit Communications Society (Inuvik), Northern Native Broadcasting, Yukon (Whitehorse), the OkalaKatiget Society (Labrador), Taqramiut Nipingat Incorporated (Northern Quebec), the Native Communications Society of the Western N.W.T. (Yellowknife), the Government of the Northwest Territories, Yukon College, and the National Aboriginal Communications Society. Associate Members are CBC Northern Service, Kativik School Board (Quebec), Labrador Community College, Northern Native Broadcasting, Terrace, Telesat Canada, and Wawatay Native Communications Society (Sioux Lookout). Services extend to Labrador, Arctic Quebec, Nunavut (formerly the Inuit regions of the Northwest Territories), Western Northwest Territories, and Yukon. Programming is produced in at least seven aboriginal languages, as well as in English and French. The TVNC's mission statement elaborates its goals:

> Television Northern Canada shall be (is) a dedicated northern satellite distribution system, for the primary benefit of aboriginal people in the North, by which residents of communities across northern Canada may distribute television programming of cultural, social, political and educational importance to each other, increasing communications access and promoting dialogues among their remote and underserved homelands (TVNC, 1 March 1993).

As a primary level of service for the North, TVNC spans five time zones and covers an area of over 4.3 million kilometers (one-third of Canada's territory). The organizations involved broadcast approximately 100 hours per week to 94 communities. TVNC is not a programmer, but a distributor of its members' programming. Core Northern programming consists of:

- 38 hours per week of aboriginal language and cultural programming;

- 23 hours per week of formal and informal educational programming;

- 12 hours per week of produced and acquired children's programming, over half of which is in aboriginal languages.

As seen from the number of programming hours scheduled, the Native Communications Societies cannot afford to produce enough materials for round-the-clock broadcast. Consequently, TVNC is wrapped around with *Environment Canada* weather forecasts, as well as newstexts from *Broadcast News.*

Despite funding cutbacks, TVNC is the only aboriginal television network in the world which broadcasts such a high volume of programming from indigenous sources. As a Pan-Arctic distribution undertaking, it is theoretically in a position to forge connections with Inuit and aboriginal groups in other countries, such as Greenland, Alaska, Finland, Russia (Siberia), Australia, New Zealand, Brazil, and Bolivia (among others). This could be achieved through program exchanges and uplink-downlink satellite arrangements, but TVNC administration faces two problems: incompatible video/electrical standards and financial barriers. What TVNC does do is offer Northern viewers limited access to programming about activities of indigenous peoples from around the globe, when this is feasible.

In May 1995, TVNC applied to be placed on the Revised Lists of Eligible Satellite Services to be picked up by cable operators throughout Canada. Approval was granted in November 1995. Availability of Northern-produced programming on Southern channels expands Northern broadcasting to a new dimension. It represents the completion of the Canadian broadcasting mandate—permitting broadcasting to move in all directions from the South to the North, North-to-North, and North-to-South. This is a leap forward from the unidirectional importation of Southern culture to the North which began in 1973. By permitting TVNC to be broadcast in the South, the Canadian regulatory body is attempting to ensure that all Canadians have an opportunity to acquire a more coherent understanding of the North and its residents.

In December 1995, the CRTC approved a deal between Arctic cooperatives Limited (ACL) and NorthwesTel to split up the northern cable TV market between them. At the same time, the CRTC noted that it expects, but does not require, ACL and NorthwesTel to pay 55 cents per cable TV subscriber into a special programming fund to be administered by TVNC. The money is intended to pay for the development and distribution of First Peoples television programs. Hopefully, these expectations will be met without resistance.

The most current initiative of TVNC is their participation with ACL and Northwestern in the creation of a new company for the purpose of constructing an affordable and accessible high-speed communications network in Northern Canada.

Satellite up-link for Northern Native People's Television
Photo courtesy of Telecine Multimedia Inc.

The granting of a license to TVNC and the integration of TVNC in the Northern information highway infrastructure represents Canada's recognition of the importance of

Northern-based control over the distribution of its own native and Northern programming and telecommunications services. TVNC is the vehicle through which First Peoples are able to represent themselves and their concerns across Canada's expansive territories. First Peoples are no longer restricted by the geography of diffusion technology to local or regional self representation and identity-building. In this sense TVNC constitutes a de facto recognition of the communication rights of First Peoples in Canada's North.

—Lorna Roth

FURTHER READING

Bell, Jim. "Co-ops Northwestern Split the North Between Them." *Nunatsiaq News,* 22 December 1995.

Jackson, Kristy. "North Switches Channels to TVNC." *Above and Beyond,* Summer 1992.

Meadows, Michael. "Ideas from the Bush: Indigenous Television in Australia and Canada." *Canadian Journal of Communication* (Saskatoon, Canada), Spring 1995.

Roth, Lorna. "Northern Voices and Mediating Structures: The Emergence and Development of First Peoples' Television Broadcasting in the Canadian North" (doctoral dissertation. Montreal: Concordia University, 1994).

TVNC. *Television Northern Canada: A Proposal for a Shared Television Distribution Service in Northern Canada.* 1 June 1987.

Young, Pamela. "Southern Reflections on The Rural North." *Maclean's* (Toronto, Canada), 28 March 1988.

CRTC DOCUMENTS AVAILABLE TO THE PUBLIC

Public Notice CRTC 1985-274. Northern Native Broadcasting. 1985.

Public Notice CRTC 1990-12. Review of Native Broadcasting: A Proposed Policy. 1990.

Public Notice CRTC 1990-89. Native Broadcasting Policy. 1990.

Decision CRTC 91-826. Television Northern Canada Incorporated. Ottawa. 28 October 1991.

TVNC. Application to Add PvNC Television Network Signal to the Lists of Eligible Satellite Services Pursuant to the Cable Television Regulations, 1986. Iqaluit. 31 May 1995.

Public Notice CRTC 1995-129. Call for Comments on a Request to Add the Service of Television Northern Canada Incorporated to the Lists of Eligible Satellite Services. Ottawa. 28 July 1995.

Public Notice CRTC 1995-189. Revised Lists of Eligible Satellite Services. Ottawa. 6 November 1995.

PUBLIC COMMENTS AVAILABLE AT CRTC OFFICES

Canadian Cable Television Association. Ottawa. August 31, 1995.

Northwest Territories Minister of Education, Culture and Employment. Yellowknife. 31 August 1995.

See also First People's Television Broadcasting in Canada

TELEVISION STUDIES

Television studies is the relatively recent, aspirationally disciplinary name given to the academic study of television. Modeled by analogy with longer established fields of study, the name suggests that there is an object, "television", which, in courses named, for example, "Introduction to Television Studies", is the self-evident object of study using accepted methodologies. This may be increasingly the case, but it is important to grasp that most of the formative academic research on television was inaugurated in other fields and contexts. The "television" of television studies is a relatively new phenomenon, just as many of the key television scholars are employed in departments of sociology, politics, communication arts, speech, theatre, media and film studies. If it is now possible, in 1996, to speak of a field of study, "television studies" in the anglophone academy, in a way in which it was not in 1970, the distinctive characteristics of this field of study include its disciplinary hybridity and continuing debate about how to conceptualise the object of study "television." These debates, which are and have been both political and methodological, are further complicated in an international frame by the historical peculiarities of national broadcasting systems. Thus, for example, the television studies that developed in Britain or Scandinavia, while often addressing U.S. television programmes, did so within the taken-for-granted dominance of public service models. In contrast, the U.S. system is distinguished by the normality of advertising spots and breaks. In the first instance then, television studies signifies the contested, often nationally inflected, academic address to television as primary object of study—rather than, for example, television as part of international media economies or television as site of drama in performance.

There have been two prerequisites for development of television studies in the "West"—and it is primarily a western phenomenon, which is not to imply that there is not, for example, a substantial literature on Indian television (cf. Krishnan and Dighe, 1990). The first was that television as such be regarded as worthy of study. This apparently obvious point is significant in relation to a medium which has historically attracted distrust, fear and contempt. These responses, which often involve the invocation of television as both origin and symptom of social ills, have, as many scholars have pointed out, homologies with responses to earlier popular genres and forms such as the novel and the cinema. The second prerequisite was that television be granted, conceptually, some autonomy and specificity as a medium. Thus television had to be regarded as more than simply a transmitter of world, civic or artistic events and as distinguishable from other of the "mass media". Indeed, much of the literature of television studies could be characterised as attempting to formulate accounts of the specificity of television, often using comparison with, on the one hand, radio (broadcast, liveness, civic address) and on the other, cinema (moving pictures, fantasy), with particular attention, as discussed below, to debate about the nature of the television text and the television audience. Increasingly significant also are the emergent histories of television whether it be the autobiographical accounts of insiders, such as Grace Wyndham Goldie's history of her years at the BBC, *Facing the Nation*, or the painstaking archival research of historians such as William Boddy with his history of the quiz scandals in 1950s U.S. television or Lynn Spigel with her pioneering study of the way in which television was "installed" in the U.S. living room in the 1950s, *Make Room for TV*.

Television studies emerges in the 1970s and 1980s from three major bodies of commentary on television: journalism, literary/dramatic criticism and the social sciences. The first, and most familiar, was daily and weekly journalism. This has generally taken the form of guides to viewing and reviews of recent programmes. Television reviewing has, historically, been strongly personally voiced, with this authorial voice rendering continuity to the diverse topics and programmes addressed. Some of this writing has offered formulations of great insight in its address to television form—for example the work of James Thurber, Raymond Williams, Philip Purser or Nancy Banks-Smith—which is only now being recognised as one of the origins of the discipline of television studies. The second body of commentary is also organised through ideas of authorship, but here it is the writer or dramatist who forms the legitimation for the attention to television. Critical method here is extrapolated from traditional literary and dramatic criticism, and the television attracts serious critical attention as an "home theatre". Indicative texts here would be the early collection edited by Howard Thomas, *Armchair Theatre* (1959) or the later, more academic volume edited by George Brandt, *British Television Drama* (1981). Until the 1980s, the address of this type of work was almost exclusively to "high culture": plays and occasionally series by known playwrights, often featuring theatrical actors. Only with an understanding of this context is it possible to see how exceptional Raymond William's defence of television soap opera is in *Drama in Performance* (1968), or Horace Newcomb's validation of popular genres in *TV: The Most Popular Art* (1974).

Both of these bodies of commentary are mainly concerned to address what was shown on the screen, and thus conceive of television mainly as a text within the arts humanities academic traditions. Other early attention to television draws, in different ways, on the social sciences to address the production, circulation and function of television in contemporary society. Here, research has tended not to address the television text as such, but instead to conceptualise television either through notions of its social *function* and *effects*, or within a governing question of *cui bono?* (whose good is served?). Thus television, along with other of the mass media, is conceptualised within frameworks principally concerned with the maintenance of social order; the reproduction of the status quo, the relationship between the state, media ownership and citizenship, the constitution of the public sphere. With these concerns, privileged areas of inquiry have tended to be non-textual: patterns of international cross-media ownership; national and international regulation of media production and distribution; professional ideologies; public opinion; media audiences. Methodologies here have been greatly contested, particularly in the extent to which Marxist frameworks, or those associated with the critical sociology of the Frankfurt School have been employed. These debates have been given further impetus in recent years by research undertaken under the loose definition of cultural studies. The privileged texts, if attention has been directed at texts, have been news and current affairs, and particularly special events such as elections, industrial disputes and wars. It is this body of work which is least represented in "television studies", which, as an emergent discipline, tends towards the textualisation of its object of study. The British journal *Media, Culture and Society* provides an exemplary instance of media research—in which television plays some part—in the traditions of critical sociology and political economy.

Much innovatory work in television studies has been focused on the definition of the television text. Indeed, this debate could be seen as one of the constituting frameworks of the field. The common-sense view points to the individual programme as a unit, and this view has firm grounding in the way television is produced. Television is, for the most part, made as programmes or runs of programmes: series, serials and miniseries. However, this is not necessarily how television is watched, despite the considerable currency of the view that it is somehow better for the viewer to choose to watch particular programmes rather that just having the television on. Indeed, BBC television in the 1950s featured "interludes" between programmes, most famously, "The Potter's Wheel", a short film showing a pair of hands making a clay pot on a wheel, to ensure that viewers did not just drift from one programme to another. It is precisely this possible "drifting" through an evening's viewing that has come to seem, to many commentators, one of the unique features of television watching, and hence something that must be attended to in any account of the television text.

The inaugural formulation is Raymond William's argument, in his 1974 book, *Television: Technology and Cultural Form,* that "the defining feature of broadcasting" is "planned flow". Williams developed these ideas through reflecting on four years of reviewing television for the weekly periodical *The Listener,* when he suggests that the separating of the television text into recognisable generic programme units, which makes the reviewer's job much easier, somehow misses "the central television experience: the fact of flow" (1974). Williams's own discussion of flow draws on analysis of both British and U.S. television and he is careful to insist on the national variation of broadcasting systems and types and management of flow, but his attempt to describe what is specific to the watching of television has been internationally generative, particularly in combination with some of the more recent empirical studies of how people do (or don't) watch television.

If Williams's idea of flow has been principally understood to focus attention on television viewing as involving more viewing and less choosing than a critical focus on individual programmes would suggest, other critics have picked up the micro-narratives of which so much television is composed. Thus John Ellis approached the television text using a model ultimately derived from film studies, although he is precisely concerned, in his book *Visible Fictions,* to differentiate cinema and television. Ellis suggests that the key unit of the television text is the "segment", which he defines as "small, sequential unities of images and sounds whose maximum duration seems to be about five minutes" (1982). Broadcast television, Ellis argues, is composed of different types of combination of segment: sometimes sequential, as in drama series, sometimes cumulative, as in news broadcasts and commercials. As with Williams's "flow", the radical element in Ellis's "segment" is the way in which it transgresses common sense boundaries like "programme" or "documentary" and "fiction" to bring to the analyst's attention common and defining features of broadcast television as a medium.

However, it has also been argued that the television text cannot be conceptualised without attention to the structure of national broadcasting institutions and the financing of programme production. In this context, Nick Browne has argued that the U.S. television system is best approached through a notion of the "super-text". Browne is concerned to address the specificities of the U.S. commercial television system in contrast to the public service models—particularly the British one—which have been so generative a context for formative and influential thinking on television such as that of Raymond Williams and Stuart Hall. Browne defines the "super-text" as, initially, a television programme and all introductory and interstitial material in that programme's place in a schedule. He is thus insisting on an "impure" idea of the text, arguing that the programme as broadcast at a particular time in the working week, interrupted by ads and announcements, condenses the political economy of television. Advertising, in Browne's schema, is the central mediating institution in U.S. television, linking programme schedules to the wider world of production and consumption.

The final concept to be considered in the discussion about the television text is Newcomb and Hirsch's idea of the "viewing strip" (1987). This concept suggests a mediation between broadcast provision and individual choice, attempting to grasp the way in which each individual negotiates his or her way through the "flow" on offer, putting together a sequence of viewing of their own selection. Thus different individuals might produce very different "texts"— viewing strips—from the same nights viewing. Implicit within the notion of the viewing strip— although not a pre-requisite—is the remote control device, allowing channel change and channel surfing. And it is this tool of audience agency which points us to the second substantial area of innovatory scholarship in television studies, the address to the audience.

The hybrid disciplinary origins of television studies are particularly evident in the approach to the television audience. Here, particularly in the 1980s, we find the convergence of potentially antagonistic paradigms. Very simply, on the one hand, research traditions in the social sciences focus on the empirical investigation of the already existing audience. Research design here tends to seek representative samples of particular populations and/or viewers of a particular type of programming (adolescent boys and violence; women and soap opera). Research on the television audience has historically been dominated, particularly in the U.S., by large-scale quantitative surveys, often designed using a model of the "effects" of the media, of which television is not necessarily a differentiated element. Within the social sciences, this "effects" model has been challenged by what is known as the "uses and gratifications" model. In James Halloran's famous formulation, "we should ask not what the media does to people, but what people do to the media." (Halloran, 1970). Herta Herzog's 1944 research on the listeners to radio daytime serials was an inaugural project within this "uses and gratifications" tradition, which has recently produced the international project on the international decoding of the U.S. prime time serial, *Dallas* (Liebes and Katz, 1990).

This social science history of empirical audience investigation has been confronted, on the other hand by ideas of a textually-constituted "reader" with their origins in literary and film studies. This is a very different conceptualisation of the audience, drawing on literary, semiotic and psychoanalytic theory to suggest—in different and disputed ways—hat the text constructs a "subject position" from which it is intelligible. In this body of work, the context of consumption and the social origins of audience members are irrelevant to the making of meaning which originates in the text. However—and it is thus that we seen the potential convergence with social science "uses and gratifications" models— literary theorists such as Umberto Eco (1979) have posed the extent to which the reader should be seen as active in meaning-making. It is, in this context, difficult to separate the development of television studies, as such, from that of cultural studies, for it is within cultural studies that we begin to find the most sophisticated theorisations and empirical investigations of the complex, contextual interplay of text and "reader" in the making of meaning.

The inaugural formulations on television in the field of cultural studies are those of Stuart Hall in essays such as "Encoding and Decoding in Television Discourse" (1974)

(Hall, 1997) and David Morley's audience research (1980). However this television specific work cannot theoretically be completely separated from other cultural studies work conducted at Birmingham University in the 1970s such as the work of Dick Hebdige and Angela McRobbie which stressed the often oppositional agency of individuals in response to contemporary culture. British cultural studies has proved a successful export, the theoretical paradigms there employed meeting and sometimes clashing with those used, internationally, in more generalised academic re-orientation towards the study of popular culture and entertainment in the 1970s and 1980s. Examples of influential scholars working within or closely related to cultural studies paradigms would by Ien Ang and John Fiske. Ang's work on the television audience ranges from a study of *Dallas* fans in the Netherlands to the interrogation of existing ideas of audience in a postmodern, global context. John Fiske's work has been particularly successful in introducing British cultural studies to a U.S. audience, and his 1987 book, *Television Culture* was one of the first books about television to take seriously the feminist agenda that has been so important to the recent development of the field. For if television studies is understood as a barely established institutional space, carved out by scholars of television from, on the one hand, mass communications and traditional marxist political economy, and on the other, cinema, drama and literary studies, the significance of feminist research to the establishment of this connotationally feminized field cannot be underestimated, even if it is not always recognised. E. Ann Kaplan's collection, *Regarding Television,* with papers from a 1981 conference gives some indication of early formulations here.

The interest of new social movements in issues of representation, which has been generative for film and literary studies as well as for television studies, has produced sustained interventions by a range of scholars, approaching mainly "texts" with questions about the representation of particular social groups and the interpretation of programmes such as, for example, *thirtysomething, Cagney and Lacey, The Cosby Show,* or various soap operas. Feminist scholars have, since the mid-1970s, tended to focus particularly on programmes for women and those which have key female protagonists. Key work here would include Julie D'Acci's study of *Cagney and Lacey* and the now substantial literature on soap opera (Seiter et al., 1989). Research by Sut Jhally and Justin Lewis has addressed the complex meanings about class and "race" produced by viewers of *The Cosby Show,* but most audience research in this "representational" paradigm has been with white audiences. Jacqueline Bobo and Ellen Seiter argue that this is partly a consequence of the "whiteness" of the academy which makes research about viewing in the domestic environment potentially a further extension of surveillance for those ethnicized by the dominant culture.

Television studies in the 1990s, then, is characterised by work in four main areas. The most formative for the emergent discipline have been the work on the definition and interpretation of the television text and the new media ethnographies of viewing which emphasise both the contexts and the social relations of viewing. However, there is a considerable history of "production studies" which trace the complex interplay of factors involved in getting programmes on screen. Examples here night include Tom Burn's study of the professional culture of the BBC (1977), Philip Schlesinger's study of "The News" (1978) or the study of MTM co-edited by Jane Feuer, Paul Kerr and Tise Vahimagi (1984). Increasingly significant also is the fourth area, that of television history. Not only does the historical endeavour frequently necessitate working with vanished sources—such as the programmes—but it has also involved the use of material of contested evidentiary status. For example, advertisements in women's magazines as opposed to producer statements. This history of television is a rapidly expanding field, creating a retrospective history for the discipline, but also documenting the period of nationally regulated terrestrial broadcasting—the "television" of "television studies"—which is now coming to an end.

—Charlotte Brunsdon

FURTHER READING

Ang, Ien. *Desperately Seeking the Audience.* London: Routledge, 1993.

———. *Living Room Wars.* London: Routledge, 1995.

———. *Watching Dallas.* London: Methuen, 1985.

Bobo, Jacqueline, and Seiter, Ellen. "Black Feminism and Media Criticism." *Screen* (London), 1991; reprint in Brunsdon, Charlotte, J. D'Acci and L. Spigel, editors. *fem.tv.* Oxford: Oxford University Press, 1996.

Boddy, William. *Fifties Television: The Industry and its Critics.* Urbana: University of Illinois Press, 1990.

Brandt, George. *British Television Drama.* Cambridge: Cambridge University Press, 1981.

Browne, Nick. "The Political Economy of the Television (Super) Text." *Quarterly Review of Film Studies* (Los Angeles), 1984.

Burns, Tom. *The BBC: Public Institution, Private World.* London: Macmillan, 1977.

D'Acci, Julie. *Defining Women: Television and the Case of Cagney and Lacey.* Chapel Hill: University of North Carolina Press, 1994.

Eco, Umberto. *The Role of the Reader.* Bloomington: University of Indiana Press, 1979.

Ellis, John. *Visible Fictions.* London: Routledge and Kegan Paul, 1982.

Feuer, Jane, Paul Kerr, and Tise Vahimagi. *MTM: "Quality Television."* London: British Film Institute, 1984.

Fiske, John. *Television Culture.* London: Methuen, 1987.

Fiske, John, and John Hartley. *Reading Television.* London: Methuen, 1978.

Goldie, Grace Wyndham. *Facing the Nation: Television and Politics, 1936-1976.* London: Bodley Head, 1978.

Hall, Stuart. *Early Writings on Television.* London: Routledge, 1997.

Hall, Stuart, Dorothy Hobson, Andrew Lowe, and Paul Willis, editors. *Culture, Media, Language*. London: Hutchinson and the Centre for Contemporary Cultural Studies, 1980.

Halloran, James. *The Effects of Television*. London: Panther, 1970.

Hebdige, Dick. *Subculture and the Meaning of Style*. London: Methuen, 1978.

Herzog, Herta. "What Do We really Know about Daytime Serial Listeners." In, Lazersfeld, Paul, and Frank Stanton, editors. *Radio Research 1942-43*. New York: Duell, Sloan and Pearce, 1944.

Jhally, Sut, and Justin Lewis. *Enlightened Racism: The Cosby Show and the Myth of the American Dream*. Boulder, Colorado: Westview Press, 1992.

Kaplan, E. Ann. *Regarding Television*. Los Angeles: American Film Institute, 1983.

Krishnan, Prabha, and Anita Dighe. *Affirmation and Denial: Construction of Femininity on Indian Television*. New Delhi: Sage, 1990.

Liebes, Tamar, and Elihu Katz. *The Export of Meaning*. Oxford: Oxford University Press, 1990.

McRobbie, Angela. *Feminism and Youth Culture*. Basingstoke, England: Macmillan, 1991.

Morley, David. *Television, Audiences and Cultural Power*. London: Routledge, 1992.

———. *The Nationwide Audience*. London: British Film Institute, 1980.

Newcomb, Horace. *TV: The Most Popular Art*. New York: Doubleday, 1974.

Newcomb, Horace, and Paul Hirsch. "Television as a Cultural Forum: Implications for Research." In, Newcomb, Horace, editor. *Television: The Critical View*. New York: Oxford, 1994.

O'Connor, Alan, editor. *Raymond Williams on Television*. London: Routledge, 1989.

Purser, Philip. *Done Viewing*. London: Quartet, 1992.

Schlesinger, Philip. *Putting "Reality" Together*. London: Constable, 1978.

Seiter, Ellen, with others. *Remote Control*. London: Routledge, 1989.

Spigel, Lynn. *Make Room for TV*. Chicago: University of Chicago Press, 1993.

Thomas, Howard, editor. *The Armchair Theatre*. London: Weidenfeld and Nicolson, 1959.

Thurber, James. *The Beast in Me and other Animals*. New York: Harcourt Brace, 1948.

Williams, Raymond. *Drama in Performance*. London: C.A. Watts, 1968.

———. *Television, Technology and Cultural Form*. London: Fontana, 1974.

See also Audience Research; Audience Research: Cultivation Analysis; Audience Research: Effects Analysis; Audience Research: Reception Analysis; Hood, Stuart; Television Criticism (Journalistic); Williams, Raymond

TELEVISION TECHNOLOGY

The technology that makes television work is a complex subject, explained here in a basic, introductory fashion. Though television seems a thoroughly modern invention, available only since mid-20th century, the concept of recreating moving images electrically was developed much earlier than is generally thought. It can be traced at least to 1884 when Paul G. Nipkow created the rotating scanning disk which provided a way of sending a representation of a moving image over a wire using varying electrical signals created by mechanically scanning that moving image.

Mechanical scanning of an image involved a spinning disk, with a spiral grouping of holes, located at both the sending and receiving ends. At the sending end, a photocell-like device varied the strength of an electrical signal at a rate representing the amount of light hitting the cell through the holes in the disk; the greater the amount of light, the greater the strength of the electrical signal. At the receiving end, a source of light varied in intensity at the rate of the electrical signal it received and could be seen through the holes in the rotating disc, thereby recreating a crude copy of the image scanned at the sending end. Today, moving images are scanned electronically as described below and the varying electronic signal representing the scanned images can be transmitted or sent through wire to be recreated at the receiver or monitor.

The earliest practical mechanical scanning and transmitting of moving images occurred in the mid-1920s, and by the early 1930s electronic scanning had generally replaced the mechanical scanning methods. At first the images were crude, little more than shadow-pictures, but as the potential for television as a profit-making medium became apparent, more money and effort went into television experimentation and improvements continued through the 1930s.

By 1941, technical standards for the scanning and transmission of television images in America had been agreed upon and these standards have, in general, been maintained ever since. The American standard, known as National Television System Committee (NTSC), utilizes 525-line, 60-field, 30-frame, interlaced scanning. This means that images are scanned in the television camera and reproduced in the television receiver or monitor 30 times each second. Each full image, or frame, is scanned by dividing the image into 525 horizontal lines, and then sequentially scanning first all the even lines (every other line) from top to bottom, creating one field, and then scanning the odd numbered lines in the same manner, creating a second field. The two fields, when combined (interlaced), create one frame. Therefore, 30 complete images or frames, each made up of two fields, are created each second. Because it is not possible to perceive individual changes in light and image happening so quickly, the 30-times-per-second scanned images are perceived as continuous movement, a trait known as "persistence of vision," similar to motion picture viewing. The NTSC standard is used in Canada, parts of Asia, including Japan, and much of Latin America as well as in the United States. But there are two other "standards" in common use today. The PAL systems, a 25-frame-per-second standard with a number of variants, are used throughout most of Western Europe and India, as well as other areas. The SECAM 25-frame-per-second standard is used in many parts of the world, including France, Russia and most of Eastern Europe. Countries that use 60 Hertz (cycles per second) A.C. (alternating current) power have adopted a 30-frame-per-second television system. Countries that utilize a 50 Hertz (Hz.) power system have a 25-frame-per-second television system. In all these television systems the frame-per-second rate is equal to half the A.C. power frequency.

The aspect ratio of the television screen, the ratio of the horizontal dimension to the vertical dimension, is 4:3. For instance, if a TV receiver screen is 16 inches wide, the screen will be 12 inches high. (TV picture tubes are defined by their diagonal measurement, so in this example the screen would be described as a 20-inch TV). Often, motion pictures are shown on television in a "letter-box" format. Because motion pictures are usually shot in an aspect ratio greater than 4:3, it is necessary to leave a black space at the top and bottom of the television screen so that the film can be viewed in a form resembling its theatrical dimensions, without cutting off the sides. "High Definition" television also utilizes a greater aspect ratio, generally 16:9.

The television camera consists of a lens to focus an image onto the front surface of one or more pick-up-devices, and, within the camera housing, the pick-up-device(s) and the electronics to make the camera work. A viewfinder to monitor the camera's images is normally mounted in or on the camera. The pick-up-device, either a camera tube or charge-coupled device (CCD), reads the focused visual image and converts the image into a varying electronic signal that represents the image. On high quality cameras, three pick-up-devices are often utilized; one to pick up each of the three primary colors (blue, green and red) that make up the color image.

The face of the camera tube has a photoemissive material that gives off electrical energy when exposed to light. The stronger the light at any given point, the more energy created. By reading the amount of energy on the surface of the camera tube at each point, an electronic representation of the visual image can be created. The camera tube "reads" the amount of energy created on its surface by the focused image by scanning the image, both horizontally and vertically, with a moving electronic beam. The scanning occurs because of precise magnetic deflection of the beam.

The CCD replaces the camera tube in most modern cameras, commonly called chip cameras. This solid-state device measures the energy at the points on its surface, known as pixels, converts and stores this information and then sends out this varying electronic signal that represents the image. CCD image pick-up-devices are becoming more popular due to their small size, long life, greater sensitivity and light tolerance, minimal power requirements, less image distortion, and ruggedness.

In the receiver's, or monitor's, picture tube, the camera tube process is essentially reversed. The face of the picture tube is coated with a phosphor-like material that glows when struck by a beam of electrons. The glow lasts long enough to make the scanned image visible to the viewer. An electron gun shoots the thin beam of electrons at the face of the screen from within the picture tube. The beam's direction is varied in a precise manner by magnetic deflection in a way that matches or synchronizes with the original image scanned by the television camera. Color picture tubes can have one electron gun, such as in the Trinitron, or three guns, one for each primary color. One major difference between a receiver and a monitor should be mentioned here. A receiver is able to tune in a television station frequency and show the images being transmitted. A monitor does not have a tuning component and can receive video signals by wire only.

At a television station, the electronic signal from a television camera can be combined or mixed with video signals from other devices such as video tape players, computers, film chains or telecines (motion picture and slide projector units whose outputs have been converted to video signals) using what is known as a switcher. The switcher is also used to create various special visual effects electronically. The video output from the switcher can then be recorded, sent to another studio or master control room, or sent directly to a transmitter.

The complete video signal sent to a transmitter or through wire to a monitor consists of signals representing the picture (luminance), color (chrominance), and synchronization. Synchronizing signals force the receiver to correctly lock onto (sync-up) and reproduce the original image correctly. Otherwise, for example, the receiver might begin to scan an image that begins half-way down the screen.

Television stations are assigned a specific transmitting frequency and operating power. In the United States, VHF (very-high frequency) television, channels 2–13, occupies a portion of the frequency spectrum from 54–216 MHz (million Hertz or cycles-per-second). Channels 2–6 are located between 54 and 88 MHz. The F.M. radio band, 88–108 MHz, is located between television channels 6 and 7. Channels 7–13 are located between 174 and 216 MHz. UHF (ultra-high frequency) television, originally channels 14 to 83, was assigned the frequency range from 470 to 890 MHz. In 1966 the Federal Communications Commission (FCC) discontinued issuing licenses for UHF television stations above channel 69. In 1970 the FCC took away the frequency range from 807 to 890 MHz for other communication uses and so the UHF band now consists of channels 14–69, from 470 to 806 MHz. The upper end of this current range is being coveted for other frequency spectrum uses and it appears that the number of available channels in the UHF band will continue to decrease. Each television channel has a frequency bandwidth of 6 MHz. So, for instance, channel 2 has a frequency bandwidth from 54 MHz to 60 MHz. Within its assigned band each station transmits the video signal as described earlier, an audio signal, and specialized signals such as closed-captioning information.

In the television transmitter a carrier wave is created at an assigned frequency. This carrier wave travels at the speed of light through space with specific transmission or propagation characteristics determined by the individual frequency. The video signal is piggy-backed onto the much higher frequency carrier wave using a process known as modulation. Modulation, in the simplest terms, means that the carrier wave is modulated, or varied slightly, at the rate of the signal being piggy-backed. In a television transmission the video signal varies the amplitude or strength of the carrier wave at the rate of the video signal. This is known as amplitude modulation (A.M.) and is similar to the method used to transmit the audio of an A.M. radio station. However, the television station audio signal is piggy-backed onto the carrier wave using frequency modulation (F.M.). With television audio the carrier wave's frequency (instead of its amplitude) is varied slightly at the rate of the audio signal.

The modulated television station carrier wave is sent from the transmitter to an antenna. The antenna then radiates the signal out into space in a pattern determined by the physical design of the transmitting antenna. Traditionally the transmitter and antenna were terrestrially located, but now television signals can be radiated or delivered by transmitters and antennas located on satellites in orbit around the earth. In this case the television signal is transmitted to the satellite on a specific frequency and then retransmitted at a different frequency by the satellite's transmitter back to the earth.

Besides delivery by carrier wave transmission, television is often sent through cable directly to homes and businesses. These signals are delivered by satellite, over-the-air from terrestrial antennas, and sometimes directly from video players to the distribution equipment of cable television (CATV) service providers for feeding directly into homes. The signals are sent at specific carrier wave frequencies (sometimes called R.F., or radio frequencies) as chosen by the cable service provider.

A television receiver picks up the transmitted television signals sent over-the-air or by cable or satellite, removes the necessary video and audio signals that had been piggy-backed on the carrier wave, discards the carrier wave, and amplifies and converts the video and audio signals into picture and sound. A television monitor accepts direct video and audio signals to provide pictures and sound. As mentioned above, a monitor cannot receive carrier waves.

From primitive experimentation in the 1920s and 1930s through the advent of commercial television in the late 1940s, to color television as the standard by the mid- 1960s, television has grown quickly to become perhaps the most important single influence on society today. From a source of information and entertainment to what some have dubbed the real "soma" of Alex Huxley's *Brave New World,* television has become the most influential medium of the second half of the twentieth century. While the medium continues to evolve and change, its importance, influence and pervasiveness appear to continue unabated. How will new technology change the face of television? Once the realm of science fiction, we are now seeing new delivery systems, on-call access, a greater number of available channels, two-way interaction, and the coupling of television and the computer. We are in the process of experiencing better technical quality including improved resolution, HDTV, the convenience of flatter and lighter television receivers, and digital processing and transmission. And yet, the basic standard for television broadcast technology has been with us, with only minor changes and improvements, for well over 50 years.

—Steve Runyon

FURTHER READING

Benson, K. Blair. *Television Engineering Handbook.* New York: McGraw-Hill, 1986.

Hartwig, Robert. *Basic TV Technology: A Media Manual.* Boston: Focal Press, 1990; 2nd edition, 1995.

Inglis, A.F. *Video Engineering.* White Plains, New York: Knowledge Industry Publications, 1992.

Mazda, Fraidoon, editor. *Telecommunications Engineer's Reference Book.* Stoneham, Massachusetts: Butterworth-Heinemann, 1993.

Shrader, Robert L. *Electronic Communication.* New York: McGraw-Hill, 1959; 6th edition, 1990.

Townsend, Boris, and Kenneth Jackson, editors. *TV and Video Engineer's Reference Book.* Stoneham, Massachusetts: Butterworth-Heinemann, 1991.

See also All Channel Law; Color Television; High-Definition Television; Low Power Television; Microwave; Steadicam; Translator; United States: Cable; United States: Networks; Video Editing; Videotape

TERRORISM

"Terroism" is a term that cannot be given a stable defintion; to do so forstalls any attempt to examine the major feature of its relation to television in the contemporary world. As the central public arena for organising ways of picturing and talking about social and political life, TV plays a pivotal role in the contest between competing defintions, accounts and explanations of terrorism.

Politicians frequently try to limit the terms of this competition by asserting the primacy of their preferred versions. Jeanne Kirkpatrick, former U.S. representative to the United Nations, for example, had no difficulty recognising "terrorism" when she saw it, arguing that "what the terrorist does is kill, maim, kidnap, torture. His victims may be schoolchildren.... industrialists returning home from work, political leaders or diplomats". Television journalists, in contrast, prefer to work with less elastic defintions. The BBC's News Guide, for example, advises reporters that "the best general rule" is to use the term "terrorist" when civilians are attacked and "guerrillas" when the targets are members of the official security forces.

Which term is used in any particular context is inextricably tied to judgements about the legitimacy of the action in question and of the political system against which it is directed. Terms like "guerrilla," "partisan," or "freedom fighter" carry positive connotations of a justified struggle against an occupying power or an oppressive state; to label an action as "terrorist" is to consign it to illegitimacy.

For most of the television age, from the end of World War II to the collapse of the Soviet Union, the deployment of positive and negative political labels was an integral part of Cold War politics and its dualistic view of the world. "Terrorism" was used extensively to characterise enemies of the United States and its allies, as in President Reagan's assertion in 1985, that Libya, Cuba, Nicaragua and North Korea constituted a "confederation of terrorist states" intent on undermining American attempts "to bring stable and democratic government" to the developing world. Conversely, friendly states, like Argentina, could wage a full scale internal war against "terrorism", using a defintion elastic enough to embrace almost anyone who criticised the regime or held unacceptable opinions, and attract comparatively little censure despite the fact that this wholesale use of state terror killed and maimed many more civilians than the more publicised incidents of "retail" terror—assasinations, kidnappings and bombings.

The relations between internal terrorism and the state raise particularly difficult questions for liberal democracies. By undermining the state's claim to a legitimate monopoly of force within its borders, acts of "retail" terror pose a clear threat to internal security. And, in the case of subnational and separatist movements which refuse to recognise the integrity of those borders, they directly challenge its political legitimacy. Faced with these challenges, liberal democracies have two choices. Either they can abide by their own declared principles, permit open political debate on the underlying causes and claims of terrorist movements, uphold the rule of law, and respond to insurgent violence through the proceedures of due process. Or they can curtail public debate and civil liberties in the name of effective security. The British state's response to the conflict in Northern Ireland, and to British television's attempts to cover it, illustrate this tension particularly well.

Television journalism in Britain has faced a particular problem in reporting "the Irish Question" since the Republican movement has adopted a dual strategy using both the ballot box and the bullet, pursuing its claim for the ultimate

The seige of the Munich Olympic Village, 1972
Photo courtesy of AP/Wide World Photos

reunification of Ireland electorally, through the legal political party, Sinn Fein, and militarily, through the campaign waged by the illegal Irish Republican Army. Added to which, the British state's response has been ambiguous. Ostensibly, as Prime Minister Thatcher argued in 1990, although "they are at war with us....we can only fight them with the civil law." Then Home Secretary Douglas Hurd admitted in 1989 that, in his view "with the Provisional IRA...it is nothing to do with a political cause any more. They are professional killers....No political solution will cope with that. They just have to be extirpated". Television journalists' attempts to explore these contradictions produced two of the bitterest peacetime confrontations between British broadcasters and the British state.

Soon after British troops were first sent to Northern Ireland in the early 1970s, there were suspicions that the due process of arrest and trial was being breached by a covert but officially sanctioned shoot-to-kill campaign against suspected members of Republican paramilitary groups. In 1988, three members of an IRA active service unit were shot dead by members of an elite British counter terrorist unit in Gibraltar. Contrary to the initial official statements, they were later found to be unarmed and not in the process of planting a car bomb as first claimed. One of the leading commercial television companies, Thames Television, produced a documentary entitled *Death on the Rock*, raising questions about the incident. It was greeted with a barrage of hostile criticism from leading Conservative politicians, including Prime Minister Thatcher. The tone of official condemnation was perfectly caught in an editorial headline in the country's best-selling daily paper *The Sun*, claiming that the programme was "just IRA propaganda."

The representation of the Provisional IRA was at the heart of the second major conflict, over a BBC documentary entitled *At the Edge of the Union*. This featured an extended profile of Martin McGuiness of Sinn Fein, widely thought to be a leading IRA executive responsible for planning bombings. The programme gave him space to explain his views and showed him in his local community and at home with his family. Then Home Secretary Leon Brittan (who had not seen the film) wrote to the chairman of the BBC's Board of Govenors urging them not to show it, arguing that "Even if [it] and any surrounding material were, as a whole, to present terrorist organisations in a wholly unfavourable light, I would still ask you not to permit it to be broadcast". The governors convened an emergency meeting and decided to cancel the scheduled screening. This very public vote of no confidence in the judgement of the corporation's senior editors and managers was unprecedented and was met with an equally unprecedented response from BBC journalists. They staged a one-day strike protesting against government interference with the Corporation's independence.

In his letter, Brittan had claimed that it was "damaging to security and therefore to the public interest to provide a boost to the morale of the terrorists and their apologists in this way." Refusing this conflation of "security" with the "public interest" is at the heart of television journalism's struggle to provide an adequate information base for a mature democracy. As the BBC's assistant director general put it in 1988, "it is necessary for the maintenance of democracy that unpopular, even dangerous, views are heard and thoroughly understood. The argument about the 'national interest' demanding censorship of such voices is glib and intrinsically dangerous. Who determines the 'national interest?' How far does the 'national interest' extend?" His argument was soundly rejected by the government. In the autumn of 1988, they instructed broadcasters not to transmit direct speech from members of eleven Irish organisations, including Sinn Fein. This ban has since been lifted, but its imposition illustrates the permanent potential for conflict between official conceptions of security and the national interest and broadcasters' desire to provide full information, rational debate and relevant contextualisation on areas of political controversy and dispute. As the BBC's former director general, Ian Trethowan, pointed out, the basic dilemma posed by television's treatment of terrorism is absolutely "central to the ordering of a civilised society: how to avoid encouraging terrorism and violence while keeping a free and democratic people properly informed."

Television's ability to strike this balance is not just a question for news, current affairs and documentary production however. The images and accounts of terrorism offered by televsion fiction and entertainment are also important in orchestrating the continual contest between the discourse of government and the state, the discourses of legitimated opposition groups, and the discourses of insurgent movements. This struggle is not simply for visibility—to be seen and heard. It is also for credibility—to have one's views discussed seriously and one's case examined with care. The communicative weapons in this battle are unevenly distributed however.

As the saturation coverage that the U.S. news media gave to the Shi'ite hijacking of a TWA passenger jet at Beirut in 1985 demonstrated very clearly, spectacular acts of retail terror can command a high degree of visibility. But the power to contextualise and to grant or withold legitimacy lies with the array of offical spokespeople who comment on the event and help construct its public meaning. As the American political scientist, David Paletz, has noted, because television news "generally ignores the motivations, objectives and long-term goals of violent organisations" it effectively prevents "their causes from gaining legitimacy with the public." This has led some commentators to speculate that exclusion from the general process of meaning making is likely to generate ever more spectacular acts designed to capitalise on the access provided by the highly visible propoganda of the deed.

Bernard Lewis, one of America's leading experts on the Arab world noted in his comments on the hijacking of the TWA airliner, that those who plotted the incident "knew that they could count on the American press and television to provide them with unlimited publicity and perhaps even some form of advocacy," but because the coverage ignored the political roots of the action in the complex power struggles within Shi'ite Islam, it did little to explain its causes or to foster informed debate on appropriate responses. As the televsion critic of the *Financial Times* of London, put it; "There is a criticism to be made of the coverage of these events, but it is not that television aided and abetted terrorists. On the contrary, it is that television failed to convey, or even to consider, the reasons for what President Reagan called 'ugly , vicious, evil terrorism.'"

News is a relatively closed form of television programming. It priviledges the views of spokespeople for governments and state agencies and generally organises stories to converge around officially sanctioned resolutions. Other programme forms, documentaries, for example, are potentially at least more open. They may allow a broader spectrum of perspectives into play, including those that voice alternative or oppositional viewpoints, they may stage debates and pose awkward questions rather than offering familiar answers. Television in a democratic society requires the greatest possible diversity of open programme forms if it is to address the issues raised by terrorism in the complexity they merit. Whether the emerging forces of technological change, in production and reception, channel proliferation, increased competition for audiences and transnational distribution, will advance or block this ideal is a question well worth examining.

—Graham Murdock

FURTHER READING

Alali, A. Odasu, and Gary W. Byrd. *Terrorism and the News Media: A Selected, Annotated Bibliography.* Jefferson, North Carolina: McFarland, 1994.

Alali, A. Odasu, and Kenoye Kelvin Eke, editors. *Media Coverage of Terrorism: Methods of Diffusion.* Newbury Park, California: Sage, 1991.

Alexander, Yonah, and Robert G. Picard, editors. *In the Camera's Eye: News Coverage of Terrorist Events.* Washington D.C.: Brassey's, 1991.

Dobkin, Bethami A. *Tales of Terror: Television News and the Construction of the Terrorist Threat.* New York: Praeger, 1992.

Livingston, Steven. *The Terrorism Spectacle.* Boulder, Colorado: Westview, 1994.

Miller, Abraham, editor. *Terrorism, the Media, and the Law.* Dobbs Ferry, New York: Transnational, 1982.

Nacos, Brigitte Lebens. *Terrorism and the Media: From the Iran Hostage Crisis to the World Trade Center Bombing.* New York: Columbia University Press, 1994.

O'Neill, Michael J. *Terrorist Spectaculars: Should TV Coverage Be Curbed.* New York: Priority Press, 1986.

Paletz, David L., and Alex Peter Schmid, editors. *Terrorism and the Media.* Newbury Park, California: Sage, 1991.

Picard, Robert G. *Media Portrayals of Terrorism: Functions and Meanings of News Coverage.* Ames: Iowa State University Press, 1993.

Schaffert, Richard W. *Media Coverage and Political Terrorists: A Quantitative Analysis.* New York: Praeger, 1992.

Signorielli, Nancy, and George Gerbner. *Violence and Terror in the Mass Media: An Annotated Bibliography.* New York: Greenwood Press, 1988.

Weimann, Gabriel, and Conrad Winn. *The Theater of Terror: Mass Media and International Terrorists.* New York: Longman, 1994.

See also *Death on the Rock*

THAT GIRL

U.S. Situation Comedy

That Girl was one of the first television shows to focus on the single working girl, predating CBS' *Mary Tyler Moore Show* by four years. This situation comedy followed heroine Ann Marie's adventures as she struggled to establish herself on the New York stage while supporting herself with a variety of temporary jobs.

That Girl was reputedly inspired by the life of its star, Marlo Thomas. The daughter of famous television comedian Danny Thomas wanted success on her own merits, so she moved to England where her father was unknown. After five years struggling, she won acclaim in Mike Nichol's 1965 London version of *Barefoot in the Park.* Returning home, she starred in an ABC pilot, *Two's Company,* about a model married to a photographer. Although it was not picked up, ABC head Ed Sherick offered Thomas other roles, including the lead in *My Mother, the Car.* She rejected these parts and instead approached the network with an idea for a show called *Miss Independence* centered on the life of a young, single career girl. ABC was interested, but wanted some kind of chaperone as a regular character.

Like *The Patty Duke Show, Peyton Place,* and *Gidget, That Girl* was one of many shows ABC targeted at the young, female audience during the mid- to late-1960s. The network had successfully turned to this up-and-coming demographic as early as 1963, capitalizing on the nascent women's movement and youth revolution. Like most of these shows, *That Girl* followed an already established trend, offering a diluted and sanitized version of the glamorized single girl lifestyle popularized by the likes of Helen Gurley Brown, Mary McCarthy, and Jacqueline Susann. Unlike their heroines, though, Ann Marie remained, at the behest of network standards and practices offices, chaste. The executives even wanted her to marry steady boyfriend, magazine executive, Don Hollinger (whom she met in the first episode) but Thomas resisted, consenting only to a September 1970 engagement.

While it focused on a self-supporting woman, *That Girl* did not center on the workplace (unlike *The Mary Tyler Moore Show*), largely because Ann's employment was essentially itinerant. Instead, her efforts to succeed revealed the

That Girl
Photo courtesy of Marlo Thomas

merger of public and private life. The erratic nature of her employment undermined everyday routines of working life, positioning her independence as highly precarious—particularly when contrasted to the steady rituals of Don's career. Ann's temporary jobs presented comedic opportunities as she struggled to retain her dignity in the face of often demeaning circumstances while foregrounding her continued reliance on her parents and Don. Female independence was thus presented as a site of struggle—both against the restrictions of the male-dominated workplace and the social and familial pressures for marriage. Meanwhile, her very choice of profession—the stage—undermined her desire for success, casting it in terms of fantasy. This lack of realism was evident from the start. Even Thomas noted that her struggling actress heroine never changed or developed. This refusal of change ultimately led to the show's 1971 cancellation: despite good ratings Thomas announced that she could not face playing the same character for eternity.

—Moya Luckett

CAST

Ann Marie	Marlo Thomas
Don Hollinger	Ted Bessell
Lou Marie	Lew Parker
Helen Marie (1966–70)	Rosemary DeCamp
Judy Bessemer (1966–67)	Bonnie Scott
Dr. Leon Bessemer (1966–67)	Dabney Coleman
Jules Benedict	Billie De Wolfe
Jerry Bauman	Bernie Kopell
Ruth Bauman (1967–69)	Carolyn Daniels
Ruth Bauman (1969–71)	Alice Borden
Harvey Peck (1966–67)	Ronnie Schell
George Lester (1966–67)	George Carlin
Seymour Schwimmer (1967–68)	Don Penny
Margie "Pete" Peterson (1967–68)	Ruth Buzzi
Mary	Reva Rose
Gloria	Bobo Lewis
Jonathan Adams	Forest Compton
Bert Hollinger	Frank Faylen
Mildred Hollinger	Mabel Albertson
Sandi Hollinger	Cloris Leachman
Nino	Gino Conforti
Mr. Brantano	Frank Puglia
Mrs. Brantano	Renata Vanni
Sandy Stone	Morty Gunty

PRODUCERS Bill Persky, Sam Denoff, Bernie Orenstein, Saul Turteltaub, Jerry Davis

PROGRAMMING HISTORY 136 Episodes

• ABC

September 1966–April 1967	Thursday 9:30-10:00
April 1967–January 1969	Thursday 9:00-9:30
February 1969–September 1970	Thursday 8:00-8:30
September 1970–September 1971	Friday 9:00-9:30

FURTHER READING

Douglas, Susan J. *Where the Girls Are: Growing Up Female with the Mass Media.* New York: Times Books, 1994.

See also *Mary Tyler Moore Show*

THAT WAS THE WEEK THAT WAS

British Satirical Review

The idea for *That Was the Week That Was* (which familiarly became known as *TW3*) came partly from the then director general of the BBC, Hugh Greene, who wanted to "prick the pomposity of public figures"—but it was the team of Ned Sherrin, Alasdair Milne and Donald Baverstock that was responsible for developing its successful format. The trio had previously worked on the BBC's daily early-evening news magazine show *Tonight* (1957–65—revived and revamped version, 1975–79) and the light-hearted style and wide-ranging brief of that show often allowed certain items to be covered in a tongue-in-cheek, irreverent or even satirical way. *TW3*, in its late-night Saturday slot, moved those elements a stage further, and, taking a lead from the increased liberalism of theatre and cinema in Britain, was able to discuss and disect the week's news and newsmakers using startlingly direct language and illustration. Whereas *Tonight* was gentle, *TW3* was savage, unflinching in its devotion to highlight cant and hypocrisy and seemingly fearless in its near libellous accusations and inuendos. It

became an influential, controversial and ground-breaking satire series, which pushed back the barriers of what was acceptable comment on television. Complaints poured in, but so did congratulations, and, despite enormous political pressure, Hugh Greene—determined in his quest to see a modern, harder BBC through the 1960s—stood by his brainchild.

Stylistically the show broke many rules: although it was commonplace on "live" shows of the 1950s (like the rock 'n' roll show *6-5 Special*) to see the cumbersome cameras being pushed from one set to the next, *TW3* went beyond that. A camera mounted high up in the studio would offer a bird's-eye view of the entire proceedings, showing the complete studio set-up with the flimsy sketch sets, the musicians, backroom personnel, the audience, other cameras, etc. It seemed to indicate that the viewing audience was to be treated as equals—that both creator and viewer knew it was a studio, knew the sketches weren't really set in a doctor's waiting-rooms but in a three walled mock-up, knew that

That Was the Week That Was

make-up girls would wait in the wings with powder and paint—so why hide it? The format of the show was simple, rigid enough to keep it all together, flexible enough to let items lengthen or shorten or disappear altogether, depending on time. Millicent Martin (the only permanent female member of the team) would sing the title song (music by Ron Grainer with Caryl Brahms providing a new set of lyrics each week relating to the news of the past few days) then David Frost, as host, would introduce the proceedings and act as link man between the items and often appearing throughout in sketches or giving monologues. (Originally John Bird was to be host but declined; Sherrin saw Frost at a club, doing an act where he gave a press conference as Prime Minister Harold MacMillan, and offered him the role of co-host with Brian Redhead who dropped out after doing the untransmitted pilot.) Bernard Levin interviewed people in the news or with strongly held views and his acid wit added an edge which occasionally produced flare-ups both verbal and physical. (A member of the studio audience once punched him, rather ineffectually, following a scathing review he had written.) Lance Percival acted in sketches and

sang topical calypsos (a device used on *Tonight*) many of which were ad libbed. David Kernan was a resident singer whose strength was his ability to parody other singers and styles, Timothy Birdsall drew cartoons, Al Mancini pulled faces and the engine room was provided by Willie Rushton, Kenneth Cope and Roy Kinnear who fleshed out the sketches and comic chatter. The show occassionally featured guest artistes, most famously comedian Frankie Howerd whose popularity had waned somewhat. His one appearance on *TW3* managed to dramatically resurrect his career, as his humour seemed to work for both traditionalists and this new, younger, harder generation.

The writing credits for the show read like a Who's Who of the sharp young talent of the time: John Albery, John Antrobus, Christopher Booker, Malcolm Bradbury, John Braine, Quentin Crewe, Brian Glanville, Gerald Kaufman, Herbert Kretzmer, David Nathan and Dennis Potter, David Nobbs, Peter Shaffer, Kenneth Tynan, Stephen Vinaver, Keith Waterhouse and Willis Hall—plus contributions from the show's creative staff: Sherrin, Frost, and Levin.

Memorable moments from the series include Gerald Kaufman's list of silent MPs which highlighted politicians who hadn't spoken in the House of Commons in ten or fifteen years. The sketch caused a furore when it was read out by the team, despite the fact that the information was readily available. Kenneth Cope's "confession" monologue (written by John Braine) featured a figure, hidden in shadows, who confesses to being heterosexual and relates the misery it can cause. Frost's scathing profile of Home Secretary Henry Brooke insinuated, amongst other things, that his intractability in an immigration case had led to the murder of the subject. Millicent Martin sang with black-faced minstrels about racism in the Southern States. And most memorable of all was the truly serious edition immediately following President Kennedy's assasination. The whole show was given over to the subject, tackling the shock felt and the implications of the shooting with rare solemnity and dignity. (That episode was lodged at the Smithsonian Institute.)

A U.S. version of the series (also featuring Frost) debuted 10 January 1964 on NBC and ran until May 1965. Singer Nancy Ames took the Millicent Martin role and Buck Henry, Pat Englund and Alan Alda were among the regulars. The show proved equally groundbreaking in the United States, and, like the British version, was no stranger to controversy.

—Dick Fiddy

CAST

David Frost
Millicent Martin
Bernard Levin
Lance Percival

Roy Kinnear
William Rushton
Timothy Birdsall
John Wells
Kenneth Cope
David Kernan
Al Mancini
John Bird
Eleanor Bron
Roy Hudd

PRODUCER   Ned Sherrin

PROGRAMMING HISTORY   36 50-Minute Episodes; 1 150-Minute Special; 1 100-Minute Special

• BBC

29 September 1962	150 Minute Special
24 November 1962–27 April 1963	23 Episodes
28 September 1963–21 December 1963	13 Episodes
28 December 1963	100 Minute Special

FURTHER READING

Campey, George, J.T. Archer, and Ian Coates. *The BBC Book of That Was the Week That Was.* London: British Broadcasting Corporation, 1963.
Frost, David. *That Was the Week That Was.* London: W.H. Allen, 1963.

See also British Programming; Frost, David

THAW, JOHN

British Actor

A versatile and successful British actor, John Thaw has worked in television, theatre, and cinema. But the small screen has guaranteed him almost continual employment throughout his exceptional career.

After training at the Royal Academy of Dramatic Art, at a 1960 stage debut he was "discovered" and promoted by Granada TV. His first TV outing was in 1961; since then he has taken the lead role in an impressive array of series. He has had parts ranging from *The Avengers* to *Z Cars*, and the lead in the series *Redcap* before his big break in *The Sweeney* (1974-78), a landmark in the police-action genre. Thaw played rough-mannered detective Jack Regan of the Flying Squad. *The Sweeney* was described as a U.S.-influenced imitation of West Coast shows, and was prominent in debates about the levels of violence and bad language on television, criticised for glamorising guns and car chases. Its superiority over standard violent fare, however, owed much to Thaw's performance, along with the growing rapport between his and Dennis Waterman's characters and the show's constant originality.

For years after *The Sweeney*, Thaw found it difficult to throw off the Jack Regan image, but in 1987 he began another long-running detective series for which he is perhaps best known. *Inspector Morse* was remarkably popular with critics and audiences internationally. Its ITV ratings in Britain were second only to those of *Coronation Street*. Again, the show owed much of its success to Thaw's central BAFTA-winning performance. He holds together Morse's eccentricities—the irascible, world-weary, and introspective crossword and classical music lover. Julian Mitchell, writer of several episodes of *Morse*, sees Thaw as the consummate TV actor: "His technique is perfect, and by seeming to do very little he conveys so much." In this way he suggests hidden depths to Morse, and conveys his troubled morality. The tranquility and gentle English manner associated with *Morse* are a far cry from *The Sweeney*. It has gained fans as an antidote to violent American TV, and has established Thaw as the thinking woman's crumpet.

Audiences are accustomed to Thaw's downbeat manner in gloomy roles, but he claims to prefer doing comedy. He played the lead in the sitcom *Home to Roost*, appeared with *Sweeney* partner Waterman in the 1976 *Morecambe and Wise Christmas Show*, and starred in the widely derided *A Year in Provence*, which lost a record ten million viewers during one series.

Despite this hiccup, Thaw remained a very bankable star. *Kavanagh QC*, a part written especially for him after *Provence*, was another big hit. He was back on familiar territory as a barrister reconciling principle and his working-class roots with a lucrative law practice.

Thaw sees himself as a "jobbing actor, no different from a plumber." Part of his success may be his ability to play everyman roles that people can relate to easily. Despite a distinctly unclassical repertoire, he has continued to act on stage whenever his busy TV career has allowed, latterly in "special guest star" roles. He has also appeared in several feature films, including two *Sweeney* films, and *Cry Freedom*.

—Guy Jowett

JOHN THAW. Born in Manchester, Lancashire, England, 3 January 1942. Attended Dulcie Technical High School, Manchester; Royal Academy of Dramatic Art (Vanbrugh Award, Liverpool Playhouse Award). Married: 1) Sally Alexander (divorced); child: Abigail; 2) Sheila Hancock, 1973; children: Joanna and stepdaughter Melanie. Stage debut, Liverpool Playhouse, 1960; London debut, Royal Court Theatre, 1961; became widely familiar to television audiences in *The Sweeney* and subsequently as star of the *Inspector Morse* series. Companion of the Order of the British Empire, 1993. Recipient: British Academy of Film and Television Arts Award for Best Television Actor, 1993. Address: John Redway Associates, 5 Denmark Street, London WC2H 8LP, England.

TELEVISION SERIES

1965–66	*Redcap*
1974	*Thick As Thieves*
1974–78	*The Sweeney*
1983	*Mitch*
1985–89	*Home to Roost*
1987–93	*Inspector Morse*
1991	*Stanley and the Women*
1992	*A Year in Provence*
1995–	*Kavanagh QC*

MADE-FOR-TELEVISION MOVIE

1981	*Drake's Venture*

TELEVISION SPECIALS

1974	*Regan*
1984	*The Life and Death of King John*
1992	*Bomber Harris*
1993	*The Mystery of Morse*
1994	*The Absence of War*

John Thaw

FILMS

Nil Carborundum, 1962; *The Loneliness of the Long Distance Runner*, 1962; *Five to One*, 1963; *Dead Man's Chest*, 1965; *The Bofors Gun*, 1968; *Praise Marx and Pass the Ammunition*, 1970; *The Last Grenade*, 1970; *The Abominable Dr. Phibes*, 1971; *Dr. Phibes Rises Again*, 1972; *The Sensible Action of Lieutenant Holst*, 1976; *The Sweeney*, 1977; *The Sweeney II*, 1978; *Dinner at the Sporting Club*, 1978; *The Grass is Singing*, 1981; *Asking for Trouble*, 1987; *Business As Usual*, 1987; *Cry Freedom*, 1987; *Charlie*, 1992.

STAGE

A Shred of Evidence, 1960; *The Fire Raisers*, 1961; *Women Beware Women*, 1962; *Semi-Detached*, 1962; *So What About Love?*, 1969; *Random Happenings in the Hebrides*, 1970; *The Lady from the Sea*, 1971; *Collaborators*, 1973; *Absurd Person Singular*, 1976; *Night and Day*, 1978; *Sergeant Musgrave's Dance*, 1982; *Twelfth Night*, 1983; *The Time of Your Life*, 1983; *Henry VIII*, 1983; *Pygmalion*, 1984; *All My Sons*, 1988; *The Absence of War*, 1993.

See also *Sweeney*

THIRTYSOMETHING

U.S. Drama

Winner of an Emmy for best dramatic series in 1988, *thirtysomething* (ABC, 1987–1991) represented a new kind of hour-long drama, a series which focused on the domestic and professional lives of a group of young urban professionals—a socio-economic category of increasing interest to the television industry. The series attracted a cult audience of viewers who strongly identified with one or more of its eight central characters, a circle of friends living in Philadelphia. And its stylistic and story-line innovations led critics to respect it for being "as close to the level of an art form as weekly television ever gets," as *The New York Times* put it. When the series was canceled due to poor ratings, a *Newsweek* eulogy reflected the baby boomers' sense of losing a rendezvous with their mirrored lifestyle: "the value of the Tuesday night meetings was that art, even on the small screen, reflected our lives back at us to be considered as new." Hostile critics, on the other hand, were relieved that the self-indulgent whines of yuppiedom had finally been banished from the schedules.

The show *thirtysomething* spearheaded ABC's drive to reach a demographically younger and culturally more capital-rich audience. Cover stories in *Rolling Stone* and *Entertainment Weekly* explored the parallels between the actors' and characters' lives, as well as the rapport generated with the audience, who were seen as sharing their inner conflicts. Michael Steadman, an advertising copywriter struggling with the claims of his liberal Jewish background, and his wife Hope, a part-time social worker and full-time mother are the "settled" couple. The Steadmans were offset against Elliot, a not-really-grown-up graphic artist who was Michael's best friend at Penn, and his long-suffering wife Nancy, an illustrator who separated from him and developed breast cancer in subsequent seasons. Three unmarried friends also date back from college days: Ellyn is a career executive in city government; Gary teaches English at a liberal arts college; and Melissa, a freelance photographer, is Michael's cousin. While the two couples wrestle with their marriages and raising their children, the three others have a series of love affairs with outsiders to the circle. For Gary, after a quasi-incestuous relation with Melissa, fate holds a child out of wedlock with temperamental feminist Susanna, the college's denial of tenure, his life as a househusband, and finally—in one of the most publicized episodes—sudden death in a cycling accident.

The title, referring to the age of the characters, was written as one word (togetherness) and in lower case (ee cummings and the refusal of authority). "Real life is an acquired taste" was the network promo for the series, as its makers explored the boundaries between soap operatics and verisimilitude, between melodrama and realism. Co-creators Edward Zwick and Marshall Herskovitz (who had met at the American Film Institute) claimed a "mandate of small moments examined closely", dealing with "worlds of incre-

thirtysomething

mental change", loosely modeled on their own lives and those of their friends. Central to their sense of this fictional world was a high degree of self-consciousness and media awareness. "Very *Big Chill*", as one character put it, referring to Lawrence Kasdan's 1983 film. The movie was often seen as a progenitor of the series, defining a generation through their nostalgia for their fancy-free days before adulthood. *The Big Chill* focus on a "reunion of friends" in turn refers to the small budget *Return of the Secaucus Seven* made by John Sayles in 1980. And yet another cinematic touchstone for the ciné-literate makers was *It's a Wonderful Life* (Capra, 1946), the perennial favorite of American movie-goers, to which homage was paid in the production company ("Bedford Falls") logo. Capra's political liberalism emerged in the series in the distaste for patriarchal and capitalist power (embodied in Miles, the ruthless CEO of the advertising company), while the film aesthetic carried over into the cinematography, intertextual references and ambitious story-lines, which occasionally incorporated flashback, daydream and fantasy sequences. This complex mixture of cinematic and cultural antecedents can be summed by suggesting that in many ways *thirtysomething's* four seasons brought the sophistication of Woody Allen's films to the small screen.

Although in the vanguard for centering on "new" (post-feminist) men, for privileging "female truth," and dealing with touchy issues within sexual relations, with disease and death, the series never really challenged gender roles. While the problem of the domestication of men, of defining them within a familial role without lessening their desirability and their sense of self-fulfillment was one of its key preoccupations, in the end the traditional sexual division of labor was ratified. Although it was the first series to show a homosexual couple in bed together, it posed very gingerly any alternative to the heterosexual couple. Nevertheless, the prominence of a therapeutic discourse, the negotiation of identity in our postmodern era, won it an accolade from professional psychologists.

The series was occasionally criticized, too, for its social and political insularity, for not dealing with problems outside the affluent lifestyle and 1960s values of its characters. Zwick and Herskovitz described it as "a show about creating your own family. All these people live apart from where they grew up, and so they're trying to fashion a new sense of home—one made up of friends, where holidays, job triumphs, illnesses, and gossip all take on a kind of bittersweet significance."

The series' influence was evident long after it moved to syndication on the Lifetime cable network and its creators moved on to feature film careers. That influence was evident in everything from the look and sound of certain TV advertisements, to other series with feminine sensibilities and preoccupations with the transition from childhood to maturity (*Sisters*), to situation comedies about groups of friends who talk all the time (*Seinfeld*). *My So-Called Life* (ABC, 1994), a later and less successful series produced by many of the same personnel, even extended the subjectivity principle to a teenage girl caught between her family and school friends. That series was perhaps an indication of a new shift in the targeting of "generational audiences," the new focus now on "twentysomethings", as television searched for a way to reach the offspring of the baby boomers.

—Susan Emmanuel

CAST

Michael Steadman . . . . Ken Olin

Hope Murdoch Steadman . . . . Mel Harris
Janey Steadman . . . . Brittany and Lacey Craven
Elliot Weston . . . . Timothy Busfield
Nancy Weston . . . . Patricia Wettig
Ethan Weston . . . . Luke Rossi
Brittany Weston . . . . Jordana "Bink" Shapiro
Melissa Steadman . . . . Melanie Mayron
Ellyn . . . . Polly Draper
Professor Gary Shepherd . . . . Peter Horton
Miles Drentell (1989–91) . . . . David Clennon
Susannah Hart (1989–91) . . . . . . Patricia Kalember
Billy Sidel (1990–91) . . . . Erich Anderson

PRODUCERS Edward Zwick, Marshall Herkovitz, Scott Winant

PROGRAMMING HISTORY 85 Episodes

• ABC

September 1987–September 1988   Tuesday 10:00-11:00
December 1988–May 1991   Tuesday l0:00-11:00
July 1991–September 1991   Tuesday 10:00-11:00

FURTHER READING

Heide, Margaret J. "Mothering Ambivalence: The Treatment of Women's Gender Role Conflicts Over Work and Family on *thirtysomething*." *Women's Studies: An Interdisciplinary Journal* (Claremont, California), 1992.

Joyrich, Lynne. "All That Heaven Allows: TV Melodrama, Postmodernism And Consumer Culture." *Camera Obscura* (Berkeley, California), January 1988.

"*thirtysomething*: A Chronicle of Everyday Life." *New York Times*, 24 February 1988.

Torres, Sasha. "Melodrama, Masculinity and the Family: *thirtysomething* as Therapy." In, Penley, Constance, and Sharon Willis, editors. *Male Trouble*. Minneapolis: University of Minnesota Press, 1993.

See also Family on Television; Gender and Television; Melodrama

THIS HOUR HAS SEVEN DAYS

Canadian Public Affairs Series

This Hour Has Seven Days has repeatedly been cited as the most exciting and innovative public affairs television series in the history of Canadian broadcasting. It was certainly the most popular, drawing more than three million viewers at the time of its controversial cancellation by CBC management, which was unable to withstand the cries of outrage from offended guardians of public morality and the growing insurgence of the *Seven Days* production team. The creation of two young producers, Patrick Watson and Douglas Leiterman, the series debuted on 4 October 1964 and

came to its well-publicized end after 50 episodes on 8 May 1966.

Watson and Leiterman had worked together as co-producers on two previous public affairs series, *Close-Up* and *Inquiry*. Given the go-ahead to create a new public affairs series, they envisioned a show that would be stimulating and exciting for the Canadian public, and that would develop a wider and more informed audience than previous public affairs shows. Both producers were deeply committed to the importance of public service broadcasting and to the importance of pushing the boundaries of television journalism to reflect the techniques of investigation and advocacy more prevalent in print journalism. Leiterman in particular argued against the prevailing ideology of CBC journalistic practice that called for adhering to the strict tenets of objectivity and "studious neutrality." Watson brought a more intellectual approach to the show, having studied English literature and linguistics in undergraduate and graduate school.

The show was launched with great fanfare in the fall of 1964 with a relatively large budget by the CBC of over $30,000 per show, about twice the average of other public affairs programs. The first year's shows were co-hosted by John Drainie, Laurier LaPierre, an academic historian turned TV talent, and Carole Simpson, soon replaced by Dinah Christie. The role of the women was limited primarily to songs or satire. Upon Drainie's illness at the start of the second year, Watson was persuaded to abandon his producer role to join the on-air team in a move that CBC management thought would reduce the controversial style of the program. A very talented and energetic young team of producers, reporters, interviewers, and filmmakers was recruited. They included some of the prime future talent in Canadian documentary film and television, such as Beryl Fox, Donald Brittain, Allan King, Daryl Duke, Peter Pearson, Alexander Ross, and Larry Zolf.

Clearly inspired by the earlier British satirical review of the news, *That Was the Week That Was, Seven Days* utilized a one-hour, magazine format that combined satirical songs and skits with aggressive "bear pit" style interviews, investigative reports and mini-documentaries. On an irregular basis the entire show would be devoted to an in-depth documentary film under the title "Document." Several important award-winning films were produced and shown. One of the most noted was Beryl Fox's "Mills of the Gods," a moving examination of life for American soldiers and Vietnam peasants during the Vietnam War. A distinct point of view, which was new to public affairs TV, was often very present in these productions.

A concrete example of one show's line-up might best illustrate the basic elements of the magazine format and explain why the series made CBC executives nervous and upset the more traditional journalists and members of the public. The episode for 24 October 1965 opened with a satirical and irreverent song by Christie about the Ku Klux Klan, followed by preview cuts of later show segments, credits and a welcome of the live studio audience by LaPierre. (Live audiences were a staple of the program, contributing to its actuality impact.) The first story was a filmed report on the funeral for a Sudbury, Ontario, policeman, including an interview with his family and a colleague. It underscored the important role of the unrecognized policemen across Canada. The second story focused on the current Federal election featuring sometimes irreverent street interviews from Toronto and Vancouver, and finishing with a shot of an empty chair and the question of whether the party leaders will show up to be questioned. The next segment was a satirical sketch portraying Harold Wilson, then prime minister of England, in conversation with Lester Pearson, then prime minister of Canada running for reelection. The fourth story was

This Hour Has Seven Days

a short feature on *Penthouse* magazine with pictures, interviews with the publisher and two British clergy, and commentary about the objectification of women. The fifth story was an on-location interview of Orson Welles by Watson. The sixth story was a filmed, almost lyrical, portrait of the Canadian boxer George Chuvalo. Running almost twenty-two-and-a-half minutes was the final story on the Ku Klux Klan (KKK). After an introduction by Christie, a satire of the Ku Klux Klan appearance before the U.S. House Un-American Activities Committee, and a short film of the civil rights struggle in the United States, two members of the Klan were invited into the "hot seat" to be interviewed in full costume. About halfway through the interview and after a question as to whether the Klansmen would shake hands with a black man, a black civil rights leader from the United States was invited to join the interview. There was some exchange of views, then the interviewer tried to get the KKK members to shake hands with the black leader, at which time they stood up and left the set. The show closed with a request for feedback and a reprise of the Christie song.

The fast pace, the topicality of many of the segments, the portrayal and incitement of conflict, the irreverence of songs and skits, and the occasional emotionalism of the on-air team members, all added to the popularity and the controversy that built around *Seven Days*. LaPierre was once shown wiping away a tear after one filmed interview—a gesture that the then CBC President Ouimet remembered angrily years later as one more affront to appropriate journalistic practice. The production team was proud of its non-traditional approaches to portraying the news, selecting guests, and even the way it gathered material for the show. At different times "regular" journalists accused *Seven Days* reporters of stealing material or of poaching on their territory. One of the final straws for the program was going behind the scencs of a "Miss Canada Pageant" to film and interview contestants in their hotel rooms and bedclothes, despitc the fact that the rival CTV network had an exclusive coverage contract with the pageant. This and other journalistic "improprieties" led to a memo from Bud Walker, vice president of the CBC that foreshadowed the demise of the series.

The cancellation of *Seven Days* and the firing of Watson and LaPierre in the spring of 1966 (Leiterman was later forced out) was met with a large public outcry, probably the largest in Canadian history for any TV program, and certainly for any public affairs program. Partly orchestrated by Watson, Leiterman and LaPierre, there were public demonstrations, thousands of letters and phone calls, indignant editorials, threats to resign by CBC staff, and calls for Parliamentary inquiries. As a result, a Parliamentary committee hearing that favorably featured the *Seven Days* team stretched over several weeks. Prime Minister Pearson appointed a special investigator, which kept the program in the news for several more weeks. The final reports seemed to chastise both sides in the dispute, but was harshest with the CBC for its heavy handedness and bureaucratic timidity. Watson, Leiterman and LaPierre were public heroes for a time. Several members of management resigned, at least two in protest at the handling of the show and its principals. Vice president Walker lost his job ostensibly for the way he handled the dispute but also as a demonstration to politicians that the CBC had gotten the message.

Despite its non-traditional approaches, *Seven Days* usually dealt with mainstream concerns and issues, taking a slightly left-leaning perspective on social issues. It might have challenged members of the Canadian elite but it rarely went outside the frame of dominant beliefs. It was often creative in the way that it visualized stories originating in studio, considering the available technology; further, it imaginatively took advantage of the recent breakthroughs in hand-held cameras and portable sound recording in its filmed stories and documentaries. Watson, Leiterman, and the *Seven Days* team often seemed to achieve the goal of involving the viewer in the emotion and actuality of television while innovating on and stretching the conventions of TV journalism. It is also clear that the team was often seduced by the power of television to embarrass guests or sensationalize issues through manipulative set-ups like the KKK interview. The series often entertained, perhaps more than it informed, foreshadowing the current concern and debate over the line between news and entertainment. While the program demonstrated ways to attract, provoke, and stimulate a mass audience for current affairs, the conflict and ultimate sanction that resulted made it difficult for television journalists to experiment or take on controversial issues for several years after. In the years since *Seven Days* aired it has taken on the mythic mantle of "that was the way it was in the good old days" of Canadian TV journalism. While much of that reputation is deserved, the series also needs to be appreciated with a critical eye and ear.

—William O. Gilsdorf

HOSTS
Laurier LaPierre, John Drainie, Patrick Watson, Dinah Christie, Carole Simpson, and others

PRODUCERS   Patrick Watson, Douglas Leiterman, Bill Hogg, Reeves Haggan, Hugh Gauntlett, Robert Hoyt, Ken Lefolii

DIRECTOR
David Rushkin

PROGRAMMING HISTORY

• CBC

| 4 October 1964–8 May 1966 | Sunday 10:00-11:00 |

FURTHER READING

Koch, Eric. *Inside Seven Days: The Show that Shook the Nation.* Scarborough, Ontario: Prentice-Hall, 1986.

Nash, Knowlton. *The Microphone Wars: A History of Triumph and Betrayal at the CBC.* Toronto: McClelland and Stewart, 1994.

Peers, Frank W. *The Public Eye: Television and the Politics of Canadian Broadcasting, 1952–1968.* Toronto: University of Toronto Press, 1979.

Rutherford, Paul. *Prime Time Canada.* Toronto: University of Toronto Press, 1990.

Stewart, Sandy. *Here's Looking at Us.* Toronto: CBC Enterprises, 1986.

See also Canadian Programming in English; Watson, Patrick

THIS IS YOUR LIFE

U.S. Biography Program

This Is Your Life, which was broadcast from 1952 to 1961, is one of the best remembered television series from the 1950s. The format of *This Is Your Life* was based on a rather simple principle—guests were surprised with a presentation of their past life in the form of a narrative read by host Ralph Edwards and reminiscences by relatives and friends. But the format was also quite shrewd in its exploitation of television's capacity for forging intimacy with viewers through live transmission and on-air displays of sentimentality.

This Is Your Life was the creation of host Ralph Edwards, who was also the host of radio's popular *Truth or*

Consequences. In a 1946 radio broadcast of the latter program, Edwards presented a capsule narrative of the past life of a disabled World War II veteran who was having difficulties adjusting to post-war life. Edwards received such positive feedback from this show that he developed the formula for a separate radio program called *This Is Your Life.* It began airing on radio in 1948, and became a live television program in 1952, running on the NBC network until 1961, and reappearing in syndicated versions briefly in the early 1970s and 1980s (during this last period, it was hosted by actor Joseph Campanella).

In its network television years, *This Is Your Life* alternated in presenting the life stories of entertainment personalities and "ordinary" people who had contributed in some way to their communities. Edwards always insisted that the theme of "Love thy neighbor" was clear no matter who was the subject of a particular program. The host was often quoted as saying that the lives under examination must represent something "constructive," must have been "given a lift 'above and beyond the call of duty' and. . .in turn, he or she has passed on the help to another." For that reason, the emotion expressed by the guest, who having first been surprised by Edwards with the on-air announcement "this is your life!" and then with the appearance of people from his or her past, was justified as a source for audience inspiration rather than voyeurism.

Entertainment personalities who were subjects of the program ranged from broadcast journalist Lowell Thomas (who displayed obvious anger and embarrassment over the "surprise") to singer Nat "King" Cole, from the famous silent film star Gloria Swanson to contemporary movie favorite Debbie Reynolds. While Edwards claimed that there were few "leaks" to the subjects about the show (if there were leaks, that subject was immediately dropped), there were several notable occasions when guests were informed in advance of their tributes—for example, Eddie Cantor was told because his heart trouble worried producers regarding the show's "surprise factor," and singer-actress Lillian Roth and actress Frances Farmer were told because their well-known troubled pasts were considered subjects too delicate (and perhaps unpredictable) for the program's usual spectacle of surprise.

When *This Is Your Life* reviewed the lives of "ordinary people," Edwards and the show staff relied on help from the individual's community. In some ways the program's cover-age of individuals who achieved despite handicaps was ahead of its time when indicating how the subject had surmounted societal bigotry. Not surprisingly, the show shared with its times a Cold War fervor towards conformity and patriotism. For example, in a 1958 program featuring a Japanese-American druggist who had been sent to an internment camp during World War II, the life narrative recounts his struggle to establish a pharmacy practice in a bigoted community. But Edwards praises the subject's behavior in the internment camp when he squelched a camp uprising protesting forced labor. At the end of the show, members from his most recent community embrace him and Edwards announces that Richard Nixon has donated an American flag and Ivory soap has donated money for a flag pole for the town which has overcome racial prejudice.

In the late 1980s, Edwards and his production company made many of the episodes featuring Hollywood celebrities available for re-broadcasting. American Movie Classics cable network channel aired these for several years to accompany screenings of movies from studio-era Hollywood.

—Mary Desjardins

HOST
Ralph Edwards

ANNOUNCER
Bob Warren

PRODUCERS   Axel Greenberg, Al Pascholl, Richard Gottlieb, Bill Carruthers, Jim Washburn

PROGRAMMING HISTORY

• NBC

October 1952–June 1953	Wednesday 10:00-10:30
June 1953–August 1953	Tuesday 9:30-10:00
July 1953–June 1958	Wednesday 9:30-10:00
September 1958–September 1960	Wednesday 10:00-10:30
September 1960–September 1961	Sunday 10:30-11:00

FURTHER READING

Balling, Fredda. "The World is His Neighbor." *TV-Radio Mirror* (New York), June 1959.
Hall, Gladys. "Four Magic Words." *TV-Radio Mirror* (New York), 1954.

THOMAS, DANNY

U.S. Comedian/Actor

Danny Thomas was one of television's most beloved and enduring entertainers. His comedic talents were surpassed only by his shrewd production activities and his well-known philanthropy. Thomas began his career as the stand-up comic Amos Jacobs, developing his story-telling shtick into a familiar routine of lengthy narratives peppered with a blend of Irish, Yiddish, Lebanese and Italian witticisms. Quite often these routines tended toward sentimentality, only to be rescued in the end by what Thomas called the "treacle cutter," a one-liner designed to elevate the maudlin bathos into irony.

Like many early television comics, Thomas developed his routines touring in a variety of clubs. Restricted mostly to his home environs of the midwest, he secured a three-year deal at Chicago's 5100 Club, where he was spotted by the powerful head of the William Morris Agency. "Uncle" Abe Lastfogel was to become Danny's mentor, overseeing his New York night-club appearances, arranging a USO tour for him with Marlene Dietrich and landing him a part on Fanny Brice's radio show. By 1945, he was declared "best new-comer in radio" by the trade papers and Joe Pasternak cast him in his film, *The Unfinished Dance*. Refusing the advice of three different studio heads to surgically alter his trade-mark nose, Thomas' film career was short-lived, but fairly respectable. In the early 1950s, he left the film industry to good reviews for his title role appearance in the 1951 Warner Brothers release of *The Jazz Singer*, and his co-starring role in the Doris Day vehicle *I'll See You in My Dreams.*

Meanwhile, tired of the night-club circuit, Thomas was anxiously pursuing a television series. His first television appearance was on NBC's *Four Star Revue*, where he co-starred with Jimmy Durante, Jack Carson and Ed Wynn. The variety show format, with its fast-paced, three-minute sketches, was ill-suited to Danny's comedic style which depended upon expository monologues and lengthy narra-tives. For the series' second season, the network ordered a format change wherein the four rotating hosts were replaced by a procession of headliners. With all but Ed Wynn's departure, the program became the *All Star Revue.*

Thomas obtained his own program when agent Abe Lastfogel pressured fledgling network ABC into accepting Thomas as part of their terms for acquiring the much-cov-eted Ray Bolger. ABC, familiar with Thomas' previously ill-received television performances, insisted upon a sitcom. It was during a prolonged brainstorming session with pro-ducer Lou Edleman and writer Mel Shavelson that Thomas inadvertently came up with the autobiographical premise that was to become *Make Room for Daddy*. As the three worked futilely into the night, Thomas grew impatient and pleaded that he simply wanted a series so that he could stay put with his family for awhile. The result was *Make Room for Daddy*, a show which revolved around the absentee-fa-ther dilemmas of traveling singer-comic "Danny Williams." The title was suggested by Thomas' real-life wife, Rose Marie, who, during Danny's frequent tours, allowed their children to sleep with her. Upon her husband's return, the children would have to empty dresser drawers and leave the master bed to, quite literally, "make room for Daddy."

Incorporating Thomas' singing and story-telling tal-ents, the program was a blend of domestic comedy and variety program (during Danny's fictionalized "nightclub engagements"). It became one of television's most successful comedies, winning numerous awards, including best new show for the 1952–53 season. Despite its success, the pro-gram underwent a number of transformations, most notably when Jean Hagen, who played the part of wife Margaret, left the series to attend to her film and stage careers. For the

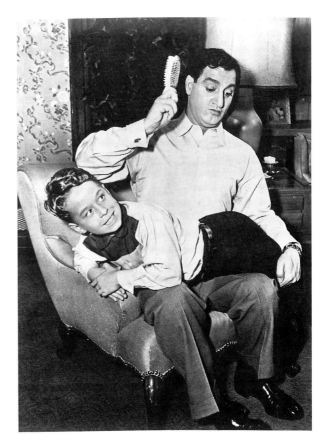

Danny Thomas with Rusty Hamer

fourth season, Danny played a widower, and a succession of guest stars appeared as potential replacement wives. In the 1956 season finale, Danny proposed to guest star Marjorie Lord who, along with child star Angela Cartwright, joined the Williams' family for the program's remaining seven years. The start of the 1957 television season also saw the program on a new network (CBS) when ABC president (and Hagen ally) Robert Kintner lost interest in the series. The newly titled *Danny Thomas Show* slid into the spot formally occupied by CBS' mega-hit *I Love Lucy*, where it remained in the top ten until voluntarily leaving the network when the performers sought new avenues of creative expression.

While starring in *Make Room for Daddy*, Thomas met Sheldon Leonard, a former gangster-type actor with aspira-tions for directing. Leonard took over as director for the program midway into its first season, eventually becoming executive producer. Together, Thomas and Leonard estab-lished Thomas-Leonard Productions, a powerhouse produc-tion company, based on the Desilu lot that was responsible for a multitude of successful series, including *The Real McCoys, The Andy Griffith Show, The Joey Bishop Show, The Bill Dana Show* and *The Dick Van Dyke Show*. In 1965 when Leonard left to develop *I Spy*, Thomas continued indepen-dently, producing *The Danny Thomas Hour*, an anthology series for NBC, and joining with Aaron Spelling to create and produce *The Mod Squad* and other programs. While a

1967 attempt to buy Desilu from Lucille Ball was unsuccessful, Thomas continued to create and produce programs under the banner of Danny Thomas Productions.

Thomas had an enormous positive impact upon the growing medium. The off-camera stand-up routines he performed for the in-studio audience just prior to filming each episode of *Make Room for Daddy* were imitated on other programs and institutionalized as the now commonplace "warm-up." *The Andy Griffith Show* was the first real spin-off for network television, originating in a 1960 episode of *The Danny Thomas Show*. As a producer Thomas read scripts and supervised a plethora of top rated programs and was personally responsible for casting Mary Tyler Moore as Laura Petrie in *The Dick Van Dyke Show*. His influence as producer continued not only in his own projects but through the work of his children, notably daughter Marlo, who became a renowned actress-producer-director, and his son Tony, who with partners Susan Harris and Paul Junger Witt is responsible for a veritable catalogue of 1970s and 1980s hit programs, including *Soap* and *The Golden Girls*.

Thomas' personal integrity is as well known as his acting and producing talents. In the 1950s he successfully protected two blacklisted writers who continued to write for his television series under assumed names. And, in 1983, he was rewarded with the Congressional Medal of Honor for his work in establishing the St. Jude's Children's Research Hospital, a cause he continued to promote and support until his death in 1991.

—Nina C. Leibman

DANNY THOMAS. Born Muzyad Yakhoob in Deerfield, Michigan, U.S.A., 6 January 1914. Married: Rose Marie Cassaniti, 1936; two daughters and one son. Began career in radio, Detroit, 1934; worked as master of ceremonies in night club, 1938–40; appeared on Chicago radio, 1940; worked as master of ceremonies, 5100 Club, Chicago, 1940–43; developed own radio, television programs, performed in clubs and theaters worldwide, through 1940s; performed overseas during World War II with Marlene Dietrich and company, and solo; performed with Fanny Brice on radio, 1944; made motion picture debut in *The Unfinished Dance*, 1946; starred in long-running television series, *Make Room for Daddy*; produced successful television series such as *The Dick Van Dyke Show*. Honorary degrees: L.H.D., Belmont Abbey, International College, Springfield, Massachusetts; Christian Brothers College, Memphis; LL.D., Loyola University, Chicago; D. Performing Arts, Toledo University, Loyola-Marymount University, Los Angeles. Founder, St. Jude Children's Research Hospital, 1962, Memphis. Recipient: Emmy Award, 1954; Layman's Award from the American Medical Association; Better World Award from the Veterans of Foreign Wars, 1972; Michelangelo Award from Boys Town of Italy, 1973; Humanitarian Award from Lions International, 1975; Father Flanagan–Boys Town Award, 1981; Murray-Green-Meany Award, AFL-CIO, 1981; Hubert H. Humphrey Award, Touchdown Club, 1981; American Education Award, 1984; Humanitarian Award, Variety Clubs International, 1984; Congressional Medal of Honor, 1984; Sword of Loyola Award, Loyola University, Chicago, 1985; decorated Knight Malta; knight commander with star, Knights of Holy Sepulchre, Pope Paul VI. Died in Los Angeles, 6 February 1991.

TELEVISION SERIES

1950–52	*All Star Revue*
1953–57	*Make Room for Daddy*
1957–64	*The Danny Thomas Show*
1964–68	*Danny Thomas Specials*
1967–68	*The Danny Thomas Hour*
1970	*Make Room for Granddaddy*
1976–77	*The Practice*
1980–81	*I'm a Big Girl Now*
1986	*One Big Family*

MADE-FOR-TELEVISION MOVIE

1988	*Side By Side*

FILMS

The Unfinished Dance, 1946; *The Big City*, 1947; *Call Me Mister*, 1948; *I'll See You in my Dreams*, 1951; *The Jazz Singer*, 1951.

PUBLICATION

Make Room for Danny, with Bill Davidson. New York: Putnam, 1991.

FURTHER READING

Gelbart, Larry. "The Man Could Make a Short Story Long." *The New York Times*, 17 February 1991.

Grote, David. *The End of Comedy: The Sit-com and the Comedic Tradition*. Hamden, Connecticut: Archon Books, 1983.

Hamamoto, Darrell Y. *Nervous Laughter: Television Situation Comedy and Liberal Democratic Ideology*. New York: Praeger, 1989.

Javna, John. *The Best of TV Sitcoms: Burns and Allen to The Cosby Show, The Munsters to Mary Tyler Moore*. New York: Harmony Books, 1988.

Jones, Gerard. *Honey, I'm Home!: Sitcoms, Selling the American Dream*. New York: Grove Weidenfeld, 1992.

Leibman, Nina. *Living Room Lectures: The Fifties Family in Film and Television*. Austin: University of Texas Press, 1995.

Marc, David. *Comic Visions: Television Comedy and American Culture*. Boston, Massachusetts: Unwin Hyman, 1989.

Mitz, Rick. *The Great TV Sitcom Book*. New York: R. Marek, 1980.

Waldron, Vince. *Classic Sitcoms: A Celebration of the Best of Prime-time Comedy*. New York: Macmillan, 1987.

See also *Andy Griffith Show; Dick Van Dyke Show*

THOMAS, TONY

U.S. Producer

Tony Thomas, a native of California and member of one of U.S. television's leading families, began his own TV career as an associate producer at Screen Gems and moved from that position to become a producer at Spelling/Goldberg Productions. These associations brought Thomas into early contact with his future partner, Paul Junger Witt, who also started his career at Screen Gems. Indeed, their first significant venture together was the award-winning made-for-television movie, *Brian's Song*, which Witt produced. The Academy of Television Arts and Sciences recognized *Brian's Song* with six Emmys, including one for Outstanding Single Program.

In 1975 Thomas and Witt formed their own company, Witt/Thomas Productions, and a year later the two men joined with the talented writer, Susan Harris, to form a second entity, Witt/Thomas/Harris. The three launched their first series in 1977, the highly acclaimed *Soap*. Brutally attacked by a reviewer for *Newsweek*, who had not even seen the show, *Soap* quickly drew fire from uninformed conservative religious leaders who threatened to boycott the ABC comedy. As Thomas recalls, it was very close to the time of the first broadcast before a full complement of sponsors was assembled. And sponsorship was a continuing difficulty for the network. The producers credit Fred Silverman of ABC for standing firmly behind their creation in spite of the attacks.

There followed a string of successes, including *Empty Nest, Benson*, and *The Golden Girls*, for which Thomas, along with Witt and Harris, received Emmys in 1985-86 and 1886-87. In the 1996-97 season, Witt/Thomas began its fourth year of producing *The John Larroquette Show*, and introduced two new series, *Pearl* and *Common Law*.

Through the company, Thomas also began producing feature films with Witt. Working with Touchstone Pictures, they produced the Oscar-winning film *Dead Poets' Society*. Their feature work also included the 1992 release *Final Analysis*, and continued in 1996 with three films in production in association with Warner Brothers.

Tony Thomas is active in fundraising efforts on behalf of St. Jude's Hospital, founded by his father Danny in 1961. It is the world's largest childhood cancer research center.

—Robert S. Alley

TONY (ANTHONY C.) THOMAS. Born in Los Angeles, California, U.S.A., 7 December 1948. Educated at the University of San Diego. Assistant to the producer, *Young Rebels* television series, 1970; associate producer, *Getting Together* television series, 1971; associate producer, *Brian's Song*, 1972: associate producer and producer for numerous other television series; with Paul Junger Witt formed Witt/Thomas production company, 1975; later, with Susan Harris, formed Witt/Thomas/Harris production company, 1976. Recipient: several Emmy Awards. Address: Witt/Thomas Productions, Witt/Thomas/Harris Productions, 1438 North Gower Street, Hollywood, California 90028, U.S.A.

Tony Thomas
Photo courtesy of Tony Thomas

TELEVISION SERIES (selection)

1970–71	*Young Rebels* (assistant to the producer)
1971–72	*Getting Together* (associate producer)
1976-77	*The Practice*
1977	*Loves Me, Loves Me Not*
1977–81	*Soap*
1979–86	*Benson*
1982–83	*It Takes Two*
1983	*Just Married*
1985–92	*The Golden Girls*
1987–90	*Beauty and the Beast*
1988–95	*Empty Nest*
1991–93	*Nurses*
1991	*Good and Evil*
1991–95	*Blossom*
1991–93	*Herman's Head*
1991–93	*Nurses*
1993	*Whoops*
1993-	*The John Larroquette Show*
1995	*Muscle*
1996	*Local Heroes*

1996   *Pearl*

MADE-FOR-TELEVISION MOVIES (selection)
1972   *Brian's Song* (associate producer)
1973   *Blood Sport*

FILMS
Firstborn, 1984; *Dead Poets' Society,* 1988; *Final Analysis,* 1992; *Mixed Nuts,* 1994.

See also *Benson, Golden Girls,* Harris, Susan; *Soap,* Witt, Paul Junger

THE THORN BIRDS

U.S. Miniseries

The miniseries *The Thorn Birds,* based on Colleen McCullough's 1977 best-selling novel, was broadcast on ABC for 10 hours between 27 and 30 March 1983. Set primarily on Drogheda, a fictional sheep station in the Australian outback, the melodrama focused on the multi-generational Cleary family, and spanned the years from 1920 to 1962.

At the outset, the family—patriarch Paddy Cleary (Richard Kiley), his wife, Fiona (Jean Simmons), and children—moved from New Zealand to Australia to help run Drogheda, owned by Paddy's wealthy sister, Mary Carson (Barbara Stanwyck). Over the years, numerous deaths and disasters—fire, a drowning, a goring by a wild boar— were to befall the family.

While the saga recounted the story of the entire Cleary clan, it focused primarily on the lone Cleary daughter, Meggie (Rachel Ward) and her relationship with Father Ralph de Bricassart (Richard Chamberlain). Although they met when she was just a child, Meggie grew up and fell in love with the handsome, young Catholic priest who had been banished to the outback for a previous disobedience. Father Ralph was torn between his own love for Meggie, his love for God, and his ambition to rise in the Catholic hierarchy. Spurred on by the spiteful Mary Carson—who was herself attracted to the priest—Father Ralph was forced to choose between his own advancement in the church and his love for Meggie. He chose the former, and soon found himself at the Vatican. As Father Ralph rose quickly through the hierarchy of the Catholic Church (eventually becoming a cardinal), Meggie married a sheep shearer named Luke O'Neill (Bryan Brown), bore a daughter (played as an adult by Mare Winningham), and ended up working as a maid in Queensland.

Years later, de Bricassart returned to Australia and to Meggie, who eventually left her husband. In the controversial third episode, the two consummated their relationship in what *Newsweek*'s Harry F. Waters called "the most erotic love scene ever to ignite the home screen," but de Bricassart still was unable to give up the church. Unbeknownst to him, Meggie gave birth to his son (played as an adult by Philip Anglim), who in an ironic twist of fate himself became a priest before dying in a drowning accident. As in McCullough's novel, the key underlying message of this miniseries was that each generation is doomed to repeat the missteps and failures of the previous generation.

While winning the 1983 Golden Globe Award for Best Miniseries, *The Thorn Birds* was not without its controversy. The subject matter—a priest breaking his vow of celibacy—was contestable enough, but the fact that ABC chose to broadcast the program beginning on Palm Sunday and running through Holy Week raised the ire of the United States Catholic Conference. In response, McDonald's Corporation initially requested that its franchisees not advertise during the broadcasts. In the end, however, the company simply advised its franchisees to advertise only before Father Ralph and Meggie consummated their relationship.

Despite its controversial subject matter (or perhaps because of it), *The Thorn Birds* garnered an average 41 rating and 59 share over the course of its four-night run, making it then the second highest rated miniseries ever, second only to *Roots* (1977). Its controversial third episode, in which Meggie and Father Ralph consummated their relationship, was at the time the fourth highest rated network entertainment show of all time (preceded only by the final episode of *M*A*S*H,* the "Who Shot JR?" episode of *Dallas,* and the eighth episode of *Roots*). In the end, an estimated 110 million to 140 million viewers saw all or some of the miniseries. *TV Guide,* in fact, has listed *The Thorn Birds* as one of the top 20 programs of the 1980s.

Produced for an estimated $21 million, *The Thorn Birds* appeared during the heyday of the network television miniseries, from the late 1970s to the mid-1980s, when the form was seen as "the salvation of commercial television." In this context *The Thorn Birds* stood out for both its controversial qualities and its success. Like *Roots* and *The Winds of War* before it, *The Thorn Birds* exemplified the miniseries genre—family sagas spanning multiple generations, featuring large, big-name casts, and laden with tales of love, sex, tragedy, and transcendence that kept the audience coming back night after night. In 1996 ABC broadcast a sequel to *The Thorn Birds* in which Father Ralph and Meggie are again separated, and again struggle with their passion and their consciences. Though widely promoted, the program received far less attention from both critics and audiences.

—Sharon R. Mazzarella

The Thorn Birds

The Thorn Birds

CAST

Father Ralph de Bricassart	Richard Chamberlain
Meggie Cleary (as a girl)	Sydney Penny
Meggie Cleary (adult)	Rachel Ward
Mary Carson	Barbara Stanwyck
Fiona Cleary	Jean Simmons
Archbishop Contini-Verchese	Christopher Plummer
Rainer Hartheim	Ken Howard
Justine O'Neill	Mare Winningham
Anne Mueller	Piper Laurie
Paddy Cleary	Richard Kiley
Luddie Mueller	Earl Holliman
Luke O'Neill	Bryan Brown
Sarah MacQueen	Antoinette Bower
Stuart Cleary	Dwier Brown
Alastair MacQueen	John de Lancie
Angus MacQueen	Bill Morey
Stuart Cleary (as a boy)	Vidal Peterson
Miss Carmichael	Holly Palance
Judy	Stephanie Faracy
Dane O'Neill	Philip Anglim
Frank Cleary	John Friedrich
Mrs. Smith	Allyn Ann McLerie
Harry Gough	Richard Venture
Pete	Barry Corbin
Jack Cleary	Stephen Burns
Bob Cleary	Brett Cullen
Annie	Meg Wylie
Sister Agatha	Nan Martin
Barker at the fair	Wally Dalton
Arne Swenson	Chard Hayward
Doc Wilson	Rance Howard
Martha	Lucinda Dooling
Phaedre	Aspa Nakopolou

PRODUCERS   David Wolper, Edward Lewis, Stan Margulies

PROGRAMMING HISTORY   4 Episodes

• ABC

27 March–30 March 1983

FURTHER READING

Bawer, Bruce. "Grand Allusions Sacred and Profane." *Emmy Magazine* (Los Angeles), March/April 1982.

Morris, Gwen. "An Australian Ingredient in American Soaps: *The Thorn Birds* by Colleen McCullough." *Journal of Popular Culture* (Bowling Green, Ohio), Spring 1991.

Waters, Harry F. "Sex and Sin in the Outback." *Newsweek* (New York), 28 March 1983.

See also Adaptations; Miniseries

THREE'S COMPANY

U.S. Situation Comedy

Three's Company, an enormously popular yet critically despised sitcom farce about a young man living platonically with two young women, aired on ABC from 1977 to 1984. After a spring try-out of six episodes beginning Thursday, 15 March 1977, *Three's Company* ranked number 11 among all TV shows for the entire 1976–77 season—at that time, an unheard-of feat. The next year *Three's Company* moved to Tuesdays behind ABC powerhouses *Happy Days* and *Laverne and Shirley,* which it also followed that year as number three in the ratings. In 1978 and 1979, *Three's Company* nudged out *Happy Days* for the number two spot, and late in this season moved its caustic landlords onto their own short-lived spin-off, *The Ropers* (which ranked number eight among all network shows after a spring tryout of six episodes, but was cancelled in 1980 after a dismal second season). In 1979 and 1980, *Three's Company* shot past both of its lead-ins to become the highest-rated TV comedy in America. That summer, ABC ran back-to-back reruns of the show in its daytime line-up, foreshadowing huge success in syndication, which the series entered in 1982, two years before its network demise.

 Three's Company entered the television scene in the midst of TV's "jiggle era" that began in 1976 with ABC's *Charlie's Angels,* and was the medium's response to the sexual revolution and the swinging single. *Three's Company,* though otherwise apolitical in content, was the first sitcom to address the sexual implications and frustrations of co-ed living, which in 1977 was still somewhat taboo. In the minds of many, male-female cohabitation was anything but innocent and, apparently, would lead only to the evils of premarital sex. *Three's Company* toyed with this dilemma in its premise, an Americanized version of the 1973-76 British TV comedy *Man About the House.*

 Set in Santa Monica, California, the series chronicled the innuendo-laden, slapstick-prone misadventures of the affably klutzy bachelor Jack Tripper (played by John Ritter) and the two single, attractive women—one a cute, down-to-earth brunette named Janet Wood (Joyce DeWitt), the other a sexy, dimwitted blonde named Christmas "Chrissy" Snow (Suzanne Somers). The three shared an apartment in order to beat the high cost of living, but Jack was also present to provide "manly protection." Though he never broke his vow of keeping a "strictly platonic" relationship with his roommates (the three were really best friends who always looked after each other), the series was rife with double entendre suggesting they were doing much naughtier stuff. Antagonists in this domestic farce were the the trio's downstairs landlords—first the prudish Stanley Roper, an Archie Bunker-type played by Norman Fell, and later the comically swaggering "ladies man" Ralph Furley, played by Don Knotts. The landlords were so suspicious of the "threesome" arrangement that they would not permit it until after Jack told them he was gay, a "lifestyle" against which, ironically,

Three's Company
Photo courtesy of DLT Entertainment, Ltd.

neither discriminated by refusing housing. Though Jack was a heterosexual with many girlfriends, he masqueraded as an effeminate "man's man" around the near-sighted Roper, who called him "one of the girls," and Furley, who often tried to "convert" him; this comic device played heavily at first but was toned down considerably by the show's fourth season. When out of Roper's and Furley's reach, Jack and his upstairs buddy, Larry Dallas (Richard Kline), leered at and lusted after every female in sight, including, in early episodes, Janet and Chrissy. Chrissy, especially, was prone to bouncing around the apartment braless in tight sweaters when she wasn't clad in a towel, nightie, short-shorts or bathing suit. The irony here was that even though sex was so ingrained in the *Three's Company* consciousness, nobody on the show ever seemed to be doing it—not even the show's only married characters, the sex-starved Helen Roper (Audra Lindley) and her impotent handyman husband, Stanley, the butt of numerous faulty plumbing jokes.

 Three's Company's sexiness and libidinal preoccupation helped gain the show tremendous ratings and media exposure. A February 1978 *Newsweek* magazine cover story on "Sex and TV" featured the trio in a sexy, staged shot. *60 Minutes* presented an interview with Somers, who, in the tradition of *Charlie's Angels'* Farrah Fawcett, became a sex symbol and magazine cover-girl with top-selling posters, dolls and other merchandise. TV critics and other intellectuals rallied against the show, calling its humor sophomoric, if not insulting. Feminists objected to what they called exploitative portrayals of women (namely Chrissy) as bubble-brained sexpots. And while *Three's Company* was not as harshly condemned among conservative educators and religious organizations as its ABC counterpart *Soap* (a more satirical comedy with a shock value so high ABC almost delayed its premiere in fall 1977), it received low marks from the Parent-Teacher Association and was targeted in a sponsor blacklisting by the Reverend Donald Wildmon's National Federation for Decency.

Though *Three's Company* would become notorious as titillation television, its origins are that of British bedroom farce and America's "socially relevant" sitcoms. In 1976, *M*A*S*H* writer/producer Larry Gelbart penned an initial *Three's Company* pilot script borrowing scenario and characterizations from Thames Television's *Man About the House*. But that pilot, with Ritter, Fell, Lindley and two other actresses, didn't sell. Fred Silverman, programming chief at ABC, requested a revamped pilot for a show he believed would be a breakthrough in sexiness the same way CBS' *All in the Family* was in bigotry. So show owners Ted Bergman and Don Taffner commissioned *All in the Family* Emmy-winning head writers and *Jeffersons* producers Don Nicholl, Michael Ross and Bernie West to rewrite the pilot. The roommates, in Gelbart's script an aspiring filmmaker and actresses, took on more bourgeois jobs in the new pilot—Jack became a gourmet cooking student, Janet a florist and Chrissy an office secretary. The female leads were recast (DeWitt was added for the second pilot, and Somers for a third), the chemistry clicked and ABC bought the series.

Most critics called *Three's Company* an illegitimate attempt to use the TV sitcom's new openness for its own cheap laughs. But Gerard Jones, author of *Honey, I'm Home: Sitcoms: Selling the American Dream*, notes that the minds behind *Three's Company* intelligently responded to the times. He suggests that producers Nicholl, Ross and West recognized that even the highly praised work of producer Norman Lear's shows "had always been simple titillation." The producers simply went a step further. They "took advantage of TV's new hipness" to present even more titillation "in completely undemanding form," thus creating "an ingenious trivialization that the public was waiting for."

Though *Three's Company* jiggled beneath the thin clothing of titillation, the show was basically innocent and harmless, a contradiction that annoyed some critics. Its comedy, framed in the contemporary trapping of sexual innuendo, was basically broad farce in the tradition of *I Love Lucy,* very physical and filled with misunderstandings. (Lucille Ball loved *Three's Company* and Ritter's pratfalls so much she hosted the show's 1982 retrospective special). As fast-paced, pie-in-your-face farce, *Three's Company* spent little time on characterization. But underlying themes of care and concern among the roommates often fueled the comedy and occasionally led to a tender resolve by episode's end.

Behind the scenes three was company until fall 1980, when Somers and her husband/manager, Alan Hamel, asked for a raise from $30,000 per episode to $150,000 per episode plus 10% of the show's profits. Co-stars Ritter and DeWitt, confused and angry, refused to work with Somers, whose role was reduced to a phone-call from a separate soundstage at the end of each episode (Chrissy had been sent to take care of her ailing mother in Fresno). For the remainder of the 1980-81 season, Jenilee Harrison performed as a "temporary" roommate, Chrissy's clumsy cousin Cindy Snow. By fall 1981, Somers was officially fired, and Priscilla Barnes was cast as a permanent replacement, playing nurse Terri Alden, a more sophisticated blonde (Harrison's character moved out to attend a university but occasionally visited through spring 1982). Though viewership dropped when Somers left, *Three's Company* remained very popular, focusing more on Ritter's physical abilities and his character's transition from cooking student to owner of Jack's Bistro, a French cuisine restaurant.

Three's Company, weathering key cast changes and America's waning interest in sitcoms, remained a top ten hit through the 1982–83 season. But in 1984, after 174 episodes, a final People's Choice Award as Favorite Comedy Series and an eighth, embattled season in which it dropped out of the top thirty in the face of competition from NBC's comically violent *The A-Team, Three's Company* changed its format. A final one-hour episode saw Janet get married, Terri move to Hawaii and Jack fall in love and move in with his new girlfriend. Ritter, who won an Emmy for Outstanding Male Lead in a Comedy in 1984, was the only *Three's Company* cast member to remain when production resumed in the fall with a new cast and new title. Recycling much of its parent show's comic formula, *Three's a Crowd* focused on Jack Tripper's consummated relationship with his live-in girlfriend (Mary Cadorette), whose disapproving father (*Soap*'s Robert Mandan) became their landlord. This incarnation lasted one season.

Three's Company, though later considered tame television, pushed the proverbial envelope in the late 1970s, opening the door for sexier, if not sillier, comedies offering audiences both titillation and mindless escape.

—Chris Mann

CAST

Jack Tripper	John Ritter
Janet Wood	Joyce DeWitt
Chrissy Snow (1977–81)	Suzanne Somers
Helen Roper (1977–79)	Audra Lindley
Stanley Roper ((1977–79)	Norman Fell
Larry Dallas (1978–84)	Richard Kline
Ralph Furley (1979–84)	Don Knotts
Lana Shields (1979–80)	Ann Wedgeworth
Cindy Snow (1980–82)	Jenilee Harrison
Terri Alden (1981–84)	Priscilla Barnes
Mike, the Bartender (1981–84)	Brad Blaisdell

PRODUCERS Don Nicholl, Michael Ross, Bernie West, Budd Gossman, Bill Richmond, Gene Perret, George Burdit, George Sunga, Joseph Staretski

PROGRAMMING HISTORY   164 Episodes

• ABC

March 1977–April 1977	Thursday 9:30-10:00
August 1977–September 1977	Thursday 9:30-10:00
September 1977–May 1984	Tuesday 9:00-9:30
May 1984–September 1984	Tuesday 8:30-9:00

FURTHER READING

Hamamoto, Darrell Y. *Nervous Laughter: Television Situation Comedy and Liberal Democratic Ideology.* New York: Praeger, 1989.

Javna, John. *The Best of TV Sitcoms: Burns and Allen to the Cosby Show, The Munsters to Mary Tyler Moore.* New York: Harmony, 1988.

Jones, Gerard. *Honey, I'm Home! Sitcoms, Selling the American Dream.* New York: Grove Weidenfeld, 1992.

Waldron, Vince. *Classic Sitcoms: A Celebration of the Best of Prime-Time Comedy.* New York: MacMillan, 1987.

See also Comedy, Domestic Settings

TIANANMEN SQUARE

Tiananmen Square will forever be remembered as a political rally that turned into a bloody massacre viewed on live television. The square in Beijing, China, was the site of a pro-democracy student demonstration in the spring of 1989, a demonstration violently crushed by the Chinese military. Scenes of the brutal crackdown were broadcast throughout the world. These images embittered the international public toward the Chinese government and had profound impact on subsequent foreign policy decisions. The demonstrations presented the media with an opportunity for a telegenic foreign story that was also easy for viewers to identify with. All of the major American networks and news organizations from many other countries had previously stationed prime-time news anchors and camera crews in Beijing in order to provide live broadcasts of Soviet President Mikhail Gorbachev's visit to the city. That visit marked a step toward rapprochement between China and the Soviet Union.

Thousands of students comprising China's pro-democracy movement also planned to use the state visit and the obligatory media coverage for their purposes. They had assembled and camped in the Square for two weeks in late May and early June. Among their demands were the rights to free speech and a free press, and they erected a symbolic Statue of Liberty named the "Goddess of Democracy." Their cause and the images they employed were very familiar to Americans and to other audiences around the world.

However, this hopeful demonstration came to a sudden and horrifying end. On the night of 3 June and into the early morning hours of 4 June, the army launched an assault on the unarmed civilians in the square. They stormed the area with tanks and machine guns, firing into the crowd at random. Hundreds of young students were killed and thousands wounded in the attack. Scenes of brutality and chaos were broadcast from Tiananmen Square, and there were reports of students and civilians being imprisoned in other parts of China.

The fear inspired by the government's crackdown was so powerful that, almost immediately, students and demon-

stration organizers stopped talking to the media. The excitement and generous spirit with which interviews had been granted just two days before had eerily disappeared. An official news blackout was imposed, and in addition to sources drying up, reporters and crews themselves were being threatened and interrogated. In a tragic distortion of intentions, the televised interviews and pictures were also used by Chinese officials to identify and incarcerate many of the students involved. The Chinese people outside of Beijing never really saw or heard the true horror of what happened. They received "official" versions from the state-run news organization. These broadcasts described scenes of violent student protesters and angry dissidents attacking innocent government authorities.

The Western media, however, was not so easily manipulated. Even though human rights violations were thought to be commonplace under the Communist Party rule, the topics received little consistent or significant mention in the mainstream press. Never before had television so graphically exposed the abuse of individual rights and disregard for human life that took place there. Tiananmen Square received continuous coverage during the first day of the massacre, representing one of the earliest efforts by U.S. news media to devote non-stop air-time to a breaking International news event. But in one of the most dramatic moments of the event audiences were able to watch a Chinese government official physically unplug the satellite transmitter carrying CBS' broadcast. As *CBS Evening News* anchor Dan Rather stood by, registering his protest, television screens suddenly carried nothing but blurred static until New York transmission opened its own feed to network affiliate stations.

China experienced nearly three years of economic sanctions and scorn from the international community after the massacre, yet the Chinese government continued its hard-line policies toward all civilian dissent. On subsequent anniversaries of the military attack, Beijing has maintained an official position of denial and repression. A heavy police presence stifles the city each year on 4 June

The student uprising in Tiananmen Square
Photo courtesy of AP/ Wide World Photos

and international news broadcasts commemorating the event are interrupted and blocked. Hotels have all been instructed to unplug their satellite connections to CNN.

Despite the government's attempts at censorship, the images broadcast from Tiananmen Square cannot be erased from public memory. Few who watched the coverage will ever forget the sight of a lone student standing defiantly against a column of army tanks, or of soldiers clubbing demonstrators until they were bloody and lifeless, or the panic-stricken faces of the people in the Square. Although the Chinese government would like to strike Tiananmen Square from the record books, television has insured that its lessons will be taught for many years to come.

—Jennifer Moreland

FURTHER READING

Calhoun, Craig J. *Neither Gods for Emperors: Students and the Struggle for Democracy in China.* Berkeley: University of California Press, 1994.

"China: The Weeks Of Living Dangerously." *Broadcasting* (Washington, D.C.) 12 June 1989.

Li, Peter, Steven Mark, and Marjorie H. Li, editors. *Culture and Politics in China: The Anatomy of Tiananmen Square.* New Brunswick, New Jersey: Transaction Books, 1991.

Salisbury, Harrison Evans. *Tiananmen Diary: Thirteen Days in June.* Boston: Little Brown, 1989.

Tonetto, Walter, editor. *Earth Against Heaven: A Tiananmen Square Anthology.* Wollongong, New South Wales, Australia: Five Islands Press, 1990.

Watson, Trevor. *Tremble and Obey: An ABC Correspondent's Account of the Bloody Beijing Uprising.* Crows Nest, New South Wales, Australia: ABC Enterprises for the Australian Broadcasting Corporation, 1990.

Yang, L.Y., and Marshall L. Wagner, editors. *Tiananmen: China's Struggle for Democracy: Its Prelude, Development, Aftermath, and Impact.* Baltimore: University of Maryland, 1990.

See also News, National; Satellite

TILL DEATH US DO PART

British Situation Comedy

One of the first British shows to take a serious and sustained interest in race themes was *Till Death Us Do Part,* originally broadcast in the mid-1960s on BBC1. Five weeks into the first series, the show had already toppled its immediate competitor, *Coronation Street,* in the ratings war. Although the idea for the series had been in the mind of its creator, Johnny Speight, for several years, it wasn't until Frank Muir took over comedy at the BBC that production began, initially as a pilot but subsequently as a fully-fledged series. The comedy centred on the Garnett family, with the main "star" of the show in the person of the patriarch "Alf," sometimes known as "Chairman Alf" for his ready willingness to engage in scurrilous diatribes against the Conservative party. The other significant target of his rantings were black people and it is for the extreme views expressed by Alf on issues of race that the programme is most remembered (and denounced).

Although Alf's creator argued at the time of the original broadcasts (and since) that his intention was to expose racist bigotry through the exaggerated utterances of Alf, such an intention has back-fired for many commentators. The enormous popularity of the show signified that there was something about it which appealed to a significant proportion of the viewing public. Wherever the series has been shown—in Great Britain or in the United States or Germany (the last two in local adaptations)—the effects have by no means always been what the author intended. Alf's rhetoric was not always seen as the voice of the ignorant bigot, but often as the stifled cry of the authentic (white) working class. While the Garnett family, and Alf in particular, were clearly represented as disgraceful and abject characters, extreme even as caricatures, many critiques of the show suggest that part of its fascination for the audience was the kernel of truth buried in the lunatic wailings. Thus, the crucial difference between Alf's grotesque soliloquies and the viewers' beliefs was that Alf was simply too stupid to understand that racist sentiment must be concealed beneath a sheen of respectability: the persuasive and polished performance of Alessandra Mussolini in her Italian political career is more credible than Alf's degenerate ramblings but contains much the same message.

The inflammatory and controversial subject matter of the show and its American counterpart, *All in the Family,* ensured that they both became the focus of academic enquiry. Research findings were mixed, some suggesting that such shows had a neutral effect on viewers while others claimed that viewers identified heavily with the xenophobic ravings of Alf/Archie. It is likely that many British viewers, worried by the alleged "immigrant avalanche" constantly reported in the media during the 1960s and fueled by Irish Protestant leader Enoch Powell's rabid jingoism, found a certain resonance in the racist bigotry espoused by Alf. Although Alf was challenged in his more ludicrous diatribes by his daughter Rita and son-in-law Mike, with the odd wry observation from his long-suffering wife "Old Moo", Warren Mitchell's powerful performance as Alf relegated the rest to mere bit players, as deserving butts of his wild wit.

Through Alf, a cascade of fear and prejudice was given unique prime-time exposure and articulated with such passion that during its transmission, 12 million viewers (then half the adult British population) tuned in to watch. It is highly unlikely that all these viewers were laughing at rather than with Alf, that they were all making wholly satirical readings of Alf's obscene racism and applauding Speight's clever exposition as they cackled at the "jokes". Looking again at the show with a 1990s sensibility, the virulent racism stands out as extraordinary and its nature and extent have never been repeated on British television. *Till Death Us Do Part* may have been written as brave social commentary, but thirty years later, it looks seriously flawed and gives the lie to the notion that what the writer intends is always "correctly" interpreted and understood by her/his audience.

There is little evidence to support the claim of programme producers and writers that mixing humour with bigotry will automatically underline the stupidity of the latter through the clever device of the former. If bigots do not perceive such programmes as satire, and much of the research effort so far seems to indicate that a satirical reading is by no means universal, then they are unlikely to become less prejudiced as a result of watching these shows. At the end of the 1980s, an Alf Garnett exhibition was staged at the Museum of the Moving Image in London, where visitors pressed buttons representing particular social problems and Alf appeared on video to opine on the selected subject. It is a strange idea and exemplifies the ease with which TV characters can make the transition from one medium to another, in this instance mutating from demon to sage in one easy movement. If it is a little too glib, from the smug security of the 1990s, to label *Till Death Us Do Part* as a straightforwardly racist text, it is nonetheless instructive to consider the limits of acceptability which prevail in any given decade and to continue the campaign for equality and respect while at the same time supporting the radical take.

—Karen Ross

CAST

Alf Garnett	Warren Mitchell
Else Garnett	Dandy Nichols
Rita	Una Stubbs
Mike	Anthony Booth

PRODUCERS   Dennis Main Wilson, David Croft, Graeme Muir

PROGRAMMING HISTORY   52 Half-hour episodes; 1 45-minute special

• BBC

July 1965	Comedy Playhouse (pilot)
June 1966–August 1966	7 episodes
December 1966–February 1967	10 episodes
January 1968–February 1968	7 episodes
September 1972–October 1972	6 episodes
December 1972	Christmas Special
January 1974–February 1974	7 episodes
December 1974–February 1975	7 episodes
November 1975–December 1975	6 episodes

FURTHER READING

Cantor, Muriel G. *Prime-Time Television: Content and Control.* Beverly Hills, California, and London: Sage, 1980; 2nd edition, with Joel Cantor, 1992.

Daniels, Therese, and Jane Gerson. *The Colour Black: Black Images in British Television.* London: British Film Institute, 1989.

Hood, Stuart. *On Television.* London: Pluto Press, 1980.

Ross, Karen. *Black and White Media: Black Images in Popular Film and Television.* Cambridge: Polity, 1995.

Vidmar, Neil, and Milton Rokeach. "Archie Bunker's Bigotry." *Journal of Communication* (Philadelphia, Pennsylvania), 1974.

Till Death Us Do Part
Photo courtesy of BBC

Woll, Allen, and Randall Miller. *Ethnic and Racial Images in American Film and Television: Historical Essays and Bibliography.* New York and London: Garland, 1987.

See also *All in the Family*; Speight, Johnny

TILLSTROM, BURR

U.S. Puppeteer

Burr Tillstrom, the creative talent behind the extraordinarily successful *Kukla, Fran and Ollie*, was one of television's earliest pioneers and a principal participant in a number of television "firsts." In the late 1930s, Tillstrom joined the RCA Victor television demonstration show for a tour throughout the Midwest. At the completion of the tour, he was invited to present his Kuklapolitan Players at the 1939 New York World's Fair, where he demonstrated the new medium at the RCA Victor exhibit. In the spring of 1940, RCA sent Tillstrom to Bermuda to do the first ship-to-shore telecasts. The Kuklapolitans were also featured on the 1941 premiere broadcast of the Balaban and Katz station WBKB in Chicago. By drawing large audiences for television puppetry, Tillstrom opened the door for future puppeteers and their puppets, such as Paul Winchell and Jerry Mahoney, Shari Lewis and Lamb Chop, and Jim Henson and the Muppets.

Tillstrom demonstrated his improvisational talents at an early age when he entertained neighborhood children using teddy bears, dolls, and any other objects that he could animate to mimic performances and film stories. Following one year of college during the mid-1930s, he joined the Chicago Parks District's puppet theater, created under the auspices of the Works Progress Administration (WPA), and developed his own puppets and characters after work. Kukla,

the puppet who was the first member of the Kuklapolitan Players, was actually designed and constructed by Tillstrom for a friend in 1936, but Tillstrom found he couldn't part with his creation. The character remained nameless until a chance meeting with Russian ballerina Tamara Toumanova who, upon seeing the puppet, called him "kukla" (Russian for "doll" and a term of endearment).

The format for *Kukla, Fran and Ollie* had its roots in Tillstrom's work at the 1939 World's Fair. His puppets, who served as an entr'acte for another marionette group, made comments to the audience, and interacted with actresses and models (spokespersons for the new medium) invited onto the stage. Tillstrom performed more than 2,000 shows at the fair, each performance different because he disliked repetition.

Tillstrom continued to hone his craft by performing with other marionette troupes and managing the puppet theatre at Marshall Field's department store in Chicago. He performed benefits for the USO during World War II and at local hospitals for the Red Cross. During a bond-selling rally in Chicago, Tillstrom met a young radio singer and personality, Fran Allison, who joined his troupe for a trial 13-week local program, a trial that lasted for many years and attracted millions of fans.

Tillstrom created each puppet on *Kukla, Fran and Ollie* by hand and was the sole manipulator and voice for 15 characters. He shifted easily—usually with only a momentary pause—among characters and created unique personalities and voices for each "kid" (as he referred to his creations), ranging from the sweet voice of Kukla, a baritone singing voice for Ollie, the flirtatious Buelah Witch, to the indistinguishable gibberish of Cecil Bill. Standing behind the small stage, Tillstrom could observe the on-stage action through the use of a small monitor, a technique that was later adopted and expanded by Jim Henson for *The Muppet Show.*

Although he is most closely identified with *Kukla, Fran and Ollie,* Tillstrom was featured on the series *That Was the Week That Was* (*TW3*) in 1964 without the Kuklapolitans. He won a special Emmy Award for a hand-ballet symbolizing the emotional conflicts caused by the Berlin Wall crisis. His work on *TW3* was cited by the George Foster Peabody committee which, in 1965, decided to recognize distinguished individual achievements rather than general program categories after chiding the radio and television industry for "a dreary sameness and steady conformity" in its programming.

Following his success on television in various reincarnations and syndicated specials of *Kukla, Fran and Ollie,* including a Broadway production, annual holiday productions at Chicago's Goodman Theater, and a sound recording (for which he was nominated for a Best Recording for Children Grammy in 1972), Tillstrom brought his characters to the printed page in his 1984 work *The Dragon Who Lived Downstairs.* A generous spirit who enjoyed sharing his knowledge and experience with future performers, Tillstrom served as an artist-in-residence at Hope College in Holland, Michigan. At the time of his death in December 1985, he was working on a musical adaptation of his story for television. In March 1986, he was inducted posthumously into the Hall of Fame of the Academy of Television Arts and Sciences for his significant contributions to the art of television.

—Susan R. Gibberman

BURR TILLSTROM. Born in Chicago, Illinois, U.S.A., 13 October 1917. Attended the University of Chicago, 1935. Puppeteer, from the early 1930s; created the puppet Kukla, 1936; manager of the puppet exhibits and marionette theater, Marshall Field and Company, Chicago, 1938; joined the RCA Victor television demonstration show, 1939; produced television show on Chicago television station WBKB with his "Kuklapolitans," 1947; program picked up by NBC, 1948–52; show moved to ABC, 1954–57; revived for PBS, 1969; staged a Broadway production with the Kuklapolitans, 1960; host, CBS Children's Film Festival, 1970s; appeared on NBC series *That Was the Week that Was,* 1964–65. Recipient: more than 50 entertainment awards, including five Emmys. Died in Palm Springs, California, 6 December 1985.

TELEVISION SERIES

1948–52, 1954–57,
  1961–62, 1969–71,
    1975–76   *Kukla, Fran and Ollie*
1964–65   *That Was the Week That Was*

Burr Tillstrom

PUBLICATION

The Kuklapolitan Players Present the Dragon Who Lived Downstairs. New York: William Morrow, 1984.

FURTHER READING

Brown, Joe. "Burr Tillstrom's Intimate World." *Washington* (D.C.) *Post,* 22 December 1982.
"Burr Tillstrom Dies at 68; Creator of 'Kukla, Fran, Ollie.'" *Variety* (Los Angeles), 11 December 1985.
Christiansen, Richard. "Burr Tillstrom: An Innovator Who Had His Hand in Here-And-Now Projects." *Chicago Tribune,* 15 December 1985.
Corren, Donald. "Kukla, Me, and Ollie: Remembering Chicago's Legendary Puppeteer, Burr Tillstrom." *Chicago Magazine,* July 1986.
Crimmins, Jerry. "Burr Tillstrom, 68, Originated 'Kukla.'" *Chicago Tribune,* 7 December 1985.
"The Dragon Who Lived Downstairs" (book review). *Publishers Weekly* (New York), 24 February 1984.
Nix, Crystal. "Burr Tillstrom, Puppeteer, Dies." *The New York Times,* 8 December 1985.
Shales, Tom. "A Troupe's Worth of Talent Died with Burr Tillstrom." *Chicago Tribune,* 18 December 1985.

See also Allison, Fran; Chicago School of Television; Children and Television; Henson, Jim; *Kukla, Fran, and Ollie*

TIME SHIFTING

The practice of recording a television show onto video tape with a video recorder (VCR) for the purpose of playing the tape back later at a more convenient time for the viewer is known as time shifting. By law, with few exceptions, a person is not permitted to make an unauthorized copy of a copyrighted work like a television show. One exception to this is the concept of "fair use." Fair use allows copying and using copyrighted material for certain non-profit, educational and/or entertaining purposes.

The VCR was introduced into the home television market in the United States during the mid-1970s. As the sale of VCRs increased in the early 1980s, more and more viewers began taping programs off-the-air. Program producers and other copyright owners went to court to stop what they believed to be infringement of their copyrights. Universal Studios sued Sony Corporation, the inventor and patent holder of the Betamax VCR, in hopes of stopping home taping of television programs, or of charging royalties for such copying. A U.S. Court of Appeals ruled in Universal's favor, but the matter went to the U.S. Supreme Court which issued its famous "Betamax" decision in 1984. Central to that decision was the granting of permission to home television viewers to record television shows for purposes of viewing them later at a more convenient time (i.e. time shifting.) The high court ruled that such copying constituted fair use, and would not hurt the market value of the programming itself to program pro-

ducers. The court's decision was vague on the issue of warehousing tape copies. For example, if a viewer is a fan of a soap opera such as *As The World Turns*, and makes copies of each and every episode with the intention of building a library of the entire program series for repeated playback in the future, that would be warehousing. The court may have left this matter deliberately vague, however, because it would be virtually impossible to enforce a ban on such warehousing without violating a person's right to privacy.

The unauthorized copying issue is raised again each time a new electronic media technology is introduced to the public. The courts are likely to continue to support the concept of time shifting and other, similar personal uses of these technologies in the future.

—Robert G. Finney

FURTHER READING

Levy, Mark R., and Barrie Gunter. *Home Video and the Changing Nature of the Television Audience.* London: Libbey, 1988.

Levy, Mark R., editor. *The VCR Age: Home Video and Mass Communication.* Newbury Park, California: Sage, 1989.

See also Betamax Case; Copyright Law and Television; Sony Corporation; Videocassette; Videotape

TIME WARNER

U.S./International Media Conglomerate

This vast media conglomerate was created in 1989 when Warner Communication and Time, Inc., merged. Through the 1990s it ranked as the largest media company in the world, owning assets in excess of $20 billion, and yearly generating revenues also measured in billions of dollars. In late 1995, in the stormy climate of corporate media mergers begun when the Disney Company bought Capital Cities/ABC, Time Warner set out to create another huge merger with the purchase of the Turner Broadcasting Company. Eclipsed in size by the Disney deal, the combination with Turner would once again make Time Warner the largest media conglomerate in the world.

Even before the latest mergers, however, Time Warner has functioned as a major player in the television business. Its Warner Brothers' studio produces a vast array of TV programs and distributes them around the world. Its cable television division counts millions of subscribers, and owns and operates leading networks such as HBO and Cinemax. Each year Time Warner also sells millions of home video recordings. Fully half its massive revenues come from television-related subsidiaries. The rest flow from moviemak-

ing, owing and operating one of the top six major music labels, and publishing a string of magazines including *Time, Fortune,* and *Money.*

Time Warner is everywhere in the television business of the 1990s. The conglomerate owns 23% of the Turner Broadcasting Company and holds three memberships on the board of that organization. Additional representatives sit on the board of WTBS-the SuperStation, CNN, and TNT, as well as the boards of Black Entertainment Television, and the Comedy Central. In 1995 it counted more than eleven million cable subscribers to its own systems, and planned for more expansion. In short, as the 20th century drew to a close, Time Warner could count rival Tele-Communications, Inc., as its only true competitor in the cable television industry and only the other five major Hollywood studios as serious rivals in making television shows.

The Warner of the company side entered television production first. During the mid-1950s Warner Brothers served as the site for the creation of such early hits as *Cheyenne* and *77 Sunset Strip.* Since then, in various configurations, the company has been involved in the production, ownership and syndication of an imposing list of television shows that includes *Martin, Living Single, Jenny Jones, Love Connection, Dukes of Hazard, Eight Is Enough, Knots Landing, The Waltons, Wonder Woman, Alf, Family Matters, Full House, Head of the Class, My Favorite Martian, Murphy Brown,* and *Perfect Strangers*. Dozens of made-for-TV movies also stream out of the Warner studio each year. Time, Inc., came to cable TV in the 1960s and became a leader on the strength of its innovation of HBO a decade later.

The 1989 merger linked these assets and pushed Time Warner toward becoming the biggest television company in world history. The organization truly hit its corporate stride when refashioned in the 1970s into Warner Communication under Steven J. Ross.

Ross created Time Warner, and as much as a single person can be responsible for merging a multi-billion dollar world-wide media conglomerate, the credit has to go to him. Ross sought to create an American company that could stand up to Japanese conglomerates Sony and Matsushita, then buying into Hollywood and taking over other media businesses in the United States. The merger was advertised as a combination of equals and at first Ross and J. Richard Monro of Time, Inc., were listed as Co-Chief Operating Officers. But this "sharing of power" proved short-lived; within a year Ross stood alone atop his media colossus.

When Ross died in December 1992, the actual day-to-day running of Time Warner fell to his protege, Gerald M. Levin. Levin inherited and thereafter expanded some of Ross' bold experiments. In September 1992, for example, Ross initiated *New York 1 News*, a 24-hour local cable channel for one million subscribers living on Manhattan Island. Ross knew such a proposition would be expensive, but New York City was his home town and he planned for journalistic giant, Time, Inc., to help make a 24-hour local cable TV news operation profitable. *New York 1 News* represents an experiment on a grand scale only Time Warner could attempt. Here the largest media company in the world, in the largest media market in the United States, under the close glare of Madison Avenue's advertising experts, tries to make "the future" profitable now. With the close involvement of *The New York Times, New York 1 News* represents one significant case by which future "information superhighway" watchers will judge the success (or failure) of the new age of mass communications.

Another telling experiment is Time Warner's Orlando, Florida, trial, providing full service, 500-channel cable television to 4000 homes. The offering is complete with video games and movies on demand, multiple interactive shopping channels, and information including news and reference guides. All these options are available to individual homes as desired.

Technical innovation and corporate synergy stand at the center of Time Warner's future. Can the company truly bring the world of 500 channels to the 100 million living rooms in the United States and then to billions more around the world? Can Time Warner's book and magazine divisions really generate new television shows and make long promised corporate synergy profitable? And can Gerald Levin negotiate these future forays while still paying off the billions of dollars of corporate debt accumulated as part of the Time Warner merger? Most importantly, can the company successfully implement still greater expansion? The 1996 merger with Turner Broadcasting propels Time-Warner into yet another level of corporate complexity, filled with potential for even greater involvement in the construction of world-wide media systems—and with potential financial disaster.

The jury is still out. The future of Time Warner is unclear as the 20th century ends. Gerald M. Levin, the hand picked successor, will need all his skills to navigate the new promised land of 500 cable TV channels. He has announced that Time Warner will offer telephone service to businesses in New York City and has formed an alliance with "Baby Bell" U.S. West to offer both cable and telephone service in the Rocky Mountain states. Other possible investments in foreign cable TV businesses and new television networks have also been mentioned. Levin will be judged on whether these ventures turn profitable.

In the meantime we must judge Time Warner as a bold corporate experiment in progress. Is it a venture successful only in building huge debt or a farsighted sequence of mergers which resulted in a new type of media conglomerate that redefined the television industry?

—Douglas Gomery

FURTHER READING

Bruck, Connie. *Master of the Game: Steve Ross and the Creation of Time Warner*. New York: Simon and Schuster, 1994.

———. "Jerry's Deal." *The New Yorker* (New York), 19 February 1996.

Byron, Christopher. *The Fanciest Dive: What Happened when the Giant Media Empire of Time/Life Leaped without Looking into the Age of High-tech*. New York: Norton, 1986.

Corliss, Richard. "Time Warner's Head Turner." *Time* (New York), 11 September 1995.

Clurman, Richard M. *To the End of Time: The Seduction and Conquest of a Media Empire*. New York: Simon and Schuster, 1992.

Fabrikant, Geraldine. "Government Review of Turner-Time Deal." *The New York Times*, 10 October 1995.

Greenwald, John. "Hands Across the Cable: The Inside Story of How Media Titans Overcame Competitors and Egos to Create a $20 Billion Giant." *Time* (New York), 2 October 1995.

Lander, Mark. "Time Warner and Turner Seal Merger: After 5 Weeks, a $7.5 Billion Stock Deal." *The New York Times*, 23 September 1995.

Oneal, Michael. "The Unlikely Mogul: Can Jerry Levin Keep a Grip on Time Warner? Don't Count Him Out Yet." *Business Week* (New York), 11 December 1995.

Roberts, Johnnie L. "An Urge to Merge: Foxes in the Chicken Coop." *Newsweek* (New York), 11 September 1995.
———. "Time's Uneasy Pieces." *Newsweek* (New York), 2 October 1995.
United States Congress, House Committee on the Judiciary, Subcommittee on Economic and Commercial Law.

Time-Warner Merger: Competitive Implications. Hearing before the Committee. Washington, D.C.: United States Government Printing Office, 1989.

See also Levin, Gerald; Turner Broadcasting Systems; Turner, Ted

TINKER, GRANT

U.S. Producer/Media Executive

While Grant Tinker's career in television spans more than 30 years and a number of positions in network programming and production, he is best known for his work in the 1970s and 1980s as founder and president of MTM Enterprises, and as "the man who saved NBC" when he served as the network's chair and chief executive officer from 1981 to 1986. Throughout his career, he has been associated with literate, sophisticated programming usually referred to as "quality television."

Tinker and wife Mary Tyler Moore formed MTM Enterprises in 1970 to produce *The Mary Tyler Moore Show* when Moore was offered a 13-episode series commitment from CBS. Tinker put into practice his philosophy of hiring the best creative people and letting them work without interference from executives at the networks or at MTM. He built MTM into a "writers' company" that produced some of the most successful and award-winning series of the 1970s and 1980s. Beginning with the writer-producer team of James Brooks and Allan Burns, who created *The Mary Tyler Moore Show*, Tinker and MTM nurtured the talents of a host of top writers and producers whose work would go on to dominate network television schedules and the Emmy Awards through the 1990s. The staff included Gary David Goldberg, Steven Bochco, Bruce Paltrow, Mark Tinker, Hugh Wilson, Joshua Brand, and John Falsey. MTM's early hits were primarily sitcoms in the *Mary Tyler Moore* mold (including spin-offs *Rhoda* and *Phyllis*) as well as *The Bob Newhart Show* and *WKRP in Cincinnati.* Beginning in the late 1970s and 1980s, however, MTM produced a number of network television's most successful and innovative dramas, including *Lou Grant, The White Shadow, Remington Steele, Hill Street Blues,* and *St. Elsewhere,* shows which benefited from Tinker's combination of benign neglect in creative matters and tenacious support in dealing with the networks.

In 1981, Tinker left MTM to become chair and chief executive officer of NBC, the perennial last-place network. With no shows in the Nielsen top ten, and only two in the top 20, NBC had suffered through a season of dismal profits (one-sixth the level of ABC's or CBS') and affiliate defections. Based on the belief that good-quality programming makes a strong network, Tinker worked with programming chief Brandon Tartikoff to revitalize NBC's prime-time schedule. They allowed low-rated but promising series to remain on the schedule until they built an audience, and courted the best producers to supply the network with

programs. Under this philosophy, NBC recovered first the upscale urban audience prized by advertisers, then industry approval with more Emmy Awards than CBS and ABC combined, and finally rose to first place in the ratings with blockbusters like the famed Thursday night lineup—*Cosby, Family Ties, Cheers, Night Court,* and *Hill Street Blues*—billed as "the best night of television on television." That his programming strategy relied heavily on work from MTM (*Hill Street Blues, St. Elsewhere,* and *Remington Steele*) and MTM alumni (Goldberg's *Family Ties,* Charles Burrows and Glen and Les Charles' *Cheers*) eventually cost Tinker his share of MTM, when NBC's parent company RCA ordered him to sell in the early 1980s. In any case, NBC's turnaround helped shore up the network system in an era when new

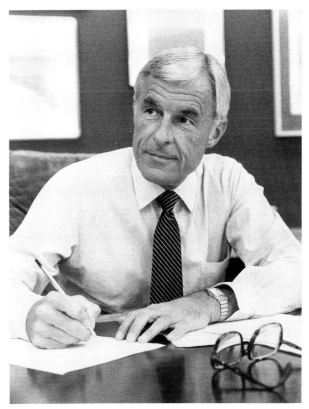

Grant Tinker
Photo courtesy of Grant Tinker

programming alternatives such as cable and VCRs had begun eroding the once-monolithic network audience. Tinker left NBC in 1986, shortly after it had been acquired by General Electric.

His stint as chair and chief executive officer was not Tinker's first experience with NBC. In 1949, after graduation from Dartmouth, he became the network's original executive trainee, learning about each of its departments before settling into a job in the station's night operations. He left the network in 1951 for employment in a series of production and programming jobs in radio, television, and advertising. He served as director of program development at McCann Erickson in the early 1950s, when advertisers were responsible for producing much of the networks' schedules, and at Warwick and Legler, where he rehabilitated Revlon's corporate image after it had been tarnished in the quiz show scandals. He also served as Benton and Bowles' vice president in charge of programs, where he was involved in developing Proctor and Gamble's *The Dick Van Dyke Show*, and where he met his second wife, Mary Tyler Moore.

Tinker returned to NBC in the early 1960s as West Coast head of programs, with responsibility for program development of a number of popular series, including *Bonanza, I Spy, Dr. Kildaire,* and *The Man from U.N.C.L.E.* After returning to New York to serve as the network's vice president in charge of programs, he left NBC to work as a production executive at Universal (where he was instrumental in birthing *It Takes a Thief* and *Marcus Welby, M.D.,* as well as *The ABC Movie of the Week*) and 20th Century-Fox, before forming MTM in 1970.

After serving as NBC chair and chief executive officer, Tinker tried to repeat the success of MTM Enterprises by forming GTG (Grant Tinker-Gannett) Entertainment with the communications giant Gannett, producer of the syndicated news-magazine *USA Today on TV* and the dramatic program *WIOU*, which aired for a short time on CBS. The partnership was dissolved in 1990.

—Susan McLeland

GRANT TINKER. Born in Stamford, Connecticut, U.S.A., 11 January 1926. Educated at Dartmouth College. Married: 1) Ruth Byerly (divorced); one daughter and three sons; 2) Mary Tyler Moore, 1963 (divorced, 1981). Worked in radio program department, NBC, 1949–54; TV department, McCann-Erickson Advertising Agency, 1954–58; Benton and Bowles Advertising Agency, 1958–61; vice president of programs, West Coast, NBC, 1961–66; vice president in charge of programming, West Coast, NBC, New York City, 1966–67; vice president, Universal TV, 1968–69; vice president, 20th Century-Fox, 1969–70; president, Mary Tyler Moore (MTM) Enterprises, Inc., 1970-81; chair of the board and chief executive officer, NBC, Burbank, California, 1981–86; independent producer, Burbank, since 1986; president, GTG Entertainment, Culver City, California, 1986–90.

PUBLICATIONS

"With NBC Still Rated No. 3, Grant Tinker Ponders His Own Decisions—And the Audience's" (interview). *People Weekly* (New York), 14 May 1984.
Tinker in Television: From General Sarnoff to General Electric, with Bud Rukeyser. New York: Simon and Schuster, 1994.

FURTHER READING

Auletta, Ken. *Three Blind Mice: How the TV Networks Lost Their Way.* New York: Random House, 1991.
Coe, Steve. "Tinker Writes the Book on Television; Former NBC Chairman Looks at 40 Years inside the Magic Box." *Broadcasting and Cable* (Washington, D.C.), 5 September 1994.
"NBC's Tortoise Overtakes the Hares." *Broadcasting* (Washington, D.C.), 5 November 1984.

See also *Dick Van Dyke Show; Mary Tyler Moore Show;* Moore, Mary Tyler; National Broadcasting Company; United States: Networks

TINKER TAILOR SOLDIER SPY

British Thriller/Miniseries

When first broadcast in September 1979, *Tinker Tailor Soldier Spy* was greeted with opposing voices as "turgid, obscure, and pretentious" or as "a great success." It is in keeping with the ambiguous nature of John Le Carré's narratives that one can simultaneously agree with both formulations without contradiction. As Roy Bland, paraphrasing Scott Fitzgerald, observes: "An artist is a bloke who can hold two fundamentally opposing views and still function." The obscurity is a consequence of the themes of deception and duplicity at the centre of the narrative: to those who, like Sir Hugh Greene, prefer the moral certainties of

Buchan's version of British Intelligence, Le Carré's world will not only be difficult to follow but morally perplexing. On the other hand, the success of the serial was not only demonstrated by good audience ratings but by general critical acclaim for the acting, a judgment ratified by subsequent BAFTA awards for best actor (Alec Guinness) and for the camerawork of Tony Pierce-Roberts. Ambiguity persisted in America where the serial won critical acclaim when shown on PBS but failed to be taken up by the networks.

Although Le Carré published his first novel, *Call For the Dead,* in 1961, and his first major success, *The Spy Who*

Came in from the Cold (1963), was turned into a film in 1966, *Tinker Tailor Soldier Spy* was his first venture into television. He rejected the project of turning it into a film because of the compression, but felt the space afforded by TV serialization would do justice to his narrative. He was also impressed with the skill of Arthur Hopcraft's screenplay which extensively reordered the structure of the novel to clarify the narrative for a television audience without violating its essential character (Hopcraft for example begins the narrative with the debacle in Czechoslovakia which only begins to be treated in the novel in chapter 27). Le Carré was even more taken by the interpretation of Smiley provided by Alec Guinness, so much so that as he was writing *Smiley's People* he found himself visualizing Guinness in the role and incorporated some of the insights afforded by the actor in the sequel to the trilogy. A trivial example will stand for many. During the production of *Tinker Tailor,* Guinness complained that the characterizing idiosyncrasy of Smiley, polishing his glasses with the fat end of his tie, cannot be done naturally because the cold weather in London means that Smiley will be wearing a three-piece suit, thus a handkerchief has to be substituted. At the end of *Smiley's People* Le Carré includes a teasingly oblique rejoinder:

> From long habit, Smiley had taken off his spectacles and was absently polishing them on the fat end of his tie, *even though he had to delve for it among the folds of his tweed coat* [emphasis added].

The story of *Tinker Tailor* has an archetypal simplicity reminiscent of the *Odyssey:* the scorned outsider investigates the running of the kingdom, tests the loyalty of his subjects and kin by means of plausible stories before disposing of the usurpers and restoring right rule. In Le Carré's modern story the elements are transposed onto the landscape of conflicted modern Europe in the throes of Cold War.

A botched espionage operation in Czechoslovakia ensures that Control (Head of British Intelligence) and his associates are discredited. Shortly after, Control dies, George Smiley his able lieutenant is retired and the two are succeeded by Tinker Tailor Soldier Spy: Percy Alleline, Bill Haydon, Roy Bland and Toby Esterhaze. Six months later, Riki Tarr, a maverick Far Eastern agent, turns up in London with a story suggesting there is a mole (a deeply concealed double agent) in the Circus (intelligence HQ, located at Cambridge Circus). Lacon of the Cabinet Office entices Smiley out of retirement to investigate the story. Smiley gradually pieces together the story by analyzing files, interrogating witnesses and trawling through his own memory and those of other retired Circus personnel, notably Connie Sachs (a brilliant cameo role played by Beryl Reid), until he finally unmasks the mole "Gerald" at the heart of the Circus.

The mood of the story, however, is far from simple. Duplicity and betrayal, personal as well as public (Smiley's upperclass wife is sexually promiscuous, betraying him to "Gerald"), informs every aspect of the scene. While the traitor is eventually unmasked the corrupt nature of the intelligence service serves as a microcosm of contemporary England: secretive, manipulative, class-ridden, materialistic and emotionally sterile. Thus, if the Augean stables have been cleaned, they will be soon be soiled again. This downbeat tone accounts for the serial not being taken up by the American networks and marks it off from the charismatic spy adventures of James Bond, but it also accounts for its particular appeal to British middle-brow audiences.

The spy genre is virtually a British invention: although other countries produce spy writers, the centrality of the genre to British culture is longstanding and inescapable: John Buchan, Somerset Maugham, Graham Greene, Ian Fleming, Frederick Forsyth, Len Deighton, as well as John Le Carré, have all achieved international success for their spy stories—not to mention television dramas by Dennis Potter (*The Blade on the Feather*) and Alan Bennett (*An Englishman Abroad* and *A Question of Attribution*). To account for this obsession with spies, we only have to consider the political circumstances of Britain in the twentieth century: a declining Imperial power, whose overseas possessions have to be ruled and defended more by information than by outright physical force; an offshore island of a divided Europe, seeing itself threatened by German, then Soviet, military ambitions. Perhaps even more significant than these external threats are those from within. A ruling class which maintains its grip on power by exclusion—a public school and Oxbridge educated elite hold a disproportionate share of positions of power in Cabinet, Whitehall, the BBC and government institutions—is liable to marginalize or demonize those who openly challenge its assumptions. The result is liable to be subversion from within—a tactic fostered by the duplicitous jockeyings for power of rival gangs in the enclosed masculine world of the public schools. The symbolic and emotional link between the world of the public school and that of the circus is established in *Tinker Tailor* by Jim Prideaux. The injured and betrayed agent teaches at a prep school after his failed Czech mission and enlists the aid of a hero-worshipping pupil as his watcher. Thus the fictions that Le Carré invented have their counterpart in the real world and tap familiar English fears and obsessions. In the same year, 1979, that saw the serialization of *Tinker Tailor*, the BBC also produced two documentary series *Public School* and *Spy,* reinforcing the connections with Le Carré's work. "The Climate of Treason" concerned itself with speculating about the Fourth Man of the Burgess, MacLean, Philby double agents within MI5. On 15 November 1979 Margaret Thatcher identified Sir Anthony Blunt, adviser of the Queen's Pictures and Drawings, as the Fourth Man who had been recruited by the Russians in the 1930s. Le Carré's novel was read as a fictionalized version of these events.

The success of *Tinker Tailor* lies in the realism, not only of character portrayal—and the acting of Alec Guinness has achieved as definitive a performance as Olivier's *Richard III* or Edith Evans as Lady Bracknell—but of the way in which intelligence institutions work. But the claim for realism must not be pressed too far: Le Carré has admitted that the vocabulary used was invented: babysitters, lamplighters, the Circus, the nursery, moles—though he was also amused to discover that real agents had begun to appropriate some of his vocabulary once the stories were published. Moreover, much intelligence work is bureaucratic and boring: Smiley's reflections turn the drudgery of reading files into a fascinating intellectual puzzle which, unlike the real experience, always produces significant information.

At the symbolic level, however, the portrayal of the workings of bureaucracy *is* authentic: bureaucracies serve those who govern by gathering, processing and controlling access to information. In a world increasingly governed by means of information, those who control it have power and wealth, so that the resonance of Le Carré's story will carry beyond the cold war setting that is its point of departure.

—Brendan Kenny

CAST

George Smiley	Alec Guiness
Annie Smiley	Sian Phillips
Tinker (Percy Alleline)	Michael Aldridge
Soldier (Roy Bland)	Terence Rigby
Poor Man (Toby Esterhaze)	Bernard Hepton
Peter Guillam	Michael Jayston
Lacon	Anthony Bate
Control	Alexander Knox

PRODUCER   Jonathan Powell

PROGRAMMING HISTORY   7 50-minute episodes

• BBC

10 September 1979–22 October 1979

FURTHER READING

Bloom, Harold. *John Le Carré*. New York: Chelsea, 1987.
Bold, Alan Norman. *The Quest for Le Carré*. New York: St. Martin's, 1988.
Lewis, Peter. *John le Carré*. New York: Ungar, 1985.
Monaghan, David John. "Le Carré and England: A Spy's-Eye View." *Modern Fiction Studies* (West Lafayette, Indiana), Autumn 1983.

See also British Programming; Spy Programs

TISCH, LAURENCE

U.S. Media Mogul

In 1986 Laurence Tisch, a fabled Wall Street investor, took control of CBS, often considered the crown jewel of American broadcasting. Tisch ran the TV network, the owned and operated television stations, and other corporate properties until 1995. Throughout the decade, he manipulated and modified CBS, looking to cash in with an eventual sale of the property. In 1995 the deal came through. Westinghouse offered $5 billion for CBS; Tisch made an estimated $2 billion.

In the view of many television critics and media industry observers, Tisch badly mismanaged the former "Tiffany network" with policies that caused ratings to drop, earnings to fall, and affiliates to defect. In a stunning pair of 1994 deals, fellow mogul Rupert Murdoch contracted broadcasting rights for the National Football League (NFL) and tempted a number of CBS affiliates to switch to the FOX Broadcasting Company. CBS was further embarrassed when Tisch demoted Connie Chung from her position as co-anchor position with Dan Rather. And media pundits lambasted Tisch for CBS' Sunday afternoon golf coverage in May 1995 when ABC and NBC carried President Bill Clinton's address to the mourners of the Oklahoma City bombing. CBS opted to stay with the golf tournament to save $1 million in advertising.

Andy Rooney, long a fixture on CBS' highest rated show, *60 Minutes,* stated openly what many in the industry felt about Tisch's negative impact on CBS' long-fabled news division. On rival network ABC's *Primetime Live* Rooney castigated Tisch for allowing CBS to slip: "We need a hero in the business. I don't see why someone like Larry Tisch . . . doesn't say: 'I've got all this money, why don't I just make the best news division in the world.'"

Tisch's relations with CBS had not begun on such a rancorous note. During the mid-1980s, when Ted Turner tried to make a make a hostile bid for CBS, longtime CBS chief William S. Paley looked for a "white knight" to save his beloved company. In October 1985, Paley and his hand-chosen corporate directors asked Tisch to join the CBS board and thwart Turner. Before his takeover, Tisch had simply been another faceless New York City multimillionaire, making money in tobacco, insurance, and hotels. His rescue of CBS made him a media celebrity.

After serving in the U.S. Army's intelligence office during World War II, Tisch joined forces with his younger brother Bob and began his rise to corporate power and profit with the 1949 purchase of Laurel-in-the-Pines, a New Jersey hotel. For the next decade the brothers bought and sold hotels, particularly in Miami Beach and Atlantic City. In 1959, the Tisch brothers bought the Loews theater chain from Metro-Goldwyn-Mayer, and changed the name of their company to Loews Corporation.

From this base they continued to expand their investment efforts and by the mid-1980s Loews Corporation ranked as a multibillion dollar conglomerate success story. Loews was built by acquiring other companies through tender offers, beginning with the takeover of Lorillard, a tobacco products company, in 1968. In early 1974 Loews announced it had acquired just over 5% of an insurance subsidiary CNA Financial, then an independent company. Before the end of that year Loews had successfully completed a hostile tender offer for the company's stock, and CNA became the principal source of Loews' success. In the case of both Lorillard and CNA the Tisch brothers reversed the fortunes of ailing companies and made millions in the process.

Privately, Laurence Tisch then began to undertake philanthropic causes. He managed the investments of the Metropolitan Museum of Art, provided endowment and buildings for New York University, and led fund-raising for the United Jewish Appeal.

In 1986 William S. Paley stepped aside and Tisch became not only CBS' major stockholder, but chief executive officer as well. To no one's surprise, Tisch restructured the company into a "lean and mean" operation. Within months, he had launched the biggest single staff and budget reduction in network TV history. When the dust had settled, hundreds had lost their long secure jobs, news bureaus had been shuttered, and CBS was a shell of its former self.

On a larger corporate scale, Tisch systematically began to sell every CBS property not connected to television. First sold was CBS' educational and professional publishing, which included Holt, Rinehart and Winston, one of the country's leading publishers of textbooks; and W.B. Saunders, a major publisher of medical tomes. CBS picked up $500 million in the deal.

But that sum proved small change compared to the $2 billion paid by Sony Corporation of Japan for the CBS Music Group. One of the world's dominant record and compact disc companies, CBS Music boasted a stable of stars that then included Bruce Springsteen, Michael Jackson, the Rolling Stones, Billy Joel, Cindy Lauper, Paul McCartney, and James Taylor. This single 1987 sale enabled the new CBS to earn a substantial profit that year.

With the layoffs, budget cuts, and sales of CBS properties completed, Tisch faced the need to improve TV programming. This proved difficult, and speculation began about precisely when Tisch would cash in his CBS stock. Potential buyers for the network included MCA/Universal Pictures, Disney, Viacom, and QVC, a television home-shopping company. Throughout the early 1990s, Tisch quietly engineered stock repurchases by CBS, and by selling much of his own stock back to the corporation he covered his original investment. Whatever he would receive for his remaining 18% of the company would be pure profit. Thus the 1995 Westinghouse deal moved Tisch from the status of a multimillionaire to a multibillionaire. In television history, however, Laurence Tisch would be remembered for how he had decimated the once dominant television network.

—Douglas Gomery

LAURENCE ALAN TISCH. Born in New York City, New York, U.S.A., 15 March 1923. Educated at New York University,

Laurence Tisch
Photo courtesy of Broadcasting and Cable

B.Sc. cum laude 1942; University of Pennsylvania, M.A. in industrial engineering 1943; studied at Harvard Law School, 1946. Married: Wilma Stein, 1948; four sons. Served in U.S. Army, Office of Strategic Services, during World War II. President, Tisch Hotels, Inc., New York City, 1946–74; chair of the board and chief executive officer, Loews Corp., New York City, from 1960, co-chief executive officer, from 1988; president, chief executive officer, CBS Inc., New York City, from 1987, chair, president, chief executive officer, board of directors, from 1990. Chair and member of board of directors, CNA Finance Corporation, Chicago. Board of directors: Bulove Corporation, New York City; ADP Corporation; Petrie Stores Corporation; R.H. Macy and Company; United Jewish Appeal Federation; chair, board of trustees, New York University. Trustee: Metropolitan Museum of Art, New York City; New York Public Library; Carnegie Corporation.

FURTHER READING

Auletta, Ken. *Three Blind Mice: How the TV Networks Lost Their Way.* New York: Random House, 1991.
———. "Network for Sale?" *The New Yorker,* 25 July 1994.
Diamond, Edwin. "The Tisching of CBS; The New Chief Takes Control—But Where Is He Going?" *The New York Times* 16 May 1988.

Klein, Edward. "Eye of the Storm." *The New York Times,* (New York), 15 April 1991.

"Larry Tisch and the New Realism at CBS" (interview). Broadcasting (Washington, D.C.), 27 October 1986.

Loftus, Jack. "Tisch Is Keeping it Simple: Bets the Farm on CBS-TV" (interview). *Television-Radio Age* (New York), 28 September 1987.

McCabe, Peter. *Bad News at Black Rock: The Sell-Out of CBS News.* New York: Arbor House, 1987.

Peretz, Martin. "Media Moguls." *The New Republic* (Washington, D.C.), 4 September 1995.

"The Perils of Networking." *The Economist* (London), 10 December 1994.

See also Columbia Broadcasting System; Paley, William S.

TODMAN, BILL See GOODSON, MARK, AND BILL TODMAN

THE TOMMY HUNTER SHOW

Canadian Country Music Program

Known as "Canada's Country Gentleman", Tommy Hunter was for many years one of Canada's most popular and well-known television personalities. He became a fixture on Canadian television as the host of *The Tommy Hunter Show,* one of North America's longest-running variety shows, and is also one of the few figures in Canadian popular music to have evolved through television rather than through recording and radio airplay. He has received numerous awards for his role in television, in country music, and in Canadian cultural life.

The Ontario native's career in television started when he was 19 years old on *Country Hoedown,* a weekly country music program produced and aired at the Canadian Broadcasting Corporation (CBC), where Hunter would spend the rest of his television career. The show was an on-stage revue with a house band and featured various musical guests both from Canada and the United States. Starting out as a rhythm guitarist, Hunter soon became a featured performer on the show, leading to his own daily noontime CBC radio program, *The Tommy Hunter Show;* it became a television series in 1965.

Much of *Country Hoedown*'s format and tone were carried over into *The Tommy Hunter Show.* Over its 27-year run on CBC (1965–92)—rerun three times a week on the Nashville Network between 1983 and 1991—the show was noted for nurturing Canadian country music, which it showcased alongside big-name American country stars. Hunter wanted to break with the hokey, country-hick feel characterized by shows like *Hee Haw,* though, and tried to present country music as "respectable". The result was a program that some labelled a country version of Lawrence Welk's show. Inspired by television variety-show hosts such as Johnny Carson and Perry Como, Hunter felt that the host should have a relaxed, comfortable style, establishing a certain rapport with the audience. By sticking to his country purist approach, he was able to establish such a rapport, building up an intensely loyal fan base which planned its Saturday evenings around *The Tommy Hunter Show.* Over the years, Hunter sustained an ongoing battle with CBC

producers who wanted to rely on demographics and "slickify" the show. He maintained that targeted programming precluded establishing a real relationship with the audience. His show relied upon the on-stage revue format, which mixed various musical sequences with dance and other coun-

Tommy Hunter
Photo courtesy of the Country Music Foundation

try entertainment. Despite attempts to alter the program by incorporating other styles and sensibilities, Hunter persevered in maintaining the show's traditional country tone. It was this purist approach that would ultimately sound the show's death knell, however, and a lack of younger viewers and slipping audience ratings led to its cancellation in 1992.

As a long-running music television program, *The Tommy Hunter Show* demonstrated how television's imbrication with popular music dates back long before the rise of MTV and the music video. Hence, while it provided country music fans with entertainment each week, it also helped to rearticulate a brand of country music many associated with Nashville as a Canadian popular music genre, in a period which saw the rise of an anti-American Canadian cultural nationalism. Indeed, through the program's year-in, year-out presence on the CBC, the state-owned broadcaster and self-styled "national network," the country music of *The Tommy Hunter Show* became a national symbol for many Canadians, and Tommy Hunter a figure of Canadianness. This ability of television to reach around the generic division of popular music into record or radio formats, then, helped shape a "Canadian country music" genre which would combine the traditional music of Canadian folk performers with the country music of artists like Tommy Hunter.

As much as *The Tommy Hunter Show* displayed how television intervenes into other areas of popular culture such as popular music, though, it also threw into relief the tensions that arise between them. The behind-the-scenes conflict between CBC television workers and Tommy Hunter, a country musician, derived from their emergence from two separate cultural formations: on the one hand, the "world" of television production, with its own sensibilities and its own priorities; and on the other hand, the "world" of country music, with its internal organization and logic. Thus, CBC personnel wanted to target specific demographic ranges by "updating" the show with natty set designs and a wider variety of musical styles. But Hunter's desire for austere sets and traditional country music, and his concern with providing family entertainment for a country audience, derived from the emphases on "sincerity" and "authenticity" which underpin country music as a genre and define fundamental aspects of the country music "world". Indeed, the conflicts behind *The Tommy Hunter Show* foreshadowed a later reticence towards music videos on the part of the country music industry as a whole, wary of the videoclip format's "slickness" that is so antithetical to country music's "authenticity".

The privileged role played by authenticity in country music, with its accompanying stress on "ordinary people", was central to *The Tommy Hunter Show*. Although based in Toronto, the show went on the road frequently, playing to sold-out audiences across Canada. Hunter's insistence that the set in each city reflect the locale of the taping illustrated his constant striving to reinsert a local feel into the globalizing pull of television. A harsh critic of the television industry even as a television star, Hunter felt that TV programmers had little understanding of country music audiences; for Hunter, the institutional imperatives of a mass-mediated country music

compromised his audience's position. These views carried over to his recording career. Hunter preferred to record albums independently rather than with major record labels, reasoning that this would allow him to aim at pleasing country audiences, rather than radio stations. And in 1992, following cancellation of *The Tommy Hunter Show*, he toured Canada with a stage version of the show, playing to sold-out audiences, meeting his fans from the other side of the television screen.

The only program to survive a wave of rural, family-oriented CBC programming in the 1950s and 1960s that included shows like *Don Messer's Jubilee*, *The Tommy Hunter Show* was a country show produced in an urban environment. It was a family-oriented show in an age of splintering demographics. It made a country singer into a television star. And in the process it had a profound impact on the Canadian popular music landscape. By the end of the show's run, Hunter had won three Juno Awards as Canada's best male country singer (1967–1969) and become the fifth Canadian to be inducted into the Country Music Hall of Fame's Walkway of Stars (1990) for his music; he received an award from the Broadcast Executive Society as well as a Gemini Award for best Canadian variety show (1991); and was named to the Order of Canada for his part in Canadian cultural life.

—Bram Abramson

REGULAR PERFORMER
Tommy Hunter

PRODUCERS   Dave Thomas, Bill Lynn, David Koyle, Les Pouliot, Maurice Abraham, Joan Toson, and others

PROGRAMMING HISTORY

• CBC

| 1965–1970 | Half hour weekly during fall/winter season |
| 1970–1992 | One hour weekly |

FURTHER READING

Abramson, Bram. "'A Country of One's Own': Rita MacNeil, Infomercials, and Canadian Country Music." In, Anderson, C., editor. *Working Papers in Canadian Studies: Proceedings of the "Instituting Cultures/Cultural Institutions" Conference*. Montreal: McGill Institute for the Study of Canada, 1995.

Conrad, Charles. "Work Songs, Hegemony, and Illusions of Self." *Critical Studies in Mass Communication* (Annandale, Virginia), Spring 1988.

Fenster, Mark. "Country Music Video." *Popular Music* (Detroit, Michigan), 1988.

Hunter, Tommy, with Liane Heller. *My Story*. Agincourt, Ontario, Canada: Methuen, 1985.

Lacey, Liam. "Canada's Country Gentleman." *The Globe and Mail* (Toronto), 24 October 1987.

Marquis, Greg. "Country Music: The Folk Music of Canada." *Queen's Quarterly* (Kingston, Ontario, Canada), Summer 1988.

See also Canadian Programming in English

TONIGHT

British Magazine Programme

Tonight was a 40-minute topical magazine programme which went out every week-day evening between 6:00 P.M. and 7:00 P.M., and was first broadcast by the BBC in February 1957. The programme was produced under the aegis of the BBC's Talks Department by Alasdair Milne and edited by Donald Baverstock, who later went on to occupy a senior position within the BBC. It was presented by Cliff Michelmore, who had already collaborated with Baverstock and Milne on *Highlight,* a shorter, less ambitious version of *Tonight.* With *Tonight,* Michelmore quickly acquired status as a broadcaster, picking up an award for artistic achievement and was twice named Television Personality of the Year. Indeed *Tonight* was significant for its ability to attract and cultivate new broadcasting talent and over its eight-year run managed to launch a number of notable careers including those of Alan Whicker, Ned Sherrin, Julian Pettifer and Trevor Philpott.

The programme was conceived by the BBC as their response to the ending of the "toddlers' truce"—the hour in the evening when television closed down to allow parents to see their children off to bed. As such, *Tonight* went out to a new and untried audience, an audience who, at this time of the evening, would be quite active rather than settled, who would be busy preparing food, putting kids off to bed or getting ready to go out. *Tonight* was designed around the needs of this audience and its style reflected this: the tone was brisk and informal, mixing the light with the serious and items were kept short allowing audiences to "dip in" at their convenience. This emphasis on the needs of the audience was something of a departure for the BBC who had tended to adopt a paternal tone with its viewers, giving them not what they wanted but what they should want. *Tonight* was going to be different. It wasn't to talk down to the viewer, but would, as the *Radio Times* put it, "be a reflection of what you and your family talk about at the end of the day" (Watkins: 29). In Baverstock's words, *Tonight* would "celebrate communication with the audience", and indeed the programme came across not as the institutional voice of the BBC but as the voice of the people.

Tonight was recognised by many to be evidence of the BBC's fight back against the new Independent Television companies who were quickly gaining ground and by 1957

Tonight
Photo courtesy of BBC

had overtaken the BBC with a 72% share of the audience. But if *Tonight* was largely a result of competition and the breaking of the monopoly, which in effect forced the BBC to adopt a more populist programming philosophy, the style and content of the programme also reflected broader social and cultural changes. *Tonight* seemed to capture an emerging attitude of disrespect and popular scepticism towards institutions and those in authority. Furthermore, the adjectives which were often used to describe the programme at the time, such as "irreverent", "modern", and "informal", could have easily described the mood that was beginning to inform other areas of the arts and popular culture.

Tonight introduced a number of innovations to British television. It was one of the first programmes to editorialise and adopt a point of view, flaunting the Public Service demands of balance and impartiality. The programme also introduced a new (some might say aggressive) style of interviewing where guests would be pushed and harassed if it was thought they were being evasive or dishonest. *Tonight* eschewed the carefully prepared question and answer type format that had prevailed in current affairs programming until then. Furthermore, broadcasters had tended to fetishise the production process, concealing the means of communication and carefully guarding against mistakes and technical breakdowns which threatened to demystify the production. *Tonight,* though, kept in view such things as monitors and telephones. Its interviews were kept unscripted and any technical faults or mistakes were skillfully incorporated into the programme flow giving *Tonight* an air of spontaneity and immediacy.

Tonight was meant to be a temporary response to the ending of the "toddlers' truce" and was initially given a three-month run. It quickly proved popular however and within a year was drawing audiences of over 8 million. In addition the programme won critical acclaim, receiving the Guild of Television Producers Award for best factual programme in 1957 and 1958. The programme generated other material as well, including feature length documentaries and was the inspiration behind *That Was the Week That Was,* a show that stepped up *Tonight's* irreverent, hard-hitting approach for a late night adult audience.

Baverstock left *Tonight* in 1961 to become assistant controller of programmes and his place was taken by Alasdair Milne. Milne proved to be a capable editor and indeed oversaw a number of innovations including the feature length documentaries.

However, the programme would not be the same without Baverstock whose leadership and vision had made *Tonight* something of an individual success. By 1962 it was felt the programme had become rigid and stale. As is the case with many innovative and ground breaking enterprises, the programme could not sustain the pace of its initial inventiveness. The final edition went out in June 1965. Nevertheless in its eight-year run it had established a format for current affairs programming which mixed the light with the serious, which blurred distinctions between education and entertainment and which managed in the process to soften the image of the BBC, transforming it, as Watkins has noted, from an "enormous over-sober responsible corporation", to something that looked "more like a man and a brother".

—Peter McLuskie

ANCHOR
Cliff Michelmore

FIELD REPORTERS
Derek Hart, Geoffrey Johnson Smith, Alan Whicker, Fyfe Robertson, Trevor Philpott, Macdonald Hastings, Julian Pettifer, Kenneth Allsop, Brian Redhead, Magnus Magnusson

PRODUCER   Donald Baverstock

DIRECTOR
Alisdair Milne

PROGRAMMING HISTORY

• BBC

1957–1965	Weekdays 6:00-7:00 P.M.

FURTHER READING

Briggs, Asa. *The History of Broadcasting in the United Kingdom, Volume 5: Competition.* Oxford: Oxford University Press, 1995.

Corner, John, editor. *Popular Television in Britain.* London: British Film Institute, 1991.

Goldie, Grace Wyndham. *Facing the Nation: Broadcasting and Politics,* 1936–76. London: Bodley Head, 1977.

Watkins, G., editor. *BFI Dossier 15: Tonight.* London: British Film Institute, 1982.

See also British Programming

THE TONIGHT SHOW

U.S. Talk/Variety Show

A long-running late-night program, *The Tonight Show* was the first, and for decades the most-watched, network talk program on television. Since 1954 NBC has aired a number of versions of the show which has, as of the mid-1990s, seen four principle hosts and one consistent format except for a brief diversion in its early days. What started out as a music, comedy and talk program first hosted by Steve Allen became, for a time, a magazine-type program,

Steve Allen

broadcasting news and entertainment segments from various correspondents located in different cities nationally. That short-lived format, however, lacked the appeal of a comedy-interview show revolving around one dynamic host. From mid-1957 until the present, Jack Paar, Johnny Carson and Jay Leno have all three followed Allen's lead and hosted a show of celebrity interviews, humor and music, each host leading his show with signature style. Late night talk in the first three decades of television was dominated by *The Tonight Show*, and for the majority of that time by Johnny Carson. However, during the 1980s and early 1990s the late-night landscape began to change as more talk shows took to the air. Change was accelerated by the appeal of David Letterman and a combination of other factors, including inexpensive production, audience interest in celebrity and entertainment gossip, and an overall increased reliance on the talk show as forum for information and debate about the important as well as unimportant issues of the day. The late-night talk genre expanded as network competitors and

comrades sought the kind of success that was originally the province of *The Tonight Show*.

Each of *The Tonight Show* principal hosts brought his own unique talent and title to the program. All of the shows featured an opening monologue, a sidekick or co-host, in-house musicians and cadre of guest hosts. Steve Allen's *Tonight!* featured his musical talents and penchant for unique comedy. He was well known for performing his own musical numbers on the piano and for humorous antics such as on-the-street improvisations and bantering with the audience, both of which were forerunners to the kinds of comedy stunts that became a staple much later on *Late Night with David Letterman*, also on NBC. In 1957 Allen left *Tonight!* to concentrate on another variety show he hosted on Sunday evenings. Allen's version of the show was immediately followed by the unsuccessful magazine format, *Tonight: America After Dark*, which lasted only a few weeks. That show was led by Jack Lescoulie, but he was never the central figure Allen had been. Essentially,

Jack Paar (right) with Hugh Downs

Lescoulie introduced the segments and correspondents around the nation.

In July 1957 Jack Paar took over as new host of *The Jack Paar Tonight Show*. Paar brought the show back to its in-studio interview format. More a conversationalist than comedian, audiences were drawn to Paar's show because of the interesting guests he brought on, from entertainers to politicians, and for the controversy that occasionally erupted there. Paar did not shy away from politics or confrontation, and often became emotionally involved with his subject matter and guests. He had a few stormy run-ins, both on camera and off, and finally left the show following controversy surrounding his broadcast from the Berlin Wall in 1962. With another change in hosts came a complete change in tone and style.

In October 1962 Johnny Carson took over as host of *The Tonight Show Starring Johnny Carson*. Carson was more emotionally detached and less political than Paar. He, like Allen, was a comic. Named the king of late night, Carson hosted the show for thirty years, from 1962 to 1992. During that time the show moved from New York City to Burbank, California. Carson was known for his glib sense of humor and his middle-American appeal, and quickly recognized his increasing popularity as well as the strain of doing comedy and talk five nights a week. He threatened to leave the show, but was lured back with a generous offer that included a huge salary increase and more time off. Guest hosts during Carson's tenure included comedians Joan Rivers, Jay Leno and David Letterman.

When Carson retired, Jay Leno was appointed the next principal host of *The Tonight Show with Jay Leno*. Leno, a well-known stand-up comedian, brought to the show his own writers and comic style, showcasing it in his opening monologues and banter with guests.

Changes in Leno's show reflected other major changes in television since its earlier days. By the late 1980s late-night talk had become slightly less a white male domain. Joan Rivers hosted her own talk show for a short time, and popular black comedian Arsenio Hall had his own show which enjoyed a wide following, attracting mostly a young black audience, a segment previously ignored in late-night talk. The first leader of Jay Leno's late-night studio band was the accomplished black jazz musician Branford Marsalis. The second band leader and Leno sidekick was Kevin Eubanks, also black. A big change for *The Tonight Show* during Leno's tenure was its first serious competition.

Starting in the mid- to late-1980s, television talk shows, both daytime and late-night, multiplied in number. The in-studio talk program was inexpensive to produce and audiences were increasingly drawn to the sensationalism and celebrity showcased each day and night on television. Some late-night talk shows—including those hosted by Joan Rivers, Chevy Chase and Pat Sajak on the FOX network—came and went quickly. Arsenio Hall's show was on the air for several years before cancellation. Especially successful in late night was the up-and-coming David Letterman. *Late Night with David Letterman* started out on NBC, airing immediately after *The Tonight Show*

Johnny Carson (center with Doc Severinsen and Ed McMahon)

from 1982 until 1993. Passed over for the host position on *The Tonight Show* when Leno was chosen for the post, Letterman moved to CBS where his new show ran in direct competition with Leno.

For the first time *The Tonight Show* shared the late-night spotlight. The two host/comedians, Leno and Letterman, were polished performers with large audiences. They became, as Carson had been, the gauge by which mainstream entertainment and politics were measured. On both programs comedy was delivered—and guests and issues of day treated—the same way, as gossip and light entertainment. After four decades *The Tonight Show* was still outlining and defining, even when not at the forefront of, the essence of contemporary televised culture.

—Katherine Fry

THE TONIGHT SHOW
September 1954–January 1957

HOST
Steve Allen
Ernie Kovacs (1956–57)

REGULAR PERFORMERS
Gene Rayburn
Steve Lawrence
Eydie Gorme
Pat Marshall (1954–55)
Pat Kirby (1955–57)
Hy Averback (1955)
Skitch Henderson and His Orchestra
Peter Handley (1956–57)
Maureen Arthur (1956–57)
Bill Wendell (1956–57)
Barbara Loden (1956–57)
LeRoy Holmes and Orchestra (1956–57)

TONIGHT! AMERICA AFTER DARK
28 January 1957–26 July 1957

HOST
Jack Lescoulie (January–June)
Al "Jazzbo" Collins (June–July)

THE JACK PAAR SHOW
July 1957–March 1962

HOST
Jack Paar

REGULAR PERFORMERS
Hugh Downs
Jose Melis and Orchestra
Tedi Thurman (1957)
Dody Goodman (1957–58)

THE TONIGHT SHOW
2 April 1962–28 September 1962

ANNOUNCER
Hugh Downs
John Haskell
Ed Herlihy

REGULAR PERFORMERS
Skitch Henderson and His Orchestra

THE TONIGHT SHOW STARRING JOHNNY CARSON
October 1962–May 1992

HOST
Johnny Carson

REGULAR PERFORMERS
Ed McMahon
Skitch Henderson (1962–66)
Milton Delugg (1966–67)
Doc Severinsen (1967–92)
Tommy Newsom (1968–92)

THE TONIGHT SHOW WITH JAY LENO
May 1992–

HOST
Jay Leno

REGULAR PERFORMERS
Branford Marsalis (1992–1995)
Kevin Eubanks (1995–)

PROGRAMMING HISTORY

• NBC

Jay Leno

September 1954–October 1956
              Monday-Friday 11:30-1:00 A.M.
October 1956–January 1957
              Monday-Friday 11:30-12:30 A.M.
January 1957–December 1966
              Monday-Friday 11:15-1:00 A.M.
January 1965–September 1966
              Saturday or Sunday 11:15-1:00 A.M.
September 1966–September 1975
              Saturday or Sunday 11:30-1:00 A.M.
January 1967–September 1980
              Monday-Friday 11:30-1:00 A.M.
September 1980–August 1991
              Monday-Friday 11:30-12:30 A.M.
September 1991
              Monday-Friday 11:35-12:35 A.M.

FURTHER READING

Carter, Bill. *The Late Shift: Letterman, Leno, and the Network Battle for the Night.* New York: Hyperion, 1994.
Cox, Stephen. *Here's Johnny!: Thirty Years of America's Favorite Late-Night Entertainment.* New York: Harmony, 1992.
De Cordova, Frederick. *Johnny Came Lately: An Autobiography.* New York: Simon and Schuster, 1988.
Metz, Robert. *The Tonight Show.* New York: Playboy, 1980.
Munson, Wayne. *All Talk: The Talkshow in Media Culture.* Philadelphia, Pennsylvania: Temple University Press, 1993.
Smith, Ronald L. *Johnny Carson: An Unauthorized Biography.* New York: St. Martin's, 1987.
Tynan, Kenneth. *Show People.* New York: Simon and Schuster, 1979.

See also Allen, Steve; Carson, Johnny; Downs, Hugh; Leno, Jay; Letterman, David; National Broadcasting Company; Paar, Jack; Talk Shows

TOP OF THE POPS

British Music Programme

Top of the Pops is Britain's longest running pop music programme. It was first broadcast in January 1964 and since then has occupied a prime-time slot on BBC television. Its primary value has been in introducing generations of youngsters to the pleasures and excitement of pop music, while for older people the show has become a reassuringly familiar item in the television schedules.

The key to the show's success lay in its revolutionary new format. Before 1964 (and to a large extent after), pop shows tended to respond to emerging trends and fashions. Earlier shows such as *The Twist* and *The Trad Fad* were a response to current dance and music styles while the highly popular *Ready Steady Go* was largely a Mod programme and tended to showcase Mod lifestyles and tastes. The problem with this type of show was that its life cycle was bound to the fashion or style that it reflected: when it passed so did the show. What was unique about the *Top of the Pops* format was that it was based round the top 20 music chart—expanding to the top 40 in 1984. This meant that the show was not associated with a fashion or a trend; it had no angle on pop music but was merely responding objectively to whatever was popular at that moment. In this way *Top of the Pops* was always going to be current, it was always going to be at the cutting edge of pop music.

The format of the chart "countdown", coupled with the policy of only featuring records moving up the charts, provided the show with a certain structure and dynamism. Unlike many other pop shows *Top of the Pops* contained the narrative ingredients of development, anticipation and closure: with each episode, as the countdown commenced, the audience would be kept in suspense by the big question, "who will be top of the pops this week?"

In many respects the *Top of the Pops* format was informed by radio, the medium that had been closer to the pulse of teen tastes and pop trends. The top 20 format was already an established feature of radio and *Top of the Pops* presenters were nearly always radio DJ's. To this end early episodes of the programme tended to show a DJ putting the disc on the turntable with a fade to the performer miming to the song. The programme was about records and hits, and even when the performer was unavailable for the show the record would still go on, a policy that sometimes meant using improvised, and often innovative, visual effects to cover the absence of the performer.

Another factor contributing to the show's continuing popularity is its accessibility: while ostensibly aimed at a fairly small teenage audience, *Top of the Pops* has nevertheless always thought of itself as a family show. Indeed, audience research carried out in the 1980s found that the majority of the viewing constituency was over 25 years old. This appeal to a wider family audience has no doubt contributed to the show's continuing success and buoyant ratings; however it has also left the show open to charges of conservatism and

Top of the Pops
Photo courtesy of the British Film Institute

policing standards in musical taste; proof of this is usually offered by pointing to the show's infamous banning of the Sex Pistols and Frankie Goes to Hollywood.

Top of the Pops has been an important actor in the music business, with immense ability to make or break a performer. An appearance on the show could almost guarantee an immediate leap up the charts. Similarly pop music retailers have found that their sales often peak the day after the show is broadcast. There is no doubt therefore that *Top of the Pops* has functioned as a powerful gatekeeper to the industry and performers and promoters continue to clamour for a spot on the show.

Although the basic format of the chart countdown has remained constant over the years, the show has introduced many changes to keep itself up to date. Innovations such as the video chart, the "breakers" spot, Europarade and the introduction of live broadcasts have all functioned to keep *Top of the Pops* in step with new audiences and a changing music scene.

The programme's high point was the mid-1970s when audience figures regularly reached 16 million. This undoubtedly reflected trends in the music industry which saw record sales peak at roughly the same period. However, the acts that were appearing on the show were peculiarly televisual and complemented perfectly the medium's newly acquired colour: the dominance of television inspired novelty acts such as the Goodies and the Wombles plus the emergence of Glam Rock with its theatricality and glitz, seemed to return pop music to the values of showbiz and entertainment.

Viewing figures have steadily declined since the mid-1970s. Some blamed the initial shock of Punk music which lacked the kind of "razzmatazz" that *Top of the Pops* thrived on. Punk re-introduced notions of authenticity and its anti-commercial stance sat uneasily with the show's emphasis on glamour and entertainment. Even though the 1980s saw the return of flamboyant pop performers, led by New Romanticism and the New Pop, the decline nevertheless continued. This was partly to do with a decline in the singles market and an increase in television channels dedicated to the music scene. This, combined with the general competitiveness of the television industry in the 1980s, has led to a severe drop in viewing figures.

By the early 1990s audience figures had fallen to 5 million. Nevertheless, *Top of the Pops* has continued to fend off all rivals, and competitors have found the show to be an immovable fixture in the schedules. The history of British television has seen a host of music shows come and go, but while they often achieved fleeting success, none of them has been able to match the staying power or the popularity of *Top of the Pops.*

—Peter McLuskie

PRODUCERS   Johnny Stewart, Robin Nash

PROGRAMMING HISTORY

• BBC

January 1964–

FURTHER READING

Blacknell, S. *The Story of Top of the Pops.* London: Patrick Stephens, 1985.

Cubitt, S. "Top of the Pops; The Politics of the Living Room." In, Masterman, L., editor. *Television Mythologies.* London: Comedia, 1986.

See also British Programming

A TOUR OF THE WHITE HOUSE WITH MRS. JOHN F. KENNEDY

U.S. Documentary

On the night of 14 February 1962 three out of four television viewers tuned to CBS or NBC to watch a *A Tour of the White House with Mrs. John F. Kennedy.* Four nights later, ABC rebroadcast the program to a sizable national audience before it then moved on to syndication in more than fifty countries around the globe. In all, it was estimated that hundreds of millions of people saw the program, making it the most widely viewed documentary during the genre's so called golden age. But the White House tour is also notable because it marked a shift in network news strategies, since it was the first primetime documentary to explicitly court a female audience.

Between 1960 and 1962 most network documentaries focused on major public issues such as foreign policy, civil rights, and national politics. These domains were overwhelmingly dominated by men and the programs were exclusively hosted by male journalists. Yet historians of the period have shown that many American women were beginning to express dissatisfaction with their domestic roles and their limited access to public life. Not only did women's magazines of this period discuss such concerns, but readers seemed fascinated by feature articles about women who played prominent roles in public life. Jacqueline Kennedy was an especially intriguing figure as she accompanied her husband on diplomatic expeditions and was seen chatting with French President De Gaulle, toasting with Khrushchev, and delivering speeches in Spanish to enthusiastic crowds in

Jacqueline Kennedy with Charles Collingwood
Photo courtesy of the John F. Kennedy Library

Latin America. She even jetted off to India on her own for a quasi-official good will visit. Kennedy quickly became a significant public figure in popular media, her every move closely followed by millions of American women.

Consequently, Jacqueline Kennedy's campaign to re-decorate the White House with authentic furnishings and period pieces drew extensive coverage. Taking the lead in fundraising and planning, she achieved her goals in a little over a year and, as the project neared completion, she acceded to requests from the networks for a televised tour of the residence. It was agreed that CBS producer Perry Wolff, Hollywood feature film director Franklin Schaffner, and CBS correspondent Charles Collingwood would play leading roles in organizing the program, but that the three networks would share the costs and each would be allowed to broadcast the finished documentary. The weekend before the videotaping, nine tons of equipment were put in place by fifty-four technicians and cutaway segments were taped in advance. The segments featuring Jacqueline Kennedy were recorded during an eight hour session on Monday.

The final product, though awkward in some regards, effectively represents changing attitudes about the public and private roles of American women. For here was Jacqueline Kennedy fulfilling her domestic duty by providing visitors a tour of her home. Yet she also was performing a public duty as the authoritative voice of the documentary: providing details on her renovation efforts, informing the audience about the historical significance of various furnishings, and even assuming the position of voice-over narrator during extended passages of the program. In fact, this was the first prime-time documentary from the period in which a woman narrated large segments of the text. Kennedy's authoritative status is further accentuated by her position at the center of the screen. This framing is striking in retrospect because correspondent Charles Collingwood, who "escorts" Kennedy from room to room, repeatedly walks out of the frame leaving her alone to deliver descriptions of White House decor and its national significance. Only at the very end of the program, when President Kennedy "drops in" for a brief interview, is Jacqueline repositioned in a subordinate role as wife and mother. Sitting quietly as the two men talk, she listens attentively while her husband hails her restoration efforts as a significant contribution to public awareness of the nation's heritage.

The ambiguities at work in this program seem to be linked to widespread ambivalence about the social status of the American woman at the time of this broadcast. Jacqueline Kennedy takes a national audience on a tour of her home, which is at once a private and public space. It is her family's dwelling, but also a representation of the nation's home. Furthermore, she is presented both as a mother—indeed, the national symbol of motherhood—and as a modern woman: a patron of the arts, an historical preservationist, and a key figure in producing the nation's collective memory. In these respects, she might be seen as symbolic of female aspirations to re-enter the public sphere, and this may help to explain the documentary's popularity with female viewers.

The White House tour was soon joined by a number of similar productions, each of which drew prime-time audiences as large as those for fictional entertainment. For example, *The World of Sophia Loren* and *The World of Jacqueline Kennedy* each drew a third of the nightly audience, while *Elizabeth Taylor's London* drew close to half. In general, elite television critics reviewed these programs skeptically, noting that entertainment values were privileged at the expense of a more critical assessment of their subject matter. Yet the appeal of these programs may have had less to do with the dichotomy between entertainment and information per se than with the way in which they tapped into women's fantasies about living a more public life while largely maintaining their conventional feminine attributes. As numerous feminist scholars have argued, one of the fundamental appeals of television programming is the opportunity it affords for the viewer to fantasize about situations and identities which are not part of one's everyday existence. In the early 1960s, such fantasies may have been important not only for women who chafed at the constraints of domesticity, but also for women who were imagining new possibilities.

—Michael Curtin

FURTHER READING

Curtin, Michael. *Redeeming the Wasteland.* New Brunswick, New Jersey: Rugters Univesity Press, 1995.
Watson, Mary Ann. *The Expanding Vista.* New York: Oxford University Press, 1990.

See also Documentary; Secondari, John

TRADE MAGAZINES

The television industry is analyzed and reported on by a variety of trade magazines reflecting the perspectives of producers, advertisers, media buyers, networks, syndicators, station owners, and new technology developers. The general television trade press is complemented by coverage of television in the advertising and entertainment industry trade press. Additional specialty magazines cover cable television, satellites, newsgathering, and religious programming. The advent of satellite distribution and the expansion of transnational media corporations has led to a growing internationalization of television industry press coverage, especially in the television trade press of Canada and Great Britain.

Broadcasting and Cable, subtitled *The Newsweekly of Television and Radio*, covers top stories of general industry interest, including regulatory issues, ratings, company and personnel changes, advertising and marketing strategies, and programming trends. Aimed at broadcast executives, *Broadcasting and Cable*'s concise journalistic coverage has been recognized as an authoritative source for industry news. Originating as a radio trade paper named *Broadcasting* in 1931, the weekly eventually expanded its coverage into the media of television and cable. Along the way, it was also known as *Broadcasting-Telecasting* (1945–57), and absorbed other important trade publications such as *Broadcast Advertising* (in 1936) and *Television* (in 1968). Currently, *Broadcasting and Cable* consists of sections that cover the top weekly stories, broadcasting, cable, and technology. Additional columns treat federal lawmaking, personnel moves, and station sales. Recently, *Broadcasting and Cable* has expanded coverage of new media technologies; its "Telemedia Week" section covers the World Wide Web, interactive media, CD-ROMs, and Internet developments. *Broadcasting and Cable International*, a companion magazine, provides more international television industry coverage.

Electronic Media (1982–), a tabloid-size weekly, covers American visual electronic media (television, cable, and video). Aimed at managerial executives, *Electronic Media* reports on technological changes, advertising strategies, management methods, regulatory developments, and programming. *Electronic Media* often draws on perspectives from throughout the industry when it covers such debates as the relative effectiveness of advertising on network television versus cable television, or the appropriateness of talk show subjects. With its regular features such as "The Insider," "Who Is News," and sections on international television, technology, ratings, and finance, *Electronic Media* is an excellent source for tracking current events in the television industry.

Advertising industry trades *Advertising Age* (1930–) and *Adweek* (1960–) cover television from the perspective of media buyers. Advertising agencies buy time on television for their clients' commercials and thus seek up-to-date and accurate information on ratings, programming strategies, schedule shifts, regulatory changes, and personnel moves. Pertinent articles in both weeklies concern specific commercial campaigns, sponsorship issues, demographic research, effectiveness of network versus cable television advertising, advertising agency activities, production company news, and ratings information. Since media buyers are customers of station managers and network executives, the editorial opinions of *Advertising Age* and *Adweek* sometimes differ from those of *Broadcasting and Cable* and *Electronic Media*.

The long-lived show business trade periodicals *Variety*, *Daily Variety*, and *Hollywood Reporter* also report on the television industry. The tabloid-size weekly *Variety* has covered entertainment industries such as vaudeville, films, television, radio, music, and theater since 1905. In addition to extensive hard news coverage of show business and insider "Buzz," *Variety* is renowned for its headline style (for example, "Comcast Ladles Beans for Frank" heads an article on the hiring of a new chief executive named Rich Frank by a cable production company).

Variety's television section includes news about program production, ratings, regulatory issues, syndication deals, Nielsen ratings, and network and cable company activities. *Variety*'s "World View" section also includes articles on international broadcasting. Additionally, in-depth television program reviews provide production information, analysis of production values, and predictions of a program's potential success or failure. *Daily Variety*, the daily counterpart to the weekly *Variety*, has covered the entertainment business since 1933, reporting production news, personality news, and entertainment stock prices.

Hollywood Reporter has been a daily newspaper for the entertainment industry since 1930, reporting on production deals, program budgets, distribution arrangements, personalities in show business, entertainment stocks, and upcoming entertainment industry events. Additionally, *Hollywood Reporter*'s television program reviews include behind-the-scenes production information. Like *Variety*, *Hollywood Reporter* is well-known for its journalistic coverage and commentary on the business end of show business.

As cable television has developed into a major competitor for network television audiences, trade publications devoted to the cable industry also have grown. *Cablevision* (1975–) serves cable television managers by providing articles on federal cable regulation, technological developments, original cable programming, pay-per-view programming, and customer service. *Cablevision* is especially concerned with cable operators' relations with local governments and increasing advertiser interest in cable.

Since 1980 the tabloid-size weekly *Multichannel News* has sought to provide breaking news to cable industry management. Covering cable industry conventions, regional cable systems, and regulatory changes, *Multichannel News* is also a clearinghouse for cable industry employment. The semimonthly *Cable Television Business* (1982–) reports on cable programming trends and financial strategies, profiles cable executives, and covers cable industry associations and conventions.

Cable World, a weekly since 1989, emphasizes the international nature of cable television financing, construction, programming, and management. *Cable World* editors have noted that U.S. personnel manage certain European cable systems, that Belgians financed the cable system in Hong Kong, and that Cable News Network is carried internationally. Aimed at the cable executive with little spare time, *Cable World* provides concise news sections on cable operations, technology, financing, advertising, and programming.

Cable and broadcast networks' reliance on satellite delivery systems has increased interest in satellite news. The monthly *Via Satellite* (1986–) covers the applications of satellite technology to international broadcasting. In addition to satellite company and personnel news, articles in *Via Satellite* address the financial and technological issues of satellite broadcasting, the changing policy and regulatory environments worldwide, and potential future applications of satellite broadcasting. Likewise, *Satellite Week* (1979–) reports on the satellite broadcasting industry, its changing international markets and regulatory environments.

Other trade publications address specific television fields. For information on the broadcast news business, *Communicator* (1988–), published monthly by the Radio-Television News Directors Association, offers coverage of television news production, personnel moves, network/station relations, and local news markets. *Religious Broadcasting* (1969–) addresses the concerns of religious broadcasters such as Focus on the Family and the Christian Broadcasting Network. In addition to "Opinion" features, *Religious Broadcasting* provides "Industry Information," including religious programming strategies, personnel training, international religious broadcasting, and news analysis.

For historical research purposes, several now-defunct trade publications offer much information on the earlier decades of the television industry. In addition to *Broadcasting-Telecasting* mentioned above, *Sponsor* (1946–68) and *Television* (1944–68) are excellent sources for articles on evolving programming strategies, regulatory issues, financing, advertising techniques, and intra-industry competition. Early issues of *Television* include many "how to" articles, often designed for the advertising agencies then in charge of much program production. Likewise, early issues of *Sponsor*, which was subtitled *Buyers of Broadcast Advertising*, trace the attitudes of advertisers and sponsors toward the decline of national network radio and the rise of network television, reflecting shifts in programming strategies and increased network control of television programming.

The biweekly *Television/Radio Age*, which originated as *Television Age* in 1953, provided analytic coverage of television industry issues until 1989. Arguing that few other industries had grown as rapidly or faced as many problems as television, the magazine's editors sought to provide in-depth analysis with which to address the television industry's regulatory, financial, and programming concerns. In addition to publishing articles written by major broadcasting executives, many *Television/Radio Age* articles closely examine specific advertising campaigns, ratings trends and techniques, network programming strategies, and Wall Street financing.

The discontinued *Channels* (1981–90) is also a good source for analytic articles on the television industry of the 1980s. Originally subtitled "of Communications" and edited by well-known television journalist Les Brown, *Channels* was later subtitled *The Business of Communications*, and sought to analyze the expanding role of television in society while reporting on the regulatory environment, production deals, programming strategies, and media markets.

Trade publications in Canada, Australia, and Great Britain not only cover national television industries but also report on the international aspects of the television industry. The Canadian monthly *Broadcaster* (1942–) often addresses issues such as how to develop and sustain Canadian-produced programming that can be competitive with well-financed and well-distributed programming from the United States. Aimed at broadcast managers, *Broadcaster* reports on developments in technology, financing,

advertising, and programming, in addition to news about the state-owned Canadian Broadcasting Corporation. Information about the Canadian cable television industry can be found in *Cable Communications Magazine*, which began as a journal on telephony in 1934. Designed for cable managers, *Cable Communications Magazine* offers international news while emphasizing Canadian issues. *Cablecaster* (1989–) covers the management, technology, regulation, and programming of Canadian cable television. A more technical perspective on Canadian broadcasting is provided by *Broadcast Technology* (1975–), also known as *Broadcast + Technology*. Although originally designed for technicians, *Broadcast Technology* has expanded into business reporting and includes articles on programming, marketing, and personnel changes.

The Australian television industry is covered by *Encore*, which reports on all audiovisual production industries in Australia. *Encore* emphasizes production news, including stories on new program series and financing arrangements, but it also covers new technology developments and regulatory issues. *B and T* (1950–), formerly known as *Broadcasting and Television*, covers Australian media markets, ratings, new productions, network strategies, and media personnel moves, as does the more advertising-trade oriented *AdNews*.

British television trade press maintains a strong international slant and is a useful source for news about European television industries. The weekly *Broadcast* (1973–), formerly known as *Television Mail*, covers British television and cable programming, regulation, financing, technology, and ratings, in addition to articles on the international scope of trends in programming and technology. *Screen Digest* (1971–) provides summaries of world news of the film, television, video, satellite, and consumer electronics industries. *Screen Digest* covers industry events and conventions, publications, and market research data for "screen media worldwide." *TBI* (or *Television Business International*, 1988) covers international broadcast, cable, and satellite markets for the broadcast executive, including articles in English, German, and Japanese. *TV World* (1977–), subtitled *Award Winning International Magazine for the Television Industry*, focuses on programming, usually profiling the trends in a particular country for a section of each issue, in addition to reviewing specific productions and festivals. Designed for executives in broadcast production and distribution, as well as those in governmental broadcast organizations, *TV World* also covers the technological developments in satellite and cable delivery systems, the shifting alliances among transnational media companies, and international coventions such as NATPE and VIDCOM. *TV World*'s truly international scope makes it an excellent source for information on the television industry worldwide.

The diversity of these trade magazines reflects the multifaceted nature of today's television industry. Since its beginnings the television industry has been closely tied to the film and advertising industries. Now television has expanded beyond broadcasting into cable, satellites, and interactive technologies. An examination of trade publications reflecting these different perspectives should provide the reader with insights into the history and future of the rapidly changing international television industry.

—Cynthia Meyers

FURTHER READING

Advertising Age (Available on-line)
Crain Communications
220 E. 42nd Street
New York, NY 10017
USA

AdNews
G.P.O. Box 606
Sydney NSW
Australia

Adweek (Available on-line)
1515 Broadway
New York, NY 10036
USA

B and T
P.O. Box 815
Strawberry Hills
NSW 2012
Australia

Broadcast
International Thomson Business Publishing
7 Swallow Place
London W1R 7AA
England

Broadcast + Technology
Diversified Publications
6 Farmers Lane
Box 420
Bolton, Ontario L0P 1A0
Canada

Broadcaster
Southam Business Communications
1450 Don Mills Rd.
Don Mills, Ontario M3B 2X7
Canada

Broadcasting and Cable (Available on-line)
Cahners Publishing Co.
1705 DeSales Steet, NW
Washington, DC 20036
USA

Broadcasting and Cable International
Cahners Publishing Co.
475 Park Avenue South
New York, NY 10016
USA

Cable Communications Magazine
Ter-Sat Media Publications
1421 Victoria Street, N.
Kitchener, Ontario
Canada

Cable Television Business
Cardiff Publishing Company
6300 S. Syracuse Way, Suite 650
Englewood, CO 80111
USA

Cable World
Cable World Associates
1905 Sherman Street, Suite 1000
Denver, CO 80203
USA

Cablecaster
1450 Don Mills Rd.
Don Mills, Ontario M3B 2X7
Canada

Cablevision
Capital Cities Media
825 Seventh Avenue
New York, NY 10019
USA

Communicator
Radio-Television News Directors Association
1000 Connecticut Avenue, NW, Suite 615
Washington, DC 20036
USA

Daily Variety
5700 Wilshire Boulevard, Suite 120
Los Angeles, CA 90036
USA

Electronic Media (Available on-line)
Crain Communications
740 N. Rush Street
Chicago, IL 60611
USA

Encore
P.O. Box 1377
Darlinghurst
NSW 2010
Australia

Hollywood Reporter (Available on-line)
5055 Wilshire Boulevard
Los Angeles, CA 90036
USA

Multichannel News
Diversified Publishing
825 Seventh Avenue
New York, NY 10019
USA

Religious Broadcasting
National Religious Broadcasters
7839 Ashton Avenue
Manassas, VA 22110
USA

Satellite Week (Available on-line)
2115 Ward Ct., NW
Washington, DC 20037
USA

Screen Digest (Available on-line)
   37 Gower Street
   London WC1 6HH
   England
TV World
   Emap Media Ltd.
   7 Swallow Place
   London W1R 7AA
   England

Variety
   Reed Publishing
   5700 Wilshire Boulevard, Suite 120
   Los Angeles, CA 90036
   USA
Via Satellite
   1201 Seven Locks Road
   Potomac, MD 20854
   USA

TRANSLATORS

Television translators are broadcast devices that receive a transmitted signal from over the air, automatically convert the frequency, and re-transmit the signal on a separate channel. Closely related are TV boosters, that amplify the incoming channel and re-transmit it, but without translating from one frequency to another.

In the United States, television stations originally were assigned to specific channels and communities, in a pattern designed to distribute service as widely as possible to all communities. The distribution plan adopted by the Federal Communications Commission in 1952 utilized a highly simplified model of physical terrain, and predicted desired coverage in a fairly smooth radius outward from the transmitter location. In reality, an obstacle such as a 9,000 foot peak would completely block any reception.

TV boosters began as a practical self-help solution to this problem wherever the terrain was mountainous, but especially in the inter-mountain West from the Front Range of the Rockies to the Cascades and through the Sierra and Coastal ranges of California. Typically, a local TV repairman or appliance salesman offering the latest in console TV sets would install a sensitive receiver on the other side of the ridge, bring the signal to the near side, and boost the signal on channel from high above the community into the valley floor.

The first booster probably was built by Ed Parsons in 1948, to extend the reach of his cable system in Astoria, Oregon. Other boosters in the Pacific Northwest soon followed. In 1954, an FCC inspector went out to Bridgeport, Washington, and ordered the local booster shut down, because it was operating without a license. It soon was returned to extra-legal operation, under the auspices of the Bridgeport Junior Chamber of Commerce. The FCC issued a cease and desist order, and on appeal, the Circuit Court of Appeals for the D.C. Circuit refused to enforce the order, holding that the FCC had a statutory duty to make provision for the use of broadcast channels, and had been remiss in not devising a means for boosters to be licensed (*C. J. Community Services v. FCC*. D.C. Cir., 1957).

In Colorado, Governor Ed Johnson began issuing state "licenses," appointing the local operators to his communication "staff," and ordering them to continue their efforts to boost television signals on channel. By 1956, there already were some 800 unlicensed boosters and translators known to be in operation. The first stirrings of cable television, or community antenna television, as it was then known, were in the same interval after 1948. As an alternate delivery mechanism, cable was the natural competitor of boosters and translators. Where cable gained initial inroads, as in Pennsylvania, it had the advantage that each home user was connected and could be charged a monthly fee. The boosters were typically supported by donations, and were a broadcast service with no toll-keeper. As cable took its initial steps as a fledgling industry, it sought protection from the FCC, urging that translators and boosters be restricted or outlawed.

Because of this early rivalry, and especially because the FCC was wedded to its pre-conceived plan for the orderly development of television in accord with the assignments it issued, the FCC refused to approve boosters and authorized translators in 1956, only to the virgin territory of UHF Channels 10-83. Power was limited to 10 watts. The rural residents essentially ignored this action, and continued to offer VHF service on Channels 2 through 13, increasingly moving away from the primitive booster, in favor of cleaner translator technology.

In 1958 the FCC announced that it was stepping up enforcement efforts, intending to get the extra-legals off the air in 90 days. Congress was deluged with protests of this action, and the Senate Committee on Interstate and Foreign Commerce conducted field hearings during 1959 in Montana, Idaho, Colorado, Utah, and Wyoming. In July 1960 Congress amended the Communications Act to waive operator requirements and otherwise authorize booster and translator operations, including those already on the air. Three weeks later, the FCC authorized VHF translators for the first time.

Translators continue to be an important component of rural TV delivery, especially in the West. As of 31 December 1995 the FCC reported 4,844 licensed translators, slightly over one-half operating on UHF. All of these re-broadcast a primary TV station. In 1982, the FCC made provision for them to originate their own programs, as low power television stations, and an additional 1,787 LPTV's have been licensed.

—Michael Couzens

FURTHER READING

Cox, Kenneth A. "The Problem of Television Service for Smaller Communities." *Staff Report to the Senate Committee on Interstate and Foreign Commerce.* Washington, D.C.: 26 December 1958.

Federal Communications Commission. *Report and Recommendations in the Low Power Television Inquiry* (BC Docket No. 78-153), 9 September 1980.

U.S. Senate, Committee on Interstate and Foreign Commerce. *Report to Accompany Senate 1886.* 86th Congress, 1st Session. Washington, D.C., 4 September 1959.

See also Low Power Television; United States: Cable

TRODD, KENITH

British Producer

Few television producers ever gain name recognition beyond their industry, but Kenith Trodd is arguably one who has. Described as the most successful of all British television drama producers, he is the winner of countless awards for the many one-off plays and films he has shepherded to the screen, and a figure seen as indispensable to the health of the Drama Department of the BBC, out of which he has worked almost continuously for over 30 years. Trodd's career is also unusual in that it has spanned the history of British television drama—from its golden age of experimentation in the 1960s to today's more hard-nosed era of cost-efficiency and ratings imperatives.

Trodd is perhaps best known for his work with the doyen of television playwrights, Dennis Potter. Both came from similar working-class and Christian fundamentalist backgrounds. (The son of a crane driver, Trodd was brought up as a member of the Plymouth Brethren). Both did National Service as Russian-language clerks at Whitehall where, during the height of the Cold War, they became firm friends with shared left-wing convictions. It was only at Oxford, from 1956 to 1959, that they found a convenient outlet for their political views, rising to become stars of a radical network of working-class students which gained national media coverage and taught them about the value of courting public controversy.

Originally, Trodd had intended to become an academic, and it was only after returning from a stint of teaching in Africa in 1964 that he received an offer from another ex-Oxford friend, Roger Smith, that would change his life. Smith had been appointed story editor of the innovative *Wednesday Play* slot and desperately needed two assistants to help him recruit as many new writers to television as possible. Along with Tony Garnett, Trodd joined the BBC just at the time the single television play was entering a radical phase of experimentation and permissiveness, as a new generation of talent began to make its presence felt. Working as a story editor on *The Wednesday Play* and *Thirty Minute Theatre* (a shorter experimental play slot), Trodd became central to this wave of innovation in the 1960s, nurturing writers such as Potter, Jim Allen, and Simon Gray.

In 1968, Trodd gained his chance to become drama producer when, along with Tony Garnett, he was lured to the rival commercial company, London Weekend Television (LWT), on the promise of forming an autonomous collective within the organisation. Notable as the first independent drama production company in British TV, Kestrel scored some successes during its two-year association with LWT, but the arrangement ended in acrimony, with Trodd eventually decamping back to the BBC, where he became producer of the *Play for Today* slot throughout the 1970s.

Never any stranger to trouble, he returned to a Drama Department in political turmoil, as managers cracked down on the freedoms programme-makers had enjoyed during the 1960s. While producing some of Dennis Potter's most controversial work, Trodd often had to make a public fuss to defend the writer's freedom, most notably in 1976 when *Brimstone and Treacle* was banned. He also found himself blacklisted by the BBC as a suspected Communist sympathiser for his support of a range of radical left-wing practitioners.

Though these difficulties were eventually resolved, Trodd continued to campaign for greater independence within the BBC, particularly after the success of his Potter serial, *Pennies from Heaven*, in 1978. In marked contrast to Potter, he became a passionate advocate for TV drama filmed on location rather than recorded in the studio (the dominant practice up to that time). This drive for change came to a head in 1979, when he again left the BBC for LWT, as part of a deal involving the formation of an independent production company with Potter. Once more, the arrangement ended in acrimony. Trodd returned to the BBC, but this time on the eve of the foundation of Channel Four, the network that would do so much to legitimise the concept of the independent producer in British television.

In the early 1980s, Trodd became chair of the Association of Independent Producers as one of the new breed of "independents," although he continued to work within the very heart of institutional television at the BBC. Under his influence, however, things were changing there, too. He had finally achieved his goal of remaining within the corporation while being able to produce independent projects as well. This ideal soon became accepted practice, as did his campaign for shooting on film.

In 1984, Trodd formed part of a BBC working party convened to examine how the corporation should respond to the feature film-making for TV and theatrical release that

Channel Four had pioneered. The outcome was the abandonment of the old concept of the studio Play for Today and the introduction of new BBC film slots, *Screen One* and *Screen Two*, with Trodd helping to oversee the first batch of films in 1985.

Despite the success of his campaigning, Trodd's recent career raises uncomfortable questions about whether he has not made himself somewhat redundant by the changes he helped bring about in the 1980s. The decline in the annual number of single drama slots due to the increased costs of film-making, plus the corresponding decline in writers and directors required to fill these slots, indicates a much tougher and more competitive environment than the one which allowed him to experiment with new ideas and untried talent in the 1960s. Nor, despite the success of a few of his BBC "single films," such as *After Pilkington* (1987) and *She's Been Away* (1989), has there been anything like the constant stream of outstanding material that secured his reputation in the 1970s. A rift with Potter in the late 1980s (not healed until the writer's death in 1994) also did not help matters in this respect. Certainly, Trodd's function has changed from the days when, as a BBC tyro, he filled his many play slots with a motley crew of young writers and directors—the question is whether for the best.

—John Cook

KENITH TRODD. Born in Southampton, Hampshire, England. Educated at Oxford. Began television career as story-editor, *The Wednesday Play*, 1964; producer, London Weekend Television, 1968-70; producer, BBC Drama Department, 1970–79; producer, London Weekend Television, and partner with playwright Dennis Potter, 1979; BBC Drama Department and independent film producer, from 1980. Recipient: Royal Television Society Silver Medal, 1986–87; British Academy of Film and Television Arts Alan Clarke Award, 1993.

TELEVISION PLAYS (selection)

1969	*Faith and Henry*
1976	*Double Dare*
1976	*Brimstone and Treacle*
1978	*Pennies from Heaven*
1978	*Dinner at the Sporting Club*
1979	*Blue Remembered Hills*
1980	*Shadows on Our Skin*
1980	*Caught on a Train*
1980	*Blade on the Feather*
1980	*Rain in the Roof*
1980	*Cream in My Coffee*
1981	*A United Kingdom*
1986	*The Singing Detective*
1987	*After Pilkington*
1988	*Christabel*
1989	*She's Been Away*

See also British Programming; Channel Four; *Film on Four*; Garnett, Tony; Loach, Ken; *Pennies from Heaven*; Potter, Dennis; *Wednesday Play*

TROUGHTON, PATRICK

British Actor

Patrick Troughton was the second actor to take on the mantle of British television's Doctor Who in the long-running science-fiction series of the same name, playing the role for three years, from 1966 to 1969. This was by no means the only part he played on television, and he also had a full and varied career as an actor in the theater and in the cinema, but it is for his flamboyant and quixotic portrayal of BBC's celebrated Time Lord that he is usually remembered.

Troughton followed William Hartnell as Doctor Who after his predecessor, suffering from multiple sclerosis and disillusioned with the changing character of the programme (which had originally been intended to have a strong educational content), withdrew from the series. Troughton determined at once that his Doctor would be in marked contrast to the white-haired dotty professor-type depicted by Hartnell, and in his hands the Doctor became a colourfully whimsical and capricious penny-whistle-playing eccentric who could be testy, courageous, and downright enigmatic as the mood took him. Such a radical change in character was made possible within the confines of the programme through the introduction of the concept that the Doctor underwent a mysterious regenerative metamorphosis at various stages of his centuries-long existence.

Troughton settled quickly into the role, and children throughout Britain cowered behind the sofa as his Doctor did weekly battle with such fearsome alien foes as the Daleks and the Cybermen. After three years, he finally passed the responsibility for playing television's famous Time Lord on to Jon Pertwee.

By the time he was selected to play Doctor Who, Troughton had long established his reputation as a performer in a wide range of roles and productions, being particularly well regarded as a Shakespearean actor. Among the most acclaimed of his previous appearances had been his performance as Hitler in the play *Eva Braun* at Edinburgh's Gateway Theatre in 1950 and supporting roles in Laurence Olivier's Shakespearean film epics *Hamlet* and *Richard III*. On television he had made appearances in such enduringly popular series as *Coronation Street*, in which he was George Barton, and *Doctor Finlay's Casebook*. Notable among his later credits on the small screen were the series *The Six Wives* of Henry VIII, in which he was cast as the Duke of Norfolk, the World War II prison camp drama Colditz, and the sitcom The Two of Us, in which he gave his usual good value as Nicholas Lyndhurst's grandfather Perce (after Troughton's death, Tenniel Evans took over the role). Always a jobbing actor who was ready to turn his hand to a variety of roles of contrasting sizes, his familiar face would pop up in all manner of series, and he guested on Special Branch, The Protectors, The Goodies, Churchill's People, Minder, and Inspector Morse, to name but a few.

But it was with Doctor Who that Troughton's name was destined to remain indelibly linked in the last years of his life. His death occurred while he was actually attending a Doctor Who convention in the United States.

—David Pickering

PATRICK GEORGE TROUGHTON. Born in London, England, 25 March 1920. Attended schools in London; Embassy School of Acting, London; Leighton Rollin's Studio for Actors, Long Island, New York. Married three times; children: Joanna, Jane, Jill (stepdaughter), David, Michael, Peter, Mark, and Graham (stepson). Served in Royal Navy during World War II. Joined Bristol Old Vic, concentrating on Shakespeare productions, 1946; made film and television debuts, 1948; achieved fame as central character in television's Doctor Who, 1966. Died 28 March 1987.

Patrick Troughton as Doctor Who

TELEVISION SERIES

1962–63	Man of the World
1966–69	Doctor Who
1970–71	The Six Wives of Henry VIII
1972–74	Colditz
1982–84	Foxy Lady
1986–87	The Two of Us

TELEVISION SPECIALS

1950	Toad of Toad Hall
1953	Robin Hood
1955	The Scarlet Pimpernel
1960	The Splendid Spur
1987	Knights of God

FILMS

Hamlet, 1948; Escape, 1948; Cardboard Cavalier, 1949; Badger's Green, 1949; Waterfront, 1950; Treasure Island, 1950; Chance of a Lifetime, 1950; The Woman with No Name, 1950; White Corridors, 1951; The Franchise Affair, 1951; The Black Knight, 1954; Richard III, 1955; The Curse of Frankenstein, 1957; The Moonraker, 1958; Misalliance, 1959; Phantom of the Opera, 1962; Jason and the Argonauts, 1963; The Gorgon, 1964; Frankenstein and the Monster from Hell, 1973; The Omen, 1976; Sinbad and the Eye of the Tiger, 1977.

STAGE (selection)

Eva Braun, 1950.

FURTHER READING

Haining, Peter. Doctor Who, The Key to Time: A Year-by-Year Record. London: W.H. Allen, 1984.

Peel, John, and Terry Nation. The Official Doctor Who and the Daleks Book. New York: St. Martin's, 1988.

Tulloch, John, and Henry Jenkins. Science Fiction Audiences: Watching Doctor Who and Star Trek. London; New York: Routlege, 1995.

Tulloch, John, and Manuel Alvarado. Doctor Who: The Unfolding Text. New York: St. Martin's, 1983.

See also Doctor Who

TURNER, TED

U.S. Media Mogul

Ted Turner is one of the entrepreneurs responsible for rethinking the way we use television, especially cable television in the 1970s, 1980s and 1990s. But Ted Turner is known, loved, or hated as much for his unique personal style as for any particular accomplishment. He is a flamboyant Southern businessman in industries normally run from New York and Los Angeles. Turner's penchant for wringing every possible use from his corporations has enabled him to establish a corporate empire that touches virtually every area of the entertainment industry. In 1995, in what could be the most significant personal and financial deal of his career, he agreed to merge his holdings with those of international media conglomerate Time Warner. Turner linked his corporation to an unusually powerful managing partner.

Turner's career in broadcasting began in 1970, when Turner Communications, a family billboard company, merged with Rice Broadcasting and gained control of WTCG, Channel 17, in Atlanta. WTCG succeeded under Turner's ownership losing $900,000 before the merger in 1970, to making $1.8 million in revenue in 1973. Turner made WTCG cable's first "superstation," broadcast by satellite to cable households around the United States. Renamed WTBS (for Turner Broadcasting System) in 1979, the station remained one of the most popular basic cable options through the growth in cable households in the 1980s. The program schedule featured a mixture of movies and series produced by Turner subsidiaries, reruns from Turner's vast entertainment libraries, broadcasts of Turner-owned Atlanta Braves' and Hawks' games, and shows related to Turner's interest in the environment, such as explorer *Jacques Cousteau's Undersea Adventures* and Audobon Society specials.

Turner's second great innovation in cable, the Cable News Network (CNN), was launched in 1980. Turner's personal involvement in CNN appeared to handicap the network from the start, since WTBS's joke-filled late-night news program and CNN's shoestring budget suggested that Turner would not commit to serious journalism. But CNN's 24-hour news programming gained viewer loyalty and industry respect as it challenged—and often surpassed—the major networks' authority in reporting breaking events such as the Persian Gulf War. Turner, as well, refashioned himself as a global newsman as CNN expanded into new markets (by 1995, it reached 156 million subscribers in 140 countries around the world), banning the word "foreign" from CNN newscasts in favor of "international." And following Turner's philosophy of finding as many outlets for his products as possible, the CNN franchise has grown to include CNN International, CNN Headline News, CNN Radio and CNN Airport Network, as well as a variety of computer on-line services.

Turner's holdings are not limited to cable networks, although he also owns Turner Network Television (TNT), Turner Classic Movies, Sportsouth, and the Cartoon Network. His Turner Entertainment Company manages one of

Ted Turner
Photo courtesy of Turner Broadcasting System, Inc.

the world's largest film libraries, including the MGM library, with licensing rights for Hollywood classics such as *Gone with the Wind, The Wizard of Oz,* and *Citizen Kane.* Production companies include New Line Cinema, Castle Rock Entertainment (which produced *Seinfeld*), Hanna-Barbera Cartoons, and Turner Pictures Worldwide; all provide programming sources for his cable and broadcast outlets. His Turner Home Entertainment manages the video release of titles from the Turner library, as well as overseeing a publishing house, educational services company, and a division devoted to exploring ways to bring Turner titles on-line. And throughout his career, Turner has endeavored to purchase one of the three major networks, targeting each for takeover as it has become financially vulnerable.

Turner's possessions cannot begin to capture the essence of the personality which has made him one of the entertainment industry's most recognizable figures. He earned the nickname "Captain Outrageous" during his yachting days (capturing America's Cup in 1977 and losing it in 1980), but his reputation for eccentric behavior has not been limited to the sporting arena. When his efforts to "colorize" films from his extensive black-and-white movie library—thereby broadening the films' appeal to audiences who prefer color—raised the hackles of film lovers and prompted congressional hearings on the authorship and ownership of cinematic texts, Turner threatened to add color to *Citizen Kane*, the 1941 Orson Welles classic which has been lauded as the greatest film ever made. (Although Turner owns the film, he didn't.)

Turner has actively sought publicity both for himself and for a number of causes he supports, such as the environmental movement and world peace, especially when they have been associated with Turner's media or sports holdings. Two examples are WTBS's *Captain Planet* environmental cartoon and the Goodwill Games between United States and Soviet athletes to which Turner has broadcasting rights. And with his third wife, former actress, fitness guru, political activist, and multimedia mogul Jane Fonda, Turner has added support for Native American causes (including a series of original films on TNT) to atone for his earlier racist promotions of the Atlanta Braves. Long accustomed to his role as "captain of his own fate," it remains to be seen how he will arrange a position in a corporate structure he arranged for but does not control.

—Susan McLeland

TED (ROBERT EDWARD) TURNER. Born in Cincinnati, Ohio, U.S.A., 19 November 1938. Educated at Brown University. Married: 1) Judy Nye, 1960 (divorced); one daughter and one son; 2) Jane Shirley Smith, 1965 (divorced, 1988); one daughter and two sons; 3) Jane Fonda, 1991. Account executive, Turner Advertising Company, Atlanta, Georgia, 1961–63, president and chief operating officer, 1963–70; president and chair of the board, Turner Broadcasting System, Inc., Atlanta, since 1970; chair of the board, Better World Society, Washington, 1985–90. Honorary degrees: D.Sc. in Commerce, Drexel University, 1982; LL.D., Samford University, Birmingham, Alabama, 1982, Atlanta University, 1984; D. Entrepreneurial Sciences, Central New England College of Technology, 1983; D. in Public Administration, Massachusetts Maritime Academy, 1984; D. in Business Administration, University of Charleston, 1985. Board of directors: Martin Luther King Center, Atlanta. Recipient: America's Cup in his yacht *Courageous*, 1977; named yachtsman of the year four times; outstanding Entrepreneur of the Year Award, *Sales Marketing and Management Magazine*, 1979; National Cable Television Association President's Award, 1979 and 1989; National News Media Award, Veterans of Foreign Wars (VFW), 1981; Special Award, Edinburgh International Television Festival, Scotland, 1982; Media Awareness Award, United Vietnam Veterans Organization, 1983; Special Olympics Award, Special Olympics Committee, 1983; World Telecommunications Pioneer Award, New York State Broadcasters Association, 1984; Golden Plate Award, American Academy of Achievement, 1984; Silver Satellite Award, American Women in Radio and Television; Lifetime Achievement Award, New York International Film and Television Festival, 1984; Tree of Life Award, Jewish National Fund, 1985; Golden Ace Award, National Cable Television Academy, 1987; Sol Taishoff Award, National Press Foundation, 1988; Chairman's Award, Cable Advertising Bureau, 1988; Directorate Award NATAS, 1989; Paul White Award, Radio and Television News Directors Association Award, 1989; numerous other awards.

FURTHER READING

Adler, Jerry. "Jane and Ted's Excellent Adventure." *Esquire* (New York), February 1991.

Bibb, Porter. *It Ain't as Easy as it Looks: Ted Turner's Amazing Story*. New York: Crown, 1993.

Bruck, Connie. "Jerry's Deal." *The New Yorker*, 19 February 1996.

Carter, Bill. "Ted Turner's Time of Discontent" (interview). *The New York Times*, 6 June 1993.

Chakravarty, Subrata N. "What New Worlds to Conquer?" *Forbes* (New York), 4 January 1993.

Dawson, Greg. "Ted Turner: Let Others Tinker with the Message: He Transforms the Medium Itself" (interview). *American Film* (Washington, D.C.), January-February 1989.

Fabrikant, Geraldine. "Government Review of Turner-Time Warner Deal." *The New York Times*, 10 October 1995.

Fahey, Alison. "'They're not so Big'; Turner Pleased with Battle against the Networks." *Advertising Age* (New York), 28 October 1991.

Goldberg, Robert, and Gerald Jay Goldberg. *Citizen Turner: The Wild Rise of an American Tycoon*. New York: Harcourt Brace, 1995.

Greenwald, John, and John Moody. "Hands Across the Cable: The Inside Story of How Media Titans Overcame Competitors and Egos to Create a $20 Billion Giant." *Time* (New York), 2 October 1995.

"Hear, O Israel." *The Economist* (London), 4 November 1989.

Henry, William A., III. "History as it Happens" ("Man of the Year" cover story). *Time* (New York), 6 January 1992.

Lanham, Julie. "The Greening of Ted Turner." *The Humanist*, November-December 1989.

"Neither Broke nor Broken: The Ever-resurgent Ted Turner" (interview). *Broadcasting* (Washington, D.C.), 17 August 1987.

Painton, Priscilla. "The Taming of Ted Turner" ("Man of the Year" cover story). *Time* (New York), 6 January 1992.

"Prince of the Global Village" ("Man of the Year" cover story). *Time* (New York), 6 January 1992.

Scully, Sean. "Turner Backs Violence Guidelines." *Broadcasting and Cable* (Washington, D.C.), 28 June 1993.

"Signoff" ("Talk of the Town" column). *The New Yorker* (New York), 12 September 1988.

Stern, Christopher. "FTC Puts TW/Turner Under Microscope." *Broadcasting and Cable* (Washington, D.C.), 29 April 1996.

Stutz, Bruce. "Ted Turner Turns it on" (interview). *Audubon*, November-December 1991.

"Ted Turner's Quantum Leap" (interview). *Broadcasting* (Washington, D.C.), 31 March 1986.

Vaughan, Roger. *The Grand Gesture: Ted Turner, Mariner, and the America's Cup.* Boston: Little, Brown, 1975.

Walley, Wayne. "Ted Turner Hones His Midas Touch" (interview). *Advertising Age* (New York), 11 December 1989.

Whittemore, Hank. *CNN, The Inside Story.* Boston: Little, Brown, 1990.

Williams, Christian. *Lead, Follow or Get out of the Way: The Story of Ted Turner.* New York: Times Books, 1981.

Zoglin, Richard. "The Greening of Ted Turner: As his once Shaky Ventures Thrive He Turns into a Liberal Activist." *Time* (New York), 22 January 1990.

See also Cable Networks; Cable News Network; Colorization; Superstation; Time Warner; Turner Broadcasting Systems; United States: Cable

TURNER BROADCASTING SYSTEM

U.S. Media Conglomerate

Over the course of three decades, Turner Broadcasting System (TBS) has grown from a regional outdoor advertising firm into one of the world's largest and most successful media conglomerates. Beginning in the late 1960s, Ted Turner changed his father's company, Turner Advertising, first into Turner Communications Company and then into Turner Broadcasting System. Each name change represented a stage in the building of an empire that would come to encompass broadcast television and radio, cable program services, movie and television production companies, home video, and sports teams.

TBS began with Turner's purchase of failing Atlanta UHF station, WJRJ, in 1968. He immediately renamed the station WTCG (for Turner Communications Group) and began to look for programming. What Turner found were old movies and syndicated television series, many of which he purchased outright with a view toward unrestricted future showings. He used these to counterprogram the network affiliates, going after such audience segments as children and people who did not watch the news. By the early 1970s, WTCG also offered local sports programming—first professional wrestling and then Atlanta Braves baseball, Atlanta Hawks basketball, and Atlanta Flames hockey. In 1976, Turner purchased the Braves, securing long-term access to his single most critical source of programming.

The old movies and TV programs combined with the sports coverage proved to be a formula for success. By 1972, WTCG boasted a 15% share of the Atlanta audience, and the station's signal had begun to be carried by microwave to cable systems in the Atlanta region. When Turner heard about Home Box Office's groundbreaking satellite debut in 1975, he quickly began preparations to use the same technology to extend WTCG's signal. Through a series of adroit negotiations, Turner set up (as a business separate from Turner Communications) a company called Southern Satellite Systems, Inc. to uplink WTCG's signal to an RCA communications satellite. In 1976, WTCG became the second satellite-delivered cable program service and the first satellite superstation.

The superstation was renamed WTBS in the late 1970s. In 1981 Cable News Network (CNN), the first of Turner Broadcasting System's cable-only program services, was launched. Throughout the following decade CNN branched into specialized news services, including CNN Radio, CNN International, CNN Headline News, and CNN Airport Network.

During the 1980s, strategic programming acquisitions led to more new cable ventures for Turner Broadcasting. In 1986 Turner added the entire MGM film library to his existing stock of old movies. Two years later Turner Network Television (TNT), a general-interest cable program service that features many movies, was launched. The Turner film library also supplies Turner Classic Movies, launched in 1994. Turner's 1991 acquisition of Hanna-Barbera Cartoons, both the production studio and the syndication library, ensured a continuous supply of programming for both the TBS superstation and the Cartoon Network, launched in 1992. Several foreign-language versions of the Cartoon Network either exist or are being developed. Finally, in addition to the TBS superstation's established market position as a sports programming outlet, Turner Broadcasting also owns Sportsouth, a regional sports programming service.

Other Turner holdings include New Line Cinema, Castle Rock Entertainment, Turner Entertainment Company, Turner Pictures Worldwide, Turner Home Entertainment, Turner Publishing, Turner Educational Services, Turner Interactive, and the Atlanta Hawks.

From the earliest efforts to revamp WTCG, much of Ted Turner's television success has lay in his and his employees' ability to acquire innovative and inexpensive sources of programming and to make that programming available through as many outlets as possible. Thus Turner Broadcasting System's current holdings represent both program material—in the form of film and television libraries, production houses, and sports teams—and the means of distributing that programming.

In 1995 TBS began what may yet be its most significant negoiations when it entered into an agreement to become part of the Time Warner media conglomerate. If approved by courts and regulatory agencies TBS would add its resources, its staff—and Ted Turner to one of the largest media organizations in the world.

—Megan Mullen

FURTHER READING

Amdur, Meredith. "The Boundless Ted Turner: Road to Globalization." *Broadcasting and Cable* (Washington, D.C.), 11 April 1994.

Bibb, Porter. *It Ain't as Easy as it Looks: Ted Turner's Amazing Story.* New York: Crown, 1993.

Carter, Bill. "Ted Turner is Not Afraid to Turn His Personal Vision into a Multimillion-Dollar Project for His Company." *New York Times,* 22 November 1993.

———. "Ted Turner's Time of Discontent" (interview). *New York Times,* 6 June 1993.

Corliss, Richard. "Time Warner's Head Turner." *Time* (New York), 11 September 1995.

"Europe Plan by Turner." *The New York Times,* 9 March 1993.

Fabrikant, Geraldine. "Government Review of Turner-Time Deal." *The New York Times,* 10 October 1995.

Goldberg, Robert, and Gerald J. Goldberg. *Citizen Turner: The Wild Rise of an American Tycoon.* New York: Harcourt Brace, 1995.

Lander, Mark. "Time Warner and Turner Seal Merger: After 5 Weeks, a $7.5 Billion Stock Deal." *The New York Times,* 23 September 1995.

Roberts, Johnnie L. "An Urge to Merge: Foxes in the Chicken Coop." *Newsweek* (New York), 11 September 1995.

———. "Time's Uneasy Pieces." *Newsweek* (New York), 2 October 1995.

Taub, James. "Reaching for Conquest." *Channels of Communication* (New York), July-August 1983.

See also Cable Networks; Cable News Network; Superstation; Turner, Ted; United States: Cable

THE 20TH CENTURY

U.S. Historical Documentary Program

From the one-hour premiere episode "Churchill, Man of the Century" (20 October 1957) to its last episode *The 20th Century* unit produced 112 half-hour historical compilation films and 107 half-hour "originally photographed documentaries" or contemporary documentaries. Narrated by Walter Cronkite, the series achieved critical praise, a substantial audience, and a dedicated sponsor, the Prudential Insurance Company of America, primarily with its historical compilation films. The compilation documentaries combined film footage from disparate archival sources—national and international, public and private—with testimony from eyewitnesses, to represent history. Programs averaged 13 million viewers a week, but periodically reached 20 million with action-oriented installments. The series foreshadowed the production and marketing strategies of weekly compilation and documentary series that populate cable television today.

Irving Gitlin, CBS vice president of public affairs programming, originally conceived the series as broad topic compilations based on Mark Sullivan's writings, *Our Times*. Burton Benjamin, whose career at CBS news began as the series' producer and progressed to executive producer, radically revised the concept. He stressed compilations focused on one man's impact on his times or an event ("Patton and the Third Army," "Woodrow Wilson: The Fight for Piece"). These were to be interspersed with more traditional biographical sketches of individual lives ("Mussolini," "Gandhi," and "Admiral Byrde"). Benjamin also added a mix of "back of the book" stories, or historical episodes receiving scant attention in history texts and unfamiliar to the general public. These "essays" dealt with individuals, such as

Mustafa Kemal Ataturk ("The Incredible Turk"), and topics, such as the Kiska campaign ("The Frozen War"), and the Danish resistance movement ("Sabotage"). The series' researchers, both literary and film, were instructed to pursue detailed factual information that would add the unknown to the familiar. Information such as the $8.50 price levied on those who wished to watch Goering's wedding parade or the details of Rommel's visit to his family on D-Day surrounded primary story elements. With the assistance of associate producer Isaac Kleinerman, editor and film researcher for *Victory at Sea* (NBC, 1952-53) and *Project XX* (NBC, 1954-73), the series established a successful formula by stressing pivotal dramatic incidents in battles, conflicts, political uprisings, and the repercussions of actions by great male leaders. Accounting for the many battle-oriented programs, Benjamin admitted that the series was "as much a show biz show as any dramatic half-hour." But when the availability of dramatic and unusual footage of personalities existed for an historical period or event, such as "Paris in the Twenties" and "The Olympics," the unit produced broad-canvas compilation films. On a weekly basis audiences stayed with the series, expecting the unique and unfamiliar even in recognizable topics.

When the series started to look familiar, Benjamin revised. In the third season the series shifted to the individual in history and more contemporary topics. The biographical form slowly expanded to contemporary men in the arts and sciences, law, and politics while giving "eyewitnesses" a more complex role in the compilation films. The successful use of German Captain Willi Bratgi in "The Remagen Bridge," dramatically describing how an American shell changed

The 20th Century
Photo courtesy of Wisconsin Center for Film and Theater Research

history's course by accidentally severing a detonation cable, led the production team to search out figures with strong emotional and informational ties to the past. From 1961 through the series' end, the most innovative compilations used central, compelling personalities to weave a dramatic structure. These included Countess Nina Von Stauffenberg and Captain Axel Von Dem Bussche in "The Plots Against Hitler," and Mine Okubo, author of *Citizen 13360* in "The Nisei: The Pride and the Shame." But as the series progressed, contemporary documentaries gradually outnumbered compilation films. Contemporary documentaries depicted the enduring value of democracy's struggle against Communism, the modernization of America, and the pioneering human spirit facing adversity.

Although accepted by the public, 28 contemporary documentaries over the nine years were greeted with criticism. These depicted U.S. military defense systems and hardware, and functioned as publicity releases for the Department of Defense by equating liberty with technology. By filming documentaries such as "Vertijet" and "SAC: Aloft and Below," the producers received extraordinary military assistance in declassifying footage in government archives for the compilation films. Still, Benjamin strove for journalistic integrity in a politicized atmosphere, even canceling biographies on General MacArthur and Curtis Le May when the military requested final script approval.

Social and political change overseas dominated the list of contemporary subjects. Although evident in the compilation films, the series' anti-Communist ideology and commitment to democratic modernization was blatant in programs such as "Poland on a Tightrope" and "Sweden: Trouble in Paradise." Periodically, the producers sought new approaches to the contemporary documentary, in response to waning critical reception and audience desire for the dramatic. When Sam Huff was outfitted with a microphone and transmitter, in "The Violent World of Sam Huff," the landscape of television documentary shifted. Other experiments in quasi-cinema-verite documentaries such as "Rhodes Scholar" and "Duke Ellington Swings through Japan" illustrated new approaches for television. But strong diversions from the series' dominant form and content, such as the grim Appalachian conditions depicted in "Depressed Area, U.S.A.," were rare and usually came from freelance film directors such as Willard Van Dyke and Leo Seltzer.

CBS executives admired the series' meticulous production process. The producers allocated 24 weeks for a program's production, with each stage such as literary research, film research, location shooting, editing, script writing, and music allocated a specific time parameter on a flow chart. By the sixth season the series ran itself, allowing Benjamin to work simultaneously on other CBS news projects. Into this production mechanism, Benjamin periodically added the attraction of established journalists and historians, including John Toland, Robert Shaplen, Sidney Hertzberg, and Hanson Baldwin. Although Alfredo Antonini composed music for 50% of the programs, Franz Waxman, Glen Paxton, George Kleinsinger, George Antheil, and others contributed original scores, working with Antonini and the CBS Orchestra within strict time limitations. This would be the last time a documentary series turned consistently to talent outside a network.

Prudential supported the series' use of these film, literary, and musical figures, but became a restraint on the series' creative potential. The company approved and prioritized each year's topics, submitted by Benjamin and Kleinerman, and admitted not wanting controversial programs on social and religious topics. The sponsor—and the Department of Defense—also expected a conservative and uncritical repre-

sentation of military activity, past and present. Certain subjects such as gambling, the labor movement, and U.S. relations with Canada were rejected by Prudential. Even though Benjamin was aware of the corporate perspective, he fought several years for the approval to air biographies on "Lenin and Trotsky" and "Norman Thomas." Prudential directly limited the boundaries of subjects and investigation of any issue potentially upsetting to a large audience. Unknown to many, the series, particularly the compilation films, were tools for insurance agents who screened them at conventions and community events. Prudential withdrew sponsorship after the ninth season, when sports programming reduced the number of available time slots to 18, and the production unit's value to new directions in news and documentary could not assure Prudential the recognizable and dramatic compilation film and documentary subjects deemed suitable for its audience.

—Richard Bartone

NARRATOR
Walter Cronkite

PROGRAMMING HISTORY   219 Episodes

• CBS

October 1957–May 1958	Sunday 6:30-7:00
September 1958–August 1961	Sunday 6:30-7:00
September 1961–August 1966	Sunday 6:00-6:30
January 1968–October 1968	Sunday 6:00-6:30
January 1969–September 1969	Sunday 6:00-630
January 1970	Sunday 6:00-6:30

FURTHER READING

Bartone, Richard C. *The Twentieth Century (CBS, 1957–1966) Television Series: A History and Analysis* (Ph.D. dissertation, New York University, New York, 1985).

Benjamin, Burton. *The Documentary: An Endangered Species* (paper no. 6, Gannett Center for Media Studies). New York: Columbia University, October 1987.

———. "The Documentary Heritage." *Television Quarterly* (New York), February 1962.

———. "TV Documentarian's Dream in Challenging World." *Variety* (Los Angeles), 4 January 1961.

———. "From Bustles to Bikinis—And All That Drama." *Variety* (Los Angeles), 27 July 1960.

Bluem, William A. *Documentary in American Television: Form, Function, Method.* New York: Hastings House, 1977.

Burns, Bret. "The Changing Techniques for TV Documentaries." *National Observer* (New York), 17 December 1962.

Cronkite, Walter. "Television and the News." In, Shayon, Robert Lewis, editor. *The Eighth Art.* New York: Holt, Rinehart and Winston, 1962.

Crosby, John. "The Right Kind of Documentary." *New York Herald Tribune*, 4 November 1957.

Gates, Gary Paul. *Air Time: The Inside Story of CBS News*. New York: Harper and Row, 1978.

Higgins, Robert. "First Eight Years of *The Twentieth Century*." *TV Guide* (Radnor, Pennsylvania), 5-11 June 1964.

"History is Terribly Dramatic." *Christian Science Monitor* (Boston, Massachusetts), 17 October 1959.

Kleiner, Dick. "Bud and Ike—CBS Film Detectives." *New York World Telegram*, 14 March 1959.

Kleinerman, Isaac. "Shooting and Searching Behind the Iron Curtain." *Variety* (Los Angeles), 6 January 1960.

Krisher, Bernard. "They Find the World in Different Shape." *New York Telegram*, 28 January 1961.

Leyda, Jay. *Films Beget Films*. New York: Hill and Wang, 1964.

Patureau, Alan. "The Man Behind *20th Century* Success." *Newsday* (Hempstead, New York), 22 August 1963.

Reed, Dena. "Isaac Kleinerman: Recorder of World Facts." *American Red Cross Journal* (Washington, D.C.), January 1961.

Sapinsley, Barbara. "We Get Letters." *Television Quarterly* (New York), Winter 1964.

Shanley, John P. "Japanese Agent Returns to Naval Base for *Twentieth Century* Telecast." *The New York Times*, 26 November 1961.

U.S. Congress, House Committee on Interstate and Foreign Commerce. *Television Network Program Procurement* (testimony by Henry M. Kennedy, Second Vice President, Public Relations and Advertising, Prudential). 88th Congress, 1st session, 1963.

See also Columbia Broadcasting System; Cronkite, Walter

TWILIGHT ZONE

U.S. Science-Fantasy Anthology

The *Twilight Zone* is generally considered to be the first real "adult" science-fantasy anthology series to appear on American television, introducing the late 1950s TV audience to an entertaining and at the same time thought-provoking collection of human condition stories wrapped within fantastic themes. Although the series is usually labeled a science-fiction program, its true sphere was fantasy, embracing elements of the supernatural, the psychological, and "the almost-but-not-quite; the unbelievable told in terms that can be believed" (Rod Serling).

During the show's five-year, 155-episode run on CBS (1959–64) the program received three Emmy Awards (Rod Serling, twice, for Outstanding Writing Achievement in Drama, and George Clemens for Outstanding Achievement in Cinematography), three World Science Fiction Convention Hugo Awards (for Dramatic Presentation: 1960, 1961, 1962), a Directors Guild Award (John Brahm), a Producers Guild Award (Buck Houghton for Best Produced Series), and the 1961 Unit Award for Outstanding Contributions to Better Race Relations, among numerous other awards and presentations.

The brain-child of one of the most successful young playwrights of his time (with such "Golden Age" TV successes as "Patterns" and "Requiem for a Heavyweight"), Rod Serling's *The Twilight Zone* began life as a story called "The Time Element," which Serling had submitted to CBS, where it was produced as part of the *Westinghouse-Desilu Playhouse* anthology. Although it was little more than a simple time-warp tale, starring William Bendix as a man who believes he goes back in time to the attack on Pearl Harbor, the TV presentation received an extraordinary amount of complementary mail and prompted CBS to commission a *Twilight Zone* pilot for a possible series. With his "Time Element" script already used, Serling prepared another story which would be the pilot episode for the series. "Where Is Every-body?" opened *The Twilight Zone* on 2 October 1959, and featured a riveting one-man performance by Earl Holliman as a psychologically stressed Air Force man who hallucinates that he is completely alone in a deserted but spookily "lived in" town while actually undergoing an isolation experiment. It was this hallucinatory human stress situation placed in a could-be science-fantasy landscape, complete with an O. Henry-type "snapper ending", that was to become the standard structure of *The Twilight Zone*. "Here's what *The Twilight Zone* is," explained Serling to *TV Guide* magazine in November 1959. "It's an anthology series, half hour in length, that delves into the odd, the bizarre, the unexpected. It probes into the dimension of imagination but with a concern for taste and for an adult audience too long considered to have IQs in negative figures."

Serling's contract with the network stipulated that he would write 80% of the first season's scripts which would be produced under Serling's own Cayuga Productions banner. The prolific Serling, of course, ended up writing well over 50% of the entire show's teleplays during its five years on the air. This enormous output was for the most part supported by two other writers of distinction in the science-fantasy genre: Richard Matheson and Charles Beaumont. Matheson's literary and screenplay work before and during the series ran parallel to that of Beaumont, not suprisingly, since they were personal friends and often script-writing collaborators during their early days in television. Matheson's early writing had included the short story collection, *Born of Man and Woman,* and a novel, *I Am Legend* (both published 1954), and later the screenplays for *The Incredible Shrinking Man* (1957; from his own novel), *House of Usher* (1960), and *The Pit and the Pendulum* (1961). Beaumont's work included similar science-fiction and horror-fantasy writings, with the short story collections *Shadow Play* (published 1957) and *Yonder* (1958), as well

The Twilight Zone: The After Hours

The Twilight Zone: In Praise of Pip

The Twilight Zone: Night of the Meek

The Twilight Zone: Time Enough at Last

as screenplays for *Premature Burial* (1962) and *The Haunted Palace* (1963), alongside others in a similar vein. Their individual scripts for *The Twilight Zone* were perhaps the nearest in style and story flavor to Serling's own work. George Clayton Johnson was another young writer who, emerging from Beaumont's circle of writer friends, produced some outstanding scripts for the series, including the crackling life-or-death bet story "A Game of Pool", featuring excellent performances from Jack Klugman and Jonathan Winters. Earl Hamner, Jr., later to be creator and narrator of the long-running *The Waltons*, supplied eight scripts to the series, most of which featured good-natured rural folk and duplicitous city slickers. The renowned science-fiction author Ray Bradbury was asked by Serling to contribute to the series before the show had even started, but due to the richness of Bradbury's written work, he contributed only one script, "I Sing the Body Electric", based on his own short story.

As an anthology focusing on the "dimension of imagination" and using parable and suggestion as basic techniques, *The Twilight Zone* favored only a dozen or so story themes. For instance, the most recurring theme appeared to be Time, involving time warps and accidental journeys through time: a World War I flier lands at a modern jet air base (Matheson's "The Last Flight"), a man finds himself back in 1865 and tries to prevent the assassination of President Lincoln (Serling's "Back There"), three soldiers on National Guard maneuvers in Montana find themselves back in 1876 at the Little Big Horn (Serling's "The 7th Is Made Up of Phantoms"). Another theme explored the Confrontation with Death/the Dead: a girl keeps seeing the same hitchhiker on the road ahead, beckoning her toward a fatal accident (Serling's "The Hitchhiker", from Lucille Fletcher's radio play), an aged recluse, fearing a meeting with Death, reluctantly helps a wounded policeman on her doorstep and cares for him overnight before she realizes that he is Death, coming to claim her (Johnson's "Nothing in the Dark"). Expected science-fiction motifs regarding Aliens and Alien Contact, both benevolent and hostile, provide another story arena: a timid little fellow accustomed to being used as a doormat by his fellow man is endowed with super-human strength by a visiting scientist from Mars (Serling's "Mr. Dingle, the Strong"), visiting aliens promise to show the people of earth how to end the misery of war, pestilence and famine until a code clerk finally deciphers their master manual for earth and discovers a cook book (Serling's "To Serve Man", from a Damon Knight story). Other themes common to the series were Robots, with Matheson's excellent "Steel" a standout; the Devil, Beaumont's "The Howling Man"; Nostalgia, Serling's "Walking Distance" and "A Stop at Willoughby"; Machines, Serling's "The Fever"; Angels, Serling's poetic "A Passage for Trumpet"; and "Premonitions/Dreams/Sleep," Beaumont's "Perchance to Dream". The general tone of many *Twilight Zone* stories was cautionary, that humans can never be too sure of anything that appears real or otherwise.

In 1983 Warner Brothers, Steven Spielberg and John Landis produced *Twilight Zone—the Movie*, a four segment tribute to the original series presenting pieces directed by Landis (also written by Landis), Spielberg (written by George Clayton Johnson, Richard Matheson, Josh Rogan, based on the original 1962 episode "Kick the Can"), Joe Dante (written by Matheson, based on the original 1961 episode "It's a Good Life"), and George Miller (written by Matheson from his own story and original 1963 episode "Nightmare at 20,000 Feet"). From 1985 onwards CBS Entertainment produced a new series of *The Twilight Zone*. Honored science-fiction scribe Harlan Ellison acted as creative consultant under executive producer Philip DeGuere; the series is particularly noted for the participating name directors, such as Wes Craven, William Friedkin, and Joe Dante. In more recent times, *Twilight Zone: Rod Serling's Lost Classics* presented a two-hour TV movie based on two unproduced works discovered by the late writer's widow and literary executor, Carol Serling: Robert Markowitz directed both "The Theater" (scripted by Matheson from Serling's original story) and "Where the Dead Are" (from a completed Serling script).

With its subtext of escape from reality, a nostalgia for more simple times, but generally a hunger for other-worldly adventures, it seems appropriate that the original *The Twilight Zone* series appeared at about the right time to take viewers away, albeit briefly, from the contemporary real-life fears of the Cold War, the Berlin Wall, the Cuban Missile Crisis, and, eventually, the tragic events of Dallas. That *The Twilight Zone*, directly or indirectly, inspired such later fantasy and science-fiction anthologies as *Thriller* (1960–62), with its dark Val Lewtonesque atmosphere, and, following that, the superb *The Outer Limits* (1963–64), a delicious tribute to 1950s science-fiction cinema when it was at its most imaginative, remain testimony to both Rod Serling and his *Twilight Zone*'s spirit of poetry and principle.

—Tise Vahimagi

HOST
Rod Serling (1959–65)

NARRATORS
Charles Aidman (1985–87)
Robin Ward (1987–88)

PRODUCERS   Rod Serling, Buck Houghton, William Froug, Herbert Hirschman

PROGRAMMING HISTORY   134 Half-hour Episodes; 17 One-hour Episodes

• CBS

October 1959–September 1962	Friday 10:10:30
September 1961–September 1964	Friday 9:30-10:00
January 1963–September 1963	Thursday 9:00-10:00
May 1965–September 1965	Sunday 9:00-10:00

September 1985–April 1986	Friday 8:00-9:00
June 1986–September 1986	Friday 8:00-9:00
September 1986–October 1986	Saturday 10:00-11:00
December 1986	Thursday 8:00-8:30
July 1987	Friday 10:00-11:00
1987–1988	First Run Syndication

FURTHER READING

Boddy, William. "Entering the *Twilight Zone.*" *Screen* (London), July–October, 1984.

Javna, John. *The Best of Science Fiction TV: The Critics' Choice: From Captain Video to Star Trek, from The Jetsons to Robotech.* New York: Harmony, 1987.

Lentz, Harris M. *Science Fiction, Horror and Fantasy Film and Television Credits: Over 10,000 Actors, Actresses, Directors.* Jefferson, North Carolina: McFarland, 1983.

———. *Science Fiction, Horror and Fantasy Film and Television Credits, Supplement 2, Through 1993.* Jefferson, North Carolina: McFarland, 1994.

Rothenberg, Randall. "Synergy of Surrealism and *The Twilight Zone.*" *The New York Times,* 2 April 1991.

Sander, Gordon F. *Serling: The Rise and Twilight of Television's Last Angry Man.* New York: Dutton, 1992.

Schumer, Arlen. *Visions from the Twilight Zone.* San Francisco: Chronicle, 1990.

Zicree, Marc Scott. *The Twilight Zone Companion.* Toronto and New York: Bantam, 1982.

Ziegler, Robert E. "Moving Out of Sight: Fantastic Vision in *The Twilight Zone.*" *Lamar Journal of the Humanities* (Beaumont, Texas), Fall 1987.

See also Science-fiction Programs; Serling, Rod

TWIN PEAKS

U.S. Serial Drama

Scheduled to appear as a limited-run, mid-season replacement series on ABC, *Twin Peaks* attracted considerable critical attention even before its premiere in spring 1990. Both the network and national critics aggressively publicized the show as an unprecedented form of television drama, one that promised to defy the established conventions of television narrative while also exploring a tone considerably more sinister than previously seen in the medium. In short, critics promoted the series as a rare example of television "art," a program that publicists predicted would attract a more upscale, sophisticated, and demographically desirable audience to television. Upon its premiere, the series generated even more critical admiration in the press, placed higher than expected in the ratings, and gave Americans the most talked-about television enigma since "Who Shot J.R.?"

The "artistic" status of *Twin Peaks* stemmed from the unique pedigrees of the series' co-creators, writer-producer Mark Frost and writer-director David Lynch. Frost was most known for his work as a writer and story editor for the highly acclaimed *Hill Street Blues,* where he had mastered the techniques of orchestrating a large ensemble drama in a serial format. Lynch, meanwhile, had fashioned one of Hollywood's more eccentric cinematic careers as the director of the cult favorite *Eraserhead* (1978), the Academy Award-winning *The Elephant Man* (1980), the epic box-office flop *Dune* (1984), and the perverse art-house hit *Blue Velvet* (1986). A prominent American auteur, Lynch was already well-known for his oblique narrative strategies, macabre mise-en-scenes, and obsessive thematic concerns.

Twin Peaks combined the strengths of both Frost and Lynch, featuring an extended cast of characters occupying a world not far removed from the sinister small town Lynch had explored in *Blue Velvet.* Ostensibly a murder mystery, the series centered on FBI agent Dale Cooper and his investigation of a murder in the northwestern community of Twin Peaks, a town just a few miles from the Canadian border. The victim, high-school prom queen Laura Palmer, is found wrapped in plastic and floating in a lake. Cooper gradually uncovers an ever more baroque network of secrets and mysteries surrounding Laura's death, all of which seem to suggest an unspeakable evil presence in the town. Quickly integrating himself into the melodramatic intrigues of the community, Cooper's search for Laura's murderer eventually leads him to track "Killer Bob," a malleable and apparently supernatural entity inhabiting the deep woods of the Pacific Northwest.

Although the enigma of Laura's killer was pivotal to the series' popularity, so much so that *TV Guide* featured a forum of popular novelists offering their own solutions to the murder mystery, *Twin Peaks* as an avowedly "artistic" text was in many ways more about style, tone, and detail than narrative. Many viewers were attracted to the series' calculated sense of strangeness, a quality that led *Time* to dub Lynch as "the czar of bizarre." As in Lynch's other work, *Twin Peaks* deftly balanced parody, pathos, and disturbing expressionism, often mocking the conventions of television melodrama while defamiliarizing and intensifying them. The entire first hour of the premiere episode, for example, covered only a single plot point, showing the protracted emotional responses of Laura's family and friends as they learned of her death. This slow yet highly overwrought story line was apparently considered so disruptive by ABC that the network briefly discussed airing the first hour without commercial interruption (although this could have been a strategy designed to promote the program as "art"). Throughout the run of the series, the story line accommodated many such directorial set-pieces, stylistic tours-de-force that allowed the "Lynchian" sensibility to make its artistic presence felt most acutely. The brooding synthesizer score and dreamy jazz interludes provided by composer Angelo Badalamenti, who had worked previously with Lynch, also greatly enhanced the eerie, bizarre, and melancholy atmosphere.

As the series progressed, its proliferation of sinister enigmas led the viewer deeper into ambiguity and continually frustrated any hope of definitive closure. Appropriately, the first season ended with a cliffhanger that left many of the major characters imperiled, and still provided no clear solution to Laura Palmer's murder. Perhaps because of the series' obstinate refusal to move toward a traditional resolution, coupled with its escalating sense of the bizarre, the initially high ratings dropped over the course of the series' run. Despite such difficulties, and in the face of a perhaps inevitable critical backlash against the series, ABC renewed the show for a second season, moving it to the Saturday schedule in an effort to attract the program's quality demographics to a night usually abandoned by such audiences. After providing a relatively "definitive" solution to the mystery of Laura's killer early in the second season, the series attempted to introduce new characters and enigmas to reinvigorate the story line, but the transition from what had essentially been an eight-episode miniseries in the first season to an open-ended serial in the second had a significant, and many would say negative, impact on the show. The series attempted to maintain its sense of mystery and pervasive dread, but having already escalated its narrative stakes into supernatural and extraterrestrial plotlines, individual episodes increasingly had to resort to either absurdist comedy or self-reflexive commentary to sustain an increasingly convoluted world. After juggling the troubled series across its schedule for several months, ABC finally packaged the season's concluding two episodes together as a grand finale, and canceled the series after just 30 total episodes.

Exported in slightly different versions, *Twin Peaks* proved to be a major hit internationally, especially in Japan. In the United States, the brief but dramatic success of *Twin Peaks* inspired a cycle of shows that attempted to capitalize on the American public's previously untested affinity for the strange and bizarre. Series as diverse as *Northern Exposure* (CBS), *Picket Fences* (CBS), *The X-Files* (FOX), and *American Gothic* (CBS) have all been described in journalistic criticism as bearing the influence of *Twin Peaks*. The series also spawned a devoted and appropriately obsessed fan culture. In keeping with the program's artistic status, fan activity around the show has concentrated on providing ever closer textual readings of the individual episodes, looking for hidden clues that will help clarify the series' rather obtuse narrative logic. This core audience was the primary target of a cinematic prequel to the series released in 1993, *Twin Peaks: Fire Walk with Me*. Again directed by Lynch, *Fire Walk with Me* chronicled Laura Palmer's activities on the days just before her death. Freed from some of the constraints of network standards and practices, Lynch's cinematic treatment of *Twin Peaks* was an even more violent,

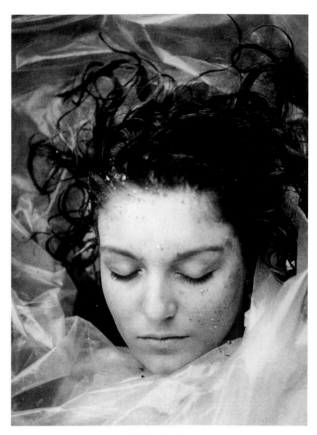

Twin Peaks

disturbing, and obsessive reading of the mythical community, and provided an interesting commentary and counterpoint to the series as a whole.

—Jeffrey Sconce

CAST

Dale Cooper	Kyle MacLachlan
Sheriff Harry S. Truman	Michael Ontkean
Shelly Johnson	Maedchen Amick
Bobby Briggs	Dana Ashbrook
Benjamin Horne	Richard Beymer
Donna Marie Hayward	Sara Flynn Boyle
Audrey Horne	Sherilyn Fenn
Dr. William Hayward	Warren Frost
Norma Jennings	Peggy Lipton
James Hurley	James Marshall
"Big Ed" Hurley	Everett McGill
Pete Martell	Jack Nance
Leland Palmer	Ray Wise
Catherine Packard Martell	Piper Laurie
Montana	Rick Giolito
Midge Loomer	Adele Gilbert
Male Parole Board Officer	James Craven
Female Parole Board Member #2	Mary Chalon
Emory Battis	Don Amendolia
The Dwarf	Michael J. Anderson
Jeffrey Marsh	John Apicella

Ronette Pulaski . . . . Phoebe Augustine
Johnny Horne . . . . Robert Bauer
Mrs. Tremond . . . . Frances Bay
Ernie Niles . . . . James Booth
Mayor Dwyane Milford . . . . John Boylan
Richard Tremayne . . . . Ian Buchanan
Blackie O'Reilly . . . . Victoria Catlin
Josie Packard . . . . Joan Chen
The Log Lady/Margaret . . . . . . Catherine E. Coulson
Herself . . . . Julee Cruise
Sylvia Horne . . . . Jan D'Arcy
Leo Johnson . . . . Eric DaRe
Maj. Garland Briggs . . . . Don S. Davis
Eileen Hayward . . . . Mary Jo Deschanel
DEA Agent Dennis/Denise Bryson . . . . David Duchovny
Agent Albert Rosenfield . . . . Miguel Ferrer
Deputy Andy Brennan . . . . Harry Goaz
Nancy O'Reilly . . . . Galyn Gorg
Annie Blackburn . . . . Heather Graham
Vivian Smythe . . . . Jane Greer
Nicolas "Little Nicky" Needleman . . . . Joshua Harris
Mike Nelson . . . . Gary Hershberger
Deputy Tommy "Hawk" Hill . . . . . . Michael Horse
Jerry Horne . . . . David Patrick Kelly
Madeleine Ferguson/Laura Palmer . . . . . . Sheryl Lee
Lana Budding . . . . Robyn Lively
Malcolm Sloan . . . . Nicholas Love
Pierre Tremond . . . . Austin Jack Lynch
Agent Gordon Cole . . . . David Lynch
Diane, Cooper's secretary . . . . . . Carol Lynley
Caroline Powell Earle . . . . Brenda E. Mathers
Evelyn Marsh . . . . Annette McCarthy
Hank Jennings . . . . Chris Mulkey
Andrew Packard . . . . Dan O'Herlihy
Jones . . . . Brenda Strong
RCMP Officer Preston King . . . . . Gavan O'Herlihy
Jacques Renault . . . . Walter Olkewicz
The Giant . . . . Carel Struycken
Jonathan Kumagai . . . . Mak Takano
Jean Renault . . . . Michael Parks
Lucy Moran . . . . Kimmy Robertson
Janek Pulaski . . . . Alan Ogle
Doctor Lawrence Jacoby . . . . Russ Tamblyn
Nadine Hurley . . . . Wendy Robie
Bob . . . . Frank Silva
Suburbis Pulaski . . . . Michelle Milantoni
Elizabeth Briggs . . . . Charlotte Stewart
Harold Smith . . . . Lenny Von Dohlen
Trudy . . . . Jill Rogosheske
Philip Michael Gerard/Mike/
    The One-Armed Man . . . . Al Strobel
Harriet Hayward . . . . Jessica Wallenfells
Bartender . . . . Kim Lentz
Thomas Eckhardt . . . . David Warner
Swabbie . . . . Charlie Spradling
Windom Earle . . . . Kenneth Welsh

Joey Paulson . . . . brett Vadset
Bernard Renault . . . . Clay Wilcox
Emerald/Jade . . . . Erika Anderson
Roger Hardy . . . . Clarence Williams III
Chet . . . . Lance Davis
Mrs. Tremond . . . . Mae Williams
Jared . . . . Peter Michael Goetz
The Room-Service Waiter . . . . Hank Worden
Tojamura . . . . Fumio Yamaguchi
Sarah Palmer . . . . Grace Zabriskie
John Justice Wheeler . . . . Billy Zane
Gwen Morton . . . . Kathleen Wilhoite
Female Parole Board Member #1 . . . . Mary Bond Davis
Einar Thorson . . . . Brian Straub
Heba . . . . Mary Stavin
Theodora Ridgely . . . . Eve Brent
Jenny . . . . Lisa Ann Cabasa
Decker . . . . Charles Hoyes
Tim Pinkle . . . . David L. Lander
Gersten Hayward . . . . Alicia Witt
Mr. Neff . . . . Mark Lowenthal
Eolani Jacoby . . . . Jennifer Aquino

PRODUCERS   David Lynch, Mark Frost, Gregg Fienberg, David J. Latt, Harley Peyton

PROGRAMMING HISTORY   30 Episodes

• ABC

8 April 1990	Sunday 9:00-11:00
April 1990–May 1990	Thursday 9:00-10:00
August 1990–February 1991	Saturday 10:00-11:00
March 1991–April 1991	Thursday 9:00-10:00
10 June 1991	Monday 9:00-11:00

FURTHER READING

Carrion, Maria M. "*Twin Peaks* and the Circular Ruins of Fiction: Figuring (Out) the Acts of Reading." *Literature/Film Quarterly* (Salisbury, Maryland), 1993.

Carroll, Michael. "Agent Cooper's Errand in the Wilderness: *Twin Peaks* and American Mythology." *Literature/Film Quarterly* (Salisbury, Maryland), 1993.

Davenport, Randi. "The Knowing Spectator of *Twin Peaks*: Culture, Feminism, and Family Violence." *Literature/Film Quarterly* (Salisbury, Maryland), 1993.

Deutsch, Helen. "'Is It Easier to Believe?' Narrative Innocence from Clarissa to *Twin Peaks*." *Arizona Quarterly: A Journal of American Literature, Culture, and Theory* (Tucson, Arizona), 1993.

Giffone, Tony. "*Twin Peaks* as Post-Modernist Parody: David Lynch's Subversion of the British Detective Narrative." *The Mid-Atlantic Almanac: The Journal of the Mid-Atlantic Popular/American Culture Association* (Greencastle, Pennsylvania), 1992.

Horne, Philip. "Henry Hill and Laura Palmer." *London Review of Books*, 20 December 1990.

Huskey, Melynda. "*Twin Peaks*: Rewriting the Sensation Novel." *Literature/Film Quarterly* (Salisbury, Maryland), 1993.

Kimball, Samuel. "'Into the Light, Leland, into the Light': Emerson, Oedipus, and the Blindness of Male Desire in David Lynch's *Twin Peaks*." *Genders* (Austin, Texas), 1993.

Lavery, David, editor. "Peaked Out." Special issue, *Literature/Film Quarterly* (Salisbury, Maryland), 1993.

Ledwon, Lenora. "*Twin Peaks* and the Television Gothic." *Literature/Film Quarterly* (Salisbury, Maryland), 1993.

Nickerson, Catherine. "Serial Detection and Serial Killers in *Twin Peaks*." *Literature/Film Quarterly* (Salisbury, Maryland), 1993.

Nochimson, Martha. "Desire under the Douglas Firs: Entering the Body of Reality in *Twin Peaks*." *Film Quarterly* (Berkeley, California), 1992.

Pollard, Scott. "Cooper, Details, and the Patriotic Mission of *Twin Peaks*." *Literature/Film Quarterly* (Salisbury, Maryland), 1993.

Shoos, Diane, Diana George, and Joseph Comprone. "*Twin Peaks* and the Look of Television: Visual Literacy in the Writing Classroom." *Journal of Advanced Composition* (Moscow, Idaho) Fall 1993.

Stevenson, Diane. "Family Romance, Family Violence: David Lynch's *Twin Peaks*." *Boulevard* (Victoria, British Columbia), Spring 1993.

Zaniello, Tom. "Hitched or Lynched: Who Directed *Twin Peaks*?" *Studies in Popular Culture* (Louisville, Kentucky), October 1994.

See also Movie Professionals and Television

227

U.S. Domestic Comedy

The show *227*, initially aired in September 1985, played five seasons on NBC before its final episode in July 1990. Based on a play of the same name, this situation-comedy was set primarily around an apartment building (number 227) located in a racially-mixed neighborhood of Washington, D. C. Featuring an ensemble cast that included such noted African-American television personalities as Marla Gibbs, Hal Williams, Alaina Reed Hall, and Jackee (Harry), *227* succeeded in becoming a top-rated television program. Surviving criticisms and early comparisons to other television programs with predominantly African-American principals, *227* proved a successful comedy, humorously portraying the everyday lives of apartment building 227.

The original play, *227*, had been written by Christine Houston of Chicago, and performed by Marla Gibbs' own Cross Roads Academy, a local community theater troupe in Los Angeles. After its successful theatrical debut, *227* was soon adapted and produced for television by Lorimar. In its earliest episodes, *227* was criticized as being too much like *The Cosby Show*, another highly successful, predominantly African-American situation-comedy broadcast on NBC at the same time. However, even in its first year *227* proved successful in its own right, earning top ratings that opening season. While Cosby portrayed an image of upper-middle class success, *227* supporters argued, *227* depicted a more working-class image of the same strong community and family values.

Most episodes taking place within and around the apartment building, from the front steps, to the laundry room, to the individual apartments, *227* invited the viewer within the most mundane and personal aspects of its characters' lives. The Jenkins, Mary and Lester, were one of the families struggling day by day to survive their various duties and commitments. Mary, played by Marla Gibbs

whose eleven seasons as the feisty, verbally aggressive maid Florence on *The Jeffersons* no doubt prepared her for this similarly outspoken character, was a mother of one, juggling the numerous responsibilities of household, family and personal life with invariably humorous results. Lester, played by Hal Williams, was a father and small-time contractor struggling to stay on top of his own family and job responsibilities. Together, Mary and Lester had their hands full with daughter Brenda, a studious, talented, and mostly well behaved young woman just beginning adolescence.

Other important characters included Rose Holloway, Mary's confidante and cohort in gossip, portrayed by Alaina Reed. Rose, the landlady of building 227, often sat with Mary on the front steps, the two laughing and gossiping about various other residents. In particular, Rose and Mary enjoyed discussing and berating sexually outspoken tenant Sandra Clark, the building's resident vamp. Played by Jackee, the one-named wonder who made Sandra, and herself, famous, Sandra's whining voice and wiggling, tight-dressed body became staple features of *227*. Her many men friends and sexually oriented antics a source of constant humor, Sandra sauntered through episode after episode, occasionally eliciting help from Mary for some dilemma she was experiencing. Another frequent front porch gossip was Pearl Shay, an older woman who often leaned out her front window to comment on Rose and Mary's discussions. The grandmother of young Calvin Dobbs, the burgeoning love interest of Brenda Jenkins, Pearl's time was frequently spent scolding and disciplining this gangly adolescent grandson.

Successful in depicting the everyday aspects of its many characters' lives, *227* offered an interesting working class version of African-American values and images. The program brought the viewer within its characters' lives, providing a personal look within this entertaining apartment complex.

—Brent Malin

CAST

Mary Jenkins	Marla Gibbs
Lester Jenkins	Hal Williams
Rose Lee Holloway	Alaina Reed-Hall
Sandra Clark	Jackee (Harry)
Brenda Jenkins	Regina King
Tiffany Holloway (1985–86)	Kia Goodwin
Pearl Shay	Helen Martin
Calvin Dobbs	Curtis Baldwin
Alexandria DeWitt (1988–89)	Countess Vaughn
Eva Rawley (1989–90)	Toukie A. Smith
Julian C. Barlow (1989–90)	Paul Winfield
Dylan McMillan (1989–90)	Barry Sobel
Travis Filmore (1989–90)	Stoney Jackson
Warren Merriwether (1989–90)	Kevin Peter Hall

PRODUCERS Bill Boulware, Bob Myer, Bob Young

PROGRAMMING HISTORY 116 Episodes

• NBC

September 1985–March 1986	Saturday 9:30-10:00
April 1986–June 1986	Saturday 9:30-10:00
June 1986–May 1987	Saturday 8:30-9:00
June 1987–July 1987	Saturday 8:00-8:30
July 1987–September 1988	Saturday 8:30-9:00
October 1988–July 1989	Saturday 8:00-8:30
September 1989–February 1990	Saturday 8:30-9:00
April 1990–May 1990	Sunday 8:30-9:00
June 1990–July 1990	Saturday 8:00-8:30

227

FURTHER READING

Collier, Aldore. "Jackee Harry: How Her TV Role is Ruining Her Love Life." *Ebony* (Chicago), June 1987.

Dates, Jannette, and William Barlow, editors. *Split Images: African Americans in the Mass Media.* Washington, D.C.: Howard University Press, 1990.

Randolph, Laura B. "Who is Toukie Smith and Why Are People Talking About Her?" *Ebony* (Chicago), May 1990.

Sanders, Richard. "After Mopping Up as the Maid on *The Jeffersons*, Marla Gibbs Polishes Her Image as the Star of *227.*" *People Weekly* (New York), 25 November 1985.

"Thurston Sees *227* Sales Growth with Affiliates Staying with Sitcoms." *Television-Radio Age* (New York), 24 July 1989.

Whitaker, Charles. "Brassy, Sassy Jackee Is on a Roll." *Ebony* (Chicago), January 1988.

See also Comedy, Domestic Settings; Racism, Ethnicity, and Television

U

THE UNCOUNTED ENEMY: A VIETNAM DECEPTION

U.S. Documentary

The *CBS Reports* documentary *The Uncounted Enemy: A Vietnam Deception*, which aired on 23 January 1982, engendered one of the most bitter controversies in television history. The 90-minute program spawned a three-year ordeal for CBS, including disclosures by *TV Guide* that the report violated CBS News standards, an internal investigation by Burton (Bud) Benjamin, and an unprecedented $120 million libel suit by retired U.S. Army General William C. Westmoreland.

Westmoreland sued producer George Crile III, correspondent Mike Wallace, and others for alleging that Westmoreland participated in a conspiracy to defraud the American public about progress in the Vietnam War. The suit was dropped, however, before reaching the jury, with CBS merely issuing a statement saying the network never meant to impugn the general's patriotism.

CBS subsequently lost its libel insurance. The controversy was also drawn into the debate over repeal of the financial interest and syndication rules. CBS chair Tom Wyman twice admonished his news division in 1984 for hindering broadcast deregulation. In part as a result of the controversies, fewer CBS documentaries were produced than ever before.

The lawsuit generated an abundance of literature, as well as soul-searching among broadcast journalists regarding ethics, First Amendment protection, libel law, and the politicization of TV news. Unlike the case for a similar, but lesser, controversy over *The Selling of the Pentagon*, *The Uncounted Enemy* failed to uplift TV news, and instead contributed to the documentary's decline.

The program states that the 1968 Tet Offensive stunned Americans because U.S. military leaders in South Vietnam arbitrarily discounted the size of the enemy that was reflected in CIA reports. Former intelligence officers testify that field command reports withheld information from Washington and the press, ostensibly under orders from higher military command, and that a 300,000-troop ceiling was imposed on official reports to reflect favorable progress in the war. This manipulation of information was characterized as a "conspiracy" in print ads and at the top of the broadcast.

The first part of the documentary chronicles the CIA-MACV dispute over intelligence estimates. Part two reports that prior to Tet, infiltration down the Ho Chi Minh Trail exceeded 20,000 North Vietnamese per month. Again, the report alleges, these figures were discounted. The last segment charges that intelligence officers purged government databases to hide the deception.

The most provocative scene features correspondent Mike Wallace interviewing Westmoreland. An extreme close-up captures the general trying to wet his dry mouth as Wallace fires questions. The visual image in conjunction with other program material suggests that Westmoreland engineered a conspiracy, and, as viewers can see, he appears guilty. Westmoreland publicly rebuked these claims and demanded forty-five minutes of open airtime to reject *The Uncounted Enemy* assertions. CBS refused the request.

In the spring of 1982, a CBS News employee disclosed to *TV Guide* that producer George Crile had violated network standards in making the program. The 24 May story by Sally Bedell and Don Kowet, "Anatomy of a Smear: How CBS News Broke the Rules and 'Got' Gen. Westmoreland," stipulated how the production strayed from accepted practices. Significantly, *TV Guide* never disputed the premise of the program. The writers attacked the journalistic process, pointing out, for instance, that Crile screened interviews of other participants for one witness and then shot a second interview, that he avoided interviewing witnesses who would counter his thesis, and that answers to various questions were edited into a single response.

CBS News president Van Gordon Sauter, who was new to his position, appointed veteran documentary producer Burton Benjamin to investigate. His analysis, known as the "Benjamin Report," corroborated *TV Guide*'s claims.

According to a report in *The American Lawyer*, several conservative organizations, such as the Richard Mellon Scaife Foundation, the Olin Foundation, and the Smith Richardson Foundation, financed Westmoreland's suit in September 1982. One goal of the Smith Richardson Foundation was to kill *CBS Reports*. Another was to turn back the 1964 *New York Times vs. Sullivan* rule, which required that public officials prove "actual malice" to win a libel judgment. The *Westmoreland* case went to trial two years later and was discontinued in February 1985.

One of the significant by-products of the controversy is the "Benjamin Report." Benjamin's effort remains widely respected within the journalistic community for revealing unfair aspects of the program's production. Some observers, however, criticized the report for having a "prosecutorial tone," for failing to come to terms with the producer's purpose, and for measuring fairness and balance by a math-

General William C. Westmoreland cross-examined by David Boles. Mike Wallace is at right.
Courtesy of Marilyn Church

ematical scale. In his conclusion, Benjamin acknowledges the enduring value of the documentary: "To get a group of high-ranking military men and former Central Intelligence Agents to say that this is what happened was an achievement of no small dimension." The production flaws, however, overshadowed the program's positive attributes.

The Uncounted Enemy helps explain an aspect of Tet and gives voice to intelligence officers who were silenced during the war. But the program tried unsuccessfully to resolve a complex subject in ninety minutes, and it fails to convey the context of national self-delusion presented in lengthier treatments, such as the thirteen-hour PBS series, *Vietnam: A Television History* or Neil Sheehan's book *A Bright Shining Lie*. CIA analyst George Allen, who was interviewed in the documentary, explained in a letter to Burton Benjamin in June 1982 his belief that the intelligence dispute was "a symptom of a larger and more fundamental problem, i.e. the tendency of every American administration from Eisenhower through Nixon toward self-delusion with respect to Indochina." Allen reasserted his support for *The Uncounted Enemy* as a valid illustration of

the larger issue and subsequently used the program as a case study in politicized intelligence.

Although many works disprove the conspiracy charge, General Westmoreland did subsequently acknowledge the potential significance of a public disclosure of intelligence information prior to Tet. Appearing on the NBC *Today* show in May 1993, Westmoreland explained: "It was the surprise element, I think, that did the damage. And if I had to do it over again, I would have called a press conference and made known to the media the intelligence we had."

—Tom Mascaro

CORRESPONDENT
Mike Wallace

PRODUCER   George Crile III

PROGRAMMING HISTORY

• CBS

23 January 1982

FURTHER READING

Bedell, Sally, and Dan Kowet. "Anatomy of a Smear: How CBS News Broke the Rules and 'Got' Gen. Westmoreland." *TV Guide,* 24 May 1982.

Benjamin, Burton. *Fair Play.* New York: Harper and Row, 1988.

The CBS Benjamin Report. Washington, D.C.: Media Institute, 1984.

Loftus, Jack. "Goldwater Points a Loaded Gun at CBS." *Variety* (Los Angeles), 25 August 1982.

Schneir, Walter, and Miriam Schneir. "The Right's Attack on the Press." *The Nation* (New York), 30 March 1985.

VIETNAM: A Documentary Collection—Westmoreland v. CBS. Microfiche. New York: Clearwater, 1985.

See also Columbia Broadcasting System; Documentary; Stanton, Frank; Wallace, Mike

UNIONS/GUILDS

The television industry is one of the more highly organized, or, unionized, in the United States. Qualified candidates are numerous for a few available jobs. Producing and airing programs lend themselves to odd working hours, location shoots, holidays, weekends, long working days and often short-term temporary employment. Such conditions would normally permit management to exploit employees by offering low wages, few fringe benefits, and no job security to employees. Historically, unionization in U. S. industry began to eliminate such exploitation, and the television industry is no exception.

Although some of the unions in television and film today grew out of earlier creative guilds like Actors' Equity and the Dramatists' Guild, the primary reference point for effective unionization of the industry was passage of the National Labor Relations Act in 1935. Known as the Wagner Act, in honor of its congressional sponsor, it was a major piece of "New Deal" legislation passed during the Franklin D. Roosevelt administration. The NLRA made it legal for workers to form unions. It set up the National Labor Relations Board as an arm of government to enforce it. Unions could bargain for wages and working conditions.

Today, unions and guilds representing employees in television and film bargain with networks and production companies for minimum wage scales, pension funds and other fringe benefits. A major bargaining issue in recent years between producers and creative guilds has been residuals. Residuals is the term used to describe royalties paid to actors, directors, and writers for airing programs originally and in subsequent replays and re-runs, and for cassette sales and rentals.

The degree of unionization in television today varies considerably by geographic region. Television stations and cable systems in most of the larger media markets, like New York City, Los Angeles, and Chicago, are almost totally unionized. Local television stations and cable systems in small markets, however, may not be unionized. Networks and major production companies are all unionized, whereas small independent producers tend not to be.

The term "union" in the television industry describes labor organizations that represent technical personnel, and are referred to as "below-the-line" unions. The term "guild" describes labor organizations that represent creative personnel, and are referred to as "above-the-line" unions. These designations result from their actual position on the pages of production budgets in which "creative" and "technical" costs are divided by a line. In a typical television show production budget, below-the-line costs are fixed, whereas above-the-line costs are flexible. For example, the budget for a one-hour drama enters a camera operator's wages below-the-line because there is a standard wage scale in the union contract with management for camera operators shooting a one-hour drama. The salary for the show's leading actor is entered above-the-line because there is considerable disparity between a relatively unknown actor's salary and the salary of a major TV star like Tim Allen or Angela Lansbury.

Four very large unions represent most below-the-line technical personnel in television and cable today: the National Association of Broadcast Employees and Technicians (NABET), the International Brotherhood of Electrical Workers (IBEW), the International Alliance of Theatrical and Stage Employees (IATSE), and the Communication Workers of America (CWA).

NABET began as a union of engineers at NBC in 1933. It is the only union among the four devoted exclusively to representing workers employed in broadcasting, film, recording, and allied industries. Today it is the exclusive bargaining agent for below-the-line personnel at the ABC, NBC, FOX, and PBS networks, as well as at many local independent television stations in large cities.

IBEW is one of the largest unions in the United States and represents workers in construction, manufacturing, and utilities, in addition to below-the-line personnel at CBS, Disney, independent TV stations, and some cable companies.

IATSE was founded in New York City in 1893 as the National Alliance of Theatrical State Employees. Today, it is organized primarily along craft lines with over 800 local chapters, each representing specialized occupations within the union's overall national membership of more than 70,000 workers. In the Los Angeles area alone, some of the occupations represented by separate local chapters are: set designers-model makers, illustrators-matte artists, costumers, makeup artists-hair stylists, film editors, film cartoonists, script supervisors, film set painters, studio electricians, stagehands, and story analysts. IATSE represents almost every below-the-line occupation at the major production studios and many independent production companies that produce shows on film for theaters, television, and cable.

CWA, historically, has represented workers in the telephone industry and other common carrier fields. In recent years, it has increased its membership and influence in the cable television industry, and represents below-the-line personnel in cable multiple system operators, cable networks, and local cable companies.

There are many above-the-line guilds representing creative workers in television. The major guilds with the most influence are: the American Federation of Television and Radio Artists (AFTRA), the Screen Actors Guild (SAG), the Directors Guild of America (DGA), the Writers Guild of America (East and West; known as WGAE and WGAW), and the American Federation of Musicians (AFM). Most members of these unions do not work full time or regularly, and those who do almost never work for minimum wage scale.

AFTRA grew out of the American Federation of Radio Artists, founded in 1937. It added television performers and "television" to its name in 1952. Today, AFTRA represents over 70,000 performers nationally who appear on television or cable programs that are produced on videotape or broadcast live. In addition to actors this number includes many performers such as announcers, dancers, newspersons, sportscasters, game show emcees, and talk show hosts, stunt people, and sound effects artists. AFTRA has about 30,000 members in its Los Angeles area alone, a small percentage of whom earn their living primarily from performing on radio, cable, or television. Most television performers work other jobs to support themselves while seeking occasional temporary employment as a television, cable, film or radio performer.

SAG represents performers who appear on television or cable programs produced on film. These include feature films produced for theatrical release and later aired on television in addition to film programs produced expressly for television exhibition. Related to SAG is the Screen Extras Guild (SEG), which represents bit performers who appear in programs produced on film. Most celebrities and successful performers belong to both AFTRA and SAG, so they are not limited from performing in all three production modes of live, tape, or film.

The DGA was organized originally in 1936 as the Screen Directors Guild by a group of famous film directors, including King Vidor and Howard Hawks. Television directors were admitted in 1950, and the name Directors Guild of America was adopted in 1960. Today, it has a West chapter in Hollywood and an East chapter in New York City. It represents directors, associate directors, unit production managers, stage managers, and production assistants in television, and directors, assistant directors, and stage managers in film. Both chapters work cooperatively to represent their members regardless of the location of a production or shoot. The East chapter, for example, represents most play directors, and the West chapter represents most film directors.

The WGAE (East) and the WGAW (West) are incorporated separately because of differing laws of incorporation in New York and California. WGAE is located in New York City, and WGAW is located in Los Angeles. Though incorporated separately, they function as a single organization that represents the interests of over 8,000 members nationally, although the WGAE has only half the membership of the WGAW, and has a significant number of playwrights among its membership, whereas WGAW is dominated by screenwriters. In 1962, WGA also joined with sister guilds in Great Britain, Canada, Australia, and New Zealand to form an international union alliance among these English-speaking nations.

The AFM began in 1896, and represents musicians, including vocalists and instrumentalists who perform live or on film, tape, record, or disk. It has local chapters throughout the United States that bargain with local television stations and cable systems in geographic regions they cover.

With computers, satellites, and digital technology globalizing electronic communication, unions and guilds will continue to add new occupational groups to their membership and become increasingly more international in scope. In a democratic society like the United States, viable unions remain necessary to provide oversight of big business and management policies and practices toward their employees.

—Robert G. Finney

FURTHER READING

Bielby, William T., and Denise D. Bielby. *The 1989 Hollywood Writers' Report: Unequal Access, Unequal Pay.* West Hollywood, California: Writers Guild of America, West, 1989.

Block, Alex Ben. "Hollywood's Labor Pains; Lower Revenues and Rising Costs Are Spawning a Tough New Attitude to Unions by TV Networks and Producers." *Channels: The Business of Communications* (New York), July-August 1988.

Directors Guild of America. *Constitution and Bylaws.* Hollywood, California, 1991.

The Journal. Los Angeles, California: Writers Guild of America, 1982.

The Journal of the Producers Guild of America. Beverly Hills, California, 1982–.

Prindle, David F. *The Politics of Glamour: Ideology and Democracy in the Screen Actors Guild.* Madison, Wisconsin: University of Wisconsin Press, 1988.

Schwartz, Nancy Lynn. *The Hollywood Writers' Wars.* New York: Knopf, 1982.

The Story of the Screen Actors Guild. Hollywood, California: Screen Actors Guild, 1966.

See also Director, Television; Writing for Television

UNITED KINGDOM SEE BRITISH TELEVISION, IRELAND, SCOTLAND, WALES

UNITED STATES: CABLE TELEVISION

In its short history, cable television has redefined television in many ways. It became a cultural force that profoundly altered news, sports and music programming with services such as Cable News Network (CNN), C-SPAN (Cable Satellite Public Affairs Network), ESPN (Entertainment and Sports Network), and Music Television (MTV). It spawned a huge variety of "narrowcast" programming services as well as new broad appeal services, including 94 basic and 20 premium services by 1994. It altered the structure of the programming industry by developing new markets for both very old and very new program types. It became an entertainment service that contributed to changed viewing practices, suggested by the proliferating use of remote controls to "surf" along the now extensive channel lineup. And it began an important debate concerning the ability of citizens to control, and contribute to, local media. Cable's organizational development, economic relationships, and regulatory status profoundly altered the video landscape in ways entirely unforeseen, and in the course of its growth and development many accepted notions about First Amendment rights of speakers and listeners or viewers, and about the functions and obligations of communication industries, have been challenged. The cable television industry eclipsed broadcasting's asset and revenue values by the late 1980s as it created moguls and empires that joined the largest media firms in the United States. The first of many communication systems to stretch the meanings and boundaries established in the Communication Act of 1934, cable television has had a pivotal role in altering conceptions about television.

Now the dominant multi-channel provider in the United States, cable television contributed to the substantial drop in the broadcast network viewing from 1983 to 1994 when weekly broadcast audience shares dropped from 69 to 52 while basic cable networks' shares rose from 9 to 26 during the same period, according to A. C. Nielsen as of 1995. Cable television service is available to 95% of all television households in the United States, and about two-thirds of all television households subscribe to it. Most of those systems offer at least 30 channels (57% have between 30 to 53 channels and 13% have 54 or more channels according to the 1995 *Television and Cable Factbook*). Even with this number of channels, however, broadcast fare carried over cable is still among the most heavily viewed, and most viewers regularly spend their time with only from five to nine of those many cable channels.

Cable service comprises a collection of several industries. Primary among them are the distributors of video product called operators or sometimes "multiple system operators" (MSOs). Cable operators establish and own the physical system that delivers television signals to homes using coaxial cable, although in the 1990s optical fiber began to replace much of the traditional coaxial cable in portions of the network. Programming services produce or compile programming and also sell their services to cable as well as to direct broadcast satellite (DBS) operators. Other entities and institutions connected to the cable industry include investors underwriting distribution or production efforts, the creative community, and loosely coupled groups such as advertisers, local community groups and producers, recording companies, equipment suppliers, satellite and terrestrial microwave relay companies, and telephone companies.

Cable service relies on three fundamental operations. The first is signal reception, using satellite, broadcast, microwave and other receivers, at a "headend" where signals are processed and combined. Second, signals are distributed from that headend to the home using coaxial cable or optical fiber or microwave relays, abetted by amplifiers and other electronic devices that insure quality of signal to households. Third, components at the home or near the home such as converters must change cable signals into tunable television images, descramblers must be able to decode encrypted programming, and still other equipment may be used to allow for delivery of services on demand, a process called "addressability." Cable television's traditional tree-and-branch system network design typifies one-way delivery services, in contrast to telephone services' star design which maximizes interconnection. Its huge and always-growing channel capacity or bandwidth enables cable television to support a variety of programming services and has always left it favorably positioned to expand into other service areas, such as high definition television, compressed video, and pay-per-view channels. However, the tree-and-branch network limits its interactive potential, a factor that became significant in the 1990s as interactive services were explored more intensively.

Programming on cable television began with retransmitted broadcast fare, but evolved to services unique to cable, some targeted at specialized audience groups such as children, teenagers, women, or ethnic groups, and some providing only one type of programming—weather, news, or sports for example. Such narrowcast programming that appeals to specific demographic groups rather than to broadcast television's wide audience attracts advertisers who require more targeted approaches.

Traditionally cable operators organized their programming into "tiers," with different subscriber charges accruing at different levels. At the base was the least expensive "basic tier" which includes retransmitted broadcast channels. Moving up leads to special cable-only packages of channels often called "expanded basic." And on the most expensive tier are single-channel premium services such as Home Box Office (HBO), Showtime, Disney or Playboy with separate fees. Programming in each of these levels has expanded because cable television's surplus of channel space and low costs helped to spawn several new formats after the early 1970s, including infomercials, 24-hour news and weather services, music video services, home shopping channels, arts channels, and a host of other narrowly targeted programming. Federal regulations of the 1970s that required cable operators to support community access channels dedicated to public, educational and governmental programming likewise led in many cases to distinctive public service programming. Although cable systems have always been engineered as predominantly one-way delivery systems, they have some capability to provide limited two-way services and could be designed to offer more interactivity. Future cable systems will focus on developing two-way services, even though one-way programming has been the foundation service.

Because cable systems must lay cable in the ground or string it along telephone or electric poles, they must negotiate for the use of poles and rights of way. This is the crux of cable television's dependence on municipalities since cities and towns control their own rights-of-way and in many cases also own the utility poles used by cable companies. Cable operators must negotiate franchises with municipalities that entitle them to use rights-of-way in exchange for fees, capped at 5% of revenues, to the city. A conventional franchise lasts for 15 years. Several aspects of cable television resemble those of traditional utilities: it uses public rights-of-way and deploys a capital intensive network; it conveys but does not create content; it bills subscribers on a monthly basis. These utility-like aspects encouraged communities to treat it as a utility in early years: generally only one cable company has been franchised in a single municipality, effectively rendering it a monopoly. Rates charged to subscribers (and sometimes even rates of return) have been regulated differently at different points in time. And service quality is monitored. One source of long-standing friction between cities and cable companies often develops in the area of franchise conditions, particularly the designation of specific services a municipality may expect a cable operator to provide (e.g., specialized channels or funds for public, educational or government access). These controversies have been attributable, in part, to cable television's common carrier or utility characteristics.

Cable television, like home video, taps viewers' willingness to pay directly for programs, a source of revenue untouched by traditional broadcasters. Subscribers pay a monthly fee for programming to the operators, and the operators in turn pay programming networks such as ESPN or MTV for the right to use the services. The price of the programming depends on the specific programming (ESPN is more expensive generally, for example, than the Learning Channel) and the size (subscribership) of the MSO or operator, although the very largest MSOs take advantage of their economies of scale to obtain smaller unit prices on programming. Most basic programming services carry advertisements, and also allow local cable operators to insert ads (called "ad avails") during designated programming segments. Advertising revenues, both national and local, were slow to develop for programming services, awaiting significant subscriber levels and solid ratings data that could indicate viewer levels. Nevertheless, ad revenues grew steadily and have proved to be an important part of programming services' revenues. Premium services such as HBO, Showtime, and the Disney Channel eschew ads and instead rely on higher, separate subscription fees assessed to subscribers.

Cable television's development was very dependent on the regulatory treatment and economic models developed for predecessor systems of telephony and broadcasting. As a hybrid communications system unanticipated in the Communication Act of 1934, cable television challenged regulators' conceptions of what it should be, how it should operate in a landscape already dominated by broadcasters, and how it might take advantage of its delivery system and capacity. The consequences of this uncertainty included some dramatic shifts in ideas of cable obligations to the public and to the communities it serves, and in the scope of cable television's First Amendment rights. The changing shape of cable television has four distinct phases. The first slow growth period, from cable television's inception through 1965, predates any major regulatory efforts. During the second phase, from 1965 to roughly 1975, the Federal Communications Commission (FCC) attempted to restrict cable television to non-urban markets and to mold it into a local media service. In the third phase, from 1975 to 1992, a series of judicial, legislative and regulatory acts including the Cable Communications Policy Act of 1984 catalyzed cable television's expansion across the country and promoted dozens of new satellite-delivered programming services. The fourth phase, signaled by the Cable Television Consumer Protection and Competition Act of 1992 and the Telecommunications Act of 1996, re-regulated certain aspects of cable television and deregulated others, even as new competitors to the service appeared in the form of MMDS, direct broadcasting satellites and new telephone company ventures into video media. As cable television moves into a more competitive environment in which many different delivery systems can duplicate its services, its separate identity is fading as very large, merged telephone-cable-entertainment conglomerates move into video programming and transmission.

I. Rural Roots and Slow Growth

Although cable television systems are now present in many regions of the globe, they began in the rural areas of North America. A product of both the geographic inaccessibility of terrestrial broadcast signals and a television spectrum allocation scheme that favored urban markets, cable systems, also called "community antenna television" or CATV, grew out of simple amateur ingenuity. Retransmission apparatuses such as extremely high antenna towers or microwave repeater stations, often erected by television repair shops or citizens groups, intercepted over-the-air signals and redelivered them to households that could not receive them using regular VHF or UHF antennas. The earliest cable television systems, established in 1948, are usually credited to Astoria, Oregon, or Mahoney City, Pennsylvania, both mountainous, rural communities. Such retransmission systems spread across remote and rural America throughout 1950s and 1960s. According to *Television Factbook 1980–81,* there were 640 systems with 650,000 subscribers in 1960. By 1970 these numbers had grown to 2,490 systems with 4,500,000 subscribers. The systems were generally "Mom-and-Pop" operations with 12 channels at best, although the MSO form of cable system ownership, in which one company owned several cable distribution systems in different communities, was already spreading.

When cable systems began importing signals from more distant stations using microwave links, broadcasters' objections to the new service escalated. Many broadcasters had never been happy with cable service, claiming that such systems "siphoned" their programming since cable operators had no copyright liability and therefore never paid for the programming. In 1956 broadcasters petitioned the FCC to generate a policy regarding cable television. The Commission initially declined; it did not possess clear regulatory authority over CATV because the technology did not use the airwaves. The agency reconsidered, however, and finally asserted jurisdiction over cable television in 1962 in the *Carter Mountain Transmission Corporation v. FCC* case. Its rationale for regulating CATV focused on cable's impact on broadcasters: to the extent that cable television's development proved injurious to broadcasting—an industry the FCC was obligated to sustain and promote—cable television required regulation. While this justification sustained the FCC's position throughout the second phase of cable television's development, it later crumbled under judicial scrutiny.

II. Restricted Expansion and Localism, 1965–1975

While the case addressed only the microwave—and hence over-the-air—portion of CATV service, the FCC eventually extended its authority to all aspects of cable television, and issued two major policy statements, the *First Cable Television Report and Order,* 1965, and *Second Cable Television Report and Order,* 1966. In these orders the FCC, hoping to prevent any deleterious effects on broadcasting, required cable operators to carry local broadcast signals under "must-carry" rules. With its ruling on "nonduplication" the Commission required cable companies to limit imported programming that duplicated anything on local broadcast. A set of 1969 rules deliberately kept cable television from growing toward urban markets or from attaining the capital or benefits of entrenched industries by placing ownership prohibitions or limitations on television and telephone companies and by preventing cable television from entering the top 100 markets. Programming mandates instituted channels for local public access and created a prohibition on showing movies

less than 10 years old and sporting events that had been on broadcast television within the previous five years. These rules were intended to promote cable's local identity and prevent it from obtaining programming that might interest or compete with broadcasters.

Although cable operators continued to press for limitations on the FCC's ability to impose such program obligations, the courts rebuffed their claims. For example, when Midwest Video Corporation challenged the FCC's requirement that it originate local programming, the Supreme Court found that such a rule was "reasonably ancillary" to the FCC's broadcasting jurisdiction (*U.S. v. Midwest Video Corp.,* 440 U.S. 689, 1972).

The net effect was to severely constrain the programming options for cable television operators, and in particular to diminish opportunities for a pay television service that would show movies or sports. During the 1960s the FCC conceived of cable television as an alternative to broadcasting and promulgated the must-carry, nonduplication, and other rules with the intention of enhancing cable television's community presence and possibilities and at the same time protecting broadcasters from competition with the new delivery system. The agency positioned cable television as a hybrid common carrier- broadcasting service, one limited to mandatory channels (the must-carry rules, local access channels, constrained non-local programming) with regulated rates. This fettered opportunities for networking, for national distribution, and for direct competition with broadcasters.

By the late 1960s and early 1970s, more public interest in cable television fueled by a coalition of community groups, educators, cable industry representatives, and think tanks such as Rand Corp. heralded cable television's potential for creating a wide variety of social, educational, political and entertainment services beneficial to society. These constituencies objected to the FCC's policies because they seemed to inhibit the promise of the "new technology." Ralph Lee Smith's 1972 book, *Wired Nation,* captured many people's imaginations with its scenarios of revolutionary possibilities cable television could offer if only it were regulated in a more visionary fashion, particularly one that supported developing the two-way capabilities of cable and moving it toward more participatory applications. The discourse of cable as a cornucopia, as progress, as an electronic future captivated many.

In 1970 and 1971 the White House's Office of Telecommunication Policy spearheaded a series of meetings among cable, programming and broadcast companies that culminated in the FCC revising its cable rules. This *1972 Cable Television Report and Order* issued new rules that softened some of the restrictions on cable television's expansion to new markets, particularly with respect to importing distant signals ("leapfrogging"). However, it continued several rules and standards that the industry found onerous, such as mandatory two-way cable service in certain markets and local origination rules requiring operators to generate programs. Still more programming restrictions on movies and sporting events adopted in 1975 chafed at the cable industry's desires to offer something new and appealing to subscribers.

III. Deregulation, National Networks, Rapid Development, 1975–1992

Nevertheless, in the wake of the *1972 Report and Order*, as cable delivered more than just local broadcast signals to viewers by importing programs from distant markets via microwave, its attractiveness and profitability grew. Two significant events spurred even more growth in the late 1970s. First, HBO became a national service in 1975 by using a communications satellite to distribute its signal, at once demonstrating the ability to bypass telephone companies' expensive network carriage fees (commercial television networks depended on AT and T's lines for their national transmissions) and the possibility for many new program services to cost-effectively form national networks. Second, a series of judicial decisions sanctioned the cable industry's rights to program as it pleased, to enter the top television markets, and to offer new services. This third phase was cable television's highest growth period.

As early as 1972 HBO had offered, on the East coast, event programming such as sports on a "pay cable" basis using a microwave relay, but with satellite feeds it could reach cable operators across the country. HBO wanted to switch from microwave relays to the new RCA satellite Satcom I, which would take its signal across the entire country once the satellite launched in 1975. There were two major impediments to this plan. First, the FCC required each cable operator to use large, nine meter dish antennas to receive a satellite feed, and these receiver dishes were expensive. Second, the restrictive FCC programming rules still prevented cable services from acquiring certain types of programming. HBO helped pay for the receiving dishes cable operators needed to receive its signal, and became Satcom I's first television customer. Just two years later the service was being taken by 262 systems around the nation, yet the best programming, current movies and sporting events, was still off limits. HBO then took the commission to court, claiming that the FCC had exceeded its jurisdiction in limiting programming options. Supporting HBO's position in *HBO v. FCC*, the District of Columbia Court of Appeals concluded that the FCC's broadcast protectionism was unjustifiable and, perhaps more important, that cable television service resembled newspapers more than broadcasting and consequently deserved greater First Amendment protections. This reasoning paved the way for the cable industry to argue against other government rules, which fell aside one by one after the strong message sent by the HBO case to the FCC.

Even as the agency stripped away federal syndicated exclusivity rules, reduced the size and consequently the cost of allowable satellite dishes, and eliminated remaining distant signal importation rules, the courts underscored cable television's rights to expand as it wished and to use any programming it desired. On the heels of the HBO case, the 1979

United States v. Midwest Video Corp. decision found that the FCC's rules imposed unacceptable obligations on cable operators, undermining the earlier Midwest Video decision. Insofar as those rules required cable operators to function as common carriers with the access channels—operators had no control over the content of access channels and they had to carry community programs on a first-come, first served basis—and insofar as they prescribed a minimum number of channels, they violated cable's First Amendment rights. The industry claimed these court decisions affirmed its status as an electronic publisher, and has continued its fight against regulatory obligations under this banner ever since. The cable industry has advanced its electronic publisher label to underscore its First Amendment status: like print publishers, cable television selects and packages materials for exhibition, and like print, should be under no obligation to exhibit material that regulatory powers prescribe.

With the regulatory barriers to entry now reduced, cable systems experienced huge growth from the late 1970s through the early 1980s: The 3,506 systems serving nearly 10 million subscribers in 1975 leaped to 6,600 systems serving nearly 40 million subscribers just ten years later. Programming services likewise emerged. Ted Turner's UHF station WTCG, renamed superstation WTBS (and later just TBS), followed HBO's lead in national satellite delivery in 1976, as did Christian Broadcast Network's CBN Cable (later the Family Channel). The Showtime movie service and sports service Spotlight followed suit in 1978. Two other superstations, New York's WOR and Chicago's WGN, began around the same time. Warner launched the children's service Nickelodeon and the Movie Channel in 1979, while Getty Oil began the Sports Programming Network (later called ESPN). Turner's Cable News Network launched in 1980, to the jeers of broadcast network news operations who dubbed it the "Chicken Noodle Network" and claimed an upstart like Turner could not do justice to the news. Other programmers rushed to satellite distribution, so that by 1980 there were 28 national programming services available, according to National Cable Television Association records.

These programming innovations affected broadcasting and related industries in several ways. For example, Turner's CNN, though it lost money for about five years before moving into profitability, had a substantial audience even in its earliest years. In fact, many network affiliates contracted with Turner for late night news in the early 1980s, prompting the broadcast networks to launch their own competing late night news shows such as NBC's *News Overnight*, CBS' *Nightwatch*, and *The CBS Early Morning News*. MTV, a popular music program service which began in 1981, prompted copycat programming on the part of the broadcast networks as well, and even episodes of popular NBC police drama *Miami Vice* were likened to one long music video. Music videos also assumed a new and critical role in establishing popular hits for the music industry. Program competition between broadcasting and cable drove up the cost of certain program categories, especially sports, and cable networks eventually outbid broadcasters for certain offerings even as they developed cost-effective ways to deliver regional sporting events to local audiences. New cable networks provided an after-market for broadcast series reruns and for series such as *The Paper Chase* that did not succeed in network broadcasting. Some networks repackaged older, mainstream broadcast series to render them more appealing. For example, Nickelodeon's Nick at Nite relies on popular series from the 1950s, 1960s and 1970s repositioned as trendy, tongue-in-cheek fare.

As programmers developed new channels to view, cable operators moved quickly to claim new markets in suburban and urban areas. Their systems finally had something new to offer these urban areas already used to several over-the-air broadcast signals, and they sought to wire the most lucrative areas as fast as possible. The MSO ownership form bought out many independent cable systems even as they sought new territories to wire. The period of time between roughly 1978 and 1984, often called the "franchise war" era, saw cable companies competing head to head with each other in negotiating franchises with communities, often promising very high capacity, two-way cable systems in order to win contracts, only to renege on these promises later. Warner Amex's QUBE system, a highly publicized but actually very limited, two-way cable service that the company promised to develop in many of its markets, was one such casualty, as were security systems, special two-way institutional networks called I-Nets, and a host of other cost-inefficient services, including public access channels. Most large, urban markets were franchised at this time, and several were promised 100 channel systems with two-way capabilities plus extensive local access facilities although few ended up with such amenities. Companies such as Time's American Television and Communications Corporation, Warner-Amex, TelePrompTer, Jones Intercable, Times-Mirror, Canada-based Rogers, Cablevision Systems, Cox, United, Viacom, Telecommunications, Inc. (TCI), and other large MSOs garnered much of these franchises. In spite of their historically harsh rhetoric against cable television, broadcasters too became convinced of cable television's profitability. They invested in transmission systems and ultimately made substantial investments in programming as well, with ABC's acquisition of ESPN a notable early success and CBS Cable, launched in 1981, a notable failure.

Expanded markets and new programming services abetted by favorable judicial decisions contributed to the cable industry's power to lobby for more favorable treatment in other domains. The industry's pleas met favorable response within the Reagan administration, and Mark Fowler, the Reagan-appointed chair of the FCC from 1981 to 1987, supported a marketplace approach to media regulation that would essentially put cable on a more equal footing with broadcasting.

The Cable Communications Policy Act of 1984 addressed the two issues that still hindered cable television's growth and profitability: rate regulation and the relative uncertainty surrounding franchise renewals. Largely the result of extensive negotiation and compromise between the cable industry's national organization, the National Cable Television Association, and the League of Cities representing municipalities franchising cable systems, the act pro-

vided substantial comfort to the cable industry's future. Its major provisions created a standard procedure for renewing franchises that gave operators relatively certain renewal, and it deregulated rates so that operators could charge what they wanted for different service tiers as long as there was "effective competition" to the service. This was defined as the presence of three or more over-the-air signals, a very easy standard that over 90% of all cable markets could meet. The act also allowed cities to receive up to 5% of the operator's revenues in an annual franchise fee and made some minor concessions in mandating "leased access" channels to be available to groups desiring to "speak" via cable television. Other portions of the act legalized signal scrambling, required operators to provide lock boxes to subscribers who wanted to keep certain programming from children, and provided subscriber privacy protections. When in the following year must-carry rules were overturned in *Quincy Cable TV v. FCC* (1985), the cable industry's freedom from most obligations and regulatory restraints seemed final.

With rate deregulation and franchise renewal assured, the cable industry's value soared, and its organization, investments, and strategies changed. MSOs consolidated, purchasing more independent systems or merging, even as they expanded into new franchises, with large MSOs getting even bigger. The growth of TCI, shepherded by John Malone to become the largest MSO for many years, garnered a great deal of criticism. Several systems changed hands as large MSOs sought to "cluster" their systems geographically so they could reap the benefits of economies of scope by having several systems under regional management. More finances poured into the industry after 1984 since its future seemed assured, and the industry's appetite for expansion made it a leader in the use of junk bonds and highly leveraged transactions, questionable financial apparatuses that later received Congressional scrutiny. Many of the largest companies such as Time (later Time Warner), TCI, and Viacom acquired or invested in programming services, leading to a certain degree of vertical integration. The issues both of size and vertical integration became the subject of Congressional inquiries in the late 1980s, but resulted only in warnings to the industry. Investments in programming, operators argued, justified higher rates, and after 1984 rates jumped tremendously—according to Government Accounting Office surveys, an average of 25% to 30% from 1986 to 1988 alone, vastly greater than the inflation rate. Subscription charges increased so much so quickly that a backlash among consumer groups grew. As the industry's market penetration and control over programming escalated, its growth strategies targeted new markets, predominantly in Europe and Latin America, and also focused on thwarting new domestic competitors such as direct broadcasting satellites, multipoint distribution service (MDS) and its offspring system called multichannel-multipoint distribution service (MMDS). The multichannel capabilities of MMDS and direct broadcast satellites could provide real competition to cable television.

In this profitable decade many new programming services launched and flourished. The 28 national networks in 1980 grew to 79 in 1990. New systems were built, bringing cable television to 60 million television households by 1990; channel capacity expanded, making the 54-channel system common (in about 70% of all systems). Although pay service subscriptions leveled off as most American households purchased videocassette recorders (VCRs), and although offerings such as pay-per-view—single programs or events subscribers could order for a premium fee on a one-time basis—never worked well technologically or economically, cable services quietly grew, so that by 1992 they were in over 60% of all American households.

However, several issues simmered on throughout the 1980s. One concerned the rate increases that many consumers and policy makers felt escalated too rapidly. Another was the availability of reasonably priced programming to rural viewers who expected to receive them using their own satellite dishes; that such newly scrambled services (after the 1984 act that legalized scrambling) were unavailable to them or only available at what they considered very high prices created an especially heated exchange in Congress. The size and vertical integration of several MSOs worried some policy makers, who felt the companies had undue opportunities to exercise their power over a captive market. Broadcasters continued their cry for remuneration for the three major network channels carried by cable television. Even though most cable subscribers still spent much of their viewing time with network channels, operators paid nothing for that programming. Moreover, as cable operators' power grew, concerns rose about the convention of municipalities authorizing only one cable system for a given territory, thus creating a *de facto* monopoly. One company, TCI, for example, was singled out for criticism because its systems served more than half of all television households in some states, a situation some critics felt conceded too much power to large cable operators. Finally, the growing deregulation of telephone companies made cable television services a target of their expansion desires.

IV. Re-regulation, 1992 and Beyond

The cumulative weight of these criticisms swung back the regulatory pendulum when the Cable Television Consumer Protection and Competition Act of 1992 attempted to resolve some of these issues. The act re-regulated rates for basic and expanded services, and required that the FCC generate a plan(called must-carry/retransmission consent), by which broadcasters would receive compensation for their channels. The retransmission consent portion of this legislation was the culmination of years of lobbying by the broadcast industry, and effectively forced cable operators to financially acknowledge the importance of broadcast programming on their tiers. The act called for new definitions of effective competition and for supervised costing mechanisms for other aspects of cable service such as installation charges, and it decreed that programming services must be available to third-party distributors such as satellite systems and MMDS providers. However, portions of this legislation, the only legislation during Presi-

dent Bush's administration to command an override of his veto, ultimately succumbed to the considerable momentum behind reducing government regulation and promoting marketplace forces in industries such as telephony and its growing family of related services.

The Telecommunications Act of 1996, though primarily focused on restructuring the telephone industry, also affected the cable industry. Not only did it designate a new service category, called "open video systems", that allows telephone companies to provide video programming, it also relaxed some of the 1992 Cable Act's rules; significantly, it determined that by 1999 rate regulation would once again be eliminated for all cable services except those in the basic tier. Rate regulation of small cable operators was available immediately. The 1996 act recognizes the convergent capabilities of the many media systems that historically had been viewed as very separate and consequently were regulated differently. A product of strong industry pressure and with scant input from citizen groups, the Telecommunications Act of 1996 was landmark deregulatory legislation.

With growing competition from the new multi-channel providers such as MMDS and direct broadcast satellite services, and with telephone companies entering the video entertainment marketplace, cable television's future appears far less certain as the 21st century begins. Major deregulation initiatives, legislative and judicial, for telephone companies in the 1990s enable them to move into new home information and entertainment services. As the "other wire" entering homes, telephone systems are well positioned to compete with cable television, although they may choose to collaborate with cable television by buying cable systems rather than competing with them: telcos are very interested in joining with both computer and cable companies to mold a new service capable of Internet-style interactivity as well as video programming. An attempted 1993 merger between the largest cable operator, TCI, and a major Bell operating company, Bell-Atlantic, was symptomatic of the cable industry's scramble to forge strategic partnerships with media systems that may eclipse its technological capabilities. The cable television industry's key advantages are that its 1980s-built plant is already in most American homes, its lines could serve fully 95% of U.S. households, and its channel capacity is considerable. Since beginning to experiment with video compression and upgrading coaxial cable to fiber, cable operators are poised to continue to expand signal carriage capacity and to offer competitive one-way video. Additionally, the extensive vertical integration among many operators and programmers appears to guarantee that the cable industry will maintain a favored position with regard to the critical resource of programming.

Whether the cable industry will be able to keep abreast of peoples' desires for programming and interactive services epitomized by the Internet remains uncertain. The 1990s have been marked by consolidation among operators and programmers and other entertainment companies as a dominant organizational response to regulatory and technological opportunity. As

Time Warner merges with the Ted Turner empire, and as Disney merges with Capital Cities/ABC, the large, vertically-integrated and multi-faceted company with international holdings seems to be the new industry template for survival. The cable industry remade the television world of the "Big Three" networks, upsetting their hold on programming and viewers and initiating a 24-hour, tumultuous and changeable video domain. As the larger video media industry changes, the cable industry's boundaries, roles and influences will likewise be reshaped, but the historical legacy of its accomplishments will surely continue to be felt.

—Sharon Strover

FURTHER READING

Blumler, J. *The Role of Public Policy in the New Television Marketplace.* Washington, D.C.: Benton Foundation, 1989.

Cabinet Committee on Communications. *Cable: Report to the President.* Washington, D.C., 1974.

Fowler, M., and D. Brenner. "A Marketplace Approach to Broadcast Regulation." *Texas Law Review* (Austin), 1982.

Garay, Ronald. *Cable Television: A Reference Guide to Information.* New York: Greenwood, 1988.

Horwitz, Robert. *The Irony of Regulatory Reform.* New York: Oxford University Press, 1989.

Le Duc, Don. *Cable Television and the FCC: A Crisis in Media Control.* Philadelphia, Pennsylvania: Temple University Press, 1973.

———. *Beyond Broadcasting: Patterns in Policy and Law.* New York: Longman, 1987.

Owen, B., and S. Wildman. *Video Economics.* Cambridge, Massachusetts: Harvard University Press, 1992.

Sloan Commission on Cable Communications. *On The Cable.* New York: McGraw-Hill, 1971.

Smith, R. L. *The Wired Nation: The Electronic Communication Highway.* New York: Harper and Row, 1972.

Streeter, T. "The Cable Fable Revisited: Discourse, Policy and the Making of Cable Television." *Critical Studies in Mass Communication* (Annandale, Virginia), 1987.

Waterman, D. "The Failure of Cultural Programming on Cable TV: An Economic Interpretation." *Journal of Communication* (Philadelphia, Pennsylvania), 1986.

Whiteside, T. "Cable I, II, III." *The New Yorker*, 20, 27 May, 3 June 1985.

U.S. Department of Commerce, National Telecommunication Information Agency. *Video Program Distribution and Cable Television: Current Policy Issues and Recommendations.* Washington, D.C.: Government Printing Office, (NTIA Report 88-233), 1988.

COURT CASES, LEGISLATION AND FCC ACTIONS

Cable Communications Policy Act of 1984. Pub. L. No. 98-549, 98 Stat. 2779 (codified as amended at 47 U.S.C. section 521).

Cable Television Consumer Protection and Competition Act of 1992. Pub. L. No. 102-385, 106 Stat. 1460 (codified at 47 U.S.C. section 533).

Cable Television Report and Order, 36 FCC 2d 143 (1972).

Carter Mountain Transmission Corporation, 32 FCC 459 (1962), affirmed 321 F. 2d 359 (D.C. Circuit), cert. denied, 375 U.S. 951 (1963).

Home Box Office v. Federal Communications Commission, 567 F. 2d (D.C. Cir. 1977). Cert. denied, 434 U.S. 829 (1977).

Quincy Cable TV v. FCC, 768 F. 2d 1434 (D.C. Cir. 1985), cert. denied, 106 S. Ct. 2889 (1986).

Telecommunications Act of 1996. Public L. No. 104-104, 110 Stat. 56 (1996).

United States v. Midwest Video Corp., 440 U.S. 689 (1979). (Midwest Video Case II).

United States v. Midwest Video Corp., 406 U.S. 649 (1972). (Midwest Video Case I).

See also Association of Independent Television Stations; Cable Networks; Cable News Network; Canadian Cable Television Association; Distant Signal; Dolan, Charles F.; Financial Interest and Syndication Rules; Home Box Office; Malone, John; Narrowcasting; National Cable Television Association; Malone, John; Microwave; Midwest Video Case; Pay Cable; Pay-Per-View Cable; Pay Television; Public Access Television; Reruns/Repeats; Satellite; Superstation; Syndication; Telcos; Televison Technology; Translator; Turner, Ted; Turner Broadcasting Systems; U.S. Policy: Telecommunication Act of 1996

UNITED STATES: NETWORKS

Networks are organizations that produce or acquire the rights to programs, distribute these on systems of interconnection, and secure uniform scheduled broadcast on a dispersed group of local outlets. In commercial broadcasting, "networking" was recognized at an early date as the clearest path to profitability, because the costs of program production were—and are—fixed, and revenue turned on securing the maximum degree of efficient distribution and exposure to mass audiences.

In the United States, the number of broadcast networks existing at a particular time, and the prospects for entry by new networks, have always been the combined result of the current state of technology, in tension with an extensive role for government regulation. Television broadcasting, tentatively begun prior to the American entry to World War II in 1941, was suspended for the duration of the war, and did not resume until the first wave of station activations in 1946 through 1948. By then, the dynamics of technology and regulation in radio broadcasting already had shaped the possibilities for television networks.

Beginning in 1920, radio entrepreneurs had developed an array of informational and entertainment fare, originated in live performances at local stations, and increasingly at network studios in New York City, from which feeds to stations could be disseminated in real time over telephone lines. Commercials, like other copy, were read and performed live. Strong local stations prospered in this system, but the highest return was enjoyed by two major networks, Columbia Broadcasting System (CBS), and the National Broadcasting Company (NBC) unit of a premier radio equipment manufacturer, Radio Corporation of America (RCA). RCA operated dual networks, the Red and Blue. In radio, as was to be the case in television, industry leadership was exercised by a charismatic executive and founder, Robert Sarnoff at NBC, William S. Paley at CBS, Allen B. DuMont and a few others.

The first comprehensive radio law, the Radio Act of 1927, did not confer on government any express power to regulate networks directly, but empowered it to regulate stations engaged in "chain broadcasting." This served to consolidate industry control by the network organizations already underway. The law mandated that radio broadcasting stations be allotted in a manner that equitably served the various states and localities, but withheld actual station ownership of broadcast channels, in favor of renewable licenses for limited times. It also prohibited the licensing of a person or entity that had been convicted of unfair competition or monopolization. These precepts carried over with the Communications Act of 1934, and shaped the relationship among stations, networks and the government throughout the emergence of television.

At the eve of American entry to the war, the Federal Communications Commission (FCC), acting under its powers to investigate and regulate stations, concluded a probe of "chain broadcasting" and announced a series of prohibited practices in radio. These included contracts that permitted networks to command and resell advertising time for their own account, or to option time. The rulings also prohibited the specific ownership of dual networks by a single entity, NBC being the singular example. The Supreme Court's decision upholding these actions in 1943 prompted the divestiture of NBC Blue, acquired that year by Lifesaver magnate Edward J. Noble, and renamed the American Broadcasting Companies (*National Broadcasting Co. v. U.S.*, 319 U.S. 190, 1943).

After 1945, as Americans turned to peace time pursuits, including the realization of television, commercial radio already was settled into a pattern with program fare dominated by two or, generously, perhaps three networks, each of them fortified against hard times by the ownership of a handful of highly-profitable local stations in the largest trading areas. The critical determinant of the number of networks that could be supported was—as it is today—the number of local outlets that could be assured for network audience, by ownership or by contract.

By 1945 the FCC preliminarily had allotted some 19 VHF Channels, 1 through 19, for television broadcasting. Almost immediately Channels 14 through 19 were reallocated to the military, and Channel 1 was put aside for two-way radio. By the end of 1946, seven stations were broadcasting (all on Channels 2 through 6), and approximately 5,000 household receivers were in use. From that point, and even in the absence of detailed technical standards to guard against mutual interference, or other standards, applications for new stations poured in. The FCC imposed a freeze on new applications on 30 September 1948. Virtually all pre-freeze filers actually built broadcasting facilities, so that by the time the freeze was lifted on 13 April 1952, some 107 VHF stations had been activated in 63 markets, and receivers in use had grown to 15.5 million. Denver led the list of many important markets that had no television at all. During the freeze, NBC moved aggressively to apply for and activate stations in the top markets. CBS got a late start, and proceeded to acquire its first stations by purchase. ABC and a fourth network, DuMont Laboratories, participated actively in the FCC proceedings, but were unable or unwilling to initiate major station investment, pending resolution of the knotty regulatory issues.

The framework adopted by the FCC in 1952 allotted television channels to specific communities throughout the United States, roughly in proportion to market size. VHF Channels 2 through 13 and UHF channels 14 to 83 were utilized, but as of 1952, virtually all TV sets were capable of VHF reception only. The first UHF set-top converter was introduced in March 1952. The decision also sacrificed efficiency, and reduced the potential number of stations, by grandfathering the existing 107 outlets, helter-skelter wherever they had started. Practically speaking, the FCC's allocations provided only enough VHF outlets to provide two-channel service to about 90% of the population, and third-channel service to substantially less. NBC and CBS, each emerging with five powerful owned-and-operated stations, and program offerings spun off from their popular radio fare, quickly expanded affiliations.

The Emmy Awards, first presented on 25 January 1949, were an accurate barometer of network emergence. A local station, KTLA in Los Angeles, dominated the awards for year 1948, with the most popular program (*Pantomime Quiz Time*), most outstanding personality (Shirley Dinsdale and her puppet, Judy Splinters), and the station award. By the second year, with KTLA still prominent, NBC cracked the line-up, jointly with its New York flagship KNBH, winning best kinescope show (*Texaco Star Theater*) and personality (Milton Berle). A network spot for Lucky Strike won best commercial. In the third presentation, for 1950, Alan Young and Gertrude Berg were best actor and actress, for CBS jointly with Los Angeles independent KTTV, and their co-produced *Alan Young Show* was recognized for best variety show. Outstanding personality was NBC/KNBH's Groucho Marx. By the end of the FCC's freeze these networks had unqualified leadership of program origination.

In the complex fight over regulation DuMont Laboratories had advocated a plan with a minimum of four VHF's allotted to each of the 140 largest trading areas. Rebuffed at the FCC, DuMont never achieved more than 10 primary or full schedule network affiliates. As the few UHF operators incurred mounting losses, DuMont folded its network in 1955. These by-products of the freeze and subsequent FCC decision to grandfather incumbent stations and intermix VHF and UHF channels have been harshly criticized.

Throughout this period, ABC was barely operating, and Noble stated that he had never declared a dividend nor taken a salary through 1952. In 1953, however, ABC received FCC approval to merge with United Paramount Theaters. The chain had been spun off from Paramount Pictures Corp., under court decree that followed the Supreme Court's antitrust decision of 1948, upholding divestment of theatrical production from exhibition. The significance of government involvement could not be more clear, with ABC's very existence jeopardized by one government action, and resolved favorably by another. ABC used its Hollywood connections adroitly, teaming with a studio to co-venture a break-through program, to that date the most expensively produced in history, *Disneyland*.

Collectively the networks could have only so many affiliates as there were stations on the air. Commercial VHF stations grew from 233 in 1954 to 458 in 1962. Commercial UHF stations stood at 121 in 1954, and struggled against the lack of UHF receivers. Many UHFs went dark and returned their licenses for cancellation, and by 1962 their numbers had shrunk to 83. In total, the commercial station universe as it grew roughly from 350 to 550 was adequate to support approximately two-and-a-half national networks. Local stations, in the enviable position of having multiple suitors, frequently left ABC with no local outlet. Congress enacted a law in 1962 mandating that all receivers be capable of UHF tuning, but it was only by the mid-1970s that local stations were plentiful enough for ABC to achieve full comparability.

As the networks consolidated their control of station time during the 1950s, a broad shift occurred in their relationship with the sponsor, enhancing their control even further. In the early part of the decade, shows typically were produced by the sponsor live, or contracted for by the sponsor and delivered to the network on expensive film or kinescope. Production centered in New York. With the introduction by Ampex of quadruplex videotape recording in 1956, it became possible for programs to be produced and recorded anywhere, and the new orders for entertainment fare shifted to the concentration of expertise in Hollywood studios. Increasingly, the network replaced the sponsor in development, acquisition, and revision to final programming form. From the 1950s can be charted the realization of core concepts in prime time programming, including the ensemble situation comedy, cop shows, westerns, and regularly scheduled newscasts. The interval often is referred to as the Golden Age of television, perhaps precisely because of its experimental flavor. But while major market stations achieved immediate and impressive profitability, networking was still a gamble, the program performance remained uneven, and in 1961 critic-for-a-day Newton N. Minow derided the totality as a "vast wasteland."

The true golden age of three-network hegemony probably traces from 1963, when each network inaugurated a half-hour prime-time newscast, and network television drew the entire nation together in grief after the assassination of President Kennedy. From 1963 until the late 1970s, the networks created a refracted version of the significant events of the day that was shared by all. This cohesion intensified with expanding use of color transmissions and color set sales during the 1960s. One nation resonated with the networks' triune voice, in a manner unparalleled in the past, and likely never again to be seen in the future. ABC, gradually shoring up its group of strong affiliates, and hiring a visionary programmer in Fred Silverman, finally took the Summer Olympics to its first full-season ratings victory in 1976-77. The "third network's" potential had been clear for years, but several attempts to acquire ABC during the 1960s were rebuffed, and an attempted buyout by IT and T foundered in 1968, after criticisms were vetted during two years of FCC proceedings.

The membership quota for this elite club of three networks, however, was eventually dismantled by a technology developing quietly during these same years—cable television. The FCC's original framework of 1952 did not assure three-network or *any* network service, to all households, and was particularly deficient where terrain obstacles degraded reception over the air. Community antenna television (CATV) was a local self-help response, tying hilltop repeaters to wires into the home. Because cablers did not utilize the broadcast spectrum, the government was uncertain of its jurisdiction until a Supreme Court decision came down in favor of a broad authority to regulate, *U.S. v Southwestern Cable Co.*, 392 U.S. 157 (1968). Thereafter broadcasters, well aware of the potential competition, leaned on the FCC to retard cable, specifically by forbidding the importation of distant signals that were not available in the local market over-the-air. By 1970, a regime of anti-cable regulation was firmly in place and for ten years it served to retard competition and preserve the networks' position. A newer technological device again led to significant change in this arrangement.

Domestic communications satellites were authorized in 1972, and by 1975 RCA and Western Union had space satellites launched and working. In 1975 RCA sold time on its Satcom I for Home Box Office, the first program service designed to bypass conventional delivery channels, and offer a unified program lineup directly to cable systems and thus to the home—in the true sense a network. The following year, uncertainties surrounding the re-sale of broadcast programs to cable were resolved, with passage of a new Copyright Act, requiring broadcasters to license to cablers under certain conditions, at below-market rates to be established through a bureaucratic process.

The opportunity presented by the resolution of the two knottiest issues—distribution and rights—was first recognized by Ted Turner, not a cabler but a broadcaster, operator of WTCG in Atlanta (later, WTBS), an independent UHF on Channel 17. By 1978, the FCC had been having second thoughts about the heavy hand it had placed on cable development. Turner approached the agency with a plan to offer Channel 17 to a common carrier he created for the purpose, Southern Satellite Systems. In turn, Southern would deliver the station by satellite to cable head-ends, charging five cents per household per month. Because imbedded in FCC common carrier regulation was the idea of nondiscriminatory rates, for large and small customers (or cable systems) alike, Southern needed a waiver to charge by the number of local subscribers. Astonishingly, the FCC said yes. The debut of Channel 17 as the first "super station" in 1980 assured, year by year, that the three-network share of the program universe would continue to shrivel inexorably. By 1981 the FCC also was in process of a cable "deregulation," abandoning its 10-year folly of attempting to re-bottle the genie of cable program origination. The networks, barred by FCC rules from owning cable systems, began to invest in new cable program services side-by-side with cable companies, Turner, and others.

With President Ronald Reagan taking office in 1981, the deregulatory thrust continued. The former actor, when he thought about such matters, was willing to favor Hollywood studios in their primordial battles with the television networks, and to endorse the expansion of channels for program delivery. A cable television bill, passed in 1984, pre-empted local rate regulation, and so gave the cable industry working capital to continue its strides as program creator and distributor.

These strides were being matched with the opening of a wholly new channel into the home. Sony had introduced a practical, consumer videotape player recorder, the Beta VCR, in 1976, at suggested retail of $1,295. Recording time was one hour. Sony's Japanese rival, Matsushita, which markets under the name Panasonic, followed shortly with an incompatible format that eventually became standard, VHS. Hollywood studios, led by Universal Pictures and Disney, promptly brought a challenge in Federal Court, claiming that the device inherently was useful only for stealing copyrighted material. The issue oscillated in court until 1984, when the U.S. Supreme Court ruled that home taping for home use was not an infringement of copyright (*Sony Corp. v. Universal City Studios, Inc.,* 464 U.S. 417 [1984], called the "Betamax case"). From that date, sales of home recorders and the rental of tapes exploded. The studios have come to enjoy greater revenue from cassette sales and rentals than from theatrical exhibition, and must look back in wonder at their temporary insanity when the player-recorders first were sighted in

North America. But for the networks, this technology presents long-term problems. The rating services have assumed so far that programs can be credited as viewed if they are recorded, but it may become apparent in time that the facts of actual audience behavior are otherwise. VCRs in their most typical use occupy the household's attention for non-network fare such as movies, just coming off their initial theatrical run.

As cable and cassettes continued to splinter the market, Reagan's FCC abolished many of the rules and policies that had stood in the background of television broadcasting also. In 1984, the rule restricting each television network to the ownership of a maximum five VHF stations, and seven VHF plus UHF, was replaced with a quota of up to twelve VHF so long as the station grouping did not exceed 25% of all TV households. While this liberalization was still at the discussion stage at the FCC, Thomas S. Murphy, chairman of the Capital Cities station group, approached ABC about a merger. Once the rule was finalized, Capital Cities in 1986 announced the acquisition of the much larger network, for $3.5 billion, with financing from Warren E. Buffett and Berkshire Hathaway, Inc.

By 1986, RCA was a diminished echo of the industrial giant of the post-war. It had departed the computer mainframe business in the early 1970s with massive losses, and its equipment markets had been overtaken by Japanese manufacturers. Its television network remained competitive and highly successful, but in no position to refurbish from working capital for the intensified program battles ahead. RCA and its NBC network were sold to General Electric in 1986 for $6.3 billion. General Electric had been instrumental in creating RCA in the 1920s before David Sarnoff, and now closed the circle in an era more receptive to combinations.

CBS entered this period smarting from a lengthy battle with General William C. Westmoreland over the *CBS Reports* documentary, *The Uncounted Enemy: A Vietnam Deception.* The advocacy group, Accuracy in Media, and Senator Jesse Helms in 1985 were urging their constituencies to take over CBS by stock purchase, with the ultimate goal to fire Dan Rather. Seeing an opportunity, Ted Turner announced the intention to do his own hostile take-over, to be financed with junk bonds. The network beat back this effort with a $1 billion stock repurchase, but was left with more debt, little working capital, and a reduced stock valuation. The board and the aging founder, Paley, passed leadership, and thereafter effective control of the stock, to Loews Corp. and its proprietor, Lawrence Tisch. By a combination of ill-fated acquisitions and divestments, and under competitive pressure, CBS had to focus on cost containment. The news division, successors of Edward R. Murrow, was pruned by 230 people. In 1987 CBS dropped to third place in the season ratings for the first time.

Ever since the sputtering start for UHF in the first two decades of television, FCC commissioners had spoken longingly of the desire, first to assure three-network service to every corner of the nation, and next to somehow realize the dream of a fourth network. By the time the fourth network arrived, conditions had so changed as to raise the question, four of what? In any given household at any given time, two or three television sets could be in use, watching network fare; or independent stations; or movies, sports, original fare and reruns on cable; or first-run movies on premium, pay cable; or movies or exercise videos on cassette; or possibly computer games. Nevertheless, the fabled fourth network did indeed come in 1990, when an Australian publisher, naturalized as a U.S. citizen, Rupert Murdoch, acquired the strong major-market grouping of Metro Media stations, and placed them under the same roof with the Twentieth Century-Fox studio. Murdoch eschewed ABC's original 1950s approach—programming mostly cannon fodder against its rivals on a full seven nights—instead making a staged entry with two nights, then three, four. The FOX network finally attained a full-time run, and in less than five years from launch, FOX could be seen actually winning a time period here and there. In 1994 FOX purchased rights to the National Football Conference, building from sports, and luring affiliates in NFC territories, all exactly as dictated by the ancient scripting of ABC.

During the 1990s the FCC continued to chip away at rules intended to adjust the playing field between Hollywood studios and other program suppliers on one hand, the networks on the other. Rules had evolved that imposed a quota on network self-produced fare, that forbade the networks to own rights for secondary distribution of the programs they originated (called the Fin-Sin Rules), and that kept an hour of prime time out of the hands of networks, reserved for local stations to program (the Prime Time Access Rule). Because FOX combined a network with a studio, some of these rules had the perverse effect of thwarting development of a fourth network, and for a time the FCC liberally accorded waivers to FOX alone.

Other rules no longer served a purpose in the multi-channel environment. By 1994, the liberalization of ground rules emboldened three more Hollywood studios to try their hand at networking directly. Warner Brothers launched a network in its own name, and Universal, which had grown to eminence as a prime source for NBC, teamed with Paramount, proud source of the inexhaustible *Star Trek*, to form UPN (United Paramount Network). The aspirations of these mini-networks remain *ad hoc*, choosing their nights and their time periods with care. It is obvious that there are too few local television broadcast stations to support five, six or more full-time networks. The immediate impact of these entrants was to drive up the prices of television stations, as too many programs chased too few outlets. In time, a new wave of consolidations is assured.

In 1995, Westinghouse, a strong group owner, acquired CBS, in a transaction that echoed the Cap Cities take-over of ABC ten years earlier. Also in 1995, Capital Cities/ABC

agreed to be acquired by Walt Disney Studios for $19 billion in cash and stock. In the long view, the CBS sale is likely to appear as but one more episodic reorganization. The Disney combination with CapCities prefigures a new level of competition among few great communications trusts equipped to provide multiple channels of information, entertainment and merchandising in coordinated fashion throughout the world. Such networks are difficult to describe, because none yet exists. The largest multiple system cable operator, TCI, which has diverse program interests, is poised to be one in the future. Viacom, as owner of Paramount, impresario of UPN, owner of Blockbuster Video, and cable programmer in other capacities, may be another.

In 1996 Congress passed and the president signed a new telecommunications act. It reduces or eliminates historic barriers that have separated telephone long distance companies and the regional Bell operating companies from the local cable television companies. In broadcasting, it abolishes the numerical limit on television stations in common ownership, and provides a liberalized cap of 35% of national audience for any one station owner. It abolishes the "dual network" ban that divested NBC Blue in 1941, and invites the FCC to undertake proceedings, looking to the authorization of more than one local TV station in common ownership (now forbidden). Since the advent of television in 1941, there never has been a regulatory change—permitting combinations not previously allowed—that did not trigger moves by the affected parties, to the full, lawful outer limits. The turn of the century is bound to witness the additional three networks (now a college of four) dropping below that point where they own even so much as a majority of prime time viewing attention. But that development, in steady process for thirty years, will be overshadowed by the emergence of new network forms, rendering the classical shape of the three no longer recognizable.

—Michael Couzens

FURTHER READING

Auletta, Ken. *Three Blind Mice: How the TV Networks Lost Their Way.* New York: Random House, 1991.

Bagdikian, Ben H. *The Media Monopoly.* Boston, Massachusetts: Beacon, 1992.

Barnouw, Erik. *The Image Empire: A History of Broadcasting in the United States*, vol. 3. New York: Oxford University Press, 1970.

———. *Tube of Plenty: The Evolution of American Television.* New York: Oxford University Press, 1975; revised edition, 1990.

Bedell, Sally. *Up the Tube: Prime Time TV in the Silverman Years.* New York: Viking, 1981.

Block, Alex Ben. *Outfoxed: Marvin Davis, Barry Diller, Rupert Murdoch, Joan Rivers, and the Inside Story of America's Fourth Television Network.* New York: St. Martin's, 1990.

Boddy, William. *Fifties Television: The Industry and Its Critics.* Urbana: University of Illinois Press, 1990.

Brown, Les. *Televi$ion: The Business behind the Box*. New York: Harcourt, Brace, 1971.

Castleman, Harry, and Walter J. Podrazik. *The TV Schedule Book: Four Decades of Network Programming from Sign-on to Sign-off*. New York: McGraw-Hill, 1984.

————. *Watching TV: Four Decades of American Television*. New York: McGraw Hill, 1982.

Cooper, R.B., Jr. "The Infamous Television Allocation Freeze of 1948." *Community Antenna Television Journal*, March 1975.

Inglis, Andrew F. *Behind the Tube: A History of Broadcasting Technology and Business*. Boston, Massachusetts: Focal, 1990.

Kiernan, Thomas. *Citizen Murdoch*. New York: Dodd Mead, 1986.

MacDonald, J. Fred. *One Nation under Television*. New York: Pantheon, 1990.

Metz, Robert. *CBS: Reflections in a Bloodshot Eye*. Chicago: Playboy, 1975.

Paul, Michael, and James Robert Parish. *The Emmy Awards: A Pictorial History*. New York: Crown, 1970.

Sloan Commission on Cable Communications. *On the Cable: The Television of Abundance*. New York: McGraw-Hill, 1971.

Wilk, Max. *The Golden Age of Television: Notes From the Survivors*. New York: Delacourte Press, 1976.

GOVERNMENT STUDIES

Federal Communications Commission. Network Inquiry Special Staff. *Final Report: New Television Networks: Entry, Jurisdiction, Ownership and Regulation*. Volume I, Final Report; Volume I, Background Reports. Washington, D.C., October, 1980.

Federal Communications Commission. *Report on Chain Broadcasting*. Commission Order No. 37, Docket No. 5060, May 1941.

Federal Communications Commission. *Second Interim Report of the Office of Network Study: Television Network Program Procurement, Part I*. Washington, D.C., 1965.

U.S. House of Representatives (88th Congress, 1st Session). *Television Network Program Procurement*. House Report No. 281. Washington, D.C., 8 May 1963.

U.S. House of Representatives (85th Congress, 2nd Session). Committee on Interstate and Foreign Commerce. *Network Broadcasting ("The Barrow Report")*. House Report No. 281. Washington, D.C., 8 May 1963.

U.S. House of Representatives (97th Congress, 1st Session) Committee on Energy and Commerce. *Telecommunications in Transition: the Status of Competition in the Telecommunications Industry*. Commission Print 97-V, Report by the Majority Staff. Washington, D.C., 8 November 1981.

UNIVISION

U.S. Network

Univision (in Spanish, *Univisión*), the largest Spanish language television network in the United States with more than 600 affiliates, has historical roots in Mexican broadcasting. Since 1992, Univision has been owned by a consortium headed by Jerry Perenchino, an entertainment financier who once owned a New Jersey Spanish language television station. Twenty-five% of the network is owned by Venevision, a Venezuelan media company, another 25% by the Mexican entertainment conglomerate, Televisa, the largest producer of Spanish language television programming in the world.

This structural configuration is often viewed as but a marginal variation in Televisa's long-standing domination of U.S. Spanish language television. The majority of Univision programming is produced in Mexico, by Televisa, as it has been since the first Spanish language television stations were established in the United States in 1961. The network was then called SIN, the Spanish International Network. In 1986 the Federal Communications Commission (FCC) found SIN to be in violation of the U.S. law that prohibits foreign ownership of U.S. broadcast stations. Televisa was ordered to divest itself of its U.S. subsidiary and SIN was sold to Hallmark Cards of Kansas City, Missouri, and renamed Univision.

Courtesy of Univision

Under Hallmark ownership, about half of Univision programming was Televisa rebroadcasts (*telenovelas* or soap operas, sports, movies and variety programming), and half was produced in the United States. The U.S. produced programming, which included a *telenovela*, a situation comedy and greatly expanded national U.S. news and public affairs programming, proved popular with U.S. Latino audiences. Nonetheless, between 1986 and 1992, Hallmark, which had financed its purchase of the Spanish language network with junk bonds, was unable to recover its initial investment in Univision. In 1992 Hallmark sold the network to the Perenchino group, which prominently featured Televisa. Among the new owners' first moves was the firing of about a third of the network's Miami based staff. This resulted in the cancellation of most of the U.S. produced programs, and the recreation of a broadcast day largely comprised of Televisa programs.

Univision has been at the forefront of the creation of a national "Hispanic Market," the notion that U.S. Latinos are an attractive, commercially viable market segment, and so an audience that advertisers should attempt to reach. Previous to the mid-1980s the Hispanic population was configured as three markets: Puerto Rican in the eastern United States, Cuban in south Florida and Mexican in the southwest. Advertising agencies, accordingly, produced three separate Spanish language advertising campaigns. Univision's extensive market and audience research persuaded Madison Avenue that these three audiences should be considered one national audience. This effort was given a major boost by the Hispanic Nielsen Survey, a specially designed methodology for measuring U.S. Spanish language television audiences, commissioned by Univision and Telemundo, and implemented by the A.C. Nielsen Company in the early 1990s. This new audience measurement system found a U.S. Spanish language television audience 30 to 40% larger than had previously been identified.

Network research conducted by Univision shows that most of its audience are recent Latin American immigrants. Another group is made up of those who have lived in the United States for years, who, because of a myriad of factors, prefer to view television in the Spanish language. Most of these immigrant audience members are from Mexico, though an increasing proportion are Central American. A smaller portion of the Univision audience are more acculturated, bilingual U.S. Latinos, a generally wealthier group much sought after by network planners. Overall, Univision research shows that about 70% of the Univision audience is Mexican or Mexican American, 10% each Puerto Rican and Cuban American, with the remainder from other Latin American countries.

The most watched Univision programs are Televisa *telenovelas*, serialized melodramas which, in contrast to U.S. soap operas, usually end after two or three months. Also, notably present in the Univision top ten (at number six) is the nightly U.S. national newscast, the *Noticiero Univisión*. Apparently the Univision immigrant audience, while maintaining its links to "the old country" through the traditional *telenovelas*, is also seeking out knowledge of its adopted U.S. home. Each year the U.S. Spanish speaking audience has more television programs to choose amongst. Telemundo, another U.S. Spanish language television network founded in 1986, has grown to several hundred affiliates. Galavision and Showtime en Espanol, two premium cable channels, as well as several regional Spanish language cable networks, including Spanish language ESPN and MTV, are challenging Univision's previously uncontested hold on U.S. Spanish language television.

—America Rodriguez

FURTHER READING

Aguirre, Adalberto, Jr. "Critical Notes Regarding the Dislocation of Chicanos by the Spanish-language Television Industry in the United States." *Ethnic and Racial Studies* (New York), January 1993.

Andrews, Edmund L. "FCC Clears Hallmark Sale of Univision TV Network." *New York Times*, 1 October 1992.

Constantakis-Valdes, Patricia E. *Spanish-language Television and the 1988 Presidential Elections: A Case Study of "Dual-Identity" of Ethnic Minority Media.* (Ph.D. Dissertation, University of Texas at Austin, 1993).

Foisie, Geoffrey. "Nielsen Expands Hispanic TV Ratings." *Broadcasting and Cable* (Washington, D.C.), 8 August 1994.

Fisher, Christy. "Ratings Worth $40 Million? Networks Have No Doubts." *Advertising Age* (New York), 24 January 1994.

"Hispanic Networks Building Up Steam; Ad Rates Seen Rising." *Television-Radio Age* (New York), 27 November 1989.

Mendoza, Rick. "The Year Belongs to Univision." *Hispanic Business* (Santa Barbara, California), December 1994.

Rodriguez, America. *Made in the USA: The Constructions of Univision News* (Ph.D. Dissertation. University of California, San Diego, 1993).

Ruprecht, Gustavo. "Tuning In To Cultural Diversity." *Americas* (Washington, D.C.), July-August 1990.

Silva, Samuel. "The Latin Superchannels." *World Press Review* (New York), November 1991.

Waters, Harry F. "The New Voice of America." *Newsweek* (New York), 12 June 1989.

See also Telemundo; Telenovela

THE UNTOUCHABLES

U.S. Crime Series

Based on the 1947 novel by Eliot Ness and Oscar Fraley, *The Untouchables* was the first dramatic series created at Desilu Productions, the studio owned by Desi Arnaz and Lucille Ball, and famous for providing situation comedies to U.S. television. Airing on ABC from 1959 to 1963, the series was panned for what critics at the time deemed "excessive and senseless violence." But it was enormously popular with audiences and made names for producer Quinn Martin and actor Robert Stack.

The series centered on a greatly embellished version of the real-life Eliot Ness, played by Robert Stack, and his incorruptible treasury agents whom Chicago newspapers had dubbed "the Untouchables." Their battles against organized crime served as the source material for the television series. While the fictional Ness and his Untouchables were somewhat lifeless characters, the back-stories and motivations established for the series' criminals were incredibly well-defined. This was due, in large part, to the talented actors, including Robert Redford, William Bendix, Lloyd Nolan, J. Carroll Naish and Peter Falk, guest actors who played the series' criminal kingpins. This, of course, led to one of the basic problems of the series—the criminals appeared more human than the heroes.

The series began as a two-hour made-for-television movie documenting Ness's fight against Chicago-mob leader Al Capone. The movie, and its episodic counterpart, maintained an earthy grittiness with its stark sets and dark, studio backlot exterior sequences. A realistic mood was added by narrator Walter Winchell (who had, incidentally, a few years before, broken the real-life scandal of Lucille Ball's alleged communist ties during the McCarthy-era blacklisting period). Winchell's staccato delivery of introductory background material set the stage for each week's episode.

ABC justified the series' violence on grounds of historical accuracy, yet the network often violated the same rule by having their fictional Ness responsible for nabbing mob leaders such as George "Bugsy" Moran and Ma Barker, figures with whom he had no actual dealings. Indeed, a number of FBI agents complained about their real-life victories being credited to the fictionalized Ness. Such pressure eventually forced ABC to create additional FBI characters to more accurately portray the people involved in the show's historically-based cases.

The Untouchables also drew controversy for its stereotyped ethnic characters. The Italian-American community protested the series' use of Italian names for criminal characters. The Capone family also brought a million-dollar lawsuit against producer Desi Arnaz for using the Capone likeness for profit. This was particularly upsetting for Arnaz, a classmate and friend of Al Capone's son.

The show was tremendously successful in its second season, but its popularity rapidly declined when NBC countered with the musical variety program *Sing Along with Mitch*. Pro-

The Untouchables

ducer Quinn Martin built his *Untouchables* success into an impressive string of cop-based dramatic hits, including *The FBI* (1965) and *The Streets of San Francisco* (1972). Robert Stack became a popular TV actor and has since starred in other successful dramas in which he has played similar crime fighters and adventurers. Since 1988 he has been most visible as the host of *Unsolved Mysteries*, a popular "reality" program. *The Untouchables* inspired two revivals—a 1980s movie version as well as a 1990s syndicated series.

—Michael B. Kassel

NARRATOR
Walter Winchell

CAST

Eliot Ness	Robert Stack
Agent Martin Flaherty (1959–60)	Jerry Paris
Agent William Youngfellow	Able Fernandez
Agent Enrico Rossi	Nick Georgiade
Agent Cam Allison (1960)	Anthony George
Agent Lee Hobson (1960–63)	Paul Picerni
Agent Jack Rossman (1960–63)	Steve London
Frank Nitti	Bruce Gordon

Al Capone	Neville Brand
"Bugs" Moran	Lloyd Nolan
Dutch Schultz	Lawrence Dobkin
"Mad Dog" Coll	Clu Gulager

PRODUCERS Quinn Martin, Jerry Thorpe, Leonard Freeman, Howard Hoffman, Alan A. Armer, Alvin Cooperman, Lloyd Richards, Fred Freiberger, Charles Russell

PROGRAMMING HISTORY 114 Episodes

• ABC

October 1959–October 1961	Thursday 9:30-10:30
October 1961–September 1962	Thursday 10:00-11:00
September 1962–September 1963	Tuesday 9:30-10:30

FURTHER READING

Arnaz, Desi. *A Book.* New York: Warner, 1976.

Boddy, William. *Fifties Television: The Industry and Its Critics.* Urbana: University of Illinois Press, 1990.

Powers, Richard Gid. *G-Men, Hoover's F.B.I. in American Popular Culture.* Carbondale: Southern Illinois University Press, 1983.

See also Arnaz, Desi; Martin, Quinn; Police Programs; *Westinghouse-Desilu Playhouse*

UPSTAIRS, DOWNSTAIRS

British Serial Drama

Upstairs, Downstairs, originally produced in England by Sagitta Productions for London Weekend Television (LWT), became one of the most popular programs in the history of *Masterpiece Theatre* on the U.S. Public Broadcasting Service and was beloved throughout much of the world. The series presents the narrative of the upper-class Bellamy family and their servants during the turbulent first third of this century in Britain. Their stories, focused individually but always illustrative of complex and intertwined relationships, unfold chronologically, highlighting members of both the upstairs biological family and the downstairs "work family" of servants.

The series accurately represented and mirrored the societal milieu of its time and has been greatly acclaimed for the producers' and authors' meticulous attention to accurate period detail. Historical events served as the context for the characters' situations and actions in a narrative that carried them from 1903 Edwardian England, through World War I and the political upheavals of the 1920s, to a conclusion set soon after the stock market crash in the summer of 1930. *Upstairs, Downstairs* captured and held a rapt television audience through 68 episodes in Britain and 55 in America. It was the most extensive series on *Masterpiece Theatre* and brought a new and refreshing image of British television to many Americans whose only perception of British programming, not necessarily correct, was of ponderous adaptations of dated British literature. In so doing, the series brought a great many new viewers to PBS and *Masterpiece Theatre*.

According to long-time *Masterpiece Theatre* host Alistair Cooke, quoted in Terrence O'Flaherty's *Masterpiece Theatre*, "I loved *Upstairs, Downstairs*. When I first saw it, my reaction was, 'I'll be amazed if this thing doesn't really hit the headlines. It's marvelous. It allows you to identify with the downstairs people while vicariously enjoying the life of the upstairs people.'" Followed closely episode by episode, the upstairs and downstairs families became a part of "our" family. The audience genuinely cared about the characters, came to know them intimately and developed a strong empathy for them.

The Bellamys and their staff of domestic servants resided in a five-story townhouse at 165 Eaton Place, Belgravia, in London, an address well known to the series' many fans. The upstairs family includes Lord Richard Bellamy (David Langton), his first wife Lady Marjorie (Rachel Gurney) who dies tragically on the Titanic, their two children James (Simon Williams) and Elizabeth (Nicola Pagett), Richard's second wife Virginia (Hannah Gordon), James' wife Hazel (Meg Wynn Owen) who dies in a flu epidemic, and cousin to James and Elizabeth Georgina Worsley (Lesley-Anne Down). Among the most memorable of the downstairs staff are Hudson the butler (Gordon Jackson), Mrs. Bridges the cook (Angela Baddeley), Rose (Jean Marsh), Ruby (Jenny Tomasin), Edward (Christopher Beeny) and Daisy (Jacqueline Tong). Among the many other characters who appeared in a number of episodes, perhaps Sarah (Pauline Collins), Watkins (John Alderton), Sir Geoffrey the family solicitor (Raymond Huntley), and Lady Pru (Joan Benham) are the most fondly remembered by viewers. The large cast, only partially noted here, is considered to include some of the best actors from British stage, film and television. The series earned the respect of professional peers as well as that of the audience. Its cast won numerous awards, both in Britain and America, including eight Emmys, Writers Guild of Great Britain Awards, American Drama Critics Circle Awards, Golden Globe Awards, and a Peabody Award. Angela Baddeley (Mrs. Bridges) received the C.B.E. (Commander of the British Empire), awarded in the Queen's 1975 New Year's Honours List. According to Queen Elizabeth II, *Upstairs, Downstairs* was her favorite program in 1975 and Mrs. Bridges her favorite character. In addition, Gordon Jackson (Hudson) received the coveted Queen's Order of the British Empire Award.

Upstairs, Downstairs
Photo courtesy of Goodman Associates

The idea for the series came from actresses Jean Marsh (who played the role of house-parlour maid Rose) and Eileen Atkins. The series was developed by John Hawkesworth, whose long and distinguished career in film and television extends from art director on the film *The Third Man* to producer of the well regarded Sherlock Holmes series featuring Jeremy Brett. This was the first program from LWT to be purchased for *Masterpiece Theatre* and only the second non-BBC program to be scheduled. *Upstairs, Downstairs* was one of the first series of its type to be produced on videotape rather than film (though certain scenes, mainly exteriors and

location shots, were shot on film). It was one of the first series on *Masterpiece Theatre* that was not biographical or based on a written work. It was created purely for television. As originally produced for British television each episode in the series was written in three acts. On *Masterpiece Theatre* each episode was shown without interruption.

Significant confusion was created when the series was shown on American television because thirteen episodes of the first 26 produced for British television were not shown. This created a rather bizarre lack of continuity. Six of the first original British episodes had been taped in black and white due to a strike. *Masterpiece Theatre* only wanted episodes in color and so the first episode ("On Trial") was revised and reshot in color for American television. Of the first 26 original episodes shot for British TV, Episodes 2 through 9, 11 and 12, 16, 19 and 20 were not shown on American television. These "lost" episodes were not made available for American viewing until 1989. The original black-and-white version of Episode One has never been made available to American television.

Upstairs, Downstairs was first shown on British television in 1971 and continued through four series of 13 episodes each (two Edwardian series, a later pre-war series, and a World War I series) and a fifth series of 16 episodes (post-war), making a total of 68 episodes produced and broadcast. On *Masterpiece Theatre* the original 26 Edwardian period episodes, pared down to 13, were first shown 6 January to 31 March 1974. From 3 November 1974 to 26 January 1975, the post-Edwardian pre-war series of 13 episodes was broadcast. The 13 World War I episodes were shown 1 January to 28 March 1976. The final series of 16 post-war episodes was broadcast 16 January to 1 May 1977 making, in all, 55 episodes shown on *Masterpiece Theatre*. The 55 episodes were later repeated on *Masterpiece Theatre* and selected episodes were shown as a part of a "10th Anniversary Season Festival of Favorites" and as a part of the "Twentieth Anniversary Favorites" series early in 1991. *Upstairs, Downstairs* was the inspiration for the short-lived CBS television series *Beacon Hill* that concerned a well-to-do Boston family and their domestic staff during the 1920s (broadcast fall 1975).

Upstairs, Downstairs is one of the highest rated programs in the history of PBS. The series has been syndicated to both commercial and non-commercial stations in America and is one of the most successful and watched dramatic series in television history. It is estimated that approximately one billion people in over 40 countries have enjoyed *Upstairs, Downstairs* and the series is still in active syndication.

—Steve Runyon

CAST

Lady Marjorie Bellamy	Rachel Gurney
Richard Bellamy	David Langton
James	Simon Williams
Elizabeth	Nicola Pagett
Hudson	Gordon Jackson
Mrs. Bridges	Angela Baddeley
Rose	Jean Marsh
Sarah	Pauline Collins
Emily	Evin Crowley
Alfred	George Innes
Roberts	Patsy Smart
Pearce	Brian Osborne
Edward	Christopher Beeny
Laurence	Ian Ogilvy
Ruby	Jenny Tomasin
Watkins	John Alderton
Hazel	Meg Wynn Owen
Daisy	Jacqueline Tong
Georgina Worsley	Lesley-Anne Down
Virginia	Hannah Gordon
Alice	Anne Yarker
William	Jonathan Seely
Frederick	Gareth Hunt
Lily	Karen Dotrice

PRODUCERS   Rex Firkin, John Hawkesworth

PROGRAMMING HISTORY   68 50-minute episodes

- ITV

10 October 1971–5 March 1972
22 October 1972–19 January 1973
27 October 1973–19 January 1974
14 September 1974–7 December 1974

FURTHER READING

Cooke, Alistair. *Masterpieces (A Decade of Masterpiece Theatre)*. New York: VNU Books International-Alfred A. Knopf, 1981.

Floyd, Patty Lou. *Backstairs with Upstairs, Downstairs*. New York: St. Martin's, 1988.

Hardwick, Mollie. *The World of Upstairs, Downstairs*. New York: Hold, Rinehart and Winston, 1976.

———. *Upstairs, Downstairs III: The Years of Change* (a novel based on the third series of thirteen episodes). New York: Dell, 1987.

Hawkesworth, John. *Upstairs, Downstairs* (a novel based on the first series of thirteen episodes). New York: Dell, 1973.

———. *Upstairs, Downstairs II: In My Lady's Chamber* (novel based on the second series of thirteen episodes). New York: Dell Publishing, 1987.

O'Flaherty, Terrence. *Masterpiece Theatre: A Celebration of 25 Years of Outstanding Television*. San Francisco, California: KQED Books, 1996.

See also British Programming; Jackson, Gordon; Miniseries

U.S. CONGRESS AND TELEVISION

The first effort to link the United States Congress and broadcasting occurred in 1922 when Representative Vincent M. Brennan introduced a bill to allow radio coverage of U.S. House proceedings. The bill failed, and not until the late 1940s was the idea revived. Television, having arrived as a mass medium by then, was allowed in 1948 to

Live television coverage of the United States House of Representatives
Photo courtesy of C-SPAN

cover hearings of the Senate Armed Services Committee. Since few Americans had television receivers in 1948, it was not until the early 1950s that televised congressional hearings generated any viewer interest.

Two televised Senate hearings during the 1950s caused a sensation. Hearings conducted by the Senate Special Committee to Investigate Organized Crime in Interstate Commerce brought the faces and words of notorious mobsters into millions of U.S. homes via coast-to-coast network television. A short time later, Americans once more were drawn to their television screens to watch the hearings of a Senate Committee on Government Operations subcommittee investigate alleged communist infiltration of the U.S. Armed Forces. The hearings were better known as the Army-McCarthy Hearings, identified closely with subcommittee chairperson, Senator Joseph McCarthy.

Two decades later, in 1973, the Senate Select Committee on Presidential Campaign Activities conducted what became known as the Watergate Hearings. Evidence of

misdeeds by President Richard Nixon led the next year to House Judiciary Committee hearings on articles of presidential impeachment. Nearly all public deliberations of both of these committees were televised gavel-to-gavel.

Serious attention to allowing television coverage of actual congressional floor proceedings arose once more with the 1973 formation of the Joint Committee on Congressional Operations. The Committee's charge was to examine means by which Congress could better communicate with the American public. The Committee's subsequent recommendation that television be allowed in the U.S. House and Senate chambers met with resistance in the latter body, but House members seemed more receptive. As a result, House Speaker Thomas (Tip) O'Neill, Jr., ordered testing of a House television system to begin in March 1977. Remote controlled cameras placed at strategic locations in the House chamber were to be used so as not to disrupt House decorum. The television test proved a success. However, full implementation of House television coverage awaited a decision from the House Rules

Committee on who would finally control the television cameras—the House itself or television networks who would remain independent of House authority. The Rules Committee decision that the House would best be served by retaining such control was approved by a vote of 235-to-150 in June 1978. Nine months later, on 19 March 1979, the House television system was fully in place and live telecasts of House floor deliberations began.

Television from the U.S. Senate chambers would have to wait still longer. Although a number of senators supported the idea of Senate television, a powerful block opposed it. Senate television opponents saw cameras as disruptive to Senate decorum and unable to present a favorable image of Senate debate to the American public. Senate television proponents nonetheless prevailed, and a Senate Rules Committee recommendation to allow testing of a Senate chamber television system was approved by a vote of 67-to-21 on 27 February 1986. The tests were satisfactory enough to convince members of the Senate to vote on 29 July 1986, to allow gavel-to-gavel coverage of Senate floor proceedings.

Both the U.S. Senate and House include rules for television coverage among their general procedural rules for committee and chamber conduct. Concern over protecting witness privacy and due process rights led the Senate to allow individual committee chairpersons to adopt television rules most appropriate for their particular committee. Such rules generally require that television coverage be prohibited at the request of a committee witness; that television cameras, lights and microphones be unobtrusive; that television personnel conduct themselves in an orderly fashion inside the hearing room; and that no commercial sponsorship of committee hearings be allowed. House rules are similar to Senate rules regarding the conduct of televised hearings. However, House rules require that television be allowed to cover House committee hearings only upon a majority vote of the committee members.

The manner by which House floor proceedings are televised is entirely under the authority of the House speaker. The speaker decides when and if proceedings will be televised and who will be authorized to distribute the television signals to the public. House rules originally required that television cameras focus only on House members as they spoke from lecterns or in the well of the House. Cameras were not to pan the House chamber to show what oftentimes was a sea of empty chairs. Rules prohibiting such panning were abolished by the Speaker in 1984.

Senate rules for televising chamber proceedings fall under the authority of the Senate Rules and Administration Committee. Most rules are similar to those in the House, save for the prohibition on panning the chamber that remains in effect (except during roll-call votes) for the Senate chamber.

Whether television has improved public debate in either the House or Senate is uncertain. Some observers argue that television has led to more grandstanding and contentious rhetoric on the House floor, whereas Senate debate appears more disciplined and more substantive. However, there is general agreement that persons who view televised House and Senate proceedings are introduced to a vast array of issues and debates unimagined before television arrived.

—Ron Garay

FURTHER READING

Blanchard, Robert, editor. *Congress and the News Media.* New York: Hastings House, 1974.

Congressional Research Service. *Congress and Mass Communications: An Institutional Perspective,* prepared for the Joint Committee on Congressional Operations, 93rd Congress. Second Session, 1974.

Crain, W. Mark, and Brian Goff. *Televised Legislatures: Political Information Technology and Public Choice.* Boston, Massachusetts: Kluwer Academic Publishers, 1988.

Garay, Ronald. *Congressional Television: A Legislative History.* Westport, Connecticut: Greenwood, 1984.

Hess, Stephen. *Live from Capitol Hill! Studies of Congress and the Media.* Washington, D.C.: Brookings Institution, 1991.

Schlesinger, Arthur M., and Roger Bruns, editors. *Congress Investigates: A Documented History, 1792–1974.* New York: Chelsea House, 1975.

Shuman, Samuel I. *Broadcasting and Telecasting of Judicial and Legislative Proceedings.* Ann Arbor, Michigan: University of Michigan Legislative Research Center, 1956.

Straight, Michael. *Trial by Television.* Boston, Massachusetts: Beacon Press, 1954.

Summers, Robert E. *The Role of Congressional Broadcasting in a Democratic Society* (Ph.D. Dissertation, Ohio State University, 1955).

Twentieth Century Fund Task Force on Broadcasting and the Legislature. *Openly Arrived At.* New York: Twentieth Century Fund, 1974.

U.S. Congress. Joint Committee on Congressional Operations. *A Clear Message to the People,* 94th Congress. First Session, 1975.

U.S. Congress. Joint Committee on Congressional Operations. *Broadcasting House and Senate Proceedings,* 93rd Congress. Second Session, 1974.

U.S. Congress. Joint Committee on Congressional Operations. *Congress and Mass Communications,* 93rd Congress. Second Session, 1974.

U.S. House. House Committee on Rules. *Broadcasting the Proceedings of the House,* 95th Congress. Second Session, 1978.

See also Parliament and Television; Political Processes and Television; U.S. Presidency; Watergate

U.S. POLICY: COMMUNICATIONS ACT OF 1934

U.S. Communications Policy Legislation

This legislative act remains the cornerstone of American television policy six decades after its initial passage. Though often updated through amendments, and itself based on the pioneering Radio Act of 1927, the 1934 legislation which created the Federal Communications Commission has endured remarkably well through an era of dramatic technical and social change.

Congress first specifically regulated broadcasting with its 1927 Radio Act which created a Federal Radio Commission designed to regulate in "the public interest, convenience, or necessity." But federal regulation of communications was shared by the Department of Commerce and the Interstate Commerce Commission. By 1934 pressure to consolidate all telecommunication regulation for both wired and wireless services prompted new legislation with a broader purpose.

President Franklin Roosevelt's message requesting new legislation was published in January 1934, the Senate held hearings on several days in March while the House held a single day of hearings in April, a conference report melding the two differing bills together appeared in early June, and the act was passed on 19 June. Given the act's subsequent longevity, it generated little controversy at the time it was considered. Few proposed substantial alteration of the commercially-based broadcast system encoded in the 1927 law. Some critics expressed concern about educational radio's survival—and though Congress mandated the new FCC to consider setting aside some frequencies for such stations, this only occurred in 1941 with approval of FM service.

Running some 45 pages in the standard government printed version as originally passed, the act is divided into several dozen numbered sections of a paragraph or more which were originally divided into six parts called titles (a seventh was added in 1984 concerning cable television). The first title provides general provisions on the FCC, the second is devoted to common carrier regulation, the third deals with broadcasting (and is of primary concern here), the fourth with administrative and procedural matters, the fifth with penal provisions and forfeitures (fines), and the sixth with miscellaneous matters.

The act has been updated through amendment many times—chiefly with creation of public television in 1967 (provisions on the operation and funding of the Corporation for Public Broadcasting expanded title III), and the cable act of 1984 (which created a new title VI devoted to cable regulation, sections of which were expanded in cable legislation of 1992).

Attempts to substantially update or totally replace the act have arisen in Congress several times, most notably during a series of "rewrite" bills from 1977 to 1982, and again in the mid-1990s. Such attempts are driven partly by frustration with legislation based upon analog radio and telephone technology still in force in a digital era of convergence. They are driven as well by increasing rivalries among competing industries—broadcast, cable, telephone and others. They are also driven by political ideology that argues government should no longer attempt to do all things for all people—and by economic constraints that force government to operate more efficiently. The 1934 act, despite its many amendments, is increasingly seen as an anachronism needing replacement to match today's needs.

—Christopher H. Sterling

FURTHER READING

Berry, Tyler. *Communications by Wire and Radio: A Treatise on the Law.* Chicago: Callaghan, 1937.

"Communications Act of 1934: 50th Anniversary Supplement." *Federal Communications Law Journal* (Los Angeles), January 1985.

Federal Communications Commission. *Annual Report.* Washington, D.C.: Government Printing Office, 1934—(issues for 1935–56 reprinted, New York: Arno Press, 1971).

McChesney, Robert W. *Telecommunications, Mass Media and Democracy: The Battle for Control of U.S. Broadcasting, 1928–1935.* New York: Oxford University Press, 1993.

McMahon, Robert S. *Regulation of Broadcasting: Half a Century of Government Regulation of Broadcasting and the Need for Further Legislative Action.* 85th Congress, Second Session, Subcommittee Print: Washington, D.C., 1958.

Paglin, Max D. *A Legislative History of the Communications Act of 1934.* New York: Oxford University Press, 1989.

Rosen, Philip T. *The Modern Stentors: Radio Broadcasters and the Federal Government, 1920–1934.* Westport, Connecticut: Greenwood, 1980.

See also Allocation; Educational Television; "Freeze" of 1948; License; Ownership; Public Interest, Convenience, and Necessity; U.S. Policy: Telecommunication Act of 1996

U.S. POLICY: TELECOMMUNICATIONS ACT OF 1996

U.S. Communications Policy Legislation

The Telecommunications Act of 1996, the first successful attempt to rewrite the sixty-two-year old Communications Act of 1934, was passed on 1 February 1996. The act refocuses federal communications policymaking after

years of confused, multi-agency and intergovernmental attempts to regulate and make sense of a burgeoning telecommunications industry. The bill relies on increased competition for development of new services in broadcasting and cable, telecommunications, information and video services while it reasserts Congress' leadership role as the dominant communications policymaker.

Portions of the act became effective immediately after President Clinton signed the bill into law on 8 February 1996. Other sections of the act will be implemented as the Federal Communications Commission (FCC) promulgates new rules and regulations to meet provisional requirements of the act. Noting the historic nature of the bill, President Clinton stated that the legislation would "stimulate investment, promote competition, provide open access for all citizens to the Information Superhighway." However, many public interest groups are concerned that the act undermines public interest values of access. The act includes several highly controversial provisions that various interests groups claim restrict speech or violate constitutional protections. One section of the bill prohibits the transmission of indecent and obscene material when the material is likely to be seen or read by a minor, and another provision requires broadcasters to formulate a ratings scheme for programs. After nearly four years of work, the bill's passage was eagerly awaited by government and industry leaders alike. Public interest and various industry groups, upset with provisions that would restrict First Amendment rights of telecommunications users vowed to challenge the constitutionality of those provisions in court. Within hours of the bill's passage, a number of civil liberties groups led by the ACLU sought an injunction against provisions of the act.

The Telecommunications act of 1996 is a complex reform of American communication policymaking that attempts to provide similar ground rules and a level playing field in virtually all sectors of the communications industries. The act's provisions fall into five general areas:

- radio and television broadcasting
- cable television
- telephone services
- Internet and on-line computer services
- telecommunications equipment manufacturing

The act abolishes many of the cross-market barriers that prohibited dominant players from one communications industry, such as telephone, from providing services in other industry sectors such as cable. New mergers and acquisitions, consolidations and integration of services previously barred under FCC rules, antitrust provisions of federal law, and the "Modified Final Judgment," the ruling governing 1984 "break-up" of the AT and T telephone monopoly, will be allowed for the first time, illustrating the belief by Congress that competition should replace other regulatory schemes as we enter a new century.

Radio and Television Broadcasting

The act incorporates numerous changes to the rules dealing with radio and television ownership under the Communications Act of 1934. Notably broadcasters have substantial regulatory relief from old and sometimes outmoded federal restrictions on station ownership requirements. Broadcast ownership limits on television stations have been lifted. Group owners can now purchase television stations with a maximum service area cap of 35% of the U. S. population, up from the previous limit of 25% established in 1985. Limits on the number of the radio stations that may be commonly owned have been completely lifted, though the bill does provide limits on the number of licenses that may be owned within specific markets or geographical areas. Also amended are previous restrictions on foreign ownership of stations.

Terms of license for both radio and television have been increased to eight years and previous rules allowing competing applications for license renewals have been dramatically altered in favor of incumbent licensees. New provisions under the act prevent the filing of a competing application at license renewal time unless the FCC first finds that a station has *not* served the public interest or has committed other serious violations of agency or federal rules. This provision will make it increasingly difficult for citizen's groups to mount a license challenge against a broadcast station. The act requires licensees to file a summary of comments and suggestions received from the public while prohibiting the commission from requiring licensees to file information not directly pertinent to the renewal question. However, the bill gives the FCC no guidance as to how it should interpret service in the "public interest" in light of the new legislative mandates. Public interest groups who oppose relaxing ownership provisions claim that the combined effect of the new rules will be to accelerate current trends toward increased control of most media outlets by a few communications conglomerates.

The Telecommunications Act of 1996 makes significant changes in FCC rules regarding station affiliations and cross-ownership restrictions. Stations may choose affiliation with more than one network. Though broadcasting networks are barred from merging or buying out other networks, they may start new program services. For the first time, broadcasters will be allowed to own cable television systems, but television licensees are still prohibited from owning newspapers in the same market. The act affirms the continuation of local marketing agreements (LMAs) and waives the previous restrictions on common control of radio and television stations in the top fifty markets, the one-to-a-market rule.

While broadcasters won new freedoms in licensing and ownership, the act mandates that the industry develop a ratings system to identify violent, sexual and indecent or otherwise objectionable programming. The Communications Decency Act of 1996, embedded in the Telecommunications Act, requires the FCC to devise a rating system if the industry fails to develop such a system within one year of passage of the act. However, early indicators appear to signal a desire on the part of the industry to develop its own ratings system rather than allow government to define program standards. Although development of a ratings system is required under the act, application of the system is voluntary. In conjunction with the establishment of a ratings system, the Telecommunications Act requires television set manufacturers to install a blocking device, called the V-chip, in television receivers larger than 13 inches in screen size by 1998. Recognizing the potential for constitutional challenges of these provisions, the Act allows for accelerated judicial review by a special three-judge federal district court panel. Other provisions of the Communication Decency Act require programmers to limit minors' exposure to objectionable material by scrambling channels depicting explicit sexual behavior and blocking access channels that might contain offensive material.

Perhaps the biggest concession to the broadcast industry centers around provisions for allowing, but not mandating, the FCC to allocate extra spectrum for the creation of advanced television (ATV) and ancillary services. Eligibility for advanced television licenses is limited to existing television licensees, insuring current broadcasters a future in providing digital and enhanced television services. However, Senate Majority Leader Bob Dole (R-Kansas) expressed reservations about giving broadcasters extra spectrum without requiring payment for the new spectrum. Thus, the bill includes a provision that allows Congress to revisit this issue before the FCC awards any digital licenses. Broadcasters vehemently oppose the notion of paying for spectrum, but the act includes provisions that would allow the Commission to impose spectrum fees for any ancillary (non-broadcast) services that broadcasters may provide with these new allocations.

Generally, though the act provides for new possibilities for broadcasters and calls for the FCC to eliminate unnecessary oversight rules, a substantial portion of regulation implemented since the passage of the 1934 Act remains. Thus, while FCC Chairman Reed Hundt issued a statement that claimed that the ubiquitous world of telecommunications had changed forever, analysts and industry experts, remind us that the act amends, but does not replace, the Communications Act of 1934.

Cable Television

Dramatic changes in rate structures and oversight contained within the Telecommunications Act of 1996 are meant to provide new opportunities and flexibility as well as new competition for cable service providers. Under the provisions of the act, *uniform rate structure* requirements will no longer apply to cable operators where there is effective competition from other service providers including the telephone company, multichannel video, direct broadcast satellites and wireless cable systems. However, for the new effective competition standards to apply, comparable video programming services would have to be available to the franchise community. For smaller cable companies, programming tier rates and basic tier rates would be deregulated in franchise areas where there are fewer than 50,000 sub-

scribers. Additionally, states and local franchise authorities are barred from setting technical standards, or placing specific requirements on customer premise equipment and transmission equipment. Sale or transfer of licenses are expedited under the act. Franchise authorities are required to act upon requests for approval to sell or transfer cable systems within 120 days. Failure to comply with the 120 window will provide an "automatic" approval of the sale unless interested parties agree to an extension.

Common carriers and other operators that utilize radio communications to provide video programming will not be regulated under cable rules if the services are provided under a common carriage scheme. Common carriers who choose programming for their video services will be regulated as cable operators unless the services are provided under the "open video systems" provision of the Telecommunications Act. Open video systems operators can apply to the Commission for certification under section 653 of the act which will provide the operator with reduce regulatory burdens. Local Exchange Carriers (LECs) can provide video services under the open video provisions. Further, LECs are not required to make space on their open video systems available on a non-discriminatory basis. Joint ventures and partnerships between local exchange carriers and cable operators are generally barred unless the services qualify under provisions for rural exemptions, or LECs are purchasing a smaller cable system in a market with more than one cable provider, or the systems are not in the top 25 markets.

In an attempt to spur competition between cable operators and local exchange carriers, Congress provided incentives for cable operators to compete with local telecommunications companies. Under the act, cable systems operators are not required to obtain additional franchise approval for offering telecommunications services.

Telephone Services

The Telecommunications Act of 1996 contains sweeping provisions that will restructure the telephone industry in the United States. As noted, LECs can offer video programming services themselves or carry other video programming services under the "open video systems" provisions of the act. In addition to allowing telephone companies to offer video services, important structural barriers erected under the Modified Final Judgment (MFJ) have been swept away. The act allows the seven regional Bell operating companies to offer long-distance telephone service for the first time since the 1984 breakup of AT and T. At the same time, long distance companies and cable operators are allowed to provide local exchange service in direct competition with the regional Bell operating companies, but the act prohibits cross subsidies from non-competitive services to competitive services. Representative Thomas Bliley (R-Virginia) stated, "we have broken up two of the biggest government monopolies left: the monopolies in local telephone service and in cable television." While investors and legislators hailed a new era of competition in the telephone industry, it now becomes the task of the FCC to work out details of the act with state public utilities commissions (PUCs) to ensure a smooth transition of services. The act preempts all previous state rules that restrict or limit competition in telephone services for both local and long distance services.

The act requires regional telephone companies (regional Bell operating companies) to undertake a series of reforms designed to open competition in their service areas. Companies must implement these reforms in order to "qualify" for providing long distance service outside their regional areas. LECs are also required to interconnect new telecommunications service providers and to "unbundle" their networks to provide for exchange access, information access, and interconnection to their systems. In order to provide customers continuity of service, LECs must provide number "portability" by allowing customers to keep their telephone numbers when switching from one service provider to another. The FCC has the task of assessing whether RBOCs and LECs have met the necessary requirements in order to offer long distance services while state public utilities commissions (PUCs) are charged with implementing local telephone competition.

Section 254 of the act defines the nature of "universal service" as "an evolving level of telecommunications services" that take into account telecommunications service advancements. The FCC and a working group of PUC officials are charged with designing policies to promote universal service, especially among rural, high cost and low-income telecommunications users. Also included in the act is a provision that directs the FCC to create discounted telecommunications services for schools and libraries.

Regional telephone companies are now free to manufacture telephone equipment once the FCC qualifies and approves their applications for long distance services. The act prohibits Bellcore, the research arm of the RBOCs from manufacturing as long as it is owned by one or more regional operating companies.

Internet and On-line Computer Services

The Telecommunication Act of 1996 includes Title V, called the Communications Decency Act of 1996 (CDA). The inclusion of the CDA culminates more than a year of debate by members of Congress over the degree to which government could regulate the transmission of objectionable material over computer networks. It creates criminal penalties for anyone who knowingly transmits material that could be construed as indecent to minors. The act criminalizes the intentional transmission of "any comment, request, suggestion, image, or other communications which is obscene, lewd, lascivious, filthy, or indecent...." Enforcement of the CDA includes the filing of criminal charges against any person who uses the computer network for such a transmission. Additionally, the CDA establishes an "anti-flame" provision by prohibiting any computer network transmission for the purpose of annoying or harassing the

recipients of messages. If enforced, penalties under the CDA could range as high as $250,000 for each violation.

The act exempts commercial on-line services that engage in "blocking" from prosecution if they have demonstrated "good faith, reasonable, effective and appropriate" actions to restrict or prevent access by minors. In addition, the CDA contains provisions for a "Good Samaritan Defense" against civil liability for on-line service providers who voluntarily restrict access or availability of material that the provider considers "obscene, lewd, lascivious, excessively violent or otherwise objectionable." The act does not authorize the FCC to enforce the statutory requirements as written.

Various free speech advocates and First Amendment scholars claim that the language in the Communications Decency Act of 1996 is overly broad. Computer experts express concern over whether government should regulate the flow of information on the Internet and other computer-based networks. On the day the President signed the bill into law, the ACLU and other plaintiffs filed suit against Attorney General Janet Reno seeking to enjoin the enforcement of the provisions of Title V on the grounds that it was unconstitutional. Judge Ronald L. Buckwater, a federal judge in Philadelphia, ruled that the language in the law regarding indecent material was unconstitutionally vague but upheld parts of the law regulating obscene and patently offensive information. The Justice Department has stated that it would not prosecute anyone under the law until the challenges mounted against the act were resolved in court. This suit and a companion suit filed by the American Library Association may ultimately go to the Supreme Court for resolution.

The Telecommunications Act of 1996 has garnered substantial praise as a pro-competitive bill designed to allow anyone to enter any communications business and to let any communications business to compete in any market against other competitors. Supporters of the bill predict job creation and lower telecommunications costs as two benefits likely to accrue as a result of its passage. Other experts say the Telecommunications Act will allow smaller telephone companies to successfully compete with larger companies for telephone, paging and cellular services. Manufacturers of cable modems and network connectivity devices should benefit from rapid advances as a result of increased competition.

Critics of the act claim its extensive deregulatory provisions coupled with relaxed restrictions on concentration of media ownership dilute the public responsibility guarantees build into the Communications Act of 1934 and tilt the preference in favor of private market forces. Critics claim that in many areas of the country which are not likely to see real competition, the cost of telecommunications and video services are likely to rise dramatically. Other critics oppose giving broadcasters extra spectrum at a time when the government could reap hundreds of millions of dollars for those frequencies through spectrum auction.

At this time it is too early to predict the outcomes of the Telecommunications Act of 1996. Analysts and financial experts views are mixed but they predict market shake-outs and consolidations are likely to radically transform the telecommunications industry in the next few years as a result of the implementation of the act.

—Fritz J. Messere

FURTHER READING

Andrews, E.L. "Congress Votes to Reshape Communications Industry Ending a 4-Year Struggle." *New York Times,* 2 February 1996.

———. "President Signs Telecommunications Bill." *Cyber Times Extra, New York Times,* 9 February 1996.

———. "What the Bill Already Did." *New York Times,* 2 February 1996.

Carter, B. "The Networks See Potential for Growth." *New York Times,* 2 February 1996.

Clinton, W. "Remarks by the President in Signing Ceremony for the Telecommunications Act Conference Report" (press release). Washington, D.C.: Library of Congress, 8 February 1996.

———. "Statement by the President" (press release). Washington, D.C.: Office of the Press Secretary, U.S. White House, 1 February 1996.

Hinden, S. "The Greatest Telecommunications Show in Years: The 1996 Legislation Leaves Fund Managers Wondering: Which Company's Act is a Winner?" *The Washington Post,* 24 March 1996.

Jones, K. "Net Access Providers Worried as FCC Rethinks On-Line Regulation." *Cyber Times Extra, New York Times,* 29 February 1996.

Landler, M. "For Telephone Companies, Excitement Over New Markets." *New York Times,* 2 February 1996.

Lewis, P. H. "Internet Courtroom Battle Gets Cyberspace." *Cyber Times Extra, New York Times,* 20 March 1996.

Raysman, R., and P. Brown. "Liability of Internet Access Provider Under Decency Act." *New York Law Journal,* 12 March 1996.

United States Congress, House of Representatives. *104—H.R. 1555, The Telecommunications Act of 1995.* 104th Congress, 1st Session. Washington, D.C.: introduced 3 May 1995.

United States Congress, Senate. *S.652—The Telecommunications Act of 1996.* 104th Congress, 2nd Session. Washington, D.C.: 3 January 1996.

United States Congress. *Joint Explanatory Statement of the Committee of Conference on S.652—The Telecommunications Act of 1996.* 104th Congress, 2nd Session. Washington, D.C.: 3 January 1996.

See also Allocation; Cable Networks; License; Ownership; Telcos; United States: Cable; U.S. Policy: Communication Act of 1934

U.S. PRESIDENCY AND TELEVISION

Ten dates, some momentous, some merely curious, tell the story of presidential television. In its own way, each date sheds light on the complex relationship between the U.S. presidency and the American television industry. Over the years, that relationship has grown complex and tempestuous (virtually every president from Harry Truman through Bill Clinton has become disaffected with the nation's press). More than anything else, however, this relationship has been symbiotic—the president and the press now depend upon one another for sustenance. Ten dates explain why:

September 23, 1952 — Richard Nixon's "Checkers" Speech

Oddly, it was Richard Nixon who discovered the political power of the new medium. Richard Nixon, who was pilloried by the press throughout his career, nonetheless discovered the salvific influence of television. Imaginatively, aggressively, Nixon used television in a way it had never been used before to lay out his personal finances and his cultural virtues and, hence, to save his place on the Republican national team (and, ultimately, his place in the American political pantheon). That same year, 1952, also witnessed the first televised coverage of a national party convention and the first TV advertisements. But it was Nixon's famous speech that turned the tide from a party-based to a candidate-controlled political environment. By using television as he did—personally, candidly, visually (his wife Pat sat demurely next to him during the broadcast)—Nixon single-handedly created a new political style.

January 19, 1955 — Dwight Eisenhower's Press Conference

When he agreed to let the television cameras into the White House for the first time in American history, Dwight Eisenhower changed the presidency in fundamental ways. Until that point, the White House press corps had been a cozy outfit but very much on the president's leash or, at least, the lesser partner in a complex political arrangement. Television changed that. The hue and cry let out by the deans of U.S. print journalism proved it, as did television's growing popularity among the American people. More proof awaited. It was not long after Dwight Eisenhower opened the doors to television that American presidents found themselves arranging their work days around network schedules. To have a political announcement receive top billing on the nightly news, after all, meant that it had to be made by 2:00 P.M., Eastern Standard Time. If the

U.S. Presidents George Bush, Ronald Reagan, Jimmy Carter, Gerald Ford and Richard Nixon

news to be shared was bad news, the slowest news days—Saturday and Sunday—would be chosen to carry the announcement. These may seem like small expediencies but they presaged a fundamental shift of power in Washington, D.C. After Eisenhower, television was no longer a novelty but a central premise in all political logic.

January 25, 1961 — John Kennedy's Press Conference

Before Ronald Reagan and Bill Clinton there was John F. Kennedy. No American president has better understood television than these three. By holding the first live press conference in the nation's history, Kennedy showed that boldness and amiability trump all suits in an age of television. In his short time in office Kennedy also showed: (1) that all communication, even presidential communication, must be relational; (2) that the substance of one's remarks is irrelevant if one cannot say it effortlessly; (3) that being "on line" and "in real time" bring a special energy to politics. Prescient as he was, Kennedy would therefore not have been surprised to learn that 50% of the American people now find television news more believable and more attractive than

print news (which attracts a mere quarter of the populace). Kennedy would also not be surprised at the advent of CNN, the all-news, all-day channel, nor would he be surprised to learn that C-SPAN (Congress' channel) has also become popular in certain quarters. Being the innovator he was, Kennedy fundamentally changed the temporal dimensions of American politics. Forever more, his successors would be required to perform the presidency during each moment of each day they held office.

February 27, 1968 — Walter Cronkite's Evaluation of the Vietnam War

Lyndon Johnson, we are told, knew he had lost the Vietnam War when CBS news anchor Walter Cronkite declared it a quagmire during an evening documentary. To be sure, Cronkite's hard-hitting special was nuanced and respectful of the presidency, but it also brought proof to the nation's living rooms that the president's resolve had been misplaced. Cronkite's broadcast was therefore an important step in altering the power balance between the White House and the networks. CBS' Dan Rather continued that trend, facing-down Richard Nixon during one cantankerous press conference and, later, George Bush during an interview about the Iran-Contra scandal. Sam Donaldson and Ted Koppel of ABC News also took special delight in deflating political egos, as did CNN's Peter Arnett who frustrated George Bush's efforts during the Gulf War by continuing to broadcast from the Baghdad Hilton even as U.S. bombs were falling on that city. Some attribute the press's new aggressiveness to their somnolescence during the Watergate affair, but it could also be credited to the replacement of politics' old barter system, which featured material costs and rewards, by an entertainment-based celebrity system featuring personal achievements and rivalries. In this latter system, it is every man for himself, the president included.

November 25, 1968 — Inauguration of the White House's Office of Communication

One of Richard Nixon's first acts as president was to appoint Herb Klein to oversee a newly enlarged unit in the White House that would coordinate all out-going communications. This act, perhaps more than any other, signalled that the new president would be an active player in the persuasion game and that he would deal with the mass media in increasingly innovative ways. Perhaps Nixon sensed the trends scholars would later unearth: (1) that citizens who see a political speech in person react far more favorably than those who see it through television reporters' eyes; (2) that the average presidential "soundbite" has been reduced to 9.8 seconds in the average nightly news story; and (3) that negative news stories about the president have increased over time. This is the bad news. The good news is that 97% of CBS' nightly newscasts feature the president (usually as the lead story) and that 20% of a typical broadcast will be devoted to comings and goings in the White House. In other words, the president is the fulcrum around which television reportage pivots; hence, he is well advised to monitor carefully the information he releases (or refuses to release).

September 17, 1976 — Gerald Ford's Pasadena Speech

Neither Gerald Ford's address nor the occasion were memorable. His was a standard stump speech, this time at the annual reception of the Pasadena Golden Circle. The speech's sheer banality signalled its importance: Ford spoke to the group not because he needed to convince them of something but because their predictable, on-camera applause would certify his broader worthiness to the American people. Ford gave some 200 speeches of this sort during the 1976 campaign. Unlike Harry Truman, who spoke to all-comers on the village green during the 1948 election, Ford addressed such "closed" audiences almost exclusively during his reelection run. In addition, Ford and his successors spoke in ritualistic settings 40% of the time since bunting, too, photographs well. The constant need for media coverage has thereby turned the modern president into a continual campaigner and the White House into a kind of national booking agency. It is little wonder, then, that the traditional press conference, with its contentiousness and unpredictability, has become rare.

January 20, 1981 — Inauguration of Ronald Reagan

Ronald Reagan and television have become American cliches. Reagan grew up with television and television with him. By the time he became president, both had matured. Reagan brought to the camera what the camera most prized: a strong visual presence and a vaunted affability. Reagan was the rare kind of politician who even liked his detractors and television made those feelings obvious. Reagan also had the ability to concretize the most abstract of issues—deficits, territorial jurisdictions, nuclear stalemates. By finding the essential narrative in these matters, and then by humanizing those narratives, Reagan produced his own unique style. Television favors that style since it is, after all, the most intimate of the mass media, with its ability to *show* emotion and to do so in tight-focus. So it is not surprising that political advertising has now become Reaganesque—visual, touching, elliptical, never noisy or brash. Like Reagan, modern political advertising never extends its stay; it says in thirty seconds all that needs to be said and then it says no more.

January 16, 1991 — George Bush's Declaration of the Gulf War

From the beginning, George Bush was determined not to turn the Gulf War into another Vietnam. His military commanders shared that determination. But what, exactly, are the lessons of Vietnam? From the standpoint of television they are these: (1) make it an air war, not a ground war, because ground soldiers can be interviewed on camera; (2) make it a short war, not a long war, because television has a short attention span; and (3) make it a technical war, not a political war, because Americans love the technocratic and fall out with one another over ends and means. Blessedly, the Gulf War was short, and, via a complex network of satellite feeds, it entertained the American people with its visuals: SCUD missiles exploding, oil-slicks spreading, yellow ribbons flying. Iraq's Saddam Hussein fought back—on television—in avuncular poses with captured innocents and by staying tuned to CNN from his bunker. The Gulf War therefore marked an almost postmodern turn in the history of warfare, with the texts it produced now being better remembered than the deaths it caused. What such a turn means for the presidency, or for humankind, has yet to be determined.

October 25, 1992 — Richmond, Virginia Debate

Several trends converged to produce the second presidential debate of 1992. In the capital of the Old South, Bush, Clinton and Perot squared off with one another in the presence of two hundred "average Americans" who questioned them for some ninety minutes. The debate's format, not its content, became its headline: the working press had been cut out of the proceedings and few seemed to mourn their passing. The president of the United States face-to-face with the populace—here, surely, was Democracy Recaptured. The 1992 campaign expanded upon this theme, with the candidates repairing to the cozy studio (and cozy questions) of talk-show host Larry King. Thereafter, they made the rounds of the morning talk-over-coffee shows. The decision to seek out these friendly climes followed from the advice politicians had been receiving for years: choose your own audience and occasion, forsake the press, emphasize your humanity. Coupled with fax machines, E-mail, cable specials, direct-mail videos, and the like, these "alternative media formats" completed a cycle whereby the president became a rhetorical entrepreneur and the nation's press an afterthought.

April 20, 1993 — Bill Clinton's MTV Appearance

Not a historic date, perhaps, but a suggestive one. It was on this date that Bill Clinton discussed his underwear with the American people (briefs, not boxers, as it turned out). Why would the leader of the free world unburden himself like this? Why not? In television's increasingly postmodern world, all texts—serious and sophomoric—swirl together in the same discontinuous field of experience. To be sure, Clinton made his disclosure because he had been asked to do so by a member of the MTV generation, not because he felt a sudden need to purge himself. But in doing so Clinton exposed several rules connected to the new phenomenology of politics: (1) because of television's celebrity system, presidents are losing their distinctiveness as social actors and hence are often judged by standards formerly used to assess rock singers and movie stars; (2) because of television's sense of intimacy, the American people feel they know their presidents as persons and hence no longer feel the need for party guidance; (3) because of the medium's archly cynical worldview, those who watch politics on television are increasingly turning away from the policy sphere, years of hyper-familiarity having finally bred contempt for politics itself.

For good and ill, then, presidential television grew apace between 1952 and the present. It began as a little-used, somewhat feared, medium of exchange and transformed itself into a central aspect of American political culture. In doing so, television changed almost everything about life in the White House. It changed what presidents do and how they do it. It changed network programming routines, launched an entire subset of the American advertising industry, affected military strategy and military deployment, and affected how and why voters vote and for whom they cast their ballots. In 1992, Ross Perot of Dallas, Texas, tested the practical limits of this technology by buying sufficient airtime to make himself an instant candidate as well as an instantly serious candidate. History records that Perot failed to achieve his goal. But given his billions and given television's capacity to mold public opinion, Perot, or someone like him, may succeed at some later time. This would add an eleventh important date to the history of presidential television.

—Roderick P. Hart and Mary Triece

FURTHER READING

Allen, Craig. "'Robert Montgomery Presents': Hollywood Debut in the Eisenhower White House." *Journal of Broadcasting and Electronic Media* (Washington, D.C.), Fall 1991.

Benjamin, Louise M. "Broadcast Campaign Precedents from the 1924 Presidential Election." *Journal of Broadcasting and Electronic Media* (Washington, D.C.), Fall 1987.

Brody, Richard A. *Assessing the President: The Media, Elite Opinion, and Public Support.* Stanford, California: Stanford University Press, 1991.

Devlin, L. Patrick. "Contrasts in Presidential Campaign Commercials of 1992." *American Behavioral Scientist* (Princeton, New Jersey), November-December 1993.

Grossman, Michael B., and Martha J. Kumar. *Portraying the President: The White House and the News Media.* Baltimore, Maryland: Johns Hopkins University Press, 1981.

Hart, Roderick P. *Seducing America: How Television Charms the Modern Voter.* New York: Oxford University Press, 1994.

———. *The Sound of Leadership: Presidential Communication in the Modern Age.* Chicago: University of Chicago Press, 1987.

Hinckley, Barbara. *The Symbolic Presidency: How Presidents Portray Themselves.* New York: Routledge, 1990.

Iyengar, Shanto, and Donald Kinder. *News That Matters: Television and American Opinion.* Chicago: University of Chicago Press, 1987.

Jamieson, Kathleen H. *Eloquence in an Electronic Age: The Transformation of Political Speechmaking.* New York: Oxford University Press, 1988.

Jamieson, Kathleen H., and David S. Birdsell. *Presidential Debates: The Challenge of Creating an Informed Electorate.* New York: Oxford University Press, 1988.

Kaid, Lynda Lee. "Political Argumentation and Violations of Audience Expectations: An Analysis of the Bush-Rather Encounter." *Journal of Broadcasting and Electronic Media* (Washington, D.C.), Winter 1990.

———. "Television News and Presidential Campaigns: The Legitimization of Televised Political Advertising." *Social Science Quarterly* (Austin, Texas), June 1993.

Lemert, James B. "Do Televised Presidential Debates Help Inform Voters?" *Journal of Broadcasting and Electronic Media* (Washington, D.C.), Winter 1993.

Lowry, Dennis T. "Effects of TV 'Instant Analysis and Querulous Criticism:' Following the Bush-Dukakis Debate." *Journalism Quarterly* (Urbana, Illinois), Winter 1990.

MacNeil, Robert. "Taking Back the System." *Television Quarterly* (New York), Spring 1992.

Maltese, John A. *Spin Control: The White House Office of Communications and the Management of Presidential News.* Chapel Hill, North Carolina: University of North Carolina Press, 1992.

Meyrowitz, Joshua. *No Sense of Place: The Impact of Electronic Media on Social Behavior.* New York: Oxford University Press, 1985.

Mickelson, Sig. *From Whistle Stop to Sound Bite: Four Decades of Politics and Television.* New York: Praeger, 1989.

Miller, Arthur H., and Bruce E. Gronbeck, editors. *Presidential Campaigns and American Self Images.* Boulder, Colorado: Westview Press, 1994.

Morreale, Joanne. *The Presidential Campaign Film: A Critical History.* Westport, Connecticut: Praeger, 1993.

Patterson, Thomas E. *Out of Order.* New York: A. Knopf, 1993.

Ranney, Austin. *Channels of Power: The Impact of Television on American Politics.* New York: Basic Books, Inc., 1983.

Rosenstiel, Tom. *Strange Bedfellows: How Television and the Presidential Candidates Changed American Politics.* New York: Hyperion, 1993.

Sabato, Larry. *Feeding Frenzy: How Attack Journalism Has Transformed American Politics.* New York: Free Press, 1991.

Smith, Carolyn D. *Presidential Press Conferences: A Critical Approach.* New York: Praeger, 1990.

Smith, Craig Allen, and Kathy B. Smith. *The White House Speaks: Presidential Leadership as Persuasion.* Westport, Connecticut: Praeger, 1994.

Stone, David M. *Nixon and the Politics of Public Television.* New York: Garland, 1985.

Tulis, Jeffrey K. *The Rhetorical Presidency.* Princeton, New Jersey: Princeton University Press, 1987.

Watson, Mary Ann. "How Kennedy Invented Political Television." *Television Quarterly* (New York), Spring 1991.

West, Darrell M. *Air Wars: Television Advertising in Electronic Campaigns, 1952–1992.* Washington D.C.: n.p. 1993.

See also Political Processes and Television; Presidential Nominating Conventions and Television; Press Conference; Reagan, Ronald; U.S. Congress; Watergate

V

THE VALOUR AND THE HORROR

Canadian Documentary

Aired on the publicly owned Canadian Broadcasting Corporation (CBC), *The Valour and the Horror* is a Canadian-made documentary about three controversial aspects of Canada's participation in World War II. This three-part series caused a controversy almost unprecedented in the history of Canadian television. Canadian veterans, outraged by what they considered an inaccurate and highly biased account of the war, sued Brian and Terence McKenna, the series directors, for libel. An account of the controversy surrounding *The Valour and the Horror* with statements by the directors, the CBC ombudsman and an examination of the series by various

historians can be found in Bercuson and Wise's *The Valour and the Horror Revisited.*

The Valour and the Horror consists of three separate two-hour segments aired on consecutive Sunday evenings in 1992. In the first, "Savage Christmas Hong Kong 1941," the McKennas explore the ill-preparedness of the Canadian troops stationed in Hong Kong, the loss of the city to the Japanese, and the barbarous treatment of Canadian troops interned in slave labour camps for the duration of the war. Arguably the most moving of the three episodes, it was the least controversial. The eyewitness testimony of two surviving veterans, combined with archival photographs and

The Valour and the Horror
Photo courtesy of Galafilm, Inc.

reenactments of letters written by prisoners of war, testifies to the strength of emotion which can be generated by television documentary.

The second episode, "Death by Moonlight: Bomber Command," proved to be the most controversial of the three episodes. It details the blanket bombing of German cities carried out by Canadian Lancaster bombers, including the firestorm caused by the bombings of Dresden and Munich. The McKennas claim that the blanket bombing, which caused enormous casualties among both German civilians and Canadian aircrews, did nothing to hasten the end of the war, and was merely an act of great brutality with little military significance. In particular British commander Sir Arthur "Bomber" Harris is cited for his bloodthirstiness.

"In Desperate Battle: Normandy 1944," the third episode, deals primarily with the massive loss of Canadian troops at Verrieres Ridge during the assault on Normandy, citing the incompetence and inexperience of Canadian military leadership as the cause for the high casualty rate. This episode also accuses the Canadian forces of war crimes against German soldiers—war crimes which were never prosecuted after the war.

All three episodes consist of black-and-white archival footage of the war, combined with present-day interviews with both allied and enemy veterans. Each episode has a voice-over narration by Brian McKenna, and is accompanied by music taken from Gabriel Faure's Requiem of 1893. The sections taken from the Requiem are those sung primarily by young boys. The accompaniment was perhaps chosen because the McKennas emphasize, throughout each episode, the youth of the combatants, and the terrible but preventable waste of Canada's young men.

The youthfulness of the soldiers is also emphasized in some very controversial reenactments in which actors speak lines taken from the letters and diaries of Canadian and British military personnel. Although these reenactments are well marked as such, the veterans have claimed that they are misleading and extremely selective about what they include. Reenactments, which are more characteristic of "Reality TV," like *America's Most Wanted* and *Rescue 911*, are problematic in conventional documentary practice. As Bill Nichols argues in *Representing Reality*, "documentaries run some risks of credibility in reenacting an event: the special indexical bond between image and historical event is ruptured." Certainly reenactments are more conventional in television than in cinematic documentary.

The battle which ensued over *The Valour and the Horror* is a battle over the interpretation of history and the responsibilities of publicly funded television. The McKennas have argued, in the tradition of investigative journalism, that they wished to set aside the official account of the war and examine events from the point of view of the participants.

They have also argued that the real story has never been told, and that their own research has shown gross incompetence, mismanagement and cover-ups on the part of the Canadian government. Historians and veterans have argued that *The Valour and the Horror* is a revisionist history which is both historically inaccurate and poorly researched.

The major complaints against *The Valour and the Horror* by historians are its lack of context, poor research, and bias which led to misinterpretation and inaccuracy. The McKennas, in defending themselves, have to a degree been their own worst enemies. By claiming that their series is fact, and contains no fiction, and also claiming that their research is "bullet proof," they have set themselves up for all kinds of attacks—attacks which have also affected the status of publicly funded television in Canada. Publicly funded institutions are particularly vulnerable to attacks by powerful lobbies, whose animosity can and does jeopardize their financial stability. *The Valour and the Horror* can be seen as a particularly acrimonious chapter in the continuing battle between a publicly funded institution and the taxpayers who support it. In this, it is not unlike the battle waged in the United States between veterans and the Smithsonian Institute over the representation of the bombing of Hiroshima and Nagasaki.

—Jeannette Sloniowski

PRODUCERS   Arnie Gelbart, André Lamy

NATIONAL FILM BOARD OF CANADA PRODUCER   Adam Symansky

CANADIAN BROADCASTING CORPORATION PRODUCER
Darce Fardy

DIRECTORS
Brian McKenna, Terence McKenna

WRITERS
Brian McKenna, Terence McKenna

PROGRAMMING HISTORY
January 1992   3 Parts

FURTHER READING

Bercuson, David J., and S.F. Wise. *The Valour and the Horror Revisited.* Montreal: McGill-Queen's University Press, 1994.

Nichols, Bill. *Representing Reality.* Bloomington: Indiana University Press, 1991.

———. *Blurred Boundaries.* Bloomington: Indiana University Press, 1994.

See also Canadian Programming in English

VAN DYKE, DICK

U.S. Actor

Dick Van Dyke's entertainment career began during World War II when he participated in variety shows and worked as an announcer while serving in the military. That career has continued, with five decades of work as an actor on network and local television, the stage and in motion pictures. The television work started with his role as host of variety programs in Atlanta, Georgia, and his first foray into network television came in 1956 as the emcee of CBS Television's *Cartoon Theatre*.

It was his role as Rob Petrie on the classic CBS situation comedy *The Dick Van Dyke Show* that insured his place in television history. He was cast by series creator Carl Reiner and series producer Sheldon Leonard in the role of a television comedy writer (Reiner himself played this role in the series pilot *Head of the Family*). He was selected over another television pioneer, Johnny Carson. Plucked from a starring role on the Broadway stage in *Bye Bye Birdie*, Van Dyke used his unique talent for physical comedy, coupled with his ability to sing and dance, to play Robert Simpson Petrie, the head writer of the *Alan Brady Show*. Complementing Van Dyke was a veteran cast of talented comedic actors including Rose Marie, Morey Amsterdam, Jerry Paris, Carl Reiner (as Alan Brady), as well as a newcomer to television, Mary Tyler Moore, who played Rob's wife Laura Petrie.

In many ways *The Dick Van Dyke Show* broke new ground in network television. The series created quite a stir when, in the early 1960s, husband and wife, though still sleeping in separate beds, were shown to actually have a physical relationship. Mary Tyler Moore was even shown wearing Capri pants, unheard of at the time. But the quintessential example of the innovations offered by *The Dick Van Dyke Show* occurred when, after the network rejected the script, only an appeal from Sheldon Leonard himself secured permission to film the episode "That's My Boy??" In this episode, Rob (Van Dyke) is convinced that the baby he and Laura brought home from the hospital was not theirs, but a baby belonging to another couple, the Peters. Constant mix-ups with flowers and candy at the hospital, caused by the similarity in names (Petrie and Peters), convinced Rob that the babies were somehow switched, and he decided to confront the Peters family. Only when the Peters show up at Rob and Laura's house does Rob learns that the Peters are African American. Some have speculated that the overwhelming positive reaction by audiences to this episode led Sheldon Leonard to eventually cast another future television megastar, Bill Cosby, in *I Spy*.

Dick Van Dyke won three Emmy Awards for his role in *TDVDS*, and the series received four Emmy Awards as outstanding comedy series. The series, which began in 1961, ended its network television run in 1966, although audiences have enjoyed the program through its extended life in syndication.

Although Dick Van Dyke went on to star in such feature films as *Chitty Chitty Bang Bang, Mary Poppins* and *The Comic*, he has continued to be a staple on network television with *The New Dick Van Dyke Show, Van Dyke and Company* (for which he received his fourth Emmy) and a critically-acclaimed and Emmy-nominated dramatic performance in the made-for-television movie *The Morning After*. In his fifth decade in television, Van Dyke has been seen in the 1990s prime-time series *Diagnosis Murder* for CBS, in which he co-starred with his son Barry Van Dyke.

—Thomas A. Birk

DICK VAN DYKE. Born in West Plains, Missouri, U.S.A., 13 December 1925. Married: Marjorie Willett, 1948; three daughters and two sons. Served in U.S. Army Air Corps, during World War II. Founded advertising agency with Wayne Williams, Danville, Illinois, 1946; appeared with Phillip Erickson in pantomime act The Merry Mutes, Eric and Van, 1947–53; television master of ceremonies, *The Music Shop*, Atlanta; hosted television variety show *The Dick Van Dyke Show*, New Orleans; master of ceremonies, *The Morning Show*, CBS, 1955, and *Cartoon Theatre*, 1956; hosted weekly television show *Flair*, ABC, 1960; performed on Broadway in *Bye Bye Birdie*, 1960–61; starred in weekly television sitcom *The Dick Van Dyke Show*, CBS, 1961–66;

Dick Van Dyke
Photo courtesy of Dick Van Dyke

performed in such films as *Mary Poppins*, 1965, and *Chitty Chitty Bang Bang*, 1968; returned to television series format with *Diagnosis Murder*, 1994; chair, Nick at Nite, since 1992. Recipient: Theater World Award, 1960; Antoinette Perry Award, 1961; four Emmy Awards.

TELEVISION SERIES

1955	*The Morning Show*
1956	*Cartoon Theatre*
1958–59	*Mother's Day*
1959	*Laugh Line*
1960	*Flair*
1961–66	*The Dick Van Dyke Show*
1971–74	*The New Dick Van Dyke Show*
1976	*Van Dyke and Company*
1988	*The Van Dyke Show*
1994–	*Diagnosis Murder*

MADE-FOR-TELEVISION MOVIES

1974	*The Morning After*
1977	*Tubby the Tuba* (voice only)
1982	*Drop-Out Father*
1983	*Found Money*
1987	*Ghost of a Chance*

FILMS

Bye Bye Birdie, 1963; *What a Way To Go*, 1964; *Lt. Robin Crusoe, USN*, 1965; *Mary Poppins*, 1965; *Divorce American Style*, 1967; *Never a Dull Moment*, 1967; *Chitty Chitty Bang Bang*, 1968; *The Comic*, 1969; *Some Kind of Nut*, 1969; *Cold Turkey*, 1971; *The Runner Stumbles*, 1979; *Drop-Out Father*, 1982; *Dick Tracy*, 1990; *Freddie Goes to Washington* (voice only), 1992.

STAGE

The Girls Against the Boys, 1959; *Bye Bye Birdie*, 1960–61.

PUBLICATIONS

Faith, Hope, and Hilarity. Garden City, New York: Doubleday, 1970.

Karlen, Neal. "A Familiar Face Introduces himself to a New Generation" (interview). *The New York Times*, 21 October 1992.

FURTHER READING

Grote, David. *The End of Comedy: The Sit-com and the Comedic Tradition*. Hamden, Connecticut: Archon Books, 1983.

Hamamoto, Darrell Y. *Nervous Laughter: Television Situation Comedy and Liberal Democratic Ideology*. New York: Praeger, 1989.

Javna, John. *The Best of TV Sitcoms: Burns and Allen to The Cosby Show, The Munsters to Mary Tyler Moore*. New York: Harmony Books, 1988.

Jones, Gerard. *Honey, I'm Home!: Sitcoms, Selling the American Dream*. New York: Grove Weidenfeld, 1992,

Leibman, Nina. *Living Room Lectures: The Fifties Family in Film and Television*. Austin: University of Texas Press, 1995.

Marc, David. *Comic Visions: Television Comedy and American Culture*. Boston, Massachusetts: Unwin Hyman, 1989.

Mitz, Rick. *The Great TV Sitcom Book*. New York: R. Marek, 1980.

Waldron, Vince. *Classic Sitcoms: A Celebration of the Best of Prime-time Comedy*. New York: Macmillan, 1987.

———. *The Official Dick Van Dyke Show Book: The Definitive History and Ultimate Viewer's Guide to Television's Most Enduring Comedy*. New York: Hyperion, 1994.

Weissman, G., and C. S. Sanders. *The Dick Van Dyke Show: Anatomy of a Classic*. New York: St. Martin's Press, 1983.

See also Comedy, Domestic Settings; Comedy, Workplace Settings; *Dick Van Dyke Show*; Moore, Mary Tyler; Reiner, Carl

VARIETY PROGRAMS

Variety programs were among the most popular prime-time shows in the early years of American television. *Texaco Star Theater* starring Milton Berle was so popular for its first two or three years in the late 1940s and early 1950s that restaurants closed the night it was on, water usage plummeted during its hour, and in 1949, almost 75% of the television audience watched it every week. Whether emphasizing musical performance or comedy, or equal portions of each, the variety genre provided early television with the spectacular entertainment values television and advertising executives believed was important to its growth as a popular medium.

Variety shows almost always featured musical (instrumental, vocal, and dance) performances and comedy sketches, and sometimes acrobatics, animal or magic tricks, and dramatic recitations. Some had musical or comedy stars as hosts, often already known from radio or the recording industry, who displayed their talents solo or with guest performers. Others featured personalities, such as Ted Mack or Ed Sullivan, who acted as emcees and provided continuity for what was basically a series of unrelated acts. This genre was produced by both networks and local television stations. Some of the most popular musical variety programs, such as *The Lawrence Welk Show* and *The Liberace Show*, began as local productions for Los Angeles stations. The form has its heritage in 19th-century American entertainment—minstrel, vaudeville, and burlesque shows—and the 20th-century nightclub and Catskills resorts revues (where such

This Is Tom Jones

Perry Como's Kraft Music Hall

The Sonny and Cher Comedy Hour

The Judy Garland Show

talents as Sid Caesar, Imogene Coca, and Carl Reiner were found).

These forms of entertainment emphasized presentational or performative aspects—immediacy, spontaneity, and spectacle—over storyline and character development. Performers might develop a "persona," but this character mask would usually represent a well-known stereotype or exhibit a particular vocal or dance talent, rather than embody a fleshed-out character growing within the context of dramatic situations. The vaudeville show, which had achieved a middle-class following by the 20th century, presented a series of unrelated acts, featured stars or "headliners," in addition to supporting acts. Many of the form's most important stars made the transition to radio or films in the 1920s and 1930s, and some of these, such as Ed Wynn, were also among the stars of television's first variety shows. Two of the most significant "headliners" of vaudeville and stars of radio, Jack Benny and Burns and Allen made a successful transition to television, but while their shows retained aspects of vaudeville and variety (especially Benny's program with movie star guests and the regularly featured singer Dennis Day), they also combined those elements with the narrative features of situation comedy. A less successful radio comedian, Milton Berle, brought vaudeville back in a much bigger way (his and other television variety-vaudeville shows were called "vaudeo") because his performances emphasized the visual spectacle of the live stage impossible on radio.

The spontaneous, rowdy antics and adult humor of Milton Berle, or of Sid Caesar and company on *Your Show of Shows*, were most popular on the east coast, where they could be aired live (before the co-axial cable was laid across the country), and where an urban population might be familiar with their styles from nightclubs and resorts. As demographics and ratings from other parts of the country became more important to advertisers and networks, as telefilm programming (usually sitcoms and western dramas) became more successful, and as moral watchdog groups and cultural pundits criticized the genre for its "blue" jokes, some comedy-variety shows fell out of favor. The gentle, child-like humor of Red Skelton became more popular than the cross-dressing of Berle, just as the various comic "personas" of Jackie Gleason (such as the Poor Soul, Joe the Bartender, Ralph Kramden) proved more acceptable to wide audiences than the foreign movie spoofs performed by Caesar and company. While Berle and Caesar stayed on the air for most of the 1950s, it was these other comics and their variety hours that made the transitions into the 1960s.

Variety shows emphasizing music, such as *The Dinah Shore Show, The Perry Como Show, The Tennessee Ernie Ford Show, The Lawrence Welk Show, Your Hit Parade, The Bell Telephone Hour* and *The Voice of Firestone* (the latter two emphasizing classical music performance), had long runs and little controversy. Nat "King" Cole, the first major black performer to have a network variety series, had a great difficulty securing sponsors for his show when it debuted in 1956 and most of the important black musical stars of the time—and many of the white ones as well—appeared for reduced fees to help save the show. NBC cancelled it a little over a year after its debut.

Besides several of the above mentioned shows, *The Smothers Brothers Show, The Carol Burnett Show*, and *The Ed Sullivan Show* (which would leave the air in 1971 after 23 years) found success in the 1960s, even as the prime-time schedule became more and more filled with dramatic programs and situation comedies. *The Smothers Brothers Show* caused some controversy with its anti-Vietnam war jokes, and the brothers tangled with CBS over Pete Seeger's singing of "Waist Down in the Big Muddy." Ed Sullivan stayed popular by booking rock acts, such as the Beatles and the Rolling Stones, and Carol Burnett continued the delicious spoofing of film that *Your Show of Shows* had started. But for the most part, the cultural changes in the late 1960s and 1970s overtook the relevance of the variety form. *The Glen Campbell Goodtime Hour, The Sonny and Cher Comedy Hour, Tony Orlando and Dawn*—all shows featuring popular music stars with a youth culture following—achieved some popularity in the 1970s. *Rowan and Martin's Laugh-In*, a different type of variety program, prefigured the faster, more culturally literate and irreverant style that would survive, in limited form, into the 1990s. Clearly more oriented toward satire and sketch comedy than to the music-variety form of other programs, *Laugh-In* in its way recalled the inventiveness of *Your Show of Shows*.

Only one other show from the 1970s, with the focus on the youth demographic, has lasted into the 1990s—NBC's *Saturday Night Live*. This program, mainly emphasizing satirical comedy and featuring a different host and musical guest or group every week, captured the teen, college, and young adult crowd with a late-night airing (11:30 P.M. Eastern and Pacific time). Although periodically critics cry for its demise as the quality of writing waxes and wanes, the show has created film and television stars out of many of its regular performers. Although this network variety show hangs on into the 1990s, the lack of the genre on television despite the proliferation of cable channels, perhaps suggests its permanent eclipse. Now, the viewer with a remote control can create his or her own variety show, switching from stand-up comedy on A and E or the Comedy Central to ballet and opera on PBS or Bravo, from rock and roll on MTV to country music on the Nashville Network (TNN).

—Mary Desjardins

FURTHER READING

Krolik, Richard. "Here's to the Musical Variety Show!" *Television Quarterly* (New York), Spring 1992.

Marc, David. *Comic Visions: Television Comedy and American Culture*. Boston, Massachusetts: Unwin Hyman, 1989.

———. "Carol Burnett: The Last of the Big-time Comedy-variety Stars." *Quarterly Review of Film and Video* (Chur, Switzerland), July 1992.

Shulman, Arthur, and Roger Youman. *How Sweet It Was.* New York: Bonanza, 1966.

Spector, Bert. "A Clash of Cultures: The Smothers Brothers vs. CBS Television." In, O'Connor, John E., editor. *American History/American Television: Interpreting the Video Past.* New York: Ungar, 1987.

Spigel, Lynn. *Make Room for TV: Television and the Family Ideal in Postwar America.* Chicago: University of Chicago Press, 1992.

Wertheim, Arthur Frank. "The Rise and Fall of Milton Berle." In, O'Connor, John E., editor. *American History/American Television: Interpreting the Video Past.* New York: Ungar, 1987.

Wilk, Max. *The Golden Age of Television: Notes from the Survivors.* New York: Delta, 1976.

See also Burnett, Carol; *Carol Burnett Show; Ed Sullivan Show; Original Amateur Hour;* Sullivan, Ed; Special/Spectacular

VICTORY AT SEA

U.S. Compilation Documentary

Victory at Sea, a 26-episode series on World War II, represented one of the most ambitious documentary undertakings of early network television. The venture paid handsomely for NBC and its parent company RCA, however, in that it generated considerable residual income through syndication and several spinoff properties. It also helped establish compilation documentaries, programs composed of existing archival footage, as a sturdy television genre.

The series premiered on the last Sunday of October 1952, and subsequent episodes played each Sunday afternoon through May 1953. Each half-hour installment dealt with some aspect of World War II naval warfare and highlighted each of the sea war's major campaigns: the Battle of the North Atlantic, the attack on Pearl Harbor, the Battle of Midway, antisubmarine patrol in the South Atlantic, the Leyte Gulf campaign, etc. Each episode was composed of archival footage originally accumulated by the United States, British, Japanese or German navies. The footage was carefully edited and organized to bring out the drama of each campaign. That drama was enhanced by the program's sententious voice-over narration and by Richard Rogers's stirring musical score.

Victory at Sea won instant praise and loyal viewers. Television critics greeted it as breakthrough for the young television industry: an entertaining documentary series that still provided a vivid record of recent history. *The New York Times* praised the series for its "rare power"; *The New Yorker* pronounced the combat footage "beyond compare"; and *Harper's* proclaimed that "*Victory at Sea* [has] created a new art form." It eventually garnered 13 industry awards, including a Peabody and a special Emmy.

The project resulted from the determination of its producer Henry Salomon and from the fact the NBC was in a position to develop and exploit a project in compilation filmmaking. Salomon had served in the U.S. Navy during the war and was assigned to help historian Samual Eliot Morison write the Navy's official history of it combat operations. In that capacity, Salomon learned of the vast amounts film footage the various warring navies had accumulated. He left military service in 1948, convinced that the footage could be organized into a comprehensive historical account of the conflict. He eventually broached the idea to his old Harvard classmate Robert Sarnoff, who happened to be the son of RCA Chairman David Sarnoff and a rising executive in NBC's television network. The younger Sarnoff was about to take over the network's new Film Division as NBC anticipated shifting more of its schedule from live to filmed programming. A full documentary series drawn entirely from extant film footage fit perfectly with plans for the company's Film Division.

Production began in 1951 with Salomon assigned to oversee the enterprise. NBC committed the then-substantial sum of $500,000 to the project. Salomon put together a staff of newsreel veterans to assemble and edit the footage. The research took them to archives in North America, Europe, and Asia through 1951 and early 1952. Meanwhile Salomon received the full cooperation of U.S. Navy, which expected to receive beneficial publicity from the series. The crew eventually assembled 60 million feet of film, roughly 11,000 miles. This was eventually edited down to 61,000 feet. Salomon scored a coup when musical celebrity Richard Rogers agreed to compose the program's music. Rogers was fresh from several Broadway successes, and his name added prestige to the entire project. More important, it offered the opportunity for NBC's parent company RCA to market the score through its record division.

When the finished series was first broadcast, it did not yet have sponsorship. NBC placed it in the line-up of cultural programs on Sunday afternoon. The company promoted it as a high-prestige program, an example of history brought to life in the living room through the new medium of television. In so doing, the company was actually preparing to exploit the program in lucrative residual markets. As a film (rather than live) production, it could be rebroadcast indefinitely. And the fact that *Victory at Sea* dealt with a historical subject meant that its information value would not depreciate as would a current-affairs documentary.

Victory at Sea went into syndication in May 1953 and enjoyed a decade of resounding success. It played on 206

Victory at Sea

local stations over the course of ten years. It had as many as 20 reruns in some markets. This interest continued through the mid-1960s when one year's syndication income equalled the program's entire production cost. NBC also aggressively marketed the program overseas. By 1964, *Victory at Sea* had played in 40 foreign markets. Meanwhile, NBC recut the material into a 90-minute feature. United Artists distributed the film theatrically in 1954, and it was subsequently broadcast in NBC's prime-time schedule in 1960 and 1963. The Richard Rogers score was sold in several record versions through RCA-Victor. By 1963, the album version had grossed four million dollars, and one tune from the collection, "No Other Love," earned an additional $500,000 as a single.

The combination of prestige and residual income persuaded NBC to make a long-term commitment to the compilation documentary as a genre. NBC retained the *Victory at Sea* production crew as Project XX, a permanent production unit specializing in prime-time documentary

specials on historical subjects. The unit continued its work through the early 1970s, producing some 22 feature-length documentaries for the network.

Victory at Sea demonstrated the commercial possibilities of compilation documentaries to other networks as well. Such programs as *Air Power* and *Winston Churchill: The Valiant Years* directly imitated the *Victory at Sea* model, and the success of CBS' long-running historical series *The 20th Century* owed much to the example set by Salomon and his NBC colleagues. The fact that such programs still continue to play in syndication in the expanded cable market demonstrates the staying power of the compilation genre.

—Vance Kepley Jr.

NARRATOR
Leonard Graves

PRODUCER   Henry Salomon

Victory at Sea

MUSIC COMPOSER
Richard Rogers

PROGRAMMING HISTORY   26 Episodes

• NBC
October 1952–April 1953   Sunday Non-Prime time

FURTHER READING
Bluem, William. *Documentary in American Television.* New York: Hastings, 1965.
Kepley, Vance, Jr. "The Origins of NBC's Project XX in Compilation Documentaries." *Journalism Quarterly* (Urbana, Illinois), 1984.
Leyda, Jay. *Films Beget Film.* New York: Hill and Wang, 1964.

See also Music on Television; National Broadcasting Company; Sarnoff, Robert; War on Television

VIDEO EDITING

Three broad historical phases characterize the development of video editing that followed: physical film/tape cutting, electronic transfer editing, and digital non-linear editing. Even before the development of a successful videotape recording format in 1956 (the Ampex VR-1000), time zone requirements for national broadcasting required a means of recording and transporting programs. Kinescopes, filmed recordings of live video shows for delayed west coast airing, were used for this practice. Minimal film editing of these kinescopes was an obligatory part of network television.

Once videotape found widespread use, the term "stop-and-go recording" was used to designate those "live" shows that would be shot in pieces then later edited together. Physically splicing the two-inch quad videotape proved cumbersome and unforgiving, however, and NBC/Burbank developed a system in 1957 that used 16mm kinescopes—not for broadcasting—but as "work-prints" to rough-cut a show before physically handling the videotape. Audible cues on the film's optical sound track allowed tape editors to match-back frame for frame each cut. Essentially, this was the first "offline" system for video. Known as ESG, this system of rough-cutting film and conforming on tape (a reversal of what would become standard industry practice in the 1990s), reached its zenith in 1968 with *Rowan and Martin's Laugh-In.* That show required 350 to 400 tape splices and 60 hours of physical splicing to build up each episode's edit master.

A cleaner way to manipulate prerecorded video elements had, however, been introduced in 1963 with Ampex's all electronic "Editec." With VTRs (videotape recorders) now controlled by computers, and in- and out-points marked by audible tones, the era of electronic "transfer editing" had begun. Original source recordings were left unaltered, and discrete video shots and sounds were re-recorded in a new sequence on a second generation edit master. In 1967, other technologies added options now commonplace in video editing studios. Ampex introduced the HS-100 videodisc recorder (a prototype for now requisite slow motion and freeze frame effects) that was used extensively by ABC in the 1968 Olympics. "Helical-scan" VTRs (which threaded and recorded tape in a spiral pattern around a rotating head) appeared at the same time, and ushered in a decade in which technological formats were increasingly miniaturized (enabled in part by the shift to fully transitorized VTRs like the RCA TR-22 in 1961). New users and markets opened up with the shift to helical: educational, community activist, and cable cooperatives all began producing on the half-inch EIAJ format that followed; producers of commercials and industrial video made the three-quarter inch U-matic format pioneered by Sony in 1973 its workhorse platform for nearly two decades; newsrooms jettisoned 16mm newsfilm (along with its labs and unions) for the same videocassette-based format in the late 1970s; even networks and affiliates replaced venerable two-inch quad machines with one-inch helical starting in 1977.

The standardization of "time-code" editing, more than any other development, made this proliferating use viable. Developed by EECO in 1967, time-code was awarded an Emmy in 1971, and standardized by SMPTE shortly thereafter. The process assigned each video frame a digital "audio address," allowed editors to manage lists of hundreds of shots, and made frame accuracy and rapidly cut sequences a norm. The explosive growth of non-network video in the 1970s was directly tied to these and other refinements in electronic editing.

Nonlinear digital editing, a third phase, began in the late 1980s both as a response to the shortcomings of electronic transfer editing, and as a result of economic and institutional changes (the influence of music video, and the merging of film and television). To "creative personnel" trained in film, state-of-the-art online video suites had become little more than engineering monoliths that prevented "cutting-edge" directors from working intuitively. In linear time-code editing, for example, changes made at minute 12 of a program meant that the entire program after that point had to be re-edited to accomodate the change in program duration. Time code editing, which made this possible, also

essentially "quantified" the process, so that the "art" of editing meant merely managing "frame in/out" numbers for shots on extensive edit decision lists (EDLs). With over 80% of primetime television still shot on film by the end of the 1980s, the complicated abstractions and obsolescence that characterized these linear video formats also meant that many Hollywood television producers simply preferred to deliver programs to the networks from film prints—cut on flatbeds and conformed from negatives. The capital intensive nature of video post-production, also segregated labor in the suites. Directors were clients who delegated edit rendering tasks to house technicians and DVE artists. Online linear editing was neither spontaneous nor user-friendly.

Nonlinear procedures rejected videotape entirely and attacked the linear "straight-jacket" on several fronts. Systems were developed to "download" or digitize (rather than record) film/video footage onto video discs (CMX 6000) or computer hard-drive arrays (Lightworks, the Cube). This created the possible of random access retrieval as an "edited" sequence. Yet nonlinear marked an aesthetic and methodological shift as much as a technological breakthrough. Nonlinear technologies desegregated the editing crafts; synthesized post-production down to the "desktop" level, the personal computer scale; allowed users to intervene, rework, and revise edited sequences without recreating entire programs; and enabled editors to render and recall for clients at will numerous stylistic variations of the same show. Directors and producers now commonly did their own editing—in their own offices. The trade journals marvelled at the Avid's "32 levels of undo," the ability to restore extensive changes to various previous states. Nothing was locked in stone.

This openness allowed for a kind of presentational and formal "volatility" perfectly suited for the stylistic excesses that characterized contemporary television in the late 1980s and 1990s. When systems like the Avid and the Media 100 were upgraded to "online" mastering systems in the 1990s—complete with on-command digital video effects—the anything-can-go-anywhere premise made televisual embellishment an obligatory user challenge. The geometric growth of hard-disc memory storage, the pervasive paradigm of desktop publishing, and the pressure to make editing less an engineering accomplishment than a film artist's intuitive statement sold nonlinear procedures and technologies to the industry.

Video editing faces a trajectory far less predictable than that in the 1950s, when an industrial-corporate triumvirate of Ampex/RCA/NBC controlled technology and use. The future is open largely because editing applications have proliferated far beyond those developed for network oligopoly. Video is everywhere. Nonlinear established its beachhead in the production of commercials and music videos, not in network television. Still, by 1993 mainstream ATAS (Academy of Television Arts and Sciences) had lauded Avid's nonlinear system with an Emmy. By 1995, traditional television equipment manufacturers like Sony, Panasonic, and Grass Valley were covering their bets by selling user-friendly, non-linear Avid-clones even as they continued slugging it out over digital tape-based electronic editing systems. At the same time, program producing factories like Universal/MCA Television continued to use a wide range of editing systems for their series—film, linear, and nonlinear.

Hollywood's obsession with "digital interactivity" in the 1990s, means that sequencing video imagery in "post-production" will remain central to the fabrication of entertainment "software." Storage formats (film, tape, video disc) will, clearly, continue to change. Yet industry forays into the "information superhighway," now suggest a prototype for interactive editing that is closer in spirit to television's historic paradigm of multi-source "switching." Many now envision the "video server"—networked by wide bandwidth fiber-optic cable—as a bottomless, digitized, motion picture storage pit, as an image-sound repository that does not even need to reside in the sequencing platform of the digital video editor. If this server-network model survives, the role of the nonlinear digital editor might then stand as the very model

for all video-on-demand consumers in the domestic sphere as well. Viewers will become their own editors.

—John Thornton Caldwell

FURTHER READING

Anderson, Gary H. *Video Editing and Postproduction.* White Plains, New York: Knowlege Industry Publications, 1988.

Browne, Stephen E. *Videotape Editing: A Postproduction Primer.* Boston: Focal, 1989.

Caldwell, John Thornton. *Televisuality: Style, Crisis, and Authority in American Television.* New Brunswick, New Jersey: Rutgers University Press, 1995.

Scheieder, Arthur. *Electronic Post-Production and Videotape Editing.* Boston: Focal, 1989.

Zettl, Herb. *Television Production Handbook.* Belmont, California: Wadsworth Press, 1992.

See also Computers in Television; Videotape

VIDEOCASSETTE

In 1956, the Ampex company announced that it had developed a new device: the videotape machine. This large reel-to-reel tape machine used four record heads (and was for this reason given the name "quad") and two-inch wide tape. The invention was quickly embraced by the broadcasting community, and on 30 November 1956, CBS broadcast the first program using videotape. Videotape is very similar in composition to audiotape. Most videotape consists of a Mylar backing, a strong, flexible plastic material, that provides a base for a thin layer of ferrous oxide. This oxide is easily magnetized and is the substance that stores the video and audio information.

In 1969, Sony introduced its EIAJ-standard three-quarter-inch U-Matic series, a videocassette system. Although there were earlier attempts to establish a standard cassette or cartridge system, the U-Matic format was the first to become solidly accepted by educational and industrial users. Similar in construction and function to the audiocassette, the videocassette is a plastic container in which a videotape moves from supply reel to take-up reel, recording and playing back short program segments through a videocassette recorder (VCR). This form of construction emerged as a distinct improvement on earlier, reel-to-reel videotape recording and playback systems. The cassette systems, especially after they were integrated with camera and sound systems, enabled ease of movement and flexible shooting arrangements. The new devices helped create a wave of video field production ranging from what is now known as "electronic news gathering" to the use of video by political activist groups, educators, and home enthusiasts.

This last group was always perceived by video hardware manufacturers as a vast opportunity for further sales. After several abortive attempts to establish a consumer market with a home cartridge or cassette system, Sony finally succeeded with its Betamax format. Sony's success with

Betamax was followed closely by other manufacturers with VHS (the "video home system"), a consumer-quality 1/2-inch videocassette system introduced by JVC. Although the VHS format still dominates the home entertainment field, several competing formats are vying for both the consumer market and the professional field. The greatly improved Super-VHS (S-VHS) format has technical specifications that equal broadcast and cable TV quality. The S-VHS system is in turn being challenged by two 8mm cassette formats—Video 8 (a consumer-grade video format developed by Sony that uses eight-millimeter-wide tape) and Hi8 (an improvement on Sony's Video-8 format that uses metal

Various professional quality videocassettes
Photo courtesy of 3M

particle tape and a higher luminance bandwidth). Other formats that are competing for the professional market include the 1/2-inch Betacam and Betacam SP systems, the 1/2-inch M-formats (M and M-II), 3/4-inch U-matic SP, and the even more recent digital formats (D-1 and D-2).

It is safe to say that the development of videocassette systems has transformed many aspects of televisual industries and more general experience with television. The innovations within news services, the rapid expansion of home video systems that transformed the financial base of the film industry, and the acceptance of "video" as an everyday aspect of contemporary experience all rely to a great extent on the videocassette.

—Eric Freedman

FURTHER READING

Browne, Steven E. *Video Editing: A Postproduction Primer.* Boston: Focal, 1989; 2nd edition, 1993.

Burrows, Thomas D., Donald N. Wood, and Lynne Schafer Gross. *Television Production: Disciplines and Techniques.* Dubuque, Iowa: Brown, 1978; 5th edition, 1992.

Zettl, Herbert. *Television Production Handbook.* London: Pitman, 1961; 5th edition, Belmont, California: Wadsworth, 1992.

See also Ancillary Markets; Betamax Case; Camcorder; Home Video; Sony Corporation; Videotape

VIDEODISC

Videodiscs are records that play high-fidelity sound and pictures through conventional television receivers. The dominant videodisc technology is the LaserDisc (LD), a replay-only video disc system based upon the same laser-read optical disc technology used by the compact disc digital audio format. LaserDisc has also been referred to by the terms LaserVision, DiscoVision, and CD-Video. The competing format, known as Capacitance Electronic Disc (CED), has become obsolete in the home market.

Videodiscs produce a picture with 400 horizontal lines, imparting a clearer, sharper image than the 240 lines displayed by conventional videotape. The LaserVision system has two speeds: Constant Linear Velocity discs play for 60 minutes per side, and Constant Angular Velocity discs play for only 30 minutes per side. Both CLV and CAV discs can be played on all LV machines, but the CLV format does not support freeze-frame and other special effects. The obsolete CED system employed discs with a capacity of one hour per side.

The first consumer videodisc players were developed in Britain during the late 1920s by John Logie Baird. Baird's system, known as "phonovision", had only 30 lines of resolution. The capacitance system was developed in the 1960s and was used in commercial broadcasting applications prior to the development of videotape. Capacitance systems were able to play full bandwidth images by means of a stylus riding in the grooves of the videodisc that translated variations in electrical capacitance into video and audio signals. Laser optical disc technology, which uses a laser beam rather than a stylus to play back sound and video images, was developed jointly by MCA and N.V. Philips in the early 1970s. Their collaboration resulted in the DiscoVision system under the Magnavox label.

Initially, videodisc players failed to be widely adopted by consumers. This was due in part to the small number of prerecorded titles offered that could be played on the systems, and in part to the competing technology of video cassette players which allowed consumers the additional ability to record video as well as play back prerecorded products. Their recent resurgent popularity may be traced to improvements in videodisc players, large screen television sets, and improved home sound systems, as well as by increased demand by consumers for a better quality picture. Film buffs and collectors are also attracted to the longer product life of videodiscs: since the audio and video information is protected under an acrylic shield and no stylus or head makes physical contact with the laser disc surface, it is less subject to wear and tear than conventional videocassettes. There are just under one million videodisc players in home use in the United States (compared to 85,000,000 VCRs), and over 2 million units in Japan.

Although the LD videodisc's shiny acrylic surface resembles that of digital audio compact discs (CDs), the laser disc differs in that it may be encoded with both analog and digital data. Most videodisc players are thus able to playback both digital and analog sound. Many videodiscs released now incorporate a digital-audio soundtrack which uses exactly the same standard used by compact discs. However, most players still support the analog soundtracks of older discs released from between 1978 and 1986. Many LD format videodisc players are also able to play conventional digital audio compact discs. Videodisc players often have features similar to those found on CD players, such as track numbers (known as "chapters" in videodisc terminology), real-time counters and rapid random access or direct access to any chapter on the side by chapter number or by specific time.

In comparison with North America and Japan, the market for videodisc players in PAL territories is still small. Since videodisc players must store a complete video signal, they are engineered in accordance with one of two incompatible formats, either the 525-line NTSC system or the 625-line PAL system. Fortunately the further sub-division of the 625-line based systems into the PAL and SECAM color systems was avoided: color information that is recorded on the disc in the PAL color system is internally decoded into SECAM by players meant for this market.

NTSC videodisc players are the norm in many Southeast Asian countries despite their use of the PAL standard for local television broadcasts. However, domestic videodisc releases for the PAL color system are made in Europe and Australia, but this small market is not well developed and fewer titles are available on disc. This has encouraged the videodisc player manufacturers to begin the production of dual standard players.

—Aviva Rosenstein

FURTHER READING

Lenk, John D. *Lenk's Laser Handbook: Featuring CD, CDV, and CD-ROM Technology.* New York: McGraw Hill, 1992.

Pratt, Douglas. *The Laser Video Disc Companion.* New York: Zoetrope, 1988.

Rovin, Jeff. *The Laserdisc Film Guide: Complete Ratings for the Best and Worst Movies Available on Disc.* New York: St. Martin's, 1993.

VIDEOTAPE

By the late 1990s videotape was familiar to most television viewers in developed countries. The videocassette was a central product throughout the home video market and in various formats was widely used as a consumer item for home recording. Despite these widespread and common uses, however, videotape is of relatively recent origin. Its immediate antecedent is, of course, audiotape.

The processes of recording audiotape and videotape work on the same principle. An audio or video recording head is a small electromagnet containing two coils of wires separated by a gap. An electrical current passing through the wires causes a magnetic charge to cross the gap. When tape, coated with metal particles, passes through the gap, patterns are set on the material. On audiotape, each syllable, musical note, or sneeze sets down its own distinct pattern. For videotape, which carries several hundred times as much information as audiotape, each image has its own pattern.

In the late 1940s and early 1950s the explosive growth of television created an enormous demand for a way to record programs. Until links could be established through television lines or microwave broadcast relay, a blurry kinescope was the only means by which a network program could be recorded and replayed on different local television stations. As a result, "television" programs were unstable, ephemeral events. Once transmitted electronically they were, for the most part, lost in time and space, unavailable for repeated use as either aesthetic, informational, or economic artifacts.

In 1951, engineers at Bing Crosby Enterprises demonstrated a black-and-white videotape recorder that used one-inch tape (tape size refers to tape width) running at 100 inches per second. At that rate a reel of tape three feet in diameter held about fifteen minutes of video. Crosby continued to fund the research, driven not only by a sense of commercial possibilities for videotape, but reportedly also by his wish to record television programs so that he could play golf without being restricted to live performances. Two years later RCA engineers developed a recorder which reproduced not only black-and-white but color pictures. However, tape ran past the heads at a blinding 360 inches per second, which is 20 miles per hour. Neither machine produced pictures of adequate quality for broadcast. It simply was not possible to produce a stable picture at such a high tape speed.

During this same period, Ampex, a small electronics firm in California, was building a machine on a different principle, spinning the recording head. They succeeded in 1956 with a recorder the size of two washing machines. Four video heads rotated at 14,400 revolutions per minute, each head recording one part of a tape that was two inches wide. One of the engineers on the project, Ray M. Dolby, later became famous for his tape noise reduction process.

The quality of Ampex recordings was such an improvement over fuzzy kinescope images that broadcasters who saw the first demonstration, presented at a national convention, actually jumped to their feet to cheer and applaud. The television industry responded so enthusiastically that Ampex could not produce machines fast enough. It was the true beginning of the video age.

West Coast television stations could now, without sacrificing picture quality, delay live East Coast news and entertainment broadcasts for three hours until evening prime time, when most viewers reached their homes after work. By 1958 the networks were recording video in color and by 1960 a recorder was synchronized with television studio electronics for the familiar film editing techniques of the "dissolve" and "wipe."

Large "two-inch" reel-to-reel Ampex machines survived for a generation before they were replaced by more compact and efficient "one-inch" reel-to-reel machines and "three-quarter-inch" cassette machines. By 1990 most of the bigger recorders had been retired.

While American companies were manufacturing two-inch, four-head, quadruplex scan machines, Japanese engineers were building the prototype of a helical scan machine that employed a single spinning head. Toshiba introduced the first helical scan VTR machine in 1959. JVC soon followed. The picture quality produced by these machines would remain inferior to "quad" machines for another ten years, unsuitable for the broadcast industry. But the smaller, more "user-friendly" helical scan machines, costing a fraction of the price of larger machines, quickly dominated the industrial and educational markets.

In 1972 Sony introduced the "Port-a-pak" black-and-white video recorder, weighing less than 10 pounds. The tape had to be threaded by hand, but the "Port-a-Pak" was an important step on the way to electronic news gathering, known in the television industry as ENG. The next big step, Sony's U-matic three-quarter inch tape machine which played tape cassettes, eliminated physical handling of tape. CBS-TV News sent a camera team equipped with an Ikegami video camera and a U-Matic tape recorder to cover President Richard Nixon's trip to Moscow. News stories were soon being microwaved back to stations for taping or live feeds. Prior to these developments the visual portion of news broadcasts had been produced on film. Videotape was the far superior medium for news. It needed no developing time, was reusable, and was more suited to the television's sense of immediacy. With the coming of videotape, television news editors replaced razor blades with electronic editing devices.

With broadcasting, educational and industrial markets in hand, Japanese video companies turned their attention to the potentially vast home market. Hobbyists had already shown the way. With slightly modified portable reel-to-reel machines, they were taping television programs at home to play again later.

Sony, whose research was led by Nobutoshi Kihara, had considered the home market from the start. Recognizing that not only television stations but viewers ought to be able to time-shift programs, Sony president Akio Morita said, "People do not have to read a book when it's delivered. Why should they have to see a TV program when it's delivered?" Sony introduced its half-inch Betamax machine in 1975. A year later rival Japanese companies, led by JVC, brought out VHS machines, a format incompatible with Betamax. VHS gradually captured the home market. People at home could simply and inexpensively record television programs, and could buy or rent tapes. At last it was possible to go to the movies without leaving home.

Tape renting began when businessman Andre Blay made a deal to buy cassette production rights to fifty Twentieth Century-Fox movies. Blay discovered that few customers wanted to buy his tapes, but everyone wanted to rent them.

The motion picture industry considered the videodisc a better way to bring a movie into the home, pointing out its sharper picture image, stereo sound, lower cost, and copy protection. However, the public wanted recording capability, not so much to copy rented films illegally as to record movies and television programs off the air for later playback. Videodisc players could not match the flexibility of videocassette recorders for time-shifting. In the battle over competing disc and tape formats, VHS tapes emerged the clear winner.

The simplicity, flexibility, low cost, and high quality of tape technology created new worlds of visual production. In the final decade of the twentieth century, one hundred years after motion pictures were invented, millions of users could

"make a movie." Video cameras found their way into schools as learning tools. The high-school library is now often referred to as "the media center," and the video yearbook has joined the printed version. Even in elementary schools, curious fingers are pushing camera buttons.

Videotape has also introduced specific changes at a very different level, expanding the production community in the professional arena. It is possible to produce a motion picture of technically acceptable quality at modest cost. The phrase desktop video has become part of our language, often in relation to desktop publishing.

Videotape has had wide impact everywhere on earth, including remote villages, where inexpensive tapes bring information and entertainment. A truck carrying a videotape player, a television set, and a portable generator is not an uncommon sight in many parts of the world. Peoples living as far from urban centers as the Kayapo of the Brazil rain forest and the Inuit of northern Canada have been introduced to video, and have themselves produced tapes to argue for political justice and to record their cultural heritage.

Several Third World governments have actively promoted videotape programs for adult education. For example, the Village Video Network in several countries provides an exchange for tapes on such subjects as farming, nutrition, and population control. International groups have given some villages video cameras and training to produce their own films, which are later shown to other villages.

Another result of video diffusion has been a widening of video journalism capability. The taping of the Rodney King beating was just one example of how ordinary citizens are making a difference not only in news coverage but in the course of events. The potential for a "video vigilantism" by "visualantes" has not gone unnoticed, with its effects not only on journalism but on law enforcement itself.

Far less significant uses of videotape technology have also developed. Replacing the traditional matchmaker, for example, is the video dating club. Participants tell a video camera of their interests, their virtues, and the type of person they would like to meet. They look at other videotapes and their videotape is shown to prospects.

Serious social and legal problems are also directly related to the easy use of this technology. Video piracy is rampant. A vast underground network feeds millions of illegal copies of videotape movies throughout the world. The national film industries of a number of countries have been battered both by the pirating of their own films and by the influx of cheap illegal copies of Western films.

Some of these issues may be resolved with the development of still newer technologies. For both the video and computer industries, the future of information storage and retrieval may lie not with tape but with such optical media as CD-ROM and CD-I, which offer the advantages of high density, random access and no physical contact between the storage medium and the pickup device. As with the earlier videotape "revolution," the television and film industries are now shifting their investments and altering their industrial practices to deal with the newer, digitally-based devices. The results of these changes for consumers, educators, and journalists are not easily predicted, yet there is no question but that all these groups will experience alteration in media use akin to that caused by the introduction of videotape.

—Irving Fang

FURTHER READING

Alvarado, Manuel. *Video World-Wide.* Paris: UNESCO, 1988.

Dobrow, Julia R., editor. *Social and Cultural Aspects of VCR Use.* Hillsdale, New Jersey: Lawrence Erlbaum, 1990.

Ennes, Harold E. *Television Broadcasting: Tape Recording Systems.* Indianapolis, Indiana: Howard W. Sams, 1979.

Lardner, James. *Fast Forward.* New York: W.W. Norton, 1987.

Levy, Mark, editor. *The VCR Age.* Newbury Park, California: Sage, 1989.

See also Betamax Case; Home Video; Reruns/Repeats; Sony Corporation; Video Editing; Videocassette

VIDEOTEXT/ONLINE SERVICES

The term "videotext" refers to any interactive electronic system which allows users to send and receive data from either a personal computer or a dedicated terminal. The term "videotext" is often used interchangeably with appellations such as "online service" or "interactive network." Videotext systems deliver information and transactional services such as banking and shopping. These systems differ from broadcast media delivery systems due to the special qualities of interactivity engendered by the technology which allow the user to personalize his media use rather than act as a passive member of an aggregate audience.

Traditionally, videotext systems displayed information only in text format, but as color monitors became more commonplace during the early 1990s, these services began to offer graphical user interfaces (GUIs) which incorporated sound and visually striking computer graphic displays. Although users connected to early videotext systems on dedicated terminals, most online services can now be accessed by the user via a phone line and a personal computer equipped with a modem or Ethernet connection. Videotext users may pay a per-use charge or a monthly subscription fee to access the service.

The first videotext systems were developed in Europe in the 1970s by government-owned telephone companies. The world's largest videotext service is the French Teletel system, which boasts approximately eight million users. This system was launched in the early 1980s as part of an economic plan aimed at making France a leader in information technology. Free "Minitel" terminals were distributed by the French government (in lieu of paper versions of phone directories) and the service was promoted widely as a matter of public policy. Smaller videotext systems in Italy, Ireland, and other countries have made use of the French technology, whereas Germany, Japan, Korea, Britain and other nations have chosen to develop their own videotext technologies.

In the United States, videotext systems were initially launched by the newspaper publishers who made news and advertisements through special terminals hooked up to television monitors, but most of these services met with little commercial success. However, the increased diffusion of personal computers into the home eventually enabled consumer oriented videotext systems to succeed in the mass marketplace. By the mid-1990s, more than four million households had subscribed to one or more of the largest consumer-oriented U.S. videotext systems: America Online, Prodigy, CompuServe, and Genie. Currently these providers are incorporating gateways to Internet applications within their services, including World Wide Web browsers, Usenet newsgroups and electronic mail.

Television broadcasters are making increasing use of online information services to promote their programming; furthermore, several information services aimed at providing services related to the broadcasting industry have been sponsored by the major service providers mentioned above. On CompuServe, for example, users may access the Hollywood Hotline, which provides news and information about the entertainment industry, or they may obtain daily summaries of soap operas or printed transcripts of selected television shows. The CompuServe Broadcast Professionals Area contains information about publications and trade associations related to broadcast engineering, programming and television production. On America Online several networks and cable services have sponsored areas where fans can get information, register their opinions, or obtain sound samples or photos from their favorite programs.

Services provided by videotext fall into one of three areas: information retrieval services such as obtaining stock prices or weather forecasts, transactional message services which enable the purchasing of merchandise over the network, and interpersonal message exchanges which may include conferencing, chat channels or electronic mail. This last application has been the most successful, indicating that consumers are more interested in using the services to talk to other people than to retrieve information.

New developments in broadband television delivery, with its ability to display high-quality video and its incorporation of stereo sound, has encouraged some developers to experiment with providing videotext services over interactive TV systems.

—Aviva Rosenstein

FURTHER READING

Branscomb, Anne W. "Videotext: Global Progress and Comparative Policies." *Journal of Communication* (New York), Winter 1988.

Cutler, Blayne. "The Fifth Medium." *American Demographics* (Ithaca, New York), 1990.

Mayer, R. "The Growth of the French Videotex System and Its Implications for Consumers." *Journal of Consumer Policy* (Dordrecht, Holland), 1988.

Mosco, Vincent. *Pushbutton Fantasies: Critical Perspectives on Videotex and Information Technology.* Norwood, New Jersey: Ablex, 1982.

VIETNAM: A TELEVISION HISTORY

U.S. Compilation Documentary

Vietnam: A Television History was the most successful documentary produced by public television at the time it aired in 1983. Nearly 9% of all U.S. households tuned in to watch the first episode, and an average of 9.7 million Americans watched each of the 13 episodes. A second showing of the documentary in the summer of 1984 garnered roughly a 4% share in the five largest television markets.

Before it was aired in the United States, over 200 high schools and universities nationwide paid for the license to record and show the documentary in the classroom as a television course on the Vietnam War. In conjunction with this educational effort, the Asian Society's periodical, *Focus on Asian Studies*, published a special issue entitled, "Vietnam: A Teacher's Guide" to aid teachers in the use of this documentary in the classroom.

The roots of the documentary reach back to 1977 when filmmaker Richard Ellison and foreign correspondent Stanley Karnow first discussed the project. Karnow had been a journalist in Paris during the 1950s and a correspondent in French Indochina since 1959. Karnow and Ellison then signed on Dr. Lawrence Lichty, professor at the University of Wisconsin at the time, as director of media research to help gather, organize and edit media material ranging from audio and videotape and film coverage, to still photographs and testimonial. As a result, *Vietnam: A Television History* became a "compilation" documentary relying heavily on a combination of fixed moments (photographs, written text) as well as fluid moments (moving video and film).

The final cost of the project totaled approximately $4.5 million. At the time of its broadcast in 1983, it was one of the most expensive ventures ever undertaken by public television. While the initial funding came from WGBH-TV Boston and the National Endowment for the Humanities, the Corporation for Public Broadcasting refused financial support. Ellison and Karnow sought additional backing abroad, gaining support from Britain's Associated Television (later to become Central Independent Television). Co-production with French Television (Antenne-2) enabled access to important archives from the French occupation of the region. Antenne-2 produced the earliest episodes of the documentary, and Associated Television partially produced the fifth episode.

Karnow and Ellison saw the documentary as an opportunity to present both sides of the Vietnam War story, the American perspective and the Vietnamese perspective. Throughout the 1960s and 1970s, documentaries and films on the Vietnam War tended to look solely at American involvement and its consequences both at home and in the region. Karnow and Ellison sought a more comprehensive historical account that traced the history of foreign invasion and subsequent Vietnamese cultural development over sev-

Vietnam: A Television History
Photo courtesy of Wisconsin Center for Film and Theater Research

Vietnam: A Television History
Photo courtesy of Wisconsin Center for Film and Theater Research

eral hundred years. Both producers believed that to gain a more comprehensive view of Vietnam would enable the documentary to become a vehicle for reconciliation as well as reflection.

The series aired first in Great Britain to good reviews, although it did not receive the high ratings it achieved in the United States. At the time of its broadcast in the United States in the fall of 1983, the documentary received very positive reviews from *The New York Times*, *The Washington Post* and *Variety*. Furthermore, both *Time* magazine and *Newsweek* hailed the series as fair, brilliant, and objective.

Still, other critics of the documentary were less complimentary and viewed it as overly generous to the North Vietnamese. The organization Accuracy in Media (AIM) produced and aired a response to the documentary seeking to "correct" the inaccurate depiction of Vietnam in the series. PBS's agreement to air the two-hour show, entitled *Television's Vietnam: The Real Story*, was seen by many liberal critics as bowing to overt political pressure. PBS's concession to air AIM's response to the documentary (its own production) was rare, if unprecedented, in television history.

The controversy surrounding *Vietnam: A Television History* and the response to it, *Television's Vietnam: The Real Story*, raise the important question concerning bias in documentary production. Bias in the interpretation of historical events has fueled, and continues to fuel, rigourous debates among historians, politicians and citizens. The experience

Karnow and Ellison had in creating this documentary underscores the sense that the more "producers" involved in a project, the more difficult the task of controlling for bias becomes. The episodes prepared by the British and French teams were noticeably more anti-American in tone.

Despite the controversy, *Vietnam: A Television History* remains one of the most popular history documentaries used in educational forums. It inspired Stanley Karnow's best-selling book, *Vietnam: A History*, which was billed as a "companion" to the PBS series. The book also remains one of the top history texts used in college courses concerning the war and its controversy, both in the United States and around the world.

—Hannah Gourgey

FURTHER READING

Banerian, James, editor. *Losers Are Pirates: A Close Look at the PBS Series "Vietnam: A Television History."* Phoenix, Arizona: Sphinx, 1985.

Bluem, A. William. *Documentary in American Television: Form, Function, and Method.* New York: Hastings House, 1965.

Broyles, W. "Vietnam: A Television History." *Newsweek* (New York), 10 October 1983.

Henry, W. A. "Vietnam: A Television History." *Time* (New York), 3 October 1983.

Karnow, Stanley. *Vietnam: A History.* Middlesex, England: Penguin, 1983.

Vietnam: A Television History
Photo courtesy of Wisconsin Center for Film and Theater Research

Lichty, Lawrence. "Vietnam: A Television History: Media Research and some Comments." In, Rosenthal, Alan, editor. *New Challenges For Documentary.* Berkeley: University of California Press, 1988.

Maurer, Marvin. "Screening Nuclear War and Vietnam." *Society* (New Brunswick, New Jersey), November-December 1985.

McGrory, Mary. "The Strategy of Stubbornness and the Policy all too Familiar." *The Washington Post,* 22 December 1983.

O'Connor, John E. *Teaching with Film and Television.* Washington, D.C.: American Historical Association, 1987.

Renov, Michael, editor. *Theorizing Documentary.* New York, London: Routledge, 1993.

Rhodes, Susan, editor. "Vietnam: A Teacher's Guide." *Focus on Asian Studies* (New York), Fall 1983.

Springer, Claudia. "Vietnam: A Television History and the Equivocal Nature of Objectivity." *Wide Angle* (Athens, Ohio), 1985.

Toplin, Robert Brent. "The Filmmaker as Historian." *American Historical Review* (Washington, D.C.), December 1988.

Walkowitz, Daniel. "Visual History: The Craft of the Historian Filmmaker." *Public Historian* (Santa Barbara, California), Winter 1985.

See also Documentary; Vietnam on Television; War on Television

VIETNAM ON TELEVISION

Vietnam was the first "television war." The medium was in its infancy during the Korean conflict, its audience and technology still too limited to play a major role. The first "living-room war," as Michael Arlen called it, began in mid-1965, when Lyndon Johnson dispatched large numbers of U.S. combat troops, beginning what is still surely the biggest story television news has ever covered. The Saigon bureau was for years the third largest the networks maintained, after New York and Washington, with five camera crews on duty most of the time.

What was the effect of television on the development and outcome of the war? The conventional wisdom has generally been that for better or for worse it was an anti-war influence. It brought the "horror of war" night after night into people's living rooms and eventually inspired revulsion and exhaustion. The argument has often been made that any war reported in an unrestricted way by television would eventually lose public support. Researchers, however, have quite consistently told another story.

A reporter records battlefield activity in Vietnam for ABC News

There were, to be sure, occasions when television did deliver images of violence and suffering. In August 1965, after a series of high-level discussions which illustrate the unprecedented character of the story, CBS aired a report by Morley Safer which showed Marines lighting the thatched roofs of the village of Cam Ne with Zippo lighters, and included critical commentary on the treatment of the villagers. This story could never have passed the censorship of World War II or Korea, and it generated an angry reaction from Lyndon Johnson. In 1968, during the Tet offensive, viewers of NBC news saw Col. Nguyen Ngoc Loan blow out the brains of his captive in a Saigon street. And in 1972, during the North Vietnamese spring offensive, the audience witnessed the aftermath of errant napalm strike, in which South Vietnamese planes mistook their own fleeing civilians for North Vietnamese troops.

These incidents were dramatic, but far from typical of Vietnam coverage. Blood and gore were rarely shown. Just under a quarter of film reports from Vietnam showed images of the dead or wounded, most of these fleeting and not particularly graphic. Network concerns about audience sensibilities combined with the inaccessibility of much of the worst suffering to keep a good deal of the "horror of war"

off the screen. The violence in news reports often involved little more than puffs of smoke in the distance, as aircraft bombed the unseen enemy. Only during the 1968 Tet and 1972 Spring offensives, when the war came into urban areas, did its suffering and destruction appear with any regularity on TV.

For the first few years of the "living room war" most of the coverage was upbeat. It typically began with a battlefield roundup, written from wire reports based on the daily press briefing in Saigon—the "Five O'Clock Follies," as journalists called it—read by the anchor and illustrated with a battle map. These reports had a World War II feel to them—journalists no less than generals are prone to "fighting the last war"—with fronts and "big victories" and a strong sense of progress and energy.

The battlefield roundup would normally be followed by a policy story from Washington, and then a film report from the field—typically about five days old, since film had to be flown to the United States for processing. As with most television news, the emphasis was on the visual and above all the personal: "American boys in action" was the story, and reports emphasized their bravery and their skill in handling the technology of war. A number of reports directly countered Morley Safer's Cam Ne story, showing the burning of huts, which was a routine part of many search-and-destroy operations, but emphasizing that it was necessary, because these were Communist villages. On Thursdays, the weekly casualty figures released in Saigon would be reported, appearing next to the flags of the combatants, and of course always showing a good "score" for the Americans.

Television crews quickly learned that what New York wanted was "bang-bang" footage, and this, along with the emphasis on the American soldier, meant that coverage of Vietnamese politics and of the Vietnamese generally was quite limited. The search for action footage also meant it was a dangerous assignment: nine network personnel died in Indochina, and many more were wounded.

Later in the war, after Tet and the beginning of American troop withdrawals in 1969, television coverage began to change. The focus was still on "American boys," to be sure, and the troops were still presented in a sympathetic light. But journalists grew skeptical of claims of progress, and the course of the war was presented more as an eternal recurrence than a string of decisive victories. There was more emphasis on the human costs of the war, though generally without graphic visuals. On Thanksgiving Day 1970, for example, Ed Rabel of CBS reported on the death of one soldier killed by a mine, interviewing his buddies, who told their feelings about his death and about a war they considered senseless. An important part of the dynamic of the change in TV news was that the "up close and personal style" of television began to cut the other way: in the early years, when morale was strong, television reflected the upbeat tone of the troops. But as withdrawals continued and morale declined, the tone of field reporting changed. This shift was paralleled by developments on the "home front." Here, divisions over the war received increasing air time, and the anti-war movement, which had been vilified as Communist-inspired in the early years, was more often accepted as a legitimate political movement.

Some accounts of television's role regarding this war assign a key role to a special broadcast by Walter Cronkite wrapping up his reporting on the Tet Offensive. On 27 February 1968, Cronkite closed "Report from Vietnam: Who, What, When, Where, Why?" by expressing his view that the war was unwinnable, and that the United States would have to find a way out. Some of Lyndon Johnson's aides have recalled that the president watched the broadcast and declared that he knew at that moment he would have to change course. A month later Johnson declined to run for reelection and announced that he was seeking a way out of the war; David Halberstam has written that "it was the first time in American history a war had been declared over by an anchorman."

Cronkite's change of views certainly dramatized the collapse of consensus on the war. But it did not create that collapse, and there were enough strong factors pushing toward a change in policy that it is hard to know how much impact Cronkite had. By the fall of 1967, polls were already showing a majority of Americans expressing the opinion that it had been a "mistake" to get involved in Vietnam; and by the time of Cronkite's broadcast, two successive secretaries of defense had concluded that the war could not be won at reasonable cost. Indeed, with the major changes in television's portrayal of the war still to come, television was probably more a follower than a leader in the nation's change of course in Vietnam.

Vietnam has not been a favorite subject for television fiction, unlike World War II, which was the subject of shows ranging from action-adventure series like *Combat* to sitcoms like *Hogan's Heroes*. During the war itself it was virtually never touched in television fiction—except, of course, in disguised form on *M*A*S*H*. After Hollywood scored commercially with *The Deer Hunter* (1978), a number of scripts were commissioned, and NBC put one pilot, *6:00 Follies*, on the air. All fell victim to bad previews and ratings, and to political bickering and discomfort in the networks and studios. Todd Gitlin quotes one network executive as saying, "I don't think people want to hear about Vietnam. I think it was destined for failure simply because I don't think it's a funny war." World War II, of course, wasn't any funnier. The real difference is probably that Vietnam could not plausibly be portrayed either as heroic or as consensual, and commercially successful television fiction needs both heroes and a sense of "family" among the major characters.

An important change did take place in 1980, just as shows set in Vietnam were being rejected. *Magnum, P.I.* premiered that year, beginning a trend toward portrayals of Vietnam veterans as central characters in television fiction. Before 1980 vets normally appeared in minor roles, often portrayed as unstable and socially marginal. With *Magnum, P.I.* and later *The A-Team, Riptide, Airwolf* and others, the veteran emerged as a hero, and in this sense the war experience, stripped of the contentious backdrop of the war itself, became suitable for television. These characters drew their strength from their Vietnam experience, including a preserved war-time camaraderie which enabled them to act as a team. They also tended to stand apart from dominant social institutions, reflecting the loss of confidence in these institutions produced by Vietnam, without requiring extensive discussion of the politics of the war.

Not until *Tour of Duty* in 1987 and *China Beach* in 1988 did series set in Vietnam find a place on the schedule. Both were moderate ratings successes; they stand as the only major Vietnam series to date. The most distinguished, *China Beach*, often showed war from a perspective rarely seen in post-World War II popular culture: that of the women whose job it was to patch up shattered bodies and souls. It also included plenty of the more traditional elements of male war stories, and over the years it drifted away from the war, in the direction of the traditional concern of melodrama with personal relationships. But it does represent a significant Vietnam-inspired change in television's representation of war.

—Daniel C. Hallin

FURTHER READING

Anderegg, Michael A. *Inventing Vietnam: The War in Film and Television.* Philadelphia, Pennsylvania: Temple University Press, 1991.

Berg, Rick. "Losing Vietnam: Covering the War in an Age of Technology." In, Rowe, John Carlos, and Rick Berg, editors. *The Vietnam War and American Culture.* New York: Columbia University Press, 1991.

Braestrup, Peter. *Big Story: How the American Press and Television Reported and Interpreted the Crisis of Tet 1968 in Vietnam and Washington.* Boulder, Colorado: Westview Press, 1977.

Gibson, James William. "American Paramilitary Culture and the Reconstruction of the Vietnam War." In, Walsh, Jeffrey, and James Aulich, editors. *Vietnam Images: War and Representation.* New York: St. Martin's, 1989.

Hallin, Daniel C. *The "Uncensored War": The Media and Vietnam.* New York: Oxford University Press, 1986.

Hammond, William M. *Public Affairs: The Military and the Media, 1962-1968.* Washington, D.C.: Center of Military History, U.S. Army, 1988.

Heilbronn, Lisa M. "Coming Home a Hero: The Changing Image of the Vietnam Vet on Prime Time Television." *Journal of Popular Film and Television* (Washington, D.C.), Spring 1985.

Martin, Andrew. "Vietnam and Melodramatic Representation." *East-West Film Journal* (Honolulu, Hawaii), June 1990.

Rollins, Peter C. "Historical Interpretation or Ambush Journalism? CBS vs. Westmoreland in *The Uncounted Enemy: A Vietnam Deception.*" *War, Literature, and the Arts* (U.S. Air Force Academy, Colorado Springs, Colorado), 1990.

———. "The Vietnam War: Perceptions through Literature, Film, and Television." *American Quarterly* (Washington, D.C.), 1984.

Rowe, John Carlos. "'Bringing it all back Home': American Recyclings of the Vietnam War." In, Armstrong, Nancy, and Leonard Tennenhouse, editors. *The Violence of Representation: Literature and the History of Violence.* London: Routledge, 1989.

———. "From Documentary to Docudrama: Vietnam on Television in the 1980s." *Genre* (Norman, Oklahoma), Winter 1988.

Rowe, John Carlos, and Rick Berg, editors. *The Vietnam War and American Culture.* New York: Columbia University Press, 1991.

Springer, Claudia. "Vietnam: A Television History and the Equivocal Nature of Objectivity." *Wide Angle* (Athens, Ohio), 1985.

Trotta, Liz. *Fighting for Air: In the Trenches with Television News.* New York: Simon and Schuster, 1991.

Turner, Kathleen J. *Lyndon Johnson's Dual War: Vietnam and the Press.* Illinois: University of Chicago Press, 1985.

See also *China Beach*; Documentary; *Selling of the Pentagon*; *60 Minutes*; *Uncounted Enemy*; *Vietnam: A Television History*; Wallace, Mike

VIOLENCE AND TELEVISION

Underlying concern for the level of violence in society has lead authorities in several countries to set up investigative bodies to examine the portrayal of violence on television. In 1969 the U.S. Surgeon General was given the task of exploring evidence of a link between television and subsequent aggression. The research that was a product of this inquiry attempted to find a "scientific" answer to the issue of whether television violence causes aggressive behavior, in much the way an earlier investigation had examined the link between cigarettes and lung cancer. The conclusions of the report were equivocal, and while some saw this as reflecting vested interests in the membership of the committee, research over the following 20 years has not silenced the debate. While in 1985 the American Psychological Association stated a belief that the overwhelming weight of evidence supports a causal relation, there is not unanimity even among American psychologists for this position. Not only the specific conclusions but the whole "scientific" framework of what has become known as effects research has been challenged. Reports by the British Broadcasting Standards Council and the Australian Broadcasting Tribunal investigation into TV Violence in Australia, in the late 1980s to early 1990s reflect a very different set of questions and perspectives.

The traditional question of whether viewing violence can make audiences more aggressive has been investigated by a variety of techniques. As social science, and psychology in particular, attempted to emulate the rigorous methods of the physical sciences, the question of television and violence was transferred to careful laboratory experiments. Inevitably the nature of the issue placed practical and ethical constraints on scientific inquiry. A range of studies found evidence that subjects exposed to violent filmed models were subsequently more aggressive (Bandura, 1973). Questions have been raised, however, as to what extent these findings can be generalised to natural viewing situations. What did participants understand about the task they were given? What did they think was expected of them? Can the measures of aggression used in such studies, such as hitting dolls or supposedly inflicting harm by pushing buttons be compared to violent behavior in real world settings? Are these effects too short term to be of practical concern?

One strategy to overcome some of these problems was to conduct studies in natural settings: preschools, reform

Buck Rogers in the 25th Century

Hunter

The Lawman

The Untouchables

homes, etc. Children watched a diet of violent or non-violent television over a period of several weeks and the changes in their behavior were monitored. Such studies resemble more closely the context in which children normally watch television and measure the kinds of aggressive behavior that create concern. Results, however, have been varied and the practical difficulties of controlling natural environments over a period of time mean that critics have been quick to point to flaws in specific studies.

From time to time researchers have been able to capitalise on naturally occurring changes, gathering data over the period when television is first introduced to a community. A Canadian study compared children in two communities already receiving television to those in a community where television was introduced during the course of the study. Increases in children's aggressive behavior over time were found to accompany the introduction of television. A similar conclusion was drawn from a major study into the effects of the introduction of television in South Africa.

An alternative to manipulating or monitoring group changes in exposure to violence, is simply to measure the amount of television violence children view and relate it to their level of aggressive behavior. While many studies have found a clear association between higher levels of violent viewing and more aggressive behavior, proving that television caused the aggression is a more complex issue. It is quite possible that aggressive children choose to watch more violent programs or that features of their home, socioeconomic or school background explain both their viewing habits and their aggression. Attempts to test these alternative models have involved complex statistical techniques, and perhaps most powerfully, studies of children over extended periods of time, in some cases over many years. Studies by Huesmann and his colleagues have followed children in a variety of different countries. They argue that the results of their research demonstrate that the extent of viewing TV in young children is an independent source of later aggression. They also suggest that aggressive children chose to watch more violent programs which in turn stimulates further aggression. The research group gathered data from a range of countries which indicates that the relationship can be found even in countries where screen violence is much lower than the United States. A comparison of Finland with the United States found, however, no relationship between violent viewing and aggressive behavior in Finnish girls. This suggests that the impact of television has to be understood in a cultural context and involves social expectations about appropriate gender roles.

An alternative technique to the longitudinal study was used by Belson in London. He selected, from a large sample of adolescent boys, two small groups that differed in the extent to which they viewed violent television but were very carefully matched on socioeconomic and other variables. Belson concluded from his comparison that greater viewing, particularly of realistic violent drama, was associated with more aggressive behavior.

Critics of these attempts to relate viewing and aggression have questioned both the accuracy with which reports of television habits and preferences were gained, either from parents or by retrospective recall, and the measures used to demonstrate aggression. In reviewing debates on research findings, it becomes clear that any study can be flawed by those taking an opposing position. The majority of researchers who have used the techniques described here believe the evidence does indicate a causal link between violence on television and violent behavior and point to the mutual support provided by the variety of empirical techniques employed.

Even among researchers who are convinced of a causal link between television and violence, explanations of when and why this occurs are varied. One of the simplest ideas is that children imitate the violence they see on television. Items associated with violence through television viewing can serve as cues to trigger aggressive behavior in natural settings. The marketing of toys linked to violent programs taps into these processes. Children are more likely to reenact the violence they have seen on television when they have available products which they have seen being used in violent scenarios. The challenge for social learning theorists has been to identify under what conditions modelling occurs. Does it depend on viewers' emotional state, for instance a high level of frustration, or on a permissive social environment? Is it important whether the violence is seen to be socially rewarded or punished? It has also been claimed that high levels of exposure to violent programs desensitise children making them more tolerant of and less distressed by violence. Thus children who had been watching a violent program were less willing to intervene and less physiologically aroused when younger children whom they had been asked to monitor via a television screen were seen fighting, than those children who had watched a non-violent program. Alternatively, high arousal itself has been suggested as an instigator of violence. The significance of such an explanation is that it does not focus on violence as such; other high action, faster cutting programs may stimulate aggression. It is evident that once focus shifts from proving causation to identifying processes, the characteristics of particular violent programs become important because programs vary in many ways besides being classifiable as violent or non-violent.

The traditional violence effects approach has been criticised as employing a hypodermic model, where the link between television violence and viewer aggression was seen as automatic. Such an approach not only ignored the complexity of television programs, but how responses to television are mediated by characteristics of viewers, their thoughts and values. As psychology has become more concerned with human thinking, there has been greater interest in how viewers, particularly children, interpret the television they watch. Research has shown that children's judgements of violent actions relate to their understanding of the plot. This in turn may be influenced by issues like plot complexity, the presence and placement of commercial breaks, the age of the child, etc. Rather than seeing violence as a behavior pattern that children internalise and reproduce on cue, children are seen to develop schematic understanding of violence. The values they attach to such behavior may depend on more complex issues, such as the extent to which they identify with a violent character, the apparent justifiability of their actions, and the rewards or punishments perceived for acting aggressively.

It has often been feared that children are particularly vulnerable to violence on television because their immature cognitive development does not enable them to discriminate between real and fictional violence. In a detailed study of children's responses to television and cartoons in particular, Hodge and Tripp (1986) found that children could make what they termed "modality judgements" as young as six years old. They were well aware that the cartoon was not real. What developed at a later stage was an understanding of certain programs as realistic, building the links between television and life experience. Such research demonstrates a coming together of psychological and cultural approaches to television. Researchers interested in the structure of program meanings and in children's psychological processes can collaborate to increase our knowledge of how children actively interpret a violent cartoon.

Another dimension of the television violence debate has been a concern that frequent viewing of violence on television makes people unrealistically fearful of violence in their own environment. Gerbner's "enculturation" thesis appeared supported by evidence that heavier viewers of television believed the world to be more violent than those who watched television less. Alternative explanations have been offered for these findings both in terms of social class (heavy viewers may actually live in more dangerous areas), and personality variables. It has also been suggested that those fearful of violence may chose to watch violent programs such as crime dramas, where offenders are caught and punished. Again viewers are seen as actively responding to violence on television, rather than simply being conditioned by it. Gerbner presents a valuable description of the violent content on television: who are portrayed as attackers and who are the victims in our television world. Yet Greenberg has argued against a cumulative drip-drip-drip view of how television affects viewers' perceptions of the world. Instead he poses a "drench" hypothesis that single critical images can have powerful effects, presumably for good or ill.

Traditional television violence effects research employed simple objective criteria for determining the extent of violence in a program. A feature of this approach has been the development of objective definitions of violence that have enabled researchers to quantify the extent of violence on our screens (80% of prime-time American television contains at least one incident of physical violence). From this perspective cartoons are just as violent as news footage and a comic cartoon like *Tom and Jerry* is among the most violent on television. Such judgements do not accord with public perceptions and in recent years there has been an interest in discovering what the public consider violent. A carefully

controlled study of audience perceptions of violence was conducted in Britain by Barrie Gunter. He found that viewers rated a similar action as more violent, if the program was closer to their life experience than if it was a cartoon, western or science fiction drama. He also found that ratings of violence were linked in complex ways to characteristics of the attacker, victim and setting and to the personality of the rater. This focus on what audiences found violent and disturbing and what they believed would disturb children has provided a rather different framework for considering issues of violence on television.

Research for the Australian investigation of violence on television, in contrast to the U.S. Surgeon General's report, was not concerned with establishing causal links but on finding how audience groups reacted to specific programs. The aim was to improve the quality of guidelines to programmers and the information provided for prospective audiences. The research concluded that the most important dimension for viewers in responding to violence was whether the subject matter was about real life. The interest in public perceptions of violence of television has stimulated new research techniques. British researchers have asked their subjects to take editing decisions as to what cuts are appropriate before material is put to air. Docherty has argued that certain material, both fiction and non-fiction can elicit strong emotional reactions which he has termed "deep play." People's cuts to a horror movie like *Nightmare on Elm Street* appeared largely a question of taste. In contrast a docudrama about football hooliganism provoked polarised and intense reactions. Some viewers felt the violent material was important and should not be cut, others reacted with great hostility to a portrayal of violence that challenged their sense of social order.

The issue of the appropriate level of televised violence arises not just with fictional violence but with the televising of news footage. Here the problem for reporters is a balance between reporting what is occurring in the world and making the violence they cover palatable for the living room. Reporters have put themselves at risk attempting to film savage violence in a way that can tell their story but not overwhelm the viewers. The violence of the Vietnam War played out nightly in American living rooms has been seen as a major factor in generating the anti-war movement. More recently, coverage of the Gulf War indicates how use of the media, especially television, has become part of wartime strategy. Research on the role of the media in the Gulf War suggests that viewers were often happy to be spared the details of the war as long as their side was winning. It is not perhaps surprising that despite concern expressed about the impact of such a violent crisis on impressionable children, the news image that evoked most anger and sadness in British children was on the plight of sea birds covered in oil.

The portrayal of the war, the sanitised images of high technology, the frequently employed analogy of the video game, the absence of blood and gore are also issues about violence and television. The fact that the political debates about violence on television have focused so strongly on the potential harm to children may act to divert attention away from the way certain violence is censored in the interests of the state. An excessive focus on screen violence can deflect attention from the complex issues of state and interpersonal violence that exist in our world.

Until recently the potential of television to challenge viewers to think about issues of violence has been largely ignored. A study by Tulloch and Tulloch of children's responses to violence in a series of programs has found young people more disturbed by a narrative about a husband's violent assault on his wife than the objectively more serious violence of a Vietnam War series. Their research has demonstrated clearly that the meanings children attach to violence on television is a function of their age, gender and social class. Not only does this confirm other findings that relate the perception of violence to personal significance, it points to the potential educative effects of violence on television. Once the portrayal of violence is not seen as necessarily increasing violence, the ways programs can work towards the promotion of non-violence can be investigated.

—Marian Tulloch and John Tulloch

FURTHER READING

Australian Broadcasting Tribunal. *TV Violence in Australia.* Sydney: Commonwealth of Australia, 1990.

Bandura, A. *Aggression: A Social Learning Analysis.* Englewood Cliffs, New Jersey: Prentice Hall, 1973.

Belson, W. A. *Television Violence and the Adolescent Boy.* Farnborough, England: Saxon House, 1978.

Cumberbatch, D., and D. Howitt. *A Measure of Uncertainty: The Effects of the Mass Media.* Broadcasting Standards Council Research Monograph Series:1. London: John Libbey, 1989.

Friedlich-Cofer, L., and A.C. Huston. "Television Violence and Aggression: The Debate Continues." *Psychological Bulletin* (Washington, D.C.), 1986.

Friedman, J. L. "Telelevision Violence and Aggression: A Rejoinder." *Psychological Bulletin* (Washington, D.C.), 1986.

Greenberg, B. S., and W. Gantz, editors. *Desert Storm and the Mass Media.* Cresskill, New Jersey: Hampton Press, 1993.

Gunter, B. *Dimensions of Television Violence.* Aldershot: Gower Press, 1985.

Gunter, B., and J. McAleer. *Children and Television: The One Eyed Monster?* London: Routledge, 1990.

Hodge, R, and D. Tripp. *Children and Television: A Semiotic Approach.* Cambridge: Polity, 1986.

Huesman, L.R., and L.D. Eron, editors. *Television and the Aggressive Child: A Cross National Comparison.* Hillsdale New Jersey: Erlbaum, 1986.

Oskamp, S., editor. "Television as a Social Issue." *Applied Social Psychology Annual 8.* Newbury Park, California: Sage, 1988.

Tulloch, J. C., and M.I. Tulloch. "Discourses About Violence: Critical Theory and the 'TV Violence' Debate." *Text* (The Hague, Netherlands), 1992.

Wober M., and B. Gunter. *Television and Social Control.* Aldershot: Gower Press, 1988.

See also Audience Research: Effects Analysis; Audience Research: Industry and Market Market Analysis; Broadcasting Standards Council; Children and Television; Detective Programs; Police Programs; Standards and Practices; Terrorism; War on Television; Westerns

THE VOICE OF FIRESTONE

U.S. Music Program

One of network television's preeminent cultural offerings, *The Voice of Firestone* was broadcast live for approximately twelve seasons between 1949 and 1963. With its forty-six piece orchestra under the direction of Howard Barlow, this prestigious award winning series offered viewers weekly classical and semi-classical concerts featuring celebrated vocalists and musicians. This series is also highly representative of the debate that still rages over the importance of ratings and mass-audience appeals as opposed to cultural-intellectual appeals targeted to comparatively small audiences in the development of network television schedules.

Sponsored throughout its history by the Firestone Tire and Rubber Company, *The Voice of Firestone* began as a radio offering in December 1928, and transferred to television as an NBC simulcast on 5 September 1949. Long on musical value but often short on television production value, the show was faulted occasionally for its somewhat stilted visual style, its pretentious nature and its garish costume choices.

In time, however, the series drew critical praise and a consistent audience of two to three million people per broadcast.

Notwithstanding its "small" viewership, the Firestone series vigorously maintained its classical/semi-classical format adding only an occasional popular music broadcast with stars from Broadway, night clubs and the recording industry and an occasional theme show developed around various topics of interest, e.g., 4-H clubs, highway safety and the United Nations. The program attracted the great performers of the day for nominal fees with Rise Stevens setting the record for program appearances at forty-seven. In his *Los Angeles Times* feature of 1 November 1992, Walter Price observed that the Metropolitan Opera star "had the face, figure and uncanny sense of the camera to tower above the others in effect."

In 1954, *The Voice of Firestone*'s audience size became a major issue. Citing low ratings and the negative effect of those ratings on other programs scheduled around it, NBC demanded a time change. Historically, the show had been broadcast in a Monday, 8:30–9:00 P.M., prime-time period. As an alternative, NBC officials suggested leaving the Monday evening radio program in its established time but moving the television version to Sunday at 5:30 P.M. or to an earlier or later slot on Monday. Firestone officials, considering the millions their company had spent for air time and talent fees over the previous twenty-six years, refused to budge.

Determined to lure viewers away from *Arthur Godfrey's Talent Scouts*, CBS' highly rated competition for the time period, NBC exercised control of its schedule and canceled both the radio and television versions of *The Voice of Firestone* effective 7 June 1954. The following week, the simulcast reappeared on ABC in its traditional day and time where it remained until June 1957. In that month, the radio portion was dropped but, after a summer hiatus, the television show returned on Monday evenings at 9:00 P.M. In June 1959, despite more popular music in its format, poor ratings again forced the show's cancellation in favor of the short-lived detective series *Bourbon Street Beat*.

Amid numerous critical outbursts, threats of Federal Communications Commission (FCC) action and a joint resolution by the National Education Association and National Congress of Parents and Teachers lamenting its cancellation, all three networks offered *Voice of Firestone* fringe time slots which the Firestone Company rejected. ABC officials indicated that the series was simply the victim of the greater attention paid to television ratings. In radio, critics pointed out, audience delivery to program adjacencies was never considered as important as it was in television and concert music programs in prime time were regarded as too weak to hold ratings through the evening schedule. Condemning the loss of the Firestone program, Norman Cousins wrote in his 9 May 1959, *Saturday Review* editorial, that stations were now pursuing a policy designed to eliminate high quality programs "even if sponsors are willing to pay for them." Cousins decried the fact that station managers measured program weakness through ratings, and a "'weak spot' in the evening programming . . . must not be allowed to affect the big winners."

The Voice of Firestone was brought back to ABC on Sunday evenings, 10:00–10:30 P.M., in September 1962. However, despite numerous commendations, positive critical reviews and a star-studded rotation of musical conductors and performers, the audience remained at two and a half million people. *The Voice of Firestone* left the air for its third and final time in June 1963. With its passing, the American public lost an alternative form of entertainment whose long heritage was one of quality, good taste and integrity.

—Joel Sternberg

NARRATOR
John Daly (1958–1959)

REGULAR PERFORMERS
Howard Barlow and the Firestone Concert Orchestra

PROGRAMMING HISTORY

• NBC

September 1949–June 1954   Monday 8:30-9:00

• ABC

June 1954–June 1957   Monday 8:30-9:00
September 1957–June 1959   Monday 9:00-9:30
September 1962–June 1963   Sunday 10:00-10:30

FURTHER READING

Adams, Val. "Firestone Show to End on June 1." *New York Times*, 15 April 1959.

Brooks, Tim, and Earle Marsh. *The Complete Directory to Prime Time Network TV Shows 1946–Present.* New York: Ballantine, 1979; 5th edition, 1992.

Cousins, Norman. "The Public Still Owns the Air." *Saturday Review* (New York), 9 May 1959.

"Firestone at 8:30." *Newsweek* (New York), 21 June 1954.

"Firestone Loses Its Voice." *Business Week* (New York), 22 May 1954.

"Firestone's Voice Silenced by N.B.C." *New York Times*, 15 May 1954.

Gould, Jack. "Radio-TV in Review." *New York Times*, 26 May 1954.

———. "Victim of Ratings." *New York Times*, 19 April 1959.

"Old Opera House now Big TV Show." *New York Times*, 15 June 1954.

"Paramount Seeks to House TV Show." *New York Times*, 3 June 1954.

Price, Walter. "Before MTV, There was Opera." *Los Angeles Times*, 1 November 1992.

Shanley, John P. "Television: Death at 31." *New York Times*, 2 June 1959.

Shepard, Richard F. "Firestone Series Returning to TV." *New York Times*, 28 March 1962.

Spalding, John Wendell. *An Historical and Descriptive Analysis of "The Voice of Firestone" Radio and Television Program, 1928–1959* (Ph.D. dissertation, University of Michigan, 1961).

See also Advertising, Company Voice; Music on Television

VOICE-OVER

Voice-over (VO or V/O) is the speaking of a person or presenter (announcer, reporter, anchor, commentator, etc.) who is not seen on the screen while her or his voice is heard. Occasionally, a narrator may be seen in a shot but not be speaking the words heard in the voice-over.

Voice-over has diverse uses in a variety of television genres. Like other forms of television talk, it aims at being informal, simple and conversational. However, except for on-the-spot reporting such as sports events, voice-over is often less spontaneous than the language of talk shows; it is heavily scripted especially in genres such as the documentary. Voice-over is not simply descriptive; it also contextualizes, analyses and interprets images and events. Commentaries have the power to reverse the significance of a particular visual content. Voice-over is, therefore, an active intervention or mediation in the process of generating and transmitting meaning. However, viewers are rarely aware or critical of the scope of mediation in part because the visual image itself confers credibility and authenticity on the voice-over. But voice is at times more credible than vision; it is an integral part of a person's identity. This was experienced in the 1988 British government ban on broadcast interviews with representatives of eleven Irish organizations, including Sinn Fein, the political wing of the Irish Republican Army. Broadcasters were allowed, however, to voice-over or caption a banned representative's words.

Voice-over is used as a form of language transfer or translation. Viewers of news programs are familiar with the use of voice-over translation of statements or responses of interviewees who do not speak in the language of the viewing audience. Inherited from radio, this form of language trans-fer allows the first and last few words in the original language to be heard, and then fades them down for revoicing a full translation. The voice-over should be synchronous with the speaker's talk, except when a still picture is used to replace footage or live broadcast. Usually gender parity between the original and revoiced speakers is maintained.

As a form of language transfer, voice-over is not limited to the translation of brief monologues; sometimes it is used to cover whole programs such as parliamentary debates, conferences or discussions. Its production is usually less expensive than dubbing and subtitling. Some countries, such as Poland and the Balkan states, use voice-over as the main method of revoicing imported television programs. Usually, the revoicing is done without much performance or acting, even when it involves drama genres.

—Amir Hassanpour

FURTHER READING

Collins, Richard. "Seeing Is Believing: The Ideology of Naturalism." In, Corner, John, editor. *Documentary and the Mass Media.* London: Edward Arnold, 1986.

Dries, Josephine. "Breaking Language Barriers Behind the Broken Wall." *Intermedia* (London), December/January 1994.

Murdock, Graham. "Patrolling the Border: British Broadcasting and the Irish Question in the 1980s." *Journal of Communication* (New York), 1991.

Wilson, Tony. *Watching Television: Hermeneutics, Reception and Popular Culture.* Cambridge, England: Polity Press, 1993.

See also Dubbing; Language and Television; Subtitling

W

WAGON TRAIN

U.S. Western

Wagon Train, a fusion of the popular Western genre and the weekly star vehicle, premiered on Wednesday nights, 7:30-8:30 P.M. in September 1957 on NBC. The show took its initial inspiration from John Ford's 1950 film, *The Wagonmaster*. NBC and Revue productions, an MCA unit for producing telefilms, conceived of the program as a unique entry into the growing stable of Western genre telefilm, combining quality writing and direction with weekly guest stars known for their work in other media, primarily motion pictures. Each week, a star such as Ernest Borgnine (who appeared in the first episode, "The Willie Moran Story"), Shelly Winters, Lou Costello, or Jane Wyman would appear along with series regulars Ward Bond and Robert Horton. The show, filmed on location in California's San Fernando Valley, had an impressive budget of one hundred thousand dollars per episode, at a time when competing hour-long Westerns, such as ABC's *Sugarfoot,* cost approximately seventy thousand dollars per episode.

Star presence enticed viewers; powerful writing and directing made the show a success. Writers with experience in other Westerns, such as *Gunsmoke* and *Tales of Wells Fargo,* developed scripts that eventually became episodes, Western novelist Borden Chase and future director Sam Peckinpah among them. Directors familiar with the Western telefilm contributed experience, as did personnel who had been involved with *GE Theatre*, a program influential in the conception of *Wagon Train*'s use of stars. Promotional materials suggested that motion picture directors John Ford, Leo McCarey, and Frank Capra had expressed interest in directing future episodes; whether wishful thinking or real possibility, *Wagon Train*'s producers envisioned their Western as television on a par with motion pictures.

Each episode revolved around characters and personalities who were traveling to California by wagon train caravan from St. Joseph, Missouri. Series regulars conducted the train through perils and adventures associated with the landscapes and inhabitants of the American West. The star vehicle format worked in tandem with the episodic nature of series television, giving audiences a glimpse into the concerns of different pioneers and adventurers from week to week. Returning cast members gave the show stability: audiences expected complaints and comedy from Charlie Wooster, the train's cook; clashes of experience with exuberance in the relationship between the wagonmaster and his dashing frontier scouts. The recurring cast's interrelationships, problems, and camaraderie contributed greatly to the sense of "family" that bound disparate elements of the series together.

Wagon Train lasted eight seasons, moving from NBC to ABC in September of 1962. In 1963, its format expanded to 90 minutes, but returned to hour length for its final run from 1964–65. It survived several cast changes: Ward Bond (Major Adams), the original wagonmaster, died during filming in 1960, and was replaced by John McIntyre (Chris Hale); Robert Horton (Flint McCullogh) left the series in 1962 and was replaced as frontier scout by Robert Fuller (Cooper Smith). Only two characters survived the eight-year run in their original positions: Frank McGrath, as comical

Wagon Train

cook Charlie Wooster, and Terry Wilson's assistant wagonmaster Bill Hawks.

The show's ability to survive a network switch and periodic cast changes during its eight-year-run attests to the popularity of the program. In the fall of 1959, two years after its inception, the show was number one in Great Britain; of seven Westerns in the Nielsen top ten in the United States, *Wagon Train* was in constant competition with *Gunsmoke* for supremacy. By 1959, the show was firmly ensconced in the top twenty five programs in the country, bouncing as high as number one in the spring of 1960, and maintaining its number one position over *Gunsmoke* throughout the 1961–62 season. In a field awash with Westerns, *Wagon Train* established a unique style reminiscent of the anthology drama, but indelibly entrenched in Western traditions.

—Kathryn C. D'Alessandro

CAST

Major Seth Adams (1957–61)	Ward Bond
Flint McCullough (1957–62)	Robert Horton
Bill Hawks	Terry Wilson
Charlie Wooster	Frank McGrat
Duke Shannon (1961–64)	Scott Miller
Chris Hale (1961–65)	John McIntire
Barnaby West (1963–65)	Michael Burns
Cooper Smith (1963–65)	Robert Fuller

PRODUCERS   Howard Christie, Richard Lewis

PROGRAMMING HISTORY   442 Episodes

• NBC

September 1957–September 1962   Wednesday 7:30-8:30

• ABC

September 1962–September 1963	Wednesday 7:30-8:30
September 1963–September 1964	Monday 8:30-9:30
September 1964–September 1965	Sunday 7:30-8:30

FURTHER READING

Brauer, Ralph. *The Horse, the Gun and the Piece of Property: Changing Images of the TV Western.* Bowling Green, Ohio: Popular Press, 1975.

Cawelti, John. *The Six-Gun Mystique.* Bowling Green, Ohio: Popular Press, 1984.

MacDonald, J. Fred. *Who Shot The Sheriff? The Rise and Fall of the Television Western.* New York: Praeger, 1987.

Morrison, C. "Ward Bond and Wagon Train." *Look* (New York), 27 October 1959.

West, Richard. *Television Westerns: Major and Minor Series, 1946-1978.* Jefferson, North Carolina: MacFarland, 1987.

Yoggy, Gary A. *Riding the Video Range: The Rise and Fall of the Western on Television.* Jefferson, North Carolina: McFarland, 1994.

See also *Cheyenne; Gunsmoke; Have Gun, Will Travel; Warner Brothers Presents;* Westerns

WALES

As a small but culturally and linguistically distinct nation within the United Kingdom, Wales offers an enlightening case study of the role of television in constructing cultural identity. Broadcasting in Wales has played a crucial role in ensuring the survival of the Welsh language, one of the oldest languages spoken on a daily basis in Europe. Coupled with recent educational policies, which include Welsh-language instruction as either a core or secondary subject in all Welsh schools, and European-wide recognition of the cultural and linguistic rights of indigenous speakers, the nation has seen a slight increase in the percentage of Welsh-speakers. Welsh television is currently comprised of BBC-1 Wales and BBC-2 Wales, the independent television (ITV) commercial franchise holder, Harlech Television (HTV Wales), and Sianel Pedwar Cymru ([S4C] Channel Four Wales), the Welsh equivalent of Britain's commercial Channel Four. BBC-1 Wales, BBC-2 Wales, and HTV Wales broadcast entirely in English, whereas S4C's schedules contain a mix of locally-produced Welsh-language and English-language Channel 4 United Kingdom programs. Welsh-language television is the progeny of battles over the national and cultural rights of a linguistic minority who, from the outset of television in Britain, lobbied hard for Welsh-language programming. Of the 2.7 million population of Wales, 20% speak Welsh, and since 1 November 1982, the bilingual minority have been able to view Welsh-language programs on S4C during the lunch and prime-time periods, seven days a week.

From the outset of television in Wales, the mountainous topography of the country presented broadcasters with transmission problems; despite the construction of new and more powerful transmitters, there were gaps in service as late as the 1980s. At the time of the opening of the first transmitter in Wales, 36,236 households had a combined radio and television license, a number that more than doubled to 82,324 by September 1953, in anticipation of the televising of the coronation of Queen Elizabeth II. By 1959, 50% of Welsh households had a television set (450,720 licenses); 70% of these viewers received their broadcasts from the Welsh transmitter (Wenvoe), which also reached an identical viewing base in South-west England. However, 10% of the Welsh population could still not receive television, and 20% received their programs from transmitters located in England.

A key player in early Welsh-language television was Alun Oldfield-Davies (senior regional BBC controller from 1957 to 1967), who persuaded the BBC in 1952 to allow Welsh-language programs to be occasionally transmitted from the Welsh transmitter outside network hours. Oldfield-Davies went on to become an inveterate campaigner for Welsh-language television, and stepped up his lobbying with the introduction of commercial television in Wales in 1956. The first television program broadcast entirely in Welsh was transmitted on St. David's Day (Wales' patron saint day) on 1 March 1953, and featured a religious service from Cardiff's Tabernacle Baptist Chapel. The first Welsh-language feature program was a portrait of the Welsh bibliophile Bob Owen; despite replacing only the test card, the program antagonized English viewers, who complained about the incomprehensible language. This reaction was to intensify in later years, when English programs were substituted by Welsh-language productions.

The Broadcast Council for Wales (BCW) was established as an advisory body in 1955, although its presence had little impact on the tardy appearance of full production facilities in Cardiff, the last regional center in the United Kingdom to be adequately equipped for production in 1959. (The BBC expanded the Broadway Methodist Chapel in Cardiff, a site that had functioned as a drive-in studio since 1954). The first program filmed before a live audience in Wales took place in 1953, while the first televised rugby match and Welsh-language play, *Cap Wil Tomos* (*Wil Tomos' Cap*) were both transmitted in January 1955. (The first televised English-language play produced in Wales, *Wind of Heaven*, was broadcast in June 1956). However, despite these important breakthroughs in Welsh television, the number of programs locally produced for both bilingual and English-speaking audiences remained small; for example, in 1954, only two hours and 40 minutes of English programming and one hour and 25 minutes of Welsh-language programming were broadcast each week. The first regular Welsh-language program, *Cefndir* (Background), aired in February 1957; introduced by Wyn Roberts, the show adopted a magazine format featuring topical items.

The BBC's monopoly in British broadcasting was broken with the launch of ITV, which could first be received by the inhabitants of North-east Wales (and many in Northwest Wales) in 1956, following the launch of Granada television in Manchester. South Wales did not receive ITV until Television Wales West (TWW) was awarded a franchise in 1958, and opened a transmitter in the South which also served the South-west of England. More than a little complacent that the commercial imperatives of ITV would preclude Welsh-language ITV broadcasts, the BBC was stunned when the ITV Granada studios in Manchester launched a series of twice-weekly 60-minute Welsh-language programs, greatly overshadowing the BBC's weekly provision of half an hour. As a result, the political stakes

involved in addressing the interests of Welsh-language viewers were raised, although both the BBC and ITV recognized the low ratings generated by such programs, given the minority status of Welsh-language- speakers. Gwynfor Evans, who went on to play a pivotal role in the emergence of S4C in the early 1980s, joined the BCW in 1957, and along with Plaid Cymru (the Welsh Nationalist Party), vigorously lobbied for an increase in Welsh-language broadcasting. The issue of Welsh-language programming for children also assumed a greater urgency in the late 1950s. The broadcasting demands of the campaigners were given institutional recognition in 1960 with the publication of the findings of the Pilkington Committee—the first broadcasting inquiry mainly concerned with television—which argued that "the language and culture of Wales would suffer irreparable harm" if Welsh-language production were not increased.

A second ITV franchise, Television Wales West and North (TWWN, known in Wales as Teledu Cymru [Welsh Television]), began broadcasting in Wales in September 1962. Initially transmitting 11 hours of Welsh-language and Welsh-interest programming a week, TWWN obtained half of its programs from TWW. However, TWWN's future as a broadcaster was short-lived; facing bankruptcy, it was taken over by TWW in September 1963. At this time, the BBC and ITV reached an agreement over the scheduling of Welsh-language programs, requiring that each broadcaster's schedules be exchanged so as to avoid a clash of Welsh-language programs (leaving non-Welsh-speakers no alternative broadcast during this time slot). By and large, the policy worked, although some overlapping did occur.

In 1963, the BBC in Wales broadcast three hours of programming for Welsh viewers per week, and occasionally produced programs exclusively for the network. *Heddiw* (*Today*), a long-running Welsh-language weekday news bulletin, was broadcast outside network hours from 1:00 to 1:25 P.M., while its English-language equivalent, *Wales Today*, occupied an early evening slot between 6:10 and 6:25 P.M. TWW also had its own Welsh-language magazine program called *Y Dydd* (*The Day*).

BBC Wales was launched in February 1964 when it received its own wavelength for television broadcasting (Channel 13). Oldfield-Davies was central in orchestrating the move, and oversaw its implementation (television sets had to be converted in order to receive Channel 13). Up to this point, most Welsh-language programs had been transmitted during non-network hours; the introduction of BBC Wales meant that Wales would opt out of the national service for a prescribed number of hours per week—8.9 hours per week in 1964—in order to transmit locally-produced English- and Welsh-language programs. However, the arrival of BBC Wales meant that non-Welsh-speaking viewers whose aerials received BBC Wales from Welsh transmitters had no way of opting out of this system, unless they could also pick up the national BBC service by pointing their aerials towards English transmitters. The inclusion of a small number of Welsh-language programs on the television schedules at this time thus incensed some English-speaking Welsh viewers, who claimed that they were more poorly served by the BBC than other English-speaking national minorities such as the Scots, and resented losing programs to Welsh-language productions. By the fall of 1984, 68% of Welsh people received programs from transmitters offering BBC Wales, a number that increased to 75% by June 1970. BBC-2, the first BBC service transmitted on UHF, was launched in South-east England in 1962, reaching South Wales and South-west England in 1965. By the early 1970s, it was available to 90% of Welsh television homes. The first color program produced by BBC Wales was transmitted on 9 July 1970 and consisted of coverage of the Llangollen Eisteddfod.

As pressure for more Welsh-language programs increased, TWW's franchise was successfully challenged in 1968 by John Morgan and Lord Harlech. Commencing in March 1958, HTV pledged to address the "particular needs and wishes of Wales," and a ten-member committee was established to consider a range of topics affecting broadcasting in Wales. These issues were addressed more forcefully in a 1969 booklet published by Cymdeithas yr Iaith Gymraeg (the Welsh Language Society), entitled "Broadcasting in Wales: To Enrich or Destroy Our National Life?" Facing a wall of silence from BBC Wales following publication of the document, three members of the society embarked upon a campaign of civil disobedience, and in May 1970 interrupted a program broadcast from Bangor in North Wales. The following year a small group of men unlawfully gained entry into the Granada television studios in Manchester and caused limited damage to television equipment; television masts were also climbed, parliament was interrupted, and roads were blocked. In addition to these high-profile disturbances, hundreds of people were prosecuted for not paying their television license fees. In the fall of 1970, the society submitted a document to the Welsh Broadcasting Authority (WBA) which contained the first proposal for a fourth Welsh channel; an interim scheme proposed by the society suggested that the unalloted fourth UHF channel in Wales should transmit 25 hours of Welsh-language programming a week and should be jointly administered by a BBC Wales and HTV committee. Soon after, ITV made a formal submission requesting that the fourth channel be used as a second ITV service broadcasting all HTV's current Welsh-language programming and making HTV Wales an all-English channel. The battle for a Welsh fourth channel had begun in earnest.

Against a backdrop of ongoing campaigns by the Welsh Language Society in the early and mid-1970s, the Crawford Committee on Broadcast Coverage examined patterns of rural reception in Wales and explored the possibility of using the fourth channel for Welsh-language programming. Those in favor of retaining the current system of integration argued that a separate Welsh-language channel would ghettoize the language and culture (a view supported by the 1977 Annan Report commissioned by the Labor government); they also drew attention to the fact that English-speaking viewers would still be deprived of English programs broadcast on the U.K. fourth channel, and questioned whether there was a solid enough economic and cultural base in Wales to maintain a fourth channel. An average of 12 hours a week of Welsh- and English-language programs, seven and five hours respectively, were broadcast on BBC Wales between 1964 and 1974, with almost half the time taken up with news and current-affairs programs such as *Heddiw* (*Today*), *Cywain* (*Gathering*), *Wales Today*, and *Week in Week out*.

Welsh-language television up to this point had gained a reputation of being quite high-brow, often consisting of nonfiction programs examining major Welsh institutions and traditions. However, the enormous popularity of sport, especially the national game of rugby, always guaranteed representation and high ratings on the schedules; moreover, the 1974 launch of the hugely successful Welsh-language soap opera entitled *Pobol y Cwm* (*People of the Valley*) did even more to shift the balance toward popular programming. *Pobol y Cwm's* 20-minute episodes are currently broadcast five days a week; the continuing serial is the highest rated program on S4C, attracting an average viewership of 180,000. English subtitles are available on teletext on daily episodes, and the five episodes are repeated on Sunday afternoon with open subtitles.

Welsh-speaking comedic stars also made their mark in light entertainment during the 1970s; these included Ryan Davies, who enjoyed widespread fame with his partner Ronnie Williams in the 1971 show *Ryan a Ronnie*, and in the first Welsh sitcom *Fo a Fe* (*Him and Him*—derived from North and South Walean dialects for "him") written by Rhydderch Jones. Stand-up comedian Max Boyce also became a household name with his own 1978 one-man series. Religious programming was still popular with audiences (as it had been in radio), and a BBC Sunday half-hour hymn-singing program entitled *Dechrau Canu, Dechrau Canmol* (*Begin Singing, Begin Praising*) drew large audiences (it has continued through the 1990s). Two successful English-language programs made for the BBC network in the mid-1970s included a seven-hour miniseries on the life of Welsh politician David Lloyd George (1977) and an animated children's cartoon entitled *Ivor the Engine* (1976). One of the most successful English-language dramas of the 1970s, a program regularly repeated on Welsh television, was *Grand Slam* (1975), which hilariously documented the exploits of a group of Welsh rugby fans traveling to Paris for an international match.

Meanwhile, political lobbying for a fourth Welsh language channel intensified as the Welsh Language Society organized walking tours, petitions, leaflet distribution, and the public burning of BBC television licenses. Published in November 1975, the government-sponsored Siberry Report recommended that the Welsh fourth channel should broadcast 25 hours a week of Welsh-language programs, with the BBC and HTV each responsible for three-and-a-half days a week. Welsh MP's also argued that the seven hours of programming on BBC

Wales opened up by the transfer of Welsh-language programs to a fourth channel should be filled with BBC Wales programs in English rather than BBC network material. In their 1979 general election manifestos, both Labor and Conservative parties pledged support for a fourth Welsh channel; however, facing resistance to the plan from the independent broadcasting authority (IBA) and HTV, Conservative Party Home Secretary William Whitelaw repudiated the Welsh fourth channel in a speech given at Cambridge University in September 1979. Welsh reaction was swift; at Plaid Cymru's annual conference in October, a fund was established into which supporters opposed to Whitelaw's decision could deposit their television license fee (2,000 protesters pledged support and a number received prison sentences the following spring). Noted political and academic figures in Wales also joined the campaign and were arrested for civil disobedience. It was, however, the intervention of Plaid Cymru MP Gwynfor Evans that had the most profound effect on public and political opinion. In May 1980, Evans announced that he would go on a hunger strike on 5 October and continue with the protest until the government restored their earlier promise of giving Wales a fourth Welsh-language channel. In the wake of public demonstrations during visits to Wales by Prime Minister Margaret Thatcher and Welsh Secretary Nicholas Edwards, Cledwyn Evans (Labor's ex-Foreign Secretary) led a deputation to Whitelaw's office in London demanding that the decision be reversed. The government finally backed down on 17 September, stating that a Welsh Fourth Channel Authority would be formed (provisions were incorporated into the 1980 Broadcasting Bill through a House of Lords amendment). The BBC would be responsible for providing ten hours per week and HTV and independent companies eight hours per week. S4C had finally arrived.

Funded by an annual budget from the Treasury which is based on a rate of 3.2% of the net advertising revenue of all terrestrial television in the United Kingdom, S4C is a commissioning broadcaster rather than a program producer with program announcements and promotions the only material produced in-house. By the mid-1990s, S4C was transmitting approximately 1,753 locally-produced hours of programming in Welsh, and 5,041 hours in English per annum; the English-language broadcasts were rescheduled U.K. C4's output. These figures translate into roughly 30 hours of programming a week in Welsh and 93 hours in English. S4C reaches a target share of approximately 20% of Welsh-speaking viewers, although its remit also includes targeting both Welsh-learners and English-speakers through the use of teletext services that enable participating viewers to call-up English subtitles for most Welsh programs. Some 75% of all local advertisers produce campaigns in both Welsh and English on S4C, while a number of multinational companies, such as McDonald's and Volvo, have also advertised in Welsh.

Of the 30 hours of Welsh-language programming shown on S4C each week, ten hours come from BBC Wales; the remaining 20 come from HTV Wales and independent producers. BBC Wales also produces ten hours of English-

language programming for viewers living in Wales which is broadcast on BBC-1 and BBC-2. The BBC's Royal Charter charges the BBC to provide services reflecting "the cultures, tastes, interests, and languages of that country," and via the BCW, the service is regularly reviewed to ensure that programs meet the requirements set down in the Royal Charter. HTV Wales produced 588 hours of English-language programs for Wales during 1995, a figure that amounted to approximately 25 hours per week.

Since 1 January 1993, S4C has been responsible for selling its own advertising (previously overseen by HTV); this has meant that revenues can now be ploughed directly back into program production. S4C provides a wide range of program genres, including news and current affairs, drama, games, and quizzes, and youth and children's programming. The main S4C news service, *Newyddion* (*News*), is provided by BBC Wales; S4C also has two investigative news shows, *Taro Naw* (*Strike Now*) and *Yr Byd ar Bedwar* (*The World on Four*), as well as documentaries exploring the diverse lives of Welsh men and women: *Hel Straeon* (*Gather Stories*), *Cefn Gwlad* (*Countryside*), and *Filltir Sgwar* (*Square Mile*). Recent comedy series have included *Nosan Llawen* (*Folk Evening of Entertainment*), *Licyris Olsorts* (*Licorice Allsorts*), and the satirical show *Pelydr X* (*X-Ray*). Series examining contemporary issues through the lens of popular drama have ranged from *Hafren*, a hospital drama; *Halen yn y Gwaed* (*Salt in the Blood*), which followed the lives of a ferry crew sailing between Wales and Ireland; *A55*, a hard-hitting series about juvenile crime; and *Pris y Farchnad* (*Market Price*), which examined the lives of a family of auctioneers. Children and teenage viewers are catered to via *Sali Mali*, *Rownd a Rownd* (*Round and Round*), which looks at the exploits of a paper round, and *Rap*, a magazine program for Welsh-learners.

Non-Welsh-speaking viewers receive their local news from BBC Wales' *Wales Today* and HTV Wales' *Wales This Week*. Other recent nonfiction programs have included *Grass Roots*, *The Really Helpful Show*, *The Once and Future Valleys*, and *The Infirmary*, from HTV Wales, and *Between Ourselves*, *All Our Lives*, and *Homeland*, produced by BBC Wales.

Thanks to S4C, Wales now has a thriving independent production sector centered in Cardiff (where 46% of the Welsh media industry is located) and Caernarfon. Welsh television's success in the field of children's animation has continued, with *Wil Cwac Cwac* and *SuperTed* making their first appearance in 1982 (both have appeared on the Disney Channel in the United States), followed by *Fireman Sam* and *Toucan Tecs*. By the early 1990s, Cardiff boasted five animation houses, 45 independent production companies, and a pool of approximately 150 professional animators. Animation co-productions from the mid-1990s have included *Shakespeare: The Animated Tales*, *Operavox: The Animated Operas*, *Testament: The Bible in Animation*, *The Little Engine That Could*, and *The Legend of Lochnagar*. Over 90 of S4C's programs have been exported to almost 100 countries worldwide, and co-productions have been negotiated with production companies in France, Italy, Germany, Australia, and the United States.

Finally, it is important to point out that the political advocacy which secured the rights of Welsh-speakers within a broadcasting system for Wales ultimately benefited both Welsh- and English-speakers, since the language campaign fostered the production of more English-language programs for Wales as a whole. The current system of Welsh broadcasting would certainly never have existed had it not been doggedly pursued by Welsh-language activists. Recent audience research into the penetration levels of S4C indicate that in the mid-1990s, between 80 and 85% of Welsh-speakers watch S4C at some point each week, and between 65 and 70% of all viewers (English- and Welsh-speaking) tune in to S4C some time each week. The S4C model in Wales has been emulated by several other European linguistic minorities, including the Basque channel Euskal Telebista 1 in Spain (launched in 1982), and a Catalan channel which started in 1983.

—Alison Griffiths

FURTHER READING

Annual Reports of the National Broadcasting Council for Wales.

BBC Wales Annual Review 1994/95. Cardiff, Wales: BBC Wales, 1995.

Bevan, David. "The Mobilization of Cultural Minorities." *Media, Culture and Society* (London), 1984.

Blanchard, Simon, and David Morley, editors. *What's This Channel Four?* London: Comedia, 1982.

Browne, Donald R. *Electronic Media and Indigenous Peoples: A Voice of Our Own?* Ames: Iowa University Press, 1996.

Cooke, Philip, and Carmel Gahan. "The Television Industry in Wales." Regional Industrial Research/Igam Ogam Research: S4C, BBC Wales, and WDA, 1988.

Curtis, Tony, editor. *Wales: The Imagined Nation. Essays in Cultural and in National Identity*. Bridgend, United Kingdom: Poetry Wales Press, 1986.

Davies, John. *Broadcasting and the BBC Wales in Wales*. Cardiff: University of Wales Press, 1994.

Griffiths, Alison. "National and Cultural Identity in a Welsh Language Soap Opera." In, Allen, Robert C., editor. *To Be Continued: Soap Operas Around the World*. London: Routledge, 1995.

Howell, W.J., Jr. "Bilingual Broadcasting and the Survival of Authentic Culture in Wales and Ireland." *Journal of Communication* (Philadelphia, Pennsylvania), Autumn 1982.

————. "Britain's Fourth Channel and the Welsh Language Controversy." *Journal of Broadcasting* (Washington, D.C.), Spring 1981.

HTV Annual Report and Accounts 1995. Cardiff: Westdale Press, 1996.

S4C Annual Report and Accounts. Cardiff: S4C, 1995.

Williams, Eyrun Ogwen. "The BBC and the Regional Question in Wales." In, Harvey, Sylvia, and Kevin Robbins. *The Regions, the Nations and the BBC*. London: British Film Institute, 1993.

See also First People's Broadcasting in Canada; Language and Televison

WALLACE, MIKE

U.S. Broadcast Journalist

Although he spent many years in broadcasting before turning to journalism, Mike Wallace became one of America's most enduring and prominent television news personalities. Primarily known for his work on the long-running CBS magazine series *60 Minutes*, he developed a reputation as an inquisitorial interviewer, authoritative documentary narrator, and powerful investigative reporter. While his journalistic credentials and tactics have been questioned at times, his longevity, celebrity, and ability to land big interviews have made him one of the most important news figures in the history of television.

Wallace's early career differed from those of his well-known peers at CBS News. Edward R. Murrow, Walter Cronkite, Eric Sevareid, Andy Rooney, and others worked as wartime radio and print correspondents before moving to television. Wallace, however, studied broadcasting at the University of Michigan and began an acting and announcing career in 1939. Throughout the 1940s, he performed in a variety of radio genres—quiz shows, talk shows, serials, commercials, and news readings. After service in the Navy, the baritone-voiced radio raconteur landed a string of early television jobs in Chicago. As early as 1949, "Myron" Wallace acted in the police drama *Stand by for Crime*, and later appeared on the CBS anthology programs, *Suspense* and *Studio One*. He emceed local and network TV quiz and panel shows while also working in radio news for CBS throughout 1951–55. Wallace's move into interviewing at the network level came in the form of two husband-and-wife talk shows, *All Around the Town* and *Mike and Buff*, which CBS adapted from a successful Chicago radio program. With his wife Buff Cobb, Wallace visited various New York locations and conducted live interviews with celebrities and passers-by. In 1954, after a three-season run on CBS, Wallace had a brief stint as a Broadway actor, but immediately returned to broadcasting.

In 1955, Wallace began anchoring nightly newscasts for the DuMont network's New York affiliate. The following year his producer, Ted Yates, created the vehicle that brought Wallace to prominence. *Night-Beat* was a live, late-night hour of interviews in which Wallace grilled a pair of celebrity guests every week night. Armed with solid research and provocative questions, the seasoned announcer with a flair for the dramatic turned into a hard-hitting investigative journalist and probing personality reporter. With the nervy Wallace as its anchor, *Night-Beat* developed a hard edge lacking in most television talk. Using only a black backdrop and smoke from his cigarette for atmosphere, Wallace asked pointed, even mischievous questions that made guests squirm. Most were framed in tight close-up, revealing the sweat elicited by Wallace's barbs and the show's harsh klieg lights.

After a successful first season during which Wallace interviewed such celebrities as Norman Mailer, Salvador Dalí, Thurgood Marshall, Hugh Hefner, William Buckley, and prominent politicians, the program moved to ABC as a half-hour prime-time show called *The Mike Wallace Interview.* Promoted as "Mike Malice" and "the Terrible Torquemada of the TV Inquisition," Wallace continued to talk to prominent personalities about controversial issues. But ABC executives, particularly after brushes with libel suits, proved wary of the brinkmanship practiced by Wallace and his guests. The show lasted only through 1958, turning more cerebral in its final weeks when the Ford Foundation became its sponsor. Intellectuals such as Reinhold Neibuhr, Aldous Huxley, and William O. Douglas replaced the Klansmen, ex-mobsters, movie stars, and more sensational interviewees seen before.

For the next five years, Wallace continued to parlay his celebrity into odd jobs on New York and network TV as quiz master, pitch man for cigarettes, chat show host (*PM East*, 1961–62), and news reader. But he began to sharpen his focus on mainstream journalism as well. He anchored *Newsbeat* (1959–61), one of the first half-hour nightly news programs, for an independent New York station, and also began working as host for David L. Wolper's TV documentary series *Biography*, narrating 65 episodes of the syndicated program. (His distinctive voice continues to be heard in many such educational productions, including the 1995 A and E cable series *The 20th Century.*) Increasingly he became a field correspondent. After a chain of Westinghouse-owned stations hired Wallace to cover the 1960 political conventions, he started traveling extensively, supplying them with daily radio and TV reports from across the country (*Closeup U.S.A.,* 1960) and abroad (*Around the World in 40 Days*, 1962).

The following year, as he described in this 1984 autobiography, Wallace decided to "go straight," giving up higher-paying entertainment jobs for a career exclusively devoted to news. In 1963 (a year in which the networks expanded their news divisions), the *CBS Morning News with Mike Wallace* premiered. Wallace remained on the show for three years before resuming full-time reporter's duties. Although seen frequently on other CBS News assignments

Mike Wallace
Photo courtesy of Mike Wallace

(Vietnam, the Middle East), Wallace's beat was the Richard Nixon comeback campaign. A confessed Nixon apologist, he nevertheless rejected an offer in 1968 to be a press secretary for the candidate.

Instead, that fall Wallace began regular duties for *60 Minutes,* a prime-time news magazine for which he and Harry Reasoner had done a pilot in February 1968. To contrast the mild-mannered Reasoner, producer Don Hewitt cast Wallace in his usual role as the abrasive, tough-guy reporter. While he could be charming when doing softer features and celebrity profiles, Wallace maintained his reputation as a bruising inquisitor who gave his subjects "Mike fright." With his personal contacts in the Nixon (and later Reagan) circles, he proved an expert reporter on national politics, particularly during Watergate. Throughout his run on *60 Minutes,* he consistently landed timely and exclusive interviews with the most important newsmakers of the day.

As *60 Minutes* was becoming a mainstay of TV news, Wallace developed its most familiar modus operandi: the ambush interview. Often using hidden cameras and one-way mirrors, Wallace would confront scam artists and other wrong-doers caught in the act. Field producers did most of the investigative work, but Wallace added the theatrical panache as he performed his on-camera muckraking. His tactics were both praised and criticized. While he has won

numerous awards as a sort of national ombudsman, a reporter with the resources and ability to expose corruption,
some critics have judged his methods too sensational, unfair,
and even unethical.

Twice Wallace was entangled in landmark libel cases.
His *60 Minutes* report, "The Selling of Colonel Herbert"
(1973), questioned a whistleblower's veracity about war
crimes. Herbert sued Wallace's producer. Although the
news team was exonerated, the Supreme Court ruled in
Herbert v. Lando (1979) that the plaintiff had the right to
examine the materials produced during the editorial process. A far bigger case followed when Wallace interviewed
General William Westmoreland for the *CBS Reports* documentary *The Uncounted Enemy: A Vietnam Deception*
(1982). When *TV Guide* and CBS' own in-house investigation charged that the producers had violated standards of
fairness, Westmoreland sued the network. The charges
Wallace aired—conspiracy to cover-up the size of Viet
Cong troop strength—were substantiated by trial evidence,
but CBS' editorial tactics proved suspect. Early in 1985,
just before Wallace was to testify, CBS issued an apology
and Westmoreland dropped the suit.

Despite such occasional setbacks, Wallace continued
his signature style of globetrotting reports and "make-'em-
sweat" interviews throughout the 1980s and 1990s. A CBS
News special, *Mike Wallace, Then and Now* (1990), offered
a retrospective of his 50 years in broadcasting, but the senior
correspondent of American television journalism continued
his *60 Minutes* work unabated.

—Daniel G. Streible

MIKE (MYRON LEON) WALLACE. Born in Brookline, Massachusetts, U.S.A., 9 May 1918. Educated at the University of
Michigan, B.A. 1939. Married: 1) Norma Kaphan, 1940
(divorced, 1948); 2) Buff Cobb, 1949 (divorced, 1955); 3)
Lorraine Perigord, 1955 (separated, 1988); children: Peter
(deceased), Christopher, and Pauline. Served in U.S. Navy,
1943–46. Newscaster, announcer, and continuity writer,
radio station WOOD WASH, Grand Rapids, Michigan,
1939–40; newscaster, narrator, announcer, WXYZ Radio,
Detroit, Michigan, 1940–41, on such shows as *The Lone
Ranger* and *The Green Hornet*; freelance radio worker, Chicago, Illinois, announcer for the soap opera *Road of Life*,
1941–42, *Ma Perkins*, and *The Guiding Light*; acted in *The
Crime Files of Flamon*; news radio announcer, *Chicago Sun's
Air Edition*, 1941–43, 1946–48; announced radio programs
such as *Curtain Time*, *Fact or Fiction*, and *Sky King*; host,
Mike and Buff, with his wife, New York City, 1950–53; host,
various television and radio shows and narrator, various
documentaries, 1951–59; star, Broadway comedy *Reclining
Figure*, 1954; organized news department for DuMont's
WABD-TV, 1955; anchor in newscasts and host for various
interview shows, 1956–63; CBS News staff correspondent,
since 1963; co-editor and co-host of *60 Minutes,* since 1968.
Member: American Federation of Television and Radio
Artists; Academy of Television Arts and Sciences (executive

vice president, 1960–61). Recipient: 18 Emmy Awards;
Peabody Awards, 1963 and 1971; duPont Columbia Journalism Awards, 1971 and 1983. Address: CBS News, *60
Minutes*, 555 West 57th Street, New York, New York
10019, U.S.A.

TELEVISION SERIES (selection)

1951–53	*Mike and Buff*
1951–52	*All Around Town*
1953–54	*I'll Buy That*
1956–57	*The Big Surprise*
1956–57	*Night-Beat*
1957–58	*The Mike Wallace Interview*
1961–62	*PM East*
1963–66	*CBS Morning News with Mike Wallace*
1968–	*60 Minutes*

STAGE

Reclining Figure (actor), 1954.

PUBLICATIONS

*Mike Wallace Asks: Highlights from 46 Controversial Inter-
views*. New York: Simon and Schuster, 1958.

A Mike Wallace Interview with William O. Douglas. New
York: American Broadcasting Company in association
with Fund for the Republic, 1958.

Close Encounters, with Gary Paul Gates. New York: Morrow,
1984.

"*60 Minutes* Into the 21st century!" *Television Quarterly*
(New York), Winter, 1990.

"5 Badfellas: In a Lifetime of Interviewing, It's Not the
Heads of State You Remember But the Guys Named
'Lunchy.'" *Forbes* (New York), 23 October 1995.

FURTHER READING

Adler, Renata. *Reckless Disregard: Westmoreland v. CBS et al.,
Sharon v. Time*. New York: Knopf, 1986.

Bar-Illan, David. "*60 Minutes* and the Temple Mount."
Commentary (New York), February 1991.

Barron, James. "Wallace Is Rebuked for Taping Interview
on Hidden Camera." *The New York Times*, 17 November 1994.

Benjamin, Burton. *Fair Play: CBS, General Westmoreland,
and How a Television Documentary Went Wrong*. New
York: Harper and Row, 1988.

Brewin, Bob, and Sydney Shaw. *Vietnam on Trial: Westmoreland vs. CBS*. New York: Atheneum, 1987.

Brown, Rich. "Wallace Blasts Trend to 'Sensationalism.'"
Broadcasting (Washington, D.C.), 30 September 1991.

CBS News. "*60 Minutes*" *Verbatim*. New York: Arno
Press/CBS News, 1980.

Clark, Kenneth R. "Getting Good Being Bad: Chicago's
Salute Brings Mike Wallace Back to His Roots." *Chicago Tribune*, 20 September 1989.

Colford, Paul D. "The Very Demanding Mr. Wallace."
Newsday (Hempstead, New York), 13 October 1988.

Darrach, Brad. "Mike Wallace: The Grand in at 75." *Life* (New York), June 1993.

Griffith, Thomas. "Water-torture Journalism." *Time* (New York), 23 May 1983.

Hall, Jane. "The Frustrations of Tough Guy Mike Wallace." *Los Angeles Times,* 26 September 1990.

Kaplan, Peter W. "For Mike Wallace: A Welcome Pause from Trial." *The New York Times,* 21 December 1984.

King, Susan. "Q and A: Mike Wallace: 40 Years of Asking." *Los Angeles Times,* 23 September 1990.

Lardner, James. "Up Against the Wallace." *Washington* (D.C.) *Post,* 18 September 1977.

Moore, Mike. "Divided Loyalties: Peter Jennings and Mike Wallace in No-Man's-Land." *The Quill* (Chicago), February 1989.

"Myron Leon Wallace: Fifth Estater." *Broadcasting* (Washington, D.C.), 9 March 1992.

Rosenthal, Donna. "Mike without Malice." *San Francisco Chronicle,* 23 September 1990.

"*60 Minutes*: A Candid Conversation about Hard News, Muckraking and Showbiz with the Creator and Correspondents of America's Most Trusted Television Show." *Playboy* (Chicago), March 1985.

Vietnam: A Documentary Collection—Westmoreland v. CBS. New York: Clearwater, 1985.

Weintraub, Joanne. "Mike Wallace Takes a Look at the 20th Century." *St. Louis* (Missouri) *Post Dispatch,* 12 July 1995.

See also Anchor; *60 Minutes*; Talk Shows; *Uncounted Enemy*

WALSH, MARY

Canadian Performer

Mary Walsh can be credited with single-handedly bringing Newfoundland culture to the rest of Canada through the medium of television. As the creator and co-star of *This Hour Has 22 Minutes,* Walsh has won three Gemini Awards, Canada's television honours. The bitingly satirical show has become a favourite, skewering politics in general, Toronto in particular, and anything else that strikes Walsh's fancy. No topic is taboo. The show takes its title from the outrageously controversial newsmagazine show *This Hour Has Seven Days,* which ran on CBC from 1964 to 1966.

A Canadian precursor to Britain's Tracey Ullman, the 43-year-old Walsh has introduced Canadian audiences over the years to a range of wacky Newfoundland archetypes, including the sharp-tongued, purple-housecoated know-it-all, Marg Delahunty, and the slovenly rooming-house owner, Mrs. Budgell. Her co-stars, fellow Newfoundlanders Cathy Jones, Greg Thomey, and Rick Mercer, all write their own characters as well.

Walsh's off-the-wall but pointed humour results in part from her unusual upbringing in St. John's, the capital of Newfoundland. One of eight siblings, at the age of eight months she contracted pneumonia and was dispatched next door to live with a still-beloved maiden aunt. She thus grew up away from her own troubled and hard-drinking family, feeling abandoned. She was also influenced by the strict rules of a convent education in the overwhelmingly Roman Catholic province of Newfoundland.

After taking acting classes at Ryerson Polytechnical Institute in Toronto and working a summer job at CBC radio in St. John's, Walsh began acting at Theatre Passe Muraille in Toronto. It was there that she met Cathy Jones, Dyan Olsen, Greg Malone, and Tommy Sexton; together they would become the comedy troupe CODCO, named after the fish which has, until recently, supported the Newfoundland culture and economy for hundreds of years. Their first production, *Cod on a Stick* (1973), was a play based on the experiences of Newfoundlanders in Toronto. It was a time of "Newfie jokes," Canada's equivalent of the racist "Polack jokes." But CODCO turned the tables on Torontonians, forcing them to laugh at themselves.

After touring the play successfully throughout Newfoundland, CODCO stayed in their home province and continued to develop the wickedly satirical sketches and characters, which they soon parlayed into the CBC television series *CODCO.* The half-hour show lasted seven seasons, from 1987 to 1993, reaching a nationwide audience.

Politicians are a particular target of the left-wing Walsh's wrathful humour: referring to Preston Manning, the conservative leader of the Reform Party, she put these words in the mouth of Marg Delahunty: "I've always enjoyed Mr. Manning's speeches. And I'm sure they're even more edifying in the original German." About a right-wing media figure, she has this to say: "That's typical of those people: they want everything—all the power and the money, and the right to call themselves victims too." Of the ongoing one-way rivalry between Newfoundland and Toronto, she has said: "I forgive Toronto and all the people in it. Toronto was the first large city I ever went to and I thought every large city was like that—cold and icy, like being in Eaton's [department store] all the time. But then I realized…it's very much a part of being specifically Toronto. It is just its outward style." She also jabs at the United States, describing her short stay in Colorado after high school and her exasperation at some Americans' misguided belief that they defeated Canada in the War of 1812.

Walsh, who is actively involved in social issues through her work in the theatre, won the Best Supporting Actress Award at the Atlantic Film Festival in 1992 for her performance in *Secret Nation,* and has guest-starred on the children's show *The Adventures of Dudley the Dragon.* She also starred as Molly Bloom at Ottawa's National Arts Centre, as well as in Eugene O'Neill's *A Moon for the Misbegotten,* in London, Ontario. In 1992 she directed Ann-Marie MacDonald's *Goodnight Desdemona, Good Morning Juliet* at Montreal's Centaur Theatre.

—Janice Kaye

MARY WALSH. Born in St. John's, Newfoundland, Canada, 1952. Studied at Ryerson Polytechnical Institute, Toronto. Began career at CBC radio, St. John's, Newfoundland; began acting career at Theatre Passe Muraille, Toronto; co-founder, *CODCO* performance group, 1973; toured Canada with *CODCO,* 1970s-80s; with *CODCO* television program, 1987–93; in film, from 1991. Recipient: Best Supporting Actress, Atlantic Film Festival, 1992; numerous Gemini Awards.

TELEVISION SERIES

1987–93	*CODCO*
1993–	*This Hour Has 22 Minutes*

TELEVISION MINISERIES

1993	*The Boys of St. Vincent*

FILMS

Secret Nation, 1991; *Buried on Sunday,* 1993.

STAGE

A Moon for the Misbegotten; Goodnight Desdemona, Good Morning Juliet (director).

See also Canadian Broadcasting in English; *CODCO*

WALT DISNEY PROGRAMS (VARIOUS TITLES)

U.S. Cartoons/Films/Children's Programming

Walt Disney was not only one of the most important producers in motion picture history, but one of the most important producers in American television history as well. He pioneered a relationship between the motion picture industry and the fledgling television industry, helped ensure the success of a third television network, promoted the transition from live broadcasts to film, and championed the conversion to color television in the mid-1960s.

Although Disney was quoted in the 1930s as having no interest in television, that opinion had changed drastically by the early 1950s, when television burst onto the American social scene. On Christmas Day in 1950 for NBC and again in 1951 for CBS, Disney produced hour-long specials that employed a number of clips from various Disney films and short subjects. Both specials achieved excellent ratings, and soon all three networks were wooing Disney to create an entire series for them.

Disney's interest in television was stimulated by his attempts to construct the Disneyland theme park in Anaheim, California. Encountering difficulty in financing the project, Walt offered network executives a television series in return for a substantial investment in the park. ABC, trailing substantially behind NBC and CBS, had just merged with United Paramount Theatres in 1953, and used this new influx of cash to fulfill Disney's request. The resultant anthology series, appropriately named *Disneyland,* premiered in late 1954, quickly becoming the first ABC program to crack the Nielsen Top 20.

Disney's relationship with ABC contradicted the strategy espoused by the rest of the film industry. During this period Hollywood studios viewed television as a competitor to motion pictures, and attempted to crush the medium. Walt, on the other hand, quickly saw TV's potential as a promotional tool. The first two specials combined old foot-age with promotions for upcoming theatrical releases such as *Alice in Wonderland* (1951). Disney's first Emmy would be awarded for an hour-long *Disneyland* episode about the filming of *20,000 Leagues Under the Sea* (1954) titled "Op-

Annette Funicello of The Mickey Mouse Club
Photo courtesy of the Walt Disney Company

eration Undersea," but humorously known within the industry as "The Long, Long Trailer." The series also worked to advertise the park, with individual episodes devoted specifically to its construction.

Soon, other studios were attempting to duplicate Disney's success. Series such as *The MGM Parade* and *Warner Brothers Presents* quickly appeared promoting their latest releases. They disappeared almost as quickly, mainly because Disney and his studio had constructed a unique image for themselves as producers of family entertainment. With a backlog of animated features and shorts, Disney came to television already known for entertaining children around the world (knowing the value of this backlog, Disney held onto the television rights to all of his films, at a time when all of the other studios were raising revenue by selling off the permanent television rights to their entire pre-1948 film catalog). From years of marketing towards children, Disney understood how children could influence their parents to buy products. After *Disneyland*'s "Davy Crockett" episodes created a merchandising phenomenon, *The Mickey Mouse Club*, a daily afternoon series, was introduced. One of the first attempts by television programming to target children, advertisers now conceived of children as a marketable group and initiated a tradition of weekday afternoon programming oriented toward younger audiences.

The studio's background in film production led to the decision to film the episodes, allowing for higher production values, rather than performing them live. The high-quality look of the series (and the subsequent involvement of other film studios in television production) helped shift television programming from live broadcasts to filmed entertainment. Long before color television technology became regulated and promoted, *Disneyland* episodes were filmed in color. Disney would promote the conversion to color when the anthology series, renamed *Walt Disney's Wonderful World of Color,* moved in 1961 to NBC, which was beginning color broadcasts.

Disney's importance to television as a producer of programming is incalculable. His success had an enormous effect on decisions by motion picture studios to enter into television production, thus guaranteeing programming for the fledgling medium. Yet, Disney is also important as a television icon as well. Working as host for the anthology series until the end of his life in 1966, Walt Disney quickly became identified by most children as "Uncle Walt." With an easy-going manner and a warm smile, Walt spoke to viewers in a Midwestern twang, enthusiastically demonstrating how certain special effects were created for his films, explaining the latest advances in space technology, or narrating a beloved fairy tale accompanied by scenes from his animated features. Usually filmed in a set that looked like his studio office, Walt gave the impression that he would drop all business to spend some time with his audience or engage in banter with cartoon characters Mickey Mouse and Donald Duck (who "magically" interacted with Walt as if they actually existed in the same space

with him). More than in any other way, Walt's presence and persona helped represent his company as promoters of American family values, and television itself as a "family medium." Even after his death, the company's television productions and subsequent cable channel reinforce that image of wholesome family entertainment.

—Sean Griffin

EXECUTIVE PRODUCER/HOST   Walt Disney (1954–1966), Michael Eisner (1986–1990)

PROGRAMMING HISTORY

- ABC

October 1954–September 1958	Wednesday 7:30-8:30
September 1958–September 1959	Friday 8:00-9:00
September 1959–September 1960	Friday 7:30-8:30
September 1960–September 1961	Sunday 6:30-7:30

- NBC

September 1961–August 1975	Sunday 7:30-8:30
September 1975–September 1981	Sunday 7:00-8:00

- CBS

September 1981–January 1983	Saturday 8:00-9:00
January 1983–February 1983	Tuesday 8:00-9:00
July 1983–September 1983	Saturday 8:00-9:00

- ABC

February 1986–September 1987	Sunday 7:00-9:00
September 1987–September 1988	Sunday 7:00-8:00

- NBC

October 1988–July 1989	Sunday 7:00-8:00
July 1989	Sunday 8:00-9:00
August 1989–May 1990	Sunday 7:00-8:00
May 1990–July 1990	Sunday 7:00-9:00
July 1990–August 1990	Sunday 8:00-9:00
August 1990–September 1990	Sunday 7:00-8:00

FURTHER READING

Anderson, Christopher. *Hollywood/TV.* Austin, Texas: University of Texas Press, 1994.

Flower, Joe. *Prince of the Magic Kingdom: Michael Eisner and the Re-Making of Disney.* New York: Wiley, 1991.

Greene, Katherine, and Richard Greene. *The Man Behind the Magic: The Story of Walt Disney.* New York: Viking, 1991.

Holliss, Richard, and Brian Sibley. *The Disney Studio Story.* New York: Crown, 1988.

Schickel, Richard. *The Disney Version.* New York: Simon and Schuster, 1968.

Smoodin, Eric, editor. *Disney Discourse: Producing the Magic Kingdom.* New York: Routledge, 1994.

See also Cartoons; Disney, Walt; Eisner, Michael

WALTERS, BARBARA

U.S. Broadcast Journalist

Although Barbara Walters would later downplay her relationship with the feminist movement, her early career is marked by a number of moves that were in part responsible for breaking down the all-male facade of U.S. network news. A *Today* show regular for 15 years, including two years as the first official female co-host, she was originally a visible presence in the program's "feature" segments, and then went on to cover "hard news"—including President Richard Nixon's historic visit to the People's Republic of China in 1972, when she was part of the NBC News team. Her most controversial breakthrough involved her decision in 1976 to leave *Today* to co-anchor the *ABC Evening News* with Harry Reasoner, the first time a woman was allowed the privileged position of network evening anchor, for a record-breaking seven-figure salary. Public reaction to both her salary and approach to the news, which critics claimed led to the creeping "infotainment" mentality which threatens traditional (male) reporting, undercut ABC News ratings, and she was quickly bumped from the anchor desk.

After this public-relations disaster, Walters undertook a comeback on ABC with *The Barbara Walters Special*, an occasional series of interviews with heads of state, newsmakers, sports figures, and Hollywood celebrities that have consistently topped the ratings and made news in themselves. In 1977, she arranged the first joint interview with Egypt's President Anwar Sadat and Israel's Prime Minister Menachem Begin; she has since interviewed six U.S. presidents, as well as political figures as diverse as British Prime Minister Margaret Thatcher, U.S. presidential contender Ross Perot, and Russian Federation President Boris Yeltsin. In 1984, ABC returned her to an anchor desk as co-host (with Hugh Downs) of the newsmagazine *20/20*.

Despite her status as both national celebrity and the recipient of numerous awards from journalists, television broadcasters, and women's groups, public reaction to Walters has remained ambivalent, perhaps as a result of changing notions of the nature of "news" in the television era. Walters' interviews have not been limited to figures embroiled in the matters covered by "hard news" subjects like politics and war; many of her more popular specials (and *20/20* segments) have been celebrity interviews and chats with more tawdry news figures. Certain memorable moments—such as the time she asked actress Katherine Hepburn what kind of tree she would like to be—have worked to undercut her image as a serious journalist. The late Gilda Radner's classic parody of Walters' distinctive style as "Baba Wawa" on *Saturday Night Live* remains popular as a timeless critique of the cult of personality in television journalism.

Walters began her career in broadcast journalism as a writer for CBS News. She also served as the youngest producer with NBC's New York station, WNBC-TV, before joining *Today*. After less than a year as a writer for *Today*, she was promoted to reporter-at-large (or, as then-host Hugh Downs described her, "the new '*Today* girl'"), although gender politics at the time severely constrained her role. According to Walters, she was not allowed to write for the male correspondents or to ask questions in "male-dominated" areas such as economics or politics, and she was forbidden to interview guests on-camera until all of the men on *Today* had finished asking their questions. Thanks in part to Walters's contributions, these commandments no longer apply.

—Susan McLeland

BARBARA WALTERS. Born in Boston, Massachusetts, U.S.A., 25 September 1931. Educated at Sarah Lawrence College, Bronxville, New York, B.A. in English 1953. Married: 1) Robert Katz (annulled), child: Jacqueline Dena; 2) Lee Guber, 1963 (divorced, 1976); 3) Merv Adelson, 1986 (divorced, 1992). Worked as a secretary at an advertising agency; assistant to the publicity director, NBC's WRCA-TV, New York; producer and writer, WRCA; writer and producer, WPIX Radio and CBS-TV; worked for a theatrical public-relations firm; hired for NBC's *Today* show, 1961, regular panel member, 1964–74, co-host, 1974–76; moderator of the syndicated program *Not for Women Only*, 1974–76; newscaster, *ABC Evening News*, 1976–78; host, *The Barbara Walters Special*, since 1976; co-host, ABC-TV news

Barbara Walters
Photo courtesy of Barbara Walters

show *20/20,* since 1984. L.H.D.: Ohio State University, 1971, Marymount College, 1975, and Wheaton College, 1983. Recipient: National Association of Television Program Executives Award, 1975; International Radio and Television Society's Broadcaster of the Year, 1975; Emmy Awards, 1975, 1980, 1982, and 1983; Lowell Thomas Award, 1990; International Women's Media Foundation Lifetime Achievement Award, 1992; Academy of Television Arts and Sciences Hall of Fame, 1990. Address: *20/20,* 147 Columbus Avenue, 10th Floor, New York City, New York 10023, U.S.A.

TELEVISION

1961–76	*Today* (co-host from 1974–76)
1974–76	*Not for Women Only*
1976–78	*ABC Evening News* (co-anchor)
1976–	*The Barbara Walters Special*
1984–	*20/20*

RADIO

Emphasis, early 1970s; *Moderator,* early 1970s.

PUBLICATION

How to Talk to Practically Anybody about Practically Anything. New York: Doubleday, 1970.

FURTHER READING

Carter, Bill. "Tender Trap." *The New York Times Magazine,* 23 August 1992.
Gelfman, Judith S. *Women in Television News.* New York: Columbia University Press, 1976.
Gottlieb, Martin. "Dangerous Liaisons: Journalists and Their Sources." *Columbia Journalism Review* (New York), July-August 1989.
Miller, Mark Crispin. "Barbara Walters's Theater of Revenge." *Harper's Magazine* (New York), November 1989.
Oppenheimer, Jerry. "The Barbara: 20/20 Vision Reveals Ms. Walters as Queen B." *Washington Journalism Review,* May 1990.
Paisner, Daniel. *The Imperfect Mirror: Inside Stories of Television Newswomen.* New York: Morrow, 1989.
Reed, Julia. "Woman in the News." *Vogue* (New York), February 1992.
Sanders, Marlene, and Marcia Rock. *Waiting for Prime Time: The Women of Television News.* Urbana: University of Illinois Press, 1988.
Wulf, Steve. "Barb's Wired." *Time* (New York), 6 November 1995.

See also Anchor; Gender and Television; News, Network

THE WALTONS

U.S. Drama

The Waltons was a highly successful family drama series of the 1970s, which portrayed a sense of family in sharp contrast to the problem-ridden urban families of the "socially relevant" sitcoms such as *All in the Family, Maude* or *Sanford and Son,* which vied with it for top billing in the Nielsen ratings. Set in the fictitious rural community of Walton's Mountain, Virginia, during the 1930s, the episodic narrative focused upon a large and dignified, "salt-of-the-earth" rural white family consisting of grandparents, parents and seven children. Based upon the semi-autobiographical writings of Earl Hamner, Jr., much of the early narrative was enunciated from the perspective of the oldest son, John Boy, an aspiring writer. The series was based on Hamner's novel *Spencer's Mountain,* which had been made into a feature film of the same name and subsequently adapted as a CBS-TV holiday special, *The Homecoming,* in 1971. The initial public reaction to the special was so overwhelming that executives Lee Rich and Bob Jacks of the newly-formed Lorimar Productions convinced CBS to continue it as a series, with Hamner as co-executive producer and story editor.

Lorimar executives constructed the series to emphasize both the locale (the Blue Ridge Mountains) and the historical period (the Great Depression), hoping to evoke a nostalgia for the recent past. They proposed to walk that fine line between "excessive sentimentality and believable human warmth," and took care not to caricature the mountain culture of the family, desiring to portray them as descendants of pioneer stock rather than stereotypical "hillbillies." Production notes in the Hamner papers emphasized the respect to be afforded the family and its culture: "That the Waltons are poor should be obvious, but there should be no hint of squalor or debased living conditions usually associated with poverty." Producers also stressed that *The Waltons* would not be like earlier wholesome family series *Father Knows Best* or *I Remember Mama* transplanted to the Blue Ridge Mountains of Virginia, but instead would be "the continuing story of a seventeen-year old boy who wants to be a writer, growing up during the Depression in a large and loving family."

Premiering in the fall of 1972, the hour-long dramatic series was scheduled in what was considered a "suicidal" time slot against two popular Thursday-night shows, ABC's *The Mod Squad* and NBC's top-rated *The Flip Wilson Show.* By its second season, *The Waltons* achieved the valedictory rank in the overall ratings, and stayed in the top 20 shows for the next several years. During its first season, the series garnered Emmy Awards for Outstanding Drama Series, Best Dramatic Actor (Richard Thomas) and Actress (Michael Learned), Best Supporting Actress (Ellen Corby) and Best Dramatic Writing (John McGreevey), and continued to receive Emmys for acting and/or writing for the next half a decade. The series endured until 1981, with the extended family maturing and changing—surviving the loss of some characters, the addition of new supporting characters, and the socio-historical changes as the community weathered the Depression era and entered that of World War II. The cast has reunited for a number of holiday and wedding specials in the nearly 15 years since the series ended, and the Walton family has endured in America's mythic imagination as well as in ratings popularity.

The Walton family was portrayed as a cohesive and nearly self-sufficient social world. The family members operated as a team, full of collective wisdom and insight, yet always finding narrative (and physical) space for their individuality. In addition to the continuing narrative development of each regular character and of the family dynamics over the course of the series, each episode frequently dealt with a conflict or tension introduced by an outsider who happened into the community (Ziegler described these characters as "foreigners, drifters, fugitives, orphans, and others just passing through") bringing their own problems which were potentially disruptive influences upon the harmony and equilibrium of the Walton's Mountain community. The narrative of each episode worked through the resolution of these tensions within the household, as well as the healing or spiritual uplift achieved by the outsider characters as they assimilated the values of the family and learned their lessons of love and morality.

The series was critically praised as being bittersweet, "wholesome", emotion-laden viewing. Reviewers noted that the series conveyed a vivid authenticity of both historical time and cultural place, as well as an emotional verisimilitude regarding the portrayal of a certain type of family life rooted in that time and place. Devoted viewers besieged the network, producers and cast members with fan letters praising the show and expressing their degree of emotional identification with many aspects of the series. Many considered the series to be the epitome of television's capacity for romantic, effective and moving storytelling in its evocation of childhood and its ability to tap into a deep desire for a mythicized community and family intimacy.

Yet the series also had its detractors, who complained that *The Waltons* was too sweet, sappily sentimental, and exploitative of viewers' emotions. Crowther remarked that its "homey wisdom and Sunday school platitudes have been known to make me gag"; others labeled it an "obviously corny, totally unreal family" with characters too good to be true. Many recognized in the show an "intolerable wistfulness" for a romanticized past constructed through the creation of false memory and hopeless longing. Some critics noted that such a romanticized image of the era could make viewers forget the real nature of rural poverty. "The Depression was not a time for the making of strong souls" or

The Waltons

healthy, well-nourished bodies, according to Roiphe, who criticized the series for associating poverty with elevated moral values and neutralizing the social, economic and political upheavals of the 1930s "behind a wall of tradition, goodness and good fortune." Roiphe noted how skillfully the media producers were able to design and articulate myths of American happiness and innocence during the historical period the series portrayed; however, the viewers who admired the series also eagerly participated in that construction of a mythical past. Other critics have noted that despite its embrace of liberal humanitarian values (against racism, etc.), *The Waltons'* inherent conservatism has made it ripe for appropriation by right-wing "family values" religious groups. Indeed, it has become a benchmark series for the Family Channel, the media outlet for Pat Robertson's Christian Coalition, which has held exclusive syndication rights for the series since 1991.

—Pamela Wilson

NARRATOR
Earl Hamner, Jr.

CAST

John Walton	Ralph Waite
Olivia Walton (1972–80)	Michael Learned
Zeb (Grandpa) Walton (1972–78)	Will Geer
Esther (Grandma) Walton (1972–79)	Ellen Corby
John Boy Walton (1972–77)	Richard Thomas

John Boy Walton (1979–81) . . . . . Robert Wightman
Mary Ellen Walton Willard . . . . Judy Norton-Taylor
Jim-Bob Walton . . . . David W. Harper
Elizabeth Walton . . . . Kami Cotler
Jason Walton . . . . Jon Walmsley
Erin Walton . . . . . . Mary Elizabeth McDonough
Ben Walton . . . . Eric Scott
Ike Godsey . . . . Joe Conley
Corabeth Godsey (1974–81) . . . Ronnie Claire Edwards
Sheriff Ep Bridges . . . . John Crawford
Mamie Baldwin . . . . Helen Kleeb
Emily Baldwin . . . . Mary Jackson
Verdie Foster . . . . Lynn Hamilton
Rev. Matthew Fordwick (1972–77) . . . . . John Ritter
Rosemary Hunter Fordwick (1973–77)   Mariclare Costello
Yancy Tucker (1972–79) . . . . Robert Donner
Flossie Brimmer (1972–77) . . . . Nora Marlowe
Maude Gormsley (1973–79) . . . . Merie Earle
Dr. Curtis Willard (1976–78) . . . . Tom Bower
Rev. Hank Buchanan (1977–78) . . . . Peter Fax
J. D. Pickett (1978–81) . . . . Lewis Arquette
John Curtis Willard (1978–81)[Alternating]
    . . . . Marshall Reed/Michael Reed
Cindy Brunson Walton (1979–81) . . . . Leslie Winston
Rose Burton (1979–81) . . . . Peggy Rea
Serena Burton (1979–80) . . . . Martha Nix

Jeffrey Burton (1979–80) . . . . Keith Mitchell
Toni Hazleton (1981) . . . . Lisa Harrison
Arlington Wescott Jones (Jonesy) (1981)   Richard Gilliland

PRODUCERS   Lee Rich, Earl Hamner, Jr., Robert L. Jacks, Andy White, Rod Peterson

PROGRAMMING HISTORY   178 Episodes

• CBS
September 1972–August 1981   Thursday 8:00-9:00

FURTHER READING
Crowther, Hal. "Boxed In." *The Humanist* (Buffalo, New York), July/August 1976.
Hamner, Earl, Papers. The Waltons Mountain Museum in Schuyler, Virginia.
Roiphe, Anne. "*The Waltons.*" *New York Times Magazine*, 18 November, 1973.
"Wholesome Sentiment in the Blue Ridge." *Life* (New York), 13 October 1972.
Ziegler, Robert E. "Memory-spaces: Themes of the House and the Mountain in *The Waltons.*" *Journal of Popular Culture* (Bowling Green, Ohio), 1981.

See also Family on Television; Melodrama

THE WAR GAME

British Drama

Over thirty years since its production, *The War Game* remains the most controversial and, perhaps, the most telling television film on nuclear war. Directed by the young Peter Watkins for the BBC, its depiction of the impact of Soviet nuclear attack on Britain caused turmoil at the corporation and in government. Although it went on to win an Oscar for Best Documentary Feature in 1966, it was denied transmission until 1985. Announcing the decision to hold back *The War Game* in 1965, the BBC explained that the film was too horrifying for the medium of broadcasting, expressing a particular concern for "children, the very old or the unbalanced."

But BBC internal documents, and newly released Cabinet papers of the period, reflect the high degree of political anxiety generated by the film, and suggest that although the BBC was keen to assert its independence and its liberalism, *The War Game* was indeed the victim of high-level censorship. The popular press of the day, for their part, largely approved the ban, often reading the film as propaganda for the youthful Campaign for Nuclear Disarmament.

The film imagines a period of some four months from the days leading up to nuclear attack. In a show of solidarity with the Chinese invasion of South Vietnam, the Russian and East German authorities have sealed off all access to

Berlin, and have threatened to invade the western sector of the city unless the United States withdraws its threat to use tactical nuclear weapons against the invading Chinese. When two NATO divisions attempt to reach Berlin, they are overrun by communist forces, triggering the U.S. president's release of nuclear warheads to NATO. The U.S.S.R. calls NATO's bluff, leading to a preemptive strike by the Allies and, in a self-protective measure, the Soviet launch against Britain.

The War Game, shot in newsreel-style black and white, and running just over three-quarters of an hour, works on a number of levels. The main discourse is that of the documentary exposition itself, chronicling and dramatising the main stages and the key features of the countdown to attack and its immediate consequences. A second discourse, also playing on the relationship between documentary and drama, takes the form of two types of *vox pop* interviews which punctuate the text: interviews which illustrate the contemporary public's consciousness of the issues, exposing wide-spread ignorance; and clearly fictional interviews with (imaginary) key figures as the attack scenario itself develops and extends.

Further elements go some way to suggesting contexts for the public's failure to perceive the realities of nuclear war.

The War Game
Photo courtesy of the British Film Institute

One strand of the film highlights the pathetically inadequate information purveyed by the official Civil Defence self-help manual (cover-price: nine old pence). A fourth level of comment, provided by inter-titles, exposes the bankruptcy of statements on the nuclear threat emerging from religious sources such as the Ecumenical Council of the Vatican.

The film concentrates on the southeast England, and, in particular, the town of Rochester in Kent. It bleakly illustrates the social chaos of the period before attack, focusing on the personal and ideological conflicts likely to arise from the enforced evacuation of large numbers of the urban population, and the impracticality of building viable domestic shelters—as the price of basics such as planks and sandbags in any case escalates—against the power of the nuclear bomb. It depicts the immediate horrors of a nuclear explosion by invoking memories of the firestorms of Dresden and Hiroshima, the earthquakes and the blinding light, thirty times more powerful than the midday sun, which is capable of melting upturned eyeballs from many miles away.

The remainder of the film concentrates on the rapid disintegration of the social fabric in the aftermath of the attack, as civilisation disappears. In images of chilling and provocative power, policemen are depicted as executioners

of the terminally ill and of minor criminals. The effects of radiation sickness are explained and illustrated, along with the psychological devastation which would befall survivors and the dying in a mute and apathetic world. There is a good chance of all this happening, the film suggests, by 1980.

The film's enduring power thus derives from a variety of sources. These include its cool articulation of momentary images—a child's eyes burned by a distant nuclear air-burst as the film itself goes into negative; a bucketful of wedding rings collected as a register of the dead, a derelict building which has become an impromptu furnace for the incineration of bodies too numerous to bury; "Stille Nacht" playing on a gramophone which, in the absence of electricity, must be turned by hand.

At a structural level the film achieves its overall rhetorical power through both its mixture and its separation of documentary and dramatic modes. It does not, for example, offer the purely "dramatic" spectacle of later TV nuclear dramas such as the U.S. *The Day After* (1983) or the British *Threads* (1984), with their more traditional identifications around character and plot. Nor does it simply document the drama in the manner of Watkins' previous *Culloden* (1965), in which the television camera revisits the battlefield of 1746 and interviews participants, or of *Cathy Come Home* (1966), Ken Loach's similar merging of the domains of documentary and drama in accounting for the rising problem of homelessness in 1960s Britain.

The War Game, on the contrary, confuses and yet demarcates the two modes. The "dramatic" sequences, with their highly "documentary" look, are retained as fragmentary and discontinuous illustrations of an ongoing documentary narrative which itself disorientingly moves back and forth between statements and assumptions that this is "really happening" before our eyes, and other types of proposition and warning that this is how it "could be" and "might look."

The British television audience was deprived of *The War Game* for two decades, until a moment in history ironically close to the events in Eastern Europe which canceled the particular Cold War scenario which underpins the film. Its banning nonetheless made the film a *cause celebre* and its notoriety grew in the troubled later 1960s as the film reached significant minority audiences in art-house cinemas and through the anti-nuclear movement. Introducing the

1985 broadcast, Ludovic Kennedy estimated that, by then, the film had already reached as many as six million viewers.
—Phillip Drummond

PRODUCER   Peter Watkins

PROGRAMMING HISTORY   Produced in 1965

• BBC-1
31 July 1985

FURTHER READING

Gomez, Joseph A. *Peter Watkins*. Boston, Massachusetts: Twayne, 1979.
Watkins, Peter. *The War Game: An Adaptation of the BBC Documentary*. New York: Avon, 1967.
Welsh, James Michael. *Peter Watkins: A Guide to References and Resources*. Boston, Massachusetts: G.K. Hall, 1986.

See also War on Television; Watkins, Peter

WAR ON TELEVISION

War on television has been the subject of both fictional accounts and extensive, often compelling, news coverage. War and other bellicose activities have inspired television documentaries, docudramas, dramatic series and situation comedies. Fictional accounts of war and documentary accounts of historical wars, however, are not discussed in this article, which focuses instead on televised coverage of contemporary warfare and related military actions.

The first noteworthy war to occur in the television age was the Korean War (1950–53). Television was, of course, in its infancy as a mass medium at the time and, as a consequence, the Korean conflict is not widely thought of as a televised war. Not only did relatively few viewers have access to television sets, but, because satellite technology was unavailable, television film had to be transported by air to broadcasters. By the time such film arrived its immediacy was much diminished; often, therefore, newspapers and radio remained the media of choice. Nonetheless, in August 1950, a CBS television news announcer reported an infantry landing as it was in-progress, and the controversy caused by this possible security breach reflects conflicts that would long continue between military authorities waging war and television reporters covering that warfare.

In some national contexts, concern about security has sometimes led to formal legal censorship of television war coverage, although, as frequently, physical or technological obstacles inherent to television broadcasting from theaters of war or erected by military personnel at the scene of a conflict served the same censorship purpose. Debates about censorship raged during many of the post-War European military campaigns to maintain control over the many col-

A soldier in Vietnam

"The Troubles" of Northern Ireland
Photo courtesy of AP/Wide World Photo

onies that would eventually achieve national independence. Informal censorship was frequent, however, as when during the 1956 Suez expedition British media were requested to refrain from reporting certain information, but were not forced to do so under penalty of law.

Television coverage also inspired controversy during the Vietnam War (1962–1975). Despite clear evidence that the war effort was less than successful in objective terms, popular opinion and much expert military opinion regard the Vietnam War as one that could have been won on the battlefield but was lost in the living room (where viewers watched their television sets). Reporters who covered the war in the early 1960s remember, however, that most of that early coverage was laudatory and that, in the words of Bernard Kalb who would later join the Cable News Network (CNN), there was "an awful lot of jingoism . . . on the part of the press in which it celebrated the American involvement in Vietnam." Methodical scholarly accounts of tele-

vised coverage also uniformly discover that television coverage was inclined overall to highlight positive aspects of the Vietnam War and that viewers exposed to the most televised coverage were also most inclined to view the military favorably. Nevertheless, domestic social schisms blamed on the Vietnam War and the war's ultimate failure to sustain a non-Communist regime in Vietnam are often blamed on television and other media.

Whether the public turned against the Vietnam War because television, in particular, and the media, in general, presented it unfavorably, or whether the public turned against the war because media accurately depicted its horrors and television did so most graphically remains an open and hotly contested question in the public debate. There is, however, no historic evidence to prove that a graphic portrayal of war disinclines a viewing public to engage in a war. Some critics suggest that the opposite may be the case when a public considers a war justified and is exposed to images of its side enduring great suffering.

Despite a less than definitive understanding of television coverage and its impact on popular support for war efforts, military strategists began to integrate domestic public relations strategy and overall military strategy during the Vietnam War. As the war progressed, military analysts continued to debate whether it was appropriate for the military to attempt to influence civilian public policy through such efforts. Within military circles and in the wake of the Vietnam War, most such debates were left behind and media relations strategies went far beyond censorship and toward a full-fledged engagement (some say co-optation) of televised media.

During 1976 naval conflicts between Britain and Iceland over fishing rights, strategies to influence televised coverage were used by the Icelandic side to depict Britain as the aggressive party, while the British Navy still even refused to allow television crews on its ships. As late as the 1982 Falklands/Malvinas War, during which Britain successfully regained control of the South Atlantic islands that Argentina's military government had invaded, British military strategists had yet to develop a comprehensive media strategy. Although, the British Navy did allow television and other media personnel to travel aboard its ships to the geographically isolated Falklands/Malvinas Islands, the British did not control the content of the war coverage by systematically influencing television media.

The following year when the United States invaded Grenada, concerns regarding less than favorable television coverage prompted military planners to exclude civilian in favor of military television camera crews. Sensitivity to unfavorable television coverage was heightened at this time by the deaths of 230 U.S. Marine and 50 French peacekeepers in a bomb attack during operations in Beirut. But in 1989, when the United States invaded Panama, the exclusion of civilian television crews was not possible and thanks to satellite technology and round-the-clock CNN coverage, television viewers were able to watch the progress of military operations with much immediacy. As had been the case during the early Vietnam War, the television media was generally inclined to stress the salutary aspects of the Panama Invasion, and U.S. military planners also did a more effective job of controlling the public perception of the invasion.

The short-lived nature of the Panama, Grenada, and Falklands/Malvinas operations may have also forestalled adverse public reactions among the civilian populations who watched their governments wage war on television. Some argue that television coverage makes short-lived military engagements more likely. Yet, despite many short-lived military endeavors, long-term warfare is still possible in the television age. Still, some observers suggest that lack of widely available independent television coverage is what makes long-term warfare palatable to the international community in contemporary times. The Iran-Iraq War (1980–88), for example, received often negligible international television coverage. Yet, the recent Civil Wars in former Yugoslavia (1991–) have continued at varying levels of intensity despite often extensive international coverage. Other extended or particularly brutal border conflicts, terrorist campaigns, coups d'état, civil wars and genocidal endeavors have also received sometimes varying levels of television coverage. Such latter-day wars have been waged in Algeria, Armenia, Azerbaijan, Cambodia, Chad, Chechnya, El Salvador, Ethiopia, Georgia, Guatemala, Liberia, Nigeria, Peru, Rwanda, the Sudan, Yemen and in other places far too numerous to mention.

Both the 1992 U.S.-led occupation of Somalia and the 1994 U.S.-led occupation of Haiti may have, however, failed to create much domestic opposition because of their short duration. The 1992 Somalia operation did, nonetheless, feature one of the most surreal interactions between military personnel and television film crews. This occurred when the first U.S. occupation forces landing on Somali beaches at night found their landings illuminated by the television lights of international news organizations. Criticism of the security risk this illumination entailed harks back to similar criticism of the 1950 CBS report on the infantry landing in Korea.

By far the most noteworthy recent interaction between military and television was occasioned not by a localized conflict but by the U.S.-led, internationally sponsored 1991 Gulf War against Iraq. In the aftermath of this war, television and other media were criticized for having failed to provide a balanced and complete account of the war. Some critics, most notably Douglass Kellner in the *The Persian Gulf TV War,* argue that television and other media failed to provide a balanced and complete account of the war because the corporate owners of commercial networks felt it was not in their business interest to do so. Other critics suggest that television coverage simply reflects popular prejudices. To a great extent, however, during the actual war, as in previous wars, the various national media had to rely on the military forces for access to events and for access to their broadcast networks. According to the *Wall Street Journal's* John Fialka, the importance of military cooperation is seen in this: that U.S. Marines, despite their smaller role in the war, received more U.S. news coverage than the U.S. Army, in part, because U.S. Marines were more dedicated to opening the lines of communication between reporters in their operations area and the reporters' news organizations back home. Overall, however, British television coverage—benefiting from access policies put in place after the Falklands/Malvinas War—featured the timeliest reports on front-line action. The British military forces were the only ones to allow satellite up-links near the front lines.

Military cooperation with the media also made possible the most notable television innovation during the 1991 Gulf War. This was the access broadcast television had to the closed-circuit video images that emanated from camera-equipped high-tech weaponry directed against Iraqi targets. Thanks to this access, television viewers were literally able to see from the viewpoint of missiles and other weapons as these bore down on Iraqi civilian and military targets—mostly vehicles, buildings and other inanimate infrastructure. Significantly, also according to the *Journal's* Fialka, videotape from cameras mounted on U.S. Army Apache helicopter-gunships "showing Iraqi soldiers being mowed down by the gunship's Gatling gun" were seen by a *Los Angeles Times* reporter but were suppressed thereafter and made unavailable for television broadcast.

Trejo Delarbe argues that sophisticated efforts to control television coverage were also attempted by Mexico's Zapatista Army of National Liberation during its uprising (1994–) against the central government—a particularly well-televised war in contrast to many listed above. Such efforts to control televised imagery have, indeed, been attempted as part of other military actions, guerrilla movements and terrorist campaigns, but a military's having actual control of the point-of-view of televised imagery is a phenomenon thus far almost unique to the Gulf War.

Indeed, lack of control sometimes seems to work in unexpected ways. This has often seemed to be the case in the present conflict in the former Yugoslavia. It has not been uncommon to see military actions from multiple perspectives, interviews with political and military leaders from all factions, human interest stories from within every combat zone, and analyses of the aftermath of battles and shelling from civilian as well as combatant or diplomatic points of view. And when a particularly bloody mortar attack on Sarajevo came at a time of tense diplomatic activity—apparent diplomatic failure to reach a settlement of the conflict—televised images and stories seemed to provide justification for increased military action by NATO forces in an attempt to force the parties to the settlement table.

In spite of such apparently random and opportunistic events that often define warfare, the control of televised imagery is, nevertheless, a logical consequence of military planners' increasing willingness to control the media relations aspects of warfare as if exercising this control were just another aspect of military strategy. Moreover, the ability to control televised imagery is also a consequence of the evolution of military technology. Far from the contentious early days, when most military organizations considered television coverage a mere nuisance or a possible security risk, cutting-edge military planners today use many aspects of television to prosecute wars or to prepare for them. As writers for *Wired* point out, today television technology is used to provide military personnel in training with images of war conditions or maneuvers and the next step in military technological development is said to include "virtual warfare". During such warfare military personnel will be safely ensconced at distant locations as televised imagery and other telemetry allows them to direct weaponry against remote targets. Such a prospect may well signify that, as media guru Marshall McLuhan wrote in 1968, "television war (will have) meant the end of the dichotomy between civilian and military."

—Donald Humphreys

FURTHER READING

Adams, Valerie. *The Media and the Falklands Campaign.* Basingstoke: Macmillan, 1986.

Arlen, Michael J. *Living-room War.* New York: Penguin Books, 1982.

Arnett, Peter. *Live From The Battlefield: From Vietnam to Baghdad, 35 Years in the World's War Zones.* New York: Simon and Schuster, 1994.

Bennett, W. Lance, and David L. Paletz, editors. *Taken By Storm: The Media, Public Opinion, and US Foreign Policy in the Gulf War.* Chicago: University of Chicago Press, 1994.

Braestrup, Peter. *Big Story: How the American Press and Television Reported and Interpreted the Crisis of Tet 1968 in Vietnam and Washington.* Boulder, Colorado: Westview, 1977.

Cumings, Bruce. *War and Television.* London and New York: Verso, 1992.

Dennis, Everette E., with others. *The Media at War: The Press and the Persian Gulf Conflict: A Report of The Gannett Foundation.* New York: Gannett Foundation Media Center, 1991.

Dennis, Everette E., George Gerbner, and Yassen N. Zassoursky, editors. *Beyond the Cold War: Soviet and American Media Images.* Newbury Park, California: Sage, 1991.

Denton, Robert E., Jr., editor. *The Media and the Persian Gulf War.* Westport, Connecticut: Praeger, 1993.

Der Derian, James. "Cyber-Deterrence: The US Army Fights Tommorrow's War Today." *Wired 2.09* (San Francisco, California), September, 1994.

Fialka, John J. *Hotel Warriors: Covering the Gulf War.* Washington, D.C.: Woodrow Wilson Center Press; 1992.

Gitlin, Todd. *The Whole World Is Watching: Mass Media in the Making and Unmaking of the New Left.* Berkeley: University of California Press, 1980.

Glasgow University Media Group. *War and Peace News.* Philadelphia, Pennsylvania: Open University Press, 1985.

Hallin, Daniel C. *The "Uncensored War": The Media and Vietnam.* New York: Oxford University Press, 1986.

Harris, Robert. *Gotcha! The Media, the Government, and the Falklands Crisis.* London, Boston: Faber and Faber, 1983.

Hammond, William M. *Public Affairs: The Military and the Media, 1962–1968.* Washington, D.C.: Center of Military History, U.S. Army, 1988.

Hofstetter, C. Richard, and David W. Moore. "Watching TV News and Supporting the Military." *Armed Forces and Society* (Chicago, Illinois), February 1979.

Jeffords, Susan, and Lauren Rabinovitz. *Seeing Through the Media: The Persian Gulf War.* New Brunswick, New Jersey: Rutgers University Press, 1994.

Kellner, Douglas. *The Persian Gulf TV War.* Boulder, Colorado: Westview, 1992.

Kennedy, William V. *The Military and the Media: Why the Press Cannot Be Trusted to Cover a War.* Westport, Connecticut and London: Praeger, 1993.

Lee, Martin, and Tiffany Devitt. "Gulf War Coverage: Censorship Begins at Home." *Newspaper Research Journal* (Memphis, Tennessee), Winter 1991.

Lefever, Ernest W. *TV and National Defense: An Analysis of CBS News, 1972-73.* Boston: Institute for American Strategy Press, 1974.

MacArthur, John R. *Second Front: Censorship and Propaganda in the Gulf War.* New York: Hill and Wang, 1992.

Magistad, Mary Kay. "Journalists Take a Look Back at Vietnam to Mark the 20th Anniversary of the End of the Vietnam War." *Morning Edition.* National Public Radio, 27 April 1995, Transcript # 1594–5.

McLuhan, Marshall. *War and Peace in the Global Village.* New York: McGraw-Hill, 1968.

Morrison, David E., and Howard Tumber. *Journalists at War: The Dynamics of News Reporting During the Falklands Conflict.* London: Sage, 1988.

Mowlana, Hamid, George Gerbner, and Herbert I. Schiller, editors. *Triumph of the Image: The Media's War in the Persian Gulf, A Global Perspective.* Boulder, Colorado: Westview, 1992.

Rubin, David M., and Ann Marie Cunningham, editors. *War, Peace and the News Media: Conference Proceedings, March 18 and 19, 1983.* New York: New York University, 1983.

Small, Melvin. *Covering Dissent: The Media and the Anti-Vietnam War Movement.* New Brunswick, New Jersey: Rutgers University Press, 1994.

Smith, Hedrick, editor. *The Media and the Gulf War.* Washington, D.C.: Seven Locks Press, 1992.

Taylor, Philip M. *War and the Media: Propaganda and Persuasion in the Gulf War.* Manchester and New York: Manchester University Press; St. Martin's, 1992.

Thompson, Loren B., editor. *Defense Beat: The Dilemmas of Defense Coverage.* New York: Lexington Books, 1991.

Thompson, Mark. *Forging War: The Media in Serbia, Croatia and Bosnia-Hercegovina.* London: Article 19, 1994.

Turner, Kathleen J. *Lyndon Johnson's Dual War: Vietnam and the Press.* Chicago: University of Chicago Press, 1985.

Twentieth Century Fund. *Battle Lines: Report of the Twentieth Century Fund Task Force on the Military and the Media.* New York: Priority Press, 1985.

Young, Peter R., editor. *Defense and the Media in Time of Limited War.* Portland, Oregon: Frank Cass, 1992.

See also Cable News Network; *China Beach*; Terrorism; *Vietnam: A Television History*; Vietnam on Television; *War Game*; Watkins, Peter

WARNER BROTHERS PRESENTS

U.S. Dramatic Series

Warner Brothers Presents, the first television program produced by Warner Brothers Pictures, appeared on ABC during the 1955–1956 season. Hosted by Gig Young, the series featured an omnibus format with weekly episodes drawn from three rotating series based loosely on the Warner Brothers movies *King's Row, Casablanca*, and *Cheyenne*. Although a one-hour series, each weekly episode reserved the final ten minutes for a segment titled "Behind the Cameras at Warner Brothers" This segment featured behind-the-scenes footage, revealing the inner workings of a major movie studio and promoting the studio's recent theatrical releases.

This short-lived series was a hit with neither critics nor viewers, and yet it still stands as a milestone because it marked the introduction of the major Hollywood studios into television production. The 1955-56 season saw the television debut not only of *Warner Brothers Presents*, but also of the *Twentieth Century-Fox Hour* on CBS and *MGM Parade* on ABC. The common inspiration for these programs was the success of *Disneyland*, which had premiered the previous season on ABC and had given Walt Disney an unprecedented forum for publicizing the movies, merchandise, and amusement park that carried the Disney trademark. Following Disney, Warner Brothers executives saw television as a vehicle for calling attention to their motion pictures. They were much less interested in producing for television than in using the medium to increase public awareness of the Warner Brothers trademark.

ABC had its own vested interests in acquiring a Warner Brothers series. By recruiting one of Hollywood's most venerable studios to television, ABC scored a valuable coup in its bid for respectability among the networks. As the perennial third-place network, ABC welcomed the glamour and prestige associated with a major Hollywood studio. The opening credits for *Warner Brothers Presents* pointedly reminded viewers of the studio's moviemaking legacy. As the screen filled with the trademark Warner Brothers logo superimposed over a soaring aerial shot of the studio, an announcer exclaimed, "From the entertainment capital of the world comes *Warner Brothers Presents.* The hour that presents Hollywood to you. Made for television by one of the great motion picture studios." Marketing the Warner Brothers' reputation, ABC signed contracts with several sponsors who had never before advertised on the network, including General Electric and the tobacco company Liggett and Myers, two of the largest advertisers in broadcasting.

The alternating series of *Warner Brothers Presents* were seen by both studio and network as an ongoing experiment in an effort to gauge the public taste for filmed television drama. *King's Row* was a pastoral melodrama about a small-town doctor (Jack Kelly) who returns home following medical school to aid the community members and play a role in various soothing tales of moral welfare. *Casablanca* reprised the Academy Award-winning movie, with Charles McGraw in the role made famous by Humphrey Bogart. Rick's Cafe Americain became the setting for tales of star-crossed romance and, to a much lesser extent, foreign intrigue. The only series to make a significant impression in the ratings was *Cheyenne*, a rough-and-tumble Western starring Clint Walker as a wandering hero who dispenses justice while riding through the Old West.

Since the studio's objective was to reach viewers with its promotional messages, the "Behind the Cameras" segments provided a fascinating glimpse into the production process at a movie studio. They introduced viewers to the various departments at the studio, demonstrating the role played by editing, sound, wardrobe, lighting, and so forth in the production of a motion picture. Each segment featured exclusive footage and interviews with top movie stars and directors. On the set of *Giant* a wry James Dean demonstrated rope tricks and, in a rather macabre twist given his untimely death, talked about traffic safety. A gruff John Ford commanded the Monument Valley location of *The Searchers*. Director Billy Wilder and Jimmy Stewart explained how they recreated Lindbergh's legendary flight in *The Spirit of St. Louis*.

When the series failed to find an audience, however, the advertisers balked at the studio's emphatic self-promotion in these segments, particularly when the studio seemed unable to create dramatically compelling episodes. Critics, sponsors, and network executives agreed that the dramatic episodes were formulaic in their writing and perfunctory in their production. In part, this reflected the economics of early telefilm production. The entire $3 million budget that ABC paid for thirty-nine hour-long episodes of *Warner Brothers Presents* represented only a fraction of the budget for a single studio feature like *Giant* or *The Searchers*. Consequently, episodes of *Warner Brothers Presents* were written, produced, and edited on minuscule budgets at a frenetic pace unseen at the studio since the B-grade movies of the 1930s.

After considerable tinkering—including the recycling of scripts from several of the studio's western movies—*Cheyenne* emerged as the sole hit among the *Warner Brothers Presents* series. Had its ratings been calculated separately, it would have finished the season among the twenty highest-rated series. Observing the success of the bluntly conflict-driven *Cheyenne*, ABC asked the studio to heighten the dramatic tension in both *King's Row* and *Casablanca*, fearing, in the words of ABC President Robert Kintner, that neither series was "lusty and combative" enough to appeal to viewers. New scripts were written for both series, introducing murderous kidnappers and mad bombers, but neither series found an audience, and they were both canceled before the end of the season. In their place, *Warner Brothers Presents* substituted an anthology series, *Conflict*, which alternated with *Cheyenne* for the remainder of the season and for the next.

Due to the difficulties in gearing up for the rapid pace of television production, Warner Brothers lost more than a half-million dollars on *Warner Brothers Presents*. But the studio also achieved two lasting benefits. First, with the production of this initial series Warner Brothers crossed the threshold into television production where, in just four years, it would become the largest producer of network series. Second, it launched the studio's first hit series, *Cheyenne*, which went on to have an eight-year run on ABC.

—Christopher Anderson

HOST

Gig Young

PROGRAMMING HISTORY

• ABC

September 1955–September 1956   Tuesday 7:30-8:30

CASABLANCA (September 1955–April 1956)

CAST

Rick Jason	Charles McGraw
Capt. Renaud	Marcel Dalio
Sasha	Michael Fox
Sam	Clarence Muse
Ludwig	Ludwig Stossel

CHEYENNE (See separate entry)

KING'S ROW (September 1955–January 1956)

CAST

Dr. Parris Mitchell	Jack Kelly
Randy Monaghan	Nan Leslie
Drake McHugh	Robert Horton
Dr. Tower	Victor Jory
Grandma	Lillian Bronson
Dr. Gordon	Robert Burton

FURTHER READING

Anderson, Christopher. *Hollywood TV: The Studio System in the Fifties.* Austin: University of Texas Press, 1994.

Balio, Tino, editor. *Hollywood in the Age of Television.* Boston: Unwin, Hyman, 1990.

Woolley, Lynn, Robert W. Malsbary, and Robert G. Strange, Jr. *Warner Brothers Television: Every Show of the Fifties and Sixties, Episode by Episode.* Jefferson, North Carolina: McFarland, 1985.

See also *Cheyenne*; Westerns

WATCH MR. WIZARD

U.S. Children's Science Program

Watch Mr. Wizard, one of commercial television's early educational efforts, was highly successful in making science exciting and understandable for children. Presenting scientific laboratory demonstrations and information in an interesting, uncomplicated and entertaining format, this long-running series was a prime example of the Chicago School of Television and of quality education in a visual format. Created and hosted by Don Herbert, the show's low key approach, casual ad lib style, and resourceful, often magic-like demonstrations led to rapid success and brought Herbert instant recognition and critical acclaim as an innovative educational broadcaster and as a teacher of science.

Donald Jeffry Herbert, a general science and English major at LaCrosse State Teachers College in Wisconsin, had originally planned to teach dramatics. Following his graduation in 1940, he acted in summer and winter stock and then traveled to New York with an eye toward Broadway. World War II interrupted his career and the young actor entered the Army Air Forces as a private. As a B-24 bomber pilot, he flew 56 missions with the Fifteenth Air Force and subsequently participated in the invasion of Italy. Discharged as a Captain in 1945, Herbert had earned the Distinguished Flying Cross and the Air Medal with three oak leaf clusters.

After the war, Herbert accepted offers of radio work in Chicago. He acted in such children's programs as *Captain Midnight, Jack Armstrong* and *Tom Mix* and sold scripts to *Dr. Christian, Curtain Time* and *First Nighter*. In October 1949, as co-producer of the documentary health series *It's Your Life*, he was able to combine his interests in science and drama. Most importantly, his idea for *Mr. Wizard* began to take form. He became fascinated with general science experiments and studied television as a medium of presentation.

Herbert sold his idea for *Mr. Wizard* to WNBQ-TV, the Chicago outlet for NBC, and the series premiered on 3 March 1951, with Herbert as the Wizard and Bruce Lindgren as the first of his young assistants. Produced in cooperation with the Cereal Institute, Incorporated, the 30-minute show was targeted at pre-teenagers and initially broadcast on Saturdays from 4:00 to 4:30 P.M.

Within four months the series had climbed to third place among children's programs in ARB ratings and its audience was growing. Chicago's Federated Advertising Club created an award especially for the show and the Voice of America entered a standing order for recorded transcripts of each program. Within two years, approximately 290 schools were using the series as required homework. In its quiet way, wrote *Variety* on 10 September 1952, "this cleverly contrived TV tour into the world of science probably adds as much to NBC's prestige as some of the network's more highly touted educational ventures."

By 1954, *Watch Mr. Wizard* was seen live on 14 stations and via kinescope on an additional 77. The National Science

Watch Mr. Wizard
Photo courtesy of Don Herbert

Foundation (NSF) cited Herbert and his show for promoting interest in the sciences, and the American Chemical Society presented him their first citation ever awarded for "important contributions to science education." Three years into his network run, there were more than 5,000 Mr. Wizard Science Clubs across North America with a membership totaling in excess of 100,000.

Sensing the decline of Chicago as a production center, Herbert moved his show to New York in 1955. During this time, he would win a number of national awards including the prestigious Peabody Award and three Thomas Alva Edison National Mass Media Awards. The total number of *Mr. Wizard* fan clubs would increase nearly tenfold to 50,000. Notwithstanding these accomplishments, NBC canceled the series on 5 September 1965.

Herbert's abilities as a teacher-producer of quality televised science education led him to the National Educational Television network where he produced a series of shows under the title *Experiment* (1966). He also produced films for junior and senior high schools, wrote a number of books on science and developed the Mr. Wizard Science Center outside of Boston. On 11 September 1971 NBC revived *Watch Mr. Wizard* but Herbert's old leisurely pace of the 1950s seemed outdated and the show left the air on 2 September 1972.

Undaunted by his second cancellation, and challenged by the NSF to create an awareness of science in children, in the early 1970s Herbert and his wife Norma developed *Mr. Wizard Close-Ups* for broadcast on NBC's daily morning schedule. At the end of the decade, the husband and wife team also developed traveling elementary school assembly programs featuring young performers and live science demonstrations. By 1991, these tours were annually presenting programs to approximately 3,000 schools and 1.2 million students.

With NSF and General Motors financial backing, in 1980 Herbert began production of *How About*—a long-running series of 80-second reports on developments in science and technology to be used as inserts in local news programs across the country. In time, the series would earn special praise from the American Association for the Advancement of Science-Westinghouse Science Journalism awards committee. Not content to rest on his laurels, in 1984 Herbert developed an updated and faster-paced *Mr. Wizard's World* that was seen three times a week on Nickelodeon, the children's cable network.

In 1991, Herbert received the Robert A. Millikan award from the American Association or Physics Teachers for his "notable and creative contributions to the teaching of physics." Three years later, in his late 70s, he developed another new series, *Teacher to Teacher with Mr. Wizard*—a series of NSF sponsored 15-minute programs airing on Nickelodeon and highlighting exemplary elementary science teachers and projects. In addition, the seemingly indefatigable Herbert created, among others, Mr. Wizard Science Secrets kits with clips from *Watch Mr. Wizard* and a Mr. Wizard Science Video Library with 20 videos from the *Mr. Wizard's World* series.

In March, 1984, Herbert told *Discovery* magazine his purpose in life was not to teach but to have fun. "I just restrict myself to fun that has scientific content." Fortunately, for generations of children and adults attracted to his Mr. Wizard persona, this soft-spoken, Minnesota-born personality had the ability to communicate and inspire in others his passion for the "fun" to be had with science.

—Joel Sternberg

HOST (as Mr. Wizard)
Don Herbert

PRODUCERS   James Pewolar (1955–65); Del Jack (1971-72)

PROGRAMMING HISTORY

• NBC

May 1951–February 1952	Saturday 6:30-7:00
March 1952–February 1955	Saturday 7:00-7:30
1955–1965	Various Times
September 1971–September 1972	Various Times

FURTHER READING

"AAPT Recognizes Herbert, Creators of Mr. Wizard TV Series. *Physics Today* (New York), November 1991.

Bolstad, Helen. "Mr. and Mrs. Wizard." *Radio-Tv Mirror* (New York), July 1954.

Cole, K.C. "Poof! Mr. Wizard Makes a Comeback." *Discover* (Los Angeles), March 1984.

Dismuke, Diane. "Meet: Don 'Mr. Wizard' Herbert." *NEA Today* (Washington, D.C.), April 1994.

Fischer, Stuart. *Kid's TV: The First 25 Years.* New York: Facts on File Publications, 1983.

Kramer, Carol. "His Wizardry Makes Aerodynamics Snappy." *Chicago Tribune,* 31 October 1971.

Margulies, Lee. "Mr. Wizard's Science Reports for Adults." *Chicago-Sun Times,* 26 March 1980.

"Mr. Wizard." *Variety* (Los Angeles), 7 March 1951.

"Mr. Wizard." *Variety* (Los Angeles), 10 September 1952.

"NBC-TV 'Wizard's' Wizardry Clinches 54-Station Ride." *Variety* (Los Angeles), 18 March 1953.

"The Robert A. Millikan Medal." *The Physics Teacher* (Stony Brook, New York), November 1991.

"'Wizard' Hot on Kinnies." *Variety* (Los Angeles), 13 January 1954.

See also Children and Television

WATCH WITH MOTHER

British Children's Programme

Watch with Mother, the general title of a series of five individual programmes, formed a central element in making television a domestic and family medium in Britain. Although the title *Watch with Mother* did not come into existence until 1952, *Andy Pandy*, the mainstay of the series, was first broadcast in July 1950. Two years later it was joined by *The Flowerpot Men* and later in the 1950s these shows were scheduled alongside *Rag, Tag and Bobtail* in 1953, and *Picture Book* and *The Woodentops* in 1955. Initially, *Andy Pandy* was shown in the afternoon between 3:45 P.M. and 4:00 P.M. at the end of the women's programme *For Women*. But in the 1960s *Watch with Mother* was scheduled at lunch time. The different programmes within the series were shown on specific days of the week: *Picture Book* on Monday,

Andy Pandy on Tuesday, *The Flowerpot Men* on Wednesday, *Rag, Tag and Bobtail* on Thursday and *The Woodentops* on Friday. The series was eventually taken off-air and replaced by *See-Saw* in 1980.

Watch with Mother was the first television programme series which specifically addressed a preschool child audience and, along with BBC radio's *Listen with Mother*, which began in 1950, it represented a shift in BBC policy to make programmes, both on radio and television, for this very young audience. Until this time, the BBC had made occasional radio programmes for the very young; however, in the words of Derek McCulloch ("Uncle Mac"), director of *Children's Hour* radio, they did not think that the young should be "catered for deliberately". This audience, according to McCulloch, came "into no real category at all". An earlier programme, *Muffin the Mule,* which was originally shown from 1946 on BBC children's television, had all the appearances of a preschool children's programme but was in fact addressed to all children and was popular with adults as well.

In the planning stages of *Andy Pandy* there was clearly some reticence about the introduction of a television programme for very young children and the BBC had a special panel to advise them, consisting of representatives of the Ministry of Education, the Institute of Child Development, the Nursery Schools' Association, and some educational child psychologists. There was particular concern about children watching television on their own, letting the "mother" free to do other things. As a result of these concerns about the development of the child and the responsibilities of the mother, *Andy Pandy,* and the later programmes, needed to be imagined in such a way as to allay these fears. The textual form of the programme and its scheduling are important in this respect.

Andy Pandy was created by Freda Lingstrom, who was head of Children's Television Programmes at the BBC between 1951 and 1956, and her long-standing friend, programme-maker Maria Bird, as a programme specifically directed at the preschool audience. Lingstrom, while assistant head of BBC School's Broadcasting, had been responsible for *Listen with Mother* and was asked to make a television equivalent on music and movement lines. *Andy Pandy* had no linear narrative structure. Instead, it presented a series of tableaux with no apparent overarching theme. For example, in one programme Andy starts by playing on a swing, accompanied by Maria Bird singing, "Swinging high, swinging low." He is joined by Teddy. The camera then focuses on Teddy, who enacts the movements to the nursery rhyme "Round and round the garden." Finally, after a scene with Andy and Teddy playing in their cart and a scene with Looby Loo singing her song, "Here we go Looby Loo," the two male characters return to their basket and wave good-bye and Maria Bird sings "Time to go home." Lingstrom argued that the tempo was slow and there was no story, so that the action could move from one situation to another in a way totally acceptable to the very young child.

The programme was designed to bring three year olds into a close relationship with what was seen on the screen. *Andy Pandy* was intended to provide a friend for the very young viewer, and, as a three-year-old actor was out of the question, a puppet was the obvious answer. The characters took part in simple movement, games, stories, nursery rhymes and songs. The use of nursery rhymes was seen as particularly important as it worked both to establish a relationship between the mother and the development of the child and also to connect the child to a tradition and community of preschool childhood. The children were invited, not only to listen and to watch the movements of the puppets, but also to respond to invitations to join in by clapping, stamping, sitting down, standing up and so forth.

Andy Pandy drew upon the language of play in order to make itself, and also television, homely. Mary Adams, head

Watch with Mother
Photo courtesy of the British Film Institute

of Television Talks at the BBC, argued that the puppet came to the child in the security of its own home and brought nothing alarming or contradictory to the safe routines of the family. In *Andy Pandy,* and also in *The Flowerpot Men,* the fictional world of preschool childhood was presented within the confines of the domestic. Andy, Teddy and Looby Loo were always presented within the garden or the living room. Likewise, in *The Flowerpot Men,* the characters were presented within the garden and in close proximity to the little house which was pictured at the beginning of each programme opening its doors to the diegetic space. In *Andy Pandy* we hear nothing of the outside world. And in *The Flowerpot Men* the only off-screen character we hear about is the gardener, whose character, never seen or heard, signified the limits of this imaginary world.

Watch with Mother was never scheduled within the main bulk of children's programmes between 5:00 P.M. and 6:00 P.M. When, in September 1950, there was discussion that *Andy Pandy* should be shown with the rest of children's programmes, Richmond Postgate, acting head of Children's Television Programmes at the BBC, firmly responded by stating that at 5:00 P.M. three year olds should be thinking of bed. The programme was designed to fit into the routines of both mothers and small children and it was scheduled at different times during its early history. However, changes to its scheduling caused minor revolts widely reported in the press. For example, when in 1963 the BBC planned to show *Watch with Mother* at 10:45 A.M. the *Daily Sketch* declared that "for most small children 10:45 is a time to 'Watch Without Mother'. And there's not much joy in that." However, although the timing of the programme was intended to provide a space especially for mother and small child, it is clear that some viewers saw it as a means to do other things.

In the 1960s and 1970s a new stream of programmes were invented for the series (e.g. *Pogles' Wood, Trumpton,* and *Mary, Mungo and Midge*). There was still significant emotional investment in the older programmes. For example, there was much concern in 1965 when viewers thought that *Camberwick Green* was to replace *Andy Pandy* and *The Flowerpot Men*. Doreen Stephens, head of Family Programmes, reassured the audience, stating that the familiar shows would be shown, which they were, although less frequently until 1970. It was no surprise that when a number of the older programmes were released on a *Watch with Mother* video in 1986, it became a best-seller and topped the BBC's video charts.

—David Oswell

SERIES CREATOR AND PRODUCER   Freda Lingstrom

ANDY PANDY

WRITER-COMPOSER
Maria Bird

SINGER
Gladys Whitred

PUPPETEERS
Audrey Atterbury, Molly Gibson

THE FLOWERPOT MEN

WRITER-COMPOSER
Maria Bird

PUPPETEERS
Audrey Atterbury, Milly Gibson

VOICES AND SOUND EFFECTS
Peter Hawkins, Gladys Whitred, Julia Williams

RAG, TAG, AND BOBTAIL

STORY NARRATOR
Charles E. Stidwell

STORY WRITER
Louise Cochrane

GLOVE PUPPETEERS
Sam and Elizabeth Williams

THE WOODENTOPS

SCRIPTS AND MUSIC
Maria Bird

PUPPETEERS
Audrey Atterbury, Molly Gibson

VOICES
Eileen Brown, Josephina Ray, Peter Hawkins

PICTURE BOOKS

STORYTELLERS
Patricia Driscoll, Vera McKechnie

PROGRAMMING HISTORY

• BBC
Various Times

FURTHER READING
Oswell, David. "Watching with Mother in the early 1950s." In, Bazelgette, Cary, and David Buckingham, editors. *In Front of the Children*. London: British Film Institute, 1995.

See also British Programming; Children and Television

WATERGATE

"Watergate" is synonymous with a series of events that began with a botched burglary and ended with the resignation of a U.S. president. The term itself formally derives from the Watergate building in Washington, D.C., where, on the night of 17 June 1972, five burglars were arrested in the Democratic National Committee offices. Newspaper reports from that point began revealing bits and pieces of details that linked the Watergate burglars with President Richard Nixon's 1972 reelection campaign. The president and his chief assistants denied involvement, but as evidence of White House complicity continued to grow, the U.S. Congress was compelled to investigate what role the Watergate matter might have played in subverting or attempting to subvert the electoral process.

The U.S. Senate, by a 77-to-0 vote, approved a resolution on 7 February 1973, to impanel the Senate Select Committee on Presidential Campaign Activities to investigate Watergate. Known as the Ervin Committee for its chairperson, Senator Sam Ervin, the committee began public hearings on 17 May 1973, that shortly came to be known as the "Watergate Hearings."

Television cameras covered the Watergate hearings gavel-to-gavel, from day one until 7 August. 319 hours of television were amassed, a record covering a single event. All three commercial television networks then in existence—NBC, CBS, and ABC—devoted an average of five hours per day covering the Watergate hearings for their first five days.

The networks devised a rotation plan that, beginning on the hearing's sixth day, shifted coverage responsibility from one network to another every third day. Any of the three networks remained free to cover more of the hearings than required by their rotation agreement, but only once did the networks choose to exercise their option. All three networks elected to carry the nearly 30 hours of testimony by key witness and former White House counsel John Dean.

The non-commercial Public Broadcasting Service (PBS) aired the videotaped version of each day's Watergate hearing testimony during the evening. Many PBS station managers who were initially reluctant to carry such programming found that as a result of the carriage, station ratings as well as financial contributions increased.

As the Ervin Committee concluded its initial phase of Watergate hearings on 7 August 1973, the hearing's television audience had waned somewhat, but a majority of viewers continued to indicate a preference that the next hearing phase, scheduled to begin on 24 September, also be televised. The networks, however, felt otherwise. The Ervin Committee continued the Watergate hearings until February 1974 but with only scant television coverage.

Television viewers were attracted to the Watergate hearings in impressive numbers. One survey found that 85% of all U.S. households had tuned in to at least some portion of the hearings. Such interest was not universal, however. In fact, Special Prosecutor Archibald Cox had argued that television's widespread coverage of Watergate testimony could endanger the rights of witnesses to a fair trial and in doing so, could deprive Americans of ever hearing the full story of Watergate. The Ervin Committee refused Cox's request to curtail coverage, saying that it was important that television be allowed to carry Watergate testimony to the American public firsthand.

On 6 February 1974, a new phase of Watergate began when the U.S. House of Representatives voted 410-to-4 to authorize the House Judiciary Committee to investigate whether sufficient grounds existed to impeach President Nixon. If so, the committee was authorized to report necessary articles of impeachment to the full House.

The Judiciary Committee spent late February to mid-July 1974 examining documents and testimony accumulated during the Senate's Watergate hearings. When this investigatory phase ended, the Judiciary Committee scheduled public deliberations for 24-27, 29 and 30 July to debate what, if any, impeachment recommendations it would make to the House. Three articles of impeachment eventually were approved by the Committee, recommending that the House begin formal impeachment proceedings against President Richard Nixon.

The decision to televise Judiciary Committee meetings was not immediate nor did it meet with overwhelming approval. Only after several impassioned pleas from the floor of the U.S. House that such an extraordinary event should be televised to the fullest extent did the House approve a resolution to allow telecast of the Judiciary Committee's

Richard Nixon's televised resignation speech
Photo courtesy of the National Archives (Uinted States)

impeachment deliberations. The committee itself had final say on the matter and voted 31-to-7 to concur with the decision of their House colleagues. One major requirement of the Judiciary Committee was that television networks covering the committee not be allowed to break for a commercial message during deliberations.

The Judiciary Committee began its televised public debate on the evening of 24 July. The commercial networks chose to rotate their coverage in the same manner as utilized during the Senate Watergate hearings. What's more, the commercial networks telecast only the evening portions of Judiciary Committee deliberations, while PBS chose to telecast the morning and afternoon sessions as well. As a result, television viewers were provided nearly 13 hours of coverage for each of the six days of Judiciary Committee public deliberations.

Eventually, the full House and Senate voted to allow television coverage of impeachment proceedings in their respective chambers, once assurances were made that the presence of television cameras and lights would not interfere with the president's due process rights. Final ground rules were being laid and technical preparations for the coverage were underway when President Nixon's resignation on 9 August 1974, brought the impeachment episode to an end.

—Ronald Garay

FURTHER READING

Garay, Ronald. *Congressional Television: A Legislative History*. Westport, Connecticut: Greenwood, 1984.

Hamilton, James. *The Power to Probe*. New York: Random House, 1976.

Kurland, Philip B. "The Watergate Inquiry, 1973." In, Schlesinger, Arthur M., and Roger Bruns, editors. *Congress Investigates: A Documented History, 1792–1974*. New York: Chelsea House, 1975.

U.S. House. House Committee on the Judiciary. *Impeachment of Richard M. Nixon, President of the United States*, 93d Cong., 2d sess., 1974, H. Rept. 93-1305.

U.S. Senate. Senate Select Committee on Presidential Campaign Activities. *The Final Report*, 93d Cong., 2d sess., 1974, S. Rept. 93-981.

See also Political Processes and Television; U.S. Congress

WATERMAN, DENNIS

British Actor

Dennis Waterman has the distinction of being well known to the British television public, somewhat known in Australia, and almost completely unknown to the North American audience. As a screen character, Waterman is heavily dependent on a strong partner; in comedy, especially, he usually acts as a straight figure to the comic excesses of his counterparts. When he does play solo, as in the recent thriller *Circle Of Deceit*, he shows himself to lack colour and charisma.

Born in London in 1947, Waterman became a child actor, appearing in the feature film *Night Train to Inverness* (1958) and in a West End production of the musical *The Music Man*. In 1961, he landed the title role of William in the children's television series *William*, produced by the BBC. This 13 half-hour episode series was based on the very popular childrens' books by Richmal Crompton, adapted by writer C.E. Webber.

Waterman spent the following year in Hollywood working on the CBS situation comedy *Fair Exchange*. He was one of four British actors imported for the series, which concerned two families, one from New York and the other from London, who arranged to swap teenage daughters. Waterman played a younger boy in the London family who suddenly had to contend with a teenage American "sister." The series was unusual only because it had extended the situation comedy format to hour-long episodes. However, it provoked only lukewarm interest and was dropped after three months. It was briefly revived in half-hour episodes but fared no better.

Waterman's voice broke, his appearance changed, and the child actor faded. In 1976, he landed the role of Detective Sergeant George Carter in the British police crime series *The Sweeney*, produced by Thames Television's Euston Films. *The Sweeney* was premised on a fictional version of Scotland Yard's Flying Squad, a police car unit concerned with major crimes such as armed robberies. (The series title came from Cockney rhyming slang: Sweeney Todd-The Flying Squad). *The Sweeney* was the best British police series of the 1970s. It was well made, carrying excellent action scenes, good stories, and fine acting from leads John Thaw as Detective Inspector Jack Regan, Waterman as his assis-

tant, and Garfield Morgan as their boss, Detective Chief Inspector Hoskins.

The Sweeney offered Waterman not only considerable fame but also a second career. As a child actor, his accent had been middle-class and he had projected sensitivity and vulnerability. In *The Sweeney* he conveyed energy, toughness, and a gritty Cockney sense of how the world really worked. Although his character played second-fiddle to Jack Regan, Waterman still managed to infuse Carter with considerable colour and guts.

Waterman's career was boosted even further by his next series, the enormously popular *Minder*. This program,

Dennis Waterman
Photo courtesy of the British Film Institute

which introduced the character of Arthur Daley, a shady London car dealer, and Terry McCann, his ex-convict bodyguard and partner, has been described as a perfect blend of dark humour and colourful characterization. *Minder* was built around the inspired casting of George Cole as Arthur and Waterman as Terry. Cole was a veteran of British cinema, who had created a memorable forerunner to Arthur Daley in the figure of the Cockney spiv, Flash Harry, in three very funny St. Trinian films in the 1950s and 1960s. Drawing partly from the figure of Carter in *The Sweeney*, Waterman's Terry was tough and Cockney streetwise. What was new was that Waterman was playing comic straight-man as the often hapless Terry, who was usually no match for Arthur. Although *Minder* was named after the figure of Terry, it was Arthur who was the mainstay of the series, a fact underlined by its revival in 1991, some six years after Waterman's departure, with Gary Webster filling the minder role.

In 1986. Waterman's on-screen woman troubles began with BBC 2's four-hour miniseries, *The Life and Loves of a She Devil*. A gruesome black comedy which combined outrageous fantasy with close-to-the-bone social comment, *She Devil* was an enormous popular success. The series concerned an unfaithful husband (Waterman) whose ex-wife, the figure of the title, wreaks a truly memorable set of punishments on her hapless mate. In portraying Waterman as a womaniser who is finally unable to control the feminine forces that he has unleashed, *She Devil* added an interesting new dimension to the actor's screen persona.

In 1989, Waterman returned to comedy-drama with the series *Stay Lucky* for Yorkshire Television. The title, which referred to nothing in particular, was somewhat indicative of the series' problems as a whole. Like *The Sweeney* and *Minder*, *Stay Lucky* concerned a partnership, although in this instance one that was romantic as well as professional. Set aboard a houseboat, the series concerned a set of predictable oppositions between male and female leads—Waterman as Thomas and Kay Francis as Sally. As a Cockney, he was nuggety, streetwise, and realistic; as a Northener, she was glamorous, sophisticated, and headstrong.

Stay Lucky attempted to mix the comedy of the sexes with the darker world of London crime and poverty but the mixture did not quite jell. However, the series was at its strongest when it gravitated to the former theme, with Waterman usually generating solid comic exasperation, not at the outrageous schemes of an Arthur Daley, but at the outlandish stratagems of a willful, attractive woman.

Waterman's most recent series has been the BBC 1 situation comedy serial, *On the Up*. Altogether 18 half-hour episodes were made between 1990 and 1992, and the comedy/drama blend was much more successful. The series concerned the Cockney selfmade millionaire Tony (Waterman), who was less successful running both his marriage—to a beautiful, headstrong, upper-class woman—and a household of servants and friends.

—Albert Moran

DENNIS WATERMAN. Born in London, England, 24 February 1948. Attended Corona Stage School. Married: 1) Penny (divorced); 2) Patricia Maynard (divorced); children: Hannah and Julia; 3) Rula Lenska. Stage debut, at the age of 11, 1959; by the age of 16 had spent a season with the Royal Shakespeare Company, Stratford-upon-Avon, and worked in Hollywood; star, *William* TV series and other productions, 1962; star, *The Sweeney* and the *Minder* series; later appeared mainly in comedy parts; has also had some success as a singer. Address: ICM, 76 Oxford Street, London W1N 0AX, England.

TELEVISION SERIES

1962	*William*
1962	*Fair Exchange*
1975–78	*The Sweeney*
1979–85, 1988–93	*Minder*
1986	*The Life and Loves of a She-Devil*
1989–91, 1993	*Stay Lucky*
1990–92	*On the Up*
1995	*Match of the Seventies* (presenter)
1996	*Circle of Deceit*

MADE-FOR-TELEVISION MOVIE

1985	*Minder on the Orient Express*

TELEVISION SPECIALS

1959	*Member of the Wedding*
1960	*All Summer Long*
1974	*Regan*
1982	*The World Cup—A Captain's Tale* (also co-producer)

FILMS

Pirates of Blood River, 1961; *Up the Junction*, 1967; *My Lover, My Son*, 1969; *A Smashing Bird I Used to Know*, 1969; *A Promise of Bed*, 1969; *I Can't... I Can't/ Wedding Night*, 1969; *The Scars of Dracula*, 1970; *Fright*, 1970; *Man in the Wilderness*, 1971; *Alice's Adventures in Wonderland*, 1972; *The Belstone Fox*, 1973; *The Sweeney*, 1977; *The Sweeney II*, 1978; *A Dog's Day Out; Cold Justice; Father Jim.*

RECORDINGS

Night Train to Inverness, 1958; *I Could Be So Good for You*, 1980; *What Are We Gonna Get 'er Indoors*, 1983; *Down Wind with Angels; Waterman.*

STAGE

The Music Man; Windy City; Cinderella; Same Time Next Year.

FURTHER READING

Watergate: Chronology of a Crisis. Washington, D.C.: Congressional Quarterly, Inc., 1975.

See also British Programming; *Minder; Sweeney*

WATERS, ETHEL

U.S. Actor

Ethel Waters, one of the most influential jazz and blues singers of her time, popularised many song classics, including "Stormy Weather." Waters was also the first African-American woman to be given equal billing with white stars in Broadway shows, and to play leading roles in Hollywood films. Once she had established herself as one of America's highest-paid entertainers, she demanded, and won, dramatic roles. Single-handedly, Waters shattered the myth that African-American women could perform only as singers. In the early 1950s, for example, she played a leading role in the stage and screen versions of Carson McCullers' *The Member of the Wedding*. Waters played a Southern mammy, but demonstrated with a complex and moving performance that it was possible to destroy the one-dimensional Aunt Jemima image of African-American women in American theater and cinema.

In a career that spanned almost 60 years, there were few openings for an African-American woman of Waters' class, talent, and ability. She appeared on television as early as 1939, when she made two experimental programmes for NBC: *The Ethel Waters Show* and *Mamba's Daughters*. But it was her regular role as the devoted, cheerful maid in ABC's popular situation comedy *Beulah* (1950-53) that established her as one of the first African-American stars of the small screen.

Waters' dramatic roles on television were also stereotyped. Throughout the 1950s, she made appearances in such series as *Favorite Playhouse, Climax, General Electric Theater, Playwrights '56,* and *Matinee Theater.* Without exception, Waters was typecast as a faithful mammy or suffering mother. In 1961, she gave a memorable performance in a *Route 66* episode, "Good Night, Sweet Blues," as a dying blues singer whose last wish is to be reunited with her old jazz band. Consequently, Waters became the first black actress nominated for an Emmy Award. She later appeared in *The Great Adventure* ("Go Down Moses"), with Ossie Davis and Ruby Dee in 1963; *Daniel Boone* ("Mamma Cooper") in 1970; and *Owen Marshall, Counselor at Law* ("Run, Carol, Run") in 1972. But, says African-American film and television historian Donald Bogle in *Blacks in American Films and Television* (1988): "Waters' later TV appearances lack the vitality of her great performances (she has little to work with in these programs and must rely on her inner resources and sense of self to get by), but they are part of her evolving image: now she's the weathered, ailing, grand old woman of film, whose talents are greater than the projects with which she's involved."

In the late 1950s, ill health forced Waters into semi-retirement. A deeply religious woman, most of her public appearances were restricted to Billy Graham's rallies. She died in 1977 at the age of 80.

—Stephen Bourne

ETHEL WATERS. Born in Chester, Pennsylvania, U.S.A., 31 October 1896. Married: 1) Merritt Pernsley, c. 1910;

Ethel Waters

2) Clyde Matthews, c. 1928. Worked numerous maid, dishwasher, and waitressing jobs, 1903–17; sang and toured vaudeville circuit, 1917–30s; appeared in numerous theatrical productions, 1919–56; appeared in numerous films, 1929–63; appeared in numerous television programs, including the series *Beulah,* 1950–52; worked for the Billy Graham Crusade from the late 1950s. Recipient: New York Drama Critics Award for performance in *The Member of the Wedding,* 1950; U.S. Postal Service commemorative stamp, 1994. Died in Chatsworth, California, 1 September 1977.

TELEVISION SERIES

1950–53   *Beulah*

TELEVISION SPECIAL (selection)

1939   *The Ethel Waters Show*

FILMS

On with the Show, 1929; *Rufus Jones for President,* 1933; *Bubblin' Over,* 1934; *Tales of Manhattan,* 1941; *Cairo,* 1942; *Stage Door Canteen,* 1943; *Cabin in the Sky,* 1943; *Pinky,* 1950; *The Member of the Wedding,* 1952; *Carib Gold,* 1955; *The Sound and the Fury,* 1959.

STAGE

Rhapsody in Black, 1931; *As Thousands Cheer*, 1933; *At Home Abroad*, 1935; *Mamba's Daughters*, 1939; *Cabin in the Sky*, 1940-41.

PUBLICATION

His Eye Is on the Sparrow, with Charles Samuels. Garden City, New York: Doubleday, 1951.

FURTHER READING:

Bogle, Donald. *Brown Sugar—Eighty Years of America's Black Female Superstars*. Prospect, Kentucky: Harmony Books, 1980.

MacDonald, J. Fred. *Blacks and White TV: Afro-Americans in Television since 1948*. Chicago: Nelson-Hall, 1983.

See also *Beulah*, Racism, Ethnicity, and Television

WATKINS, PETER

British Director

George Bernard Shaw wrote of fin-de-siecle novelist Samuel Butler, "England does not deserve great men." Much the same might be said at the end of another century of one of the most singular, committed, and powerful directors of the last 40 years. Peter Watkins' prize-winning experimental documentaries, *Diary of an Unknown Soldier* (1959), and *The Forgotten Faces* (1961), reconstructing respectively World War I and the Hungarian uprising of 1956, earned screenings and a job at the BBC, which he used to make the remarkable *Culloden*, a Brechtian deconstruction of documentary technique in an account of the bloody defeat of the 1742 Jacobin rebellion in Scotland. *Culloden* exhibited hallmark techniques: hand-held camera, direct-to-camera address from historical and fictional characters, and interviews with them, though the near surrealism of placing a modern on-camera reporter on the battlefield was a humourous touch rarely paralleled in his later work. Using, as he has throughout his oeuvre, the heightened naturalism of amateur actors, the programme contrasted the effete figure of Bonnie Prince Charlie, actually a European adventurer, with the impoverished and still feudally-bound Gaelic-speaking peasantry of the Highlands, a cruel indictment of both Scottish patriotism and the brutal British reprisals on the Highlanders. Another work, *The War Game*, "preconstructed" the effects of a nuclear attack on southern England. Perhaps it was not just Watkins' deadpan voice-over, nor the matter-of-fact delivery of official prognostications of casualties and security measures, but his comparison of nuclear firestorms with the ever-sensitive British bombing of Dresden in 1945 (subject of two later banned programmes in the United Kingdom) that saw the film banned. Reduced to fund-raising shows for nuclear disarmament groups, the programme has rarely been discussed in terms other than those of its subject and its political fate. But its pathbreaking and still-powerful juxtaposition of interview, reconstruction, graphics, titles, and the collision of dry data with images of horror still shock, the grainy black-and-white imagery and use of telephoto, sudden zooms, and wavering focus creating an atmosphere of immediacy unique in British television. Fifty minutes that shook the world, it was banned for 25 years by the BBC amid storms of controversy which were reopened when it finally made British TV screens in a Channel 4 season of banned titles.

The War Game took the 1966 Best Documentary Oscar, opening the door to Hollywood. Universal bankrolled the feature film *Privilege*, about a pop messiah in a near-future police state, but pulled the plug on an ambitious reconstruction of the Battle of The Little Big Horn and the subjugation of the Native American Indians. From the late 1960s, Watkins' career was marked by projects cut, abandoned, or suppressed: Watkins himself listed 14 of them in a document seeking support for his 1980s film *The Journey*. *The Gladiators*, made for Swedish TV, about popular acquiescence in militarism, used the device of a fictional television programme, "The Peace Game," in which generals played games of strategy, and the savage 16mm allegory of Nixon's America, *Punishment Park*, in which "deviants" were given their chance to survive in a nightmarish outlaw zone, both saw broadcast and theatrical release, though limited. These two titles extended Watkins' repertoire of effects by their focus on individual characters caught up in evil times, though the use of montage cutting and extreme naturalism

Peter Watkins
Photo courtesy of the British Film Institute

in performances combined to minimise identification, and increased the intellectual engagement of the viewer with the narrative. Closer in technique to Brecht's practice than his theory, Watkins failed to benefit either from the vanguardism of contemporary film theory or the political clout of less challenging auteurs like Ken Loach and Dennis Potter.

Other completed projects like *70s People* (on suicide and the failures of social democracy) and *Evening Land* (a terrorist kidnap contrasted with the quelling of a strike in a military shipyard), both for Danish TV, were suppressed. Only the biopic of a Norwegian painter, *Edvard Munch,* has had major distribution, though mainly as theatrical film, rather than the three-part series it was originated as. *Munch's* passion derived not only from the subject and Watkins' handling of it, but from the identification between director and derided artist. The series was distinguished again by direct-cinema techniques, but also by complex editing around motifs, especially faces and floras, and by multi-tracked sound design layering the characters' past, present, and future into a rich montage. Like his earlier documentaries, *Munch* added voice-over to the sound mix, sometimes even over blank screens, to connect the narrative with worldwide events and political analysis. Carrying the use of natural light pioneered in his BBC projects into colour, the film achieved a profoundly affecting image of a consumptive society unable to credit those who warn of its demise until it is too late. It was its political analysis and, stylistically, its use of sophisticated montage editing, that distinguished *Munch* and its predecessors from the hand-held stylistics of some recent U.S. cop shows.

In 1982, an attempt to remake *The War Game* with Central TV fell through, and Watkins devoted the following three years to accruing donations and help to make *The Journey,* perhaps his greatest achievement. Running at over 14 hours, the film was a rarely screened account, shot in over a dozen nations, of nuclear war and its effects. It has yet to be broadcast. Watkins' peripatetic life, spent developing and trying to complete projects in cinema and TV, and his occasional embittered polemics in print, are all that is certain. Rumours circulate of an international shoestring production on ecological disaster, and about further failed projects with production houses. Watkins' intelligence, passion, and skill have been consistently masked by controversy: he is the most neglected and perhaps the most significant major British director of his generation.

—Sean Cubitt

PETER WATKINS. Born in Norbiton, Surrey, England, 29 October 1935. Attended Christ College, Grecknockshire; studied acting at Royal Academy of Dramatic Art, London. Served with East Surrey Regiment. Began career as assistant producer of television short subjects and commercials, 1950s; assistant editor and director of documentaries, BBC, 1961; director, *The War Game,* banned by the BBC, 1966; director, feature film, *Privilege,* 1967; moved to Sweden, 1968; worked in United States, 1969–71; resides in Sweden.

TELEVISION

1964   *Culloden*

FILMS

The Web, 1956; *The Field of Red,* 1958; *Diary of an Unknown Soldier,* 1959; *The Forgotten Faces,* 1961; *Dust Fever,* 1962; *The War Game,* 1966; *Privilege,* 1967; *The Gladiators,* 1969; *Punishment Park,* 1971; *Edvard Munch,* 1974; *The Seventies People,* 1975; *The Trap,* 1975; *Evening Land,* 1977; *The Journey,* 1987.

PUBLICATIONS

Blue, James, and Michael Gill. "Peter Watkins Discusses His Suppressed Nuclear Film *The War Game.*" *Film Comment* (New York), Fall 1965.

The War Game: An Adaptation of the BBC Documentary. New York: Avon, 1967

"Left, Right, Wrong." *Films and Filming* (London), March 1970.

"Peter Watkins Talks about the Suppression of His Work within Britain." *Films and Filming* (London), February 1971.

"*Punishment Park* and Dissent in the West." *Literature/Film Quarterly* (Salisbury, Maryland), 1976.

"*Edvard Munch*: A Director's Statement." *Literature/Film Quarterly* (Salisbury, Maryland), Winter 1977.

"Interview with S. MacDonald." *Journal of the University Film Association* (Carbondale, Illinois), Summer 1982.

FURTHER READING

Cunningham, Stuart. "Tense, Address, Tendenz: Questions of the Work of Peter Watkins." *Quarterly Review of Film Studies* (Los Angeles), Fall 1980.

Gomez, Joseph A. *Peter Watkins.* Boston: Twayne, 1979.

Kawin, Bruce. "Peter Watkins: Cameraman at World's End." *Journal of Popular Film* (Washington, D.C.), Summer 1973.

MacDonald, Scott. "From Zygote to Global Cinema via Su Friedrich's Films." *Journal of Film and Video* (Chicago), Spring-Summer, 1992.

———. "Filmmaker as Global Circumnavigator: Peter Watkins' *The Journey* and Media Critique." *Quarterly Review of Film and Video* (Chur, Switzerland), August 1993.

Nolley, Ken. "Narrative Innovation in *Edvard Munch.*" *Literature/Film Quarterly* (Salisbury, Maryland), 1987.

Welsh, James M. "The Dystopian Cinema of Peter Watkins." *Film Criticism* (Edinboro, Pennsylvania), Fall 1982.

———. "The Modern Apocalypse: *The War Game.*" *Journal of Popular Film and Television* (Washington, D.C.), Spring, 1983.

———. *Peter Watkins: A Guide to References and Resources.* Boston, Massachusetts: G.K. Hall, 1986.

See also Director, Television; *War Game*; War on Television

WATSON, PATRICK

Canadian Producer/Host

Patrick Watson has played a key role in the development of Canadian television, starting as producer, and then host, of many of the Canadian Broadcasting Corporation's (CBC) groundbreaking public-affairs series. In 1989, he was named chair of the CBC board of directors, a position from which he resigned in June 1994. His career in Canadian broadcasting, with several short detours into U.S. television, has been recognized by many for its innovative and substantive contribution to television journalism. He currently holds two honorary degrees, and is an Officer of the Order of Canada for his journalistic efforts. At the same time, his career has been distinguished by well-publicized struggles with CBC management and a number of Canadian politicians, both as producer and board chair. Lending substance to his television journalism has been his wide-ranging interest in the arts and social affairs.

Watson's first broadcast experience was as a radio actor in 1943, in a continuing CBC children's dramatic series called *The Kootenay Kid*. He has maintained his interest in dramatic television production by performing in several CBC dramas, and by producing and performing in his two dramatized series of fictional encounters with great historical figures: *Titans* and *Witness to Yesterday*. In 1983, he wrote and acted in a one-man stage version of the Old Testament's *The Book of Job*.

Canadian television received its bilingual launch on Saturday, 6 September 1952, on CBFT, a CBC station in Montreal. Watson's involvement with television started in those early years, first as a freelancer in 1955, then as producer of *Close-Up*, 1957–60, and the national-affairs series, *Inquiry*, 1960–64. Both shows were noted for their hard-hitting, sometimes confrontational interviews with the Canadian elite. *Inquiry* established an exciting and stimulating public-affairs television show that would attract a larger audience than the typical narrow, well-educated one.

Watson's next project, which attracted the largest audience for a public-affairs program in Canadian history, also proved to be the most controversial series of its kind. *This Hour Has Seven Days* was the creation of Watson and his co-producer from *Close-Up* and *Inquiry*, Douglas Leiterman. Broadcast before a live audience on Sunday nights from the fall of 1964 to the spring of 1966, this public-affairs show became the darling of over three million Canadians until its demise at the hands of CBC management, who could no longer withstand the criticism from parliament or the insubordination of the *Seven Days* team. Shows featured satire of politicians in song and skit mixed with "bear pit" interviews, probing film documentaries, on-location stakeouts and street interviews—all dealing with important, but often ignored, social and political issues. Critics hailed it for its freshness and probing investigations and condemned it for its sometimes sensational and "yellow" journalism. Watson was the co-producer for the first season of *Seven Days*,

Patrick Watson
Photo courtesy of CBC/Fred Phipps

and became the on-air co-host and interviewer in the second year in a move that the CBC management thought would curb some of the more controversial ideas and methods of the series. Watson and the extraordinary team of producers and writers assembled for the program (many of whom became influential documentarians and producers through the 1960s and 1970s) became even more innovative and "in-your-face" with their journalism, daring the CBC management to take action. In a later interview Watson admitted to the arrogance of those days, inciting his crew to "make people a little bit angry, frustrate them...come socking out of the screen." The management took the dare and cancelled the show to the outrage of many, some of it orchestrated by the *Seven Days* team to try and save the show. There was an avalanche of calls and letters, public demonstrations, a parliamentary committee hearing, and a special investigation by an appointee of the prime minister—quite a response to the cancellation of a TV show. The series has taken on mythic proportions in the history of television journalism. It certainly pushed the boundaries of what was considered appropriate journalism, predated the current concern over the fine line between news and entertainment, and created a very chilly environment for CBC producers of public affairs for many years.

Because of his highly visible contribution to *Seven Days* and the aftermath to its cancellation, Watson was popularly touted for president of the CBC. He let it be known that he was interested, but was not to reach high administrative office in the CBC until 25 years later. In the intervening years he turned his attention to a number of creative projects in and out of television. In addition to those already mentioned, he wrote, produced, hosted, and directed for *The Undersea World of Jacques Cousteau*, *The Watson Report*, *The Canadian Establishment*, *Lawyers*, and *The Fifty-first State* (for PBS Channel 13,

New York), among others. In 1989, before being named chair of the CBC, he created, produced, and hosted the ten-part international co-production television series, *The Struggle for Democracy*. It was the first documentary ever to appear simultaneously in French and English on the CBC's two main networks with the same host. Researched in depth and reflecting the dominant values of western democracy, this substantive and ambitious series took the viewer across the world and into history, to the sites of many experiments, successes, and failures of the democratic effort. In the years after *Seven Days*, Watson was frequently and deservedly praised for his skills as a host and interviewer.

Watson's years as chair of the CBC board of directors were difficult ones for him and the corporation. The CBC had to face many severe budget cuts, subsequent layoffs, and the closing of regional outlets. Watson was dealing with a board becoming stacked with Tory appointees, several of whom advocated the privatization of the CBC. He was expected to both manage the board and lobby parliament. Though he toured the country speaking up for public television, he was seen by many CBC staffers and some of the public as less than effective in his efforts. In his last year, the CBC was hit with a new controversy over a public-affairs series on the Canadian effort in World War II, called *The Valour and the Horror*. This program challenged many standard versions of World War II history by critically examining the actions and the fallibility of military and political leaders. While the series won awards and was praised by many, it was vilified by veterans' groups and conservative politicians. After intense pressure, including a senate hearing controlled by the critics of the program, the CBC issued an ombud's report, supported by statements from the president of the CBC and the board, that essentially chastised the show's producers for their research, methods of presentation, and conclusions. As chair of the board, Watson was criticized for not speaking out publicly in support of the journalists and for not resigning. Insiders, including the producers of the show, credit Watson for moderating the board's and the president's response and mediating the dispute with CBC management.

Ironically, it seemed that Patrick Watson's career had circled back in on him. He began by pushing the boundaries of TV public affairs, stretching the limits of management tolerance, and establishing a precedent for interpretative journalism that would eventually challenge his own authority, role, and accountability. Throughout, he has remained dedicated to the important role television can play in creating an informed public. His influential career has both reflected and addressed a number of the issues and tensions facing Canadian broadcasting and society.

—William O. Gilsdorf

PATRICK WATSON. Born in Toronto, Ontario, Canada, 1929. Educated at University of Toronto, B.A, M.A.; studied linguistics at University of Michigan. Married: 1) Beverly (divorced, 1983), three children; 2) Caroline Bamford.

Joined CBC-TV, early 1950s; founder, Patrick Watson Enterprises, 1966; co-founder, Immedia Inc., 1967; helped pioneer CBS Cable Network during 1980s; chair, CBC, 1989–94; first North American filmmaker to film in the People's Republic of China. Officer of the Order of Canada. Recipient: Bruxelles Festival Award, 1984; 12 Junos; two ACTRA Awards.

TELEVISION

1957–60	*Close-Up* (co-producer)
1960–64	*Inquiry* (producer and director)
1964–66	*This Hour Has Seven Days* (executive producer and co-host)
1967	*Search in the Deep* (producer)
1967	*The Undersea World of Jacques Cousteau* (producer)
1968	*Science and Conscience* (host)
1973–75	*Witness to Yesterday* (interviewer and writer)
1975–81	*The Watson Report* (interviewer)
1977	*The Fifty-first State* (editor and anchor)
1978	*Flight: The Passionate Affair* (host and writer)
1980	*The Canadian Establishment* (host and contributing writer)
1981	*The Chinese* (host, narrator, and contributing writer)
1981–82	*CBS Cable Service* (host)
1981	*Titans* (interviewer and writer)
1985	*Lawyers* (host)
1989	*The Struggle for Democracy* (10 parts; writer, host, and executive editor)

TELEVISION SPECIAL

1983–86	*Live from Lincoln Center* (host)

FILMS

Bethune (actor), 1963; *The 700 Million* (producer and director), 1964; *The Terry Fox Story* (actor),1982; *Countdown to Looking Glass* (actor), 1984; *The Land That Devours Ships* (co-producer), 1984.

RADIO

The Kootenay Kid (actor), 1943.

STAGE

The Book of Job (writer and performer).

FURTHER READING

Borowski, Andrew. "The Watson Report" (interview). *Canadian Forum* (Toronto), June/July 1991.
"CBC Chairman Resigns Post." *Montreal Gazette*, 15 June 1994.
"CBC 'Consultant' Watson Criticized over Sale of *Democracy* Episode." *Globe and Mail* (Toronto), 25 May 1990.
"The Future of the CBC" (interview). *Policy Options* (Montreal), January/February 1994.

Johnson, Brian D. "Making 'Democracy'." *Maclean's* (Toronto), 16 January 1989.

———. "A Televisionary." *Maclean's* (Toronto), 16 January 1989.

Koch, Eric. *Inside Seven Days.* Scarborough, Canada: Prentice-Hall, 1986.

"The New Men in Charge of the CBC." *Globe and Mail* (Toronto), 27 September 1989.

Rutherford, Paul. *When Television Was Young: Primetime Canada 1952–1967.* Toronto: University of Toronto Press, 1990.

Stewart, Sandy. *Here's Looking at Us: A Personal History of Television in Canada.* Toronto: CBC Enterprises, 1986.

"Watson's Sights Set on *Democracy* and the CBC." *Globe and Mail* (Toronto), 24 December 1988.

See also Canadian Programming in English; *This Hour Has Seven Days*

WAYNE AND SHUSTER

Canadian Comedy Act

Wayne and Shuster, who won international acclaim for their distinctive gentle satiric sketches, were the founding fathers of English Canadian TV comedy. Appearing fairly regularly on CBC radio and television from the 1940s until Wayne's death in 1990, they helped to pave the way for such successful Canadian acts as the *Royal Canadian AirFarce* and *Kids in the Hall*. At the same time, however, their near-monopoly on the CBC's commitment to TV comedy for many years may have hindered the growth of other comedic talent in Canada. During their early years, they wrote all their own material, but later made use of other writers as well.

On television, initially, they were a bigger sensation in the United States than in Canada. They made a record-setting 67 appearances on *The Ed Sullivan Show,* and edited versions of their many specials for CBC TV were highly popular in U.S. syndication. Over the years, they also made frequent appearances on the BBC and won numerous awards, including the illustrious Silver Rose of Montreux.

The eldest of seven children of a successful clothing manufacturer who spoke several languages, Johnny Wayne was born John Louis Weingarten on 28 May 1918, in the heart of downtown Toronto. Though also born in Toronto, on 5 September 1916, Frank Shuster grew up in Niagara Falls, Ontario, where his father ran a small theatre called the Colonial. Most evenings of his childhood were spent watching silent movies (and learning to read the intertitles), until his father was put out of business by a larger operation down the street. Failing to join other relatives in the United States (Frank's first cousin, Joe, who drew the *Superman* comic strip, lived in Cleveland), the family returned to Toronto.

The future comics first met in Grade 10 at Harbord Collegiate—seated in the same class alphabetically, S happened to be close to W. Under the influence of Charles Girdler, who taught ancient history at Harbord and set up the Oola Boola Club to teach students how to do sketches and variety, they wrote a series of comedy dramas for the school's dramatic guild. One of Wayne's long-standing characters, Professor Waynegartner, originated in a geometry lesson written by Girdler poking fun at one of the other teachers. To take the sting out it, Girdler suggested that it be done with a German accent.

Wayne and Shuster
Photo courtesy of the National Archives of Canada

Both men completed degrees in English at the University of Toronto where they wrote, produced, and starred in a number of variety shows. They also edited and wrote for the university newspaper, the *Varsity*. In 1941, they began a show on Toronto radio station CFRB called *Wife Preserves*, which paid them $12.50 each per week to dispense household hints for women over a network of Ontario stations. They were then contracted to write and perform on the *Shuster and Wayne* (sic) comedy show on the CBC's Trans-Canada Network for one year.

In 1942, they left the CBC to join the infantry, and were soon writing and performing for the big Army Show. They toured military bases across Canada and later, when the show was split into smaller units, took the Invasion Review into Normandy after D-Day. Later they wrote a 52-week series for veterans and spent six weeks entertaining the Commonwealth Division in Korea.

In 1946, they returned to CBC Radio on the *Wayne and Shuster Show*, broadcast live at 9:30 P.M. Thursdays. It was one of the few Canadian programs to compete successfully against American imports. Among their radio creations were the undefeated Mimico Mice, who competed against the Toronto

Maple Leafs. Legendary radio sports announcer Foster Hewitt did the play-by-play using the names of real Leaf players, but only Wayne and Shuster played for the Mice.

Although they began appearing as guests on various American TV programs as early as 1950, their biggest television success came in 1958 when Ed Sullivan, whose ratings had slipped, invited them to appear on his Sunday night variety show. He insisted that they stick to the kind of comedy they were doing in Canada, and gave them a one-year contract with complete freedom to decide on the length, frequency, content, sets, and supporting cast of all their sketches. Jack Gould of *The New York Times* described them as "the harbingers of literate slapstick." Sullivan, who became very fond of them both personally and professionally, said they were his biggest hit in ten years. In fact, his ratings shot up whenever they performed and their contract was renewed again and again. So too was their CBC contract, which had been on the verge of being canceled before their American success.

In 1961, Wayne and Shuster unwisely agreed to do a dreadful 13-week sitcom called *Holiday Lodge*, written by others as a summer replacement for Jack Benny on CBS. But they soon returned to the sophisticated sketches they did best, and in 1962 and 1963 were ranked as the best comedy team in America in polls by *Motion Picture Daily* and *Television Today*.

Fearing overexposure, they avoided doing a weekly show for CBC TV, and instead contracted for a certain number of hour-long specials each year. Their style, which consisted of a mixture of slapstick, pantomime, and groan-inducing jokes, depended heavily, at times excessively, on sets and props. Many of their early sketches were take-offs on classic situations, such as putting Shakespearean blank verse into the mouths of baseball players. In their first appearance on *Ed Sullivan*, Wayne played a Roman detective investigating the murder of Julius Caesar in "Rinse the Blood Off My Toga." His use of "martinus" as the singular of "martini" quickly became a catchphrase (some New York bars began advertising "Martinus Specials"), as did the line "I told him, 'Julie, don't go,'" uttered several times by actress Sylvia Lennick playing Caesar's wife. Even Marshall McLuhan complimented them on their word games, as when the hero of their western version of *Hamlet* refused a drink from the bar and ordered "the unkindest cut of all."

Some of the most memorable moments on their TV shows for CBC arose from tricks of the camera—they would walk down an apparently infinite number of stairs or defy gravity as painters on the Tower of Pisa. Although Shuster tended to play the straight man, both portrayed a variety of characters. In general, their comedy was literate, middle-brow, and up-beat. They always disdained cruel humor, preferring the "send-up" to the put-down. Wayne thought that the best description of their style was the phrase "innocent merriment" from Gilbert and Sullivan's *Mikado*.

By the late 1970s, some Canadian critics were complaining that the comic duo were merely going through the motions, that their comedy was hopelessly out of date, more sophomoric than sophisticated, and often embarrassingly bad. It was suggested that they had become too comfortable with the world, that they had lost the anger or frustration necessary for good comedy. There was also some criticism of their decision to do commercials for U.S.-owned Gulf Oil. Nonetheless, they remained quite popular, especially among the under-30 and over-55 age groups. The syndication of 80 half-hour specials in the United States, South Africa, and half a dozen other countries in 1980 was the CBC's largest dollar sale of programming to that date.

Despite several enticing offers from the United States, Wayne and Shuster always chose to stay in Toronto. In addition to giving Canadians the confidence to do their own comedy, they spoke passionately on behalf of Canadian cultural sovereignty. In 1978, for example, Wayne told a joint luncheon of the Ottawa Men's and Women's Clubs that "an imbalanced television system has made us a nation of American watchers, totally ignorant of our own way of life. We are being robbed of our national identity. We've put Dracula in charge of the blood bank."

—Ross A. Eaman

FURTHER READING

Rutherford, Paul. *When Television Was Young: Primetime Canada 1952–1957*. Toronto: University of Toronto Press, 1990.

See also Canadian Programming in English; *Ed Sullivan Show*; *Kids in the Hall*; *Royal Canadian Air Farce*

WEARING, MICHAEL

British Producer

Michael Wearing is one of Britain's most well-respected and successful producers of quality drama, responsible for developing a string of award-winning short series in the 1980s, including *Boys from the Blackstuff*, one of the landmarks in British television drama. His career in television began in 1976 when he was appointed script editor to the BBC's English Regions Drama Department in Birmingham, set up to encourage new writing from the regions.

From 1980 Wearing was producing both single plays and series for the unit; in 1981, he achieved a major success with the serialisation of *The History Man*.

In the development of single plays BBC producers have enjoyed considerable autonomy and, following the trend in contemporary theatre, Wearing was keen to commission socially challenging material. However, by the early 1980s single plays were being squeezed out of the schedules and their potential to create a social stir had diminished accordingly. Wearing's contribution to TV drama hinges on his success in carrying over the progressive tendencies of the single play into the short series—an altogether more difficult format to negotiate with management because of the higher costs.

In Britain, the most celebrated of these programmes was Alan Bleasdale's *Boys from the Blackstuff* (1982), a five-part play series which explored the impact of unemployment on a gang of Liverpudlian asphalt workers. The hard-hitting programme coincided with rocketing unemployment and gave a voice to the despair of the three million people in Britain who were forced to claim the dole at the time. The series touched a vital nerve and stimulated a national debate on a major social issue like few other dramas before it. Wearing then moved to London and was soon producing *The Edge of Darkness,* a nuclear thriller series by Troy Kennedy-Martin. Once again the moment was highly opportune, as the programme's transmission in 1985 coincided with widespread anxiety about the nuclear issue in the wake of Chernobyl and the deployment of cruise missiles. Subsequently the programme was sold to 26 countries and proved to be one of the BBC's most successful exports to North America. Other award-winning programmes followed, including Peter Flannery's *Blind Justice* series in 1988, which exposed the inadequacies in the British criminal justice system.

Though these series were all writer-originated projects, Wearing was the driving force behind their success. Indeed, it is entirely due to his strategic skills and tenacity in negotiating the budgets and mustering the resources that they found their way through the convoluted production process onto the screen.

Wearing has been a head of department in the BBC since 1988 and is currently head of Drama Serials. The BBC Serials product, much more than a conventional miniseries, is required to contribute to the prestige of the corporation. In the bureaucratic turmoil of the early 1990s when the corporation was attempting to secure its charter renewal, there was considerable reappraisal as to how drama might best contribute. Under Wearing's stewardship the classic serial was reintroduced, and the adaptations of *Middlemarch* and *Pride and Prejudice* were to enjoy international success. However, in the finest tradition of Sydney Newman before him, Wearing has also managed to preserve the space for more socially controversial contemporary programmes, such as *The Buddha of Suburbia, Family,* and most recently, the ambitious *Our Friends in the North.*

—Bob Millington

MICHAEL WEARING. Theatre director; script editor, BBC's English Regions Drama Department, Birmingham, 1976–81; produced *Boys from the Blackstuff,* 1982; moved to BBC's London departments; head of department, BBC, since 1988; currently head of Drama Serials. Address: Head of Drama Serials, BBC Television, Television Centre, Wood Lane, London WI2 7RJ, England.

Michael Wearing
Photo courtesy of Michael Wearing

TELEVISION (selection)

1981	*The History Man*
1982	*Boys from the Blackstuff*
1982	*Bird of Prey*
1985	*The Edge of Darkness*
1988	*Blind Justice*
1993	*The Buddha of Suburbia*
1996	*The Final Cut*
1996	*Our Friends in the North*

FURTHER READING

Brandt, George. *British Television Drama in the Nineteen Eighties.* Cambridge: Cambridge University Press, 1993.

Millington, Bob, and Robin Nelson. "*Boys from the Blackstuff*": The Making of TV Drama. London: Comedia, 1986.

Paterson, Richard, editor. *BFI Dossier 20: "Boys from the Blackstuff".* London: British Film Institute, 1983.

See also British Programming; *Boys from the Blackstuff*

WEAVER, SYLVESTER (PAT)

U.S. Media Executive/Programmer

Sylvester (Pat) Weaver enjoys a well-deserved reputation as one of network television's most innovative executives. His greatest impact on the industry came during his tenure as programming head at NBC in the late 1940s and early 1950s. There he developed programming and business strategies the other networks would imitate for years to come. He is also remembered for supporting the idea that commercial television could educate as well as entertain, and he championed cultural programming at NBC under a policy he labeled "Operation Frontal Lobes."

Weaver studied philosophy and classics at Dartmouth, graduating magna cum laude. After military service in World War II, he worked in advertising at the Young and Rubicam agency. At that time, advertisers owned the programs that were broadcast on network radio and television, and Weaver worked on program development for the agency's clients. This experience prepared him to make the move to network television.

Weaver joined NBCTV in 1949 to help the company develop its new television network, and held several top-management positions culminating in his appointment as chair of the board in 1956. During that time he maintained close control over television programming at the network and shaped NBC's entire programming philosophy.

To promote growth in the fledgling network, Weaver commissioned a series of specials he called "spectaculars." These heavily-promoted, live specials were designed to generate interest in the NBC schedule in particular and the television medium in general. He hoped that families would purchase their first television sets specifically to watch such events and would then develop regular viewing habits. The strategy especially promised to benefit NBC's parent company RCA, which controlled most patents on new receiver sets. Programming events such as Mary Martin's *Peter Pan* and the 1952 Christmas Eve broadcast of *Amahl and the Night Visitors,* the first opera commissioned for television, resulted from this plan.

While overseeing NBC's growth, Weaver also worked to enhance its power in relation to advertisers. His experience at Young and Rubicam convinced him that sponsors rather than network programmers actually ran the television industry. Because sponsors owned shows outright, the networks had minimal control over what was broadcast through their services. Some sponsors could even dictate when a show would appear in the weekly schedule. Weaver moved to shift this power to the networks by encouraging NBC to produce programs and then to offer blocks of time to multiple sponsors. He developed certain programs such as *Today* and *The Tonight Show* to provide vehicles for this practice. Advertisers could buy the right to advertise in particular segments of such shows but could not control program content. Weaver called this the "magazine concept" of advertising, comparing it to the practice in which print adver-

Pat Weaver
Photo courtesy of Pat Weaver

tisers bought space in magazines without exercising editorial control over the articles. His ambition was for NBC to develop a full schedule of programs and then persuade advertisers to purchase commercial time here and there throughout that schedule. Any given program would carry commercials of several different sponsors. Other networks eventually followed the NBC model, and by the 1960s it had become the television industry standard, commonly known as "participation advertising."

Weaver took pride in his classical education, and he championed the idea that commercial television had an educational mission. He proposed a series of cultural and public-affairs programs for NBC which he promoted under the banner "Operation Frontal Lobes." The goal, Weaver announced in 1951, was "the enlargement of the horizon of the viewer." The campaign included a number of prime-time documentary specials. For example, Project XX was a full-time documentary production unit which made feature-length documentaries on historical events. The *Wisdom* series consisted of interviews with major artists and intellectuals (Edward Steichen, Margaret Mead). Weaver even required that educational material be mixed into the entertainment schedule. For example, the popular comedy/variety program *Your Show of Shows* might include a performance of a Verdi aria among its normal array of comic monologues and Sid Caesar skits.

Weaver left NBC in 1956 when it became clear that the network could no longer follow his philosophy of program variety and innovation. His successor, Robert Kintner, pushed the network schedule toward more standardized series formats. Weaver's last major effort at television innovation came in the early 1960s when he headed Subscription Television, Inc., an early venture into the pay cable industry. His effort to set up a cable service in California was blocked by a referendum initiated by traditional broadcasters. Weaver challenged them in court, and the U.S. Supreme Court subsequently ruled the referendum unconstitutional. STV, however, was bankrupted by the process. Although Weaver's cable venture failed, the case helped remove certain barriers to the eventual development of cable television.

—Vance Kepley, Jr.

PAT WEAVER. Born Sylvester Laflin Weaver, Jr. in Los Angeles, California, U.S.A., 21 December 1908. Educated at Dartmouth College, B.A. magna cum laude 1930. Married: Elizabeth Inglis (Desiree Mary Hawkins), 1942; children: Trajan Victor Charles and Susan (Sigourney). Served in the U.S. Navy, 1942–45. Worked for Young and MacCallister, an advertising and printing firm; announcer, writer, producer, director, actor, and salesman, radio station KHJ, Los Angeles, 1932; program manager, station KFRC, San Francisco, 1934; worked for NBC and the United Cigar Company, 1935; joined Young and Rubicam advertising agency, 1935; supervisor of programs, Young and Rubicam's radio division, 1937; advertising manager, American Tobacco Company, 1938–46; associate director of communications, Office of the Coordination of Inter-American Affairs, 1941; vice president in charge of radio and television for Young and Rubicam, also serving on executive committee, 1947–49; vice president, vice chair, president, then chair, NBC, 1949–1956; chair, McCann Erickson, 1958–63; president, Subscription TV, Los Angeles, 1963-66; chair, American Heart Association, 1959–63; member, board of directors, Muscular Dystrophy Association, since 1967; president, Muscular Dystrophy Association, since 1975. Member: Phi Beta Kappa. Recipient: Peabody Award, 1956; Emmy Award, 1967; named to Television Hall of Fame, 1985. Address: 818 Deerpath Road, Santa Barbara, California 93108, U.S.A.

PUBLICATION

The Best Seat in the House: The Golden Years in Radio and Television. New York: Knopf, 1994.

FURTHER READING

Baughman, James. "Television in the 'Golden Age': An Entrepreneurial Experiment." *The Historian* (Kingston, Rhode Island), 1985.

Boddy, William. "'Operation Frontal Lobes' Versus the Living Room Toy." *Media, Culture and Society* (London), 1987.

Kepley, Vance, Jr. "The Weaver Years at NBC." *Wide Angle* (Athens, Ohio), 1990.

———. "From 'Frontal Lobes' to the 'Bob-and-Bob' Show: NBC Management and Programming Strategies, 1949-1965." In, Balio, Tino, editor. *Hollywood in the Age of Television.* Boston: Unwin Hyman, 1990.

See also Advertising; Advertising Agency; National Broadcasting Company; Sarnoff, David; Special/Spectacular; *Tonight Show*

WEBB, JACK

U.S. Actor/Producer

Although he will be remembered most for his physically rigid portrayal of the morally rigid cop Joe Friday on *Dragnet,* Jack Webb had one of the most varied and far-reaching careers in television history. In his four decades in broadcasting, Webb performed nearly every role imaginable in the industry: actor, director, producer, writer (under the pseudonym John Randolph), editor, owner of an independent production company, and major studio executive. Webb's importance stems not only from his endurance and versatility, but also from his innovation and success.

Webb entered broadcasting as a radio announcer in 1945. After leading roles in radio dramas such as *Pat Novak for Hire,* he conceived of his own police program based on discussions with Los Angeles police officers about the unrealistic nature of most "cop" shows. *Dragnet* began on NBC radio in 1949, based on "actual cases" from the files of the L.A.P.D., and featuring Webb as director, producer, co-writer, and star in the role of the stoic Sergeant Joe Friday. Webb broke the traditional molds of both "true story" crime dramas and "radio *noir*" by de-emphasizing violence, suspense, and the personal life of the protagonists; he instead strove for maximum verisimilitude by using police jargon, showing "business-only" cops following dead-end leads and methodical procedures, and sacrificing spectacle for authenticity. Webb's personal ties to the L.A.P.D. (which approved scripts and production for every *Dragnet* episode) and his own admitted "ultra-conservative" political beliefs tinted his version of "reality" in all of his productions, where good always triumphed over evil and the law always represented the best interests of all members of society at large.

Dragnet was a huge success, moving to television in 1951, where it became the highest-rated crime drama in broadcast history. The television version featured more Webb innovations, including passionless dialogue and acting (obtained by forcing actors to read dialogue "cold" from cue-cards) and using camera and editing techniques taken from a film model. The show's success fueled Webb's career as an independent producer and director of both television and feature films. His Mark VII Limited production company produced *Dragnet* throughout its run on television, including its four-year return in the late 1960s. He also produced numerous other shows with varied degrees of success, including *Adam-12, Emergency,* and *General Electric True,* but all Mark VII productions featured Webb's special blend of heightened realism, rapid-fire emotionless dialogue, and conservative politics. In 1954, *Dragnet* spawned one of the first in a long line of successful television-inspired films. Webb directed and produced more feature films throughout the 1950s, most notably an acclaimed version of *Pete Kelly's Blues* in 1955.

Webb's least successful venture was his brief tenure as a studio executive. Webb, whose association with Warner Brothers ran back to his mid-1950s film projects, was named head of production at Warner Brothers Television in early 1963. Although his previous successes created high expectations, he was only able to sell one show to a network (NBC's short-lived western, *Temple Houston*), and his singular style was incompatible with Warner's only other series on the air, *77 Sunset Strip.* This "ultra-hip" crime show was created in direct opposition to the grim procedural quality of *Dragnet,* but Webb pushed the already waning show in a new direction—toward the stark realism of his previous work. *77 Sunset Strip* was canceled at the end of the season, but Webb didn't last as long—he was fired in December 1963, ending a failed ten-month tenure.

Upon Webb's death in 1982, most reports and coverage focused on Joe Friday. His performance style has been parodied since his emergence in the 1950s, but Webb's impact on television has never been properly assessed. Always anomalous and bucking the tide of televisual convention, Webb's style lives on in syndicated episodes of *Dragnet,* but his innovations and creations are consistently being copied or forsaken on every crime show today.

—Jason Mittell

JACK WEBB. Born in Santa Monica, California, U.S.A., 2 April 1920. Educated at Belmont High School. Married: 1) Julie Peck (London), 1947 (divorced, 1954); children: Stacy and Lisa; 2) Dorothy Thompson, 1955 (divorced, 1957); 3) Jackie Loughery, 1958 (divorced, 1964). Served with the U.S. Army Air Force during World War II, 1942–45. Radio announcer, star, and producer, 1945–61; television producer, director, and actor, from 1951; star and director of motion pictures, from 1948; founder, production company Mark VII, Ltd., and music publishing firms of Mark VII Music and Pete Kelly Music; executive in charge of television production, Warner Brothers Studios, 1963. Member: Screen Actors Guild; Screen Directors Guild; American Society of Cinematographers; American Federation of Television and Radio Artists; United Cerebral Palsy Association (honorary chair). Recipient: Academy of Television Arts and Sciences' Best Mystery Show, 1952–54; over 100 commen-

Jack Webb

dations of merit awarded by radio and television critics. Died in Los Angeles, California, 23 December 1982.

TELEVISION (executive producer)

1951–59	*Dragnet* (actor, producer, and director)
1968–70	*Adam 12* (creator and producer)
1970–71	*The D.A.*
1970–71	*O'Hara, U.S. Treasury*
1971–5	*Emergency!*
1973	*Escape* (narrator only)
1973	*Chase*
1974–75	*The Rangers*
1975	*Mobile Two*
1977	*Sam*
1978	*Project U.F.O.*
1978	*Little Mo*

FILMS (selection; actor)

He Walked by Night, 1948; *Sunset Boulevard,* 1950; *The Men,* 1950; *Halls of Montezuma,* 1950; *You're in the Navy Now,* 1951; *Dragnet* (also director), 1954; *Pete Kelly's Blues* (also director), 1955; *The D.I.* (also director), 1957; *The Last Time I Saw Archie* (also director), 1961.

RADIO

Pat Novak for Hire, 1946; *Johnny Modero Pier 23,* 1947; *Dragnet* (creator, director, producer, and star), 1949-55; *Pete Kelly's Blues* (creator), 1951; *True Series* (creator), 1961.

FURTHER READING

Anderson, Christopher. *Hollywood TV.* Austin: University of Texas Press, 1994.

Collins, Max Allan, and John Javna. *The Best of Crime and Detective TV.* New York: Harmony, 1988.

MacDonald, J. Fred. *Don't Touch That Dial.* Chicago: Nelson-Hall, 1979.

Meyers, Richard. *TV Detectives.* San Diego: Barnes, 1981.

Varni, Charles A. *Images of Police Work and Mass-Media Propaganda: The Case of "Dragnet"* (Ph.D. dissertation, Washington State University, 1974).

See also Detective Programs; *Dragnet;* Police Programs

THE WEDNESDAY PLAY

British Anthology Series

The *Wednesday Play* is now nostalgically looked back upon as part of the legendary past of British television drama—a halcyon time in the 1960s when practitioners had the luxurious freedom of exploring the creative possibilities of the medium through the one-off television play, egged on by broadcasters and audiences alike. To many writers and directors today, it stands as a wistful beacon, a symbol of the possible, as they gaze enviously at the apparent freedoms of their forebears from the seemingly ratings-led, series-dominated wasteland of their TV dramatic present.

As with any legend, there is more than a grain of truth to this view of the past, but also a considerable amount of misty idealisation. *The Wednesday Play* arose, in fact, not as a benign gift of liberal broadcasters but as a desperate attempt by the head of BBC TV drama, Sydney Newman, to save the single play from being axed from the BBC's premier channel (BBC-1), due to poor ratings. Newman, who had been impressed by Scots director James MacTaggart's work on the earlier experimental play strands, *Storyboard* (1961) and *Teletale* (1963), hired him as producer of the new BBC-1 play slot, handing him a brief to commission a popular series of plays.

Newman's stipulations were significant. He wanted a play slot that would be relevant to the lives of a mainstream popular audience, and that would reflect the "turning points" of society: the relationship between a son and a father; a parishioner and his priest; a trade union official and his boss. He also wanted plays that would be fast, not only telling an exciting narrative sparely rather than building up mood, but also hooking the audience's attention by way of an intriguing pre-titles "teaser" sequence. Borrowing from the techniques of the popular series that was threatening to displace the single play in the schedules, Newman wanted the slot to have a recognisable "house style," so that audiences knew that if they tuned in each week, they could expect to see a certain type of show. Finally, mimicking his own success in commercial television several years earlier (on ITV's *Armchair Theatre* slot), Newman prioritised a search for material that would more accurately reflect the experience of the audience, by instituting a system of story editors whose task it was to bring fresh new writers to television.

MacTaggart absorbed Newman's guidelines but translated them in his own way, not least by appointing as his story editor a young writer and actor with whom he had worked on *Teletale*: Roger Smith. It was with Smith's help that the play slot soon came to acquire the reputation for "controversy" and "outrage" that would mark its subsequent history. The script commissioned for MacTaggart and Smith's very first *Wednesday Play* outing in January 1965 set the seal for what would follow. Written by a convicted murderer (James O'Connor) and depicting the cynical progress of a villain from gangster to baronet, *A Tap on the Shoulder* marked a conscious break with the conventions of the polite, "well-made" TV play.

Its determination to break new ground came to characterise *The Wednesday Play* ethos as a whole—from the first crucial season in 1965 to the last in 1970. The slot also acted as a showcase for new talent, in keeping with Newman's original vision. Many well-known practitioners gained their first big break on *The Wednesday Play*, including Tony Garnett and Kenith Trodd (recruited by Smith as assistant story editors), Dennis Potter, and Ken Loach, *A Tap's* director, whose contributions to the slot eventually numbered some of the most seminal TV plays of the 1960s: the "docudramas" *Up the Junction* (1965) and *Cathy Come Home* (1966).

As *The Wednesday Play* developed, shifts in emphasis, however, took place. Under the first season of MacTaggart and Smith, the plays were much more "expressionist" in style, and concerned with exploiting the resources of the television studio, as the earlier *Teletale* had done. It is significant that the slot's first non-naturalistic dramas, from writers like Dennis Potter and David Mercer, were commissioned at this time. In later seasons, though, after MacTaggart and Smith had departed and Tony Garnett was named chief story editor, many of the plays became noticeably more "documentary," reflecting a determination to transcend the confines of the TV studio in order to record more faithfully the rapidly changing character of life in 1960s Britain. Having gained access to lightweight 16mm filming equipment, Garnett and his collaborator Loach abandoned the studio for location shooting, and their form of filmed documentary realism became one of the most familiar hallmarks of *The Wednesday Play.*

The Loach-Garnett documentary style also became one of the most controversial, and was accused both outside and within the BBC of unacceptably blurring the distinctions between fictional drama and factual current affairs. Meanwhile, the play slot itself came under attack from some

The Wednesday Play, "The Lump"
Photo courtesy of the British Film Institute

quarters for its general "filth" and "squalor." "Clean-Up TV" campaigner Mary Whitehouse harried it for what she saw as its gross sexual immorality, though the effect of her attacks was simply to boost publicity and the all-important ratings. Audiences climbed from one to eight million, as people tuned in each week to see for themselves the latest play trailed as "controversial" in the press. For one of the very few times in TV history, Newman's dream of a popular series of plays became reality. By the end of the 1960s, however, it was clear the slot had become a victim of its own past reputation: a reaction had set in against its perceived "permissiveness" and anti-establishment bias amongst significant proportions of the audience who were now deliberately not tuning in. Accordingly, Newman's successor as head of drama, Shaun Sutton, tried to win new audiences by giving the BBC's contemporary play slot a new time and title. In 1970, he mutated it into *Play for Today*, thereby

inadvertently creating the legend of the lost golden age which *The Wednesday Play* has become.

—John Cook

PROGRAMMING HISTORY

• BBC

January 1965–1970   Anthology

FURTHER READING

Cook, John R. *Dennis Potter: A Life on Screen.* Manchester and New York: Manchester University Press, 1995.

Kennedy-Martin, Troy. "Nats Go Home: First Statement of a New Drama for Television." *Encore* (London), March-April 1964.

———. "Up the Junction and After." *Conrast* (London), Winter-Spring 1965–66.

Madden, Paul, editor. *Complete Programme Notes for a Season of British Television Drama, 1959-73.* London: British Film Institute, 1976.
Shubik, Irene. *Play for Today: The Evolution of Television Drama.* London: Davis-Poynter, 1975.

Williams, Raymond. "A Lecture on Realism." *Screen* (London), Spring 1977.

See also British Programming; *Cathy Come Home,* Garnett, Tony; Loach, Ken; Mercer, David; Potter, Dennis; Trodd, Kenith

WEINBERGER, ED

U.S. Writer/Producer

Ed Weinberger is one of television's most respected writer-producers who, along with James L. Brooks, David Davis, Allan Burns, and Stan Daniels, comprised the heart of the MTM creative team. Weinberger has received many awards for his contributions to a number of successful or critically acclaimed series for both MTM and the John Charles Walters Company, of which he was a partner.

Weinberger's early TV experience included writing for *The Dean Martin Show,* where he was teamed with Stan Daniels, who eventually became Weinberger's writing partner at MTM. Weinberger had also been a writer for Bob Hope, traveling with him to Vietnam. In the late-1960s, Weinberger wrote a screenplay about a divorced woman who was struggling to make it on her own. Although it was never produced, *Mary Tyler Moore Show* creators James L. Brooks and Allan Burns saw a copy of the script and hired Weinberger during the series' second season.

In addition to his Emmy Award-winning work on *The Mary Tyler Moore Show,* Weinberger, along with Daniels, created and produced the MTM sitcoms *Phyllis, Doc,* and *The Betty White Show.* In 1977, Weinberger, with Brooks, Davis, and Daniels, were wooed away by Paramount, which was looking to finance other independent production companies for ABC programming. The MTM alumni welcomed the change, if only because the cozy MTM atmosphere was being gradually replaced by a growing bureaucracy that hampered creativity. Brooks, Davis, Daniels, and Weinberger formed the John Charles Walters Company, which produced its most famous sitcom, *Taxi,* in 1978.

In *Taxi,* Weinberger and the other members of the new creative team were able to successfully echo the quality television that had become synonymous with MTM. Much like an MTM show, *Taxi* was a sophisticated example of humor derived from carefully-crafted character exploration. *Taxi* also pursued the "work-place as family" theme so prominent in the best of MTM sitcoms. Canceled in 1982 by ABC, *Taxi* was picked up by NBC for a continuing season. Thus, Weinberger helped deliver a second-generation of quality television that extended into the 1980s.

In 1983, after NBC also canceled *Taxi,* Weinberger seemed to take a giant step backward when he co-produced *Mr. Smith,* a sitcom featuring a talking chimp for which Weinberger provided the voice. This was not the first time Weinberger had used his voice-over talents; the sigh in the John Charles Walters Company end credit logo is Weinberger's as well. In 1984, Weinberger was back on the quality track when he co-wrote the Emmy Award-winning pilot episode for *The Cosby Show.* Weinberger's later production credits also include the disappointing-yet-wildly successful series *Amen,* as well as the critically-acclaimed-yet-unpopular sitcom *Dear John.*

—Michael B. Kassel

ED WEINBERGER. Attended Columbia University, New York City. Married: Carlene Watkins. Writer for nightclub comedians, monologues for Johnny Carson, Bob Hope in Vietnam, Dick Gregory in Mississippi, and Dean Martin specials; creator, writer, and producer of television comedy, since 1970s, working with Stan Daniels for the early part of his career. Recipient: nine Emmy Awards.

TELEVISION SERIES (selection)
1965–74	*The Dean Martin Show* (writer)
1970–77	*The Mary Tyler Moore Show* (writer and producer)
1975–76	*Doc* (producer)
1975–77	*Phyllis* (writer and producer)
1977–78	*The Betty White Show* (producer)
1978–83	*Taxi* (creator, writer, and producer)
1983	*Mr. Smith* (creator and producer)
1984–92	*The Cosby Show* (co-creator and writer)
1986–91	*Amen* (creator and producer)
1989–91	*Dear John* (producer)
1991–92	*Baby Talk* (producer)

MADE-FOR-TELEVISION MOVIE
1978   *Cindy* (co-writer)

FILM
The Lonely Guy , 1984 (co-writer)

FURTHER READING
Feuer, Jane, Paul Kerr, and Tise Vahimagi, editors. *MTM: "Quality Television."* London: British Film Institute, 1984.

See also *Amen; Cosby Show; Mary Tyler Moore Show; Taxi*

WELDON, FAY

British Writer

Most widely known in Britain and abroad as an irreverent novelist usually concerned with women's issues, Fay Weldon has also pursued a wide variety of projects for television, radio, and the stage. The daughter of a novelist, granddaughter of a *Vanity Fair* editor, and a niece to novelist-screenwriter-radio and television dramatist Selvyn Jepson, Weldon's first published novel in 1967 simply expanded upon her 1966 teleplay for *The Fat Woman's Tale*. The teleplay had been written while Weldon was working as a highly successful copywriter for English print and television advertising; her previous work included the still-remembered "Get to work on an egg" campaign. Weldon remained in advertising until the 1970s, yet she still produced teleplays for productions such as *A Catching Complaint* (1966) and *Poor Cherry* (1967).

While Weldon's real progress as a writer has often been traced back to the mid-1960s, it was in the early 1970s that she began fully to establish both her name and public voice. Where Weldon fit in British culture was another matter. *The Fat Woman's Tale* had told a decidedly proto-feminist story of a housewife's anger toward her philandering husband, yet Weldon's public espousal of domestic joys and the use of "Mrs." seemed to mark her as an opponent to the growing British women's rights movement. But as David Frost learned in 1971, Weldon's relation to feminism is not always what it might seem: invited onto Frost's television program to rebut feminist activists, she instead surprised everyone by publicly embracing their complaints. That same year Weldon won the best series script award from the Writers Guild of Great Britain for "On Trial," the first episode of *Upstairs, Downstairs*. She wrote only one other episode, and in many ways the series' sober, understated visual style was quite different from the satiric, reflexive, often fantastic surfaces of much of Weldon's other work, including her sedate, but still barbed television adaptation of *Pride and Prejudice* (1980).

Perhaps it is no coincidence that the imagined recipient of Weldon's *Letters to Alice: On First Reading Jane Austen,* 1984, is a punk-haired but literary niece; that juxtaposition of texts and attitudes, together with Weldon's own later televised comments on the (mis)teaching of Austen, led some critics to accuse Weldon of unjustly attacking Austen's work.

Yet the melodramatic pleasures of both *Upstairs, Downstairs* and *Pride and Prejudice* run through nearly all of Weldon's work and inform her understanding of gender. She not only won a prestigious Booker Prize nomination for *Praxis* (1978), but also chaired the prize's 1983 panel. Yet Weldon has never divorced her "serious" literary work from her own enjoyment of what she calls "that whole women's magazine area, the communality of women's interests, and the sharing of the latest eye-shadow." With such an attitude,

Fay Weldon
Photo courtesy of Fay Weldon/Isolde Ohlbaum

Weldon penned the polemical prison docudrama *Life for Christine* (1980), polished the script for Joan Collins' *Sins* miniseries (1985), and turned a critical eye toward pastoral life in *The Heart of the Country* (1987).

Despite her willingness to adapt the work of others, Weldon has been protective of the rights to her own work. Nevertheless, she has been most notably represented on television in Britain and abroad not through her own scripts, but through two popular multi-part adaptations from her novels: *The Life and Loves of a She-Devil* (1983, televised 1986), which sharply satirized conventions of both heterosexual romance and the romance novel, and *The Cloning of Joanna May* (1989, televised 1991), a slightly more genteel version of *She-Devil's* antics, this time as practiced by a devilish husband. The same creative team (including writer Ted Whitehead, director Philip Saville, and star Patricia Hodge) helmed both adaptations, but it is the highly praised *The Life and Loves of a She-Devil* which remains the strongest evocation of Weldon's own ethos, despite the intervening memory of Susan Seidelman's limp Americanized film adaptation (*She Devil*, 1990).

Oddly enough, Seidelman's film omitted Weldon's most visually rich and outrageous portion, the fantastic surgical reconstruction of the She-Devil into her nemesis, the physical form of female romantic perfection. This excision removed what is most remarkable throughout much of Weldon's work, her Mary Shelley-like coupling of deliberately excessive Gothic fantasy with sharp feminist perception.

Weldon has not been alone in the use of such fantastic elements. Indeed, as Thomas Elsaesser (1988) has suggested, Weldon and "New Gothic" companion Angela Carter (*The Magic Toyshop*, 1986) may present a female-centered television parallel to the male-centered and often fantastic films of Peter Greenaway, Derek Jarman, and other directors prominent in the 1980s "New British Cinema." If these filmmakers were "learning to dream" again (to quote the familiar title of James Park's study), then Weldon has been one of British television's more prominent instructors in the same task.

—Robert Dickinson

FAY WELDON. Born Fay Birkinshaw in Alvechurch, Worcestershire, England, 22 September 1931. Grew up in New Zealand. Attended University of St. Andrews, M.A. in economics and psychology 1954. Married: 1) Ron Weldon, 1962 (died, 1994); 2) Nick Fox, 1995; four sons. Writer for Foreign Office and *Daily Mirror*, London, late 1950s; worked in advertising; author of television and radio plays, dramatizations and series, and novels and stage plays. Chair, Booker McConnell Prize judges' panel, 1983. Recipient: Writers Guild Award, 1973; Giles Cooper Award, 1978; Society of Authors traveling scholarship, 1981; Los Angeles *Times* Award, 1989. Address: Giles Gordon, Anthony Sheil Associates, 43 Doughty Street, London WC1N 2LF, England.

TELEVISION SERIES

1980	*Pride and Prejudice*
1986	*The Life and Loves of a She-Devil*
1987	*Heart of the Country*

TELEVISION PLAYS (selection)

1966	*The Fat Woman's Tale*
1966	*A Catching Complaint*
1967	*Poor Cherry*
1972	*Splinter of Ice*
1980	*Life for Christine*
1991	*The Cloning of Joanna May*
1991	*Growing Rich*

FILM
She-Devil, 1990.

RADIO
Spider, 1973; *Housebreaker*, 1973; *Mr. Fox and Mr. First*, 1974; *The Doctor's Wife*, 1975; *Polaris*, 1978; *Weekend*, 1979; *All the Bells of Paradise*, 1979; *I Love My Love*, 1981.

STAGE
A Small Green Space, 1989 (libretto).

PUBLICATIONS (selection)

The Fat Woman's Joke (novel). London: MacGibbon and Kee, 1967; as *...and the Wife Ran Away*. New York: McKay, 1968.

Down among the Women (novel). London: Heinemann, 1971; New York: St. Martin's, 1972.

Female Friends (novel). London: Heinemann, and New York: St. Martin's, 1975.

Remember Me (novel). London: Hodder and Stoughton, and New York: Random House, 1976.

Words of Advice. New York: Random House, 1977; as *Little Sisters*, London: Hodder and Stoughton, 1978.

Praxis (novel). London: Hodder and Stoughton, and New York: Summit, 1978.

Puffball (novel). London: Hodder and Stoughton, and New York: Summit, 1980.

Watching Me, Watching You (short stories). London: Hodder and Stoughton, and New York: Summit, 1981.

The President's Child (novel). London: Hodder and Stoughton, 1982; New York: Doubleday, 1983.

The Life and Loves of a She-Devil (novel). London: Hodder and Stoughton, 1983; New York: Pantheon, 1984.

Letters to Alice: On First Reading Jane Austen. London: Joseph, 1984; New York: Taplinger, 1985.

Polaris and Other Stories. London: Hodder and Stoughton, 1985; New York: Penguin, 1989.

Rebecca West. London and New York: Viking, 1985.

The Shrapnel Academy (novel). London: Hodder and Stoughton, 1986; New York: Viking, 1987.

The Heart of the Country (novel). London: Hutchinson, 1987; New York: Viking, 1988.

The Hearts and Lives of Men (novel). London: Heinemann, 1987; New York: Viking, 1988.

The Rules of Life (novella). London: Hutchinson, and New York: Harper, 1987.

Leader of the Band (novel). London: Hodder and Stoughton, 1988; New York: Viking, 1989.

The Cloning of Joanna May (novel). London: Collins, 1989; New York: Viking, 1990.

Darcy's Utopia (novel). London: Collins, and New York: Viking, 1990.

Growing Rich (novel). London: Harper Collins, 1992.

Life Force (novel). London: Harper Collins, 1992.

Natural Love (novel). London: Harper Collins, 1993.

Affliction (novel). London: Harper Collins, 1994.

Splitting (novel). London: Flamingo, 1995.

Wicked Women (short stories). London: Flamingo, 1995.

FURTHER READING

Brandt, George W. *British Television in the 1980s*. Cambridge and New York: Cambridge University Press, 1993.

Elsaesser, Thomas. "Games of Love and Death, or an Englishman's Guide to the Galaxy." *Monthly Film Bulletin* (London), 1988.

Pearlman, Mickey, editor. *Listen to Their Voices: Twenty Interviews with Women Who Write*. New York: Norton, 1993.

See also British Programming

WELLAND, COLIN

British Actor/Writer

Colin Welland is widely respected both as an actor and writer for television, the cinema, and the stage. Rotund and unfailingly good-humoured, he has given invaluable support in a range of plays and serials.

Welland first became a familiar face on British television when he landed the role of Constable David Graham, one of the original characters based at Newtown police station in the long-running police serial *Z Cars* in the 1960s. The series broke new ground, introducing a fresh realism to police dramas, and the regular stars all became household names. Welland stayed with the show for some time, as PC Bert Lynch's second partner on the beat, before eventually leaving for new pastures. He reappeared, together with other stars from the early years of the show, when the last episode was filmed in 1978.

Thus established in television as a performer, Welland went on to star in various plays and television movies, often also contributing the scripts (he was voted Best TV Playwright in Britain in 1970, 1973, and 1974). True to his Lancashire roots, his plays often had an earthy northern humour and dealt with themes accessible to the working-class "man in the street." He also enjoyed huge success as a writer for the cinema, notably with his screenplays for *Yanks* and *Chariots of Fire*, an Oscar-winning smash that was heralded (somewhat prematurely) as signalling a new golden era in British moviemaking. Welland himself picked up an Academy Award for Best Screenplay. Among subsequent films that have garnered their share of praise have been *A Dry White Season*, a drama dwelling on the cruelties imposed by the policy of apartheid in South Africa (co-written with Euzhan Palcy), and *The War of the Buttons*, delving into the often dark and violent world of children. Also much admired were his appearances in such films as *Kes*, in which he played the sympathetic Mr. Farthing, and Willy Russell's *Dancing Through the Dark*, which was set in familiar north-western territory, in the bars and clubs of Liverpool.

Perhaps the most memorable image from Welland's lengthy career as a television actor came in 1979, when he was one of a first-class cast that was chosen to appear in Dennis Potter's award-winning play *Blue Remembered Hills*, which recalled the long-lost days of his own childhood. In company with Helen Mirren, Michael Elphick, Colin Jeavons, and John Bird, among others, all of whom were adults playing the roles of young children, Welland cavorted gleefully around woods and fields, his bulk grotesquely crammed into a pair of boy's shorts. Potter's brilliantly realised play, exposing the native cruelty beneath the outwardly innocent world of children, was hailed as a masterpiece and Welland himself, not for the first time in his distinguished career, was singled out for special praise.

—David Pickering

COLIN WELLAND. Born Colin Williams in Leigh, Lancashire, England, 4 July 1934. Attended Newton-le-Willows Grammar School; Bretton Hall College; Goldsmith's College, London, Teacher's Diploma in Art and Drama. Married: Patricia Sweeney, 1962; children: Genevieve, Catherine, Caroline, and Christie. Art teacher, 1958–62; joined Library Theatre, Manchester, 1962–64; established popular fame as PC Graham in *Z Cars*, 1962–65; has since worked as writer and actor for film, television, and theatre. Recipient: Best Television Writer and Best Supporting Film Actor, British Academy of Film and Television Arts Award, 1970; Best Television Playwright, Writers Guild, 1970, 1973, and 1974; Academy Award, 1981; *Evening Standard* Award, 1981; Broadcasting Press Guild Awards,

Colin Welland
Photo courtesy of Colin Welland

1973, 1981. Address: Peters, Fraser and Dunlop, Fifth Floor, The Chambers, Chelsea Harbour, Lots Road, London SW10 0XF, England.

TELEVISION SERIES

1962-65	Z Cars (actor)

MADE-FOR-TELEVISION MOVIES (actor)

1976	Machine Gunner
1979	Blue Remembered Hills
1990	The Secret Life of Ian Fleming/Spymaker: The Secret
1993	Femme Fatale (also writer)

TELEVISION PLAYS (writer)

1969	Bangelstein's Boys (also actor)
1970	Say Goodnight to Grandma
1970	Roll on Four O'Clock (also actor)
1973	Kisses at Fifty (also actor)
1974	Leeds United (also actor)
1974	The Wild West Show
1974	Jack Point
1976	Your Man from Six Counties (also actor)

1977	Bank Holiday
1994	Bambino Mio

FILMS (actor)

Kes, 1969; Straw Dogs, 1971; Villain, 1971; Sweeney!, 1977; Dancing Through the Dark, 1990.

FILMS (writer)

Yanks, 1979; Chariots of Fire, 1981; Twice in a Lifetime, 1985; A Dry White Season, 1989; War of the Buttons, 1994; The Yellow Jersey.

STAGE (writer)

Say Goodnight to Grandma, 1973; Roll on Four O'Clock, 1981.

PUBLICATIONS

A Roomful of Holes (play). London: Davis-Poynter, 1972.
Say Goodnight to Grandma (play). London: Davis-Poynter, 1973.
Northern Humour, 1982.

See also ZCars

WENDT, JANA

Australian Broadcast Journalist

Jana Wendt is Australian television's best-known female current affairs reporter and presenter. She is also widely regarded as one of Australian commercial television's most skilled interviewers.

The daughter of Czech immigrants, Melbourne-born Wendt began her career in journalism researching documentaries for the government-funded Australian Broadcasting Commission in 1975. After completing an arts degree at Melbourne University, she accepted a job in commercial television, joining Ten Network as an on-camera news reporter in their Melbourne newsroom. Shortly after moving into the role of news presenter at Ten Network, Wendt was offered a position as a reporter on Nine Network's new prime-time current affairs show, Sixty Minutes.

Under the guidance of executive producer Gerald Stone, an American with broad experience in both Australian and U.S. news and current affairs programming, Sixty Minutes proceeded to set the standard for quality commercial current affairs in Australia both in terms of content and production values. The youngest correspondent to join the Sixty Minutes team, Wendt quickly established a reputation for her aggressive interviewing style and glamorous, ice-cool on-camera demeanour. It was this combination of acuity and implacability which earned Wendt her nickname "the perfumed steamroller".

In 1988, Wendt left Sixty Minutes to anchor another Nine Network program, the nightly prime-time half-hour current affairs show, A Current Affair, where she cemented her journalistic reputation with a series of incisive and revealing interviews with national and international political figures. Her subjects included Libya's Colonel Gaddafi, U.S. Vice President Dan Quayle, former U.S. Secretary of State Henry Kissinger, former Philippines President Ferdinand Marcos, and media barons Rupert Murdoch and Conrad Black. In 1994, Wendt returned to Sixty Minutes to fill the newly-created role of anchor.

Wendt's departure from A Current Affair the previous year followed accelerating criticism of the program for its increasingly tabloid accent. The trend, evidenced for critics by A Current Affair's frequent use of hidden cameras, walk-up interviews, and stories with a voyeuristic, sexual theme, was at odds with Wendt's image as a guarantor of dispassionate investigative reporting. While she declined to criticise the program on her departure, she did register her general professional objections to the tabloidisation of Australian current affairs on her return to Nine Network in 1994. The first Sixty Minutes she hosted was an hour-long studio debate on journalistic ethics and the tabloidisation of news and current affairs.

A traditionalist who endorses the notions of journalistic objectivity and the watchdog role of the media in the public sphere, Wendt is an icon of an era many media analysts believe to be passing in Australian commercial current affairs television. The approach of pay television, as well as the debt burdens many network owners inherited in the 1980s, caused Australian broadcast networks to look carefully at their production budgets and demand that news and current affairs divisions show increasing profitability. The result has been an attempt to move the focus of such programs away from public sphere issues like politics, economics, and science and concentrate on domestic matters such as relationships, consumer issues, sexuality, and family life. In many instances, this shift in focus has been accompanied by a more melodramatic, emotional approach on the part of journalists and hosts. It is a trend which Wendt has consistently resisted, and which has led her to become a respected, but somewhat isolated figure in the commercial current affairs landscape of the 1990s.

—Catharine Lumby

JANA WENDT. Born in Melbourne, Australia, 1956. Educated at Melbourne University. Researcher, Australian Broadcasting Corporation (ABC) TV documentaries 1975–77; field reporter, ATV-10 News, 1979, co-anchor, 1980; reporter, Nine Network's *Sixty Minutes*, 1982–87; host, Nine's *A Current Affair*, 1987–92; host and reporter, *Sixty Minutes*, 1994; host, Seven Network's current affairs program *Witness*, from 1995.

TELEVISION

1972–	*A Current Affair*
1979–	*Sixty Minutes*

See also Australian Programming

Jana Wendt
Photo courtesy of TCN Channel Nine

WESTERNS

The western has always been a dusty rear-view mirror for reflecting back on the U.S. experience. Whether celebrating the pioneering spirit of the Scotch-Irish invading class or lamenting the genocidal whitewashing of the continent under the banner of "Manifest Destiny," the western has operated as an instrument for navigating through the fog of contemporary political, social, and cultural anxieties by reinterpreting and rewriting the nation's mythic past. In the 1930s, during the most desperate days of the Great Depression, singing cowboys sporting white hats offered hopeful visions of good guys finishing first to a nation starved for optimism; during the dawning of the Cold War era, Hollywood's "A" westerns provided relatively safe vehicles for commenting on McCarthyism (*High Noon*) and American apartheid (*The Searchers*); prime-time westerns in the 1960s often addressed, though allegorically and indirectly, the generational discord of the decade, as well as the con-

flicting frustrations over U.S. involvement in an undeclared war; and in the 1980s and 1990s, revisionist westerns have taken multicultural angles on the Western Expansion (*Dances with Wolves*) or libertarian spins on the genre's long-standing infatuation with law and order (*The Unforgiven*). The western is, in other words, best understood as a "hindsight" form—a form that deploys the rich imagery of the Old West in an ongoing rewriting of the pride and shame of what it means be American.

This rewriting and reinterpreting of the American experience is even evident in the first "modern" western novel, *The Virginian* by Owen Wister. Published in 1902, Wister's classic cowboy novel sparked something of a range war in the heartland of popular literature. According to contemporary literary critics, Wister's novel and the rise of the cowboy hero represented a masculinist and secular reaction to the so-called "sentimental novel" that had been so popular in

The Big Valley

Dr. Quinn, Medicine Woman

Rawhide

The Virginian

the late 19th century. In the tradition of *Uncle Tom's Cabin* and *Little Women*, the sentimental novel celebrated feminine moral authority, domesticity, and religion. The 20th-century western, in stark contrast, denounced the civilized world of women and flaunted, instead, rugged images of courageous men free from the constraints of family. Ultimately, these taciturn men were more given to flirtation with death than with women, and more attached to their horses and six-shooters than they were to their mothers, sisters, sweethearts, wives, or daughters.

Although rooted in the novel, the first westerns appearing on television were more directly connected to Hollywood's mass-produced version of the genre. In television's infancy, recycled "B" westerns from marginal production companies like Mascot, Monogram, PRC, Lonestar, and Republic played a prominent role in transforming television into a mass medium, by stimulating much of the initial enthusiasm for the medium especially among youngsters and rural audiences. Formulaic features and serials displaying the exploits of familiar names like Ken Maynard, Bob Steele, Hoot Gibson, and Tex Ritter were telecast locally, usually during juvenile viewing hours, in showcases with names such as *Six-Gun Playhouse*, *Sage-Brush Theater*, and *Saddle and Sage Theater*. Thanks to such scheduling, a survey of the programming preferences of children in New York City conducted in April 1949 ranked westerns at the top of the list, a full two percentage points ahead of *Howdy Doody*.

The astute marketing of William Boyd's Hopalong Cassidy was by far the most profitable repackaging of a "B" western hero in television's infancy. Performing as a romantic leading man in silent films, Boyd had trouble even mounting a horse when he first landed the role of Hopalong Cassidy in 1935. However, by 1948, after completing 66 western features, Boyd was not only at home in the saddle, but also savvy enough to secure the TV rights to his Hoppy films. In 1949, as a weekly series on NBC, *Hopalong Cassidy* ranked number seven in the Nielsen ratings—and Boyd quickly cashed in on his popularity through product endorsements that included Hoppy roller skates, soap, wristwatches, and, most notably, jackknives (of which one million units were sold in ten days). Clearly influenced by the *Hopalong Cassidy* phenomenon, the first wave of made-for-TV westerns was targeted specifically at the juvenile market, which was a particularly appealing and expansive demographic segment because of the post-war baby boom. Some of the first western series produced expressly for television, most notably *The Gene Autry Show* and *The Roy Rogers Show*, recycled prominent stars of the "B" western. Others, like *The Cisco Kid* and *The Lone Ranger*, were more familiar as radio series. All featured squeaky-clean heroes who modeled what was considered positive roles for their prepubescent fans.

Perhaps the most self-conscious moralist of television's first western stars was Gene Autry, who in the early 1950s authored the Cowboy Code:

1. A cowboy never takes unfair advantage, even of an enemy.

2. A cowboy never betrays a trust.
3. A cowboy always tells the truth.
4. A cowboy is kind to small children, to old folks, and to animals.
5. A cowboy is free from racial and religious prejudice.
6. A cowboy is always helpful, and when anyone's in trouble, he lends a hand.
7. A cowboy is a good worker.
8. A cowboy is clean about his person, and in thoughts, word, and deed.
9. A cowboy respects womanhood, his parents, and the laws of his country.
10. A cowboy is a patriot.

With its emphasis on the work ethic and patriotism, the Cowboy Code adequately captures the seemingly-benign, though unapologetically sexist values animating the juvenile westerns of America's Cold-War culture. But "Thou Shall Not Kill" is noticeably missing from Autry's Ten Commandments—and this omission would later come to be the source of much public concern.

In the mid-1950s, as major powers in Hollywood stampeded into the television industry, a second wave of made-for-TV westerns would elevate the production values of juvenile programs and, more importantly, introduce the first of the so-called adult western series. On the kiddie frontier, Screen Gems, the TV subsidiary of Columbia Pictures, blazed the trail for tinsel town with *The Adventures of Rin Tin Tin* which premiered on ABC in October 1954. Walt Disney Productions ventured into the territory of TV westerns with three hour-long installments of the *Disneyland* anthology show that presented Fess Parker's clean-cut portrayal of an American legend: *Davy Crockett, Indian Fighter* (first telecast on 15 December 1954); *Davy Crockett Goes to Congress* (26 January 1955); and *Davy Crockett at the Alamo* (23 February 1955). The merchandising hysteria that accompanied the initial broadcasting of the Crockett trilogy even surpassed the earlier Hopalong frenzy as Americans consumed around $100 million in Crockett products, including 4 million copies of the record, "The Ballad of Davy Crockett" and 14 million Davy Crockett books. In the fall of 1957, Disney would branch out into series production with *Zorro* which celebrated the heroics of a masked Robin-Hood figure who was fond of slashing the letter "Z" onto the vests of his many foes.

On the adult frontier, four series premiering in September 1955 would start a programming revolution: *Gunsmoke* on CBS, *Frontier* on NBC, and on ABC, *Cheyenne* and *The Life and Legend of Wyatt Earp*. While *Cheyenne* is notable for being part of Warner Brothers Studio's first foray into television production, the most important and enduring of the original adult westerns is, without a doubt, *Gunsmoke*. Adapted from a CBS radio series in which the rotund William Conrad provided the mellifluous voice of Marshall Matt Dillon, the television version recast the taller, leaner, and more telegenic James Arness in the starring role. Destined to become one of the longest running prime-time series in network television history, the premiere episode of *Gunsmoke* was introduced by none other than John Wayne. Positioned behind a hitching post, Wayne directly addressed the camera, telling viewers that *Gunsmoke* was the first TV western in which he would feel comfortable appearing. Linking the program to Hollywood's prestigious, big-budget westerns, Wayne's endorsement was obviously a self-conscious attempt by CBS to legitimize *Gunsmoke* by setting it apart from typical juvenile fare.

The impact of the adult western was stunning and immediate. In the 1958–59 television season, there were 28 prime-time westerns crowding the network schedule. That year seven westerns (*Gunsmoke, Wagon Train, Have Gun, Will Travel, The Rifleman, Maverick, Tales of Wells Fargo,* and *The Life and Legend of Wyatt Earp*) ranked among the top ten most-watched network programs. But the extraordinary commercial success of the television western was not without its detractors. Although adult westerns displayed characters with more psychological complexity and plots with more moral ambiguity than their juvenile counterparts, the resolution of conflict still involved violent confrontations that left saloons, main streets, and landscapes littered with the dead and dying. The body count attracted the scorn of a number of concerned citizens—but by far the most powerful and threatening figure to speak out against such violence was Newton Minow. On 9 May 1961, soon after being appointed the chair of the Federal Communications Commission (FCC) by President John F. Kennedy, Minow delivered his "vast wasteland" speech to a meeting of the National Association of Broadcasters. In this famous harangue, the FCC chairman singled out the TV western for special denunciation. After roundly condemning the "violence, sadism, murder, western badmen, western good men" on television, Minow rebuked westerns as a hindrance in the not-so-cold propaganda war with the Soviet bloc. "What will the people of other countries think of us when they see our western badmen and good men punching each other in the jaw in between the shooting?" Minow asked. "What will the Latin American or African child learn from out great communications industry? We cannot permit television in its present form to be our voice overseas."

In part because of such criticism from high places, and in part because of burn-out in the mass audience, the western would, once again, be rewritten in the 1960s. As the networks attempted to deemphasize violence, the domestic western emerged as a kinder, gentler programming trend. In contrast to action-oriented westerns dealing with the adventures of law officers (*The Deputy*), bounty hunters (*Have Gun, Will Travel*), professional gunmen (*Gunslinger*), scouts (*Wagon Train*), cow punchers (*Rawhide*), gamblers (*Maverick*), and trail-weary loners (*The Westerner*), the domestic western focused on the familial. The patriarchal Murdoch Lancer and his two feuding sons in *Lancer*, the matriarchal Victoria Barkley and her brood in *The Big Valley*, and the Cannon clan in *The High Chaparral*—all were ranching families in talky melodramas that attempted to

replicate the success of the Cartwrights of *Bonanza* fame (Lorne Greene's Ben, Pernell Roberts' Adam, Dan Blocker's Hoss, and Michael Landon's Little Joe). Television's most distinguished domestic western—and the first western series to be televised in color—*Bonanza* ranked among the top ten TV shows for 10 of its 14 seasons and for three consecutive years from 1964 to 1967 was the nation's most watched program.

Unfortunately, this gloss of the western cannot do justice to all of the interesting wrinkles in the genre. The innovations of series like *Branded* and *Kung Fu* are lost in such a brief accounting—and comedic westerns like *The Wild, Wild West* and *F Troop* can only be mentioned in passing. It is also impossible to catalog the accomplishments and contributions of the many talented artists who brought the western to life on television—whether working behind the camera (Lewis Milestone, Sam Fuller, Robert Altman, and Sam Peckinpah, for instance), or in front of it (Amanda Blake, Ward Bond, Richard Boone, Robert Culp, Clint Eastwood, Linda Evans, James Garner, Steve McQueen, Hugh O'Brian, Barbara Stanwyck, and Milburn Stone, to name a few). Suffice it to say that this dinosaur of a programming form once attracted many of television's most creative storytellers and most compelling performers.

In fact, no one was really surprised in 1987 when J. Fred MacDonald wrote the TV western's obituary in his book, *Who Shot the Sheriff?* Declaring that the western was "no longer relevant or tasteful," MacDonald noted the irony that "the generation [baby boomers] that once made the western the most prolific form of TV programming has lived to see a rare occurrence in American popular culture: the death of a genre." Indeed, between 1970 and 1988, fewer than 28 new westerns in total were introduced as regular network series. The last time a western made the top ten list of weekly prime-time programs was in 1973 when *Gunsmoke* was ranked eighth. With the exception of the strange popularity in the early 1980s of made-for-TV movies starring singer Kenny Rogers in the role of "The Gambler," the thunder of the western has been silenced in prime time.

Even so, after the publication of MacDonald's book, the TV western would have at least one more moment of glory when the adaptation of Larry McMurtry's epic western novel, *Lonesome Dove*, became the television event of the 1988-89 season. The highest rated miniseries in five years, *Lonesome Dove* documented the final days of a lifelong partnership between two characters who represent distinctly different models of manhood: Woodrow Call and Augustus "Gus" McCrae. Call enacted the strong, silent tradition of the western hero. Like John Wayne's characters in *Red River* (Tom Dunson) and *The Searchers* (Ethan Edwards), Call was a powerful, tireless, generally humorless leader who outwardly feared no enemy, though his rugged individualism drove him toward the misery of self-imposed isolation. Call was masterfully portrayed by Tommy Lee Jones—but it was Robert Duval's performance of McCrae that stole the show. Where Call's outlook was utilitarian, Gus's was romantic. In some ways, Gus resembled the funny, spirited sidekicks of westerns past: Andy Devine in *Stagecoach*, Walter Brennan in *Red River*, Pat Brady in *The Roy Rogers Show*, or Dennis Weaver and Ken Curtis in *Gunsmoke*. But in *Lonesome Dove*, the eccentric sidekick achieved equal status with the strong silent hero—and as a counterpoint to Call, Gus rewrote the meaning of the western hero. Valuing conversation, irony, the personal, and the passionate, Gus openly shed tears over the memory of a sweetheart. In a genre marred by misogyny since the publication of *The Virginian* in 1902, Gus was no woman-hater. Instead, Gus actively sought the company of women, not merely for sexual gratification, but for their conversation and civilization: he was as comfortable around women as he was around men. The rewriting of the western hero in the Gus character, then, goes a long way toward explaining why *Lonesome Dove* attracted a mammoth audience in which the women viewers actually outnumbered the men. For a story in a genre that has traditionally been written almost exclusively by men for men, this was no small accomplishment.

At the end of *Lonesome Dove*, Call returns to Texas after leading the first cattle drive to Montana. The quest for untamed land beyond the reach of bankers, lawyers, and women has been costly for Call. Narrow graves scattered along the trail north contain the remains of men who served with Call in the Texas Rangers, who worked with him in the Hat Creek Cattle Company, and who looked to him for friendship, leadership and discipline. As Call surveys the ruins of the forlorn settlement that he once called his head-quarters, he is approached by a young newspaper reporter from San Antonio. An agent of the expanding civilization that Call has spent a lifetime loathing and serving, the reporter presses the uncooperative Call for an interview.

"They say you are a man of vision," says the reporter. Reflecting with anguish on the deaths of his friends (including Gus whose dying words were "What a party!"), Call replies, "A man of vision, you say? Yes, a hell of a vision."

As the final words of the miniseries, "hell of a vision" spoke to Call's disillusionment with the dream of Montana

as "Cattleman's Paradise"—a vision that inspired the tragic trail drive. Defeated and alone, his invading heart had, finally, been chastened. But in punctuating what appears to be the great last stand of the cowboy on the small screen, "hell of a vision" takes on even more profound connotations as an epitaph—an epitaph for the television western.

—Jimmie L. Reeves

FURTHER READING

Barabas, SuzAnne, and Gabor Barabas. *Gunsmoke: A Complete History and Analysis of the Legendary Broadcast Series with a Comprehensive Episode-By-Episode Guide to Both the Radio and Television Programs.* Jefferson, North Carolina: McFarland, 1990.

Buscombe, Edward, editor. *The BFI Companion to the Western.* New York: Atheneum, 1988.

Jackson, Ronald. *Classic TV Westerns: A Pictorial History.* Seacaucus, New Jersey: Carol, 1994.

MacDonald, J. Fred. *Who Shot The Sheriff?: The Rise and Fall of the Television Western.* New York: Praeger, 1987.

Marsden, Michael T., and Jack Nachbar. "The Modern Popular Western: Radio, Television, Film and Print." In, *A Literary History of the American West.* Fort Worth: Texas Christian University Press, 1987.

Peel, John. *Gunsmoke Years: The Behind-the-Scenes Story: Exclusive Interviews with the Writers and Directors: A Complete Guide to Every Episode Aired: The Longest Running Network Television Drama Ever!* Las Vegas, Nevada: Pioneer, 1989.

West, Richard. *Television Westerns: Major and Minor Series, 1946–1978.* Jefferson, North Carolina: McFarland, 1987.

Yoggy, Gary A. *Riding the Video Range: The Rise and Fall of the Western on Television.* Jefferson, North Carolina: McFarland, 1994.

See also Cheyenne; Gunsmoke; Wagon Train; Warner Brothers Presents; Walt Disney Programs; Zorro; Westinghouse-Desilu Playhouse

WESTINGHOUSE-DESILU PLAYHOUSE

U.S. Anthology Series

Westinghouse-Desilu *Playhouse*, an anthology series broadcast on CBS between 1958 and 1960, never received the critical acclaim of *Playhouse 90* or *Studio One*, nor did it last as long as those two dramatic programs. However, among the episodes in its brief run were two productions that, in effect, served as pilots for *The Twilight Zone* and *The Untouchables*, two of the most memorable (and most widely syndicated in reruns) television shows of the 1960s.

Westinghouse-Desilu Playhouse was produced by Desilu, a telefilm production company owned by Desi Arnaz and Lucille Ball that owed its genesis and initial success to a single series—*I*

Love Lucy (CBS, 1951–57). By the late 1950s, the company was producing, through a variety of financial arrangements (wholly owning, co-producing, leasing of facilities and person-nel), several situation comedies and western dramas. *Desilu Playhouse* was to be the realization of Arnaz's dream to make Desilu the most significant telefilm production company and to give himself the opportunity for creative play and control beyond his role as producer and actor on *I Love Lucy* and *The Lucy-Desi Comedy Hour* (an hour-long comedy series with the cast and characters of *I Love Lucy* that aired once a month during the 1957–58 television season). Departing from the standard practice of networks committing to series only after a sponsor had agreed to bankroll production costs, CBS bought *Desilu Playhouse* on the strength of the Desilu track record and with a promise that *The Lucy-Desi Comedy Hour* would be among the planned package of dramas, comedies, and musical spectaculars.

Westinghouse committed to sponsorship a month after the sale to CBS in early 1958, agreeing to a record of $12 million production cost outlay. The company was already sponsor of the prestigious anthology series *Studio One*, but this show was canceled shortly after the deal with Desilu. Historians as well as former personnel of Desilu and Westinghouse suggest that it was Westinghouse president Mark Cresap's love of *I Love Lucy* and the persuasiveness of the charming Arnaz—who promised Cresap that the series would double Westinghouse's business in the first year—that encouraged the company to lay out so much money for the telefilmed anthology series.

The first episode of the *Westinghouse-Desilu Playhouse*, aired in October 1958, was "Lucy Goes to Mexico," a *Lucy-Desi Hour* with guest star Maurice Chevalier. The following week the first dramatic hour premiered, "Bernadette" (a biography of Saint Bernadette, the young girl claiming visitation from the Virgin Mary in 19th century Lourdes, France), starring Pier Angeli. Despite Arnaz's claim that the series would never show anything offensive to children, its highest rated telecasts were the two hours of *The Untouchables*, featuring Robert Stack as Eliot Ness, leader of the crack FBI team who pursued Al Capone and other gangsters during the Prohibition. When *The Untouchables* became a regular series on ABC in 1959, it was the subject of great controversy because of its violence and allegedly negative stereotypes of Italian-Americans.

Westinghouse-Desilu Playhouse did not survive long for a variety of reasons—the inability to attract big-star guests every week, the waning power of the anthology series form due to cost and subject matter, the growing popularity of other dramatic programming (such as westerns and cop shows), and the divorce of Ball and Arnaz, which ended their partnership as Lucy and Ricky Ricardo as well. Although *Westinghouse-Desilu Playhouse* did prove Desilu to be multifaceted at telefilm production, Desi Arnaz did not get a chance to expand his acting range, and the musical spectaculars he had envisioned producing for the series fell short of the quantity and quality promised to Westinghouse. The legacy of the series lies in its launching of *The Twilight Zone* and *The Untouchables*, and its continuation of *The Lucy-Desi Hour*, which still appears regularly in syndicated reruns.

—Mary Desjardins

HOST
Desi Arnaz

WESTINGHOUSE SPOKESPERSON
Betty Furness

PRODUCERS   Desi Arnaz, Bert Granet

PROGRAMMING HISTORY   48 Episodes

• CBS
October 1958–September 1959   Monday 10:00-11:00
October 1959–June 1960   Friday 9:00-10:00

FURTHER READING

Anderson, Christopher. *Hollywood TV: The Studio System in the Fifties.* Austin: University of Texas Press, 1994.
Andrew, Bart. *The "I Love Lucy" Book.* New York: Doubleday, 1985.
Sanders, Coyness Steven, and Tom Gilbert. *Desilu: The Story of Lucille Ball and Desi Arnaz.* New York: William Morrow, 1993.

See also Anthology Dramas; Arnaz, Desi; Ball, Lucille

WEYMAN, RON

Canadian Producer

The story of Ron Weyman is the story of the beginning of film drama on Canadian national television in the 1960s and early 1970s, a time when there were no full-length dramatic features being made on a regular basis in Canada. In Weyman's own words, "I was in the business of getting home-town (i.e., Canadian) writers to write films, which would in fact be feature pictures. They could then break through the artificial relationship (as I saw it) between television and the screen."

Weyman, an executive producer of film drama, took on this mission in the midst of a varied career. In the 1950s, he spent a number of years with the National Film Board of Canada as producer, director, writer, and editor of over 20 films. He traveled extensively and learned the craft of shooting film on location, a skill which he eventually brought back to the Canadian Broadcasting Corporation (CBC), where he was responsible for moving the CBC into the production of filmed series, and encouraging a corporate commitment to dramatic film production.

Several years earlier, when technologies had improved and business had changed to the point that the U.S. model of the filmed series obliterated the live-television anthology genre, Weyman had begun to explore the possibilities offered by film in a form new to Canada—the serial. Serials were still studio-bound in Canada, but Weyman put film crews out on locations across the land to film sequences for insertion into the stories. The response was remarkable. Viewers loved to see where they lived—and other places in their sprawling country—on television. At the same time, with Weyman's support, Philip Keately was producing four or five stories in his limited series *Cariboo Country*—on film, on location in the Chilcotin.

The relationship between the National Film Board (NFB) and the CBC was characterised at this time by uneasy and intermittent cooperation. Opinions on the relationship are divided. It is clear that as far as the medium of film—as opposed to kinescope copies of "live" or "live to tape"—productions were concerned, the two agencies were rivals in some areas. As in many other countries, film was considered to be the paramount medium in a hierarchy of entertainment that excluded theatre but included radio and television. When the question of television drama on film was raised, the perceived wisdom was that this was the NFB's job. When both agencies were urged to co-produce fictional films for the centennial year (1967), the premise was that CBC director/producers understood actors and NFB producers and directors (their roles were separate in film but not in television) understood film. Inevitably this led to internal conflicts and overspent budgets. The result was three rather ordinary dramas on film, broadcast on the CBC flagship Sunday night anthology *Festival*. The one remarkable colour film from that period, *The Paper People*, did not involve the NFB. Physically removed from the working headquarters of the CBC (English) language division, Weyman and his crews and editors were free from middle management's interference—and were seen as a drama production unit of their own.

The result of this freedom was the hit series *Wojeck* (a concept which was run through the Hollywood blender to emerge as the bland *Quincy*) and *Corwin,* a medical series. Meanwhile, with David Gardner, Weyman also produced another hit series, *Quentin Durgens, M.P.,* about an idealistic member of parliament. This program was shot on tape, but still went on location for part of each episode, and made a star of actor Gordon Pinsent. Weyman also produced a half-hour comedy program set in an 1837 pioneer settlement, *Hatch's Mill,* and *McQueen: The Actioneer,* a series about a newspaper columnist.

The common thread in all of these works, even *Hatch's Mill,* was engagement with topical social issues, an examination of the uses and abuses of power, and questions of individual and communal responsibility. Most episodes raised uncomfortable questions for the audience and often chose not to present the easy answers supplied by most television drama at that time. Within the series form, Weyman fused the documentary style and spirit of inquiry with the personalised focus of continuing characters, who were supplied with literate dialogue, and the subtext, nuance, and structural freedoms of fiction.

Weyman's influence continues to be felt in the work of producer Maryke McEwan, who began with the docudramas of *For the Record,* shaped the series *Street Legal,*

Ron Weyman
Photo courtesy of Peter Weyman/Herb Nott and Company

and then returned to documentary and docudrama specials. Many successful series in Canada still reflect the blend of documentary and drama which Weyman and Keately created 30 years ago.

—Mary Jane Miller

RON WEYMAN. Born in Kent, England, 1915. Studied briefly at Art Students' League in New York, U.S.A. Married: Giovanna; two sons. Served as lieutenant-commander RCNVR RN, on destroyer escort duty in North Atlantic, aboard landing craft at Normandy during the D-Day invasion, and in Southeast Asia, 1940–45. Producer, writer, and director, over 20 films, National Film Board of Canada, 1946–54; director and producer, CBC, 1954–80; author of books, since 1980. Recipient: Venice Film Festival First Award; Canadian Film Awards First Award.

TELEVISION SERIES (selection)

1965	The Serial
1966, 1968	*Wojeck*
1966–69	*Quentin Durgens, M.P.*
1969–71	*Corwin*
1969–70	*McQueen: The Actioneer*
1970–71	*The Manipulators*

See also *Quentin Durgens, M.P.; Wojeck*

WHELDON, HUW

British Producer/Media Executive

Sir Huw Wheldon was one of the leading figures among BBC television program makers in the 1960s and a top BBC administrator in the 1970s. A man of profound intellect and understanding, he inspired great loyalty among those who had the privilege of working with him.

After a distinguished war career, Wheldon became the arts council director for Wales, and was awarded an OBE for his contributions to the Festival of Britain. Joining the BBC publicity department in 1952, he quickly established himself as a gifted television presenter with the children's program *All Your Own*. Wheldon's greatest contribution to modern television in Britain was his editorship of the arts program *Monitor* from 1958 to 1964. He both produced the program and appeared as its principal interviewer and anchor, surrounding himself with a brilliant team of young directors which included David Jones, Ken Russell, and Melvyn Bragg. Wheldon was a wonderful encourager. He made a major contribution to the work of young directors like Ken Russell, whose career was boosted by his *Monitor* film on the life of Edward Elgar.

Wheldon made *Monitor* the seminal magazine program of the arts. As interviewer, he guided his audience by his readiness to learn and to inquire rather than to pontificate. His sensitivity to language and his skilled use of film sequences made *Monitor* the outstanding arts program of its day. Though some criticised his editorship as promoting a "middle culture" which was neither high art nor pop art, *Monitor* captured and held a large and varied audience. Wheldon described this group as "a small majority, the broad section of the public well-disposed to the arts."

The second part of Wheldon's career was as a manager and administrator. He became head of documentary programs in 1962, a post that was enlarged the following year to head of music and documentary programs. He proved himself a good administrator who could detect and promote real talent. At that time Wheldon believed it was difficult to find superior documentary makers outside the department, and he seldom used freelances. Three years later, however, when he became controller of programs, he accepted the value of the BBC's employing brilliant freelance film makers such as Jack Gold, Ken Russell, and Patrick Garland. In 1968 Wheldon succeeded Kenneth Adam as director of BBC television.

The post was later redesignated as managing director and, in that position, Wheldon was committed to three conflicting objectives: to maintain and enhance standards; to secure at least half of the viewing audience in competition with ITV; and to contain costs in an era of inflation. Wheldon easily maintained and enhanced standards, but the challenge of competitive scheduling was formidable. His published paper, *The British Experience in Television*, revealed how the BBC television audience as a whole suffered because the ITV companies ran very popular programs such as *Coronation Street* and *Emergency Ward 10* at 7:30 P.M.,

thus winning the audience in the early evening and keeping it. Wheldon's solution was to fight like with like, pitting film against film, current affairs against current affairs. He wrote, "Both BBC-1 and ITV had to adopt broadly competitive policies if they were to remain, each of them in a 50-50 position. Neither could afford to be in a 20-80 position...A 50-50 position was achieved in the sixties, and broadly speaking, has prevailed ever since."

Containing costs was an ever harder task; the BBC employed the management consultants McKinsey to make recommendations, and as a result of their report, the corporation, through the efforts of Wheldon and others, introduced a system of total costing. Under this system, individual programs were charged a true proportion of the overheads. The prospect of employment casualization worried the broadcasting unions; every time Wheldon imposed cutbacks the unions became restive. Wheldon believed that 70% of the program staff should be on permanent budget, and the other 30% on temporary or short-term contracts.

Sir Ian Trethowan, who succeeded Wheldon as managing director of television, described Wheldon's style of leadership as tending toward the flamboyant and inspirational. Wheldon was also a shrewd professional broadcaster, with a passion for the public service role of the BBC. He believed it was the BBC's organisational foundation that made it possible to work well and achieve excellence. For Wheldon, the singularity of the BBC lay in its privileged position. Supported by the license fee, and armed with all the radio channels and two television channels, it could afford excellence.

Huw Wheldon was perhaps the last great leader in BBC television; none of his successors measured up to his achievements. He was described as the "last of the great actor-managers," but such a judgment underestimates a man who was much more than a performer. It is fascinating to speculate what would have happened if age had not debarred him from succeeding Charles Curran as director general. Instead the job went to his immediate successor as managing director of television, Ian Trethowan. It was Wheldon's misfortune that his luck ran out just when he could have made his greatest contribution to the fortunes of the BBC as director general.

—Andrew Quicke

HUW WHELDON. Born in Wales, 1916. Attended schools in Wales and Germany. Served in armed forces during World War II; Military Cross, 1944. Publicity officer, BBC, 1952; producer and presenter, various children's programmes; editor and presenter, arts programme *Monitor*, 1957–64, commissioning first films from Ken Russell, John Schlesinger, and Humphrey Burton; head of documentary and music programmes, 1963–65, and controller of programmes, 1965–68; managing director of television, BBC, 1968, deputy director general, BBC, 1976; after retirement from senior posts at the BBC, continued to work as a writer and presenter. President, Royal Television Society, 1979–85. Officer of the Order of the British Empire, Died 1986.

TELEVISION SERIES (presenter)

1954	*All Your Own*
1958–64	*Monitor* (also editor)
1977	*Royal Heritage* (also co-writer)

PUBLICATIONS

Monitor: An Anthology. London: MacDonald, 1962.
"British Traditions in a World Wide Medium." London: BBC, 1973.

"The Achievement of Television: A Lecture." London: BBC, 1975.
"The British Experience in Television." London: BBC, 1976.

FURTHER READING

Bakewell, Joan, with Nicholas Garnham. *The New Priesthood: British Television Today.* London: Allen Lane, 1970.
Ferris, Paul. *Sir Huge: The Life of Huw Wheldon.* London: Joseph, 1990.

See also British Television; Russell, Ken

WHICKER, ALAN

British Broadcast Journalist

Alan Whicker is a globe-trotting television commentator without equal. For some 40 years, on behalf of both the BBC and independent British television networks, he has roamed far and wide in search of the eccentric, the ludicrous, and the socially-revealing aspects of everyday life as lived by some of the more colourful of the world's inhabitants.

Since the late 1950s, when the long-running *Whicker's World* documentary series was first screened, Whicker—a former journalist and reporter for television's *Tonight* programme (he was once reported dead while working as a war correspondent in Korea)—has probed and dissected the often secretive and unobserved private worlds of the rich and famous, rooting out the most implausible and sometimes ridiculous characters after gaining admittance to the places where they conduct their leisure hours. These have ranged from fabulously appointed cruise ships and the Orient Express to cocktail parties, world tours, health spas, and gentlemen's clubs. His focus has been truly international, with series from Australia, the Indian continent and Hong Kong, as well Britain and the United States.

Whicker's satire is so subtle it is often almost undetectable. The objects of his interest are allowed to condemn or recommend themselves and their way of life almost entirely through their own words and appearances, with often little more than the odd encouraging question or aside from Whicker himself. With long-practiced ease and studied diffidence he infiltrates the most select clubs and institutions and moves almost invisibly from person to person, seeking out the most promising individuals and, generally, being more than amply rewarded with the results. Never aggressive in his questioning and carefully cultivating the image of the relaxed but politely interested ex-patriot ready to accept the world as it comes, he has lured countless individuals into allowing him a privileged glimpse of sometimes extraordinary lives.

On occasion over the years Whicker has concentrated his attention upon a single individual, usually someone of immense influence or prestige who is rarely seen in the public eye. Attracted by the air of mystery surrounding such personages, he has drawn general conclusions about the problems and privileges

of living with wealth and power through his detailed portraits of such enigmatic and sometimes deeply disturbed (and disturbing) figures as billionaire John Paul Getty, Paraguay's General Stroessner, and Haiti's greatly feared dictator "Papa Doc" Duvalier. Sometimes the tone is openly critical, but more often the viewer is allowed to draw his own conclusions.

Whicker's World, over the years, has consistently claimed a place in the top ten ratings and Whicker himself has been widely recognized for his talents as a social commentator, winning numerous major awards.

—David Pickering

Alan Whicker
Photo courtesy of Alan Whicker

ALAN DONALD WHICKER. Born in Cairo, Egypt, 2 August 1925. Attended Haberdashers' Aske's School, London. Served as captain in Devonshire Regiment, World War II; director, Army Film and Photo Section with British 8th Army and U.S. 5th Army. Newspaper war correspondent in Korea; foreign correspondent, novelist, writer, and radio broadcaster; joined BBC television, 1957, and presented nightly film reports from around the world for *Tonight*, as well as studio interviews and outside broadcasts; participated in first Telstar two-way transmission at opening of United Nations, 1962; host, *Whicker's World*, BBC, 1959–60; helped launch Yorkshire Television, 1967; left BBC, 1968; producer and host, numerous television specials and documentaries and further series of *Whicker's World*; worked for BBC, 1982–92; returned to ITV, 1992. Fellow, Royal Society of Arts, 1970. Recipient: numerous awards, including Screenwriters Guild Best Documentary Script Award, 1963; Guild of Television Producers and Directors Personality of the Year, 1964; Royal Television Society Silver Medal, 1968; University of California DuMont Award, 1970; Hollywood Festival of TV Best Interview Program Award, 1973; British Academy of Film and Television Arts Dimbleby Award, 1978; *TV Times* Special Award, 1978; Royal Television Society Hall of Fame, 1993. Address: Le Gallais Chambers, St. Helier, Jersey, England.

TELEVISION SERIES

1957–65	*Tonight*
1959–60	*Whicker's World*
1961	*Whicker Down Under*
1962	*Whicker on Top of the World!*
1963	*Whicker in Sweden*
1963	*Whicker in the Heart of Texas*
1963	*Whicker Down Mexico Way*
1964	*Alan Whicker Report Series: The Solitary Billionaire (J. Paul Getty)*
1965–67	*Whicker's World*
1968	*General Stroessner of Paraguay*
1968	*Count von Rosen*
1968	*Papa Doc—The Black Sheep*
1969	*Whicker's New World*
1969	*Whicker in Europe*
1970	*Whicker's Walkabout*
1971	*World of Whicker*
1972	*Whicker's Orient*
1972	*Broken Hill—Walled City*
1972	*Gairy's Grenada*
1972	*Whicker within a Woman's World*
1973	*Whicker's South Seas*
1973	*Whicker Way Out West*
1974–77	*Whicker's World*
1976	*Whicker's World—Down Under*
1977	*Whicker's World: U.S.*
1978	*Whicker's World: India*
1979	*Whicker's World: Indonesia*
1980	*Whicker's World: California*
1980	*Peter Sellers Memorial Programme*
1982	*Whicker's World Aboard the Orient Express*
1982	*Around Whicker's World in 25 Years*
1982	*Whicker's World—The First Million Miles*
1984	*Whicker's World—A Fast Boat to China*
1984	*Whicker!*
1985	*Whicker's World—Living with Uncle Sam*
1987–88	*Whicker's World—Living with Waltzing Matilda*
1990	*Whicker's World—Hong Kong*
1992	*Whicker's World—A Taste of Spain*
1992	*Around Whicker's World—The Ultimate Package!*
1992	*Whicker's World—The Absolute Monarch*
1993	*Whicker's Miss World*
1993	*Whicker's World—The Sun King*
1994	*Whicker's World Aboard the Real Orient Express*
1994	*Whicker—The Mahatir Interview*
1994	*Pavarotti in Paradise*

FILM

The Angry Silence, 1960.

RADIO

Start the Week (chair); *Whicker's Wireless World*, 1983.

PUBLICATIONS (selection)

Within Whicker's World: An Autobiography. London: Elm Tree, 1982.

Whicker's New World. London: Weidenfeld and Nicolson, 1985.

Whicker's World Down Under. London: Collins, 1988.

See also British Programming; *Tonight*

WHITE, BETTY

U.S. Actor

One of television's most beloved, talented actresses, Betty White began as a local TV "personality" and then, defying convention, became star and producer of her own nationally broadcast sitcom. But it was later that she obtained her greatest fame. In a pair of very different roles on sitcom hits, in the 1970s and 1980s, her skillful acting as part of an ensemble and her way with a comic line earned her acclaim and a loving following; a following that has made her a legend.

Early on, White played leads at Beverly Hills High. After graduation, she took on stage roles at the Bliss-Hayden Little

Theater Group. She began to work as a radio actress as well; local TV quickly followed since it was a natural "option for someone just starting." In 1949, Los Angeles TV personality Al Jarvis called White and gave her her first regular TV assignment. Jarvis took to the airwaves six days a week on KLAC to act as a "disc jockey," to play records just like on radio. Between selections, he delivered commercials, performed in sketches and conducted interviews. White was hired as his on air "girl Friday" to do much of the same. Jarvis left in 1952 and soon after White took over full hosting duties.

While still appearing on daily Los Angeles television, White, with two male partners, co-founded Bandy Productions in 1952 to produce her own self-starring situation comedy. A direct out-growth of some of White's daytime sketches, *Life with Elizabeth* told the story of married couple Elizabeth and Alvin (played by Del Moore). It was an unusual program in several respects, not the least of which was its twenty-eight year old co-creator, producer, and star. White was one of only two women in the early days of television (Gertrude Berg being the other) to wield creative control both in front of and behind the camera. A second distinctive feature of the program were its non-linear stories—each episode consisted of three vignettes, three different plots. Leisurely paced, *Elizabeth*'s stories had a ring of *I Love Lucy* about them. While Elizabeth never launched any outrageous schemes, the comic conflicts often grew out of husband Alvin's disapproval of her logic.

Originally, *Elizabeth* aired only in the Los Angeles area, but by 1953 Guild Films began to syndicate the series nationally and the program was in production until 1955. Afterward, the show's three act format made it possible for each episode to be divided up and marketed to stations as fillers. As ten minute segments *Elizabeth* ran successfully and profitably for many years. Betty White earned her first Emmy in 1952 for *Life with Elizabeth*.

While *Elizabeth* was still in production, White moved to NBC and to her own daily daytime variety show. Bandy Production's *The Betty White Show* premiered February 1954. White would appear in the two programs simultaneously for a year. The NBC daytime show ended in early 1955 and White filled the next two years working, primarily, for game show packagers Goodman and Toddson.

In 1957, White co-created the prime time sitcom *A Date with the Angels*. She played Vicki Angel and Bill Williams starred as her husband Gus. More typical in its format and stories than *Life with Elizabeth*, the Angels were newlyweds and were seen fumbling through their first year of wedded bliss. The program aired on ABC for six months before the network retooled it into the comedy-variety vehicle *The Betty White Show*. Lackluster ratings, which inspired the revamping, lingered and that program ended in April 1958.

Over the next several years, White concentrated on guest work. She was a regular visitor to *The Jack Paar Show* where her funny, slightly risqué remarks made her an audience favorite. She also was a frequent visitor to daytime, as a game show panelist.

It was on *Password* in 1961 that White met her husband, host Allen Ludden. They were married in Las Vegas in 1963.

Betty White
Photo courtesy of Betty White

The Luddens were good friends of actress Mary Tyler Moore and her producer husband Grant Tinker, the two powerhouses behind the hit *The Mary Tyler Moore Show*. When script #73 for the series came along it called for an "icky sweet Betty White type" and the show's casting director eventually decided to call the genuine article. Though usually thought of as a series regular, White did not make her first appearance on *The Mary Tyler Moore Show* until the program's fourth year and in her most active season she appeared in only twelve of twenty-six show regularly scheduled episodes. Nevertheless, she made herself an integral part of that show's family and dynamic. As Sue Ann Nivens, the host of "The Happy Homemaker," White created a sparkling presence. Satirizing her own image, White threw herself into the role of a catty, man-chaser who hid her true self behind a gooey shell of sugar. White won Emmys in the 1974–75 and 1975–76 seasons for Best Supporting Actress. She was part of *The Mary Tyler Moore Show*'s final episode in 1977.

After its end White began her own series. The sitcom *The Betty White Show* premiered in 1977 on CBS. Critically acclaimed and co-starring such pros as John Hillerman and Georgia Engel, the program faced tough competition on Monday nights and CBS did not wait for the show to build an audience. It was canceled in early 1978.

In 1983, White joined the small, exclusive group of women to have hosted a daytime game show. *Just Men!* had White as host and seven male guest stars who tried to help two female contestants win cars. Though the program lasted only six months, White proved funny and unflappable as "femcee" and won the Emmy for best game show host that year. She remains, to date, the only female winner of that top honor. Back on prime time she took guest roles on *St. Elsewhere* and other shows.

In 1985, White, at age 63, began the biggest hit of her career. *The Golden Girls*, from Disney, reunited three of TV's greatest comediennes: White, Beatrice Arthur, and Rue McClanahan. (From the New York stage it imported Estelle Getty.) A highly anticipated show, it was the biggest hit of NBC's new fall season. At the end of the first year, all three lead actresses were nominated for Emmys. White won, for her innocent, adorably ignorant Rose Nylund whose nature bespoke of a more optimistic and trusting time. In some ways Rose brought Betty White full circle: Elizabeth of *Life with Elizabeth* was sweet and a little naive and so was Rose.

Golden Girls ran for seven years. The program was repackaged, without Arthur, for CBS the following season. *Golden Palace*, with White, McClanahan and Getty running a Florida hotel, aired for one year. Then, for White, it was on to *Bob*, Bob Newhart's third series, for a few months in early 1994. There she played Sylvia, the no nonsense head of a greeting card company. After *Bob*, White did several guest spots and some television commercials.

White's eagerly awaited autobiography, *Here We Go Again: My Life in Television*, was published that summer not long after it was announced that she would return to series TV. *Maybe This Time*, a Disney-produced sitcom co-starring actress/singer Marie Osmond premiered in the fall of 1995. That same year saw White's induction into the Academy of Television Arts and Sciences' Hall of Fame. Inducted along with Dick Van Dyke, Bill Moyers and Jim McKay, among others, White was the tenth woman so honored.

It has been a long, highly diverse career. From early TV "DJ" to producer/actress to game-show regular to Emmy-winning ensemble player—from "girl Friday" to "Golden Girl." White has said her longevity is based on her "familiarity" to audiences: the generation who knew her as Elizabeth stayed with her up through Rose. Subsequent generations have discovered her, like a shiny new penny, along the way. Each incarnation of Betty White has brought with it a new set of fans.

But whether as herself or as a character (and in her career she has shown a range greater than that of most actors) Betty White always connects with her audience through her honesty and genuineness. And that quality, intimate and comfortable, makes some TV performers truly unique and long-lasting—legendary.

—Cary O'Dell

BETTY WHITE. Born in Oak Park, Illinois, U.S.A., 17 January 1922. Attended public schools in Beverly Hills, California. Married: Allen Ludden, 1963 (died). Began career with appearances on radio shows; has appeared as star, regular and guest in various television series, from 1950s. Recipient: Emmy Awards, 1952, 1975, 1976, and 1986. Inductee, Television Academy Hall of Fame, 1996.

TELEVISION SERIES

1953–55	*Life With Elizabeth*
1954–58	*The Betty White Show*
1957–58	*A Date with the Angels*
1970–77	*The Mary Tyler Moore Show*
1971	*The Pet Set*
1977–78	*The Betty White Show*
1979	*The Best Place to Be*
1980	*The Gossip Columnist*
1985–92	*The Golden Girls*
1992–93	*The Golden Palace*
1993	*Bob*
1995–96	*Maybe This Time*

TELEVISION SPECIALS

1982	*Eunice*
1986	*Walt Disney World's 15th Birthday Celebration* (co-host)
1991	*The Funny Women of Television* (co-host)

FILM

Advice and Consent, 1962.

STAGE (selection)

Summer stock presentations from late 1960s: *Guys and Dolls*; *Take Me Along*; *The King and I*; *Who Was That Lady?*; *Critic's Choice*; *Bells Are Ringing*.

PUBLICATIONS

Betty White in Person. Garden City, New York: Doubleday, 1987.

Here We Go Again: My Life in Television. New York: Scribners, 1995.

FURTHER READING

O'Dell, Cary. *Women Pioneers in Television*. Jefferson, North Carolina: McFarland, 1996.

See also *Golden Girls*; *Mary Tyler Moore Show*

WHITFIELD, JUNE

British Comedy Actor

June Whitfield is a durable comedy actor whose entire career has been spent providing excellent support to virtually every major British comedian on radio and television. In the 1950s she became a radio favourite, playing the perennially engaged "Eth" in the famous Jimmy Edwards comedy series, *Take It from Here*, but her lasting stardom was due to a remarkable succession of television appearances supporting Britain's best-loved comedians, and her long-running sitcom series, *Terry and June*. The list of male comedians with whom Whitfield has worked reads like a *Who's Who* of British comedy talent, and includes Benny Hill, Tony Hancock, Frankie Howerd, Morecambe and Wise, and Dick Emery. However, she was most closely associated with Jimmy Edwards, with whom she co-starred in a number of comedy playlets under the generic title *Faces of Jim* (*Seven Faces of Jim*, 1961; *Six More Faces of Jim*, 1962; and *More Faces of Jim*, 1963; all BBC). She also appeared in many series with Terry Scott, including *Scott On...* (1964–74, BBC), and *Terry and June* (1979–87, BBC), which was a continuation of an earlier series, *Happy Ever After* (1974–78, BBC).

Whitfield made her debut on television in 1951 in *The Passing Show* (BBC), and later appeared as support to Bob Monkhouse and Derek Goodwin in *Fast and Loose* (1954, BBC). After guesting in various sitcoms for 12 years, she landed a starring role in *Beggar My Neighbour* (1966–68), a show about ill-matched neighbours.

Terry and June was Whitfield's most famous vehicle, and her portrayal of a typical long-suffering wife (June Fletcher) with a perennially adolescent husband (Terry Fletcher, played by Terry Scott), while not stretching her talent as an actor, nevertheless demonstrated her amazing consistency and willingness to bring the best out of any material. Throughout the 1980s and into the 1990s, she also reestablished herself as a radio star, working with comedian Roy Hudd in *The News Huddlines*, where she demonstrated a hitherto unknown talent for impersonation, particularly for her "Margaret Thatcher."

The "new wave" of comedy which began to make serious inroads into British television in the 1980s provided Whitfield with further opportunities. Comediennes Dawn French and Jennifer Saunders used the actor in their sketch show, *French and Saunders* (1987-, BBC), and Jennifer Saunders later chose her for the role of "Mother" in *Absolutely Fabulous* (1992–95, BBC).

Absolutely Fabulous was a groundbreaking British sitcom of the 1990s, with a dazzling mix of politically incorrect, outrageousness, and savage wit. The clever casting of Whitfield as "Mother" allowed Saunders to utilise the actor's housewife persona in a subversive way, employing dialogue and plot to investigate areas of the character never glimpsed in *Terry and June*.

Absolutely Fabulous and similar shows written by and starring women are no longer rarities on British television, but the majority of Whitfield's career has been spent supporting male comedians who dominated the medium, with most of the programmes on which she worked bearing the name of the male star (*The Benny Hill Show* and *The Dick Emery Show*, among others). She is not the only funny woman of British television to have had such a comedy-support career, but she is arguably one of the busiest. One can only lament that it was never considered viable in British television to produce *The June Whitfield Show*.

—Dick Fiddy

June Whitfield
Photo courtesy of June Whitfield

JUNE ROSEMARY WHITFIELD. Born in London, England, 11 November 1925. Attended Streatham High School; Royal Academy of Dramatic Art, diploma 1944. Married: Timothy John Aitchison, 1955; child: Suzy. Has appeared in revue, musicals, pantomime, films, radio, and television, from 1950s; formed long-running situation comedy partnership with Terry Scott, 1969–88. Officer of the Order of the British Empire, 1985. Freeman, City of London, 1982. Recipient: British Comedy Awards' Lifetime Achievement Award, 1994. Address: April Young, 11 Woodlands Road, Barnes, London SW13 0JZ, England.

TELEVISION SERIES

1954	*Fast and Loose*
1961–63	*Faces of Jim*
1964–74	*Scott On...*
1966–68	*Beggar My Neighbour*
1967	*Hancock's Hour*
1969	*The Best Things in Life*
1969	*The Fossett Saga*
1974–78	*Happy Ever After*
1979–87	*Terry and June*
1990	*Cluedo*
1992–95	*Absolutely Fabulous*
1994–95	*What's My Line?*

FILMS (selection)

Carry on Nurse, 1959; *The Spy with a Cold Nose*, 1966; *Carry on Abroad*, 1972; *Bless This House*, 1972; *Carry on Girls*, 1973; *Carry on Columbus*, 1992.

RADIO

Take It from Here, 1953–60; *The News Huddlines*, 1984–; *JW Radio Special*, 1992; *Murder at the Vicarage*, 1993; *A Pocketful of Rye*, 1994; *At Bertram's Hotel*, 1995.

STAGE (selection)

A Bedful of Foreigners; Not Now, Darling; An Ideal Husband, 1987; *Ring Round the Moon*, 1988; *Over My Dead Body*, 1989; *Babes in the Wood*, 1990, 1991, 1992; *Cinderella*, 1994.

See also *Absolutely Fabulous;* British Programming

WIDOWS

British Crime Drama

Widows, a drama series with six 52-minute episodes written by Lynda La Plante, was first broadcast on British television in the spring of 1983. The series had a simple, effective conceit, which was initially condensed into the opening credits, in which we saw a carefully planned robbery of a security van go badly wrong, with the apparent death of all participants. The widows of the title are the three women left alone by this catastrophe which has befallen Harry's gang. They decide, under the leadership of Harry's widow, Dolly (Ann Mitchell), to follow through the already laid plans for the next robbery—which they will conduct themselves after recruiting another recently widowed woman, Bella (Eva Mottley). This simple variation on a traditional crime story formula—the gang of robbers planning and carrying out a raid under the surveillance of the police—offered a series of pleasures for both male and female viewers in what is traditionally a men's genre. The production company, Euston Films, a wholly owned subsidiary of Thames Television, set up in 1971 to make high quality films and film series for television, had a strong track record with the crime genre, being responsible for *Special Branch, The Sweeney, Out*, and *Minder*. Characteristics of the Euston series included London location shooting in a "fast" realist style, working-class and often semi-criminal milieux and sharp scripts. *Widows* offered these familiar pleasures, but also engaged with changing ideas of appropriate feminine behaviour by audaciously presenting the widows of the title tutoring themselves in criminality so they could be agents not victims. In this sense the series, which had Verity Lambert as executive producer and Linda Agran as producer, was clearly a Euston product; it also must be understood in relation to earlier shows which had tried to insert women into the crime genre—such as *Cagney and Lacey, The Gentle Touch* and *Juliet Bravo*. The difference with *Widows* was that the women were on the wrong side of the law.

Following the success of the first series—which had six episodes and a continuous narrative—a second series was commissioned and the two were broadcast together in 1985. Again, the narrative was continuous over the two series, and at the end of *Widows II* the central character, Dolly Rawlins, was imprisoned. Some years later, in 1995, Lynda La Plante, the writer of the first series, produced the final part to what had become a trilogy, *She's Out*, in which Dolly returns. *She's Out* reprises *Widows I* to some extent in that its climax was a carefully planned train robbery—conducted, spectacularly, by women on horseback—but the general critical consensus was that neither of the sequels quite matched *Widows I*.

Retrospectively, *Widows* is now perhaps most interesting as Lynda La Plante's first successful foray into a territory she has made peculiarly her own, the hard world of women in the television crime genre. Her subsequent projects, which include the internationally successful *Prime Suspect*, in which Helen Mirren plays a chief inspector on a murder case, and *The Governor*, in which Janet McTeer plays an inexperienced governor given a prison to run, have tended to place their central female characters within a male hierarchy and visual repertoire. Here they must both confront the prejudice of their colleagues and successfully inhabit and wield power in the context of law enforcement and criminal justice. In contrast, *Widows*, the first of La Plante's "women in a man's world" dramas, was set explicitly within a criminal milieu with the women attempting to support themselves through robbery, rather than learning

Widows
Photo courtesy of the British Film Institute

how to occupy masculine positions of power. This had a series of interesting consequences.

First, the representation of female criminality in the crime series is strongly focused around the figures of the prostitute and the shop-lifter, not the ambitious and successful bank robbers we find here. So the series shook up expectations about what women in crime series can do. Second, because the women are having to learn to perform as men, femininity is "made strange" and becomes a way of behaviour that the women consciously turn on when they need to escape detection. Finally, it should be noted that the heroes of this series, three white, one black, were all working class in origin—although Dolly, well-off from the proceeds of Harry's crimes, listens to opera—and the series thus has a place in the history of honourable endeavour by both Euston Films and Lynda La Plante to depict working class life as diverse and contradictory—and more than comic.

—Charlotte Brunsdon

CAST

Dolly . . . . Ann Mitchell
Bella . . . . Eva Mottley

PRODUCERS   Verity Lambert, Linda Agran

PROGRAMMING HISTORY   6 52-minute episodes
16 March 1983–20 April 1983

FURTHER READING

Alvarado, Manuel, and John Stewart, editors. *Made for Television: Euston Films Limited*. London: British Film Institute, 1985.

Baehr, Helen, and Gillian Dyer, editors. *Boxed In: Women and Television*. London: Pandora, 1987.

See also British Programming; La Plante, Lynda; *Prime Suspect*

WILD KINGDOM

U.S. Wildlife/Nature Program

Mutual of Omaha's Wild Kingdom (also titled *Wild Kingdom*) was one of television's first wildlife/nature programs, and stands among the genre's most popular and longest-running examples. *Wild Kingdom* premiered in a Sunday afternoon time slot on NBC in January 1963, and remained a Sunday afternoon staple until the start of the 1968–69 television season, when it was moved to Sunday evenings. NBC dropped *Wild Kingdom* from its regular series lineup altogether in April 1971 as part of the programming changes and cutbacks each of the three networks were making at that time in response to the newly-created Prime-Time Access Rule. Interestingly, *Wild Kingdom* found its largest audience as a prime-access syndicated program, playing to an estimated 34 million people on 224 stations by 1974, and beating out the likes of *The Lawrence Welk Show* and *Hee Haw* to top the American Research Bureau ratings for syndicated series in October of that year. Though a good number of the episodes aired after 1971 were repackaged reruns from earlier network days, new episodes continued to be produced and included in the syndicated program packages as well. *Wild Kingdom* was produced and distributed in first-run syndication until the fall of 1988.

The perennial host and figurehead of *Wild Kingdom* was zoologist Marlin Perkins. Perkins began his zoological career as reptile curator at the St. Louis Zoo in 1926, and then became director of the Buffalo Zoo in the late 1930s and early 1940s, the Lincoln Park Zoo (Chicago) through the 1950s, and finally the St. Louis Zoo in 1962, a position he held until his death on 14 June 1986. Throughout his career, Perkins was drawn to the medium of television as a means of promoting a conservationist ethic and popular-

izing a corresponding understanding of wildlife and the natural world.

Marlin Perkins

Perkins initiated his involvement in the production of nature programming in 1945, when television itself was only beginning to work its way into the fabric of American life. Having recently been named director of Chicago's Lincoln Park Zoo, Perkins began hosting a wildlife television program on a small local Chicago station, WBKB. He then became the host of *Zoo Parade* in 1949, which began its eight-year run on Chicago station WNBQ before becoming an NBC network show early in 1950. A precursor of sorts to the regularly-featured animal segments on *The Tonight Show* and other late-night talk shows, *Zoo Parade* was a location-bound production (filmed in the reptile house basement) during which Perkins would present and describe the life and peculiarities of Lincoln Park Zoo animals. Soon after his move to the St. Louis Zoo in 1962, Perkins and *Zoo Parade's* producer-director Don Meier were convinced by representatives of the Mutual of Omaha Insurance Company to create *Wild Kingdom*. Perkins remained involved with the production of *Wild Kingdom* until a year before his death on 14 June 1986.

Unlike *Zoo Parade*, *Wild Kingdom* was shot on film almost entirely in the field, and featured encounters with wildlife in their natural habitat. Indeed, one of the program's signature features was the footage of Marlin Perkins, or his assistants Jim Fowler and later Stan Brock, pursuing and at times physically engaging with the wildlife-of-the-week, whether that meant mud-wrestling with alligators, struggling to get free from the vice-like grip of a massive water snake, running from unexpectedly awakened elephants or seemingly angered sea lions, or jumping from a helicopter onto the back of an elk in the snows of Montana. Edited to emphasize the dangerous, dramatic, or comedic interplay between man and beast, accompanied by the appropriate soundtrack mix of music and natural sound, and always punctuated by the familiar voice-overs of Marlin or Jim, the popular narrative conceit of *Wild Kingdom* at times was criticized by some zoologists and environmentalists for putting entertainment values before those of ecological education. Yet *Wild Kingdom* reflected in precisely these ways many of the dominant ecophilosophical and ecological tenets of its day. Set "out in nature," as one reviewer put it, and structured around the actions of protagonists who have left the ordered world of the zoo to explore the unpredictable and often alien landscape of nature, *Wild Kingdom* echoed the conservationist idea of the natural world and the human world as, at best, separate but equal kingdoms.

Many wildlife/nature series since *Wild Kingdom* have developed different and less human-centered narrative strategies with which to represent the natural world, strategies which may themselves reflect a contemporary shift away from the anthropocentric essence of conservationism toward a more ecocentrically-defined environmentalism. In their day, however, Marlin Perkins and Jim Fowler were, in the words of Charles Seibert, "television's cowboy naturalists," and their weekly rides proved to be among the most popular in television history.

—Jim Wehmeyer

HOSTS

Marlin Perkins

Jim Fowler

Stan Brock

PROGRAMMING HISTORY

- NBC

January 1963–December 1968	Sunday Non-prime time
January 1968–June 1968	Sunday 7:00-7:30
January 1969–June 1969	Sunday 7:00-7:30
September 1969–June 1970	Sunday 7:00-7:30
September 1970–April 1971	Sunday 7:00-7:30
First Run Syndication 1971–1988	

FURTHER READING

Cimons, Marlene. "It's Not Easy to Deceive a Grebe." *TV Guide* (Radnor, Pennsylvania), 26 October 1974.

"How to Capture a Live Fur Coat." *TV Guide* (Radnor, Pennsylvania), 15 February 1964.

Kern, J. "Marlin Perkins' Wild Wild Kingdom." *TV Guide* (Radnor, Pennsylvania), 20 April 1963.

"Marlin Perkins." *Variety Obituaries*, Vol. 10. New York: Garland, 1988.

Rouse, Sarah, and Katharine Loughrey, compilers. *Three Decades of Television: The Catalog of Television Programs Acquired by the Library of Congress, 1949-1979*. Washington, D.C.: Library of Congress, 1989.

Siebert, Charles. "The Artifice of the Natural." *Harper's* (New York), February, 1993.

Walsh, Patrick. "Television's Dr. Dolittle Returns to the Air." *TV Guide* (Radnor, Pennsylvania), 17 February 1968.

See also Wildlife and Nature Programs

WILDLIFE AND NATURE PROGRAMS

Television has capitalized on a cultural fascination with the non-human, the mysterious, the unknown, the exotic, and the remote aspects of the natural world in the form of programs devoted to the study and presentation of wildlife, geography, and other features of the biological universe. Watching such offerings, viewers can "go" to locations normally inaccessible because of physical and fiscal limitations. While there is certainly an entertainment value to such programs, they also play an important educational role. And, like all such offerings, while entertaining and educating, they also construct their own interpretation of "nature" or "the wild" or "the animal kingdom." Indeed, wildlife and nature presentations are among the most prominent in emphasizing television's capacity for "framing" and "constructing" particular points of view, while omitting others.

Most wildlife and nature programs are documentary in format. They can be classified roughly under three related categories; tourism, scientific discovery, and environmental preservation. Of these categories, the first may be distinguished from purely educational or scientific inquiry because of its commercial connection. The last is also distinct because of its political motivation.

Since most documentaries are shot on location production costs are relatively high and grants or sponsorship of some kind are necessary to sustain them. On location, film crews are kept small and efficient to minimize costs. The director often doubles as stand up and voice over narrator. Equipment usually consists of a single camera, microphone, sound recorder and lighting kit, where necessary.

Wildlife and nature programming first appeared on U.S. television in 1948 with the success of a fifteen-minute science program called *The Nature of Things*. The series' success lasted until 1954 and paved the way for a host of nature programs to follow. From the start, the introduction of nature and wildlife programming attracted audiences as a "great escape." These programs were fun and exhilarating to watch, and had viewers on the edge of their seats waiting for the commercial-breaks to end and the show to resume. Programs such as *Zoo Parade* (1950–57) a half-hour Sunday afternoon series which looked at animals and animal behavior, included travel footage from such locations as the Amazon jungles. Another such program was *Expedition* (1960–63) which documented journeys to the various remote regions of the world. Known for presenting exciting and sometimes controversial places around the globe, one episode presented a tribe in New Guinea, ruled by Tambaran—the cult of the ghost which venerated the sweet potato. In another episode, *Expedition* presented an aboriginal Indian tribe who had never before seen a white man.

Following the success of adult-oriented programs, such as *Zoo Parade* and *Expedition*, nature and wildlife shows changed strategies and focused attention on attracting younger audiences. Programs were often set up in a format designed to "introduce" the phenomena of wildlife and nature. *Exploring* (1962–66) targeted children ages five to eleven by using methods such as storytelling, mathematics, music, science, and history. *Discovery* (1962–71) searched the world over for natural wonders, as did *Zoo Parade* and *Expedition,* but with the aim of attracting a younger audience. The *Discovery* series was designed to stimulate the cultural, historical, and intellectual curiosity of seven to twelve-year olds regarding nature. Young people were piloted through a spectrum of wonders including how animals use their tails, dramatized

Walt Disney's Wonderful World of Color:
A Country Coyote Goes Hollywood
Photo courtesy of the Walt Disney Company

essays on the history of dance, the voyage of Christopher Columbus, a visit to a Texas ranch, and were introduced to the desert Native Americans. In keeping with the same format, *First Look: Wonders of the World* (1965–66) was designed to provide young children with an introduction to natural history, science, and the various inventions of the world. *First Look*'s topics varied from exploring sea life to experiencing a simulated prehistoric expedition of the dinosaur period.

From the 1960s through the 1970s, wildlife and nature programming introduced a new format designed to give audiences an "untamed" and "dangerous" view into the world of nature. Programs became more "adventurous" in their presentational style. Perhaps the best known and successful of such a series was *Wild Kingdom* (1963–88), sponsored by Mutual of Omaha and hosted for most of its duration by Marlin Perkins. *Wild Kingdom* traveled to out-of-the way places in Africa, South America, the Arctic, Alaska, across the United States, Canada, and the Soviet Union in search of unusual creatures and wild adventures. The series covered such diverse topics as animal survival in the wilds, treatment of animals in captivity, and the lives and habitats of animals and primitive people and their struggle for survival. Similar documentary series followed which focused on animals and their struggle for survival, including *The Untamed World* (1969–71); *Wild, Wild World of Animals* (1973–76); *The World of Survival* (1971); *Safari To Adventure* (1971–73); and *Animal World (Animal Kingdom)* (1968–80). Another such program was *Jane Goodall and the World of Animal Behavior* (1973–74). ABC aired several nature documentaries featuring Miss Goodall, who came to national attention as a scientist who lived among the apes. Here the scientist as "adventurer-hero" became a central narrative focus. Two successful efforts in her ABC series were "The Wild Dogs of Africa" (1973) and the "Baboons of Gombe" (1974) which attracted audiences with their "realism" and detailed an intimate visual portraits.

In order to give audiences an alternative to the harsh realities of nature, wildlife programs added a "sophisticated approach" with the airing of such programs as the *National Geographic Specials* (1965–). Produced in cooperation with the National Geographic Society, this long running series of specials on anthropology, explorations, biological, historical and cultural subjects was first aired on CBS (1965–73), then on ABC (1973–74) and currently can be seen on PBS (1975–). The *National Geographic Specials*, in keeping with the traditions of the journal and the society that stand behind them, are noted for exceptional visual qualities. Another such program was *Animal Secrets* (1966–68), which disclosed the mysteries of wildlife behavior in an appealing nature series and explored such phenomena as how bees buzz, how fish talk, and why birds migrate. A program on "The Primates," filmed in Kenya, presented a study of baboons; their social order and living patterns were observed to find clues to the development of man. The high-quality film series *Nova* (1974–) also relied on detailed productions with exceptional production values. *Nova* is noted for examining complex scientific questions in a manner comprehensible to laymen and in a relatively entertaining fashion. For the most part, the series concerns itself with the effects on nature and society of new developments in science. The close connection of this program with the Public Broadcasting Service (PBS) has almost reached "brand" identification, and the program is often cited as an example of what PBS is and can do.

For a short period of time, wildlife documentaries added a new frontier to the nature of inquiry by examining "oceans and marine worlds." With the appearance of such programs as *Water World* (1972–75) and the very popular *Undersea World of Jacques Cousteau* (1967–76), a new market was opened and added to the previous audience. The *Undersea World* centered around the scientific expeditions of Captain Jacques Cousteau and the crew of his specially equipped vessel, the *Calypso*. The first show began on ABC in 1968 and continued for nearly eight years. ABC dropped the series in 1976, but it continued on PBS with underwriting by the Atlantic Richfield Corporation. Since 1981 Cousteau's environmental series and specials have been produced for Turner Broadcasting System (TBS) in a number of short series.

As the decade of the 1970s closed there was a movement towards bringing back traditional methods of presenting wildlife and nature programming—as if "reintroducing" the areas would stir up an interest in the subject. One such program, *Animals, Animals, Animals* (1976–81), explored the relationship of humans and other animals in order to help youngsters and inquiring adults understand various wildlife phenomena and the interrelated scheme of nature. An entertainment focus was combined with an introduction to the world of science, zoology and biology, and each episode focused on a particular animal in an exciting, yet simplistic manner. By the 1980s a few wildlife and nature programs such as *Nature* (1982–) and *Wild America* (1982–) sustained the "adventurous" format that marked the era of the 1960s and 1970s. For the most part, however, 1980s programming appeared to make greater strides when the focus was on ecology and "saving the planet." During this period programs such as *Universe (Walter Cronkite's Universe)* (1980–82) and *Life on Earth* (1982) often focused on space—the solar system and beyond—in order to understand the phenomena of nature and society.

Another major advancement for wildlife and nature programming occurred in 1985 when the Discovery Channel, an all-documentary cable network, was launched into homes across the nation. This network was devoted chiefly to presenting documentaries on nature, science-technology, travel, history, and human adventure—finally, there was something for everyone. By 1990, the Discovery Channel's penetration passed the 50 million mark, making it one of the fastest growing cable networks of all time. Today, the Discovery Channel has become an alternative outlet for the kinds of nature and wildlife programs that in the 1980s had to depend on public television for exposure. With the success of Discovery Channel, another cable network has joined the nature campaign. *Nickelodeon* (1979–), a children's programming network, recently teamed with Sea World of Florida to educate young people about the importance of conserving the earth's natural resources, the protection of endangered species, pollution prevention, and the importance of recycling. In the 1990s, "Nickelodeon's Cable in the Classroom" service and "Sea World's Shamu TV: Sea World Video Classroom" service provides a hands-on program for audiences from preschoolers to college postgraduates about sea-life and the ecology.

A number of programs focused on nature and wildlife have stepped beyond the most common U.S. television goals of entertaining and informing. They have not only attempted to support the preservation of species and environments but to hold corporations and governmental agencies accountable for acts of pollution and destruction. Films of this type often record dramatic confrontations between those who seek to conserve and those who seek to exploit the environment. The environmental activist group Greenpeace, for example, adopts as part of its policy the need to identify and protest callous indifference toward animals and the environment, and has used such films with great advantage. It remains to be seen whether or not television will eventually be used in a similar manner, or whether "nature" will continue to be presented either as an entertaining commodity or as an exotic topic for popular education.

—Nannetta Durnell and Richard Worringham

FURTHER READING

Bovet, Susan F. "Teaching Ecology: A New Generation Influences Environmental Policy." *Public Relations Journal* (New York), April 1994.

Brooks, Tim, and Earle Marsh. *The Complete Directory to Prime Time Network TV Shows: 1946–Present.* New York: Ballantine, 1979.

Brown, Les. *Les Brown's Encyclopedia of Television.* New York: Times Books, 1977; reprint, Detroit: Gale, 1992.

McNeil, Alex. *Total Television: A Comprehensive Guide to Programming from 1948 to the Present.* New York: Penguin, 1984.

Woolery, George W. *Children's Television: The First Thirty-Five Years, 1946–81.* New Jersey: Scarecrow Press, 1985.

See also Cousteau, Jacques Yves; *Wild Kingdom*

WILDMON, DONALD

U.S. Minister/Media Reformer

As social mores have evolved in the United States in recent years, increasing concern over the role of the media, particularly that of television, has come from outspoken "media reformers" such as Donald Wildmon. Wildmon is regarded by some as a self-appointed censor. To others, he is a minister whose congregation crosses the nation and is comprised of followers upset with the kinds of material seen on television. His ministry comes not so much from preaching, but from the publication of the *American Family Association Journal*, which boasts a readership of nearly two million and income of over $10 million. The central theme of the publication and of Wildmon's work is the advocation of effecting change in television content through boycotting the advertisers of programs which present language and themes that he believes to be anti-Christian.

Wildmon, a soft-spoken fundamentalist Methodist minister from Tupelo, Mississippi, graduated from Emory University's Divinity School. He has spoken often of the roots for his current cause: in the mid-1970s, when his family of young children were gathered around the TV set, he found nothing but sex and violence, adultery and swearing. He vowed to his family that he would do something about it.

At the time, he was the pastor of a Methodist church in Mississippi. He asked his congregation to go without television for one week and found such a striking reaction to the content of programming and to this action taken against the medium that he soon formed the National Federation for Decency, in Tupelo. From that time he never re-entered the regular ministry.

Early on, Wildmon discovered that preaching to the network chiefs, advertisers and programmers was not an easy task. By 1980 he joined with the Reverend Jerry Falwell, leader of the Moral Majority, to form the Coalition for Better Television (CBTV). Members began to observe and record, with a form of "content analysis," the numbers of sexual references, or episodes ridiculing Christian characters, and other aspects of programming deemed offensive. Armed with statistics that, to him, demonstrated the erosion of Christian principles by television programs, Wildmon visited corporate heads. On one occasion, he convinced the chairman of Proctor and Gamble to withdraw advertising from approximately 50 TV shows.

Disputes between Wildmon and Falwell broke up CBTV and Wildmon started another group, Christian Leaders for Responsible Television (CLEAR-TV). His concern spread from television to movies to the distribution of adult magazines. He targeted movie studios such as MCA–Universal, distributor of *The Last Temptation of Christ*, with its "blasphemous depiction" of biblical accounts. He organized campaigns against retail chains—7-Eleven, and K Mart, parent company of Waldenbooks—where adult magazines were sold. And he protested against hotel chains such as Holiday Inns for carrying adult movies on in-house cable systems.

Wildmon's boycotting strategies have been both direct, going to the heads of companies requesting avoidance of anti-Christian materials, and indirect, asking media users to avoid buying those products advertised on questionable programs. In some cases, he seems to have been successful. Pepsico was persuaded to cancel commercials in which Madonna's uses of religious imagery appear. Mazda Motor of America withdrew advertising from NBC's *Saturday Night Live* because of its "indecent vulgar and offensive" nature. And when Burger King was found advertising on TV shows containing "sex, violence, profanity and anti-Christian bigotry," it was induced to run a newspaper ad, an "Open Letter to the American People," declaring its support of "traditional American family values on TV." Some of Wildmon's critics question whether such persuasion by Wildmon is a form of censorship. Others, including Wildmon, insist that such boycotting and public pressure is "as American as the Boston tea party."

Donald Wildmon
Photo courtesy of Broadcasting and Cable

Past issues of Wildmon's *American Family Association Journal* have carried names of members of Congress, together with phone and fax numbers, suggesting active consumer participation in the law-making process. A typical issue of the journal includes "TV Reviews" focused on demonstrating the presence of themes not in keeping with Wildmon's perspectives of traditional values, but there is also text that highlights "The Good Stuff." The journal also regularly presents a list of "troublesome" TV programs and identifies the advertisers supporting the shows. And this is followed by a column listing the "Action Index" or the more emphatic "Boycott Box," listing names of corporations, the chief executive officers, addresses, and phone numbers. Articles cover a number of topics, such as NEA's funding of "anti-Christian" art. Advertisements offer related items such as the video, *MTV Examined*, described as a "comprehensive—and sometimes shocking—look at the destructive effects of MTV and how the programming often crosses the line from entertainment to promotion of illicit sex, violence, drug abuse, immorality, profanity and liberal politics."

More liberal forms of media have been outspoken in reacting to these efforts. *Playboy* has regularly lashed out against Wildmon, presumably because of his attacks on retail outlets that sell the magazine. Other media simply ignore him.

In 1994 Wildmon's attacks hit a crescendo and gained national attention when he brought to public attention, before its airing on ABC, the controversial cop show, *NYPD Blue*. The show's producer, Steven Bochco, had indicated that he would push the frontier of what would be seen on prime-time TV with a series that included controversial language, adult situations, perhaps even brief nudity. This would be television akin to what might be seen in R-rated movies. Wildmon called for a boycott, loudly. With Bochco's promotions and Wildmon's protests, the show attracted viewers and received good ratings, as well as many positive critical notices. A number of ABC affiliates chose not to carry the show, however, and there was some controversy surrounding its advertisers. But the viewing public soon became acclimated; the show did not seem strikingly indecent to many and it continued unchanged through the season. Wildmon later conceded that his loud protests against the show probably attracted attention to it.

While the idea of consumer activism and consumer boycotting came from liberals in the 1960s and 1970s, in ensuing decades such causes and tactics came from the political right. Donald Wildmon, as leader of the forces attacking the media and television in particular, brought to many people the idea that they were not helpless in countering media influences. In doing so, he has taken a prominent place in a long line of advocates addressing the social and cultural role of television.

—Val E. Limburg

DONALD WILDMON. Born in Dumas, Mississippi, U.S.A., 18 January 1938. Attended Mississippi State University, graduated from Millsaps College, Jackson, Mississippi, 1960; Emory University, Atlanta, Georgia, Master of Divinity. Married: Lynda Lou Bennett, 1961; two daughters and two sons. Served in U.S. Army, 1961–63. Ordained as minister, 1964; quit pastorate to protest pornography and violence in media, 1977; founded National Federation for Decency, 1977 (changed name to American Family Association, 1987); founded Coalition for Better Television, 1981 (disbanded, 1982); organized Christian Leaders for Responsible Television, 1982; widened scope of protests by submitting lists of sellers of pornographic magazines and books to Attorney General Edwin Meese's commission on pornography, 1986; convinced Federal Communications Commission to issue warning to radio personality Howard Stern, 1987; protested release of film *The Last Temptation of Christ*, 1988; protested video for and advertising use of Madonna's song "Like a Prayer," 1989; protested National Endowment for the Arts policies, since 1989.

PUBLICATIONS (selection)

Thoughts Worth Thinking. Tupelo, Mississippi: Five Star, 1968.

Practical Help for Daily Living. Tupelo, Mississippi: Five Star, 1972.

Stand Up to Life. Nashville, Tennessee: Abingdon Press, 1975.

The Home Invaders. Wheaton, Illinois: Victor Books, 1985.

FURTHER READING

Kinsley, Michael. "Sour Grapes." *The New Republic* (Washington, D.C.), 10 December 1990.

Lafayette, Jon. "Protesting via Post Card: Wildmon Group Reminds Agencies of Product Boycotts." *Advertising Age* (New York), 15 July 1991.

LaMarche, Gara. "Festival Furore." *Index on Censorship* (London), July-August 1992.

Mendenhall, Robert Roy. *Responses to Television from the New Christian Right: The Donald Wildmon Organizations* (Ph.D. dissertation, University of Texas at Austin, 1994).

"*Playboy*, Others take Offense Against Product Boycotters." *Broadcasting* (Washington, D.C.), 6 November 1989.

Stafford, Tim. "Taking on TV's Bad Boys." *Christianity Today* (Carol Stream, Illinois), 19 August 1991.

"Weighing the Wildmon Effect (editorial)." *Advertising Age* (New York), 28 August 1989.

See also Advertising; Censorship; Religion on Television

WILLIAMS, RAYMOND

British Media Critic

Raymond Williams was to become one of Britain's greatest post-war cultural historians, theorists and polemicists. He was a distinguished literary and social thinker in the Left-Leavisite tradition. He was concerned to understand literature and related cultural forms not as the outcome of an isolated aesthetic adventure, but as the manifestation of a deeply social process that involved a series of complex relationships between authorial ideology, institutional process, and generic/aesthetic form. Pioneering in the context of the British literary academy, these concerns are heralded in the brief-lived post-war journal *Politics and Letters*, which he co-founded. They are perhaps best summarised in *Culture and Society 1780–1950*, his critical panorama of literary tradition from the Romantics to Orwell, predicated on the key terms "industry", "democracy", "class", "art" and "culture". This ideological sense of cultural etymology became the basis of his influential pocket dictionary *Keywords: A Vocabulary of Culture and Society*.

Marked by a commitment to his class origins and his post-war experiences of adult education, his expansion of the traditional curriculum for English also entailed an early engagement with the allied representational pressures of film and cinema, in books such as *Preface to Film, Drama from Ibsen to Eliot, Drama in Performance, Modern Tragedy*, and *Drama from Ibsen to Brecht*. His perception of the links between film and drama remains evident in his 1977 *Screen* essay on the politics of realism in Loach's TV film *The Big Flame* and in his historical introduction to Curran's and Porter's *British Cinema History* (1983).

His preoccupation with the relationships between ideology and culture, and the development of socialist perspectives in the communicative arts, were to continue in such works as *The Long Revolution, May Day Manifesto 1968, The English Novel from Dickens to Lawrence, The Country and the City, Marxism and Literature, Problems in Materialism and Culture, Culture, Writing in Society, Towards 2000, Resources of Hope, The Politics of Modernism*, and *Politics, Education, Letters. Politics and Letters: Interviews with 'New Left Review'* provides a useful retrospective.

In the 1960s Williams' work was to take on new dimensions. He published his first, autobiographical novel, *Border Country*, which was to be followed by *Second Generation, The Volunteers* and *The Fight for Manod*. At the beginning of the decade, he was to write his first book directly addressing the new world of contemporary mass media, *Communications*, an informative and influential volume in the early history of media studies in Great Britain and internationally. He was to move to the centre of left cultural politics, in the crucible of 1968, with his chairmanship of the Left National Committee and his edition of the *May Day Manifesto 1968*.

Throughout the 1960s he was participating in what he remembered as innumerable TV discussion programmes as the young medium found its style. Two of his novels became

TV plays, now sadly lost—a "live" version of *A Letter from the Country* (1966) and *Public Inquiry* (1967), filmed in his native Wales.

From 1968 to 1972 he contributed a weekly column on TV to the BBC magazine *The Listener*. Now collected as *Raymond Williams on Television: Selected Writings*, these illustrate Williams' response to a wide range of TV themes and pleasures—from an enthusiasm for television sport to a distrust in the medium's stress on "visibility", to arguments about the economic and political relationships between production and transmission.

He went on to develop these ideas more formally in the book *Television: Technology and Cultural Form*, one of the first major theoretical studies of the medium, largely written on a visiting professorship at Stanford in 1972. There he soaked up American TV, almost inevitably developed his influential concept of TV "flow", and encountered the newly emerging technologies of satellite and cable.

In 1970 he had contributed a personal documentary, *Border Country*, to the BBC series *One Pair of Eyes*, which was to be followed, at the end of the decade, by *The Country and The City: A Film with Raymond Williams*, the last of five programmes in the series *Where We Live Now: Five Writers Look at Our Surroundings* (1979). In the 1980s he contributed to a trio of Open University/BBC programmes—*Language in Use: "The Widowing of Mrs Holroyd"* (1981), *Society, Education and the State: Worker, Scholar and Citizen* (1982) and *The State and Society in 1984* (1984). He also appeared in *Identity Ascendant: The Home Counties* (1988), an episode in the HTV/Channel 4 series *The Divided Kingdom*, and in *Big Words, Small Worlds* (1987), Channel 4's record of the Strathclyde *Linguistics of Writing* Conference.

Williams' contribution to cultural thinking was that of the Cambridge professor who never forgot the Welsh village of his childhood. He was a theorist of literature who himself wrote novels; an historian of drama who was also a playwright; and a commentator on TV and the mass media who himself regularly contributed to the medium in a variety of ways. For him, unlike so many academics, the medium of television was a crucial cultural form, as relevant to education as the printed word. When Channel 4 began transmission in Great Britain in 1982, it was entirely appropriate that this innovative channel's opening feature film should be *So That You Can Live*, Cinema Action's elegy for the industrial decay of the Welsh valleys, explicitly influenced by the work of Williams, from whose work the film offers us readings.

The Second International Television Studies Conference, held in London in 1986, was honoured to appoint him as its co-president alongside Professor Hilde Himmelweit. But it was a gathering, eventually, that he could not join, and by the time the next event came round in 1988 the conference sadly honoured not his presence, but his passing. The breadth of his impact in the U.K. cultural arena can be gauged from the British Film Institute monograph, *Raymond Williams: Film/TV/Cinema* (1989), produced to accompany a Williams memorial season at the National Film Theatre and containing a contribution by his widow.

—Phillip Drummond

RAYMOND (HENRY) WILLIAMS. Born in Llanfihangel Crocorney, Wales, 31 August 1921. Attended Abergavenny Grammar School, 1932–39; Trinity College, Cambridge, M.A. 1946. Served in Anti-Tank Regiment, Guards Armoured Division, 1941–45. Married: Joyce Marie Dalling, 1942; children: one daughter and two sons. Editor, *Politics and Letters*, 1946–47; extra-mural tutor in literature, Oxford University, 1946–61; fellow, Jesus College, Cambridge, from 1961; reader, Cambridge University, 1967–74; professor of drama, Cambridge University, 1974–83; visiting professor of political science, Stanford University, 1973; general editor, *New Thinkers Library*, 1962–70; reviewer, *The Guardian*, from 1983; adviser, John Logie Baird Centre for Research in Television and Film, from 1983; president, Classical Association, 1983–84. Litt.D.: Trinity College, Cambridge, 1969; D.Univ.: Open University, Milton Keynes, 1975; D.Litt.: University of Wales, Cardiff, 1980. Member: Welsh Academy. Died in Cambridge, 26 January 1988.

TELEVISION PLAYS

1966	*A Letter from the Country*
1967	*Public Inquiry*
1979	*The Country and the City*

PUBLICATIONS

Reading and Criticism. London: Miller, 1950.

Drama from Ibsen to Eliot. London: Chatto and Windus, 1952.

Drama in Performance. London: Watts, 1954.

Preface to Film, with Michael Orram. London: Film Drama, 1954.

Modern Tragedy. London: Verso, and California: Stanford University Press, 1958.

Culture and Society, 1780–1950. London and New York: Columbia University Press, 1958.

Border Country (novel). London: Chatto and Windus, 1960.

The Long Revolution. London and New York: Columbia University Press, 1961.

Communications. London: Penguin, 1962; 3rd edition, 1976.

Second Generation (novel). London: Chatto and Windus, 1964.

May Day Manifesto 1968, editor. London: Harmondsworth Penguin, 1968.

Drama from Ibsen to Eliot. New York: Oxford University Press, 1968.

The Pelican Book of English Prose: From 1780 to the Present Day, editor. London: Harmondsworth Penguin, 1970.

The English Novel from Dickson to Lawrence. London: Chatto and Windus, 1970.

Orwell. London: Fontana, 1971.

D. H. Lawrence on Education, editor, with Joy Williams. London: Harmondsworth Penguin, 1973.

The Country and the City. London: Chatto and Windus, 1973.

Television: Technology and Cultural Form. London: Fontana, 1974.

George Orwell: A Collection of Critical Essays, editor. Englewood Cliffs, New Jersey: Prentice Hall, 1974.

Keywords: A Vocabulary of Culture and Society. London: Fontana, 1975.

English Drama: Forms and Development: Essays in Honour of Muriel Clara Bradbrook, editor, with Marie Axton. Cambridge: Cambridge University Press, 1977.

Marxism and Literature. London and New York: Oxford University Press, 1977.

The Volunteers (novel). London: Eyre Methuen, 1978.

The Fight for Manod (novel). London: Chatto and Windus, 1979.

Politics and Letters: Interviews with "New Left Review". London and New York, 1979.

Problems in Materialism and Culture: Selected Essays. London and New York: Verso, 1980.

Contact: Human Communication and Its History, editor. London: Thames and Hudson, 1981.

Culture. London: Fontana, 1981.

The Sociology of Culture. New York: Schocken, 1982.

Cobbett. London and New York: Oxford University Press, 1983.

Towards 2000. London: Chatto and Windus, 1983.

Writing in Society. London: Verso, 1983.

Gabriel Garcia Marquez. Boston: Twayne, 1984.

Loyalties. London: Chatto and Windus, 1985.

People of the Black Mountains. London: Chatto and Windus, 1989.

The Politics of Modernism: Against the New Conformists, edited by Tony Pinkney. London and New York: Verso, 1989.

Raymond Williams on Television: Selected Writings. New York: Routledge, 1989.

Resources of Hope: Culture, Democracy, Socialism, edited by Robin Gale. London and New York: Verso, 1989.

What I Came to Say. London: Hutchinson, 1989.

FURTHER READING

Eagleton, Terry, editor. *Raymond Williams: Critical Perspectives*. Boston: Northeastern University Press, 1989.

Ethridge, J.E.T. *Raymond Williams: Making Connections*. New York: Routledge, 1994.

Gorak, Jan. *The Alien Mind of Raymond Williams*. Columbia: University of Missouri Press, 1988.

Inglish, Fred. *Raymond Williams*. New York: Routledge, 1995.

O'Connor, Alan. *Raymond Williams: Writing, Culture, Politics*. Oxford and New York: Oxford University Press, 1989.

Pinkney, Tony, editor. *Raymond Williams*. Bridgend, Mid Glamorgan, England: Sern Books, 1991.

Stevenson, Nick. *Culture, Ideology, and Socialism: Raymond Williams and E.P. Thompson*. Aldershot, England: Avebury, 1995.

See also Television Studies

WILSON, FLIP

U.S. Comedian

Flip Wilson was among a group of rising black comics of the early 1970s, of such notoriety as Bill Cosby, Nipsey Russell, and Dick Gregory. He is best remembered as the host of *The Flip Wilson Show*, the first variety show bearing the name of its African-American host, and for his role in renewing stereotype comedy.

With a keen wit developed during his impoverished youth, Wilson rose quickly to fame as a stand-up comic and television show host. Under the stage name Flip, inherited from Air Force pals who joked he was "flipped out," Wilson began performing in cheap clubs across the United States. His early routines featured black stereotypes of the controversial Amos 'n' Andy-type. After performing in hallmark black clubs such as the Apollo in Harlem and the Regal in Chicago, Wilson made a successful appearance on *The Ed Sullivan Show*. Recommended by Redd Foxx, Wilson also performed on *The Tonight Show* to great accolades, becoming a substitute host.

After making television guest appearances on such shows as *Love, American Style* and *That's Life*, and starring in his own 1969 NBC special, Wilson was offered an hour-long prime-time NBC show, *The Flip Wilson Show,* which saw a remarkable four-year run. Only Sammy Davis, Jr. had enjoyed similar success with his song-and-dance variety show; comparatively, shows hosted by Nat "King" Cole and Bill Cosby were quickly canceled, due to lack of sponsorship and narrow appeal. At the show's high point, advertising rates swelled to $86,000 per minute, and by 1972, *The Flip Wilson Show* was rated the most popular variety show, and the second-most popular show overall in the United States.

Wilson's television success came from his unique combination of "new" stereotype comedy and his signature stand-up form. His style combined deadpan delivery and dialect borrowed from his role models, Redd Foxx and Bill Cosby, but replaced their humorous puns with storytelling. His fluid body language, likened to that of silent-screen actor Charlie Chaplin, gave Wilson's act a dynamic and graceful air. The show benefited from his intensive production efforts, unprecedented for a black television performer; he wrote one-third of the show's material, heavily edited the work of writers, and demanded a five-day work week from his staff and guests to produce each one-hour segment. Audiences appreciated the show's innovative style risks, such as the intimate theater-in-the-round studio, and medium-long shots which replaced close-ups, to fully capture Wilson's expressive movements.

Wilson altered his club act for television to accommodate family viewing, relying on descriptive portraits of black characters and situations rather than ridicule. Still, his show offended many African-Americans and civil rights activists who believed Wilson's humor depended on race. A large multi-ethnic television audience, however, found universal humor in the routines, and others credited Wilson with subtly ridiculing the art of stereotyping itself. Wilson denied

Flip Wilson
Photo courtesy of Flip Wilson

this claim, strongly denouncing suggestions that his race required that his art purport anti-bias messages.

These divergent interpretations in fact reflect the variety among Wilson's characters. Some were easily offensive, such as the money-laundering Reverend Leroy and the smooth swinger, Freddy the Playboy. Others, such as Sonny, White House janitor and the "wisest man in Washington," were positive black portraits. The show's most popular character, Geraldine, exemplifies Wilson's intention to produce race-free comedy. Perfectly coifed and decked out in designer clothes and chartreuse stockings, Geraldine demanded respect and, in Wilson's words, "Everybody knows she don't take no stuff." Liberated yet married, outspoken yet feminine, ghetto-born yet poised, Geraldine was neither floozy nor threat. This colorful black female image struck a positive chord with viewers; her one-liners—"The devil made me do it," and "When you're hot, you're hot"—became national fads. Social messages were imparted indirectly through Wilson's characters; the well-dressed and self-respecting Geraldine, for example, countered the female-degrading acts of other popular stand-up comics. Through Geraldine, Wilson also negotiated race and class bias by positively characterizing a working-class black female, in contrast to the absence of female black images on 1970s television, with the exception of the middle- class black nurse of the 1969 sitcom *Julia*. Wilson addressed race more directly through story and theme; one skit, for example, featured Native American women discourteously greeting Christopher Columbus and crew on their arrival in North America. Such innovative techniques enabled Wilson's humorous characters and themes to suggest racial and gender tolerance.

Wilson's career lost momentum when his show was canceled in 1974. Though the recipient of a 1970 Emmy Award for outstanding writing and a 1971 Grammy for best comedy record, Wilson's career never rekindled. He continued to make television specials and TV guest appearances, debuted in Sidney Poitier's successful, post-blaxploitation film, *Uptown Saturday Night*, and performed in two subsequent unsuccessful films. His 1985 television comeback, *Charlie and Company*—a sitcom following *The Cosby Show*'s formula—had a short run.

Wilson saw himself first as an artist, and humor was more prominent than politics in his comic routines. This style, however, allowed him to successfully impart occassional social messages into his act. Moreover, he achieved unprecedented artistic control of his show, pressing the parameters for black television perfomers and producers. Through Geraldine, Wilson created one of a few respectful television images for black women, who were generally marginalized by both the civil rights and women's movements of that era. Finally, though no regular black variety show took up where Wilson left off, its success paved the way for the popularity of later sitcoms featuring middle- and working-class black families, situations, and dialect, shows such as *Sanford and Son*, *The Jeffersons*, and *Good Times*.

—Paula Gardner

FLIP WILSON. Born in Jersey City, New Jersey, U.S.A., 8 December 1933. Married: 1957 (divorced, 1967); four children. Served in U.S. Air Force, 1950–54. Bellhop and part-time entertainer, Manor Plaza Hotel, San Francisco, 1954; travelled country performing in night clubs, late 1950s; regular act at New York City's Apollo Theater, early 1960s; appearances on *The Tonight Show*, from 1965; appeared in numerous television shows, including *Rowan and Martin's Laugh-In*, 1967–68; recorded comedy records, 1967–68; star, *The Flip Wilson Show*, 1970–74; appeared in films, from 1970s; appeared in television series *Charlie and Company*, 1985–86. Recipient: Emmy Award, 1970; Grammy Award, 1971. Address: c/o Triad Artists Inc., 16th Floor, 10100 Santa Monica Boulevard, Los Angeles, California 90067, U.S.A.

TELEVISION SERIES

| 1970–74 | *The Flip Wilson Show* |
| 1985–86 | *Charlie and Company* |

TELEVISION SPECIALS

1974	*Flip Wilson...Of Course*
1974	*The Flip Wilson Special*
1975	*The Flip Wilson Special*
1975	*The Flip Wilson Special*
1975	*Travels With Flip*
1975	*The Flip Wilson Comedy Special*

FILMS

Uptown Saturday Night, 1974; *Skatetown, U.S.A.*, 1979; *The Fish That Saved Pittsburgh*, 1979.

RECORDINGS

Cowboys and Colored People, 1967; *Flippin'*, 1968; *Flip Wilson, You Devil You*, 1968.

FURTHER READING

Davidson, B. "Many Faces of Flip." *Good Housekeeping* (New York), 1971.
Robinson, Louie. "The Evolution of Geraldine." *Ebony* (Chicago), 1970.
"When You're Hot You're Hot." *Time* (New York), 1972.

See also *Flip Wilson Show;* Racism, Ethnicity, and Television

WINDSOR, FRANK

British Actor

Frank Windsor is one of the most well-known stalwarts of British police drama serials, having co-starred in several such productions since the 1960s. His career as a television performer started in radically different shows from those with which he was destined to become most closely associated, with appearances in the Shakespearean anthology *An Age of Kings* and subsequently in the science-fiction series *A for Andromeda*, in which he played scientist Dennis Bridger. In 1962, however, he made his debut in the role with which he became virtually synonymous—that of Newtown's Detective Sergeant John Watt. As one of the crime-busting team crewing *Z Cars*, Watt was right-hand man to Detective Inspector Barlow (Stratford Johns), and was often placed in the role of the "nice guy" to Stratford John's more aggressive, often bullying, senior officer. The two actors formed a dynamic, absorbing partnership that survived well beyond their departure from the series in 1965.

The two stars resumed the same screen personas in their own follow-up series, *Softly, Softly*, a year after leaving the Newtown force. With Barlow raised to the rank of detective chief superintendent and Watt detective chief inspector, the pair continued to hunt down criminals in their "nice and nasty" partnership, though now based in the fictional region of Wyvern, which appeared to be somewhere near Bristol.

Three years into the series the pair were relocated to Thamesford Constabulary's CID Task Force, and the programme itself was retitled *Softly, Softly—Task Force*. Barlow disappeared from the series in 1969, when he left for his own series, *Barlow at Large*, leaving Watt to continue the battle with new partners for another seven years.

Barlow and Watt were brought together again in 1973, when they disinterred the case files connected with the real-life "Jack the Ripper" murders of the 1880s. They pored over the various theories concerning the identity of the murderer, including the possibility that he might have been a member of the royal family, but in the end even television's two most celebrated police detectives could draw no firm conclusion. Along similar lines was *Second Verdict*, another short series in which the two characters investigated unsolved murder cases from real life.

The extent to which Windsor became linked to just one role has subsequently militated against his taking parts that would challenge public perceptions of his original persona. He has, however, appeared as a guest in supporting roles in a number of established series (including *All Creatures Great and Small*, *Boon*, and *Casualty*), has participated in quiz shows, and has also accumulated a number of film and stage credits.

—David Pickering

FRANK WINDSOR. Born in Walsall, Staffordshire, England, 12 July 1927. Attended St. Mary's School, Walsall. Married: Mary Corbett; children: Amanda and David. Began career as performer on radio; founding member, Oxford and Cambridge Players, later the Elizabethan Players; acted classical roles on British stage; television actor as Detective Sergeant Watt in the series *Z Cars*; has since appeared in further police series and other productions. Address: Scott Marshall, 44 Perryn Road, London W3 7NA, England.

TELEVISION SERIES

1960	*An Age of Kings*
1961	*A for Andromeda*
1962–65	*Z Cars*
1966–70,	
1970–76	*Softly, Softly*
1976	*Second Verdict*

MADE-FOR-TELEVISION MOVIES (selection)

1981	*Dangerous Davies—The Last Detective*
1982	*Coming out of the Ice*

Frank Windsor
Photo courtesy of Frank Windsor

FILMS
This Sporting Life, 1963; *Spring and Port Wine*, 1970; *Sunday, Bloody Sunday*, 1971; *Hands of the Ripper*, 1971; *The Dropout*, 1973; *Barry MacKenzie Holds His Own*, 1974; *Assassin*, 1975; *Who is Killing the Great Chefs of Europe?*, 1978; *The London Connection*, 1979; *Night Shift*, 1979; *The Shooting Party*, 1984; *Revolution*, 1985; *Oedipus at Colonus*, 1986; *First Among Equals*, 1987; *Out of Order*, 1987.

STAGE (selection)
Androcles and the Lion; *Brand*; *Travesties*; *Middle-age Spread*; *Mr. Fothergill's Murder*.

FURTHER READING
Corner, John, editor. *Popular Television in Britain: Studies in Cultural History*. London: British Film Institute, 1991.
See also British Programming; *Z Cars*

WINFREY, OPRAH

U.S. Talk Show-Host

Oprah Winfrey, known primarily as the nationally and internationally syndicated American talk-show host of *The Oprah Winfrey Show*, has successfully charted and navigated a career that has built on the television industry as

a form of public therapy. The proliferation of talk-show programs in 1980s and 1990s that have been constructed around the public airing of private trials can be directly attributed to the success of Oprah Winfrey and, a decade earlier, Phil Donahue. It is a genre of television that blends the private and the public into a public confessional. On *Oprah Winfrey* both ordinary people and guest celebrities are there to reveal their inner truths. And it is these revelations which create in the audience the dual sentiments that have been critical to the success of Oprah: there is a voyeuristic pleasure in hearing about what is normally hidden by others, and there is the cathartic sensation that the public revelation will lead to social betterment.

One of the key features of Oprah Winfrey's television persona is that her own private life has been an essential element of her talk-show format of public therapy. Her poor black background and her past and current problems with child abuse, men, and weight have made Oprah an exposed public personality on television and have allowed her loyal audience to feel that they "know" her quite well. This televisual familiarity is part of the power of Oprah Winfrey.

Winfrey was born in Kosciusko, Mississippi, in 1954 and was raised solely by her paternal grandmother for her first six years on a rural pig farm. Her now famous name Oprah was in fact a misspelling of the Biblical name Orpah. Throughout her childhood and adolescence she moved between her father's residence in Nashville and her mother's in Milwaukee. By her early teens she had settled more permanently in Nashville and it was there that she developed her first contacts with broadcasting.

Her path into the profession was partially connected to her success in two beauty pageants. At 16, Winfrey was the first black Miss Fire Prevention for Nashville. From that position and her obvious and demonstrated abilities in public speaking, she was invited to be the newsreader on a local black radio station, WVOL. Later, she maintained her public profile by winning the Miss Black Tennessee and gained a scholarship to Tennessee State University. In her final year of studying speech, drama and English, Winfrey was offered a position as co-anchor on the television news program of the CBS affiliate, WVTF. She has described her early role model for news broadcasting as Barbara Walters.

Although not entirely comfortable with her role as news journalist/anchor, Winfrey gained a more lucrative co-anchor position at WJZ, the ABC affiliate in Baltimore in 1977. She struggled for several months in the position—her greatest weaknesses derived from not reading the newscopy before airtime and from her penchant for extensive ad libbing. She was pulled from the anchor position and given the role of co-hosting a morning chat show, *People are Talking.* Able to be relaxed and natural on air, Winfrey excelled in this position. By the end of her run, her local morning talk show had transformed into a program dealing with more controversial issues and Winfrey's presence helped the show outdraw *Donahue,* the nationally syndicated talk show in the local Baltimore market.

In 1983, she followed her associate producer Debra Di Maio to host *A.M. Chicago,* a morning talk show on Chicago station WLZ–TV. By 1985 the name was changed to *The Oprah Winfrey Show* and again the program was drawing a

Oprah Winfrey
Photo courtesy of Oprah Winfrey

larger audience than Donahue in the local Chicago market. Winfrey also gained a national presence through her Oscar-nominated role in Steven Spielberg's *The Color Purple* (1985). The large television program syndicator King World, realizing the earning potential of Winfrey, took over production of her show in 1986 and reproduced the daily program for the national market. Within weeks of the launch in September 1986, *The Oprah Winfrey Show* became the most watched daytime talk show in the United States.

The deal struck with King World in 1986 instantly made Winfrey the highest paid performer in the entertainment industry with estimated earnings from the program of $31 million in 1987. She has continued to be one of the wealthiest women in the entertainment industry and has used that power to establish her own production company, Harpo Productions. Harpo's presence on television has been evident in a number of arenas. First, in dramatic programming, Harpo produced the miniseries *The Women of Brewster Place* (1989) and the follow-up situation-drama comedy *Brewster Place* (1990). Winfrey both starred in and produced these programs. She has produced and hosted several prime-time documentaries, one specifically on children and abuse. In recent years, she has supplanted Barbara Walters in securing one-off interviews with key celebrities. Her prime time interview of Michael Jackson in February 1993 (ABC) succeeded at garnering a massive television audience both nationally and internationally. Similarly her interview with basketball star Michael Jordan in October 1993 reaffirmed Winfrey's omnipresence and power in television.

The centrepiece of both her wealth and public presence continues to be her daily talk show, which is also broadcast successfully internationally. Borrowing the "run and microphone thrust" device from Phil Donahue, she makes the television audience part of the performance. With this and other techniques, Winfrey has managed to create an interesting public forum that transforms the feminist position that "the personal is political" into a vaguely political television program. Themes range from the bizarre, ("Children Who Abuse Parents") to the titillating ("How Important is Size in Sex?"), from the overtly political ("Women of the Ku Klux Klan") to the personal trials and tribulations of her own weight loss/gain and the "problems" of fellow celebrities. The sensational quality to the topics has often been cited to discount the seriousness of her show and others. But Winfrey has been a central part of this televisual transformation of public debate in the United States. Partly through her own public revelations of her private battles and her capacity to move from the serious to the humorous, Oprah Winfrey has aided in the expansion of television as public therapy.

—P. David Marshall

OPRAH WINFREY. Born in Kosciusko, Mississippi, U.S.A., 29 January 1954. Educated at Tennessee State University, B.A. in Speech and Drama, 1987. Began career as news reporter for WVOL Radio, Nashville, 1971–72; reporter, news anchorperson, WTVF-TV, Nashville, 1973–76; news anchorperson, WJZ-TV, Baltimore, 1976–77; host, morning talk show, *People Are Talking*, 1977–83; host, talk show, WLS-TV, Chicago, 1984; host, *The Oprah Winfrey Show*, locally broadcast in Chicago, 1985–86, nationally syndicated, since 1986; received Oscar and Golden Globe nominations for dramatic film debut in *The Color Purple*, 1985; owner and producer, Harpo Productions, since 1986; moved to television acting with *Brewster Place* miniseries on ABC, 1990; host, series of television specials, including *Oprah: Behind the Scenes*, from 1992. Recipient: Woman of Achievement Award, National Organization of Women, 1986; Emmy Awards, 1987, 1991, 1992, 1993, 1994, and 1995; named Broadcaster of the Year, International Radio and TV Society, 1988; America's Hope Award, 1990; Industry Achievement Award, Broadcast Promotion Marketing Executives/Broadcast Design Association, 1991; Image Awards, National Association for the Advancement of Colored People (NAACP), 1989, 1990, 1991, and 1992; Entertainer of the Year Award, NAACP, 1989; CEBA Awards, 1989, 1990, and 1991. Address: Oprah Winfrey Show, 110 North Carpenter Street, Chicago, Illinois 60607, U.S.A.

TELEVISION SERIES

1977–83	*People Are Talking*
1986–	*The Oprah Winfrey Show*
1990	*Brewster Place*

MADE-FOR-TELEVISION MOVIES

| 1989 | *The Women of Brewster Place* |
| 1992 | *Overexposed* (executive producer) |

TELEVISION SPECIALS

1991-93	*ABC Afterschool Special* (host, supervising producer)
1992	*Oprah: Behind the Scenes* (host, supervising producer)
1993	*Michael Jackson Talks . . . to Oprah: 90 Prime-Time Minutes with the King of Pop*

FILMS

The Color Purple, 1985; *Native Son*, 1986.

FURTHER READING

Barthel, Joan. "Here Comes Oprah! From the Color Purple to TV Talk Queen." *Ms.* (New York), August 1986.

Gillespie, Marcia Ann. "Winfrey Takes All." *Ms.* (New York), November 1988.

Haag, Laurie L. "Oprah Winfrey: The Construction of Intimacy in the Talk Show Setting." *Journal of Popular Culture* (Bowling Green, Ohio), Spring 1993.

Harrison, Barbara Grizzuti. "The Importance of Being Oprah." *The New York Times Magazine* (New York), 11 June 1989.

King, Norman. *Everybody Loves Oprah!: Her Remarkable Life Story.* New York: Morrow, 1987.

Mair, George. *Oprah Winfrey: The Real Story.* Secaucus, New Jersey: Carol, 1994.

Marshall, P. David. *Celebrity and Power.* Minneapolis: University of Minnesota Press, 1996.

Mascariotte, Gloria-Jean. "'C'mon Girl': Oprah Winfrey and the Discourse of Feminine Talk." *Genders* (Austin, Texas), Fall 1991.

Randolph, Laura B. "Oprah Opens Up . . . ." *Ebony* (Chicago), October 1993.

Waldron, Robert. *Oprah!* New York: St. Martin's Press, 1988.

White, Mimi. *Tele-Advising: Therapeutic Discourse in American Television.* Chapel Hill: University of North Carolina Press, 1992.

Zoglin, Richard. "Lady with a Calling . . . ." *Time* (New York), 8 August 1988.

See also Talk shows; *Women of Brewster Place*

WINTERS, JONATHAN

U.S. Comedian

Jonathan Winters began his career in radio as a disk jockey on station WING (Dayton, Ohio), and then moved to television at WBNS (Columbus, Ohio), where he hosted a local program for three years. He moved to New York in the 1950s and performed in night clubs on Broadway. But it is TV that has made Winters both famous and familiar to a huge and grateful U.S. audience for more than four decades. Known for his numerous characters and voices, his stream-of-consciousness humor has influenced countless other performers, a prime example being the contemporary comic actor Robin Williams.

Winters' first network television appearances came during the 1950s with enormously successful guest spots on talk/variety shows such as *The Jack Paar Show, The Steve Allen Show,* and *The Tonight Show.* He went on to appear in many television programs, including *Omnibus* (where he was the show's first stand-up comedian) *Playhouse 90, Twilight Zone,* and *Here's the Show* (a summer replacement for *The George Gobel Show*). *The NBC Comedy Hour,* originally designed as a Sunday showcase for new talent, was revamped to feature Gail Storm as the hostess and Jonathan Winters as the show's comedian. He also hosted his own program, *The Jonathan Winters Show,* in 1956–57. Aired on NBC from 7:30-7:45 P.M. to fill a 15-minute spot following the NBC evening news, the show was structured around Winters' sketches, blackouts, and monologues. The program was revived by CBS in a one-hour format for two seasons beginning in December 1967, and featured the famous Maude Frickert, as well as the character Willard "From the Couple up the Street" sketch. In some ways, these shows indicated that Winters' comedy was almost too unpredictable for conventional network television, and he was allowed more freedom in *The Wacky World of Jonathan Winters,* a syndicated program that focused on Winters' bravura improvisations.

Younger viewers may remember Winters from *Mork and Mindy,* where he played the role of Mork and Mindy's son. Paired with Robin Williams in his Mork role, Winters was wildly inventive. The comedy in this show was at times truly explosive, with one improvisational genius playing off the other. In the more conventional sitcom, *Davis Rules,* Winters was confined to a character, yet somehow managed to work many of his other personae into the stories. His performance earned an Emmy for best supporting actor in a comedy. In addition to on-camera roles, Winters frequently provides the voice for commercials and cartoons. These performances are usually wedded to his distinctive style, allowing audiences the pleasure of recognition for yet another Jonathan Winters moment.

—William Richter

JONATHAN WINTERS. Born in Dayton, Ohio, U.S.A., 11 November 1925. Educated at Kenyon College, Gambier, Ohio,

Jonathan Winters
Photo courtesy of Jonathan Winters

1946; Dayton Art Institute, B.F.A. 1950. Married: Eileen Schauder, 1948; one daughter and one son. Served in U.S. Marine Corps Reserve, 1943-46. Began career at radio station WING, Dayton, Ohio, 1949; disc jockey, station WBNS-TV, Columbus, Ohio, 1950-53; nightclub comedian, New York, 1953; successful in film and as author and painter; recorded 12 albums for "Verve." Honorary chair, National Congress of American Indians. Recipient: Emmy Award, 1991. Address: c/o George Spota Productions, Inc., 11151 Ophir Drive, Los Angeles, California 90024, U.S.A.

TELEVISION SERIES (selection)

1956–57	*The Jonathan Winters Show*
1967–69	*The Jonathan Winters Show*
1972–74	*The Wacky World of Jonathan Winters*
1975–80	*Hollywood Squares*
1978–82	*Mork and Mindy*
1991–92	*Davis Rules*

MADE-FOR-TELEVISION MOVIES

1968	*Now You See It, Now You Don't*
1980	*More Wild, Wild West*
1985	*Alice in Wonderland*
1987	*The Little Troll Prince* (voice only)

TELEVISION SPECIALS (selection)

1964	*The Jonathan Winters Special*
1965	*The Jonathan Winters Show*
1965	*The Jonathan Winters Show*
1967	*Guys 'n' Geishas*
1970	*The Wonderful World of Jonathan Winters*
1976	*Jonathan Winters Presents 200 Years of American Humor*
1977	*Yabba Dabba Doo! The Happy World of Hanna-Barbera* (co-host)
1986	*King Kong: The Living Legend* (host)
1991	*The Wish that Changed Christmas* (voice)

FILMS

It's a Mad, Mad, Mad, Mad World, 1963; *The Loved One,* 1964; *The Russians Are Coming! The Russians Are Coming!,* 1966; *Penelope,* 1967; *The Midnight Oil,* 1967; *8 On the Lam,* 1967; *Oh Dad, Poor Dad, Mama's Hung You in the Closet and I'm Feeling So Sad,* 1968; *Viva Max,* 1969; *The Fish That Saved Pittsburgh,* 1979; *The Longshot,* 1986; *Say Yes,* 1986; *Moon Over Parador,* 1988; *The Shadow,* 1994; *The Flintstones,* 1994.

PUBLICATIONS

Mouse Breath, Social Conformity and Other Ills. Indianapolis: Bobbs-Merrill, 1965.

Winters' Tales: Stories and Observations for the Unusual. New York: Random House, 1987.

Hang Ups: Paintings by Jonathan Winters. New York: Random House, 1988.

FURTHER READING

Adir, Karin. *The Great Clowns of American Television.* Jefferson, North Carolina: McFarland, 1988.

Sanoff, Alvin P. "The Stand-Up Art of Jonathan Winters." *U.S. News and World Report* (Washington, D.C.), 5 December 1988.

See also Talk Shows

WISEMAN, FREDERICK

U.S. Documentary Filmmaker

Frederick Wiseman is arguably the most important American documentary filmmaker of the past three decades. A law professor turned filmmaker in 1967, Wiseman, in his most dramatically powerful documentaries, has poignantly chronicled the exercise of power in American society by focusing on the everyday travails of the least fortunate Americans caught in the tangled webs of social institutions operating at the community level. An underlying theme of many of these documentaries is the individual's attempt to preserve his or her humanity and dignity while struggling against laws and dehumanizing bureaucratic systems. Wiseman functions as producer, director, and editor of the films, which numbered 29 by 1996. The documentaries have all been broadcast on public television in the United States, presented by New York station WNET, and have regularly marked the opening of the new PBS season. Wiseman's documentaries have won numerous awards, including two Emmys, and a Dupont Award. Wiseman was awarded the prestigious MacArthur Prize Fellows Award in 1982, and received a Peabody Award for his contribution to documentary film.

Wiseman's aesthetic falls squarely in the "direct cinema" tradition of documentary filmmaking, which emphasizes continued filming, as unobtrusively as possible, of human conversation and the routines of everyday life, with no music, no interviews, no voice-over narration, and no overt attempt to interpret or explain the events unfolding before the camera.

Wiseman calls his films "reality-fictions," reflecting his tight thematic structuring of the raw footage in the editing process. Eschewing "leading characters," Wiseman skillfully interweaves many small stories to provide contrast and thematic complexity.

Wiseman's debut as a documentarian was both auspicious and highly controversial. His first film, *Titicut Follies* (1967), was shot in the Massachusetts State Hospital for the Criminally Insane at Bridgewater. Here we see the impact of a social institution—a publicly-funded mental hospital—on society's rejects. Often described as an "expose" (a description Wiseman rejects), *Titicut Follies* chronicled the indignities suffered by the inmates, many of whom were kept naked and force-fed through nasal tubes. *Titicut Follies*

Frederick Wiseman
Photo courtesy of Frederick Wiseman/John Goodman

caused a public outcry and demands for institutional reform. The film was officially barred from general public showings until 1993 by order of a U.S. court on grounds that it violated an inmate's privacy.

A succession of critically-acclaimed documentaries quickly followed. In *High School* (1968), Wiseman examined a largely white and middle-class Philadelphia high school and the authoritarian, conformist value system inculcated in students by teachers and administrators. The official ideology reflected in the educational power structure was largely seen as an expression of the value framework of the surrounding community.

Law and Order (1969) was filmed in Kansas City, Missouri. Here, Wiseman cast his gaze on the daily routine of police work in the Kansas City police department. Most of the sequences were filmed in the black district of the city. Examples of police brutality and insensitivity were juxtaposed with other examples of sympathetic patrol officers attempting to assist citizens with a variety of minor, and sometimes humorous, problems. On the whole, however, police behavior was depicted as symptomatic of deeper social crisis, including racism, poverty, and the resultant pervasive violence in the inner city.

His next film, *Hospital* (1969), for which Wiseman won two Emmys for Best News Documentary, was set in the operating room, emergency ward and out-patient clinics of New York City's Metropolitan Hospital. As in *Law and Order*, Wiseman used an institutional setting to examine urban ills. Stabbing and drug overdose victims, abused children, the mentally disturbed, and the abandoned elderly pass through the public hospital. But unlike the authority figures in *Titicut Follies*, the doctors, nurses, and orderlies at Metropolitan come off as much more humane, responding to patients with sympathy and understanding.

In *Juvenile Court* (1973), as in *Hospital*, Wiseman reveals the compassionate side of authority. The court officials in the Memphis, Tennessee, juvenile court discuss, with evident concern, the futures of young offenders accused of crimes such as child abuse and armed robbery.

Welfare (1975) is one of the most provocative and understated of Wiseman's institutional examinations. Shot in a New York City welfare office, the documentary, in seemingly interminable shots, chronicles the frustration and pain of abject welfare recipients who spend their time sitting and waiting, or being shunted from office to office, as the degrading milieu of the welfare system grinds on. Welfare bureaucrats are largely seen as agents of dehumanization.

The Store (1983), Wiseman's first color film, at first glance appears to depart from the typical "weighty" subject matter of most of his previous films. That, however, is deceptive. For while the institution under scrutiny, the world-famous Neiman-Marcus department store in Dallas, Texas, may seem to be light-weight material, Wiseman's treatment of the activities of store employees and the mostly wealthy customers ultimately reveals the shallow lives of America's economic elite and those who

service them. Conspicuous consumption is everywhere in evidence. The clientele while away days in the store's dressing rooms, trying on expensive gowns and furs. A compliant group of saleswomen are led in smile exercises as they prepare to meet their condescending customers. The bourgeoisie and proletariat are complicit in this sordid dance of money and unproductive leisure. *The Store* stands in stark and powerful contrast to the despair depicted in *Welfare*.

The ethics of Wiseman's filmmaking has been criticized by some as invading the privacy of its subjects (*Titicut Follies* is the clearest case-in-point). Wiseman's response is unequivocal. He argues that if an institution receives public tax support, citizens are entitled to observe its operation. Reportorial access, Wiseman adds, is a constitutional right with regard to public institutions. In his early documentaries, if any subject objected at the time of shooting to being filmed, Wiseman eliminated the footage in question from the final cut. Later, however, he denied subjects veto rights. Some subjects, while initially pleased with their portrayals, later became upset with others' negative reactions to those portrayals. This may be one of Wiseman's major contributions to the documentary form, to permit subjects to examine their own behavior—to confront the consequences of their own social actions—as seen through the eyes of others.

—Hal Himmelstein

FREDERICK WISEMAN. Born in Boston, Massachusetts, U.S.A., 1 January 1930. Educated at Williams College, Williamstown, Massachusetts, B.A. 1951; LL.B., Yale University, New Haven, Connecticut, 1954. Worked as law professor; turned to television documentary filmmaking, 1967. Recipient: Emmy Awards, 1969 (twice), and 1970; John Simon Guggenheim Memorial Foundation Fellowship, 1980–81; John D. and Catherine T. MacArthur Fellowship, 1982–87; International Documentary Association, Career Achievement Award, 1990; Peabody Award, Personal Award, 1991. Address: Zipporah Films, Inc., One Richdale Avenue, Unit #4, Cambridge, Massachusetts 02140, U.S.A.

TELEVISION DOCUMENTARIES (all as producer, director, and editor)

1967	*Titicut Follies*
1968	*High School*
1969	*Law and Order*
1970	*Hospital*
1971	*Basic Training*
1972	*Essene*
1973	*Juvenile Court*
1974	*Primate*
1975	*Welfare*
1976	*Meat*
1977	*Canal Zone*
1978	*Sinai Field Mission*

1979	*Manoeuvre*
1980	*Model*
1982	*Seraphita's Diary*
1983	*The Store*
1985	*Racetrack*
1986	*Blind*
1986	*Deaf*
1986	*Adjustment and Work*
1986	*Multi-Handicapped*
1987	*Missile*
1989	*Near Death*
1989	*Central Park*
1991	*Aspen*
1993	*Zoo*
1994	*High School II*
1995	*Ballet*
1996	*La Comedie Francaise*

FURTHER READING

Arlen, Michael J. *The Camera Age: Essays on Television.* New York: Penguin, 1982.

Atkins, Thomas R. "Frederick Wiseman's America: Titicut Follies to Primate." In, Jacobs, Lewis, editor. *The Documentary Tradition.* New York: Hopkinson and Blake, 1971; 2nd edition, New York: Norton, 1979.

Barnouw, Erik. *Documentary: A History of the Non-fiction Film.* New York: Oxford University Press, 1974; revised edition, 1983.

Denby, David. "Documentary in America." *Atlantic Monthly* (New York), March 1970.

Rifkin, Glenn. "Wiseman Looks at Affluent Texans." *New York Times,* 11 December 1983.

Rosenthal, Alan, editor. *New Challenges in Documentary.* Berkeley: University of California Press, 1988.

WITT, PAUL JUNGER

U.S. Producer

Native New Yorker Paul Junger Witt took his first television position with Screen Gems in Los Angeles immediately following his graduation from the University of Virginia. At Screen Gems, one of Hollywood's most active television production companies, he worked as an associate producer and director of *The Farmer's Daughter* and *Occasional Wife.* In 1971, Witt produced the enormously successful and influential—and Emmy-winning—made-for-television movie, *Brian's Song.* On that project he worked for the first time with his future partner, Tony Thomas. He then assumed producer-director duties on *The Partridge Family.*

In 1971, he moved on to become a producer with Spelling-Goldberg Productions, where he was involved in several films. A year later, he joined Danny Thomas Productions as president, serving as executive producer of five movies for television and two series, including *Fay*, which was created and written by Susan Harris.

In 1975, Witt joined with Tony Thomas, son of the legendary comedian, Danny Thomas, to form Witt/Thomas Productions. A year later, the two men teamed up with Susan Harris to form Witt/Thomas/Harris Productions. Their first venture, *Soap*, was both a critical and popular success, although it was roundly attacked by religious and cultural conservatives. Witt found the criticisms particularly disturbing since no one in the groups making the attacks had ever seen the series. Yet several ABC affiliates responded to the critiques and either refused to air *Soap* or relegated it to late hours. It is Witt's belief that the unfair depictions of the show by those bent upon removing it from the air continued to have a chilling effect on advertisers for all the remaining years that the program was on ABC.

A unique television event, *Soap* set in motion a long string of major television hits for the three partners, includ-

Paul Junger Witt
Photo courtesy of Witt/Thomas/Harris Productions

ing *Benson, The Golden Girls,* and *Empty Nest.* Of these series, *Soap* and *Golden Girls* reflected a continuing emphasis on strong female characters. The company also produced at least five other shows with modest success that focused upon women. In addition, Witt/Thomas produced *Beauty and the Beast, Blossom,* and *The John Larroquette Show.*

The huge success of the company solidified Witt/Thomas/Harris as a powerful force in the television industry. Witt observed that their reputation gave them significant access to network time slots. In 1984, Witt/Thomas also began production of feature films with *Firstborn, Dead Poets' Society,* and *Final Analysis.*

—Robert S. Alley

PAUL JUNGER WITT. Born in New York City, New York, U.S.A., 20 March 1941. Educated at University of Virginia, Charlottesville, Virginia, B.A. in fine arts 1963. Married: Susan Harris; one son, Oliver; one daughter and two sons from a previous marriage. Associate producer and director, Screen Gems, Hollywood, California, 1965–67, producer and director, 1967–71; producer, Spelling/Goldberg Productions, Hollywood, 1971–73; president and executive producer, Danny Thomas Productions, Hollywood, 1973–74; founder and executive producer, Witt/Thomas/Harris Productions, Hollywood, 1976–81, executive producer, since 1975; executive producer, Witt/Thomas/Harris Productions, Witt/Thomas Productions, Witt/Thomas Films, 1992. Recipient: Emmy Awards, 1972, 1985, and 1986. Member: Board of Directors, Environmental Defense Fund; National Board, Medicine Sans Frontiers; Directors Guild of America; Writers Guild of America. Address: Witt/Thomas Productions, Witt/Thomas/Harris Productions, 1438 North Gower Street, Hollywood, California 90028, U.S.A.

TELEVISION SERIES (selection)

1972–76	*The Rookies*
1977–81	*Soap*
1980–81	*I'm a Big Girl Now*
1980–82	*It's a Living*
1982–83	*It Takes Two*
1983	*Condo*
1985	*Hail to the Chief*
1985–92	*The Golden Girls*
1987–90	*Beauty and the Beast*
1988–95	*Empty Nest*
1991–95	*Blossom*
1991	*Good and Evil*
1991–93	*Herman's Head*
1991–93	*Nurses*
1993	*Whoops*
1993–	*The John Larroquette Show*
1995	*Muscle*
1996	*Local Heroes*
1996–	*Pearl*

MADE-FOR-TELEVISION MOVIES

1972	*Brian's Song*
1972	*No Place to Run*
1972	*Home for the Holidays*
1973	*A Cold Night's Death*
1973	*The Letters*
1973	*Bloodsport*
1974	*Remember When*
1974	*The Gun and the Pulpit*
1975	*Satan's Triangle*
1976	*Griffin and Phoenix*
1976	*High Risk*
1980	*Trouble in Big Timber Country*
1996	*Radiant City*

FILMS

Firstborn, 1984; *Dead Poets' Society,* 1988; *Final Analysis,* 1992; *Mixed Nuts,* 1994.

See also *Benson; Golden Girls;* Harris, Susan; *Soap;* Thomas, Tony

WOJECK

Canadian Drama Series

First aired on the anglophone network of the Canadian Broadcasting Corporation for two seasons (1966 and 1968), *Wojeck* was a magnificent aberration: a popular, homegrown dramatic series made for the pleasure of English-Canadian viewers. Early on, francophone producers in Montreal had developed a particular genre of social melodrama, known as *téléromans,* that did captivate the imagination of French-Canadian viewers. Not so their anglophone counterparts. The record of domestic dramatic series in English Canada had been short and dismal, a collection of failures, or at best partial successes, usually modeled on

American hits but lacking either the inspiration or the funding necessary to succeed. Audiences much preferred watching the originals, the stories Hollywood had made—until *Wojeck* arrived. Early in its first season, *Wojeck* was purportedly attracting more viewers than many American imports, and it received even higher ratings when rebroadcast in the summer of 1967.

Part of the success of *Wojeck* rested upon its visual style. It was the first time the CBC had produced a filmed dramatic series for its national audience. Executive producer Ronald Weyman drew upon his experience at the National

Film Board to deliver stories which had the look of authenticity. This was especially true in the first season, when each episode was in black and white, and scenes were sometimes shot with a hand-held camera, giving the productions a gritty, realistic quality that at times suggested the news documentary. The look of authenticity was less apparent in the second season, when the series was shot in color.

Success, however, had as much to do with the subject, the script, and above all the acting. *Wojeck* created stories around a big city coroner and his quest for justice. The character and setting were novel twists on the very popular 1960s American genre of work-place dramas that focused on the exploits of such professionals as lawyers, doctors, and even teachers and social workers. The notion of a crusading coroner would become much more familiar to North American audiences because of the hit U.S. *Quincy*, of course, which began its long run on NBC in 1976. But at the time, *Wojeck* was an original, possibly inspired by the much-publicized exploits of an actual coroner of the city of Toronto.

The show did conform nonetheless to the formula of such American hits as *Ben Casey* (1961–66) and *Mr. Novak* (1963-65). All of the episodes (written in the first season by Philip Hersch) centered on the seamy side of life: racism, ageism, discrimination (one program dealt with male prostitution and homosexuality), and other species of injustice. Often the "heavy" was society itself, whose indifference or intolerance had bred evil. *Wojeck* was a kind of "edutainment," since viewers were supposed to absorb some sort of moral lesson about the country's social ills while enjoying their hour of diversion. The first show, an outstanding episode entitled "The Last Man in the World," looked at why an Indian committed suicide in the big city, exposing "Canada's shame"—its mistreatment of its native peoples.

Wojeck featured a strong male lead, Dr. Steve Wojeck, superbly played by John Vernon, who was backed up by a "team" that included his wife (the understanding helpmate), an assistant (efficient but unobtrusive), and a sometimes reluctant crown attorney (the well-meaning bureaucrat). Wojeck was emphatically masculine: big and rough, aggressive, short-tempered, and domineering. These qualities were most apparent when he dealt with the police and other authorities. He was easily moved to anger and moral outbursts, but was much more understanding when he dealt with society's outcasts. Wojeck was the engaged liberal: an advocate for the powerless committed to reforming the practices of the system so that it ensured justice for all. Like his Hollywood counterparts, Wojeck embodied the 1960s myth of the professional as hero who would turn his talents and skills to making our sadly flawed world a better place.

Wojeck had no real successors. Weyman and others did produce a number of forgettable dramas in the next few years, but none could match the appeal of the imports. Ironically, the very success of *Wojeck* had spelled trouble for CBC's drama department. John Vernon was lured away to Hollywood, where he came to specialize in playing villains. Indeed, Weyman later

Wojeck
Photo courtesy of the National Archives of Canada

claimed that much of the talent which had contributed to the appeal of *Wojeck* was drawn away to the greener pastures down south. The memory of that brief, glorious moment was sufficient to justify replaying some of the episodes of *Wojeck* on the CBC network over 20 years later.

—Paul Rutherford

CAST

Dr. Steve Wojeck	John Vernon
Marty Wojeck	Patricia Collins
Crown Attorney Bateman	Ted Follows
Byron James	Carl Banas

PRODUCER   Ronald Weyman

PROGRAMMING HISTORY   20 Episodes

• CBC

September 1966–November 1966	Tuesday 9:00-10:00
January 1968–March 1968	Tuesday 9:00-10:00

FURTHER READING

Miller, Mary Jane. *Turn up the Contrast: CBC Television Drama Since 1952.* Vancouver: University of British Columbia Press/CBC Enterprises, 1987.

Rutherford, Paul. *When Television Was Young: Primetime Canada 1952–1967.* Toronto: University of Toronto Press, 1990.

Wolfe, Morris. *Jolts: The TV Wasteland and the Canadian Oasis.* Toronto: James Lorimer, 1985.

See also Canadian Programming in English; Weyman, Ron

WOLPER, DAVID L.

U.S. Producer

David L. Wolper is arguably the most successful independent documentary producer to have ever worked in television. Through a career span of nearly fifty years, this prolific filmmaker has left his imprint with documentary specials, documentary series, dramatic miniseries, movies made for theatrical release, movies made for television, television sitcoms, entertainment specials and entertainment special events.

Wolper began his career in the late 1940s by selling B-movies, English-dubbed Soviet cartoons, and film serials, including *Superman*, to television stations. Interested in producing television documentaries, in 1958 he established Wolper Productions. Working with exclusive Russian space program footage and NASA cinematography of American missile launches, within two years, his first film, *The Race for Space*, was completed and had attracted a sponsor. Wolper offered the film to all three networks but an unofficial rule of the time dictated that only news programs and documentaries produced by network personnel were allowed on the air. Not to be discouraged, the young producer fell back on his sales experience and syndicated the film to 104 local stations across the United States—the overwhelming majority of these stations network affiliates willing to preempt other programming for the Wolper show. For the first time in television history a non-network documentary special achieved near-national audience coverage. Having been released to theaters prior to television, *The Race for Space* also received an Academy Award nomination in the best documentary category—another first for a television film.

Wolper's notoriety helped to launch a significant number of documentary projects that found their way to network time slots. Utilizing a basic compilation technique, these early films consisted of editing photo stills and film clips to narration and music, with occasional recreations of footage, minimal editorial viewpoint and high-information, high-entertainment value. Increasingly successful, within four years of establishing Wolper Productions, Wolper's method would place him on a level with NBC and CBS as one of the three largest producers of television documentaries and documentary specials.

A major turning point in Wolper's career occurred in 1960 when he bought the rights to Theodore H. White's book, *The Making of the President*. Aired on ABC, Wolper's potentially controversial film presented an incisive look at the American political process, won four Emmy Awards including 1963 Program of the Year and guaranteed Wolper's celebrity.

In 1964, Wolper sold his documentary production unit to Metromedia but stayed on as the company's chief of operations. With this media giant's backing, Wolper's projects grew in scope and substance. He became a regular supplier of documentary programs to all three commercial

David L. Wolper
Photo courtesy of the Wolper Organization, Inc.

networks creating such memorable series as *The March of Time*, in association with Time, Inc., and a series of nature specials in collaboration with the National Geographic Society. For the latter, he introduced American audiences to French oceanographer Jacques Cousteau. This in turn led to the first ever documentary spin-off, *The Undersea World of Jacques Cousteau*.

Breaking away from Metromedia in 1967, Wolper continued his documentary work but also tried his hand at theatrical release motion pictures. He created a number of unexceptional films including *The Bridge at Remagen* (1968), *If It's Tuesday, This Must be Belgium* (1969), and *Willy Wonka and the Chocolate Factory* (1971). In fiction television, he found more success with regularly scheduled television series that included *Get Christie Love!* (1974–75), featuring the first black policewoman character in television history, *Chico and the Man* (1974–78) and *Welcome Back, Kotter* (1975–79).

Perhaps Wolper's most significant accomplishment was his developmental work with the television non-fiction drama miniseries. In the mid-1970s, after bypass heart surgery and sale of his company to Warner Brothers, he helped to invent the docudrama genre with his award-winning production of Alex Haley's acclaimed family saga, *Roots*. Reconstructing history in an unprecedented twelve-hour film, the series was broadcast in one- and two-hour segments over an eight-day period in January 1977. Contrary to initial concerns over the high risk nature of the venture, the series brought ABC a 44.9 rating and 66% share of audience to set viewership records that place it among the most watched programs in the history of television.

In 1984, Wolper stepped out of his usual role as film producer to orchestrate the opening and closing ceremonies for the Summer Olympics in Los Angeles. The first ever to be staged by a private group, the ceremonies received a 55% share of audience outranking all other Olympic coverage. For his efforts, Wolper was rewarded with a special Emmy and the Jean Hersholt Humanitarian Award at the Oscar ceremony in 1985. The following year he was recruited to produce the Liberty Weekend hundredth anniversary celebration for the Statue of Liberty. The four-day event was viewed by 1.5 billion people worldwide.

As a producer, filmmaker, entrepreneur, historian and visionary, David Wolper's career has been one of taking risks and continually breaking new ground. Most importantly, through his more than 600 films his innovative and creative spirit has educated and entertained millions.

—Joel Sternberg

DAVID L(LOYD) WOLPER. Born in New York City, New York, U.S.A., 11 January 1928. Studied at Drake University, 1946; University of Southern California, 1948. Married: 1) Margaret Davis Richard, 1958 (divorced, 1969); one daughter and two sons; 2) Gloria Diane Hill, 1974. Began career as vice president, then treasurer, Flamingo Films, TV sales company, 1948–50; vice president, West Coast Operations, 1954–58; chair and president, Wolper Productions, Los Angeles, since 1958; president, Fountainhead International, since 1960; president, Wolper TV Sales Company, since 1964; vice president, Metromedia, Inc., 1965-68; president and chair, Wolper Pictures Limited, since 1968; consultant and executive producer, Warner Brothers, Inc., since 1976. Member: U.S. Olympic Team Benefit Committee; advisory committee, National Center for Jewish Film; Academy of Motion Picture Arts and Sciences; Producers Guild of America; Caucus for Producers, Writers and Directors. Trustee: Los Angeles County Museum of Art, 1984; American Film Institute; Los Angeles Thoracic and Cardiovascular Foundation. Board of directors: Amateur Athletic Association of Los Angeles, 1984; Los Angeles Heart Institute; Southern California Committee for the Olympic Games, 1977; Academy of Television Arts and Sciences Foundation, 1983; University of Southern California Cinema/Television Department. Recipient: Award for documentaries, San Francisco International Film Festival, 1960; Distinguished Service Award, U.S. Junior Chamber of Commerce; Monte Carlo International Film Festival Award, 1964; Cannes Film Festival Grand Prix for TV Programs, 1964;

Oscar Award: Jean Hersholt Humanitarian Award, 1985; named to TV Hall of Fame, 1988; Medal of Chevalier, French National Legion of Honor, 1990; Lifetime Achievement Award, Producers Guild of America, 1991; 8 Globe Awards; 5 Peabody Awards; 40 Emmy Awards; numerous other awards. Address: Wolper Organization, Inc., 4000 Warner Boulevard, Burbank, California 91522, U.S.A.

TELEVISION SERIES (selection)

1961–64, 1979	*Biography*
1962–65	*Story of . . .*
1963–64	*Hollywood and the Stars*
1965–66	*March of Time*
1965–76	*National Geographic*
1968–76	*The Undersea World of Jacques Cousteau*
1971–73	*Appointment with Destiny*
1972–73	*Explorers*
1974–78	*Chico and the Man*
1974–75	*Get Christie Love*
1975–79	*Welcome Back, Kotter*

TELEVISION MINISERIES (selection)

1976	*Victory at Entebbe*
1977	*Roots*
1979	*Roots: The Next Generations*
1983	*The Thorn Birds*
1985	*North and South Book I*
1986	*North and South Book II*
1987	*Napoleon and Josephine*

MADE-FOR-TELEVISION MOVIES (selection)

1973	*500 Pound Jerk*
1974	*Men of the Dragon*
1974	*Unwed Father*
1974	*The Morning After*
1974	*Get Christie Love*
1976	*Brenda Starr*
1982	*Agatha Christie Movie: Murder Is Easy*
1983	*Agatha Christie Movie: Sparkling Cyanide*
1984	*Agatha Christie Movie: Caribbean Mystery*
1989	*The Plot to Kill Hitler*
1989	*Murder in Mississippi*
1990	*Dillinger*
1990	*When You Remember Me*
1991	*Bed of Lies*
1992	*Fatal Deception: Mrs. Lee Harvey Oswald*
1993	*The Flood: Who Will Save our Children?*
1994	*Without Warning*

TELEVISION SPECIALS (selection)

1958	*The Race for Space*
1959	*Project: Man in Space*
1960	*Hollywood: The Golden Years*
1960, 1964, 1968	*The Making of the President*

1961	*Biography of a Rookie*
1961	*The Rafer Johnson Story*
1962	*D-Day*
1962	*Hollywood: The Great Stars*
1963	*Hollywood: The Fabulous Era*
1963	*Escape to Freedom*
1963	*The Passing Years*
1963	*Ten Seconds That Shook the World*
1963	*Krebiozen and Cancer*
1963	*December 7: Day of Infamy*
1963	*The American Woman in the 20th Century*
1964	*The Legend of Marilyn Monroe*
1964	*The Yanks Are Coming*
1964	*Berlin: Kaiser to Khrushchev*
1964	*The Rise and Fall of American Communism*
1964	*The Battle of Britain*
1964	*Trial at Nuremberg*
1965	*France: Conquest to Liberation*
1965	*Korea: The 38th Parallel*
1965	*Prelude to War*
1965	*Japan: A new Dawn Over Asia*
1965	*007: The Incredible World of James Bond*
1965	*Let my People Go*
1965	*October Madness: The World Series*
1965	*Race for the Moon*
1965	*The Bold Men*
1965	*The General*
1965	*The Teenage Revolution*
1965	*The Way Out Men*
1965	*In Search of Man*
1965	*Mayhem on a Sunday Afternoon*
1966	*The Thin Blue Line*
1966	*Wall Street: Where the Money Is*
1966	*A Funny Thing Happened on the Way to the White House*
1967	*China: Roots of Madness*
1967	*A Nation of Immigrants*
1967	*Do Blondes Have More Fun*
1968	*The Rise and Fall of the Third Reich*
1968	*On the Trail of Stanley and Livingstone*
1970	*The Unfinished Journey of Robert F. Kennedy*
1970–72	*George Plimpton*
1971	*Say Goodbye*
1971	*They've Killed President Lincoln*
1971–73	*Appointment With Destiny*
1972	*They've Killed President Lincoln*
1973–74	*American Heritage*
1973–75	*Primal Man*
1974	*Judgment*
1974	*The First Woman President*
1974–75	*Smithsonian*
1975–76	*Sandburg's Lincoln*
1976	*Collision Course*
1980	*Moviola*
1984	*Opening and Closing Ceremonies, 1984 Olympic Games*
1986	*Liberty Weekend*
1987	*The Betty Ford Story*
1988	*What Price Victory*
1988	*Roots: The Gift*

FILMS (selection)

Four Days in November, 1964; *Devil's Brigade*, 1967; *The Bridge at Remagen*, 1968; *If It's Tuesday, This Must Be Belgium*, 1968; *I Love my Wife*, 1970; *The Helstrom Chronicle*, 1971; *Willy Wonka and the Chocolate Factory*, 1971; *King, Queen, Knave!*, 1972; *One Is a Lonely Number*, 1972; *Wattstax*, 1973; *Visions of Eight*, 1973; *Birds Do It...Bees Do It...*, 1974; *The Animal within*, 1974; *Victory at Entebbe*, 1976; *The Man who Saw Tomorrow*, 1980; *This is Elvis*, 1981; *Imagine: John Lennon*, 1988; *Murder in the First*, 1994; *Surviving Picasso*, 1996.

FURTHER READING

Angelo, Bonnie. "Liberty's Ringmaster of Ceremonies." *Time* (New York), 7 July 1986.

Arar, Yardena. "And the Show Goes On." *Los Angeles Daily News*, 2 May 1990.

Bluem, A. William. *Documentary in American Television: Form, Function, Method.* New York: Hastings House, 1965.

Berlin, Joey. "David Wolper's 'Imagine' Takes a Documentary Approach." *New York Post*, 5 October 1988.

Goldenson, Leonard H., with Martin J. Wolf. *Beating the Odds: The Untold Story Behind the Rise of ABC: The Stars, Struggles and Egos That Transformed Network Television By the Man Who Made It Happen.* New York: Scribner's, 1991.

Harvey, Alec. "Tragedy is Tale of Hope, Says Wolper." *Birmingham (Alabama) News*, 2 February 1990.

"Hollywood Fights Back." In, Cole, Barry, editor. *Television Today: A Close-Up View Readings from TV Guide.* Oxford: Oxford University Press, 1981.

Marc, David, and Robert J. Thompson. *Prime Time, Prime Movers: From "I Love Lucy" to "L.A. Law"—America's Greatest TV Shows and the People Who Created Them.* Boston: Little, Brown, 1992.

O'Connor, Colleen, with Martin Kasindorf. "Wolper: Impresario of the Big Event." *Newsweek* (New York), 7 July 1986.

"Wolper, David L." In, Monush, Barry, editor. *1993 International Television and Video Almanac.* New York: Quigley, 1993.

"Wolper, David L(loyd)." In, Moritz, Charles, editor. *Current Biography Yearbook.* New York: H. W. Wilson, 1987.

"Wolper Performs Hat Trick Again: Documentaries on All 3 Webs." *Variety* (Los Angeles), 11 May 1966.

"Young King David." *Newsweek* (New York), 23 November 1964.

See also Documentary; *Roots*

THE WOMEN OF BREWSTER PLACE

U.S. Miniseries

The Women of Brewster Place, a miniseries based on the novel by Gloria Naylor, was produced in 1989 by Oprah Winfrey's firm Harpo, Inc. Winfrey served as executive producer and starred along with noted actors, Mary Alice, Jackee, Lynn Whitfield, Barbara Montgomery, Phyllis Yvonne Stickney, Robin Givens, Olivia Cole, Lonette McKee, Paula Kelly, Cicely Tyson, Paul Winfield, Moses Gunn and Douglas Turner Ward. The story, spanning several decades, includes a cast of characters that depict the constant battles fought by African-American women against racism, poverty, and sexism. Interpersonal struggles and conflicts also pepper the storyline, often revolving around black men who may be fathers, husbands, sons, or lovers.

The Winfrey character, Mattie, opens the drama. Her road to Brewster Place began when she refused to reveal the name of her unborn child's father to her parents (Mary Alice and Paul Winfield). Milestones for Mattie included living in the home of Eva Turner (Barbara Montgomery) until she died and willed the house to Mattie; then forfeiting the house when her son, Basil, jumped bail after Mattie used their home as collateral for his bond. The other characters' journeys to the tenement on Brewster Place were just as unpredictable and crooked. Kiswana, portrayed by Robin Givens, moved to the neighborhood to live with her boyfriend. They worked to organize the neighbors, to plan special activities for the neighborhood, and to protest their excessive rent. One of the most powerful scenes in the drama occurs between Kiswana and her mother, Mrs. Browne (Cicely Tyson). When Tyson comes for a visit, she and Givens begin a conversation that progresses into a heated argument regarding Kiswana's name change. Mrs. Browne reveals why she named her daughter Melanie (after her grandmother), and in a powerful soliloquy tells the story of that grandmother's strength and fearlessness when facing a band of angry white men.

Other women from the building reveal bruises inflicted either by the men in their lives, or by the world in general. Cora Lee (Phyllis Stickney) continues to have children because she wants the dependency of infants; once they become toddlers, her interest in them falters. By the end of the series, however, she begins to see the importance of all her children, and after being prodded by Kiswana, she attends the neighborhood production of an African-American adaptation of a Shakespearean play. Through this experience and her children's reaction to it, the audience sees a change in Cora Lee.

Miss Sophie (Olivia Cole), an unhappy woman and the neighborhood busybody, spreads vicious gossip about her neighbors in the tenement. Etta Mae (Jackee), Mattie's earthy, flamboyant and loyal childhood friend, moved to Brewster Place for refuge from her many failed romances. Lucielia Louise Turner, housewife and mother (Lynn Whitfield), lived a somewhat happy life with her husband Ben (Moses Gunn) and daughter Serena in one of the tenement apartments until Ben lost his job and left home. Lucielia then aborted their second child and her daughter Serena was electrocuted when she used a fork to chase a roach into a light socket. Theresa and Lorraine (Paula Kelly and Lonette McKee) decided to reside on Brewster Place because, as lesbians, they were seeking some place where they could live without ridicule and torment. Their relationship, soon discovered by their neighbors, became the backdrop for the drama's finale.

Criticism of the miniseries began before the drama aired. The National Association for the Advancement of Colored People requested review of the scripts before production to determine whether the negative images of the African-American male, present in the Naylor book, appeared in the television drama. This request was denied, but Winfrey, also concerned with the image of black men in the novel, altered several of their roles. Ben Turner, the tenement's janitor and a drunk in Naylor's novel, was revamped for the teleplay, and in a scene created for especially for the series, explains why he felt pressed into desertion. The producers also attempted to cast actors who could bring a level of sensitivity to the male roles and create characters who were more than one-dimensional villains.

Still, newspaper columnist Dorothy Gilliam criticized the drama in a two-part series for the *Washington Post,* as one of the most stereotype-ridden polemics against black men ever seen on television, a series which, she claimed, trotted out nearly every stereotype of black men that had festered in the mind of the most feverish racist. In spite of such criticism the series won its time period Sunday and Monday nights against heavy competition, *The Wizard of Oz* on CBS and a *Star Wars* installment, *Return of the Jedi,* on NBC.

Though criticized for its portrayal of African-American men and women, *The Women of Brewster Place* offered its audience a rare glimpse of America's black working class and conscientiously attempted to probe the personal relationships, dreams, and desires of a group of women who cared about their children and friends, who worked long hours at jobs they may have hated in order to survive, and who moved forward despite their disappointments. A spin-off of the miniseries titled *Brewster Place,* also produced by Harpo, Inc., aired for a few weeks in 1990 on ABC, but was canceled because of low ratings.

—Bishetta D. Merritt

CAST

Mattie Michael . . . . Oprah Winfrey
Etta Mae Johnson . . . . Jackee
Mrs. Browne . . . . Cicely Tyson
Kiswana Browne . . . . Robin Givens

The Women of Brewster Place

Lorraine	Lonette McKee
Cora Lee	Phyllis Stickney
Ben	Moses Gunn
Butch Fuller	Clark Johnson
Ciel	Lynn Whitfield
Basil	Eugene Lee
Mattie's Father	Paul Winfield
Mattie's Mother	Mary Alice
Eva Turner	Barbara Montgomery
Reverend Wood	Douglas Turner Ward
Miss Sophie	Olivia Cole

PRODUCERS   Oprah Winfrey, Carole Isenberg

PROGRAMMING HISTORY

• ABC

19-20 March 1989	9:00-11:00

FURTHER READING

Bobo, Jacqueline, and Ellen Seiter. "Black Feminism and Media Criticism: *The Women of Brewster Place.*" *Screen* (Glasgow, Scotland), Autumn 1991.

Kort, Michele. "Lights, Camera, Affirmative Action." *Ms.* magazine (New York), November 1988.

See also Racism, Ethnicity, and Television; Winfrey, Oprah

THE WONDER YEARS

U.S. Domestic Comedy

The Wonder Years, a gentle, nostalgic look at Baby Boom youth and adolescence, told stories in weekly half-hour installments presented entirely from the point of view of the show's main character. Young Kevin Arnold, portrayed on screen in youth by fresh-faced Fred Savage,

provided the center of the action. Adult Kevin, whose voice was furnished by unseen narrator Daniel Stern, commented on the events of his youth with grownup wryness twenty years after the fact. The series traced Kevin's development in suburban America from 1968, when he was 11 years old, until the summer of 1973, his junior year in high school.

A typical week's plot involved Kevin facing some rite of passage on the way to adulthood. His first kiss, fleeting summer love, first day at high school, the struggle to get Dad to buy a new, color TV—these were the innocuous narrative problems of *The Wonder Years*. The resolutions seemed simple but often were surprising. Kevin the narrator always conveyed the unsettling knowledge that, in our struggle toward maturity, we make decisions that prevent us from going back to the comfortable places of youth. For example, when pubescent Kevin stood up to his mother's babying, he took pride in his new independence. But his victory was bittersweet—he realized that he had hurt Norma by reacting harshly to her well-meaning mothering, and he had lost a piece of the relationship forever.

Mundane situations that would resonate with most Americans' youth experiences were shaded by the backdrop of everyday life in the late 1960s or early 1970s. Hip hugger pants, Army surplus gear, and toilet-paper-strewn yards helped to place the show in the collective memory of the baby boomers who were watching it (and whose dollars advertisers were vigorously seeking). Attention to period detail was often thorough, but occasional anachronisms managed to slip through, such as the use of a television remote control device in the Arnold home in about 1970. The program often opened with TV news clips from the era—showing a war protest, President Nixon waving good-bye at the White House, or some other instantly recognizable event—accompanied by a classic bit of rock music. Joe Cocker's rendition of "I Get By With a Little Help from My Friends" was the show's theme song, played over a montage of home movie clips depicting a harmonious Arnold family and Kevin's friends, Paul and Winnie.

Much of the series' historical identification had to do with oblique connections with hippie counterculture and the Vietnam War. Kevin's older sister, Karen, was a hippie, but Kevin was not, and his observation of the counterculture was from the sideline. While Karen struggled to define her identity against the grain of her parents' traditions, Kevin, for the most part, accepted the world around him. He was portrayed as an average kid, personally uninvolved with most of the larger cultural events swirling about him. One serious treatment of the Vietnam War did intrude in Kevin's personal experience, however, when Brian Cooper, older brother of his neighbor and girlfriend, Winnie, was killed. Kevin struggled to support Winnie, first in the loss of her brother and, later, after her parents' separation resulted from the brother's death.

Episodes of *The Wonder Years* were often based on challenges in Kevin's relationship with a family member, friend, authority figure, or competitor. Kevin's father, Jack; mother, Norma; sister, Karen; brother, Wayne; neighborhood best friend, Paul Pfeiffer; and childhood sweetheart, Winnie Cooper, were heavily involved in the storyline. Much of the action took place in and around the middle-

The Wonder Years
Photo courtesy of New World Television

class Arnold home or at Kevin's school (Robert F. Kennedy Junior High and, later, William McKinley High School).

While each episode was self-contained, Kevin's struggles and changes were evident as the series developed. In one episode, Kevin's older sister became estranged from their father because of her involvement in the hippie culture. Other episodes reflected that estrangement, and, in a later season, the program depicted Karen's reconciliation with her father. Kevin's observations and feelings, of course, remained central to exploring such issues. Although episodes sometimes showed how characters' perspectives shifted, the emphasis was on Kevin's own observation of his world. This acknowledgment of the character's egocentrism melded with a major program theme—adolescent self-involvement.

Sometimes, the primary point of the program was the effect of another character's struggle on the egocentric Kevin. He watched as father Jack quit a stultifying middle-manager's job at the Norcom corporation and as frustrated homemaker Norma enrolled in college classes and launched her own career. Often, Kevin spent much of his time reacting to the personal impact of such events, then feeling guilty about expressing his selfish thoughts. At the end of each episode, relations, although marked by change, became harmonious once again.

As an example of a "hybrid genre," the half-hour dramedy, *The Wonder Years* never amassed the runaway ratings of a show such as *Cheers* (though it did wind up in the Nielsen Top Ten for two of its five seasons). After a time, it was apparent to producers and the television audience that

Kevin Arnold's wonder years were waning. Creative differences between producers and ABC began to spring up from such instances as Kevin's touching a girl's breast during the 8:00 hour usually reserved for "family viewing." Economic pressures, including rising actor salaries and the need for more location shooting after Kevin acquired a driver's license, also helped to end the show. During its 115-episode run, however, *The Wonder Years* generated intensely loyal fans and collected important notice.

The final episode on 12 May 1993 exercised a luxury few ending series have: tying up loose ends. Bob Brush, executive producer of the show after creators Neal Marlens and Carol Black left in the second season, took a cue from sagging ratings when the last episode was shot. In it, Kevin quit his job working in Jack Arnold's furniture store and struck out on his own. Sadly, for some viewers, he and Winnie Cooper did not wind up together. Unfortunately, the show's resolution occurred in the summer following Kevin's junior year in high school, so the formal finality of graduation, a rite of passage so familiar to much of the audience, was missing.

Among the awards bestowed on *The Wonder Years* were an Emmy for best comedy series in 1988—after only six episodes had aired—and the George Foster Peabody Award in 1990. *TV Guide* named the show one of the 1980s' 20 best.

—Karen E. Riggs

CAST

Kevin Arnold (age 12)	Fred Savage
Kevin (as adult; voice only)	Daniel Stern
Wayne Arnold	Jason Hervey
Karen Arnold	Olivia d'Abo
Norma Arnold	Alley Mills
Jack Arnold	Dan Lauria
Paul Pfeiffer	Josh Saviano
Winnie (Gwendolyn) Cooper	Danica McKellar
Coach Cutlip	Robert Picardo
Becky Slater	Crystal McKellar
Mrs. Ritvo (1988–89)	Linda Hoy
Kirk McCray (1988–89)	Michael Landes
Carla Healy (1988–90)	Krista Murphy
Mr. DiPerna (1988–91)	Raye Birk
Mr. Cantwell (1988–91)	Ben Stein
Doug Porter (1989–)	Brandon Crane
Randy Mitchell (1989–)	Michael Tricario
Craig Hobson (1989–90)	Sean Baca
Ricky Halsenback (1991–93)	Scott Nemes
Jeff Billings (1992–93)	Giovanni Ribisi
Michael (1992)	David Schwimme

PRODUCERS   Neal Marlens, Carol Black, Jeffrey Silver, Bob Brush

PROGRAMMING HISTORY   115 Episodes

• ABC

March 1988–April 1988	Tuesday 8:30-9:00
October 1988–February 1989	Wednesday 9:00-9:30
February 1989–August 1990	Tuesday 8:30-9:00
August 1990–August 1991	Wednesday 8:00-8:30
August 1991–February	Wednesday 8:30-9:00
March 1992–September 1993	Wednesday 8:00-8:30

FURTHER READING

Blum, David. "Where Were You in '68?" *The New York Times*, 27 February 1989.

Gross, Edward A. *The Wonder Years*. Las Vegas, Nevada: Pioneer Books, 1990.

Kaufman, Peter. "Closing the Album on *The Wonder Years*." *The New York Times*, 9 May 1993.

Kinosian, Janet. "Fred Savage: Having Fun." *Saturday Evening Post* (Indianapolis, Indiana), January-February 1991.

See also Comedy, Domestic Settings

WOOD, ROBERT

U.S. Media Executive

Robert Wood moved network prime-time programming out of TV's adolescent phase into adulthood. In 1971 he broke with patterned success by jettisoning long-lived popular shows in order to attract younger audiences coveted by advertisers. At the same time he set aside traditional standards of gentle and slightly vacuous comedy for "in your face" dialog and contemporary situations that delighted masses, offended some, and pulled network entertainment into the post-assassination/civil rights/Vietnam era.

His strategy in 1970 was to cancel rural- and older-skewed classic series (*Green Acres, Beverly Hillbillies, Petticoat Junction*, and *Hee Haw*), and veteran stars (Red Skelton, Jackie Gleason, Ed Sullivan, and Andy Griffith), in favor of more contemporary, urban-oriented programming. He scheduled the challenging comedy *All in the Family*, developed by producer Norman Lear, which ABC had twice rejected. After a weak initial half season, in the spring of 1971, the series built a strong viewership during summer reruns and became a sensation by the fall season. Massive audience popularity, including sought-after younger adults, and critical praise helped Lear's production company add to CBS' schedule *The Jeffersons, Maude, Good Times, Sanford and Son, One Day at a Time* and other series. Rather than farcical situation comedies (sitcoms), these shows built on issues affecting characters as interacting persons, thus becoming "character comedies" instead.

Wood presided over the entertainment revolution that changed the tenor of America's living rooms during evening television viewing. Other networks emulated the move, sometimes outpacing CBS' entries in teasing audience acceptability with double entendre. But the nation's TV screens had moved to a new plateau (some cynics would claim a lower one) with Wood's determined risk-taking. TV and cable in the following decades pushed forward dramatic and comedic themes from that position.

Wood was energetic, optimistic, thoughtful, and shrewd. But his strategies never undercut people as he formed policies for the stations he managed (KNXT, Los Angeles; the CBS television stations division of owned-and-operated outlets) and the network he led (CBS-TV) from 1969 to 1976. He was the longest-lived and last executive totally in command of the national television fortunes of CBS Inc.

As the industry grew more complex, he advocated shifting the programming department from network headquarters in New York City to the West Coast where most entertainment programming was developed. After he retired from the network, his position was eventually divided into several presidencies, including: TV network, entertainment (programming—on the west coast), sports, affiliate relations, sales, and marketing. Competing networks had already begun splitting network executives' responsibilities, after Wood had proposed such a structure within CBS.

Wood was the rare network executive who was respected and liked, often with genuine affection, by broadcast colleagues, executives, staff members, local station managers, program producers, and talented actors. He dealt with each graciously and with good cheer, caring for those he worked with, not taking himself too seriously, and was totally committed to his top management responsibilities which he handled skillfully and with enormous success. After a brief stint as an independent producer he became president of Metromedia Producers Corporation in 1979. He died in 1986.

—James A. Brown

ROBERT WOOD. Born in Boise, Idaho, U.S.A., 17 April 1925. Educated at the University of Southern California, B.S. in advertising 1949. Married: Nancy Harwell, 1949; children: Virginia Lucile and Dennis Harwell. Served in U.S. Naval Reserve, 1943–46. Worked as sales service manager, KNXT, Hollywood, California, 1949; account executive, KTTV, c.1950–51; account executive, CBS owned-and-operated KNXT-TV station, Los Angeles, 1952–54; account executive, CBS television stations division's national sales department, 1954; general sales manager, KNXT-TV, 1955–60; vice president and general manager, KNXT-TV, 1960–66; executive vice president,

Robert Wood
Photo courtesy of Broadcasting and Cable

CBS television stations division, 1966–67; president, CBS television stations division, 1967–69; president, CBS, 1969–1976; later headed own TV production company. Died in Santa Monica, California, 20 May 1986.

FURTHER READING

Halberstam, David. *The Powers that Be.* New York: Knopf, 1979.
Metz, Robert. *CBS: Reflections in a Bloodshot Eye.* Chicago: Playboy Press, 1983.
Paley, William S. *As It Happened: A Memoir.* Garden City, New York: Doubleday, 1979.
Paper, Lewis J. *Empire: William S. Paley and the Making of CBS.* New York: St. Martin's, 1987.
Slater, Robert. *This...Is CBS: A Chronicle of 60 Years.* Englewood Cliffs, New Jersey: Prentice-Hall, 1988.
Smith, Sally Bedell. *In all his Glory: The Life of William S. Paley, the Legendary Tycoon and his Brilliant Circle.* New York: Simon and Schuster, 1990.

See also Columbia Broadcasting System; Demographics

WOOD, VICTORIA

British Comedy Actor/Writer/Singer

Victoria Wood is a talented comedy actor/writer/singer who has built up a national reputation following a string of self-written TV plays, films, and sketch shows. Born in 1953 in Lancashire in Northern England, she first had small-screen exposure on the TV talent search show *New Faces* when she sang comedy songs of her own composition. Accompanying herself on the piano, she scored heavily with viewing audiences with her jaunty tunes, which often belied her sharp, poignant lyrics. Her regular themes of unrequited love, tedium, mismatched couples, and suburban living, as well as her ability to find humour in the minutiae of modern life, stood her in good stead when she moved into writing plays for the stage and later for television.

Talent, her first play adapted for television (Granada, 5 August 1979), reunited her with Julie Walters, whom she had met at Manchester Polytechnic. Their partnership would launch both their careers. *Talent* dealt with a mismatched couple: the ambitious would-be cabaret singer Julie Stephens (Walters) and the eternally sniffing Maureen, her plump, dull, but loyal friend (played by Wood), who had accompanied Julie to a talent contest. The bittersweet comedy explored themes of desperation, dashed hopes, lost ambition, and hopeless romances. The fact that *Talent* managed to be both funny and truthful demonstrated Wood's skill as a writer and the pair's acting ability. A sequel, *Nearly a Happy Ending* (Granada, 1 June 1980), appeared the following year. This time the couple were going out for a night on the town, pausing en route at a slimming club. Wood was then quite portly, and occasionally her material dealt with what being overweight meant to oneself and others. Later in her career, she slimmed down considerably.

Following *Nearly a Happy Ending,* Wood and Walters appeared in a one-off special, *Wood and Walters: Two Creatures Great and Small* (Granada, 1 January 1981), which led to the series *Wood and Walters* (Granada, 1982). It was the series *Victoria Wood: As Seen on TV* (BBC), however, that truly established Wood as a major TV star. A sketch show introduced by a stand-up routine from Wood, the program also featured a musical interlude. Julie Walters, Patricia Routledge, Susie Blake, Duncan Preston, and Celia Imrie provided strong support, and one favourite section of the show was "Acorn Antiques," a spoof of cheaply-made soap operas.

As Walters' film career blossomed, Wood's comedic talent continued to mature, and by the end of the 1980s she was a big draw on the live circuit. Her stand-up routine relied on observational humour as she drew laughs from finding the idiosyncrasies of normal modern life. She followed a long line of (male) northern comedians with her style of taking her story lines into surreal areas, as well as her character inventions, especially the gormless "Maureen." On television she remained determined to try something new

Victoria Wood
Photo courtesy of Victoria Wood

and not merely revamp winning ideas. To this end, she wrote and starred in a number of half-hour comedy playlets under the generic title *Victoria Wood* (BBC, 1989), her first series not to attract universal acclaim. She also appeared in a number of solo stand-up shows, and in a one-off spoof of early morning television news magazine programs, *Victoria Wood's All Day Breakfast* (BBC, 31 December 1992).

The feature-length TV film *Pat and Margaret* (BBC, 11 September 1994), Wood's most ambitious project to date, was her most accomplished reworking of her mismatched couple theme. In this context, Pat (Julie Walters) was a successful English actor in a hit U.S. soap (a la Joan Collins) who was reunited with her sister Margaret (Wood) on a TV chat show. The pair hadn't been in touch for 27 years, and neither was happy about the meeting. Once again, bittersweet themes of escape and despair were explored; once again, despite this tone, Wood's comedic ability triumphed.

—Dick Fiddy

VICTORIA WOOD. Born in Prestwich, Lancashire, England, 19 May 1953. Attended Bury Grammar School for Girls; University of Birmingham, B.A. in drama and theatre arts. Married: Geoffrey Durham, 1980; one son and one daughter. Worked on regional television and radio, 1974=-78; theater writer; formed television comedy partnership with Julie Walters; star of her own series and one-woman stage shows, writing her own material; appeared in numerous television series; author of several books. D.Litt: University of Lancaster, 1989, University of Sunderland, 1994. Recipient: Pye Colour Television Award, 1979; Broadcasting Press Guild Award, 1985; British Academy of Film and Television Arts Awards, 1985 (twice), 1986, 1987, and 1988 (twice); Variety Club BBC Personality of the Year Award, 1987; Writers Guild Award, 1992; Broadcasting Press Guild Award, 1994; Monte Carlo Best Single Drama Critics' Award, 1994; Monte Carlo Nymphe d'Or Award, 1994. Address: Richard Stone Partnership, 25 Whitehall, London SW1A 2BS, England.

TELEVISION SERIES

1976	*That's Life!*
1981–82	*Wood and Walters*
1984, 1986	*Victoria Wood: As Seen on TV*
1989	*Victoria Wood*
1994	*Victoria Wood Live in Your Own Home*

MADE-FOR-TELEVISION MOVIE

| 1994 | *Pat and Margaret* |

TELEVISION SPECIALS (selection)

1979	*Talent*
1988	*An Audience with Victoria Wood*
1992	*Victoria Wood's All Day Breakfast*

STAGE (selection)

Talent, 1980; *Good Fun*, 1980; *Funny Turns*, 1982; *Lucky Bag*, 1984; *Victoria Wood*, 1987; *Victoria Wood Way Up West*, 1990.

PUBLICATIONS (selection)

Up to You, Porky. London: Methuen, 1985.
Good Fun and Talent. London: Methuen, 1988.
Mens Sana in Thingummy Doodah. London: Methuen, 1990.

See also British Programming

WOODWARD, EDWARD

British Actor

Edward Woodward has enjoyed a long and varied career since he first became a professional performer in 1946. A graduate of the Royal Academy of Dramatic Art, he has acted in England, Scotland, Australia and the United States, on both London and Broadway stages, and has appeared in a wide range of productions from Shakespeare to musicals. Despite being known for dramatic roles, he can also sing and has made over a dozen musical recordings. In recent years, his distinctive, authoritative voice has narrated a number of audio books.

Although he has played supporting roles in prestigious films like *Becket* (1964) and *Young Winston* (1972), Woodward is best known for two hit television series, *Callan* in Britain and *The Equalizer* in the United States. Despite the fact that the series were made over a decade apart, Woodward played essentially the same character in each—a world-weary spy with a conscience.

Woodward's definitive screen persona of an honorable gentleman struggling to maintain his own personal morality in an amoral, even corrupt, world was prefigured in two motion pictures in which the actor starred, *The Wicker Man* (1974) and *Breaker Morant* (1980). In *The Wicker Man*, Woodward was a priggish Scottish policeman investigating a child's disappearance; he stumbles upon an island of modern-day pagans led by Christopher Lee. In *Breaker Morant*, Woodward starred as the title character, a British Army officer well-respected by his men, who is arrested with two other soldiers for war crimes and tried in a kangaroo court during the Boer War. In both cases, Woodward's character's life is sacrificed, a victim of larger hostile social and political forces he is too decent to understand or control.

Callan, an hour-long espionage series which ran in Britain on Thames Television from 1967 to 1973, starred

Edward Woodward as The Equalizer

Woodward as David Callan, an agent who carried a license to kill, working for a special secret section of British Intelligence. The section's purpose was "getting rid of" dangerous or undesirable people through bribery, blackmail, frame-ups, or, in the last resort, death. Described in one episode as "a dead shot with the cold nerve to kill," Callan was the section's best operative and indeed, killing seemed to be his main occupation. The character paid a high moral and emotional price for his expertise: he was brooding, solitary, and friendless except for a grubby petty thief named Lonely (Russell Hunter), and his only hobby was collecting toy soldiers. Callan also had two personal weaknesses: he was rebellious and he cared. Although he always did what his bosses told him, he inevitably argued or defied them first, and more importantly, he often became concerned or involved with those whose paths he crossed during the course of his assignments. Despite its bleak subject matter, *Callan* was a hit in Britain. It spawned both a theatrical film (*Callan*, 1974), and later a television special (*Wet Job*, 1981), in which loyal viewers learned of Callan's ultimate fate.

On one *Callan* episode, "Where Else Could I Go?", a psychiatrist working for British Intelligence says that Callan is "brave, aggressive, and can be quite ruthless when he believes in the justice of his cause." This description could also be applied to Robert McCall, the lead character of *The Equalizer*, which ran in the United States on CBS from 1985 to 1989. McCall was a retired espionage agent who'd been working for an American agency (probably the CIA). After forcing the agency to let him go, he decided to use his professional skills to aid helpless people beset by human predators in the urban jungle, usually free of charge. His ad running in the New York classifieds read: "Got a problem? Odds against you? Call the Equalizer." Although McCall's clients came from all walks of life, they shared one thing in common: they all had a problem that conventional legal authorities, such as the courts and the police, could not handle. McCall had an ambivalent relationship with his ex-superior, Control (Robert Lansing), but often borrowed agency personnel (Mickey Kostmayer, played by Keith Szarabajka, was a frequent supporting player) to assist in the "problem-solving."

In a time of rising crime rates, *The Equalizer* was a potent paranoiac fantasy, made more so because Woodward as McCall cut a formidable figure. He seemed the soul of decency, always polite and impeccably dressed, but one could also detect determination in his steely-eyed gaze and danger in his rueful laugh. To many critics familiar with *Callan*, McCall seemed to be just an older, greyer, version of the same character. However, there were significant differences. Like Callan, McCall was suffering from a crisis of conscience, but unlike the earlier character, he had found a way to expiate his sins. While Callan was the instrument and even the victim of his superiors, McCall was the master of his fate.

A year after *The Equalizer*'s run, Woodward starred in another detective drama, *Over My Dead Body*. An attempt

by producer William Link to create a male version of his successful *Murder, She Wrote*, the show paired Woodward as a cranky crime novelist with a young reporter turned amateur sleuth played by Jessica Lundy. Unfortunately, there was a lack of chemistry between the stars and the series lasted barely a season.

Afterward, Woodward returned to England to lend his authoritative voice and presence to a real-life crime series called *In Suspicious Circumstances*, a sort of British version of the American show, *Unsolved Mysteries*. In 1995, Woodward was back on U.S. television screens in a TV movie, *The Shamrock Conspiracy*, playing a retired Scotland Yard inspector who tangles with IRA terrorists. The film, reportedly the first of a series starring Woodward as the inspector, was shot in Toronto, Canada.

In addition to his series work, Woodward has appeared in several other television movies both in Britain and the United States. His roles have been offbeat, to say the least, including most notably Merlin in *Arthur the King*, a strange version of the Camelot legend told by way of Lewis Carroll, and the Ghost of Christmas Present in the very fine 1984 production of *A Christmas Carol*, starring George C. Scott as Scrooge.

—Cynthia W. Walker

EDWARD WOODWARD. Born in Croydon, Surrey, England, 1 June 1930. Attended Eccleston Road and Sydenham Road School, Croydon; Elmwood School, Wallingford; Kingston College; Royal Academy of Dramatic Art. Married: 1) Venetia Mary Collett, 1952 (divorced); children: Sarah, Tim and Peter; 2) Michele Dotrice, 1987; child: Emily Beth. Began career as stage actor at the Castle Theatre, Farnham, 1946; worked in repertory companies throughout England and Scotland; first appeared on the London stage, 1955; continued stage work in London over next four decades, occasionally appearing in New York as well; has appeared in numerous films and in over 2,000 television productions, including *Callan*, 1967–73, and *The Equalizer*, 1985–89; has recorded twelve albums of music (vocals), three albums of poetry and fourteen books on tape. Officer of the Order of the British Empire, 1978. Recipient: Television Actor of the Year, 1969, 1970; Sun Award for Best Actor, 1970, 1971, 1972; Golden Globe Award; numerous others. Address: Ginette Chalmers, Peters, Fraser and Dunlop, 503–04 The Chambers, Chelsea Harbour, London SW10 0XF, England.

TELEVISION SERIES

1967	*Sword of Honour*
1967–73, 1981	*Callan*
1972	*Whodunnit?* (host)
1977–78	*1990*
1978	*The Bass Player and the Blonde*
1981	*Winston Churchill: The Wilderness Years*
1981	*Nice Work*
1985–89	*The Equalizer*
1987	*Codename Kyril*

1990	*Over My Dead Body*
1991	*In Suspicious Circumstances*
1991–92	*America at Risk*
1994	*Common as Muck*

MADE-FOR-TELEVISION MOVIES

1983	*Merlin and the Sword*
	(U.S. title, *Arthur the King*)
1983	*Love Is Forever*
1984	*A Christmas Carol*
1986	*Uncle Tom's Cabin*
1988	*The Man in the Brown Suit*
1990	*Hands of a Murderer*
1995	*The Shamrock Conspiracy*

TELEVISION SPECIALS

1969	*Scott Fitzgerald*
1970	*Bit of a Holiday*
1971	*Evelyn*
1979	*Rod of Iron*
1980	*The Trial of Lady Chatterley*
1981	*Wet Job*
1980	*Blunt Instrument*
1986	*The Spice of Life*
1988	*Hunted*
1990	*Hands of a Murderer, or The Napoleon of Crime*
1991	*In My Defence*

1994	*Harrison*
1995	*Cry of the City*
1995	*Gulliver's Travels*

FILMS

Where There's a Will, 1955; *Inn For Trouble*, 1960; *Becket*, 1964; *File on the Golden Goose*, 1968; *Incense for the Damned*, 1970; *Charley One-Eye*, 1972; *Young Winston*, 1972; *Hunted*, 1973; *Sitting Target*, 1974; *The Wicker Man*, 1974; *Callan*, 1974; *Three for All*, 1975; *Stand Up Virgin Soldiers*, 1977; *Breaker Morant*, 1980; *The Appointment*, 1981; *Comeback*, 1982; *Who Dares Wins*, 1982; *Merlin and the Sword*, 1982; *Champions*, 1983; *King David*, 1986; *Mister Johnson*, 1990; *Deadly Advice*, 1993; *A Christmas Reunion*, 1994.

STAGE (selection)

Where There's a Will, 1955; *Romeo and Juliet*, 1958; *Hamlet*, 1958; *Rattle of a Simple Man*, 1962; *Two Cities*, 1968; *Cyrano de Bergerac*, 1971; *The White Devil*, 1971; *The Wolf*, 1973; *Male of the Species*, 1975; *On Approval*, 1976; *The Dark Horse*, 1978; *The Beggar's Opera* (also director), 1980; *Private Lives*, 1980; *The Assassin*, 1982; *Richard III*, 1982; *The Dead Secret*, 1992.

FURTHER READING

Jefferson, Margo. "The Equalizer." *Ms.* magazine (New York), September 1986.

WOODWARD, JOANNE

U.S. Actor

Joanne Woodward has been recognized as an exceptional television performer from the beginning of her career in 1952 when she appeared on *Robert Montgomery Presents* in a drama entitled "Penny." She performed in over a dozen live New York productions from 1952 to 1958, and was also active on the stage during that period, a vocation she has pursued throughout her career. In those early years Woodward made appearances on *Goodyear Playhouse*, *Omnibus*, *Philco Television Playhouse*, *Studio One*, *Kraft Televison Theatre*, *U.S. Steel Hour*, *Playhouse 90* and *The Web*, in which she played opposite Paul Newman in 1954. Woodward remembers those experiences as "marvelous days."

In 1957 Woodward was cast in her first starring role in a feature film, *The Three Faces of Eve*, for which she received an Academy Award as Best Actress. Since then, Woodward has been recognized primarily as a feature film actress; however, her television roles have been numerous and highly memorable.

Woodward received an Emmy Award for her starring performance in *See How She Runs* on CBS in 1978. In 1985 she won a second Emmy for her role in *Do You Remember Love?*, a provocative and moving drama about the impact of Alzheimer's disease. In 1990 she received her third Emmy

Joanne Woodward

Award for producing and hosting a PBS special *American Masters*. In addition, she has been nominated three times for other performances on television.

Her roles in television drama over the past twenty years have frequently addressed social issues. Her 1981 performance as Elizabeth Huckaby in the CBS drama *Crisis at Central High* is an example of her unique ability to draw the audience into the character by becoming that character.

—Robert S. Alley

JOANNE GIGNILLIAT WOODWARD. Born in Thomasville, Georgia, U.S.A., 27 February 1930. Attended Louisiana State University, 1947-49; graduated from Neighborhood Playhouse Dramatic School, New York City. Married: Paul Newman, 1958; three daughters. Made first television appearance in "Penny," for *Robert Montgomery Presents*, 1952; numerous appearances in specials and television movies; appeared in numerous stage plays and films. Recipient: Kennedy Center Honors for Lifetime Achievement in the Performing Arts (with Paul Newman); Academy Award, 1957; Foreign Press Award, 1957; Cannes Film Festival Award, 1972; New York Film Critics Award, 1968, 1973, and 1990; Emmy Awards, 1978, 1985, 1990.

MADE-FOR-TELEVISION MOVIES (selection)

1952	*Robert Montgomery Presents*: "Penny"
1976	*All the Way Home*
1976	*Sybil*
1977	*Come Back, Little Sheba*
1978	*See How She Runs*
1979	*Streets of L.A.*
1980	*The Shadow Box*
1981	*Crisis at Central High*
1985	*Do You Remember Love?*
1989	*Foreign Affairs*
1993	*Blind Spot*
1994	*Hallmark Hall of Fame*: "Breathing Lessons"

TELEVISION SPECIALS

1989	*Broadway's Dreamers:* "The Legacy of the Group Theater"
1990	*American Masters*
1996	*Great Performances:* "Dance in America: A Renaissance"

FILMS (selection)

Count Three and Pray, 1955; *A Kiss Before Dying*, 1956; *The Three Faces of Eve*, 1957; *No Down Payment*, 1957; *Rally Round the Flag Boys*, 1958; *The Long Hot Summer*, 1958; *The Sound and the Fury*, 1959; *The Fugitive Kind*, 1960; *Paris Blues*, 1961; *The Stripper*, 1963; *A New Kind of Love*, 1963; *A Big Hand for the Little Lady*, 1965; *A Fine Madness*, 1965; *Rachel, Rachel*, 1968; *Winning*, 1969; *WUSA*, 1970; *They Might Be Giants*, 1971; *The Effect of Gamma Rays on Man-in-the-Moon Marigolds*, 1972; *Summer Wishes, Winter Dreams*, 1973; *The Drowning Pool*, 1975; *The End*, 1978; *Harry and Son*, 1984; *The Glass Menagerie*, 1987; *Mr. and Mrs. Bridge*, 1990; *Philadelphia*, 1993; *The Age of Innocence* (narrator/voice only), 1993.

FILMS (director)

Come Along With Me, 1982; *The Hump Back Angel*, 1984.

STAGE

Picnic (understudy), 1953; *Baby Want a Kiss*, 1964; *Candida*, 1982; *The Glass Menagerie*, 1985; *Sweet Bird of Youth*, 1988.

FURTHER READING

McGillivray, David. "Joanne Woodward." *Films and Filming* (London), October 1984.

Morella, Joe, and Edward Z. Epstein. *Paul and Joanne: A Biography of Paul Newman and Joanne Woodward*. New York: Delacorte, 1988.

Netter, Susan. *Paul Newman and Joanne Woodward*. London: Piatkus, 1989.

Stern, Stewart. *No Tricks in My Pocket: Paul Newman Directs*. New York: Grove Press, 1989.

WORKPLACE PROGRAMS

U.S. television, from its earliest years, has developed prime-time programs which focus on the workplace. This trend is understandable enough, given TV's essential investment in the "American work ethic" and in consumer culture, although it also evinces TV's basic domestic impulse. By the 1970s and 1980s, in fact, TV's most successful workplace programs effectively merged the medium's work-related and domestic imperatives in sitcoms like *The Mary Tyler Moore Show*, *M*A*S*H*, *Taxi*, and *Cheers*, and in hour-long dramas like *Hill Street Blues*, *St. Elsewhere*, and *LA Law*. While conveying the working conditions and the professional ethos of the workplace, these programs also depicted co-workers as a loosely knit but crucially interdependent quasi-family within a "domesticated" workplace. This strategy was further refined in 1990s sitcoms like *Murphy Brown* and *Frasier*, and even more notably in hour-long dramas like *ER*, *NYPD Blue*, *Picket Fences*, *Chicago Hope*, and *Homicide: Life in the Streets*. These latter series not only marked the unexpected resurgence of hour-long drama in prime time, but in the view of many critics evinced a new "golden age" of American television.

This integration of home and work was scarcely evident in 1950s TV, when the domestic arena and the work-

ER

Dr. Kildare

Law and Order

The Bold Ones

place remained fairly distinct. The majority of workplace programs were male-dominant law-and-order series which generally focused less on the workplace itself than on the professional heroics of the cops, detectives, town marshals, bounty hunters, who dictated and dominated the action. *Dragnet*, TV's prototype cop show, did portray the workaday world of the L.A. police, albeit in uncomplicated and superficial terms. The rise of the hour-long series in the late 1950s brought a more sophisticated treatment of the workplace in courtroom dramas like *Perry Mason*, detective shows like *77 Sunset Strip*, and cop shows like *Naked City* (which ran as a half-hour show in the 1958–59 season and then returned as an hour-long drama in 1960). More than simply a "home base" for the protagonists, the workplace in these programs was a familiar site of personal and professional interaction.

The year 1961 saw three new important hour-long workplace dramas: *Ben Casey*, *Dr. Kildare*, and *The Defenders*. The latter was a legal drama whose principals spent far less time in the courtroom and more time in the office than did Perry Mason. And while Mason's cases invariably were murder mysteries, with Mason functioning as both lawyer and detective, *The Defenders* treated the workaday legal profession in more direct and realistic terms. Both *Ben Casey* and *Dr. Kildare*, meanwhile, were medical dramas set in hospitals, and they too brought a new degree of realism to the depiction of the workplace setting—and to the lives and labors of its occupants. As *Time* magazine noted in reviewing *Ben Casey*, the series "accurately captures the feeling of sleepless intensity of a metropolitan hospital."

Another important and highly influential series to debut in 1961 was a half-hour comedy, *The Dick Van Dyke Show*, which effectively merged the two dominant sitcom strains—the workplace comedy with its ensemble of disparate characters, and the domestic comedy centering on the typical (white, middle-class) American home and family. At the time, most workplace comedies fell into three basic categories: school-based sitcoms like *Mr. Peepers* and *Our Miss Brooks*; working-girl sitcoms like *Private Secretary* and *Oh Suzanna*; and military sitcoms like *The Phil Silvers Show* and *McHale's Navy*. The vast majority of half-hour comedies were domestic sitcoms extolling (or affectionately lampooning) the virtues of home and family. These occasionally raised work-related issues—via working-stiffs like Chester Riley (*The Life of Riley*) lamenting an American Dream just out of reach, for instance, or on an "unruly" housewife like Lucy Ricardo (*I Love Lucy*) comically resisting her domestic plight. And some series like *Hazel* centered on "domestic help" (maids, nannies, etc.), thus depicting the home itself as a workplace.

The Dick Van Dyke Show created a hybrid of sorts by casting Van Dyke as Rob Petrie, an affable suburban patriarch and head writer on the fictional Alan Brady Show. Setting the trend for workplace comedies of the next three decades, *The Dick Van Dyke Show* featured a protagonist who moved continually between home and work, thus

creating a format amenable to both the domestic sitcom and the workplace comedy. The series' domestic dimension was quite conventional, but its treatment of the workplace was innovative and influential. The work itself involved television production (as would later workplace sitcoms like *The Mary Tyler Moore Show*, *Buffalo Bill*, and *Murphy Brown*), and thus the program carried a strong self-reflexive dimension. More importantly, *The Dick Van Dyke Show* developed the prototype for the domesticated workplace and the work-family ensemble—Rob and his staff writers Buddy (Morey Amsterdam) and Sally (Rose Marie); oddball autocrat Alan Brady (Carl Reiner, the creator and executive producer of *The Dick Van Dyke Show*); and Alan's producer and brother-in-law, the ever-flustered and vaguely maternal Mel (Richard Deacon). Significantly, Rob was the only member of the workplace ensemble with a stable and secure "home life," and thus he served as the stabilizing, nurturing, mediating force in the comic-chaotic and potentially dehumanizing workplace.

The influence of *The Dick Van Dyke Show* on TV's workplace programs was most obvious and direct in the sitcoms produced by MTM Enterprises in the early 1970s, particularly *The Mary Tyler Moore Show* and *The Bob Newhart Show*. While these and other MTM sitcoms featured a central character moving between home and work, *The Mary Tyler Moore Show* was the most successful in developing the workplace (the newsroom of a Minneapolis TV station, WJM) as a site not only of conflict and comedic chaos but of community and kinship as well. And although Moore, who had played Rob's wife on *The Dick Van Dyke Show*, was cast here as an independent single woman, her nurturing instincts remained as acute as ever in the WJM newsroom.

While the MTM series maintained the dual focus on home and work, another crucial workplace comedy from the early 1970s, *M*A*S*H*, focused exclusively on the workplace—in this case a military surgical unit in war-torn Korea of the early 1950s (with obvious pertinence to the then-current Vietnam War). Alan Alda's Hawkeye Pierce was in many ways the series' central character and governing sensibility, especially in his caustic disregard for military protocol and his fierce commitment to medicine. Yet *M*A*S*H* was remarkably "democratic" in its treatment of the eight principal characters, developing each member of the ensemble as well as the collective itself into a functioning work-family. While ostensibly a sitcom, the series often veered into heavy drama in its treatment of both the medical profession and the war; in fact, the laugh track was never used during the scenes set in the operating room. And more than any previous workplace program, whether comedy or drama, *M*A*S*H* was focused closely on the professional "code" of its ensemble, on the shared sense of duty and commitment which both defined their medical work and created a nagging sense of moral ambiguity about the military function of the unit—that is, patching up the wounded so that they might return to battle.

A domestic sitcom hit from the early 1970s, *All in the Family*, also is pertinent here for several reasons. First, in Archie Bunker (Carroll O'Connor), the series created the most endearing and comic-pathetic working stiff since Chester Riley. Second, parenting on the series involved two grown "children," with the generation-gap squabbling between Archie and son-in-law Mike (Rob Reiner) frequently raising issues of social class and work. Moreover, their comic antagonism was recast in other generation-gap sitcoms set in the workplace, notably *Sanford and Son* and *Chico and the Man*. And third, *All in the Family* itself evolved by the late 1970s into a workplace sitcom, *Archie Bunker's Place*, with the traditional family replaced by a work-family ensemble.

The trend toward workplace comedies in the early 1970s was related to several factors both inside and outside the industry. One factor, of course, was the sheer popularity of the early-1970s workplace comedies, and their obvious flexibility in terms of plot and character development. These series also signaled TV's increasing concern with demographics and its pursuit of "quality numbers"—the upscale urban viewers coveted by sponsors. Because these series often dealt with topical and significant social issues, they were widely praised by critics, thus creating an equation of sorts between quality demographics and "quality programming." And in a larger social context, this programming trend signaled the massive changes in American lifestyles which accompanied a declining economy and runaway inflation, the sexual revolution and women's movement, the growing ranks of working wives and mothers, rising divorce rates, the aging of the baby-boom generation, and so on.

Thus, the domestic sitcom with its emphasis on traditional home and family all but disappeared from network schedules in the late 1970s and early 1980s, replaced by workplace comedies like *Alice*, *Welcome Back, Kotter*, *WKRP in Cincinnati*, *Taxi*, *Cheers*, *Newhart*, and *Night Court*. The domestic sitcom did rebound in the mid-1980s with *The Cosby Show* and *Family Ties*, and by the 1990s the domestic and workplace sitcoms had formed a comfortable alliance—with series like *Murphy Brown*, *Coach*, and *Frasier* sustaining the MTM tradition of a central, pivotal character moving between home and the workplace.

TV's hour-long workplace dramas underwent a transformation as well in the 1970s, which was a direct outgrowth, in fact, of MTM's workplace sitcoms. In 1977, MTM Enterprises retired *The Mary Tyler Moore Show* and created a third and final spin-off of that series, *Lou Grant*, which followed Mary's irascible boss (Ed Asner) from WJM-TV in Minneapolis to the *Los Angeles Tribune*, where he took a job as editor. *Lou Grant* was created by two of MTM's top comedy writer-producers, James Brooks and Allan Burns, along with Gene Reynolds, the executive producer of *M*A*S*H*. It marked a crucial new direction for MTM not only as an hour-long drama, but also because of its primary focus on the workplace (a la *M*A*S*H*) and its aggressive treatment of "serious" social and work-related issues. In that era of Vietnam, Watergate, and *All the President's Men*, *Lou Grant* courted controversy week after week, with Lou and his work-family of investigative journalists not only pursuing the "Truth" but agonizing over their personal lives and professional responsibilities as well.

MTM's hour-long workplace dramas hit their stride in the 1980s with *Hill Street Blues* and *St. Elsewhere*, which effectively revitalized two of television's oldest genres, cop show and doc show. Each shifted the dramatic focus from the all-too-familiar heroics of a series star to an ensemble of co-workers and to the workplace itself—and not simply as a backdrop but as a social-service institution located in an urban-industrial war zone with its own distinctive ethos and sense of place. Each also utilized serial story structure and documentary-style realism, drawing viewers into the heavily populated and densely plotted programs through a heady, seemingly paradoxical blend of soap opera and cinema verite. Documentary techniques—location shooting, hand-held camera, long takes and reframing instead of cutting, composition in depth, and multiple-track sound recording—gave these series (and the workplace itself) a "look" and "feel" that was utterly unique among police and medical dramas.

Hill Street Blues and *St. Elsewhere* also emerged alongside prime-time soap operas like *Dallas* and *Falcon Crest*, and shared with those series a penchant for "continuing drama." While this serial dimension enhanced both the Hill Street precinct and St. Eligius hospital as a "domesticated workplace," the genre requirements of each series (solving crimes, healing the sick) demanded action, pathos, jeopardy, and a dramatic payoff within individual episodes. Thus, a crucial component of MTM's workplace dramas was their merging of episodic and serial forms. The episodic dimension usually focused on short-term, work-related conflicts (crime, illness), while the serial dimension involved the more "domestic" aspects of the characters' lives—and not only their personal lives, since most of the principals were "married to their work," but also the ongoing interpersonal relationships among the co-workers.

Hill Street co-creator Steven Bochco left MTM in the mid-1980s and developed *LA Law*, which took the ensemble workplace drama "upscale" into a successful big-city legal firm. While a solid success, this focus on upscale professionals marked a significant departure from *Hill Street* and *St. Elsewhere*—and from most workplace dramas in the 1990s as well. Indeed, prime-time network TV saw a remarkable run of MTM-style ensemble dramas in the 1990s, notably *ER*, *Homicide*, *Law and Order*, *Chicago Hope*, and another Bochco series, *NYPD Blue*. Most of these were set, like *Hill Street* and *St. Elsewhere*, in decaying inner cities, and they centered on co-workers whose commitment to their profession and to one another was far more important than social status or income. Indeed, a central paradox in these programs is that their principal characters, all intelligent, well-educated professionals, eschew material rewards to work in under-funded social institutions where commitment outweighs income, where the work is never finished nor the conflicts satisfactorily resolved, and where the work itself, finally, is its own reward.

Despite these similarities to *Hill Street* and *St. Elsewhere*, the 1990s workplace dramas differed in their emphasis on workaday cops and docs. Those earlier MTM series carried a strong male-management focus, privileging the veritable "patriarch" of the work-family—Captain Frank Furillo and Dr. Donald Westphall, respectively—whose role (like Lou Grant before them) was to uphold the professional code and the familial bond of their charges. The 1990s dramas, conversely, concentrated mainly on the workers in the trenches, whose shared commitment to one another and to their work defines the ethos of the workplace and the sense of kinship it engendered.

More conventional hour-long workplace programs have been developed alongside these MTM-style dramas, of course, from 1970s series like *Medical Center, Ironside*, and *Baretta* to more recent cop, doc, and lawyer shows like *Matlock, T.J. Hooker*, and *Quincy*. In the tradition of *Dragnet* and *Marcus Welby*, the lead characters in these series are little more than heroic plot functions, with the plots themselves satisfying the generic requirements in formulaic doses and the workplace setting as mere backdrop. Two recent hour-long dramas more closely akin to the MTM-style workplace programs are *Northern Exposure* and *Picket Fences*. Both are successful ensemble dramas created by MTM alumni who took the workplace form into more upbeat and off-beat directions—the former a duck-out-of-water doc show set in small-town Alaska which veered into magical realism, the latter a hybrid cop-doc-legal-domestic drama set in small-town Wisconsin. But while both are effective ensemble dramas with an acute "sense of place," they are crucially at odds with urban-based medical dramas like *ER* and *Chicago Hope*, and police dramas like *Homicide* and *NYPD Blue*, whose dramatic focus is crucially wed to the single-minded professional commitment of the ensemble and is deeply rooted in the workplace itself.

Indeed, *ER* and *Homicide* and the other MTM-style ensemble dramas posit the workplace as home and work itself as the basis for any real sense of kinship we are likely to find in the contemporary urban-industrial world. As Charles McGrath writes in *The New York Times Magazine*, "The Triumph of the Prime-Time Novel," such shows appeal to viewers because "they've remembered that for a lot of us work is where we live more of the time; that, like it or not, our job relationships are often as intimate as our family relationships, and that work is often where we invest most of our emotional energy." McGrath is one of several critics who view these workplace dramas as ushering in a renaissance of network TV programming, due to their Dickensian density of plot and complexity of character, their social realism and moral ambiguity, their portrayal of workers whose heroics are simply a function of their everyday lives and labors.

The workplace in these series ultimately emerges as a character unto itself, and one which is both harrowing and oddly inspiring to those who work there. For the characters in *ER* and *NYPD Blue* and the other ensemble workplace dramas, soul-searching comes with the territory, and they know the territory all too well. They are acutely aware not only of their own limitations and failings but of the inadequacies of their own professions to cure the ills of the modern world. Still, they maintain their commitment to one another and to a professional code which is the very life-blood of the workplace they share.

—Thomas Schatz

FURTHER READING

Brooks, Tim, and Earle Marsh. *The Complete Directory to Prime Time Network TV Shows*. New York: Ballantine, 1979; 5th edition, 1992.

Feuer, Jane, Paul Kerr, and Tise Vahimagi, editors. *MTM: 'Quality Television.'* London: British Film Institute, 1984.

Gitlin, Todd. *Inside Prime Time*. New York: Pantheon, 1983.

McGrath, Charles. "The Triumph of the Prime-Time Novel." *New York Times Magazine,* 22 October 1995.

Schatz, Thomas. "*St. Elsewhere* and the Evolution of the Ensemble Series." In, Newcomb, Horace, editor. *Television: The Critical View*. New York: Oxford University Press, 1976; 4th edition, 1987.

Williams, Betsy. "'North to the Future': *Northern Exposure* and Quality Television." In, Newcomb, Horace, editor. *Television: The Critical View*. New York: Oxford University Press, 1976; 5th edition, 1994.

See also Comedy, Workplace Settings; Detective Programs; Police Programs

WORLD IN ACTION

British News Documentary

W*orld in Action*, Britain's long-running and most illustrious current affairs programme, goes out in prime-time on ITV (the main commercial channel) and is produced by Granada Television, a company with a reputation for innovation and "quality" programming. First launched in 1963, with Tim Hewat, an ex-*Daily Express* reporter, as its editor, *World in Action* was the first weekly current affairs programme in Britain to pioneer pictorial journalism on film and to risk taking an independent editorial stance. In comparison with *Panorama*, the BBC's rival current-affairs programme, which was studio-based and featured several items, *World in Action* was, in the words of Gus McDonald, "born brash." It devoted each half-hour episode to a single issue and, abandoning the studio and presenter, put the story itself up-front. The lightweight film equipment gave the production team the mobility to follow up the stories first hand and to bring raw images of the world into the living room. A conspicuous and influential style evolved with interviewees framed in close-up talking directly to camera, cross-cut with fast-edited observation of relevant action and environmental detail. The hard-hitting approach compelled attention and made complex social issues accessible to a mass audience for the first time.

Having firmly established the idea of picture journalism on TV, *World in Action* consolidated its position in 1967 under David Plowright when an investigative bureau was set up, and it is on the quality of its investigative journalism that the programme's reputation chiefly rests. Award-winning episodes have included "The Demonstration" (1968) observing the mass protest outside the U.S. embassy against the bombing of North Vietnam; "Nuts and Bolts of the Economy" (1976), a series exploring different aspects of the world economy; and an investigation into "The Life and Death of Steve Biko"(1978). The programme has been equally wide-ranging with domestic topics, covering stories such as the exposure of police corruption in "Scotland Yard's Cocaine Connection" (1985), revealing the British Royal Family's tax loop-hole (1991), and investigating the dangers of different types of contraceptive pill (1995). Over the years the programme has fearlessly and impartially pursued the truth, exposing injustice and falsehood, and frequently running at odds with the powers that be. In this respect the programme's long-standing, but eventually successful, fight to secure the release of the six men wrongfully convicted for the IRA pub bombing in Birmingham provides the outstanding example.

World in Action stands as one of the finest achievements of public service television in Britain—of programming driven by the desire to inform and educate viewers as much as to entertain them. In the course of its long run it has provided the training round for some of the most distinguished names in British broadcasting, as well as pioneering innovative programme approaches such as under-cover and surveillance work, and drama documentary. How it will continue to fare in the more competitive broadcast market following deregulation remains to be

World in Action: The Man Who Wouldn't Keep Quiet
Photo courtesy of the British Film Institute

seen. However, it is possible that to maintain its prime time slot the emphasis will shift away from costly long-term investigations and international stories to focus on populist health and consumer issues which can be guaranteed to deliver large audiences.

—Judith Jones and Bob Millington

PROGRAMMING HISTORY

• ITV

1963–1965
1967–

FURTHER READING

Corner, John. *The Art of Record.* Manchester: Manchester University Press, 1996.

Granada: The First Twenty-Five Years. London: British Film Institute, 1988.

See also British Programming

WORRELL, TRIX

British Writer

Trix Worrell has lived in Britain for most of his life, having moved there from St. Lucia when he was five. When he began his acting career, he also started writing because there were so few good parts for black actors to play. As a teenager, Worrell worked with the Albany Theatre in South London, where he wrote and directed his first play, *School's Out,* in 1980. Eventually, he enrolled at the National Film and Television School, initially as a producer, but soon decided to concentrate on writing and directing. Even before his NFTS course, he had achieved recognition as a writer.

In 1984, Worrell won Channel 4 Television's "Debut New Writers" competition with his play *Mohicans,* which was broadcast on Channel 4 as *Like a Mohican* in 1985. At that time, the young Worrell was a more modest individual, and it was a colleague rather than Worrell himself who sent in the script to the competition. When he won, his pleasure was somewhat dulled when he realized that despite his success, the small print of the competition meant that Channel 4 did not actually have to broadcast his work. Showing the determination which would stand him in good stead for subsequent battles with commissioning editors, Worrell fought to have his play broadcast, and successfully challenged Channel 4's insistence that single dramas were too expensive to produce. Having leapt that first hurdle, he then argued forcefully for the play to keep its original language, including the ubiquitous swearing which is an intrinsic part of polyglot London's authentic voice. Fortunately, his persistence paid off, and after this success he went on to co-author (with Martin Stellman) the feature film *For Queen and Country* (1989) before returning again to the small screen.

In the late 1980s, Channel 4 was interested in commissioning a new sitcom, and Worrell contacted the producer Humphrey Barclay with a view to working up an idea. Though he had never written television comedy before, he had penned various satirical works for the theatre and felt confident, if slightly anxious, about entering this extremely difficult terrain. Worrell tells the story that he was on his way to meet Barclay to talk through possibilities when his bus pulled up at a traffic light and he saw a barber shop with three barbers peering through the shop window to ogle the women going past: suddenly he had found his comedy situation. The subsequent show, *Desmond's,* was one of Channel 4's most successful programmes, producing seven series in five years, from 1989 to 1994. As with all good sitcoms, *Desmond's* was organised around a particular location, in this case, the inside of the barber shop, with occasional shoots in the world outside or scenes set in the flat over the shop, which served as home for the eponymous Desmond and his family.

Although this was not the first British comedy series about a black family, Worrell was keen to work through a number of complex issues and important features of black migrant experiences in Britain in ways which would make sense to both black and white viewers. *Desmond's* was always intended for a mixed audience, and Worrell wanted to expose white audiences to an intact black family whose members experienced precisely the same problems and joys as those of white families. At the same time, he wanted to reflect a positive and realistic black family for black viewers as an antidote to the routinely stereotypical portraits which more usually characterise programmes about black people in Britain.

In talking about the production of *Desmond's,* Worrell has revealed the considerable antagonisms he faced from black colleagues who regarded writing sitcoms as an act of betrayal, or at the very least as a soft-option sell-out. But this type of criticism misses the point: powerful sentiment and subversive commentary can be made by comedy characters precisely because their comedic tone and domesticated milieu are unthreatening—the viewer is invited to laugh and empathise *with* the characters, not to scorn them. In later episodes of *Desmond's,* programme narratives were pushed into more controversial areas such as racism because identification and loyalty had already been secured from the audience and more risks could be taken.

Worrell is very aware of the limited opportunities which exist for black writers wanting to break into television. By the third series of *Desmond's,* he had brought together a new team to work on the show, enabling him to concentrate more on directing as well as providing valuable production experience to a cohort of black writers, many of whom were women. Despite the considerable success of *Desmond's,* Worrell believes he still has to fight much harder than white colleagues to get new programme ideas accepted. There are significant problems in trying to negotiate new and challenging territory which questions the cosy prejudices of the status quo, and British broadcasters now tend towards the conservative rather than the innovative in their relentless battle to retain market share. While there is a continued interest in series which reflect the assumptions and preconceptions that white editors have about black communities, Worrell is keen to explore the diversities of life as it is actually lived by Britain's blacks. His work breaks out of the suffocating straightjacket of dismal (racist) stereotypes, instead examining the complex realities of black experiences, which are as much about living, loving, and working within strongly multicultural environments as about the hopeless crackheads, pimps, and villains who inhabit London's ghetto slums. There is no one story—there are many.

In late 1994, Worrell teamed up with Paul Trijbits to create the film and TV production company, Trijbits-Worrell. With corporate backing from the Dutch-based Hungry Eye Entertainment Group, Worrell is currently working up a number of new ideas for both film and television. Two TV projects which are already considerably advanced are *Quays to the City,* a post-yuppie soap opera for the BBC set in London's Docklands, and *Saturday Dad,* a sitcom about a group of single fathers who only see their children at weekends. Although Worrell is quite pessimistic about the future for black writers, producers, and directors trying to penetrate the industry, the continued success of his own work ensures that there is at least one act to follow.

—Karen Ross

TRIX WORRELL. Born in St. Lucia; immigrated to Britain at the age of five. Educated at the National Film and Television School, London. Writer and actor, Albany Youth Theatre, Deptford, South East London; winner of Channel 4's Debut '84 New Writers competition for *Mohicans, Like a Mohican* aired on Channel 4, 1985; writer and director, *Desmond's,* Channel 4 situation comedy, 1989–94; executive producer, science-fiction film, *Hardware,* 1990; co-founder, with Paul Trijbits, Trijbits-Worrell, film and television production company, 1994.

TELEVISION SERIES
1989–94   *Desmond's*

TELEVISION PLAY
1985   *Like a Mohican*

FILMS

For Queen and Country (with Martin Stellman), 1989; *Hardware* (executive producer), 1990.

STAGE

School's Out, 1980.

FURTHER READING

Pines, Jim, editor. *Black and White in Colour: Black People in British Television since 1936.* London: British Film Institute, 1992.

Ross, Karen. *Black and White Media: Black Images in Popular Film and Television.* Cambridge: Polity Press, 1995.

See also British Programming; *Desmond's*

WRATHER, JACK

U.S. Media Executive/Producer

Born in Amarillo, Texas, Jack Wrather became an oil wildcatter who eventually rose to be president of an oil company founded by his father. He later expanded his resources into real estate, hotels, motion pictures, and broadcast properties. Following service in the U.S. Marine Corps during World War II, Wrather relocated to California, where he diversified his holdings in the movie business, creating Jack Wrather Pictures, Inc. and Freedom Productions. Between 1946 and 1955, Wrather produced feature films for Eagle Lion, Warner Brothers, Allied Artists, and United Artists, including *The Guilty, High Tide, Perilous Waters, Strike It Rich, Guilty of Treason, The Lone Ranger and the Lost City of Gold, The Magic of Lassie,* and *The Legend of the Lone Ranger.*

A true entrepreneur, Wrather established television syndication services during the 1950s such as Television Programs of America and Independent Television Corporation. He was also co-owner of television stations licensed to Wrather-Alvarez Broadcasting Company in Tulsa, San Diego, and Bakersfield.

Perhaps Wrather is most noted for several of the television series he produced: *The Lone Ranger, Lassie,* and *Sergeant Preston of the Yukon.* These programs, which were standards among early syndicated television offerings, served stations affiliated with networks as well as independent stations, and demonstrated that formulaic, filmed entertainment could provide both audiences and a resalable product. In many ways, Wrather's operations foreshadowed some of the most significant developments in the economic support structure for the next generation of television, a fact he obviously recognized.

After paying three million dollars to George W. Trendle for rights to *The Lone Ranger,* Wrather considered his purchase an important part of American history. The 221-episode half-hour western series, licensed through the years to ABC, CBS, and NBC, remains in syndication today. In the 1950s Wrather also produced the popular weekly *Lassie* adventure series and 78 episodes of *Sergeant Preston.*

Among other Wrather holdings were the *Queen Mary,* and Howard Hughes' transport aircraft, the *Spruce Goose.* He also owned Disneyland Hotel, and served as board director or board chair for Continental Airlines, TelePromTer, Muzak, Inc., and the Corporation for Public Broadcasting.

Wrather was among several prominent business executives who became members of Ronald Reagan's original transition committee when Reagan became president in 1981. Jack Wrather died of cancer in 1984 at age 66.

–Denis Harp

JACK WRATHER. Born John Devereaux Wrather, Jr. in Amarillo, Texas, U.S.A., 24 May 1918. Educated at the University of Texas at Austin, B.A. 1939. Married: Bonita Granville, 1947; children: Molly, Jack, Linda, and Christopher. Served in U.S. Marine Corps Reserves, 1942–53. Independent oil producer in Texas, Indiana, and Illinois; president, Evansville Refining Company, 1938–40, Overton Refining Company, Amarillo Producers, and Refiners

Jack Wrather
Photo courtesy of Broadcasting and Cable

Corporation, Dallas, 1940–49; owner, Jack Wrather Pictures, Inc., 1947–49, and Freedom Productions Corporation, from 1949; president, Western States Investment Corporation, from 1949; president, Wrather Television Productions, Inc., from 1951; Wrather-Alvarez Broadcasting, Inc., Lone Ranger, Inc., Lassie, Inc., Disneyland Hotel, Anaheim, California; owner, KFMB, KERO, and KEMB-TV in San Diego; owner, KOTY-TV in Tulsa, Oklahoma; part-owner, WNEW, New York City; chair, Muzak, Inc., Independent Television Corporation and Television Programs of America, Inc., Stephens Marine, Inc.; president and chair, Wrather Corporation; director, TelePrompTer Corporation, Continental Airlines, Transcontinent Television Corporation, Jerold Electronics Corporation, Capitol Records, Inc.; board of directors, Community Television of Southern California, Corporation for Public Broadcasting, 1970. Member: development board, University of Texas; board of counselors for performing arts, University of Southern California; Independent Petroleum Association of America; International Radio and Television Society; Academy of Motion Picture Arts and Sciences; National Petroleum Council, 1970. Died in Santa Monica, California, 12 November 1984.

TELEVISION SERIES (producer)

1949–57	*The Lone Ranger*
1957–74	*Lassie*
1955–58	*Sergeant Preston of the Yukon*

FILMS (producer)

The Guilty, 1946; *High Tide*, 1947; *Perilous Water*, 1947; *Strike It Rich*, 1948; *Guilty of Treason*, 1949; *The Lone Ranger and the Lost City of Gold*, 1958; *The Magic of Lassie*, 1978; *The Legend of the Lone Ranger*, 1981.

See also *Lassie*, *Lone Ranger*, Syndication

WRIGHT, ROBERT C.

U.S. Media Executive

Robert C. Wright succeeded the legendary Grant Tinker as president of NBC in 1986 when "the peacock network" was acquired by General Electric (GE) for $6.3 billion. Under General Electric chief executive officer Jack Welch, Wright immediately began to reshape a new NBC, moving it out of radio altogether and headlong into cable television. In 1988 Wright allied with Cablevision Systems, Inc., in a $300 million deal which led in the following year to the start up of a 24-hour cable network, CNBC (Consumer News and Business Channel). He also acquired shares of Visnews, an international video news service, and immediately initiated selling NBC News products to hundreds of clients overseas, and of the cable channel, CourtTV.

The first half of the 1990s were equally busy for Wright. The Australian Television Network became NBC's first overseas affiliate. In 1991, NBC bought out CNBC's chief rival, the Financial News Network, for well in excess of $100 million, closed it down, and merged its core components into CNBC. He invested in the Super Channel, an advertising-supported satellite service based in London, England; began NBC Asia; and poured millions into NBC's News Channel, a TV wire service based in Charlotte, North Carolina. But the biggest deal during the first half of the 1990s came when Wright and Bill Gates announced a multimillion dollar alliance of NBC and Microsoft to create an all-news channel, MSNBC, to rival CNN around the world.

Wright, under the tutelage of Jack Welch, remade NBC within 10 years, and served as the longest reigning NBC head since David Sarnoff. And like mentor Welch, Wright came from a Catholic household (from suburban Long Island), was the son of an engineer, had not gone to an Ivy league college, was devoted to GE, and was no fan of television. Wright had entered the GE corporate ladder as a staff attorney, but quickly moved to the decision-making

Robert C. Wright

side, running GE's plastic sales division (1978–1980), working as the head of the housewares and audio equipment division (1983–1984), and promoted to the presidency of GE Financial Services (1984–86).

Wright's first ten years at NBC were not without failure. Most notably he led NBC to well in excess of $50 million in losses by way of its pay-per-view venture "Triplecast" during the 1992 Olympics. Still, he is credited with transforming NBC, and manouvering NBC through a key intersection of the technological, economic, political, social, and cultural forces that helped shape television in the United States, at the end of the 20th century.

—Douglas Gomery

ROBERT C(HARLES) WRIGHT. Born in Hempstead, New York, U.S.A., 23 April 1943. Holy Cross College, B.A. in history 1965; University of Virginia, LL.B. 1968. Married: Suzanne Werner, 1967; children: Kate, Christopher and Maggie. Served in U.S. Army Reserve. Admitted to Bar: New York, 1968; Virginia, 1968; Massachusetts, 1970; New Jersey, 1971. Attorney, General Electric Company, 1969–70, and 1973–79; general manager, plastics sales department, 1976–80; law secretary to chief judge, U.S. District Court, New Jersey, 1970–73; president, Cox Cable Communications, Atlanta, 1979–83; executive vice president, Cox Communications, 1980–83; vice president and general manager, GE housewares, electronics and cable TV operations, 1983–84; president and chief executive officer, GE Financial Services, Inc., 1984–86; president and chief executive officer, NBC, New York City, from 1986. Address: NBC, 30 Rockefeller Plaza, New York, New York 10112, U.S.A.

FURTHER READING

Auletta, Ken. *Three Blind Mice: How the TV Networks Lost their Way.* New York: Random House, 1991.

Goldberg, Robert, and Gerald Jay Goldberg. *Anchors: Brokaw, Jennings and Rather and the Evening News.* New York: Birch Lane, 1990.

Tichy, Noel M., and Stratford Sherman. *Control Your Own Destiny or Someone Else Will: How Jack Welch Is Making General Electric the World's Most Competitive Corporation.* New York: Doubleday, 1993.

See also National Broadcasting Company; United States: Networks

WRITER IN TELEVISION

A commonplace in the television industry is that "it all begins with the script." In part, this notion recognizes the centrality of writers in the early days of live television, when authors such as Reginald Rose, Paddy Chayevsky and Rod Serling established the medium as an arena for the exploration of character, psychology, and moral complexity in close intimate settings. With the television industry's move to Hollywood in the 1950s, and its increasing reliance on filmed, formulaic, studio factory productions, writers were often reduced to "hack" status, churning out familiar material that was almost interchangeable across genres. This week's western could be reformatted for next week's crime drama. This view oversimplifies, of course, and ignores extraordinary work in television series such as *Naked City*, *The Defenders*, *Route 66*, and others. But it does capture conventional assumptions and expectations.

In the 1970s, with the rise of socially conscious situation comedy often identified with producer Norman Lear and the "quality" comedies associated with MTM Productions, writers once again moved to positions of prominence. Lear himself was a writer-producer, one of the many "hyphenates" who would follow into positions of authority and control. And Grant Tinker, head of MTM, sought out strong writers and encouraged them to create new shows—and new types of shows—for television. Indeed, the legacy of MTM stands strong in today's television industry. Names such as James Brooks, Allan Burns, Steven Bochco, David Milch, and others can trace their careers to that company.

At the present time almost every major producer in American television is also a writer. Writers oversee series development and production, create new programs, and see to the coordination and conceptual coherence of series in progress. Their skills are highly valued and, for the very successful few, extremely highly rewarded. Never the less, the role of the writer is affected by many other issues, and despite new respect and prominence, remains a complex, often conflicted position within the television industry.

The film and television industries, for example, have been, until quite recently, very separate entities. Even in the early years of television writers were recruited not from film but from radio and the theater. In many ways, the environment for writers in television still remains distinct from that of the film industry. TV writers are quick to remark that it is nearly impossible to start out in television and move on to film, but that there are no barriers to moving in the other direction—it is, rather, a fact that writers in the film industry will not write television "unless they are starving." This belief summarizes a power relationship in which writers are clearly identified as *either* "television" or "film," or even by genre, early in their careers. One important difference lies in the common perception that writers in television have more clout, simply because there is a well-defined career path by which writers can move up through the ranks of a production company to become a senior producer and therefore control their work in ways typically denied to film scriptwriters.

An interesting aspect of writing for television is the hierarchical organization of the profession. Many production companies now employ "staff writers," although most TV writers work as freelancers competing for a diminishing number of assignments. At the bottom of the pyramid are the outside freelancers who may write no more than two or three episodes a season for various shows. At the top are the producers and executive producers. In between are readers, writer's assistants, a handful of junior staff writers (with contracts of varying lengths), and assistant and associate producers. Producer titles are often given to writers and are usually associated with seniority and supervisory responsibilities for a writing team. The desirable career path, then, involves moving from freelancer, to staff writer, to associate producer, to supervising producer to executive producer. Executive producers are given sole responsibility for controlling a television series, are usually owners or part owners of the series, and may work on several series at once.

Writers usually become executive producers by creating their own series. But this generally occurs only after writing successfully in other positions, and after being recognized by studio and network executives as someone with the potential to create and control a series. Only in the rarest of circumstances are new program ideas purchased or developed from freelancers or beginning writers.

Readers are a critical element in a freelance television writer's working life, because they control whether or not one's work reaches senior staff with hiring authority. Readers analyze samples of a writer's work and evaluate the appropriateness of a writer's skills, experience, and background for the series, and they are used routinely as a "first cut" mechanism throughout the industry. The criteria used by readers is often very specific, sometimes seemingly arbitrary, but because of their importance TV writers learn to "write to the reader" in order to advance to the next assessment level. An entire subordinate industry exists in Los Angeles to educate writers about the process and criteria reviewers employ, even though readers describe themselves as without significant influence.

Agents are also a fact of a television writer's life because production companies and their readers generally will not consider any work from a writer unless it is submitted by an agent, preferably an agent known to that production company. A common frustration for writers is that agents refuse to represent writers without credits but credits cannot be earned without agent representation.

The Writers Guild Of America (WGA) founded in 1912 is the official trade union and collective bargaining unit for writers in the film and television industries and actively monitors working conditions for writers. The WGA has warned that contemporary writers face a hostile environment with ageism and sexism a common complaint. Hollywood is enamored with youth culture and consequently producers and network executives often seek creative talent they feel will be capable of addressing that audience. According to WGA statistics, a definite bias toward younger writers has emerged in the industry. In addition, the WGA and another organization, Women in Film, recently released reports showing that although women comprise 25% of the Hollywood writing pool they receive a smaller share of assignments proportional to their number. Although there are several prominent female writers and producers in television, many industry observers believe that there are structural and cultural barriers to the advancement of women throughout the industry that cannot be easily removed.

Because the production of most television shows (prior to syndication sales) must be "deficit-financed" (network payment for the rights to the series is less than the cost to produce the episodes), writers often bear the brunt of the resulting financial insecurity, taking less cash upfront in salary or per-episode fees and hoping for healthy residuals if the series becomes successful. Although the WGA sets minimum payments for each type of writing assignment, writers are often seen at the popular "Residuals Bar" in Van Nuys where a residuals check for $1 or less earns the bearer a free drink. 70% of television writers earn less than $50,000 a year through their efforts in this field. In spite of this harsh reality, hundreds of aspiring writers write thousands of new scripts each year, hoping for the chance to write the next huge hit.

In other television systems writers continue to enjoy a similar sort of prestige. Television authors such as Dennis Potter and Lynda La Plante have offered audiences outstanding, formally challenging work for this medium. Because of their work as well as because of the American system's financial and aesthetic rewards, television writing is now perhaps recognized as a truly legitimate form of creativity, and has taken its place alongside the novel, the stage play, and the film screenplay as one of the most significant expressive forms of the age.

—Cheryl Harris

FURTHER READING

Berger, Arthur Asa. *Scripts: Writing for Radio and Television.* Newbury Park, California: Sage, 1990.

Bielby, William T., and Denise D. Bielby. *The 1989 Hollywood Writers' Report: Unequal Access, Unequal Pay.* West Hollywood, California: Writers Guild of America, West, 1989.

Blum, Richard A. *Television Writing: From Concept to Contract.* Boston: Focal Press, 1995.

Brady, Ben. *The Understructure of Writing for Film and Television.* Austin: University of Texas Press, 1988.

Di Maggio, Madeline. *How to Write for Television.* New York: Prentice Hall, 1990.

DiTillio, Lawrence G. "'I Hate Stories.' Script Is easy. Story Is hard." *Writer's Digest* (Cincinnati, Ohio), June 1994.

———. "Scripting a Sample 'Seinfeld.'" *Writer's Digest* (Cincinnati, Ohio), December 1993.

Macak, Jim. "How Writers Survive." *The Journal* (Los Angeles), Writers Guild of America West, January 1994.

Potter, Dennis. *Seeing the Blossom: Two Interviews, a Lecture, and a Story.* London and Boston: Faber and Faber, 1994.

Root, Wells. *Writing the Script: A Practical Guide for Films and Television.* New York: Holt, Rinehart, and Winston, 1980.

Stempel, Tom. *Storytellers to the Nation: A History of American Television Writing.* New York: Continuum, 1992.

Straczynski, J. Michael. "The TV Commandments." *Writer's Digest* (Cincinnati, Ohio), April 1992.

Walter, Richard. *Screenwriting: The Art, Craft, and Business of Film and Television Writing.* New York: New American Library, 1988.

See also Chayefsky, Paddy; Bochco, Steven; Huggins, Roy; La Plante, Lynda; Mercer, David; Potter, Dennis; Rose, Reginald; Serling Rod; Silliphant, Sterling; Tarses, Jay

WYMAN, JANE

U.S. Actor/Producer

Jane Wyman is one of the few Hollywood movie stars to have had an equally successful television career. She was at the height of her film career in the mid-1950s when she launched her first television series, *Jane Wyman Theater.* Modeled after the successful *The Loretta Young Show,* the prime-time filmed anthology series presented a different drama each week, with Wyman as host, producer, and sometimes actress. Between 1958 and 1980, Wyman appeared occasionally as a guest star on television series and in made-for-TV movies. Then, in 1981, she scored another series success with her portrayal of ruthless matriarch Angela Channing on CBS' prime-time soap opera, *Falcon Crest.*

Wyman broke into movies in the early 1930s as a Goldwyn Girl and continued to play chorus girls until the mid-1940s. By 1948, when she won the Best Actress Academy Award for *Johnny Belinda,* her image was that of a capable, dramatic actress. In the early 1950s, her success continued with romantic comedies like *Here Comes the Groom* (1951) and melodramas like *Magnificent Obsession* (1954). She was considered a "woman's star," mature yet glamorous, a woman with whom middle-class, middle-aged women could identify. Amid speculation as to why a currently successful film star would want to do series television, Wyman started work on her own anthology drama series. According to her, television seemed like the right thing to do at that time. The movie industry was changing, and she wanted to try the new medium. Moreover, film roles for fortyish female stars were in short supply.

Procter and Gamble's *Fireside Theater,* a filmed anthology series, had been a fixture on NBC since 1949 but, by the end of the 1954–55 season, ratings had slipped. The show was overhauled in 1955 and became Wyman's series. Her production company, Lewman Productions (co-owned with MCA's Revue Productions), produced the series. As host, she was glamorous Jane Wyman. As producer, she chose the stories. As actress, she chose her occasional roles. Presentations were dramas or light comedies, with Wyman acting in about half of the episodes. The series carried on the tradition established by *Fireside Theater* and *The Loretta Young Show*—filmed, half-hour anthology dramas that attracted substantial audiences while critics praised live, 60- and 90-minute anthology dramas like *Studio One* and *Playhouse 90.*

Wyman's series was initially titled *Jane Wyman Presents the Fireside Theater,* but was later shortened to *Jane Wyman Theater.* (It was called *Jane Wyman Presents* when ABC aired reruns in 1963.) Like *The Loretta Young Show,* Wyman's series was rerun on network daytime schedules (to target women audiences) and in syndication. (An aspiring writer,

Jane Wyman
Photo courtesy of Jane Wyman

Aaron Spelling, found work with *Jane Wyman Theater* and later became one of television's most successful producers.) Wyman also hosted a summer series that featured teleplays originally shown on other anthology dramas. This 1957 program was called *Jane Wyman's Summer Playhouse.*

In the years following the cancellation of *Jane Wyman Theater,* Wyman guest-starred on television programs, made a few feature films (with starring roles in two Disney films), and appeared in a made-for-TV movie. In 1971, Wyman guest starred on an episode of *The Bold Ones* as Dr. Amanda Fallon. This production provided the basis for a series pilot, but never became a series. In 1979, she received attention for her supporting role in the made-for-TV movie *The Incredible Journey of Dr. Meg Laurel.* She then made appearances on two of Aaron Spelling's series, *The Love Boat* and *Charlie's Angels.* The spotlight really returned in 1981—for two different reasons.

As the ex-wife of the newly-elected President Ronald Reagan, Wyman was being sought out by the media. Her publicity value did not escape Lorimar Productions' Earl Hamner and CBS. Seeking to capitalize on their success with *Dallas* and *Knots Landing,* Lorimar and CBS launched *Falcon Crest* in 1981 with Wyman starring as a female version of *Dallas's* ruthless and manipulative J.R. Ewing. For nine seasons, she portrayed Angela Channing, the powerful matriarch of a wealthy, wine-making family. Wyman had made a successful return to series television, but in a role quite different from her earlier work. As Angela Channing, she was not the likable, clean-cut woman she had so often portrayed in the past, but she played the part to perfection. In 1984, she won a Golden Globe Award for her *Falcon Crest* performances, and was reported to be the highest paid actress on television at that time. Jane Wyman's television career began in the mid-1950s, after she had already achieved stardom in the movies. Like Loretta Young and Lucille Ball, she was one of the few film stars, one of fewer women, to have her own successful television series. She also was one of the few women to star in her own anthology drama series. Thirty years later, in the 1980s, Wyman accomplished something even more unusual: as an actor of old Hollywood and early television, she starred in another, even more successful series, *Falcon Crest.*

—Madelyn Ritrosky-Winslow

JANE WYMAN. Born Sarah Jane Fulks in St. Joseph, Missouri, U.S.A., 4 January 1916. Attended the University of Missouri, Colombia, 1935. Married: 1) Myron Futterman, 1937 (divorced, 1939); 2) Ronald Reagan, 1940 (divorced, 1948); children: Maureen and Michael; 3) Freddie Karger, 1952 (divorced, 1955). Actress in films, from 1932; debuted as Sarah Jane Fulks in *The Kid From Spain;* radio singer under the name of Jane Durrell; contract with Warner Brothers, 1936–49; host and actor in television series *The Jane Wyman Theater,* 1955-58; starring role in *Falcon Crest,* 1981–90. Recipient: Best Actress Academy Award, 1948; Golden Globe Award, 1984. Address: c/o Michael Mesnick, 500 South Sepulveda Boulevard, Los Angeles, CA 90049, U.S.A.

TELEVISION SERIES

1955-58	*The Jane Wyman Theater*
1957	*Jane Wyman's Summer Playhouse*
1981-90	*Falcon Crest*

MADE-FOR-TELEVISION MOVIES

1971	*The Failing of Raymond*
1979	*The Incredible Journey of Dr. Meg Laurel*

FILMS

(as Sarah Jane Fulks) *The Kid from Spain,* 1932; *Elmer the Great,* 1933; *College Rhythm,* 1934; *Rumba,* 1935; *All the King's Horses,* 1935; *Stolen Harmony,* 1935; *King of Burlesque,* 1936; *Anything Goes,* 1936; *My Man Godfrey,* 1936; (as Jane Wyman) *Stage Struck,* 1936; *Cain and Mabel,* 1936; *Polo Joe,* 1936; *Smart Blonde,* 1936; *Gold Diggers of 1937,* 1937; *Ready, Willing, and Able,* 1937; *The King and the Chorus Girl,* 1937; *Slim,* 1937; *The Singing Marine,* 1937; *Mr. Dodd Takes the Air,* 1937; *Public Wedding,* 1937; *The Spy Ring,* 1938; *Fools for Scandal,* 1938; *She Couldn't Say No,* 1938; *Wide Open Faces,* 1938; *The Crowd Roars,* 1938; *Brother Rat,* 1938; *Tail Spin,* 1939; *Private Detective,* 1939; *The Kid from Kokomo,* 1939; *Torchy Plays with Dynamite,* 1939; *Kid Nightingale,* 1939; *Brother Rat and a Baby,* 1940; *An Angel from Texas,* 1940; *Flight Angels,* 1940; *My Love Came Back,* 1940; *Tugboat Annie Sails Again,* 1940; *Gambling on the High Seas,* 1940; *Honeymoon for Three,* 1941; *Bad Men of Missouri,* 1941; *You're in the Navy Now,* 1941; *The Body Disappears,* 1941; *Larceny, Inc.,* 1942; *My Favorite Spy,* 1942; *Footlight Serenade,* 1942; *Princess O'Rourke,* 1943; *Make Your Own Bed,* 1944; *Crime By Night,* 1944; *The Doughgirls,* 1944; *Hollywood Canteen,* 1944; *The Lost Weekend,* 1945; *One More Tomorrow,* 1946; *Night and Day,* 1946; *The Yearling,* 1946; *Cheyenne,* 1947; *Magic Town,* 1947; *Johnny Belinda,* 1948; *A Kiss In the Dark,* 1949; *The Lady Takes a Sailor,* 1949; *It's a Great Feeling,* 1949; *Stage Fright,* 1950; *The Glass Menagerie,* 1950; *Three Guys Named Mike,* 1951; *Here Comes the Groom,* 1951; *The Blue Veil,* 1951; *Starlift,* 1951; *The Story of Will Rogers,* 1952; *Just for You,* 1952; *Let's Do It Again,* 1953; *So Big,* 1953; *Magnificent Obsession,* 1954; *Lucy Gallant,* 1955; *All That Heaven Allows,* 1955; *Miracle in the Rain,* 1956; *Holiday for Lovers,* 1959; *Pollyanna,* 1960; *Bon Voyage,* 1962; *How to Commit Marriage,* 1969; *The Outlanders.*

FURTHER READING

Bawden, J. "Jane Wyman: American Star Par Excellence." *Films in Review* (New York), April 1975.

Morella, Joe, and Edward Z. Epstein. *Jane Wyman: A Biography.* New York: Delacorte, 1985.

Parish, James Robert, and Don E. Stanke. *The Forties Gals.* Westport, Connecticut: Arlington House, 1980.

Quirk, Lawrence J. *Jane Wyman: The Actress and the Woman.* New York: Dembner, 1980.

See also *Fireside Theater;* Gender and Television; Melodrama; Young, Loretta

Y

YES, MINISTER

British Situation Comedy

Yes, Minister, a classic situation comedy exposing the machinations of senior politicians and civil servants in Great Britain, was first broadcast by the BBC in 1980. Such was the standard of scripts and performance and the accuracy of the satire that the programme became required viewing for politicians, journalists, and the general public alike, and both the initial three-season series and the two-season sequels that were made in the 1980s under the title *Yes, Prime Minister,* were consistently among the top-rated shows.

The idea for the series was developed by writer Antony Jay and former *Doctor in the House* star Jonathan Lynn while both were on the payroll of the video production company set up by John Cleese in the mid-1970s. The BBC bought

Yes, Minister
Photo courtesy of the British Film Institute

the rights to the pilot episode and work on a full series finally got under way in 1979.

The humour of each episode revolved around the maneuverings of the Right Honourable James Hacker, MP, the idealistic and newly installed minister for Administrative Affairs (and ultimately prime minister), and his cynical and wily permanent under-secretary, Sir Humphrey Appleby, who was committed to seeing that his ministerial charge never meddled too much in the business of the department and that the real power remained securely in the hands of the civil service. Every time Hacker conceived some notion aimed at reform of the ministry, Sir Humphrey Appleby and Private Secretary Bernard Woolley were there to thwart him by various ingenious means. If Hacker inquired too closely into the reasons why he was not going to get his way about something, Sir Humphrey Appleby was more than able to throw up a smokescreen of obfuscation and technical jargon, which as often as not discouraged further questioning and persuaded the civil servant that his charge was now nearly "house-trained". This was not to say that Sir Humphrey always got his way, however: sometimes a last-minute development would deliver him into the minister's hands, leaving the civil servant speechless with rage and indignation.

The script of *Yes, Minister*, was both perceptive and hugely funny, and the casting of the main roles was perfect. Paul Eddington was completely convincing as the gullible and idealistic Hacker, while Nigel Hawthorne was masterly as the machiavellian Sir Humphrey, assisted by Derek Fowlds as the genial Bernard Woolley. The show was an immediate success and was showered with numerous awards.

Among its many devotees were such distinguished figures as Margaret Thatcher, who named it as her favourite programme and saw to it that writer Antony Jay received a knighthood (Eddington and Hawthorne both got CBEs in the 1986 New Year's Honours list). Also connected with the programme, providing invaluable insights into the operations of Whitehall behind the scenes, was Harold Wilson's one-time secretary, Lady Marcia Falkender.

—David Pickering

CAST

Rt. Hon. James Hacker	Paul Eddington
Sir Humphrey Appleby	Nigel Hawthorne
Bernard Woolley	Derek Fowlds

PRODUCERS   Stuart Allen, Sydney Latterby, Peter Whitmore

PROGRAMMING HISTORY   37 Half-hour episodes; 1 Special

• BBC2

February 1980–April 1980	7 Episodes
February 1981–April 1981	7 Episodes
November 1982–December 1982	7 Episodes
17 December 1984	Christmas Special
January 1986–February 1986	8 Episodes
December 1987–January 1988	8 Episodes

See also British Programming

YOUNG, LORETTA

U.S. Actor

Loretta Young was one of the first Hollywood actors to move successfully from movies to a television series. She made that transition in 1953 with *Letter to Loretta* (soon retitled *The Loretta Young Show*), an anthology drama series. Anthology dramas were a staple of 1950s programming, presenting different stories with different characters and casts each week. Young hosted and produced the series, and acted in over half the episodes as well. Capitalizing on her glamorous movie star image, her designer fashions became her television trademark. The show's success spurred other similar series, but Young's was the most successful. She was one of the few women who had control of her own successful series, the first woman to have her own dramatic anthology series on network television, and the first person to win both an Academy Award and an Emmy Award.

Loretta Young began her acting career with bit parts as a child extra in silent films. By the mid-1930s, fashion and glamour were important components of her star image. By 1948, after more than twenty years in films, she was recognized for her acting when she won the Best Actress Academy Award for her performance in *The Farmer's Daughter*, a romantic comedy. In 1952, she made her last feature film and jumped eagerly into television. For older movie actors, television offered new opportunities and at forty Young was considered "older" when she began her series. Following her lead with prime-time anthology dramas were actors Jane Wyman, June Allyson, and Barbara Stanwyck.

As a movie star and as a woman, Young realistically had two options for a television series in 1953. CBS, the situation comedy network, home of Lucille Ball and *I Love Lucy*, suggested a sitcom. NBC offered an anthology drama. Not a zany comedienne like Ball or Martha Raye (who appeared in comedy-variety shows), Young went for the anthology drama. In doing so, she would follow film actor Robert Montgomery (*Robert Montgomery Presents*) to prime-time success as host and actor in her own dramatic anthology series. She wanted—and the anthology format afforded—acting variety, a format for conveying moral messages, and a showcase for her glamorous, fashionable movie star image. Though many anthology dramas were broadcast live,

Young—like most movie stars trying series TV—chose telefilm production, a mode that could bring future profit through syndication.

Young and husband Thomas Lewis (who was instrumental in setting up Armed Forces Radio during World War II and developed numerous radio programs) created Lewislor Enterprises to produce the series. Lewis initially served as executive producer, but left the show by the end of the third season. Young became sole executive producer. When her five-year contract with NBC was up, Young formed a new company, Toreto Enterprises, which produced the series' last three seasons.

Religious and moral questions had long concerned Young. Known for her religious faith and work on behalf of Catholic charities, the stories she selected for production in her series carried upbeat messages about family, community, and personal conviction, and every story was summed up with a quotation from the Bible or some other recognized source. Concerned about postwar changes in American society, Young advocated TV entertainment with a message. Scripts hinged on the resolution of moral dilemmas. Numerous civic and religious groups honored her for this. She also won three Emmys, the first in 1955 as best dramatic actress in a continuing series.

Fashion had also been an important component of Young's star image, and was central to her television program. Indeed, fashion may be the most memorable feature of *The Loretta Young Show*. Every episode opened with a swirling entrance that showcased her designer dresses, a move that became her television trademark. Many of the dresses she wore on the show were designed by Dan Werle, and some were marketed under the label Werle Originals. Young's strong feelings about fashion were publicized again in the early 1970s when she won a suit against NBC for allowing her then-dated fashion introductions to be shown in syndication. While this emphasis on fashion actually served Young's conviction that women had to maintain their femininity, as a star she epitomized a supposed paradox: she was beautiful and feminine, but she was also a strong-willed woman with a career.

While the star and her fashions often attracted reviewers, some complained that Young and her show were sentimental, low-brow women's entertainment, a typical criticism of women's fiction, where stories focus on the relationships and emotions comprising women's traditional sphere of home and family. The criticism was also typical of a 1950s conceit that filmed television series were inferior to prestigious live anthology dramas such as *Studio One* and *Philco Television Playhouse*.

Young's anecdotal and philosophical book, *The Things I Had To Learn*, was published in 1961, the same year her prime-time series went off the air. Her philosophies about life, success, and faith were the basis of the book, just as they had been for *The Loretta Young Show*.

She returned to series television in the 1962–63 season with *The New Loretta Young Show*, a situation comedy, and formed LYL Productions to produce the series. The story originally centered on her as a widowed writer-mother, but her

Loretta Young

character was married by the end of the season. This new series lasted only one season and Young did not return to television again until 1986, when she appeared in a made-for-TV movie, *Christmas Eve*. She won a Golden Globe Award for that performance. Her most recent television appearance was in another made-for-TV movie, *Lady in the Corner* (1989), in which she played the publisher of a fashion magazine.

Loretta Young is probably most important to television's history as a woman who blazed a path for other women as both an actor and a producer, who succeeded with her own prime-time show in a format that was not a situation comedy, and who was able to transfer success in film to success in television. Few film stars have made this transition.

—Madelyn Ritrosky-Winslow

LORETTA YOUNG. Born Gretchen Michaela Belzer in Salt Lake City, Utah, U.S.A., 6 January 1914. Attended Immaculate Heart College, Hollywood, California. Married: 1) Grant Withers, 1930 (divorced, 1931); child: Judy; 2) Thomas H.A. Lewis, 1940; children: Chistopher Paul and Peter. Debuted as an extra in *The Only Way*, 1919; contract with First National film company, late 1920s; contract with 20th Century-Fox, 1933–40; host and occasionally actor in anthology series, *The Loretta Young Show*, 1953–61; star of series *The New Loretta Young Show*, 1962–63. Recipient: Emmy Awards, 1955, 1956, 1959; Special Prize, Canne Film Festival; Academy Award, 1947; Golden Globe Award, 1986. Address: c/o Lewis, 1705 Ambassador Avenue, Beverly Hills, California 90210-2720, U.S.A.

TELEVISION SERIES

1953–61   *The Loretta Young Show*
1962–63   *The New Loretta Young Show*

MADE-FOR-TELEVISION MOVIES

1986   *Christmas Eve*
1989   *Lady in the Corner*

FILMS

The Only Way, 1919; *Sirens of the Sea,* 1919; *The Son of the Sheik,* 1921; *Naughty But Nice,* 1927; *Her Wild Oat,* 1928; *The Whip Woman,* 1928; *Laugh, Clown, Laugh,* 1928; *The Magnificent Flirt,* 1928; *The Head Man,* 1928; *Scarlett Seas,* 1928; *The Squall,* 1929; *The Girl in the Glass Cage,* 1929; *Fast Life,* 1929; *The Careless Age,* 1929; *The Show of Shows,* 1929; *The Forward Pass,* 1929; *The Man from Blankley's,* 1930; *The Second-Story Murder,* 1930; *Loose Ankles,* 1930; *Road to Paradise,* 1930; *Kismet,* 1930; *The Truth About Youth,* 1930; *The Devil to Pay,* 1930; *Bea Ideal,* 1931; *The Right of Way,* 1931; *Three Girls Lost,* 1931; *Too Young to Marry,* 1931; *Big Business Girl,* 1931; *I Like Your Nerve,* 1931; *Platinum Blonde,* 1931; *The Ruling Voice,* 1931; *Taxi,* 1932; *The Hatchet Man,* 1932; *Play Girl,* 1932; *Weekend Marriage,* 1932; *Life Begins,* 1932; *They Call It Sin,* 1932; *Employee's Entrance,* 1933; *Grand Slam,* 1933; *Zoo in Budapest,* 1933; *The Life of Jimmy Dolan,* 1933; *Midnight Mary,* 1933; *Heroes for Sale,* 1933; *The Devil's in Love,* 1933; *She Had to Say Yes,* 1933; *A Man's Castle,* 1933; *The House of Rothschild,* 1934; *Born to Be Bad,* 1934; *Bulldog Drummond Strikes Back,* 1934; *Caravan,* 1934; *The White Parade,* 1934; *Clive of India,* 1935; *Shanghai,* 1935; *Call of the Wild,* 1935; *The Crusades,* 1935; *The Unguarded Hour,* 1936; *Private Number,* 1936; *Ramona,* 1936; *Ladies in Love,* 1936; *Love Is News,* 1937; *Café Metropole,* 1937; *Love Under Fire,* 1937; *Wife, Doctor, and Nurse,* 1937; *Second Honeymoon,* 1937;

Four Men and a Prayer, 1938; *Three Blind Mice,* 1938; *Suez,* 1938; *Kentucky,* 1938; *Wife, Husband, Friend,* 1939; *The Story of Alexander Graham Bell,* 1939; *Eternally Yours,* 1939; *The Doctor Takes a Wife,* 1940; *The Lady from Cheyenne,* 1941; *The Men in Her Life,* 1941; *Bedtime Story,* 1942; *A Night to Remember,* 1943; *China,* 1943; *Ladies Courageous,* 1944; *And Now Tomorrow,* 1944; *Along Came Jones,* 1945; *The Stranger,* 1946; *The Perfect Marriage,* 1947; *The Farmer's Daughter,* 1947; *The Bishop's Wife,* 1947; *Rachel and the Stranger,* 1948; *The Accused,* 1949; *Mother Is a Freshman,* 1949; *Come to the Stable,* 1949; *Key to the City,* 1950; *Cause for Alarm,* 1951; *Half Angel,* 1951; *Paula,* 1952; *Because of You,* 1952; *It Happens Every Thursday,* 1953.

STAGE

An Evening with Loretta Young, 1989.

PUBLICATION

The Things I Had To Learn, as told to Helen Ferguson. Indianapolis, Indiana: Bobbs-Merrill, 1961.

FURTHER READING

Atkins, J. "Young, Loretta." In, Thomas, N., editor. *International Dictionary of Films and Filmmakers, Volume 3: Actors and Actresses.* Detroit, Michigan: St. James, 1992.

Bowers, R.L. "Loretta Young: Began as a Child-extra and Exuded Glamor for Forty Years." *Films in Review* (New York), 1969.

Morella, Joe, and Edward Z. Epstein. *Loretta Young: An Extraordinary Life.* New York: Delacorte, 1986.

Siegel, S., and B. Siegel. *The Encyclopedia of Hollywood.* New York: Facts on File, 1990.

See also Anthology Drama; Gender and Television; *Loretta Young Show;* Wyman, Jane

YOUNG, ROBERT

U.S. Actor

Robert Young came to television out of film and radio, and for nearly 30 years he was revered as television's quintessential father-figure. With his roles as Jim Anderson in the domestic melodrama *Father Knows Best,* and as the title character in the long-running medical drama, *Marcus Welby, M.D.,* he was admired as a strict, but benevolent patriarch. Gentle, moralistic, and highly interventionist, Young's television character corrected and guided errant behavior initially in a family setting, then as an omnipotent doctor, and perhaps most self-consciously, when he portrayed "himself" in a decade-long series of decaffeinated coffee commercials. With a simple raised eyebrow and a tilt of the head, Young's character convinced even the most hedonistic of co-stars to relinquish their selfish ways for a greater noble purpose.

Young began his career as a second lead in Hollywood films. Displaying a generally unrecognized versatility, Young portrayed villains, best buddies and victims with equal aplomb, and performed for many of Hollywood's finest directors, including Alfred Hitchcock, Frank Borzage, and Edward Dmytryk. Frustrated with his secondary status (he described his parts as those refused by Robert Montgomery), Young ventured into radio, in 1949, where with his good friend and business partner, Eugene Rodney, he co-produced and starred in a family comedy, *Father Knows Best?* Running for five years, the program was a soft-hearted look at a family in which the benevolent head of the family was regarded with love but skepticism and in which mother generally supplied the wisdom. At the time, most family comedies were characterized by wise-cracking moms and inept fathers. Young took the role on the condition that the father, in his words, not be "an idiot. Just make it so he's unaware. He's not running the ship, but he thinks he is."

In 1954, Young and Rodney were approached by Screen Gems to bring the program to television. While Young was hesitant at first, a promise of joint ownership in the program convinced him to make the move. Upon network insistence, the question mark was dropped (they thought it demeaning) and *Father Knows Best* premiered on CBS, under the sponsorship of Kent cigarettes. Because of advertising and network time-franchises, the program was placed too late in the evening to attract a family audience, and quickly died in the ratings. A fan-letter campaign and the personal intervention of Thomas McCabe, president of the Scott Paper Company, resurrected the program, which was to become an NBC staple for the next five years.

The television series was quite different from the radio version. Most significantly, radio's ambivalence about the father's wisdom was removed and replaced by an emphatic belief that Jim Anderson was the sole possessor of knowledge and child-rearing acumen. Although the original head writer, Roswell Rogers, remained with the program, most of the radio scripts had to be re-written or completely scrapped for the visual television medium. With the exception of Robert Young, the Anderson family was completely re-cast, with Jane Wyatt signing on after a year-long search. Many of the episodes were based on the real-life exploits of Young's daughter Kathy, while Wyatt was described as an amalgamation of the wives of Young, Rodney, and Rogers.

The program was heralded by the popular press and audiences alike as a refreshing change from "dumb Dad" shows. With near-irritating consistency, Jim Anderson resolved his family's dilemmas through a pattern of psychic intimidation, guilt and manipulation, causing the errant family member to recant his or her selfish desires, and put the good of the community, family and society ahead of personal pleasure. The wife and the three children, played by Elinor Donahue, Billy Gray, and Laurin Chapin, were lectured with equal severity by the highly exalted father, whose virtues were often the focus for episodic tribute.

The program won numerous awards, and spawned a host of domestic melodramas that were to dominate the television schedule (including *The Donna Reed Show,* and *Leave It to Beaver).* So popular was the program and so powerful its verisimilitude that viewers came to believe the Anderson family really existed. Women wrote to star Jane Wyatt with questions about cooking and advice about home decorating or child-rearing. Young was named Mt. Sinai "father of the year," and gathered similar honors throughout the series' run. In one of the stranger blends of fact and fiction, the producers were approached to do a U.S. Savings Bond benefit for the American Federation of Labor and the Treasury Department. "24 Hours in Tyrant Land" depicted the Anderson's fictional Springfield community caught in the clutches of a tyrannical despot. Never aired on television, the episode toured the country's town halls and churches.

Robert Young

By 1960, the personal difficulties of both Young and the teenage cast members, and the creative fatigue of Rogers, prompted the producers to cease first-run production, although re-runs continued to air in prime time on ABC for two more years.

Despite a couple of television films, Young's career was basically dormant during the 1960s until the highly acclaimed television movie, *Marcus Welby, M.D.* The pilot film, revolving around the heroic efforts of a kindly general practitioner and his "anti-establishment" young assistant (played by James Brolin), became a hit television series that was to air on ABC for the next seven years. Each phenomenally slow-moving episode, featured Welby, his partner Dr. Steven Kiley, and the friendly (but usually confused) nurse, Consuella, treating a single patient whose disease functioned as some sort of personal or familial catastrophe. Even for the 1970s, the program was anachronistic—Welby practiced out of his well-appointed Brentwood home, and both he and Kiley made housecalls. Significantly, the show did try to bring public attention to current health crises or recent medical discoveries. Thus, episodes dealt with Tay-Sach's disease, amniocenteses, abortion rights (when abortion was still illegal). With kindly didacticism, Welby would lecture the guest star (and the television viewer) on the importance of consistent medical care, early detection, immunization and the like.

By the mid-1970s, Young grew weary of the program, and this, coupled with Brolin's career ambitions, and a post-Watergate viewership hostile toward elderly male authority figures, contributed to the program's demise. With the end of the program, Young continued to work in television, starring in a couple of *Welby* movies, and a *Father Knows Best* reunion. He gained critical acclaim in a television film dealing with Alzheimer's disease and euthanasia.

His bitterness towards Hollywood casting practices never diminished however, and in the early 1990s Young attempted suicide, revealing a vulnerability and despair totally at odds with his carefully constructed patriarchal persona.

—Nina C. Leibman

ROBERT (GEORGE) YOUNG. Born in Chicago, Illinois, U.S.A., 22 February 1907. Attended Lincoln High School, Los Angeles. Married: Elizabeth Louise Henderson, 1933; children: Carol Anne, Barbara Queen, Elizabeth Louise, and Kathleen Joy. Earned living as clerk, salesman, reporter, and loan company collector during four years of studies and acting with the Pasadena Playhouse; toured with stock company production *The Ship*, 1931; contract with MGM, 1931–45; on radio program *Good News of 1938*, and on *Maxwell House Coffee Time*, 1944; co-founder, with Eugene Rodney, of Cavalier Productions, 1947; star of radio series *Father Knows Best?*, 1949–54; star of television version of same, 1954–61; star of *Marcus Welby, M.D.*, 1969–76. Recipient: Emmy Awards: 1956, 1957. Address: c/o Herb Tobias, 1901 Avenue of the Stars, Suite 840, Los Angeles, California 90067, U.S.A.

TELEVISION SERIES

1954–60	*Father Knows Best*
1961–62	*The Window On Main Street*
1969–76	*Marcus Welby, M.D.*
1979	*Little Women*

MADE-FOR-TELEVISION MOVIES

1969	*Marcus Welby, M.D.: A Matter of Humanities*
1971	*Vanished*
1972	*All My Darling Daughters*
1973	*My Darling Daughters' Anniversary*
1977	*The Father Knows Best Reunion*
1978	*Little Women*
1984	*The Return of Marcus Welby, M.D.*
1987	*Mercy or Murder?*
1989	*Conspiracy of Love*

FILMS

The Black Camel, 1931; *The Sin*, 1931; *The Guilty Generation*, 1931; *The Wet Parade*, 1931; *New Morals for Old*, 1932; *Unashamed*, 1932; *Strange Interlude*, 1932; *The Kid from Spain*, 1932; *Men Must Fight*, 1933; *Today We Live*, 1933; *Hell Below*, 1933; *Tugboat Annie*, 1933; *Saturday's Children*, 1933; *The Right to Romance*, 1933; *La Ciudad de Carton*, 1933; *Carolina*, 1934; *Spitfire*, 1934; *The House of Rothschild*, 1934; *Lazy River*, 1934; *Hollywood Party*, 1934; *Whom the Gods Destroy*, 1934; *Paris Interlude*, 1934; *Death On the Diamond*, 1934; *The Band Plays On*, 1934; *West Point of the Air*, 1935; *Vagabond Lady*, 1935; *Calm Yourself*, 1935; *Red Salute*, 1935; *Remember Last Night*, 1935; *The Bride Comes Home*, 1935; *Three Wise Guys*, 1936; *It's Love Again*, 1936; *The Bride Walks Out*, 1936; *Secret Agent*, 1936; *Sworn Enemy*, 1936; *The Longest Night*, 1936; *Stowaway*, 1936; *Dangerous Number*, 1937; *I Met Him in Paris*, 1937; *Married Before Breakfast*, 1937; *The Emperor's Candlesticks*, 1937; *The Bride Wore Red*, 1937; *Navy Blue and Gold*, 1937; *Paradise For Three*, 1938; *Josette*, 1938; *The Toy Wife*, 1938; *Three Comrades*, 1938; *Rich Man—Poor Girl*, 1938; *The Shining Hour*, 1938; *Honolulu*, 1939; *Bridal Suite*, 1939; *Miracles For Sale*, 1939; *Maisie*, 1939; *Northwest Passage*, 1940; *Florian*, 1940; *The Mortal Storm*, 1940; *Sporting Blood*, 1940; *Dr. Kildare's Crisis*, 1940; *The Trial of Mary Dugan*, 1941; *Lady Be Good*, 1941; *Unmarried Bachelor*, 1941; *H.M.Pulham, Esq.*, 1941; *Joe Smith—American*, 1942; *Cairo*, 1942; *Journey For Margaret*, 1942; *Slightly Dangerous*, 1943; *Claudia*, 1943; *Sweet Rosie O'Grady*, 1943; *The Canterville Ghost*, 1944; *The Enchanted Cottage*, 1945; *Those Endearing Young Charms*, 1945; *Lady Luck*, 1946; *The Searching Wind*, 1946; *Claudia and David*, 1946; *They Won't Believe Me*, 1947; *Crossfire*, 1947; *Relentless*, 1948; *Sitting Pretty*, 1948; *Adventure In Baltimore*, 1949; *Bride for Sale*, 1949; *That Forsyte Woman*, 1949; *And Baby Makes Three*, 1949; *The Second Woman*, 1951; *Goodbye, My Fancy*, 1951; *The Half Breed*, 1952; *Secret of the Incas*, 1954.

RADIO

Good News of 1938; *Father Knows Best?*, 1949–53.

PUBLICATION

"How I Won the War of the Sexes By Losing Every Battle." *Good Housekeeping* (New York), January 1962.

FURTHER READING

Parish, James Robert, and Gregory W. Mank. *The Hollywood Reliables*. Westport, Connecticut: Arlington House, 1980.

See also *Father Knows Best*; *Marcus Welby, M.D.*

YOUR HIT PARADE

U.S. Music Variety

Your Hit Parade was a weekly network television program that aired from 1950 to 1959. The program enjoyed some popularity but was never as successful as its radio predecessor which began in 1935 and ran for fifteen years before moving to television. Both the radio and television versions featured the most popular songs of the previous week as determined by a national "survey" of record and sheet music sales. The methodology behind this survey was never revealed but most audience members were willing to accept the tabulations without question. Both the TV and radio versions were sponsored by the American Tobacco Company's Lucky Strike cigarettes.

Original cast members for the TV program included Eileen Wilson, Snooky Lanson, Dorothy Collins and a wholesome array of young fresh-scrubbed "Hit Parade Singers and Dancers." Gisele MacKenzie joined the cast in 1953.

The TV version featured the top seven tunes of the week and several Lucky Strike extras. These extras were older, more established popular songs that were very familiar to audiences. The top seven tunes were presented in reverse order not unlike the various popular music countdowns currently heard on radio. The top three songs were presented with an extra flourish and audience members would speculate among themselves as to which tunes would climb to the top three positions and how long they would stay there.

The continuing popularity of certain songs over a multiple-week period had never been a problem for the radio version of the program with its Top Ten list. Regular listeners were willing to hear a repeat performance of last week's songs perhaps with a different vocalist than the previous week to provide variation. The television *Hit Parade* attempted to dramatize each song with innovative skits, elaborate sets, and a large entourage of performers. Creating new skits for longer running popular songs proved much more difficult on television, particularly when we recall such hits from the period as "How Much Is that Doggie in the Window," and "Shrimp Boats Are Coming."

A much more serious problem facing the program was the changing taste in American popular music. Rock 'n' roll was displacing the syrupy ballads that had been the mainstay of popular music during the 1930s and 1940s. The earlier music had a multi-generational appeal and the radio version of *Your Hit Parade* catered to a family audience. The rock music of the 1950s was clearly targeted to younger listeners and actually thrived on the disdain of its older critics.

Further, much of the popularity of the faster paced rock hits was dependent on complex instrumental arrangements and the unique styling of a particular artist or group. Rock music's first major star, the brooding, sensuous Elvis Presley, was a sharp contrast to the sedate styles of Snooky Lanson and Dorothy Collins. As rock (and Presley) gained in popularity, the ratings for *Your Hit Parade* plummeted. The cast was changed in 1957, the show temporarily canceled in 1958, but revived under new management with Dorothy Collins and Johnny Desmond in 1953. Despite these changes, the program was simply out of touch with the current musical scene and the last program was broadcast on 24 April 1959.

—Norman Felsenthal

ANNOUNCERS
Andre Baruch (1950–57)
Del Sharbutt (1957–58)

Your Hit Parade

VOCALISTS
Eileen Wilson (1950–52)
Snooky Lanson (1950–57)
Dorothy Collins (1950–57,1958–59)
Sue Bennett (1951–52)
June Valli (1952–53)
Russell Arms (1952–57)
Gisele MacKenzie (1953–57)
Tommy Leonetti (1957–58)
Jill Corey (1957–58)
Alan Copeland (1957–58)
Virginia Gibson (1957–58)
Johnny Desmond (1958–59)
Kelly Garrett (1974)
Chuck Woolery (1974)
Sheralee (1974)

DANCERS
The Hit Paraders (chorus and dancers) (1950–58)
Peter Gennaro Dancers (1958–59)
Tom Hansen Dancers (1974)

ORCHESTRA
Raymond Scott (1950–57)
Harry Sosnik (1958–59)
Milton Delugg (1974)

PRODUCERS   Dan Lounsberry, Ted Fetter

PROGRAMMING HISTORY

• NBC

July 1955–August 1950	Monday 9:00-9:30
October 1950–June 1958	Saturday 10:30-11:00

• CBS

October 1958–April 1959	Friday 7:30-8:00

August 1974   Friday 8:00-8:30

FURTHER READING

Williams, John R. *This Was Your Hit Parade.* Camden, Maine: n.p., 1973.

See also Music on Television

YOUTH TELEVISION

Canadian Youth Cable Network

Youth TV (YTV) is a Canadian cable television network aimed at young people up to the age of 18. Since its launch in September 1988, YTV has proven remarkably successful, far surpassing even its most optimistic economic and audience projections. An important part of YTV's success is predicated upon its ownership structure. Although investors include numerous producers with a long-standing commitment to children's television, over 50% of YTV is owned by just two cable firms, CUC Ltd. and Rogers Communications, the latter being Canada's largest cable operator. Their financial interest has helped make YTV available in the over 85% cabled homes in Canada. This high rate of cable penetration has in turn made YTV an attractive advertising vehicle for products and services aimed at a youth demographic.

Additionally, YTV has been able to insert itself into a traditional area of Canadian programming strength. Canadian production companies have long excelled at children's and young people's programming for three main reasons: (a) children's programming was relatively inexpensive, (b) it could easily be exported, and (c) it tended to be neglected by more powerful American production companies. As a result, YTV has been able to draw upon a considerable catalogue of Canadian children's programming and to provide opportunities for the expansion of this traditional expertise.

Finally, YTV has proven very successful in attracting its target audience. It engages in extensive polling of young people to determine their aspirations and concerns, buying patterns, political views, and to spot trends. As a result, YTV has crafted a schedule which mixes old, familiar shows with new, highly-targeted programs. YTV has therefore very rapidly emerged not only as a leading showcase but also an important producer of children's programming. It has produced or co-produced such shows as *Maniac Mansion, The Adventures of the Black Stallion, Deke Wilson's Mini-Mysteries*, and *StreeNOISE*, some of which have received wide international distribution. Indeed, YTV regularly exceeds its Canadian content production requirements by very wide margins.

YTV has also emerged as a socially conscious broadcaster which contributes to numerous charities and fund raisers (Children's Wish Foundation, Muscular Dystrophy, Kids Help Phone, etc.) and which provides educational grants. In its few years, YTV has already received several national and international awards for excellence in programming, for promoting international human rights, for aiding the cause of literacy, and for work in other areas of social concern.

Ironically, YTV's greatest problems have come not from the marketplace or from viewers but from the Canadian Radio-Television and Telecommunication Commission (CRTC). In its concern that YTV not appeal to audience members or age groups beyond its mandated audience, and thereby threaten the market of established broadcasters, the CRTC instituted the "protagonist clause" also known as the "Little Joe" rule. This clause requires that 100% of YTV's drama programming broadcast in the evening feature "a major protagonist that is a child, youth under the age of 18 years, puppet, animated character or creature of the animal kingdom."

The clause acquired its nickname when YTV discovered that Little Joe, a main character of *Bonanza* which it had purchased to strip in prime time, actually celebrated his 19th birthday in one of the early episodes. The CRTC ordered *Bonanza* off the air and YTV has since lobbied to have the clause removed or altered.

YTV complains that the protagonist clause prevents it from showing material which legitimately appeals to its

Courtesy of YTV

target audience—characters such as Superman, Batman, and Robin Hood, who are all well over 18—hockey superstar Wayne Gretzky, works of classic literature such as *Great Expectations* in which the hero starts as a child but grows past 18, the life stories of most musical groups, and so on. YTV claims that it is difficult to co-produce or sell internationally if a major protagonist must be "a puppet, animated character or creature of the animal kingdom."

YTV's efforts met with some success when the CRTC amended the protagonist clause in 1992 to include comic book characters, folk and superheroes, and classical or historical heroes. Nonetheless, YTV has generally managed to reach a loyal audience, to produce hundreds of hours of original content, and to ensure its financial success while also meeting public service and social responsibility objectives.

—Paul Attallah

Z

Z CARS

British Police Series

Z Cars was the innovative, long-running BBC police series of the 1960s, which has programmed more episodes (667) than any other weekly crime programme on British television. Created by Troy Kennedy-Martin and Elwyn Jones, and produced by David Rose, the series brought a new realism to the genre as it featured day-to-day policing in Newtown, a fictitious town to the north of Liverpool. At the spearhead of operations were four police constables: "Jock" Weir, "Fancy" Smith, Bob Steele and Bert Lynch. They occupied the two radio crime cars called Z-Victor 1 and Z-Victor 2, from which the series gained its title. Supervising operations via a VHF radio operator in the station, and securing prosecutions in the interrogation room, were Detective Sergeant Watt and the formidable Detective Inspector Barlow. Watched by nearly 14 million viewers in its first season, Z Cars rapidly captured the public imagination, and the leading characters became household names. Though in later seasons new characters might be brought in as replacements and the crime cars up-dated, the same basic formula applied. Bert Lynch, played by James Ellis, remained throughout the programme's run. Promoted to station sergeant in 1966 he was still in place at the desk when the doors were finally closed down on the cars in 1978.

In terms of programme aesthetics, Z Cars attempted to counter the film appeal of early U.S. cop programmes, such as Highway Patrol, with "gritty" realism. This was achieved by close attention to authentic police procedure, observation of working-class behaviour and, most especially, the adoption of regional speech. "Northern" working-class subject matter was prominent in 1960s culture, exemplified in feature films like Saturday Night and Sunday Morning and A Taste of Honey. However, Z Cars had more in common with the dialogue-led drama and actor-centred performances of ATV's Armchair Theatre and the early years of Granada's Coronation Street. Though later series were able to make more use of film and locations, the look of Z Cars was constructed almost entirely in the television studio. The 50 minutes of continuous recorded performance provided the space for displays of male comradeship and teamwork, sharp verbal exchanges with members of the community, and, most characteristic of all, intense drama in the interrogation room as Barlow bullied and coaxed confessions from his suspects.

Overall, Z Cars succeeded in presenting a more human and "down to earth" image of the police than had been previously created on British television. Major crime remained at the periphery of the series and the emphasis was placed instead on domestic and juvenile crime. The programme adopted the social-democratic view of society so prevalent in 1960s Britain, and at times the PCs behaved more like social workers than policemen, as criminal behaviour was explained in terms of social deprivation. The liberal approach, however, was showing signs of exhaustion. Barlow upheld the law with a fierce authoritarianism in the station, and the PCs needed all their ingenuity and skill to enforce it effectively in the community. An on-going theme is the personal cost of securing law and order, and most of the police characters have unsatisfactory family relationships. In one episode, for instance, Watt was shown agreeing to a divorce and in another Steele beats up his wife. The image of policemen as fallible human beings created some controversy and for a time the chief inspector of Lancashire withdrew his support from the programme, apprehensive that it might undermine public confidence in the police.

In the course of its long run the programme established the reputations of many production participants, including actors such as Stratford Johns, Frank Windsor, Colin Welland, Brian Blessed and James Ellis, producers and directors such as Shaun Sutton, David Rose and John McGrath, and writers such as Troy Kennedy-Martin, John Hopkins, Alan Plater and Allan Prior. Z Cars has been a major influence on the course of TV police fiction in Britain. The long-running C.I.D. series Softly Softly (1966–75) was a direct spin-off from it, achieved by promoting Barlow to the rank of chief inspector, transferring him to a regional crime squad and replacing the squad car with a dog-handling unit. Recent British programmes about community policing as different as The Bill and Heartbeat continue to draw from the Z Cars idea. One of the most interesting reworkings of the programme's basic format was BBC's Juliet Bravo (1980–88) which, in keeping with 1980s gender politics, transferred the power from male C.I.D. officers to a uniformed female inspector.

—Bob Millington

CAST

Charlie Barlow	Stratford Johns
John Watt	Frank Windsor
Bert Lynch	James Ellis
Fancy Smith	Brian Blessed

Z Cars
Photo courtesy of the British Film Institute

Jock Weir	Joseph Brady
Bob Steele	Jeremy Kemp
Sgt. Twentyman	Leonard Williams
Ian Sweet	Terence Edmond
Insp. Dunn	Dudley Foster
David Graham	Colin Welland
Sgt. Blackitt	Robert Keegan
Sally Clarkson	Diane Aubrey
Insp. Bamber	Leonard Rossiter
PC Robbins	John Philips
Insp. Millar	Leslie Sands
Ken Baker	Geoffrey Whitehead
Arthur Boyle	Edward Kelsey
PC Foster	Donald Webster
PC Boland	Michael Grover
Ray Walker	Donald Gee
Sam Hudson	John Barrie
Tom Stone	John Slater
Steve Tate	Sebastian Breaks
Alec May	Stephen Yardley
Owen Calshaw	David Daker

Jane Shepherd	Luanshya Greer
Insp. Brogan	George Sewell
PC Newcombe	Bernard Holley
Insp. Todd	Joss Ackland
PC Jackson	John Wreford
Insp. Witty	John Woodvine
PC Roach	Ron Davies
PC Bannerman	Paul Angelis
Insp. Goss	Derek Waring
Joe Skinner	Ian Cullen
Mick Quilley	Douglas Fielding
PC Culshaw	John Challis
Sgt. Moffat	Ray Lonnen
Jill Howarth	Stephanie Turner
PC Covill	Jack Carr
PC Lindsay	James Walsh
PC Scatliff	Geoffrey Hayes
PC Render	Alan O'Keefe
PC Hicks	Godfrey James
PC Logie	Kenton Moore
PC Birch	John Woodnutt

Sgt. Hagger	John Collin
WPC Cameron	Sharon Duce
Insp. Connor	Gary Watson
PC Yates	Nicholas Smith
WPC Bayliss	Alison Steadman
DC Braithwaite	David Jackson
Sgt. Knell	John Dunn-Hill
PC Preston	Michael Stirrup
Sgt. Chubb	Paul Stewart
DC Bowker	Brian Grellis
Insp. Maddan	Tommy Boyle
WPC Beck	Victoria Plucknett

PRODUCERS David Rose, Colin Morris, Ronald Travers, Richard Benyon, Ron Craddock, Roderick Graham

PROGRAMMING HISTORY 291 50-minute Episodes; 376 25-minute Episodes

• BBC

January 1962–July 1962	31 Episodes
September 1962–July 1963	42 Episodes
September 1963–June 1964	42 Episodes
September 1964–June 1965	43 Episodes
October 1965–December 1965	12 Episodes
March 1967–April 1971	334 Episodes
August 1971–March 1972	28 Episodes of 25 Minutes
	1 Episode of 50 Minutes
April 1972–August 1972	14 Episodes of 25 Minutes
	11 Episodes of 50 Minutes
September 1972–July 1973	40 Episodes
October 1973–June 1974	28 Episodes
September 1974–May 1975	31 Episodes
January 1976–March 1976	12 Episodes
April 1977–July 1977	13 Episodes
June 1978–September 1978	13 Episodes

FURTHER READING

"Allen Prior and John Hopkins Talking About the *Z Cars* Series." *Screen Education* (London), September-October 1963.

Casey, A. "Blood Without Thunder." *Screen Education* (London), September-October 1962.

Hurd, Geoffrey. "The Television Presentation of the Police." In, Bennet, Tony, with others, editors. *Popular Television and Film.* London: British Film Insitute, 1981.

Kennedy-Martin, T. *Z Cars.* London: May Fair Books, 1962; Severn House, 1975.

Laing, Stuart. "Banging in Some Reality: The Original *Z Cars.*" In, Corner, John, editor. *Popular Television in Britain, Studies in Cultural History.* London: British Film Institute, 1991.

Vahimagi, Tise, editor. *British Television: An Illustrated Guide,* London: Oxford, 1994.

"*Z Cars* and their Impact: A Conference Report." *Screen Education* (London), September-October 1963.

See also British Programming; Police Programs; Welland, Colin; Windsor, Frank

ZAPPING

Zapping is the use of a remote control device (RCD) to avoid commercials by switching to another channel. The process is often paired with "zipping," fast-forwarding through the commercials in recorded programs. Although zapping and zipping have received much attention recently, viewers have always avoided commercials by changing channels, leaving the viewing area or simply shifting their attention away from the set. But as the penetration of RCDs increased to about 90% and videocassette recorders (VCRs) to over 77% of U.S. households by 1993, advertiser concern over zapping and zipping has accelerated. RCDs and VCRs, combined with a multitude of viewing options on cable and home satellite systems, have led to the zapping or zipping of 10 to 20% of all commercials, according to some industry studies. Cable networks specializing in short form programming (music videos, news stories, comedy shorts) are well suited to filling commercial breaks. Thus, the once "captive" audience of television is exercising its option to zap or zip boring or annoying commercials. Indeed, several studies of RCD gratifications have consistently identified commercial avoidance as a major motivation to use remote control devices.

In the 1980s, RCDs and VCRs proliferated, while the advertising and television industries debated the relative impact of zapping and zipping. Advertisers argued that program ratings did not reflect decreasing audience attention to commercials, while broadcasters cited studies that minimized the increase in channel changing during commercials. Several studies showed that the content of a commercial greatly affected the degree of zapping, encouraging many advertisers to restructure their television commercials by focusing on more entertaining content, fast-paced editing, or high quality special effects. When research showed that commercials placed during sports programming were particularly susceptible to zapping, some advertisers responded with commercials that combined both program and advertising elements. For example, IBM's "you make the call" commercials inserted an advertising message between question and answer segments of a sports quiz. Advertisers also tried to thwart the RCD's impact through more careful audience targeting and by reducing the length of some commercials. As the decade wore on, advertisers increased their use of place-based advertising and integrated marketing to replace the ad exposures lost to zapping and zipping.

Although some observers see RCD enhanced zapping as a modest intensification of the television audience's long standing urge to avoid bad commercials, others have argued that zapping will lead to gradual structural changes in the commercial television industry. Refinements in RCDs and VCRs may make zapping and zipping even easier. Thus, as these two sources of commercial avoidance decrease the value of commercially sponsored programming, advertisers may continue to shift resources to other advertising media and marketing approaches, or begin to offer compensation to viewers for simply watching commercials. Program providers may need to seek other revenue streams such as pay-per-view and subscriber fees to replace the lost revenue from advertisers. The result of these structural changes may be fewer viewing options for those unable or unwilling to pay these new charges and a wider gap between the information and entertainment haves and have not's.

—James R. Walker

FURTHER READING

Ainslie, Peter. "Commercial Zapping: TV Tapers Strike Back; VCR Owners Are Skipping Station Breaks, and Advertisers Are Getting Worried." *Rolling Stone* (New York), 28 February 1985.

Bellamy, Jr., R.V., and J.R. Walker. *Grazing on a Vast Wasteland: The Remote Control and Television's Second Generation.* New York: Guilford, 1995.

Walker, J.R., and R.V. Bellamy, Jr., editors. *The Remote Control in the New Age of Television.* Westport, Connecticut: Praeger, 1993.

See also Remote Conrol Device

ZIV TELEVISION PROGRAMS, INC.

U.S. Production and Syndication Company

As the most prolific producer of programming for the first-run syndication market during the 1950s, Ziv Television Programs occupies a unique niche in the history of U.S. television. Bypassing the networks and major national sponsors, Ziv rose to prominence by marketing its series to local and regional sponsors, who placed them on local stations, generally in time slots outside of prime time. Using this strategy, Ziv produced several popular and long-lived series, including *The Cisco Kid* (1949–56), *Highway Patrol* (1955-59), and *Sea Hunt* (1957–61).

Frederick W. Ziv, the company's founder, was born in Cincinnati, Ohio, in 1905. The son of immigrant parents, he attended the University of Michigan, where he graduated with a degree in law. Returning to his native Cincinnati, Ziv chose not to practice the legal profession, but instead opened his own advertising agency. His corporate strategies and his vision of the broadcasting business developed from this early experience in the Midwest.

During the radio era, Cincinnati was a surprisingly active regional center for radio production. Clear-channel station WLW, owned by the local Crosley electronics firm, broadcast a powerful signal that could be heard over much of the Midwest. Due to its regional influence, WLW became a major source of radio programming that offered local stations an alternative to network-originated programming. Cincinnati was also home to Procter and Gamble, the most influential advertiser in the radio industry at a time when most radio programming was produced by sponsors. Consequently, Procter and Gamble was directly responsible for developing many of radio's most lasting genres, including the soap opera.

Ziv's small advertising agency gained valuable experience in this fertile regional market. Ziv produced several programs for WLW, where he met John L. Sinn, a writer who would become his right-hand man. In 1937, the two men launched the Frederick W. Ziv Company into the business of program syndication. From his experience in a regional market, Ziv recognized that local and regional advertisers could not compete with national-brand sponsors because they could not afford the budget to produce network-quality programs. In an era dominated by live broadcasts, Ziv produced pre-recorded programs, "transcriptions" recorded onto acetate discs, bypassing the networks and selling his programs directly to local advertisers on a market-by-market basis. Programs were priced according to the size of each market; this gave local sponsors a chance to break into radio with affordable quality programming that could be scheduled in any available slot on a station's schedule.

Ziv produced a wide range of programming for radio, including sports, music, talk shows, soap operas, anthology dramas, and action-adventure series such as *Boston Blackie, Philo Vance,* and *The Cisco Kid.* By 1948, he was the largest packager and syndicator of radio programs—the primary source of programming outside the networks.

In 1948, Ziv branched into the television market by creating the subsidiary, Ziv Television Programs. His fortunes in television were entirely tied to the market for first-run syndication, which grew enormously during the first half of the 1950s before going into a steep decline by the end of the decade. In the early years of U.S. television, local stations needed programming to fill the time slots outside of prime time that were not supplied by the networks. More importantly, local and regional sponsors needed opportunities to advertise their products on television. As in radio, Ziv supplied this market with inexpensive, pre-recorded programs that could be scheduled on a flexible basis. In 1948, the first Ziv series, *Yesterday's Newsreel* and

Frederick W. Ziv (right)
Photo courtesy of Wisconsin Center for Film and Theater Research

Sports Album, featured 15-minute episodes of repackaged film footage.

In 1949, Ziv branched into original programming with his first dramatic series, *The Cisco Kid,* starring Duncan Renaldo as the Cisco Kid and Leo Carillo as his sidekick, Pancho. Ziv's awareness of the long-term value of filmed programming was signaled by his decision to shoot *The Cisco Kid* in color several years before color television sets were even available. *The Cisco Kid* remained in production until 1956, but its 156 episodes had an extraordinarily long life span in syndication thanks to the decision to shoot in color. In its first decade of syndication, the series grossed $11 million.

Ziv produced more than 25 different series during the 1950s, all of which were half-hour dramas based on familiar male-oriented, action-adventure genres. His output included science-fiction series such as *Science Fiction Theater* (1955–57), *Men into Space* (1959–60), and *The Man and the Challenge* (1959–60), westerns such as *Tombstone Terri-*

tory (1957–60), *Rough Riders* (1958–59), and *Bat Masterson* (1958–61), and courtroom dramas such as *Mr. District Attorney* (1954–55) and *Lockup* (1959–61).

In order to carve out a unique market niche, Ziv tried to spin variations on these familiar genres. In the crime genre, for instance, he produced few series that could be considered typical cop shows. His most notorious crime series, *I Led Three Lives* (1953–56), featured Richard Carlson as Herbert Philbrick, an undercover FBI agent sent to infiltrate Communist organizations throughout the United States. While the major networks generally avoided the subject of the Red Scare, preferring to blacklist writers and performers while barely alluding to the perceived Communist threat in their programming, Ziv attacked the issue with an ultra-conservative zeal. By organizing the series around Philbrick's fight against the menace of Communism, the series implied that Communism was every bit as threatening and ubiquitous as urban crime.

Another crime series, *Highway Patrol*, starring Broderick Crawford, moved the police out of the familiar urban landscape, placing them instead on an endless highway—an important symbolic shift in a postwar America obsessed with automobile travel as a symbol of social mobility. *Sea Hunt*, which was produced for Ziv by Ivan Tors (who would go on to produce *Flipper* and *Daktari*), took the crime series onto the sea, where star Lloyd Bridges as Mike Nelson solved crimes and found adventure under the ocean's surface. The underseas footage added a touch of low-budget spectacle to the crime genre.

The market for first-run syndication swelled through the mid-1950s, and Ziv rode the wave with great success. The watchword for Ziv productions was economy, and the company even formed a subsidiary called Economee TV in 1954. Production budgets were held to $20,000 to $40,000 per episode, which were generally shot in two to three days. As the demand for syndicated programming grew, Ziv expanded rapidly. In 1953, Ziv opened an international division to sell its series overseas. The operation proved to be such a success in England that Ziv found itself with revenues frozen by protectionist British legislation designed to force American companies to spend their profits in Great Britain. In order to make use of these frozen funds, in 1956-57, Ziv produced two series in England: *The New Adventures of Martin Kane* and *Dial 999*.

With production at the studio booming, Ziv stopped leasing space from other studios, and purchased its own Hollywood studio in 1954. By 1955, the company's annual revenues were nearly doubling every year. Ziv was then producing more than 250 half-hour TV episodes annually, with a production budget that exceeded $6 million—a figure that surpassed virtually every other television producer in Hollywood.

But the tide was turning in the market for first-run syndication. By 1956, the networks had begun to syndicate reruns of their older prime-time programs. Since these off-network reruns—with their established audience appeal—had already earned money during the initial run in prime time, networks were able to sell them to local markets at deep discounts. As a consequence, the market for first-run syndication began to shrink dramatically. In 1956, there were still 29 first-run syndicated series on television, with the number dropping to ten by 1960. By 1964, there was only one such series left on the air.

As the networks extended their influence beyond prime time and the market for first-run syndication dwindled, Ziv began to produce series specifically for network use—a decision that the company had actively avoided for over two

decades. Ziv's first network series was *West Point* (1956-57) for CBS, followed by four other network programs: *Tombstone Territory*, *Bat Masterson*, *Men into Space*, and *The Man and the Challenge*.

In 1959, Ziv elected to sell 80% of his company to an alliance of Wall Street investment firms for $14 million. "I sold my business," he explained, "because I recognized the networks were taking command of everything and were permitting independent producers no room at all. The networks demanded a percentage of your profits, they demanded script approval and cast approval. You were just doing whatever the networks asked you to do. And that was not my type of operation. I didn't care to become an employee of the networks."

In 1960, United Artists purchased Ziv Television Programs, including the 20% share still held by chair of the board, Frederick Ziv, and president, John L. Sinn, for $20 million. The newly merged production company was renamed Ziv-United Artists. United Artists had never been very successful in television, having placed only two series in prime time, *The Troubleshooters* (1959–60) and *The Dennis O'Keefe Show* (1959–60). This pattern continued after the merger. Ziv-UA produced 12 pilots during the first year and failed to sell any of them. In 1962, the company phased out Ziv Television operations and changed its name to United Artists Television. Frederick Ziv left the board of directors at this time to return to Cincinnati, where he spent his retirement years.

—Christopher Anderson

FURTHER READING

Balio, Tino. *United Artists: The Company that Changed the Film Industry*. Madison: University of Wisconsin Press, 1987.

Boddy, William. *Fifties Television: The Industry and Its Critics*. Urbana: University of Illinois Press, 1990.

Moore, Barbara. "The Cisco Kid and Friends: The Syndication of Television Series from 1948 to 1952." *The Journal of Popular Film and Television* (Washington, D.C.), Spring 1980.

Rouse, Morleen Getz. *A History of the F. W. Ziv Radio and Television Syndication Companies, 1930-1960*. (Ph.D. dissertation, University of Michigan, 1976).

See also Syndication

ZNAIMER, MOSES

Canadian Media Producer/Executive

Moses Znaimer, an internationally known Canadian broadcaster and producer, is the executive producer and president of Citytv, one of Canada's leading commercial media production organizations. There he guides program services such as MuchMusic, Bravo!, and MusiquePlus.

Znaimer's work in forging a distinctive style of television within Canada, and internationally, identifies him as a clear *auteur* within television production and he can rightfully claim that he is the visionary of Canadian television. His early work in broadcasting was as a co-creator and producer of the CBC national radio program, *Cross-Country Check-up* in the 1960s (a first in the world), and in television as a co-host and producer of the CBC afternoon talk show *Take-Thirty* with Adrienne Clarkson. After being denied the opportunity to remake the radio phone-in program into a national television program, Znaimer quit the CBC and launched into private broadcasting. With no VHF licenses available, Znaimer began Toronto's first UHF station, Channel 57, known as Citytv, on a limited budget in offices on Queen Street in Toronto in 1972. The unique programming of Citytv has been Znaimer's central contribution to the world of broadcasting. The station originally created a sensation in the 1970s for its late-night, soft-core porn movie stripping, Baby Blue Movies which shocked Toronto. But, its inner-city focus, its celebration of a cosmopolitan ethnic diversity in its choice of personalities and reporters, its transformation of news into something that was decidedly less formal, more identifiably urban and generally more positive, and its programming mix of just news, movies and music all clearly made the station distinctive. Indeed, Znaimer and his small UHF station served as the real-life starting point for David Cronenberg's dystopic film *Videodrome* (1983).

Through the platform of Citytv, Znaimer has successfully produced a number of programs, many of which have gained national and international distribution. *The New Music* (1978–), designed as a *Rolling Stone*-style magazine of the air, was widely sold within Canada and internationally. More recently, Znaimer has broadcast and distributed two fashion related programs, *Fashion Television* and *Ooh-La-La*, both nationally and internationally. *Movie Television*, an interview and news program about Hollywood in particular, has also been well syndicated throughout Canada's independent stations. The success of Citytv under Znaimer's direction allowed the company that bought the station in 1981, CHUM Limited, to launch Canada's first satellite to cable music specialty channel MuchMusic. What was clear about the look of MuchMusic was that it emulated Citytv. Its style was irreverent, its use of hand-held cameras at often canted angles was unending, its dependence on the liveness of television and its possibility for spontaneity and its transformation of the studio "backstage" into the foreground were signatures of Znaimer's work as executive producer.

Znaimer has contributed specific forms of television which celebrate the potential spontaneity of the medium. His Toronto ChumCity building (1987), the home of Citytv, MuchMusic and Bravo! is described as the first studio-less television station. With complete cabling and wiring through 35 exposed "hydrants", any part of the building can be converted into an exhibition site for broadcast. Several conceptual approaches to television have been registered trademarks developed by Znaimer. The building itself is trademarked as the "Streetfront Studio-less Television Operating System" and is marketed internationally. The Vox populi box at the front of the building is trademarked "The Speaker's Corner," where anyone by dropping

Moses Znaimer
Photo courtesy of Moses Znaimer/Citytv

a dollar into the slot can speak on any issue and the message will be broadcast.

Recent ventures of Znaimer, both nationally and internationally, have met with more circumscribed success. His involvement with a 1992 bid to set up a similar inner-city style of television for Britain (along with Thames Television and Time-Warner) for the proposed Channel Five was in the end not accepted. His recent launch of another specialty channel, Bravo!, which rebroadcasts past Canadian television programs and films, has had limited appeal and financial viability. Znaimer was involved in setting up a third television network in New Zealand, which once again built on his tried programming flow strategies developed at Citytv. His launch of a Spanish version of MuchMusic, MuchMusica, in Buenos Aires, Argentina, in 1994 has gained access to over 1.5 million via cable and thousands of others via satellite in South America. The launch of MuchMusic into the U.S. cable market in 1994 has also produced access to a further 4 million viewers.

Znaimer's versatility within the arts has occasionally led to on-camera performances. He has been an on-and-off actor over the last two decades with film credits including *Atlantic City* (1980) and, more regularly, an on-air narrator/interviewer in a number of programs, most notably *The Originals*. His most recent-large scale production for the CBC is a clear acknowledgment of his role in pioneering a unique style of television.

A four-part series entitled *TVTV: The Television Revolution* (1995) was hosted and produced by Znaimer.

Znaimer's style of television represents a unique contribution to broadcasting. He has developed a localized style with up to 40 hours a week of local content that because of its connection to the particular urban landscape has gained a certain resonance and exportability to other urbanized cultures. In addition, Znaimer has emphasised the concept of the flow of television in various formats. Rather than a focus on narrative conclusion, Znaimer's programming style identifies how television can attempt to capture—however partially—the becoming aspect of contemporary life. He has been able to achieve this vision of interactive, urban, hip television through repeated financial success in Toronto, generally recognized as one of the most competitive television markets in North America. The apparent cost of his studio-less studio is roughly one-quarter that of regular television stations. Portions of this style have been copied throughout North American television and to a lesser degree internationally.

—P. David Marshall

MOSES ZNAIMER. Born in Kulab, Tajikistan, 1942; family fled to Shanghai, arrived in Canada in 1948 and settled in Montreal. Educated at McGill University, Montreal, B.A. in philosophy and politics; Harvard University, M.A. in government. Joined the Canadian Broadcasting Corporation as radio and TV producer/director/host of several shows, from 1965 to 1969; vice president, T'ang Management Ltd. and Helix Investments; co-founder, president, chief executive officer and executive officer, Citytv, 1972, MuchMusic, 1984, Musique Plus, 1986, and Bravo!, 1995.

TELEVISION SPECIAL

1995   *TVTV: The Television Revolution*

RADIO

Cross-Country Checkup (co-producer), 1960s.

FURTHER READING

"Access Boys Ready for Opposition to Moses Znaimer's Access Network Deal with Alberta Government." *Calgary* (Canada) *Herald*, 12 March 1995.

"The Gospel According to Moses: the Bad Boy of Canadian Broadcasting." *Maclean's* (Toronto), 8 May 1995.

"Looking for Meaning in TV's 'Flow'." *Globe and Mail* (Toronto), 3 May 1995.

"The Masque of Moses Znaimer's Medium—Drainie." *Globe and Mail* (Toronto, Canada), 13 April 1995.

"Moses Disposes Formal Launch of Bravo!" *Globe and Mail* (Toronto), 28 March 1995.

Robins, Max "Toronto's Citytv to Export its Savvy." *Variety* (Los Angeles), 26 July 1993.

"Wholly Moses: Znaimer Takes on TV." *Montreal* (Quebec) *Gazette*, 7 April 1995.

See also Canadian Programming in English; Citytv; MuchMusic

ZORRO

U.S. Western

The television version of *Zorro*, like its previous movie incarnations, was based on stories written by Johnston McCulley. These stories recounted exploits of the swashbuckling alter-ego of Don Diego de la Vega in colonial California.

The most popular and recognizable TV version of *Zorro* was the Disney Studios production for ABC. The two organizations had entered into a joint production agreement in 1954, an agreement which bore immediate fruit with *Disneyland* and *The Mickey Mouse Club*. Walt Disney had purchased the rights to the Zorro stories in the early 1950s but pilot production stalled while Walt focused on construction of his Disneyland theme park. *Zorro* went into production in 1957 and enjoyed immense popularity on ABC for two years, from October 1957 to September 1959.

Guy Williams played Zorro, the mysterious hero who righted wrongs perpetrated on the common people by the evil Captain Monastario (Britt Lomond), commandant of the Fortress de Los Angeles. Don Diego's father, Don Alejandro (George J. Lewis), persuaded his son to return to California from Spain and do his utmost to foil Monastario and his dimwitted underling, Sergeant Garcia (Henry Calvin). Zorro's true identity was known only to his deaf-mute servant Bernardo (Gene Sheldon). Depending on the situation Zorro rode one of two trusty mounts, one black (Tornado) and one white (Phantom). Each episode began with Zorro sticking a message on the Commandant's door, "My sword is a flame to right every wrong, so heed well my name—Zorro."

Though it used almost all Caucasian actors, the story of Zorro stands out in the television landscape of 1957 for featuring an Hispanic hero figure. Roles and role models for Hispanic-Americans were absent from the television productions of the era and this acknowledgement of the Hispanic culture and the heroism of many of its constituents was considered a forward step.

Yet the characters were broadly drawn and often stereotypical. The conflict in *Zorro* was a simple distillation: a decadent, militaristic monarchy which exercised a corrupt, greedy rule over simple, God-loving folk versus the mysterious, altruistic defender of honesty and virtue. The archetypal characters of Monastario, Garcia, and Zorro provided easy markers of good and evil for the children of *Zorro's* target audience. Evil was effeminate, devious, slovenly, and doltish. Good was decisive and (in the words of another Disney Studios product), "brave, truthful, and unselfish." Even as the prime-time western genre was approaching the end of its cycle by reinventing itself as "adult," the western genre for children remained a comfortable and predictable haven of values championed by Walt Disney and, in turn, the middle class.

By the late 1950s and early 1960s, the relationship between ABC and the Disney Studios had soured. *The Mickey Mouse Club* was dropped after its fourth season. Though the network claimed this was due to flagging sponsorship, Walt Disney believed it was because of excessive commercial minutes. *Zorro*, still quite popular, was also cancelled. ABC now owned enough shows to make the purchase of programs from independent producers less necessary. To make matter worse, ABC forbade the Disney Studios from selling its product to a competing network, and while legal wrestling changed that restriction, it was clear that the Disney Studios had become a casualty of the fledgling network's success.

But *Zorro* also serves as an early example of what can happen to the popularity of a show when it is extensively merchandised. Because it was a Disney Studios product, *Zorro* had the benefit of the studio's massive merchandising machinery. During the run of the show, and for many years thereafter, *Zorro* spawned a huge number of items—hats, knives, masks, capes, pencil and lunch boxes—sold with the *Zorro* logo. The original theme was recorded for the opening of the show by Henry Calvin, who played Sergeant Garcia, and made into a popular hit record by the musical group called the Chordettes. During the two years that *Zorro* ran on ABC, the Disney merchandising juggernaut generated millions in additional income and kept the profile of the program high, especially with children. Even years after the popularity of the Disney Studios and ABC's *Zorro* had waned, the merchandising continued. When *Zorro* became a children's cartoon in the 1970s, a PEZ dispenser capped with Zorro's masked visage enjoyed healthy sales.

In some ways, *Zorro* serves as a model for much that is right and much that is wrong with children's television. It often propounded positive values and altruistic behavior, but it was ultimately one of the first of a long line of productions used solely to deliver a huge number of children to advertisers.

The image of Zorro remains prevalent today. From McCulley's original stories, through the movie with Tyrone Power and the serial with Clayton Moore, the Disney version for ABC, the Saturday morning cartoon, and the cable remake on the Family Channel in 1988, Zorro still has appeal. Even today, colorized versions of the original black-and-white episodes shot by Disney are cablecast on the Disney Channel, introducing the next wave of children to "a horseman known as Zorro."

—John Cooper

Zorro
Photo courtesy of the Walt Disney Company

CAST

Don Diego de la Vega ("Zorro")	Guy Williams
Don Alejandro	George J. Lewis
Bernardo	Gene Sheldon
Captain Monastario	Britt Lomond
Sergeant Garcia	Henry Calvin
Nacho Torres	Jan Arvan
Elena Torres	Eugenia Paul
Magistrate Galindo	Vinton Hayworth
Anna Maria Verdugo (1958-1959)	Jolene Brand
Senor Gregorio Verdugo (1958-1959) . . .	Eduard Franz
Corporal Reyes (1958-1959)	Don Diamond

PRODUCERS   Walt Disney, William H. Anderson

PROGRAMMING HISTORY
• ABC
October 1957–September 1959   Thursday 8:00-8:30

FURTHER READING

Hollis, Richard. *The Disney Studio Story*. London: Octopus Books, 1989.

Schnider, Cy. *Children's Television*. Chicago: NTC Books, 1987.

West, Richard. *Television Westerns: Major and Minor Series, 1946–1978*. Jefferson, North Carolina: McFarland, 1987.

See also Walt Disney Programs; Westerns

ZWORYKIN, VLADIMIR

U.S. Inventor

For his fundamental and crucial work in creating the iconoscope and the kinescope, inventor Vladimir Zworykin is often described as "the father of television". These basic technologies revolutionized television and led to the worldwide adoption of electronic television rather than mechanical television, a device which used synchronized moving parts to generate rudimentary pictures.

At the Petersburg Institute of Technology, Zworykin studied electrical engineering with Boris Rosing, who believed cathode ray tubes would be useful in television's development because they could shoot a steady stream of charged particles. After graduating from St. Petersburg in 1912, he studied X-ray technology with well-known French physicist Paul Langevin at the College de France in Paris. Both experiences influenced Zworykin's later work after he emigrated to the United States in 1919.

In 1920 Zworykin joined Westinghouse to work on the development of radio tubes and photocells. While there, he earned his Ph.D. in physics at the University of Pittsburgh and wrote his dissertation on improving photoelectric cells. But electronic television's development captured his attention, and in December 1923 he applied for a patent for the iconoscope, which produced pictures by scanning images. Within the year he applied for a patent for the kinescope, which reproduced those scanned images on a picture tube. Electronic television was now possible. After demonstrating his new system to Westinghouse executives, they decided not to pursue his research.

He found a more receptive audience in 1929 at the Radio Corporation of America (RCA), where he was hired as associate research director for RCA's electronic research laboratory in Camden, New Jersey. This same year, he filed his first patent for color television. Reportedly, Zworykin told RCA president David Sarnoff that it would take $100,000 to perfect television. Sarnoff later told the *New York Times*, "RCA spent $50 million before we ever got a penny back from TV."

In 1930, Zworykin's experiments with G.A. Morton on infrared rays led to the development of night-seeing devices. He also began to apply television technology to microscopy, which led to RCA's development of the electron microscope. His work also led to text readers, electric eyes used in security systems and garage door openers, and electronically-controlled missiles and vehicles. During World War II he advised several defense organizations, and immediately after the war, he worked with Princeton professor John von Neumann to develop computer applications for accurate weather forecasting.

After retiring from RCA in 1954, he was named an honorary vice president and its technical consultant. He was also appointed director of the Medical Electronics Center at Rockefeller Institute, and worked on electronically based medical applications.

Zworykin received numerous awards related to these inventions, especially television. They included the Institute of Radio Engineers' Morris Liebmann Memorial prize in 1934; the American Institute of Electrical Engineers' highest honor in 1952, the Edison Medal; and the National Academy of Sciences' National Medial of Science in 1967.

—Louise Benjamin

VLADIMIR K(OZMA) ZWORYKIN. Born in Mourom, Russia, 30 July 1889. Degree in engineering from St. Petersburg Institute of Technology (Russia), 1912; attended College de France, 1912–14; University of Pittsburgh, Pennsylvania, U.S.A., Ph.D. 1926. Married 1) Tatiana Vasilieff, 1916 (divorced); two children; 2) Katherine Polevitsky, 1951. Served in Signal Corps, Russian Army, World War I. Immigrated to U.S., 1919; naturalized, 1924. Bookkeeper, financial agent, Russian Embassy, Washington, D.C., 1919–20; electronics researcher, Westinghouse Electric and Manufacturing Company, Pittsburgh, 1920, 1922, 1923–29; researcher, electronics development firm, Kansas, 1922–23; filed first of 120 patents, for electronic camera tube called an "iconoscope," 1923; patented kinescope, 1924; patented color television, 1929; director of electronics re-

Vladimir Zworykin
Photo courtesy of Broadcasting and Cable

search lab, Radio Corporation of America (RCA), Camden, New Jersey, 1929–42; sponsored development of early version of electron microscope, 1940; associate research director, RCA Labs, Princeton, New Jersey, 1942–45, director of electronic research, 1946–54; vice president, from 1947; honorary vice president and consultant, 1954–82; director, Medical Electronics Research Center, Rockefeller Institute (now Rockefeller University), New York City, from 1954; developed radio endosonde, 1957; developed ultraviolet color-translating television microscope, 1957; researcher, Princeton University, 1970s; visiting professor, Institute for Molecular and Cellular Evolution, University of Miami, 1970–82; contributed numerous papers concerning electronics to scientific journals. National chair, Professional Group on Medical Electronics, Institute of Radio Engineers; founder and president, International Federation for Medical Electronics and Biological Engineering; officer of the Academy, French Ministry of Education; governor, International Institute for Medical Electronics and Biological Engineering, Paris. Fellow: American Association for the Advancement of Science; American Institute of Physics; American Physical Society; Institute of Electrical and Electronics Engineers. Member: American Academy of Arts and Sciences; American Philosophical Society; charter member, Electron Microscope Society of America; National Academy of Engineering; National Academy of Sciences; charter member, Society of Television Engineers; charter member, Society of Television Pioneers; Sigma Xi. Honorary fellow: Institute Internazionale delle Comunicazione, Italy; Television Society, England. Honorary member: British Institute of Radio Engineers; Société Francaise des Électriciens et des Radioélectriciens; Television Engineers of Japan. Eminent member, Eta Kappa Nu Association. Recipient: Liebman Memorial Prize, 1934; Overseas Award, 1939; National Association of Manufacturers Modern Pioneer Award, 1940; American Academy of Arts and Sciences Rumford Medal, 1941; U.S. War Department Certificate of Appreciation, 1945; U.S. Navy Certificate of Commendation, 1947; Franklin Institute Potts Medal, 1947; Presidential Certificate of Merit, 1948; chevalier, Légion d'Honneur, 1948; American Institute of Electrical Engineers (AIEE) Lamme Medal, 1949; Poor Richard Club Gold Medal of Achievement, 1949; Society of Motion Picture and Television Engineers Progress Medal, 1950; Medal of Honor, 1951; establishment of Television Prize in his name by the Institute of Radio Engineers, 1952; AIEE Edison Medal, 1952; Union Francaise des Inventeurs Gold Medal, 1954; University of Liege Trasenster Medal, 1959; Christoforo Columbo Award and Order of Merit, Italy, 1959; Broadcast Pioneers Award, 1960; American Society of Metals Sauveur Award, 1963; University of Liege Medical Electronics Medal, 1963; British Institution of Electrical Engineers Faraday Medal, 1965; DeForest Audion Award, 1966; National Medal of Science, 1966; American Academy of Achievement Golden Plate Award, 1967; National Academy of Engineering Founders Medal, 1968; named to National Inventor's Hall of Fame, 1977; Eduard Rhein Foundation ring, 1980. Died in Princeton, New Jersey, 29 July 1982.

PUBLICATIONS

Photocells and Their Applications, with E.D. Wilson. New York: Wiley, 1930.

Television: The Electronics of Image Transmission, with G.A. Morton. New York: Wiley, 1940.

Electron Optics and the Electron Microscope, with G.A. Morton, E.G. Ramberg, and others. New York: Wiley, 1945.

Photoelectricity and its Application, with E.G. Ramberg. New York: Wiley, 1949.

Television: The Electronics of Image Transmission in Color and Monochrome, with G.A. Morton. New York: Wiley, 1954.

Television in Science and Industry, with E.G. Ramberg, and L.E. Flory. New York: Wiley, 1958.

FURTHER READING

Abramson, Albert. *Zworykin, Pioneer of Television*. Urbana: University of Illinois Press, 1995.

Cheek, Dennis W., and A. Kim. "Vladimir Zworykin." In McMurray, Emily J., editor. *Notable Twentieth-Century Scientists*, Volume 4. Detroit, Michigan: Gale, 1995.

Parker, Sybil P., editor. *McGraw-Hill Modern Scientists and Engineers*, Volume 3. New York: McGraw Hill, 1980.

Thomas, Robert M., Jr. "Vladimir Zworykin, Television Pioneer, Dies at 92." *New York Times Biographical Service*, August 1982.

See also Television Technology

NOTES ON
CONTRIBUTORS

NOTES ON CONTRIBUTORS

ABBOTT, Gina. Graduate student, Communication, University of Houston, United States.

ABRAMSON, Bram. Doctoral candidate, Department of Communication, University of Montreal, Canada. Author, "The Birth of a Nation: Genealogy and Quebec," in *Border/lines 36*, 1995; "'A Country of One's Own': Rita Mae Neil, Infomercials, and Canadian Country Music," in *Working Papers in Canadian Studies: Proceedings of the "Instituting Cultures/Cultural Institutions" Conference*, 1995. Member, editorial board, *Mediatribe*, 1994–95.

ACLAND, Charles. Assistant professor, Communications, University of Calgary, Canada. Author, *Youth, Murder, Spectacle: The Cultural Politics of Youth in Crisis*, 1995; articles in *Wide Angle, Communication*, and *Canadian Journal of Film Studies*. Member, editorial board, *Cultural Studies*.

ALDRIDGE, Henry B. Professor, Telecommunications and Film, Eastern Michigan University, United States. Co-author, *Television, Cable, and Radio: A Communications Approach*, 1992; *Audio/Video Production: Theory and Practice*, 1990. Chair, Telecommunications and Film Program, Eastern Michigan University, 1995–98, 1973–88.

ALEXANDER, Alison. Professor and head, Department of Telecommunications, University of Georgia, United States. Author, *Taking Sides*, 1995; *Media Economics*, 1993.

ALLEN, Erika Tyner. Doctoral candidate, Speech Communication, University of Texas at Austin, United States.

ALLEN, Robert C. James Logan Godfrey Professor, American Studies and History, University of North Carolina, United States. Author, *Horrible Prettiness: Burlesque and American Culture*, 1992; *Speaking of Soap Operas*, 1985. Editor, *To Be Continued...Soap Operas around the World*, 1995; *Channels of Discourse: Reassembled*, 1992.

ALLEY, ROBERT S. Professor Emeritus, Humanities, University of Richmond, United States. Co-author, *Murphy Brown: Anatomy of a Sitcom*, 1990; *Love is All Around: The Making of the Mary Tyler Moore Show*, 1989; *The Producer's Medium: Conversations with Creators of American TV*, 1983; author, *Television: Ethics for Hire*, 1977; also numerous books and essays on the relation of American religion to the First Amendment, free speech, the presidency, and the Supreme Court.

ALLOR, Martin. Associate professor, Communication, Concordia University, Canada. Co-author, *L'Etat de Culture: Généalogie Discursives des Politiques Culturelles Québécoises*, 1994. Co-director, Research Group on Cultural Citizenship.

ALVARADO, Manuel. Research fellow, Media Arts, Luton University, England; editorial director, John Libbey Media Publications. Co-author, *Made for Television: Euston Films Limited*, 1985; *Doctor Who: The Unfolding Text*, 1983; *Hazell: The Making of a TV Series*, 1978. Co-editor, *Media Education: An Introduction: The Workbook*, 1992; *East of Dallas: The European Challenge to American Television*, 1988; editor, *Video Worldwide: An International Study*, 1988.

ALVEY, Mark. Ph.D. in Radio-TV-Film 1995, University of Texas at Austin, United States.

AMIN, Hussein Y. Associate professor, Journalism and Mass Communication, American University in Cairo, Egypt. Co-author, "Global TV News in Developing Countries: CNN's Expansion to Egypt," in *Ecquid Novi, Journal of Journalism in Southern Africa*, 1993; "The Impact of Home Video Cassette Recorder on Egyptian Film and Television Consumption Pattern," in *The European Journal of Communication*, 1993. Member, board of trustees, Egyptian Radio and Television Union; General Assembly, Egyptian Ministry of Information; advisory board, *Journal of International Communication*.

ANDERSON, Christopher. Associate professor, Telecommunications, Indiana University, United States. Author, *Hollywood TV: The Studio System in the Fifties*, 1994.

ARON, Danielle. Ph.D. candidate, London School of Economics, England.

ASHLEY, Laura. Graduate student, Communication, University of Houston, United States.

ATTALLAH, Paul. Associate director, Journalism and Communication, Carleton University, Toronto, Canada. Author, "Broadcasting and Narrowcasting: VCRs, Home Video, DBS," in *Cultural Studies: Into the 21st Century*, 1995; "Reconstructing

North American Television Audiences," in *New Researcher*, 1992; "Trends and Developments in Canadian Television," in *Media Information Australia*, 1991; *Theories de la Communication: Sens, Sujets, Savoirs*, 1991; *Theories de la Communication: Histoire, Contexte, Pouvoir*, 1989. Member, editorial board, *MIA* from 1991; *La Revue Communication*, from 1987.

AUFDERHEIDE, Patricia. Associate professor, American University, United States. Author, "Tongues Untied: Controversy in Public Television," in *Journalism Quarterly*, Autumn 1994; "Public Television and the Public Sphere," in *Critical Studies in Mass Communication*, 1991; "A Funny Thing's Happening to TV's Public Forum," in *Columbia Journalism Review*, November/December 1991. Senior editor, *American Film*, 1982-83; cultural/senior editor, *In These Times*, from 1978.

AUSTER, Albert. Teacher, Media and Communication Studies, Fordham University, United States. Author, *Tune in...Turn on...Television and Radio in the U.S.A.*, 1994; *How the War was Remembered: Hollywood and Vietnam*, 1988.

AUTER, Philip J. Assistant professor, Communication, University of South Alabama, United States. Author of a number of articles on media use, new communication technology, and television history.

AVERY, Robert K. Professor, Communication, University of Utah, United States. Author, *Public Service Broadcasting in a Multichannel Environment*, 1993; *Critical Perspectives on Media and Society*, 1991; articles in *Journal of Communication, Journal of Broadcasting and Electronic Media, Communication Education* and *Western Journal of Communication*. Founding editor, *Critical Studies in Mass Communication*.

BABE, Robert. Professor, Communication, University of Ottawa, Canada. Author, *Communication and the Transformation of Economics*, 1995; *Telecommunications in Canada: Technology, Industry and Government*, 1990; co-author, *Broadcasting Policy and Copyright Law*, 1983. Contributing editor, *Information and Communication in Economics*, 1994.

BAL, Vidula V. Doctoral candidate, Speech Communication, University of Texas at Austin, United States.

BALIO, Tino. Professor, Communication Arts, University of Wisconsin, Madison, United States. Author, *Grand Design: Hollywood as a Modern Business Enterprise, 1930-1939*, 1993. Editor, *Hollywood in the Age of Television*, 1990.

BARAN, Stanley J. Professor, Radio-TV, San Jose State University, California, United States. Author, *Mass Communication Theory: Foundations, Ferment, Future*, 1995; *Television Criticism: Reading, Writing, and Analysis*, 1994; *The Known World of Broadcast News*, 1991.

BAREISS, Warren. Doctoral candidate, Telecommunications, Indiana University, United States.

BARRERA, Eduardo. Assistant professor, Communication, El Colegio de la Frontera Norte, Mexico and University of Texas at El Paso.

BARTONE, Richard. Assistant Professor, Communication, William Patterson College of New Jersey, United States. Author of articles on film, television documentary programs, and news programs.

BEASLEY, Vanessa B. Visiting assistant professor, Texas A and M University, United States. Article, "The Logic of Power in the Hill-Thomas Hearings: A Rhetorical Analysis," in *Political Communication II*, 1994.

BECHELLONI, Giovanni. Professor, Faculty of Political Science; director, M.A. in Media Studies, University of Florence, Italy. Author of numerous books and articles on journalism, mass media, and television in Italy.

BELLAMY, Robert V., Jr. Associate professor, Communication, Duquesne University, United States. Co-author, *Grazing on a Vast Wasteland: The Remote Control and Television's Second Generation*, 1996; articles and chapters in *Journal of Communication, Journal of Broadcasting and Electronic Media, Journalism Quarterly, The Cable Networks Handbook*, and *Media, Sports, Society*. Co-editor, *The Remote Control and the New Age of Television*, 1993. Member, editorial board, *Journal of Sport and Social Issues* and *NINE*.

BENJAMIN, Louise. Associate director, George Foster Peabody Awards, and assistant professor, School of Journalism and Mass Communication, University of Georgia, United States. Author of articles on communication history and law and policy in the *Journal of Broadcasting and Electronic Media, Free Speech Yearbook, Journalism Quarterly*, and the *Historical Journal of Film, Radio and Television*.

BENSHOFF, Harry M. Film and video teacher in the Los Angeles area, California, United States. Author, *Monsters in the Closet: Homosexuality and the Horror Film*, forthcoming.

BERGER, Arthur Asa. Professor, Broadcast and Electronic Communication Arts, San Francisco State University, California, United States. Author, *Cultural Criticism: A Primer of Key Concepts*, 1995; *An Anatomy of Humor*, 1993; *Popular Culture Genres*, 1992; *Media Analysis Techniques*, 1982, revised edition, 1992. Film and television review editor, *Society*; editor, *Classics in Communication*; consulting editor, *Humor*.

BERNARDI, Daniel. Editor, *Race and the Emergence of American Film*, 1996.

BIRD, J.B. Freelance writer. M.A. in Radio-TV-Film 1994, University of Texas at Austin, United States.

BIRD, William L., Jr. Curator, Political History Collection, National Museum of American History, Smithsonian Institution, Washington, D.C., United States. Author, "Television in the Ice Age," in *Hail to the Candidate*, 1992; "From the Fair to the Family," in New Museum of Contemporary Art, *From Receiver to Remote Control: The TV Set*, 1990; "Enterprise and Meaning: Sponsored Film, 1939–49," in *History Today*, December 1989. Curator, Smithsonian exhibit, "American Television from the Fair to the Family, 1939–1989."

BIRK, Thomas A. Assistant professor, Communication, Southern Illinois University, United States.

BLASINI, Gilberto M. Doctoral candidate, Critical Studies, University of California, Los Angeles, United States.

BLEICHER, Joan. Research team member, television program research, University of Hamburg, Germany. Author of several articles on television aesthetics and the history of German television.

BLUMLER, Jay G. Professor Emeritus, Communication, University of Leeds, England, and University of Maryland, United States. Research advisor to the Broadcast Standards Council, England. Co-author, *The Crisis of Public Communication*, 1995; *Television in Politics: Its Uses and Influences*, 1968. Co-editor, *Comparatively Speaking: Communication and Culture Across Space and Time*, 1992; *Television and the Public Interest: Vulnerable Values in West European Broadcasting*, 1992; *The Uses of Mass Communication: Current Perspectives on Gratifications Research*, 1974.

BODROGHKOZY, Aniko. Lecturer, Communication Studies, Concordia University, Montreal, Canada. Author, "'Is This What You Mean by Color TV?' Race, Gender and Contested Meanings in NBC's *Julia*," in *Private Screenings: Television and the Female Consumer*, 1992; "'We're the Young Generation and We've Got Something to Say': A Gramscian Analysis of Entertainment Television and the Youth Rebellion of the 1960s," in *Critical Studies in Mass Communication*, June 1991.

BORTHWICK, Stuart. Lecturer, Communications, Liverpool John Moore's University, England.

BOUNDS, J. Dennis. Assistant professor, Cinema-Television, Regent University, Virginia Beach, Virginia, United States. Author, *Perry Mason: The Authorship and Reproduction of a Popular Hero*, 1996; "Noble Counselor: Perry Mason and Its Impact on the Legal Profession," in *The Lawyer in Popular Culture*, 1993.

BOURNE, Stephen. British author and television archivist. Contributor, *Black and White in Colour—Black People in British Television Since 1936*, 1992; *The Colour Black—Black Images in British Television*, 1989.

BREEN, Myles P. Professor, Communication, Charles Stuart University, Bathurst, Australia. Author of articles in *Communication Monographs, Journalism Quarterly, Journal of Communication, Media Information, Australia* and *Washington Journalism Review*, and electronic journal, *APEX-J (Asia-Pacific Exchange Journal)*.

BROOKS, Carolyn N. Assistant specialist, Colleges of Arts and Sciences, University of Hawaii at Manoa, United States. Editor, *The Velvet Light Trap*, 1983–86.

BROWER, Sue. Freelance writer and editor. Ph.D. in Radio-TV-Film 1990, University of Texas at Austin, United States. Author, "Fans as Tastemakers: Viewers for Quality Television," in *The Adoring Audience: Fan Culture and Popular Media*, 1992.

BROWN, James A. Associate professor, Telecommunication and Film Department, University of Alabama, United States. Co-author, *Radio-Television-Cable Management*, 1995; author, *Television "Critical Viewing Skills" Education: Major Media Literacy Projects in the United States and Selected Countries*, 1991. Chair of broadcasting/telecommunication departments, University of Alabama, 1982–88; University of Southern California, 1973–74; University of Detroit, 1967–70; consultant to CBS TV; Bertlesmann Foundation; Aruba's government commercial TV.

BRUNSDON, Charlotte. Lecturer, Film and Television Studies, University of Warwick, England. Author, *Screen Tastes*, 1997; "Television: Aesthetics and Audiences," 1990; "Writing About Soap Opera," 1984; "*Crossroads*: Notes on Soap Opera," 1981; co-author, *Everyday Television: Nationwide*, 1978. Co-editor, *Feminist Television Criticism*, 1997; Editor, *Films for Women*, 1986.

BRYANT, Steve. Keeper of Television, National Film and Television Archive, United Kingdom. Author, "The Television Heritage," in *The Broadcasting Debate*, No. 4, 1989.

BURNS, Gary. Associate professor, Communication Studies, Northern Illinois University, United States. Author of articles in *Journal of Popular Film and Television, Journal of Film and Video* and *Wide Angle*. Co-editor, *Making Television: Authorship and the Production Process*, 1990; *Television Studies: Textual Analysis*, 1989; editor, *Popular Music and Society*.

BUSCOMBE, Edward. Former head, Publishing, British Film Institute, London, England. Author, *Stagecoach*, 1992; co-author, *Hazell: The Making of a Television Series*, 1978. Editor, *The B.F.I. Companion to the Western*, 1988.

BUTCHER, Margaret Miller. Adjunct Professor, Kansas Newman College, United States.

BUTLER, Jeremy G. Associate professor, Communication, University of Alabama, United States. Author, *Television: Critical Methods and Applications*, 1994; articles in *Cinema Journal, Journal of Film and Video*, and *Jump Cut*. Editor, *Star Texts: Image and Performance in Film and Television*, 1991.

BUTSCH, Richard. Professor, Sociology, Rider University, New Jersey, United States. Author, "Bowery B'hoys and Matinee Ladies: The Re-gendering of Nineteenth Century American Theatre Audiences," in *American Quarterly*, September 1994; "Class and Gender in Four Decades of Television Situation Comedies: Plus ça Change Plus C'est Le Meme Chose," in *Critical Studies in Mass*

Communication, December 1992; *For Fun and Profit: The Trans-formation of Leisure into Consumption*, 1990.

BUXTON, Rodney A. Assistant professor, Mass Communication and Journalism Studies, University of Denver, Colorado, United States. Author, "'After it Happened...': The Battle to Present AIDS in Television Drama," in *Television: the Critical View*, 1994; "The Late-Night Talk Show: Humor in Fringe Television," in *Television Criticism: Approaches and Applications*, 1991; "Dr. Ruth Westheimer: Upsetting the Normalcy of the Late Night Talk Show," in *Gay People, Sex, and the Media*, 1991.

CALDWELL, John Thornton. Chair, Radio-TV-Film, California State University, Long Beach, United States. Author, *Televisuality: Style, Crisis, and Authority in American Television*, 1995; articles in *American Television: New Directions in History and Theory, Cinema Journal, Jump Cut*, and *Genre*. Broadcast productions on WTTW, WGBH, WNED, and the SBS-Network in Australia.

CATRON, Christine R. Head, Media Studies, Department of English, St. Mary's College, San Antonio, Texas, United States.

CHMIELEWSKI, Jaqui. Doctoral candidate, Radio-TV-Film, University of Texas at Austin, United States. Research assistant in political communication at the University of Texas at Austin, 1994; Institute of Social Research, University of Michigan, Ann Arbor, 1993; senior researcher in the fields of public opinion and marketing at the Gallup Institute in Argentina, 1988–91.

CHORBA, Frank J. Professor, Communications, Washburn University, Topeka, Kansas, United States.

CIRKSENA, Kathryn. Director, Communications Program, Russell Sage College, Troy, New York, United States. Author of articles in *Women Making Meaning, Journal of Communication Inquiry*, and *Romanian Journal of Sociology*. Fulbright lecturer, University of Timisoara, Romania, 1991-92; chair, Feminist Scholarship Division of the International Communication Association.

CLARK, Kevin A. Assistant instructor, Speech Communication, University of Texas at Austin, United States.

CONKLIN, George C. Director, Global Ecumenical Newsroom, Berkeley, California, United States.

CONSTANTAKIS-VALDES, Patricia. Ph.D. in Radio-TV-Film 1993, University of Texas at Austin, United States. Visiting scholar, Academic Systems Corporation, a firm specializing in interactive instructional media for diverse and underrepresented students.

COOK, John. Lecturer, Media Studies, De Montfort University, Leicester, England. Author, *Dennis Potter: A Life on Screen*, 1995.

COOPER, John. Assistant professor, Telecommunications and Film, Eastern Michigan University, United States. Author, *The Fugitive: A Complete Episode Guide*, 1994.

CORNER, John. Professor, Politics and Communication Studies, University of Liverpool, England. Author, *Television Form and Public Address*, 1995. Co-editor, *Communication Studies*, fourth edition, 1993; editor, *Media Culture and Society*.

COUZENS, Michael. Communications attorney, Oakland, California, United States. Federal Communications Commission staff attorney, 1978–81; Network Inquiry special staff and Low Power Television Inquiry staff.

CRAIG, Robert. Professor, Broadcast and Cinematic Arts, Central Michigan University, United States.

CUBITT, Sean. Reader, Video and Media studies, Liverpool John Moore's University, Liverpool, England. Author, *Videography: Video Media as Art and Culture*, 1993; *Timeshift: On Video Culture*, 1991. Member, editorial board, *Screen* and *Third Text*.

CULLUM, Paul. Freelance writer, Los Angeles, California, United States, specializing in screenwriting, fiction, and critical approaches to film and popular culture.

CUNNINGHAM, Stuart O. Associate professor, Media and Communication, Queensland University of Technology, Australia. Co-author, *Away from Home: Australian Television and International Mediascapes*, 1995; *Contemporary Australian Television*, 1994; author, *Framing Culture: Criticism and Policy in Australia*, 1992; *Featuring Australia: The Cinema of Charles Chauvel*, 1991. Co-editor, *New Patterns in Global Television*, 1995; *The Media in Australia: Industries, Texts, Audiences*, 1993.

CURTHOYS, Ann. Professor, History, Australian National University. Author, *For and against Feminism: A Personal Journey into Theory and History*, 1988.

CURTIN, Michael. Assistant professor, Telecommunications, Indiana University, United States. Author, *Redeeming the Wasteland: Television Documentary and the Cold War*, 1995; articles in *Cinema Journal, Journalism Monographs*, and *Journal of Broadcasting and Electronic Media*.

D'ALESSANDRO, Kathryn C. Assistant professor, Media Arts, Jersey City State College, New Jersey, United States.

DAYAN, Daniel. Fellow, Laboratoire Communication et Politique, Centre National de la Recherche Scientifique, Paris, France, and External Professor of Communication, University of Oslo, Norway. Co-author, *Media Events: The Live Broadcasting of History*, 1992; author, *Western Graffiti: Jeux D'images et Programmation du Spectateur dans La Chevauchee Fantastique de John Ford*, 1983. Co-editor of the French communication studies journal, *Hermes*.

de la GARDE, Roger. Professor, Communication, Laval University, Quebec City, Canada. Co-author, "Cultural Development: State of the Question and Prospects for Quebec," in *Canadian Journal of Communication*, 1994; "To speak one's culture," in *The London Journal of Canadian Studies*, 1990. Co-editor, *Small Nations, BIG Neighbour: Denmark and Quebec/Canada Compare Notes on American Popular Culture*, 1993; *Les pratiques culturelles de grande consommation. Le marche francophone*, 1992; editor, *Communication*.

DEANE, Pamala S. Lecturer and writer-researcher of broadcast and film history, specializing in the black experience in America. Former military historian.

DESJARDINS, Mary. Assistant professor, Radio-TV-Film, University of Texas at Austin, United States. Author of articles in *Film Quarterly, Quarterly Review of Film and Video, The Velvet Light Trap*, and *The Spectator*.

DESSART, George. Professor and deputy chair, Graduate Studies, Television and Radio, and executive director, Center for the Study of World Television, Brooklyn College, City University of New York, United States. Former vice president, Program Practices, CBS Broadcast Group. Author, "Of Tastes and Times: What Happened to Broadcast Standards?" *Television Quarterly*, 1994; *More Than You Want to Know About PSA's: A Guide to Production and Placement of Effective Public Service Announcements on Radio and Television*, 1982. Publisher and editor, *ALMANAC: The Annual of the International Council, National Academy of Television Arts and Sciences*.

DICKINSON, Robert. Instructor, University of Southern California, Los Angeles, United States. Author, "Hearing the Body of Bruce Weber's Work: *Let's Get Lost* and *Broken Noses*," in *The Spectator*, Spring 1993; "The Unbearable Weight of Winning: Garci's Trilogy of Melancholoy and the Foreign Language Oscar," in *The Spectator*, Spring 1991.

DOCKER, John. Australian Research Council, Research Fellow, Humanities Research Centre, Australian National University, Australia. Author, *Postmodernism and Popular Culture: A Cultural History*, 1994.

DOHERTY, Thomas. Chair, Film Studies, Brandeis University, United States. Author, *Projections of War: Hollywood, American Culture, and World War II*, 1993; *Teenagers and Teenpics: The Juvenilization of American Movies in the 1950s*, 1988.

DONNELLY, David F. Assistant professor, Communication, University of Houston, United States. Author of articles in *Communication Research, Telematics and Informatics, International Teleconferencing Association Yearbook*, and others.

DOWLER, Kevin. Assistant professor, Carleton University, Toronto, Canada.

DOWNING, John. John T. Jones, Jr., Centennial Professor, Radio-TV-Film, University of Texas at Austin, United States.

Author, "The Cosby Show and American Racial Discourse," in *Discourse and Discrimination*, 1988; "Trouble in the Backyard: Soviet Media Coverage of Afghanistan," in *Journal of Communication*, 1988; *Film and Politics in the Third World*, 1987. Co-editor, *Questioning The Media*, 1995.

DRUMMOND, Phillip. Faculty of Culture, Communication and Societies, Institute of Education, University of London. Co-editor, *National Identity and Europe: The Television Revolution*, 1993; *Television and its Audiences: International Research Perspectives*, 1988; *Television in Transition: Papers from the First International Television Studies Conference*, 1985.

DUNN, J.A. Slavonic Languages, University of Glasgow, Scotland. Author, "A Lot of Boiling Milk," in *Rusistika*, 8, 1993; "The Rise, Fall and Rise (?) of Soviet Television," in *Rusistika*, 4, 1991.

DURNELL, Nannetta. Assistant professor, Communication, Florida Atlantic University, United States. Author, "Dragon Slayer: National News Magazines Portrayal of Jesse Jackson as a Mythical Hero During the 1988 Presidential Campaign," in *Black Religious Leaders in America*, 1996.

EAMAN, Ross A. Associate professor, Carleton University, Toronto, Canada. Author, *Channels of Influence: CBC Audience Research and the Canadian Public*, 1994.

EDGERTON, Gary R. Professor and chair, Communication and Theatre Arts, Old Dominion University, Virginia, United States. Author, *Film and the Arts in Symbiosis*, 1988; *American Film Exhibition*, 1983; articles in *Film and History, Television Quarterly, Journal of American Culture*, and *Critical Studies in Mass Communication*.

ELMER, Greg. Doctoral candidate, Communication, University of Massachusetts-Amherst, United States. Editor, *CommOddities*.

EMANUEL, Susan. Translator. Contributor to British publications on British and American television. Former member, editorial board, *Screen*. Former lecturer in film and television studies.

EPSTEIN, Michael. Practicing attorney; doctoral candidate in American Culture, University of Michigan, United States. Regular contributor to *Television Quarterly* and for popular magazines.

ERLER, Robert. Librarian, Long Island University, New York, United States.

EVERETT, Anna. Doctoral candidate, Critical Studies, University of Southern California, Los Angeles, United States. Author, "Recolonizing Africa for the Twentieth Century," in *Ufahamu*, 1993.

EVERETT, Robert. University assistant secretary, York University, Ontario, Canada. Co-author, "Political Communication in Canada," in *Communications in Canadian Society*, 1995; "TV in the 1988 Campaign: Did it Make a Difference," in *Politics: Canada*, 1991.

FANG, Irving. Professor, University of Minnesota, United States. Author, *Pictures*, 1993; *The Computer Story*, 1988; *Television News, Radio News*, 1985.

FELSENTHAL, Norman. Professor, Communications and Theater, Temple University, Philadelphia, United States. Author, *Mass Communication*, 1981; articles in *Journal of Broadcasting*.

FERGUSON, Robert. Course leader, M.A. Media Programme, Culture, Communication, and Societies, Institute of Education, University of London.

FIDDY, Dick. British freelance television researcher and scriptwriter. Writer-researcher, Channel 4 Television, T.V. Heaven, 1992; *1001 Nights of Television*, 1991, and writer-creator, *The A to Z of Television*, 1990. Editor, *The Virgin Television Year Book*, 1985; *Primetime: The Television Magazine*, 1980s.

FINNEY, Robert G. Professor, Radio-TV-Film, California State University, Long Beach, United States. Associate editor, *ACA Journal*, from 1992; editor, *Feedback*, 1976-82.

FLETCHER, Frederick J. Professor, Political Science and Environmental Studies, York University, Ontario, Canada. Co-author, "Political Communication in Canada," in *Communications in Canadian Society*, 1995; "The Mass Media: Private Ownership, Public Responsibilities," in *Canadian Politics in the 1990s*, 1995; author, "Media, Elections and Democracy," in *Canadian Journal of Communication*, 1994; co-author and editor, *Media, Elections and Democracy*, 1991.

FLETCHER, James E. Professor, Telecommunications, University of Georgia, United States. Author of articles in *Quarterly Journal of Speech, Western Speech, Journal of Broadcasting, Jewish Social Studies, Journal of Advertising Research, Tobacco Control,* and *Psychophysiology*. Co-editor, *Music and Program Research*, 1986; *Broadcast Research Methods*, 1985; editor, *Handbook of Radio-TV: Research Procedures*, 1981; editor, *Feedback*.

FLEW, Terry. Lecturer, Communications, Faculty of Social Sciences, University of Technology, Sydney, Australia. Co-editor, "The Policy Moment," in *Media Information Australia*, 1994. Member, editorial board, *Metro*.

FOSTER, Nicola. Researcher, Electric Pictures Video, London, England.

FOX, Jeanette. Doctoral candidate, Media Studies, University of Iowa, United States.

FREEDMAN, Eric. Doctoral candidate, Critical Studies, Cinema-Television, University of Southern California, Los Angeles, United States. Public access producer and independent video artist. Author, "From Excess to Access: Televising the

Subculture," in *The Spectator*, Fall 1993. Editor, *The Spectator*, Fall 1994.

FRY, Katherine. Assistant professor, Television and Radio, Brooklyn College, City University of New York, United States. Author, "Regional Consumer Magazines and the Ideal White Reader," in *The American Magazine: Research Perspectives and Prospects*, 1994.

GANZ-BLAETLLER, Ursula. Researcher, Mass Media Department, University of Zurich, Switzerland. Author of articles on serial fiction in television, and on translation and sub-titling in television.

GARAY, Ronald. Professor and associate dean, Undergraduate Studies and Administration, Manship School of Mass Communication, Louisiana State University, United States. Author, *Gordon McLendon: The Maverick of Radio*, 1992; *Cable Television: A Reference Guide to Information*, 1988; *Congressional Television: A Legislative History*, 1984.

GARDNER, Paula. Doctoral candidate, Communication, University of Massachusetts-Amherst, United States. Editorial board, *CommOddities*.

GATEWARD, Frances K. Assistant professor, Communications, American University, Washington, D.C., United States. Independent film and video maker.

GIBBERMAN, Susan R. President, SRG Research Services, United States. Author, *Star Trek: An Annotated Guide to Resources on the Development, the Phenomenon, the People, the Television Series, the Films, the Novels and the Recordings*, 1991. Former researcher, Walt Disney Company.

GIBSON, Mark. Associate lecturer, Communication and Media Studies, Central Queensland University, Australia. Author, "A Centre of Flux: Japan in the Australian Business Press," in *Continuum*, 1994; "Eastern Contexts/Western Theory—Cultural Studies and Japan," in *South East Asian Journal of Social Science*, 1994. Member, editorial board, *Social Semiotics*.

GILSDORF, William O. Chair and professor, Communication Studies, Concordia University, Montreal, Canada. Author, "Questioning Balance: Struggles over Broadcasting Policies and Content," in *Canadian Journal of Communication*, Winter 1992. Co-editor, *Small Nations, BIG Neighbor: Denmark and Quebec/Canada Compare Notes on American Popular Culture*, 1993.

GLENNON, Ivy. Associate professor, Speech Communication, Eastern Illinois University, United States.

GLYNN, Kevin. Teaches in the Department of American Studies, University of Canterbury, Christchurch, New Zealand.

GODDARD, Peter. Tutor/Researcher, Communication Studies, University of Liverpool, England. Author, "Hancock's Half-

Hour: A Watershed in British Television Comedy," in *Popular Television in Britain*, 1991. Member, Liverpool Public Communications Group.

GODFREY, Donald G. Associate professor, Walter Cronkite School of Journalism and Telecommunication, Arizona State University, United States. Author, *Reruns of File: A Guide to Electronic Media Archives*, 1992. Video review editor, *Journalism History*.

GOMERY, Douglas. Professor, Journalism, University of Maryland, United States. Senior researcher, Media Studies Project of the Woodrow Wilson International Center for Scholars, 1988–92. Author, *The Future of News*, 1992; *Shared Pleasures*, 1992; *American Media*, 1989; and *The Hollywood Studio System*, 1986; over 400 articles in other sources including *Modern Maturity*.

GOURGEY, Hannah. Doctoral candidate, Speech Communication, University of Texas at Austin, United States.

GRANT, August. Associate professor, Radio-Television-Film, University of Texas at Austin. Author of numerous articles on new media technologies, television audiences, and issues in broadcasting. Editor, *Communication Technology Update*, 1996.

GRAY, Herman. Professor, Sociology, University of California at Santa Cruz, United States. Author, *Watching Race: Television and the Struggle for "Blackness*, 1995; *Producing Jazz: The Experience of an Independent Record Company*, 1988; numerous articles on African Americans, race, and ethnicity relating to mass media.

GRIFFIN, Sean. Doctoral candidate, Cinema-Television, University of Southern California, Los Angeles, United States. Associate editor, *The Spectator*.

GRIFFITHS, Alison. Doctoral candidate, Cinema Studies, New York University, United States. Author of articles in *To Be Continued: Soap Operas around the World*, 1995; *Continuum*, *Society for Visual Anthropology Newsletter*, 1994; *National Identity and Europe*, 1993.

GROSS, Lynne Schafer. Author, *The International World of Electronic Media*, 1995; *Television Production: Techniques and Disciplines*, 1995; *Telecommunications*, 1995; *Programming for TV, Radio, and Cable*, 1994; *Electronic Moviemaking*, 1994; *Radio Production Worktext*, 1994.

GUNZERATH, David. Doctoral candidate, Communication Studies, University of Iowa, United States. Columnist, *TV Guide*, 1987–91; editorial and management positions, *TV Guide*, 1984–93.

HAGINS, Jerry. Freelance writer/researcher. M.A. in Radio-TV-Film 1995, University of Texas at Austin, United States.

HALLIN, Daniel C. Professor, Communication, University of California, San Diego, United States. Author, *We Keep America*

on Top of the World: Television Journalism and the Public Sphere, 1994; *The 'Uncensored War': The Media and Vietnam*, 1986.

HAMMILL, Geoffrey. Associate professor, Telecommunications and Film, Eastern Michigan University, United States.

HAMOVITCH, Susan. Independent producer. Author, "The Talking Dream: A Case Study Using Video in the Public Schools," in *Television Quarterly*, 1997.

HAMPSON, Keith C. Doctoral candidate and tutor, University of Queensland, Australia.

HARP, Denis. Associate Director, School of Mass Communication, Texas Tech University, United States.

HARRIS, Cheryl. Assistant professor, California State University, United States. Executive director of Media Research Center. Author, *Theorizing Fandom: Fans, Subculture and Identity*, 1995; *An Internet Education*, 1995.

HART, Roderick. P.F.A. Liddell Professor of Communication, and professor of government, University of Texas at Austin, United States. Author, *Seducing America: How Television Charms the Modern Voter*, 1994; *Modern Rhetorical Criticism*, 1990; *The Sound of Leadership*, 1987; *Verbal Style and the Presidency*, 1984; articles in *Quarterly Journal of Speech, Presidential Studies Quarterly, Philosophy and Rhetoric, Critical Studies in Mass Communication*, and the *Journal of Communication*. Professor of the Year for the State of Texas, National Council for the Advancement and Support of Education, 1991.

HARTLEY, John. Professor, Mass Communication, University of Wales College of Cardiff, Wales. Author, *Popular Reality: Journalism, Modernity, Popular Culture*, 1996; *Teleology: Studies in Television*, 1992; *The Politics of Pictures: The Creation of the Public in the Age of Popular Culture*, 1992; *Understanding News*, 1982; co-author, *Reading Television*, 1978.

HASSANPOUR, Amir. Communication Studies, Concordia University, Montreal, Canada. Author, "The Internationalization of Language Conflict: The Case of Kurdish," in *Language Contact-Language Conflict*, 1993; *Nationalism and Language in Kurdistan, 1918–1985*, 1992; "The Pen and the Sword...," in *Knowledge, Culture and Power: International Perspectives on Literacy as Policy and Practice*, 1992.

HAWKINS-DADY, Mark. British freelance editor. Contributor, *John Osborne: A Casebook*, 1996; *New Theatre Quarterly*. Editor, *Reader's Guide to Literature in English*, 1996; series editor, *International Dictionary of Theatre*, 1992–95.

HAY, James. Associate professor, Communication, University of Illinois, United States. Author, *Popular Film Culture in Fascist Italy*, 1987. Co-editor, *The Audience and its Landscape*, 1996.

HAYNES, Richard. Research fellow, Film and Media Studies, University of Stirling, Scotland. Author, *The Football Imagina-*

tion: *The Rise of Football Fanzine Culture*, 1995; book chapter, *The Passion and the Fashion: Football Fandom in the New Europe*, 1993.

HILMES, Michele. Associate professor, Communication Arts, University of Wisconsin-Madison, United States. Author, *Hollywood and Broadcasting: From Radio to Cable*, 1990.

HIMMELSTEIN, Hal. Professor, Television and Radio, Brooklyn College-City University of New York, United States. Author, *Television Myth and the American Mind*, 1994; *On the Small Screen*, 1981. Member, editorial board, *Critical Studies in Mass Communication*, 1990–92.

HOERSCHELMANN, Olaf. Doctoral candidate/associate instructor, Telecommunications, Indiana University, United States.

HONG, Junhao. Assistant professor, Communications, State University of New York, Buffalo, United States. Author of articles in *Gazette, Media Development, Telecommunications Policy, Asian Survey, Media Asia, Media Information Australia,* and *Intercultural Communication Studies*.

HOOVER, Stewart M. Associate professor, Journalism and Mass Communication, University of Colorado, United States. Author, *Mass Media Religion*, 1988.

HUGETZ, Ed. Director, University of Houston System, Texas, United States. Co-host and co-producer, *The Territory*, a television program devoted to presentation and discussion of new film and video.

HUMPHREYS, Donald. Ph.D. candidate, Speech Communication, University of Texas at Austin, United States. Author, "Lessons Unlearned: An Historical Sketch of Governmental English-Spanish Translation and Interpretation in South-Central Texas," in *Vistas: Proceedings of the Thirty-Fifth Annual Conference of the American Translators Association*, 1994.

HUNT, Darnell M. Assistant professor, Sociology, University of Southern California, Los Angeles, United States. Author, *Screening the Los Angeles "Riots": Race, Seeing, and Resistance*, 1996.

JACKA, Elizabeth. Senior lecturer, Media and Communications, Macquarie University, Sydney, Australia. Author, *The ABC of Drama*, 1990; co-author, *The Screening of Australia*, Vols. I and II, 1987, 1988. Editor, *Continental Shift: Globalisation and Culture*, 1993; editor and book review editor, *Media Information Australia*, from 1991; co-editor, *The Imaginary Industry*, 1988.

JACKSON, Matt. Doctoral candidate, Indiana University, United States. Author of articles in *Journal of Broadcasting and Electronic Media* and *Communication Law and Policy*. Associate editor, *Federal Communications Law Journal*, from 1995.

JACOBS, Jason J. Lecturer, Film and TV, University of Warwick, England.

JARVIS, Sharon. Doctoral candidate, Speech Communication Department, University of Texas at Austin.

JENKINS, Henry. Director, Film and Media Studies, Massachusetts Institute of Technology, Cambridge, United States. Co-author, *Science Fiction Audiences Watching Doctor Who and Star Trek*, 1995; author, *Classical Hollywood Comedy*, 1995; *Textual Poachers: Television Fans and Participatory Culture*, 1992; *What Made Pistachio Nuts?: Early Sound Comedy?* and *The Vaudeville Aesthetic*, 1992.

JENNINGS, Ros. Teaching assistant, Film and Television, University of Warwick, England. Author, "Desire and Design: Ripley Undressed," in *Immortal, Invisible*, 1995.

JONES, Clifford A. Adjunct professor of law and lecturer, United States. Oklahoma City University, 1980; University of Oklahoma College of Law, 1979. President, Clifford A. Jones and Associates, Norman, Oklahoma. Author of articles in *Oklahoma Bar Journal*: "New Campaign Finance Rules: Issues Affecting Lawyers and Judges," 65, 1994; "Antitrust and Patent Licensing Problems: Are the Nine 'No-Nos' the Nine 'Maybes'?", 53, 1982; "Franchise Disclosure Requirements," 53, 1982; "Muddling Through: The Standing Requirement in Private Antitrust Actions," 52, 1981. Editor, *The Cost of Democracy: Continuing Education Materials and Compliance Workshop Materials*, 1994.

JONES, Jeffrey P. Assistant instructor/doctoral candidate, Radio-TV-Film, University of Texas at Austin, United States.

JONES, Judith. Lecturer, Liverpool John Moores University, Liverpool, England. Freelance production assistant.

JOWETT, Garth. Professor, Communication, University of Houston, United States. Co-author, *Propaganda and Persuasion*, 1991; *Movies as Mass Communication*, 1989. Editor, *Foundations of Popular Culture Series*, co-editor, *History of Mass Communication Series*.

JOWETT, Guy. Researcher, British Film Institute, London, England.

KAID, Lynda Lee. Professor, Communication, and director, Political Communication Center, University of Oklahoma, United States. Co-author, *Political Advertising in Western Democracies: Parties and Candidates on Television*, 1995; articles in *Political Communication, World Communication, European Journal of Communication, Journal of Communication Studies,* and *Social Science Quarterly*. Member, editorial board, *Journal of Broadcasting and Electronic Media*, from 1991; *Social Science Quarterly*, from 1982.

KARIITHI, Nixon K. Doctoral candidate, University of Houston, United States. Research fellow, Freedom Forum Media Studies Center, New York, 1994–95. Author of two books and articles on African development issues. Founding assistant editor, *The Economic Review*, a weekly news magazine in Kenya.

KASSEL, Michael B. Doctoral candidate, Michigan State University, United States. Freelance media and feature writer. Author, *America's Favorite Radio Station: WKRP in Cincinnati*, 1993.

KAYE, Janice. Doctoral candidate, Critical Studies, Cinema-Television, University of Southern California, Los Angeles, United States.

KEARNEY, Mary C. Doctoral candidate, Critical Studies, Cinema-Television, University of Southern California, Los Angeles, United States. Co-author, "The History and Representation of Ida Lupino as Director of Television," in *Queen of the B's: Ida Lupino Behind the Camera*, 1995.

KEIRSTEAD, Phillip O. Professor, Florida A and M University, United States. Co-author, *ENG: Television News*, 3rd edition, 1995; *The World of Telecommunication*, 1990. Correspondent, *International Broadcasting/Multi-Channel Media*, London.

KELLNER, C.A. Professor, Communication, Marshall University, United States., retired. Retired executive, Arbitron Corporation.

KELLNER, Douglas. Professor, Philosophy, University of Texas at Austin, United States. Author, *Media Culture: Cultural Studies, Identity, and Politics Between the Modern and the Postmodern*, 1995; *Camera Politica: The Politics and Ideology of Contemporary Hollywood Film*, 1988; co-author, *Critical Theory, Marxism, and Modernity*, 1989; co-author, *Jean Baudrillard: From Marxism to Postmodernism and Beyond*, 1989.

KENNY, Brendan. Senior lecturer, English, Social and Historical Studies, University of Portsmouth, England. Author of articles in *Literature, Society, and the Sociology of Literature; Browning Institute Studies*.

KEPLEY, Vance, Jr. Professor, Communication Arts, University of Wisconsin-Madison, United States. Author, "The Weaver Years at NBC," in *Wide Angle*; "From 'Frontal Lobes' to the 'Bob-and-Bob Show': NBC Management and Programming, 1949–65," in *Hollywood in the Age of Television*, 1990.

KIM, Lahn S. Doctoral candidate, Film and Television, University of California, Los Angeles, United States.

KIM, Won-Yong. Associate professor, Mass Communication and Journalism, Sung Kyun Kwan University, Seoul, Korea. Co-author, *Revolution vs. Modernization*, 1992; author, *Theories of Critical Communication*, 1991. Member, Presidential Commission of Policy Planning.

KLEIMAN, Howard M. Associate professor, Mass Communication, Miami University, Oxford, Ohio, United States. Author, "Content Diversity and the FCC's Minority and Gender Licensing Policies," in *Journal of Broadcasting and Electronic Media*, 1991. Fellow, Cable News Network, 1994.

KUBEY, Robert. Associate professor, Communication, Rutgers University, United States. Co-author, *Television and the Quality of Life*, 1990.

LACKAMP, J. Jerome. Former director, Communications, Archdiocese of Cleveland, Ohio, United States.

LAMONTAGNE, Manon. Writer and researcher, National Film Board of Canada, Montreal, Quebec, Canada.

LANE, Christina. Doctoral candidate, Radio-TV-Film, University of Texas at Austin, United States. Coordinating editor, *The Velvet Light-Trap*.

LAPASTINA, Antonio C. Doctoral candidate, Radio-TV-Film, University of Texas at Austin, United States.

LEACH, Jim. Professor, Film and Communication Studies, Brock University, St. Catharines, Canada. Author, *A Possible Cinema: The Films of Alain Tanner*, 1984; articles in *Cinema Canada, Wide Angle, Literature/Film Quarterly, Dalhousie Review, Film Criticism,* and *Journal of Canadian Studies*.

LEE, Stephen. Associate professor and director, Television Emphasis, Santa Clara University, California, United States. Author of articles in *Popular Music* and *Journal of Media Economics*.

LEIBMAN, Nina C. Author, *Living Room Lectures: The Fifties Family in Film and Television*, 1995; "Leave Mother Out: The Family in Film and TV," in *Wide Angle*; "Decades and Retro-decades: Historiography in the Case of *Easy Rider* and *Shampoo*," in *Mid-Atlantic Almanac*; "The Family Spree of Film Noir," in *Journal of Popular Film and Television*.

LEMIEUX, Debra A. M.A., College of William and Mary, United States.

LEMIEUX, Robert. Ph.D., University of Georgia, United States. Author of articles in *Women's Health Care Campaigns: The Rhetoric of Reproduction*, 1995; *Communication Research*.

LEWIS, Lisa Anne. Independent scholar and film producer. Author, *Gender Politics and MTV: Voicing the Difference*, 1990. Editor, *The Adoring Audience: Fan Culture and Popular Media*, 1992.

LIEBES, Tamar. Head, Smart Institute of Communication Research, Hebrew University of Jerusalem, Israel. Co-author, "The Structure of Family and Romantic Ties in the Soap Opera: An Ethnographic Approach," in *Communication Research*, 1994; *The Export of Meaning: Cross-Cultural Readings of Dallas*, 1993. Editor, *Narrativization of the News*, 1994; member, editorial board, *Political Communication, Communication Theory,* and *Journal of Broadcasting and Electronic Media*. Member, Plenary Council of Public Broadcasting.

LIGGETT, Lucy A. Professor, Telecommunications and Film, Eastern Michigan University, United States. Co-author,

Audio/Video Production: Theory and Practice, 1990; author, *Ida Lupino as Film Director, 1940–1953*, 1979. Executive secretary, Michigan Association of Speech Communication, 1988–94.

LIMBURG, Val E. Associate professor, Edward R. Murrow School of Communication, Washington State University, United States. Author, *Electronic Media Ethics*, 1994; *Mass Media Literacy*, 1988; articles in *Journalism Quarterly*, *Judicature*, and *Journalism of Mass Media Ethics*.

LIVINGSTONE, Sonia. Lecturer, Social Psychology, London School of Economics, England. Co-author, *Talk on Television*, 1994; author, *Making Sense of Television*, 1990. Member, editorial board, *Journal of Communication* and *European Journal of Communication*.

LOGAN, Pamela. Assistant keeper of Television, National Film and Television Archive, England. Author, *Jack Hylton Presents*, 1995.

LOMETTI, Guy E. Dean, Communication and Arts, Marist College, United States. Author of chapters in *New Research Toward Better Understanding of Children as Consumers*, 1993; *Television and Nuclear Power: Making the Public Mind*, 1992; articles in *Journal of Broadcasting and Electronic Media*, *Society*, *Human Communication Research*, *Communication Research*, *Journalism Quarterly*, *The Public Perspective*, and *Television and Children*.

LOOMIS, Amy W. Assistant professor, Santa Clara University, California, United States. Author of articles in *Visual Communication Quarterly*, *CommOddities*, and *Moving Images*. Editorial assistant, *Journal of Broadcasting and Electronic Media*.

LOVDAL, Lynn T. Lecturer, Ohio University, United States. Author, "Sex-Role Messages in Television Commercials: An Update," in *Sex Roles*, 1989.

LUCKETT, Moya. Assistant professor, Radio-TV, University of Houston, United States. Author of articles in *The Velvet Light Trap*, 1995; *Disney Discourse*, 1994. Editor, *The Velvet Light Trap*, 1991–93.

LUMBY, Catharine. Lecturer, Communications, Macquarie University, Australia. Feature writer, *The Sydney Morning Herald*. Author, "Tabloid Television," in *Not Just Another Business*, 1994.

MAGDER, Ted. Director, Mass Communication Programme, York University, Toronto, Canada. Author, *Canada's Hollywood: The Canadian State and Feature Films*, 1993; "Public Discourse and the Structures of Communication," in *Building on the New Canadian Political Economy*.

MALIK, Sarita. Ph.D. researcher, British Film Institute, England.

MALIN, Brent. Graduate student, Media Studies, University of Iowa, United States.

MANN, Chris. Newspaper writer-editor, *The Tulsa World*, *The Tulsa Sentinel*, *The University of Tulsa Collegian*, and *The New York Times*. Editor and publisher, *The Roomie Report*, and quarterly *Three's Company* fanzine.

MARC, David. U.S. freelance writer. Author, *Bonfire of the Humanities*, 1995; *Comic Visions*, 1989; *Demographic Vistas*, 1984; co-author, *Prime Time, Prime Movers*, 1992.

MARSHALL, P. David. Director, Media and Cultural Studies Centre, English Department, University of Queensland, Australia. Author, *Celebrity and Power*, 1995; "Panic Television: Bush Fires and the Media Future," in *Meanjin*, 1994; "At Last a Co-Production We Can all Enjoy: The Australian Canadian Co-Production of Black Robe and Golden Fiddles," in *Media Information Australia*, 1992.

MARTIN, William. Professor, Sociology, Rice University, Houston, United States. Author, *My Prostate and Me: Dealing with Prostate Cancer*, 1994; *A Prophet with Honor: The Billy Graham Story*, 1991.

MASCARO, Tom. Ph.D. 1994, Wayne State University. Author of articles in *Television Quarterly* and *Electronic Media*.

MASHON, Mike. Curator, Broadcast Pioneers Library, University of Maryland, United States. Author of articles in *Wide Angle* and *Historical Journal of Film, Radio, and Television*. Former editor, *The Velvet Light Trap*.

MASSEY, Kimberly B. Associate professor, Radio/Television, San Jose State University, California, United States. Co-author, *Introduction to Radio: Production and Programming*, 1995; *Television Criticism: Reading, Writing, and Analysis*, 1994.

MAXWELL, Richard. Assistant professor, Radio-TV-Film, Northwestern University, United States. Author, *The Spectacle of Democracy: Spanish Television, Nationalism, and Political Transition*, 1995.

MAZZARELLA, Sharon R. Assistant professor, Television-Radio, Ithaca College, Ithaca, United States. Co-author of articles in *Journal of Broadcasting and Electronic Media* and *Communication Research*.

MCALLISTER, Matthew P. Assistant professor, Communication Studies, Virginia Polytechnic Institute and State University, United States. Author, *The Commercialization of American Culture: New Advertising, Control, and Democracy*, 1996; articles in *Journal of Communication*, *Journal of Popular Culture*, and *Journal of Popular Film and Television*.

MCCARTHY, Anna. Assistant professor, Communication Studies, University of North Carolina, United States. Author of articles in *Consoling Passions Anthology on Cultural History*, 1996; *Cinema Journal*, 1995; *The Velvet Light Trap*, 1993. Pre-doctoral fellow, Smithsonian Institution, National Museum of American History.

MCCOURT, Tom. Lecturer, Radio-Television-Film, University of Texas at Austin.

MCDERMOTT, Mark R. Freelance writer. M.A. in Popular Culture, Bowling Green State University, Ohio, United States. Five-time *Jeopardy!* champion and semi-finalist in the 1993 10th Anniversary Tournament.

MCKEE, Alan. Lecturer, Media Studies at Edith Cowan University, Australia. Author of articles in *Screen, Continuum,* and *Cultural Studies.* Co-editor, *Telling Both Stories: Indigenous Australia and the Media.*

MCKINNON, Lori Melton. Assistant professor, Advertising and Public Relations, University of Alabama, United States. Author of articles in *Journalism and Mass Communication Quarterly, Electronic Journal of Communication/La Revue Electronique de Communication,* and *Argumentation and Advocacy;* chapter in *The American Magazine: Research Perspectives and Prospects,* 1995.

MCLELAND, Susan. Doctoral candidate, Radio-TV-Film, University of Texas at Austin, United States. Author, "Re-shaping the Grotesque Body: Roseanne, *Roseanne,* Breast Reduction and Rhinoplasty," in *The Spectator,* 1993. Co-coordinating editor, *The Velvet Light Trap,* Spring 1995.

MCLUSKIE, Peter. Teaching assistant, University of Warwick, England.

MERRITT, Bishetta D. Associate professor and chair, Radio-Television-Film, Howard University. Author, "The Debbie Allen Touch," in *Women and the Media: Content, Careers, and Criticism,* 1995; "Illusive Reflections: African American Women on Primetime Television," in *Our Voices: Essays in Culture, Ethnicity, and Communications,* 1993; "Black Family Imagery and Interactions on Television," in *Journal of Black Studies,* 1993; "Bill Cosby: TV Auteur?" *Journal of Popular Culture,* 1992.

MESSERE, Fritz J. Faculty fellow, Annenberg Washington Program, and associate professor, State University of New York, Oswego, United States. Co-author, *Telecommunications: An Introduction to Electronic Mass Media.*

MEYERS, Cynthia. Doctoral candidate, Radio-TV-Film, University of Texas at Austin, United States.

MILLER, Mary Jane. Professor, Film and Dramatic Literature, Brock University, St. Catharines, Canada. Author, *Rewind and Search: Conversations with Makers and Decision-Makers in Television Drama,* 1996; *Turn Up the Contrast: CBC Television Drama since 1952,* 1987; articles and chapters in *Canadian Drama* and *Beyond Quebec: Taking Stock of Canada.* Book review editor, *Theatre History in Canada.*

MILLER, Toby. Associate professor, Cinema Studies, New York University, United States. Co-author, *Contemporary Australian Television,* 1994; author, *The Well-Tempered Self: Citizenship, Culture, and the Postmodern Subject,* 1993; articles in

Media Information Australia, Social Text, Continuum, Australian Journal of Communication, Media Culture and Society, and *Social Semiotics.*

MILLINGTON, Bob. Senior lecturer, School of Media, Critical and Creative Arts, Liverpool John Moores University, Liverpool, England. Author, "Myths of Creation: Theatre on the Southbank Show," in *Boxed Sets,* 1996; co-author, *Boys from the Black Stuff: The Making of TV Drama,* 1986.

MITTELL, Jason. Graduate student, Telecommunications, University of Wisconsin, Madison, United States.

MONTGOMERIE, Margaret. Senior lecturer, Film and Television, University of Derby, United Kingdom.

MOODY, Nickianne. Freelance researcher, Liverpool, England. Currently conducting an oral history of the Boots Booklovers' Library.

MORAN, Albert. Senior research fellow, Griffith University, Brisbane, Australia. Author, *Public Voices, Private Interests,* 1995; *Film Policy,* 1994; *Moran's Guide to Australian TV Series,* 1993; *Stay Tuned,* 1992; *Projecting Australia,* 1991.

MORAN, James. Doctoral candidate, Cinema-Television, University of Southern California, Los Angeles, United States. Author, "Wedding Video and Its Generation," in *Resolutions: Essays on Contemporary Video Practice and Theory,* 1996; co-author, "Ida Lupino as Director of Television," in *Queen of the B's: Ida Lupino Behind the Camera,* 1995.

MORELAND, Jennifer. Freelance writer, Chicago, United States. M.A. in Radio-TV-Film 1995, University of Texas at Austin.

MOREY, Anne. Doctoral candidate, Radio-TV-Film, University of Texas at Austin, United States. Author, "The Judge Called Me an Accessory: Women's Prison Films, 1950–1962," in *Journal of Popular Film and Television,* Summer 1995; "A Whole Book for a Nickel? L. Frank Baum as Filmmaker," in *Children's Literature Association Quarterly,* Winter 1995.

MORGAN, Michael. Professor, Communication, University of Massachusetts-Amherst, United States. Co-author, *Democracy Tango: Television, Adolescents, and Authoritarian Tensions in Argentina,* 1995; numerous articles on television and cultivation theory.

MORLEY, David. Reader, Communication, Goldsmiths' College, University of London. Author, *Television, Audiences, and Cultural Studies,* 1992; *Family Television: Cultural Power and Domestic Leisure,* 1986; *The Nationwide Audience,* 1978; co-author, *Spaces of Identity: Global Media, Electronic Landscapes, and Cultural Boundaries,* 1995; *Everyday Television: Nationwide,* 1978.

MORSE, Margaret. Assistant professor, Theater Arts-Film-Video, University of California, Santa Cruz, United States. Author, "What Do Cyborgs Eat? Oral Logic in an Information

Society," in *Culture on the Brink: Ideologies of Technology*, 1994. Member, advisory board, *Convergence* and *Quarterly Review of Film and Video*.

MULLEN, Megan. Assistant professor, Communication, University of New Hampshire, United States. Author of articles in *The Velvet Light Trap* and *Quarterly Review of Film and Video*.

MURDOCK, Graham. Reader, Sociology of Culture, Loughborough University of Technology, England. Co-author, *Televising 'Terrorism,'* 1983. Co-editor, *Communicating Politics*, 1986; editorial board, *New Formations* and *The Communication Review*.

MURRAY, Matthew. Doctoral candidate, Communication Arts, University of Wisconsin-Madison, United States. Author of articles in *The Velvet Light Trap, Journal of Popular Film and Television*, and *Society and Space*.

MURRAY, Sue. Doctoral candidate, Radio-TV-Film, University of Texas at Austin, United States. Contributor, *The Independent Film and Video Monthly*. Freelance editor and researcher.

NEGRA, Diane M. Visiting assistant professor, Radio, TV, Film, University of North Texas, United States. Author of articles and book reviews in *The Velvet Light Trap, Literature and Film Quarterly*. Coordinating editor, *The Velvet Light Trap*.

NEWCOMB, Horace. Professor, Radio-Television-Film, University of Texas at Austin. Author, *TV: The Most Popular Art*, 1974; co-author, *The Producer's Medium: Conversations with Makers of American TV*, 1983. Editor, *Television: The Critical View*, 1976, revised edition, 1994.

NICKS, Joan. Associate professor and chair, Film Studies, Dramatic and Visual Arts, Brock University, St. Catharines, Canada. Author, "Sex, Lies and Landscape: Meditations on Vertical Tableaus in Joyce Wieland's *The Far Shore* and Jean Beaudin's *J.A. Martin, Photographe*," in *Canadian Journal of Film Studies*; "Crossing into Eden's Storehouse," in *Textual Studies in Canada 2*.

NIELSEN, Poul Erik. Assistant professor, University of Aarhus, Denmark. Author, "Film og TV-Produktion I USA og Danmark," in *Mediekultur*, 1995; "Bag Hollywoods Dromme Fabrik," in *Foreningen af Nye Medier*, 1993.

NILL, Dawn Michelle. Ph.D. in Communication 1996, University of Missouri-Columbia, United States.

NOYES, Gayle. Instructor, Communication Studies, Virginia Polytechnic Institute and State University, Blacksburg, Virginia; instructor, Media Studies, Radford University, Radford, Virginia.

O'DELL, Cary. Archives director, Museum of Broadcast Communications, Chicago, Illinois, United States. Author,

Women Pioneers in Television: Biographies of Fifteen Industry Leaders, 1996.

ORLIK, Peter B. Professor, Broadcast and Cinematic Arts, Central Michigan University, United States. Author, *Broadcast/Cable Copywriting*, 5th edition, 1994; *Electronic Media Criticism*, 1994; *The Electronic Media: An Introduction to the Profession*, 1992; *Critiquing Radio and Television Content*, 1988; numerous articles and book chapters.

OSWELL, David. Lecturer, Communication and Information Studies, Brunel University, England. Author, "True Love in Queer Times," in *Guns, Roses and Fatal Attractions*, 1996; "Watching with Mother in the 1950s," in *In Front of the Children*, 1995. Co-editor, *Pleasure Principles: Explorations in Politics, Sexuality and Ethics*, 1993.

PACK, Lindsy E. Assistant professor, Communication and Theatre Arts, Frostburg State University, Maryland, United States.

PARKS, Lisa. Doctoral candidate, Communication Arts, University of Wisconsin-Madison, United States. Member, editorial board, *The Velvet Light Trap*.

PATERSON, Chris. Assistant professor, Georgia State University, United States. Author, "Who Owns TV Images in Africa?," *Issue*, 1994; "Remaking South African Broadcasting in America's Image," in *FAIR Extra*, January/February 1994; "Television News from the Frontline States," in *Africa's Media Image*, 1992.

PEARSON, Tony. Senior lecturer, Film and Television Studies, University of Glasgow, Scotland. Author, "Re-chartering the BBC," in *Media Information Australia*, 1992; "Soviet Television in Transition," in *Coexistence*, 1992; "Britian's Channel Five: at the Limits of the Spectrum?" *Screen*, 1990.

PETTITT, Lance. Author of articles, *Irish Studies Review* and *The Sunday Tribune* (Dublin). Editor, British Association of Irish Studies, *Newsletter*.

PICKERING, David. British freelance writer and editor of arts and general reference books. Editor, *Dictionary of Abbreviations*, 1995; *Dictionary of Superstitions*, 1995; *International Dictionary of Theatre 3*, 1995; *Brewer's 20th Century Music*, 1994; *Encyclopedia of Pantomime*, 1993. Assistant editor, Market House Books, Ltd., 1984–92.

PLOEGER-TSOULOS, Joanna. Teaching assistant, University of Georgia, United States.

POHL, Gayle M. Assistant professor, University of Northern Iowa, United States. Author, *Public Relations: Designing Effective Communication*, 1995; articles in *Communication Reports, Popular Culture Review*, and *International Society of Exploring Teaching Alternatives Reports*.

POPELNIK, Rodolfo B. Professor, Communication, University of Puerto Rico, United States. Author, "The Problem of the Social Problem Film: A Review and a Proposal,"; *Memorias del Congreso Internacional de Cine y Literatura*, 1994. Co-editor, *IMAGINES*, 1985–90.

PORTER, Vincent. Professor, Communication and Information Studies, University of Westminster, England. Author, "The Copyright Protection of Compilations and Pseudo-Literary Works in EC Member States," in *Journal of Business Law*, 1993; *Beyond the Berne Convention, Copyright Broadcasting and the Single European Market*, 1991. Member, editorial board, *Journal of Media Law and Practice*.

PRINCE, Julie. Freelance writer, Corvalis, Oregon, United States.

QUICKE, Andrew. Associate professor, Cinema-Television, Regent University, Virginia, United States. Author, *Hidden Agendas: The Politics of Religious Broadcasting in Britain: 1987–1991*, 1992.

RABOY, Marc. Professor, Communication, University of Montreal, Quebec, Canada. Author, *Media, Crisis and Democracy*, 1992; *Missed Opportunities: The Story of Canada's Broadcasting Policy*, 1990; *Movements and Messages: Media and Radical Politics in Quebec*, 1984. Co-editor, *Communication for and against Democracy*, 1989.

REEVES, Jimmie L. Assistant professor, Mass Communication, Texas Tech University, Lubbock. Author of numerous articles on television and American culture; co-author, *Cracked Coverage: Television News, The Anti-Cocaine Crusdade, and the Reagan Legacy*, 1994.

REVELL, Jane. Trainer, Neuro-Linguistic Programming, and hypnotherapist, London, England. M.A. in Media Studies, Institute of Education, University of London.

RICHARDS, Jef. Associate professor, Advertising, University of Texas at Austin, United States. Author, *Deceptive Advertising: Behavioral Study of a Legal Concept*, 1990; articles in *Journal of Public Policy and Marketing, Journal of Consumer Affairs, Advertising Law Anthology, American Business Law Journal*, and *Boston University Law Review*.

RICHTER, William. Assistant professor, Communication, Lenoir-Rhyne College, United States.

RIDGMAN, Jeremy. Senior lecturer and convener, Film and Television Studies, Roehampton Institute, London, England. Editor, *Boxed Sets: Television Representations of Theatre*.

RIGGS, Karen E. Assistant professor, Mass Communication, University of Wisconsin-Milwaukee, United States. Co-author, "Televised Sports and Ritual: Fan Experiences," in *Sociology of Sport Journal*, 1994; "Manufactured Conflict in the 1992 Win-

ter Olympics: The Discourse of Television and Politics," in *Critical Studies in Mass Communication*, 1993.

RING, Trudy. Senior news reporter, Lambda Publications, Chicago, Illinois, United States. Freelance editor and writer.

RITROSKY-WINSLOW, Madelyn. Doctoral candidate, Telecommunications, Indiana University, United States.

RODRIGUEZ, America. Assistant professor, Radio-Television-Film, University of Texas at Austin, United States. Author, "Made in the USA: The Production of the Noticiero Univision," in *Journalism Monographs*, Winter 1995; "The Control of National Newsmaking," in *Questioning the Media*, 1995. Correspondent, National Public Radio, 1982–87.

ROSENSTEIN, Aviva. Doctoral candidate, Radio-TV-Film, University of Texas at Austin, United States.

ROSS, Karen. Research officer, Centre for Mass Communication Research, University of Leicester, England. Author, *Black and White Media: Black Images in Popular Film and Television*, 1995; "Women and the News Agenda," in *Journal of Gender Studies*, 1995; "Gender and Party Politics," in *Media, Culture and Society*, 1995; *Television in Black and White*, 1992.

ROTH, Lorna. Assistant professor, Communication Studies, Concordia University, Montreal, Canada. Author, "Media and the Commodification of Crisis," in *Media, Crisis and Democracy: Essays on Mass Communication and the Disruption of Social Order*, 1992; "The Role of CBC Northern Service in the Federal Election Process," in *Election Broadcasting in Canada*, 1991; co-author, "Aboriginal Broadcasting in Canada: A Case Study in Democratization," in *Communication For and Against Democracy*, 1989.

ROTHENBUHLER, Eric. Associate professor, Communication Studies, University of Iowa, United States. Author of books and journal articles on research on media institutions, popular culture, sociology of the audience, and communication theory.

RUNYON, Steve. Director, Media Studies, and KUSF (FM), University of San Francisco, United States.

RUTHERFORD, Paul. Professor, History, University of Toronto, Canada. Author, *The New Icons?: The Art of Television Advertising*, 1994; *When Television Was Young: Primetime Canada 1952–1967*, 1990.

SAENZ, Michael. Assistant professor, Communication Studies, University of Iowa. Author of articles on television and culture and media and ethnic groups.

SCHAEFER, Eric. Assistant professor, Emerson College, Boston, United States. Author of articles in *Film History, The Journal of Film and Video*, and *The Velvet Light Trap*. Co-editor, book series, *Inside Popular Film*.

SCHATZ, Thomas. Professor, Radio-TV-Film, University of Texas at Austin, United States. Author, *The Genius of the System*, 1988; *Hollywood Genres*, 1981.

SCHWOCH, James. Associate professor, Radio-TV-Film, Northwestern University, United States. Co-author, *Media Knowledge*, 1992; author, *The American Radio Industry and Its Latin American Activities, 1900–1939*, 1990.

SCODARI, Christine. Associate professor, Communication, Florida Atlantic University, United States. Author, "Possession, Attraction, and the Thrill of the Chase: Gendered Myth-Making in Film and Television Comedy of the Sexes," in *Critical Studies in Mass Communication*; "Operation Desert Storm as 'Wargames': War, Sport, and Media Intertexuality," in *Journal of American Culture.*

SCONCE, Jeffrey. Assistant professor, Radio-TV-Film, University of Wisconsin at Oshkosh, United States. Author of articles in *The Studio System*, 1995; *Film Theory Goes to the Movies*, 1993; *Screen and Wide Angle.* Editor, *The Velvet Light Trap*, 1991–93.

SEATON, Beth. Assistant professor, York University, Canada. Editorial collective, *Borderlines: Canada's Journal of Cultural Studies.*

SEEL, Peter B. Assistant professor, San Francisco State University, United States. Author, *Television Wars: The Local Effects of Competition Between Multinational Telecommunications Corporations*, 1993.

SHAPIRO, Mitchell E. Professor, University of Miami, Florida, United States. Author, chapter in "Network Non-prime Time Programming," in *Broadcast/Cable Programming*, 1993; *Television Network Weekend Programming 1959–1990*, 1992; *Television Network Daytime and Late Night Programming 1959–1989*, 1990; articles in *Journalism Quarterly, Television Quarterly, Newspaper Research Journal, Mass Communication Review, Current Research in Film*, and *Florida Communication Journal.*

SHELTON, Marla L. Doctoral candidate, Cinema-Television, University of Southern California, Los Angeles, United States.

SHERMAN, Barry. Professor, Telecommunications, University of Georgia, United States. Director, George Foster Peabody Awards. Author, *Telecommuniction Management: Broacasting/Cable and the New Technologies*, 1995; co-author, *Broadcasting, Cable, and beyond: An Introduction to Modern Electronic Media*, 1990.

SHIRES, Jeff. Graduate student, University of Missouri-Columbia, United States.

SHUMATE, Robbie. Teaching assistant, University of Georgia, United States.

SILLARS, Jane. Lecturer, University of Stirling, Scotland.

SILVO, Ismo. Director, European Audiovisual Observatory, Strasbourg, France. Author of numerous studies and reports on television in Finland. Former Head of Research, Finnish Broadcasting Company.

SIMON, Ron. Curator, Museum of Television and Radio, New York. Author of numerous articles and exhibition notes related to television.

SINHA, Nikhil. Assistant professor, Radio-Television-Film, University of Texas at Austin. Author of numerous articles on the political economy of telecommunications in developing nations.

SLONIOWSKI, Jeannette. Assistant professor, Film Studies, Brock University, St. Catharines, Canada.

SMITH, B. R. Associate professor, Broadcast and Cinematic Arts, Central Michigan University, United States.

SOUKUP, Paul A. Associate professor, Communication, Santa Clara University, California, United States. Author, *Christian Communication*, 1989. Co-editor, *Mass Media and the Moral Imagination*, 1994; *Media, Consciousness, and Culture*, 1991.

SPARKS, Colin. Centre for Communication and Information Studies, University of Westminster, London, England. Editor, *Media, Culture and Society.*

SPIGEL, Lynn. Associate professor, University of Southern California, United States. Author, *Make Room for TV: Television and the Family Ideal in Postwar America*, 1992; *Private Screenings*, 1992; *Close Encounters*, 1991; articles in *Screen, Critical Studies in Mass Communication, Camera Obscura*, and numerous anthologies. Co-editor, *Feminist Television Criticism*, 1997.

SRAGOW, Mike. Senior writer, *Seattle Weekly*, United States. Contributor, *The New Yorker* and *The Atlantic.* Editor, *Produced and Abandoned: The National Society of Film Critics Write on the Best Films You've Never Seen*, 1990.

STAIGER, Janet. Professor, Radio-Television-Film, University of Texas at Austin, United States. Director, Senior Fellows Honors Program, College of Communication. Author, *Bad Women*, 1995; *Interpreting Films*, 1992; co-author, *The Classical Hollywood Cinema*, 1985. Editor, *The Studio System*, 1995.

STERLING, Christopher H. Associate dean, Graduate Affairs, George Washington University, Columbian College, and Graduate School of Arts and Sciences, United States. Professor, National Center for Communication Studies, from 1982. Co-author, *Broadcasting in America*, 1994; *Stay Tuned: A Concise History of American Broadcasting*, 1990. Editor, *Decision to Divest: Major Documents in United States v. A T and T 1974–1988*, 4 vols., 1988, 1985; editor, *Communication Booknotes.*

STERNBERG, Joel. Associate professor, Saint Xavier University, Chicago, United States. President, Sternberg Communications, Inc., from 1984.

STRANGE, Nicola. Production co-ordinator, Independent Television Productions, United Kingdom. M.A, Film and Television Studies, University of Warwick, England.

STRAUBHAAR, Joseph. Professor and director, Communication Research Center, Brigham Young University, United States. Author, "Brazil," in *International World of Electronic Media*, 1994; "Beyond Media Imperialism: Asymmetrical Interdependence and Cultural Proximity," in *Critical Studies in Mass Communication*, 1991; co-author, "The Role of Television in the 1989 Brazilian Presidential Elections," in *Television, Politics and the Transition to Democracy in Latin America*, 1993; contributor, *Critical Studies in Mass Communication* and *Journal of Communication*.

STREIBLE, Daniel G. Assistant professor, Radio-TV-Film, University of Wisconsin, Oskosh, United States. Author of articles in *Film History*, *The Velvet Light Trap*, *Libraries and Culture*, *Arachne*, and *Screen*.

STROVER, Sharon. Associate professor, Radio-Television-Film, University of Texas at Austin, United States. Co-author, *Electronic Byways*, 1993. Co-editor, *Telecommunications and Rural Development*, 1991; *The New Urban Infra-structure*, 1990.

STULLER-GIGLIONE, Joan. Doctoral candidate, Radio-TV-Film, University of Texas at Austin, United States. Journalism instructor, Los Angeles Valley College, California.

TAN, Zoe. Director of research, SRT Nielsen Media Services, Taipei, Taiwan. Author, "The Role of Mass Media in Insurgent Terrorism: Issues and Perspectives," 1989, *Gazette* (Amsterdam); "Mass Media and Insurgent Terrorism: A Twenty-year Balance Sheet," 1988, *Gazette* (Amsterdam).

TCHOUNGUI, Gisele. M.A. 1955, Laval University, Canada

TEDESCO, John C. Doctoral candidate, University of Oklahoma, United States. Author of articles in *Argumention and Advocacy*, *Journal of Communication Studies*, and *Informatologia*.

TETZLAFF, David J. Assistant professor, Communication, University of San Francisco, United States. Author of essays in *The Madonna Connection* and *Culture and Power*. Member, editorial board, *The Velvet Light Trap*, 1994–95.

THOMPSON, Robert J. Associate professor, Newhouse School of Communication, Syracuse University, United States. Author of several books, including *Television's Second Golden Age*, 1996; co-author, *Prime Time, Prime Movers*. Book series editor, Syracuse University Press.

THORBURN, David. Professor, Literature; Director, Program in Cultural Studies, Massachusetts Institute of Technology, United States. Author, "Television Melodrama," *Television as a Cultural Force*, 1976; *Conrad's Romanticism*, 1974. Co-editor,

John Updike: A Collection of Critical Essays, 1979; *Romanticism: Vistas, Instances, Continuities*, 1973.

TIMBERG, Bernard M. Associate professor, Media Studies, Radford University, Virginia, United States. Author, "The Unspoken Rules of Television Talk," in *Television: The Critical View*, 1994; co-author, "Encounters with the Television Image: Thirty Years of Encoding Research," in *Communication Yearbook*, 1991.

TORRE, Paul J. Doctoral candidate, Cinema-Television, University of Southern California, Los Angeles, United States.

TOVARES, Raul D. Assistant professor, University of North Dakota, United States. Author, "Ganging up on the Gang," in *Chicanos in Contemporary Society*, 1995; "The Legalization of Cocaine and Marijuana," in *War on Drugs: Opposing Viewpoints*, 1990.

TREVIÑO, Liza. M.A. in Radio-Television-Film 1995, University of Texas at Austin, United States.

TRIANTAFILLOU, Soti. Freelance writer, teaches film and television studies at the Hellenic Cinema and Television School Lykourgus Stavrakos, Athens, Greece.

TRIECE, Mary. Doctoral candidate, Speech Communication, University of Texas at Austin, United States.

TULLOCH, John. Professor, Cultural Studies, Charles Stuart University, Bathurst, Australia. Co-author, *Science Fiction Audiences: Watching Doctor Who and Star Trek*, 1995; author, *Television Drama: Agency, Audience and Myth*, 1990. Co-editor, *Australian Television: Programs, Pleasures and Politics*, 1987.

TULLOCH, Marian. Senior lecturer, Psychology, Charles Stuart University, Bathurst, Australia. Author, "Evaluating Aggression: School Student's Responses to Television Portrayals of Institutionalized Violence," in *Journal of Youth and Adolescence*, 1995; co-author, "Understanding TV Violence: A Multifaceted Cultural Analysis," in *Nation, culture, text: Australian Cultural and Media Studies*, 1993.

TURNER, J.C. Associate Professor, Communications, University of Northern Iowa, United States.

TUROW, Joseph. Professor, Annenberg School for Communication, University of Pennsylvania, United States. Author, *Playing Doctor: Television, Storytelling, and Media Power*, 1989; *Entertainment, Education, and the Hard Sell: Three Decades of Network Children's Television*, 1981; numerous articles and book chapters.

VAHIMAGI, Tise. Film and television researcher, British Film Institute, England. Co-editor, *British Television*, 1994; *MTM: Quality Television*, 1984.

VANDE BERG, Leah R. Professor, Communication, California State University, Sacramento, United States. Co-author,

Organizational Life on Television, 1989. Co-editor, *Television Criticism: Approaches and Applications*, 1991.

WAITE, Clayland H. Professor, Media Studies, Radford University, Virginia, United States. Research interests, multimedia, electronic journalism, corporate and professional communication.

WALKER, Cynthia W. Doctoral candidate, Information and Library Science, Rutgers University, United States. Adjunct professor, Media Arts, Jersey City State College, New Jersey. Author, *The Teacher's Guide to Star Trek*, 1994. Former journalist, *The News Tribune*, New Jersey.

WALKER, James R. Associate professor, Mass Communications, Saint Xavier University, Chicago, United States. Co-author, *Grazing on a Vast Wasteland: The Remote Control and The Second Generation of Television*, 1995; *The Remote Control in the New Age of Television*, 1993; articles in *Journalism Quarterly* and *Journal of Broadcasting and Electronic Media*. Member, editorial board, *The Journal of Broadcasting and Electronic Media*.

WALSH, Kay. Assistant professor, Communication and Theatre Arts, Frostburg State University, Maryland, United States.

WARNER, Charles. Goldenson Professor, University of Missouri School of Journalism, United States. Co-author, *Broadcast and Cable Selling*, 1993. Editor, *Media Management Review*, 1995.

WATSON, Mary Ann. Professor, Telecommunications and Film, Eastern Michigan University, United States. Author, "From *My Little Margie* to *Murphy Brown*: Women's Lives on the Small Screen," in *Television Quarterly*, 1994; *The Expanding Vista: American Television in the Kennedy Years*, 1990.

WEHMEYER, James. Assistant professor, English, Fort Lewis College, Colorado, United States.

WEISBLAT, Tinky "Dakota". U.S. freelance writer and editor. Reviews and commentaries in *American Quarterly*, *The Boston Globe*, *Film Quarterly*, *The Hartford Courant*, *Remember*, and *The Velvet Light Trap*.

WHITE, Mimi. Associate professor, Radio-TV-Film, Northwestern University, United States. Author, *Tele:Advising: Therapeutic Discourse in American Television*, 1992; co-author, *Media Knowledge: Popular Culture, Pedagogy, and Critical Citizenship*, 1992; articles in *Screen*, *Cinema Journal*, *Cultural Studies*, and *Wide Angle*.

WIGGINS, D. Joel. Assistant instructor, Speech Communication, University of Texas at Austin, United States.

WILDING, Derek. School of Media and Journalism, Queensland University of Technology, Australia.

WILKINS, Karin Gwinn. Assistant professor, Radio-Television-Film, University of Texas at Austin, United States. Author, "Gender, News Exposure and Political Cynicism: Public Opinion of Hong Kong's Future Transition," in *International Journal of Public Opinion Research*, 1995.

WILLIAMS, Carol Traynor. Professor, Humanities, Roosevelt University, United States. Author, *"It's Time for My Story": Soap Opera, Sources, Structure, and Responses*, 1992; *The Dream Beside Me: The Movies and the Children of the Forties*, 1980.

WILLIAMS, Mark. Assistant professor, Film Studies, Dartmouth College, New Hampshire. Author of numerous articles on film and television, with special interest in the history of local television programming in the United States.

WILLIAMS, Suzanne Hurst. Associate professor, Communication, Trinity University, San Antonio, United States. Author, "Bugs Bunny Meets He-Man: A Historical Comparison of Values in Animated Cartoons," in *Television Criticism: Approaches and Applications*, 1991; articles in *Southwestern Mass Communication Journal* and *Feedback*; editor, *Society for Animation Studies Newsletter*.

WILSON, Pamela. Doctoral candidate, University of Wisconsin-Madison. Author, "Upscale Feminine Angst: Molly Dodd, the Lifetime Cable Network, and Gender Marketing," in *Camera Obscura*, 1995; "All Eyes on Montana: Regional Television Reception and Native American Cultural Politics in the 1950s," in *Quarterly Review of Film Studies*, 1995; and various articles on television, ethnicity, and gender.

WINSTON, Brian. Professor and Director, Mass Communication, University of Wales, College of Cardiff, Wales. Author, *Claiming the Real*, 1995; *Misunderstanding Media*, 1986.

WORRINGHAM, Richard. Chair, Media Studies Department, Radford University, United States.

ZAJACZ, Rita. M.A. candidate, Telecommunications, Indiana University, United States.

ZECHOWSKI, Sharon. Doctoral candidate, Telecommunications, Ohio University, United States.

ZUBERI, Nabeel. Lecturer, Department of Film, Television, and Media Studies, University of Auckland, New Zealand.

INDEX

Listings are arranged in alphabetical order. Page numbers in italics indicate entries on the subject.